Fundamental Theories of Physics

Volume 188

The international monograph series "Fundamental Theories of Physics" aims to stretch the boundaries of mainstream physics by clarifying and developing the theoretical and conceptual framework of physics and by applying it to a wide range of interdisciplinary scientific fields. Original contributions in well-established fields such as Quantum Physics, Relativity Theory, Cosmology, Quantum Field Theory, Statistical Mechanics and Nonlinear Dynamics are welcome. The series also provides a forum for non-conventional approaches to these fields. Publications should present new and promising ideas, with prospects for their further development, and carefully show how they connect to conventional views of the topic. Although the aim of this series is to go beyond established mainstream physics, a high profile and open-minded Editorial Board will evaluate all contributions carefully to ensure a high scientific standard.

More information about this series at http://www.springer.com/series/6001

Klaas Landsman

Foundations of Quantum Theory

From Classical Concepts to Operator Algebras

Springer Open

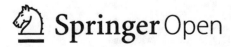

Klaas Landsman
IMAPP
Radboud University
Nijmegen
The Netherlands

ISSN 0168-1222 ISSN 2365-6425 (electronic)
Fundamental Theories of Physics
ISBN 978-3-319-84738-2 ISBN 978-3-319-51777-3 (eBook)
DOI 10.1007/978-3-319-51777-3

Softcover reprint of the hardcover 1st edition 2017

Printed on acid-free paper

This Springer imprint is published by Springer Nature
The registered company is Springer International Publishing AG
The registered company address is: Gewerbestrasse 11, 6330 Cham, Switzerland

To Jeremy Butterfield

Preface

'Der Kopf, *so* gesehen, hat mit dem Kopf, *so* gesehen, auch nicht die leiseste Ähnlichkeit (...) Der Aspektwechsel. "Du würdest doch sagen, dass sich das Bild jetzt gänzlich geändert hat!" Aber was ist anders: mein Eindruck? meine Stellungnahme? (...) Ich *beschreibe* die Änderung wie eine Wahrnehmung, ganz, als hätte sich der Gegenstand vor meinen Augen geändert.' (Wittgenstein, *Philosophische Untersuchungen* II, §§127, 129).[1]

As the well-known picture above is meant to allegorize, some physical systems admit a dual description in either classical or quantum-mechanical terms. According to Bohr's "doctrine of classical concepts", measurement apparatuses are examples of such systems. More generally—as hammered down by decoherence theorists—the classical world around us is a case in point. As will be argued in this book, the measurement problem of quantum mechanics (highlighted by Schrödinger's Cat) is *caused* by this duality (rather than *resolved* by it, as Bohr is said to have thought).

[1] 'The head seen in *this* way hasn't even the slightest similarity to the head seen in *that* way (...) The change of aspect. "But surely you'd say that the picture has changed altogether now! But what is different: my impression? my attitude? (...) I *describe* the change like a perception; just as if the object has changed before my eyes.' Translation: G.E.M. Anscombe, P.M.S. Hacker, & J. Schulte (Wittgenstein, 2009/1953, pp. 205–206).

The aim of this book is to analyze the foundations of quantum theory from the point of view of classical-quantum duality, using the mathematical formalism of operator algebras on Hilbert space (and, more generally, C*-algebras) that was originally created by von Neumann (followed by Gelfand and Naimark). In support of this analysis, but also as a matter of independent interest, the book covers many of the traditional topics one might expect to find in a treatise on the foundations of quantum mechanics, like pure and mixed states, observables, the Born rule and its relation to both single-case probabilities and long-run frequencies, Gleason's Theorem, the theory of symmetry (including Wigner's Theorem and its relatives, culminating in a recent theorem of Hamhalter's), Bell's Theorem(s) and the like, quantization theory, indistinguishable particle, large systems, spontaneous symmetry breaking, the measurement problem, and (intuitionistic) quantum logic. One also finds a few idiosyncratic themes, such as the Kadison–Singer Conjecture, topos theory (which naturally injects intuitionism into quantum logic), and an unusual emphasis on both conceptual and mathematical aspects of limits in physical theories.

All of this is held together by what we call ***Bohrification***, i.e., the mathematical interpretation of Bohr's classical concepts by *commutative* C*-algebras, which in turn are studied in their quantum habitat of *noncommutative* C*-algebras.

Thus the book is mostly written in mathematical physics style, but its real subject is *natural philosophy*. Hence its intended readership consists not only of mathematical physicists, but also of philosophers of physics, as well as of theoretical physicists who wish to do more than 'shut up and calculate', and finally of mathematicians who are interested in the mathematical and conceptual structure of quantum theory.

To serve all these groups, the native mathematical language (i.e. of C*-algebras) is introduced slowly, starting with finite sets (as classical phase spaces) and finite-dimensional Hilbert spaces. In addition, all advanced mathematical background that is necessary but may distract from the main development is laid out in extensive appendices on Hilbert spaces, functional analysis, operator algebras, lattices and logic, and category theory and topos theory, so that the prerequisites for this book are limited to basic analysis and linear algebra (as well as some physics). These appendices not only provide a direct route to material that otherwise most readers would have needed to extract from thousands of pages of diverse textbooks, but they also contain some original material, and may be of interest even to mathematicians.

In summary, the aims of this book are similar to those of its peerless paradigm:

> 'Der Gegenstand dieses Buches ist die einheitliche, und, soweit als möglich und angebracht, mathematisch einwandfreie Darstellung der neuen Quantenmechanik (. . .). Dabei soll das Hauptgewicht auf die allgemeinen und prinzipiellen Fragen, die im Zusammenhange mit dieser Theorie entstanden sind, gelegt werden. Insbesondere sollen die schwierigen und vielfach noch immer nicht restlos geklärten Interpretationsfragen näher untersucht werden.' (von Neumann, *Mathematische Grundlagen der Quantenmechanik*, 1932, p. 1).[2]

---

[2] 'The object of this book is to present the new quantum mechanics in a unified presentation which, so far as it is possible and useful, is mathematically rigorous. (. . .) Therefore the principal emphasis shall be placed on the general and fundamental questions which have arisen in connection with this theory. In particular, the difficult problems with interpretation, many of which are even now not fully resolved, will be investigated in detail.' Translation: R.T. Beyer (von Neumann, 1955, p. vii).

Two other quotations the author often had in mind while writing this book are:

> 'And although the whole of philosophy is not immediately evident, still it is better to add something to our knowledge day by day than to fill up men's minds in advance with the preconceptions of hypotheses.' (Newton, draft preface to *Principia*, 1686).[3]

> 'Juist het feit dat een genie als DESCARTES volkomen naast de lijn van ontwikkeling is blijven staan, die van GALILEI naar NEWTON voert (...) [is] een phase van den in de historie zoo vaak herhaalden strijd tusschen de bescheidenheid der mathematisch-physische methode, die na nauwkeurig onderzoek de verschijnselen der natuur in steeds meer omvattende schemata met behulp van de exacte taal der mathesis wil beschrijven en den hoogmoed van het philosophische denken, dat in één genialen greep de heele wereld wil omvatten (...).' (Dijksterhuis, *Val en Worp*, 1924, p. 343).[4]

Acknowledgements

1. Research underlying this book has been generously supported by:
    - Radboud University Nijmegen, partly through a sabbatical in 2014.
    - The Netherlands Organization for Scientific Research (NWO), initially by funding various projects eventually contributing to this book, and most recently by paying the Open Access fee, making the book widely available.
    - The Templeton World Charity Foundation (TWCF), by funding the Oxford–Princeton–Nijmegen collaboration *Experimental Tests of Quantum Reality*.
    - Trinity College (Cambridge), by appointing the author as a *Visiting Fellow Commoner* during the Easter Term 2016, when the book was largely finished.
2. The author was fortunate in having been surrounded by outstanding students and postdocs, who made essential contributions to the insights described in this book. In alphabetical order these were Christian Budde, Martijn Caspers, Ronnie Hermens, Jasper van Heugten, Chris Heunen, Bert Lindenhovius, Robin Reuvers, Bas Spitters, Marco Stevens, and Sander Wolters. Those were the days!
3. The author is indebted to Jeremy Butterfield, Peter Bongaarts, Harvey Brown, Dennis Dieks, Siegfried Echterhoff, Aernout van Enter, Jan Hamhalter, Jaap van Oosten, and Bas Terwijn for comments on the manuscript. In addition, through critical feedback on a Masterclass at Trinity, Owen Maroney and Fred Muller indirectly (but considerably) improved Chapter 11 on the measurement problem.
4. Angela Lahee from Springer thoughtfully guided the publication process of this book from the beginning to the end. Thanks also to her colleague Aldo Rampioni.

Finally, it is a pleasure to dedicate this book to Jeremy Butterfield, in recognition of his ideas, as well as of his unrelenting support and friendship over the last 25 years.

---

[3] Newton (1999), p. 61.

[4] 'The very fact that a genius like Descartes was completely sidelined in the development leading from Galilei to Newton (...) represents a phase in the struggle—that has so often been repeated throughout history—between the modesty of the approach of mathematical physics, which after precise investigations attempts to describe natural phenomena in increasingly comprehensive schemes using the exact language of mathematics, and the haughtiness of philosophical thought, which wants to comprehend the entire world in one dazzling grasp.' Translation by the author.

Contents

Introduction

After 25 years of confusion and even occasional despair, in March 1926 physicists suddenly had *two* theories of the microscopic world (Heisenberg, 1925; Schrödinger, 1926ab), which hardly could have looked more differently. Heisenberg's ***matrix mechanics*** (as it came to be called a bit later) described experimentally measurable quantities (i.e., "observables") in terms of discrete quantum numbers, and apparently lacked a state concept. Schrödinger's ***wave mechanics*** focused on unobservable continuous matter waves apparently playing the role of quantum states; at the time the only observable within reach of his theory was the energy. Einstein is even reported to have remarked in public that the two theories excluded each other.

Nonetheless, Pauli (in a letter to Jordan dated 12 April 1926), Schrödinger (1926c) himself, Eckart (1926), and Dirac (1927) argued—it is hard to speak of a complete argument even at a heuristic level, let alone of a mathematical *proof* (Muller, 1997ab)— that in fact the two theories were equivalent! A rigorous equivalence proof was given by von Neumann (1927ab), who (at the age of 23) was the first to unearth the mathematical structure of quantum mechanics as we still understand it today. His effort, culminating in his monograph *Mathematische Grundlagen der Quantenmechanik* (von Neumann, 1932), was based on the abstract concept of a ***Hilbert space***, which previously had only appeared in examples (i.e. specific realizations) going back to the work of Hilbert and his school on integral equations.

The novelty of von Neumann's abstract approach may be illustrated by the advice Hilbert's former student Schmidt gave to von Neumann even at the end of the 1920s:

> 'Nein! Nein! Sagen Sie nicht Operator, sagen Sie Matrix!" (Bernkopf, 1967, p. 346).[5]

Von Neumann proposed that observables quantities be interpreted as (possibly unbounded) self-adjoint operators on some Hilbert space, whilst pure states are realized as rays (i.e. unit vectors up to a phase) in the same space; finally, the inner product provides the probabilities introduced by Born (1926ab). In particular, Heisenberg's observables were operators on $\ell^2(\mathbb{N})$, whereas Schrödinger's wave-functions were unit vectors in $L^2(\mathbb{R}^3)$. A unitary transformation between these Hilbert spaces then provided the mathematical equivalence between their competing theories.

[5] 'No! No! You shouldn't say operator, you should say matrix!'

This story is well known, but it is worth emphasizing (cf. Zalamea, 2016, §I.1) that the most significant difference between von Neumann's mathematical axiomatization of quantum mechanics and Dirac's heuristic but beautiful and systematic treatment of the same theory (Dirac, 1930) was not so much the lack of mathematical rigour in the latter—although this point was stressed by von Neumann (1932, p. 2) himself, who was particularly annoyed with Dirac's δ-function and his closely related assumption that every self-adjoint operator can be diagonalized in the naive way of having a basis of eigenvectors—but the fact that Dirac's approach was *relative* to the choice of a (generalized) basis of a Hilbert space, whereas von Neumann's was *absolute*. In this sense, as a special case of his (and Jordan's) general transformation theory, Dirac showed that Heisenberg's matrix mechanics and Schrödinger's wave mechanics were related by a (unitary) transformation, whereas for von Neumann they were two different realizations of his abstract (separable) Hilbert space. In particular, von Neumann's approach *a priori* dispenses with a basis choice altogether; this is precisely the difference between an *operator* and a *matrix* Schmidt alluded to in the above quotation. Indeed, von Neumann's abstract approach (which as a co-founder of functional analysis he shared with Banach, but not with his mentor Hilbert) was remarkable even in mathematics; in physics it must have been dazzling.

It is instructive to compare this situation with special relativity, where, so to speak, Dirac would write down the theory in terms of inertial frames of reference, so as to subsequently argue that due to Poincaré-invariance the physical content of the theory does not depend on such a choice. Von Neumann, on the other hand (had he ever written a treatise on relativity), would immediately present Minkowski's space-time picture of the theory and develop it in a coordinate-free fashion.

However, this analogy is also misleading. In special relativity, all choices of inertial frames are genuinely equivalent, but in quantum mechanics one often does have preferred observables: as Bohr would argue from his Como Lecture in 1927 onwards (Bohr, 1928), these observables are singled out by the choice of some experimental context, and they are jointly measurable iff they commute (see also below). Though not necessarily developed with Bohr's doctrine in mind, Dirac's approach seems tailor-made for this situation, since his basis choice is equivalent to a choice of "preferred" physical observables, namely those that are diagonal in the given basis (for Heisenberg this was energy, while for Schrödinger it was position).

Von Neumann's abstract approach can deal with preferred observables and experimental contexts, too, though the formalism for doing so is more demanding. Namely, for reasons ranging from quantum theory to ergodic theory via unitary group representations on Hilbert space, from 1930 onwards von Neumann developed his theory of "rings of operators" (nowadays called ***von Neumann algebras***), partly in collaboration with his assistant Murray (von Neumann, 1930, 1931, 1938, 1940, 1949; Murray & von Neumann, 1936, 1937, 1943). For us, at least at the moment the point is that Dirac's diagonal observables are formalized by ***maximal commutative von Neumann algebras*** A on some Hilbert space. These often come naturally with some specific realization of a Hilbert space; for example, on Heisenberg's Hilbert space $\ell^2(\mathbb{N})$ on has $A_d = \ell^\infty(\mathbb{N})$, while Schrödinger's $L^2(\mathbb{R}^3)$ is host to $A_c = L^\infty(\mathbb{R}^3)$, both realized as multiplication operators (cf. Proposition B.73).

Although the second (1931) paper in the above list shows that von Neumann was well aware of the importance of the commutative case of his theory of operator algebras, he—perhaps deliberately—missed the link with Bohr's ideas. As explained in the remainder of this Introduction, providing this link is one of the main themes of this book, but we will do so using the more powerful formalism of ***C*-algebras***. Introduced by Gelfand & Naimark (1943), these are abstractions and generalizations of von Neumann algebras, so abstract indeed that Hilbert spaces are not even mentioned in their definition. Nonetheless, C*-algebras remain very closely tied to Hilbert spaces through the GNS-construction originating with Gelfand & Naimark (1943) and Segal (1947b), which implies that any C*-algebra is isomorphic to a well-behaved algebra of bounded operators on some Hilbert space (see §C.12).

Starting with Segal (1947a), C*-algebras have become an important tool in mathematical physics, where traditionally most applications have been to quantum systems with infinitely many degrees of freedom, such as quantum statistical mechanics in infinite volume (Ruelle, 1969; Israel, 1979; Bratteli & Robinson, 1981; Haag, 1992; Simon, 1993) and quantum field theory (Haag, 1992; Araki, 1999).

Although we delve from the first body of literature, and were at least influenced by the second, the present book employs C*-algebras in a rather different fashion, in that we exploit the unification they provide of the commutative and the noncommutative "worlds" into a single mathematical framework (where one should note that as far as physics is concerned, the commutative or classical case is not purely C*-algebraic in character, because one also needs a Poisson structure, see Chapter 3). This unified language (supplemented by some category theory, group(oid) theory, and differential geometry) gives a mathematical handle on Wittgenstein's *Aspektwechsel* between classical and quantum-mechanical modes of description (see Preface), which in our view lies at the heart of the foundations of quantum physics. This "change of perspective", which roughly speaking amounts to switching (and interpolating) between commutative and noncommutative C*-algebras, is *added* to Dirac's transformation theory (which comes down to switching between generalized bases, or, equivalently, between maximal commutative von Neumann algebras).

The central conceptual importance of the *Aspektwechsel* for this book in turn derives from our adherence to Bohr's ***doctrine of classical concepts***, which forms part of the ***Copenhagen Interpretation*** of quantum mechanics (here defined strictly as a body of ideas shared by Bohr and Heisenberg). We let the originators speak:

> 'It is decisive to recognize that, however far the phenomena transcend the scope of classical physical explanation, the account of all evidence must be expressed in classical terms. The argument is simply that by the word *experiment* we refer to a situation where we can tell others what we have done and what we have learned and that, therefore, the account of the experimental arrangements and of the results of the observations must be expressed in unambiguous language with suitable application of the terminology of classical physics.' (Bohr, 1949, p. 209)

> 'The Copenhagen interpretation of quantum theory starts from a paradox. Any experiment in physics, whether it refers to the phenomena of daily life or to atomic events, is to be described in the terms of classical physics. The concepts of classical physics form the language by which we describe the arrangement of our experiments and state the results. We cannot and should not replace these concepts by any others.' (Heisenberg 1958, p. 44)

The last quotation even opens Heisenberg's only systematic presentation of the Copenhagen Interpretation, which forms Chapter III of his Gifford Lectures from 1955; apparently this was the first occasion where the name "Copenhagen Interpretation" was used (Howard, 2004). In our view, several other defining claims of the Copenhagen Interpretation appear to be less well founded, if not unwarranted, although they may have been understandable in the historical context where they were first proposed (in which the new theory of quantum mechanics needed to get going even in the face of the foundational problems that all of the originators—including Bohr and Heisenberg—were keenly aware of). These spurious claims include:

- The emphatic rejection of the possibility to analyze what is going on during measurements, as expressed in typical Bohr parlance by claims like:

  > 'According to the quantum theory, just the impossibility of neglecting the interaction with the agency of measurement means that every observation introduces a new uncontrollable element.' (Bohr, 1928, p. 584),

  or, with similar (but somehow less off-putting) dogmatism by Heisenberg:

  > 'So we cannot completely objectify the result of an observation' (1958, p. 50).

- The closely related interpretation of quantum-mechanical states (which Heisenberg indeed referred to as "probability functions") as mere catalogues of the probabilities attached to possible outcomes of experiments, as in:

  > 'what one deduces from observation is a probability function, a mathematical expression that combines statements about possibilities or tendencies with statements about our knowledge of facts' (Heisenberg 1958, p. 50),

In addition, there are two ingredients of the avowed Copenhagen Interpretation Bohr and Heisenberg actually seem to have disagreed about. These include:

- The ***collapse of the wave-function*** (i.e., upon completion of a measurement), which was introduced by Heisenberg (1927) in his paper on the uncertainty relations. As we shall see in Chapter 11, this idea was widely adopted by the pioneers of quantum mechanics (and it still is), but apparently it was never endorsed by Bohr, who saw the wave-function as a "symbolic" expression (cf. Dieks, 2016a).
- Bohr's doctrine of ***Complementarity***, which—though never precisely articulated—he considered to be a revolutionary philosophical insight of central importance to the interpretation of quantum mechanics (and even beyond). Heisenberg, on the other hand, regarded complementary descriptions (which Bohr saw as *incompatible*) as mathematically *equivalent* and at best paid lip-service to the idea. The reason for this discord probably lies in the fact that Heisenberg was typically guided by (quantum) *theory*, whereas Bohr usually started from *experiments*; Heisenberg once even referred to his mentor as a 'philosopher of experiment'. Therefore, Heisenberg was satisfied that for example position and momentum were related by a unitary operator (i.e. the Fourier transform), whereas Bohr had the incompatible experimental arrangements in mind that were required to measure these quantities. Their difference, then, contrasted theory and experiment.

Let us now review the philosophical motivation Bohr and Heisenberg gave for their mutual doctrine of classical concepts. First, Bohr (in his typical convoluted prose):

> 'The elucidation of the paradoxes of atomic physics has disclosed the fact that the unavoidable interaction between the objects and the measuring instruments sets an absolute limit to the possibility of speaking of a behavior of atomic objects which is independent of the means of observation. We are here faced with an epistemological problem quite new in natural philosophy, where all description of experience has so far been based on the assumption, already inherent in ordinary conventions of language, that it is possible to distinguish sharply between the behavior of objects and the means of observation. This assumption is not only fully justified by all everyday experience but even constitutes the whole basis of classical physics. (...) As soon as we are dealing, however, with phenomena like individual atomic processes which, due to their very nature, are essentially determined by the interaction between the objects in question and the measuring instruments necessary for the definition of the experimental arrangement, we are, therefore, forced to examine more closely the question of what kind of knowledge can be obtained concerning the objects. In this respect, we must, on the one hand, realize that the aim of every physical experiment—to gain knowledge under reproducible and communicable conditions—leaves us no choice but to use everyday concepts, perhaps refined by the terminology of classical physics, not only in all accounts of the construction and manipulation of the measuring instruments but also in the description of the actual experimental results. On the other hand, it is equally important to understand that just this circumstance implies that no result of an experiment concerning a phenomenon which, in principle, lies outside the range of classical physics can be interpreted as giving information about independent properties of the objects.'

This text has been taken from Bohr (1958, p. 25), but very similar passages appear in many of Bohr's writings from his famous Como Lecture (Bohr, 1928) onwards. In other words, the (supposedly) unavoidable interaction between the objects and the measuring instruments, which for Bohr represents *the* characteristic feature of quantum mechanics (and which we would now express in terms of entanglement, of which concept Bohr evidently had an intuitive grasp), threatens the objectivity of the description that is characteristic of (if not the defining property of) of classical physics. However, this threat can be countered by describing quantum mechanics through classical physics, which (or so the argument goes) restores objectivity. Elsewhere, we see Bohr also insisting on the need for classical concepts in *defining* any meaningful theory whatsoever, as these are the only concepts we really understand (though, as he always insists, classical concepts are at the same time challenged by quantum theory, as a consequence of which their use is necessarily limited).

Although Heisenberg's arguments for the necessity of classical concepts start similarly, they eventually take a conspicuously different direction from Bohr's:

> 'To what extent, then, have we finally come to an objective description of the world, especially of the atomic world? In classical physics science started from the belief—or should one say from the illusion?—that we could describe the world or at least parts of the world without any reference to ourselves. This is actually possible to a large extent. We know that the city of London exists whether we see it or not. It may be said that classical physics is just that idealization in which we can speak about parts of the world without any reference to ourselves. Its success has led to the general ideal of an objective description of the world. Objectivity has become the first criterion for the value of any scientific result. Does the Copenhagen interpretation of quantum theory still comply with this ideal? One may perhaps say that quantum theory corresponds to this ideal as far as possible. Certainly quantum theory does not contain genuine subjective features, it does not introduce the mind

> of the physicist as a part of the atomic event. But it starts from the division of the world into the object and the rest of the world, and from the fact that at least for the rest of the world we use the classical concepts in our description. This division is arbitrary and historically a direct consequence of our scientific method; the use of the classical concepts is finally a consequence of the general human way of thinking. But this is already a reference to ourselves and in so far our description is not completely objective. (...)
>
> The concepts of classical physics are just a refinement of the concepts of daily life and are an essential part of the language which forms the basis of all natural science. Our actual situation in science is such that we do use the classical concepts for the description of the experiments, and it was the problem of quantum theory to find theoretical interpretation of the experiments on this basis. There is no use in discussing what could be done if we were other beings than we are. (...)
>
> Natural science does not simply describe and explain nature; it is a part of the interplay between nature and ourselves; it describes nature as exposed to our method of questioning.' (Heisenberg, 1958, p. 55–56, 56, 81)

The well-known last part may indeed have been the source of the crucial 'I'm the one who knocks' episode in the superb tv-series *Breaking Bad* (whose criminal main character operates under the cover name of "Heisenberg"). This is worth mentioning here, because Heisenberg (and to a lesser extent also Bohr) displays a puzzling mixture between the hubris of claiming that quantum mechanics has restored Man's position at the center of the universe and the modesty of recognizing that nonetheless Man has to know his limitations (in necessarily relying on the classical concepts he happens to be familiar with at the current state of evolution and science).

Our own reasons for favoring the doctrine of classical concepts are threefold. The first is closely related to Heisenberg's and may be expressed even better by the following passage from a book by the renowned Dutch primatologist Frans de Waal:

> '*Die Verwandlung* [i.e., *The Metamorphosis* by Franz Kafka, in which Gregor Samsa famously wakes up to find himself transformed into an insect], published in 1915, was an unusual take-off for a century in which anthropocentrism declined. For metaphorical reasons, the author had picked a repulsive creature, forcing us from the first page onwards to feel what it would be like to be an insect. Around the same time, the German biologist Jakob von Uexküll drew attention to the fact that each particular species has its own perspective, which he called its *Umwelt*. To illustrate this new idea, Uexküll took his readers on a tour through the worlds of various creatures. Each organism observes its environment in its own peculiar way, he argued. A tick, which has no eyes, climbs onto a grass blade, where it awaits the scent of butyric acid off the skin of mammals that pass by. Experiments have demonstrated that ticks may survive without food for as long as 18 years, so that a tick has ample time to wait for her prey, jump on it, and suck its warm blood, after which she is ready to lay her eggs and die. Are we in a position to understand the *Umwelt* of a tick? Its seems unbelievably poor compared to ours, but Uexküll regarded its simplicity rather as a strength: ticks have set themselves a narrow goal and hence cannot easily be distracted. Uexküll analysed many other examples, and showed how a single environment offers hundreds of different realities, each of which is unique for some given species. (...) Some animals merely register ultraviolet light, others live in a world of odors, or of touch, like a star nose mole. Some animals sit on a branch of an oak, others live underneath the bark of the same oak, whilst a fox family digs a hole underneath its roots. Each animal observes the tree differently.' (De Waal, 2016, pp. 15–16. Translation by the author).

Indeed, it is hardly an accident that De Waal preceded this passage by a quotation from Heisenberg almost identical to the last one above.

A second argument in favour of the doctrine lies in the possibility of a peaceful outcome of the Bohr–Einstein debate, or at least of an important part of it; cf. Landsman (2006a), which was inspired by earlier work of Raggio (1981, 1988) and Bacciagaluppi (1993). This debate initially centered on Einstein's attempts to debunk the Heisenberg uncertainty relations, and subsequently, following Einstein's grudging acceptance of their validity, entered its most famous and influential phase, in which Einstein tried to prove that quantum mechanics, although admittedly *correct*, was *incomplete*. One could argue that both antagonists eventually lost this part of the debate, since Einstein's goal of a local realistic (quantum) physics was quashed by the famous work of Bell (1964), whereas against Bohr's views, deterministic versions of quantum mechanics such as Bohmian mechanics and the Everett (i.e. Many Worlds) Interpretation turned out to be at least logical possibilities.

However incompatible the views of Einstein and Bohr on physics and its goals may have been, unknown to them a common battleground did in fact exist and could even have led to a reconciliation of at least the epistemological views of the great adversaries. The common ground referred to concerns the problem of ***objectification***, which at first sight Bohr and Einstein approached in completely different ways:

- Bohr objectified a *quantum* system through the specification of a classical experimental context, i.e. by looking at it through appropriate classical glasses.
- Einstein objectified *any* physical system by claiming its independent existence:

> 'The belief in an external world independent of the perceiving subject is the basis of all natural science.' (Einstein, 1954, p. 266).

On a suitable mathematical interpretation, these conditions for the objectification of the system turn out to be equivalent! Namely, identifying Bohr's apparatus with Einstein's perceiving subject, calling its algebra of observables A, and denoting the algebra of observables of the quantum system to be objectified by B, our reading of the doctrine of classical concepts (to be explained in more detail below) is simply that A be commutative. Einstein, on the other hand, insists that the system under observation has its own state, so that there must be no entangled states on the tensor product $A \otimes B$ that describes the composite system. Equivalently, every pure state on $A \otimes B$ must be a product state, so that both A and B have states that together determine the joint state of $A \otimes B$. This is the case if and only if A or B is commutative, and since B is taken to be a quantum system, it must be A (see the notes to §6.5 for details). Thus Bohr's objectification criterion turns out to coincide with Einstein's!

Thirdly, the doctrine of classical concepts describes all known applications to date of quantum theory to experimental physics; and therefore we simply have to use it if we are interested in understanding these applications. This is true for the entire range of empirically accessible energy and length scales, from molecular and condensed matter physics (including quantum computation) to high-energy physics (in colliders as well as in the context of astro-particle physics). So if people working in a field like quantum cosmology complain about the Copenhagen Interpretation then perhaps they should ask themselves if their field is more than a chimera.

Given its clear empirical relevance, it is a moot point whether the doctrine of classical concepts is as necessary as Bohr and Heisenberg claimed it was:

> 'In their attempts to formulate the general content of quantum mechanics, the representatives of the Copenhagen School often used formulations with which they do not merely say how things *are* in their opinion, but beyond that, they say that things *must* be thus and so (...) They chose formulations for the mere communication of an item in which at the same time the inevitability of what is communicated is asserted. (...) The assertion of the necessity of a proposition adds *nothing* to its content.' (Scheibe, 2001, pp. 402–403)

The doctrine of classical concepts implies in particular that the measuring apparatus is to be described classically; indeed, along with its coupling to the system undergoing measurement, it is its classical description which turns some device—which *a priori* is a quantum system like anything else—into a measuring apparatus. This point was repeated over and over by Bohr and Heisenberg, but in our view the clearest explanation of this crucial point has been given by Scheibe:

> 'It is necessary to avoid any misunderstanding of the buffer postulate [i.e., the doctrine of classical concepts], and in particular to emphasize that the requirement of a classical description of the apparatus is not designed to set up a special class of objects differing fundamentally from those which occur in a quantum phenomenon as the things examined rather than measuring apparatus. This requirement is essentially epistemological, and affects this object only *in its role as apparatus*. A physical object which may act as apparatus may in principle also be the thing examined. (...) The apparatus is governed by classical physics, the object by the quantum-mechanical formalism.' (Scheibe, 1973, p. 24–25)

Thus it is essential to the Copenhagen Interpretation that one can describe at least some quantum-mechanical devices classically: those for which this is possible include the candidate-apparatuses (i.e. measuring devices). In view of its importance for their interpretation of quantum mechanics, it is remarkable how little Bohr, Heisenberg, and their followers did to seriously address this problem of a dual description of at least part of the world, although they were clearly aware of this need:

> 'In the system to which the quantum mechanical formalism is to be applied, it is of course possible to include any intermediate auxiliary agency employed in the measuring process. Since, however, all those properties of such agencies which, according to the aim of measurements have to be compared with the corresponding properties of the object, must be described on classical lines, their quantum mechanical treatment will for this purpose be essentially equivalent with a classical description.' (Bohr, 1939, pp. 23–24; quotation taken from Camilleri & Schlosshauer, 2015, p. 79)

In defense of this alleged equivalence, we read almost circular explanations like:

> 'the necessity of basing the description of the properties and manipulation of the measuring instruments on purely classical ideas implies the neglect of all quantum effects in that description.' (Bohr, 1939, p. 19)

Since it delineates an appropriate regime, the following is slightly more informative:

> 'Incidentally, it may be remarked that the construction and the functioning of all apparatus like diaphragms and shutters, serving to define geometry and timing of the experimental arrangements, or photographic plates used for recording the localization of atomic objects, will depend on properties of materials which are themselves essentially determined by the quantum of action. Still, this circumstance is irrelevant for the study of simple atomic phenomena where, in the specification of the experimental conditions, we may to a very high degree of approximation disregard the molecular constitution of the measuring instruments.

> If only the instruments are sufficiently heavy compared with the atomic objects under investigation, we can in particular neglect the requirement of the [uncertainty] relation as regards the control of the localization in space and time of the single pieces of the apparatus relative to each other. (Bohr, 1948, pp. 315–316).

Even Heisenberg restricted himself to very general comments like:

> 'This follows mathematically from the fact that the laws of quantum theory are for the phenomena in which Planck's constant can be considered as a very small quantity, approximately identical with the classical laws. (Heisenberg, 1958, pp. 57).

Notwithstanding these vague or even circular explanations, the connection between classical and quantum mechanics was at the forefront of research in the early days of quantum theory, and even predated quantum mechanics. For example, Jammer (1966, p. 109) notes that already in 1906 Planck suggested that

> 'the classical theory can simply be characterized by the fact that the quantum of action becomes infinitesimally small.'

In fact, in the same context as Planck, namely his radiation formula, Einstein made a similar point already in 1905. Subsequently, Bohr's ***Correspondence Principle***, which originated in the context of atomic radiation, suggested an asymptotic relationship between quantum mechanics and classical electrodynamics. As such, it played a major role in the creation of quantum mechanics (Bohr, 1976, Jammer, 1966, Mehra & Rechenberg, 1982; Hendry, 1984; Darrigol, 1992), but the contemporary (and historically inaccurate) interpretation of the Correspondence Principle as the idea that all of classical physics should be a certain limiting case of quantum physics seems of much later date (cf. Landsman, 2007a; Bokulich, 2008).

Ironically, the possibility of giving a dual classical–quantum description of measurement apparatuses, though obviously crucial for the consistency of the Copenhagen Interpretation, simply seems to have been taken for granted, whereas also the more ambitious problem of explaining at least the appearance of the classical world (i.e. beyond measurement devices) from quantum theory—which is central to current research in the foundations of quantum mechanics—is not to be found in the writings of Bohr (who, after all, saw the explanation of experiments as his job).

Perhaps Heisenberg could have used the excuse that he regarded the problem as solved by his 1927 paper on the uncertainty relations; but on both technical and conceptual grounds it would have been a feeble excuse. One of the few expressions of at least some dissatisfaction with the situation from within the Copenhagen school—if phrased ever so mildly—came from Bohr's former research associate Landau:

> 'Thus quantum mechanics occupies a very unusual place among physical theories: it contains classical mechanics as a limiting case, yet at the same time it requires this limiting case for its own formulation.' (Landau & Lifshitz, 1977, p. 3)

In other words, the relationship between the (generalized) Correspondence Principle and the doctrine of classical concepts needs to be clarified, and such a clarification should hopefully also provide the key for the solution of the grander problem of deriving the classical world from quantum theory under appropriate conditions.

As a first step to this end, Bohr's conceptual ideas should be interpreted within the formalism of quantum mechanics before they can be applied to the physical world, an intermediate step Bohr himself seems to have considered superfluous:

> 'I noticed that mathematical clarity had in itself no virtue for Bohr. He feared that the formal mathematical structure would obscure the physical core of the problem, and in any case, he was convinced that a complete physical explanation should absolutely precede the mathematical formulation.' (Heisenberg, 1967, p. 98)

Fortunately, von Neumann did not return the compliment, since beyond its brilliant mathematical content, his *Mathematische Grundlagen der Quantenmechanik* from 1932 devoted considerable attention to conceptual issues. For example, he gave the most general form of the Born rule (which is the central link between experimental physics and the Hilbert space formalism), he introduced density operators for quantum statistical mechanics (which are still in use), he conceptualized projection operators as yes-no questions (paving the way for his later development of quantum logic with Birkhoff, as well as for Gleason's Theorem and the like), in his analysis of hidden variables he introduced the mathematical concept of a state that became pivotal in operator algebras (including the algebraic approach to quantum mechanics), *en passant* also preparing the ground for the theorems of Bell and Kochen & Specker (which exclude hidden variables under physically more relevant assumptions than von Neumann's), and, last but not least, his final chapter on the measurement problem formed the basis for all serious subsequent literature on this topic.

Nonetheless, much as Bohr's philosophy of quantum mechanics would benefit from a precise mathematical interpretation, von Neumann's mathematics would be more effective in physics if it were supplemented by sound conceptual moves (beyond the ones he provided himself). Killing two birds with one stone, we implement the doctrine of classical concepts in the language of operator algebras, as follows:

The physically relevant aspects of the noncommutative operator algebras of quantum-mechanical observables are only accessible through commutative algebras.

Our ***Bohrification program***, then, splits into two parts, which are distinguished by the precise relationship between a given noncommutative operator algebra A (representing the observables of some quantum system, as detailed below) and the commutative operator algebras (i.e. classical contexts) that give physical access to A.

While delineated mathematically, these two branches also reflect an unresolved conceptual disagreement between Bohr and Heisenberg about the status of classical concepts (Camilleri, 2009b). According to Bohr—haunted by his idea of Complementarity—only one classical concept (or one coherent family of classical concepts) applies to the experimental study of some quantum object at a time. If it applies, it does so exactly, and has the same meaning as in classical physics; in Bohr's view, any other meaning would be undefined. In a different experimental setup, some other classical concept may apply. Examples of such "complementary" pairs are particle versus wave (an example Bohr stopped using after a while), space-time description versus "causal description" (by which Bohr means conservation laws), and, in his later years, one "phenomenon" (i.e., an indivisible unit of a quantum object plus an experimental arrangement) against another. For example:

'My main purpose (...) is to emphasize that in the phenomena concerned we are (...) dealing with a rational discrimination between essentially different experimental arrangements and procedures which are suited either for an unambiguous use of the idea of space location, or for a legitimate application of the conservation theorem of momentum (...) which therefore in this sense may be considered as *complementary* to each other (...) Indeed we have in each experimental arrangement suited for the study of proper quantum phenomena not merely to do with an ignorance of the value of certain physical quantities, but with the impossibility of defining these quantities in an unambiguous way. (Bohr, 1935, p. 699).

Heisenberg, on the other hand, seems to have held a more relaxed attitude towards classical concepts, perhaps inspired by his famous 1925 paper on the quantum-mechanical reinterpretation (*Umdeutung*) of mechanical and kinematical relations, followed by his equally great paper from 1927 already mentioned. In the former, he introduced what we now call *quantization*, in putting the observables of classical physics (i.e. functions on phase space) on a new mathematical footing by turning them into what we now call operators (initially in the form of infinite matrices), where they also have new properties. In the latter, Heisenberg tried to find some operational meaning of these operators through measurement procedures. Since quantization applies to all classical observables at once, all classical concepts apply simultaneously, but approximately (ironically, like most research on quantum theory at the time, the 1925 paper was inspired by Bohr's Correspondence Principle).

To some extent, then, Bohr's view on classical concepts comes back mathematically in ***exact Bohrification***, which studies (unital) commutative C*-subalgebras C of a given (unital) noncommutative C*-algebra A, whereas Heisenberg's interpretation of the doctrine resurfaces in ***asymptotic Bohrification***, which involves asymptotic inclusions (more specifically, deformations) of commutative C*-algebras into noncommutative ones. So the latter might have been called *Heisenbergification* instead, but in view of both the ugliness of this word and the historical role played by Bohr's Correspondence Principle just alluded to, the given name has stuck.

The precise relationship between Bohr's and Heisenberg's views, and hence also between exact and asymptotic Bohrification, remains to be clarified; their joint existence is unproblematic, however, since the two programs complement each other.

- *Exact* Bohrification turns out to be an appropriate framework for:
  - *The Born rule* (for single case probabilities).
  - *Gleason's Theorem* (which justifies von Neumann's notion of a state as a positive linear expectation value, assuming the operator part of quantum theory).
  - *The Kochen–Specker Theorem* (excluding non-contextual hidden variables).
  - *The Kadison–Singer Conjecture* (concerning uniqueness of extensions of pure states from maximal commutative C*-subalgebras of the algebra $B(H)$ of all bounded operators on a separable Hilbert space H to $B(H)$).
  - *Wigner's Theorem* (on unitary implementation of symmetries of pure states with transition probabilities, and its analogues for other quantum structures).
  - *Quantum logic* (which, if one adheres to the doctrine of classical concepts, turns out to be intuitionistic and hence distributive, rather than orthomodular).
  - *The topos-theoretic approach* to quantum mechanics (which from our point of view encompasses quantum logic and implies the preceding claim).

- *Asymptotic* Bohrification, on the other hand, provides a mathematical setting for:
  - *The classical limit of quantum mechanics.*
  - *The Born rule* (for probabilities measured as long-run frequencies).
  - *The infinite-volume limit of quantum statistical mechanics.*
  - *Spontaneous symmetry breaking* (SSB).
  - *The Measurement Problem* (highlighted by Schrödinger's Cat).

On the philosophical side, the limiting procedures inherent in asymptotic Bohrification may be seen in the light of the (alleged) phenomenon of ***emergence***. From the philosophical literature, we have distilled two guiding thoughts which, in our opinion, should control the use of limits, idealizations, and emergence in physics and hence play a paramount role in this book. The first is ***Earman's Principle***:

> 'While idealizations are useful and, perhaps, even essential to progress in physics, a sound principle of interpretation would seem to be that no effect can be counted as a genuine physical effect if it disappears when the idealizations are removed.' (Earman, 2004, p. 191)

The second is ***Butterfield's Principle***, which in a sense is a corollary to Earman's Principle, and should be read in the light of Butterfield's own definition of emergence as 'behaviour that is novel and robust relative to some comparison class', which among other virtues removes the reduction-emergence opposition:

> "there is a weaker, yet still vivid, novel and robust behaviour that occurs before we get to the limit, i.e. for finite N. And it is this weaker behaviour which is physically real."
> (Butterfield, 2011, p. 1065)

Indeed, the link between theory and reality stands or falls with an adherence to these principles, for real materials (like a ferromagnet or a cat) are described by the *quantum* theory of *finite* systems (i.e., $\hbar > 0$ or $N < \infty$, as opposed to their idealized limiting cases $\hbar = 0$ or $N = \infty$), and yet they do display the remarkable phenomena that strictly speaking are only possible in the corresponding limit theories, like symmetry breaking, or the fact that cats are either dead or alive, as a metaphor for the fact that measurements have outcomes. This simple observation shows that any physically relevant conclusion drawn from some idealization must be foreshadowed in the underlying theory already for positive values of $\hbar$ or finite values of N.

Despite their obvious validity, it is remarkable how often idealizations violate these principles. For example, all rigorous theories of spontaneous symmetry breaking in quantum statistical mechanics (Bratteli & Robinson, 1981) and in quantum field theory (Haag, 1992) strictly apply to infinite systems only, since ground states of finite quantum systems are typically unique (and hence symmetric), whilst thermal equilibrium states of such systems are even always unique (see also Chapter 10). As explained in Chapter 11, the "Swiss" approach to the measurement problem based on superselection rules faces a similar problem, and must be discarded for that reason. Bohr's doctrine of classical concepts is particularly vulnerable to Earman's Principle, since classical physics (in whose language we are supposed to express the account of all evidence) is not realized in nature but only in the human mind, so to speak. This necessitates great care in implementing this doctrine.

Interestingly, in his famous lecture "Über das Unendliche", in which he expounded his finitary program intended to save mathematics against the devilish intuitionist challenge of L.E.J. Brouwer, Hilbert (1925) expressed similar principles controlling the use of infinite idealizations in mathematics:

> "Und so wie bei den Grenzprozessen der Infinitesimalrechnung das Unendliche im Sinne des Unendlichkleinen und des Unendlichgroßen sich als eine bloße Redensart erweisen ließ, so müssen wir auch das Unendliche im Sinne der Unendlichen Gesamtheit, wo wir es jetzt noch in den Schlußweisen vorfinden, als etwas bloß scheinbaren erkennen. Und so wie das Operieren mit dem Unendlichkleinen durch Prozesse im Endlichen ersetzt wurde, welche ganz dasselbe leisten und zu ganz denselben eleganten formalen Beziehungen führen, so müssen überhaupt die Schlußweisen mit dem Unendlichen durch endliche Prozesse ersetzt werden, die gerade dasselbe leisten, d.h. dieselben Beweisgänge und dieselben Methoden der Gewinning von Formeln und Sätzen ermöglichen." (Hilbert, 1925, p. 162).[6]

In addition, asymptotic Bohrification has three rather more technical roots:

1. A new approach to quantization theory developed in the 1970s under the name of ***deformation quantization*** (Berezin, 1975; Bayen et al, 1978), where the noncommutative algebras characteristic of quantum mechanics arise as deformations of Poisson algebras. In Rieffel's (1989, 1994) approach to deformation quantization, further developed in Landsman (1998a), the deformed algebras are C*-algebras, and hence the apparatus of operator algebras and noncommutative geometry (Connes, 1994) becomes available. Deformation quantization gives a mathematically precise and physically relevant meaning to the limit $\hbar \to 0$, and shows that quantization and the classical limit are two sides of the same coin.
2. The mathematical analysis of the BCS-model of superconductivity initiated by Bogoliubov (1958) and Haag (1962), which, in the more general setting of mean-field models of solid state physics, culminated in the work of Bona (1988, 2000), Raggio & Werner (1989), and Duffield & Werner (1992). These authors showed that in the macroscopic limit $N \to \infty$, non-commutative algebras of quantum-mechanical observables (which are typically tensor powers of matrix algebras $M_n(\mathbb{C})$) converge to some commutative algebra (typically consisting of all continuous functions on the state space of $M_n(\mathbb{C})$), at least for macroscopic averages.
3. The role of low-lying states and the ensuing instability of ground states under tiny perturbations in the two limits at hand, discovered by Jona-Lasinio, Martinelli, & Scoppola (1981) for the classical limit $\hbar \to 0$, and by Koma & Tasaki (1994) for the macroscopic limit $N \to \infty$. In combination with the previous items, this led to a new approach to the measurement problem (Landsman & Reuvers, 2013) and to spontaneous symmetry breaking and emergence (Landsman, 2013), which in particular addresses these issues in the framework of asymptotic Bohrification.

---

[6] 'Just as in the limit processes of the infinitesimal calculus, the infinite in the sense of the infinitely large and the infinitely small proved to be merely a figure of speech, so too we must realize that the infinite in the sense of an infinite totality, where we still find it in deductive methods, is an illusion. Just as operations with the infinitely small were replaced by operations with the finite which yielded exactly the same results and led to exactly the same elegant formal relationships, so in general must deductive methods based on the infinite be replaced by finite procedures which yield exactly the same results, i.e., which make possible the same chains of proofs and the same methods of getting formulas and theorems.' (Benaceraff & Putnam, 1983, p. 184).

This book is organized into two parts. Rather than following the partition of our approach into exact and asymptotic Bohrification, these parts reflect the (mathematical) sophistication of the material, starting with finite sets, and ending with a combination of C*-algebras and topos theory. Part I, called *$C_0(X)$ and $B(H)$*, gives a mathematical introduction to both classical and quantum mechanics from an operator-algebraic point of view, in which these theories are kept separate, whilst mathematical analogies are stressed whenever possible. This part emphasizes the notion of symmetry, and includes some of the main abstract mathematical results about quantum mechanics (i.e., those not involving the study of Schrödinger operators and concrete models), such as the Born rule, the theorems of Gleason and Kochen & Specker already mentioned, the one of Wigner (on symmetries) and its numerous derivatives, including a new one on unitary implementability of symmetries of the poset $\mathscr{C}(B(H))$ of unital commutative C*-subalgebras of $B(H)$, and Stone's Theorem on unitary implementability of time evolution in quantum mechanics. This part may also serve as a reference for such fundamental theorems about quantum mechanics. An unusual ingredient of this part is our discussion of the Kadison–Singer Conjecture, included because of its fit into (exact) Bohrification. Also elsewhere, results are (re)phrased in a language appropriate to this ideology.

Experts in the C*-algebraic approach to quantum mechanics will be able to read the second part independently of the first (which they might therefore skip if they find it to be too elementary), but the spirit of Bohrification will only be instilled in the reader if (s)he reads the entire book; indeed, it is this very spirit that keeps the two parts together and turns the book into a whole. Part II, entitled *Between $C_0(X)$ and $B(H)$*, starts with a survey of some known results on the grey area between classical and quantum, such as Bell's Theorem(s) and the so-called Free Will Theorem. It then embarks on the asymptotic Bohrification program, including (deformation) quantization and the classical limit (including a small excursion into indistinguishable particles), large systems and their (thermodynamic) limit, and the Born rule (revisited). This part centers on a somewhat idiosyncratic treatment of spontaneous symmetry breaking (SSB) and the closely related measurement problem of quantum mechanics, which is given an unusual but technically precise formulation in the spirit of the Copenhagen Interpretation, and hence is meant to be relevant to actual experimental physics (which is what the Copenhagen Interpretation covers).

Our treatment of both quantization and SSB relies mathematically on continuous bundles of C*-algebras, while the principles of Earman and Butterfield provide philosophical guidance. This is also true for our approach to the measurement problem, which combines elements of quantization and SSB. Although experiments and detailed theoretical models are lacking so far, this powerful combination of mathematical and philosophical tools leads to a compelling scenario for solving the measurement problem, harboring the hope of finally laying this problem to rest. Like dynamical collapse models that require modifications of quantum mechanics, our scenario looks at the wave-function realistically, and hence describes measurement as a physical process, including the collapse that settles the outcome (as opposed to reinterpretations of the uncollapsed state, as in modal or Everettian interpretations). However, in our approach collapse takes place within unitary quantum theory.

Insolubility theorems for the measurement problem are circumvented, because these rely on the counterfactual that *if* ψ_n *were* the initial state, then *for each n* it *would* evolve (linearly) according to the Schrödinger equation with *given* Hamiltonian h, whereas *if* the initial state *were* $\sum_n c_n \psi_n$, also then it *would* evolve according to the *same* Hamiltonian h. However, Butterfield's Principle implies that this counterfactual is inapplicable precisely in the measurement situations it is meant for, because the dual description of the apparatus as both classical and quantum-mechanical causes extreme sensitivity of the wave-function to even the tiniest perturbations of the Hamiltonian. Indeed, such perturbations dynamically enforce some particular outcome of the measurement. Our scenario also rejects the typical way of looking at measurement as a two-step process (going back to von Neumann himself and widely adopted in the literature ever since), i.e., of firstly a transition of a pure state to a mixed one (this is his ill-fated "process 1"), followed by the registration of a single outcome. In real measurements (like elsewhere), pure states remain pure! If our scenario is correct, the mistaken impression that quantum theory seems to imply the irreducible randomness of nature, then arises because measurement outcomes are merely unpredictable "for all practical purposes", indeed they are unpredictable in a way that dwarfs even the apparent randomness of classical chaotic systems.

The final chapter on topos theory and quantum logic elaborates on ideas originating with Isham and Butterfield. It centers on the poset $\mathscr{C}(A)$ of all unital commutative C*-subalgebras of a unital C*-algebra A, ordered by inclusion; with some goodwill, one might call $\mathscr{C}(A)$ the mathematical home of Complementarity (although the construction applies even when A itself is commutative). The power of this poset is already clear in Part I, where the special case $A = B(H)$ leads to a new version of Wigner Theorem on unitary implementability of symmetries. Hamhalter's Theorem, which is a far-reaching generalization of this version, then shows that $\mathscr{C}(A)$ carries at least as much information about A as the pure state space. Furthermore, $\mathscr{C}(A)$ enforces a (new) notion of quantum logic that turns out to be *intuitionistic* in being distributive but denying the law of the excluded middle (on which both classical logic and the non-distributive quantum logic of Birkhoff–von Neumann are based). Finally, $\mathscr{C}(A)$ gives rise to a quantum phase space (which is lacking in the usual formalism), on which observables are functions and states are probability measures, just like in classical physics (but now "internal" to a particular topos, i.e., a mathematical universe alternative to set theory, in which logic is typically intuitionistic).

About a third of the book is devoted to mathematical appendices. Those on functional analysis and operator algebras give thorough introductions to these subjects, sparing the reader the effort to study books like Bratteli & Robinson (1981), Conway (2007), Dudley (1989), Kadison & Ringrose (1983, 1986), Lance (1995), Pedersen (1989), Reed & Simon (1972), Schmüdgen (2012), and Takesaki (2002, 2003). The appendices on logic, category theory, and topos theory, on the other hand, are far from exhaustive (though self-contained): they provide a shortcut to the necessary parts of e.g. Johnstone (1987), Mac Lane (1998), and Mac Lane & Moerdijk (1992), or, alternatively, of Bell & Machover (1977) and Bell (1988). Though primarily meant to support the main body of the book, these appendices may also be of some interest by themselves, especially to philosophers, but even to mathematicians.

As a "Quick Start Guide" for readers in a hurry, we now summarize the main definitions in the theory of operator algebras. A ***C*-algebra*** is an associative algebra (over $\mathbb{C}$) equipped with an involution (i.e., a real-linear map $a \mapsto a^*$ such that

$$a^{**} = a, \ (ab)^* = b^* a^*, \ (\lambda a)^* = \overline{\lambda} a^*,$$

for all $a, b \in A$ and $\lambda \in \mathbb{C}$), as well as a norm in which A is complete (i.e., a Banach space), such that algebra, involution, and norm are related by the axioms

$$\|ab\| \leq \|a\| \|b\|;$$
$$\|a^* a\| = \|a\|^2.$$

The two main classes of C*-algebras are:

- The space $C_0(X)$ of all continuous functions $f : X \to \mathbb{C}$ that vanish at infinity (i.e., for any $\varepsilon > 0$ the set $\{x \in X \mid |f(x)| \geq \varepsilon\}$ is compact), where X is some locally compact Hausdorff space, with pointwise addition and multiplication, involution

  $$f^*(x) = \overline{f(x)},$$

  and a norm

  $$\|f\|_\infty = \sup_{x \in X}\{|f(x)|\}.$$

  It is of fundamental importance for physics and mathematics that $C_0(X)$ is *commutative*. Conversely, Gelfand & Naimark (1943) proved that every commutative C*-algebra is isomorphic to $C_0(X)$ for some locally compact Hausdorff space X, which is determined by A up to homeomorphism (X is called the *Gelfand spectrum* of A). Note that $C_0(X)$ has a unit (i.e. the function 1_X that is equal to 1 for any x) iff X is compact.
- Norm-closed subalgebras A of the space $B(H)$ of all bounded operators on some Hilbert space H for which $a^* \in A$ iff $a \in A$; this includes the case $A = B(H)$. Here one uses the standard operator norm

  $$\|a\| = \sup\{\|a\psi\|, \psi \in H, \|\psi\| = 1\},$$

  the algebraic operations are the natural ones, and the involution is the adjoint. If $\dim(H) > 1$, $B(H)$ is a *non-commutative* C*-algebra. An important special case is the C*-algebra $B_0(H)$ of all *compact* operators on H, which has no unit whenever H is infinite-dimensional (whereas $B(H)$ is always unital). In their fundamental paper, Gelfand & Naimark (1943) also proved that every C*-algebra is isomorphic to $A \subset B(H)$ for some Hilbert space space X.

These classes are related as follows: in the commutative case $A = C_0(X)$, take

$$H = L^2(X, \mu),$$

where the support of the measure μ is X, on which $C_0(X)$ acts by multiplication operators, that is, $m_f \psi = f\psi$, where $f \in C_0(X)$ and $\psi \in L^2(X, \mu)$.

As already noted, C*-algebras were introduced by Gelfand & Naimark (1943), generalizing the rings of operators studied by von Neumann during 1930–1949, partly in collaboration with Murray (von Neumann, 1930, 1931, 1938, 1940, 1949; Murray & von Neumann, 1936, 1937, 1943). These rings are now called *von Neumann algebras*, and arise as the special case where a C*-algebra $A \subset B(H)$ satisfies

$$A = A'',$$

in which for any subset $S \subset B(H)$ the *commutant* of S is defined by

$$S' = \{a \in B(H) \mid ab = ba \forall b \in S\},$$

in terms of which the *bicommutant* of S is given by $S'' = (S')'$. Equivalently, a C*-algebra is a von Neumann algebra M iff it is the dual of some Banach space M_* (which is unique, and contains the so-called *normal states* on M).

Generalizing von Neumann's concept of a state on $B(H)$, a *state* on a C*-algebra A (as first defined by Segal in 1947) is a linear map

$$\omega : A \to \mathbb{C}$$

that is *positive* in that

$$\omega(a^*a) \geq 0$$

for each $a \in A$, and *normalized* in that, noting that positivity implies boundedness,

$$\|\omega\| = 1,$$

where $\|\cdot\|$ is the usual norm on the Banach dual A^*. If A has a unit 1_A, then in the presence of positivity, the above normalization condition is equivalent to

$$\omega(1_A) = 1.$$

The Riesz–Radon representation theorem in measure theory gives a bijective correspondence between states ω on $A = C_0(X)$ and probability measures μ on X, viz.

$$\omega(f) = \int_X d\mu\, f,$$

for any $f \in C_0(X)$. At the other end of the operator-algebraic world, if $A = B(H)$, then any density operator ρ on H gives a state ω on $B(H)$ by

$$\omega(a) = \mathrm{Tr}(\rho a),$$

but if H is infinite-dimensional there are other states, which cannot be normal. Such "singular" states are the C*-algebraic analogues of improper eigenstates for eigenvalues in the continuous spectrum of some self-adjoint operator (think of position or momentum), and hence they make perfect sense physically. Singular states play an important role also mathematically, especially in the Kadison–Singer Conjecture.

Let me close this Introduction with a small personal note on the way this book came into being. Of the three disciplines relevant to the foundations of physics, namely mathematics, physics, and philosophy, my expertise has always been located within the first two, more specifically in mathematical physics. Nonetheless, my interest in the foundations of physics was triggered already at school, notably by books like *The Dancing Wu-Li Masters* by Gary Zukav, *The Tao of Physics* by Fritjof Capra (both of which may appear suspicious in hindsight), and especially by Werner Heisenberg's fascinating (though historically unreliable) autobiography *Physics and Beyond* (called *Der Teil und das Ganze* in German). The second autobiography that made a huge impression on me at the time was Bertrand Russell's, which in particular made me want to go to Cambridge and become a so-called Apostle (i.e. a member of an elitist secret conversation society that once included such illustrious members as Moore, Keynes, Hardy, and Russell himself); the first dream was eventually realized (see below), about the second I have to remain silent.

My interest in foundations was reinforced by two books on general relativity which I read as a first-year physics student, namely *Raum · Zeit · Materie* by Weyl (1918) and *The Mathematical Theory of Relativity* by Eddington (1923). Although these were beyond my grasp at the time, they were clearly written in the spirit of Newton's *Principia*, in that they were primarily treatises in natural philosophy, for which mathematical physics just provided the technical underpinning. Nonetheless, despite an unforgettable seminar by Jan Hilgevoord on the Heisenberg uncertainty relations in 1984, reporting on his recent joint work with Jos Uffink, foundations remained dormant during my undergraduate and PhD years (1981–1989).

As a postdoc in Cambridge from 1989 onwards, I initially attended all seminars in any subject related to mathematics and/or physics I found remotely interesting, including the so-called *Sigma Club*, which at the time was organized by Michael Redhead. Michael was surrounded by a group of people I began to increasingly like, although I was and still am worried by their deification of John Bell (one speaker even asked his audience to stand whilst he was reading a passage from *Speakable and Unspeakable in Quantum Mechanics*). In any case, I was very kindly invited to speak at the Sigma Club on my recent paper on superselection rules and the measurement problem (whose approach I now eschew, since it violates Earman's Principle, see above as well as Chapter 11 below), followed by a private dinner in the posh Riverside Restaurant with Michael (who asked my opinion about David Lewis, whom I unfortunately had never heard of). Indeed, the generosity of inviting an absolute beginner in the philosophy of physics to speak in such a prestigious seminar endeared me even further to both the subject and the community.

My main business remained mathematical physics, but, reinforcing the earlier spark I had got from reading Weyl and Eddington (and later also from von Neumann as well as Newton), two people (unfortunately no longer with us) made it clear to me that the goal of this discipline may include not only mathematics and physics, but also foundations, i.e., natural philosophy. These were Rob Clifton, who was a PhD student of Redhead and Butterfield, and Rudolf Haag, in whose group I had the honour to work during my year at Hamburg (1993-1994) as an Alexander von Humboldt Fellow (this was Haag's last active year at the university, cf. Haag, 2010).

My first book in 1998, which I wrote during my last two years at Cambridge, when the prospect of having to leave Academia and hence the urge to leave a permanent record loomed large, did not yet reflect this attitude. But my lengthy article on the classical-quantum interface in the *Handbook of the Philosophy of Physics* edited by Butterfield and Earman already did, and so does the present book.

There is an inherent danger in a mathematical physics approach to foundations:

> 'I'm guided by the beauty of our weapons' (Leonard Cohen)

Our mathematical weapons, that is; this book is predicated on the idea that operator algebras provide the right language for quantum theory. If they don't—for example, if path integrals are really its essence, as researchers especially in quantum gravity seem to believe, and there turns out to be a difference between the two toolkits—the mathematical underpinning of Bohrification would fall. Since our conceptual program is closely linked to this mathematical language, it would presumably collapse, too. Even if operator algebras stand, once some noncommutative alien gets direct access to the quantum world in defiance of Bohr's doctrine of classical concepts, the conceptual framework behind Bohrification (and with it much of this book) would tremble. So far there has been no evidence for any of this, and as long as physics remains an empirical science I offer this book to the reader both as an introduction to modern mathematical methods in physics (in so far as these are relevant to foundational questions), and also as an alternative to various interpretations of quantum mechanics that seem to philosophize the physics of the problems away.

Notes

Each chapter is followed by a section called *Notes*, in which background and credits for the results in the given chapter are given. Such information is therefore absent in the main text (expect when—typically famous—theorems are named after their discoverers, like Gleason, Wigner, and the like). This Introduction, which anomalously contains some references, is an exception, but we still provide some notes to it.

Since this book is not an exegesis of Bohr but rather an exposition of some mathematical ideas partly inspired by his work (with no claim to retroactive endorsement by Bohr or his followers), we hardly relied on the secondary literature on his philosophy, except, as already mentioned, on Scheibe (1973) and Beller (1999), both of which are pretty critical of Bohr. For a more balanced picture, one might consult monographs like Folse (1985), Murdoch (1987), McEvoy (2001), Brock (2003), the collection of essays edited by Faye & Folse (2017), as well as Dieks (2016a) and Zinkernagel (2016). Secondary literature on Heisenberg's philosophy of physics is scarce, but includes Camilleri (2009b). Though irrelevant to the present book, one cannot resist mentioning Landsman (2002) on Heisenberg's controversial political war record, from which he tried to escape by writing the intriguing essay *Ordnung der Wirklichkeit*, published 50 years later as Heisenberg (1994).

A propos, notes on von Neumann and operator algebras follow §C.25.

Strictly speaking, no previous knowledge of quantum mechanics is needed to understand this book, but it is hard to imagine readers of this book without such a background. Beyond standard undergraduate physics courses, for mathematically serious introductions to quantum mechanics—further to von Neumann (1932), which founded the subject—we recommend Bongaarts (2015), Gustafson & Sigal (2003), Hall (2013), Takhtajan (2008), and Thirring (2002). No previous acquaintance with the philosophy of quantum theory is required either, but once again it might be expected that typical readers of the present book have at least some awareness of this field. In fact, the author himself has only read a few such books from cover to cover, including Heisenberg (1958), Jammer (1966, 1974), Scheibe (1973), Earman (1986), van Fraassen (1991), Bub (1997), Beller (1999), and Wallace (2012).

From these books, apart from its obvious source Heisenberg (1958), Bohrification (at least in its 'exact' variant) is conceptually akin to the program of Bub (1997), which was based on Clifton & Bub (1996); the past tense seems appropriate here, since Bub has meanwhile abandoned this program in favour of foundations based on information theory (Bub, 2004). Anyway, given some preferred observable $a \in B(H)_{\mathrm{sa}}$ and pure state $e \in \mathscr{P}_1(H)$ (i.e., a one-dimensional projection on H), the Bub–Clifton approach looks for the largest C*-subalgebra A of $B(H)$ on which one may define something like a hidden variable compatible with the Born probabilities emanating from the *given* state e (the emphasis on some given e comes form the modal interpretation(s) of quantum mechanics). For generic states e and observables a, this typically allows A to be noncommutative, which blasts the conceptual framework of exact Bohrification. Requiring compatibility with quantum mechanics for *arbitrary* states e, on the other hand, would force A to be commutative. All this relates to the Kochen–Specker Theorem; see the Notes to §6.1 for further details.

Finally, though remote from Wallace (2012) in our attempt to solve (or, in the light of the first quotation below, one should say "address") the measurement problem through physics rather than philosophy, even with this polar opposite author we share the following attitude towards the foundations of quantum mechanics:

> 'The basic thesis of this book is that there is no quantum measurement problem (...) What I mean is that there is actually no conflict between the dynamics and ontology of (unitary) quantum theory and our empirical observations. (...) [I do not] wish to be read as offering yet one more "interpretation of quantum mechanics".
>
> This book takes an extremely conservative approach to quantum mechanics (...) quantum mechanics can be taken literally (...) there is just unitary quantum mechanics.
>
> The way in which cats or tables exist is as structures within the underlying microphysics (...) [they are] emergent objects, higher-order entities.' (Wallace, 2012, pp. 1, 2, 13, 38, 40)

But although it may indeed apply to the town of Oxford, one might take issue with:

> 'It is simply false that there are alternative explanatory theories to Everett-interpreted quantum mechanics which can reproduce the predictions of quantum theory (...) The Everett interpretation is the only game in town.' (Wallace, 2012, p. 43)

Part I
$C_0(X)$ and $B(H)$

Chapter 1
Classical physics on a finite phase space

Throughout this chapter, X is a *finite set*, playing the role of the configuration space of some physical system, or, equivalently (as we shall see), of its pure state space (in the continuous case, X will be the phase space rather than the configuration space). One should not frown upon finite sets: for example, the configuration space of N bits is given by $X = \underline{2}^{\underline{N}}$, where for arbitrary sets Y and Z, the set Y^Z consists of all functions $x : Z \to Y$, and for any $N \in \mathbb{N}$ we write $\underline{N} = \{1,2,\ldots,N\}$ (although, following the computer scientists, $\underline{2}$ usually denotes $\{0,1\}$). More generally, if one has a lattice $\Lambda \subset \mathbb{Z}^d$ and each site is the home of some classical object (say a "spin") that may assume N different configurations, then $X = \underline{N}^{\Lambda}$, in that $x : \Lambda \to \underline{N}$ describes the configuration in which the "spin" at site $\mathbf{n} \in \Lambda$ takes the value $x(\mathbf{n}) \in \underline{N}$.

Although the setting is *a priori* deterministic, in that (knowing) some point $x \in X$ in its guise as a pure state at least in principle determines everything (there is to say), the mathematical language will be probabilistic. Even within the confines of classicality this allows one to do statistical physics, and as such it also sheds light on e.g. the special status of x as an extreme probability measure (see below). Furthermore, the use of this language may be motivated by the goal of describing classical and quantum mechanics as analogously as possible at this elementary level.

The following concepts play a central role in this chapter. Recall that the power set $\mathscr{P}(X)$ of X is the set of all subsets of X (for finite X, these are all measurable).

Definition 1.1. *1. An* **event** *is a subset $U \subseteq X$, i.e., $U \in \mathscr{P}(X)$.*
2. *A* **probability distribution** *on X is a function $p : X \to [0,1]$ such that $\sum_x p(x) = 1$.*
3. *A* **probability measure** *on X is a function $P : \mathscr{P}(X) \to [0,1]$ such that $P(X) = 1$ and $P(U \cup V) = P(U) + P(V)$ whenever $U \cap V = \emptyset$.*
4. *For a given probability measure P on X, and an event $V \subseteq X$ such that $P(V) > 0$, the* **conditional probability** *$P(U|V)$ of U given V is defined by*

$$P(U|V) = \frac{P(U \cap V)}{P(V)}. \tag{1.1}$$

5. *A* **random variable** *on X is a function $f : X \to \mathbb{R}$.*
6. *The* **spectrum** *of a random variable f is the subset $\sigma(f) = \{f(x) \mid x \in X\}$ of $\mathbb{R}$.*

K. Landsman, *Foundations of Quantum Theory*,
Fundamental Theories of Physics 188, DOI 10.1007/978-3-319-51777-3_1

1.1 Basic constructions of probability theory

Probability distributions p and probability measures P determine each other by

$$P(U) = \sum_{x\in U} p(x); \tag{1.2}$$

$$p(x) = P(\{x\}), \tag{1.3}$$

but this is peculiar to finite sets (in general, probability *measures* will be primary). Two special classes of probability measures and of random variables stand out:

- Each $y \in X$ defines a probability distribution p_y by $p_y(x) = \delta_{xy}$, or explicitly $p_y(x) = 1$ if $x = y$ and $p_y(x) = 0$ if $x \neq y$; for the corresponding probability measure one has $P_y(U) = 1$ if $y \in U$ and $P_y(U) = 0$ if $y \notin U$.
- Each event $U \subset X$ defines a random variable 1_U (i.e., the ***characteristic function*** of U) by $1_U(x) = 1$ if $x \in U$ and $1_U(x) = 0$ if $x \notin U$. Clearly, $\sigma(1_U) = \{0\}$ when $U = \emptyset$, $\sigma(1_U) = \{1\}$ when $U = X$, and $\sigma(1_U) = \{0,1\}$ otherwise. Note that $1_U(x) = P_x(U)$. Conversely, any random variable f with spectrum $\sigma(f) \subseteq \{0,1\}$ is given by $f = 1_U$ for some $U \subseteq X$; just take $U = \{x \in X \mid f(x) = 1\}$. Such functions may be construed as yes-no questions to the system (i.e. $f = 1$ versus $f = 0$) and will lie at the basis of the logical interpretation of the theory (cf. §1.4).

The single most important construction in probability theory is as follows.

Theorem 1.2. *A probability distribution p on X and a random variable $f : X \to \mathbb{R}$ jointly yield a probability distribution p_f on the spectrum $\sigma(f)$ by means of*

$$p_f(\lambda) = \sum_{x\in X \mid f(x)=\lambda} p(x). \tag{1.4}$$

In terms of the corresponding probability measure P on X, one has

$$p_f(\lambda) = P(f = \lambda), \tag{1.5}$$

where $f = \lambda$ denotes the event $\{x \in X \mid f(x) = \lambda\}$ in X. Similarly, the probability measure P_f on $\sigma(f)$ corresponding to the probability distribution p_f is given by

$$P_f(\Delta) = P(f \in \Delta), \tag{1.6}$$

where $\Delta \subseteq \sigma(f)$ and $f \in \Delta$ denotes the event $\{x \in X \mid f(x) \in \Delta\}$ in X.

The proof is trivial. Instead of $f = \lambda$, the notation $f^{-1}(\{\lambda\})$ might be used, and similarly, $f^{-1}(\Delta)$ is the same as $f \in \Delta$. If $\lambda \in \sigma(f)$ is non-degenerate in that there is exactly one $x_\lambda \in X$ such that $f(x_\lambda) = \lambda$, then one simply has $P(f = \lambda) = p(x_\lambda)$.

For example, combining both our special cases $P = P_y$ and $f = 1_U$ above yields

$$P_y(1_U = 1) = 1 \text{ and } P_y(1_U = 0) = 0 \text{ if } y \in U; \tag{1.7}$$

$$P_y(1_U = 1) = 0 \text{ and } P_y(1_U = 0) = 1 \text{ if } y \notin U. \tag{1.8}$$

Given some probability measure P, the ***expectation value*** $E_P(f)$ and the ***variance*** $\Delta_P(f)$ of a random variable f with respect to P are defined by, respectively,

$$E_P(f) = \sum_{x\in X} f(x)p(x); \tag{1.9}$$

$$\Delta_P(f) = E_P(f^2) - E_P(f)^2. \tag{1.10}$$

A simple calculation shows that E_P may be written directly in terms of P itself as

$$E_P(f) = \sum_{\lambda\in\sigma(f)} P(f=\lambda)\cdot\lambda. \tag{1.11}$$

Note that $\Delta_P(f) \geq 0$. The special role of the point measures P_y may now be clarified:

Proposition 1.3. *A probability measure P takes the form $P = P_y$ for some $y \in X$ iff $\Delta_P(f) = 0$ for all random variables $f : X \to \mathbb{R}$.*

Proof. For "$\Rightarrow$", we compute $E_{P_y}(f) = f(y)$, and hence $E_{P_y}(f^2) = f(y)^2$. In the opposite direction, take $f = p_y$, so that $f^2 = f$ and hence $\Delta_P(f) = p(y) - p(y)^2$. The assumption $\Delta_P(f) = 0$ for each f implies that either $p(y) = 0$ or $p(y) = 1$ for each $y \in X$. Definition 1.1.2 then implies that $p(y) = 1$ for exactly one $y \in X$. □

More generally, a collection $f_1,\dots,f_n$ of n random variables and a (single) probability distribution p on X jointly define a probability distribution $p_{f_1,\dots,f_n}$ on the product $\sigma(f_1)\times\cdots\times\sigma(f_n)$ of the individual spectra by

$$p_{f_1\cdots f_n}(\lambda_1,\dots,\lambda_n) = \sum_{x\in X\,|\,f_1(x)=\lambda_1,\dots,f_n(x)=\lambda_n} p(x). \tag{1.12}$$

Once again, this may be rewritten as

$$p_{f_1\cdots f_n}(\lambda_1,\dots,\lambda_n) = P(f_1=\lambda_1,\dots,f_n=\lambda_n), \tag{1.13}$$

where the argument of P denotes the intersection $\cap_{k=1}^n (f_k = \lambda_k)$, i.e.,

$$P(f_1=\lambda_1,\dots,f_n=\lambda_n) = \{x\in X \mid f_1(x)=\lambda_1,\dots,f_n(x)=\lambda_n\}. \tag{1.14}$$

Simple calculations then yield results for the so-called ***marginal distributions***, like

$$\sum_{\lambda_{l+1}\in\sigma(f_{l+1}),\dots,\lambda_n\in\sigma(f_n)} P(f_1=\lambda_1,\dots,f_n=\lambda_n) = P(f_1=\lambda_1,\dots,f_l=\lambda_l), \tag{1.15}$$

where $1 \leq l < n$. The above constructions also apply to the corresponding conditional probabilities: given m additional random variables $a_1,\dots,a_m$, one has

$$\sum_{\lambda_{l+1}\in\sigma(f_{l+1}),\dots,\lambda_n\in\sigma(f_n)} P(f_1=\lambda_1,\dots,f_n=\lambda_n|a_1=\alpha_1,\dots a_m=\alpha_m) \tag{1.16}$$

$$= P(f_1=\lambda_1,\dots,f_l=\lambda_l|a_1=\alpha_1,\dots a_m=\alpha_m). \tag{1.17}$$

1.2 Classical observables and states

Given a finite set X, we may form the set $C(X)$ of all complex-valued functions on X, enriched with the structure of a complex vector space under pointwise operations:

$$(\lambda \cdot f)(x) = \lambda f(x) \; (\lambda \in \mathbb{C}); \tag{1.18}$$
$$(f+g)(x) = f(x) + g(x). \tag{1.19}$$

We use the notation $C(X)$ with some foresight, anticipating the case where X is no longer finite, but in any case, since for the moment it is, every function is continuous. Moreover, the vector space structure on $C(X)$ may be extended to that of a commutative algebra (where, by convention, all our algebras are associative and are defined over the complex scalars) by defining multiplication pointwisely, too:

$$(f \cdot g)(x) = f(x)g(x). \tag{1.20}$$

Note that this algebra has a unit 1_X, i.e., the function identically equal to 1.

For finite X, this structure suffices for X to be recovered from $C(X)$, as follows.

Definition 1.4. *The* **Gelfand spectrum** $\Sigma(A)$ *of a (complex) algebra A is the set of all nonzero linear maps* $\omega : A \to \mathbb{C}$ *that satisfy* $\omega(fg) = \omega(f)\omega(g)$.

These are, of course, precisely the nonzero algebra homomorphisms from A to $\mathbb{C}$.

Proposition 1.5. *The Gelfand spectrum* $\Sigma(C(X))$ *is isomorphic (as a set) to* X.

Proof. Each $x \in X$ defines a map $\omega_x : C(X) \to \mathbb{C}$ by $\omega_x(f) = f(x)$. One obviously has $\omega_x \in \Sigma(C(X))$, so we have a map $X \to \Sigma(C(X))$, $x \mapsto \omega_x$. We show that this map is a bijection. Injectivity is easy: if $\omega_x = \omega_y$, then $f(x) = f(y)$ for each $f \in C(X)$, so taking $f = \delta_z$ for each $z \in X$ gives $x = y$ (here $\delta_z(x) = \delta_{xz}$). To prove surjectivity, we note that since $C(X)$ is finite-dimensional as a vector space, with basis $(\delta_y)_{y\in X}$, each linear functional $\omega : C(X) \to \mathbb{C}$ takes the form

$$\omega(f) = \sum_x \mu(x) f(x), \tag{1.21}$$

for some function $\mu : X \to \mathbb{C}$. For $\omega \in \Sigma(C(X))$, find some $z \in X$ for which $\mu(z) \neq 0$ (this has to exist, as $\omega \neq 0$). For arbitrary $w \in X$, imposing $\omega(\delta_w \delta_z) = \omega(\delta_w)\omega(\delta_z)$ enforces $\mu = \delta_z$ (which also shows that z is unique), and hence $\omega = \omega_z$. □

The physically relevant set $R(X)$ of all real-valued functions on X is obviously a real vector space inside $C(X)$. To recover it algebraically, we equip $C(X)$ with an ***involution***, which on an arbitrary (not necessarily commutative) algebra A is defined as an anti-linear anti-homomorphism that squares to id_A, i.e., a linear map $* : A \to A$ (written $a \mapsto a^*$) that satisfies $(\lambda a)^* = \overline{\lambda} a^*$, $(ab)^* = b^* a^*$, and $a^{**} = a$. In our case $A = C(X)$, which is commutative, the latter property simply becomes $(fg)^* = f^* g^*$. In any case, we define this involution by pointwise complex conjugation, i.e.,

$$f^*(x) = \overline{f(x)}. \tag{1.22}$$

We evidently recover the real-valued functions in the involutive algebra $C(X)$ as

$$R(X) \equiv C(X)_{\mathrm{sa}} = \{f \in C(X) \mid f^* = f\}. \tag{1.23}$$

Finally, although we do not need this yet, we note that $C(X)$ has a natural ***norm***

$$\|f\|_\infty = \sup_{x \in X}\{|f(x)|\}. \tag{1.24}$$

These structures turn $C(X)$ into a ***commutative C*-algebra*** (cf. Definition C.1).

Definition 1.6. *The* **algebra of observables** *of the physical system described by the phase space X is $C(X)$, seen as a (commutative)* C*-algebra *in the above way.*

Thence elements of $C(X)$ are called ***observables*** (a term that really should be applied only to its self-adjoint elements, i.e., those satisfying $f^* = f$).

We have thus equipped the *random variables* on X with enough structure to recover X itself, and now turn to the other side of the coin, viz. the *probability measures* on X. Here the relevant mathematical structure is that of a *compact convex set*, a concept we only need to define in the context of an ambient (real) vector space.

Definition 1.7. *A subset K of a (real or complex) vector space V is called* **convex** *if the straight line segment between any two points on K lies in K. Expressed formally, this means that whenever $v,w \in K$ and $t \in (0,1)$, one has $tv + (1-t)w \in K$.*

The following probabilistic reformulation of this notion is very useful.

Proposition 1.8. *A set $K \subset V$ is convex iff for any k, given k probabilities $(t_1,\dots,t_k)$ (i.e., $t_i \geq 0$ and $\sum_i t_i = 1$) and k points $(v_1,\dots,v_k)$ in K, one has $\sum_{i=1}^k t_i \cdot v_i \in K$.*

Proof. Taking $k = 2$ recovers Definition 1.7 from its probabilistic version. Conversely, one uses induction on k, using the identity (assuming $0 < t_k < 1$):

$$t_1 v_1 + \cdots + t_k v_k = (1-t_k)\left(\frac{t_1}{1-t_k}v_1 + \cdots + \frac{t_{k-1}}{1-t_k}v_{k-1}\right) + t_k v_k. \qquad \square$$

Any linear subspace of V is trivially convex, as is any translate thereof (i.e., any ***affine*** subspace of V). Another, much more important example is the ***convex hull*** $\mathrm{co}(S)$ of any subset $S \subset V$; noting that the intersection of any family of convex sets is again convex, $\mathrm{co}(S)$ may be defined as the intersection of all convex subsets of V that contain S, or, equivalently, as the smallest convex subset of V that contains S (whose existence is guaranteed by the previous remark). Proposition 1.8 then yields

$$\mathrm{co}(S) = \left\{ \sum_{i=1}^k t_i \cdot v_i \mid k \in \mathbb{N}, (v_1,\dots,v_k) \in S^k, t_i \geq 0, \sum_i t_i = 1 \right\}. \tag{1.25}$$

In particular, if $S = \{v_1,\dots,v_k\}$ is a finite set, then one simply has

$$\mathrm{co}(\{v_1,\dots,v_k\}) = \left\{ \sum_{i=1}^k t_i \cdot v_i \mid t_i \geq 0, \sum_i t_i = 1 \right\}. \tag{1.26}$$

The convex hull of any finite set of points in $\mathbb{R}^{n+1}$ is called a ***convex polytope***. Such convex sets are closed and bounded (since none of the $t_i \geq 0$ can walk away too far without violating the condition $\sum_i t_i = 1$), and hence are compact. In particular,

$$\Delta_n = \{x \in \mathbb{R}^{n+1} \mid x_i \geq 0, \sum_i x_i = 1\} \tag{1.27}$$

is a convex polytope called a ***simplex***. For example, Δ_1 is the line segment from $(0,1)$ to $(1,0)$ in $\mathbb{R}^2$. We would like to say that Δ_1 is "isomorphic" to the unit interval $[0,1]$, so we define two convex sets K_1, K_2 to be ***isomorphic*** (as such) if there is a bijection $f : K_1 \to K_2$ that is ***affine***, in that for $t \in (0,1)$ and $v_1, v_2 \in K_1$, we have

$$f(tv_1 + (1-t)v_2) = tf(v_1) + (1-t)f(v_2). \tag{1.28}$$

Then the function $f : \Delta_1 \to [0,1]$ given by $f(\lambda, 1-\lambda) = \lambda$, where $\lambda \in [0,1]$, will do. Similarly, $\Delta_2 \subset \mathbb{R}^3$ is isomorphic to any equilateral triangle in $\mathbb{R}^2$ with sides of unit length, whereas Δ_3 is just the tetrahedron (which is one of the five Platonic solids).

There are many other convex polytopes (cf. §B.11), but simplices are of prime importance for us, since Δ_n is isomorphic to the set $\Pr(X)$ of all probability distributions on a set $X = \{0, \ldots, n\}$ with $n+1$ points; the identification $\Pr(X) \ni p \leftrightarrow x \in \Delta_n$ is given by $x_i = p(i+1)$. In particular, we see that for any finite set X, $\Pr(X)$ is a compact convex set. This is also clear from Definitions 1.1 and 1.7 (and will even be true for general compact phase spaces X, cf. Corollary B.17 and §C.25).

Definition 1.9. *The* **state space** *of the physical system described by a (finite) space X is the set* $\Pr(X)$ *of all probability measures on X (or, equivalently, of all probability distributions on X), seen as a* compact convex set.

Thus a probability measure (or distribution) on X is often called a ***state*** (of the physical system described by X). The operation of passing from states $P, Q \in \Pr(X)$ to a new state $tP + (1-t)Q \in \Pr(X)$, where $t \in (0,1)$ as usual, or, more generally, from a (finite) family of states (P_i) and a set (t_i) of probabilities (i.e., $t_i \geq 0$ and $\sum_i t_i = 1$) to the convex sum $\sum_i t_i P_i$, is called ***mixing***.

It is possible to recover X from its associated state space $\Pr(X)$, as follows.

Definition 1.10. *The* **(extreme) boundary** $\partial_e K$ *of a convex set K consists of all points $v \in K$ satisfying the following condition:*

if $v = tw + (1-t)x$ for certain $w, x \in K$ and $t \in (0,1)$, then $v = w = x$.

Elements $v \in \partial_e K$ of the boundary are called **extreme points** *of K.*

We will now compute the boundary of $\Pr(X)$. The result may be expressed by

$$\partial_e \Delta_n = \{\mathbf{e}_1, \ldots .\mathbf{e}_{n+1}\}, \tag{1.29}$$

where $(\mathbf{e}_1, \ldots .\mathbf{e}_{n+1})$ is the standard basis of $\mathbb{R}^{n+1}$ (i.e., $\mathbf{e}_1 = (1,0,\ldots,0)$, etc.). However, we will give a direct probabilistic proof. We already noted the special probability measures P_x, $x \in X$. The association $x \mapsto P_x$ defines a map from X to $\Pr(X)$.

Proposition 1.11. *The set X is isomorphic to the boundary $\partial_e \Pr(X)$ through $x \mapsto P_x$.*

Proof. It is convenient to work with probability distributions p rather than probability measures P. First, $x \mapsto p_x$ is trivially injective from X to $\Pr(X)$: if $x \neq y$ then $p_x(x) = 1$ whereas $p_y(x) = 0$, so $p_x \neq p_y$. Second, $p_x \in \partial_e \Pr(X)$. For suppose one has $p_x = tp + (1-t)q$ for some $p, q \in \Pr(X)$ and $t \in (0,1)$. Hence $p_x(y) = tp(y) + (1-t)q(y)$. Taking $y \neq x$ yields $p(y) = q(y) = 0$, so that $p = q = p_x$. Consequently, $X \subseteq \partial_e \Pr(X)$.

The converse inclusion is (contrapositively) equivalent to the property that for any $p \neq p_x$ (for all x), there are q and r, $q \neq r$, and $t \in (0,1)$, with $p = tq + (1-t)r$. Indeed, if $p \neq p_x$, there is some $x_0 \in X$ with $0 < p(x_0) < 1$. Now define q, r, and t by $q(x_0) = 1$ and $q(x) = 0$ for all $x \neq x_0$, $r(x_0) = 0$ and $r(x) = p(x)/(1 - p(x_0))$, and finally $t = p(x_0)$. Then $p = tq + (1-t)r$ and $q \neq r$. □

The simplest example would be $X = \{0,1\}$, so that $\Pr(X) \cong [0,1]$ by mapping the distribution $p \in \Pr(X)$ to $p(1)$. Since one may directly verify that $\partial_e[0,1] = \{0,1\}$, under the above isomorphism one therefore has $\partial_e \Pr(X) \cong \{0,1\}$. Analogously, $\partial_e(0,1) = \emptyset$, so that the boundary of a convex set may apparently be empty. Hence we see that one remarkable ingredient of Proposition 1.11 lies in the claim that the convex set $\Pr(X)$ actually *has* a (nonempty) boundary! This is no accident: by the Krein-Milman Theorem (cf. §B.10), this is true for any *compact* convex set (which is consistent with the counterexample just given). For example in quantum mechanics we will encounter the case of $K = B^3$ (i.e. the closed unit ball in $\mathbb{R}^3$) as the state space of a qubit, whose (extreme) boundary is the two-sphere S^2, cf. Proposition 2.9. Something similar is true in any dimension, but beware of surprises: if $K = \Delta_2$ is an equilateral triangle in the plane, then its *extreme* boundary $\partial_e K$ consists of the *vertices* of K (whereas its *faces* form the *geometric* boundary of the triangle).

The general problem arises whether some point $v \in K$ of a compact convex set K may be written as a convex sum (or, more generally, an integral) of extreme points of K, and if so, to what extent this ***extremal decomposition***

$$v = \sum_{i \in I} t_i v_i, \; t_i \geq 0, \; \sum_i t_i = 1, \; v_i \in \partial_e K, \tag{1.30}$$

which for simplicity has been assumed to be a finite sum here, is unique. Without proof, we state a general result of convexity theory, called ***Caratheodory's Theorem***:

Theorem 1.12. *If K is a nonempty compact convex subset of $\mathbb{R}^n$, then $\partial_e K \neq \emptyset$, and each point of K is a convex sum of at most $n+1$ points in $\partial_e K$.*

If $K = \Delta_n$, then this sum generically has $n+1$ points and is unique. Probabilistically:

Proposition 1.13. *If X is finite, then any probability measure $P \in \Pr(X)$ may be written in a unique way as a finite mixture of extreme probability measures, viz.*

$$P = \sum_{x \in X} t_x P_x. \tag{1.31}$$

Proof. Take $t_x = P(\{x\})$ in the sense of Definition 1.1, or, equivalently, $t_x = E_P(\delta_x)$ in the sense of (1.9). To see that this decomposition is unique, use Proposition 1.11, i.e. $\partial_e \mathrm{Pr}(X) \cong X$, in (1.30) to force $I = X$ and apply both sides of (1.31) to δ_x. □

The state space and the algebra of observables may also be defined in terms of each other. We start with the (re)construction of states from observables, where the following definition and proposition may leave a hybrid impression. The rationale behind our approach is that for many purposes it is easier to work with the *complex* algebra $C(X)$, but on the other hand, compact convex sets are most naturally defined in terms of *real* vector spaces. Fortunately, it is easy to switch between the two: we already know how to obtain the real part $R(X)$ from $C(X)$, see (1.23), and conversely, $C(X)$ is simply the complexification of the real vector space $R(X)$.

Definition 1.14. *A* **state** *on $C(X)$ is a linear map $\omega : C(X) \to \mathbb{C}$ that satisfies:*

1. *$\omega(f^2) \geq 0$ for each $f \in C(X)$ with $f^* = f$ (**positivity**);*
2. *$\omega(1_X) = 1$ (**normalization**).*

The first condition obviously comes down to $\omega(f) \geq 0$ whenever $f \geq 0$ pointwise.

Equivalently, we may define a state on $R(X)$ as a real-linear map $\omega_{\mathbb{R}} : R(X) \to \mathbb{R}$ that satisfies the very same conditions. Indeed, a state $\omega_{\mathbb{R}}$ on $R(X)$ defines a complex-linear map $\omega : C(X) \to \mathbb{C}$ by $\omega(f + ig) = \omega_{\mathbb{R}}(f) + i\omega_{\mathbb{R}}(g)$, where $f, g \in R(X)$. This map satisfies the same conditions of positivity and normalization. Conversely, ω may be restricted to the real part $R(X)$ of $C(X)$, so that there is no real (sic) difference between ω and $\omega_{\mathbb{R}}$. Hence we will use these interchangeably, often even dropping the suffix $\mathbb{R}$ on ω. One advantage of this ability to switch is that a state ω on $C(X)$ may be regarded as an element of the *real* vector space $R(X)^*$. Doing so shows that the terminology of Definitions 1.9 and 1.14 is consistent:

Theorem 1.15. *There is a bijective correspondence between states ω on $C(X)$ and probability measures P on X, given by $\omega \leftrightarrow E_P$, cf.* (1.9) *and* (1.11). *Therefore, as a subset of the (real) vector space $R(X)^*$ of all (real-) linear maps from $R(X)$ to $\mathbb{R}$, the set $S(C(X))$ of all states on $C(X)$ coincides with the set* $\mathrm{Pr}(X)$ *of all probability measures on X. In particular, the state space $S(C(X))$ of $C(X)$ is a compact convex set in $R(X)^*$ (as a finite-dimensional vector space with its usual topology).*

Proof. Given a state ω, define a function $p : X \to \mathbb{R}$ by $p(x) = \omega(\delta_x)$. Since $\delta_x \geq 0$ pointwise, positivity of ω yields $p(x) \geq 0$. Noting that $1_X = \sum_x \delta_x$, normalization then forces $\sum_x p(x) = 1$, so that p is a probability distribution on X. Hence $P \in \mathrm{Pr}(X)$, where P is the probability measure corresponding to p. Conversely, $P \in \mathrm{Pr}(X)$ defines a map $E_P : R(X) \to \mathbb{R}$ by (1.9), which is positive and normalized. Note that compactness and convexity of the set $S(C(X))$ in $R(X)^*$ follow directly from its definition, i.e., even without knowing that it equals $\mathrm{Pr}(X)$. □

Consequently, we may refer to $S(C(X))$ as *the* ***state space*** of $C(X)$ without any ambiguity, and we will always regard state spaces of (unital) C*-algebras A (cf. Appendix C) as compact convex sets $S(A)$, where in the present case $A = C(X)$.

1.3 Pure states and transition probabilities

For any C*-algebra A (with unit), and hence in particular for $A = C(X)$, elements of the boundary $\partial_e S(A)$ are called ***pure states***, and we call

$$P(A) \equiv \partial_e S(A) \tag{1.32}$$

the ***pure state space*** of A. States that are not pure are called ***mixed***.

Theorem 1.16. *One has $P(C(X)) \cong X$, in that the following map is an isomorphism:*

$$X \to P(C(X)),\ x \mapsto \omega_x,\ \omega_x(f) = f(x). \tag{1.33}$$

Proof. Combine Proposition 1.11 and Theorem 1.15. □

For finite X this isomorphism is merely meant as a bijection between sets (and for general compact Hausdorff spaces X it will be a homeomorphism of topological spaces), but we will now introduce some additional structure on pure state spaces that will enrich Theorem 1.16 to an isomorphism of so-called ***sets with a transition probability***. This will be necessary in order to reconstruct the observables from the pure states, but it also clarifies the general probabilistic structure of physics (note that the following definition is unusual in probability theory!).

Definition 1.17. *1. A* **transition probability** *on a set X is a function*

$$\tau : X \times X \to [0,1] \tag{1.34}$$

that satisfies $\tau(x,y) = 1$ iff $x = y$ and $\tau(x,y) = \tau(y,x)$ (**symmetry**).

The simplest example of a transition probability (on any set X) is obviously

$$\tau(x,y) = \delta_{xy}. \tag{1.35}$$

The point is that this transition probability may be derived from the classical C*-algebra of observables $C(X)$ by the following formula (assuming X finite):

$$\delta_{xy} = \inf\{f(x) \mid f \in C(X), 0 \leq f \leq 1_X, f(y) = 1\}. \tag{1.36}$$

Indeed, for $x = y$ this is a tautology, whereas for $x \neq y$ the infimum (which is zero) is attained by $f = \delta_y$. In terms of the pure state space $P(C(X))$, which is *isomorphic* to but not *equal* to X, cf. Theorem 1.16, this formula may be written as

$$\delta_{xy} = \inf\{\omega_x(f) \mid f \in C(X), 0 \leq f \leq 1_{C(X)}, \omega_y(f) = 1\}. \tag{1.37}$$

Furthermore (and this is the *real* point, so that we already have to mention it here, ahead of a more detailed treatment in the context of quantum mechanics), the right-hand side of (1.37) may be generalized to any finite-dimensional C*-algebra A by

$$\tau^A(\omega, \omega') = \inf\{\omega(a) \mid a \in A, 0 \leq a \leq 1_A, \omega'(a) = 1\}, \tag{1.38}$$

where $\omega, \omega' \in P(A)$. Since (1.38) clearly generalizes (1.37), for $A = C(X)$ we have

$$\tau^{C(X)}(\omega_x, \omega_y) = \delta_{xy}. \tag{1.39}$$

Note that the symmetry property in Definition 1.17 is not obvious from (1.38), but in the classical case $A = C(C)$ it is true by computation, and the same will hold in quantum theory. To motivate these definitions, we recall that f in (1.37), and likewise a in (1.38), are yes-no question to the system, so that the transition probability $\tau^A(\omega, \omega')$ monitors to what extent the states ω and ω' may be sharply distinguished by asking such questions. If they can, there should be some question a for which $\omega'(a) = 1$ and $\omega(a) = 0$, so that $\tau^A(\omega, \omega')$ (if $\omega \neq \omega'$, of course). As we have seen, in the classical case this can always be done. However, we shall see this is no longer the case in quantum mechanics, where pure states may be thus distinguished iff they correspond to orthogonal unit vectors in Hilbert space. Further motivation for the expression (1.38) is *post hoc*, as it turns out to allow a reconstruction of the vector space of observables A, supplemented by the part of its algebraic structure that determines its logical and probabilistic structure (viz. the ability to form squares, $a \mapsto a^2$) from $P(A)$ *with its associated transition probability*. See Theorem C.179.

First, we develop some theory that puts both classical and quantum mechanics into a more general setting. Notwithstanding the formal incorporation of the former, the underlying Hilbert space thinking will be obvious throughout.

Definition 1.18. *Let (X, τ) be a set with a transition probability.*

1. *A subset $O \subset X$ is* **orthonormal** *if $\tau(x,y) = \delta_{xy}$ for all $x, y \in O$.*
2. *A* **basis** *of a set X with a transition probability τ is an orthonormal family $B \subset X$ such that for each $x \in X$ one has*

$$\sum_{u \in B} \tau(x,u) = 1. \tag{1.40}$$

A basis of a subset $S \subset X$ is an orthonormal family $B \subset S$ such that (1.40) *holds for each $x \in S$. Relative to such a basis B of S, we define $\tau_S : X \to \mathbb{R}$ by*

$$\tau_S(x) = \sum_{u \in B} \tau(x,u). \tag{1.41}$$

As a special case, for $S = \{u\}$ we write $\tau_{\{u\}} \equiv \tau_u$, so that

$$\tau_u(x) = \tau(x,u). \tag{1.42}$$

3. *The* **orthocomplement** *$S^\perp$ of some subset $S \subset X$ is defined as*

$$S^\perp = \{y \in X \mid \tau(x,y) = 0 \forall x \in S\}. \tag{1.43}$$

4. *A subset $S \subset X$ is* **orthoclosed** *if $S^{\perp\perp} = S$ (where $S^{\perp\perp} = (S^\perp)^\perp$).*
5. *A* **resolution of the identity** *in X is a family of orthogonal orthoclosed subsets $(S_j)_j$ (i.e., $\tau(x_i, x_j) = 0$ if $x_i \in S_i$, $x_j \in S_j$, and $i \neq j$), for which $\sum_j \tau_{S_j} = 1_X$.*

6. *An* **observable** *for the pair* (X,τ) *is a bounded function* $f : X \to \mathbb{R}$ *of the form*

$$f = \sum_i c_i \cdot \tau_{y_i},\ c_i \in \mathbb{R},\ y_i \in X. \tag{1.44}$$

The real vector space of such observables is called $\ell^\infty(X,\tau)$.

7. *A* **spectral resolution** *of an observable* $f \in \ell^\infty(X,\tau)$ *is a decomposition*

$$f = \sum_\lambda \lambda \cdot \tau_{S_\lambda}, \tag{1.45}$$

where $(S_\lambda)_\lambda$ *is a resolution of the identity and each* $\lambda \in \mathbb{R}$ *occurs at most once.*

In the present section X is finite, whilst in the following section on quantum mechanics on finite-dimensional Hilbert spaces at least all bases will be finite, so that there are no convergence issues. In general, B may be infinite, in which case (1.40) is defined as the least upper bound of all finite partial sums, and all sums in Definition 1.18 are defined pointwise (i.e., in x). In that case, eq. (1.45) may need to be adapted through limit constructions. Furthermore, one may worry about the basis-dependence of τ_S in (1.41), but fortunately it turns out that in all sets with a transition probability that arise as pure state spaces defined by C*-algebras according to (1.38), the function τ_S is independent of the basis B whenever S is orthoclosed. In that case, spectral resolutions exists and are unique, and one may turn the real vector space $\ell^\infty(X,\tau)$ of part 6 into a ***Jordan algebra*** by defining a product $\circ$ through

$$f^2 = \sum_\lambda \lambda^2 \cdot \tau_{S_\lambda}; \tag{1.46}$$

$$f \circ g = \tfrac{1}{4}((f+g)^2 - (f-g)^2). \tag{1.47}$$

In the classical case this yields the pointwise product (1.20), whereas in quantum mechanics it recovers the anti-commutator. Both are examples of ***Jordan products*** (cf. §C.25), i.e., commutative products $\circ$ satisfying the curious axiom (C.619).

All this trivializes if $\tau = \tau^{C(X)}$ is given by (1.35), where X need not even be finite:

1. Any subset $O \subset X$ is orthonormal.
2. The set $B = X$ itself is the only basis of (X,τ), and analogously $B = S$.
3. The orthocomplement $S^\perp$ is the set-theoretic complement $S^c \equiv X \backslash S$.
4. Hence any subset $S \subset X$ is orthoclosed.
5. Any partition $X = \bigsqcup_j S_j$ yields a resolution of the identity.
6. Any bounded function $f : X \to \mathbb{R}$ is an observable, so that when X is finite,

$$\ell^\infty(X,\tau) = R(X) \equiv C(X,\mathbb{R}); \tag{1.48}$$

7. The spectral resolution (1.45) of f is given (analogously to operator theory) by

$$f = \sum_{\lambda \in \sigma(f)} \lambda \cdot \tau_{f=\lambda}, \tag{1.49}$$

cf. Definition 1.1.5. In particular, spectral resolutions in (1.48) are unique.

1.4 The logic of classical mechanics

Whatever one's route to $C(X,\mathbb{R})$ as the algebra of observables, i.e. either as a starting point or as a derived concept as in (1.48), it determines the logical structure of classical mechanics (we here restrict ourselves to propositional logic). According to the general scheme reviewed in §D.2, apart from the usual logical connectives $\neg$, $\wedge$, $\vee$, and $\rightarrow$ for *not*, *and*, *or*, and *implies*, a propositional theory needs a set Σ_X of *atomic propositions*. These are provided by $C(X,\mathbb{R})$, and Σ_X consist of all expressions $f \in \Delta$ (we expect no confusion between this notation for both *propositions* in logic and *events* in probability theory), where $f : X \rightarrow \mathbb{R}$ is a function, and Δ is some subset of $\mathbb{R}$. As we shall see, $f \in \Delta$ is always false if $\Delta \cap \sigma(f) = \emptyset$, so we might as well assume that $\Delta \subseteq \sigma(f)$. We write $f = \lambda$ for $f \in \{\lambda\}$. From these elementary propositions, propositions are constructed inductively using the iterative rules of propositional logic (see §D.2). This produces a set $B_X \equiv B_{\Sigma_X}$ of propositions.

Of course, there are logical relations between our atomic propositions (and hence between elements of B_X). For example, if $\Delta \subset \Delta'$, then $f \in \Delta$ should imply $f \in \Delta'$. Such relations may be formulated as axioms of some propositional theory $\mathscr{T}_X$ describing the logic of classical mechanics. These axioms take the following form:

$$(f \in \Gamma) \rightarrow (g \in \Delta) \text{ iff } f^{-1}(\Gamma) \subseteq g^{-1}(\Delta). \tag{1.50}$$

This may also be formulated through the notion of ***semantic entailment***. For each $x \in X$, we define a valuation $V_x : \Sigma_X \rightarrow \{0,1\}$ (cf. §D.2) by

$$V_x(f \in \Delta) = 1 \text{ iff } f(x) \in \Delta, \tag{1.51}$$

extended to a map $V_x : B_X \rightarrow \{0,1\}$ through the recursive use of truth tables. Defining the semantic entailment relation $\models_X$ on B_X by $\alpha \models_X \beta$ iff $V_x(\alpha) = 1$ implies $V_x(\beta) = 1$ for all $x \in X$, it is easy to see that $\alpha \rightarrow \beta$ as defined in (1.50) iff $\alpha \models_X \beta$.

In order to compute the ensuing Lindenbaum algebra $L_X \equiv L_{\Sigma_X}$, we note that

$$(f \in \Gamma) \leftrightarrow (g \in \Delta) \text{ iff } f^{-1}(\Gamma) = g^{-1}(\Delta). \tag{1.52}$$

Writing $\sim_X$ for $\sim_{\mathscr{T}_X}$ (which is the equivalence relation given by $\models_X$, too), we find

$$(f \in \Delta) \sim_X (1_{f^{-1}(\Delta)} = 1), \tag{1.53}$$

where we recall that 1_A is the characteristic (or indicator) function of A. Using the truth tables for $\wedge$ and for $\neg$, we also obtain (in terms of the complement $\Delta^c = \mathbb{R}\backslash\Delta$):

$$(f \in \Gamma) \wedge (g \in \Delta) \sim_X (1_{f^{-1}(\Gamma) \cap g^{-1}(\Delta)} = 1); \tag{1.54}$$

$$(\neg f \in \Delta) \sim_X (f \in \Delta^c) \sim_X (1_{f^{-1}(\Delta^c)} = 1). \tag{1.55}$$

Finally, the truth tables yield logical (and hence semantic) equivalences like

$$\alpha \vee \beta \sim_X \neg(\neg\alpha \wedge \neg\beta), \tag{1.56}$$

Combining the specific and the general equivalences (1.53) - (1.56), we have:

Lemma 1.19. *Any proposition in B_X is logically (and semantically) equivalent (relative to X) to one of the form $1_U = 1$, for some event $U \subset X$. Furthermore,*

$$(\neg 1_U = 1) \sim_X (1_{U^c} = 1); \tag{1.57}$$
$$(1_U = 1) \wedge (1_V = 1) \sim_X (1_{U \cap V} = 1); \tag{1.58}$$
$$(1_U = 1) \vee (1_V = 1) \sim_X (1_{U \cup V} = 1). \tag{1.59}$$

Theorem 1.20. *The Lindenbaum algebra L_X is isomorphic (as a Boolean algebra) to the power set $\mathscr{P}(X)$ of X under the map $\varphi : L_X \to \mathscr{P}(X)$ induced by*

$$\varphi([f \in \Delta]_X) = f^{-1}(\Delta). \tag{1.60}$$

In particular, the logical connectives $\neg$, $\wedge$ and $\vee$ (descended to L_X) turn into set-theoretic complementation $(-)^c$, intersection $\cap$, and union $\cup$, respectively, in that

$$\varphi([\neg\alpha]_X) = \varphi([\alpha]_X)^c; \tag{1.61}$$
$$\varphi([\alpha \wedge \beta]_X) = \varphi([\alpha]_X) \cap \varphi([\beta]_X; \tag{1.62}$$
$$\varphi([\alpha \vee \beta]_X) = \varphi([\alpha]_X) \cup \varphi([\beta]_X), \tag{1.63}$$

and φ maps the partial order $\leq$ on L_X into set-theoretic inclusion $\subseteq$, i.e.,

$$[\alpha]_X \leq [\beta]_X \textit{ iff } \varphi([\alpha]_X) \subseteq \varphi([\beta]_X). \tag{1.64}$$

This is immediate from Lemma 1.19. Interestingly, the *Boolean* algebra structure just derived as the governor of the (propositional) logic of classical mechanics may be reformulated in terms of the *Jordan* algebraic structure (1.46) - (1.47) of $\ell^\infty(X\tau)$, or, when X is finite, of the C*-algebra of observables $C(X)$ itself:

- Events $U \subseteq X$ (and hence, by Theorem 1.20, logical equivalence classes of propositions) correspond bijectively to characteristic functions 1_U on X, that is, with ***yes-no questions*** (having spectrum in $\{0,1\}$). Algebraically, these are precisely the ***idempotents*** in $\ell^\infty(X,\tau)$, i.e., those functions e satisfying $e^2 - e$.
- In terms of those, the partial ordering and the logical connectives are given by

$$e \leq f \text{ iff } e \circ f = e; \tag{1.65}$$
$$\neg e = 1_X - e; \tag{1.66}$$
$$e \wedge f = e \circ f; \tag{1.67}$$
$$e \vee f = e + f - e \circ f. \tag{1.68}$$

Indeed, in this case $\circ$ is pointwise multiplication (1.20). Using $1_U \cdot 1_V = 1_{U \cap V}$ yields (1.67), (1.65) comes down to $U \subseteq V$ iff $U \cap V = U$, (1.66) is $1_X - 1_U = 1_{U^c}$, and (1.68) follows by writing its right-hand side as $1_X - (1_X - e) \wedge (1_X - f)$.

1.5 The GNS-construction for $C(X)$

As a bridge from classical to quantum mechanics (as well as a good exercise), we finally inject some Hilbert space theory into classical physics by discussing the GNS-***construction*** of C*-algebra theory for the special case of $C(X)$, where X remains finite. In general, for each state ω on a C*-algebra A, the GNS-construction canonically yields a ***Hilbert space*** H_ω (which is finite-dimensional for $A = C(X)$ with finite X) and a ***representation*** of A on H_ω, in the sense of a (complex) linear map

$$\pi_\omega : A \to B(H_\omega) \tag{1.69}$$

that satisfies

$$\pi_\omega(ab) = \pi_\omega(a)\pi_\omega(b); \tag{1.70}$$
$$\pi_\omega(a^*) = \pi_\omega(a)^*. \tag{1.71}$$

Furthermore, H_ω contains a special *unit vector* Ω_ω that is ***cyclic*** for π_ω in that

$$\pi_\omega(A)\Omega_\omega \equiv \{\pi_\omega(a)\Omega_\omega, a \in A\} = H_\omega, \tag{1.72}$$

at least in the relevant case where $\dim(H_\omega) < \infty$; otherwise, the left-hand side is merely dense in H_ω and one needs to take the (norm) closure to obtain H_ω. Furthermore, Ω_ω realizes the state ω as a quantum-mechanical expectation value by

$$\omega(a) = \langle \Omega_\omega, \pi_\omega(a)\Omega_\omega \rangle_{H_\omega}. \tag{1.73}$$

Given $\omega \in S(A)$, the GNS-construction starts with the vector spaces

$$N_\omega = \{a \in A \mid \omega(a^*a) = 0\}; \tag{1.74}$$
$$H_\omega = A/N_\omega. \tag{1.75}$$

Now, if $b \in N_\omega$ and $a \in A$, then $ab \in N_\omega$, because of the important inequality

$$\omega(b^*a^*ab) \leq \|a\|^2 \omega(b^*b). \tag{1.76}$$

This is true for any C*-algebra A, but below we prove it only for our example. Assuming (1.76) for the moment, the action of A on itself by left multiplication descends to a well-defined action on H_ω, which we call π_ω. In other words, if $b_\omega \in H_\omega$ is the image of $b \in A$ under the canonical projection $A \to A/N_\omega$, then

$$\pi_\omega(a)b_\omega = (ab)_\omega. \tag{1.77}$$

Crucially, this vector space H_ω is equipped with a canonical inner product

$$\langle a_\omega, b_\omega \rangle = \omega(a^*b). \tag{1.78}$$

Indeed, this form is well defined, and is positive definite because ω is a state.

In general, H_ω as defined by (1.75) with inner product (1.78) is merely a pre-Hilbert space, which needs to be completed in the associated norm, and it takes some effort to check that the operators defined by (1.77) are bounded. In our example, on the other hand, H_ω is finite-dimensional and hence complete. In any case, it is easy to verify the properties (1.70) - (1.73), whilst (1.72) holds with the unit $1 = 1_H$.

We now prove (1.76) for $A = C(X)$. Fom Theorem 1.15 we have $\omega = E_P$, and by (1.9) and (1.24), the inequality (1.76) comes down to the obviously correct result

$$\sum_x |f(x)g(x)|^2 \leq \|f\|_\infty^2 \sum_x |g(x)|^2. \tag{1.79}$$

Writing $N_{E_P} \equiv N_P$, we may also check directly that if $g \in N_P$ and $f \in C(X)$, then $fg \in N_P$. Indeed, in terms of the set $\text{supp}\,(P) \subseteq X$ defined by

$$\text{supp}\,(P) = \{x \in X \mid p(x) > 0\}, \tag{1.80}$$

we have

$$N_P = \{f \in C(X) \mid f(x) = 0 \forall x \in \text{supp}\,(P)\}, \tag{1.81}$$

and clearly $g = 0$ on $\text{supp}\,(P)$ implies $fg = 0$ on $\text{supp}\,(P)$. We now compute H_P and π_P. From (1.81) we have $f - g \in N_P$ and hence $f \sim g$ iff $f(x) = g(x)$ for all $x \in \text{supp}\,(P)$, where $\sim$ is the equivalence relation whose equivalence classes f_P define elements of $H_P = C(X)/N_P$. Hence f_P is simply the restriction of f to $\text{supp}\,(P)$, and

$$H_P = \ell^2(X, P) \tag{1.82}$$

is the Hilbert space that consists of these restriction, with inner product

$$\langle f_P, g_P \rangle = \sum_{x \in \text{supp}\,(P)} p(x)\overline{f(x)}g(x). \tag{1.83}$$

The representation (1.77) then trivially gives

$$\pi_P(f)g_P = f_P g_P, \tag{1.84}$$

so that $\pi_P(f)$ is the ***multiplication operator*** defined by f on $\ell^2(X, P)$. In functional analysis one often denotes elements $g_P \in \ell^2(X, P)$ by the functions g themselves, and similarly writes $\pi_P(f)$ as f, so that (1.84) simply reads $\pi_P(f)g = fg$.

The operator norm of $\pi_P(f)$ is easily computed to be

$$\|\pi_P(f)\| = \sup\{|f(x)|, x \in \text{supp}\,(P)\} = \|f_{|\text{supp}\,(P)}\|_\infty. \tag{1.85}$$

Indeed, the bound $\|\pi_P(f)\| \leq \|f_{|\text{supp}\,(P)}\|_\infty$ is immediate from the definition

$$\|\pi_P(f)\| = \sup\{\|\pi_P(f)g_P\|, g_P \in H_P, \|g_P\| = 1\}, \tag{1.86}$$

and equality in this bound follows from applying the operator $\pi_P(f)$ to the function $g = 1_U$, where $U \subset X$ is any set where $|f|$ attains its maximum $\|f_{|\text{supp}\,(P)}\|_\infty$.

Notes

§1.1. **Basic constructions of probability theory**

§1.2. **Classical observables and states**

For (advanced) treatments of convexity theory and probability theory in contexts relevant to mathematical physics we recommend Israel (1979), Alfsen & Shultz (2001), and Simon (2001).

§1.3. **Pure states and transition probabilities**

Transition probabilities (in the abstract sense meant here) were introduced by von Neumann, but his manuscript from 1937 was only published in 1981 as von Neumann (1981/1937). This remarkable paper has remained largely unused (or even unknown) in both mathematical physics and operator algebras; Mielnik (1968), Shultz (1982), and Landsman (1996, 1997) are exceptions. An extensive discussion with further references may be found in Landsman (1998a).

§1.4. **The logic of classical mechanics**

Unless one counts Boole (1847), it seems that the logical analysis of classical mechanics was initiated by the famous paper of Birkhoff & von Neumann (1936), which was primarily concerned with quantum logic (cf. §2.10). Our use of semantic implication (also in the quantum case) was inspired by Rédei (1998).

§1.5. **The GNS-construction for $C(X)$**

See §C.12 for the GNS-construction in general.

Chapter 2
Quantum mechanics on a finite-dimensional Hilbert space

The quantum analogue of a finite set X (in its role as a configuration space in classical mechanics) is the finite-dimensional Hilbert space $\ell^2(X)$, by which we mean the vector space of functions $\psi : X \to \mathbb{C}$, equipped with the inner product

$$\langle \psi, \varphi \rangle = \sum_{x \in X} \overline{\psi(x)} \phi(x). \tag{2.1}$$

There is no issue of convergence here, but later on we will use the same notation for infinite sets X, where $\ell^2(X)$ is restricted to those functions (i.e. sequences) for which $\sum_{x\in X} |\psi(x)|^2 < \infty$ (which also guarantees convergence of the sum in (2.1)).

If $X \cong \underline{n}$ as sets (i.e., $|X| = n$), we have a unitary isomorphism of Hilbert spaces

$$\ell^2(\underline{n}) \cong \mathbb{C}^n, \tag{2.2}$$

through the map $\psi \mapsto (\psi(1), \ldots, \psi(n))$, where $\mathbb{C}^n$ has the standard inner product. $\langle w, z \rangle = \sum_i \overline{w}_i z_i$. In particular, the function $\delta_k \in \ell^2(\underline{n})$, defined by $\delta_k(l) = \delta_{kl}$, is mapped to the k'th standard basis vector $u_k \equiv |k\rangle$ of $\mathbb{C}^n$, i.e., $u_1 = (1, 0, \ldots, 0)$, etc. In the special case $X = \underline{N}^\Lambda$ considered in Chapter 1, we have $|X| = N^{|\Lambda|}$ and hence

$$\ell^2(\underline{N}^\Lambda) \cong \mathbb{C}^{(N^{|\Lambda|})} \cong \left(\mathbb{C}^N\right)^{\otimes|\Lambda|} = \bigotimes_{\mathbf{n}\in\Lambda} \mathbb{C}^N_{\mathbf{n}} \equiv \bigotimes_{\Lambda} \mathbb{C}^N, \tag{2.3}$$

where $\mathbb{C}^N_{\mathbf{n}} = \mathbb{C}^N$ for each $\mathbf{n} \in \Lambda$, so that the suffix $\mathbf{n}$ merely labels which copy of $\mathbb{C}^N$ is meant (see §C.13 for tensor products of Hilbert spaces). Explicitly, a canonical unitary isomorphism $\ell^2(\underline{N}^\Lambda) \to \bigotimes_\Lambda \mathbb{C}^N$ is given by linear extension of the map

$$\delta_x \mapsto \otimes_{\mathbf{n}\in\Lambda} u_{x(\mathbf{n})}, \tag{2.4}$$

where $x : \Lambda \to \underline{N}$ and hence $u_{x(\mathbf{n})} \in \mathbb{C}^N$. Thus elements of the tensor product $\bigotimes_\Lambda \mathbb{C}^N$ may be seen as wave-functions on spin configuration space (and *vice versa*). In particular, elementary tensor products of basis vectors in $\bigotimes_\Lambda \mathbb{C}^M$ correspond to wave-functions in $\ell^2(\underline{M}^\Lambda)$ that are δ-peaked at some 'classical' spin configuration.

K. Landsman, *Foundations of Quantum Theory*,
Fundamental Theories of Physics 188, DOI 10.1007/978-3-319-51777-3_2

2.1 Quantum probability theory and the Born rule

In preparation for this chapter, the reader would do well to review Appendix A.

The probabilistic setting of quantum mechanics is given by the following counterpart of Definition 1.1 (from which conditional probabilities are lacking, though).

Definition 2.1. *Let H be a finite-dimensional Hilbert space.*

1. *A* **(quantum) event** *is a linear subspace L of H (which is automatically closed).*
2. *A* **(quantum) probability distribution** *is a* **density operator**, *i.e., a positive operator ρ on H (in that $\langle\psi, \rho\psi\rangle \geq 0$ for all $\psi \in H$) such that*

$$\mathrm{Tr}(\rho) = 1. \tag{2.5}$$

   We denote the set of all density operators on H by $\mathscr{D}(H)$.

3. *A* **(quantum) random variable** *is a self-adjoint operator a on H (i.e., $a^* = a$).*
4. *The* **spectrum** *of a self-adjoint operator a is the set $\sigma(a) \subset \mathbb{R}$ of its eigenvalues.*

Being positive, a density matrix ρ is self-adjoint, so by Theorem A.10, notably (A.40), and Definition 2.1.2 we have

$$\rho = \sum_i p_i |\upsilon_i\rangle\langle\upsilon_i|, \; p_i > 0, \; \sum_i p_i = 1, \tag{2.6}$$

where the (υ_i) form an orthonormal set in H and $|\upsilon_i\rangle\langle\upsilon_i|$ is the (orthogonal) projection on the one-dimensional subspace $\mathbb{C} \cdot \upsilon_i$. As in the classical case, one special class of density operators and one special class of random variables stand out:

- Each *unit vector* $\psi \in H$ defines a density operator

$$\rho_\psi \equiv e_\psi = |\psi\rangle\langle\psi|, \tag{2.7}$$

  i.e., the (orthogonal) projection e_ψ on the one-dimensional subspace $\mathbb{C} \cdot \psi$. A basis (which by convention always means an *orthonormal* basis) of eigenvectors of ρ_ψ consists of $\upsilon_1 = \psi$ itself, supplemented by any basis $(\upsilon_2, \ldots, \upsilon_{\dim(H)})$ of the orthogonal complement of $\mathbb{C} \cdot \psi$. The corresponding probabilities in (2.6) are evidently $p_1 = 1$ and $p_i = 0$ for all $i > 1$.
- Each *quantum event* $L \subset H$ defines the corresponding projection e_L (which is self-adjoint, i.e. a random variable): If (υ_j) is a basis of L, then $e_L = \sum_j |\upsilon_j\rangle\langle\upsilon_j|$. If $L = H$ then $e_L = 1$ with $\sigma(e_L) = \{1\}$. If $L = \{0\}$ then $e_L = 0$ with $\sigma(e_L) = \{0\}$. In all other cases, i.e. for proper subspaces L, one has $\sigma(e_L) = \{0, 1\}$. Conversely, any self-adjoint operator a with spectrum $\sigma(a) \subseteq \{0, 1\}$ is given by $a = e_L$ for some subspace $L \subseteq H$; just take $L = \{\psi \in H \mid a\psi = 1\}$. Such operators correspond to yes-no questions to the system and lie at the basis of the logical interpretation of quantum theory due to Birkhoff and von Neumann; see §2.10.

The following quantum analogue of Theorem 1.2 is based on Theorem A.10.

Theorem 2.2. *A density operator* ρ *on* H *and a self-adjoint operator* $a : H \to H$ *jointly yield a probability distribution* p_a *on the spectrum* $\sigma(a)$ *by the* **Born rule**

$$p_a(\lambda) = \mathrm{Tr}(\rho e_\lambda). \tag{2.8}$$

The associated probability measure P_a *is given at* $\Delta \subseteq \sigma(a)$ *by (cf.* (A.42)*)*

$$P_a(\Delta) = \mathrm{Tr}(\rho e_\Delta). \tag{2.9}$$

Proof. Positivity of the numbers $p_a(\lambda)$ follows by taking the trace over a basis of eigenvectors υ_i of ρ, with corresponding eigenvalues $p_i \geq 0$. This yields

$$\mathrm{Tr}(\rho e_\lambda) = \sum_i p_i \|e_\lambda \upsilon_i\|^2 \geq 0.$$

Eqs. (A.38) and (2.5) then give $\sum_\lambda p_a(\lambda) = 1$. Eq. (2.8) follows from the equality $P_a(\Delta) = \sum_{\lambda \in \Delta} p_a(\Delta)$, cf. (1.2), and (A.42). □

In particular, if $\rho = \rho_\psi$, writing p_a^ψ for the associated probability, (2.8) yields

$$p_a^\psi(\lambda) = \langle \psi, e_\lambda \psi \rangle = \|e_\lambda \psi\|^2. \tag{2.10}$$

If in addition $\lambda \in \sigma(a)$ is non-degenerate, so that $e_\lambda = |\upsilon_\lambda\rangle\langle\upsilon_\lambda|$ for some unit vector υ_λ with $a\upsilon_\lambda = \lambda\upsilon_\lambda$, then the Born rule (2.9) assumes its original form

$$p_a^\psi(\lambda) = |\langle \psi, \upsilon_\lambda \rangle|^2. \tag{2.11}$$

Specializing (2.10) to the random variable $a = e_L$ defined by an event $L \subset H$ yields

$$p_{e_L}^\psi(1) = \|e_L \psi\|^2. \tag{2.12}$$

If $L = \mathbb{C} \cdot \varphi$ is one-dimensional, too, in which case we write $p_{e_\varphi}^\psi \equiv p_\varphi^\psi$, we have

$$p_\varphi^\psi(1) = |\langle \psi, \varphi \rangle|^2; \tag{2.13}$$

note the following equality of probability distributions on $\sigma(e_\varphi) = \sigma(e_\psi) = \{0, 1\}$:

$$p_\varphi^\psi(1) = p_\psi^\varphi(1). \tag{2.14}$$

Expectation values and variances may be defined as in the classical case, viz.

$$E_\rho(a) = \mathrm{Tr}(\rho a); \tag{2.15}$$
$$\Delta_\rho(a) = E_\rho(a^2) - E_\rho(a)^2. \tag{2.16}$$

Similar to (1.11), we may also write the expectation value as

$$E_\rho(a) = \sum_{\lambda \in \sigma(a)} \lambda \cdot p_a(\lambda). \tag{2.17}$$

The special case $\rho = \rho_\psi$, for which we write $E_{\rho_\psi} \equiv E_\psi$, gives the usual formula

$$E_\psi(a) = \mathrm{Tr}\,(\rho_\psi a) = \langle \psi, a\psi \rangle. \tag{2.18}$$

As in the classical case one always has $\Delta_\rho(a) \geq 0$, but a major contrast between classical and quantum mechanics lies in the following result, cf. Proposition 1.3.

Proposition 2.3. *For each density operator ρ there exists a self-adjoint operator b such that $\Delta_\rho(b) > 0$. On the other hand, if $a^* = a$, then $\Delta_\rho(a) = 0$ iff the image of ρ lies in some fixed eigenspace of a, i.e., in terms of the spectral decomposition* (2.6) *we have $a\upsilon_i = \lambda \upsilon_i$ where λ is independent of i.*

Proof. We first prove the first claim for $H = \mathbb{C}^2$. By an appropriate choice of basis, we may assume that ρ is diagonal, i.e., $\rho = \mathrm{diag}(p_1, p_2)$, with $p_1, p_2 \in [0,1]$ and $p_1 + p_2 = 1$. Now take $b = \sigma_x$ (i.e., the first Pauli matrix), so that $\mathrm{Tr}\,(\rho b) = 0$ and $\mathrm{Tr}\,(\rho b^2) = 1$. Hence $\Delta_\rho(b) = 1$. Secondly, for general $H \cong \mathbb{C}^n$, diagonalize ρ and order the eigenvectors such that the above 2×2 case forms the upper left block, with at least one of the eigenvalues p_1, p_2 strictly positive. Take b to be σ_x in the upper left corner, and zero elsewhere. This once again yields $\Delta_\rho(b) = 1$.

For the second claim we use (2.6), and write $\rho_i \equiv \rho_{\upsilon_i}$. We note the inequality

$$\Delta_\rho(a) \geq \sum_i p_i \Delta_{\rho_i}(a), \tag{2.19}$$

with equality iff $\rho_i(a) = \rho_j(a)$ for all i, j; this follows from convexity of the function $x \mapsto x^2$. We now show that for any unit vector ψ we have $\Delta_{\rho_\psi} = 0$ iff $a\psi = \lambda\psi$. Assuming the latter gives $E_\psi(a) = \langle \psi, a\psi \rangle = \lambda$ and likewise $E_\psi(a^2) = \lambda^2$, hence $\Delta_{\rho_\psi}(a) = 0$. In the opposite direction, using $a^* = a$, elementary manipulations yield

$$\Delta_{\rho_\psi}(a) = \|(a - \langle \psi, a\psi \rangle)\psi)\|^2. \tag{2.20}$$

This clearly vanishes iff $a\psi = \langle \psi, a\psi \rangle \psi$, so $a\psi = \lambda\psi$, with $\lambda = \langle \psi, a\psi \rangle$.

Putting $\psi = \upsilon_i$ gives $\Delta_{\rho_i} = 0$ iff $a\upsilon_i = \lambda_i \upsilon_i$, and then $\Delta_{\sum_i p_i \rho_i}(a) = 0$ iff in addition $\rho_i(a) = \rho_j(a)$ for all i, j. Since $\rho_i(a) = \langle \upsilon_i, a\upsilon_i \rangle = \lambda_i$, we obtain $\lambda_i = \lambda_j$. □

As first recognized by von Neumann, Theorem 2.2 may be generalized to a family of self-adjoint operators *as long as they commute*. Thus we obtain the following counterpart of (1.12) - (1.13): a collection $a_1, \ldots, a_n$ of n *commuting* self-adjoint operators and a (single) density operator ρ on H jointly define a probability distribution $p_{a_1,\ldots,a_n}$ on the product $\sigma(a_1) \times \cdots \times \sigma(a_n)$ of the individual spectra by

$$p_{a_1,\ldots,a_n}(\lambda_1, \ldots, \lambda_n) = \mathrm{Tr}\,(\rho e^{(1)}_{\lambda_1} \cdots e^{(n)}_{\lambda_n}). \tag{2.21}$$

The proof of positivity of these numbers requires the spectral projections $e^{(i)}_{\lambda_i}$ to commute, which they do provided the a_i commute (if the a_i fail to commute, positivity of (2.21) is not guaranteed, although they do still sum op to unity; the possibility of defining joint probabilities is strictly limited to commuting random variables).

2.2 Quantum observables and states

Given a finite-dimensional Hilbert space H, the set $B(H)$ of all linear operators on H (which for $H = \mathbb{C}^n$ may be identified with the set $M_n(\mathbb{C})$ of complex $n \times n$ matrices) forms an involutive algebra under the natural (pointwise) operations

$$(\lambda \cdot a)\psi = \lambda(a\psi); \tag{2.22}$$
$$(a+b)\psi = a\psi + b\psi; \tag{2.23}$$
$$(ab)\psi = a(b\psi), \tag{2.24}$$

and finally with a^* given by the usual operator adjoint (A.15). Compare the corresponding classical expressions (1.18) - (1.20) and (1.22). Analogous to (1.24), we also have a norm on $B(H)$, defined by (A.18). It follows that *like* its classical counterpart $C(X)$, the involutive algebra $B(H)$ (or, in this case, $M_n(\mathbb{C})$) is a C*-algebra, cf. Definition C.1 in Appendix C. It crucially *differs* from $C(X)$ in that $B(H)$ is *non-commutative*. For this reason, the Gelfand spectrum, which in the classical case allowed us to reconstruct X from $C(X)$, turns out to be empty, cf. Proposition 2.10 below. Nonetheless, it makes good sense to copy Definition 1.14, *mutatis mutandis*:

Definition 2.4. *A* **state** *on $B(H)$ is a complex-linear map $\omega : B(H) \to \mathbb{C}$ satisfying:*

1. *$\omega(a^*a) \geq 0$ for each $a \in B(H)$* (**positivity**)*;*
2. *$\omega(1_H) = 1$* (**normalization**)*.*

The **state space** *$S(B(H))$ is the set of all states $\omega : B(H) \to \mathbb{C}$.*

Physicists may not like this definition, since it involves non-observable quantities. As in the classical case, we may introduce the self-adjoint (or 'real') part of $B(H)$:

$$B(H)_{\mathrm{sa}} = \{a \in B(H) \mid a^* = a\}, \tag{2.25}$$

which is a real vector space (though not a real algebra in the usual sense, cf. §C.25).

Definition 2.5. *A* **state** *on $B(H)_{\mathrm{sa}}$ is a real-linear map $\omega : B(H)_{\mathrm{sa}} \to \mathbb{R}$ satisfying:*

1. *$\omega(a^2) \geq 0$ for each $a \in B(H)$ with $a^* = a$* (**positivity**)*;*
2. *$\omega(1) = 1$* (**normalization**)*.*

The **state space** *$S(B(H)_{\mathrm{sa}})$ is the set of all states $\omega : B(H)_{\mathrm{sa}} \to \mathbb{R}$.*

Fortunately, there is no need for a fight over this point; the discussion is similar to the one below Definition 1.14 and is settled as follows.

Proposition 2.6. *The state spaces $S(B(H))$ and $S(B(H)_{\mathrm{sa}})$ may be identified: an element ω of the former defines an element $\omega_{\mathbb{R}}$ of the latter by restriction, whilst the unique decomposition $c = a + ib$ (where $a^* = a$ and $b^* = b$ are given by $a = \frac{1}{2}(c+c^*)$ and $b = -\frac{1}{2}i(c-c^*)$, respectively) gives $\omega(c) = \omega_{\mathbb{R}}(a) + i\omega_{\mathbb{R}}(b)$. Moreover,*

$$\|\omega\| = \|\omega_{\mathbb{R}}\| = 1. \tag{2.26}$$

Here the norm on the dual (Banach) space $B(H)^*_{\mathrm{sa}}$ of $B(H)_{\mathrm{sa}}$ is given by

$$\|\omega\| = \sup\{|\omega(a)|, a \in B(H)_{\mathrm{sa}}, \|a\| = 1\}. \tag{2.27}$$

This lemma holds for any Hilbert space H (cf. Theorem C.52), but it is instructive to restrict our proof to the finite-dimensional setting in which we currently work.

Proof. The first few claims are immediate from Proposition A.22. To prove (2.26), it suffices to prove that for any $a \in B(H)$ one has

$$|\omega(a)| \leq \|a\|, \tag{2.28}$$

since by normalization of states the bound is saturated by $a = 1_H$. Furthermore, even if ω is seen as an element of $B(H)^*$ rather than $B(H)^*_{\mathrm{sa}}$, eq. (2.28) needs to be shown only for self-adjoint a, for positivity of ω implies the Cauchy–Schwarz inequality

$$|\omega(a^*b)|^2 \leq \omega(a^*a)\omega(b^*b), \tag{2.29}$$

cf. (A.1), in which we may take $a = 1_H$ to find, assuming (2.28) for self-adjoint a,

$$|\omega(b)|^2 \leq \omega(b^*b) \leq \|b^*b\| = \|b\|^2, \tag{2.30}$$

where the last equality holds for any $b \in B(H)$ (turning the latter into a C*-algebra). Noting that b^*b is self-adjoint, this gives (2.28) for any a. To prove (2.28) for $a^* = a$, then, we firstly use (A.47), and secondly use Theorem 2.7 and eq. (2.6) to obtain

$$|\omega(a)| = |\mathrm{Tr}\,(\rho a)| = |\sum_i p_i\langle \upsilon_i, a\upsilon_i\rangle| \leq \sum_i p_i |\langle \upsilon_i, a\upsilon_i\rangle|. \tag{2.31}$$

Now let (ξ_j) be a basis of H consisting of eigenvectors of a, so that

$$\langle \upsilon_i, a\upsilon_i\rangle = \sum_j |\langle \upsilon_i, \xi_j\rangle|^2 \lambda_j, \ \sum_j |\langle \upsilon_i, \xi_j\rangle|^2 = 1.$$

Since $|\lambda_j| \leq \|a\|$ and $\sum_i p_i = 1$, the bound (2.28) follows from the estimate

$$\sum_i p_i |\langle \upsilon_i, a\upsilon_i\rangle| \leq \sum_i p_i \sum_j |\langle \upsilon_i, \xi_j\rangle|^2 |\lambda_j| \leq \sum_i p_i \sum_j |\langle \upsilon_i, \xi_j\rangle|^2 \|a\| = \|a\|. \tag{2.32}$$

Finally, combining (2.31) and (2.32) gives (2.28) for self-adjoint a. □

In view of this, we may work with either $S(B(H)_{\mathrm{sa}})$ or $S(B(H))$; denoting states simply by ω, the context will usually show if it is defined on $B(H)_{\mathrm{sa}}$ or on $B(H)$.

Despite its easy proof, the following result is of fundamental importance.

Theorem 2.7. *If H is finite-dimensional, there is a bijective correspondence between states ω on $B(H)$ or $B(H)_{\mathrm{sa}}$ and density operators ρ on H, given by*

$$\omega(a) = \mathrm{Tr}\,(\rho a). \tag{2.33}$$

Proof. First note that linear algebra already yields (2.33) as a bijective correspondence between complex-linear maps ω and operators ρ, for example, because

$$\langle a,b\rangle = \mathrm{Tr}\,(a^*b) \tag{2.34}$$

defines an inner product on $B(H)$. Positivity and normalization of ω then translate to the corresponding properties of ρ. □

The quantum analogue of Theorem 1.15, then, is as follows.

Theorem 2.8. *The state space $S(B(H)_{\mathrm{sa}}) = S(B(H))$ forms a compact convex set in the (real) vector space $B(H)^*_{\mathrm{sa}}$ (in its w^*-topology) and, putting the corresponding topology on $\mathscr{D}(H)$, eq. (2.33) defines an affine homeomorphism*

$$S(B(H)) \cong \mathscr{D}(H). \tag{2.35}$$

Proof. Convexity of $S(B(H))$ holds by Definition 2.4. For compactness, by Proposition 2.6 the state space $S(B(H))$ is contained in the closed unit ball B_1 of $B(H)^*_{\mathrm{sa}}$, which is compact in the w^*-topology (in the case at hand this is simply because $B(H)^*_{\mathrm{sa}}$ is finite-dimensional). It is easy to see that a convergent sequence of states actually converges to a state, since both conditions in Definition 2.4 are clearly preserved by w^* limits (in which $\omega_n \to \omega$ iff $\omega_n(a) \to \omega(a)$ for each $a \in B(H)$). □

For infinite-dimensional Hilbert spaces eq. (2.35) is false; see §4.2. At the opposite end, the case $H = \mathbb{C}^2$ provides a beautiful illustration of this theorem (and more).

Proposition 2.9. *The state space $S(M_2(\mathbb{C}))$ of the 2×2 matrices is isomorphic (as a compact convex set) to the closed unit ball $B^3 = \{(x,y,z) \in \mathbb{R}^3 \mid x^2+y^2+z^2 \leq 1\}$.*

On this isomorphism, the extreme boundary (cf. Definition 1.10)

$$\partial_e B^3 = S^2 = \{(x,y,z) \in \mathbb{R}^3 \mid x^2+y^2+z^2 = 1\} \tag{2.36}$$

corresponds to the set of all density matrices $\rho = \rho_\psi$, where $\psi \in \mathbb{C}^2$ with $\|\psi\| = 1$.

Proof. Any self-adjoint 2×2 matrix may be parametrized by $(t,x,y,z) \in \mathbb{R}^4$ as

$$\rho(t,x,y,z) = \tfrac{1}{2}\begin{pmatrix} t+z & x-iy \\ x+iy & t-z \end{pmatrix}. \tag{2.37}$$

The eigenvalues λ_i of $\rho(t,x,y,z)$, computed from its characteristic polynomial, are

$$\lambda_\pm = \tfrac{1}{2}(t \pm \sqrt{x^2+y^2+z^2}). \tag{2.38}$$

Condition (2.5) yields $t = 1$. Positivity of $\rho(1,x,y,z)$ is equivalent to positivity of its eigenvalues λ_i, which gives $x^2+y^2+z^2 \leq 1$. For the second claim, note that the ρ_ψ are just the one-dimensional projections, which in turn are the density matrices satisfying $\rho^2 = \rho$ (or require $\lambda_+ = 1$, $\lambda_- = 0$), so $x^2+y^2+z^2 = 1$. Finally, since convex sums $t\mathbf{v} + (1-t)\mathbf{w}$ in B^3 $(0 \leq t \leq 1)$ are given by straight line segments connecting $\mathbf{w}$ and $\mathbf{v}$ in $\mathbb{R}^3$, it immediately follows geometrically that $\partial_e B^3 = S^2$. □

2.3 Pure states in quantum mechanics

In classical physics, the phase space X arose both as the Gelfand spectrum $\Sigma(C(X))$ of the C*-algebra of observables $C(X)$, cf. Definition 1.4 and Proposition 1.5, and as the pure state space $P(C(X))$ of $C(X)$, see Definition 1.10 and Theorem 1.16. In particular, $\Sigma(C(X)) \cong P(C(X))$ at least as sets. Because of this, any pure state $\omega \in P(C(X))$ is dispersion-free, since as an element of $\Sigma(C(X))$ it satisfies $\omega(f^2) = \omega(f)^2$ for any $f \in C(X)$. These two definitionally different (but classically coinciding) guises of X will fall apart in quantum mechanics; cf. Proposition 2.3.

Proposition 2.10. *If $\dim(H) > 1$, the Gelfand spectrum $\Sigma(B(H))$ of $B(H)$ is empty, i.e., there are no nonzero linear maps $\omega : B(H) \to \mathbb{C}$ that satisfy $\omega(ab) = \omega(a)\omega(b)$.*

In particular, there are no nonzero linear maps $\omega : B(H) \to \mathbb{C}$ that are **dispersion-free**, *i.e., satisfy $\Delta_\omega(a) = 0$, with $\Delta_\omega(a) = \omega(a^2) - \omega(a)^2$.*

Proof. Suppose $\omega \in \Sigma(B(H))$. Multiplicativity for $b = a = a^*$ implies that ω is positive, whereas for $b = 1_H$ it implies that ω is normalized. Hence ω must be a state. Now use Theorem 2.7 and use multiplicativity for $b = a = a^*$, implying that $\Delta_\rho(a) = 0$. This contradicts Proposition 2.3. □

On the other hand, the pure state space of $B(H)$ is by no means empty, and despite Proposition 2.10, we will see that the special density operators $\rho_\psi \equiv e_\psi$ in (2.7) to some extent do play the role of the points $x \in X$. Let us write

$$\mathscr{P}_1(H) = \{e \in B(H) \mid e^2 = e^* = e, \mathrm{Tr}\,(e) = 1\} \tag{2.39}$$

for the set of all one-dimensional projections on H; note that $\mathrm{Tr}\,(e) = \dim(eH)$ for $e \in \mathscr{P}(H)$. Each $e \in \mathscr{P}_1(H)$ takes the form $e = e_\psi$ for some unit vector ψ, see (2.7).

Lemma 2.11. *A density operator ρ is an extreme point of the convex set $\mathscr{D}(H)$ of all density operators on H iff $\rho = \rho_\psi$ for some unit vector $\psi \in H$.*

Proof. The argument is similar to the proof of Proposition 1.11. To show that $\rho_\psi \in \partial_e S(B(H))$, assume $\rho_\psi = t\rho_1 + (1-t)\rho_2$ for some $t \in (0,1)$ and $\rho_1, \rho_2 \in S(B(H))$. Evaluating this equality at $a = |\varphi\rangle\langle\varphi|$, where $\varphi \perp \psi$ yields $\langle\varphi, \rho_i\varphi\rangle = 0$ for $i = 1,2$, so that $\rho_1 = \rho_2 = \rho_\psi$. Conversely, the spectral decomposition (2.6) shows that $\rho \notin \partial_e S(B(H))$ whenever $\rho \neq \rho_\psi$ for some unit vector $\psi \in H$. □

Consequently, for the moment just as sets (and even as topological spaces), one has

$$P(\mathscr{D}(H)) = \mathscr{P}_1(H); \tag{2.40}$$

$$P(B(H)) \cong \mathscr{P}_1(H), \tag{2.41}$$

where the second isomorphism is given by (2.33). Defining a state ω_ψ by

$$\omega_\psi(a) = \langle\psi, a\psi\rangle, \tag{2.42}$$

cf. (2.18), the isomorphism (2.41) is the correspondence $\omega_\psi \leftrightarrow e_\psi$, cf. (2.7).

This isomorphism becomes more interesting if we note that both spaces are naturally equipped with ***transition probabilities***. For $P(B(H))$ we canonically have

$$\tau^{B(H)}(\omega_\psi, \omega_\varphi) = \inf\{\omega_\psi(a) \mid a \in B(H), 0 \leq a \leq 1_H, \omega_\varphi(a) = 1\}, \tag{2.43}$$

as in (1.38) for $A = B(H)$. Furthermore, on $\mathscr{P}_1(H)$ we define (with some foresight)

$$\tau^{\mathscr{P}_1(H)}(e, f) = \mathrm{Tr}(ef). \tag{2.44}$$

Theorem 2.12. *The pairs* $(P(B(H)), \tau^{B(H)})$ *and* $(\mathscr{P}_1(H), \tau^{\mathscr{P}_1(H)})$ *are isomorphic as sets with a transition probability. In particular, we have, cf.* (2.13),

$$\tau^{B(H)}(\omega_\psi, \omega_\varphi) = |\langle\psi, \varphi\rangle|^2 = \mathrm{Tr}(e_\psi e_\varphi) = \tau^{\mathscr{P}_1(H)}(e_\psi, e_\varphi). \tag{2.45}$$

Proof. The last equality is a simple computation. The first follows if we can show that the infimum in (2.43) is reached at $a = e_\varphi$. To this end, we prove that for any $0 \leq a \leq 1_H$ with $\omega_\varphi(a) = 1$ we must have $\langle\psi, a\psi\rangle \geq |\langle\varphi, \psi\rangle|^2$. Indeed, the condition $\omega_\varphi(a) = \langle\varphi, a\varphi\rangle = 1$ with $\|a\| \leq 1$ (which follows from $0 \leq a \leq 1_H$) and $\|\varphi\| = 1$ imply, by Cauchy–Schwarz, that $a\varphi = \varphi$. Since $a^* = a$ (by positivity of a), we also have $a : (\mathbb{C} \cdot \varphi)^\perp \to (\mathbb{C} \cdot \varphi)^\perp$, so we may write $a = e_\varphi + a'$, with $a'\varphi = 0$ and a' mapping $(\mathbb{C} \cdot \varphi)^\perp$ to itself. Then $a \geq 0$ implies $a' \geq 0$. If $\langle\psi, a\psi\rangle < |\langle\varphi, \psi\rangle|^2$, then $\langle\psi, a'\psi\rangle < 0$, which contradicts positivity of a' (and hence of a). □

The theory of observables and spectral resolutions of the kind (1.45) may be worked out completely for the "quantum" transition probabilities in this theorem:

Proposition 2.13. *1. There is a bijective correspondence between self-adjoint operators* $a \in B(H)$ *and observables* f *on* $(\mathscr{P}_1(H), \tau^{\mathscr{P}_1(H)})$ *à la Definition 1.18.6:*

- *Given a self-adjoint operator* a*, define an observable* f_a *at* $e_\psi \in \mathscr{P}_1(H)$ *by*

$$f_a(e_\psi) = \mathrm{Tr}(e_\psi a) = \langle\psi, a\psi\rangle; \tag{2.46}$$

- *Given an observable* $f = \sum_i c_i \tau_{e_i}^{\mathscr{P}_1(H)}$*, define an operator* a_f *by*

$$a_f = \sum_i c_i e_i. \tag{2.47}$$

2. Each such observable $f = f_a$ *has a unique spectral resolution as in* (1.45)*, i.e.,*

$$f_a = \sum_{\lambda \in \sigma(a)} \lambda \cdot \tau_{S_\lambda}, \tag{2.48}$$

where S_λ *is the (automatically orthoclosed) subset of* $\mathscr{P}_1(H)$ *whose elements* e *satisfy* $eH \subseteq H_\lambda$*, where* $H_\lambda \subseteq H$ *is the eigenspace for the eigenvalue* $\lambda \in \sigma(a)$.

3. The product defined by (1.46) - (1.47) *is equal to*

$$f_a^2 = f_{a^2}; \tag{2.49}$$
$$f_a \circ f_b = f_{(ab+ba)/2}. \tag{2.50}$$

Proof. Any spectral decomposition $a = \sum_i \lambda_i |\upsilon_i\rangle\langle\upsilon_i|$ puts f_a as defined in (2.46) in the general form (1.44), with $c_i = \lambda_i$ and $y_i = e_{\upsilon_i}$. The rest should be clear. □

We now turn to the quantum counterpart of Proposition 1.13. The main difference is that although extremal decompositions of mixed states into pure ones always exist, they are no longer unique. For example, for $H = \mathbb{C}^2$, we have

$$\rho \equiv \mathrm{diag}(2/3, 1/3) = \tfrac{2}{3}\rho_{u_1} + \tfrac{1}{3}\rho_{u_2} = \tfrac{1}{2}(\rho_{\xi_1} + \rho_{\xi_2}),$$

where (u_1, u_2) is the standard basis of $\mathbb{C}^2$, and

$$\xi_1 = (\sqrt{2/3}, \sqrt{1/3}),\ \xi_2 = (\sqrt{2/3}, -\sqrt{1/3}).$$

More generally, take any basis (w_i) of $H \cong \mathbb{C}^n$, assume (2.6), and for each i for which $\sqrt{\rho}w_i \neq 0$ (where $\sqrt{\rho} = \sum_i \sqrt{p_i}|\upsilon_i\rangle\langle\upsilon_i|$), define $t_i = \|\sqrt{\rho}w_i\|^2$, as well as the unit vector $\xi_i = \sqrt{\rho}w_i/\|\sqrt{\rho}w_i\|$. Then $\rho = \sum_i t_i \rho_{\xi_i}$ is an extremal decomposition of ρ. The above example corresponds to the special case $t_1 = t_2 = 1/2$, with

$$n = 2,\ p_1 = 2/3,\ p_2 = 1/3,\ w_1 = (1/\sqrt{2}, 1/\sqrt{2}),\ w_2 = (1/\sqrt{2}, -1/\sqrt{2}).$$

One might require the ξ_i to be mutually orthogonal, but even that does not imply uniqueness of the extremal decomposition: take, for example, $\rho = (1/n) \cdot 1_n$, where 1_n is the $n \times n$ unit matrix on $H = \mathbb{C}^n$. Then any basis induces (2.6).

Nonetheless, under appropriate assumptions uniqueness does follow.

Proposition 2.14. *1. Any density operator ρ on H has an extremal decomposition*

$$\rho = \sum_{i=1}^{m} p_i \rho_{\psi_i}, \tag{2.51}$$

where $m \leq \dim(H)$, the p_i are probabilities, and the ψ_i are distinct unit vectors.
2. This decomposition can be chosen such that the ψ_i are mutually orthogonal, in which case it is unique iff each of the non-zero eigenvalues of ρ is simple.

Proof. The existence of the *extremal* decomposition (2.51) of ρ follows from its *spectral* decomposition (2.6), which also proves claim 2. If ρ has some degenerate non-zero eigenvalue, the example just given yields non-uniqueness of (2.51). For the converse direction, use uniqueness of the decomposition (2.6) under the condition that each of the non-zero eigenvalues of ρ is simple. □

In the light of Theorem 2.7, it would be interesting to reformulate Proposition 2.14 directly in terms of the states on $B(H)$; note our standing assumption $\dim(H) < \infty$!

Proposition 2.15. *1. Any state ω on $B(H)$ has an extremal decomposition*

$$\omega = \sum_{i=1}^{m} p_i \omega_i, \tag{2.52}$$

into distinct pure states $\omega_i \in P(B(H))$, where $m \leq \dim(H)$, $p_i > 0$, and $\sum_i p_i = 1$.

2. *The unit vectors* ψ_i *that correspond to the pure states* ω_i *in* (2.52) *via* (2.42) *are mutually orthogonal (and hence are part or all of a basis of* H*) iff*

$$\|\omega_i - \omega_j\| = 2 \ (i \neq j). \tag{2.53}$$

3. *Extremal decompositions* (2.52) *satisfying* (2.53) *exist and correspond bijectively to orthogonal families* (e_i) *of one-dimensional projections on* H *(i.e.,* $e_i e_j = \delta_{ij} e_i$ *and* $\mathrm{Tr}\,(e_i) = 1$*, respectively) for which* $\omega(e_i) > 0$, $\sum_i \omega(e_i) = 1$, *and*

$$\omega(ae_i) = \omega(e_i a), \ a \in B(H). \tag{2.54}$$

In terms of such a family, the decomposition (2.52) *is given by*

$$p_i = \omega(e_i); \tag{2.55}$$

$$\omega_i(a) = \frac{\omega(ae_i)}{\omega(e_i)}. \tag{2.56}$$

Hence an extremal decomposition (2.52) *with all* ω_i *mutually orthogonal in the sense of* (2.53) *is unique iff the family* (e_i) *with the above properties is.*

Proof. Claim 1 clearly follows from no. 3. To prove (2.53), assume (2.42), so that

$$\|\omega_i - \omega_j\| = \sup\{|\langle\psi_i, a\psi_i\rangle - \langle\psi_j, a\psi_j\rangle|, a \in B(H), \|a\| = 1\}. \tag{2.57}$$

Clearly, $|\langle\psi, a\psi\rangle| \leq 1$ when $\|a\| = \|\psi\| = 1$, hence $|\langle\psi_i, a\psi_i\rangle - \langle\psi_j, a\psi_j\rangle| \leq 2$, and the upper bound $\|\omega_i - \omega_j\| = 2$ in (2.57) is reached iff $|\langle\psi_1, a\psi_1\rangle| = 1$ and $\langle\psi_2, a\psi_2\rangle = -\langle\psi_1, a\psi_1\rangle$. By Cauchy–Schwarz, this holds iff $a\psi_1 = \lambda\psi_1$ as well as $a\psi_2 = -\lambda\psi_2$ for some $\lambda \in \mathbb{T}$. If $\psi_i \perp \psi_j$, then this is accomplished by the operator $a = |\psi_i\rangle\langle\psi_i| - |\psi_j\rangle\langle\psi_j|$; note that $\sigma(a) = \{-1, 1\}$ for $\dim(H) = 2$ and $\sigma(a) = \{-1, 0, 1\}$ for $\dim(H) > 2$, so indeed $\|a\| = 1$ by (A.47). If, on the other hand, $\langle\psi_i, \psi_j\rangle \neq 0$, then no a with $\|a\| = 1$ can meet these eigenvalue equations. One way to see this is to reduce to $H = \mathbb{C}^2$, since a in (2.57) can be replaced by eae, where e is the projection onto the linear span of ψ_i and ψ_j. Picking a basis of $\mathbb{C}^2$ (with say $\upsilon_1 = \psi_1$), the two eigenvalue equations for a yield a matrix representation of a, from which $\|a\|^2 = \|a^*a\|$ may be computed by calculating the eigenvalues of a^*a and using (A.47). This gives $\|a\| > 1$ unless $\langle\psi_i, \psi_j\rangle = 0$.

One direction of the proof of the third claim easily follows from Theorem 2.7: any spectral decomposition (2.6) of ρ provides the projections

$$e_i = |\upsilon_i\rangle\langle\upsilon_i| \tag{2.58}$$

of the proposition. For example, eq. (2.54) comes down to $[\rho, e_i] = 0$, which is the case iff e_i commutes with all spectral projections of ρ, which clearly holds for (2.58). Uniqueness of the e_i then corresponds to uniqueness of (2.6) and hence to non-degeneracy of the non-zero eigenvalues p_i of ρ, as in Proposition 2.14.

The opposite direction, i.e., proving that (2.58) exhausts all possibilities for (2.53) - (2.54), is based on the GNS-construction and requires an entire subsection.

2.4 The GNS-construction for matrices

The proof of Proposition 2.15 may be completed on the basis of the GNS-construction began in §1.5, which in this subsection we develop for $A = B(H)$, where, as usual, $\dim(H) < \infty$. In that case, we may use Theorem 2.7 to simplify matters.

First, to prove (1.76) we use (2.33) and cyclicity of the trace, compute the trace by summing over a basis (υ_i) of eigenvectors of a^*a, say $a^*a\upsilon_i = \mu_i\upsilon_i$, where $\mu_i \geq 0$ by positivity of a^*a, and use (A.47) (for a^*a rather than a) to obtain:

$$\begin{aligned}\omega(b^*a^*ab) &= \mathrm{Tr}\,(\rho b^*a^*ab) = \sum_i \langle \upsilon_i, b\rho b^*a^*a\upsilon_i\rangle = \sum_i \mu_i\langle \upsilon_i, b\rho b^*\upsilon_i\rangle \\ &\leq \|a^*a\|\sum_i \langle \upsilon_i, b\rho b^*\upsilon_i\rangle = \|a\|^2\mathrm{Tr}\,(\rho b^*b) = \|a\|^2\omega(b^*b),\end{aligned}$$

where we used $\langle \upsilon_i, b\rho b^*\upsilon_i\rangle = \langle b^*\upsilon_i, \rho b^*\upsilon_i\rangle \geq 0$ to justify the inequality.

We now explain all cases of interest, paying special attention to the ***commutant***

$$\pi_\omega(A)' = \{B \in B(H_\omega) \mid \pi_\omega(a)B = B\pi_\omega(a) \forall a \in A\}; \tag{2.59}$$

to distinguish operators on H from operators on H_ω, we write the latter in capitals. For simplicity we also put $H = \mathbb{C}^n$ (with the standard inner product), so that

$$B(H) = M_n(\mathbb{C}), \tag{2.60}$$

and all operators are matrices. Performing a suitable unitary transformation or change of basis if necessary, we also assume that the unit vectors υ_i in the spectral decomposition (2.6) of ρ form (all or part of) the standard basis $(\upsilon_1, \ldots, \upsilon_n)$ of $\mathbb{C}^n$. As in (1.74), we denote the null space by

$$N_\rho = \{a \in B(H) \mid \mathrm{Tr}\,(\rho a^*a) = 0\}. \tag{2.61}$$

- If $\rho = |\upsilon_j\rangle\langle\upsilon_j|$, the corresponding pure state (2.42) is $\omega(a) = \langle \upsilon_j, a\upsilon_j\rangle$, with

$$N_\rho = \{a \in A \mid a\upsilon_j = 0\}. \tag{2.62}$$

Hence $a \in N_\rho$ iff the j'th column $C_j(a)$ of a vanishes, so we have $a - b \in N_\rho$ iff $C_j(a) = C_j(b)$. Thus the equivalence class $a_\rho \in M_n(\mathbb{C})/N_\rho$ may be identified with $C_j(a)$. Consequently, we obtain

$$H_\rho = M_n(\mathbb{C})/N_\rho \cong \mathbb{C}^n, \tag{2.63}$$

under the unitary isomorphism $u : H_\rho \to \mathbb{C}^n$, $a_\rho \mapsto C_j(a)$, with inverse $u^{-1} : z \mapsto a_\rho$, $z \in \mathbb{C}^n$, where a is the matrix with $C_j(a) = z$ and zeros elsewhere (i.e., $a_{ij} = z_i$ and $a_{ik} = 0$ for all i and $k \neq j$). We likewise write $u^{-1}w = b_\rho$, with $b_{ij} = w_i$ and $b_{ik} = 0$ for all i and $k \neq j$. With $ua_\rho = z$ and $ub_\rho = w$, we obtain (beware: no sum over j!):

$$\langle a_\rho, b_\rho\rangle = \mathrm{Tr}\,(\rho a^*b) = \sum_i \overline{a_{ij}}b_{ij} = \sum_i \overline{z}_i w_i = \langle z, w\rangle_{\mathbb{C}^n} = \langle ua_\rho, ub_\rho\rangle_{\mathbb{C}^n}.$$

The GNS-representation π_ρ, originally given on H_ρ by (1.77), is accordingly transformed to $u\pi_\rho(a)u^{-1} \equiv \hat{\pi}_\rho$ on $\mathbb{C}^n$, which is given by

$$\hat{\pi}_\rho(a)w = u\pi_\rho(a)b_\rho = u(ab)_\rho = C_j(ab) = aw,$$

and the cyclic vector $u\Omega_\rho \in \mathbb{C}^n$ is just the basis vector υ_j from which we started. More generally, for a pure state (2.42) the GNS-representation $\pi_{\omega_\psi}(M_n(\mathbb{C}))$ is equivalent to the defining representation on $\mathbb{C}^n$, with canonical cyclic vector ψ. Finally, since only multiples of the unit matrix commute with all matrices, it follows that

$$\pi_{\omega_\psi}(M_n(\mathbb{C}))' \cong \mathbb{C}. \tag{2.64}$$

• The 'opposite' case occurs when ρ is *invertible*, in other words, when the sum over i in (2.6) has n nonzero terms. Hence

$$\mathrm{Tr}(\rho a^* a) = \sum_{i=1}^{n} p_i \|a\upsilon_i\|^2 \tag{2.65}$$

vanishes iff $a\upsilon_i = 0$ for each i, i.e., $a = 0$, so that $N_\rho = \{0\}$ and hence

$$H_\rho = M_n(\mathbb{C}). \tag{2.66}$$

The GNS-constructed inner product on $M_n(\mathbb{C})$, cf. (1.78), given by

$$\langle a_\rho, b_\rho \rangle = \mathrm{Tr}(\rho a^* b), \tag{2.67}$$

may be transformed into the usual one (2.34) by the following linear map:

$$u : M_n(\mathbb{C}) \to M_n(\mathbb{C}); \tag{2.68}$$

$$u a_\rho = a_\rho \rho^{1/2}. \tag{2.69}$$

This map is unitary from the Hilbert space $(M_n(\mathbb{C}), \langle \cdot, \cdot \rangle_\rho)$ to the Hilbert space $(M_n(\mathbb{C}), \langle \cdot, \cdot \rangle)$, for it is invertible, with inverse $u^{-1}a = a_\rho \rho^{-1/2}$, as well as isometric:

$$\langle u(a), u(b) \rangle = \mathrm{Tr}(\rho^{1/2} a^* b \rho^{1/2}) = \mathrm{Tr}(\rho a^* b) = \langle a_\rho, b_\rho \rangle.$$

The transformed representation $\hat{\pi}_\rho = u\pi_\rho(a)u^{-1}$ on $M_n(\mathbb{C})$ is simply given by

$$\hat{\pi}_\rho(a)b = ab, \tag{2.70}$$

and the cyclic vector $u\Omega_\rho$ in $M_n(\mathbb{C})$ becomes $\rho^{1/2}$, so that, as in (1.73),

$$\langle \rho^{1/2}, \hat{\pi}_\rho(a)\rho^{1/2} \rangle = \mathrm{Tr}(\rho a). \tag{2.71}$$

In this case, the commutant is easily computed to be

$$\hat{\pi}_\rho(M_n(\mathbb{C}))' \cong M_n(\mathbb{C}), \tag{2.72}$$

since any linear map $C : M_n(\mathbb{C}) \to M_n(\mathbb{C})$ that satisfies $C(ab) = aC(b)$ for each $a, b \in M_n(\mathbb{C})$ is of the form $C(a) = ac \equiv R_c(a)$ for some $c \in M_n(\mathbb{C})$, namely $c = C(1)$; to see this, just take $b = 1$. Since this involves *right* multiplication R_c by c, which messes up the order in that $R_c R_d = R_{dc}$, one has a choice in implementing the isomorphism (2.72) either as a *linear anti-homomorphism* (of algebras) $C \mapsto R_c$, or as an *anti-linear homomorphism* $C \mapsto R_{c^*}$ (see also Theorem C.159).

Further insight into the structure of this representation comes from the realization

$$M_n(\mathbb{C}) \cong \mathbb{C}^n \otimes \mathbb{C}^n, \tag{2.73}$$

as Hilbert spaces under the unitary map $v : a \mapsto \sum_{ij} a_{ij} \upsilon_i \otimes \upsilon_j$. This yields

$$v\hat{\pi}_\rho(a)v^* = a \otimes 1_n, \tag{2.74}$$

as an operator on $\mathbb{C}^n \otimes \mathbb{C}^n$, and indeed for any Hilbert spaces H_1, H_2 one has

$$(B(H_1) \bigotimes \mathbb{C} \cdot 1_{H_2})' = \mathbb{C} \cdot 1_{H_1} \bigotimes B(H_2). \tag{2.75}$$

• Finally, in the 'intermediate' case the sum in the spectral decomposition (2.6) has $1 < m < n$ nonzero terms. Using the ensuing (partial) basis $(\upsilon_1, \ldots, \upsilon_m)$ of $\mathbb{C}^m$ (viz. $\mathbb{C}^n$), analogously to (2.66) with (2.73) we obtain, up to unitary equivalence,

$$H_\rho \cong \mathbb{C}^n \otimes \mathbb{C}^m; \tag{2.76}$$
$$\pi_\rho(a) \cong a \otimes 1_m; \tag{2.77}$$
$$\Omega_\rho \cong \sum_{i=1}^{n} \sqrt{p_i}\, \upsilon_i \otimes \upsilon_i; \tag{2.78}$$
$$\pi_\rho(M_n(\mathbb{C}))' \cong M_m(\mathbb{C}). \tag{2.79}$$

The relevance of all this to the decomposition of states on $B(H)$ is as follows.

Proposition 2.16. *Let ω be a state on $B(H) \cong M_n(\mathbb{C})$. Then each decomposition*

$$\omega = \sum_i p_i \omega_i, \tag{2.80}$$

where the p_i are probabilities (but the states ω_i are not necessarily pure) is induced by a family (A_i) of nonzero operators in the commutant $\pi_\omega(B(H))'$ that satisfy:

$$0 \leq A_i \leq 1; \tag{2.81}$$
$$\sum_i A_i = 1. \tag{2.82}$$

Namely, given such a family of operators A_i, the decomposition (2.80) *is given by:*

$$p_i = \langle \Omega_\omega, A_i \Omega_\omega \rangle; \tag{2.83}$$
$$\omega_i(a) = \frac{\langle \Omega_\omega, \pi_\omega(a) A_i \Omega_\omega \rangle}{\langle \Omega_\omega, A_i \Omega_\omega \rangle}. \tag{2.84}$$

Proof. The claim that such a family yields (2.80) is trivial, except for the remark that automatically $p_i > 0$, since $\langle \Omega_\omega, A_i \Omega_\omega \rangle = 0$ would imply $\sqrt{A_i}\Omega_\omega = 0$ and hence

$$\sqrt{A_i} a_\omega = \sqrt{A_i} \pi_\omega(a) \Omega_\omega = \pi_\omega(a) \sqrt{A_i} \Omega_\omega = 0$$

for any $a \in B(H)$; by (1.72) this gives $\sqrt{A_i} = 0$ and therefore $A_i = \sqrt{A_i}^2 = 0$.

Conversely, each state ω_i in (2.80) defines a sesquilinear form Q_i on H_ω by $Q_i(a_\omega, b_\omega) = \omega_i(a^* b)$, which is well defined by $\omega_i(a^* a) \le \omega(a^* a)$ and (A.1), and is positive because ω_i is a state. Proposition A.23 then provides us with a positive operator A_i for which $Q_i(a_\omega, b_\omega) = \langle a_\omega, A_i b_\omega \rangle$, hence $\omega_i(a^* b) = \langle a_\omega, A_i b_\omega \rangle$. Next,

$$\langle a_\omega, A_i \pi_\omega(c) b_\omega \rangle = \langle a_\omega, A_i (cb)_\omega \rangle = \omega_i(a^* cb) = \langle (c^* a)_\omega, A_i b_\omega \rangle = \langle a_\omega, \pi_\omega(c) A_i b_\omega \rangle,$$

so $A_i \in \pi_\omega(B(H))'$. Finally, the bound (2.81) corresponds to $0 \le p_i \le 1$ in (2.80), whilst $\omega(1) = 1$, or equivalently $\sum_i p_i = 1$, yields (2.82). □

We now complete the proof of Proposition 2.15. We assume (2.33), where we initially take ρ to be invertible. We omit the hat in (2.70) as well as the suffix ω or ρ on vectors. As noted, we then have $\Omega_\rho = \rho^{1/2}$, and we also know that A_i is given by $A_i b = b a_i$ for some $a_i \in M_n(\mathbb{C})$, viz. $a_i = A_i 1_n$ (where $1_n = 1_H$ is to be distinguished from $\Omega_\rho = \rho^{1/2}$). In this case, (2.81) means $0 \le \mathrm{Tr}(b^* b a_i) \le 1$ for each b with $\mathrm{Tr}(b^* b) = 1$, which is true iff $0 \le a_i \le 1$, whereas (2.82) immediately yields $\sum_i a_i = 1$. In terms of such a family (a_i) in $M_n(\mathbb{C})$ itself, the decomposition (2.80) of $\omega = \mathrm{Tr}(\rho -)$ into *arbitrary* states ω_i follows from (2.83) - (2.84) as

$$p_i = \mathrm{Tr}(\rho a_i); \tag{2.85}$$

$$\omega_i(a) = \mathrm{Tr}(\rho_i a); \tag{2.86}$$

$$\rho_i = \frac{\rho^{1/2} a_i \rho^{1/2}}{\mathrm{Tr}(\rho a_i)}. \tag{2.87}$$

To obtain *pure and orthogonal* states ω_i, we subsequently ask when the new density matrices ρ_i are mutually orthogonal one-dimensional projections $\rho_i = |\upsilon_i\rangle\langle\upsilon_i|$.

To answer this, we use the spectral theorem (A.37) - (A.38) applied to ρ, which gives $\rho = \sum_j p_j e_j$ and hence $\rho^{1/2} = \sum_j \sqrt{p_j} e_j$, so that

$$\rho^{1/2} a_i \rho^{1/2} = \sum_{j,k} \sqrt{p_j p_k} e_j a_i e_k. \tag{2.88}$$

This can only be proportional to a one-dimensional projection if each a_i is a one-dimensional projection that commutes with all spectral projections e_j of ρ (and hence also commutes with ρ itself), and all further constraints on the a_i may then only be satisfied if $a_i = |\upsilon_i\rangle\langle\upsilon_i|$, for some basis (υ_i) of eigenvector υ_i of ρ.

A similar analysis applies to non-invertible ρ, the only new point being that projections e_i orthogonal to the range of ρ fall into the null space N_ρ, cf. (2.76) - (2.79), and hence do not contribute to (2.52), so that they may be ignored. □

2.5 The Born rule from Bohrification

The Bohrification approach to quantum mechanics studies noncommutative algebras of observables like $B(H)$ through their commutative subalgebras. In this section we show how the Born rule (2.8) emerges from that perspective. Our discussion is based on the interplay between the three kinds of (finite-dimensional) C*-algebras:

- $C(X)$ is a C*-algebra under the pointwise operations (1.18) - (1.20) and the supremum-norm (1.24); we still assume that X is finite.
- $B(H)$ is a C*-algebra under the pointwise operations (2.22) - (2.24) and the operator norm (A.18); our standard assumption remains $\dim(H) < \infty$.
- $C^*(a)$ is the C*-algebra generated by $a \in B(H)$ and 1_H (i.e., the intersection of all unital C*-algebras in $B(H)$ that contain a). If $a^* = a$, then $C^*(a)$ is commutative.

Each of these is *unital*, since $C(X)$ has a unit 1_X (i.e. the function $x \mapsto 1$), $B(H)$ has a unit 1_H (i.e. the operator $\psi \mapsto \psi$), and $C^*(a)$ shares the unit 1_H. The first two classes overlap just in case $\dim(H) = 1$ and X is a singleton (in which case $B(\mathbb{C}) = C(*) = \mathbb{C}$); otherwise, the fundamental difference between the two is that $C(X)$ is *commutative* in that $fg = gf$ for all f, g, whereas $B(H)$ is *non-commutative*. However, the system of C*-algebras $C^*(a)$ within $B(H)$, where $a \in B(H)_{\text{sa}}$ varies, to some extent bridges the gap between the commutative and the non-commutative worlds. This relatively simple situation goes to the heart of exact Bohrification.

Theorem 2.17. *Let $a^* = a \in B(H)$, where H is a finite-dimensional Hilbert space.*

1. *The commutative C*-algebra $C^*(a)$ consists of all polynomials in a.*
2. *Any element of $C^*(a)$ is a linear combination of the spectral projections e_λ of a.*
3. *For functions $f : \sigma(a) \to \mathbb{C}$, the map $f \mapsto f(a)$ defined by*

$$f(a) = \sum_{\lambda \in \sigma(a)} f(\lambda) \cdot e_\lambda. \tag{2.89}$$

gives a (necessarily unital) isomorphism of commutative C-algebras*

$$C(\sigma(a)) \cong \mathbb{C}^{|\sigma(a)|} \cong C^*(a). \tag{2.90}$$

Proof. Noting that any function on the finite subset $\sigma(a)$ of $\mathbb{R}$ is continuous, this is a restatement of Theorem A.15 for finite-dimensional Hilbert spaces. □

We now come to the main point. States on unital C*-algebras A may be defined just as in Definitions 1.14 and 2.5, i.e. as positive linear functionals $\omega : A \to \mathbb{C}$ that satisfy $\omega(1_A) = 1$ (cf. Proposition C.5). Recall Theorem 1.15 and Theorem 2.7.

Theorem 2.18. *Let ω be a state on $B(H)$, represented by a density operator ρ via* (2.33), *and let $a \in B(H)$ be a self-adjoint operator. Then the restriction of ω to $C^*(a) \subset B(H)$ is a state, which also induces a state $\omega_{|C(\sigma(a))}$ on $C(\sigma(a))$ through* (2.89) - (2.90), *i.e., $\omega_{|C(\sigma(a))}(f) = \omega(f(a))$. The probability measure on $\sigma(a)$ that corresponds to the state $\omega_{|C(\sigma(a))}$ on $C(\sigma(a))$, then, is given by the Born rule* (2.9).

Proof. First, the restriction of a state on a given unital C*-algebra to a unital C*-subalgebra remains a state. Second, isomorphisms of unital C*-algebras pull back to state spaces in that, if $\varphi : A \to B$ is an isomorphism, and ω is a state on B, then $\varphi^*\omega : A \to \mathbb{C}$ is a state on A, where $\varphi^*(a) = \omega(\varphi(a))$. We now compute

$$\begin{aligned}\omega_{|C(\sigma(a))}(f) &= \omega(f(a)) = \mathrm{Tr}\,(\rho f(a)) \\ &= \sum_{\lambda\in\sigma(a)} \mathrm{Tr}\,(\rho e_\lambda) f(\lambda) = \sum_{\lambda\in\sigma(a)} p_a(\lambda) f(\lambda) \\ &= E_{P_a}(f), \end{aligned} \tag{2.91}$$

where, from left to right, the first equality is just the definition of $\omega_{|C(\sigma(a))}$, whereas the others in turn follow from (2.33), (2.89), (2.8), and (1.9), respectively. □

Note that Theorem 2.18 implies Theorem 2.2. The simplest nontrivial illustration is:

$$H = \mathbb{C}^n; \tag{2.92}$$

$$\omega = \omega_\psi; \tag{2.93}$$

$$\psi = \sum_{i=1}^n c_i u_i; \tag{2.94}$$

$$a = \mathrm{diag}(\lambda_1, \ldots, \lambda_n) = \sum_{i=1}^n \lambda_i |u_i\rangle\langle u_i|, \tag{2.95}$$

with respect to the standard basis (u_i) of $\mathbb{C}^n$, with all $\lambda_i \in \mathbb{R}$ different, cf. (2.42). The C*-algebra $C^*(a) \cong \mathbb{C}^n$ then consists of all diagonal matrices

$$b = \mathrm{diag}(b_1, \ldots, b_n). \tag{2.96}$$

Since obviously

$$\sigma(a) = \{\lambda_1, \ldots, \lambda_n\}, \tag{2.97}$$

the isomorphism (2.90) is given by

$$f \mapsto \mathrm{diag}(f(\lambda_1), \ldots, f(\lambda_n)). \tag{2.98}$$

The computation (2.91) in the proof of Theorem 2.18 then becomes

$$\begin{aligned}\omega_{\psi|C(\sigma(a))}(f) &= \langle\psi, \mathrm{diag}(f(\lambda_1), \ldots, f(\lambda_n))\psi\rangle = \sum_{i=1}^n |c_i|^2 f(\lambda_i) \\ &= \sum_{i=1}^n p_a(\lambda_i) f(\lambda_i), \end{aligned} \tag{2.99}$$

from which the Born probabilities p_a may be read off as the familiar expressions

$$p_a(\lambda_i) = |c_i|^2. \tag{2.100}$$

For an analogous treatment of the generalized Born rule (2.21), we first refer to Definition A.16 for the the pertinent definitions, especially of the joint spectrum

$$\sigma(\underline{a}) \subseteq \sigma(a_1) \times \cdots \times \sigma(a_n) \subset \mathbb{R}^n$$

of a family $\underline{a} = (a_1, \ldots, a_n)$ of commuting self-adjoint operators. As in the case of a single operator, we define $C^*(\underline{a})$ as the smallest unital C*-subalgebra of $B(H)$ that contains each a_i. Generalizing Theorem A.15, we have:

Theorem 2.19. *Let $\underline{a} = (a_1, \ldots, a_n)$ be commuting self-adjoint operators on H. Then $C^*(\underline{a})$ is commutative, and there is a unique isomorphism of C*-algebras*

$$C^*(\underline{a}) \cong C(\sigma(\underline{a})), \tag{2.101}$$

under which $1_H \in C^*(\underline{a})$ corresponds to the unit function $1_{\sigma(\underline{a})} : \underline{\lambda} \mapsto 1$ in $C(\sigma(a))$, and $a_i \in C^*(\underline{a})$ corresponds to the projection $\pi_i : \underline{\lambda} \mapsto \lambda_i$ in $C(\sigma(\underline{a}))$.

For further discussion, see Appendix A, Theorem A.17.

Theorem 2.18 may then be generalized in the following way, with similar proof.

Theorem 2.20. *Let ω be a state on $B(H)$, represented by a density operator ρ, and let $\underline{a} = (a_1, \ldots, a_n)$ be commuting self-adjoint operators on H. Then the restriction of ω to $C^*(\underline{a}) \subset B(H)$ is a state, which induces a state $\omega_{|C(\sigma(\underline{a}))}$ on $C(\sigma(\underline{a}))$ through the isomorphism* (2.101). *Then the probability measure on the joint spectrum $\sigma(\underline{a})$ that corresponds to $\omega_{|C(\sigma(\underline{a}))}$ is given by the generalized Born rule* (2.21), *i.e.,*

$$p_{\underline{a}}(\underline{\lambda}) = \mathrm{Tr}\,(\rho e_{\underline{\lambda}}). \tag{2.102}$$

Strictly speaking, in the present context one should restrict (2.21) to $\underline{\lambda} \in \sigma(\underline{a})$, but the claim is correct even if one does not, for the (Born) probability assigned to values $\underline{\lambda} \in \sigma(a_1) \times \cdots \times \sigma(a_n)$ that do not lie in $\sigma(\underline{a})$ is simply zero.

As shown in Proposition A.19 in Appendix A, the multi-operator case is a special case of the single-operator case, in that $C^*(\underline{a}) = C^*(a)$ for a suitable self-adjoint operator a. Since the converse is obvious, Theorems 2.18 and 2.20 are equivalent. Corollary A.20 in Appendix A even shows that *any* unital commutative C*-algebra C in $B(H)$ takes the form $C = C^*(a)$ for some self-adjoint operator $a \in B(H)$. Comparing the restrictions of a state ω on $B(H)$ to C as the latter varies therefore comes down to asking how the various Born probability distributions p_a on $C^*(a)$ are related to each other as a varies. It is clear from (2.8) that if p_a and p_b come from the same density operator ρ (as the notation indicates), then for $\lambda \in \sigma(a)$ and $\mu \in \sigma(b)$,

$$e^{(a)}_{\lambda} = e^{(b)}_{\mu} \Rightarrow p_a(\lambda) = p_b(\mu). \tag{2.103}$$

Indeed, this is the only compatibility condition between p_a and p_b, showing that $p_a(\lambda)$ only depends on a and λ through the associated spectral projection $e^{(a)}_{\lambda}$. Condition (2.103) is a version of a general property of quantum mechanics called ***non-contextuality***, which in this case means that, given its spectral projection $e^{(a)}_{\lambda}$, the 'context' operator a is otherwise irrelevant for the Born probability $p_a(\lambda)$.

2.6 The Kadison–Singer Problem

It should be clear from the example in the previous section that *pure* states ω_ψ on $B(H)$ may well give rise to *mixed* states on $C^*(a)$; referring to (2.94) and (2.100), this is the case whenever $c_i \neq 0$ for more than one value of the index i. If, on the other hand, $c_i \neq 0$ for just a single value $i = j$, then $\psi = u_j$ (up to a phase), or, equivalently, $\omega_\psi(a) = \langle u_j, au_j \rangle$. In that case, the given state ω_ψ is pure both on $B(H)$ and on $C^*(a)$, and the associated probability measure $\omega_{\psi|C(\sigma(a))}$ on the spectrum $\sigma(a)$ is supported by a single point, namely $\lambda_j \in \sigma(a)$.

This example suggests a general problem (first posed in the non-trivial case where H is infinite-dimensional by Kadison and Singer in 1959) that is of great relevance for the Bohrification program. Namely, let A be a maximal commutative unital C*-algebra in $B(H)$ and let ω_A be a pure state on A. We may then ask:

1. Does ω_A have an extension to a state ω on $B(H)$ at all (i.e., $\omega_{|A} = \omega_A$)?
2. If so, is ω uniquely determined by its restriction ω_A?
3. Either way, if ω exists, can it be chosen so as to be pure (assuming ω_A is)?

If $\dim(H) < \infty$, all these questions are easy to answer at one stroke:

Theorem 2.21. *Let* $\dim(H) < \infty$ *and let* ω_A *be a pure state on a maximal commutative unital C*-algebra A in* $B(H)$*. Then* ω_A *has a unique extension to a state* ω *on* $B(H)$*, which is necessarily pure.*

Proof. As explained after the proof of Corollary A.20 in Appendix A, we may simply assume that $H = \mathbb{C}^n$ and that A consists of all diagonal matrices; call this collection $D_n(\mathbb{C})$ (for every other case is unitarily equivalent to this one). Clearly,

$$D_n(\mathbb{C}) \cong \mathbb{C}^n, \tag{2.104}$$

from which we see that if ω_A is pure, then it must be given on $b \in D_n(\mathbb{C})$ by

$$\omega_A(b) = b_j, \tag{2.105}$$

for some j, cf. (2.96). If ω exists, it is given by (2.33). Using (2.6), condition (2.105) then enforces the following constraint on the p_i and υ_i (where (u_i) is the standard basis of $\mathbb{C}^n$ and (υ_i) is an orthonormal set diagonalizing the density operator ρ):

$$\sum_i p_i |\langle u_j, \upsilon_i \rangle|^2 = 1. \tag{2.106}$$

Since $\sum_i p_i = 1$ and $|\langle u_j, \upsilon_i \rangle| \leq 1$, eq. (2.106) can only hold, for given j, if

$$|\langle u_j, \upsilon_i \rangle| = 1 \tag{2.107}$$

for all i with $p_i > 0$. Since u_j is a unit vector whilst the (υ_i) are an orthonormal set, (2.107) can only be true if there is a single i for which $p_i > 0$, namely $i = j$ (and hence $p_j = 1$), in which case υ_j must equal u_j up to a phase. Hence $\rho = |u_j\rangle\langle u_j|$, which shows that ρ exists, is unique, and is pure. □

At least in operational interpretations of quantum mechanics, this theorem implies that a *pure* quantum state (i.e., on $B(H)$) is completely determined by the outcome of a measurement of some maximal observable a, whose outcome, after all, gives one of the eigenvalues λ_j in (2.95) and hence fixes the post-measurement state to be the one given by (2.105). This is, indeed, a typical way of preparing a state.

As one might expect, this is no longer true if $A = C^*(a)$ fails to be maximal (in which case a measurement of a would not provide enough information about the quantum state). Namely, suppose $a = \sum_{\lambda\in\sigma(a)} \lambda \cdot e_\lambda$, as in (A.37); the maximal case occurs iff $\mathrm{Tr}\,(e_\lambda) = \dim(H_\lambda) = 1$ for all $\lambda \in \sigma(a)$ (equivalently, all eigenvalues λ_i in (A.37) are different). If not, suppose $\dim(H_\lambda) > 1$ for some λ. Then any unit vector $\psi \in H_\lambda$ gives rise to a pure state ω_ψ on $B(H)$, which remains pure on A (it is given by $\omega_{\psi|A}(a) = \lambda$ and hence induces the Dirac probability measure δ_λ on $\sigma(a)$).

Dropping the purity condition on ω_A loses uniqueness of the extension ω, too, even if A is maximal: take $b = \mathrm{diag}(b_1,\ldots,b_n) \in A = D_n(\mathbb{C})$, and assume that

$$\omega_A(b) = \sum_i p_i b_i \tag{2.108}$$

has more than one term (with $p_i > 0$ and $\sum_i p_i = 1$ as always), cf. (2.105). Then:

- *any* pure state ω_ψ as in (2.94), such that $|c_i|^2 = p_i$ for all i, extends ω_A;
- the "decohered" *mixed* state $\omega = \sum_i p_i |\upsilon_i\rangle\langle\upsilon_i|$ extends ω_A, too.

Further insight in the state extension problem comes from the following result.

Proposition 2.22. *Let A be any unital C*-algebra in $B(H)$ (i.e., A is not necessarily commutative) and let ω_A be a pure state on A. Then the set*

$$S_A = \{\omega \in S(B(H)) \mid \omega_{|A} = \omega_A\} \tag{2.109}$$

of all states on $B(H)$ whose restriction $\omega_{|A}$ to A is the given state ω_A, is a compact convex subspace of the total state space $S(B(H))$ of $B(H)$, whose extreme boundary $\partial_e S_A$ consist of pure states on $B(H)$, i.e., $\partial_e S_A \subset P(B(H))$. Consequently, ω_A has a unique extension to a state on $B(H)$ iff it has a unique pure extension.

Proof. Convexity and (w^*) compactness are obvious. Let $\omega \in \partial_e S_A$ and suppose $\omega = t\omega_1 + (1-t)\omega_2$ for some $t \in (0,1)$ and $\omega_1, \omega_2 \in S(B(H))$. By assumption, $\omega_A = \omega_{|A} = t\omega_{1|A} + (1-t)\omega_{2|A}$ is pure on A, so $\omega_{1|A} = \omega_{2|A} = \omega_A$, hence $\omega_1, \omega_2 \in S_A$. Since $\omega \in \partial_e S_A$, this implies $\omega_1 = \omega_2 = \omega$. Hence ω is pure on $B(H)$.

Finally, S_A is a singleton iff its boundary $\partial_e S_A$ is (since any state in S_A has a convex decomposition in terms of states in its boundary), yielding the last claim. □

This proposition remains true for infinite-dimensional H (and even for arbitrary C*-algebras), but Theorem 2.21 becomes much more complicated. As we shall see, maximal commutative unital C*-subalgebra of $B(H)$ are no longer unique up to unitary equivalence, and the validity of the claim depends on which type of maximal subalgebra is considered. Also, the proof of what then is called the ***Kadison–Singer Conjecture*** becomes extremely difficult (with questionable relevance to physics).

2.7 Gleason's Theorem

Gleason's Theorem answers the following question in the positive: given probability distributions p_a on $\sigma(a)$, for each self-adjoint operator $a \in B(H)$, satisfying (2.103), is there a single state ω on $B(H)$ inducing these probabilities through the Born rule? This question is closely related to various others that involve equivalent structures, cf. Definition 1.1. We denote the unit sphere in H by $H_1 = \{\psi \in H, \|\psi\| = 1\}$, and write $\mathscr{P}(H) = \{e \in B(H) \mid e^2 = e^* = e\}$ for the set of all projections on H.

Definition 2.23. *Let H be a finite-dimensional Hilbert space, with unit sphere H_1.*

1. *A* **probability distribution** *on $\mathscr{P}(H)$ is a map $p : H_1 \to [0,1]$ that satisfies*

$$\sum_{i=1}^{\dim H} p(\upsilon_i) = 1, \textit{ for any basis } (\upsilon_i) \textit{ of } H. \tag{2.110}$$

2. *A* **probability measure** *on $\mathscr{P}(H)$ is a map $P : \mathscr{P}(H) \to [0,1]$ that satisfies:*

$$P(e+f) = P(e) + P(f) \textit{ whenever } ef = 0 \Leftrightarrow eH \perp fH; \tag{2.111}$$

$$P(1_H) = 1. \tag{2.112}$$

Note that p is really defined on $\mathscr{P}_1(H)$, for we have $p(z\upsilon) = p(\upsilon)$ for all $z \in \mathbb{T}$ and $\upsilon \in H_1$; to see this, extend $z\upsilon$ and υ to a basis of H in the same way and use (2.110).

As in Definition 1.1, these notions of probability are equivalent, cf. (A.28):

- Given a probability measure P, one obtains a probability distribution p by

$$p(\upsilon) = P(e_\upsilon). \tag{2.113}$$

- Given a probability distribution p, Lemma 2.24 below guarantees that

$$P(e) = \sum_{i=1}^{\dim(eH)} p(\upsilon_i), \tag{2.114}$$

where (υ_i) is any basis of eH, defines a probability measure P.

Lemma 2.24. *If p is a probability distribution on $\mathscr{P}(H)$ and $L \subset H$ is a linear subspace, with basis (υ_i), then $\sum_{i=1}^{\dim(L)} p(\upsilon_i)$ is independent of this basis choice.*

Proof. Extend (υ_i) to a basis of H by adding a basis (υ'_j) of $L^\perp$. Take another basis (υ''_i) of L and complete it to a basis of H by using the same basis (υ'_j) of $L^\perp$. Then

$$\sum_i p(\upsilon_i) + \sum_j p(\upsilon'_j) = \sum_i p(\upsilon''_i) + \sum_j p(\upsilon'_j) = 1, \tag{2.115}$$

where we once again used (2.110). Hence $\sum_i p(\upsilon_i) = \sum_i p(\upsilon''_i)$. □

Clearly, a state ω on $B(H)$ induces a probability measure P on $\mathscr{P}(H)$ by

$$P(e) = \omega(e) = \mathrm{Tr}(\rho e), \tag{2.116}$$

where ρ is the density operator associated to ω, as in (2.33). Therefore, it is a natural question if any probability measure on $\mathscr{P}(H)$ is induced by some state on $B(H)$ by (2.116). This question is equivalent to the one above:

Proposition 2.25. • *A probability measure P on $\mathscr{P}(H)$ induces non-contextual probability distributions p_a on $\sigma(a)$ for each self-adjoint $a \in B(H)$ by*

$$p_a(\lambda) = P(e_\lambda^{(a)}); \tag{2.117}$$

• *Conversely, a family (p_a) of non-contextual probability distributions (i.e. satisfying (2.103)) gives rise to a probability measure P on $\mathscr{P}(H)$ by*

$$P(e) = p_e(1). \tag{2.118}$$

Proof. As defined by (2.117), p_a is a probability distribution on $\sigma(a)$: by (A.38),

$$\sum_{\lambda\in\sigma(a)} p_a(\lambda) = \sum_{\lambda\in\sigma(a)} P\left(e_\lambda^{(a)}\right) = P\left(\sum_{\lambda\in\sigma(a)} e_\lambda^{(a)}\right) = P(1_H) = 1. \tag{2.119}$$

Conversely, suppose $ef = 0$. Introduce $g = 1 - e - f$, and consider the self-adjoint operator $a = \lambda_1 e + \lambda_2 f + \lambda_3 g$, for three different real numbers $\lambda_1, \lambda_2, \lambda_3$. By (2.103),

$$P(e) = p_e(1) = p_a(\lambda_1),\, P(f) = p_f(1) = p_a(\lambda_2),\, P(g) = p_g(1) = p_a(\lambda_3).$$

Furthermore, since $\sigma(a) = \{\lambda_1, \lambda_2, \lambda_3\}$, we have $p_a(\lambda_1) + p_a(\lambda_2) + p_a(\lambda_3) = 1$ and hence $P(e) + P(f) + P(g) = 1$. Also, $P(e+f) + P(g) = P(e+f+g) = P(1_H) = 1$. The last two equations give $P(e+f) = P(e) + P(f)$. □

Suppose $(e_i)_{i=1}^N$ is a family of projections on H such that $\sum_i e_i = 1_H$ and $e_i e_j = \delta_{ij} e_i$. Such a family generates a commutative unital C*-algebra $C = C^*(e_1, \ldots, e_N)$ in $B(H)$, which coincides with $C^*(a)$ for $a = \sum_i \lambda_i e_i$, where all $\lambda_i \in \mathbb{R}$ are different, so that $\sigma(a) = \{\lambda_1, \ldots, \lambda_N\}$. All commutative unital C*-algebras in $B(H)$ arise in this way, and C is maximally abelian iff $N = \dim(H)$, i.e., iff each e_i is one-dimensional. The point is that a probability measure P on $\mathscr{P}(H)$ induces a state ω_C on each $C = C^*(e_1, \ldots, e_N)$ (or, for $C = C^*(a)$, a probability measure P_a on $\sigma(a)$):

1. if $a \in C$ is self-adjoint, then we have unique spectral resolutions (A.37), and put

$$\omega_C(a) = \sum_{\lambda\in\sigma(a)} \lambda P(e_\lambda). \tag{2.120}$$

2. if $c = a + ib \in C$ with a and b self-adjoint, we define $\omega_C(c) = \omega_C(a) + i\omega_C(b)$.

By Lemma 2.24, the map ω_C thus defined coincides with the linear extension of the map $e_i \mapsto P(e_i)$ to C, which also shows that ω_C in linear. Clearly, ω_C is a state on C.

Again by Lemma 2.24, the ensuing family of states ω_C on all commutative unital C*-algebras $C \subset B(H)$ is *non-contextual* (or, one might say *compatible*) in the sense that if $b \in C \cap C'$, then $\omega_C(b) = \omega_{C'}(b)$. In particular, if $C' \subset C$, then $\omega_{C|C'} = \omega_C$ (where $\omega_{C|C'}$ is the restriction of ω_C to C'). It is convenient to extend this non-contextual family (ω_C) of states to a well-defined map $\omega : B(H) \to \mathbb{C}$ by putting

$$\omega(a+ib) = \omega_{C^*(a)}(a) + i\omega_{C^*(b)}(b), \, a,b \in B(H), a^* = a, b^* = b. \tag{2.121}$$

Definition 2.26. *A* **quasi-state** *on $B(H)$ is a map $\omega : B(H) \to \mathbb{C}$ that is positive ($\omega(a^*a) \geq 0$) and normalized ($\omega(1_H) = 1$), cf. Definition 2.4, and otherwise:*

1. *satisfies $\omega(a) = \omega(a') + i\omega(a'')$, where $a' = \frac{1}{2}(a+a^*)$ and $a'' = -\frac{1}{2}i(a-a^*)$.*
2. *is linear on each commutative unital C*-algebra in $B(H)$.*

Note that a' and a'' are self-adjoint, so that ω is fixed by its values on $B(H)_{\mathrm{sa}}$. Hence we have $\omega(za) = z\omega(a)$, $z \in \mathbb{C}$, and $\omega(a+b) = \omega(a) + \omega(b)$ whenever $ab = ba$.

Proposition 2.27. *The map $\omega : B(H) \to \mathbb{C}$ defined by* (2.120) *and* (2.121) *is a quasi-state on $B(H)$. Any quasi-state on $B(H)$ arises in this way, giving a bijective correspondence between quasi-states on $B(H)$ and probability measures on $\mathcal{P}(H)$.*

Proof. The first claim holds by construction. Conversely, a quasi-state ω yields a probability measure P via $P(e) = \omega(e)$, cf. (2.116). □

Theorem 1.15 shows that each state on $C(X)$ is induced by a probability measure (and, trivially, also the other way round). Although Theorem 2.7 is already a quantum version of Theorem 1.15, an even better parallel would involve the probability measures of Definition 2.23. This is indeed what ***Gleason's Theorem*** achieves, *en passant* answering all versions of our lead question:

Theorem 2.28. *Let H be a finite-dimensional Hilbert space of dimension > 2. Then each probability measure P on $\mathcal{P}(H)$ is induced by a unique state ω on $B(H)$ via*

$$P(e) = \omega(e). \tag{2.122}$$

Equivalently, each probability distribution p on $\mathcal{P}(H)$ is given by

$$p(\upsilon) = \langle \upsilon, \rho \upsilon \rangle, \tag{2.123}$$

where ρ is a unique density operator on H. Hence every quasi-state is a state.

This completes the following list (of which 1–5 do not require Gleason's Theorem).

Corollary 2.29. *Let H be a finite-dimensional Hilbert space. The following notions are equivalent (i.e., there are natural bijective correspondence between):*

1. *Non-contextual families of states on commutative unital C*-algebras $C \subset B(H)$;*
2. *Non-contextual families of probability measures on spectra $\sigma(a)$, cf.* (2.103)*;*
3. *Probability distributions on $\mathcal{P}(H)$;*
4. *Probability measures on $\mathcal{P}(H)$;*
5. *Quasi-states on $B(H)$;*
6. *States on $B(H)$.*

2.8 Proof of Gleason's Theorem

The difficulty of Theorem 2.28 should already be clear from the fact that it is false if $\dim(H) = 2$: as we have seen in (2.37), a state on $M_2(\mathbb{C}) = B(\mathbb{C}^2)$ is given by three real parameters, whereas a probability measure P on $\mathscr{P}(\mathbb{C}^2)$ can assign arbitrary values $P(e)$ to one-dimensional projections e, as long as $P(1-e) = 1 - P(e)$. Equivalently, this time from the perspective of probability distributions p, each unit vector in $\mathbb{C}^2$ belongs to a unique basis (up to a phase), so that p can assign an arbitrary value to one of the two vectors in each basis and is unconstrained otherwise.

In higher dimensions, however, one-dimensional projections always belong to infinitely many orthogonal sets, whilst unit vectors belong to infinitely many bases. This constrains the possible values P or p may take, and these constraints turn out to be strong enough to enforce (2.116).

The proof of Theorem 2.28 consists of two nontrivial parts, the second of which is notoriously difficult. By exception in quantum-mechanical reasoning, both involve $\mathbb{R}^3$ as a *real* Hilbert space, whose elements $\mathbf{x} = (x,y,z)$ have standard inner product

$$\langle \mathbf{x}, \mathbf{x}' \rangle = xx' + yy' + zz', \tag{2.124}$$

with the ensuing (Pythagorean) norm and (Euclidean) notion of orthogonality.

Proposition 2.30. *If Theorem 2.28 holds for the real Hilbert space* $\mathbb{R}^3$, *then it holds for any complex finite-dimensional Hilbert space of dimension* > 2.

Proposition 2.31. *Theorem 2.28 holds for the real Hilbert space* $\mathbb{R}^3$.

Proposition 2.30 is a conjunction of two lemmas.

Lemma 2.32. *If* (2.123) *holds for* $\mathbb{R}^3$, *where* ρ *is some symmetric operator, then* (2.123) *holds for* $\mathbb{C}^3$, *where* ρ *is a self-adjoint operator.*

Neither positivity nor normalization of ρ play a role in the argument; once we have (2.123) in this more general sense, the conclusion that ρ be a density operator trivially follows from the definition of p. This also applies to the second sublemma.

Lemma 2.33. *If* (2.123) *holds for* $\mathbb{C}^3$, *then it holds for for any complex finite-dimensional Hilbert space of dimension* > 2.

It will be convenient to extend $p : H_1 \to [0,1]$ to a function $Q : H \to \mathbb{R}$ by

$$Q(0) = 0; \tag{2.125}$$

$$Q(\psi) = \|\psi\|^2 p\left(\frac{\psi}{\|\psi\|}\right) \ (\psi \neq 0), \tag{2.126}$$

so that (2.123) is evidently equivalent to the analogous expression

$$Q(\psi) = \langle \psi, \rho\psi \rangle \ (\psi \in H). \tag{2.127}$$

Given (2.127), the minimax principle for real symmetric matrices implies that Q is maximized on H_1 by $\psi \in H_1$ iff $\rho\psi = \lambda\psi$, where λ is the largest eigenvalue of ρ.

Proof of Lemma 2.32. Suppose $p : \mathbb{C}^3_1 \to [0,1]$ is a probability distribution (in the sense of Definition 2.23). The first step shows that p assumes a maximum on the unit sphere $\mathbb{C}^3_1$ (note that $\mathbb{C}^3_1$ is compact, but we do not know yet if p is continuous!). Since $0 \leq p(\upsilon) \leq 1$ for $\upsilon \in \mathbb{C}^3_1$, $M = \sup\{p(\upsilon), \upsilon \in \mathbb{C}^3_1\}$ exists, and there is a sequence (υ_n) in $\mathbb{C}^3_1$ for which $p(\upsilon_n) \to M$. Since $\mathbb{C}^3_1$ is compact, this sequence has a convergent subsequence, with limit $\upsilon_\infty \in \mathbb{C}^3_1$. Furthermore, we may assume that $\langle \upsilon_n, \upsilon_\infty \rangle \in \mathbb{R}$, for if not, we change to $\upsilon'_n = z_n \upsilon_n$ with $z_n = \langle \upsilon_\infty, \upsilon_n \rangle / |\langle \upsilon_n, \upsilon_\infty \rangle|$.

For each fixed n (with υ_n in the convergent subsequence in question), the real linear span of υ_∞ and υ_n is isomorphic to $\mathbb{R}^2$ as a Hilbert space (with standard inner product), embedded in any $\mathbb{R}^3 \subset \mathbb{C}^3$ one likes (where, once again, $\mathbb{R}^3$ is seen as a real Hilbert subspace in the sense that all inner products of vectors in $\mathbb{R}^3$ are real). By assumption, (2.123) holds on $\mathbb{R}^3$ and hence also on $\mathbb{R}^2 \subset \mathbb{R}^3$, so that, in particular,

$$
\begin{aligned}
|p(\upsilon_\infty) - p(\upsilon_n)| &= |\langle \upsilon_\infty, \rho\upsilon_\infty \rangle - \langle \upsilon_n, \rho\upsilon_n \rangle| = |\langle (\upsilon_\infty - \upsilon_n), \rho(\upsilon_\infty + \upsilon_n) \rangle| \\
&\leq \|\rho\| \|\upsilon_\infty + \upsilon_n\| \|\upsilon_\infty - \upsilon_n\| \leq 2\|\rho\| \|\upsilon_\infty - \upsilon_n\|,
\end{aligned}
$$

since $\|\upsilon_\infty + \upsilon_n\| \leq \|\upsilon_\infty\| + \|\upsilon_n\|$ and $\|\upsilon_\infty\| = \|\upsilon_n\| = 1$. Consequently,

$$
|p(\upsilon_\infty) - M| \leq |p(\upsilon_\infty) - p(\upsilon_n)| + |p(\upsilon_n - M| \leq 2\|\rho\| \|\upsilon_\infty - \upsilon_n\| + |p(\upsilon_n) - M|,
$$

so letting $n \to \infty$ makes both terms on the right-hand side vanish. Hence $p(\upsilon_\infty) = M$.

For reasons to become clear soon, we relabel $\upsilon_\infty \equiv \upsilon_1$. Take any $\upsilon_0 \in \mathbb{C}^3_1$ with $\langle \upsilon_0, \upsilon_1 \rangle = 0$ and consider the *real* Hilbert space $\mathbb{R}^2 \subset \mathbb{C}^3$ spanned by υ_1 and υ_0. By assumption, (2.127) holds, and by the minimax principle, $\rho\upsilon_1 = \lambda_1\upsilon_1 = p(\upsilon_1)\upsilon_1$, with $p(\upsilon_1) = M$. Hence for any $\upsilon = t_0\upsilon_0 + t_1\upsilon_1$, with $t_0, t_1 \in \mathbb{R}$, we have

$$
Q(\upsilon) = \langle t_0\upsilon_0 + t_1\upsilon_1, \rho(t_0\upsilon_0 + t_1\upsilon_1) \rangle = |t_0|^2 p(\upsilon_0) + |t_1|^2 p(\upsilon_1). \tag{2.128}
$$

We claim that this also holds for *complex* coefficients $t_0, t_1 \in \mathbb{C}$. Indeed, by (2.126),

$$
Q(t_0\upsilon_0 + t_1\upsilon_1) = |t_1|^2 Q\left(\frac{|t_0|}{|t_1|} \frac{|t_1|}{|t_0|} \frac{t_0}{t_1} \upsilon_0 + \upsilon_1 \right) = |t_0|^2 p(\upsilon_0) + |t_1|^2 p(\upsilon_1), \tag{2.129}
$$

where we used (2.128) with $\upsilon'_0 = (t_0/t_1)/|(t_0/t_1)|\upsilon_0$ instead of υ_0; this is still a vector orthogonal to υ_1, and we also used $Q(\upsilon'_0) = p(\upsilon'_0) = p(\upsilon_0)$.

We now repeat this analysis on the part $(\mathbb{C}^3_1)_{\perp\upsilon_1}$ of $\mathbb{C}^3_1$ that consists of all unit vectors orthogonal to υ_1, which remains compact. Thus p assumes a maximum at some unit vector $\upsilon_2 \in (\mathbb{C}^3_1)_{\perp\upsilon_1}$, and we may complete the pair (υ_1, υ_2) to a basis $(\upsilon_1, \upsilon_2, \upsilon_3)$ of $\mathbb{C}^3$. With $\upsilon_0 = t_2\upsilon_2 + t_3\upsilon_3$, the above argument (on $(\mathbb{C}^3_1)_{\perp\upsilon_1}$) gives

$$
p(\upsilon_0) = Q(\upsilon_0) = |t_2|^2 p(\upsilon_2) + |t_3|^2 p(\upsilon_3). \tag{2.130}
$$

Combined with (2.129) at $t_0 = 1$, this gives, for any coefficients $t_1, t_2, t_3 \in \mathbb{C}$,

$$Q(t_1\upsilon_1+t_2\upsilon_2+t_3\upsilon_3)=|t_1|^2p(\upsilon_1)+|t_2|^2p(\upsilon_2)+|t_3|^2p(\upsilon_3). \qquad (2.131)$$

Hence (2.127) holds on all of $\mathbb{C}^3$, with

$$\rho=p(\upsilon_1)|\upsilon_1\rangle\langle\upsilon_1|+p(\upsilon_2)|\upsilon_2\rangle\langle\upsilon_2|+p(\upsilon_3)|\upsilon_3\rangle\langle\upsilon_3|. \qquad \square$$

Proof of Lemma 2.33. Let H be a complex finite-dimensional Hilbert space of dimension ≥ 3, equipped with a probability distribution p, and define $Q:H\to\mathbb{R}$ by (2.125) - (2.126). We need to prove (2.127) for some self-adjoint operator ρ. By Propositions A.4 and A.23, this is equivalent to Q being a quadratic form. Since (A.8) evidently holds, we just need to prove (A.9). Take any three-dimensional Hilbert space $L_3\subset H$ containing v and w. By assumption, there exists a self-adjoint operator ρ_{L_3} on L_3 for which (2.127) is valid for all $\psi\in L_3$. Taking $\psi=v$, $\psi=w$, $\psi=v+w$, and $\psi=v-w$ then validates (A.9). This completes the first proof.

This lemma may also be proved without invoking Proposition A.4, as follows.

If v and w are linearly independent, they are contained in a unique two-dimensional subspace $L_2\subset H$, which in turn is contained in a (non-unique) three-dimensional subspace $L_3\subset H$. Take ρ_{L_3} as above and define a bilinear form B on L_2 by $B(v,w)=\langle v,\rho_{L_3}w\rangle$. Defining the associated quadratic form Q by (A.7), we see that (2.125) - (2.126) hold, from which we also conclude that B is independent of the choice of $L_3\supset L_2$. If v and w are linearly dependent, a similar argument shows that B is independent of the choice of the subspace L_2 containing v and w. Hence $B:H\times H\to\mathbb{C}$ is well defined, and to conclude that it is a self-adjoint form we need to check that $B(v,\lambda w+x)=\lambda B(v,w)+B(v,x)$ for all $v,w,x\in V$, $\lambda\in\mathbb{C}$, cf. Definition A.1. If v,w, and x are linearly independent, this can be done by passing to the unique three-dimensional subspace $L_3'\subset H$ containing these vectors. If they are not, we are already done by the previous step. Finally, given that B is a bilinear form, a self-adjoint operator ρ may be reconstructed from Proposition A.23, upon which (2.127) holds by construction. $\square$

Proposition 2.31 again follows from two lemmas by *modus ponens*.

Lemma 2.34. *Any probability distribution on* $\mathbb{R}^3$ *(vf. Definition 2.23) is continuous.*

Lemma 2.35. *Any continuous probability distribution in* $\mathbb{R}^3$ *satisfies* (2.127), *for some self-adjoint operator* ρ.

The operator ρ obtained by Lemma 2.35 is necessarily positive and automatically has unit trace. Another way to phrase this is to take the complex linear span of all probability distribution on the unit sphere $\mathbb{R}^3_1=S^2$ in $\mathbb{R}^3$; this yields a vector space $\mathscr{F}(S^2)$, whose elements are called ***frame functions***. These are *bounded* functions

$$f:S^2\to\mathbb{C},$$

with the property that for any basis $(\mathbf{u}_1,\mathbf{u}_2,\mathbf{u}_3)$ of $\mathbb{R}^3$ one has

$$f(\mathbf{u}_1)+f(\mathbf{u}_2)+f(\mathbf{u}_3)=w(f), \qquad (2.132)$$

where $w(f) \in \mathbb{C}$ does not depend on the basis and is called the ***weight*** of the frame function f. For a probability distribution p we obviously have $w(p) = 1$. The natural norm on $\mathscr{F}(S^2)$ is the supremum-norm inherited from $C(S^2)$, and like the latter, $\mathscr{F}(S^2)$ is closed in this norm (and hence is a Banach space in its own right, a fact that will play an important technical role in Lemma 2.40 below).

As for probability distributions, (2.132) implies a lemma that will often be used:

Lemma 2.36. *If $(\mathbf{u}_1, \mathbf{u}_2)$ is a basis of some two-dimensional linear subspace of $\mathbb{R}^3$, then $f(\mathbf{u}_1) + f(\mathbf{u}_2)$ is independent of the choice of this pair. Hence if C is some great circle in S^2 and $\mathbf{u}_1 \perp \mathbf{u}_2$ for $\mathbf{u}_1, \mathbf{u}_2 \in C$, then $f(\mathbf{u}_1) + f(\mathbf{u}_2)$ only depends on C.*

Furthermore, by similar arguments any frame function is even, i.e., $f(-\mathbf{u}) = f(\mathbf{u})$.

The proof of Lemma 2.34 will actually show that every frame function on S^2 is continuous, whilst the proof of Lemma 2.35 will establish the property that any continuous frame function on S^2 satisfies (2.127), for some self-adjoint operator ρ.

Proof of Lemma 2.34. Let $f : S^2 \to \mathbb{R}$ be a frame function (the complex-valued case follows by decomposing f into a real and an imaginary part). Since constants are frame functions, adding a constant to f if necessary we may assume

$$\inf\{f(\mathbf{x}), \mathbf{x} \in S^2\} = 0. \tag{2.133}$$

Hence for given $\varepsilon > 0$ there exists $\mathbf{p} \in S^2$ with

$$f(\mathbf{p}) < \varepsilon/2. \tag{2.134}$$

Performing a rotation if necessary, we may assume that $\mathbf{p} = (0,0,1)$ is the north pole. It is useful to introduce another frame function $g : S^2 \to \mathbb{R}^+$ by

$$g(\mathbf{x}) = f(\mathbf{x}) + f(R_z(\pi/2)\mathbf{x}), \tag{2.135}$$

where $R_z(\pi/2)$ is the (counter-clockwise) rotation around the z-axis by an angle $\pi/2$. It is easy to see that g is constant on the equator E: for $\mathbf{x} \in E$, consider the basis $(\mathbf{x}, R_z(\pi/2)\mathbf{x}, \mathbf{p})$ of $\mathbb{R}^3$, so that $g(\mathbf{x}) = w(f) - f(\mathbf{p})$ is independent of $\mathbf{x}$.

Furthermore, for any $U \subset S^2$ consider the ***oscillation*** of f at U, defined by

$$\mathrm{Osc}_U(f) = \sup_U(f) - \inf_U(f) \equiv \sup\{f(\mathbf{u}), \mathbf{u} \in U\} - \inf\{f(\mathbf{u}), \mathbf{u} \in U\}. \tag{2.136}$$

If, for given $\mathbf{x} \in S^2$, for any $\varepsilon > 0$ there is a neighbourhood $U \subset S^2$ of $\mathbf{x}$ on which $\mathrm{Osc}_U(f) < \varepsilon$, then $|f(\mathbf{x}) - f(\mathbf{u})| < \varepsilon$ for all $\mathbf{u} \in U$, so that f is continuous at $\mathbf{x}$.

The lengthier steps in the proof of Lemma 2.34 are now as follows:

Lemma 2.37. *Given that $g(\mathbf{p}) < \varepsilon$, there is an open set $U \subset S^2$ on which*

$$\mathrm{Osc}_U(g) < 3\varepsilon.$$

Lemma 2.38. *For any non-negative frame function h, if $\mathrm{Osc}_U(h) \leq \varepsilon'$ for some open U, then each point $\mathbf{x} \in S^2$ has a neighborhood V where*

$$\mathrm{Osc}_V(h) \leq 4\varepsilon'.$$

Assuming these lemmas (to be proved below), continuity of f easily follows:

1. Lemmas 2.37 and 2.38 applied to $h = g$ and $\mathbf{x} = \mathbf{p}$ yield $\mathrm{Osc}_V(g) < 12\varepsilon$ for some neighbourhood V of $\mathbf{p}$. Now $g(\mathbf{p}) < \varepsilon$, hence $\inf\{g(\mathbf{v}), \mathbf{v} \in V\} < \varepsilon$, hence
$$\sup_V(f) \leq \sup_V(g) \leq \mathrm{Osc}_V(g) + \inf_V(g) < 13\varepsilon.$$
2. Since $f \geq 0$ and hence $0 \leq \inf_V(f) \leq \sup_V(f)$, this yields $\mathrm{Osc}_V(f) < 13\varepsilon$.
3. Applying Lemmas 2.38 to $h = f$ and $U = V$ gives that each point $\mathbf{x} \in S^2$ has a neighborhood W where $\mathrm{Osc}_W(f) < 52\varepsilon$.
4. Hence $|f(\mathbf{x}) - f(\mathbf{w})| < 52\varepsilon$ for all $\mathbf{w} \in W$. Since $\varepsilon > 0$ was arbitrary, it follows that f is continuous at $\mathbf{x}$, and since $\mathbf{x}$ was arbitrary, f is continuous on all of S^2.

For $\mathbf{p} \neq \mathbf{u} \in N$, i.e., the open northern hemisphere, let $C_{\mathbf{u}}$ be the unique great circle through $\mathbf{u}$ with one (and hence both) of the following equivalent properties:

- the point of greatest latitude on $C_{\mathbf{u}}$ is $\mathbf{u}$;
- $C_{\mathbf{u}}$ cuts the equator E at two points that are both orthogonal to $\mathbf{u}$.

We write $D_{\mathbf{u}} = C_{\mathbf{u}} \cap N$, and for each $\mathbf{z} \in N$, we introduce the set
$$DD_{\mathbf{z}} = \{\mathbf{x} \in N \mid \exists \mathbf{y} \in D_{\mathbf{x}}, \mathbf{z} \in D_{\mathbf{y}}\}. \tag{2.137}$$

Geometrically, $DD_{\mathbf{z}}$ consists of the points $\mathbf{x}$ on the northern hemisphere from which $\mathbf{z}$ can be reached by "double descent", where we say that $\mathbf{y} \in N$ may be reached from some point $\mathbf{x}$ at higher latitude by (single) descent if $\mathbf{y} \in C_{\mathbf{x}}$. The proof of our lemmas relies on the following two facts from spherical geometry (stated without proof, as they have nothing to do with frame functions, though the second is easy).

Lemma 2.39. *1. The set $DD_{\mathbf{z}}$ in (2.137) has open interior.*
2. For any $\mathbf{x} \in S^2$ there exists $\mathbf{y} \in E$ such that $\mathbf{x}$ lies on the equator $E_{\mathbf{y}}$ relative to $\mathbf{y}$ regarded as the north pole (so in this terminology, $E = E_{\mathbf{p}}$).

Proof of Lemma 2.37. By definition of the infimum, for each $\varepsilon > 0$ there exists $\mathbf{z} \in N$ such that
$$\inf_N g \leq g(\mathbf{z}) \leq \inf_N g + \varepsilon. \tag{2.138}$$
The open U in question will be the interior of $DD_{\mathbf{z}}$. The crucial inequality is
$$g(\mathbf{x}) < g(\mathbf{z}) + 2\varepsilon \quad (\mathbf{x} \in DD_{\mathbf{z}}), \tag{2.139}$$
which together with (2.138) yields $\inf_N g \leq g(\mathbf{x}) \leq \inf_N g + 3\varepsilon$ for each $\mathbf{x} \in DD_{\mathbf{z}}$, whence $\mathrm{Osc}_U(g) \leq 3\varepsilon$. So we need to prove (2.139), given the assumption $g(\mathbf{p}) < \varepsilon$, which is immediate from (2.134) and (2.135).

To prove (2.139), take $\mathbf{r} \in N$ and $\mathbf{s} \in C_{\mathbf{r}} \cap E$, so $\mathbf{r} \perp \mathbf{s}$ and hence
$$g(\mathbf{r}) + g(\mathbf{s}) \leq w(g). \tag{2.140}$$

Furthermore, take $\mathbf{t}, \mathbf{u} \in E$, $\mathbf{t} \perp \mathbf{u}$, so that $(\mathbf{t}, \mathbf{u}, \mathbf{p})$ is a basis and, g being a frame function, we have

$$g(\mathbf{t}) + g(\mathbf{u}) + g(\mathbf{p}) = w(g). \tag{2.141}$$

But by construction g is constant on the equator E, so $g(\mathbf{t}) = g(\mathbf{u}) = k$, hence $2k + g(\mathbf{p}) = w(g)$, and (2.140) yields

$$g(\mathbf{r}) \leq w(g) - g(\mathbf{s}) = 2k + g(\mathbf{p}) - g(\mathbf{s}) = k + g(\mathbf{p}),$$

from which

$$k - g(\mathbf{r}) \geq -g(\mathbf{p}). \tag{2.142}$$

Furthermore, for $\mathbf{q} \in N$, $\mathbf{x}, \mathbf{r} \in D_{\mathbf{q}}$, $\mathbf{x} \perp \mathbf{r}$, there exists $\mathbf{q}' \in D_{\mathbf{q}} \cap E$ such that

$$g(\mathbf{x}) + g(\mathbf{r}) = g(\mathbf{q}) + g(\mathbf{q}') = g(\mathbf{q}) + k,$$

from which, using (2.142), we obtain

$$g(\mathbf{x}) = g(\mathbf{q}) + k - g(\mathbf{r}) \geq g(\mathbf{q}) - g(\mathbf{p}),$$

and hence

$$g(\mathbf{q}) \leq g(\mathbf{x}) + g(\mathbf{p}), \mathbf{q} \in N, \mathbf{x} \in D_{\mathbf{q}}. \tag{2.143}$$

Aplying this twice to the double descent definition domain (2.137), we find

$$g(\mathbf{x}) \leq g(\mathbf{y}) + g(\mathbf{p}) \leq g(\mathbf{z}) + 2g(\mathbf{p}), \mathbf{y} \in D_{\mathbf{x}}, \mathbf{z} \in D_{\mathbf{y}}. \tag{2.144}$$

Since (2.134) and (2.135) imply $g(\mathbf{p}) < \varepsilon$, this yields (2.139). □

Proof of Lemma 2.38. We may assume $\mathbf{p} \subset U = U_{\mathbf{p}}$. Using Lemma 2.39.2, by the argument to come we then move $U_{\mathbf{p}}$ to a neighborhood of $\mathbf{y}$ called $U_{\mathbf{y}}$, and subsequently repeat the argument so as to move $U_{\mathbf{y}}$ to $U_{\mathbf{x}} \equiv V$ as specified in the lemma.

We use spherical coordinates (ϕ, θ) for $\mathbf{x} = (x, y, z) \in S^2$, given by

$$(x = \cos\phi \sin\theta, y = \sin\phi \sin\theta, z = \cos\theta), \ \phi \in [0, 2\pi), \ \theta \in [0, \pi]. \tag{2.145}$$

Hence the north pole $\mathbf{p} = (0,0,1)$ has $\theta = 0$ and ϕ undefined (note that (ϕ, θ) are essentially (longitude, latitude), except that the latter usually starts counting downwards from $\frac{1}{2}\pi$ to $-\frac{1}{2}\pi$, with the north pole having latitude $\frac{1}{2}\pi$). Since U is open, there exists $\delta > 0$ such that all points with $0 \leq \theta < \delta$ belong to U. Pick $\mathbf{y} \in E$ as above, and define $\mathbf{r}$ as the point with the same ϕ as $\mathbf{y}$ but $\theta_{\mathbf{r}} = \theta_{\mathbf{y}} + \frac{1}{2}\delta$ (so that $\mathbf{r}$ lies a little south of $\mathbf{y}$). Then inspection of S^2 shows that one can find a neighborhood $U_{\mathbf{y}}$ of $\mathbf{y}$ with the following property: for any $\mathbf{u} \in U_{\mathbf{y}}$ there exists a great circle C through $\mathbf{r}$ and $\mathbf{u}$ that contains two further points $\mathbf{r}' \in U_{\mathbf{p}}$ and $\mathbf{u}' \in U_{\mathbf{p}}$ such that $\mathbf{r} \perp \mathbf{r}'$ and $\mathbf{u} \perp \mathbf{u}'$. Hence $h(\mathbf{r}) + h(\mathbf{r}') = h(\mathbf{u}) + h(\mathbf{u}')$. Doing this for two different points $\mathbf{u} = \mathbf{u}_1$ and $\mathbf{u} = \mathbf{u}_2$ gives

$$\begin{aligned} h(\mathbf{r}) + h(\mathbf{r}_1') &= h(\mathbf{u}_1) + h(\mathbf{u}_1'); \\ h(\mathbf{r}) + h(\mathbf{r}_2') &= h(\mathbf{u}_2) + h(\mathbf{u}_2'). \end{aligned}$$

Hence $h(\mathbf{u}_1) - h(\mathbf{u}_2) = h(\mathbf{r}_1') - h(\mathbf{r}_2') - (h(\mathbf{u}_1') - h(\mathbf{u}_2'))$, from which we obtain

$$|h(\mathbf{u}_1) - h(\mathbf{u}_2)| \leq |h(\mathbf{r}'_1) - h(\mathbf{r}'_2)| + |(h(\mathbf{u}'_1) - h(\mathbf{u}'_2)| \leq \mathrm{Osc}_U(h) + \mathrm{Osc}_U(h) \leq 2\varepsilon',$$

for by assumption, $\mathrm{Osc}_U(h) \leq \varepsilon'$. Since $\mathbf{u}_1$ and $\mathbf{u}_2$ in $U_\mathbf{y}$ were arbitrary, this gives

$$\mathrm{Osc}_{U_\mathbf{y}}(h) \leq 2\varepsilon'. \tag{2.146}$$

Repeating this with $\mathbf{y}$ as the north pole gives $\mathrm{Osc}_{U_\mathbf{x}}(h) \leq 4\varepsilon'$, i.e., the lemma. □

To prove Lemma 2.35, following Gleason himself we consider the natural action of the rotation group $SO(3)$ (with positive determinant) on $\mathbb{R}^3$, written $R : \mathbf{x} \mapsto R\mathbf{x}$. This action maps S^2 onto itself and hence induces an action U on $C(S^2)$ by pullback:

$$U(R)f(\mathbf{u}) = f(R^{-1}\mathbf{u}). \tag{2.147}$$

By Lemma 2.34 we have inclusions

$$\mathscr{F}(S^2) \subset C_e(S^2) \subset C(S^2), \tag{2.148}$$

where $\mathscr{F}(S^2)$ are the frame functions and $C_e(S^2)$ consists of the even functions in $C(S^2)$; both spaces are obviously stable under the action (2.147). The following facts, due to Weyl, which we state without proof, follow from elementary representation theory, but they are also quite easily verified by explicit computation. Let

$$\psi_\ell(x,y,z) = (x+iy)^\ell, \ell \in \mathbb{N}, \tag{2.149}$$

and restrict this function to S^2, still calling it ψ_ℓ. Let $H_\ell \subset C(S^2)$ be the vector space spanned by all transforms $U(R)\psi_\ell$, $R \in SO(3)$. This vector space:

- consists of all homogeneous polynomials of degree ℓ that are orthogonal (with respect to the inner product in $L^2(S^2)$) to any such polynomials of degree $\ell - 2$;
- has a basis consisting of the spherical harmonics Y_ℓ^m, $m = -\ell, -\ell+1, \ldots, \ell-1, \ell$;
- accordingly, has *finite* dimension equal to $\dim(H_\ell) = 2\ell+1$;
- is irreducible under the natural $SO(3)$-action (2.147).

Indeed, all (necessarily finite-dimensional) irreducible representations of $SO(3)$ arise in this way. Now $\mathscr{F}(S^2)$ is closed under the $SO(3)$-action (2.147), hence so must be $\mathscr{F}(S^2) \cap H_\ell$. Since H_ℓ is irreducible, there are merely two possibilities:

$$H_\ell \subset \mathscr{F}(S^2); \tag{2.150}$$
$$H_\ell \cap \mathscr{F}(S^2) = \{0\}. \tag{2.151}$$

Since for even/odd values of ℓ the space H_ℓ consist of even/odd functions, and $\mathscr{F}(S^2)$ only has even elements, we immediately see that (2.151) applies if ℓ is *odd*. For *even* values of ℓ, we see at once that (2.150) holds for:

- $\ell = 0$, where the constant frame function $f(x,y,z) = c = \frac{1}{3}w(f) \neq 0$ is obviously induced by the operator $\rho = c \cdot 1_3$ (where 1_3 is the 3×3 unit matrix), cf. (2.127);
- $\ell = 2$, which corresponds to frame functions f with weight $w(f) = 0$.

The latter functions are induced by operators ρ with zero trace. To see this, diagonalize ρ in $\mathbb{C}^3$ as in (2.6), without the constraints on p_i. This yields

$$f(\mathbf{x}) = \langle \mathbf{x}, \rho \mathbf{x} \rangle = \sum_{i=1}^{3} p_i |\langle \mathbf{x}, \upsilon_i \rangle|^2. \tag{2.152}$$

For $f \in H_2$, since $H_2 \perp H_0$ in $L^2(S^2)$ we must have

$$\langle 1_{\mathbb{R}^3}, f \rangle_{L^2(S^2)} = \int_{S^2} d^2\mathbf{x}\, f(\mathbf{x}) = 0. \tag{2.153}$$

For any $\upsilon \in \mathbb{C}^3$, we have

$$\int_{S^2} d^2\mathbf{x}\, |\langle \mathbf{x}, \upsilon \rangle|^2 = \frac{4\pi}{3} \|\upsilon\|^2; \tag{2.154}$$

to see this, write $|\langle \mathbf{x}, \upsilon \rangle|^2 = |\upsilon_x|^2 x^2 + |\upsilon_y|^2 y^2 + |\upsilon_z|^2 z^2$, and use the surface element $d^2\mathbf{x} = d\phi d\theta \sin\theta$ associated to the spherical coordinates (2.145) to compute

$$\int_{S^2} d^2\mathbf{x}\, x^2 = \int_{S^2} d^2\mathbf{x}\, y^2 = \int_{S^2} d^2\mathbf{x}\, z^2 = \frac{4\pi}{3}. \tag{2.155}$$

Therefore, from (2.152), noting that $\|\upsilon_i\|^2 = 1$ for each $i = 1,2,3$, we obtain

$$\int_{S^2} d^2\mathbf{x}\, f(\mathbf{x}) = \frac{4\pi}{3} \sum_{i=1}^{3} p_i = \frac{4\pi}{3} \mathrm{Tr}(\rho). \tag{2.156}$$

To settle the case $\ell \geq 4$, all we need to know about the spherical harmonics is that if ℓ is even, then, once again using spherical coordinates, one has

$$Y_\ell^m(x,y,z=0) \sim e^{im\phi} \ (m \text{ even}); \tag{2.157}$$
$$Y_\ell^m(x,y,z=0) = 0 \ (m \text{ odd}). \tag{2.158}$$

If (2.150) holds, then $Y_\ell^m \in \mathscr{F}(S^2)$ for each $m = -\ell, -\ell+1, \ldots, \ell-1, \ell$. But for any (even) $\ell \geq 4$, there are values of m for which Y_ℓ^m cannot be a frame function. To see this, take the following family of bases of $\mathbb{R}^3$, indexed by ϕ:

$$u_1 = (\cos\phi, \sin\phi, 0); \tag{2.159}$$
$$u_2 = (-\sin\phi, \cos\phi, 0); \tag{2.160}$$
$$u_3 = (0,0,1). \tag{2.161}$$

For any frame function f, the value of $f(u_1) + f(u_2) = w(f) - f(u_3)$ must therefore be independent of ϕ. However, from (2.157) - (2.158), we find

$$Y_\ell^m(u_1) + Y_\ell^m(u_2) \sim e^{im\phi} + e^{im(\phi+\pi/2)} = e^{im\phi}(1 + i^m),$$

which is independent of ϕ iff $m = 0$ or $m = 2 \pmod 4$. For $\ell = 0, 2$ these are indeed the only values that occur, but as soon as $\ell \geq 4$, the value $m = 4$ (among others) will ruin it. So (2.150) holds only for $\ell = 0$ and $\ell = 2$, whereas (2.151) is the case for all other $\ell \in \mathbb{N}$. Since H_0 and H_2 occur in $C(S^2)$ with multiplicity one, they cannot have greater multiplicity in $\mathscr{F}(S^2) \subset C(S^2)$, so the above argument suggests that

$$\mathscr{F}(S^2) = H_0 \oplus H_2, \tag{2.162}$$

which would prove the lemma. Fortunately, this is indeed the case, but to complete the argument we need the following technical results (left out by Gleason himself):

Lemma 2.40. *1. Frame functions are uniformly continuous.*

2. *The representation* (2.147) *of* $SO(3)$ *on* $\mathscr{F}(S^2)$ *is continuous (in the usual sense that the map* $(R, f) \mapsto U(R)f$ *from* $SO(3) \times \mathscr{F}(S^2)$ *to* $\mathscr{F}(S^2)$ *is continuous) with respect to the supremum-norm on* $\mathscr{F}(S^2)$.
3. *A continuous representation of a compact group G on a Banach space B is completely reducible (in that B is the closure of the direct sum of all irreducible representations of G that it contains).*

Proof. 1. The first claim follows because S^2 is compact. Another proof starts from the proof of Lemma 2.38, which has the feature that for given $\varepsilon' > 0$, if $y, y' \in E$ with $y' = R_z(\phi)$ for some angle ϕ, then $U_{y'} = R_z(\phi)U_y$ (this is immediately clear from the geometry). Similarly, as $x \in S^2$, different neighborhoods $V = U_x$ are related by a rotation. Hence the size of U_x is independent of x, so that the above proof of continuity established uniform continuity of frame functions also.

2. Let $R_n \to R$ in $SO(3)$ and $f_m \to f$ uniformly in $\mathscr{F}(S^2)$, i.e., $\|f_m - f\|_\infty \to 0$. Then, subtracting and adding a term $U(R_n)f$ and using isometricity of U, i.e.,
$$\|U(R_n)(f_m - f)\|_\infty = \|f_m - f\|_\infty,$$
we obtain the estimate
$$\|U(R_n)f_m - U(R)f\|_\infty \leq \|f_m - f\|_\infty + \|U(R_n)f - U(R)f\|_\infty,$$
cf. (2.147). As $m \to \infty$ the first term on the right-hand side vanishes by assumption, whilst the second vanishes as $n \to \infty$ by uniform continuity of f.
3. This is a Banach space version of the Peter–Weyl theorem, applied to the Banach space of frame functions equipped with the supremum-norm (see Notes). □

Something like this is necessary, because one needs to rule out the possibility that although (by the Stone–Weierstrass Theorem) the polynomial functions on $\mathbb{R}^3$, restricted to S^2, are uniformly dense in $C(S^2)$, so that the linear span of all spherical harmonics and hence of all H_ℓ is uniformly dense in $C(S^2)$, some frame functions might lie in the closure of this direct sum (or, in other words, they are given by uniformly convergent infinite sums of certain Y_ℓ^m). Lemma 2.40 clinches the proof of (2.162), since the third part implies that $\mathscr{F}(S^2)$ would contain all irreducible representations that contribute to the potential infinite sums; but we have already proved that it only contains H_0 and H_2. Thus Lemma 2.35 now also follows. □

2.9 Effects and Busch's Theorem

Gleason's Theorem is easy to state but difficult to prove; ***Busch's Theorem*** is a variation of it, which is more difficult to state but much easier to prove. Logically, Busch's Theorem is weaker than Gleason's, as the assumptions of the latter are contained in those of the former, but physically it appears to be more useful, as it covers more situations. To wit, Busch's Theorem revolves around certain generalizations of projections (which took the centre stage in Gleason's Theorem) called ***effects***: these are (necessarily self-adjoint) operators $a \in B(H)$ that satisfy $0 \le a \le 1_H$, in the sense defined after Proposition A.22. Thus $a \in B(H)$ is an effect iff

$$0 \le \langle \psi, a\psi \rangle \le 1 \ (\psi \in H). \tag{2.163}$$

The set of effects on a Hilbert space H is denoted by $\mathscr{E}(H)$ or by $[0,1]_{B(H)}$. By Theorem A.10, we have (2.163) iff $a^* = a$ and the eigenvalues λ of a lie in the interval $[0,1]$ (i.e., $\sigma(a) \subset [0,1]$). This implies that $\|a\| \le 1$, and conversely, if $a \ge 0$, using the bound $a \le \|a\| \cdot 1_H$ for any self-adjoint operator a, which easily follows from (A.47), we see that for $a \ge 0$, the condition $\|a\| \le 1$ is equivalent to $a \in \mathscr{E}(H)$. In particular, it follows that both projections and density operators are effects.

Proposition 2.41. *1. The set $\mathscr{E}(H)$ of effects on H is a compact convex subset of $B(H)$ in its σ-weak topology, with extreme boundary*

$$\partial_e \mathscr{E}(H) = \mathscr{P}(H), \tag{2.164}$$

i.e., the set of all projections on H (including 0).
2. Each $a \in \mathscr{E}(H)$ has a (typically non-unique) extremal decomposition

$$a = \sum_{i=0}^{m} t_i f_i, \tag{2.165}$$

in which $t_i \ge 0$ and $\sum_i t_i = 1$, and the f_i are projections.

The σ-weak topology on $B(H)$, defined after Corollary A.31, is the right one in this context, but if H is finite-dimensional, as we assume here, this technicality may be ignored, as the claim is even true with respect to the norm topology.

Proof. In Part 1, compactness and convexity are easily checked.

The inclusion $\partial_e \mathscr{E}(H) \subseteq \mathscr{P}(H)$ is equivalent to the claim that any $a \in \mathscr{E}(H)$, $a \notin \mathscr{P}(H)$, does not lie in $\partial_e \mathscr{E}(H)$ and hence admits a convex decomposition

$$a = ta_1 + (1-t)a_2, \ t \in (0,1), a_1, a_2 \in \mathscr{E}(H), a_1 \ne a \ne a_2, \tag{2.166}$$

or, equivalently, a has a nontrivial decomposition $a = \sum_i t_i a_i$, for certain $t_i > 0$ with $\sum_i t_1 = 1$. Indeed, the latter follows from the spectral resolution (A.37), in which the spectral projections e_λ should be rescaled if necessary to as to make the coefficients sum to unity (note that $te \in \mathscr{E}(H)$ for any projection e and any $t \in [0,1]$).

To show the opposite inclusion $\mathscr{P}(H) \subseteq \partial_e \mathscr{E}(H)$, again assume (2.166), where this time $a = e \in \mathscr{P}(H)$ is a projection. "Sandwiching" between $\psi \in H_1$, this yields

$$\langle \psi, a_1 \psi \rangle = \langle \psi, a_2 \psi \rangle = 0, \ \psi \in (eH)^{\perp}; \tag{2.167}$$
$$\langle \psi, a_1 \psi \rangle = \langle \psi, a_2 \psi \rangle = 1, \ \psi \in eH. \tag{2.168}$$

Using $0 \leq a_i \leq 1$, $i = 1, 2$, and (A.37), these equations imply that $a_1 = a_2 = e$.

The claim of part 2 is satisfied by picking the t_i and f_i in terms of the spectral data associated to a (cf. Theorem A.10), as follows: with $m = |\sigma(a)|$, order the eigenvalues $\lambda \in \sigma(a)$ according to $\lambda_1 < \cdots < \lambda_m$, and take:

$$t_0 = 1 - \lambda_m; \tag{2.169}$$
$$t_1 = \lambda_1; \tag{2.170}$$
$$t_i = \lambda_i - \lambda_{i-1} \ (i \geq 2); \tag{2.171}$$
$$f_0 = 0; \tag{2.172}$$
$$f_1 = 1_H; \tag{2.173}$$
$$f_i = \sum_{j=i}^{m} e_{\lambda_i} \ (i \geq 2). \tag{2.174}$$

The validity of (2.165) is then a trivial verification. □

Note that, in general, the extremal decomposition of *a as an effect* differs from its spectral resolutions (A.37) or (A.38) *as a self-adjoint operator*. If $a = \rho$ is a density operator, then the latter, i.e., (2.6), does provide an extremal decomposition of a construed as an effect also, which differs from the one in (2.165). This example shows that extremal decompositions in $\mathscr{E}(H)$ are not necessarily unique. Also, observe that te, for $e \in \mathscr{P}(H)$ and $t \in (0,1)$, does not lie in $\partial_e \mathscr{E}(H)$, since it admits a nontrivial decomposition $te = te + (1-t) \cdot 0$, recalling that $0 \in \mathscr{P}(H) \subset \mathscr{E}(H)$.

Busch's Theorem classifies the following objects.

Definition 2.42. *A* **probability distribution** *on $\mathscr{E}(H)$ is a function $p : \mathscr{E}(H) \to [0,1]$ that satisfies the following two conditions:*

1. *$p(1_H) = 1$;*
2. *If a (finite) family (a_i) of effects satisfies $\sum_i a_i \leq 1_H$, then*

$$p\left(\sum_i a_i\right) = \sum_i p(a_i). \tag{2.175}$$

Lemma 2.43. *if a (finite) family (a_i) of effects satisfies $\sum_i a_i = 1$, then $\sum_i p(a_i) = 1$.*

This trivial observation implies that a probability distribution on $\mathscr{E}(H)$ induces a probability distribution on $\mathscr{P}(H) \subset \mathscr{E}(H)$ by restriction, cf. Definition 2.23. Another way to see this from the perspective of probability measures is to note that any family (e_i) of projections that satisfies $\sum_i e_i \leq 1$ is automatically orthogonal.

Therefore, restricted to $\mathscr{P}(H)$, Definition 2.42 reduces to Definition 2.23.2. To see this, fix j and pick $\psi \in e_jH$. The condition $\sum_i e_i \leq 1$ gives

$$\sum_{i\neq j}\langle\psi, e_i\psi\rangle = \sum_{i\neq j}\|e_i\psi\|^2 \leq 0,$$

but since each term is positive, this implies $e_i\psi = 0$ for each $i \neq j$. Putting $\psi = e_j\varphi$, where $\varphi \in H$ is arbitrary, this gives $e_ie_j\varphi = 0$ for all φ and hence $e_ie_j = 0$.

Clearly, any state ω on $B(H)$ induces a probability distribution p_ω on $\mathscr{E}(H)$ by

$$p_\omega(a) = \omega(a). \tag{2.176}$$

Busch's Theorem shows the converse.

Theorem 2.44. *Any probability distribution p on $\mathscr{E}(H)$ takes the form $p = p_\omega$ for some state ω on $B(H)$, establishing a bijective correspondence between probability distributions on $\mathscr{E}(H)$ and states on $B(H)$.*

Proof. If $p : \mathscr{E}(H) \to [0,1]$ can be extended to a linear map $\omega : B(H) \to \mathbb{C}$, then ω is automatically a state, for normalization is assumed and positivity follows from the fact that any $0 \neq b \geq 0$ has the form $b = ra$ for some $r \in \mathbb{R}^+$ and $0 \leq a \leq 1_H$, namely with $r = \|b\|$ and $a = b/\|b\|$; then $a \geq 0$ and $\|a\| = 1$, so that, as explained earlier, a is an effect. Hence $\omega(b) = \omega(ra) = rp(a) \geq 0$. To achieve this extension:

1. We show that $p(ra) = rp(a)$ for all $r \in \mathbb{Q} \cap [0,1]$ and $0 \leq a \leq 1_H$. Indeed, for any such a and $n \in \mathbb{N}$ we write $a = (a + \cdots + a)/n$ (n terms), so that by (2.175), $p(a) = np(a/n)$. Similarly, for any $m \in \mathbb{N}$ and $0 \leq b \leq 1_H/m$, we have $p(mb) = mp(b)$. Take integers m, n such that $(m/n) \in [0,1]$ and put $b = a/n$, so that
$$p\left(\frac{m}{n}a\right) = mp\left(\frac{a}{n}\right) = \frac{m}{n}p(a). \tag{2.177}$$
2. We next prove that $p(ta) = tp(a)$ for all $t \in [0,1]$ and $0 \leq a \leq 1_H$. Positivity of p yields $p(a) \leq p(a')$ whenever $0 \leq a \leq a' \leq 1_H$. Given $t \in [0,1]$, take an increasing sequences of rationals (r_n) with $r_n \leq t$, as well as a decreasing sequence of rationals (s_n) with $t \leq s_n$, such that $r_n \uparrow t$ and $s_n \downarrow t$ in $\mathbb{R}$. With step 1, this gives
$$r_np(a) = p(r_na) \leq p(ta) \leq p(s_na) \leq s_np(a).$$
Letting $n \to \infty$, this gives $tp(a) \leq p(ta) \leq tp(a)$, and hence equality.
3. Now extend p to all $a \geq 0$, calling the extension ω, by $\omega(a) = \|a\|p(a/\|a\|)$ at $a \neq 0$ and $\omega(0) = 0$; the previous step then easily yields the compatibility property $\omega_{|[0,1]_{B(H)}} = p$ and the scaling property $\omega(ta) = t\omega(a)$ for each $t \geq 0$.
4. For $a \geq 0$ and $b \geq 0$, rescaling and (2.175) yield $\omega(a+b) = \omega(a) + \omega(b)$.
5. For general $a^* = a$ we write $a = a_+ - a_-$, with $a_\pm \geq 0$, as in Proposition A.24, and define ω on all of $B(H)_{\mathrm{sa}}$ by $\omega(a) = \omega(a_+) - \omega(a_-)$. This is well defined despite the lack of uniqueness of (A.74), for if $a = a_+ - a_- = a'_+ - a'_-$, with $a'_\pm \geq 0$, then $a_+ + a'_- = a'_+ + a_-$, whence $\omega(a_+) - \omega(a_-) = \omega(a'_+) - \omega(a'_-)$.

This argument also shows that ω remains linear on general self-adjoint a and b, since $a+b=(a_+ + b_+) - (a_- + b_-)$ is a decomposition with $(a_\pm + b_\pm) \geq 0$.

6. Finally, for general $c \in B(H)$ we (uniquely) decompose $c = a+ib$, $a^* = a$, $b^* = b$, cf. the proof of Corollary A.20, and put $\omega(c) = \omega(a) + i\omega(b)$. □

To close, we give a very brief and superficial introduction to effects as they arise from modern ("operational") quantum measurement theory. This theory associates quantum data to classical data through the concept of a ***Positive Operator Valued Measure*** or ***POVM***. Relative to some given "classical" space X (taken finite here) and Hilbert space H (assumed finite-dimensional), a POVM is defined as a map

$$\mathsf{A} : \mathscr{P}(X) \to \mathscr{E}(H) \tag{2.178}$$

that satisfies $\mathsf{A}(X) = 1_H$ as well as $\mathsf{A}(U \cup V) = \mathsf{A}(U) + \mathsf{A}(V)$ whenever $U \cap V = \emptyset$, cf. Definition 1.1. Equivalently, a POVM is a map

$$\mathsf{a} : X \to \mathscr{E}(H) \tag{2.179}$$

that satisfies

$$\sum_{x \in X} \mathsf{a}(x) = 1_H. \tag{2.180}$$

As in the classical case, these notions are trivially equivalent through

$$\mathsf{a}(x) = \mathsf{A}(\{x\}); \tag{2.181}$$

$$\mathsf{A}(U) = \sum_{x \in U} \mathsf{a}(x). \tag{2.182}$$

The motivating special case of a POVM is given by some self-adjoint operator $a \in B(H)$, which yields $X = \sigma(a)$ and $\mathsf{a}(\lambda) = e_\lambda$. In that case, each density operator ρ induces a probability distribution on $\sigma(a)$ through the Born rule (2.8). More generally, a probability distribution p on $\mathscr{E}(H)$ and a POVM (2.179) jointly determine a probability distribution p_{a} on X, given by

$$p_{\mathsf{a}}(x) = p(\mathsf{a}(x)). \tag{2.183}$$

Indeed, $p_{\mathsf{a}}(x) \geq 0$ because $a \geq 0$, and $\sum_{x \in X} p_{\mathsf{a}}(x) = 1$ by (2.180) and Lemma 2.43. The idea, then, is that a measurement of some POVM a has (classical) outcome x with probability $p_{\mathsf{a}}(x)$; this generalizes the traditional dogma that a measurement of an observable a has outcome $\lambda \in \sigma(a)$ with (Born) probability (2.8). Indeed, combined with (2.33), Busch's Theorem shows that we necessarily have

$$p_{\mathsf{a}}(x) = \mathrm{Tr}\,(\rho\mathsf{a}(x)), \tag{2.184}$$

for some density operator ρ. So nothing has been gained by introducing Definition 2.42, expect perhaps for the insight that, as in Gleason's Theorem, it is the non-contextuality of a probability distribution on $\mathscr{E}(H)$—in that $p(a(x))$ is independent of the POVM a which $a(x)$ forms part of—that eventually enforces (2.184).

2.10 The quantum logic of Birkhoff and von Neumann

In §1.4 we showed that *classical* mechanics has a *classical* logical structure, in which (equivalence classes of) propositions correspond to subsets of phase space. These subsets form a Boolean lattice in which the logical connectives $\neg$, $\wedge$, and $\vee$ fornegation, disjunction, and conjunction, respectively, are interpreted as their natural set-theoretic counterparts (i.e., complementation, intersection, and union).

In 1936, Birkhoff and von Neumann proposed a strikingly similar *quantum* logic for *quantum* mechanics, in which (closed) linear subspaces of Hilbert space play the role of (measurable) subsets of phase space, and the basic logical connectives (except implication, which is queerly lacking in this setting) are interpreted as:

$$\neg L = L^{\perp}; \tag{2.185}$$

$$L \wedge M = L \cap M; \tag{2.186}$$

$$L \vee M = L + M, \tag{2.187}$$

where $L^{\perp}$ is the orthogonal complement of L, see (A.29), $L \cap M$ is the (set-theoretic) intersection of L and M, and $L+M$ is the (closed) linear span of L and M. If $\dim(H) < \infty$, as we continue to assume, any linear subspace of H is automatically closed, and the infinite-dimensional case an attractive operator-algebraic and lattice-theoretic structure arises only if the events are taken to be *closed* linear subspaces.

Although the Brouwer–Hilbert debate on the foundations of mathematics had somewhat subsided in 1936, with hindsight it may be argued that the quantum logic of Birkhoff and von Neumann (who had been a "postdoc" *avant la lettre* with Hilbert) was predicated on their desire to preserve not only the ***law of contradiction***

$$\alpha \wedge \neg \alpha = \perp, \tag{2.188}$$

where α is any proposition and $\perp$ is the proposition that is identically false, but also, against Brouwer, the ***law of excluded middle*** (or ***tertium non datur***)

$$\alpha \vee \neg \alpha = \top, \tag{2.189}$$

where $\top$ is the proposition that is identically true. Indeed, in the Birkhoff–von Neumann model (2.185) - (2.187), where $\perp = \{0\}$ and $\top = H$, these are identities. Similarly, their model satisfies the ***law of double negation***

$$\neg\neg\alpha = \alpha, \tag{2.190}$$

which both in classical logic (where it is a tautology) and in intuitionistic logic (where it is rejected in general) is equivalent to (2.189). Also, ***De Morgan's Laws***:

$$\neg(\alpha \vee \beta) = \neg\alpha \wedge \neg\beta; \tag{2.191}$$

$$\neg(\alpha \wedge \beta) = \neg\alpha \vee \neg\beta, \tag{2.192}$$

hold in their quantum logic (despite their origin in *classical* propositional logic).

We will now derive the Birkhoff–von Neumann structure along similar lines as its classical counterpart (cf. §1.4), except that in the absence of the necessary structure for a classical propositional calculus we now rely on semantic entailment alone.

In quantum theory, the role of functions $f : X \to \mathbb{R}$ as observables in classical physics is played by self-adjoint operators $a : H \to H$ on some Hilbert space H, and hence the quantum analogue of an elementary proposition $f \in \Delta$ of classical physics is $a \in \Delta$ (where $\Delta \subset \mathbb{R}$), with special case $a = \lambda$ for $a \in \{\lambda\}$ (with $\lambda \in \mathbb{R}$).

In analogy to the points $x \in X$ of phase space, pure states ω_ψ as in (2.42), or the corresponding density operators e_ψ (where $\psi \in H$ is a unit vector), yield truth assignments to elementary propositions. To start with the simplest case, $a = \lambda$ is:

- ***true*** with respect to ω_ψ iff $p_a^\psi(\lambda) = 1$, see (2.10), or, equivalently, iff $\psi \in H_\lambda$, where $H_\lambda \subseteq H$ is the eigenspace of a for eigenvalue λ, cf. (A.36);
- ***false*** with respect to ω_ψ iff $p_a^\psi(\lambda) = 0$, or, equivalently, iff $\psi \perp H_\lambda$.

The underlying idea here is arguably that, according to some naive operational interpretation of quantum mechanics, a measurement of a in a state ω_ψ would give outcome λ with probability one (zero) iff $a = \lambda$ is true (false) with respect to ω_ψ. If $0 < p_a^\psi(\lambda) < 1$, the "truthmaker" ω_ψ actually *fails to assign a truth value* to $a = \lambda$; the *partial* nature of truthmakers marks a significant difference with the classical case, as does the closely related distinction between *false* and *not true*. Similarly, we say that an elementary proposition $a \in \Delta$ is ***true*** in some state ω_ψ iff

$$P_a^\psi(\Delta) \equiv \|e_\Delta \psi\|^2 = 1, \tag{2.193}$$

cf. (2.9) and (A.42), and ***false*** if $P_a^\psi(\Delta) = 0$. In other words, $a \in \Delta$ is true in ω_ψ iff $\psi \in H_\Delta$, and false if $\psi \perp H_\Delta$,see (A.43). Such propositions may formally be combined using the connectives $\neg$, $\wedge$, and $\vee$ (whose meaning is unfortunately far from clear in this new setting) according to the same (inductive) formation rules as in classical propositional logic. However, the classical truth tables for $\wedge$ and $\vee$ are unsound with regard to the above rules, at least if one eventually wants to arrive at (2.185) - (2.187). For example, ω_ψ may validate neither α nor β, yet it might make $\alpha \vee \beta$ true (assuming that α and β correspond to L and M, respectively, this is the case if $\psi \notin L$ and $\psi \notin M$, yet $\psi \in L + M$). Similarly, ω_ψ may render neither α nor β false, yet it may falsify $\alpha \wedge \beta$. Due to this complication, the approach of §1.4 has to be modified, as follows. Our goal remains to define a ***semantic equivalence relation*** $\sim_H$, which is predicated on an inductive definition of truth we first give.

Definition 2.45. *1. $a \in \Delta$ is* **true** *in ω_ψ iff $P_a^\psi(\Delta) = 1$, and* **false** *if $P_a^\psi(\Delta) = 0$.*
2. The negation $\neg(a \in \Delta)$ of an elementary proposition $a \in \Delta$ is given by $a \in \Delta^c$.
3. The negation $\neg\alpha$ is true iff α is false.
4. The conjunction $\alpha \wedge \beta$ is true iff both α and β are true.
5. De Morgan's Laws (2.191) - (2.192) *and the law of double negation* (2.190) *hold; in particular, the disjunction $\alpha \vee \beta$ is true iff $\neg(\neg\alpha \wedge \neg\beta)$ is true (as per 1–4).*
6. We write $\alpha \models_H \beta$ iff the truth of α implies the truth of β, for each state ω_ψ.
7. We write $\alpha \sim_H \beta$ iff $\alpha \models_H \beta$ and $\beta \models_H \alpha$.
8. If $\alpha \sim_H \beta$, then $\neg\alpha \sim_H \neg\beta$.

Lemma 2.46. *Definition 2.45 implies the following rules:*

1. *Our earlier truth attributions for the case $a \in \Delta$ with $\Delta = \{\lambda\}$. In particular, $a = \lambda$ is always false when $\lambda \notin \sigma(a)$, and so is $a \in \Delta$ whenever $\Delta \cap \sigma(a) = \emptyset$.*
2. *$a \in \Delta$ is false relative to ω_ψ iff $\psi \perp H_\Delta$.*
3. *$(a \in \Delta) \wedge (b \in \Gamma)$ is true in ω_ψ iff $\psi \in H_\Delta^{(a)} \cap H_\Gamma^{(b)}$.*
4. *$(a \in \Delta) \vee (b \in \Gamma)$ is true in ω_ψ iff $\psi \in H_\Delta^{(a)} + H_\Gamma^{(b)}$.*

Hence conjunctions behave classically, as part 3 states that $(a \in \Delta) \wedge (b \in \Gamma)$ is true iff $a \in \Delta$ and $b \in \Gamma$ are true). The proof of this lemma uses the following notation.

Definition 2.47. *If e and f are projections on a Hilbert space H, then:*

- *$e \wedge f$ is the projection onto $eH \cap fH$;*
- *$e \vee f$ is the projection onto $eH + fH$, i.e., the (closed) linear span of eH and fH.*

Note that if e and f commute, these reduce to the algebraic expressions

$$e \wedge f = ef; \tag{2.194}$$

$$e \vee f = e + f - ef. \tag{2.195}$$

Furthermore, in case of potential ambiguity we will write $e_\Delta^{(a)}$ for the spectral projection e_Δ as defined by a, and analogously $e_\Gamma^{(b)}$, etc. Similarly for $H_\Delta^{(a)}$ etc.

Proof. The first and third claims are immediate. The second one follows from the relation $e_{\Delta^c} = e_\Delta^\perp = 1 - e_\Delta$, or, equivalently, $H_{\Delta^c} = H_\Delta^\perp$. For the fourth, use Definition 2.45.6, 3, and 2 to infer that $(a \in \Delta) \vee (b \in \Gamma)$ is true iff $(a \in \Delta^c) \wedge (b \in \Gamma^c)$ is false. From the third claim, we note that

$$(a \in \Delta) \wedge (b \in \Gamma) \sim_H \left(e_\Delta^{(a)} \wedge e_\Gamma^{(b)} = 1\right), \tag{2.196}$$

so by Definition 2.45.5, $(a \in \Delta^c) \wedge (b \in \Gamma^c)$ is false iff $e_{\Delta^c}^{(a)} \wedge e_{\Gamma^c}^{(b)} = 1$ is false. Since $e_{\Delta^c}^{(a)} \wedge e_{\Gamma^c}^{(b)} = 1$ is true iff $\psi \in H_{\Delta^c}^{(a)} \cap H_{\Gamma^c}^{(b)}$, claim 2 implies $e_{\Delta^c}^{(a)} \wedge e_{\Gamma^c}^{(b)} = 1$ is false iff

$$\psi \in (H_{\Delta^c}^{(a)} \cap H_{\Gamma^c}^{(b)})^\perp = ((H_\Delta^{(a)})^\perp \cap (H_\Gamma^{(b)})^\perp)^\perp = (H_\Delta^{(a)})^{\perp\perp} + (H_\Gamma^{(b)})^{\perp\perp} = H_\Delta^{(a)} + H_\Gamma^{(b)},$$

which finishes the proof. □

Quite analogously to the classical case, Definition 2.45 implies

$$(a \in \Delta) \models_H (b \in \Gamma) \text{ iff } e_\Delta^{(a)} \subseteq e_\Gamma^{(b)}, \tag{2.197}$$

which, once again, immediately yields $(a \in \Delta) \sim_H (b \in \Gamma)$ iff $e_\Delta^{(a)} = e_\Gamma^{(b)}$. Taking $b = e_\Delta^{(a)}$ and $\Gamma = \{1\}$, analogously to (1.53), as in the above proof we have

$$a \in \Delta \sim_H e_\Delta^{(a)} = 1. \tag{2.198}$$

Furthermore, as in the proof of Lemma 2.46 we find

$$(a \in \Delta) \wedge (b \in \Gamma) \sim_H \left(e_\Delta^{(a)} \wedge e_\Gamma^{(b)} = 1\right); \tag{2.199}$$

$$(a \in \Delta) \vee (b \in \Gamma) \sim_H \left(e_\Delta^{(a)} \vee e_\Gamma^{(b)} = 1\right). \tag{2.200}$$

Consequently, we have the following counterpart of Lemma 1.19:

Lemma 2.48. *Any elementary or composite proposition is semantically equivalent (relative to H) to one of the form $e = 1$, for some projection e. Furthermore,*

$$\neg(e = 1) \sim_H \left(e^\perp = 1\right); \tag{2.201}$$

$$(e = 1) \wedge (f = 1) \sim_H (e \wedge f = 1); \tag{2.202}$$

$$(e = 1) \vee (f = 1) \sim_H (e \vee f = 1). \tag{2.203}$$

At last, the quantum version of Theorem 1.20 reads as follows:

Theorem 2.49. *The set $\mathcal{Q}(H)$ of equivalence classes $[\cdot]_H$ of propositions generated by the elementary propositions $a \in \Delta$ and the logical connectives $\neg$, $\vee$, and $\wedge$, is isomorphic to the set $\mathcal{L}(H)$ of linear subspaces of H, under the map*

$$\varphi : \mathcal{Q}(H) \overset{\cong}{\to} \mathcal{L}(H); \tag{2.204}$$

$$\varphi([a \in \Delta]_H) = e_\Delta^{(a)} H. \tag{2.205}$$

Under this isomorphism, the logical connectives $\neg$, $\wedge$ and $\vee$ turn into orthogonal complementation $(-)^\perp$, intersection $\cap$, and linear span $+$, respectively, in that

$$\varphi([\neg\alpha]_H) = \varphi([\alpha]_X)^\perp; \tag{2.206}$$

$$\varphi([\alpha \wedge \beta]_H) = \varphi([\alpha]_H) \cap \varphi([\beta]_H; \tag{2.207}$$

$$\varphi([\alpha \vee \beta]_H) = \varphi([\alpha]_H) + \varphi([\beta]_H), \tag{2.208}$$

Furthermore, if we define a partial order $\leq$ on $\mathcal{Q}(X)$ by saying that $[\alpha]_H \leq [\beta]_H$ iff $\alpha \models_H \beta$ (which is well defined), then φ maps $\leq$ into set-theoretic inclusion $\subseteq$, i.e.,

$$[\alpha]_H \leq [\beta]_H \text{ iff } \varphi([\alpha]_H) \subseteq \varphi([\beta]_H). \tag{2.209}$$

With respect to these operations, $\mathcal{L}(H)$ is a **modular lattice** *(granted that* $\dim(H) < \infty$*; otherwise, the lattice is merely* **orthomodular**, *cf. §D.1 for terminology).*

Proof. Most of this is immediate from Lemma 2.48, expect for the last claim, which follows from simple computations (and from the Amemiya–Araki Theorem). □

As in the classical case, there is an algebraic reformulation of this result, obtained from the bijective correspondence between (closed) linear subspaces L of H and projections e on H, given by $L = eH$ (see Proposition A.8).

Theorem 2.50. *The set $\mathscr{Q}(H)$ of equivalence classes $[\cdot]_H$ of propositions generated by the elementary propositions $a \in \Delta$ and the logical connectives $\neg$, $\vee$, and $\wedge$, is isomorphic to the set $\mathscr{P}(H)$ of projections on H, under the map*

$$\varphi' : \mathscr{Q}(H) \stackrel{\cong}{\to} \mathscr{P}(H); \tag{2.210}$$

$$\varphi'([a \in \Delta]_H) = e_\Delta^{(a)}, \tag{2.211}$$

where (once again) $\mathscr{P}(H)$ is the set of all projections on H.

Under this map, the logical connectives $\neg$, $\wedge$ and $\vee$ turn into (cf. Definition 2.47):

$$\varphi'([\neg\alpha]_H) = 1 - \varphi'([\alpha]_X) \tag{2.212}$$

$$\varphi'([\alpha \wedge \beta]_H) = \varphi'([\alpha]_H) \wedge \varphi'([\beta]_H); \tag{2.213}$$

$$\varphi'([\alpha \vee \beta]_H) = \varphi'([\alpha]_H) \vee \varphi'([\beta]_H), \tag{2.214}$$

Furthermore, φ' maps the partial order $\leq$ on $\mathscr{Q}(H)$ into the partial order on $\mathscr{P}(H)$ defined by $e \leq f$ iff $eH \subseteq fH$, or equivalently, iff $ef = e$.

Finally, with respect to these operations, $\mathscr{P}(H)$ is an (ortho)modular lattice.

However, unlike (1.65) - (1.68), this result is somewhat unsatisfactory in not being purely algebraic. This may partly be remedied through expressions like

$$e \wedge f = \lim_{n\to\infty} (e \circ f)^n; \tag{2.215}$$

$$e \vee f = 1 - ((1-e) \wedge (1-f)), \tag{2.216}$$

where $e \circ f = ef + fe$, and the (strong) limit in (2.215) should be taken on fixed vectors $\psi \in H$ (upon which it exists in the norm-topology of H). Even so, this specific limit still relies on the underlying Hilbert space, and in any case the expressions fail to be purely algebraic and look pretty artificial. Indeed, the same may be said about Definition 2.45, which, of course, has been fine-tuned with hindsight in order to obtain the "desired" answer in the form of Theorem 1.20, which in turn vindicates the mathematically sweet Birkhoff–von Neumann *Ansatz* (2.185) - (2.187).

In addition, there are serious *conceptual* objections to this kind of quantum logic:

1. Conjunction $\wedge$ and disjunction $\vee$ do not distribute over each other, rendering their interpretation as "and" and "or" obscure.
2. There are propositions α and β (namely those for which $\varphi'([\alpha]_H)$ and $\varphi'([\beta]_H)$ do not commute) for which the conjunction $\alpha \wedge \beta$ is physically undefined.
3. There are states in which $\alpha \vee \beta$ is true whilst neither α nor β is true.
4. There are states in which $\alpha \wedge \beta$ is false whilst neither α nor β is false.
5. In view of Schrödinger's Cat, one would expect the law of excluded middle (2.189) to *fail* in quantum mechanics, yet it *holds* in quantum logic (and this is possible because neither $\vee$ nor $\neg$ has any familiar logical meaning in it).
6. Finally, nothing is said or done about propositions that are neither true nor false.

In Chapter 12, we will therefore replace the doomed quantum logic of Birkhoff and von Neumann by the intuitionistic logic of Brouwer and Heyting.

Notes

All operator theory for this chapter may be found in Kadison & Ringrose (1983).

§2.1. **Quantum probability theory and the Born rule**

The Born rule was first stated by Born (1926b) in the context of scattering theory, following the earlier paper (Born, 1926a) in which Born omitted the absolute value squared signs (corrected in a footnote added in proof). The application to the position operator is due to Pauli (1927), who merely spent a footnote on it. The general formulation is due to von Neumann (1932, §III), following earlier contributions by Dirac (1926b) and Jordan (1927). Both Born and Heisenberg acknowledge the profound influence of Einstein on the probabilistic formulation of quantum mechanics. However, Born and Heisenberg as well as Bohr, Dirac, Jordan, Pauli and von Neumann differed with Einstein about the fundamental nature of the Born probabilities and hence on the issue of determinism. Indeed, whereas Born and the others just listed after him believed the outcome of any individual quantum measurement to be unpredictable in principle, Einstein felt this unpredictability was just caused by the incompleteness of quantum mechanics (as he saw it). See, for example, the invaluable correspondence between Einstein and Born (2005).

Mehra & Rechenberg (2000) provide a very detailed reconstruction of the historical origin of the Born rule within the context of quantum mechanics, whereas von Plato (1994) embeds a briefer historical treatment of it into the more general setting of the emergence of modern probability theory and probabilistic thinking. For the earlier history of probability see Hacking (1975, 1990). See also Landsman (2009).

§2.2. **Quantum observables and states**

Proposition 2.10 is due to von Neumann; see also Chapter 6.

§2.3. **Pure states in quantum mechanics**

This kind of thinking goes back to von Neumann (1932) and Segal (1947ab).

§2.4. **The GNS-construction for matrices**

Again, see §C.12 for the GNS-construction in general.

§2.5. **The Born rule from Bohrification**

See notes to §4.1.

§2.6. **The Kadison–Singer Problem**

The Kadison–Singer Problem was first discussed in Kadison & Singer (1959). See the Notes to §4.3 for more information.

§2.7. **Gleason's Theorem**

§2.8. **Proof of Gleason's Theorem**

Gleason's Theorem is due to Gleason (1957), whose proof we largely follow, with some simplifications due to Varadarajan (1985) and Hamhalter (2004). Lemma 2.40.3 or some analogous result is lacking from these references; it may be found in Lyubich (1988), Chapter 4, §2, Theorem. It is often claimed that Gleason's proof has been superseded by the more elementary one due to Cooke, Keane, & Moran (1985), which avoids all use of harmonic analysis. A similar proof, following up on Cooke et al but using constructive analysis only, was given by Richman & Bridges (1999). However, both because Gleason's use of rotation invariance is very natural, and also since the proof of Cooke et al has already been presented and simplified in two monographs entirely devoted to Gleason's Theorem, viz. Dvurečenskij (1993) and Hamhalter (2004), as well as in the highly efficient book by Kalmbach (1998), we prefer to return to the original source (and add some technical details).

§2.9. **Effects and Busch's Theorem**

Busch's Theorem is from Busch (2003), whose proof we follow almost *verbatim*. See also Caves et al (2004). For the use of POVM's in quantum physics see, e.g., Busch, Grabowski, & Lahti (1998), Davies (1976), Holevo (1982), Kraus (1983), Landsman (1998a, 1999), de Muynck (2002), and Schroeck (1996).

§2.10. **The quantum logic of Birkhoff and von Neumann** Our discussion is based on Rédei (1998), with some modifications though. The original source is Birkhoff & von Neumann (1936).

Chapter 3
Classical physics on a general phase space

Passing from finite phase spaces X to infinite ones yields many fascinating new phenomena, some of which even seem genuinely "emergent" in not having any finite-dimensional shadow, approximate or otherwise. Nonetheless, practically all results in the previous chapter remain valid, typically after the inclusion of some technical condition(s) that restrict the almost unlimited freedom allowed by infinite sets.

One of these restrictions is that in classical physics we assume that our phase space X is *locally compact Hausdorff*, where we recall that a space is:

- ***compact*** if every open cover has a finite subcover;
- ***locally compact*** if every point has a compact neighbourhood;
- ***Hausdorff*** (or T_2) if every pair of distinct points x,y can be separated by open sets (i.e., there are disjoint open sets U_x, U_y that contain x and y, respectively).

This combination of topological properties turns out to be very convenient; it incorporates spaces like $\mathbb{R}^k$ (and more generally all non-pathological manifolds), or lattices like $\mathbb{Z}^n$ (the price is that we exclude systems with an infinite number of degrees of freedom, such as classical field theories). A locally compact Hausdorff space X is ***regular*** in that each $x \in X$ and each closed set $F \subset X$ not containing x can be separated by open sets (i.e., there are disjoint open sets $U_x \ni x$ and $U_F \supset F$).

From the perspective of C*-algebras, the main advantage of using this particular class of spaces is that they are naturally singled out by ***Gelfand's Theorem***:

Theorem 3.1. *Every commutative C*-algebra A is isomorphic to $C_0(X)$ for some locally compact Hausdorff space X, which is unique up to homeomorphism.*

A proof may be found in Appendix C; here we just explain the notation and the main idea behind the proof (cf. Definition C.1, which we do not repeat).

First, $C_0(X)$ is the set of all continuous functions $f : X \to \mathbb{C}$ that ***vanish at infinity***, i.e., for any $\varepsilon > 0$ the set $\{x \in X \mid |f(x)| \geq \varepsilon\}$ is compact, or, equivalently, for any $\varepsilon > 0$ there is a compact set $K \subset X$ such that $|f(x)| < \varepsilon$ for all $x \notin K$. For example, if $X = \mathbb{R}$, then $f(x) = \exp(-x^2)$ lies in $C_0(\mathbb{R})$. If X is compact, then $C_0(X) = C(X)$.

Second, $C_0(X)$ is a vector space under pointwise operations (including pointwise complex conjugation as the involution), and is a Banach space in the ***sup-norm***

K. Landsman, *Foundations of Quantum Theory*,
Fundamental Theories of Physics 188, DOI 10.1007/978-3-319-51777-3_3

$$\|f\|_\infty = \sup_{x\in X}\{|f(x)|\}. \tag{3.1}$$

The space X making A isomorphic to $C_0(X)$, then, is the ***Gelfand spectrum*** $\Sigma(A)$ of A, which we already encountered (cf. Definition 1.4) as the set of nonzero algebra homomorphisms from A to $\mathbb{C}$. This set turns out to be a locally compact Hausdorff space in the topology of pointwise convergence, and the isomorphism $A \to C_0(X)$ is the **Gelfand transform** $a \mapsto \hat{a}$, where $\hat{a}(\omega) = \omega(a)$. Conversely, if X is given, then we associate the commutative C*-algebra $C_0(X)$ to it, as in Chapter 1.

Generalizing Definition 1.14, as a special case of the notion of a state we have:

Definition 3.2. *A* **state** *on $C_0(X)$ is a positive (and hence bounded) linear functional $\omega : C_0(X) \to \mathbb{C}$ with $\|\omega\| = 1$.*

If X is compact, given positivity one has $\|\omega\| = 1$ iff $\omega(1_X) = 1$, cf. Lemma C.4.

The appropriate generalization of Theorem 1.15 then reads (cf. Corollary B.21):

Theorem 3.3. *Let X be a locally compact Hausdorff space. There is a bijective correspondence between states on $C_0(X)$ and probability measures on X, namely*

$$\varphi(f) = \int_X d\mu\, f,\ f \in C_0(X). \tag{3.2}$$

Moreover, pure states correspond to Dirac measures and hence to points of X.

In particular, a nonzero linear functional $\omega : C_0(X) \to \mathbb{C}$ is multiplicative iff it is a pure state. This recovery of probability measures on phase space as states of the associated algebra of observables $C_0(X)$, and of points in phase space as the associated pure states, already familiar from the finite case, remains of great importance.

As in quantum mechanics, many interesting observables in classical mechanics fail to be bounded, let alone C_0; coordinate functions (on non-compact phase spaces) and the usual kinetic energy are a case in point. This is not a serious problem, especially not if, as we shall assume from now on, X is a (smooth) manifold (those unfamiliar with this notion may always have $X = \mathbb{R}^k$ in mind). In that case, there is a very natural class of (typically unbounded) functions on X, viz. $C^\infty(X) \equiv C^\infty(X,\mathbb{R})$, which form a commutative algebra just like $C_0(X) \equiv C_0(X,\mathbb{C})$, and provide the (algebraic) basis for the theory of symmetry and dynamics in classical physics, as we shall now show (the fact that functions in $C^\infty(X)$ may be freely added and multiplied provides a major simplification compared to unbounded operators in quantum mechanics, even self-adjoint ones, which are most easily treated by transforming them into bounded ones, as discussed in §B.21). In fact, the most natural mathematical setting of classical physics is not operator theory, or even symplectic geometry (as even mathematically minded people used to think until the 1980s), but rather the more general and flexible framework of ***Poisson geometry***, to which we now turn.

3.1 Vector fields and their flows

We do not assume familiarity with differential geometry and analysis on manifolds, so in what follows one may assume that $M = \mathbb{R}^k$ for some k. However, whenever possible we will phrase definitions and results in such a way that their more general meaning should be clear to those who *are* familiar with differential geometry etc.

An ***old-fashioned vector field*** on $X = \mathbb{R}^k$ is a map

$$\xi : \mathbb{R}^k \to \mathbb{R}^k; \tag{3.3}$$

$$\xi(x) = (\xi^1(x), \dots, \xi^k(x)), \tag{3.4}$$

which describes something like a hyper-arrow at x. However, this is a coordinate-dependent object, which is hard to generalize to arbitrary manifolds. Therefore, in a modern approach a vector field is seen as the corresponding first-order differential operator $\xi : C^\infty(X) \to C^\infty(X)$ defined by

$$\xi f(x) = \sum_{j=1}^{k} \xi^j(x) \frac{\partial f(x)}{\partial x^j}. \tag{3.5}$$

To make the idea precise that a vector field on X is essentially the same as a first-order differential operator on $C^\infty(X)$, we note that it easily follows from (3.5) that

$$\xi(fg) = \xi(f)g + f\xi(g), \tag{3.6}$$

for any $f, g \in C^\infty(X)$, where the product fg is defined pointwise, i.e.,

$$(fg)(x) = f(x)g(x). \tag{3.7}$$

Similarly, we have pointwise addition and scalar multiplication, i.e., for $s, t \in \mathbb{R}$,

$$(sf + tg)(x) = sf(x) + tg(x). \tag{3.8}$$

This turns $C^\infty(X)$ into a commutative algebra (over $\mathbb{R}$, as $C^\infty(X) \equiv C^\infty(X, \mathbb{R})$).

A ***derivation*** of an algebra A (over $\mathbb{R}$) is a linear map $\delta : A \to A$ satisfying

$$\delta(ab) = \delta(a)b + a\delta(b). \tag{3.9}$$

Thus any vector field on X defines a derivation of the algebra $C^\infty(X)$ by (3.5). Conversely, a deep theorem of differential geometry states that for any manifold X, each derivation of $C^\infty(X)$ takes the form (3.5), at least locally (and for $X = \mathbb{R}^k$ also globally). Therefore, either as a definition or as a theorem, we often simply identify vector fields on X with derivations of $C^\infty(X)$. Derivations have a rich structure:

Definition 3.4. *A (real)* **Lie algebra** *is a (real) vector space equipped with a bilinear map* $[\cdot,\cdot] : A \times A \to A$ *that satisfies* $[a,b] = -[b,a]$ *(and hence* $[a,a] = 0$*) as well as*

$$[a,[b,c]] + [c,[a,b]] + [b,[c,a]] = 0 \text{ (\textbf{Jacobi identity})}. \tag{3.10}$$

It is easy to see that the set $\mathrm{Vec}(X)$ of all old-fashioned vector fields ξ on X (i.e. in the sense (3.5)) forms a real Lie algebra under pointwise vector space operations (i.e., $(s\xi + t\eta)(f) = s\xi f + t\eta f$) and the natural bracket

$$[\xi, \eta] = \xi\eta - \eta\xi. \tag{3.11}$$

Similarly, the set $\mathrm{Der}(A)$ of all derivations on some algebra is a Lie algebra under pointwise vector space operations and Lie bracket

$$[\delta_1, \delta_2] = \delta_1 \circ \delta_2 - \delta_2 \circ \delta_1. \tag{3.12}$$

Of course, the identification of $\mathrm{Vec}(X)$ with $\mathrm{Der}(C^\infty(X))$ identifies (3.11) and (3.12).

Vector fields (or, equivalently, derivations) may be "integrated", at least *locally*, in the following sense. First, a ***curve*** through $x_0 \in X$ is a smooth map $c : I \to X$, where $I \subset \mathbb{R}$ is open and $c(t_0) = x_0$ for some $t_0 \in I$. We usually assume that $0 \in I$ with $t_0 = 0$ and hence $c(0) = x_0$. We then say that c ***integrates*** ξ near x_0 if

$$\dot{c}(t) = \xi(c(t)), \tag{3.13}$$

a somewhat symbolic equality that can be interpreted in two equivalent ways:

- Describing $c : I \to \mathbb{R}^k$ by k functions $c^j : I \to \mathbb{R}$ ($j = 1, \ldots, k$), eq. (3.13) denotes

$$\frac{dc^j(t)}{dt} = \xi^j(c^1(t), \ldots, c^k(t)),\ j = 1, \ldots, k. \tag{3.14}$$

- More abstractly, eq. (3.13) means that for any $f \in C^\infty(X)$ we have

$$\xi f(c(t)) = \frac{d}{dt} f(c(t)). \tag{3.15}$$

To pass from (3.15) to (3.14), we just have to recall (3.5), and note that

$$\frac{d}{dt} f(c(t)) = \frac{d}{dt} f(c^1(t), \ldots, c^k(t)) = \sum_{j=1}^{k} \frac{dc^j(t)}{dt} \frac{\partial f(c(t))}{\partial x^j}. \tag{3.16}$$

The theory of ordinary differential equations shows that such local integral curves exist near any point $x_0 \in X$, and that they are unique in the following sense: if two curves $c_1 : I_1 \to X$ and $c_2 : I_2 \to X$ both satisfy (3.13) with $c_1(0) = c_2(0) = x_0$, then $c_1 = c_2$ on $I_1 \cap I_2$. However, curves that integrate ξ near some point may not be defined for all t, i.e., for $I = \mathbb{R}$. This makes the concept of a ***flow*** of a vector field ξ, which is meant to encapsulate all integral curves of ξ, a bit complicated. We start with the simplest case. We say that a vector field ξ is ***complete*** if for any $x_0 \in X$ there is a curve $c : \mathbb{R} \to X$ satisfying (3.13) with $c(0) = x_0$. The simplest example of a complete vector field is $X = \mathbb{R}$ and $\xi = d/dx$, so that $\varphi_t(x) = x + t$. For an incomplete example, take $X = \mathbb{R}$ and $\xi(x) = x^2 d/dx$. It can be shown that a vector field ξ with compact support (in the sense that the set $\{x \in X \mid \xi(x) \neq 0\}$ is bounded) is complete. In particular, any vector field on a compact manifold is complete.

Definition 3.5. *Let X be a manifold and let $\xi \in \mathrm{Vec}(X)$ be a complete vector field. A* **flow** *of ξ is a smooth map $\varphi : \mathbb{R} \times X \to X$, written*

$$\varphi_t(x) \equiv \varphi(t,x), \tag{3.17}$$

that satisfies

$$\varphi_0(x) = x; \tag{3.18}$$
$$\varphi_s \circ \varphi_t = \varphi_{s+t}, \tag{3.19}$$

and that integrates ξ is the sense that for each $t \in \mathbb{R}$ and $x \in X$,

$$\xi(\varphi_t(x)) = \frac{d}{dt}\varphi_t(x). \tag{3.20}$$

As before, eq. (3.20) by definition means that for each $f \in C^\infty(X)$ we have

$$\xi f(\varphi_t(x)) = \frac{d}{dt} f(\varphi_t(x)), \tag{3.21}$$

or, equivalently, that in local coordinates, where

$$\varphi_t(x) = (\varphi_t^1(x), \ldots, \varphi_t^k(x)), \tag{3.22}$$

we have

$$\frac{d\varphi_t^j(x)}{dt} = \xi^j(\varphi_t(x)),\ j = 1, \ldots, k. \tag{3.23}$$

Indeed, the flow φ of ξ gives the integral curve c of ξ through x_0 by

$$c(t) = \varphi_t(x_0). \tag{3.24}$$

According to the Picard–Lindelöf Theorem in the theory of ordinary differential equations, any complete vector field has a unique flow. In fact, the uniqueness part of this theorem implies that (3.19) is a consequence of (3.20) with (3.18), but it is convenient to state (3.19) separately, so as to make the point that the flow of a complete vector field ξ on X is a smooth $\mathbb{R}$-action on X, as defined by conditions (3.18) - (3.19), whose orbits integrate ξ. In particular, each $\varphi_t : X \to X$ is invertible, with inverse $\varphi_t^{-1} = \varphi_{-t}$. In particular, X is a disjoint union of the integral curves of ξ, which can never cross each other because of the uniqueness of the solution of the initial-value problem (3.13) with $c(0) = x_0$).

If ξ is not complete, we do the best we can by defining the set

$$D_\xi = \{(t,x) \in \mathbb{R} \times X \mid \exists c : I \to X, c(0) = x, t \in I\} \subset \mathbb{R} \times X, \tag{3.25}$$

where it is understood that c satisfies (3.13). Obviously $\{0\} \times X \subset D_\xi$, and (less trivially) it turns out that D_ξ is open. Then a flow of ξ is a map $\varphi : D_\xi \to X$ that satisfies (3.18) for all x, eq. (3.21) for $(t,x) \in D_\xi$, as well as (3.19) whenever defined.

3.2 Poisson brackets and Hamiltonian vector fields

To obtain flows, classical mechanics requires more than a manifold structure:

Definition 3.6. *A* **Poisson bracket** *on a manifold X is a Lie bracket* $\{-,-\}$ *on (the real vector space)* $C^\infty(X)$, *such that for each* $h \in C^\infty(X)$ *the map*

$$\xi_h : f \mapsto \{h, f\} \tag{3.26}$$

is a vector field on X (or, equivalently, a derivation of $C^\infty(X,\mathbb{R})$ *with respect to its structure of a commutative algebra under pointwise multiplication). A manifold X equipped with a Poisson bracket is called a* **Poisson manifold**, $(C^\infty(X),\{\,,\})$ *is called a* **Poisson algebra**, *and* ξ_h *is called the* **Hamiltonian vector field** *of h.*

Unfolding, we have a bilinear map $\{-,-\} : C^\infty(X) \times C^\infty(X) \to C^\infty(X)$ that satisfies

$$\{g,f\} = -\{f,g\}; \tag{3.27}$$
$$\{f,\{g,h\}\} + \{h,\{f,g\}\} + \{g,\{h,f\}\} = 0; \tag{3.28}$$
$$\{f,gh\} = \{f,g\}h + g\{f,h\}. \tag{3.29}$$

Bilinearity and the abstract properties (3.27) - (3.29) imply:

Proposition 3.7. *Each Poisson bracket on X defines a Lie algebra homomorphism*

$$C^\infty(X) \to \mathrm{Der}(C^\infty(X)); \tag{3.30}$$
$$h \mapsto \delta_h, \tag{3.31}$$

or, equivalently, a Lie algebra homomorphism

$$C^\infty(X) \to \mathrm{Vec}(X); \tag{3.32}$$
$$h \mapsto \xi_h. \tag{3.33}$$

The time-honored example is $X = \mathbb{R}^{2n}$, with coordinates $x = (p,q)$ and bracket

$$\{f,g\} = \sum_{j=1}^{n} \left(\frac{\partial f}{\partial p_j}\frac{\partial g}{\partial q^j} - \frac{\partial f}{\partial q^j}\frac{\partial g}{\partial p_j} \right). \tag{3.34}$$

In that case, the Hamiltonian vector field of h is obviously given by

$$\xi_h = \sum_{j=1}^{n} \left(\frac{\partial h}{\partial p_j}\frac{\partial}{\partial q^j} - \frac{\partial h}{\partial q^j}\frac{\partial}{\partial p_j} \right). \tag{3.35}$$

The flow of ξ_h gives the motion of a system with Hamiltonian h. Writing

$$\varphi_t(p,q) = (p(t), q(t)),$$

we see from (3.23) that this flow is given by ***Hamilton's equations***

$$\frac{dp_j(t)}{dt} = -\frac{\partial h(p(t),q(t))}{\partial q^j}; \tag{3.36}$$

$$\frac{dq^j(t)}{dt} = \frac{\partial h(p(t),q(t))}{\partial p_j}. \tag{3.37}$$

Hamiltonians of the special form

$$h(p,q) = \frac{p^2}{2m} + V(q), \tag{3.38}$$

where $p^2 = \sum_j p_j^2$, give ***Newton's equation*** "$F = ma$", where $F_j = -\partial V/\partial q^j$, viz.

$$F_j(q(t)) = m\frac{d^2q^j(t)}{dt^2}. \tag{3.39}$$

Proposition 3.8. *For any vector field ξ on a manifold X, we say that a function $f \in C^\infty(X)$ is* **conserved** *if f is constant along the flow of ξ. If X is a Poisson manifold and $\xi = \xi_h$ is Hamiltonian, then f is conserved iff $\{h,f\} = 0$.*

The proof is trivial. A Poisson bracket on X may also be defined in terms of a ***Poisson tensor***. In coordinates, this is just an anti-symmetric matrix $B^{ij}(x)$ that satisfies

$$\sum_l \left(B^{li}\frac{\partial B^{jk}}{\partial x_l} + B^{lj}\frac{\partial B^{ki}}{\partial x_l} + B^{lk}\frac{\partial B^{ij}}{\partial x_l} \right) = 0, \tag{3.40}$$

for each (i,j,k). In terms of B, the Poisson bracket is then defined abstractly by

$$\{f,g\} = B(df,dg), \tag{3.41}$$

using standard notation of differential geometry, or, in coordinates, by

$$\{f,g\}(x) = \sum_{i,j} B^{ij}(x)\frac{\partial f(x)}{\partial x^i}\frac{\partial g(x)}{\partial x^j}. \tag{3.42}$$

Conversely, a Poisson bracket must come from a Poisson tensor: for any derivation δ on $C^\infty(X)$, the function $\delta(g)$ depends linearly on dg, so if $\delta_f(g) = \{f,g\}$, then $\delta_f(g) = -\delta_g(f)$, so that $\{f,g\}$ depends linearly on both df and dg. This enforces (3.42), upon which (3.41) implies (3.40). A nice example is $X = \mathbb{R}^3$, with

$$\{f,g\}(\mathbf{x}) = x\left(\frac{\partial f}{\partial y}\frac{\partial g}{\partial z} - \frac{\partial f}{\partial z}\frac{\partial g}{\partial y}\right) + y\left(\frac{\partial f}{\partial z}\frac{\partial g}{\partial x} - \frac{\partial f}{\partial x}\frac{\partial g}{\partial z}\right) + z\left(\frac{\partial f}{\partial x}\frac{\partial g}{\partial y} - \frac{\partial f}{\partial y}\frac{\partial g}{\partial x}\right);$$

$$B^{ij}(\mathbf{x}) = \sum_k \varepsilon_{kij}x^k. \tag{3.43}$$

Finally, we say that a Poisson manifold is ***symplectic*** if the corresponding Poisson tensor $B(x)$ is given by an *invertible* matrix, for each $x \in X$. This requires X to be *even-dimensional*. For example, $\mathbb{R}^{2n}$ with Poisson bracket (3.34) is symplectic.

3.3 Symmetries of Poisson manifolds

Two equivalent notions of symmetries of classical physics suggest themselves: one is based on the idea of a Poisson *manifold* (X,B), the other comes from the equivalent notion of a Poisson *algebra* $(C^\infty(X),\{\,,\})$.

Definition 3.9. *1. A symmetry of a Poisson manifold (X,B) is a diffeomorphism $\varphi : X \to X$ (that is, an invertible smooth map with smooth inverse) satisfying*

$$\varphi_* B = B. \tag{3.44}$$

2. A symmetry of a Poisson algebra $(C^\infty(X),\{\,,\})$ is an invertible linear map $\alpha : C^\infty(X) \to C^\infty(X)$ that satisfies (for each $f,g \in C^\infty(X)$):

$$\alpha(fg) = \alpha(f)\alpha(g); \tag{3.45}$$
$$\alpha(\{f,g\}) = \{\alpha(f),\alpha(g)\}. \tag{3.46}$$

Let us define the push-forward φ_* in (3.44). We do this in terms of the ***pullback*** φ^* of a smooth (i.e., infinitely often differentiable) map $\varphi : X \to X$, defined as

$$\varphi^* : C^\infty(X) \to C^\infty(X); \tag{3.47}$$
$$\varphi^* f = f \circ \varphi. \tag{3.48}$$

If φ is a diffeomorphism, the ***push-forward*** φ_* of φ, which acts on derivations, is

$$\varphi_* : \mathrm{Der}(C^\infty(X)) \to \mathrm{Der}(C^\infty(X)); \tag{3.49}$$
$$(\varphi_*\delta)(f) = \delta(\varphi^* f) \circ \varphi^{-1}; \tag{3.50}$$

this may be checked to define a derivation, as follows:

$$\begin{aligned}(\varphi_*\delta)(f \cdot g) &= (\varphi^{-1})^*\delta(\varphi^*(f \cdot g))\\ &= (\varphi^{-1})^*\delta(\varphi^*(f)\varphi^*(g))\\ &= (\varphi^{-1})^*(\delta(\varphi^*(f))\varphi^*(g) + \varphi^*(f)\delta(\varphi^*(g)))\\ &= (\varphi_*\delta)(f) \cdot g + f \cdot (\varphi_*\delta)(g).\end{aligned}$$

If, given coordinates $x = (x^1,\dots,x^k)$ on X, we now (without loss of generality) take our derivation δ to be a vector field $\xi = \sum_j \xi^j \partial/\partial x^j$, and write $\varphi(x) = (\varphi^1(x),\dots,\varphi^l(x))$, for the image $\varphi_*(\xi)$ we obtain

$$\begin{aligned}(\varphi_*\xi)(f)(x) &= (\xi(\varphi^* f))(\varphi^{-1}(x))\\ &= \sum_j \xi^j(\varphi^{-1}(x)) \left(\frac{\partial}{\partial x^j} f \circ \varphi\right)(\varphi^{-1}(x))\\ &= \sum_{j,k} \xi^k(\varphi^{-1}(x)) \frac{\partial f(x)}{\partial x^j} \frac{\partial \varphi^j}{\partial x^k}(\varphi^{-1}(x)),\end{aligned}$$

so that

$$\varphi_*\xi^j(x) = \sum_k \frac{\partial\varphi^j}{\partial x^k}(\varphi^{-1}(x))\xi^k(\varphi^{-1}(x)), \tag{3.51}$$

or, equivalently,

$$\varphi_*\xi^j(\varphi(x)) = \sum_k \frac{\partial\varphi^j}{\partial x^k}(x)\xi^k(x), \tag{3.52}$$

which only depends on $\xi(x)$, so that for each $x \in X$, φ_* may be localized to a linear map $\varphi_*(x) : T_xX \to T_{\varphi(x)}X$. This may be done even if φ is not invertible. Physicists often write this as $\varphi(x) \equiv y = y(x^1,\ldots,x^k)$, $\xi = v$, $\varphi_*\xi = v'$, so that we have a "covariant" transformation rule $(v')^i(y) = \sum_{j=1}^k \frac{\partial y^i(x)}{\partial x^j} v^j(x)$.

Taking tensor products, one obtains similar rules for higher-order tensors. For example, if $N = X$, the transformation rule for the Poisson tensor B reads

$$\varphi_*B^{ij}(\varphi(x)) = \sum_{m,n=1}^k \frac{\partial\varphi^i(x)}{\partial x^m}\frac{\partial\varphi^j(x)}{\partial x^n}B^{mn}(x), \tag{3.53}$$

so that, in coordinates, the invariance requirement (3.44) reads

$$\sum_{m,n=1}^k \frac{\partial\varphi^i(x)}{\partial x^m}\frac{\partial\varphi^j(x)}{\partial x^n}B^{mn}(x) = B^{ij}(\varphi(x)). \tag{3.54}$$

Theorem 3.10. *The two parts of Definition 3.9 are equivalent, in that:*

1. *Given a diffeomorphism $\varphi : X \to X$ satisfying* (3.44)*, the map*

$$\alpha = \varphi^*, \tag{3.55}$$

   i.e., $\alpha(f) = f \circ \varphi$, is linear, invertible, and satisfies (3.45) - (3.46).
2. *Given an invertible linear map $\alpha : C^\infty(X) \to C^\infty(X)$ that satisfies* (3.45) - (3.46)*, there is a unique diffeomorphism $\varphi : X \to X$ inducing α as in* (3.55).
3. *This correspondence defines an anti-isomorphism between the group* $\mathrm{Diff}(X,B)$ *of diffeomorphisms of X satisfying* (3.44) *and the group* $\mathrm{Aut}(C^\infty(X),\{\,,\})$ *of invertible linear maps $\alpha : C^\infty(X) \to C^\infty(X)$ that satisfy* (3.45) - (3.46).

Here an ***anti-isomorphism*** of groups is just an isomorphism that inverts the order of multiplication. This complication may be removed by writing φ^{-1} instead of φ in (3.55), but that change would make the next proposition a bit less natural.

Proof. The first claim is true by construction. The hard part is the second claim, which follows from a more general result about manifolds (note that in our terminology, manifolds are by definition assumed to be Hausdorff):

Proposition 3.11. *Let X and Y be a smooth manifolds. Then* (3.55) *establishes a bijective correspondence between linear maps $\alpha : C^\infty(X) \to C^\infty(Y)$ satisfying* (3.45) *and smooth maps $\varphi : Y \to X$.*

The proof is quite similar to a central part of the proof of Gelfand duality for commutative C*-algebras, in which (3.55) establishes a bijective correspondence between C*-homomorphisms $\alpha : C(X) \to C(Y)$ and continuous maps $\varphi : Y \to X$, where X and Y are compact Hausdorff spaces; see §C.3 and especially Proposition C.22.

For any commutative real algebra A, let $\Sigma(A)$ be the space of non-zero algebra homomorphisms $\omega : A \to \mathbb{R}$ (these are just the non-zero multiplicative linear maps), equipped with the weakest topology that makes each function $\hat{a} : \Sigma(A) \to \mathbb{R}$ continuous, where $\hat{a}(\omega) = \omega(a)$. Furthermore, if B is another commutative real algebra, then any homomorphism $\alpha : A \to B$ induces a continuous map $\alpha^* : \Sigma(B) \to \Sigma(A)$ in the obvious way, that is, by $\alpha^*\omega = \omega \circ \alpha$. In the special case $A = C^\infty(X)$ (and similarly if $A = C(X)$), one has a canonical map $\mathrm{ev}^X : X \to \Sigma(C(X))$, given by $\mathrm{ev}^X_x(f) = f(x)$. The whole point (in which the entire difficulty of the proof lies) is that this map is a bijection (see Proposition C.21), which simultaneously equips X with a smooth structure that makes ev^X a diffeomorphism (by definition of the smooth structure on $\Sigma(C(X))$. In view of all this, given a multiplicative linear map $\alpha : C^\infty(X) \to C^\infty(Y)$, we obtain a continuous map $\varphi : Y \to X$ by

$$\varphi = (\mathrm{ev}^Y)^{-1} \circ \alpha^* \circ \mathrm{ev}^X. \tag{3.56}$$

Eq. (3.55) then holds by construction. Smoothness of φ, then, is a consequence of the fact that $\alpha(f) = f \circ \varphi$ must be a smooth function on Y for any $f \in C^\infty(X)$.

Applying this to the setting of Theorem 3.10 easily yields all claims. □

In what follows, we look at smooth actions of Lie groups on (Poisson) manifolds X, in other words, at homomorphisms $\varphi : G \to \mathrm{Diff}(X)$ or $\varphi : G \to \mathrm{Diff}(X, B)$, where G is a Lie group, $\mathrm{Diff}(X)$ is the group of all diffeomorphisms of a manifold, and $\mathrm{Diff}(X, B)$ is the group of all diffeomorphisms of a Poisson manifold preserving the Poisson structure. Foregoing the underlying differential geometry, we take a pragmatic attitude and only study ***linear Lie groups***, defined as closed subgroups G of $GL_n(\mathbb{R})$ or $GL_n(\mathbb{C})$, with group multiplication given by matrix multiplication and hence group inverse being matrix inverse. Here one may think of $SU(2) \subset GL_2(\mathbb{C})$ or $SO(3) \subset GL_3(\mathbb{R})$, but also abelian Lie groups like the additive groups $\mathbb{R}^n$ fall under this scope, since one may identify $a \in \mathbb{R}^n$ with the $2n \times 2n$-matrix

$$a \equiv \begin{pmatrix} 1 & a \\ 0 & 1 \end{pmatrix}, \tag{3.57}$$

in which case matrix multiplication indeed reproduces addition. Similarly, the $2n+1$-dimensional ***Heisenberg group*** H_n is the group of real $(n+2) \times (n+2)$-matrices

$$(a,b,c) = \begin{pmatrix} 1 & a^T & c + \frac{1}{2}a^T b \\ 0 & 1_n & b \\ 0 & 0 & 1 \end{pmatrix}, \tag{3.58}$$

where $a, b \in \mathbb{R}^n$, $c \in \mathbb{R}$, and $a^T b = \langle a, b \rangle$; this gives the multiplication rule

$$(a,b,c) \cdot (a',b',c') = (a + a', b + b', c + c' - \tfrac{1}{2}(\langle a, b' \rangle - \langle a', b \rangle)). \tag{3.59}$$

If G is a linear Lie group, its ***Lie algebra*** $\mathfrak{g}$ may be defined as the vector space

$$\mathfrak{g} = \{A \in M_n(\mathbb{K}) \mid e^{tA} \in G \forall t \in \mathbb{R}\}, \tag{3.60}$$

where $\mathbb{K} = \mathbb{R}$ or $\mathbb{C}$, as determined by the embedding $G \subset GL_n(\mathbb{R}))$ or $G \subset GL_n(\mathbb{C})$. Either way, $\mathfrak{g}$ is seen as a *real* vector space, equipped with the ***Lie bracket***

$$[A,B] = AB - BA. \tag{3.61}$$

This is trivially a bilinear antisymmetric map $\mathfrak{g} \times \mathfrak{g} \to \mathfrak{g}$ satisfying the Jacobi identity

$$[A,[B,C]] + [C,[A,B]] + [B,[C,A]] = 0, \tag{3.62}$$

which in turn expresses the fact that for fixed $A \in\ g$ the map $\delta_A : \mathfrak{g} \to \mathfrak{g}$ defined by

$$\delta_A(B) = [A,B] \tag{3.63}$$

is a derivation of $\mathfrak{g}$ with respect to its Lie bracket, i.e.,

$$\delta_A([B,C]) = [\delta_A(B),C] + [B,\delta_A(C)]. \tag{3.64}$$

The ***exponential map*** $\exp : \mathfrak{g} \to G$ is then just given by its usual power series, which for matrices is norm-convergent. Conversely, one may pass from G to $\mathfrak{g}$ through

$$A = \frac{d}{dt}(e^{tA})_{|t=0}. \tag{3.65}$$

If $G = \mathbb{R}^n$, we also have $\mathfrak{g} = \mathbb{R}^n$, and eq. (3.57) implies that exp is the identity map.

For example, since $SO(3)$ is the subgroup of $GL_3(\mathbb{R})$ consisting of matrices R that satisfy $R^T R = 1_3$, its Lie algebra $\mathfrak{so}(3)$ consists of all matrices a that satisfy $a^T = -a$. As a vector space have $\mathfrak{so}(3) \cong \mathbb{R}^3$, which follows by choosing a basis

$$J_1 = \begin{pmatrix} 0 & 0 & 0 \\ 0 & 0 & -1 \\ 0 & 1 & 0 \end{pmatrix}, J_2 = \begin{pmatrix} 0 & 0 & 1 \\ 0 & 0 & 0 \\ -1 & 0 & 0 \end{pmatrix}, J_3 = \begin{pmatrix} 0 & -1 & 0 \\ 1 & 0 & 0 \\ 0 & 0 & 0 \end{pmatrix}. \tag{3.66}$$

of the 3×3 real antisymmeric matrices. The commutators of these elements are

$$[J_1,J_2] = J_3; \ [J_3,J_1] = J_2; \ [J_2,J_3] = J_1. \tag{3.67}$$

For the Lie algebra of the Heisenberg group we obtain $\mathfrak{h}_n = \mathbb{R}^{2n+1}$, with basis

$$P_i = \begin{pmatrix} 0 & 0 & 0 \\ 0 & 0 & -\mathbf{e}_i \\ 0 & 0 & 0 \end{pmatrix}, Q_j = \begin{pmatrix} 0 & \mathbf{e}_j^T & 0 \\ 0 & 0 & 0 \\ 0 & 0 & 0 \end{pmatrix}, Z = \begin{pmatrix} 0 & 0 & 1 \\ 0 & 0 & 0 \\ 0 & 0 & 0 \end{pmatrix}, \tag{3.68}$$

where $(\mathbf{e}_1, \ldots, \mathbf{e}_n)$ is the usual basis of $\mathbb{R}^n$, satisfying commutation relations

$$[P_i,Q_j] = \delta_{ij}Z; \ \ [P_i,P_j] = [Q_i,Q_j] = [P_i,Z] = [Q_j,Z] = 0. \tag{3.69}$$

3.4 The momentum map

Leaving out the Poisson structure for the moment, let X be a manifold, let G be a Lie group, and let $\varphi : G \to \mathrm{Diff}(X)$ be a homomorphism; as already mentioned, this corresponds to a smooth action $\tilde{\varphi} : G \times X \to X$, which we simply write as

$$\gamma \cdot x \equiv \varphi_\gamma(x) \equiv \tilde{\varphi}(\gamma, x).$$

In terms of the pullback $\varphi_\gamma^*(f) = f \circ \varphi_\gamma$, we then automatically have

$$\varphi_\gamma^*(fg) = \varphi_\gamma^*(f)\varphi_\gamma^*(g). \tag{3.70}$$

For each $A \in \mathfrak{g}$ we then define a map $\delta_A : C^\infty(X) \to C^\infty(X)$ by

$$\delta_A f(x) = \frac{d}{dt} f(e^{-tA} \cdot x)_{|t=0}. \tag{3.71}$$

This map is obviously linear. Moreover, it can be shown that δ is well behaved:

Proposition 3.12. *The map $\delta : \mathfrak{g} \to \mathrm{Der}(C^\infty(X))$, $A \mapsto \delta_A$ is a homomorphism of Lie algebra, i.e., each δ_A is a derivation, δ is linear in A, and, for each $A, B \in \mathfrak{g}$,*

$$[\delta_A, \delta_B] = \delta_{[A,B]}. \tag{3.72}$$

The proof relies on ***Hadamard's Lemma***, which we only need for complete vector fields, or, equivalently, for derivations with complete flow (i.e., defined for all t).

Lemma 3.13. *If δ is a derivation of $C^\infty(X)$ with complete flow φ, and $f \in C^\infty(X)$, then there is a function $g(t,x) \equiv g_t(x)$ such that for all x and t,*

$$g_0(x) = \delta f(x); \tag{3.73}$$
$$f(\varphi_t(x)) = f(x) + t g_t(x). \tag{3.74}$$

Indeed, if the flow is complete one may take

$$g_t(x) = \int_0^1 ds \dot{F}(st, x), \tag{3.75}$$

where $F(t,x) = f(\varphi_t(x))$ and (in Newton's notation) $\dot{F}$ is the time derivative of F.

Proof. To prove that δ_A is linear in A, let φ be the flow of δ_A, i.e., $\varphi_t(x) = e^{-tA}x$. For $B \in \mathfrak{g}$, Hadamard's Lemma with $\delta \rightsquigarrow \delta_A$ and $x \rightsquigarrow e^{-tB}x$ then gives us

$$f(e^{-tA}e^{-tB}x) = f(\varphi_t(e^{-tB}x)) = f(e^{-tB}x) + t g_t(e^{-tB}x);$$
$$\Rightarrow \frac{d}{dt} f(e^{-tA}e^{-tB}x)_{|t=0} = \delta_B f(x) + g_0(x) = \delta_B f(x) + \delta_A f(x). \tag{3.76}$$

On the other hand, since A and B are matrices, we may use the CBH-formula

$$e^{-tA}e^{-tB} = e^{-t(A+B)+\frac{1}{2}t^2[A,B]+O(t^3)}, \tag{3.77}$$

which gives $e^{-tA}e^{-tB} = e^{-t(A+B)}(1+O(t^2))$, and hence

$$\frac{d}{dt}f(e^{-tA}e^{-tB}x)_{|t=0} = \frac{d}{dt}f(e^{-t(A+B)}x)_{|t=0} = \delta_{A+B}f(x). \tag{3.78}$$

Comparing (3.76) with (3.78) gives $\delta_{A+B} = \delta_A + \delta_B$. The property $\delta_{sA} = s\delta_A$ is trivial. We now prove (3.72). Within the (matrix) Lie algebra $\mathfrak{g}$ we have

$$[A,B] = -\frac{d}{dt}(e^{-tA}Be^{tA})_{|t=0} = -\lim_{t\to 0}\frac{e^{-tA}Be^{tA}-B}{t}. \tag{3.79}$$

Furthermore, for any $g \in G$ one has $e^{gBg^{-1}} = ge^Bg^{-1}$, so linearity of δ gives

$$\begin{aligned}
\delta_{[A,B]}f(x) &= -\lim_{t\to 0}\frac{1}{t}(\delta_{e^{-tA}Be^{tA}}f(x) - \delta_B f(x))\\
&= \lim_{t\to 0}\frac{1}{t}\left(\frac{d}{ds}f(e^{-tA}e^{sB}e^{tA}x) - \frac{d}{ds}f(e^{sB}x)\right)\\
&= \lim_{s,t\to 0}\frac{1}{st}(f(e^{-tA}e^{sB}e^{tA}x) - f(e^{-tA}e^{tA}e^{sB}x))\\
&= \lim_{s,t\to 0}\frac{1}{st}(f\circ\varphi_t(e^{sB}e^{tA}x) - f\circ\varphi_t(e^{tA}e^{sB}x))\\
&= \lim_{s,t\to 0}\left(\frac{1}{st}(f(e^{sB}e^{tA}x) - f(e^{tA}e^{sB}x)) + \frac{1}{s}(g_t(e^{sB}e^{tA}x) - g_t(e^{tA}e^{sB}x))\right)\\
&= [\delta_A,\delta_B]f(x),
\end{aligned}$$

since in the limit $t\to 0$ the third term in the penultimate line cancels the fourth. □

Now suppose that, in addition, X is a Poisson manifold, and that each φ_γ acts on X as a Poisson symmetry, in that

$$\varphi_\gamma^* B = B, \tag{3.80}$$

cf. (3.44), or, equivalently, cf. (3.46),

$$\varphi_\gamma^*(\{f,g\}) = \{\varphi_\gamma^*(f), \varphi_\gamma^*(g)\}. \tag{3.81}$$

This implies, for each $A \in \mathfrak{g}$, and each $f,g \in C^\infty(X)$,

$$\delta_A(\{f,g\}) = \{\delta_A(f),g\} + \{f,\delta_A(g)\}. \tag{3.82}$$

Compare this with the following property δ_A already has since it is a derivation:

$$\delta_A(fg) = \delta_A(f)g + f\delta_A(g). \tag{3.83}$$

We may call a derivation $\delta : C^\infty(X) \to C^\infty(X)$ satisfying the like of (3.82), i.e.,

$$\delta(\{f,g\}) = \{\delta(f),g\} + \{f,\delta(g)\}, \tag{3.84}$$

a ***Poisson derivation***. We are already familiar with a large class of Poisson derivations: for each $h \in C^\infty(X)$, the corresponding map δ_h defined by (3.26) is a Poisson derivation (this follows from the Jacobi identity). Let us call a Poisson derivation of the kind δ_h ***inner***. This raises the question if our derivations δ_A are inner.

Definition 3.14. *A* **momentum map** *for a Lie group G acting on a Poisson manifold X is a map*

$$J : X \to \mathfrak{g}^* \tag{3.85}$$

such that for each $A \in \mathfrak{g}$,

$$\delta_A = \delta_{J_A}, \tag{3.86}$$

where the function $J_A \in C^\infty(X)$ is defined by by

$$J_A(x) = \langle J(x), A\rangle \equiv J(x)(A). \tag{3.87}$$

In other words, for each $A \in \mathfrak{g}$ and $f \in C^\infty(X)$ we must have

$$\delta_A(f) = \{J_A, f\}. \tag{3.88}$$

A Lie group action admitting a momentum map is called **Hamiltonian.**

Equivalently, a momentum map is a linear map

$$J^* : \mathfrak{g} \to C^\infty(X) \tag{3.89}$$

such that $\delta_A = \delta_{J^*(A)}$; the connection between the two definitions is given by

$$J_A = J^*(A). \tag{3.90}$$

The pullback notation J^* would suggest that it is a map $C^\infty(\mathfrak{g}^*) \to C^\infty(X)$, which is not quite the case, but it is a near miss: we embed $\mathfrak{g} \hookrightarrow C^\infty(\mathfrak{g}^*)$ by $A \mapsto \hat{A}$, where $\hat{A}(\theta) = \theta(A)$, so $J^* : \mathfrak{g} \to C^\infty(X)$ is the restriction of the pullback J^* to $\mathfrak{g}$. Another near miss would be to read J^* as the adjoint to J, which maps $\mathfrak{g}^{**} \cong \mathfrak{g}$ to the 'dual' X^*, but since X may not be a vector space, this dual cannot be defined as in linear algebra, so instead of all linear maps from X to $\mathbb{R}$ we might as well say that it consists of all smooth functions on X. Either way, the symbol J^* seems justified.

Proposition 3.15. *Let G be a connected Lie group that acts on a Poisson manifold X. If this action is Hamiltonian (i.e., if it has a momentum map), then G acts on (X,B) by Poisson symmetries (in the sense that* (3.81) *holds).*

Proof. An easy computation shows that (3.82) holds. We omit the proof of the fact that for *connected* Lie groups this "infinitesimal" property is equivalent to (3.81); this relies on the fact that G is generated by the image of the exponential map. □

The converse is not true: if G acts by Poisson symmetries, the action is not necessarily Hamiltonian. For example, take $X = \mathbb{R}^2$, with the unusual Poisson bracket

$$\{f,g\}(p,q) = p\left(\frac{\partial f}{\partial p}\frac{\partial g}{\partial q} - \frac{\partial f}{\partial q}\frac{\partial g}{\partial p}\right), \tag{3.91}$$

and let $G = \mathbb{R}$ act on $\mathbb{R}^2$ by $b \cdot (p,q) = (p, q+b)$. This action satisfies (3.81), and has a single generator $\delta = -\partial/\partial q$. But there clearly is no function $J \in C^\infty(\mathbb{R}^2)$ such that $\{J,f\} = -\partial f/\partial q$ (it should be $J(p,q) = -\log(p)$, which is singular at $p = 0$).

However, in most "everyday situations" momentum maps exist:

1. Take $X = \mathbb{R}^6 = \mathbb{R}^3 \times \mathbb{R}^3$, with coordinates $x = (\mathbf{p}, \mathbf{q})$, where $\mathbf{p} = (p_1, p_2, p_2)$ and $\mathbf{q} = (q^1, q^2, q^3)$, equipped with the canonical Poisson bracket (3.34).

   a. Let $G = \mathbb{R}^6$ act on X by

   $$(\mathbf{a}, \mathbf{b}) \cdot (\mathbf{p}, \mathbf{q}) = (\mathbf{p} + \mathbf{a}, \mathbf{q} + \mathbf{b}). \tag{3.92}$$

   This action is Hamiltonian, with momentum map

   $$J(\mathbf{p}, \mathbf{q}) = (\mathbf{q}, -\mathbf{p}). \tag{3.93}$$

   b. Let $G = SO(3)$ act on the same space X by

   $$R \cdot (\mathbf{p}, \mathbf{q}) = (R\mathbf{p}, R\mathbf{q}). \tag{3.94}$$

   Also this action is Hamiltonian, with momentum map

   $$J(\mathbf{p}, \mathbf{q}) = \mathbf{p} \times \mathbf{q}. \tag{3.95}$$

2. Let $G = SO(3)$ act on $X = \mathbb{R}^3$, equipped with the Poisson bracket (3.43), through its defining representation. This action has a momentum map

   $$J(\mathbf{x}) = \mathbf{x}, \tag{3.96}$$

   where we have identified $\mathfrak{g}$ with $\mathbb{R}^3$ by choosing the basis (3.66) of $\mathfrak{g}$, and have identified $\mathfrak{g}^*$ with $\mathfrak{g}$ (and hence with $\mathbb{R}^3$ also) by the usual inner product on $\mathbb{R}^3$.
3. The previous example is a special case of the ***Lie–Poisson structure***. Let G be a Lie group with Lie algebra $\mathfrak{g}$. Choose a basis (T_a) of $\mathfrak{g}$, with associated ***structure constants*** C^c_{ab} defined by the Lie bracket on $\mathfrak{g}$ as

   $$[T_a, T_b] = \sum_c C^c_{ab} T_c. \tag{3.97}$$

   We write θ in the dual vector space $\mathfrak{g}^*$ as $\theta = \sum_a \theta_a \omega^a$, where (ω_a) is the dual basis to a chosen basis (T_a) of $\mathfrak{g}$, i.e., $\omega_a(T_b) = \delta_{ab}$. In terms of these coordinates, the ***Lie–Poisson bracket*** on $C^\infty(\mathfrak{g}^*)$ is defined by

   $$\{f,g\}(\theta) = C^c_{ab}\theta_c \frac{\partial f(\theta)}{\partial \theta_a}\frac{\partial g(\theta)}{\partial \theta_b}. \tag{3.98}$$

   Equivalently, the Poisson bracket (3.98) may be defined by the condition

$$\{\hat{A},\hat{B}\} = \widehat{[A,B]}, \tag{3.99}$$

where $A,B \in \mathfrak{g}$ and $\hat{A} \in C^\infty(\mathfrak{g}^*)$ is the evaluation map $\hat{A}(\theta) = \theta(A)$.
Now G canonically acts on $\mathfrak{g}^*$ through the ***coadjoint representation***, defined by

$$(x\cdot\theta)(A) = \theta(x^{-1}Ax). \tag{3.100}$$

This action is Hamiltonian with respect to the Lie–Poisson bracket (3.98), the associated momentum map simply being the identity map $\mathfrak{g}^* \to \mathfrak{g}^*$, as in (3.96). In other words, we have

$$J_A = \hat{A}, \tag{3.101}$$

whose correctness may be verified from the computation

$$\begin{aligned}\delta_A\tilde{B}(\theta) &= \frac{d}{dt}\tilde{B}(e^{-tA}\cdot\theta)_{|t=0} = \frac{d}{dt}\theta(e^{tA}Be^{-tA})_{|t=0}\\ &= \theta([A,B]) = \widehat{[A,B]}(\theta) = \{\hat{A},\hat{B}\}(\theta)\\ &= \{J_A,\hat{B}\}(\theta).\end{aligned}$$

4. Let $X = T^*Q$ for some manifold Q. e.g. $Q = \mathbb{R}^n$ and hence $X = \mathbb{R}^{2n}$. We take

$$G = \mathrm{Diff}(Q), \tag{3.102}$$

i.e., the diffeomorphism group of Q. This is an infinite-dimensional Lie group (if described in the right way). The defining action of $\varphi \in G$ on Q induces an action called φ^* on T^*Q, given (in coordinates) by

$$\varphi^*(p,q) = (p',q'); \tag{3.103}$$
$$(q^i)' = \varphi^i(q); \tag{3.104}$$
$$p_i' = \sum_{j=1}^n \frac{\partial(\varphi^{-1})^j(q)}{\partial q^i}p_j. \tag{3.105}$$

This may be taken as a definition, but in the language of differential geometry this comes down to the neater prescription that if $\theta = \sum_j p_j dq^j \in T_q^*Q$, then $\varphi^*\theta \in T^*_{\varphi(q)}Q$ is the one-form that maps a vector $X \in T_{\varphi(q)}Q$ to $\theta(\varphi_*^{-1}(X))$, i.e.,

$$(\varphi^*\theta)(X) = \theta(\varphi_*^{-1}(X)), \tag{3.106}$$

where $\varphi_*^{-1}(X) = \sum_j \varphi_*^{-1}(X)^j\partial/\partial q^j$ is given componentwise by, cf. (3.52),

$$\varphi_*^{-1}X^j = \sum_j \frac{\partial(\varphi^{-1})^j(q)}{\partial q^k}X^k. \tag{3.107}$$

If $Q = \mathbb{R}^3$ and $\varphi = R \in SO(3)$, then, using $R^{-1} = R^T$, we find that (3.104) - (3.105) simply become $R^*(\mathbf{p},\mathbf{q}) = (R\mathbf{p},R\mathbf{q})$, as in (3.94).

Furthermore, if $\varphi(\mathbf{q}) = \mathbf{q}+\mathbf{b}$, then the partial derivatives in (3.105) form the identity matrix, so that $\varphi^*(\mathbf{p},\mathbf{q}) = (\mathbf{p},\mathbf{q}+\mathbf{b})$. To show that the action of Diff(Q) on T^*Q is Hamiltonian and compute its momentum map, we need to know that the Lie algebra of Diff(Q) is the space Vec(X) of all vector fields on Q, with its canonical Lie bracket (3.61)! We will not prove this, but the exponential map $\exp : \mathfrak{g} \to G$ is given through the flow φ of the vector field ξ on Q by (cf. (3.20))

$$e^{t\xi} = \varphi_t. \tag{3.108}$$

Theorem 3.16. *The action of* Diff(Q) *on* T^*Q *has momentum map*

$$J_X(p,q) = -\sum_j p_j X^j(q), \tag{3.109}$$

and hence is Hamiltonian. Moreover, this momentum map satisfies

$$\{J_\xi, J_\eta\}_\xi = -J_{[\xi,\eta]}. \tag{3.110}$$

Proof. First note that $\varphi_t^{-1} = \varphi_{-t}$, so from (3.71), (3.108), and (3.104) - (3.105),

$$\begin{aligned}\delta_\xi f(p,q) &= \frac{d}{dt} f(\varphi^*_{-t}(p,q))_{|t=0} \\ &= \sum_{i,j} \frac{\partial f}{\partial p_i}(p,q) \frac{d}{dt}\left(\frac{\partial \varphi^j_t(q)}{\partial q^i}\right)_{|t=0} p_j + \sum_i \frac{\partial f}{\partial q^i}(p,q)\frac{d}{dt}\varphi^i_{-t}(q)_{|t=0} \\ &= \sum_{i,j} p_j \frac{\partial X^j(q)}{\partial q^i}\frac{\partial f}{\partial p_i}(p,q) - \sum_j X^j(q)\frac{\partial f}{\partial q^j}(p,q).\end{aligned}$$

From this and (3.109), using the canonical Poisson bracket (3.34) we find

$$\{J_\xi, f\} = \delta_\xi f.$$

Finally, verifying (3.110) is a simple exercise. □.

Thus the momentum map is a generalization of (minus) the momentum, whence its name; the quantity in (3.95) is (minus) the angular momentum. These annoying minus signs could be removed by putting a minus sign in (3.86), but that would have other negative (*sic*) consequences. For example, with our sign choice one often has

$$\{J_A, J_B\} = J_{[A,B]}, \tag{3.111}$$

in which case the accompanying map (3.89) is a homomorphism of Lie algebras, or, equivalently, J is a morphism with respect to the given Poisson bracket on X and the Lie–Poisson bracket on $\mathfrak{g}^*$. Such a momentum map is called ***infinitesimally equivariant***, for if G is connected, (3.111) is equivalent to the equivariance property

$$J(g\cdot x) = g \cdot J(x). \tag{3.112}$$

Here the G-action on $\mathfrak{g}^*$ on the right-hand side is the coadjoint representation.

All of this is true for our examples (3.95), (3.96), (3.101), and (3.109); in the latter case we note that the Lie bracket in the Lie algebra of Diff(Q) is *minus* the commutator of vector fields. However, (3.111) does not always hold (in which case *a fortiori* also (3.112) fails). For example, it fails for (3.93): if we take the usual basis $(\mathbf{e},\mathbf{f}) \equiv (e_1,e_2,e_3,f_1,f_2,f_3)$ of $\mathfrak{g} = \mathbb{R}^6$ and relabel $e_j \equiv Q_j$ and $f_i \equiv -P_i$, then

$$J_{P_i}(\mathbf{p},\mathbf{q}) = p_i; \tag{3.113}$$

$$J_{Q_j}(\mathbf{p},\mathbf{q}) = q_j, \tag{3.114}$$

cf. (3.93), and hence, although $[P_i,P_j] = [Q_i,Q_j] = [P_i,Q_j] = 0$, we obtain

$$\{J_{P_i},J_{P_j}\} = \{J_{Q_i},J_{Q_j}\} = 0; \tag{3.115}$$

$$\{J_{P_i},J_{Q_j}\} = \delta_{ij}1_{\mathbb{R}^6}. \tag{3.116}$$

Fortunately, in cases like that one can often find a central extension G_φ of G (see §5.10 below for notation) that acts on X through its quotient group G and does have an infinitesimally equivariant momentum map. In the case at hand, the Heisenberg group H_3 does the job, whose central elements $(0,0,c)$ then act trivially on $\mathbb{R}^6$. In terms of the generators (3.68) we take J_{P_i} and J_{Q_j} as in (3.113) - (3.114), and add $J_Z = 1_{\mathbb{R}^6}$; according to (3.69) and (3.115) - (3.116) we then have (3.111), as desired.

Finally, the above formalism leads to a clean formulation of ***Noether's Theorem***, providing the well-known link between symmetries and conserved quantities:

Theorem 3.17. *Let X be a Poisson action equipped with a Hamiltonian action of some Lie group G (so that there is a momentum map $J : X \to \mathfrak{g}^*$). Suppose $h \in C^\infty(X)$ is G-invariant, in that $h(\gamma \cdot x) = h(x)$ for each $\gamma \in G$ and $x \in X$. Then for each $A \in \mathfrak{g}$, the function J_A is constant along the flow of the vector field X_h. In other words,*

$$J_A(\varphi_t(x)) = J_A(x) \tag{3.117}$$

for any $x \in X$ and any $t \in \mathbb{R}$ for which the flow $\varphi_t(x)$ of X_h is defined.

Proof. Using all assumptions as well as the definition of a flow, we compute:

$$\begin{aligned}
\frac{d}{dt}J_A(\varphi_t(x)) &= X_h(J_A)(\varphi_t(x)) = \delta_h(J_A)(\varphi_t(x)) \\
&= \{h,J_A\}(\varphi_t(x)) = -\{J_A,h\}(\varphi_t(x)) \\
&= -\delta_A(h)(\varphi_t(x)) = \frac{d}{ds}h(e^{sA}\varphi_t(x))_{|s=0} \\
&= \frac{d}{ds}h(\varphi_t(x))_{|s=0} = 0.
\end{aligned}$$

□

For example, a Hamiltonian (3.38) has conserved (angular) momentum if the potential V is translation (rotation) invariant, reflecting (3.93) and (3.95), respectively.

Notes

The traditional symplectic approach to classical mechanics, culminating in the momentum map, is exhaustively covered in Guillemin & Sternberg (1984) and Abraham & Marsden (1985). A founding paper for Poisson geometry is Weinstein (1983). The modern Poisson approach to mechanics may be found in Marsden & Ratiu (1994), from which most of the material in this chapter originates.

Our proof of Proposition 3.11 is based on Navarro González & Sancho de Salas (2003), §2.1. Burtscher (2009) is a nice survey of many similar results.

Chapter 4
Quantum physics on a general Hilbert space

In this chapter we generalize the results of Chapter 2 to infinite-dimensional Hilbert spaces. So let H be a Hilbert space and let $B(H)$ be the set of all *bounded* operators on H. Here a notable point is that linear operators on *finite-dimensional* Hilbert spaces are automatically bounded, whereas in general they are not. Thus we impose boundedness as an extra requirement, beyond linearity. This is very convenient, because as in the finite-dimensional case, $B(H)$ is a C*-algebra, cf. §C.1. At the same time, assuming boundedness involves no loss of generality whatsoever, since we can alway replace closed unbounded operators by bounded ones through the *bounded transform*, as explained in §B.21. Nonetheless, even the relatively easy setting of bounded operators leads to some technical complications we have to deal with. First, Definition 2.1 must be adjusted as follows:

Definition 4.1. *Let H be a Hilbert space.*

1. *A* **(quantum) event** *is a* closed *linear subspace L of H*
2. *A* **density operator** *is a positive* trace-class *operator ρ on H such that* $\mathrm{Tr}(\rho) = 1$*; we continue to denote the set of all density operators on H by $\mathscr{D}(H)$.*
3. *A* **(quantum) random variable** *is a* bounded *self-adjoint operator on H.*
4. *The* **spectrum** *$\sigma(a)$ of a bounded operator a is the set of all $\lambda \in \mathbb{C}$ for which the operator $a - \lambda$ is* not *invertible in $B(H)$ (cf. Definition B.80).*

As shown in Corollary B.88, if H is finite-dimensional this notion of a spectrum reduces to the set of eigenvalues of a. Even H is infinite-dimensional, the spectrum of a self-adjoint operator a is real (i.e., $\sigma(a) \subset \mathbb{R}$); this is also true if a is unbounded (see Theorem B.93). For any H, unit vectors ψ still define special density matrices e_ψ, as in (2.7); we will later see that these are pure states on $B(H)$, although the set of pure states is no longer exhausted by such density matrices. Finally, quantum events *in* H still bijectively correspond with projections *on* H; see Proposition B.76. The Born rule as well as the correspondence between density matrices and states require a separate discussion, to which we now turn.

K. Landsman, *Foundations of Quantum Theory*,
Fundamental Theories of Physics 188, DOI 10.1007/978-3-319-51777-3_4

4.1 The Born rule from Bohrification (II)

In this section we extend the characterization of the Born rule in §2.5, which was restricted to finite phase spaces X and finite-dimensional Hilbert spaces H, to the general case. Recall that a ***probability space*** is a measure space (X,Σ,μ) for which $\mu(X)=1$, and that, for compact X, a state on $C(X)$ is a positive map $\varphi : C(X) \to \mathbb{C}$ that is positive and satisfies $\varphi(1_X)=1$. Theorem B.15 and Corollary (B.17) yield:

Theorem 4.2. *Let X be a compact Hausdorff space. There is a bijective correspondence between probability measures μ on X and states ω on $C(X)$, given by*

$$\omega(f) = \int_X d\mu\, f,\ f \in C(X). \tag{4.1}$$

More precisely, the correspondence in question is between complete regular probability spaces (X,Σ,μ) and states on $C(X)$, and this is understood in what follows.

Second, we recall that if H is a Hilbert space and $a \in B(H)$, then $C^*(a)$ is the C*-algebra generated by a and 1_H (i.e., the norm-closure of the algebra of all polynomials in a). Theorems B.84, B.94, and B.93 give the following spectral theorem:

Theorem 4.3. *If $a^* = a \in B(H)$, then $C^*(a)$ is commutative, $\sigma(a) \subset \mathbb{R}$ is compact, and there is an isomorphism of (commutative) C*-algebras*

$$C(\sigma(a)) \cong C^*(a), \tag{4.2}$$

written $f \mapsto f(a)$, which is unique if it is subject to the following conditions:

1. *the unit function $1_{\sigma(a)} : \lambda \mapsto 1$ corresponds to the unit operator 1_H;*
2. *the identity function $\mathrm{id}_{\sigma(a)} : \lambda \mapsto \lambda$ is mapped to the given operator a.*

Furthermore, this **continuous functional calculus** *satisfies the rules*

$$(tf+g)(a) = tf(a)+g(a); \tag{4.3}$$
$$(fg)(a) = f(a)g(a); \tag{4.4}$$
$$f(a)^* = f^*(a). \tag{4.5}$$

Combining Theorems 4.2 and 4.3 gives a result of great importance:

Corollary 4.4. *Let H be a Hilbert space, let $a^* = a \in B(H)$, and let $\psi \in H$ be a unit vector. There exists a unique probability measure μ_ψ on the spectrum $\sigma(a)$ such that*

$$\langle \psi, f(a)\psi \rangle = \int_{\sigma(a)} d\mu_\psi\, f,\ f \in C(\sigma(a)). \tag{4.6}$$

In terms of the spectral projections $e_\Delta = 1_\Delta(a)$ (defined for Borel sets $\Delta \subseteq \sigma(a)$) constructed in (B.305) - (B.307) *and Theorem B.102, the Born measure is given by*

$$\mu_\psi(\Delta) = \|e_\Delta \psi\|^2. \tag{4.7}$$

More generally, a density operator $\rho \in \mathscr{D}(H)$ *induces a unique probability measure* μ_ρ *on* $\sigma(a)$ *for which*

$$\mathrm{Tr}(\rho f(a)) = \int_{\sigma(a)} d\mu_\rho\, f,\ f \in C(\sigma(a)),; \tag{4.8}$$

$$\mu_\rho(\Delta) = \mathrm{Tr}(\rho e_\Delta). \tag{4.9}$$

This measure on $\sigma(a)$ *is called the* **Born measure** *(defined by a and* ψ *or* ρ*).*

Proof. The point is that the map $f \mapsto \langle \psi, f(a)\psi \rangle$ defines a state on $C(\sigma(a))$:

- Linearity follows from linearity of the continuous functional calculus $f \mapsto f(a)$;
- Positivity follows because if $f \geq 0$, then $f = \sqrt{f} \cdot \sqrt{f}$, so that by (4.4) and (4.5), $\langle \psi, f(a)\psi \rangle = \|\sqrt{f}(a)\psi\|^2 \geq 0$;
- Unitality follows from Theorem 4.3.1, i.e., $\langle \psi, 1_{\sigma(a)}(a)\psi \rangle = \langle \psi, 1_H \psi \rangle = 1$.

To prove (4.7), use Lemma B.97 to approximate 1_Δ by functions $f_n \in C(\sigma(a))$ as stated. By Theorem B.13.2 (i.e., the Lebesgue Monotone Convergence Theorem), we have $\int_{\sigma(a)} d\mu_\psi f_n \to \int_{\sigma(a)} d\mu_\psi 1_\Delta = \mu_\psi(\Delta)$, whereas by (B.315) with $a_n = f_n(a)$, one has $\langle \psi, f_n(a)\psi \rangle \to \langle \psi, e_\Delta \psi \rangle = \|e_\Delta \psi\|^2$. Hence (4.7) follows from (4.6).

The proof for density operators is analogous. □

Defining the mean value $\langle a \rangle_\psi$ of a with respect to the Born measure μ_ψ by

$$\langle a \rangle_\psi = \int_{\sigma(a)} d\mu_\psi(x)x, \tag{4.10}$$

and similarly for ρ, using Theorem 4.3.2 we easily obtain

$$\langle a \rangle_\psi = \langle \psi, a\psi \rangle; \tag{4.11}$$

$$\langle a \rangle_\rho = \mathrm{Tr}(\rho a). \tag{4.12}$$

As an important special case, suppose that $\sigma(a) = \sigma_p(a)$ (i.e., each $\lambda \in \sigma(a)$ is an eigenvalue); this always happens if H is finite-dimensional. Eq. (A.57) then gives

$$\langle \psi, f(a)\psi \rangle = \sum_{\lambda \in \sigma(a)} f(\lambda) \cdot \|e_\lambda \psi\|^2,$$

where e_λ is the projection onto the eigenspace $H_\lambda = \{\psi \in H \mid a\psi = \lambda\psi\}$. Thus

$$\mu_\psi(\lambda) = \|e_\lambda \psi\|^2, \tag{4.13}$$

and using the notation $P_\psi(a = \lambda)$ for $\mu_\psi(\lambda)$, eq. (4.11) just becomes

$$\langle a \rangle_\psi = \sum_{\lambda \in \sigma(a)} \lambda \cdot P_\psi(a = \lambda). \tag{4.14}$$

It is customary to extend the Born measure on $\sigma(a) \subset \mathbb{R}$ to a (probability) measure μ'_ψ on all of $\mathbb{R}$ by simply stipulating that

$$\mu'_{\psi}(\Delta) = \mu_{\psi}(\Delta \cap \sigma(a)); \tag{4.15}$$

we will often assume this and omit the prime. This obviously implies that $\mu_{\psi}(\Delta) = 0$ for any Borel set $\Delta \subset \mathbb{R}$ disjoint from $\sigma(a)$; in particular, if $\sigma(a)$ is discrete, then μ_{ψ} is concentrated on the eigenvalues λ of a, in that

$$\mu_{\psi}(\Delta) = \sum_{\lambda \in \Delta \cap \sigma(a)} \mu_{\psi}(\lambda). \tag{4.16}$$

To state an interesting property of the Born measure we need Hausdorff's solution to the relevant special case of the famous ***Hamburger Moment Problem***:

Theorem 4.5. *If $K \subset \mathbb{R}$ is compact, then any finite measure μ on K is determined by its* **moments**

$$\alpha_n = \int_K d\mu(x) x^n. \tag{4.17}$$

Using $f(x) = x^n$ in (4.6), we therefore obtain:

Corollary 4.6. *The Born measure μ_{ψ} is determined by its moments*

$$\alpha_n = \langle \psi, a^n \psi \rangle. \tag{4.18}$$

More precisely, we need to be sure that numbers (α_n) of the kind (4.18) are the moments of some (probability) measure. This follows from the spectral theorem by running the above argument backwards, but one may also use the general solution of the Hamburger Moment Problem, which we here state without proof:

Theorem 4.7. *A sequence of real numbers (α_n) forms the moments of some measure μ on $\mathbb{R}$ iff for all $N \in \mathbb{N}$ and $(\beta_1, \ldots, \beta_N) \in \mathbb{C}^N$ one has $\sum_{n,m=0}^{N} \overline{\beta n} \beta_m \alpha^{n+m} \geq 0$. Furthermore, if there are constants C and D such that $|\alpha_n| \leq C D^n n!$, then μ is uniquely determined by its moments (α_n).*

These conditions are easily checked from (4.18).

If a is unbounded, but still assumed to be self-adjoint (in the sense appropriate for unbounded operators, cf. Definition B.70), the spectrum $\sigma(a)$ remains real (see Theorem B.93) but it is no longer compact. Nonetheless, the Born measure on $\sigma(a)$ may be constructed in almost exactly the same way as in the bounded case, this time invoking Corollary B.21 and Theorem B.158 instead of Theorems 4.2 and B.94, respectively. Corollary 4.4 then holds almost *verbatim* for the unbounded case:

Corollary 4.8. *Let H be a Hilbert space, let $a^* = a$, and let $\psi \in H$ be a unit vector. There exists a unique probability measure μ_{ψ} on the spectrum $\sigma(a)$ such that*

$$\langle \psi, f(a)\psi \rangle = \int_{\sigma(a)} d\mu_{\psi} f, \; f \in C_0(\sigma(a)). \tag{4.19}$$

Also, eqs. (4.7) and (4.9) hold, as does (4.8), with $f \in C_0(\sigma(a))$.

There is no need to worry about domains, since even if a is unbounded, $f(a)$ is bounded for $f \in C_b(\sigma(a))$, and hence also for $f \in C_0(\sigma(a))$.

The physical relevance of the Born measure is given by the ***Born rule***:

If an observable a is measured in a state ρ, then the probability $P_\rho(a \in \Delta)$ that the outcome lies in $\Delta \subset \mathbb{R}$ is given by the Born measure μ_ρ defined by a and ρ, i.e.,

$$P_\rho(a \in \Delta) = \mu_\rho(\Delta). \tag{4.20}$$

As in the finite-dimensional case, the Born measure may be generalized to families $(a_1, \ldots, a_n)$ of commuting self-adjoint operators. Assuming these are bounded, the C*-algebra $C^*(a_1, \ldots, a_n)$ is defined in the obvious way, i.e., as the smallest C*-algebra containing each a_i, or, equivalently, as the norm-closure of the algebra of all finite polynomials in the $(a_1, \ldots, a_n)$. This C*-algebra is commutative, as a simple approximation argument shows: polynomials in the a_i obviously commute, and this property extends to the closure by continuity of multiplication. However, even in the bounded case, the correct notion of a joint spectrum is not obvious. In order to motivate the following definition, it helps to recall Definition 1.4, Theorem C.24, and especially the last sentence before the proof of the latter, making the point that the spectrum $\sigma(a)$ of a single (bounded) self-adjoint operator coincides with the image of the Gelfand spectrum $\Sigma(C^*(a))$ in $\mathbb{C}$ under the map $\omega \mapsto \omega(a)$.

Definition 4.9. *1. The* **joint spectrum** $\sigma(\underline{a}) = \sigma(a_1, \ldots, a_n) \subset \mathbb{R}^n$ *of a finite family $\underline{a} = (a_1, \ldots, a_n)$ of commuting bounded self-adjoint operators is the image of the Gelfand spectrum $\Sigma(C^*(a_1, \ldots, a_n)) = \Sigma(C^*(\underline{a}))$ under the map*

$$\Sigma(C^*(a_1, \ldots, a_n)) \to \mathbb{R}^n, \;\; \omega \mapsto (\omega(a_1), \ldots, \omega(a_n)). \tag{4.21}$$

Since $\omega(a_i)$ only utilizes the restriction of ω to $C^*(a_i) \subset C^*(\underline{a})$, we have $\omega(a_i) \subset \sigma(a_i) \subset \mathbb{R}$, so that $\Sigma(C^*(\underline{a})) \subseteq \sigma(a_1) \times \cdots \times \sigma(a_n)$ is a compact subset of $\mathbb{R}^n$.

To justify this definition, we note that:

- For $n = 1$, this definition reproduces the usual spectrum, cf. Theorem C.24.
- For $n > 1$ and $\dim(H) < \infty$, we recover the joint spectrum of Definition A.16.
- For $n > 1$ and $\dim(H) = \infty$, Weyl's Theorem B.91 generalizes in the obvious way: we have $\lambda \in \sigma(a)$ iff there exists a sequence (ψ_k) of unit vectors in H with

$$\lim_{k \to \infty} \|(a_i - \lambda_i)\psi_k\| = 0, \tag{4.22}$$

for each $i = 1, \ldots, n$. The proof is similar.

One way to see the second claim is to use Proposition C.14 joined with the observation that, as in the case of $A = B(H)$ for finite-dimensional H, any pure state on a finite-dimensional C*-algebra $A \subset B(H)$ is a vector state (2.42), too. To see this, we first specialize Theorem C.133 to the finite-dimensional case (where the proof becomes elementary), so that each state on $C^*(\underline{a})$ takes the form (2.33). Subsequently, we use the spectral decomposition (2.6), and use the definition of purity: suppose $\omega(b) = \mathrm{Tr}(\rho b) = \sum_i p_i \langle \upsilon_i, b \upsilon_i \rangle \equiv \sum_i p_i \omega_{\upsilon_i}(b)$ is pure, where $b \in C^*(\underline{a})$.

Then $\omega_{\upsilon_i} = \omega$ for each i, so that ω is a vector state, say $\omega(b) = \langle \psi, b\psi \rangle$ where ψ is one of the υ_i. Once we know this, suppose $\underline{\lambda} = (\lambda_1, \ldots, \lambda_n) \in \sigma(\underline{a})$, with $\lambda_i = \omega(a_i)$. Multiplicativity of ω implies that for any finite polynomial in n real variables we have $\langle \psi, p(\underline{a})\psi \rangle = p(\underline{\lambda})$, which easily gives $a_i\psi = \lambda_i\psi$ for each i; for example, take $p(\underline{x}) = (x_i - \lambda_i)^2$, so that the previous equation gives $\|(a_i - \lambda_i)\psi\|^2 = 0$.

Conversely, if $\underline{\lambda}$ is a joint eigenvalue of $\underline{a}$, then by definition there exists a joint eigenvector ψ whose vector state $\omega(b) = \langle \psi, b\psi \rangle$ on $C^*(\underline{a})$ is multiplicative.

Using this (perhaps contrived) notion of a joint spectrum, Theorem 2.19 now holds by construction also if $\dim(H) = \infty$, where the pertinent isomorphism $f \mapsto f(\underline{a})$ is given as in the single operator case, that is, by starting with polynomials and using a continuity argument to pass to arbitrary continuous functions.

Theorem 2.18 and Corollary 4.4 then generalize to:

Theorem 4.10. *Let H be a Hilbert space, let $\underline{a} = (a_1, \ldots, a_n)$ be a finite family of commuting bounded self-adjoint operators, and let $\psi \in H$ be a unit vector. There exists a unique probability measure μ_ψ on the joint spectrum $\sigma(\underline{a})$ such that*

$$\langle \psi, f(\underline{a})\psi \rangle = \int_{\sigma(\underline{a})} d\mu_\psi f, \; f \in C(\sigma(\underline{a})), \tag{4.23}$$

or, equivalently, for special Borel sets $\underline{\Delta} = \Delta_1 \times \cdots \times \Delta_n \subseteq \sigma(\underline{a})$, where $\Delta_i \subset \sigma(a_i)$,

$$\mu_\psi(\underline{\Delta}) = \|e_{\Delta_1} \cdots e_{\Delta_n} \psi\|^2, \tag{4.24}$$

where the $e_{\Delta_i} = 1_{\Delta_i}(a_i)$ are the pertinent spectral projections (which commute).

Similarly for density operators instead of pure states.

If (some of) the operators a_i are unbounded, we use the trick of §B.21 and pass to their bounded transforms b_i, see Theorem B.152. We say that the b_i ***commute*** iff the corresponding bounded operators b_i do; this is equivalent to commutativity of all spectral projections of the a_i. We then define, in self-explanatory notation,

$$\sigma(\underline{a}) = \{\underline{\lambda}(1 - \underline{\lambda}^2)^{-1/2} \mid \underline{\lambda} \in \sigma(\underline{b}) \cap (-1,1)^n\}. \tag{4.25}$$

This leads to Born measures on $\sigma(\underline{a})$ defined either as in (4.23), with $f \in C(\sigma(\underline{a}))$ replaced by $f \in C_0(\sigma(\underline{a}))$, cf. (4.19), or as in (4.24).

For example, if $H = L^2(\mathbb{R}^n)$ and $a_i\psi(x) = x_i\psi(x)$, defined on the domain

$$D(a_i) = \{\psi \in L^2(\mathbb{R}^n) \mid \int_{\mathbb{R}^n} d^n x x_i^2 |\psi(x)|^2 < \infty\}, \tag{4.26}$$

as in (B.242), then $b_i\psi(x) = x_i(1 + x_i^2)^{-1/2}\psi(x)$, so that $\sigma(\underline{b}) = [-1,1]^n$ and hence $\sigma(\underline{a}) = \mathbb{R}^n$. For a measurable region $\underline{\Delta} \subset \mathbb{R}^n$ we then have Pauli's famous formula

$$\mu_\psi(\underline{\Delta}) = \int_{\underline{\Delta}} d^n x |\psi(x)|^2 \tag{4.27}$$

for finding the particle in the region $\underline{\Delta}$, given that the system is in a pure state ψ.

4.2 Density operators and normal states

Definition 2.4 of a state still makes good sense in the infinite-dimensional case, as it simply specializes the general definition of a state on a C*-algebra A to the case $A = B(H)$. Thus we continue to say that a state on $B(H)$ is a complex-linear map $\omega : B(H) \to \mathbb{C}$ satisfying $\omega(b^*b) \geq 0$ for each $b \in B(H)$ and $\omega(1_H) = 1$. Despite this lack of novelty in the definition of a state (i.e., compared to finite-dimensional Hilbert spaces), Theorem 2.7 no longer holds if H is infinite-dimensional: although it (almost trivially) remains true that density operators ρ on H define states on $B(H)$ through the fundamental correspondence $\omega(a) = \mathrm{Tr}\,(\rho a)$, $a \in B(H)$, cf. (2.33), there are (many) states that are *not* given in that way (see below). Fortunately, states that *do* arise through (2.33) can be characterized in a simple way.

Definition 4.11. *A state $\omega : B(H) \to \mathbb{C}$ is called* **normal** *if for each orthogonal family (e_i) of projections (i.e., $e_i^* = e_i$ and $e_i e_j = \delta_{ij} e_i$) one has*

$$\omega\left(\sum_i e_i\right) = \sum_i \omega(e_i). \tag{4.28}$$

Here $\sum_i e_i$ is defined as the projection on the smallest closed subspace K of H that contains each $e_i H$ (that is, $\sum_i e_i = \vee_i e_i$, i.e., the supremum in the poset $\mathscr{P}(H)$ of all projections on H with respect to the partial order $e \leq f$ iff $eH \subseteq fH$). Furthermore, the sum over i on the right-hand side is defined by (B.11)*, i.e., as the supremum (in $\mathbb{R}$) of the set of all sums $\sum_{i \in F} \omega(e_i)$ over finite subsets $F \subset I$ of the index set I in which i takes values. It is finite because $\sum_{i \in F} e_i \leq 1_H$ and hence, since ω is positive,*

$$\sum_{i \in F} \omega(e_i) < \omega(1_H) = 1.$$

For example, let (υ_i) be a basis of H with associated one-dimensional projections

$$e_i = |\upsilon_i\rangle\langle\upsilon_i|. \tag{4.29}$$

If ω is assumed to be a state, then the additivity condition (4.28) implies

$$\sum_i \omega(e_i) = 1, \tag{4.30}$$

or, equivalently, using Definition B.6 etc. as well as the notation $e_F \equiv \sum_{i \in F} e_i$,

$$\lim_F \omega(e_F) = 1. \tag{4.31}$$

If H is separable, any orthogonal family (e_i) of projections is necessarily countable, and (4.28) is analogous to the countable additivity condition defining a measure.

Theorem 4.12. *A state ω on $B(H)$ takes the form $\omega(a) = \mathrm{Tr}\,(\rho a)$ for some (unique) density operator $\rho \in \mathscr{D}(H)$ iff it is normal.*

Proof. First, eq. (2.33) implies (4.28). To see this, take the trace with respect to some basis (υ_j) of H that is *adapted* to the family (e_i) in the sense that for each j, either $e_i\upsilon_j = \upsilon_j$ (i.e., $\upsilon_j \in e_iH$) for one value of i, or $e_i\upsilon_j = \upsilon_j$ for all i. Then

$$\omega\left(\sum_i e_i\right) = \mathrm{Tr}\left(\rho \sum_i e_i\right) = \sum_j \langle \upsilon_j, \rho \sum_i e_i \upsilon_j\rangle = \sum_j{}' \langle \upsilon_j, \rho \upsilon_j\rangle,$$

where the sum $\sum_j'$ is over those j for which $\upsilon_j \in K \equiv \vee_i e_i H$. On the other hand, since the basis is adapted, we have $\upsilon_j \in K$ iff there is an i for which $e_i\upsilon_j = \upsilon_j$ (since otherwise $e_i\upsilon_j = 0$ and hence $\upsilon_j \perp e_iH$ for each i, so that $\upsilon_j \in K^\perp$), so

$$\sum_i \omega(e_i) = \sum_i \mathrm{Tr}\,(\rho e_i) = \sum_i \sum_j \langle \upsilon_j, \rho e_i \upsilon_j\rangle = \sup_{F\subset I} \sum_{j\in J_F} \langle \upsilon_j, \rho \upsilon_j\rangle = \sum_j{}' \langle \upsilon_j, \rho \upsilon_j\rangle,$$

where J_F consists of those j for which $\upsilon_j \in \sum_{i\in F} e_iH$. This gives (4.28).

Conversely, assume ω is normal. For the e_i in (4.28) we now take the projections (4.29) determined by some basis (υ_i). For each $a \in B(H)$ we then have

$$\omega(a) = \lim_F \omega(e_F a). \tag{4.32}$$

Indeed, using Cauchy–Schwarz for the positive semi-definite form $(a,b) = \omega(a^*b)$, as in (C.197), and using $\sum_i e_i = 1_H$ and hence $\omega(a) = \omega(\sum_i e_i a)$ we have

$$|\omega(a) - \omega(e_F a)|^2 = |\omega(e_{F^c} a)|^2 \le \omega(a^*a)\omega(e_{F^c}) \le \|a\|^2 \omega(e_{F^c}), \tag{4.33}$$

since $e_{F^c} \equiv \sum_{i\notin F} e_i$ is a projection. Since $\omega(e_F) + \omega(e_{F^c}) = \omega(1_H) = 1$, eq. (4.31) gives $\lim_F \omega(e_{F^c}) = 0$, so that (4.33) gives (4.32). For each finite $F \subset I$, the operator $e_F a$ has finite rank and hence is compact. According to Theorem B.146, the restriction of $\omega : B(H) \to \mathbb{C}$ to the C*-algebra $B_0(H)$ of compact operators on H is induced by a trace-class operator ρ, which (from the requirement that ω be a state) must be a density operator. Hence $\omega(e_F a) = \mathrm{Tr}\,(\rho e_F a)$, and we finally have

$$\omega(a) = \lim_F \omega(e_F a) = \lim_F \mathrm{Tr}\,(\rho e_F a) = \mathrm{Tr}\,(\rho a). \tag{4.34}$$

To derive the final equality, we rewrite $\mathrm{Tr}\,(\rho e_F a) = \mathrm{Tr}\,(e_F a \rho)$, cf. (A.78) and Proposition B.144, note that $a\rho \in B_1(H)$, as shown in Corollary B.147, and observe that for any $b \in B_1(H)$ we have $\lim_F \mathrm{Tr}\,(e_F b) = \mathrm{Tr}\,(b)$. To see this, simply compute the trace in the basis (υ_i) defining the projections e_i through (4.29), so that $\mathrm{Tr}\,(e_F b) = \sum_{i\in F} \langle \upsilon_i, b\upsilon_i\rangle$, and note that by Definition B.6,

$$\lim_F \sum_{i\in F} \langle \upsilon_i, b\upsilon_i\rangle = \sum_i \langle \upsilon_i, b\upsilon_i\rangle = \mathrm{Tr}\,(b).$$

Finally, suppose $\omega(a) = \mathrm{Tr}\,(\rho_1 a) = \mathrm{Tr}\,(\rho_2 a)$ for each $a \in B(H)$ and hence for each $a \in B_0(H)$. It follows from (B.476) that $\mathrm{Tr}\,(\rho a) = 0$ for all $a \in B_0(H)$ iff $\rho = 0$. Hence $\rho_1 = \rho_2$, i.e., a normal state ω uniquely determines a density operator ρ. □

If ω is normal, we may therefore use the spectral resolution (2.6) of the corresponding density operator ρ, i.e., $\rho = \sum_i p_i |\upsilon_i\rangle\langle\upsilon_i|$, where (υ_i) is some basis of H consisting of eigenvectors of ρ (which exists because ρ is compact and self-adjoint), and the corrsponding eigenvalues satisfy $p_i \geq 0$ and $\sum_i p_i = 1$; see the explanation after Definition B.148. Computing the trace in the same basis gives

$$\mathrm{Tr}\,(\rho a) = \sum_i p_i \langle \upsilon_i, a\upsilon_i \rangle. \tag{4.35}$$

We may characterize normality in a number of other ways. First note that because of the duality $B_1(H)^* \cong B(H)$ of Theorem B.146, cf. (B.477), we may equip $B(H)$ with the w^*-topology in its role as the dual of the trace-class operators $B_1(H)$, see §B.9; this means that $a_\lambda \to a$ iff $\mathrm{Tr}\,(\rho a_\lambda) \to \mathrm{Tr}\,(\rho a)$ for each $\rho \in B_1(H)$, or, equivalently, for each $\rho \in \mathscr{D}(H)$, since each trace-class operator is a linear combination of at most four density operators, as follows from Lemma C.53 with (C.8) - (C.9). The w^*-topology on $B(H)$, seen as the dual of $B_1(H)$, is called the ***σ-weak topology***. By Proposition B.46, the σ-weakly continuous linear functionals φ on $B(H)$ are just those given by $\varphi(a) = \mathrm{Tr}\,(\rho b)$ for some trace-class operator $b \in B_1(H)$.

Secondly, $B(H)$ is ***monotone complete***, in the sense that each net (a_λ) of positive operators that is bounded (i.e., $0 \leq a_\lambda \leq c \cdot 1_H$ for some $c > 0$ and all $\lambda \in \Lambda$) and increasing (in that $a_\lambda \leq a_{\lambda'}$ whenever $\lambda \leq \lambda'$) has a supremum a with respect to the standard ordering $\leq$ on $B(H)_+$, which supremum coincides with the strong limit of the net (i.e., $\lim_\lambda a_\lambda \psi = a\psi$ for each $\psi \in H$); the proof is the same as for Proposition B.98, and also here we write $a_\lambda \nearrow a$ to describe this entire situation.

Corollary 4.13. *The following conditions on a state $\omega \in S(B(H))$ are equivalent:*

1. *ω is normal, cf. Definition 4.11;*
2. *$\omega(a) = \lim_\lambda \omega(a_\lambda)$ if $a_\lambda \nearrow a$;*
3. *$\omega(a) = \mathrm{Tr}\,(\rho a)$ for some density operator $\rho \in \mathscr{D}(H)$;*
4. *ω is σ-weakly continuous.*

Proof. We have seen $1 \leftrightarrow 3 \leftrightarrow 4$, and $2 \to 1$ is obvious, so establishing $3 \to 2$ would complete the proof. To this effect, we first note that because the sum (4.35) is convergent, for $\varepsilon > 0$ we may find a finite subset $F \subset I$ for which $\sum_{i\notin F} p_i < \varepsilon/2\|a\|$ (assuming $a \neq 0$). Since $0 \leq a_\lambda \leq a$ also implies $a_\lambda \leq \|a\| \cdot 1_H$ (since $a \leq \|a\| \cdot 1_H$), we therefore have $|\sum_{i\notin F} p_i \langle \upsilon_i, (a_\lambda - a)\upsilon_i\rangle| < 2\varepsilon/3$, uniformly in λ. Moreover, since F is finite and $a_\lambda \to a$ strongly, we can find λ_0 such that for all $\lambda \geq \lambda_0$ we have

$$|\sum_{i\in F} p_i \langle \upsilon_i, (a_\lambda - a)\upsilon_i\rangle| < \varepsilon/3. \tag{4.36}$$

Consequently, for such λ,

$$|\mathrm{Tr}\,(\rho(a_\lambda - a))| \leq |\sum_{i\in F} p_i \langle \upsilon_i, (a_\lambda - a)\upsilon_i\rangle| + |\sum_{i\notin F} p_i \langle \upsilon_i, (a_\lambda - a)\upsilon_i\rangle| < \frac{2}{3}\varepsilon + \frac{1}{3}\varepsilon = \varepsilon.$$

This shows that $\lim_\lambda |\mathrm{Tr}\,(\rho(a_\lambda - a))| = 0$, so that assumption 3 implies no. 2. □

We denote the ***normal state space*** of $B(H)$, i.e., the set of all normal states on $B(H)$ by $S_n(B(H))$. It is easy to see from Definition B.148 that $S_n(B(H))$ is a convex (but not necessarily compact!) subset of the total state space $S(B(H))$.

Corollary 4.14. *The relation $\omega(a) = \mathrm{Tr}\,(\rho a)$ induces an isomorphism*

$$S_n(B(H)) \cong \mathscr{D}(H) \tag{4.37}$$

of convex sets (i.e., $\omega \leftrightarrow \rho$). Furthermore, for the corresponding pure states we have

$$P_n(B(H)) \cong \mathscr{P}_1(H), \tag{4.38}$$

i.e., any pure state ω on $B_0(H)$, as well as any normal pure state on $B(H)$, is given by $\omega = \omega_\psi$ for some unit vector $\psi \in H$, where $\omega(a) = \langle \psi, a\psi \rangle$, cf. (2.42).

The proof of (4.38) is practically the same as in the finite-dimensional case. From Theorem B.146 we obtain another characterization of $S_n(B(H))$ and hence of $\mathscr{D}(H)$:

Corollary 4.15. *If $B_0(H)$ is the C*-algebra of compact operators on H, we have*

$$S(B_0(H)) = S_n(B(H)); \tag{4.39}$$
$$P(B_0(H)) = P_n(B(H)), \tag{4.40}$$

in the sense that any (pure) state ω on $B_0(H)$ has a unique normal *extension to a (pure) state ω' on $B(H)$, given by the same density operator ρ that yields ω.*

It can be shown that any state $\omega \in S(B(H))$ has a convex decomposition

$$\omega = t\omega_n + (1-t)\omega_s, \tag{4.41}$$

where $t \in [0,1]$, ω_n is a normal state, and ω_s is called a ***singular state***. In particular, since for $t \in (0,1)$ the state ω is mixed, *a pure state is either normal or singular.*

Singular states are not as aberrant as the terminology may suggest: such states are routinely used in the physics literature and are typically denoted by $|\lambda\rangle$, where λ lies in the continuous spectrum of some self-adjoint operator (that has to be maximal for this notation to even begin to make sense, see §4.3 below). Examples of such "improper eigenstates" are $|x\rangle$ and $|p\rangle$, which many physicists regard as idealizations. However, mathematically such states are at least defined, namely as singular pure states on $B(H)$. The key to the existence of such states lies in Proposition C.15 and its proof, which should be reviewed now; we only need the case $a^* = a$.

Proposition 4.16. *Let $a = a^* \in B(H)$ have non-empty continuous spectrum, so that there is some $\lambda \in \sigma(a)$ that is not an eigenvalue of a. Then $\omega_\lambda(f(a)) = f(\lambda)$ defines a pure state on $A = C^*(a)$, whose extension to $B(H)$ by any pure state is singular.*

Proof. Normal pure states on $B(H)$ take the form $\omega_\psi(b) = \langle \psi, b\psi \rangle$, where $\psi \in H$ is a unit vector and $b \in B(H)$. We know from Proposition C.14 that ω_λ is multiplicative on $C^*(a)$. However, if some multiplicative state ω on $C^*(a)$ has the form $\omega = \omega_\psi$, then ψ must be eigenvector of a; cf. the proof of Proposition 2.3. □

4.3 The Kadison–Singer Conjecture

To obtain deeper insight into singular pure states, and as a matter of independent interest, we return to the Kadison–Singer problem, cf. §2.6. Recall that this problem asks if some abelian unital C*-algebra $A \subset B(H)$ has the ***Kadison–Singer property***, stating that a pure state ω_A on A has a *unique* pure extension ω to $B(H)$. Here the issue is uniqueness rather than existence, since at least one such extension exists: since A is necessarily unital (with $1_A = 1_H$) and ω_A is a state on A, so that in particular $\omega_A(1_A) = \|\omega_A\| = 1$, Corollary B.41 gives the existence of a bounded extension ω satisfying $\omega(1_H) = \|\omega\| = 1$, which by Proposition C.5 is a state on $B(H)$. Proposition 2.22 then gives the existence of a *pure* extension ω. As in the finite-dimensional case, the Kadison–Singer property forces A to be maximal (in the poset $\mathscr{C}(B(H))$ of all abelian unital C*-subalgebras of $B(H)$, ordered by inclusion):

Proposition 4.17. *If some abelian unital C*-subalgebra A of $B(H)$ has the Kadison–Singer property, then A is necessarily maximal.*

Proof. We use the Gelfand isomorphism $A \cong C(P(A))$, where $P(A)$ is the pure state space of A, cf. Theorem C.8 and Proposition C.14. If A has the Kadison–Singer property and $A \subseteq B \subset B(H)$, where B is an abelian unital C*-subalgebra A of $B(H)$, then ω_A has a unique pure extension ω on $B(H)$, which restricts to some state ω_B on B. The same reasoning as in the proof of Proposition 2.22 shows that ω_B is a pure state on B, so that we obtain a unique map

$$P(A) \to P(B); \tag{4.12}$$
$$\omega_A \mapsto \omega_B. \tag{4.13}$$

The inverse of this map is simply the pullback of the inclusion $A \hookrightarrow B$, i.e., $\omega_B \in P(B)$ defines $\omega_A \in P(A)$ by restriction, so that we have a bijection $P(A) \cong P(B)$, $\omega_A \leftrightarrow \omega_B$. Since for any pair of C*-algebras $A \subseteq B$ the pullback $S(B) \to S(A)$ is continuous (in the pertinent w^*-topology), the map $\omega_B \mapsto \omega_A$ is continuous. As in Lemma C.20, this implies that it is in fact a homeomorphism, so that $A \cong B$ through the inclusion $A \hookrightarrow B$. This gives $A = B$, and hence A is maximal. □

Maximality of A implies $A' = A$, so that A is a von Neumann algebra, sharing the unit of $B(H)$. To see the relevance of singular states for the Kadison–Singer problem, we first settle the normal case. We know what it means for a state on $B(H)$ to be normal (cf. Definition 4.11 and Corollary 4.13); for arbitrary von Neumann algebras $A \subset B(H)$ the situation is exactly the same: we *define* normality by (4.28) and *characterize* it by the equivalent properties in Corollary 4.13, where the σ-weak topology on A may be defined either as the one inherited from $B(H)$, or, more intrinsically, and the w^*-topology from the duality $A = A_*^*$, where the Banach space A_* is the so-called predual of A, e.g., $\ell^\infty_* \cong \ell^1$ and $L^\infty(0,1)_* = L^1(0,1)$, cf. §B.9.

Theorem 4.18. *Let H be a separable Hilbert space and let ω_A be a* normal *pure state on a maximal commutative unital C*-algebra A in $B(H)$. Then ω_A has a unique extension to a state ω on $B(H)$, which is necessarily pure and normal.*

Proof. As noted after (4.41), a pure state on $B(H)$ is either normal or singular. The possibility that ω_A is normal whereas ω is singular is excluded by Corollary 4.13.3, so ω must be normal and hence given by a density operator. The proof of uniqueness is then the same as in the finite-dimensional case, cf. Theorem 2.21. □

We now recall the classification of maximal maximal abelian *-algebras (and hence of maximal abelian von Neumann algebras) A in $B(H)$ up to unitary equivalence (cf. Theorem B.118). This classification is the relevant one for the Kadison–Singer problem, since, as is easily seen, $A \subset B(H)$ has the Kadison–Singer property iff $uAu^{-1} \subset B(uH)$ has it. The uniqueness of the finite-dimensional case will be lost:

Theorem 4.19. *If H is separable and infinite-dimensional, and $A \subset B(H)$ is a maximal abelian *-algebra, then A is unitarily equivalent to exactly one of the following:*

1. $L^\infty(0,1) \subset B(L^2(0,1))$;
2. $\ell^\infty \subset B(\ell^2)$;
3. $L^\infty(0,1) \oplus \ell^\infty(\kappa) \subset B(L^2(0,1) \oplus \ell^2(\kappa))$,

where $\ell^\infty \equiv \ell^\infty(\mathbb{N})$, $\ell^2 \equiv \ell^2(\mathbb{N})$, and κ is either $\{1,\ldots,n\}$, in which case $\ell^2(\kappa) = \mathbb{C}^n$ and $\ell^\infty(\kappa) = D_n(\mathbb{C})$, or $\kappa = \mathbb{N}$, in which case $\ell^2(\kappa) = \ell^2$ and $\ell^\infty(\kappa) = \ell^\infty$.

This classification sheds some more light on Theorem 4.18. Since $L^\infty(0,1)$ has no pure normal states and $D_n(\mathbb{C})$ has been dealt with in Theorem 2.21, the interesting case is ℓ^∞. Using Corollary 4.13.3 (or the analysis below), it is easy to check that the normal pure states on ℓ^∞ are given by $\omega_A(f) = f(x)$ for some $x \in \mathbb{N}$; these are vector state of the kind $\omega_A(f) = \langle \psi, m_f \psi \rangle$ with $\psi = \delta_x$, or, in other words, they are given by $\omega_A(f) = \mathrm{Tr}(\rho m_f)$ with $\rho = |\delta_x\rangle\langle\delta_x|$. We now invoke a fairly deep result:

Proposition 4.20. *A pure state ω on $B(H)$ is singular iff one (and hence all) of the following equivalent conditions is satisfied:*

- $\omega(a) = 0$ *for each* $a \in B_0(H)$;
- $\omega(e) = 0$ *for each one-dimensional projection* e;
- $\sum_i \omega(e_i) = 0$ *for the projections* $e_i = |\upsilon_i\rangle\langle\upsilon_i|$ *defined by some basis* (υ_i).

One direction is easy: a normal pure state certainly does not satisfy the condition in question. For example, given (2.42) one may take $a = |\psi\rangle\langle\psi|$, which as a one-dimensional projection lies in $B_0(H)$, so that $\omega_\psi(a) = 1$. We omit the other direction of the proof. We conclude from this proposition that a pure singular state on $B(\ell^2)$ cannot restrict to a normal pure state on ℓ^∞, which reconfirms Theorem 4.18.

We now study the Kadison–Singer property for each of the three cases in Theorem 4.19 (where the third will be an easy corollary of the first and the second). Since the proofs of the first two cases are formidable, we just sketch the argument.

Theorem 4.21. • *There exist (necessarily singular) pure states on $L^\infty(0,1)$ that do not have a unique extension to $B(L^2(0,1))$, and similarly for $L^\infty(0,1) \oplus \ell^\infty(\kappa)$.*
• *Any pure state on ℓ^∞ has a unique extension to $B(\ell^2)$.*

The statement about ℓ^∞ is the ***Kadison–Singer Conjecture***, which dates from 1959 but was only proved in 2013. The first claim (which was already known to Kadison and Singer themselves) is equally remarkable, however, as is the contrast between the two parts of Theorem 4.21. In particular, Dirac's notation $|\lambda\rangle$ may be ambiguous.

The key to the proof of the first claim lies in the choice of a total countable family of normal states on $L^\infty(0,1)$, from which all pure states may be constructed by a limiting operation. Here we call a (countable) family $(\omega_n)_{n\in\mathbb{N}}$ of states on some C*-algebra A ***total*** if, for any self-adjoint $a \in A$, the conditions $\omega_n(a) \geq 0$ for each n imply $a \geq 0$ (the converse is trivial). For example, the well-known ***Haar basis*** (h_n) of $L^2(0,1)$ provides such a family. The functions forming this basis are defined via some bijection β between the set of pairs (k,l) and $\mathbb{N}$, e.g., $\beta(k,l) = k + 2^l$, by

$$h_n = \chi_{\beta^{-1}(n)}, \; (n \in \mathbb{N} = \{1,2,\ldots\}); \tag{4.44}$$
$$\chi_{k,l}(x) = 2^{k/2} g(2^k x - l), \; (k \in \mathbb{N} \cup \{0\}, 0 \leq l < 2^k); \tag{4.45}$$
$$g(x) = 1_{[0,1/2)} - 1_{[1/2,1]}. \tag{4.46}$$

Basic analysis then shows that the Haar functions h_n form a basis of $L^2(0,1)$ and that the associated vector states ω_n on $L^\infty(0,1)$ form a total set, where obviously

$$\omega_n(f) = \langle h_n, m_f h_n \rangle = \int_0^1 h_n^2 f. \tag{4.47}$$

The relevance of total sets to the conjecture is explained by the following lemma.

Lemma 4.22. *If $T \subset S(A)$ is a total set of states on a unital C*-algebra A, then*

$$S(A) = \mathrm{co}(T)^-; \tag{4.48}$$
$$P(A) \subseteq T^-, \tag{4.49}$$

where $\mathrm{co}(T)^-$ is the w-closure of the convex hull of T in A^* or in $S(A)$.*

Proof. The inclusion $\mathrm{co}(T)^- \subseteq S(A)$ is obvious, since $T \subseteq S(A)$ and $S(A)$ is a compact (and hence a closed) convex set. To prove the converse inclusion, suppose $a = a^* \in A$ and $s \in \mathbb{R}$ are such that $\omega(a) \geq s$ for each $\omega \in T$. Then $\omega(a - s \cdot 1_A) \geq 0$ and hence $\omega(a) \geq s$ for each $\omega \in S(A)$. Using Theorem B.43 (of Hahn–Banach type), this property would lead to a contradiction if $S(A)$ were not contained in $\mathrm{co}(T)^-$.

The second claim, which is the one we will use, follows from the first through a corollary of the Krein–Milman Theorem B.50, stating that if $T \subset K$ is any subset of a compact convex set K such that $K = \mathrm{co}(T)^-$, then $\partial_e K \subseteq T^-$. This corollary may be proved (by contradiction) from Theorem B.43 in a similar way. □

Our next aim is to get rid of the closure in (4.49). The Haar basis yields a map

$$h : \mathbb{N} \to S(L^\infty(0,1)); \tag{4.50}$$
$$n \mapsto \omega_n, \tag{4.51}$$

with image T, i.e., the set of Haar states. Since $S(A)$ is a compact Hausdorff space (in its w^*-topology), the universal property (B.135) of the Čech–Stone compactification $\beta\mathbb{N}$ of $\mathbb{N}$ implies that h extends (uniquely) to a continuous map

$$\beta h : \beta\mathbb{N} \to S(A),$$

whose image is compact and hence closed (since $\beta\mathbb{N}$ is compact). Since $T = h(\mathbb{N}) \subset S(A)$ we have $T \subseteq \beta h(\beta\mathbb{N})$ and hence $T^- \subseteq \beta h(\beta\mathbb{N})$, so that, from (4.49),

$$P(L^\infty(0,1)) \subseteq \beta h(\beta\mathbb{N}). \tag{4.52}$$

Hence each pure state $\omega_c \equiv \omega_{L^\infty(0,1)}$ on $L^\infty(0,1)$ takes the form $\omega_c = \omega_c^{(U)}$, where

$$\omega_c^{(U)}(f) = \lim_U \omega_n(f) = \bigcap_{A\in U} \{\omega_n(f) \mid n \in A\}^-, \; f \in L^\infty(0,1), \tag{4.53}$$

and $U \in \beta\mathbb{N}$ is some ultrafilter on $\mathbb{N}$, cf. (B.136). The point of this analysis, then, is that ω_U can immediately be extended to $B(L^2(0,1))$ by the same formula, i.e.,

$$\omega^{(U)}(a) = \lim_U \omega_n(a) = \bigcap_{A\in U} \{\omega_n(a) \mid n \in A\}^-, \; a \in B(L^2(0,1)), \tag{4.54}$$

where $\omega_n(a) = \langle h_n, a h_n \rangle$. If $L^\infty(0,1)$ had the Kadison–Singer property, this were the unique extension of ω_U, and we will show that this leads to a contradiction.

Apart from the use of ultrafilters, the technically most challenging part of the argument disproving the Kadison–Singer property for $L^\infty(0,1)$ is as follows. If $A = C([0,1])$, for any $f \in A$ and any pure state $\omega \in P(A)$ there is some $x \in [0,1]$ such that $\omega(f) = f(x)$; see Propositions C.14 and C.19. For $A = L^\infty(0,1)$ the situation is not that simple due to measure zero complications. Nonetheless, it is easy to show that for each *positive* $f \in L^\infty(0,1)$ and $\omega_c \in P(L^\infty(0,1))$ and each $\varepsilon > 0$ one has

$$\mu(\{x \in (0,1) \mid f(x) \in [\omega_c(f) - \varepsilon, \omega_c(f) + \varepsilon]\}) > 0. \tag{4.55}$$

where μ is Lebesque measure on $(0,1)$. Taking the projection

$$e = 1_{\{x\in(0,1)|f(x)\in[\omega_c(f)-\varepsilon/2,\omega_c(f)+\varepsilon/2]\}},$$

it follows that for each positive $f \in L^\infty(0,1)$, $\omega \in P(L^\infty(0,1))$ and $\varepsilon > 0$ there exists a projection $e \in \mathscr{P}(L^\infty(0,1))$ with $\omega(e) = 1$ and $\|ef - e\omega_c(f)\| < \varepsilon$. Hard analysis then generalizes this property from $L^\infty(0,1)$ to $B(L^2(0,1))$, as follows:

Lemma 4.23. *If $\omega_c \in P(L^\infty(0,1))$ has a unique extension ω to $B(L^2(0,1))$ (which is necessarily pure if it is unique), then for each $a \in B(L^2(0,1))$ and $\varepsilon > 0$ there exists a projection $e \in \mathscr{P}(L^\infty(0,1))$ with $\omega_c(e) = 1$ and*

$$\|ea - e\omega(a)\| < \varepsilon. \tag{4.56}$$

To derive a contradiction between (4.54) and (4.56), we use a bijection $b : \mathbb{N} \to \mathbb{N}$ that cyclically permutes the ordered subsets $(2^k+1,\ldots,2^{k+1})$, $k = 0,1,\ldots$, that is, $(1,2)$, $(3,4)$, $(5,6,7,8)$, $(9,\ldots,16)$, etc. This bijection induces a unitary operator

$$u : L^2(0,1) \to L^2(0,1); \tag{4.57}$$
$$uh_n = h_{b(n)}, \tag{4.58}$$

which is easily shown to have the following properties:

$$\omega_n(u) = 0,\ n \in \mathbb{N}; \tag{4.59}$$
$$\|eue\| = 1,\ e \in \mathscr{P}(L^\infty(0,1)), e \neq 0. \tag{4.60}$$

To show that $L^\infty(0,1)$ fails to have the Kadison–Singer property, suppose it does, so that any $\omega_c \in P(L^\infty(0,1))$ has a unique extension $\omega \in P(B(L^2(0,1)))$. As already noted, we may then assume that $\omega_c = \omega_c^{(U)}$, as in (4.53), whilst $\omega = \omega^{(U)}$, as in (4.54). Taking $a = u$ then gives $\omega(u) = 0$, see (4.59), so that $\|eu\| < \varepsilon$ by (4.56). But this contradicts (4.60), finishing the sketch of the proof of the first claim in Theorem 4.21. The remark about $L^\infty(0,1) \oplus \ell^\infty(\kappa)$ follows from the one about $L^\infty(0,1)$.

We now pass to the (even) more difficult case of $\ell^\infty \subset B(\ell^2)$. Although this will not be used in the proof, it gives some insight to know which states on ℓ^∞ we are actually talking about, i.e., the singular pure states, and compare this with (4.53).

Theorem 4.24. *There is a bijective correspondence*

$$\omega_d(f) = \int_{\mathbb{N}} d\mu\, f \tag{4.61}$$

between states ω_d on ℓ^∞ and finitely additive probability measures μ on $\mathbb{N}$, where:

1. ω_d is normal iff μ is countably additive (and hence is a probability measure).
2. ω_d is pure iff μ corresponds to some ultrafilter U on $\mathbb{N}$, in which case: ω_d is normal iff U is principal (and hence singular iff U is free).

This follows from case no. 5 in §B.9, notably eqs. (B.153) - (B.154). In other words, the pure states ω_d on ℓ^∞ are given by ultrafilters U on $\mathbb{N}$ through

$$\omega_d^{(U)}(f) = \beta f(U) = \lim_U f(n); \tag{4.62}$$

the analogy with (4.53) is even clearer if we write $f(n) = \langle \delta_n, m_f \delta_n \rangle \equiv \omega_n(f)$. If $U = U_n$ is a principal ultrafilter, $n \in \mathbb{N}$, we thus recover the normal pure states

$$\omega_d^{(U_n)}(f) = f(n). \tag{4.63}$$

As in (4.54), we find at least one natural extension $\omega^{(U)}$ of $\omega_d^{(U)}$ to $B(\ell^2)$, namely

$$\omega^{(U)}(a) = \lim_U \omega_n(a). \tag{4.64}$$

We now show that that ℓ^∞ has the Kadison–Singer property, making $\omega^{(U)}$ the *only* extension of $\omega_d^{(U)}$. The proof relies on an extremely difficult lemma from linear algebra (formerly known as a ***paving conjecture***). We first define a linear map $D : M_n(\mathbb{C}) \to D_n(\mathbb{C})$ by $D(a)_{ii} = a_{ii}$, $i = 1, \ldots, n$, and $D(a)_{ij} = 0$ whenever $i \neq j$.

Lemma 4.25. *For any $\varepsilon > 0$ there exist $l \in \mathbb{N}$ such that for all $n \in \mathbb{N}$ and $a \in M_n(\mathbb{C})$ with $D(a) = 0$, there are l projections $(e_1, \ldots, e_l)$ in $D_n(\mathbb{C})$ such that*

$$\sum_{k=1}^{l} e_k = 1_n; \tag{4.65}$$
$$\|e_i a e_i\| \leq \varepsilon \|a\|, \; i = 1, \ldots, l. \tag{4.66}$$

Since this estimate is uniform in n, the lemma extends to ℓ^2, where $D : B(\ell^2) \to \ell^\infty$ is defined analogously, i.e., $D(a)$ is diagonal in the canonical basis (δ_n) of ℓ^2 with

$$D(a)\delta_n = \omega_n(a)\delta_n, \; n \in \mathbb{N}. \tag{4.67}$$

Lemma 4.26. *For any $\varepsilon > 0$ there exist $l \in \mathbb{N}$ such that for all $a \in B(\ell^2)$ with $D(a) = 0$, there are l projections $(e_1, \ldots, e_l)$ in ℓ^∞ such that*

$$\sum_{k=1}^{l} e_k = 1_H; \tag{4.68}$$
$$\|e_i a e_i\| \leq \varepsilon \|a\|, \; i = 1, \ldots, l. \tag{4.69}$$

Now suppose that $\omega_d \in P(\ell^\infty)$, that $\omega \in S(B(\ell^2))$ extends ω_d, and that $a \in B(\ell^2)$ has $D(a) = 0$. Let e_i be one of the projections in Lemma 4.26. Using Cauchy–Schwarz for the sesquilinear form $(a,b) = \omega(a^*b)$, we obtain (using $e_i^2 = e_i^* = e_i$)

$$|\omega(e_i a e_j)|^2 \leq \omega(e_i)\omega(e_j a^* a e_i); \tag{4.70}$$
$$|\omega(e_i a e_j)|^2 \leq \omega(a^* e_i a)\omega(e_j). \tag{4.71}$$

Since $\omega(e_i) = \omega_d(e_i)$ and ω_d is a pure state (and hence is multiplicative), we have $\omega(e_i) \in \{0, 1\}$, since e_i is a projection. Moreover, in view of (4.68) and the normalization $\omega(1_H) = 1$, there must be exactly one value of $i = 1, \ldots, l$, say $i = i_0$, such that $\omega(e_{i_0}) = 1$, and $\omega(e_i) = 0$ for all $i \neq i_0$. Eqs. (4.70) - (4.71) therefore imply that $\omega(e_i a e_j) \neq 0$ iff $i = j = i_0$. Using (4.68) once more, we see that $\omega(a) = \sum_{i,j} \omega(e_i a e_j) = \omega(e_{i_0} a e_{i_0})$, so that $|\omega(a)| \leq \|\omega\| \|e_{i_0} a e_{i_0}\| \leq 1 \cdot \varepsilon \|a\|$ by (4.66). Letting $\varepsilon \to 0$, we proved:

Lemma 4.27. *If $\omega \in S(B(\ell^2))$ extends $\omega_d \in P(\ell^\infty)$, and $D(a) = 0$, then $\omega(a) = 0$.*

Since $D^2 = D$, we have $D(a - D(a)) = 0$, so that for any $a \in B(\ell^2)$, we have

$$\omega(a) = \omega(D(a)) = \omega_d(D(a)), \tag{4.72}$$

provided that ω extends ω_d, as before. This shows that ω is determined by ω_d and hence is unique, completing the proof (sketch) of Theorem 4.21.

4.4 Gleason's Theorem in arbitrary dimension

To a large extent the thrust and difficulty of the proof of Gleason's Theorem 2.28 already lies in its finite-dimensional version, but some care is needed in the general case, and also Corollary 2.29 needs to be refined. A major point here is that Definition 2.23 has no unambiguous generalization to arbitrary Hilbert spaces.

Definition 4.28. *Let H be an arbitrary Hilbert space with unit sphere H_1.*

1. *A* **probability distribution** *on $\mathscr{P}(H)$ is a map $p : H_1 \to [0,1]$ that satisfies*
$$\sum_{i\in I} p(\upsilon_i) = 1, \textit{ for any basis } (\upsilon_i) \textit{ of } H, \tag{4.73}$$
where, as in §B.12, the sum (over a possibly uncountable index set) is meant as in Definition B.6. In particular, if H is separable and the basis is labeled and ordered by $I = \mathbb{N}$, then it is an ordinary convergent sum of the kind $\sum_{i=1}^{\infty}\cdots$.
2. *A map $P : \mathscr{P}(H) \to [0,1]$ that satisfies $P(1_H) = 1$ is called a:*
   a. **finitely additive probability measure** *if*
$$P\left(\sum_{j\in J} e_j\right) = \sum_{j\in J} P(e_j) \tag{4.74}$$
   for any finite *collection $(e_j)_{j\in J}$ of mutually orthogonal projections on H (i.e., $e_jH \perp e_kH$, or equivalently, $e_je_k = 0$, whenever $j \neq k$); this is equivalent to the condition $P(e+f) = P(e) + P(f)$ whenever $ef = 0$, cf. Definition 2.23.2.*
   b. **probability measure** *if* (4.74) *holds for any* countable *collection $(e_j)_{j\in J}$ of mutually orthogonal projections on H, where the first sum is defined in the* strong *operator topology; note that the strong sum $\sum_j e_j$ coincides with the supremum $\bigvee_j e_j$ of the given family, defined with respect to the usual ordering of projections (that is, $e \le f$ iff $eH \subseteq fH$).*
   c. **completely additive probability measure** *if* (4.74) *holds for* arbitrary *collections $(e_j)_{j\in J}$ of mutually orthogonal projections on H (the first sum again meant in the* strong *operator topology, with the same comment as above).*

Thus a probability measure is by definition σ-additive in the usual sense of measure theory; the other two cases are unusual from that perspective. However, if H is separable, then J can be at most countable, so that complete additivity is the same as σ-additivity and hence any probability measure is completely additive. Surprisingly, assuming the *Continuum Hypothesis* (CH) of set theory, it can be shown that this is even the case for arbitrary Hilbert spaces. The fundamental distinction, then, is between *finitely* additive probability measures and probability measures (which by definition are *countably* additive). As we shall see, this reflects the distinction between *arbitrary* and *normal* states on $B(H)$, respectively, cf. §4.2. In what follows, in dealing with non-separable Hilbert spaces we assume CH, in which case probability distributions on H are equivalent to probability measures on $\mathscr{P}(H)$.

The proof is the same as in finite dimension (taking into account that infinite sums over projections are defined strongly). Even without CH, Gleason's Theorem still holds for non-separable Hilbert spaces if we assume P to be completely additive, and probability distributions are equivalent to completely additive probability measures on $\mathscr{P}(H)$. For separable Hilbert spaces, CH is irrelevant and unnecessary altogether.

We then have the following generalization (and bifurcation) of Theorem 2.28.

Theorem 4.29. *Let H be a Hilbert space of dimension > 2.*

1. *Each probability measure P on $\mathscr{P}(H)$ is induced by a unique normal state on $B(H)$ via* (2.122), *i.e.,*
$$P(e) = \mathrm{Tr}\,(\rho e), \tag{4.75}$$
where ρ is a density operator on H uniquely determined by P.
Equivalently, each probability distribution p on $\mathscr{P}(H)$ is given by (2.123), *or*
$$p(\upsilon) = \langle \upsilon, \rho \upsilon \rangle. \tag{4.76}$$
Conversely, each density operator ρ on H defines a probability measure P on $\mathscr{P}(H)$ via (4.75), *as well as as a probability distribution p on $\mathscr{P}(H)$ via* (4.76).
2. *Each finitely additive probability measure P on $\mathscr{P}(H)$ is induced by a unique state ω on $B(H)$ via*
$$P(e) = \omega(e), \tag{4.77}$$
and similarly each probability distribution p on $\mathscr{P}(H)$ is given by
$$p(\upsilon) = \omega(e_\upsilon). \tag{4.78}$$
Conversely, each state ω on H defines a probability measure P on $\mathscr{P}(H)$ via (4.77), *as well as as probability distribution p on $\mathscr{P}(H)$ via* (4.78).

Proof. The proof of part 1 is practically the same as in finite dimension, except for the fact that in the proof of Lemma 2.33 the reference to Proposition A.23 should be replaced by Proposition B.79, upon which one obtains a bounded positive operator ρ for which (2.123) holds. The normalization condition (2.110) then yields $\mathrm{Tr}\,(\rho) = 1$ if the trace is taken over any basis of H, and since ρ is positive this implies $\rho \in B_1(H)$, see §B.20 (complete additivity of P is just necessary to relate it to p).

Unfortunately, the proof of part 2 exceeds the scope of this book (see Notes). □

In infinite dimension, Corollary 2.29 becomes more complicated, too; for one thing, Definition 2.26 of a quasi-state bifurcates into two possibilities. The one given still makes perfect sense and is natural from the point of view of Bohrification; to avoid confusion we call a map $\omega : B(H) \to \mathbb{C}$ satisfying the conditions in Definition 2.26 a ***strong quasi-state***. In the context of Gleason's Theorem, a slightly different notion is appropriate: a ***weak quasi-state*** on $B(H)$ satisfies Definition 2.26, except that linearity is only required on commutative C*-algebras in $B(H)$ of the form $C^*(a)$, where $a = a^* \in B(H)$ (these are *singly generated*). Since commutative unital C*-subalgebras of $B(H)$ are not necessarily singly generated, and a specific counterexample exists, weak quasi-states are not necessarily strong quasi-states.

Proposition 4.30. *The map $\omega \mapsto \omega_{|\mathscr{P}(H)}$ gives a bijective correspondence between weak quasi-states ω on $B(H)$ and finitely additive probability measures on $\mathscr{P}(H)$.*

Proof. For some finite family $(e_1, \ldots, e_n)$ of mutually orthogonal projections on H, add $e_0 = 1_H - \sum_j e_j$ if necessary and let $a = \sum_{j=0}^n \lambda_j e_j$, with all $\lambda_j \in \mathbb{R}$ different. Then $\sigma(a) = \{\lambda_0, \ldots, \lambda_n\}$, so that $C^*(a) \cong C(\sigma(a) \cong \mathbb{C}^{n+1}$ (cf. Theorem B.94) coincides with the linear span of the projections e_j. If ω is a weak quasi-state, then it is linear on $C^*(a)$ and hence also on the e_j, so that $\omega_{|\mathscr{P}(H)}$ is finitely additive.

Conversely, let μ be a finitely additive probability measure on $\mathscr{P}(H)$. If $a = a^* \in B(H)$ is given, using the notation (B.328) we symbolically define ω on a by

$$\omega(a) = \int_{\sigma(a)} d\mu(e_\lambda)\,\lambda. \tag{4.79}$$

More precisely, for any $\varepsilon > 0$ we use Corollary B.104 to define $\omega_\varepsilon(a) = \sum_{i=1}^n \lambda_i \mu(e_{A_i})$ and let $\omega(a) = \lim_{\varepsilon \to 0} \omega_\varepsilon(a)$; it follows from Lemma B.103 (or the theory underlying the Riemann–Stieltjes integral (4.79)) that this limit exists. Now let $b, c \in C^*(a)$, so that $b = f(a)$ and $c = g(a)$ for certain $f, g \in C(\sigma(a))$, and $b + c = (f+g)(a)$, cf. Theorem B.94. By (B.325) we therefore have $\omega_\varepsilon(b+c) = \sum_{i=1}^n (f+g)(\lambda_i)\mu(e_{A_i})$, which, since $(f+g)(\lambda_i) = f(\lambda_i) + g(\lambda_i)$, again by (B.325) equals $\omega_\varepsilon(b) + \omega_\varepsilon(c)$. Since this holds for every $\varepsilon > 0$, letting $\varepsilon \to 0$ we obtain $\omega(b+c) = \omega(b) + \omega(c)$, making ω linear on $C^*(a)$. It is clear that the quasi-state ω thus obtained, on restriction to $\mathscr{P}(H)$ reproduces μ, making the map $\omega \mapsto \omega_{|\mathscr{P}(H)}$ surjective. Finally, injectivity of this map follows from Corollary B.104. □.

Corollary 4.31. *If* $\dim(H) > 2$, *then each weak quasi-state on $B(H)$ (and* a fortiori *each strong quasi-state) is linear and hence is actually a state.*

This is immediate from Theorem 4.29.2. and Proposition 4.30.

Another corollary of Gleason's Theorem is the ***Kochen–Specker Theorem***, which we will explain in detail in Chapter 6, where it will also be proved in a different way.

Theorem 4.32. *If* $\dim(H) > 2$, *there are no weak quasi-states $\omega : B(H) \to \mathbb{C}$ whose restriction to each C^*-subalgebra $C^*(a) \subset B(H)$ is pure (where $a = a^* \in B(H)$).*

Equivalently, there are no nonzero maps $\omega' : B(H)_{\mathrm{sa}} \to \mathbb{R}$ that are:

- ***Dispersion-free***, i.e., $\omega'(a^2) = \omega'(a)^2$ for each $a \in B(H)_{\mathrm{sa}}$;
- ***Quasi-linear***, i.e., linear on commuting operators.

Cf. Definitions 6.1 and 6.3. To see that these conditions are equivalent to those stated in Theorem 4.32 (despite the impression that linearity on all commuting self-adjoint operators seems stronger than linearity on each $C^*(a)$), extend ω' to $\omega : B(H) \to \mathbb{C}$ by complex linearity, as in Definition 2.26.1, and note that dispersion-freeness implies positivity and hence continuity on each subalgebra $C^*(a)$ (cf. Theorem C.52 and Lemma C.4). We then see that the two conditions just stated imply that ω is multiplicative on $C^*(a)$, and hence pure, see Proposition C.14, which conversely implies that pure states on $C^*(a)$ are dispersion-free. We now prove Theorem 4.32.

Proof. If e is a projection, then $e^2 = e$, so that $\omega(e^2) = \omega(e)$. Since ω is dispersion-free (as just explained), we also have $\omega(e^2) = \omega(e)^2$, whence $\omega(e)^2 = \omega(e)$ and hence $\omega(e) \in \{0,1\}$. Furthermore, since ω is a state by Corollary 4.31, we may apply the GNS-construction, see Theorem C.88 (whose notation we use). In particular, for any projection e, using the fact that $\pi_\omega(e) = \pi_\omega(e)^*\pi_\omega(e)$, by (C.196) we have

$$\omega(e) = \langle \Omega_\omega, \pi_\omega(e)\Omega_\omega \rangle = \|\pi_\omega(e)\Omega_\omega\|^2. \tag{4.80}$$

If $\omega(e) = 0$, then $\pi_\omega(e)\Omega_\omega = 0$ from the second equality. If $\omega(e) = 1$, then $\pi_\omega(e)\Omega_\omega = \Omega_\omega$ from the first inequality and Cauchy–Schwarz (in which we have equality, so that $\pi_\omega(e)\Omega_\omega = z\Omega_\omega$ for some $z \in \mathbb{T}$, upon which (4.80) forces $z = 1$).

By the spectral theorem (e.g. in the form Corollary B.104) or the theory of von Neumann algebras, the linear span of $\mathscr{P}(H)$ is norm-dense in $B(H)$. Since Ω_ω is cyclic for $\pi_\omega(B(H))$ by the GNS-construction, it must be that $H_\omega = \mathbb{C}\cdot\Omega_\omega$, and hence $\pi_\omega(a) = \omega(a)\cdot 1_{H_\omega}$ for any $a \in B(H)$. Since $\pi_\omega(ab) = \pi_\omega(a)\pi_\omega(b)$ by the GNS-construction, this gives $\omega(ab) = \omega(a)\omega(b)$ for all $a,b \in B(H)$. However, such multiplicative statesω on $B(H)$ cannot exist if $\dim(H) > 1$. This is clear if ω is normal, cf. Proposition 2.10, so that the following argument (which also covers the normal case) is especially meant for the case where ω is singular.

1. If $\dim(H) = n < \infty$, there are n one-dimensional projections $(e_1,\dots,e_n)$ such that $\sum_j e_j = 1_H$. (indeed, we may assume that $B(H) = M_n(\mathbb{C})$ and take diagonal matrices $e_1 = \mathrm{diag}(1,0,\dots,0)$, etc.). Now for any pair (e_i,e_j) there is some $v \in B(H)$ (which by definition is a partial isometry) such that $e_i = vv^*$, $e_j = v^*v$ (in the above case e_i and e_j are thus related if $v_{ij} = 1$ and $v_{i'j'} = 0$ otherwise). Hence
$$\omega(e_i) = \omega(vv^*) = \omega(v)\omega(v^*) = \omega(v^*v) = \omega(e_j), \tag{4.81}$$
since ω is multiplicative. But ω is also additive, which implies
$$\sum_{j=1}^n \omega(e_i) = \omega\left(\sum_{j=1}^n e_j\right) = \omega(1_H) = 1. \tag{4.82}$$
Since also $\omega(e_i) \in \{0,1\}$, eqs. (4.81) - (4.82) are clearly contradictory.
2. If $\dim(H) = \infty$, separable or not, a similar contradiction arises from the ***halving lemma***, which states that there is a projection e and an operator v such that $e = vv^*$, $1_H - e = v^*v$. For example, in the separable case assume $H = \ell^2$ and take e the projection onto the closed linear span ℓ^2_e of the basis vectors (δ_x) with $x \in \mathbb{N}$ even, so that $1_H - e$ projects onto the closed linear span ℓ^2_o of the basis vectors (δ_x) with $x \in \mathbb{N}$ odd. Then $\ell^2 = \ell^2_e \oplus \ell^2_o$; take $v = 0$ on ℓ^2_e and $v : \ell^2_o \to \ell^2_e$ any unitary operator. In general, a similar method works, for if I is a set indexing some basis of H one may find a subset $E \subset I$ that has the same cardinality as its complement $I\backslash E$, upon which $\ell^2(E) \cong \ell^2(I\backslash E)$, cf. Theorem B.63.
Multiplicativity of ω then leads to similar contradiction between the properties $\omega(e) = \omega(1_H - e)$, as in (4.81), and $\omega(e) + \omega(1_H - e) = \omega(1_H) = 1$, as in (4.82): if $\omega(e) = 0$ one finds $0 = 1$, whereas $\omega(e) = 1$ implies $2 = 1$. □

Notes

§4.1. **The Born rule from Bohrification** (II)

The Born measure (and its construction along the lines of this section) is well known in functional analysis, cf. Pedersen (1989), §4.5. For the Hamburger Moment Problem see, for example, Reed, M. & Simon, B. (1975), *Methods of Modern Mathematical Physics. Vol II. Fourier Analysis, Self-adjointness* (New York: Academic Press), Theorem X.4, p. 145 and Example 4, p. 205. In fact, the proof uses spectral theory! Corollary 4.6 was suggested by the treatment of the Born rule in Hall (2013). Definition 4.9 of the joint spectrum goes back (at least) to Arens (1961) and Hörmander (1966), §3.1.13.

§4.2. **Density operators and normal states**

These are really results about von Neumann algebras and come from the pertinent literature; our proofs derive from Li (1992), §1.8 and Takesaki (2002), Ch. III.

§4.3. **The Kadison–Singer Conjecture**

As already mentioned in the notes to §2.6, the Kadison–Singer Conjecture was first discussed in Kadison & Singer (1959) and was finally proved by Marcus, Spielman, & Srivastava (2014ab), following important intermediate contributions by e.g. Anderson (1979) and Weaver (2004). For an introduction including a complete proof see Stevens (2016), and for applications of the conjecture and its proof to other areas of mathematics see Casazza et al (2005) as well as Casazza & Tremain (2016). Proposition 4.20 is due to Glimm (1960).

§4.4. **Gleason's Theorem in arbitrary dimension**

The extension of Gleason's Theorem to non-separable Hilbert space assuming complete additivity of P is due to Maeda (1980). Maeda (1990) generalizes this result to von Neumann algebras without summands of type I_2. The proof that assuming CH countable additivity implies complete additivity (and hence Gleason's Theorem) was given by Eilers & Horst (1975). Proposition 4.30 is due to Aarens (1970), whose Theorem 1 is wrong: see Aarens (1991). The proof of Theorem 4.32 is due to Döring (2004), using results of Hamhalter (1993).

Chapter 5
Symmetry in quantum mechanics

Roughly speaking, a ***symmetry*** of some mathematical object is an invertible transformation that leaves all relevant structure as it is. Thus a symmetry of a set is just a bijection (as sets have no further structure, whence invertibility is the only demand on a symmetry), a symmetry of a topological space is a homeomorphism, a symmetry of a Banach space is a linear isometric isomorphism, and, crucially important for this chapter, a symmetry of a Hilbert space H is a ***unitary operator***, i.e., a linear map $u : H \to H$ satisfying one and hence all of the following equivalent conditions:

- $uu^* = u^*u = 1_H$;
- u is invertible with $u^{-1} = u^*$;
- u is a surjective isometry (or, if $\dim(H) < \infty$, just an isometry);
- u is invertible and preserves the inner product, i.e., $\langle u\varphi, u\psi \rangle = \langle \varphi, \psi \rangle$ $(\varphi, \psi \in H)$.

The discussion of symmetries in quantum physics is based on the above idea, but the mathematically obvious choices need not be the physically relevant ones. Even in elementary quantum mechanics, where $A = B(H)$, i.e., the C*-algebra of all bounded operators on some Hilbert space H, the concept of a symmetry is already diverse. The main structures whose symmetries we shall study in this chapter are:

1. The ***normal pure state space*** $\mathscr{P}_1(H)$, i.e., the set of one-dimensional projections on H, with transition probability $\tau : \mathscr{P}_1(H) \times \mathscr{P}_1(H) \to [0,1]$ defined by (2.44).
2. The ***normal state space*** $\mathscr{D}(H)$, i.e. the convex set of density operators ρ on H.
3. The ***self-adjoint operators*** $B(H)_{sa}$ on H, seen as a Jordan algebra (see below).
4. The ***effects*** $\mathscr{E}(H) = [0,1]_{B(H)}$, seen as a convex partially ordered set (poset).
5. The ***projections*** $\mathscr{P}(H)$ on H, seen as an orthocomplemented lattice.
6. The ***unital commutative C*-subalgebras*** $\mathscr{C}(B(H))$ ***of*** $B(H)$, seen as a poset.

Each of these structures comes with its own notion of a symmetry, but the main point of this chapter will be to show these notions are equivalent, corresponding in all cases to either unitary or—surprisingly—*anti-unitary* operators, both merely defined up to a phase. The latter subtlety will open the world of *projective* unitary group representation to quantum mechanics (without which the existence of spin-$\frac{1}{2}$ particles such as electrons, and therewith also of ourselves, would be impossible).

K. Landsman, *Foundations of Quantum Theory*,
Fundamental Theories of Physics 188, DOI 10.1007/978-3-319-51777-3_5

5.1 Six basic mathematical structures of quantum mechanics

We first recall the objects just described in a bit more detail. We have:

$$\mathscr{P}_1(H) = \{e \in B(H) \mid e^2 = e^* = e, \mathrm{Tr}(e) = \dim(eH) = 1\}; \tag{5.1}$$
$$\mathscr{D}(H) = \{\rho \in B(H) \mid \rho \geq 0, \mathrm{Tr}(\rho) = 1\}; \tag{5.2}$$
$$B(H)_{\mathrm{sa}} = \{a \in B(H) \mid a^* = a\}; \tag{5.3}$$
$$\mathscr{E}(H) = \{a \in B(H) \mid 0 \leq a \leq 1_H\}; \tag{5.4}$$
$$\mathscr{P}(H) = \{e \in B(H) \mid e^2 = e^* = e\}; \tag{5.5}$$
$$\mathscr{C}(B(H)) = \{C \subset B(H) \mid C \text{ commutative C*-algebra}, 1_H \in C\}. \tag{5.6}$$

The point is that each of these sets has some additional structure that defines what it means to be a symmetry of it, as we now spell out in detail.

Definition 5.1. *Let H be a Hilbert space (not necessarily finite-dimensional).*

1. *A* **Wigner symmetry** *(of H) is a bijection*

$$\mathsf{W} : \mathscr{P}_1(H) \to \mathscr{P}_1(H) \tag{5.7}$$

that satisfies

$$\mathrm{Tr}(\mathsf{W}(e)\mathsf{W}(f)) = \mathrm{Tr}(ef),\ e, f \in \mathscr{P}_1(H). \tag{5.8}$$

2. *A* **Kadison symmetry** *is an* **affine** *bijection*

$$\mathsf{K} : \mathscr{D}(H) \to \mathscr{D}(H), \tag{5.9}$$

i.e. a bijection K that preserves convex sums: for $t \in (0,1)$ and $\rho_1, \rho_2 \in \mathscr{D}(H)$,

$$\mathsf{K}(t\rho_1 + (1-t)\rho_2) = t\mathsf{K}\rho_1 + (1-t)\mathsf{K}\rho_2. \tag{5.10}$$

3. a. *A* **Jordan symmetry** *is an invertible* **Jordan map**

$$\mathsf{J} : B(H)_{\mathrm{sa}} \to B(H)_{\mathrm{sa}}, \tag{5.11}$$

i.e., an $\mathbb{R}$-linear bijection that satisfies the equivalent conditions

$$\mathsf{J}(a \circ b) = \mathsf{J}(a) \circ \mathsf{J}(b); \tag{5.12}$$
$$\mathsf{J}(a^2) = \mathsf{J}(a)^2. \tag{5.13}$$

Here

$$a \circ b = \tfrac{1}{2}(ab + ba) \tag{5.14}$$

is the **Jordan product** *on $B(H)_{\mathrm{sa}}$, which turns the (real) vector space $B(H)_{\mathrm{sa}}$ into a* **Jordan algebra**, *cf. §C.25.*

b. *A* **weak Jordan symmetry** *is an invertible* **weak Jordan map**, *i.e., a bijection (5.11) of which the restriction $\mathsf{J}_{|C_{\mathrm{sa}}}$ is a Jordan map for each $C \in \mathscr{C}(B(H))$.*

4. *A* **Ludwig symmetry** *is an affine order isomorphism*

$$\mathsf{L}:\mathscr{E}(H)\to\mathscr{E}(H). \tag{5.15}$$

5. *A* **von Neumann symmetry** *is an order isomorphism*

$$\mathsf{N}:\mathscr{P}(H)\to\mathscr{P}(H) \tag{5.16}$$

preserving orthocomplementation, i.e. $\mathsf{N}(1-e)=1-\mathsf{N}(e)$ *for each* $e\in\mathscr{P}(H)$.

6. *A* **Bohr symmetry** *is an order isomorphism*

$$\mathsf{B}:\mathscr{C}(B(H))\to\mathscr{C}(B(H)). \tag{5.17}$$

In nos. 3 and 5–6, an ***order isomorphism*** O of the given poset is a bijection that preserves the partial order $\leq$ (i.e., if $x\leq y$, then $\mathsf{O}(x)\leq\mathsf{O}(y)$) and whose inverse O^{-1} does so, too; cf. §D.1. The names in question have been chosen for historical reasons and (except perhaps for the first and third) are not standard.

Let us note that any Jordan map has a unique extension to a $\mathbb{C}$-linear map

$$\mathsf{J}_{\mathbb{C}}:B(H)\to B(H); \tag{5.18}$$

$$\mathsf{J}_{\mathbb{C}}(a^*)=\mathsf{J}_{\mathbb{C}}(a)^*, \tag{5.19}$$

which satisfies (5.12) for all a,b, as well as

$$\mathsf{J}_{\mathbb{C}}(a+ib)=\mathsf{J}(a)+i\mathsf{J}(b), \tag{5.20}$$

with notation as in Proposition 2.6. Conversely, such a Jordan map (5.18) defines a real Jordan map (5.11) by $\mathsf{J}=\mathsf{J}_{|B(H)_{\mathrm{sa}}}$. Similarly, a weak Jordan symmetry is equivalent to a map (5.18) that satisfies (5.19), preserves squares as in (5.13), and is linear on each subspace C of $B(H)$, with $C\in\mathscr{C}(B(H))$. In other words (in the spirit of Bohrification), $\mathsf{J}_{\mathbb{C}}$ is a homomorphism of C*-algebras on each commutative unital C*-subalgebra $C\subset B(H)$. Therefore, either way J and $\mathsf{J}_{\mathbb{C}}$ are essentially the same thing, and if no confusion may arise we call it J. Note that a weak Jordan map J *a priori* satisfies (5.12) only for *commuting* self-adjoint a and b. It follows that weak (and hence ordinary) Jordan symmetries are unital: since

$$\mathsf{J}(b)=\mathsf{J}(1_H\circ b)=\mathsf{J}(1_H)\circ\mathsf{J}(b) \tag{5.21}$$

for any b, we may pick $b=\mathsf{J}^{-1}(1_H)$ to find, reading (5.21) from right to left,

$$\mathsf{J}(1_H)=\mathsf{J}(1_H)\circ 1_H=1_H. \tag{5.22}$$

The special role of unitary operators u now emerges: each such operator defines the relevant symmetry in the obvious way, namely, in order of appearance:

$$\mathsf{W}(e) = ueu^*; \tag{5.23}$$
$$\mathsf{K}(\rho) = u\rho u^*; \tag{5.24}$$
$$\mathsf{L}(a) = uau^*; \tag{5.25}$$
$$\mathsf{J}(a) = uau^*; \tag{5.26}$$
$$\mathsf{N}(e) = ueu^*; \tag{5.27}$$
$$\mathsf{B}(C) = uCu^*, \tag{5.28}$$

where $a^* = a$ in (5.26). If not, this formula remains valid also for the map $J_{\mathbb{C}}$. Furthermore, in (5.28) the notation uCu^* is shorthand for the set $\{uau^* \mid a \in C\}$, which is easily seen to be a member of $\mathscr{C}(B(H))$. Here, as well as in the other three cases, it is easy to verify that the right-hand side belongs to the required set, that is,

$$ueu^* \in \mathscr{P}_1(H),\ u\rho u^* \in \mathscr{D}(H),\ u\rho u^* \in \mathscr{E}(H), \tag{5.29}$$
$$uau^* \in B(H)_{\mathrm{sa}},\ u\rho u^* \in \mathscr{P}(H),\ uCu^* \in \mathscr{C}(B(H)), \tag{5.30}$$

respectively, provided, of course, that

$$e \in \mathscr{P}_1(H),\ \rho \in \mathscr{D}(H),\ a \in \mathscr{E}(H)\ a \in B(H)_{\mathrm{sa}},\ e \in \mathscr{P}(H),\ C \in \mathscr{C}(B(H)).$$

Indeed, if, in (5.23), $e = e_\psi = |\psi\rangle\langle\psi|$ for some unit vector $\psi \in H$, then

$$ue_\psi u^* = e_{u\psi}. \tag{5.31}$$

If $\rho \geq 0$ in that $\langle\psi, \rho\psi\rangle \geq 0$ for each $\psi \in H$, then clearly also $u\rho u^* \geq 0$, and if $\mathrm{Tr}(\rho) = 1$, then also $\mathrm{Tr}(u\rho u^*) = 1$. If $a^* = a$, then

$$(uau^*)^* = u^{**}a^*u^* = uau^*. \tag{5.32}$$

However, one may also choose u in these formulae to be *anti-unitary*, as follows:

Definition 5.2. *1. A real-linear operator $u : H \to H$ is* **anti-linear** *if*

$$u(z\psi) = \overline{z}\psi\ (z \in \mathbb{C}). \tag{5.33}$$

2. An anti-linear operator $u : H \to H$ is **anti-unitary** *if it is invertible, and*

$$\langle u\varphi, u\psi\rangle = \overline{\langle\varphi, \psi\rangle}\ (\varphi, \psi \in H). \tag{5.34}$$

The adjoint u^* of a (bounded) anti-linear operator u is defined by the property

$$\langle u^*\varphi, \psi\rangle = \overline{\langle\varphi, u\psi\rangle}\ (\varphi, \psi \in H), \tag{5.35}$$

in which case u^* is anti-linear, too. Hence we may equally well say that an anti-linear operator is anti-unitary if $uu^* = u^*u = 1_H$. The simplest example is the map

$$J : \mathbb{C}^n \to \mathbb{C}^n;$$
$$Jz = \overline{z}, \tag{5.36}$$

i.e., if $z = (z_1, \ldots, z_n) \in \mathbb{C}^n$, then $(Jz)_i = \overline{z}_i$. Similarly, one may define

$$J : \ell^2 \to \ell^2;$$
$$J\psi = \overline{\psi}, \tag{5.37}$$

and likewise on L^2, where complex conjugation is defined pointwise, that is,

$$(J\psi)(x) = \overline{\psi(x)}. \tag{5.38}$$

For any Hilbert space one may pick a basis (υ_i) and define J relative to this basis by

$$J\left(\sum_i c_i \upsilon_i\right) = \sum_i \overline{c}_i \upsilon_i. \tag{5.39}$$

For future use, we state two obvious facts.

Proposition 5.3. *1. The product of two anti-unitary operators is unitary.*
2. Any anti-unitary operator $u : H \to H$ takes the form $u = Jv$, where v is unitary and J is an anti-unitary operator on H of the kind constructed above.

It is an easy verification that (5.23) - (5.28) still define symmetries if u is anti-unitary. Note that in terms of the complexification $J_{\mathbb{C}}$, eq. (5.26) should read

$$J_{\mathbb{C}}(a) = ua^*u^*. \tag{5.40}$$

The goal of the following sections is to show that these are the only possibilities:

Theorem 5.4. *Let H be a Hilbert space, with $\dim(H) > 1$.*

1. *Each Wigner symmetry takes the form* (5.23)*;*
2. *Each Kadison symmetry takes the form* (5.24)*;*
3. *Each Ludwig symmetry takes the form* (5.25)*;*
4. *a. Each Jordan symmetry takes the form* (5.26)*;*
   b. If $\dim(H) > 2$, also each weak Jordan symmetry takes this form;
5. *If $\dim(H) > 2$, each von Neumann symmetry takes the form* (5.27)*;*
6. *Again if $\dim(H) > 2$, each Bohr symmetry takes the form* (5.28)*,*

where in all cases the operator u is either unitary or anti-unitary, and is uniquely determined by the symmetry in question up to a phase (that is, u and u' implement the same symmetry by conjugation iff $u' = zu$, where $z \in \mathbb{T}$).

As we shall see, the reason why the case $H = \mathbb{C}^2$ is exceptional with regard to weak Jordan symmetries, von Neumann symmetries, and Bohr symmetries is that in those cases the proof relies on Gleason's Theorem, which fails for $H = \mathbb{C}^2$.

To see this more explicitly, and also to prove the positive cases (i.e., nos. 1–4a) in a simple situation without invoking higher principles, before proving Theorem 5.4 in general it is instructive to first illustrate it in the two-dimensional case $H = \mathbb{C}^2$.

5.2 The case $H = \mathbb{C}^2$

We start with some background. Any complex 2×2 matrix a can be written as

$$a = a(x_0, x_1, x_2, x_3) = \frac{1}{2}\sum_{\mu=0}^{3} x_\mu \sigma_\mu \ (x_\mu \in \mathbb{C}); \tag{5.41}$$

$$\sigma_0 = \begin{pmatrix} 1 & 0 \\ 0 & 1 \end{pmatrix}, \sigma_1 = \begin{pmatrix} 0 & 1 \\ 1 & 0 \end{pmatrix}, \sigma_2 = \begin{pmatrix} 0 & -i \\ i & 0 \end{pmatrix}, \sigma_3 = \begin{pmatrix} 1 & 0 \\ 0 & -1 \end{pmatrix}, \tag{5.42}$$

i.e., the ***Pauli matrices***. Furthermore, if we equip the vector space $M_2(\mathbb{C})$ of complex 2×2 matrices with the canonical inner product (2.34), then the rescaled matrices $\sigma'_\mu = \sigma_\mu/\sqrt{2}$ form a basis ($\equiv$ orthonormal basis) of the ensuing Hilbert space.

Writing $\mathbf{x} = (x_1, x_2, x_3)$, some interesting special cases are:

- $x_0 \in \mathbb{R}$, $\mathbf{x} = i\mathbf{v}$ with $\mathbf{v} \in \mathbb{R}^3$ and $x_0^2 + v_1^2 + v_2^2 + v_3^2 = 1$, which holds iff $a \in SU(2)$;
- $x_\mu \in \mathbb{R}$ for each $\mu = 0, 1, 2, 3$, which is the case iff $a^* = a$.
- $x_0 = 1$, $\mathbf{x} \in \mathbb{R}^3$, and $\|\mathbf{x}\| = 1$, which holds iff a is a one-dimensional projection.

The first case follows because $SU(2)$ consist of all matrices of the form

$$\begin{pmatrix} \alpha & \beta \\ -\overline{\beta} & \overline{\alpha} \end{pmatrix}, \ \alpha, \beta \in \mathbb{C}, \ |\alpha|^2 + |\beta|^2 = 1. \tag{5.43}$$

The second case is obvious, and the third follows from Proposition 2.9.

Assume the third case, so that $a = e$ with $e^2 = e^* = e$ and $\mathrm{Tr}(e) = 1$. If a linear map $u : \mathbb{C}^2 \to \mathbb{C}^2$ is unitary, then simple computations show that $e' = ueu^*$ is a one-dimensional projection, too, given by $e' = \frac{1}{2}\sum_{\mu=0}^{3} x'_\mu \sigma_\mu$ with $x'_0 = 1$, $\mathbf{x}' \in \mathbb{R}^3$, and $\|\mathbf{x}'\| = 1$. Writing $\mathbf{x}' = R\mathbf{x}$ for some map $R : S^2 \to S^2$, we have

$$u(\mathbf{x} \cdot \sigma)u^* = (R\mathbf{x}) \cdot \sigma, \tag{5.44}$$

where $\mathbf{x} \cdot \sigma = \sum_{j=1}^{3} x_j \sigma_j$. This also shows that R extends to a linear isometry $R : \mathbb{R}^3 \to \mathbb{R}^3$. Using the formula $\mathrm{Tr}(\sigma_i \sigma_j) = 2\delta_{ij}$, the matrix-form of R follows as

$$R_{ij} = \frac{1}{2}\mathrm{Tr}(u\sigma_i u^* \sigma_j). \tag{5.45}$$

Define $U(2)$ as the (connected) group of all unitary 2×2 matrices (whose connected subgroup $SU(2)$ of elements with unit determinant has just been mentioned). Also, recall that $O(3)$ is the group of all real orthogonal 3×3 matrices M, a condition that may be expressed in (at least) four equivalent ways (like unitarity):

- $MM^T = M^M M = 1_3$;
- M invertible and $M^T = M^{-1}$;
- M is an isometry (and hence it is injective and therefore invertible);
- M preserves the inner product: $\langle M\mathbf{x}, M\mathbf{y} \rangle = \langle \mathbf{x}, \mathbf{y} \rangle$ for all $\mathbf{x}, \mathbf{y} \in \mathbb{R}^3$.

This implies $\det(M) = \pm 1$ (as can be seen by diagonalizing M; being a real linear isometry, its eigenvalues can only be ± 1, and $\det(M)$ is their product). Thus $O(3)$ breaks up into two parts $O_{\pm}(3) = \{R \in O(3) \mid \det(R) = \pm 1\}$, of which $O_+ \equiv SO(3)$ consists of rotations. Using an explicit parametrization of $SO(3)$, e.g., through Euler angles, or, using surjectivity of the exponential map (from the Lie algebra of $SO(3)$, which consist of anti-symmetric real matrices), it follows that $O_{\pm}(3)$ are precisely the two connected components of $O(3)$, the identity of course lying in $O_+(3)$.

Proposition 5.5. *The map $u \mapsto R$ defined by* (5.44) *is a homomorphism from $U(2)$ onto $SO(3)$. In terms of $SU(2) \subset U(2)$, this map restricts to a two-fold covering*

$$\tilde{\pi} : SU(2) \to SO(3), \tag{5.46}$$

with discrete kernel

$$\ker(\tilde{\pi}) = \{1_2, -1_2\}. \tag{5.47}$$

Proof. As a finite-dimensional linear isometry, R is automatically invertible (this also follows from unitarity and hence invertibility of u), hence $R \in O(3)$. It is obvious from (5.44) that $u \mapsto R$ is a continuous homomorphism (of groups). Since $U(2)$ is connected and $u \mapsto R$ is continuous, R must lie in the connected component of $O(3)$ containing the identity, whence $R \in SO(3)$. To show surjectivity of $\tilde{\pi}$, take some unit vector $\mathbf{u} \in \mathbb{R}^3$ and define $u = \cos(\frac{1}{2}\theta) + i\sin(\frac{1}{2}\theta)\mathbf{u} \cdot \sigma$. The corresponding rotation $R_\theta(\mathbf{u})$ is the one around $\mathbf{u}$ by an angle θ, and such rotations generate $SO(3)$.

Finally, it follows from (5.44) that $u \in \ker(\tilde{\pi})$ iff u commutes with each σ_i and hence, by (5.41), with all matrices. Therefore, $u = z \cdot 1_2$ for some $z \in \mathbb{C}$, upon which the the condition $\det(u) = 1$ (in that $u \in SU(2)$) enforces $z = \pm 1$. □

Note that the covering (5.46) is topologically nontrivial (i.e., $SU(2) \neq SO(3) \times \mathbb{Z}_2$), since $SU(2) \cong S^3$ is simply connected, whereas $SO(3)$ is doubly connected: a closed path $t \mapsto R_{2\pi t}(\mathbf{u})$, $t \in [0,1]$ in $SO(3)$ (starting and ending at 1_3) lifts to a path

$$t \mapsto \cos(\pi t) + i\sin(\pi t)\mathbf{u} \cdot \sigma$$

in $SU(2)$ that starts at the unit matrix 1_2 and ends at -1_2.

To incorporate $O_-(3)$, let $U_a(2)$ be the set of all anti-unitary 2×2 matrices. These do not form a group, as the product of two anti-unitaries is unitary, but the union $U(2) \cup U_a(2)$ is a disconnected Lie group with identity component $U(2)$.

Proposition 5.6. *The map $u \mapsto R$ defined by* (5.44) *is a surjective homomorphism*

$$\tilde{\pi}' : U(2) \cup U_a(2) \to O(3), \tag{5.48}$$

with kernel $U(1)$, seen as the diagonal matrices $z \cdot 1_2$, $z \in \mathbb{T}$. Moreover, $\tilde{\pi}'$ maps $U(2)$ onto $SO(3)$ and maps $U_a(2)$ onto $O_-(3)$.

Proof. The map $u \mapsto R$ in (5.44) sends the anti-unitary operator $u = J$ on $\mathbb{C}^2$ to $R = \mathrm{diag}(1, -1, 1) \in O_-(3)$. Since $U_a(2) = J \cdot U(2)$ and similarly $O_-(3) = R \cdot SO(3)$, the last claim follows. The computation of the kernel may now be restricted to $U(2)$, and then follows as in the last step op the proof of the previous proposition. □

We now return to Theorem 5.4 and go through its special cases one by one.

Part 1 of Theorem 5.4 is ***Wigner's Theorem***, which in the case at hands reads:

Theorem 5.7. *Each bijection* $\mathsf{W}: \mathscr{P}_1(\mathbb{C}^2) \to \mathscr{P}_1(\mathbb{C}^2)$ *that satisfies*

$$\mathrm{Tr}(\mathsf{W}(e)\mathsf{W}(f)) = \mathrm{Tr}(ef) \tag{5.49}$$

for each $e, f \in \mathscr{P}_1(\mathbb{C}^2)$ *takes the form* $\mathsf{W}(e) = ueu^*$, *where* u *is either unitary or anti-unitary, and is uniquely determined by* W *up to a phase.*

To prove, this we transfer the whole situation to the two-sphere, where it is easy:

Proposition 5.8. *The pure state space* $\mathscr{P}_1(\mathbb{C}^2)$ *corresponds bijectively to the sphere*

$$S^2 = \{(x,y,z) \in \mathbb{R}^3 \mid x^2 + y^2 + z^2 = 1\},$$

in that each one-dimensional projection $e \in \mathscr{P}_1(\mathbb{C}^2)$ *may be expressed uniquely as*

$$e(x,y,z) = \tfrac{1}{2}\begin{pmatrix} 1+z & x-iy \\ x+iy & 1-z \end{pmatrix}, \tag{5.50}$$

where $(x,y,z) \in \mathbb{R}^3$ *and* $x^2 + y^2 + z^2 = 1$. *Under the ensuing bijection*

$$\mathscr{P}_1(\mathbb{C}^2) \cong S^2, \tag{5.51}$$

Wigner symmetries W *of* $\mathbb{C}^2$ *turn into orthogonal maps* $R \in O(3)$, *restricted to* S^2.

Proof. The first claim restates Proposition 2.9. If ψ and ψ' are unit vectors in $\mathbb{C}^2$ with corresponding one-dimensional projections $e_\psi(x,y,z)$ and $e_{\psi'}(x',y',z')$ then, as one easily verifies, the corresponding transition probability takes the form

$$\mathrm{Tr}(e_\psi e_{\psi'}) = \tfrac{1}{2}(1 + \langle \mathbf{x}, \mathbf{x}' \rangle) = \cos^2(\tfrac{1}{2}\theta(\mathbf{x},\mathbf{y})), \tag{5.52}$$

where $\theta(\mathbf{x},\mathbf{y})$ is the arc (i.e., geodesic) distance between $\mathbf{x}$ and $\mathbf{y}$. Consequently, if $\mathsf{W}: \mathscr{P}_1(\mathbb{C}^2) \to \mathscr{P}_1(\mathbb{C}^2)$ satisfies (5.8), then the corresponding map $R: S^2 \to S^2$ (defined through the above identification $\mathscr{P}_1(\mathbb{C}^2) \cong S^2$) satisfies

$$\langle R(\mathbf{x}), R(\mathbf{x}') \rangle = \langle \mathbf{x}, \mathbf{x}' \rangle \;\; (\mathbf{x}, \mathbf{x}' \in S^2). \tag{5.53}$$

Lemma 5.9. *If some bijection* $R: S^2 \to S^2$ *satisfies* (5.53), *then* R *extends (uniquely) to an orthogonal linear map (for simplicity also called)* $R: \mathbb{R}^3 \to \mathbb{R}^3$.

Proof. With $(\mathbf{u}_1, \mathbf{u}_2, \mathbf{u}_3)$ the standard basis of $\mathbb{R}^3$, define a 3×3 matrix by

$$R_{kl} = \langle \mathbf{u}_k, R(\mathbf{u}_l) \rangle. \tag{5.54}$$

It follows from (5.53) that $R^{-1}(\mathbf{u}_j)_k = R_{jk}$, which implies $\langle R^{-1}(\mathbf{u}_j), \mathbf{x} \rangle = \sum_k R_{jk} x_k$, or, once again using (5.53), $R(\mathbf{x})_j = \sum_k R_{jk} x_k$. Hence the map $\mathbf{x} \mapsto \sum_{j,k} R_{jk} x_k \mathbf{u}_j$, i.e., the usual linear map defined by the matrix (5.54), extends the given bijection R. Orthogonality of this linear map is, of course, equivalent to (5.53). □

Wigner's Theorem then follows by combining Propositions 5.6 and 5.8: given the linear map R just constructed, read (5.44) from right to left, where u exists by surjectivity of the map (5.48), and the precise lack of uniqueness of u as claimed in Theorem 5.4 is just a restatement of the fact that (5.48) has $U(1)$ as its kernel. □

Kadison's Theorem is part 2 of Theorem 5.4. Explicitly, for $H = \mathbb{C}^2$ we have:

Theorem 5.10. *Each affine bijection $\mathsf{K} : \mathscr{D}(\mathbb{C}^2) \to \mathscr{D}(\mathbb{C}^2)$ is given as $\mathsf{K}(\rho) = u\rho u^*$, where u is unitary or anti-unitary, and is uniquely determined by K up to a phase.*

Proof. We once again invoke Proposition 2.9, implying that any density matrix ρ on $\mathbb{C}^2$ takes the form

$$\rho = \tfrac{1}{2}\left(1_2 + \sum_{\mu=1}^{3} x_\mu \sigma_\mu\right), \tag{5.55}$$

with $\|\mathbf{x}\| \leq 1$. Moreover, the ensuing bijection $\mathscr{D}(\mathbb{C}^2) \cong B^3$, $\rho \mapsto \mathbf{x}$, is clearly affine, in that a convex sums $t\rho + (1-t)\rho'$ of density matrices correspond to convex sums $t\mathbf{x} + (1-t)\mathbf{x}'$ of the corresponding vectors in $\mathbb{R}^3$.

Lemma 5.11. *Any affine bijection K of the unit ball B^3 in $\mathbb{R}^3$ is given by an orthogonal linear map $R \in O(3)$.*

Proof. First, K must map the boundary $\partial_e B^3 = S^2$ to itself (necessarily bijectively): if $\mathbf{x} \in S^2$ and $\mathsf{K}(\mathbf{x}) = t\mathbf{x}' + (1-t)\mathbf{x}''$, then $\mathbf{x} = tK^{-1}(\mathbf{x}') + (1-t)K^{-1}(\mathbf{x}'')$, whence

$$K^{-1}(\mathbf{x}') = K^{-1}(\mathbf{x}''), \tag{5.56}$$

since $\mathbf{x}$ is pure, whence $\mathbf{x}' = \mathbf{x}''$, so that also $\mathsf{K}(\mathbf{x})$ is pure.

Second, the basis of all further steps is the property

$$\mathsf{K}(\mathbf{0}) = \mathbf{0}. \tag{5.57}$$

This is because $\mathbf{0}$ is intrinsic to the convex structure of B^3: it is the unique point with the property that for any $\mathbf{x} \in S^2$ there exists a unique $\mathbf{x}'$ such that $\frac{1}{2}\mathbf{x} + \frac{1}{2}\mathbf{x}' = \mathbf{0}$, namely $\mathbf{x}' = -\mathbf{x}$. Thus $\mathbf{0}$ must be preserved under affine bijections. For a formal proof (by contradiction), suppose $\mathsf{K}(\mathbf{0}) \neq \mathbf{0}$, and define $\mathbf{y} = \mathsf{K}(\mathbf{0})/\|\mathsf{K}(\mathbf{0})\| \in S^2$. Then $\mathsf{K}(\mathbf{0})$ has an extremal decomposition $\mathsf{K}(\mathbf{0}) = t\mathbf{y} + (1-t)\mathbf{y}'$, with $\mathbf{y}' = -\mathbf{y}$ and $t = \frac{1}{2}(1 + \|\mathsf{K}(\mathbf{0})\|)$. Applying the affine map K^{-1} then gives

$$\|\mathsf{K}^{-1}(\mathbf{y}')\| = \|\mathsf{K}^{-1}(\mathbf{y})\| \cdot \frac{1 + \|\mathsf{K}(\mathbf{0})\|}{1 - \|\mathsf{K}(\mathbf{0})\|}.$$

Now $\mathbf{y} \in S^2$ and hence $\mathsf{K}^{-1}(\mathbf{y}) \in S^2$ by part one of this proof (applied to K^{-1}), so that $\|\mathsf{K}^{-1}(\mathbf{y})\| = 1$. But this implies $\|\mathsf{K}^{-1}(\mathbf{y}')\| > 1$, which is impossible because $\mathbf{y}' \in S^2$ and hence $\|\mathsf{K}^{-1}(\mathbf{y}')\| = 1$.

Third, for $\mathbf{x} \in B^3$ and $t \in [0,1]$ the preceding point implies that

$$\mathsf{K}(t\mathbf{x}) = \mathsf{K}(t\mathbf{x} + (1-t)\mathbf{0}) = t\mathsf{K}(\mathbf{x}) + (1-t)\mathsf{K}(\mathbf{0}) = t\mathsf{K}(\mathbf{x}). \tag{5.58}$$

The same then holds for $\mathbf{x} \in B^3$ and all $t \geq 0$ as long as $t\mathbf{x} \in B^3$: for take $t > 1$, so that $t^{-1} \in (0,1)$, and use the previous step with $\mathbf{x} \rightsquigarrow t\mathbf{x}$ and $t \rightsquigarrow t^{-1}$ to compute

$$\mathsf{K}(t\mathbf{x}) = tt^{-1}\mathsf{K}(t\mathbf{x}) = t\mathsf{K}(t^{-1}t\mathbf{x}) = t\mathsf{K}(\mathbf{x}).$$

Also, (5.58) and affinity imply that for any $\mathbf{x}, \mathbf{y} \in B^3$ for which $\mathbf{x}+\mathbf{y} \in B^3$, we have

$$\mathsf{K}(\mathbf{x}+\mathbf{y}) = 2\mathsf{K}(\tfrac{1}{2}\mathbf{x}+\tfrac{1}{2}\mathbf{y}) = 2\cdot(\tfrac{1}{2}\mathsf{K}(\mathbf{x})+\tfrac{1}{2}\mathsf{K}(\mathbf{y})) = \mathsf{K}(\mathbf{x})+\mathsf{K}(\mathbf{y}). \tag{5.59}$$

With our earlier result (5.57), this also gives

$$\mathsf{K}(-\mathbf{x}) = -\mathsf{K}(\mathbf{x}). \tag{5.60}$$

For some nonzero $\mathbf{x} \in \mathbb{R}^3$, take $s \geq \|\mathbf{x}\|$ and $t \geq \|\mathbf{x}\|$. Then by (5.58) we have

$$s\mathsf{K}(\mathbf{x}/s) = s\mathsf{K}\left(\frac{t}{s}\frac{\mathbf{x}}{t}\right) = t\mathsf{K}(\mathbf{x}/t).$$

We may therefore define a map $R : \mathbb{R}^3 \to \mathbb{R}^3$ by

$$R(\mathbf{0}) = \mathbf{0}; \tag{5.61}$$
$$R(\mathbf{x}) = s\cdot\mathsf{K}(\mathbf{x}/s)\ \ (\mathbf{x}\neq\mathbf{0}), \tag{5.62}$$

for any choice of $s \geq \|\mathbf{x}\|$. For $\mathbf{x} \in B^3$ we may take $s = 1$, so that R extends K.

To prove that R is linear, for $\mathbf{x} \in \mathbb{R}^3$ and $t \geq 0$ pick some $s \geq t\|\mathbf{x}\|$ and compute

$$R(t\mathbf{x}) = s\mathsf{K}\left(\frac{t}{s}\mathbf{x}\right) = s\mathsf{K}\left(\|\mathbf{x}\|\frac{t}{s}\frac{\mathbf{x}}{\|\mathbf{x}\|}\right) = s\cdot\|\mathbf{x}\|\frac{t}{s}\mathsf{K}\left(\frac{\mathbf{x}}{\|\mathbf{x}\|}\right) = tR(\mathbf{x}). \tag{5.63}$$

For $t < 0$, we first show from (5.60) and (5.62) that

$$R(-\mathbf{x}) = -R(\mathbf{x}), \tag{5.64}$$

upon which (5.63) gives

$$R(t\mathbf{x}) = R(|t|\cdot(-\mathbf{x})) = |t|R(-\mathbf{x}) = -|t|R(\mathbf{x}) = -tR(\mathbf{x}). \tag{5.65}$$

Furthermore, for given $\mathbf{x}, \mathbf{y} \in B^3$, pick $s' > 0$ such that $s' \geq \|\mathbf{x}\|$ and $s' \geq \|\mathbf{y}\|$, so that $s = 2s' \geq \|\mathbf{x}+\mathbf{y}\|$ by the triangle inequality, and use (5.59) to compute

$$\begin{aligned} R(\mathbf{x}+\mathbf{y}) &= s\mathsf{K}\left(\frac{\mathbf{x}+\mathbf{y}}{s}\right) = s\mathsf{K}\left(\frac{\mathbf{x}}{s}+\frac{\mathbf{y}}{s}\right) = s\mathsf{K}(\mathbf{x}/s)+s\mathsf{K}(\mathbf{y}/s) \\ &= R(\mathbf{x})+R(\mathbf{y}). \end{aligned} \tag{5.66}$$

Finally, R is an isometry by (5.62) and step one of the proof. Being also linear and invertible, R must therefore be an orthogonal transformation. □

Given step one, an alternative proof derives this lemma from Proposition 5.18 below, which shows that the transition probabilities (5.52) on S^2 are determined by the convex structure of B^3, so that affine bijections must preserve them. In other words, the boundary map $S^2 \to S^2$ defined by K preserves transition probabilities and hence satisfies the conditions of Lemma 5.9. This reasoning effectively reduces Kadison's Theorem to Wigner's Theorem, a move we will later examine in general.

In any case, Theorem 5.10 now follows from Lemma 5.11 is exactly the same way as Theorem 5.7 followed from the corresponding Lemma 5.9. □

We have given this proof in some detail, because step 3 will recur on other occasions where a given affine bijection is to be extended to some linear map.

Ludwig's Theorem is part 3 of Theorem 5.4. For $H = \mathbb{C}^2$, we have:

Theorem 5.12. *Each affine order isomorphism* $\mathsf{L} : \mathscr{E}(\mathbb{C}^2) \to \mathscr{E}(\mathbb{C}^2)$ *reads* $\mathsf{L}(a) = uau^*$, *where u is unitary or anti-unitary, and is uniquely fixed by* L *up to a phase.*

Proof. Using the parametrization (5.41), we have $a(x_0,x_1,x_2,x_3) \in \mathscr{E}(\mathbb{C}^2)$ iff each x_μ is real and $0 \leq x_0 \pm \|\mathbf{x}\| \leq 2$. In particular, we have $0 \leq x_0 \leq 2$. This easily follows from (2.38), noting that $a \in \mathscr{E}(\mathbb{C}^2)$ just means that $a^* = a$ and that both eigenvalues of a lie in $[0,1]$. Thus $\mathscr{E}(\mathbb{C}^2)$ is isomorphic as a convex set to a convex subset C of $\mathbb{R}^4$ that is fibered over the x_0-interval $[0,2]$, where the fiber C_{x_0} of C over x_0 is the three-ball $B^3_{x_0}$ with radius $\|\mathbf{x}\| = x_0$ as long as $0 \leq x_0 \leq 1$, whereas for $1 \leq x_0 \leq 2$ the fiber is $B^3_{2-x_0}$, so at $x_0 = 1$ the fiber is $C_1 = B^3 \equiv B^3_1$ (in one dimension less, this convex body is easily visualizable as a double cone in $\mathbb{R}^3$, where the fibers are disks). The partial order on C induced from the one on $\mathscr{E}(\mathbb{C}^2)$ is given by

$$(x_0,\mathbf{x}) \leq (x_0',\mathbf{x}') \text{ iff } x_0' - x_0 \geq \|\mathbf{x}' - \mathbf{x}\|, \tag{5.67}$$

which follows from (5.41) and (2.38), noting that for matrices one has $a \leq a'$ iff $a' - a$ has positive eigenvalues. A similar argument to the one proving (5.57) then shows that any affine bijection L of C must map the base space $[0,2]$ to itself (as an affine bijection), and hence either $x_0 \mapsto x_0$ or $x_0 \mapsto 2 - x_0$. The latter fails to preserve order, so L must fix x_0. Similarly, L maps each three-ball C_{x_0} to itself by an affine bijection, which, by the same proof as for Kadison's Theorem above, must be induced by some element R_{x_0} of $O(3)$. Finally, the order-preserving condition $x_0' - x_0 \geq \|\mathbf{x}' - \mathbf{x}\| \Rightarrow x_0' - x_0 \geq \|R_{x_0'}\mathbf{x}' - R_{x_0}\mathbf{x}\|$ obtained from (5.67) and the property $\mathsf{L}(x_0) = x_0$ just found can only be met if R_{x_0} is independent of x_0. □

Part 3 of Theorem 5.4 does not carry an official name; it may be attributed to Kadison, too, but the hard part of the proof was given earlier by Jacobson and Rickart. Rather than a contrived (though historically justified) name like "Jacobson–Rickart–Kadison Theorem", we will simply speak of ***Jordan's Theorem*** (for $H = \mathbb{C}^2$):

Theorem 5.13. *Each linear bijection* $\mathsf{J} : M_2(\mathbb{C})_{\mathrm{sa}} \to M_2(\mathbb{C})_{\mathrm{sa}}$ *that satisfies* (5.13) *and hence* (5.12) *takes the form* $\mathsf{J}(a) = uau^*$, *where u is either unitary or anti-unitary, and is uniquely determined by* J *up to a phase.*

Proof. First, any Jordan map (and hence *a fortiori* any Jordan automorphism) trivially maps projections into projections, as it preserves the defining conditions $e^2 = e^* = e$. Second, any Jordan automorphism J maps *one-dimensional* projections into *one-dimensional* projections: if $e \in \mathscr{P}_1(H)$, then $\mathsf{J}(e) \neq 0$ and $\mathsf{J}(e) \neq 1_2$, both because J is injective in combination with $\mathsf{J}(0) = 0$ and $\mathsf{J}(1_2) = 1_2$, respectively. Hence $\mathsf{J}(e) \in \mathscr{P}_1(H)$, since this is the only remaining possibility (a more sophisticated argument shows that this is even true for any Hilbert space H). From (5.41) and subsequent text, as in (5.44), by linearity of J we therefore have

$$\mathsf{J}\left(\sum_{j=1}^{3} x_j \sigma_j\right) = \sum_{j=1}^{3} (R\mathbf{x})_j \sigma_j, \tag{5.68}$$

from some map $R : S^2 \to S^2$, which is bijective because J is. Linearity of J then allows us to extend R to a linear map $\mathbb{R}^3 \to \mathbb{R}^3$, with matrix

$$R_{jk} = \tfrac{1}{2} \sum_{j=1}^{3} \mathrm{Tr}\,(\sigma_k \mathsf{J}(\sigma_j)), \tag{5.69}$$

cf. (5.45). By (5.69), this linear map restricts to the given bijection $R : S^2 \to S^2$, which also shows that it is isometric. Thus we have a linear isometry on $\mathbb{R}^3$, which therefore lies in $O(3)$. The proof may then be completed as in Theorem 5.7. □

The case $H = \mathbb{C}^2$ was already exceptional in the context of Gleason's Theorem, and it remains so as far as weak Jordan symmetries and Bohr symmetries are concerned.

Proposition 5.14. *The poset $\mathscr{C}(M_2(\mathbb{C}))$ is isomorphic to $\{\perp\} \cup \mathbb{RP}^2$, where the real projective plane $\mathbb{RP}^2$ is the quotient $S^2/\sim$ under the equivalence relation $\mathbf{x} \sim -\mathbf{x}$, and the only nontrivial ordering is $\perp \leq p$ for any $p \in \mathbb{RP}^2$.*

Proof. It is elementary that $M_2(\mathbb{C})$ has a single one-dimensional unital *-subalgebra, namely $\mathbb{C} \cdot 1$, the multiples of the unit; this gives the singleton $\perp$ in $\mathscr{C}(M_2(\mathbb{C}))$.

Furthermore, any two-dimensional unital *-subalgebra C of $M_2(\mathbb{C})$ is generated by a one-dimensional projection e, in that C is the linear span of e and 1_2. Hence C is also the linear span of (the projection) $1_2 - e$ and 1_2. In our parametrization of all one-dimensional projections e on $\mathbb{C}^2$ by S^2 (cf. Proposition 2.9), if e corresponds to $\mathbf{x}$, then $1 - e$ corresponds to $-\mathbf{x}$. This yields the remainder $\mathbb{RP}^2$ of $\mathscr{C}(M_2(\mathbb{C}))$.

Finally, commutative unital *-subalgebras D of $M_2(\mathbb{C})$ of dimension > 2 do not exist. For any such algebra D would contain some two-dimensional C just defined, but a simple computation (for example, in a basis were C consists of all diagonal matrices) shows that the only matrices that commute with all elements of C already lie in C (i.e., are diagonal). Hence no commutative extension of C exists. □

Bohr symmetries B for $\mathbb{C}^2$ therefore correspond to bijections of $\mathbb{RP}^2$. Similarly, weak Jordan symmetries J for $\mathbb{C}^2$ corresponds to bijections of S^2 (the difference with Bohr symmetries lies in the fact that J may also map $C = \mathrm{span}(e, 1_2)$ to itself nontrivially, i.e., by sending e to $1_2 - e$, which for B would yield the identity map). In both cases, few of these bijections are (anti-) unitarily implemented.

5.3 Equivalence between the six symmetry theorems

If $\dim(H) > 1$, the first three claims of Theorem 5.4 are equivalent; if $\dim(H) > 2$, all claims are. We will show this in some detail, if only because the proofs of the various equivalences relate the six symmetry concepts stated in Definition 5.1 in an instructive way. We will do this in the sequence Wigner $\leftrightarrow$ Kadison $\leftrightarrow$ Jordan, and subsequently Jordan $\leftrightarrow$ Ludwig, Jordan $\leftrightarrow$ von Neumann, and Jordan $\leftrightarrow$ Bohr. Consequently, in principle only one part of Theorem 5.4 requires a proof. Although redundant, we will, in fact, prove both Wigner's Theorem and Jordan's (indeed, no independent proof of the other parts of Theorem 5.4 seems to be known!). The most transparent way to state the various equivalences is to note that in each case the set of symmetries of some given kind (i.e., Wigner, ...) forms a group. In all cases, the nontrivial part of the proof is the establishment of a "natural" bijection, from which the group homomorphism property is trivial (and hence will not be proved).

Proposition 5.15. *There is an isomorphism of groups between:*

- *The group of affine bijections* $\mathsf{K} : \mathscr{D}(H) \to \mathscr{D}(H)$;
- *The group of bijections* $\mathsf{W} : \mathscr{P}_1(H) \to \mathscr{P}_1(H)$ *that satisfy* (5.8), *viz.*

$$\mathsf{W} = \mathsf{K}_{|\mathscr{P}_1(H)}; \tag{5.70}$$

$$\mathsf{K}\left(\sum_i \lambda_i e_{\upsilon_i}\right) = \sum_i \lambda_i \mathsf{W}(\upsilon_{\upsilon_i}), \tag{5.71}$$

where $\rho = \sum_i \lambda_i e_{\upsilon_i}$ *is some (not necessarily unique) expansion of* $\rho \in \mathscr{D}(H)$ *in terms of a basis of eigenvector* υ_i *with eigenvalues* λ_i, *where* $\lambda_i > 0$ *and* $\sum_i \lambda_i = 1$. *In particular,* (5.70) *and* (5.71) *are well defined.*

Proof. It is conceptually important to distinguish between $B(H)_{\mathrm{sa}}$ as a Banach space in the usual operator norm $\|\cdot\|$, and $B_1(H)_{\mathrm{sa}}$, the Banach space of trace-class operators in its intrinsic norm $\|\cdot\|_1$. Of course, if $\dim(H) < \infty$, then $B(H)_{\mathrm{sa}} = B_1(H)_{\mathrm{sa}}$ as vector spaces, but even in that case the two norms do not coincide (although they are equivalent). The proof below has the additional advantage of immediately generalizing to the infinite-dimensional case. We start with (5.70).

1. Since $\mathscr{P}_1(H) = \partial_e \mathscr{D}(H)$, by the same argument as in the proof of Lemma 5.11, any affine bijection of the convex set $\mathscr{D}(H)$ must preserve its boundary, so that K maps $\mathscr{P}_1(H)$ into itself, necessarily bijectively. The goal of the next two steps is to prove that (5.70) satisfies (5.8), i.e., preserves transition probabilities.
2. An affine bijection $\mathsf{K} : \mathscr{D}(H) \to \mathscr{D}(H)$ extends to an isometric isomorphism $\mathsf{K}_1 : B_1(H)_{\mathrm{sa}} \to B_1(H)_{\mathrm{sa}}$ with respect to the trace-norm $\|\cdot\|_1$, as follows:

   a. Put $\mathsf{K}_1(0) = 0$ and for $b \geq 0$, $b \in B_1(H)$, i.e. $b \in B_1(H)_+$, and $b \neq 0$, define

$$\mathsf{K}_1(b) = \|b\|_1 \mathsf{K}(b/\|b\|_1). \tag{5.72}$$

By construction, K_1 is isometric and preserves positivity. For $b \in B_1(H)_+$ we have $\mathrm{Tr}\,(b) = \|b\|_1$, hence $b/\|b\|_1 \in \mathscr{D}(H)$, on which K is defined.
Linearity of K_1 with positive coefficients (as a consequence of the affine property of K) is verified as in the proof of Lemma 5.11; this time, use

$$a+b = (\|a\|_1 + \|b\|_1)\cdot\left(t\frac{a}{\|a\|_1} + (1-t)\frac{b}{\|b\|_1}\right), \tag{5.73}$$

with $t = \|a\|_1/(\|a\|_1 + \|b\|_1)$. Note that if $a,b \in B_1(H)_+$, then $a+b \in B_1(H)_+$.

b. For $b \in B_1(H)_{\mathrm{sa}}$, decompose $b = b_+ - b_-$, where $b_\pm \geq 0$; see Proposition A.24 (this remains valid in general Hilbert spaces). We then define

$$\mathsf{K}_1(b) = \mathsf{K}_1(b_+) - \mathsf{K}_1(b_-). \tag{5.74}$$

To show that this makes K_1 linear on all of $B_1(H)_{\mathrm{sa}}$, suppose $b = b'_+ - b'_-$ with $b'_\pm \geq 0$. Then $b'_+ + b_- = b_+ + b'_-$, and since each term is positive,

$$\mathsf{K}_1(b'_+ + b_-) = \mathsf{K}_1(b'_+) + \mathsf{K}_1(b_-) = \mathsf{K}(b_+ + b'_-) = \mathsf{K}_1(b_+) + \mathsf{K}_1(b'_-),$$

by the previous step. Hence $\mathsf{K}_1(b'_+) - \mathsf{K}_1(b'_-) = \mathsf{K}_1(b_+) - \mathsf{K}_1(b_-)$, so that (5.74) is actually independent of the choice of the decomposition of b as long as the operators are positive. Hence for $a,b \in B_1(H)_{\mathrm{sa}}$ we may compute

$$\begin{aligned}\mathsf{K}_1(a+b) &= \mathsf{K}_1(a_+ + b_+ - (a_- + b_-)) = \mathsf{K}_1(a_+ + b_+) - \mathsf{K}_1(a_- + b_-)\\ &= \mathsf{K}_1(a_+) + \mathsf{K}_1(b_+) - \mathsf{K}_1(a_-) - \mathsf{K}_1(b_-) = \mathsf{K}_1(a) + \mathsf{K}_1((b),\end{aligned}$$

since $a_+ + b_+$ and $a_- + b_-$ are both positive.

The key point in verifying isometry of K_1 is the property $|b| = b_+ + b_-$, which follows from (A.76) or Theorem B.94. Using this property, we have

$$\begin{aligned}\|\mathsf{K}_1(b)\|_1 &= \mathrm{Tr}\,(|\mathsf{K}_1 b|) = \mathrm{Tr}\,(|\mathsf{K}_1(b_+) - \mathsf{K}_1(b_-)|) = \mathrm{Tr}\,(\mathsf{K}_1(b_+) + \mathsf{K}_1(b_-))\\ &= \mathrm{Tr}\,(b_+ + b_-) = \mathrm{Tr}\,(|b_+ - b_-|) = \mathrm{Tr}\,(|b|) = \|b\|_1.\end{aligned}$$

3. For any two unit vectors ψ, φ in H we have the formula

$$\|e_\psi - e_\varphi\|_1 = 2\sqrt{1 - \mathrm{Tr}\,(e_\psi e_\varphi)}, \tag{5.75}$$

which can easily be proved by a calculation with 2×2 matrices (since everything takes place is the two-dimensional subspace spanned by ψ and φ, expect when $\varphi = z\psi$, $z \in \mathbb{T}$, in which case (5.75) reads $0 = 0$ and hence is true also). Since K_1 is linear as well as isometric with respect to the trace-norm, we have

$$\|\mathsf{K}_1(e_\psi) - \mathsf{K}_1(e_\varphi)\|_1 = \|\mathsf{K}_1(e_\psi - e_\varphi)\|_1 = \|e_\psi - e_\varphi\|_1,$$

and hence, by (5.75), $\mathrm{Tr}\,(\mathsf{K}_1(e_\psi)\mathsf{K}_1(e_\varphi)) = \mathrm{Tr}\,(e_\psi e_\varphi)$. Eq. (5.70) then gives (5.8).

We move on to (5.71). The main concern is that this expression be well defined, since in case some eigenvalue $\lambda > 0$ of ρ is degenerate (necessarily with finite multiplicity, even in infinite dimension, since ρ is compact), the basis of the eigenspace H_λ that takes part in the sum $\sum_i \lambda_i e_{\upsilon_i}$ is far from unique. This is settled as follows:

Lemma 5.16. *Let* $\mathsf{W} : \mathscr{P}_1(H) \to \mathscr{P}_1(H)$ *be a bijection that satisfies* (5.8)*, let* $L \subset H$ *be a (finite-dimensional) subspace, and let* (υ_j) *and* (υ'_i) *be bases of* L*. Then*

$$\sum_j \mathsf{W}(e_{\upsilon_j}) = \sum_i \mathsf{W}(e_{\upsilon'_i}). \tag{5.76}$$

Proof. As usual, for projections e and f on H we write $e \leq f$ iff $eH \subseteq fH$. From (B.212) and (B.214) we have $\sum_j |\langle \upsilon_j, \psi \rangle|^2 \leq 1$ for any unit vector $\psi \in H$, with equality iff $\psi \in L$. In other words, $e_\psi \leq e_L$ iff $\sum_j \mathrm{Tr}\,(e_{\upsilon_j} e_\psi) = 1$. Furthermore, by (5.8) the images $\mathsf{W}(e_{\upsilon_j})$ remain orthogonal; hence $\sum_j \mathsf{W}(e_{\upsilon_j})$ is a projection, and $e \leq \sum_j \mathsf{W}(e_{\upsilon_j})$ iff $\sum_j \mathrm{Tr}\,(\mathsf{W}(e_{\upsilon_j})e) = 1$. By (5.8), this condition is satisfied for $e = \mathsf{W}(e_{\upsilon_i})$, so that $\mathsf{W}(e_{\upsilon'_i}) \leq \sum_j \mathsf{W}(e_{\upsilon_j})$ for each j. Since also the projections $\mathsf{W}(e'_{\upsilon_i})$ are orthogonal, this gives $\sum_i \mathsf{W}(e'_{\upsilon_i}) \leq \sum_j \mathsf{W}(e_{\upsilon_j})$. Interchanging the roles of the two bases gives the converse, yielding (5.76). □

Finally, to prove bijectivity of the correspondence $\mathsf{K} \leftrightarrow \mathsf{W}$, we need the property

$$\mathsf{K}\left(\sum_i \lambda_i e_{\upsilon_i}\right) = \sum_i \lambda_i \mathsf{K}(e_{\upsilon_i}), \tag{5.77}$$

since this implies that K is determined by its action on $\mathscr{P}_1(H) \subset \mathscr{D}(H)$. In finite dimension this follows from convexity of K, and we are done. In infinite dimension, we in addition need continuity of K, as well as convergence of the sum $\sum_i \lambda_i e_{\upsilon_i}$ not only in the operator norm (as follows from the spectral theorem for self-adjoint compact operators), but also in the trace norm: for finite n, m,

$$\|\sum_{i=n}^m \lambda_i e_{\upsilon_i}\|_1 \leq \sum_{i=n}^m |\lambda_i| \|e_{\upsilon_i}\|_1 = \sum_{i=n}^m \lambda_i,$$

since $\|e_{\upsilon_i}\|_1 = 1$. Because $\sum_i \lambda_i = 1$, the above expression vanishes as $n, m \to \infty$, whence $\rho_n = \sum_{i=1}^n \lambda_i e_{\upsilon_i}$ is a Cauchy sequence in $B_1(H)$, which by completeness of the latter converges (to an element of $\mathscr{D}(H)$, as one easily verifies).

The proof of continuity is completed by noting that K is continuous with respect to the trace norm, for it is isometric and hence bounded (see step 2 above). □

It is enlightening to give a rather more conceptual proof that $\mathsf{K}_{|\mathscr{P}_1(H)}$ satisfies (5.8), which is based on a result to be used more often in the future. In what follows, for any convex set C, the notation $A_b(K)$ stands for the real vector space of *bounded* affine functions $f : C \to \mathbb{R}$, that is, bounded functions satisfying

$$f(tx + (1-t)y) = tf(x) + (1-t)f(y), \; x, y \in C, t \in (0,1). \tag{5.78}$$

It is easily checked that $A_b(K)$ with the supremum-norm is a real Banach space.

Proposition 5.17. *For any Hilbert space H we have an isometric isomorphism*

$$A_b(\mathscr{D}(H)) \cong B(H)_{\text{sa}}, \tag{5.79}$$
$$f \leftrightarrow a; \tag{5.80}$$
$$f(\rho) = \text{Tr}(\rho a), \tag{5.81}$$

which preserves the unit (i.e., $1_{\mathscr{D}(H)} \leftrightarrow 1_H$) as well as the order (i.e, $f \geq 0$ iff $a \geq 0$).

Note that under the identification $\mathscr{D}(H) \cong S_n(B(H))$ (where in finite dimension the normal state space $S_n(B(H))$ simply coincides with the state space $S(B(H))$), where $\rho \leftrightarrow \omega$ as in (2.33), i.e., $\omega(a) = \text{Tr}(\rho a)$, the above isomorphism simply reads

$$A_b(S_n(B(H))) \cong B(H)_{\text{sa}}, \tag{5.82}$$
$$\hat{a} \leftrightarrow a; \tag{5.83}$$
$$\hat{a}(\omega) = \omega(a). \tag{5.84}$$

Proof. It is clear that for each $a \in B(H)_{\text{sa}}$ the function $f : \rho \mapsto \text{Tr}(\rho a)$ (or, equivalently, $\hat{a} : \omega \mapsto \omega(a)$) is affine as well as real-valued, and is bounded by (A.100) (supplemented, if $\dim(H) = \infty$, by Lemma B.142), noting that $\|\rho\|_1 = 1$ for $\rho \in \mathscr{D}(H)$, and in fact (B.483) yields the equality $\|f\|_\infty = \|a\|$ (or $\|\hat{a}\|_\infty = \|a\|$).

Conversely, $f \in A_b(\mathscr{D}(H))$ defines a function $Q : H \to \mathbb{R}$ by

$$Q(0) = 0; \tag{5.85}$$
$$Q(\psi) = \|\psi\|^2 f(e_{\psi/\|\psi\|}) \ (\psi \neq 0). \tag{5.86}$$

This function is clearly bounded on the unit ball of H, as in

$$|Q(\psi)| \leq \|f\|_\infty \|\psi\|^2. \tag{5.87}$$

To check that Q in fact defines a quadratic form on H, we verify the properties (A.8) - (A.9). The first is trivial. The second follows from the easily verified identity

$$te_{\frac{v+w}{\|v+w\|}} + (1-t)e_{\frac{v-w}{\|v-w\|}} = se_{\frac{v}{\|v\|}} + (1-s)e_{\frac{w}{\|w\|}}, \tag{5.88}$$

where $v, w \neq 0$, $v \neq w$, and the coefficients s, t are given by

$$t = \frac{\|v+w\|^2}{2(\|v\|^2 + \|w\|^2)}; \tag{5.89}$$
$$s = \frac{\|v\|^2}{\|v\|^2 + \|w\|^2}. \tag{5.90}$$

The affine property (5.78) then immediately yields (A.9). According to Proposition B.79, we obtain a unique operator $a \in B(H)_{\text{sa}}$ such that $Q(\psi) = \langle \psi, a\psi \rangle$, i.e.,

$$\langle \psi, a\psi \rangle = f(e_\psi), \ \psi \in H, \|\psi\| = 1. \tag{5.91}$$

Since also $\langle\psi, a\psi\rangle = \mathrm{Tr}\,(e_\psi a)$, we have established (5.81) for each $\rho = e_\psi$, where $\psi \in H$, $\|\psi\| = 1$. To extend this result to general density operators $\rho = \sum_i \lambda_i e_{\upsilon_i}$, we use (A.100) as well as convergence of the above sum in the trace norm $\|\cdot\|_1$, cf. the proof of Lemma 5.16; the details are analogous to the proof of Theorem B.146. □

Proposition 5.18. *For any unit vectors $\psi, \varphi \in H$ we have*

$$\mathrm{Tr}\,(e_\psi e_\varphi) = \inf\{f(e_\psi) \mid f \in A_b(\mathscr{D}(H)), 0 \leq f \leq 1, f(e_\varphi) = 1\}. \tag{5.92}$$

The virtue of this formula is that the expression on the left-hand side, which defines the transition probabilities on $\partial_e \mathscr{D}(H) = \mathscr{P}_1(H)$, is intrinsically given by the convex structure of $\mathscr{D}(H)$. Consequently, any affine bijection of this convex set (which already preserves the boundary) must preserve these probabilities.

Proof. By the previous proposition, eq. (5.92) is equivalent to

$$\mathrm{Tr}\,(e_\psi e_\varphi) = \inf\{\langle\psi, a\psi\rangle \mid a \in B(H)_{\mathrm{sa}}, 0 \leq a \leq 1, \langle\varphi, a\varphi\rangle = 1\}. \tag{5.93}$$

Since $\mathrm{Tr}\,(e_\psi e_\varphi) = \langle\psi, e_\varphi\psi\rangle$, we are ready if we can show that the infimum is reached at $a = e_\varphi$. Therefore, we prove that for any a as specified we must have $\langle\psi, a\psi\rangle \geq \mathrm{Tr}\,(e_\psi e_\varphi) = |\langle\varphi, \psi\rangle|^2$. To do so, we are going to find a contradiction if

$$\langle\psi, a\psi\rangle < \mathrm{Tr}\,(e_\psi e_\varphi), \tag{5.94}$$

for some such a. Indeed, $\langle\varphi, a\varphi\rangle = 1$ with $\|a\| \leq 1$ (which follows from $0 \leq a \leq 1$) and $\|\varphi\| = 1$ imply, by Cauchy–Schwarz, that $a\varphi = \varphi$. Since $a^* = a$ (by positivity of a), we also have $a : (\mathbb{C}\cdot\varphi)^\perp \to (\mathbb{C}\cdot\varphi)^\perp$, so we may write $a = e_\varphi + a'$, with $a'\varphi = 0$ and a' mapping $(\mathbb{C}\cdot\varphi)^\perp$ to itself. Then $a \geq 0$ implies $a' \geq 0$. If (5.94) holds, then $\langle\psi, a'\psi\rangle < 0$, which contradicts positivity of a' (and hence of a). □

We now turn to the equivalence between Jordan's Theorem and Kadison's Theorem.

Proposition 5.19. *There is an isomorphism of groups between:*

- *The group of affine bijections $\mathsf{K} : \mathscr{D}(H) \to \mathscr{D}(H)$;*
- *The group of Jordan automorphisms $\mathsf{J} : B(H)_{\mathrm{sa}} \to B(H)_{\mathrm{sa}}$,*

such that for any $a \in B(H)_{\mathrm{sa}}$ one has

$$\mathrm{Tr}\,(\mathsf{K}(\rho)a) = \mathrm{Tr}\,(\rho\mathsf{J}(a)) \quad (\rho \in \mathscr{D}(H)). \tag{5.95}$$

This immediately follows from the following lemma (of independent interest):

Lemma 5.20. *1. There is a bijective correspondence between:*

- *affine bijections $\mathsf{K} : \mathscr{D}(H) \to \mathscr{D}(H)$;*
- *unital positive (i.e. order-preserving) linear bijections $\alpha : B(H)_{\mathrm{sa}} \to B(H)_{\mathrm{sa}}$,*

such that for any $a \in B(H)_{\mathrm{sa}}$ one has (5.95).

2. A map $\alpha : B(H) \to B(H)$ is a unital positive linear bijection iff it is a Jordan automorphism.

Proof. 1. An affine bijection $\mathsf{K} : \mathscr{D}(H) \to \mathscr{D}(H)$ induces an isomorphism

$$\mathsf{K}^* : A_b(\mathscr{D}(H)) \to A_b(\mathscr{D}(H)); \quad (5.96)$$
$$f \mapsto f \circ \mathsf{K}, \quad (5.97)$$

which is evidently unital, positive, and isometric. Consequently, by Proposition 5.17, K^* corresponds to some isomorphism $\alpha : B(H)_{\mathrm{sa}} \to B(H)_{\mathrm{sa}}$, which necessarily shares the properties of being unital, positive, and isometric; this follows abstractly from the proposition, but may also be verified directly from (5.95). Conversely, such a map α yields a map K directly by (5.95); to see this, we identify $\mathscr{D}(H)$ with the normal state space of $B(H)$ through $\rho \leftrightarrow \omega$, as usual, cf. (2.33), and note that $\mathsf{K}\omega$ is the state defined by $(\mathsf{K}\omega)(a) = \omega(\alpha(a))$, or briefly $\mathsf{K}\omega = \omega \circ \alpha$. This is often written as $\mathsf{K} = \alpha^*$, and for future reference we write

$$\alpha^*\omega(a) = \omega(\alpha(a)). \quad (5.98)$$

2. The nontrivial direction of the proof (i.e. positive etc. $\Rightarrow$ Jordan) is based on a number of facts from operator theory:

   a. Unital positive linear maps maps on $B(H)_{\mathrm{sa}}$ preserve $\mathscr{P}(H)$, cf. (2.164).
   b. Any two projections e and f are orthogonal ($ef = 0$) iff $e + f \leq 1_H$ (easy).
   c. Any $a \in B(H)_{\mathrm{sa}}$ is a norm-limit of finite sums of the kind $\sum_i \lambda_i e_i$, where $\lambda_i \in \mathbb{R}$ and the e_i are mutually orthogonal projections (this follows from the spectral theorem for bounded self-adjoint operators in the form of Theorem B.104)
   d. Any unital positive linear map $\alpha : B(H)_{\mathrm{sa}} \to B(H)_{\mathrm{sa}}$ is continuous. Since

$$-\|a\| \cdot 1_H \leq a \leq -\|a\| \cdot 1_H \;\; (a \in B(H)_{\mathrm{sa}}), \quad (5.99)$$

   by (C.83), applying the positive map α and using $\alpha(1_H) = 1_H$ yields

$$-\|a\| \cdot 1_H \leq \alpha(a) \leq -\|a\| \cdot 1_H.$$

   This is possible only if $\|\alpha(a)\| \leq \|a\|$, and hence α is continuous with norm bounded by $\|\alpha\| \leq 1$. In fact, since a is unital we have $\|\alpha\| = 1$.

Therefore, any unital positive linear map α preserves orthogonality of projections, so if $a = \sum_i \lambda_i e_i$ (finite sum), then

$$\alpha(a^2) = \alpha\left(\sum_i \lambda_i^2 e_i\right) = \sum_i \lambda_i^2 \alpha(e_i) = \sum_{i,j} \lambda_i \lambda_j \alpha(e_i)\alpha(e_j) = \alpha(a)^2, \quad (5.100)$$

since $e_i e_j = \delta_{ij} e_j$ and by the above comment also $\alpha(e_i)\alpha(e_j) = \delta_{ij}\alpha(e_j)$. By continuity of α, this property extends to arbitrary $a \in B(H)_{\mathrm{sa}}$. Finally, since

$$a \circ b = \tfrac{1}{2}((a+b)^2 - a^2 - b^2), \quad (5.101)$$

preserving squares as in (5.100) implies preserving the Jordan product $\circ$. $\square$

We now turn to the equivalence between Ludwig symmetries and Jordan ones.

Proposition 5.21. *There is an isomorphism of groups between:*

- *The group of affine order isomorphism* $\mathsf{L} : \mathscr{E}(H) \to \mathscr{E}(H)$*;*
- *The group of Jordan automorphisms* $\mathsf{J} : B(H)_{\mathrm{sa}} \to B(H)_{\mathrm{sa}}$.

Proof. Since L is an order isomorphism, it satisfies $\mathsf{L}(0) = 0$ (as well as $\mathsf{L}(1_H) = 1_H$), since 0 is the bottom element of $\mathscr{E}(H)$ as a poset (and 1_H is its the top element). As in the proof of Lemma 5.11, one shows that this property plus convexity implies $\mathsf{L}(ta) = t\mathsf{L}(a)$ and $\mathsf{L}(a+b) = \mathsf{L}(a) + \mathsf{L}(b)$ whenever defined. Defining J by

$$\mathsf{J}(0) = 0; \tag{5.102}$$

$$\mathsf{J}(a) = s \cdot \mathsf{L}(a/s) \ (a > 0, s \geq \|a\|); \tag{5.103}$$

$$\mathsf{J}(a) = -\mathsf{J}(-a) \ (a < 0), \tag{5.104}$$

where $a > 0$ means $a \geq 0$ and $a \neq 0$, and $a < 0$ means $-a \geq 0$ and $a \neq 0$, once again the reasoning near the end of the proof of Lemma 5.11 shows that J is linear; it is a untital order-preserving bijection by construction. Hence J is a Jordan automorphism by Lemma 5.20.2 Of course, instead of (5.104) one could equivalently have defined J on general $a \in B(H)_{\mathrm{sa}}$ by $\mathsf{J}(a) = \mathsf{J}(a_+) - \mathsf{J}(a_-)$, using the (by now hopefully familiar) decomposition $a = a_+ - a_-$ with $a_\pm \geq 0$ and $a_+a_- = 0$.

Conversely, once again using Lemma 5.20.2, a Jordan automorphisms (5.11) preserves order as well as the unit, so that the inequality $0 \leq a \leq 1_H$ characterizing $a \in \mathscr{E}(H)$ is preserved, i.e., $0 \leq \mathsf{J}(a) \leq 1_H$. Thus J preserves $\mathscr{E}(H)$, where it preserves order. Convexity is obvious, since $\mathsf{L} = \mathsf{J}_{|\mathscr{E}(H)}$ comes from a linear map. □

The equivalence between Jordan's Theorem and von Neumann's Theorem (provided $\dim(H) \geq 3$) hinges on the following corollary of Gleason's Theorem (cf. §D.1).

Corollary 5.22. *Let* $\dim(H) > 2$. *Then an isomorphism* N *of* $\mathscr{P}(H)$ *as an orthocomplemented lattice has a unique extension to a linear map* $\alpha : B(H)_{\mathrm{sa}} \to B(H)_{\mathrm{sa}}$, *which is (automatically) invertible, unital, and positive.*

Proof. According to Lemma D.2, N preserves all suprema in $\mathscr{P}(H)$. Since we have $\sum_i e_i = \bigvee e_i$ for any family of mutually orthogonal projections and since N by definition preserves the orthocomplementation $e^\perp = 1 - e$ and hence preserves orthogonality of projections, we may compute

$$\mathsf{N}\left(\sum_i e_i\right) = \mathsf{N}\left(\bigvee_i e_i\right) = \bigvee_i \mathsf{N}(e_i) = \sum_i \mathsf{N}(e_i). \tag{5.105}$$

Consequently, for any normal state ω on $B(H)$, the map $e \mapsto \omega \circ \mathsf{N}(e)$ is a probability measure on $\mathscr{P}(H)$, which by Gleason's Theorem has a unique linear extension to $B(H)$ and hence *a fortiori* to $B(H)_{\mathrm{sa}}$. We use this in order to define α, as follows.

First, let $a \in B(H)_{\mathrm{sa}}$ and suppose $a = \sum_j \lambda_j f_j$ for some *finite* family (f_j) of projections (not necessarily orthogonal), and some $\lambda_j \in \mathbb{R}$. Then $\sum_j \lambda_j \mathsf{N}(f_j)$ is independent of the particular decomposition of a that has been chosen, so we may put

$$\alpha(a) = \sum_j \lambda_j \mathsf{N}(f_j). \tag{5.106}$$

To see this, put $a = \sum_{j'} \lambda'_{j'} f'_{j'}$ and hence $\alpha'(a) = \sum_{j'} \lambda'_{j'} \mathsf{N}(f'_{j'})$, and suppose $\alpha'(a) \neq \alpha(a)$. By (B.477) there exists a normal state ω such that $\omega(\alpha'(a)) \neq \omega(\alpha(a))$; indeed, each element of $B_1(H)$ is a linear combination of at most four density operators, so that each normal linear functional on $B(H)$ is a linear combination of at most four normal states. But since $\omega \circ \mathsf{N}$ is linear, this implies $\omega \circ \mathsf{N}(a) \neq \omega \circ \mathsf{N}(a)$, which is a contradiction. Hence $\alpha'(a) = \alpha(a)$ and accordingly, (5.106) is well defined. Because it is independent of the decomposition of a into projections, α is linear: if $a = \sum_j \lambda_j f_j$ and $a' = \sum_{j'} \lambda'_{j'} f'_{j'}$, then $a + a' = \sum_j \lambda_j f_j + \sum_{j'} \lambda'_{j'} f'_{j'}$, so that

$$\mathsf{N}(a+a') = \mathsf{N}\left(\sum_j \lambda_j f_j + \sum_{j'} \lambda'_{j'} f'_{j'}\right) = \sum_j \lambda_j \mathsf{N}(f_j) + \sum_{j'} \lambda'_{j'} \mathsf{N}(f'_{j'}) = \mathsf{N}(a) + \mathsf{N}(a').$$

Similarly, for any $t \in \mathbb{R}$ we have

$$\mathsf{N}(ta) = \mathsf{N}\left(\sum_j t\lambda_j f_j\right) = \sum_j t\lambda_j \mathsf{N}(f_j) = t\sum_j \lambda_j \mathsf{N}(f_j) = t\mathsf{N}(a).$$

We may now extend α to all of $B(H)_{\mathrm{sa}}$ by continuity. Indeed, according to the spectral theorem in the form (B.326), the set of all operators of the form $a = \sum_j \lambda_j f_j$ with all f_j mutually orthogonal (so that a is given by its spectral resolution) is norm-dense in $B(H)_{\mathrm{sa}}$. Applying (5.106), and noting that $\|a\| = \sup_j |\lambda_j|$, we may estimate

$$\|\alpha(a)\| = \|\sum_j \lambda_j \mathsf{N}(f_j)\| \leq \sup_j \{|\lambda_j|\} \|\sum_j \mathsf{N}(f_j)\| \leq \|a\|,$$

since the $\mathsf{N}(f_j)$ are mutually orthogonal and hence sum to some projection, which has norm 1 (unless $a = 0$). For general $a \in B(H)_{\mathrm{sa}}$, we may therefore define N by $\mathsf{N}(a) = \lim_n \mathsf{N}(a_n)$, where each a_n is of the above (spectral) form and $\|a_n - a\| \to 0$.

To prove that α is positive, we show that $\alpha(a) \geq 0$ whenever $a \geq 0$. As in the preceding step, initially suppose that $a = \sum_j \lambda_j f_j$ has a finite spectral resolution. Then $a \geq 0$ iff $\lambda_j \geq 0$ for each j, and hence $\alpha(a) \geq 0$ by (5.106), since by orthogonality of the $\mathsf{N}(f_j)$ this equation states the spectral resolution of $\alpha(a)$. Now if $a_n \geq 0$ and $a_n \to a$ (in norm), then $\langle \psi, a_n \psi \rangle \to \langle \psi, a\psi \rangle$, which must remain positive, so that $a \geq 0$. Hence positivity of α on all of $B(H)_{\mathrm{sa}}$ follows by continuity.

Finally, α inherits invertibility from N, and it is unital by (5.105), taking $e_i = |\upsilon_i\rangle\langle\upsilon_i|$ for some basis (υ_i) of H (or using the fact that it preserves $\top = 1_H$). □

Subsequently, we use Lemma 5.20 to further extend α by complex linearity to a Jordan isomorphism of $B(H)$; see Definition 5.1.

Finally, the equivalence between weak Jordan symmetries and Bohr symmetries follows from Hamhalter's Theorem 9.4, whereas Theorem 9.7 strengthens this to an equivalence between Jordan symmetries and Bohr symmetries. The proof of these theorems does not seem to simplify in the special case at hand, i.e. $A = B(H)$.

5.4 Proof of Jordan's Theorem

In view of the equivalence between the six parts of Theorem 5.4, we only need to prove one of them. In the literature, one only finds proofs of Jordan's Theorem and of Wigner's Theorem, and we present each of these (surprisingly but instructively, these proofs look completely different). We start with ***Jordan's Theorem***:

Theorem 5.23. *Any Jordan automorphism $\mathsf{J}_{\mathbb{C}}$ of $B(H)$ is given by either*

$$\mathsf{J}_{\mathbb{C}}(a) = \alpha_u(a) \equiv uau^*, \tag{5.107}$$

where u is unitary (and is determined by $\mathsf{J}_{\mathbb{C}}$ up to a phase), or by

$$\mathsf{J}_{\mathbb{C}}(a) = \alpha'_u(a) \equiv ua^*u^*, \tag{5.108}$$

where u is anti-unitary (and is determined by $\mathsf{J}_{\mathbb{C}}$ up to a phase, too).

The difficult part of the proof is Theorem C.175, which implies:

Proposition 5.24. *A Jordan automorphism α of $B(H)$ is either an automorphism or an anti-automorphism.*

Recall that an ***automorphism*** of $B(H)$ is a linear bijection $\alpha : B(H) \to B(H)$ that satisfies $\alpha(a^*) = \alpha(a)^*$ and $\alpha(ab) = \alpha(a)\alpha(b)$; an ***anti-automorphism***, on the other hand, satisfies the first property whilst the latter is replaced by $\alpha(ab) = \alpha(b)\alpha(a)$. Clearly, both automorphisms and anti-automorphisms are Jordan automorphisms. Granting this result, we may deal with the two cases separately.

Proposition 5.25. *Any automorphism $\alpha : B(H) \to B(H)$ takes the form $\alpha = \alpha_u$, see (5.107), where $u : H \to H$ is unitary, uniquely determined by α up to a phase.*

The proof uses the following lemmas. The first follows from Theorem C.62.4.

Lemma 5.26. *If $\alpha : B(H) \to B(H)$ is an automorphism and $a \in B(H)$, then*

$$\|\alpha(a)\| = \|a\|. \tag{5.109}$$

Lemma 5.27. *If $\alpha : B(H) \to B(H)$ is an automorphism and $e \in B(H)$ is a one-dimensional projection, then so is $\alpha(e)$.*

Proof. It should be obvious that automorphisms α preserve projections e (whose defining properties are $e^2 = e^* = e$). Furthermore, α preserves order, i.e., if $a \geq 0$ (in that, as always, $\langle \psi, a\psi \rangle \geq 0$ for each $\psi \in H$, or, equivalently, $a = b^*b$), then $\alpha(a) \geq 0$ (this is clear from the second way of expressing positivity). Consequently, if $a \leq b$ (in that $b - a \geq 0$), then $\alpha(a) \leq \alpha(b)$. We notice that if we define $e \leq f$ iff $eH \subseteq fH$, then $e \leq f$ iff $e \leq f$ as self-adjoint operators (in that $\langle \psi, e\psi \rangle \leq \langle \psi, f\psi \rangle$ for each $\psi \in H$); see Proposition C.170. With respect to the ordering $\leq$ the one-dimensional projections e are ***atomic***, in the sense that $0 \leq e$ (but $e \neq 0$) and if $0 \leq f \leq e$, then either $f = 0$ or $f = e$. Now automorphisms of the projection lattice $B(H)$ restrict to isomorphisms of $\mathscr{P}(H)$, which preserve atoms (as these are intrinsically defined by the partial order). □

We are now ready for the (constructive!) proof of Proposition 5.25.

Proof. For some fixed unit vector $\chi \in H$, take the corresponding one-dimensional projection e_χ and define a new unit vector φ (up to a phase) by

$$e_\varphi = \alpha^{-1}(e_\chi). \tag{5.110}$$

Now any $\psi \in H$ may be written as $\psi = a\varphi$, for some $a \in B(H)$. Attempt to define an operator u by $u\psi = \alpha(a)\chi$, i.e.,

$$ua\varphi = \alpha(a)\chi. \tag{5.111}$$

This looks dangerously ill-defined, since many different operators a may give rise to the same ψ. Fortunately, we may compute

$$\begin{aligned} \|a\varphi\|_H &= \|ae_\varphi\varphi\|_H = \|ae_\varphi\|_{B(H)} = \|\alpha(ae_\varphi)\|_{B(H)} \\ &= \|\alpha(a)\alpha(e_\varphi)\|_{B(H)} = \|\alpha(a)e_\chi\|_{B(H)} = \|\alpha(a)\chi\|_H \\ &= \|ua\varphi\|_H, \end{aligned}$$

so that if $a\varphi = b\varphi$, then $\alpha(a)\chi = \alpha(b)\chi$ and hence u is well defined. By this computation u is also isometric and since it is clearly surjective, it is unitary. The property $\alpha(a) = uau^*$ is equivalent to $ua = \alpha(a)u$, which in turn is equivalent to $uab\varphi = \alpha(a)ub\varphi$ for any $b \in B(H)$, which by definition of u is the same as

$$\alpha(ab)\chi = \alpha(a)\alpha(b)\chi. \tag{5.112}$$

But this holds by virtue of α being an automorphism. Finally, all arbitrariness in u lies in the lack of uniqueness of φ given its definition (5.110). □

Proposition 5.28. *Any antiautomorphism $\alpha : B(H) \to B(H)$ takes the form $\alpha = \alpha_u$, cf. (5.108), where $u : H \to H$ is anti-unitary, uniquely determined by α up to a phase.*

Proof. Pick an arbitrary anti-unitary operator $J : H \to H$ and define

$$\begin{aligned} \beta &: B(H) \to B(H); \\ \beta(a) &= Ja^*J^*. \end{aligned} \tag{5.113}$$

Then $\alpha \circ \beta$ is an automorphism, to which Proposition 5.25 applies, so that

$$\alpha \circ \beta(a) = \tilde{u}a\tilde{u}^*, \tag{5.114}$$

for some unitary $\tilde{u}$. Hence

$$\alpha(a) = \alpha(\beta \circ \beta^{-1}(a)) = \alpha \circ \beta(J^*a^*J) = \tilde{u}J^*a^*J\tilde{u}^*,$$

so that $\alpha(a) = ua^*u^*$ with $u = \tilde{u}J^*$.

The precise lack of uniqueness of u is inherited from the unitary case. □

5.5 Proof of Wigner's Theorem

We recall ***Wigner's Theorem***, i.e. Theorem 5.4.1:

Theorem 5.29. *Each bijection* $\mathsf{W} : \mathscr{P}_1(H) \to \mathscr{P}_1(H)$ *that satisfies*

$$\mathrm{Tr}\,(\mathsf{W}(e)\mathsf{W}(f)) = \mathrm{Tr}\,(ef), \ (e, f \in \mathscr{P}_1(H)), \tag{5.115}$$

is given by $\mathsf{W}(e) = ueu^* \equiv \alpha_u(e)$, *where the operator* u *is either unitary or anti-unitary, and is uniquely determined by* W *up to a phase.*

The problem is to lift a given map $\mathsf{W} : \mathscr{P}_1(H) \to \mathscr{P}_1(H)$ that satisfies (5.115) to either a unitary or an anti-unitary map $u : H \to H$ such that

$$\mathsf{W}(e_\psi) = e_{u\psi} = ue_\psi u^*. \tag{5.116}$$

Suppose $\mathsf{W}(e_\psi) = e_{\psi'}$. Since $e_{z\psi} = e_\psi$ for any $z \in \mathbb{T}$, and likewise for $e_{\psi'}$, this means that $u\psi = z\psi'$ for some $z \in \mathbb{T}$; the problem is to choose the z's coherently all over the unit sphere of H. There are many proofs in the literature, of which the following one—partly based on an earlier proof by Bargmann (1964)—has the advantage of making at least the construction of u explicit (at the cost of opaque proofs of some crucial lemma's). We assume $\dim(H) > 2$, since $H = \mathbb{C}^2$ has already been covered.

Fix unit vectors $\psi \in H$ and $\psi' \in \mathsf{W}(e_\psi)H$; clearly, ψ' is unique up to multiplication by $z \in \mathbb{T}$, whose choice turns out to completely determine u (i.e., the ambiguity in ψ' is the only one in the entire construction). For a modest start, we put

$$u\psi = \psi'. \tag{5.117}$$

Lemma 5.30. *If* $V \subset H$ *is a* k*-dimensional subspace (where* $k < \infty$*), then there is a unique* k*-dimensional linear subspace* $V' \subset H$ *with the following property:*

For all unit vectors $\psi \in H$, *we have* $\psi \in V$ *iff* $\mathsf{W}(e_\psi)H \subset V'$.

Proof. Pick a basis $(\upsilon_1, \ldots, \upsilon_k)$ of V and find unit vectors $\upsilon_i' \in H$ such that $\upsilon_i' \in \mathsf{W}(e_{\upsilon_i})H$, $i = 1, \ldots, k$. Then, using (5.115) we compute

$$|\langle \upsilon_i', \upsilon_j' \rangle|^2 = \mathrm{Tr}\,(e_{\upsilon_i'} e_{\upsilon_j'}) = \mathrm{Tr}\,(\mathsf{W}(e_{\upsilon_i})\mathsf{W}(e_{\upsilon_j})) = \mathrm{Tr}\,(e_{\upsilon_i} e_{\upsilon_j}) = |\langle \upsilon_i, \upsilon_j \rangle|^2 = \delta_{ij},$$

so that the vectors $(\upsilon_1', \ldots, \upsilon_k')$ form an orthonormal set and hence form a basis of their linear span V'. Now, as mentioned below (B.214), we have $\psi \in V$ iff $\sum_{i=1}^k |\langle \upsilon_i, \psi \rangle|^2 = 1$ and similarly $\psi' \in V'$ iff $\sum_{i=1}^k |\langle \upsilon_i', \psi' \rangle|^2 = 1$. Since W preserves transition probabilities, a computation similar to one just given yields

$$\sum_{i=1}^k |\langle \upsilon_i, \psi \rangle|^2 = \sum_{i=1}^k |\langle \upsilon_i', \psi' \rangle|^2, \tag{5.118}$$

so that both sides do or do not equal unity, and hence $\psi \in V$ iff $\psi' \in V'$. □

Wigner's Theorem for $H = \mathbb{C}^2$ (i.e. Theorem 5.7) implies:

Lemma 5.31. *If V and V' are related as in Lemma 5.30, and*

$$\dim(V) = \dim(V') = 2, \tag{5.119}$$

then there is a unitary or anti-unitary operator $u_V : V \to V'$ such that

$$\mathsf{W}(e) = u_V e u_V^*, \tag{5.120}$$

for any one-dimensional projection $e \in \mathscr{P}_1(V)$, where $\mathscr{P}_1(V) \subset \mathscr{P}_1(H)$ consists of all $e \in \mathscr{P}_1(H)$ with $eH \subset V$. Moreover, u_V is unique up to a phase.

Proof. A choice of basis for both V and V' gives unitary isomorphisms $u : V \xrightarrow{\cong} \mathbb{C}^2$ and $u' : V' \xrightarrow{\cong} \mathbb{C}^2$, which jointly induce a map

$$\mathsf{W}' \equiv u'\mathsf{W}u^{-1} : \mathscr{P}_1(\mathbb{C}^2) \to \mathscr{P}_1(\mathbb{C}^2). \tag{5.121}$$

This maps satisfies the hypotheses of Wigner's Theorem in $d = 2$, and so it is (anti-) unitarily induced as $\mathsf{W}' = \alpha_v$, where $v : \mathbb{C}^2 \to \mathbb{C}^2$ is (anti-) unitary. Then the operator $u_V = (u')^{-1} v u$ does the job; its lack of uniqueness stems entirely from v. □

Lemma 5.32. *Given a Wigner symmetry W, the ensuing operator u_V is either unitary or anti-unitary for all two-dimensional subspaces $V \subset H$ (simultaneously).*

Proof. We first design a "unitarity test" for W. Define a function

$$T : \mathscr{P}_1(H) \times \mathscr{P}_1(H) \times \mathscr{P}_1(H) \to \mathbb{C}; \tag{5.122}$$

$$T(e, f, g) = \mathrm{Tr}\,(efg), \tag{5.123}$$

$$T(e_{\psi_1}, e_{\psi_2}, e_{\psi_3}) = \langle \psi_1, \psi_2 \rangle \langle \psi_2, \psi_3 \rangle \langle \psi_3, \psi_1 \rangle. \tag{5.124}$$

Let $V \subset H$ be two-dimensional and pick an orthonormal basis (υ_1, υ_2). Define

$$\chi_1 = \upsilon_1,\ \chi_2 = (\upsilon_1 - \upsilon_2)/\sqrt{2},\ \chi_3 = (\upsilon_1 - i\upsilon_2)/\sqrt{2}. \tag{5.125}$$

A simple computation then shows that

$$T(e_{\chi_1}, e_{\chi_2}, e_{\chi_3}) = \tfrac{1}{4}(1 + i). \tag{5.126}$$

It follows from (5.124) that for u unitary and v anti-unitary, we have

$$T(e_{u\psi_1}, e_{u\psi_2}, e_{u\psi_3}) = T(e_{\psi_1}, e_{\psi_2}, e_{\psi_3}); \tag{5.127}$$

$$T(e_{v\psi_1}, e_{v\psi_2}, e_{v\psi_3}) = \overline{T(e_{\psi_1}, e_{\psi_2}, e_{\psi_3})}. \tag{5.128}$$

Eq. (5.126) implies that if $\mathsf{W} : V \to V'$ is (anti-) unitarily implemented, we have

$$T(\mathsf{W}(e_{\chi_1}), \mathsf{W}(e_{\chi_2}), \mathsf{W}(e_{\chi_3})) = T(e_{u\chi_1}, e_{u\chi_2}, e_{u\chi_3}) = \tfrac{1}{4}(1 \pm i), \tag{5.129}$$

with a plus sign if u is unitary and a minus sign if u is anti-unitary. Now take a second pair $(\tilde{V}, \tilde{V}')$ as above, and pick a basis $(\tilde{\upsilon}_1, \tilde{\upsilon}_2)$ of $\tilde{V}$, with associated vectors $(\tilde{\chi}_1, \tilde{\chi}_2, \tilde{\chi}_3)$, as in (5.125). Suppose $u : V \to V'$ implementing W is unitary, whereas $\tilde{u} : \tilde{V} \to \tilde{V}'$ implementing W is anti-unitary. It then follows from (5.129) that

$$T(\mathsf{W}(e_{\chi_1}), \mathsf{W}(e_{\chi_2}), \mathsf{W}(e_{\chi_3})) = T(e_{u\chi_1}, e_{u\chi_2}, e_{u\chi_3}) = \tfrac{1}{4}(1+i); \tag{5.130}$$

$$T(\mathsf{W}(e_{\tilde{\chi}_1}), \mathsf{W}(e_{\tilde{\chi}_2}), \mathsf{W}(e_{\tilde{\chi}_3})) = T(e_{\tilde{u}\tilde{\chi}_1}, e_{\tilde{u}\tilde{\chi}_2}, e_{\tilde{u}\tilde{\chi}_3}) = \tfrac{1}{4}(1-i). \tag{5.131}$$

In view of (C.637), the following expression defies a metric d on $\mathscr{P}_1(H)$:

$$d(e_\psi, e_\varphi) = \|\omega_\psi - \omega_\varphi\| = \|e_\psi - e_\varphi\|_1 = 2\sqrt{1 - |\langle\varphi, \psi\rangle|^2}, \tag{5.132}$$

with respect to which both W and T are continuous (the latter with respect to the product metric on $\mathscr{P}_1(H)^3$, of course). Let $t \mapsto (\upsilon_1(t), \upsilon_2(t))$ be a continuous path of orthonormal vectors (i.e., in $H \times H$), with associated vectors $(\chi_1(t), \chi_2(t), \chi_3(t))$, as in (5.125). Then the function $f(t) = T(\mathsf{W}(\chi_1(t)), \mathsf{W}(\chi_2(t)), \mathsf{W}(\chi_3(t)))$ is continuous, and by (5.129) it can only take the values $\frac{1}{4}(1 \pm i)$. Hence $f(t)$ must be constant. However, taking a path such that $(\upsilon_1(0), \upsilon_2(0)) = (\upsilon_1, \upsilon_2)$ and $(\upsilon_1(1), \upsilon_2(1)) = (\tilde{\upsilon}_1, \tilde{\upsilon}_2)$, gives $f(0) = \frac{1}{4}(1+i)$ and $f(1) = \frac{1}{4}(1-i)$, which is a contradiction. □

Lemma 5.33. *Wigner's Theorem holds for three-dimensional Hilbert spaces.*

Proof. Let $(\upsilon_1, \upsilon_2, \upsilon_3)$ be some basis of of H (like the usual basis of $H = \mathbb{C}^3$). We first show that if W is the identity if restricted to both $\mathrm{span}(\upsilon_1, \upsilon_2)$ and $\mathrm{span}(\upsilon_1, \upsilon_3)$, then W is the identity on H altogether. To this end, take $\psi = \sum_i c_i \upsilon_i$, initially with $c_1 \in \mathbb{R}\backslash\{0\}$. Take a unit vector $\psi' \in \mathsf{W}(e_\psi)$, with $\psi = \sum_i c_i' \upsilon_i$. By the first assumption on W we have $|\langle\upsilon, \psi'\rangle| = |\langle\upsilon, \psi\rangle|$ for any unit vector $\upsilon \in \mathrm{span}(\upsilon_1, \upsilon_2)$. Taking

$$\upsilon = \upsilon_1, \quad \upsilon = \upsilon_2, \quad \upsilon = (\upsilon_1 + \upsilon_2)/\sqrt{2}, \quad \upsilon = (\upsilon_1 + i\upsilon_2)/\sqrt{2}, \tag{5.133}$$

gives the equations

$$|c_1'| = |c_1|, \quad |c_2'| = |c_2|, \quad |c_1' + c_2'| = |c_1 + c_2|, \quad |c_1' - ic_2'| = |c_1 - ic_2|, \tag{5.134}$$

respectively. By a choice of phase we may and will assume $c_1' = c_1$, in which case the only solution is $c_2 = c_2'$ (geometrically, the solution c_2' lies in the intersection of three different circles in the complex plane, which is either empty or consists of a single point). Similarly, the second assumption on W gives $c_3 = c_3'$, whence $\psi' = \psi$. The case $c_1 = 0$ may be settled by a straightforward limit argument, since inner products (and hence their absolute values) are continuous on $H \times H$.

Given a Wigner symmetry $\mathsf{W} : \mathscr{P}_1(H) \to \mathscr{P}_1(H)$, we now construct u as follows.

1. Fix a basis $(\upsilon_1, \upsilon_2, \upsilon_3)$ with "image" $(\upsilon_1', \upsilon_2', \upsilon_3')$ under W, i.e, $\mathsf{W}(e_{\upsilon_i}) = e_{\upsilon_i'}$.
2. The unitarity test in the proof of Lemma 5.32 settles if the operators should be chosen to be unitary or anti-unitary; for simplicity we assume the unitary case.
3. Define a unitary $u_1 : H \to H$ by $u_1 \upsilon_i' = \upsilon_i$ for $i = 1, 2, 3$, and subsequently define $\mathsf{W}_1 = \alpha_{u_1} \circ \mathsf{W}$, which (being the composition of two Wigner symmetries)

is a Wigner symmetry. Clearly, $\mathsf{W}_1(e_{\upsilon_i}) = e_{\upsilon_i}$ $(i = 1,2,3)$, so that W_1 maps $\mathscr{P}_1(H_{(12)})$ to itself, where $H_{(12)} \equiv \mathrm{span}(\upsilon_1, \upsilon_2)$. Hence Lemma 5.31 gives a unitary map $\tilde{u}_1 : H_{(12)} \to H_{(12)}$ such that the restriction of W_1 to $H_{(12)}$ is $\alpha_{\tilde{u}_1}$.

4. Define a unitary $u_2 : H \to H$ by $u_2 = \tilde{u}_1^{-1}$ on $H_{(12)}$ and $u_2\upsilon_3 = \upsilon_3$, followed by the Wigner symmetry $\mathsf{W}_2 = \alpha_{u_2} \circ \mathsf{W}_1$. By construction, $\mathsf{W}_2(e_{\upsilon_i}) = e_{\upsilon_i}$ for $i = 1,2,3)$ (W_2 is even the identity on $\mathscr{P}_1(H_{(12)})$), so that W_2 maps $\mathscr{P}_1(H_{(13)})$ to itself, where $H_{(13)} \equiv \mathrm{span}(\upsilon_1, \upsilon_3)$. Hence the restriction of W_2 to $H_{(13)}$ is implemented by a unitary $\tilde{u}_2 : H_{(13)} \to H_{(13)}$, whose phase may be fixed by requiring $\tilde{u}_2\upsilon_1 = \upsilon_1$.
5. Similarly to u_2, we define $u_3 : H \to H$ by $u_3 = \tilde{u}_2^{-1}$ on $H_{(13)}$ and $u_3\upsilon_2 = \upsilon_2$, so that u_3 is the identity on $H_{(12)}$. Of course, we now define a Wigner symmetry

$$\mathsf{W}_3 = \alpha_{u_3} \circ \mathsf{W}_2 = \alpha_{u_3} \circ \alpha_{u_2} \circ \alpha_{u_1} \circ \mathsf{W}, \tag{5.135}$$

which by construction is the identity on both $\mathscr{P}_1(H_{(12)})$ and $\mathscr{P}_1(H_{(13)})$, and so by the first part of the proof it must be the identity on all of $\mathscr{P}_1(H)$. Hence

$$\mathsf{W} = \alpha_{u_1^{-1}} \circ \alpha_{u_2^{-1}} \circ \alpha_{u_3^{-1}} = \alpha_u \qquad (u = u_1^{-1}u_2^{-1}u_3^{-1}).$$ □

Lemma 5.34. *As in Lemma 5.30, if* $\dim(V) = \dim(V') = 3$, *then there is a unitary or anti-unitary operator* $u_V : V \to V'$ *such that* $\mathsf{W}(e) = u_V e u_V^*$ *for any* $e \in \mathscr{P}_1(V)$,

Proof. Given Lemma 5.33, the proof is practically the same as for Lemma 5.31. □

We now finish the proof of Wigner's Theorem. We assume that the outcome of Lemma 5.32 is that each u_V is unitary; the anti-unitary case requires obvious modifications of the argument below. The first step is, of course, to define $u(\lambda\psi) = \lambda u\psi$, $\lambda \in \mathbb{C}$ (so this would have been $\overline{\lambda}u\psi$ in the anti-unitary case). Let $\varphi \in H$ be linearly independent of ψ and consider the two-dimensional space V spanned by ψ and φ. Define $u(\varphi) = u_V\varphi$. With (5.117), this defines u on all of H. To prove that u is linear, take φ_1 and φ_2 linearly independent of each other and of ψ, so that the linear span V_3 of ψ, φ_1, and φ_2 is three-dimensional. Let V_i be the two-dimensional linear span of ψ and φ_i, $i = 1,2$. Then $u\varphi_i = u_{V_i}\varphi_i$, where the phase of u_{V_i} is fixed by (5.117). Let $w : V_3 \to V_3'$ be the unitary that implements W according to Lemma 5.33.2, with phase determined by (5.117). Since u_{V_1} and u_{V_2} and w are unique up to a phase and this phase has been fixed for each in the same way, we must have $u_{V_1} = w_{|V_1}$ and $u_{V_2} = w_{|V_2}$. Finally, we have V_{12} spanned by ψ and $\varphi_1 + \varphi_2$, and by the same token, $u_{V_{12}} = w_{|V_{12}}$. Now w is unitary and hence linear, so

$$\begin{aligned} u(\varphi_1 + \varphi_2) &= u_{V_{12}}(\varphi_1 + \varphi_2) = w(\varphi_1 + \varphi_2) = w(\varphi_1) + w(\varphi_2) \\ &= u_{V_1}(\varphi_1) + u_{V_2}(\varphi_2) = u(\varphi_1) + u(\varphi_2), \end{aligned}$$

since this is how u was defined. Since each u_V is unitary, so is u, and similarly it is easy to verify that u implements W, because each u_V does so. □

5.6 Some abstract representation theory

Since all symmetries we have considered (named after Wigner, Kadison, Jordan, Ludwig, von Neumann, and Bohr) are implemented by either unitary or anti-unitary operators, which are determined (by the given symmetry) only up to a phase $z \in \mathbb{T}$, the quantum-mechanical symmetry group $\mathscr{G}^H$ of a Hilbert space H is given by

$$\mathscr{G}^H = (U(H) \cup U_a(H))/\mathbb{T}, \tag{5.136}$$

where $U(H)$ is the group of unitary operators on H, and $U_a(H)$ is the set of anti-unitary operators on H; the latter is not a group (since the product of two anti-unitaries is unitary) but their union is. Furthermore, $\mathbb{T}$ is identified with the normal subgroup $\mathbb{T} \equiv \mathbb{T} \cdot 1_H = \{z \cdot 1_H \mid z \in \mathbb{T}\}$ of $U(H) \cup U_a(H)$ (and also of $U(H)$) consisting of multiples of the unit operators by a phase; thus the quotient $\mathscr{G}^H$ is a group.

The fact that $\mathscr{G}^H$ rather than $U(H)$ is the symmetry group of quantum mechanics has profound consequences (one of which is our very existence), which we will study from §5.10 onwards. However, this material relies on the theory of "ordinary" (i.e., non-projective) unitary representations, which we therefore review first.

Namely, let G be a group. In mathematics, the natural kind of action of G on a Hilbert space H is a ***unitary representation***, i.e., a homomorphism

$$u : G \to U(H), \tag{5.137}$$

so that $u(x)^{-1} = u(x^{-1}) = u(x)^*$ and $u(x)u(y) = u(xy)$, which imply $u(e) = 1_H$.

As to the possible continuity properties of unitary representations in case that G is a *topological* group (i.e., a group G that is also a topological space, such that group multiplication $G \times G \hookrightarrow G$ and inverse $G \to G$ are continuous), one should equip $U(H)$ with the *strong* operator topology (as opposed to the norm topology).

Proposition 5.35. *If $u : x \mapsto u(x)$ is a unitary representation of some locally compact group G on a Hilbert space H, then the following conditions are equivalent:*

1. *The map $G \times H \to H$, $(x, \psi) \mapsto u(x)\psi$, is continuous;*
2. *The map $G \to U(H)$, $x \mapsto u(x)$, is continuous in the strong topology on $U(H)$.*

Proof. Strong continuity means that if $x_\lambda \to x$ in G, then for each $\psi \in H$ we have $\|(u(x_\lambda) - u(x))\psi\| \to 0$. This is clearly implied by the first kind of continuity, giving $1 \Rightarrow 2$, so let us prove the nontrivial converse. Suppose $x_\lambda \to x$ and $\psi_\mu \to \psi$; since G is locally compact, x has a compact neighborhood K and we may assume that each $x_\lambda \in K$. If u is strongly continuous, then for any $\varphi \in H$ the set $\{u(y)\varphi, y \in K\}$ is compact in H and hence bounded. The Banach–Steinhaus Theorem B.78 gives boundedness of the corresponding operator norms, that is, $\{\|u(y)\|, y \in K\} < C_K$ for some $C_K > 0$. We now estimate

$$\|u(x_\lambda)\psi_\mu - u(x)\psi\| \leq \|u(x_\lambda)\psi_\mu - u(x_\lambda)\psi\| + \|(u(x_\lambda) - u(x))\psi\|.$$

The first term vanishes as $\psi_\mu \to \psi$ since it is bounded by $C_K\|\psi_\mu - \psi\|$, whereas the second vanishes as $x_\lambda \to x$ by the (assumed) strong continuity of u. □

Since the first kind of continuity is the usual one for group actions, this justifies the choice of strong continuity as the natural one for unitary representations (to which a pragmatic point may be added: norm continuity is quite rare for unitary representations on infinite-dimensional Hilbert spaces). Things further simplify under mild restrictions on G and H, which are satisfied in all examples of physical interest.

Proposition 5.36. *If H is separable and G is second countable locally compact (sclc), then each of the two continuity conditions in Proposition 5.35 is in turn equivalent to* **weak measurability** *of u, in that for each $\varphi, \psi \in H$ the function*

$$x \mapsto \langle \varphi, u(x)\psi \rangle$$

from G to $\mathbb{C}$ is (Borel) measurable.

Proof. This spectacular result is due to von Neumann, who more generally proved that a measurable homomorphism between sclc groups is continuous. This implies the claim: first, if H is separable, then the group $U(H)$ is sclc in its weak operator topology, so that if the map $G \to U(H)$, $x \mapsto u(x)$ is weakly measurable, then it is continuous in the weak topology on $U(H)$. Second, for any Hilbert space, weak (operator) continuity of a unitary representation implies strong continuity (so that, given the trivial converse, weak and strong continuity of unitary group representations are equivalent). We only prove this last claim: for $x, y \in G$, we compute

$$\begin{aligned} \|(u(y) - u(x))\psi\| &= \|u(x)\psi\|^2 + \|u(y)\psi\|^2 - \langle u(x)\psi, u(y)\psi\rangle - \langle u(y)\psi, u(x)\psi\rangle \\ &= 2\|\psi\|^2 - \langle \psi, u(x^{-1}y)\psi\rangle - \langle \psi, u(y^{-1}x)\psi\rangle, \end{aligned}$$

Weak continuity obviously implies that the function $x \mapsto \langle \psi, u(x)\psi\rangle$ is continuous at the identity $e \in G$, so if $y = x_\lambda \to x$, then $\|(u(x_\lambda) - u(x))\psi\| \to 0$. □

In view of this, it is hardly a restriction for a unitary representation of a locally compact group on a Hilbert space to be continuous in the sense of Proposition 5.35, so we always assume this in what follows. Furthermore, any group we consider is locally compact, so this will be a standing assumption, too. An important consequence of this assumption is the existence of a translation-invariant measure on G.

Theorem 5.37. *Each locally compact group G has a canonical nonzero (outer regular Borel) measure μ, called* **Haar measure**, *which is left-invariant in that*

$$\int_G d\mu(x)\, L_y f(x) = \int_G d\mu(x)\, f(x), \tag{5.138}$$

for each $f \in C_c(G)$ and $y \in G$, where the **left translation** *L_y of f by y is defined by*

$$L_y f(x) = f(y^{-1}x). \tag{5.139}$$

This measure is unique up to scalar multiplication. Moreover, if G is compact, then:

1. *μ is finite and hence can be normalized to a probability measure, i.e.,*

$$\mu(G) = 1. \tag{5.140}$$

2. *μ is also right-invariant in that*

$$\int_G d\mu(x)\, R_y f(x) = \int_G d\mu(x)\, f(x), \tag{5.141}$$

where the **right translation** *R_y of f by $y \in G$ is defined by*

$$R_y f(x) = f(xy). \tag{5.142}$$

3. *μ is invariant under inversion, in that*

$$\int_G d\mu(x)\, f(x^{-1}) = \int_G d\mu(x)\, f(x). \tag{5.143}$$

Existence is due to Haar and uniqueness was first proved by von Neumann. One often writes $dx \equiv d\mu(x)$ for Haar measure. Here are some examples:

- For $G = \mathbb{R}^n$, Haar measure equals Lebesgue measure μ_L (up to a constant); eqs. (5.139) and (5.141) state the familiar translation invariance of μ_L.
- For $G = \mathbb{T}$, we have

$$\int_{\mathbb{T}} d\mu(z) f(z) = \frac{1}{2\pi} \int_0^{2\pi} d\theta\, f(e^{i\theta}). \tag{5.144}$$

- For $G = GL_n(\mathbb{R})$ with $X = (x_{ij})$, we have

$$d\mu(X) = \prod_{i,j=1}^{m} dx_{ij} |\det(X)|^{-n}, \tag{5.145}$$

  which for $G = SL_n(\mathbb{R})$ of course simplifies to $d\mu(X) = \prod_{i,j} dx_{ij}$.

Definition 5.38. *A unitary representation u of a group G on a Hilbert space H is* **irreducible** *if the only closed subspaces K of H that are stable under $u(G)$ (in the sense that if $\psi \in K$, then $u(x)\psi \in K$ for all $x \in G$) are either $K = H$ or $K = \{0\}$.*

We will often need two important results about irreducibility. The first is ***Schur's Lemma***, in which the commutant S' of some subset $S \subset B(H)$ is defined by

$$S' = \{a \in B(H) \mid ab = ba \forall b' \in S\}. \tag{5.146}$$

Lemma 5.39. *A unitary representation u of a group G is irreducible iff*

$$u(G)' = \mathbb{C} \cdot 1, \tag{5.147}$$

i.e., if $au(x) = u(x)a$ for each $x \in G$ implies $a = \lambda \cdot 1_H$ for some $\lambda \in \mathbb{C}$.

This follows from Theorem C.90, of which the above lemma is a special case: take $A = u(G)'' \equiv (u(G)')'$. The second is part of the ***Peter–Weyl Theorem***.

Theorem 5.40. *Irreducible representations of compact groups are finite-dimensional.*

Proof. We first reduce the situation to the unitary case: if $\langle \cdot,\cdot,\rangle'$ is the given inner product on H, we define a new inner product $\langle \cdot,\cdot,\rangle$ by averaging with respect to Haar measure $dx \equiv d\mu(x)$, i.e.,

$$\langle \psi, \varphi \rangle = \int_G dx\, \langle u(x)\psi, u(x)\varphi \rangle. \tag{5.148}$$

Using (5.141), it is easy to verify that this new inner product makes u unitary.

So let $u : G \to u(H)$ be an irreducible unitary representation. For each unit vector $\varphi \in H$ and $x \in G$, we define the following projection and its G-average:

$$e_{u(x)\varphi} = |u(x)\varphi\rangle\langle u(x)\varphi|, \tag{5.149}$$

$$W_\varphi = \int_G dx\, e_{u(x)\varphi}. \tag{5.150}$$

The ***Weyl operator*** (5.150) is initially defined as a quadratic form by

$$\langle \psi_1, W_\varphi \psi_2 \rangle = \int_G dx\, \langle \psi_1, e_{u(x)\varphi} \psi_2 \rangle. \tag{5.151}$$

The integral exists because the integrand is continuous and bounded, defining a *bounded* quadratic form by the estimate $|\langle \psi_1, W_\varphi \psi_2 \rangle| \leq \|\psi_1\| \|\psi_2\|$, where we assumed (5.140) and used $\|e_{u(x)\varphi}\| = 1$, as (5.149) is a nonzero projection. Thus the operator W_φ may be reconstructed from its matrix elements (5.151), cf. Proposition B.79. It is easy to verify that $[W_\varphi, u(y)] = 0$ for each $y \in G$, so that Schur's Lemma yields $W_\varphi = \lambda_\varphi \cdot 1_H$ for some $\lambda_\varphi \in \mathbb{C}$. Hence $\langle \psi, W_\varphi \psi \rangle = \lambda_\varphi \|\psi\|^2$, in other words,

$$\int_G dx\, |\langle \psi, u(x)\varphi \rangle|^2 = \lambda_\varphi \|\psi\|^2. \tag{5.152}$$

If we now interchange φ and ψ and use (5.143) we find $\lambda_\varphi \|\psi\|^2 = \lambda_\psi \|\varphi\|^2$, so that, taking ψ to be a unit vector, too, since ψ and φ are arbitrary we obtain $\lambda_\varphi = \lambda_\psi \equiv \lambda$, where in fact $\lambda > 0$, as follows by taking $\psi = \varphi$ in (5.152). Finally, take n orthornormal vectors $(\upsilon_1, \ldots, \upsilon_n)$ in H, so that also $(u(x)\upsilon_1, \ldots, u(x)\upsilon_n)$ are orthonormal (since $u(x)$ is unitary), upon which Bessel's inequality (B.212) gives

$$\sum_{i=1}^n |\langle \psi, u(x)\upsilon_i \rangle|^2 \leq \|\psi\|^2. \tag{5.153}$$

Integrating both sides over G, taking $\|\psi\| = 1$, and using (5.140) gives

$$\sum_{i=1}^n \int_G dx\, |\langle \psi, u(x)\upsilon_i \rangle|^2 \leq 1. \tag{5.154}$$

On the other hand, summing (5.152) over i simply yields $n\lambda$, whence $n\lambda \leq 1$, for any $n \leq \dim(H)$. Since $\lambda > 0$ this forces $\dim(H) < \infty$. □

5.7 Representations of Lie groups and Lie algebras

We now assume that G is a ***Lie group***; as in §3.3, for our purposes we may restrict ourselves to *linear* Lie groups, i.e. closed subgroups of $GL_n(\mathbb{K})$ for $\mathbb{K} = \mathbb{R}$ or $\mathbb{C}$.

Let $u : G \to U(H)$ be a unitary representation of a Lie group G on some Hilbert space H (assumed strongly continuous). If H is finite-dimensional, the following operation is unproblematic: for $A \in \mathfrak{g}$ (i.e. the Lie algebra of G) we define an operator

$$u'(A) \,:\, H \to H; \tag{5.155}$$
$$u'(A) = \frac{d}{dt} u\left(e^{tA}\right)_{|t=0}. \tag{5.156}$$

This gives a linear map $u' : \mathfrak{g} \to B(H)$, which satisfies

$$[u'(A), u'(B)] = u'([A,B]); \tag{5.157}$$
$$u'(A)^* = -u'(A). \tag{5.158}$$

Note that physicists use Planck's constant $\hbar > 0$ and like to write

$$\pi(A) = i\hbar u'(A), \tag{5.159}$$

so that one has the following commutation relations and self-adjointness condition:

$$[\pi(A), \pi(B)] = i\hbar\pi([A,B]); \tag{5.160}$$
$$\pi(A)^* = \pi(A). \tag{5.161}$$

If one knows that $u' : \mathfrak{g} \to B(H)$ comes from $u : G \to U(H)$, one conversely has

$$u(e^A) = e^{u'(A)} = e^{-\frac{i}{\hbar}\pi(A)}. \tag{5.162}$$

More generally, we call a map $\rho : \mathfrak{g} \to B(H)$ (where $H \cong \mathbb{C}^n$ remains finite-dimensional, so that $\rho : \mathfrak{g} \to M_n(\mathbb{C})$), a ***skew-adjoint*** representation of $\mathfrak{g}$ on H if

$$[\rho(A), \rho(B)] = \rho([A,B]); \tag{5.163}$$
$$\rho(A)^* = -\rho(A). \tag{5.164}$$

The property of irreducibility of such a representation $\rho : \mathfrak{g} \to B(H)$ is defined in the same way as for groups, namely that the only linear subspaces of $H \cong \mathbb{C}^n$ that are stable under $\rho(\mathfrak{g})$ are $\{0\}$ and H. Equivalently, by Schur's Lemma, $\rho(\mathfrak{g})$ is irreducible iff the only operators that commute with all $\pi(A)$ are multiples of the unit operator. If $\rho = u'$ for some unitary representation $u(G)$, it is easy to see that u is irreducible iff u' is irreducible. In view of this, it is a reasonable strategy to try and construct irreducible unitary representations $u(G)$ by starting, as it were, from $u'(\mathfrak{g})$. More precisely, if ρ is some (irreducible) skew-adjoint representation of $\mathfrak{g}$, we may ask if there is a (necessarily irreducible) unitary representation $u(G)$ such that $\rho = u'$. Writing $\exp(\rho)$ for u, one would therefore hope that

$$u\left(e^{A}\right) \equiv e^{\rho}\left(e^{A}\right) = e^{\rho(A)}, \tag{5.165}$$

as in (5.162). Note that if G is connected, then ρ duly defines $u(x)$ for each $x \in G$ through (5.165), since by Lie theory every element x of a connected Lie group is a finite product $x = \exp(A_1) \cdots \exp(A_n)$ of exponentials of elements $(A_1, \ldots, A_n)$ of $\mathfrak{g}$.

In general, this hope is in vain, since although each operator $\exp(A)$ is unitary, the representation property $u(x)u(y) = u(xy)$ may fail for global reasons. For example, if $G = SO(3)$, then $\mathfrak{g} \cong \mathbb{R}^3$, with basis (J_1, J_2, J_3), as in (3.66). Define an *a priori* linear map $\rho : \mathfrak{g} \to M_2(\mathbb{C})$ by linear extension of

$$\rho(J_k) = -\tfrac{1}{2} i \sigma_k, \tag{5.166}$$

where $(\sigma_1, \sigma_2, \sigma_3)$ are the Pauli matrices (5.42), so that physicists would write

$$\pi(J_k) = \tfrac{1}{2} \hbar \sigma_k, \tag{5.167}$$

cf. (5.159). This is easily checked to give a skew-adjoint representation of $\mathfrak{g}$, but it does not exponentiate to a unitary representation of $SO(3)$: as already mentioned after Proposition 5.46, if $\mathbf{u}$ is a unit vector in $\mathbb{R}^3$, then a rotation $R_\theta(\mathbf{u})$ around the $\mathbf{u}$-axis by an angle $\theta \in [0, 2\pi]$ is represented by

$$u(R_\theta(\mathbf{u})) = \cos(\theta/2) \cdot 1_2 + i \sin(\theta/2) \mathbf{u} \cdot \sigma. \tag{5.168}$$

Consequently, $u(R_\pi(\mathbf{u})) = i\mathbf{u} \cdot \sigma$, so that $u(R_\pi(\mathbf{u}))^2 = -1_2$, although within $SO(3)$ one has $R_\pi(\mathbf{u})^2 = e$, the unit of $SO(3)$, so that $u(R_\pi(\mathbf{u}))^2 \neq u(R_\pi(\mathbf{u})^2)$.

However, ρ does exponentiate to a representation of $SU(2)$, which happens to be the universal covering group of $SO(3)$. This is typical of the general situation, which we state without proofs. We first need a refinement of ***Lie's Third Theorem***:

Theorem 5.41. *Let G be a connected Lie group G with Lie algebra $\mathfrak{g}$. There exists a simply connected Lie group $\tilde{G}$, unique up to isomorphism, such that:*

- *The Lie algebra of $\tilde{G}$ is $\mathfrak{g}$.*
- *$G \cong \tilde{G}/D$, where D is a discrete normal subgroup of the center of $\tilde{G}$.*
- *$D \cong \pi_1(G)$, i.e., the fundamental group of G, which is therefore abelian.*

For example, for $G = SO(3)$ we have $\tilde{G} = SU(2)$ and $D = \mathbb{Z}_2$, cf. Proposition 5.46.

Theorem 5.42. *Let G_1 and G_2 be Lie groups, with Lie algebras $\mathfrak{g}_1$ and $\mathfrak{g}_2$, respectively, and suppose that G_1 is simply connected. Then every Lie algebra homomorphism $\varphi : \mathfrak{g}_1 \to \mathfrak{g}_2$ comes from a unique Lie group homomorphism $\Phi : G_1 \to G_2$ through $\varphi = \Phi'$, where (realizing G_1 and G_2 as matrices)*

$$\Phi'(X) = \frac{d}{dt} \Phi\left(e^{tX}\right)_{|t=0}. \tag{5.169}$$

Let H be a finite-dimensional Hilbert space, so that $B(H) \cong M_n(\mathbb{C})$, where $n = \dim(H)$, and take $U(H) \cong U_n(\mathbb{C})$ to be the group of all unitary matrices on $\mathbb{C}^n$. The Lie algebra $\mathfrak{u}_n(\mathbb{C})$ of $U_n(\mathbb{C})$ consists of all skew-adjoint $n \times n$ complex matrices. Since irreducibility is preserved under the correspondence $u(G) \leftrightarrow u'(\mathfrak{g})$, we infer:

Corollary 5.43. *Let G be a simply connected Lie group with Lie algebra $\mathfrak{g}$. Any finite-dimensional skew-adjoint representation $\pi : \mathfrak{g} \to \mathfrak{u}_n(\mathbb{C})$ of $\mathfrak{g}$ comes from a unique unitary representation $u(G)$ through* (5.156)*, in which case we have*

$$e^{u'(A)} = u\left(e^A\right) \quad (A \in \mathfrak{g}). \tag{5.170}$$

Thus there is a bijective correspondence between finite-dimensional unitary representations of G and finite-dimensional skew-adjoint representations of $\mathfrak{g}$. In particular, if G is compact, this specializes to a bijective correspondence between unitary irreducible representations of G and skew-adjoint irreducible representations of $\mathfrak{g}$.

If $G \cong \tilde{G}/D$ is connected but not simply connected, then a finite-dimensional skew-adjoint representation $\rho : \mathfrak{g} \to B(H)$ exponentiates to a unitary representation $u : G \to U(H)$ iff the representation $\exp(\rho) : \tilde{G} \to U(H)$ *is trivial on D.*

For example, $G = SO(3)$, the last condition is satisfied for the irreducible representations with integer spins $j \in \mathbb{N}$ (as well as for $j = 0$), see §5.8.

A similar construction is possible when H is infinite-dimensional, except for the fact that the derivative in (5.156) may not exist. For example, $G = \mathbb{R}$ has its canonical regular representation on $H = L^2(\mathbb{R})$, defined by $u(a)\psi(x) = \psi(x-a)$, in which case (5.159) gives some multiple of the momentum operator $-i\hbar d/dx$. This operator is unbounded and hence is not defined on all of H, see also §5.11 and §5.12. As in Stone's Theorem 5.73, this problem is solved by finding a suitable domain in H on which the underlying limit, taken strongly, does exist. This is the ***Gårding domain***

$$D_G = \left\{u^{\int}(f)\psi, f \in C_c^\infty(G), \psi \in H\right\}, \tag{5.171}$$

where for each $f \in C_c^\infty(G)$ (or even $f \in L^1(G)$) the operator $u^{\int}(f)$ is defined by

$$u^{\int}(f) = \int_G dx\, f(x) u(x). \tag{5.172}$$

Like the derivative u', this integral is most easily defined weakly, i.e., the (bounded) operator $u^{\int}(f)$ is initially defined as a bounded quadratic form

$$Q(\varphi,\psi) = \int_G dx\, f(x)\langle \varphi, u(x)\psi\rangle, \tag{5.173}$$

from which the operator $u^{\int}(f)$ may be reconstructed as in Proposition B.79. Note that the function $x \mapsto \langle \varphi, u(x)\psi\rangle$ is in $C_b(G)$, so that the integral (5.173) exists.

It can be shown that D_G is dense in H, as well as ***invariant*** under $u'(\mathfrak{g})$, in the sense that if $\psi \in D_G$, then $u'(A)\psi \in D_G$ for any $A \in \mathfrak{g}$. Furthermore, for each $\varphi \in D_G$ the function $x \mapsto u(x)\varphi$ from G to H is smooth (if G is unimodular this property even characterizes D_G). The commutation relations (5.157) then hold on D_G, but the equalities (5.164) do not: one has to choose between (5.157) and (5.164), since the latter holds for the closure of each $\pi(A)$ (i.e., each $i\rho(A)$ is essentially self-adjoint on D_G), whose domain however depends on A: there is no common domain on which each $i\rho(A)$ is self-adjoint *and* the commutation relations (5.157) hold.

5.8 Irreducible representations of $SU(2)$

One of the most important groups in quantum physics is $SU(2)$, both as an internal symmetry group—e.g. of the Heisenberg model of ferromagnetism, of the weak nuclear interaction, and possibly also of (loop) quantum gravity—and as a spatial symmetry group in disguise (all projective unitary representations of $SO(3)$ come from unitary representations of $SU(2)$, preserving irreducibility, cf. Corollary 5.61). In this section we review the well-known classification and construction of its unitary irreducible representations. Since $SU(2)$ is compact, by Theorem 5.40 all its unitary irreducible representations are finite-dimensional. Since $G = SU(2)$ is also simply connected, by Corollary 5.43 its irreducible finite-dimensional (unitary) representations u bijectively correspond to the irreducible finite-dimensional skew-adjoint representations $\rho = u'$ of its Lie algebra $\mathfrak{g}$. Hence our job is to find the latter.

We already encountered the basis (3.66) of the Lie algebra $\mathfrak{so}(3) \cong \mathbb{R}^3$ of $SO(3)$; the corresponding basis of the Lie algebra $\mathfrak{su}(2)$ of $SU(2)$ is (S_1, S_2, S_3), where

$$S_k = -\tfrac{1}{2} i \sigma_k, \tag{5.174}$$

and the σ_k are the Pauli matrices given in (5.42); linear extension of the map $J_k \mapsto S_k$ defines an isomorphism between $\mathfrak{so}(3)$ and $\mathfrak{su}(2)$. These matrices satisfy

$$[S_i, S_j] = \varepsilon_{ijk} S_k, \tag{5.175}$$

where ε_{ijk} is the totally anti-symmetric symbol with $\varepsilon_{123} = 1$ etc., so that (5.175) comes down to $[S_1, S_2] = S_3$, $[S_3, S_1] = S_2$, and $[S_2, S_3] = S_1$. By linearity, finding ρ is the same as finding $n \times n$ matrices

$$L_k = i\rho(S_k) \tag{5.176}$$

that satisfy

$$[L_i, L_j] = i\varepsilon_{ijk} L_k, \tag{5.177}$$

i.e., $[L_1, L_2] = iL_3$, etc., and

$$L_k^* = L_k. \tag{5.178}$$

It turns out to be convenient to introduce the ***ladder operators***

$$L_\pm = L_1 \pm iL_2, \tag{5.179}$$

with ensuing commutation relations

$$[L_3, L_\pm] = \pm L_\pm; \tag{5.180}$$

$$[L_+, L_-] = 2L_3. \tag{5.181}$$

Furthermore, we define the ***Casimir operator***

$$C = L_1^2 + L_2^2 + L_3^2, \tag{5.182}$$

which, crucially, commutes with each L_k, i.e.,

$$[C, L_k] = 0 \ (k = 1, 2, 3). \tag{5.183}$$

By Schur's lemma, in any irreducible representation we therefore must have

$$C = c \cdot 1_H, \tag{5.184}$$

where $c \in \mathbb{R}$ (in fact, $c \geq 0$). We will also use the additional algebraic relations

$$L_+L_- = C - L_3(L_3 - 1_H); \tag{5.185}$$

$$L_-L_+ = C - L_3(L_3 + 1_H). \tag{5.186}$$

The simple idea is now to diagonalize L_3, which is possible as $L_3^* = L_3$. Hence

$$H = \bigoplus_{\lambda \in \sigma(L_3)} H_\lambda, \tag{5.187}$$

where $\sigma(L_3)$ is the spectrum of L_3 (which in this finite-dimensional case consists of its eigenvalues), and H_λ is the eigenspace of L_3 for eigenvalue λ (i.e., if $\upsilon \in H_\lambda$, then $L_3\upsilon = \lambda\upsilon$). The structure of (5.187) in irreducible representations is as follows.

Lemma 5.44. *Let $\rho : \mathfrak{su}(2) \to B(H)$ be a finite-dimensional skew-adjoint irreducible representation, so that* (5.177) *holds. Then the spectrum $\sigma(L_3)$ of the self-adjoint operator $L_3 = i\rho(S_3)$ is given by*

$$\sigma(L_3) = \{-j, -j+1, \cdots, j-1, j\}. \tag{5.188}$$

If (5.187) *is the spectral decomposition of H relative to L_3, then:*

1. *The subspace H_λ is one-dimensional for each $\lambda \in \sigma(L_3)$;*
2. *For $\lambda < j$ the operator L_+ maps H_λ to $H_{\lambda+1}$, whereas $L_+ = 0$ on H_j;*
3. *For $\lambda > -j$ the operator L_- maps H_λ to $H_{\lambda-1}$, whereas $L_- = 0$ on H_{-j}.*

Proof. For any $\lambda \in \sigma(L_3)$ and nonzero $\upsilon_\lambda \in H_\lambda$, we have:

- either $\lambda + 1 \in \sigma(L_3)$ and $L_+\upsilon_\lambda \in H_{\lambda+1}$ (as a nonzero vector);
- or $L_+\upsilon_\lambda = 0$.

Indeed, (5.180) gives $L_3(L_+\upsilon_\lambda) = (\lambda + 1)L_+\upsilon_\lambda$, which immediately yields the claim. Similarly, either $\lambda - 1 \in \sigma(L_3)$ and $L_-\upsilon_\lambda \in H_{\lambda-1}$, or $L_-\upsilon_\lambda = 0$. Now let $\lambda_0 = \min \sigma(L_3)$ be the smallest eigenvalue of L_3, and pick some $0 \neq \upsilon_{\lambda_0} \in H_{\lambda_0}$. Since H is finite-dimensional by assumption, there must be some $k \in \mathbb{N}_0 = \mathbb{N} \cup \{0\}$ such that $L_+^{k+1}\upsilon_{\lambda_0} = 0$, whereas all vectors $L_+^l\upsilon_{\lambda_0}$ for $l = 0, \ldots, k$ are nonzero (and lie in H_{λ_0+l}). With c defined as in (5.184), it then follows from (5.185) - (5.186) that

$$c - \lambda_0(\lambda_0 - 1) = 0; \tag{5.189}$$

$$c - (\lambda_0 + k)(\lambda_0 + k + 1) = 0. \tag{5.190}$$

These relations imply $\lambda_0 = -k/2$, so that by the above bullet points we also have

$$\{-k/2, -k/2+1, \ldots, k/2-1, k/2\} \subseteq \sigma(L_3). \tag{5.191}$$

To prove equality, as in (5.188), consider the vector space

$$H' = \mathbb{C} \cdot \upsilon_{\lambda_0} \oplus \mathbb{C} \cdot L_+ \upsilon_{\lambda_0} \oplus \cdots \oplus L_+^{k-1} \upsilon_{\lambda_0} \oplus L_+^k \upsilon_{\lambda_0} \subseteq H; \tag{5.192}$$

this is just the subspace of H with basis $(\upsilon_{\lambda_0}, L_+\upsilon_{\lambda_0}, \ldots, L_+^{k-1}\upsilon_{\lambda_0}, L_+^k \upsilon_{\lambda_0})$. By the previous arguments following from (5.180), we see that the operators L_+ and L_- never leave H', and the same is trivially true for L_3. Therefore, if ρ is irreducible, then we must have $H' = H$ (and conversely). All claims of the lemma are now trivially verified on H'. □

It should be clear from this proof that the actions of L_+, L_-, and L_3 (and hence of all elements of $\mathfrak{su}(2)$) on $H' = H$) are fixed, so that ρ is determined by its dimension

$$\dim(H) = 2j + 1, \tag{5.193}$$

from which it follows that j can only take the values $0, 1/2, 1, 3/2, \ldots$.

It remains to fix an inner product on H' in which ρ is skew-adjoint, i.e., in which $L_3^* = L_3$ and $L_+^* = L_-$ (which implies that $L_1^* = L_1$ and $L_2^* = L_2$, which jointly imply $\rho(X^*) = -\rho(X)$ for any $X \in \mathfrak{g}$). This may be done in principle by starting with any inner product, integrating ρ to a unitary representation of $SU(2)$, and using the construction explained at the beginning of the proof of Theorem 5.40. In practice, it is easier to just calculate: take $H = \mathbb{C}^n$ with $n = 2j+1$, standard inner product, and standard orthonormal basis (u_l), labeled as $l = 0, 1, \ldots, 2j$). Then put

$$L_3 u_l = (l - j)u_l; \tag{5.194}$$

$$L_+ u_l = \sqrt{(l+1)(n-l-1)}\, u_{l+1}; \tag{5.195}$$

$$L_- u_l = \sqrt{l(n-l)}\, u_{l-1}. \tag{5.196}$$

Note that (5.195) is even formally correct for $l = 2j$, since in that case $n - 2j - 1 = 0$, and similarly, (5.196) formally holds even for $l = 0$. The commutation relations (5.180) - (5.181) as well as the above conditions for skew-adjointness may be explicitly verified, from which it follows that for any prescribed dimension (5.193) we have found a skew-adjoint realization of ρ. Clearly, $u_l = \upsilon_{l-j}$.

In view of Theorem 5.40 and Corollary 5.43 we have therefore proved:

Theorem 5.45. *Up to unitary equivalence, any (unitary) irreducible representation of $SU(2)$ is completely determined by its dimension $n = \dim(H)$, and any dimension $n \in \mathbb{N}_0 = \mathbb{N} \cup \{0\}$ occurs. Furthermore, if j is the number in (5.188), we have*

$$n = 2j + 1. \tag{5.197}$$

Physicists typically label these irreducible representations by j (called the ***spin*** of the given representation) rather than by n, or even by $c = j(j+1)$, cf. (5.184).

Corollary 5.43 shows that one may pass from $\rho(\mathfrak{su}(2))$ to a unitary representation $u(SU(2))$, of which one may give a direct realization. For $j \in \mathbb{N}_0/2$, define H_j as the complex vector space of all homogeneous polynomials p in two variables $z = (z_1, z_2)$ of degree $2j$. A basis of H_j is given by $(z_1^{2j}, z_1^{2j-1}z_2, \ldots, z_1 z_2^{2j-1}, z_2^{2j})$, which has $2j+1$ elements. So $\dim(H_j) = 2j+1$. Then consider the map

$$D_j : SU(2) \to B(H_j); \tag{5.198}$$

$$D_j(u)f(z) = f(zu). \tag{5.199}$$

Clearly,

$$D_j(e)f(z) = f(z \cdot 1_2) = f(z)), \tag{5.200}$$

so $D_j(e) = 1$, and

$$D_j(u)D_j(v)f(z) = D_j(v)f(zu) = f(zuv) = D_j(uv)f(z),$$

so $D_j(u)D_j(v) = D_j(uv)$. Hence D_j is a representation of $SU(2)$.

We now compute $L_3 = -\frac{1}{2}iS_3$ on this space. From (5.156) with $u \rightsquigarrow D_j$, we have

$$L_3 = -\tfrac{1}{2}iD'_j\begin{pmatrix} i & 0 \\ 0 & -i \end{pmatrix} = -\tfrac{1}{2}i\frac{d}{dt}D_j\begin{pmatrix} e^{it} & 0 \\ 0 & e^{-it} \end{pmatrix}_{t=0}, \tag{5.201}$$

so that

$$L_3 f(z) = -\tfrac{1}{2}i\frac{d}{dt}f(e^{it}z_1, e^{-it}z_2)_{t=0} = \tfrac{1}{2}\left(z_1\frac{\partial f(z)}{\partial z_1} - z_2\frac{\partial f(z)}{\partial z_2}\right). \tag{5.202}$$

Similarly, we obtain

$$L_+ f(z) = z_1\frac{\partial f(z)}{\partial z_2}; \tag{5.203}$$

$$L_- f(z) = z_2\frac{\partial f(z)}{\partial z_1}. \tag{5.204}$$

Hence $f_{2j}(z) = z_1^{2j}$ gives $L_3 f_{2j} = jf_{2j}$, and $f_0(z) = z_2^{2j}$ gives $L_3 f_0 = -jf_0$. In general, $f_l(z) = z_1^l z_2^{2j-l}$ spans the eigenspace H_λ of L_3 with eigenvalue $\lambda = -j+l$. Since $l = 0, 1, \ldots, 2j$, this confirms (5.188), as well as the fact that the corresponding eigenspaces are all one-dimensional. The rest is easily checked, too, except for the unitarity of the representation, for which we refer to the proof of Theorem 5.40.

Finally, we return to $SO(3)$. Either explicit exponentiation (5.165), as done for $j = 1/2$ in (5.168), or the above construction of D_j, allows one to verify the crucial condition stated in Corollary 5.43, namely that $D_j(\delta) = 1_{H_j}$ for $\delta \in D = \mathbb{Z}_2$, which comes down to $D_j(-1_2) = 1_{H_j}$. This is easily seen to be the case iff $j \in \mathbb{N}_0$.

Corollary 5.46. *Up to unitary equivalence, each unitary irreducible representation of $SO(3)$ is completely fixed by its dimension $n = 2j+1$, where $j \in \mathbb{N}_0$ (so that $n = 1$ for spin-0, $n = 3$ for spin-1, $n = 5$ for spin-2, ...), and each such dimension occurs.*

5.9 Irreducible representations of compact Lie groups

Because of its importance for the classical-quantum correspondence (cf. §7.1) we first reformulate the main result of the previous section (i.e. the classification the irreducible representations of $SU(2)$) and on that basis generalize this result to arbitrary compact Lie groups. This gives a classification of great simplicity and beauty.

We already encountered the coadjoint representation (3.100) of a Lie group G on $\mathfrak{g}^*$, given by $(x\cdot\theta)(A) = \theta(x^{-1}Ax)$, where $x \in G$, $\theta \in \mathfrak{g}^*$, $A \in \mathfrak{g}$. The orbits under this action are called ***coadjoint orbits***. If $G = SO(3)$, we have $\mathfrak{g} \cong \mathbb{R}^3$ under the map

$$\mathbf{x}\cdot\mathbf{J} \equiv \sum_{k=1}^{3} x_x J_i \mapsto (x_1, x_2, x_3) \equiv \mathbf{x}, \tag{5.205}$$

where the matrices J_k are given in (3.66). Hence also $\mathfrak{g}^* \cong \mathbb{R}^3$ under the map

$$\theta \mapsto \left((\theta_1, \theta_2, \theta_3) : \mathbf{x} \mapsto \sum_{k=1}^{3} \theta_k x_k \right). \tag{5.206}$$

Writing $R \in SO(3)$ for a generic element $x \in G$, analogously to (5.44), we can compute the adoint action $R : A \mapsto RAR^{-1}$, seen as an action on $\mathbb{R}^3$, through

$$R(\mathbf{x}\cdot\mathbf{J})R^{-1} = (R\mathbf{x})\cdot\mathbf{J}. \tag{5.207}$$

Using the fact that the angular momentum matrices transform as vectors, i.e.,

$$RJ_iR^{-1} = \sum_j R_{ji} J_j, \tag{5.208}$$

we find that the adjoint action of $SO(3)$ on $\mathfrak{g}$, seen as $\mathbb{R}^3$, is its defining action. In general, if $\mathfrak{g} \cong \mathbb{R}^n$ and also $\mathfrak{g}^* \cong \mathbb{R}^n$ under the usual pairing of $\mathbb{R}^n$ and $\mathbb{R}^n$ through the Euclidean inner product, the coadjoint action of G on $\mathfrak{g}^*$, seen as an action on $\mathbb{R}^n$, is given by the inverse transpose of the adjoint action on $\mathfrak{g} \cong \mathbb{R}^n$. For $SO(3)$ we have $(R^{-1})^T = R$, so the coadjoint action of $SO(3)$ on $\mathbb{R}^3$ is just its defining action, too, and hence the coadjoint orbits are the 2-spheres S_r with radius $r \geq 0$.

Turning to $SU(2)$, we now make the identification of $\mathfrak{g}^*$ with $\mathbb{R}^3$ slightly differently, namely by replacing the 3×3 real matrices J_i in (5.205) by the 2×2 matrices S_i in (5.174), but the computation is similar: using (5.44) - (5.45), we find that the coadjoint action of $u \in SU(2)$ on $\mathbb{R}^3$ is given by the defining action of $\tilde{\pi}(u) \in SO(3)$, cf. (5.46). It follows that the coadjoint orbits for $SU(2)$ are the same as for $SO(3)$.

Returning to general Lie groups G for the moment, assumed connected for simplicity, we take some coadjoint orbit $\mathscr{O} \subset \mathfrak{g}^*$, fix a point $\theta \in \mathscr{O}$ (so that $\mathscr{O} = G\cdot\theta \equiv G_\theta$), and look at the stabilizer G_θ and its Lie algebra $\mathfrak{g}_\theta$. Since the derivative Ad$'$ of the adjoint action Ad of G on $\mathfrak{g}$—defined as in (5.156)—is given by

$$\mathrm{Ad}'(A) : B \mapsto [A, B], \tag{5.209}$$

it follows that the "infinitesimal stabilizer" $\mathfrak{g}_\theta$ is given by

$$\mathfrak{g}_\theta = \{A \in \mathfrak{g} \mid \theta([A,B]) = 0 \forall B \in \mathfrak{g}\}. \tag{5.210}$$

Consequently, the restriction of $\theta : \mathfrak{g} \to \mathbb{R}$ to $\mathfrak{g}_\theta \subset \mathfrak{g}$ is a Lie algebra homomorphism (where $\mathbb{R}$ is obviously endowed with the zero Lie bracket). Consider a ***character*** $\chi : G_\theta \to \mathbb{T}$, which is the same thing as a one-dimensional unitary representation of G_θ. If we regard $\mathbb{T}$ as a closed subgroup of $GL_1(\mathbb{C})$, its Lie algebra $\mathfrak{t}$ is given by $i\mathbb{R} \subset M_1(\mathbb{C}) = \mathbb{C}$. It is conventional (at least among physicists) to take $-i$ as the basis element of $\mathfrak{t}$, so that $\mathfrak{t} \cong \mathbb{R}$ under $-it \leftrightarrow t$, so that the exponential map $\exp : \mathfrak{t} \to \mathbb{T}$ (which is the usual one), seen as a map from $\mathbb{R}$ to $\mathbb{T}$, is given by $t \mapsto \exp(-it)$. Defining the derivative $\chi' : \mathfrak{g}_\theta \to \mathbb{C}$ as in (5.156), it follows that actually $\chi' : \mathfrak{g}_\theta \to i\mathbb{R}$, so that $i\chi'$ maps $\mathfrak{g}_\theta$ to $\mathbb{R}$ and is a Lie algebra homomorphism.

Definition 5.47. *Let G be a connected Lie group. A coadjoint orbit $\mathcal{O} \subset \mathfrak{g}^*$ is called* **integral** *if for some (and hence all) $\theta \in \mathcal{O}$ one has $\theta_{|\mathfrak{g}_\theta} = i\chi'$ for some character $\chi : G_\theta \to \mathbb{T}$, i.e., if there is a character χ such that for each $A \in \mathfrak{g}_\theta$ one has*

$$\theta(A) = i\frac{d}{dt}\chi\left(e^{tA}\right)_{|t=0}. \tag{5.211}$$

In the simplest case where $G = \mathbb{T}$, the coadjoint action on $\mathfrak{t}^*$ is evidently trivial, so that $G_\theta = G = \mathbb{T}$ for any $\theta \in \mathfrak{t}^* \cong \mathbb{R}$. Furthermore, any character on $\mathbb{T}$ takes the form $\chi_n(z) = z^n$, where $n \in \mathbb{Z}$, cf. (C.351). As explained above, if $\mathfrak{t} \cong \mathbb{R}$ and hence also $\mathfrak{t}^* \cong \mathbb{R}$, the identification of $\lambda \in \mathfrak{t}^*$ with $\lambda \in \mathbb{R}$ is made by $\lambda(-i) \leftrightarrow \lambda$, where $-i \in \mathfrak{t}$. If $\chi = \chi_n$, the right-hand side of (5.211) evaluated at $A = -i$ equals n, so that (5.211) holds iff $\theta = n$ for some $n \in \mathbb{Z}$. Thus the integral coadjoint orbits in $\mathfrak{t}^*$ are the integers $\mathbb{Z} \subset \mathbb{R}$. Similarly, if $G = \mathbb{T}^d$, the characters are elements of $\mathbb{Z}^d$, as in

$$\chi_{(n_1,\ldots,n_d)}(z_1,\ldots,z_d) = z_1^{n_1}\cdots z_d^{n_d}, \tag{5.212}$$

and the integral coadjoint orbits in $\mathfrak{g}^* \cong \mathbb{R}^d$ are the points of the lattice $\mathbb{Z}^d \subset \mathbb{R}^d$.

For $G = SU(2)$ we take a coadjoint orbit $S_r^2 \subset \mathbb{R}^3$ and fix $\theta_r = (0,0,r)$. If $r = 0$, then $G_\theta = G$ and (5.211) holds for the trivial character $\chi \equiv 1$, so the orbit $\{(0,0,0)\}$ is integral. Let $r > 0$. Then $G_{\theta_r} \equiv G_r$ consist of the pre-image of $SO(2)$ in $SU(2)$ under the projection $\tilde{\pi}$ in (5.46), where $SO(2) \subset SO(3)$ is the group of rotations around the z-axis. This is the abelian group

$$T = \{\mathrm{diag}(z,\bar{z}) \mid z \in \mathbb{T}\}. \tag{5.213}$$

This group is isomorphic to $\mathbb{T}$ under $\mathrm{diag}(z,\bar{z}) \mapsto z$ and hence its characters are given by $\chi_n(\mathrm{diag}(z,\bar{z})) = z^n$, where $n \in \mathbb{Z}$. The identification $\mathfrak{g}^* \cong \mathbb{R}^3$ is made by identifying $\theta \in \mathfrak{g}^*$ with $(\theta_1,\theta_2,\theta_3)$, where $\theta_1 = \theta(S_i)$. Putting $A = S_3$ in (5.211), see (5.174), therefore gives $r = n/2$ for some $n \in \mathbb{N}$. We conclude that the coadjoint orbits for $SU(2)$ are given by the two-spheres $S_r^2 \subset \mathbb{R}^3$ with $r \in \mathbb{N}_0/2$.

Similarly, for $G = SO(3)$ the stabilizer of $(0,0,r)$ is $SO(2) \cong \mathbb{T}$ itself, and putting $A = J_3$ in (5.211) one finds that the coadjoint orbits are the spheres S_r^2 with $r \in \mathbb{N}_0$.

For any (Lie) group G, let the ***unitary dual*** $\hat{G}$ be the set whose elements are equivalence classes of unitary irreducible representations of G, where we say:

Definition 5.48. *Two unitary representations $u_i : G \to U(H_i)$, $i = 1,2$, are* **equivalent** *if there is unitary $v : H_1 \to H_2$ such that $u_2(x) = vu_1(x)v^*$ for each $x \in G$.*

The examples $G = \mathbb{T}^d$ as well as for $G = SU(2)$ now suggest the following theorem:

Theorem 5.49. *If G is a compact connected Lie group, then the unitary dual $\hat{G}$ is parametrized by the set of integral coadjoint orbits in $\mathfrak{g}^*$.*

Furthermore, there is an explicit (geometric) procedure to a construct an irreducible representation $u_{\mathcal{O}}$ corresponding to such an orbit, namely by the method of *geometric quantization*. We will not explain this method, which would require some reasonably advanced differential geometry, but instead we outline the connection between coadjoint orbits and the well-known ***method of the highest weight***.

Let G be a compact connected Lie group and pick a maximal torus $T \subset G$. Let

$$W_T = N(T)/T \tag{5.214}$$

be the corresponding ***Weyl group***, where $N(T)$ is the normalizer of T in G (i.e., $x \in N(T)$ iff $xzx^{-1} \in T$ for each $z \in T$). Note that all maximal tori in compact connected Lie groups are conjugate, so that the specific choice of T is irrelevant.

For example, for $SU(2)$ we take (5.213), in which case $N(T)$ is generated by T and $\sigma_1 \in SU(2)$, so that $W \cong \mathfrak{S}_2$, i.e., the permutation group on two variables. In general the Weyl group inherits the adjoint action of $N(T)$ on T, so that W_T acts on T and hence also acts on $\mathfrak{t}$ and $\mathfrak{t}^*$; for $SU(2)$ the action of the nontrivial element of W_T, i.e., image $[\sigma_1]$ of $\sigma_1 \in N(T)$ in $N(T)/T$), on T is given by

$$[\sigma_1](\mathrm{diag}(z,\overline{z})) = \mathrm{diag}(\overline{z},z), \tag{5.215}$$

so that its action on $\mathbb{T} \cong T$ is $z \mapsto \overline{z}$, which gives rise to actions $A \mapsto -A$ of W_T on $\mathfrak{t}$ and hence $\lambda \mapsto -\lambda$ of W_T on $\mathfrak{t}^*$. This is a special case of the following bijection:

$$\mathfrak{g}^*/G \cong \mathfrak{t}^*/W_T, \tag{5.216}$$

where the G-action on $\mathfrak{g}^*$ is the coadjoint one; globally, one has $G/\mathrm{Ad}(G) \cong T/W_T$.

Indeed, for $SU(2)$ the left-hand side of (5.216) is the set of spheres S_r^2 in $\mathbb{R}^3$, $r \geq 0$, whereas the right-hand side is $\mathbb{R}/\mathfrak{S}_2$ (where $\mathfrak{S}_2$ acts on $\mathbb{R}$ by $\theta \mapsto -\theta$).

In general, a given coadjoint orbit $\mathcal{O} \subset \mathfrak{g}^*$ defines a Weyl group orbit $\mathcal{O}_W$ in $\mathfrak{t}^*$ as follows: $\mathcal{O}$ contains a point θ for which $T \subseteq G_\theta$, and we take $\mathcal{O}_W$ to be the orbit through $\theta_{|\mathfrak{t}}$. Conversely, any G-invariant inner product on $\mathfrak{g}$ induces a decomposition

$$\mathfrak{g} = \mathfrak{t} \oplus \mathfrak{t}^{\perp}, \tag{5.217}$$

which yields an extension of $\lambda \in \mathfrak{t}^*$ to $\theta_\lambda \in \mathfrak{g}^*$ that vanishes on $t^{\perp}$. Let $\Lambda \subset \mathfrak{t}^*$ be the set of integral elements in $\mathfrak{t}^*$ (as explained after Definition 5.47). Elements of Λ are called ***weights***. Theorem 5.51 below gives a parametrization

$$\hat{G} \cong \Lambda/W_T, \tag{5.218}$$

which, restricting (5.216) to the integral part $\Lambda \subset \mathfrak{t}^*$, implies Theorem 5.49.

Instead of with the quotient Λ/W_T, one may prefer to work with Λ itself, as follows: we say that $\lambda \in \mathfrak{t}^*$ is ***regular*** if $w \cdot \lambda$ for $w \in W_T$ iff $w = e$; this is the case iff $\lambda = \theta_{|\mathfrak{t}}$ with $G_\theta = T$. For $SU(2)$ all weights $\lambda \in \mathbb{Z}$ are regular except $\lambda = 0$. The set $\mathfrak{t}^*_r$ of regular elements of $\mathfrak{t}^*$ falls apart into connected components C, called ***Weyl chambers***, which are mapped into each other by W_T. For $SU(2)$ one has $\mathfrak{t}^* = (-\infty,0) \cup (0,\infty)$, so that the Weyl chambers are $(-\infty,0)$ and $(0,\infty)$.

One picks an arbitrary Weyl chamber C_d (for $SU(2)$ this is $(0,\infty)$) and forms

$$\Lambda_d = \Lambda \cap C_d^-, \tag{5.219}$$

where C_d^- is the closure of C_d in $\mathfrak{t}^*$. Elements of Λ_d are called ***dominant weights***. For each element of Λ/W_T there is a unique dominant weight representing it in Λ, so that instead of (5.218) we may also write what Theorem 5.51 actually gives, viz.

$$\hat{G} \cong \Lambda_d. \tag{5.220}$$

To explain this in some detail, we need further preparation. Any (unitary) representation $u : G \to U(H)$ on some finite-dimensional Hilbert space H restricts to T, and since T is abelian, we may simultaneously diagonalize all operators $u(z)$, $z \in T$. The operators $iu'(A)$, where $A \in \mathfrak{t}$, commute as well, so that we may decompose

$$H = \bigoplus_{\mu \in \Lambda_H} H_\mu, \tag{5.221}$$

where $\Lambda_H \subset \Lambda$ contains the weights that occur in $u_{|T}$, so that for each $\psi \in H_\mu$,

$$u(z)\psi = \chi_\mu(z)\psi \ \ (z \in T); \tag{5.222}$$

$$iu'(Z)\psi = \mu(Z)\psi \ \ (Z \in \mathfrak{t}), \tag{5.223}$$

where the character $\chi_\mu : T \to \mathbb{T}$ corresponding to the weight $\mu \in \Lambda$ is defined as in (5.212) with $\mu = (n_1, \dots, n_d)$ and $z = (z_1, \dots, z_d) \in T \cong \mathbb{T}^d$, where $d = \dim(T)$. For example, we have seen that the irreducible representations $D_j(SU(2))$ on $H_j \cong \mathbb{C}^{2j+1}$ contains weights in $\Lambda_j = \{-j, -j+1, \dots, j-1, j\}$, where $j \in \mathbb{N}_0/2$.

In particular, take $H = \mathfrak{g}_\mathbb{C}$ with some G-invariant inner product, cf. (5.148), and take $u = \mathrm{Ad}$, given by $\mathrm{Ad}(x)B = xBx^{-1}$, so that $\mathrm{Ad}'(A)(B) = [A,B]$, extended from $\mathfrak{g}$ to $\mathfrak{g}_\mathbb{C}$: we write $\mathfrak{g}_\mathbb{C} = \mathfrak{g} + i\mathfrak{g}$ and hence put $\mathrm{Ad}'(A)(B+iC) = [A,B] + i[A,C]$, where $A,B,C \in \mathfrak{g}$. We assume that the inner product $\langle \cdot,\cdot \rangle$ on $\mathfrak{g}_\mathbb{C}$ is obtained from a real inner product on $\mathfrak{g}$ by complexification. This inner product on $\mathfrak{g}$ may be restricted to $\mathfrak{t} \subset \mathfrak{g}$ and hence induces an inner product on $\mathfrak{t}^*$, also denoted by $\langle \cdot,\cdot \rangle$. For example, if G is semi-simple (like $SU(2)$), one may take the inner product on $\mathfrak{g}$ and hence on $\mathfrak{g}_\mathbb{C}$ to be the Cartan–Killing form $\langle A,B \rangle = -\frac{1}{2}\mathrm{Tr}(\mathrm{Ad}'(A)\mathrm{Ad}'(B))$, which is nondegenerate because G is semi-simple, and positive definite since G is compact. For $SU(2)$ or $SO(3)$ this gives the usual inner product on $\mathbb{R}^3$ and $\mathbb{C}^3$.

Definition 5.50. *The* **roots** *of* $\mathfrak{g}$ *are the* nonzero *weights of the adjoint representation* $u = \mathrm{Ad}$ *on* $H = \mathfrak{g}_{\mathbb{C}}$. *That is, writing* $\Delta \subset \Lambda$ *for the set of roots, we have* $\alpha \in \Delta$ *iff* $\alpha : \mathfrak{t} \to \mathbb{R}$ *is not identically zero and there is some* $E_\alpha \in \mathfrak{g}_{\mathbb{C}}$ *such that for each* $Z \in \mathfrak{t}$,

$$i[Z, E_\alpha] = \alpha(Z)E_\alpha, \tag{5.224}$$

cf. (5.223). *Furthermore, subject to the choice of a preferred Weyl chamber* C_d *in* $\mathfrak{t}_r^*$, *we say* $\alpha \in \Delta$ *is* **positive**, *denoted by* $\alpha \in \Delta^+$, *if* $\langle \alpha, \lambda \rangle > 0$ *for each* $\lambda \in C_d$.

Since $\langle \alpha, \lambda \rangle$ is real and nonzero for each $\alpha \in \Delta$ and $\lambda \in C_d$, one has either $\alpha \in \Delta^+$ or $-\alpha \in \Delta^+$, i.e., $\alpha \in \Delta^- = -\Delta^+$. Since $\mathfrak{t}$ is maximal abelian in $\mathfrak{g}$, it can also be shown that each root is nondegenerate. Writing $\mathfrak{g}_\alpha = \mathbb{C} \cdot E_\alpha$, this gives a decomposition

$$\mathfrak{g}_{\mathbb{C}} = \mathfrak{t}_{\mathbb{C}} \bigoplus_{\alpha \in \Delta^+} \mathfrak{g}_\alpha \bigoplus_{\alpha \in \Delta^-} \mathfrak{g}_\alpha. \tag{5.225}$$

For $G = SU(2)$, the single generator of $\mathfrak{t}$ is S_3, and taking $E_\pm = i(S_1 \pm iS_2)$, we see from (5.180) that $i[S_3, E_\pm] = \pm E_\pm$. Hence the roots are $\alpha_\pm$, given by $\alpha_\pm(S_3) = \pm 1$, and with $(0, \infty)$ as the Weyl chamber of choice, the root α_+ is the positive one.

We now define a partial ordering $\leq$ on Λ by putting $\mu \leq \lambda$ iff $\lambda - \mu = \sum_i n_i \alpha_i$ for some $n_i \in \mathbb{N}_0$ and $\alpha_i \in \Delta^+$. This brings us to the ***theorem of the highest weight***:

Theorem 5.51. *Let G be a connected compact Lie group. There is a parametrization* $\hat{G} \cong \Lambda_d$, *such that any unitary irreducible representation* $u_\lambda : G \to H_\lambda$ *in the class* $\lambda \in \hat{G}$ *defined by a given dominant weight* $\lambda \in \Lambda_d$ *has the following properties:*

1. H_λ *contains a unit vector* υ_λ, *unique up to a phase, such that*

$$iu'_\lambda(Z)\upsilon_\lambda = \lambda(Z)\upsilon_\lambda \ (Z \in \mathfrak{t}); \tag{5.226}$$
$$iu'_\lambda(E_\alpha)\upsilon_\lambda = 0 \ (\alpha \in \Delta^+). \tag{5.227}$$

2. Any other weight μ *occurring in H, cf.* (5.221), *satisfies* $\mu \leq \lambda$ *and* $\mu \neq \lambda$.

The crucial point is that eqs. (5.226) - (5.227) imply

$$\theta_\lambda(A) = i\langle \upsilon_\lambda, u'_\lambda(A)\upsilon_\lambda \rangle \ (A \in \mathfrak{g}), \tag{5.228}$$

where $\theta_\lambda \in \mathfrak{g}^*$ was defined after (5.217) by $\lambda \in \Lambda_d \subset \mathfrak{t}^*$. Since each operator $u_\lambda(x)$ is unitary, each vector $u_\lambda(x)\upsilon_\lambda$ is a unit vector, so we may form the G-orbit

$$\mathscr{O}'_\lambda = \{|u_\lambda(x)\upsilon_\lambda\rangle\langle u_\lambda(x)\upsilon_\lambda|, x \in G\} \tag{5.229}$$

through $|\upsilon_\lambda\rangle\langle\upsilon_\lambda|$ in the space $\mathscr{P}_1(H_\lambda)$ of all one-dimensional projections on H_λ. Denoting the coadjoint orbit $G \cdot \theta_\lambda \subset \mathfrak{g}^*$ by $\mathscr{O}_\lambda$, where $\lambda = (\theta_\lambda)_{|\mathfrak{t}}$, the map

$$x \cdot \theta_\lambda \mapsto |u_\lambda(x)\upsilon_\lambda\rangle\langle u_\lambda(x)\upsilon_\lambda|, \tag{5.230}$$

is a G-equivariant diffeomorphism (in fact, a symplectomorphism) from $\mathscr{O}_\lambda$ to $\mathscr{O}'_\lambda$. This amplifies Theorem 5.49 by making the the bijective correspondence between the set Λ_d of dominant weights and the set of integral coadjoint orbits explicit.

5.10 Symmetry groups and projective representations

Despite the power and beauty of unitary group representations in *mathematics*, in the context of e.g. Wigner's Theorem we have seen that in *physics* one should look at homomorphisms $x \mapsto \mathsf{W}(x)$, where $\mathsf{W}(x)$ is a symmetry of $\mathscr{P}_1(H)$. In view of Theorems 5.4, this is equivalent to considering a *single* homomorphism $h : G \mapsto \mathscr{G}^H$, cf. (5.136). To simplify the discussion, we now drop $U_a(H)$ from consideration and just deal with the connected component $\mathscr{G}_0^H = U(H)/\mathbb{T}$ of the identity. This restriction may be justified by noting that in what follows we will only deal with symmetries given by *connected Lie groups*, which have the property that each element is a product of squares $x = y^2$. In that case, $h(x) = h(y)^2$ is always a square and hence it cannot lie in the component $U_a(H)/\mathbb{T}$ (the anti-unitary case does play a role as soon as *discrete* symmetries are studied, such as time inversion, parity, or charge conjugation). Thus in what follows we will study continuous homomorphisms

$$h : G \to U(H)/\mathbb{T}, \tag{5.231}$$

where $U(H)/\mathbb{T}$ has the quotient topology inherited from the strong operator topology on $U(H)$, as explained above. Since it is inconvenient to deal with such a quotient, we try to lift h to some map (5.137) where, in terms of the canonical projection

$$\pi : U(H) \to U(H)/\mathbb{T}, \tag{5.232}$$

which is evidently a group homomorphism, we have

$$\pi \circ u = h. \tag{5.233}$$

This can be done by choosing a cross-section s of π, that is, a measurable map

$$s : U(H)/\mathbb{T} \to U(H), \tag{5.234}$$

or (this doesn't matter much) a map $s : h(G)/\mathbb{T} \to U(H)$, such that

$$\pi \circ s = \text{id}. \tag{5.235}$$

Given h, such a cross-section s yields a map $u : G \to U(H)$ through

$$u = s \circ h; \tag{5.236}$$

in particular, $\pi(u(x)) = h(x)$. Such a lift often loses the homomorphism property, *though in a controlled way*, as follows. Since different choices of s must differ by a phase, and h is a homomorphism of groups, there must be a function

$$c : G \times G \to \mathbb{T} \tag{5.237}$$

such that

$$u(x)u(y) = c(x,y)u(xy) \ (x, y \in G). \tag{5.238}$$

Indeed, since π and h are homomorphisms, we may compute

$$\begin{aligned}\pi(u(x)u(y)u(xy)^{-1}) &= \pi(s(h(x))\pi(s(h(y))\pi(s(h(xy)))^{-1} \\ &= h(xy)h(xy)^{-1} = h(e_G) = e_{U(H)/\mathbb{T}}.\end{aligned}$$

Hence $u(x)u(y)u(xy)^{-1} \in \pi^{-1}(e_{U(H)/\mathbb{T}}) = \mathbb{T} \cdot 1_H$, which yields (5.238), or, more directly,

$$c(x,y) \cdot 1_H = u(x)u(y)u(xy)^*. \tag{5.239}$$

Associativity of multiplication in G and the homomorphism property of h yield

$$c(x,y)c(xy,z) = c(x,yz)c(y,z), \tag{5.240}$$

and if we impose the natural requirement $u_e = 1_H$, we also have

$$c(e,x) = c(x,e) = 1. \tag{5.241}$$

Definition 5.52. *A function* $c : G \times G \to \mathbb{T}$ *satisfying* (5.240) *and* (5.241) *is called a* **multiplier** *or* **C@2-cocycle** *on* G *(in the topological case one requires* c *to be Borel measurable, and for Lie groups it should in addition be smooth* near the identity*).* *The set of such multipliers, seen as an abelian group under (pointwise) operations in* $\mathbb{T}$*, is denoted by* $Z^2(G,\mathbb{T})$*. If* c *takes the form*

$$c(x,y) = \frac{b(xy)}{b(x)b(y)}, \tag{5.242}$$

where $b : G \to \mathbb{T}$ *satisfies* $b(e) = 1$ *(and is measurable and smooth near* e *as appropriate), then* c *is called a* **2-coboundary** *or an* **exact multiplier**. *The set of trivial multipliers forms a (normal) subgroup* $B^2(G,\mathbb{T})$ *of* $Z^2(G,\mathbb{T})$*, and the quotient*

$$H^2(G,\mathbb{T}) = \frac{Z^2(G,\mathbb{T})}{B^2(G,\mathbb{T})} \tag{5.243}$$

is called the **second cohomology group** *of* G *with coefficients in* $\mathbb{T}$.

The reason 2-coboundaries and the ensuing group $H^2(G,\mathbb{T})$ are interesting for our problem is as follows. Given a map $x \mapsto u(x)$ from G to $U(H)$ with (5.238), suppose we change $u(x)$ to $u(x)' = b(x)u(x)$. The associated multiplier then changes to

$$c'(x,y) = \frac{b(x)b(y)}{b(xy)}c(x,y), \tag{5.244}$$

in that $u(x)'u(y)' = c'(x,y)u'_{xy}$. In particular, a multiplier of the form (5.242) may be removed by such a transformation, and is accordingly called ***exact***.

Proposition 5.53. *If* $H^2(G,\mathbb{T})$ *is trivial, then any multiplier can be removed by modifying the lift* u *of* h*, and the ensuing map* $u' : G \to U(H)$ *is a homomorphism and hence a unitary representation of* G *on* H*. In that case, any homomorphism* $G \to U(H)/\mathbb{T}$ *comes from a unitary representation* $u : G \to U(H)$ *through* (5.233).

This is true by construction. By the same token, if $H^2(G,\mathbb{T})$ is non-trivial, then G will have projective representations that cannot be turned into ordinary ones by a change of phase (for it can be shown that any multiplier $c \in Z^2(G,\mathbb{T})$ is realized by some projective representation). Thus it is important to compute $H^2(G,\mathbb{T})$ for any given (physically relevant) group G, and see what can be done if it is non-trivial.

To this end we present the main results of practical use. In order to state one of the main results (Whitehead's Lemma), we need to set up a cohomology theory for $\mathfrak{g}$ (which we only need with trivial coefficients). Let $C^k(\mathfrak{g},\mathbb{R})$ be the abelian group of all k-linear totally antisymmetric maps $\varphi : \mathfrak{g}^k \to \mathbb{R}$, with ***coboundary maps***

$$\delta^{(k)} : C^k(\mathfrak{g},\mathbb{R}) \to C^{k+1}(\mathfrak{g},\mathbb{R}); \tag{5.245}$$

$$(X_0,X_1,\ldots,X_k) \mapsto \sum_{i<j=1}^{k+1} (-1)^{i+j}\varphi([X_i,X_j],X_0,\ldots,\hat{X}_i,\ldots,\hat{X}_j,\ldots,X_k), \tag{5.246}$$

where the hat means that the corresponding entry is omitted. For example, we have

$$\delta^{(1)}\varphi(X_0,X_1) = -\varphi([X_0,X_1]);$$
$$\delta^{(2)}\varphi(X_0,X_1,X_2) = -\varphi([X_0,X_1],X_2)+\varphi([X_0,X_2],X_1)-\varphi([X_1,X_2],X_0).$$

These maps satisfy "$\delta^2 = 0$", or, more precisely,

$$\delta^{(k+1)} \circ \delta^{(k)} = 0, \tag{5.247}$$

and hence we may define the following abelian groups:

$$B^k(\mathfrak{g},\mathbb{R}) = \mathrm{ran}(\delta^{(k-1)}); \tag{5.248}$$
$$Z^k(\mathfrak{g},\mathbb{R}) = \ker(\delta^{(k)}); \tag{5.249}$$
$$H^k(\mathfrak{g},\mathbb{R}) = \frac{Z^k(\mathfrak{g},\mathbb{R})}{B^k(\mathfrak{g},\mathbb{R})}. \tag{5.250}$$

Note that $B^k(\mathfrak{g},\mathbb{R}) \subseteq H^k(\mathfrak{g},\mathbb{R})$ because of (5.247). In particular, for $k=2$ the group $Z^2(\mathfrak{g},\mathbb{R})$ of all ***2-cocycles*** on $\mathfrak{g}$ consists of all bilinear maps $\varphi : \mathfrak{g}\times\mathfrak{g} \to \mathbb{R}$ that satisfy

$$\varphi(X,Y) = -\varphi(Y,X); \tag{5.251}$$
$$\varphi(X,[Y,Z]) + \varphi(Z,[X,Y]) + \varphi(Y,[Z,X]) = 0, \tag{5.252}$$

and its subgroup $B^2(\mathfrak{g},\mathbb{R})$ of all ***2-coboundaries*** comprises all φ taking the form

$$\varphi(X,Y) = \theta([X,Y]),\ \theta \in \mathfrak{g}^*. \tag{5.253}$$

For example, for $\mathfrak{g} = \mathbb{R}$ any antisymmetric bilinear map $\varphi : \mathbb{R}^2 \to 0$ is zero, so that

$$H^2(\mathbb{R},\mathbb{R}) = 0. \tag{5.254}$$

This has nothing to with the fact that the Lie bracket on $\mathfrak{g}$ vanishes. Indeed, $\mathfrak{g} = \mathbb{R}^2$ does admit a unique nontrivial 2-cocycle, given by (half) the symplectic form, i.e.,

$$\varphi_0((p,q),(p',q')) = \tfrac{1}{2}(pq' - qp'). \tag{5.255}$$

Since $B^2(\mathbb{R}^2,\mathbb{R}) = 0$, this cannot be removed, hence (5.255) generates $H^2(\mathbb{R}^2,\mathbb{R})$:

$$H^2(\mathbb{R}^2,\mathbb{R}) \cong \mathbb{R}. \tag{5.256}$$

As far as cohomology is concerned, each Lie group and each Lie algebra has its own story, although in some cases a group of stories may be collected into a single narrative. As a case in point, a Lie algebra $\mathfrak{g}$ is called ***simple*** when it has no proper ideals, and ***semi-simple*** when it has no commutative ideals. A Lie algebra is semi-simple iff it is a direct sum of simple Lie algebras. If a Lie group G is (semi-) simple, then so is its Lie algebra $\mathfrak{g}$. A basic result, often called ***Whitehead's Lemma***, is:

Lemma 5.54. *If $\mathfrak{g}$ is semi-simple, then $H^2(\mathfrak{g},\mathbb{R}) = 0$.*

Proof. The key point is that $C^k(\mathfrak{g},\mathbb{R})$ is a $\mathfrak{g}$-module under the action

$$(X_0 \cdot \varphi)(X_1,\ldots,X_k) = -\sum_{i=1}^{k} \varphi(X_1,\ldots,[X_0,X_i],\ldots,X_k). \tag{5.257}$$

For $k = 2$, a simple computation shows that

$$\begin{aligned}(X_0 \cdot \varphi)(X_1,X_2) &= -\varphi([X_0,X_1],X_2) - \varphi(X_1,[X_0,X_2]) \\ &= \delta^{(2)}\varphi(X_0,X_1,X_2) - \delta^{(1)}\varphi(X_0,-)(X_1,X_2),\end{aligned} \tag{5.258}$$

where at fixed X_0, the map $\varphi(X_0,-)$ is seen as an element of $C^1(\mathfrak{g},\mathbb{R})$. This show that $\mathfrak{g}$ maps both $B^2(\mathfrak{g},\mathbb{R})$ and $Z^2(\mathfrak{g},\mathbb{R})$ onto itself. Indeed, if $\varphi = \delta^{(1)}\chi$, then the first term in (5.258) vanishes because $\delta^{(2)} \circ \delta^{(1)} = 0$, cf. (5.247), so that the right-hand side of (5.258) takes the form $\delta^{(1)}(\cdots)$ and hence lies in $B^2(\mathfrak{g},\mathbb{R})$. Similarly, if $\delta^{(2)}\varphi = 0$, then $\delta^{(2)}(X_0 \cdot \varphi) = 0$. We now use the fact that if $\mathfrak{g}$ is semi-simple, then any finite-dimensional module is completely reducible. Consequently, as a $\mathfrak{g}$-module, $Z^2(\mathfrak{g},\mathbb{R})$ must decompose as $Z^2(\mathfrak{g},\mathbb{R}) = B^2(\mathfrak{g},\mathbb{R}) \oplus V$, where V is some $\mathfrak{g}$-module. Hence if $\varphi \in V$, then $X_0 \cdot \varphi \in V$. Since $\varphi \in Z^2(\mathfrak{g},\mathbb{R})$, the first term in (5.258) vanishes, whilst the second term lies in $B^2(\mathfrak{g},\mathbb{R})$. Since $V \cap B^2(\mathfrak{g},\mathbb{R}) = \{0\}$, we therefore have $X_0 \cdot \varphi = 0$, and hence $\delta^{(1)}\varphi(X_0,-)(X_1,X_2) = 0$, which gives $\varphi(X_0,[X_1,X_2]) = 0$, for all $X_0,X_1,X_2 \in \mathfrak{g}$. At this point we use another implication of the semi-simplicity of $\mathfrak{g}$, namely $[\mathfrak{g},\mathfrak{g}] = \mathfrak{g}$. It follows that $\varphi = 0$, whence $V = \{0\}$, from which $Z^2(\mathfrak{g},\mathbb{R}) = B^2(\mathfrak{g},\mathbb{R})$, or, in other words, $H^2(\mathfrak{g},\mathbb{R}) = 0$. □

Theorem 5.55. *Let G be a connected and simply connected Lie group. Then*

$$H^2(G,\mathbb{T}) \cong H^2(\mathfrak{g},\mathbb{R}). \tag{5.259}$$

Proof. This is really a conjunction of two isomorphisms:

$$H^2(G,\mathbb{T}) \cong H^2(G,\mathbb{R}); \tag{5.260}$$
$$H^2(G,\mathbb{R}) \cong H^2(\mathfrak{g},\mathbb{R}), \tag{5.261}$$

where $\mathbb{R}$ is the usual additive group, and $Z^2(G,\mathbb{R})$, $B^2(G,\mathbb{R})$, and hence $H^2(G,\mathbb{R})$ are defined analogously to $Z^2(G,\mathbb{T})$ etc. The first isomorphism is simply induced by

$$Z^2(G,\mathbb{R}) \mapsto Z^2(G,\mathbb{T}); \tag{5.262}$$
$$\Gamma(x,y) \mapsto e^{i\Gamma(x,y)} \equiv c(x,y), \tag{5.263}$$

which preserves exactness and induces an isomorphism in cohomology (but note that (5.262) - (5.263) may not itself define an isomorphism).

The isomorphism (5.261) is induced at the cochain level, too. Given a cocycle $\varphi \in Z^2(G,\mathbb{R})$, we construct a new Lie algebra $\mathfrak{g}_\varphi$ (called a ***central extension*** of $\mathfrak{g}$) by taking $\mathfrak{g}_\varphi = \mathfrak{g} \oplus \mathbb{R}$ as a vector space, equipped though with the unusual bracket

$$[(X,v),(Y,w)] = ([X,Y],\varphi(X,Y)); \tag{5.264}$$

the condition $\varphi \in Z^2(G,\mathbb{R})$ guarantees that this is a Lie bracket. Furthermore, $\mathfrak{g}_\varphi$ is isomorphic (as a Lie algebra) to a direct sum iff $\varphi \in B^2(\mathfrak{g},\mathbb{R})$; indeed, if (5.253) holds, then $(X,v) \mapsto (X, v+\theta(X))$ yields the desired isomorphism $\mathfrak{g}_\varphi \to \mathfrak{g} \oplus \mathbb{R}$.

By Lie's Third Theorem, there is a connected and simply connected Lie group G_φ (again called a ***central extension*** of G), with Lie algebra $\mathfrak{g}_\varphi$. As a manifold, $G_\varphi = G \times \mathbb{R}$, but the group laws are given, in terms of a function $\Gamma : G \times G \to \mathbb{R}$, by

$$(x,v)\cdot(y,w) = (xy, v+w+\Gamma(x,y)); \tag{5.265}$$
$$(x,v)^{-1} = (x^{-1}, -v-\Gamma(x,x^{-1})). \tag{5.266}$$

The group axioms then imply (indeed, they are equivalent to) the condition $\Gamma \in Z^2(G,\mathbb{R})$. Furthermore, two such extensions G_φ and G'_φ are isomorphic iff the corresponding cocycles Γ and Γ' are related by (5.244), and in particular, $\Gamma \in B^2(G,\mathbb{R})$ iff G_φ is isomorphic (as a Lie group) to a direct product $G \times \mathbb{R}$, which in turn is the case iff $\varphi \in B^2(\mathfrak{g},\mathbb{R})$. Conversely, given $\Gamma \in Z^2(G,\mathbb{R})$, we define the central extension G_φ by (5.265) - (5.266), to find that the associated Lie algebra $\mathfrak{g}_\varphi$ takes the above form, defining $\varphi \in B^2(\mathfrak{g},\mathbb{R})$ through (5.264). Explicitly,

$$\varphi(X,Y) = \frac{d}{ds}\frac{d}{dt}\left[\Gamma\left(e^{tX},e^{sY}\right)\right]_{|s=t=0} - (X \leftrightarrow Y). \tag{5.267}$$

Lie's Third Theorem thus implies that the map $\varphi \leftrightarrow \Gamma$ (which is not necessarily a bijection) descends to an isomorphism $H^2(\mathfrak{g},\mathbb{R}) \to H^2(G,\mathbb{R})$ in cohomology. □

Given (5.254), Theorem 5.55 immediately gives

$$H^2(\mathbb{R},\mathbb{T}) = 0. \tag{5.268}$$

In particular, if $\mathbb{R}$ is the relevant symmetry group, which is the case e.g. with time translation, by Proposition 5.53 we may restrict ourselves to unitary representations.

Once again, this has nothing to do with abelianness or topological triviality of $\mathbb{R}$. Indeed, for $G = \mathfrak{g} = \mathbb{R}^2$, the ***Heisenberg cocycle*** (5.255) comes from the multiplier

$$c_0((p,q),(p',q')) = e^{i(pq'-qp')/2}, \tag{5.269}$$

where $\mathbb{R}^2$ is seen as the group of translations in the *phase space* $\mathbb{R}^2$ of a particle moving on $\mathbb{R}$. Accordingly, this multiplier is realized by the following projective representation of $\mathbb{R}^2$ on $L^2(\mathbb{R})$:

$$u(p,q)\psi(x) = e^{-ipq/2}e^{ixp}\psi(x-q). \tag{5.270}$$

If $\mathbb{R}^2$ is the *configuration space* of some particle, and the group $\mathbb{R}^2$ produces translations in the latter (i.e., of *position*), then the appropriate unitary representation would rather be on $L^2(\mathbb{R}^2)$ and would have trivial multiplier, viz.

$$u(q_1,q_2)\psi(x_1,x_2) = \psi(x_1-q_1,x_2-q_2). \tag{5.271}$$

Similarly, $G = \mathbb{R}^2$, now seen as generating translations of *momentum* in the phase space $\mathbb{R}^4$ of the latter example would appropriately be represented on $L^2(\mathbb{R}^2)$ as

$$u(q_1,q_2)\psi(x_1,x_2) = e^{i(x_1q_1+x_2q_2)}\psi(x_1,x_2). \tag{5.272}$$

Corollary 5.56. *Let G be a connected and simply connected semi-simple Lie group. Then $H^2(G,\mathbb{T})$ is trivial.*

Here we say that a Lie group is ***simple*** when it has no proper *connected* normal subgroups, and ***semi-simple*** if it has no proper connected normal *abelian* subgroups. For example, the "classical Lie groups" of Weyl are semi-simple, including $SO(3)$ and $SU(2)$, which are even simple (note that the latter does have a *discrete* normal subgroup, namely its center $\{\pm 1_2\} \cong \mathbb{Z}_2$). Also, products of simple Lie groups are semi-simple. However, Corollary 5.56 does not apply to $SO(3)$, which is semi-simple but not simply connected. Here the relevant general result is:

Theorem 5.57. *Let G be a connected Lie group with $H^2(\mathfrak{g},\mathbb{R}) = 0$. Then*

$$H^2(G,\mathbb{T}) \cong \widehat{\pi_1(G)}. \tag{5.273}$$

We need some background (cf. §C.15). For any abelian (topological) group A, the set

$$\hat{A} = \mathrm{Hom}(A,\mathbb{T}) \tag{5.274}$$

consists of all (continuous) homomorphisms (also called ***characters***) $\chi : A \to \mathbb{T}$; these are just the irreducible (and hence necessarily one-dimensional) unitary representations of A. This set is a group under the obvious pointwise operations

$$\chi_1\chi_2(a) = \chi_1(a)\chi_2(a); \tag{5.275}$$
$$\chi^{-1}(a) = \chi(a)^{-1}. \tag{5.276}$$

As such, the group $\hat{A}$ is called the ***(Pontryagin) dual*** of A; the ***Pontryagin Duality Theorem*** states that $\hat{\hat{A}} \cong A$. Using Theorem 5.57 and Theorem 5.41, this gives

$$H^2(SO(3),\mathbb{T}) = \mathbb{Z}_2. \tag{5.277}$$

We now use Theorem 5.41 as a lemma to prove Theorem 5.57:

Proof. We first state the map $\widehat{\pi_1(G)} \to H^2(G,\mathbb{T})$ that will turn out to be an isomorphism. Assuming Theorem 5.41, pick a (Borel measurable) cross-section

$$\tilde{s} : G \to \tilde{G} \tag{5.278}$$

of the canonical projection

$$\tilde{\pi} : \tilde{G} \to G = \tilde{G}/D. \tag{5.279}$$

As always, this means that $\tilde{\pi} \circ \tilde{s} = \mathrm{id}_G$, and $\tilde{s}$ is supposed to be smooth near the identity, and chosen such that $\tilde{s}(e_G) = e_{\tilde{G}}$, where e_G and $e_{\tilde{G}}$ are the unit elements of G and $\tilde{G}$, respectively. Given a character $\chi \in \widehat{\pi_1(G)}$, define $c_\chi : G \times G \to \mathbb{T}$ by

$$c_\chi(x,y) = \chi(\tilde{s}(x)\tilde{s}(y)\tilde{s}(xy)^{-1}). \tag{5.280}$$

This makes sense: $\tilde{\pi}$ is a homomorphism, so that (cf. the computation below (5.238))

$$\tilde{\pi}(\tilde{s}(x)\tilde{s}(y)\tilde{s}(xy)^{-1}) = \tilde{\pi}(\tilde{s}(x))\tilde{\pi}(\tilde{s}(y))\tilde{\pi}(\tilde{s}(xy))^{-1} = xy(xy)^{-1} = e_G,$$

and hence $\tilde{s}(x)\tilde{s}(y)\tilde{s}(xy)^{-1}) \in \ker(\tilde{\pi}) = D$ (where we identify D with $\pi_1(G)$, cf. Theorem 5.41). Furthermore, tedious computations show that (5.240) and (5.241) hold, so that $c_\chi \in Z^2(G,\mathbb{T})$. Different choices of s lead to equivalent 2-cocycles c, and hence by taking the cohomology class $[c_\chi]$ of c_χ we obtain an injective map

$$\widehat{\pi_1(G)} \to H^2(G,\mathbb{T}); \tag{5.281}$$
$$\chi \mapsto [c_\chi]. \tag{5.282}$$

To prove surjectivity of this map, let $c \in Z^2(G,\mathbb{T})$ and define $\tilde{c} : \tilde{G} \times \tilde{G} \to \mathbb{T}$ by

$$\tilde{c}(\tilde{x},\tilde{y}) = c(\tilde{\pi}(x),\tilde{\pi}(y)). \tag{5.283}$$

Conversely, we may recover c from $\tilde{c}$ and some cross-section $\tilde{s} : G \to \tilde{G}$ of $\tilde{\pi}$ by

$$c(x,y) = \tilde{c}(\tilde{s}(x),\tilde{s}(y)). \tag{5.284}$$

It follows that $\tilde{c} \in Z^2(\tilde{G},\mathbb{T})$. Theorem 5.55 implies that $H^2(\tilde{G},\mathbb{T})$ is trivial, so that

$$\tilde{c}(\tilde{x},\tilde{y}) = \tilde{b}(\tilde{x}\tilde{y})/\tilde{b}(\tilde{x})\tilde{b}(\tilde{y}), \tag{5.285}$$

for some function $\tilde{b} : \tilde{G} \to \mathbb{T}$ satisfying $\tilde{b}(\tilde{e}) = 1$. From (5.241), i.e., $c(e,x) = 1$, we infer that if $\tilde{x} = \delta \in D$, so that $\tilde{\pi}(\delta) = e$, then $\tilde{c}(\delta,\tilde{y}) = 1$, and hence

$$\tilde{b}(\delta\tilde{y}) = \tilde{b}(\delta)\tilde{b}(\tilde{y}). \tag{5.286}$$

Taking $\tilde{x}$ and $\tilde{y}$ both in D, we see that $\tilde{b}_{|D}$ is a character, which we call χ. Hence

$$\begin{aligned} c(x,y) &= \frac{\tilde{b}(\tilde{s}(x)\tilde{s}(y))}{\tilde{b}(\tilde{s}(x))\tilde{b}(\tilde{s}(y))} = \frac{\tilde{b}(\tilde{s}(xy))}{\tilde{b}(\tilde{s}(x))\tilde{b}(\tilde{s}(y))} \cdot \frac{\tilde{b}(\tilde{s}(x)\tilde{s}(y))}{\tilde{b}(\tilde{s}(xy))} \\ &= \frac{\tilde{b}(\tilde{s}(xy))}{\tilde{b}(\tilde{s}(x))\tilde{b}(\tilde{s}(y))} \cdot c_\chi(x,y), \end{aligned} \tag{5.287}$$

since, using (5.286) with $\delta \rightsquigarrow \tilde{s}(x)\tilde{s}(y)\tilde{s}(xy)^{-1}$ and $\tilde{y} \rightsquigarrow \tilde{s}(xy)$, we have

$$\begin{aligned} \frac{\tilde{b}(\tilde{s}(x)\tilde{s}(y))}{\tilde{b}(\tilde{s}(xy))} &= \frac{\tilde{b}(\tilde{s}(x)\tilde{s}(y)\tilde{s}(xy)^{-1}\tilde{s}(xy))}{\tilde{b}(\tilde{s}(xy))} = \tilde{b}(\tilde{s}(x)\tilde{s}(y)\tilde{s}(xy)^{-1}) \\ &= \chi(\tilde{s}(x)\tilde{s}(y)\tilde{s}(xy)^{-1}) = c_\chi(x,y). \end{aligned}$$

Thus $[c] = [c_\chi]$, and hence the map (5.281) - (5.282) is surjective. □

Definition 5.58. *In the situation and notation of Theorem 5.41, a unitary representation $\tilde{u}: \tilde{G} \to U(H)$ is called* **admissible** *if $\tilde{u}(D) \subset \mathbb{T} \cdot 1_H$.*

In that case, there is obviously a character $\chi \in \hat{D}$ such that for each $\delta \in D$ we have

$$\tilde{u}(\delta) = \chi(\delta) \cdot 1_H. \tag{5.288}$$

Unitary irreducible representations are admissible, since Schur's Lemma implies that, since D lies in the center of $\tilde{G}$, its image $\tilde{u}(D)$ consists of multiples of the unit.

If $\tilde{u}$ is admissible, we obtain a homomorphism (5.231) by means of

$$h = \pi \circ \tilde{u} \circ \tilde{s}, \tag{5.289}$$

where $\tilde{s}$ is any cross-section of $\tilde{\pi}$, cf. (5.278) - (5.279). Note that different choices $\tilde{s}, \tilde{s}'$ are related by $\tilde{s}'(x) = \tilde{s}(x)\delta(x)$, where $\delta: G \to D$ is some function, so that

$$h'(x) = \pi(\tilde{u}(\tilde{s}'(x))) = \pi(\tilde{u}(\tilde{s}(x))\tilde{u}(\delta(x))) = \pi(\tilde{u}(\tilde{s}(x)))\pi(\delta(x) \cdot 1_H) = h(x).$$

Theorem 5.59. *1. If G is a connected Lie group with $H^2(\mathfrak{g}, \mathbb{R}) = 0$, any homomorphism $h: G \to U(H)/\mathbb{T}$ as in (5.231) comes from some admissible unitary representation $\tilde{u}$ of $\tilde{G}$ by (5.289). If H is separable, then h is continuous iff $\tilde{u}$ is.*
2. Moreover, if $\tilde{u}(\tilde{G})$ is **super-admissible** *in that $\tilde{u}(\delta) = 1_H$ for each $\delta \in D$, then $u = \tilde{u} \circ \tilde{s}$ is a unitary representation of G, in which case $h = \pi \circ u$ therefore comes from a unitary representation of G itself.*

Proof. Given such a homomorphism h, pick a cross-section $s: U(H)/\mathbb{T} \to U(H)$, as in (5.234), with associated 2-cocycle c on G given by (5.239). By Theorem 5.57 and its proof, we may assume (possibly after redefining s) that there exists a character $\chi \in \hat{D}$ and a cross-section (5.278) such that $c = c_\chi$, cf. (5.280). We then define

$$\tilde{u} : \tilde{G} \to B(H); \tag{5.290}$$
$$\tilde{x} \mapsto \chi(\tilde{x} \cdot (\tilde{s} \circ \tilde{\pi}(\tilde{x}))^{-1}) u(\tilde{\pi}(\tilde{x})). \tag{5.291}$$

Simple computations then show that $\tilde{x} \cdot (\tilde{s} \circ \tilde{\pi}(\tilde{x}))^{-1} \in D$ (i.e., the center of $\tilde{G}$), that (5.288) holds, that each operator $\tilde{u}(\tilde{x})$ is unitary, that the group homomorphism properties $\tilde{u}(\tilde{x})\tilde{u}(\tilde{y}) = \tilde{u}(\tilde{x}\tilde{y})$ and $\tilde{u}(\tilde{e}) = 1_H$ hold, and that (5.289) is valid. As to the last equation, since π removes the term with χ in (5.291), and $u = s \circ h$, we have

$$\pi \circ \tilde{u} \circ \tilde{s}(x) = \pi \circ s \circ h \circ \tilde{\pi} \circ \tilde{s}(x) = h(x),$$

since $\pi \circ s = \text{id}$ (on $U(H)/\mathbb{T}$) and $\tilde{\pi} \circ \tilde{s} = \text{id}$ (on G).

If $\tilde{u}(\delta) = 1_H$ for each $\delta \in D$, then $c_\chi = 1$ from (5.280), so that $u(x)u(y) = u_{xy}$ by (5.238). If s preserves units, or, equivalently, if $h_e = 1_H$, as we always assume, we see that u is a unitary representation of G. In this case, (5.291) simply reads $\tilde{u} = s \circ h \circ \tilde{\pi}$. This immediately yields $\tilde{u} = u \circ \tilde{\pi}$, which in turn gives $u = \tilde{u} \circ \tilde{s}$.

Finally, even if h is continuous, it is *a priori* unclear if $\tilde{u}$ is, since the cross-sections s and $\tilde{s}$ appearing in the above construction typically fail to be continuous. Fortunately, since they are assumed measurable, there is no question about measurability of $\tilde{u}$, and if H is separable, continuity follows from Proposition 5.36. □

Corollary 5.60. *If G is a connected Lie group with covering group $\tilde{G}$, the formulae*

$$\tilde{u} = u \circ \tilde{\pi}; \tag{5.292}$$
$$u = \tilde{u} \circ \tilde{s}, \tag{5.293}$$

where $\tilde{s} : G \to \tilde{G}$ is any cross-section of the covering map $\tilde{\pi} : \tilde{G} \to G$, give a bijective correspondence between (continuous) super-admissible unitary representations $\tilde{u}$ of $\tilde{G}$ and (continuous) unitary representations u of G, preserving irreducibility.

Corollary 5.61. *Any homomorphism $h : SO(3) \to U(H)/\mathbb{T}$ as in (5.231) comes from an admissible unitary representation $\tilde{u}$ of $SU(2)$ by (5.289). Moreover, h comes from a unitary representation $u = \tilde{u} \circ \tilde{s}$ of $SO(3)$ itself iff $\tilde{u}$ is trivial on the center $\mathbb{Z}_2$.*

In particular, if h is irreducible, it must come from the unitary irreducible representations $\tilde{u} = D_j$, where $j = 0, \frac{1}{2}, 1, \ldots$ is the (half-) integer **spin** *label. Then $D_j(SU(2))$ is super-admissible iff j is integral, in which case it defines a unitary irreducible representation of $SO(3)$.*

Indeed, the assumption $H^2(\mathfrak{g}, \mathbb{R}) = 0$ in Theorem 5.59 is satisfied for $SO(3)$ because of Whitehead's Lemma 5.54. The case where $H^2(\mathfrak{g}, \mathbb{R}) \neq 0$ occurs e.g. for the Galilei group (cf. §7.6). It can be shown that $H^2(\mathfrak{g}, \mathbb{R})$ has finitely many generators, for which one finds pre-images $(\varphi_1, \ldots, \varphi_M)$ in $Z^2(\mathfrak{g}, \mathbb{R})$, with corresponding elements $(\Gamma_1, \ldots, \Gamma_M)$ of $Z^2(\tilde{G}, \mathbb{R})$, cf. the proof of Theorem 5.55. Of these, a subset $(\Gamma_1, \ldots, \Gamma_N)$, $N \leq M$, satisfies the relation $\Gamma_i(\delta, \tilde{x}) = \Gamma_i(\tilde{x}, \delta)$ for any $\delta \in D$ (cf. Theorem 5.41) and $\tilde{x} \in \tilde{G}$. This yields a map $\Gamma : \tilde{G} \times \tilde{G} \to \mathbb{R}^N$ given by $\Gamma(\tilde{x}, \tilde{y}) = (\Gamma_1(\tilde{x}, \tilde{y}), \ldots, \Gamma_N(\tilde{x}, \tilde{y}))$, which in turn equips the set

$$\check{G} = \tilde{G} \times \mathbb{R}^N, \tag{5.294}$$

with a group multiplication $(\tilde{x}, v) \cdot (\tilde{y}, w) = (\tilde{x}\tilde{y}, v + w + \Gamma(\tilde{x}, \tilde{y}))$. We then have the following generalization of Theorem 5.59, in which a unitary representation u of $\check{G}$ is called ***admissible*** if $u(\delta, v) \in \mathbb{T} \cdot 1_H$ for any $\delta \in D$ and $v \in \mathbb{R}^N$.

Theorem 5.62. *Let G be a connected Lie group, and H a separable Hilbert space. Then any continuous homomorphism* $h : G \to U(H)/\mathbb{T}$ *comes from some admissible continuous unitary representation* $\tilde{u}$ *of* $\check{G}$.

As we only apply this to the Galilei group (where $N = 1$), basically only for illustrative purposes, we omit the proof. The correct (and natural) notion of equivalence of projective representations is as follows: we say that two such homomorphisms $h_i : G \to U(H_i)/\mathbb{T}$, $i = 1, 2$ are ***equivalent*** if there is a unitary $w : H_1 \to H_2$ such that

$$\mathrm{Ad}_w(h_1(x)) = h_2(x), \; x \in G, \tag{5.295}$$

where $\mathrm{Ad}_w : U(H_1)/\mathbb{T} \to U(H_2)/\mathbb{T}$ is the map $[u] \mapsto [vuv^*]$, which is well defined (here $[u]$ is the equivalence class of $u \in U(H)$ in $U(H)/\mathbb{T}$ under $u \sim zu$, $z \in \mathbb{T}$).

This induces the following notion for $\check{G}$: two admissible unitary representations $\tilde{u}_1, \tilde{u}_2$ of $\tilde{G}$ on Hilbert spaces H_1, H_2 are ***equivalent*** if there is a unitary $w : H_1 \to H_2$ *and* a map $b : \check{G} \to \mathbb{T}$ such that $wu_1(\check{x})w^* = b(\check{x})u_2(\check{x})$, for any $\check{x} \in \check{G}$. It can be shown that such a map b always comes from a character $\chi : \tilde{G} \to \mathbb{T}$ through $b(\tilde{x}, v) = \chi(\tilde{x})$.

To close this long and difficult section, in relief it should be mentioned that the above theory vastly simplifies if H is finite-dimensional. By Theorem 5.40, this is true, for example, if G is compact and u is irreducible. Suppose $u : G \to U(H)$ is merely a projective unitary representation of G, so that instead of (5.157) one has

$$[u'(X), u'(Y)] = u'([X,Y]) + i\varphi(X,Y) \cdot 1_H, \tag{5.296}$$

where φ is given by (5.267). Taking the trace yields

$$\varphi(X,Y) = \frac{i}{n} \mathrm{Tr}\,(u'([X,Y])), \tag{5.297}$$

where $n = \dim(H) < \infty$. We may define a linear function $\theta : \mathfrak{g} \to \mathbb{R}$ by

$$\theta(X) = \frac{i}{n} \mathrm{Tr}(u'(X)), \tag{5.298}$$

so that $\varphi(X,Y) = \theta([X,Y])$, cf. (5.253), and hence we may remove φ by redefining

$$\tilde{u}'(X) = u'(X) + i\theta(X) \cdot 1_H, \tag{5.299}$$

which satisfies (5.157) - (5.158). Hence by Corollary 5.43 the map $\tilde{u}'$ exponentiates to a unitary representation $\tilde{u}$ of the universal covering group $\tilde{G}$ of G; it should be checked from the values of $\tilde{u}$ on D if $\tilde{u}$ also defines a unitary representation of G. This argument shows that *finite-dimensional* projective unitary representations of Lie groups always come from unitary representations of the covering group.

5.11 Position, momentum, and free Hamiltonian

The three basic operators of non-relativistic quantum mechanics are position, denoted q, momentum, p, and the free Hamiltonian h_0. Assuming for simplicity that the particle moves in one dimension, these are informally given on $H = L^2(\mathbb{R})$ by

$$q\psi(x) = x\psi(x); \tag{5.300}$$

$$p\psi(x) = -i\hbar\frac{d}{dx}\psi(x); \tag{5.301}$$

$$h_0\psi(x) = -\frac{\hbar^2}{2m}\frac{d^2}{dx^2}\psi(x), \tag{5.302}$$

where m is the mass of the particle under consideration. We put $\hbar = 1$ and $m = 1/2$.

The issue is that these operators are unbounded; see §B.13. In general, quantum-mechanical observables are supposed to be represented by self-adjoint operators, and examples like (5.300) - (5.302) show that these may not be bounded. The Hellinger–Toeplitz Theorem B.68 then shows that it makes no sense to try and extend the above expressions to all of $L^2(\mathbb{R})$, so we have to live with the fact that some crucial operators $a : D(a) \to H$ are merely defined on a dense subspace $D(a) \subset H$.

Each such operator has an ***adjoint*** $a^* : D(a^*) \to H$, whose domain $D(a^*) \subset H$ consists of all $\psi \in H$ for which the functional $\varphi \mapsto \langle\psi, a\varphi\rangle$ is bounded on $D(a)$, and hence (since $D(a)$ is dense in H) can be extended to all of H by continuity through the unique "Riesz–Fréchet vector" χ for which $\langle\psi, a\varphi\rangle = \langle\chi, \varphi\rangle$. Writing $\chi = a^*\psi$, for each $\psi \in D(a^*)$ and $\varphi \in D(a)$ we therefore have

$$\langle a^*\psi, \varphi\rangle = \langle\psi, a\varphi\rangle. \tag{5.303}$$

Assuming that $D(a)$ is dense in H, we say that a is ***self-adjoint***, written $a^* = a$, if

$$\langle a\varphi, \psi\rangle = \langle\varphi, a\psi\rangle, \tag{5.304}$$

for each $\psi, \varphi \in D(a)$ *and* $D(a^*) = D(a)$. A self-adjoint operator a is automatically ***closed***, in that its graph $G(a) = \{(\psi, a\psi) \mid \psi \in D(a)\}$ is a closed subspace of the Hilbert space $H \oplus H$ (indeed, the adjoint of any densely defined operator is closed, see Proposition B.72). In practice, self-adjoint operators often arise as closures of ***essentially self-adjoint*** operators a, which by definition satisfy $a^{**} = a^*$. Equivalently, such an operator is ***closable***, in that the closure of its graph is the graph of some (uniquely defined) operator, called the ***closure*** a^- of a, and furthermore this closure is self-adjoint, so that $a^- = a^*$. If a is closable, the domain $D(a^-)$ of its closure consists of all $\psi \in H$ for which there exists a sequence (ψ_n) in $D(a)$ such that $\psi_n \to \psi$ *and* $a\psi_n$ converges, on which we define a^- by $a^-\psi = \lim_n a\psi_n$.

The simplest case is the position operator.

Theorem 5.63. *The operator q is self-adjoint on the domain*

$$D(q) = \{\psi \in L^2(\mathbb{R}) \mid \int_{\mathbb{R}} dx\, x^2 |\psi(x)|^2 < \infty\}. \tag{5.305}$$

See Proposition B.73 for the proof. To give a convenient domain of essential self-adjointness (also for the other two operators), we need a little distribution theory.

Definition 5.64. *The* **Schwartz space** $\mathscr{S}(\mathbb{R})$ *(whose elements are* **functions of rapid decrease***) consist of all smooth function $f : \mathbb{R} \to \mathbb{C}$ for which each expression*

$$\|f\|_{n,m} = \sup\{|x^n f^{(m)}(x)|, x \in \mathbb{R}\}, \tag{5.306}$$

where $f^{(m)}$ is the m'th derivative of f, is finite. The topology of $\mathscr{S}(\mathbb{R})$ is given by saying that a sequence (or net) f_λ converges to f iff $\|f_\lambda - f\|_{n,m} \to 0$ for all $n,m \in \mathbb{N}$.

Each $\|\cdot\|_{n,m}$ happens to be a norm, but positive definiteness is nowhere used in the theory below (which therefore works for families of ***seminorms***, which satisfy the axioms of a norm expect perhaps for positive definiteness). Since there are countably many such (semi)norms defining the topology, we may equivalently say that $\mathscr{S}(\mathbb{R})$ is a metric space defined by

$$d(f,g) = \sum_{n,m=0}^{\infty} 2^{-n} \frac{\|f-g\|_{n,m}}{1+\|f-g\|_{n,m}}. \tag{5.307}$$

Indeed, $\mathscr{S}(\mathbb{R})$ is complete in this metric. A typical element is $f(x) = \exp(-x^2)$.

Definition 5.65. *A* **tempered distribution** *is a continuous linear map $\varphi : \mathscr{S}(\mathbb{R}) \to \mathbb{C}$. The space of all such maps, equipped with the topology of pointwise convergence (i.e., $\varphi_\lambda \to \varphi$ iff $\varphi_\lambda(f) \to \varphi(f)$ for each $f \in \mathscr{S}(\mathbb{R})$) is denoted by $\mathscr{S}'(\mathbb{R})$.*

It can be shown that (because of the metrizability of $\mathscr{S}(\mathbb{R})$) continuity is the same as sequential continuity, i.e., some linear map $\varphi : \mathscr{S}(\mathbb{R}) \to \mathbb{C}$ belongs to $\mathscr{S}'(\mathbb{R})$ iff $\lim_N \varphi(f_N) = \varphi(f)$ for each convergent *sequence* $f_N \to f$ in $\mathscr{S}(\mathbb{R})$. Like $\mathscr{S}(\mathbb{R})$, the tempered distributions $\mathscr{S}'(\mathbb{R})$ form a (locally convex) ***topological vector space***, that is, a vector space with a topology in which addition and scalar multiplication are continuous. The topology of $\mathscr{S}'(\mathbb{R})$ is given by a family of seminorms, namely $\|\varphi\|_f = |\varphi(f)|$, $f \in \mathscr{S}(\mathbb{R})$, and hence a simple way to prove that $\varphi \in \mathscr{S}'(\mathbb{R})$ is to find some (n,m) for which $|\varphi(f))| \leq C\|f\|_{n,m}$ for each $f \in \mathscr{S}(\mathbb{R})$, since in that case $f_N \to f$, which means that $\|f_N - f\|_{n,m} \to 0$ for *all* $n,m \in \mathbb{N}$, certainly implies that $\varphi(f_N) \to \varphi(f)$, so that φ is continuous. For example, the evaluation maps δ_x defined by $\delta_x(f) = f(x)$ are continuous (take $n = m = 0$). Similarly, each finite measure on $\mathbb{R}$ defines a tempered distribution. Taking the $(0,m)$ seminorm shows that the maps $f \mapsto f^{(m)}(x)$ for fixed $m \in \mathbb{N}$ and $x \in \mathbb{R}$ are tempered distributions.

A less obvious example (defining a so-called ***Gelfand triple***) is as follows:

Proposition 5.66. *We have continuous dense inclusions*

$$\mathscr{S}(\mathbb{R}) \subset L^2(\mathbb{R}) \subset \mathscr{S}'(\mathbb{R}), \tag{5.308}$$

where the second inclusion identifies $\varphi \in L^2(\mathbb{R})$ with the map

$$f \mapsto \langle \overline{\varphi}, f\rangle = \int_{\mathbb{R}} dx\, \varphi(x) f(x). \tag{5.309}$$

Proof. As vector spaces, the first inclusion is obvious. For $f \in \mathscr{S}(\mathbb{R})$ we estimate

$$\|f\|_2^2 = \int_{\mathbb{R}} dx\, |f(x)| \cdot |f(x)| \leq \|f\|_1 \|f\|_\infty; \tag{5.310}$$

$$\begin{aligned} \|f\|_1 &= \int_{\mathbb{R}} dx \frac{(1+x^2)|f(x)|}{1+x^2} \leq \int_{\mathbb{R}} dy \frac{1}{1+y^2} \|(1+m_{x^2})f\|_\infty \\ &\leq \pi(\|f\|_{0,0} + \|f\|_{2,0}), \end{aligned} \tag{5.311}$$

so that, noting that $\|\cdot\|_{0,0} = \|\cdot\|_\infty$, we have

$$\|f\|_2^2 \leq \pi(\|f\|_\infty + \|f\|_{2,0})\|f\|_\infty. \tag{5.312}$$

Hence $f_\lambda \to f$ in $\mathscr{S}(\mathbb{R})$, which incorporates the conditions $\|f_\lambda - f\|_{0,0} \to 0$ and $\|f_\lambda - f\|_{2,0} \to 0$, implies $\|f_\lambda - f\|_2 \to 0$. This shows that the first inclusion in (5.308) is continuous. Density may be proved in two steps. First, take some fixed positive function $h \in C_c^\infty(-1,1)$ with the property $\int dx\, h(x) = 1$, and define $h_n(x) = nh(nx)$, so that informally $h_n \in C_c^\infty(\mathbb{R})$ converges to a δ-function as $n \to \infty$. For each $\psi \in L^2(\mathbb{R})$, we consider the convolution $h_n * \psi$, where for suitable f, g,

$$f * g(x) \equiv \int_{\mathbb{R}} dy\, f(x-y)g(y). \tag{5.313}$$

Then $h_n * \psi \in C^\infty(\mathbb{R}) \cap L^2(\mathbb{R})$ and, from elementary analysis, $\|h_n * \psi - \psi\| \to 0$.

Second, for $\psi \in C_c(\mathbb{R})$, the functions $h_n * \psi$ lie in $C_c^\infty(\mathbb{R})$ and hence in $\mathscr{S}(\mathbb{R})$. Since $C_c(\mathbb{R})$ is dense in $L^2(\mathbb{R})$ by Theorem B.30, for $\psi \in L^2(\mathbb{R})$ and $\varepsilon > 0$ we can find $\varphi \in C_c(\mathbb{R})$ such that $\|\psi - \varphi\| < \varepsilon/2$, and (as just shown) find n such that $\|\varphi - \varphi_n\| < \varepsilon/2$, whence $\|\psi - \varphi_n\| < \varepsilon$. This proves that $\mathscr{S}(\mathbb{R})$ is dense in $L^2(\mathbb{R})$.

The second inclusion is continuous by Cauchy–Schwarz, which gives

$$|\varphi(f)| \leq \|\varphi\|_2 \|f\|_2,$$

to be combined with (5.312). It should be noted that also the second inclusion in (5.308) is indeed an injection, i.e., that $\varphi(f) = 0$ for each $f \in \mathscr{S}(\mathbb{R})$ implies $\varphi = 0$ in $L^2(\mathbb{R})$; this is true because $\mathscr{S}(\mathbb{R})$ is dense in $L^2(\mathbb{R})$, plus the standard fact that, in any Hilbert space H, if $\langle \varphi, f \rangle = 0$ for all f in some dense subspace of H, then $\varphi = 0$. Finally, the fact that $L^2(\mathbb{R})$ is dense in the seemingly huge space $\mathscr{S}'(\mathbb{R})$ follows from the even more remarkable fact that $\mathscr{S}(\mathbb{R})$ is dense in $\mathscr{S}'(\mathbb{R})$. On top of the functions h_n just defined, also employ a function $\chi \in C_c^\infty(\mathbb{R})$ such that $\chi(x) = 1$ on $(-1,1)$, and define $\chi_n(x) = \chi(x/n)$, so that informally $\lim_{n\to\infty} \chi(x) = 1$ (as opposed to the h_n, which converge to a δ-function as $n \to \infty$). If for any $g \in \mathscr{S}(\mathbb{R})$ and any $\varphi \in \mathscr{S}'(\mathbb{R})$ we define $g\varphi$ as the distribution that maps $f \in \mathscr{S}(\mathbb{R})$ to $\varphi(fg)$, and similarly define $g * \varphi$ as the distribution that maps f to $\varphi(g * f)$, we may define a sequence of distributions $\varphi_n = h_n * (\chi_n \varphi)$. From the point of view of (5.308), these correspond to functions $\varphi_n \in \mathscr{S}(\mathbb{R})$ in the sense that $\varphi_n(f) = \int dx\, \varphi_n(x) f(x)$, where $f \in \mathscr{S}(\mathbb{R})$. Using similar analysis as above, it then follows that for any $f \in \mathscr{S}(\mathbb{R})$ we have $\varphi_n(f) \to \varphi(f)$, so that $\varphi_n \to \varphi$ in $\mathscr{S}'(\mathbb{R})$. □

For our purposes, the point of all this is that we can define generalized derivatives of (tempered) distributions, and hence, because of (5.308), of functions in $L^2(\mathbb{R})$.

Definition 5.67. *For $\varphi \in \mathscr{S}(\mathbb{R})'$ and $m \in \mathbb{N}$, the m'th* **generalized derivative** $\varphi^{(m)}$ *is defined by*

$$\varphi^{(m)}(f) = (-1)^m \varphi(f^{(m)}). \tag{5.314}$$

The idea is that under (5.308) this is an identity if $\varphi \in \mathscr{S}(\mathbb{R})$ (partial integration). Like the constructions at the end of the proof of Proposition 5.66, this is a special case of a more general construction: whenever we have a continuous linear map $T : \mathscr{S}(\mathbb{R}) \to \mathscr{S}(\mathbb{R})$, we obtain a dual continuous linear map $T' : \mathscr{S}(\mathbb{R})' \to \mathscr{S}(\mathbb{R})'$ defined by $T'\varphi = \varphi \circ T$, i.e.,

$$(T'\varphi)(f) = \varphi(T(f)). \tag{5.315}$$

Sometimes a slight change in the definition (as in (5.314), or as in the Fourier transform below) is appropriate so that the restriction of T' to $\mathscr{S}(\mathbb{R})$ coincides with T.

Theorem 5.68. *The momentum operator $p = -id/dx$ is self-adjoint on the domain*

$$D(p) = \{\psi \in L^2(\mathbb{R}) \mid \psi' \in L^2(\mathbb{R})\}, \tag{5.316}$$

where the derivative ψ' is taken in the distributional sense (i.e., letting $\psi \in \mathscr{S}'(\mathbb{R})$).

Proof. We first show that p is symmetric, or $p \subseteq p^*$. This comes down to

$$\langle \psi', \varphi \rangle = -\langle \psi, \varphi' \rangle, \tag{5.317}$$

for each $\psi, \varphi \in D(p)$, where both derivates are "generalized". The most elegant proof (though perhaps not the shortest) uses the Sobolev space $H^1(\mathbb{R})$, which equals $D(p)$ as a vector space, now equipped, however, with the new inner product

$$\langle \psi, \varphi \rangle_{(1)} = \langle \psi, \varphi \rangle + \langle \psi', \varphi' \rangle, \tag{5.318}$$

with both inner products on the right-hand side in $L^2(\mathbb{R})$; the associated norm is

$$\|\psi\|^2_{(1)} = \|\psi\|^2 + \|\psi'\|^2. \tag{5.319}$$

Similar to the Gelfand triple (5.308), we have dense continuous inclusions

$$\mathscr{S}(\mathbb{R}) \subset H^1(\mathbb{R}) \subset \mathscr{S}'(\mathbb{R}), \tag{5.320}$$

with analogous proof. All we need for Theorem 5.68 is the first inclusion of the triple (5.320): for $\psi \in H^1(\mathbb{R})$ we now have $h_n * \psi \in C^\infty(\mathbb{R}) \cap H^1(\mathbb{R})$ as well as $h_n * \psi \to \psi$ in $H^1(\mathbb{R})$, both of which follow from the L^2-case plus the identity

$$(h_n * \psi)' = h_n * \psi'. \tag{5.321}$$

Using the same cutoff function χ as in the L^2 case, we have $\chi_n \psi \to \psi$ and $\chi_n' \psi \to 0$ in $L^2(\mathbb{R})$, so that $(\chi_n \psi)' \to \psi'$ in $L^2(\mathbb{R})$ and hence $\chi_n \psi \to \psi$ also in $H^1(\mathbb{R})$.

Furthermore, the functions $\psi_n = h_n * (\chi_n \psi)$ lie in $C_c^\infty(\mathbb{R})$ and hence in $\mathscr{S}(\mathbb{R})$; using the above facts we obtain $\psi_n \to \psi$ in $H^1(\mathbb{R})$. In sum, for each $\psi \in H^1(\mathbb{R})$ we can find a sequence (ψ_n) in $\mathscr{S}(\mathbb{R})$ such that $\psi_n \to \psi$ *and* $\psi_n' \to \psi'$ in $L^2(\mathbb{R})$. Hence

$$\langle \psi, \varphi' \rangle = \lim_n \langle \psi_n, \varphi' \rangle = -\lim_n \langle \psi_n', \varphi \rangle = -\langle \psi', \varphi' \rangle. \tag{5.322}$$

For the converse, let $\psi \in D(p^*)$, so that by definition for each $\varphi \in D(p)$ we have

$$\langle p^* \psi, \varphi \rangle = \langle \psi, p\varphi \rangle = -i\langle \psi, \varphi' \rangle. \tag{5.323}$$

Since $\mathscr{S}(\mathbb{R}) \subset D(p)$, this is true in particular for each $\varphi \in \mathscr{S}(\mathbb{R})$, in which case the right-hand side equals $-i\psi'(\varphi)$, where the derivative is distributional. But this equals $\langle p^*\psi, \varphi \rangle$ and so the distribution $-i\psi'$ is given by taking the inner product with $p^*\psi \in L^2(\mathbb{R})$. Hence $-i\psi' = p^*\psi \in L^2(\mathbb{R})$, and in particular $\psi' \in L^2(\mathbb{R})$, so that $\psi \in D(p)$. This proves that $D(p^*) \subseteq D(p)$, and since from the first step we have the oppositie inclusion, we find $D(p^*) = D(p)$ and $p^* = p$. □

For the free Hamiltonian $h_0 = -\Delta$ with $\Delta = d^2/dx^2$, we similarly have:

Theorem 5.69. *The free Hamiltonian $h_0 = -\Delta$ is self-adjoint on the domain*

$$D(\Delta) = \{\psi \in L^2(\mathbb{R}) \mid \psi'' \in L^2(\mathbb{R})\}, \tag{5.324}$$

where the double derivative ψ'' is taken in the distributional sense.

Although this may be proved in an analogous way, such proofs are increasingly burdensome if the number of derivatives gets higher. It is easier to use the Fourier transform (which also provided an alternative way of proving Theorem 5.68).

Theorem 5.70. *The formulae*

$$\hat{f}(k) = \int_{-\infty}^{\infty} \frac{dx}{\sqrt{2\pi}} e^{-ikx} f(x); \tag{5.325}$$

$$\check{f}(x) = \int_{-\infty}^{\infty} \frac{dk}{\sqrt{2\pi}} e^{ikx} f(k), \tag{5.326}$$

are rigorously defined on $\mathscr{S}(\mathbb{R})$, $L^2(\mathbb{R})$, and $\mathscr{S}'(\mathbb{R})$, and provide continuous isomorphisms of each of these spaces. Furthermore, (5.326) *is inverse to* (5.325)*, i.e.*

$$\check{\hat{f}} = \hat{\check{f}} = f, \tag{5.327}$$

so that we may (and often do) write $\hat{f} = \mathscr{F}(f)$ and $\check{f} = \mathscr{F}^{-1}(f)$, or $f = \mathscr{F}^{-1}(\hat{f})$.

In all three cases we have the identities (in a distributional sense if appropriate)

$$\mathscr{F}(x^n f^{(m)})(k) = (id/dk)^n (ik)^m \mathscr{F}(f)(k). \tag{5.328}$$

Finally, as a map $\mathscr{F} : L^2(\mathbb{R}) \to L^2(\mathbb{R})$ the Fourier transform is unitary, so that

$$\langle \hat{\psi}, \hat{\varphi} \rangle = \langle \psi, \varphi \rangle. \tag{5.329}$$

See §C.15 for further discussion. For example, we have

$$D(p) = \{\psi \in L^2(\mathbb{R}) \mid k \cdot \hat{\psi}(k) \in L^2(\mathbb{R})\}; \tag{5.330}$$
$$D(\Delta) = \{\psi \in L^2(\mathbb{R}) \mid k^2 \cdot \hat{\psi}(k) \in L^2(\mathbb{R})\}. \tag{5.331}$$

Thus we may now reformulate Theorems 5.68 and 5.69 as follows:

Theorem 5.71. *The momentum operator p is self-adjoint on the domain* (5.330). *The free Hamiltonian $h_0 = -\Delta$ is self-adjoint on the domain* (5.331).

Proof. Denoting multiplication by x^n by the symbol k^n, we have

$$p = \mathscr{F}^{-1} k \mathscr{F}; \tag{5.332}$$
$$\Delta = -\mathscr{F}^{-1} k^2 \mathscr{F}. \tag{5.333}$$

Hence the theorem follows from Proposition B.73 and unitarity of the Fourier transform $\mathscr{F}$ (plus the little observation that if $a = a^*$ on $D(a) \subset H$ and $u : H \to K$ is unitary, then $b = uau^*$ is self-adjoint on $D(b) = uD(a) \subset K$). □

Much is known about regularity properties of functions in such domains, e.g.,

$$D(p) \subset C_0(\mathbb{R}); \tag{5.334}$$
$$D(\Delta) \subset C_0^{(1)}(\mathbb{R}). \tag{5.335}$$

These are the most elementary cases of the famous ***Sobolev Embedding Theorem***.

If $\psi \in D(p)$, then $k \mapsto (1+k^2)^{1/2}\hat{\psi}(k)$ is in $L^2(\mathbb{R})$, so applying Hölder's inequality (B.15) with $p = q = 2$ to $f(k) = (1+k^2)^{1/2}\hat{\psi}(k)$ and $g(k) = (1+k^2)^{-1/2}$, which is in $L^2(\mathbb{R})$, too, gives $\hat{\psi} \in L^1(\mathbb{R})$. The Riemann–Lebesgue Lemma (see §C.15) then yields $\psi \in C_0(\mathbb{R})$. To prove (5.335), one uses $(1+k^2)$ rather than its square root.

Finally, we give a common domain of essential self-adjointness for q, p, and h_0.

Proposition 5.72. *The operators q, p, and h_0 are essentially self-adjoint on $\mathscr{S}(\mathbb{R})$.*

Proof. We see from (5.332) that the cases of p and q are similar, so we only explain the case of q. Denoting the operator of multiplication by x on the domain $\mathscr{S}(\mathbb{R})$ by q_0, as in the proof of Proposition B.73 it is easy to see that $D(q_0^*) = D(q)$. Fourier-transforming, the fact that $\mathscr{S}(\mathbb{R})$ is dense in $H^1(\mathbb{R})$ (cf. the proof of Theorem 5.68) shows that $D(q_0^-) = D(q)$,so that $D(q_0^*) = D(q_0^-)$. The actions of q_0^* and q_0^- obviously being given by multiplication by x in both cases, we have $q_0^* = q_0^-$.

The proof for h_0 is similar; in the second step we now use the fact that $\mathscr{S}(\mathbb{R})$ is dense in $H^2(\mathbb{R})$, defined as $D(\Delta)$, as in (5.324), but now seen as a Hilbert space in the inner product $\langle \psi, \varphi \rangle_{(2)} = \langle \psi, \varphi \rangle + \langle \psi'', \varphi'' \rangle$, with corresponding norm given by $\|\psi\|_{(2)}^2 = \|\psi\|^2 + \|\psi''\|^2$. This is proved just as in the case of a single derivative. □

We also say that $\mathscr{S}(\mathbb{R})$ is a ***core*** for the operators in question. For example, the canonical commutation relations $[q, p] = i\hbar \cdot 1_H$ rigorously hold on this domain.

5.12 Stone's Theorem

We now come to a central result on symmetries in quantum mechanics "explaining" the Hamiltonian. Recall that a continuous unitary representation of $\mathbb{R}$ (as an additive group) on a Hilbert space H is a map $t \mapsto u_t$, where $t \in \mathbb{R}$ and each $u_t \in B(H)$ is unitary, such that the associated map $\mathbb{R} \times H \to H$, $(t, \psi) \mapsto u_t \psi$, is continuous, and

$$u_s u_t = u_{s+t}, \; s,t \in \mathbb{R}; \tag{5.336}$$

$$u_0 = 1_H; \tag{5.337}$$

$$\lim_{t\to 0} u_t \psi = \psi \;\; (t \in \mathbb{R}, \, \psi \in H). \tag{5.338}$$

These conditions imply

$$\lim_{t\to s} u_t \psi = u_s \psi \;\; (s,t \in \mathbb{R}, \psi \in H). \tag{5.339}$$

Note that according to Proposition 5.36 continuity may be replaced by weak measurability. Probably the simplest nontrivial example is given by $H = L^2(\mathbb{R})$ and

$$u_t \psi(x) = \psi(x-t). \tag{5.340}$$

To prove (5.338), we use a routine $\varepsilon/3$ argument. We first prove (5.338) for $\psi \in C_c(\mathbb{R})$, where it is elementary in the sup-norm, i.e., $\lim_{t\to 0} \|u_t \psi - \psi\|_\infty = 0$ by continuity and hence (given compact support) uniform continuity of ψ. But then the (ugly) estimate $\|\psi\|_2^2 \leq |K| \|\psi\|_\infty$, where $K \subset \mathbb{R}$ is any compact set containing the support of ψ, also yields $\lim_{t\to 0} \|u_t \psi - \psi\|_2 = 0$. Hence for $\varepsilon > 0$ we may find $\delta > 0$ such that $\|u_t \psi - \psi\|_2 < \varepsilon/3$ whenever $|t| < \delta$. For general $\psi' \in H$, we find $\psi \in C_c(\mathbb{R})$ such that $\|\psi - \psi'\| < \varepsilon/3$, and, using unitarity of u_t, estimate

$$\begin{aligned}\|u_t \psi' - \psi'\| &\leq \|u_t \psi' - u_t \psi\| + \|u_t \psi - \psi\| + \|\psi - \psi'\| \\ &\leq \varepsilon/3 + \varepsilon/3 + \varepsilon/3 = \varepsilon.\end{aligned}$$

In the context of quantum mechanics, physicists formally write

$$u_t = e^{-ita}, \tag{5.341}$$

where a is usually thought of as the Hamiltonian of the system, although in the previous example it is rather the momentum operator. In any case, we avoid the notation h instead of a here, partly in order to rightly suggest far greater generality of the construction and partly to avoid confusion with the notation in §B.21; if h *is* the Hamiltonian, one would have $a = h/\hbar$ in (5.341). Mathematically speaking, if a is self-adjoint, eq. (5.341) is rigorously defined by Theorem B.158, where

$$e_t(x) = \exp(-itx). \tag{5.342}$$

Conversely, given a continuous unitary representation $t \mapsto u_t$ of $\mathbb{R}$ on H, one may attempt to define an operator a by specifying its domain and action by

$$D(a) = \left\{ \psi \in H \mid \lim_{s \to 0} \frac{u_s - 1}{s} \psi \text{ exists} \right\}; \tag{5.343}$$

$$a\psi = i \lim_{s \to 0} \frac{u_s - 1}{s} \psi \; (\psi \in D(a)). \tag{5.344}$$

Stone's Theorem makes this rigorous, and even turns the passage from the generator a to the unitary group $t \mapsto u_t$ (and back) into a bijective correspondence.

Theorem 5.73. *1. If $a : D(a) \to H$ is self-adjoint, the map $t \mapsto u_t$ defined by* (5.341), *which is rigorously defined by Proposition B.159 with* (5.342), *defines a continuous unitary representation of $\mathbb{R}$ on H.*

2. Conversely, given such a representation, the operator a defined by (5.343) - (5.344) *is self-adjoint; in particular, $D(a)$ is dense in H.*

3. These constructions are mutually inverse.

Proof. We use the setting of §B.21, so that b is the bounded transform of a.

1. Eqs. (5.336) - (5.337) are immediate from Theorem B.158, which also yields unitarity of each operator u_t. To prove (5.338) we first take $\varphi \in C_c^*(b)H$, which means that φ is a finite linear combinations of vectors of the type $\varphi = h(a)\psi$, where $h \in C_c(\sigma(a))$ and $\psi \in H$. Using (5.342) and (B.573), we have

$$\|u_t\varphi - \varphi\| \leq \|e_t h - h\|_\infty \|\psi\| \leq \|h\|_\infty \|e_t - 1_K\|_\infty^{(K)} \|\psi\|, \tag{5.345}$$

where K is the (compact) support of h in $\sigma(b)$. Since the exponential function is uniformly convergent on any compact set, this gives $\lim_{t \to 0} \|u_t\varphi - \varphi\| = 0$. Taking finite linear combinations of such vectors φ gives the same result for any $\varphi \in C_c^*(b)H$ (with an extra step this could have been done on $C_0^*(b)H$, too). Thus for $\varepsilon > 0$ we can find $\delta > 0$ so that $\|u_t\varphi - \varphi\| < \varepsilon/3$ whenever $|t| < \delta$. For general $\psi' \in H$, we find $\varphi \in C_0^*(b)H$ such that $\|\varphi - \psi'\| < \varepsilon/3$, and estimate

$$\begin{aligned} \|u_t\psi' - \psi'\| &\leq \|u_t\psi' - u_t\varphi\| + \|u_t\varphi - \varphi\| + \|\varphi - \psi'\| \\ &\leq \varepsilon/3 + \varepsilon/3 + \varepsilon/3 = \varepsilon, \end{aligned}$$

since $\|u_t\psi' - u_t\varphi\| = \|\psi' - \varphi\|$ by unitarity of u_t. This is equivalent to (5.338).

2. For any $\psi \in H$ and $n \in \mathbb{N}$, define $\psi_n \in H$ by

$$\psi_n = n \int_0^\infty ds\, e^{-ns} u_s \psi, \tag{5.346}$$

either as a Riemann-type integral (whose approximants converge in norm) or as a functional $\varphi \mapsto n \int_0^\infty ds\, e^{-ns} \langle u_s\psi, \varphi \rangle$, which is obviously continuous and hence is represented by a unique vector $\psi_n \in H$. Then simple computations show that

$$\lim_{s \to 0} \frac{u_s - 1}{s} \psi_n = n(\psi_n - \psi),$$

so that $\psi_n \in D(a)$. The proof that $\psi_n \to \psi$ starts with the elementary estimate

$$\|\psi_n - \psi\| \leq n \int_0^\infty ds\, e^{-ns} \|u_s\psi - \psi\|,$$

in which we split up the $\int_0^\infty$ as $\int_0^\delta \cdots + \int_\delta^\infty \cdots$, where $\delta > 0$. Using strong continuity of the map $t \mapsto u_t$, i.e., (5.338), for any n the first integral vanishes as $\delta \to 0$. In the second integral we estimate $\|u_s\psi - \psi\| \leq 2\|\psi\|$ and take the limit $n \to \infty$. Thus $\psi_n \to \psi$, so that $D(a)$ is dense in H.
To prove self-adjointness of a, we need a tiny variation on Theorem B.93:

Lemma 5.74. *Let a be symmetric. Then a is self-adjoint (i.e. $a^* = a$) iff*

$$\operatorname{ran}(a+i) - \operatorname{ran}(a-i) = H. \tag{5.347}$$

Proof. We only need the implication from (5.347) to $a^* = a$ (but the converse immediately follows from Theorem B.93). So assume (5.347). For given $\psi \in D(a^*)$ there must then be a $\varphi \in H$ such that $(a^* - i)\psi = (a-i)\varphi$. Since a is symmetric, we have $D(a) \subset D(a^*)$, so $\psi - \varphi \in D(a^*)$, and $(a^* - i)(\psi - \varphi) = 0$. But $\ker(a^* - i) = \operatorname{ran}(a+i)^\perp$, so $\ker(a^* - i) = 0$. Hence $\psi = \varphi$, and in particular $\psi \in D(a)$ and hence $D(a^*) \subset D(a)$. Since we already know the opposite inclusion, we have $D(a^*) = D(a)$. Given symmetry, this implies $a^* = a$. □

Continuing the proof of Theorem 5.73.2, symmetry of a easily follows from its definition, combined with the property $u_t^* = u_t^{-1} = u_{-t}$. Indeed, for $\psi, \varphi \in D(a)$, the weak limit $s \to 0$ below exists by definition of $D(a)$, cf. (5.343), whence:

$$\langle \varphi, a\psi \rangle = i \lim_{s\to 0} \langle \varphi, \frac{u_s - 1}{s}\psi \rangle = -i \lim_{s\to 0} \langle \frac{u_{-s} - 1}{-s}\varphi, \psi \rangle = \langle a\varphi, \psi \rangle.$$

To prove that $\operatorname{ran}(a-i) = H$, we compute $(a-i)\psi_1 = -i\psi$, with ψ_1 defined by (5.346) with $n = 1$. The property $\operatorname{ran}(-i) = H$ is proved in a similar way: now define $\tilde{\psi}_1 = \int_{-\infty}^0 ds\, e^s u_s \psi$ and obtain $(a+i)\tilde{\psi}_1 = i\psi$. Thus Lemma 5.74 applies.
3. Bijectivity has two directions: $a \mapsto u_t \mapsto a$ and $u_t \mapsto a \mapsto u_t$.

- Given a and hence (5.341) defining u_t, we change notation from a to a' in (5.343) - (5.344) and need to show that $a' = a$. Denoting the restriction of a to the domain $C_c^*(b)$ by a_0, we first show that $a_0 \subseteq a'$. The technique to prove this is similar to the argument around (5.345). We initially assume that $\varphi \in D(a_0) = C_c^*(b)H$ takes the form $\varphi = h(a)\psi$ for some $h \in C_c(\sigma(a))$ and $\psi \in H$. Just a trifle more complicated than (5.345), using (5.342), (B.573), and unitarity of u_t, we estimate:

$$\begin{aligned}\left\| \frac{u_{t+s}\varphi - u_t\varphi}{s} + ia_0 u_t \varphi \right\| &\leq \left\| \frac{e_s h - h}{s} + i \cdot \mathrm{id}_{\sigma(T)} h \right\|_\infty \|\psi\| \\ &\leq \left\| \frac{e_s - 1_K}{s} + i \cdot \mathrm{id}_K \right\|_\infty^{(K)} \|h\|_\infty \|\psi\|,\end{aligned}$$

so that by definition of the (strong) derivative we obtain

$$\frac{du_t}{dt}\varphi = \lim_{s\to 0}\frac{u_{t+s}\varphi - u_t\varphi}{s} = -iau_t\varphi, \tag{5.348}$$

initially for any φ of the said form $h(a)\psi$, and hence, taking finite sums, for any $\varphi \in D(a_0)$. The existence of this limit shows that, on the assumption $\psi \in D(a_0)$, we have $\psi \in D(a')$, and we also see that $a' = a$ on $D(a_0)$, or, in other words, that $a_0 \subseteq a'$. Since a' is self-adjoint (by part 2 of the theorem) and hence closed, we have $a_0^- \subseteq a'$. Since a_0 is essentially self-adjoint by Theorem B.159, this gives $a \subseteq a'$. Taking adjoints reverses the inclusion, and since both operators are self-adjoint this gives $a = a'$.

- Given u_t and hence (5.343) - (5.344) defining a, we change notation from u_t to u'_t in (5.341) and need to show that $u'_t = u_t$. Indeed, let

$$\psi_t = u_t\psi, \tag{5.349}$$

and similarly $\psi'_t = u'_t\psi$. If $\psi \in D(a)$, then by definition of a we have

$$i\frac{d\psi_t}{dt} = i\lim_{s\to 0}\frac{u_{t+s} - u_t}{s}\psi = i\lim_{s\to 0}\frac{u_s - 1_H}{s}u_t\psi = a\psi_t, \tag{5.350}$$

which also shows that $\psi_t \in D(a)$. Similarly, $id\psi'_t/dt = a\psi'_t$, so that ψ_t and ψ'_t satisfy the same differential equation with the same initial condition

- $$\psi^{(0)} = (\psi^{(0)})' = \psi.$$

Now consider $\hat{\psi}_t = \psi_t - \psi'_t$, which once again satisfies the same equation (i.e., $id\hat{\psi}_t/dt = a\hat{\psi}_t$), but this time with initial condition $\hat{\psi}_0 = \psi^{(0)} - (\psi^{(0)})' = \psi - \psi = 0$. The key point is that *any* solution $\hat{\psi}_t$ of this equation has the property $\|\hat{\psi}_t\| = \|\hat{\psi}_0\|$ for any $t \in \mathbb{R}$, since by symmetry of a,

$$\frac{d}{dt}\|\hat{\psi}_t\|^2 = \frac{d}{dt}\langle\hat{\psi}_t, \hat{\psi}_t\rangle = -i(\langle\hat{\psi}_t, a\hat{\psi}_t\rangle - \langle a\hat{\psi}_t, \hat{\psi}_t\rangle) = 0.$$

For our *specific* $\hat{\psi}_t$ we have $\|\hat{\psi}_0\| = 0$ and hence $\psi_t = \psi'_t$, that is, $u'_t = u_t$. □

Corollary 5.75. *With $t \mapsto u_t$ and a defined and related as in Theorem 5.73, if $\psi \in D(a)$, for each $t \in \mathbb{R}$ the vector ψ_t defined by* (5.349) *lies in $D(a)$ and satisfies*

$$a\psi_t = i\frac{d\psi_t}{dt}, \tag{5.351}$$

whence $t \mapsto \psi_t$ is the unique solution of (5.351) *with initial value $\psi^{(0)} = \psi$.*

This follows from the proof of part 3 of Theorem 5.73. With $a = h/\hbar$ (as above), this is just the famous ***time-dependent Schrödinger equation***

$$h\psi_t = i\hbar\frac{d\psi_t}{dt}. \tag{5.352}$$

Notes

§5.1. **Six basic mathematical structures of quantum mechanics**

Wigner's Theorem was first stated by von Neumann and Wigner (1928), but the first proof appeared in Wigner (1931). See Bonolis (2004) and Scholz (2006) for some history. Instead of working with $\mathscr{P}_1(H)$ with the bilinear trace form expressing the transition probabilities, one may also formulate and prove Wigner's Theorem in terms of the projective Hilbert space $\mathbb{P}H$ equipped with the Fubini–Study metric, in which case the relevant symmetries may be defined geometrically as isometries. See Freed (2012) for this proof, as well as Brody & Hughston (2001) for the underlying geometry. Kadison's Theorem may be traced back from Kadison (1965). See also Moretti (2013). Ludwig symmetries go back to Ludwig (1983); see also Kraus (1983). Our approach to von Neumann symmetries was inspired by Hamhalter (2004), and has a large pedigree in quantum logic. Bohr symmetries were introduced in Landsman & Lindenhovius (2016), where Theorem 5.4.6 was also proved.

§5.2. **The case $H = \mathbb{C}^2$**

This material is partly based on Simon (1976). The covering map (5.46) has a nice geometric description: if $\Sigma = \mathbb{C} \cup \{\infty\}$ is the Riemann sphere, we have the well-known stereographic projection

$$S^2 \stackrel{\cong}{\rightarrow} \Sigma; \tag{5.353}$$

$$(x,y,z) \mapsto \frac{x+iy}{1-z}. \tag{5.354}$$

If $u \in SU(2)$ is given by (5.43), then the associated Möbius transformation

$$z \mapsto \frac{\alpha z + \beta}{-\overline{\beta} z + \overline{\alpha}}$$

is a bijection of Σ, whose associated transformation of S^2 is the rotation $R = \tilde{\pi}(u)$.

§5.3. **Equivalence between the six symmetry theorems**

Most proofs may be also found in Cassinelli et al (2004) or Moretti (2013).

§5.4. **Proof of Jordan's Theorem**

Our proof of Jordan's Theorem is taken from Bratteli & Robinson (1987); see also Thomsen (1982) for a simplification of the purely algebraic step (which we delegated to Theorem C.175), originally proved by Jacobson & Rickart (1950).

§5.5. **Proof of Wigner's Theorem**

There are many proofs of Wigner's Theorem, none of them really satisfactory (in this respect the situation is similar to Gleason's Theorem). Our proof follows Simon (1976), who in turn relies on Bargmann (1964) and Hunziker (1972). The proof in Cassinelli et al (2004) seems cleaner, but their proof of the additivity of their operator T_ω is not easy to follow. For a geometric approach see Freed (2012).

If $\dim(H) \geq 3$, the conclusion of Wigner's Theorem follows if W merely preserves orthogonality (Uhlhorn, 1963). See also Cassinelli et al (2004). This, in turn, has been generalized in various directions, e.g. to indefinite inner product spaces (Molnár, 2002) as well as to certain Banach spaces, where one says that x is orthogonal to y if for all $\lambda \in \mathbb{C}$ one has $\|x+\lambda y\| \geq \|x\|$ (Blanco & Turnšek, 2006).

§5.6. **Some abstract representation theory**

Among numerous books on representation theory, our personal favourite is Barut & Raçka (1977), and also Gaal (1973) and Kirillov (1976) are classics at least for the abstract theory. An interesting recent paper on the unitary group on infinite-dimensional Hilbert space is Schottenloher (2013).

§5.7. **Representations of Lie groups and Lie algebras**

This section was inspired by Hall (2013) and Knapp (1988). For Lie's Third Theorem, see, for example, Duistermaat & Kolk (2000), §1.14. To obtain Theorem 5.41, consider the canonical projection $\tilde{\pi} : \tilde{G} \to G$ and define $D = \tilde{\pi}^{-1}(\{e\})$. This is a discrete normal subgroup of $\tilde{G}$, and it is an easy fact that a discrete normal subgroup of any connected topological group must lie in its center. Note that a discrete subgroup of the center of $\tilde{G}$ is automatically normal.

The exponentiation problem for skew-adjoint representations of $\mathfrak{g}$ is considerably more complicated than in finite dimension. Let H be an infinite-dimensional Hilbert space with dense subspace D and let $\rho : \mathfrak{g} \to L(D,H)$ be a linear map, where $L(D,H)$ is the space of linear maps from L to H. We say that ρ is a ***skew-adjoint representation*** of $\mathfrak{g}$ if *(i):* D is invariant under $u'(\mathfrak{g})$, *(ii):* the commutation relations (5.157) hold on D, and *(i):* each $i\rho(A)$ is essentially self-adjoint on D. For example, we have seen that if $u : G \to U(H)$ is a unitary representation, then the construction $\rho(A) = u'(A)$, defined on the Gårding domain $D = D_G$, fits the bill. Conversely, additional conditions are needed for ρ to exponentiate to a unitary representation. The best-known of those is ***Nelson's criterion***: if, given a skew-adjoint representation $\rho : \mathfrak{g} \to L(D,H)$, the ***Nelson operator*** or ***Laplacian*** $\Delta = \sum_{k=1}^{\dim(\mathfrak{g})} \rho(T_k)^2$ is essentially self-adjoint on D, then ρ exponentiates to a unitary representation of $\tilde{G}$ (with additional remarks similar to those in Corollary 5.43).

§5.8. **Irreducible representations of $SU(2)$**

§5.9. **Irreducible representations of compact Lie groups**

See e.g. Knapp (1988), Simon (1996), and Deitmar (2005), and innumerable other books. This material ultimately goes back to (É.) Cartan and Weyl.

§5.10. **Symmetry groups and projective representations**

See Varadarajan (1985), Tuynman & Wiegerinck (1987), Landsman (1998a), Cassinelli et al (2004), and Hall (2013). For different proofs of Theorem 5.59 (Bargmann, 1954) see Simms (1971) and Cassinelli et al (2004). Leaving out the anti-unitary symmetries is a pity; see e.g. Freed & Moore and Roberts (2016).

§5.11. **Position, momentum, and free Hamiltonian**

§5.12. **Stone's Theorem**

See Reed & Simon (1972), Schmüdgen (2012), Moretti (2013), Hall (2013), and many other books. Our proof of part 1 of Theorem 5.73 is original.

Part II
Between $C_0(X)$ and $B(H)$

Chapter 6
Classical models of quantum mechanics

This chapter gives an introduction to a chain of results attempting to exclude deeper layers underneath quantum mechanics that restore some form of classical physics:

> '[Such results] more or less illustrate the ways along which some opponents might hope to escape Bohr's reasonings and von Neumann's proof and the places where they are dangerously near breaking their necks.' (Groenewold, 1946, p. 454)

In so far as they are mathematically precise, such no-go results have their roots in von Neumann's 1932 book, which gave rise to two traditions that were often in polemical opposition to each other. Mathematically minded authors typically admired von Neumann's exclusion of hidden variables, yet tried to strengthen his theorem by weakening its assumptions; this sparked, for example, *Gleason's Theorem* (1957) as well as the *Kochen–Specker Theorem* (1967). Certain physicists (led by Bell), on the other hand, tried to circumvent (and later even ridicule) von Neumann's work. A high point of this tradition was *Bell's Theorem* from 1964, which was informed not only by von Neumann, but even more so by the famous Einstein–Podolsky–Rosen (EPR) paper from 1935, as well as by Bohm's deterministic pilot wave reformulation of quantum mechanics (1952). However, at the end of the day these traditions turned out to be not really divergent after all: Bell not only independently (and earlier) obtained a version of the Kochen–Specker Theorem, but, more importantly, his results from 1964 turn out to be very closely related to the culmination of the first tradition in the form of the so-called *Free Will Theorem* (FWT), which was published by Conway and Kochen during 2006–2008. Indeed, although its validity is uncontroversial, this theorem has been criticized on the following grounds:

1. Lack of novelty compared with the famous paper by Bell (1964), whose assumptions and conclusions are at least quite similar to those of the FWT (although the underlying proofs are mathematically quite distinct from those in the FWT).
2. Lack of novelty even within its own terms: versions of the FWT had actually been around for decades under less illustrious titles and authorships, e.g. Heywood & Redhead (1983), Stairs (1983), Brown & Svetlichny (1990), and Clifton (1993).
3. Circularity, in that indeterminism is presupposed (namely in the assumption that 'experimenters have a certain freedom') instead of derived.

K. Landsman, *Foundations of Quantum Theory*,
Fundamental Theories of Physics 188, DOI 10.1007/978-3-319-51777-3_6

One aim of this chapter is to clarify these matters, with the following conclusions:

1. The difference between earlier literature in the same direction and the FWT is largely one of emphasis, namely on free will (!), exemplifying a recent trend (also found elsewhere) in emphasizing free choice of the settings of experiments. Unfortunately, like Bell, Conway and Kochen even mathematically use an informal way of talking about free settings, not to speak of the complete absence of any serious philosophical analysis of free will among all three authors (for which perhaps Bell, but certainly not Conway and Kochen may be excused).
2. Granting the informal characterization of free settings, both Bell's (1964) Theorem and the FWT establish a contradiction between quantum mechanics, determinism, and locality (in the sense of Bell, which in the presence of determinism reduces to a no-signaling condition called parameter independence).
3. The technical difference between Bell's Theorem and the FWT lies in four facts:
   a. Bell's arguments rely on probability theory (whereas the FWT does not).
   b. The (optical) corner of quantum mechanics used in Bell's Theorem may be replaced by the corresponding experimental results, whereas the FWT uses uncontroversial yet untested predictions about massive spin-1 particles.
   c. The FWT must assume perfect (EPR) correlations, which are difficult to realize and hence are avoided by later versions of Bell's Theorem (i.e. through the CHSH inequalities rather than the original Bell inequalities).
   d. Like EPR, Bell and his followers focused on locality right from the beginning, and hence in Bell (1964) the inference is from locality to determinism. Conway and Kochen, on the other hand, resolve the contradiction their FWT established by inferring randomness of outcomes from freedom of settings.

We start with a very simple treatment of both von Neumann's argument against linear hidden variables and Kochen & Specker's refinement of it, in which von Neumann's controversial linearity assumption is decisively weakened so as to only apply to *commuting* operators; the Kochen–Specker Theorem excludes what are called *non-contextual quasi-linear hidden variables*. We then present what we see as a more transparent version of the FWT, whose key ingredient of replacing the non-contextuality assumption in the Kochen–Specker Theorem by a locality condition is preserved, but where this time the setting is completely deterministic. Freedom of choice then arises as a very natural independence assumption, and any threat of circularity is avoided: the conclusion is simply a contradiction between determinism, freedom of choice (i.e. of apparatus settings), locality, and quantum mechanics. Moreover, as we argue in §6.3, the philosophically precise concept of free will used in the assumptions of the FWT is what Lewis coined 'local miracle compatibilism'.

Following an interlude on the GHZ Theorem, which seamlessly fits into the given framework, we then turn to Bell's Theorems, which we compare with the FWT.

Finally, we give our own rigorous version of an argument first proposed by Colbeck and Renner to the effect that, under suitable freeness of choice and no-signaling conditions (similar to those in Bell's Theorem and the FWT), as long as they are compatible with quantum mechanics, hidden variables are at best irrelevant. In fact, this can only be proved under much stronger assumptions, obscuring the claim.

6.1 From von Neumann to Kochen–Specker

Von Neumann's Theorem 6.2 below was the first technical result excluding some class of hidden variables underneath quantum mechanics, namely (in current parlance) *linear non-contextual hidden variables*. This terminology requires some explanation. First, theorems of this kind apparently accept the mathematical structure of the observables prescribed by the usual formalism of quantum theory, i.e., observables are identified with elements of the self-adjoint part

$$H_n(\mathbb{C}) \equiv M_n(\mathbb{C})_{\mathrm{sa}} = \{a \in M_n(\mathbb{C}) \mid a^* = a\} \tag{6.1}$$

of the algebra $M_n(\mathbb{C})$ of $n \times n$ matrices (this simple case suffices to make all points of conceptual interest). Short of introducing "hidden" *observables*, hidden variable theories propose the existence of hidden *states*, which either replace or supplement the usual quantum states (which in the case at hand would be density operators). Mimicking classical (statistical) physics, such states are interpreted as probability measures on some phase space X, whose points $x \in X$ assign sharp values to quantum-mechanical observables. Naively, this is done through associated functions

$$V_x : H_n(\mathbb{C}) \to \mathbb{R}, \tag{6.2}$$

but in fact this choice already commits us to the first of two possibilities, which we pragmatically present as theories predicting measurement outcomes:

- In ***non-contextual*** deterministic theories of measurement, the outcome solely depends on the observable a that is being measured and on the (possibly 'hidden') state of the system. Theorem 6.2 below, then, rules out such theories in which values are sharp (i.e., dispersion-free), and V_x in (6.2) is *linear*. The Kochen–Specker Theorem subsequently proves the same impossibility under a weaker (and physically more reasonable) assumption called *quasi-linearity*.
- ***Contextual*** deterministic theories of measurement, on the other hand, allow the outcome of some measurement of a to depend on the ***measurement context*** (as well as on the state), which in this case is understood as the choice of possible other (compatible) observables b measured together with a (i.e., $ab = ba$). This seems a reasonable assumption, well within the spirit of quantum mechanics, though perhaps not so in the extreme form later held by Heisenberg, according to which measurement outcomes (or even "reality") are "created" by the measurement. Under a weakened non-contextuality assumption, Bell's Theorem (cf. §6.5) and the Free Will Theorem (§6.2) rule out such theories, too.

Definition 6.1. *A* **non-contextual hidden variable** *is a map $V : H_n(\mathbb{C}) \to \mathbb{R}$ that for each $a \in H_n(\mathbb{C})$, and in terms of the $n \times n$ unit matrix 1_n, satisfies*

$$V(a^2) = V(a)^2; \tag{6.3}$$
$$V(1_n) = 1. \tag{6.4}$$

That is, V is **dispersion-free** *as well as* **normalized**, *respectively.*

Theorem 6.2. *For $n \geq 2$, non-zero linear dispersion-free maps $V : H_n(\mathbb{C}) \to \mathbb{R}$ do not exist. In particular, linear non-contextual hidden variables do not exist.*

Proof. Such maps extend to complex-linear dispersion-free maps $V : M_n(\mathbb{C}) \to \mathbb{C}$ by complex linearity, so that theorem is equivalent to Proposition 2.10. □

As von Neumann perfectly well understood himself, his seemingly natural linearity assumption (given the mathematical structure of quantum mechanics unearthed by none other than he!) is unwarranted physically (and even mathematically, since eigenvalues and eigenstates, which should be the hallmark of dispersion-free states, are by no means linear in the underlying operator). This suggests the following:

Definition 6.3. *A map $V : H_n(\mathbb{C}) \to \mathbb{R}$ is called* **quasi-linear** *if for all $s,t \in \mathbb{R}$ and all $a,b \in H_n(\mathbb{C})$* that commute *(i.e., $ab = ba$) one has*

$$V(sa+tb) = sV(a)+tV(b). \tag{6.5}$$

As in the linear case, such a map uniquely extends to a map $V : M_n(\mathbb{C}) \to \mathbb{C}$ that is precisely a quasi-state in the sense of Definition 2.26. The following lemma will be useful, also showing that the above objections to linearity have been met.

Lemma 6.4. *Let $V : H_n(\mathbb{C}) \to \mathbb{R}$ be a quasi-linear non-contextual hidden variable.*

1. *For each $a \in H_n(\mathbb{C})$, the number $\lambda = V(a)$ is an eigenvalue of a.*
2. *If $(a_1,\dots,a_k)$ pairwise commute, and $b = f(a_1,\dots,a_k)$ for some polynomial f, then $V(b) = f(V(a_1),\dots,V(a_k))$.*

More generally, it follows from Theorem C.24 that if H is a Hilbert space and $V : B(H)_{\mathrm{sa}} \to \mathbb{R}$ is a quasi-linear non-contextual hidden variable (or, equivalently, its complexification $V_{\mathbb{C}} : B(H) \to \mathbb{C}$ is a dispersion-free quasi-state), then $V(a) \in \sigma(a)$ (provided $a^* = a$). This implies the above lemma, but we also provide a direct proof.

Proof. For any $b \in H_n(\mathbb{C})$ with $ab = ba$, eq. (6.3) and quasi-linearity imply that

$$V(ab) = V(a)V(b); \tag{6.6}$$

just evaluate $V((a \pm b)^2) = (V(a) \pm V(b))^2$. Taking $b = a^2$ etc. and also invoking (6.4) then yields $V(p(a)) = p(V(a))$ for any polynomial in a. If λ_i are the eigenvalues of a, its characteristic polynomial $p(a) = \prod_{i=1}^n (a - \lambda_i)$ satisfies $p(a) = 0$, so that $V(p(a)) = 0$ and hence $p(V(a)) = 0$, or $\prod_{i=1}^n (\lambda - \lambda_i) = 0$. This implies that $\lambda = \lambda_i$ for some i. The second claim is proved in a similar way. □

Theorem 6.5. *For $n \geq 3$, quasi-linear non-contextual hidden variables do not exist.*

This is the ***Kochen–Specker Theorem***. It follows from Gleason's Theorem 2.28 and von Neumann's Theorem 6.2, since according to Corollary 2.29 to the former, quasi-states on $M_n(\mathbb{C})$ are actually states (in other words, quasi-linear non-contextual hidden variables are linear). However, Kochen and Specker also gave a direct proof of their theorem, subsequently somewhat simplified along the following lines.

Proof. We prove the claim for $n = 3$, which (by restricting V to any self-adjoint subalgebra of $M_n(\mathbb{C})$ isomorphic to $H_3(\mathbb{C})$) implies the result for all $n > 3$ also. To prove Theorem 6.5 for $n = 3$, we interpret $H_3(\mathbb{C})$ as the algebra of observables of a spin-1 particle and introduce the well-known angular momentum matrices

$$J_1 = \begin{pmatrix} 0 & 0 & 0 \\ 0 & 0 & -i \\ 0 & i & 0 \end{pmatrix}, \ J_2 = \begin{pmatrix} 0 & 0 & i \\ 0 & 0 & 0 \\ -i & 0 & 0 \end{pmatrix}, \ J_3 = \begin{pmatrix} 0 & -i & 0 \\ i & 0 & 0 \\ 0 & 0 & 0 \end{pmatrix}. \tag{6.7}$$

In what follows, we will heavily use the squares

$$J_1^2 = \begin{pmatrix} 0 & 0 & 0 \\ 0 & 1 & 0 \\ 0 & 0 & 1 \end{pmatrix}, J_2^2 = \begin{pmatrix} 1 & 0 & 0 \\ 0 & 0 & 0 \\ 0 & 0 & 1 \end{pmatrix}, J_3^2 = \begin{pmatrix} 1 & 0 & 0 \\ 0 & 1 & 0 \\ 0 & 0 & 0 \end{pmatrix}, \tag{6.8}$$

each of which has eigenvalues 0 and 1. The J_i^2 commute by inspection, and satisfy

$$J_1^2 + J_2^2 + J_3^2 = 2 \cdot 1_3. \tag{6.9}$$

The (matrix-valued) angular momentum vector is given by

$$\mathbf{J} = J_1\mathbf{e}_1 + J_2\mathbf{e}_2 + J_3\mathbf{e}_3, \tag{6.10}$$

where $(\mathbf{e}_1, \mathbf{e}_2, \mathbf{e}_3)$ is the standard basis of $\mathbb{R}^3$ (seen as a vector space with the usual inner product $\langle \cdot, \cdot \rangle$), i.e., $\mathbf{e}_1 = (1,0,0)$, etc., and the angular momentum $J_{\mathbf{u}}$ along an arbitrary unit vector $\mathbf{u} = \sum_i u_i \mathbf{e}_i$ in $\mathbb{R}^3$ is given by

$$J_{\mathbf{u}} = \langle \mathbf{J}, \mathbf{u} \rangle = \sum_{i=1}^{3} J_i u_i. \tag{6.11}$$

This brings us to the crucial point: a map $V : H_3(\mathbb{C}) \to \mathbb{R}$ induces a map $\tilde{V} : S^2 \to \mathbb{R}$ on the set S^2 of all unit vectors $\mathbf{u}$ in $\mathbb{R}^3$, via

$$\tilde{V}(\mathbf{u}) = V(J_{\mathbf{u}}^2). \tag{6.12}$$

As usual, a *basis* of $\mathbb{R}^3$, denoted by $a = (\mathbf{u}_1, \mathbf{u}_2, \mathbf{u}_3)$, is always assumed *orthonormal*.

Lemma 6.6. *Let $V : H_3(\mathbb{C}) \to \mathbb{R}$ be a non-contextual quasi-linear hidden variable, with associated map $\tilde{V} : S^2 \to \{0,1\}$ given by (6.12). Then:*

1. *$\tilde{V}(-\mathbf{u}) = \tilde{V}(\mathbf{u})$ for each $\mathbf{u} \in S^2$ (so that $\tilde{V}$ is defined on the real projective plane);*
2. *If $a = (\mathbf{u}_1, \mathbf{u}_2, \mathbf{u}_3)$ is a basis, then the triple $\tilde{V}(a) \equiv (\tilde{V}(\mathbf{u}_1), \tilde{V}(\mathbf{u}_2), \tilde{V}(\mathbf{u}_3))$ must contain a single 0 and two 1's, i.e., $\tilde{V}(a)$ must be one of the triples*

$$\begin{aligned} \lambda^{(1)} &= (0,1,1); \\ \lambda^{(2)} &= (1,0,1); \\ \lambda^{(3)} &= (1,1,0). \end{aligned} \tag{6.13}$$

In Gleason-like language, $\tilde{V}$ is a $\underline{2}$-valued frame function of weight $w(\tilde{V}) = 2$.

Proof. If $a = (\mathbf{u}_1, \mathbf{u}_2, \mathbf{u}_3)$ is a basis, then $J_{\mathbf{u}_i} = uJ_iu^*$ for $i = 1,2,3$, where u is the 3×3 matrix with entries $u_{ij} = \langle \mathbf{u}_i, \mathbf{e}_j\rangle$. Since u is unitary, the matrices $J_{\mathbf{u}_i}$ and their squares have the same eigenvalues and satisfy the same relations as the J_i and their squares. Thus the eigenvalues of $J^2_{\mathbf{u}_i}$ are 0 and 1, for fixed a the squares $J^2_{\mathbf{u}_i}$ mutually commute, and they satisfy the sum rule (6.9), i.e., $J^2_{\mathbf{u}_1} + J^2_{\mathbf{u}_2} + J^2_{\mathbf{u}_3} = 2\cdot 1_3$, so $\tilde{V}(\mathbf{u}_1) + \tilde{V}(\mathbf{u}_2) + \tilde{V}(\mathbf{u}_3) = 2$. The claim then follows from Definition 6.3 and Lemma 6.4. □

Now define a ***coloring*** of $\mathbb{R}^3$ as any map $\tilde{V}: S^2 \to \{0,1\}$ satisfying the two properties in Lemma (6.6). The proof of Theorem 6.5 then reduces to the following lemma.

Lemma 6.7. *There exists no coloring of* $\mathbb{R}^3$.

Proof. Take the following unit vectors (some identical), grouped into 11 bases (for simplicity we use unnormalized vectors, e.g., $(1,0,1)$ stands for $(1/\sqrt{2}, 0, 1/\sqrt{2})$):

basis	$\mathbf{u}_1$	$\mathbf{u}_2$	$\mathbf{u}_3$
a_1	$(0,0,1)$	$(1,0,0)$	$(0,1,0)$
a_2	$(1,0,1)$	$(-1,0,1)$	$(0,1,0)$
a_3	$(0,1,1)$	$(0,-1,1)$	$(1,0,0)$
a_4	$(1,-1,2)$	$(-1,1,2)$	$(1,1,0)$
a_5	$(1,0,2)$	$(-2,0,1)$	$(0,1,0)$
a_6	$(2,1,1)$	$(0,-1,1)$	$(-2,1,1)$
a_7	$(2,0,1)$	$(0,1,0)$	$(-1,0,2)$
a_8	$(1,1,2)$	$(1,-1,0)$	$(-1,-1,2)$
a_9	$(0,1,2)$	$(1,0,0)$	$(0,-2,1)$
a_{10}	$(1,2,1)$	$(-1,0,1)$	$(1,-2,1)$
a_{11}	$(1,0,0)$	$(0,2,1)$	$(0,-1,2)$.

We will show that one cannot even color this particular finite set of vectors (let alone all unit vectors in $\mathbb{R}^3$). We denote a vector $\mathbf{u}_i$ in a basis a_μ by

$$\mathbf{u}_i^{(\mu)}, i = 1,2,3, \mu = 1,\dots,11,$$

and write e.g. $\tilde{V}(a_\mu) = (0,1,1)$ for the three conditions

$$\tilde{V}(\mathbf{u}_1^{(\mu)}) = 0,\ \tilde{V}(\mathbf{u}_2^{(\mu)}) = 1),\ \tilde{V}(\mathbf{u}_3^{(\mu)}) = 1.$$

The main point is that if some coloring $\tilde{V}$ maps a specific vector $\mathbf{u}$ to 0, then all vectors orthogonal to $\mathbf{u}$ must go to 1. In particular, two orthogonal vectors can never both be sent to 0. To find a contradiction (to the assumption that $\tilde{V}$ exists), we try to assign values $\tilde{V}(\mathbf{u}_i^{(\mu)})$ one after the other, starting in row 1. Here some specific choices will be made, but by symmetry other choices lead to similar contradictions.

1. Suppose that $\tilde{V}(a_1) = (0,1,1)$ (i.e., $\tilde{V}(\mathbf{u}_1^{(1)}) = 0$ and $\tilde{V}(\mathbf{u}_2^{(1)}) = \tilde{V}(\mathbf{u}_3^{(1)}) = 1$). In a_2 this forces $\tilde{V}(\mathbf{u}_3^{(2)}) = 1$, so that either $\mathbf{u}_1^{(2)}$ or $\mathbf{u}_2^{(2)}$ must be mapped to 0 (and the other to 1). Let $\tilde{V}(\mathbf{u}_1^{(2)}) = 0$, so that $\tilde{V}(\mathbf{u}_2^{(2)}) = 1$, i.e., $\tilde{V}(\mathbf{u}_2) = (0,1,1)$. In a_3 one has $\mathbf{u}_3^{(3)} = \mathbf{u}_2^{(1)}$, so $\tilde{V}(\mathbf{u}_3^{(3)}) = 1$. We choose $\tilde{V}(\mathbf{u}_1^{(3)}) = 0$ and hence $\tilde{V}(\mathbf{u}_2^{(3)}) = 1$, so $\tilde{V}(\mathbf{u}_2) = (0,1,1)$. In a_4, the vector $\mathbf{u}_3^{(4)}$ is orthogonal to $\mathbf{u}_1^{(1)}$, which has been mapped to zero already, so that $\tilde{V}(\mathbf{u}_3^{(4)}) = 1$. The remaining free choice is arbitrarily made as $\tilde{V}(\mathbf{u}_1^{(4)}) = 0$, so that $\tilde{V}(\mathbf{u}_2^{(4)}) = 1$ and hence $\tilde{V}(a_4) = (0,1,1)$.
2. But now everything is fixed for a_5 t/m a_{11}, as follows. From a_5, the vector $\mathbf{u}_3^{(5)}$ already occurred in $\mathbf{u}_1$, and moreover, $\mathbf{u}_2^{(5)}$ is orthogonal to $\mathbf{u}_1^{(4)}$ from a_4. Because $\tilde{V}(\mathbf{u}_1^{(4)}) = 0$, one must have $\tilde{V}(\mathbf{u}_2^{(4)}) = 1$. And so on and so forth, yielding $\tilde{V}(a_\mu) = (0,1,1)$ voor $\mu = 5,\ldots,10$ (as was the case also for $\mu = 1,2,3,4$).
3. In a_{11} one has $\mathbf{u}_1^{(11)} = \mathbf{u}_2^{(1)}$, so $\mathbf{u}_1^{(11)}$ is mapped to 1. Furthermore, $\mathbf{u}_2^{(11)}$ is orthogonal to $\mathbf{u}_1^{(4)}$, which was mapped to 0; hence $\mathbf{u}_2^{(11)}$ goes to 1. Finally, $\mathbf{u}_3^{(11)}$ is orthogonal to $\mathbf{u}_1^{(10)}$, which was mapped to 0, so that $\mathbf{u}^{(11)}$ must go to 1. Thus

$$\tilde{V}(a_{11}) = (1,1,1). \tag{6.14}$$

But $(1,1,1)$ is not an admissible value of $\tilde{V}$! So $\tilde{V}$ and hence V cannot exist. □

Corollary 6.8. *There is no function $\tilde{V}$ with the two properties stated in Lemma 6.6.*

The Kochen–Specker Theorem is often stated in the following way.

Definition 6.9. *For any finite-dimensional Hilbert space H, a* **coloring** *of the set $\mathscr{P}_1(H)$ of one-dimensional projections on H is a function*

$$W : \mathscr{P}_1(H) \to \{0,1\}$$

such that for any resolution of the identity (e_i) with $e_i \in \mathscr{P}_1(H)$, i.e.,

$$e_i e_j = \delta_{ij} e_i; \tag{6.15}$$

$$\sum_i e_i = 1_H, \tag{6.16}$$

one has

$$\sum_i W(e_i) = 1, \tag{6.17}$$

so that there is exactly one member e_i of the family such that $W(e_i) = 1$.

Note that if $e \in \mathscr{P}_1(H)$ then $e = e_\psi = |\psi\rangle\langle\psi|$ for some unit vector $\psi \in H$, so that each basis (υ_i) of H defines such a family by $e_i = |\upsilon_i\rangle\langle\upsilon_i|$, and *vice versa*, up to phase factors. The setting of Gleason's Theorem is similar, with the crucial difference that the function on $\mathscr{P}_1(H)$ in question then takes values in $[0,1]$ instead of $\{0,1\}$ and hence can be shown to exist, even amply so (as there are many states).

Theorem 6.10. *If* $\dim(H) > 2$, *there exists no coloring of* $\mathscr{P}_1(H)$.

Proof. For $H = \mathbb{C}^3$, the existence of W would yield the existence of $\tilde{V}$ through

$$\tilde{V}(\mathbf{u}) = 1 - W(e_{\mathbf{u}}), \tag{6.18}$$

where $\mathbf{u} \in \mathbb{R}^3$ is regarded as a vector in $\mathbb{C}^3$. Property 1 in Lemma 6.6 is obviously satisfied. To prove property 2, we note that for any unit vector $\mathbf{u} \in \mathbb{R}^3 \subset \mathbb{C}^3$, we have

$$J_{\mathbf{u}}^2 \mathbf{u} = 0, \tag{6.19}$$

since an explicit computation based on (6.11) shows that, with $\mathbf{u} = (u_1, u_2, u_3)$,

$$J_{\mathbf{u}}^2 = \begin{pmatrix} u_2^2 + u_3^2 & -u_1u_2 & -u_1u_3 \\ -u_1u_2 & u_1^2 + u_3^2 & -u_2u_3 \\ -u_1u_3 & -u_2u_3 & u_1^2 + u_2^2 \end{pmatrix}. \tag{6.20}$$

It follows from rotation invariance that the eigenvalues of $J_{\mathbf{u}}^2$ are the same as those of each J_i^2, cf. (6.8), i.e., $\lambda = 0$ with multiplicity one and $\lambda = 1$ with multiplicity two. Hence (6.19) gives the projection e_0 onto the eigenspace of $J_{\mathbf{u}}^2$ for $\lambda = 0$ as

$$e_0 = |\mathbf{u}\rangle\langle\mathbf{u}| \equiv e_{\mathbf{u}}. \tag{6.21}$$

Property 2 in Lemma 6.6 then follows from the assumption that W is a coloring. Since $\tilde{V}$ cannot exist by Lemma 6.7, neither can W. This proves the claim for $\mathbb{C}^3$.

We finish by induction. Suppose $\mathbb{C}^n$ contains some set $\{\mathbf{u}_k\}_{k\in K}$ of unit vectors that cannot be colored, assuming that $\mathbf{u}_0 = (1, 0, \ldots, 0)$ lies in this set. We embed each $\mathbf{u}_k$ into $\mathbb{C}^{n+1}$ by adding a zero *at the end*, calling the image $\mathbf{u}'_k$. Adding $\mathbf{v} = (0, \ldots, 0, 1)$, the only possible coloring of the set $\{\mathbf{u}'_k, \mathbf{v}\}_{k\in K}$ in $\mathbb{C}^{n+1}$ is given by $W(\mathbf{u}'_k) = 0$ for each $k \in K$ and $W(\mathbf{v}) = 1$. Indeed, if $W(\mathbf{u}'_{k_0}) = 1$ for some k_0, then, since $\mathbf{v}$ is orthogonal to each $\mathbf{u}'_k$, we must have $W(\mathbf{v}) = 0$, which means that the original set $\{\mathbf{u}_k\}_{k\in K}$ should be colorable in $\mathbb{C}^n$, but this is impossible by assumption.

We now embed each $\mathbf{u}_k$ into $\mathbb{C}^{n+1}$ by adding a zero *at the beginning*, denoting its image by $\mathbf{u}''_k$, and add $\mathbf{u}'_0 = (1, 0, \ldots, 0, 0)$. By the same token, the only coloring of the set $\{\mathbf{u}''_k, \mathbf{u}'_0\}_{k\in K}$ is given by $W(\mathbf{u}''_k) = 0$ for each $k \in K$ and $W(\mathbf{u}'_0) = 1$. But this leaves the set $\{\mathbf{u}'_k, \mathbf{u}''_k, \mathbf{v}\}_{k\in K}$ in $\mathbb{C}^{n+1}$ uncolorable, since colorability of $\{\mathbf{u}'_k, \mathbf{v}\}_{k\in K}$ gave $W(\mathbf{u}'_0) = 0$, whereas colorability of $\{\mathbf{u}''_k, \mathbf{u}'_0\}_{k\in K}$ gave $W(\mathbf{u}'_0) = 1$. □

The set thus obtained is larger than necessary. For example, already for $H = \mathbb{C}^4$ the following bases cannot be colored (again writing down unnormalized vectors):

basis	$\mathbf{u}_1$	$\mathbf{u}_2$	$\mathbf{u}_3$	$\mathbf{u}_4$
a_1	$(0,0,0,1)$	$(0,0,1,0)$	$(1,1,0,0)$	$(1,-1,0,0)$
a_2	$(0,0,0,1)$	$(0,1,0,0)$	$(1,0,1,0)$	$(1,0,-1,0)$
a_3	$(1,-1,1,-1)$	$(1,-1,-1,1)$	$(1,1,0,0)$	$(0,0,1,1)$
a_4	$(1,-1,1,-1)$	$(1,1,1,1)$	$(1,0,-1,0)$	$(0,1,0,-1)$
a_5	$(0,0,1,0)$	$(0,1,0,0)$	$(1,0,0,1)$	$(1,0,0,-1)$
a_6	$(1,-1,-1,1)$	$(1,1,1,1)$	$(1,0,0,-1)$	$(0,1,-1,0)$
a_7	$(1,1,-1,1)$	$(1,1,1,-1)$	$(1,-1,0,0)$	$(0,0,1,1)$
a_8	$(1,1,-1,1)$	$(-1,1,1,1)$	$(1,0,1,0)$	$(0,1,0,-1)$
a_9	$(1,1,1,-1)$	$(-1,1,1,1)$	$(1,0,0,1)$	$(0,1,-1,0)$

The proof is the following observation: if we present the coloring condition as

$$W(0,0,0,1)+W(0,0,1,0)+W(1,1,0,0)+W(1,-1,0,0)=1; \qquad (a_1)$$

$$\cdots \qquad (a_\bullet)$$

$$W(1,1,1,-1)+W(-1,1,1,1)+W(1,0,0,1)+W(0,1,-1,0)=1, \qquad (a_9)$$

then since there are nine such equations the sum of the right-hand sides is odd, whereas the sum of the left-hand sides is even, since each vector appears twice.

To bridge the gap between the Kochen–Specker Theorem and the Free Will Theorem, as well as the one between mathematics and physics, we now rephrase the former as a "mini FWT". We build an experiment consisting of a box containing a spin-1 particle and a device capable of measuring all of the three observables

$$(J^2_{\mathbf{u}_1}, J^2_{\mathbf{u}_2}, J^2_{\mathbf{u}_3})$$

for an arbitrary basis a of $\mathbb{R}^3$; since the operators in question commute, this simultaneous measurement is allowed by quantum theory. The choice of a is called the ***setting*** of the experiment, traditionally denoted by A (in honor of Alice, who is supposed to perform the experiment), with possible values $A=a$. In "phenomenological" notation, the observable measured in an experiment like this is called F, which in the case at hand has three components $F=(F_1,F_2,F_3)$: given the setting a, the observable F_i corresponds to $J^2_{\mathbf{u}_i}$. The notation $F=\lambda$ for $\lambda=(\lambda_1,\lambda_2,\lambda_3)$, i.e., $F_i=\lambda_i$, then expresses the fact that the outcome of a measurement of F is λ.

According to both quantum mechanics and our quasi-linear non-contextual hidden variable theory, either $\lambda_i=0$ or $\lambda_i=1$, and λ must lie in the value space

$$\Lambda=\{(0,1,1),(1,0,1),(1,1,0)\}; \qquad (6.22)$$

cf. Lemma 6.6 for the hidden variable theory, while in quantum mechanics (6.22) follows from the fact that λ must lie in the joint spectrum of the three operators $J^2_{\mathbf{u}_i}$.

This, in turn means that there must be a joint eigenvector ψ such that $J^2_{\mathbf{u}_i} = \lambda_i \psi$ for each $i = 1,2,3$. There are three such joint eigenvectors, namely $\mathbf{u}_1$, $\mathbf{u}_2$, and $\mathbf{u}_3$ (initially defined as vectors in $\mathbb{R}^3$ but now seen as vectors in $\mathbb{C}^3$), with joint eigenvalues $(0,1,1)$, $(1,0,1)$, and $(1,1,0)$, respectively.

Otherwise, quantum mechanics and our quasi-linear non-contextual hidden variable theory provide a different picture of the experiment. According to the former theory, a given spin-1 particle may be prepared in a (pure) quantum state ψ, which is a unit vector in $\mathbb{C}^3$. Quantum theory then merely predicts probabilities

$$P_\psi(F = \lambda | A = a) \equiv p_{J^2_{\mathbf{u}_1}, J^2_{\mathbf{u}_2}, J^2_{\mathbf{u}_3}}(\lambda_1, \lambda_2, \lambda_3), \tag{6.23}$$

for the possible outcomes λ, which according to the Born rule (2.21) are given by

$$P_\psi(F = \lambda^{(i)} | A = a) = |\langle \mathbf{u}_i, \psi \rangle|^2. \tag{6.24}$$

So if $\psi = \mathbf{u}_i$, then the outcome will be $\lambda = \lambda^{(i)}$ with probability one, but in a superposition $\psi = \sum_i c_i \mathbf{u}_i$ (with $\sum_i |c_i|^2 = 1$), quantum theory predicts a random sequence of outcomes $\lambda^{(i)}$, each with probability $|c_i|^2$.

Let us note that quantum mechanics is non-contextual in the following (probabilistic) sense. Alice could decide to perform just one measurement instead of three, say F_1, with setting $a_1 = \mathbf{u}_1$, or perhaps she may not know if the other two are performed. Fortunately, this does not matter, since for any unit vector $\psi \in \mathbb{C}^3$,

$$P_\psi(F_1 = \lambda_1 | A_1 = \mathbf{u}_1) = \sum_{\lambda_2, \lambda_3} P_\psi(F = \lambda | A = a), \tag{6.25}$$

so that according to quantum mechanics, it does not matter for the Born probabilities of the first measurement if the other two are performed or not.

The question now arises if some quasi-linear non-contextual hidden variable theory theory could improve on this, in that the *probabilities* quantum theory assigns to various outcomes are replaced by *predictions*. In the sprit of determinism (whilst avoiding the appearance of circularity), such a theory should also predict the settings of the experiment. Accordingly, the assumptions leading to our "mini FWT" are:

Definition 6.11. *In the context of the experiment on spin-1 particles just discussed:*

- **Determinism** firstly *means that there is a state space X with associated functions*

$$A : X \to X_A; \tag{6.26}$$
$$F : X \to \Lambda, \tag{6.27}$$

where X_A is the set of all bases in $\mathbb{R}^3$ (i.e. $a \in X_A$), and Λ is some set of possible outcomes; these functions completely describe the experiment in the sense that each state $x \in X$ determines both its settings $a = A(x)$ and its outcome $\lambda = F(x)$. Here $A = (A_1, A_2, A_3)$, where the functions $A_i : X \to S^2$ (seen as the space of unit vectors in $\mathbb{R}^3$) combine to define a basis, and $F = (F_1, F_2, F_3)$, where $F_i : X \to \mathbb{R}$.

Secondly, *there exists some set X_Z and an additional function*

$$Z : X \to X_Z, \tag{6.28}$$

such that

$$F = F(A, Z). \tag{6.29}$$

More precisely, for each $x \in X$ one has

$$F(x) = \hat{F}(A(x), Z(x)) \tag{6.30}$$

for a certain function $\hat{F} : X_A \times X_Z \to \Lambda$. Also this function is, of course, a triple $\hat{F} = (\hat{F}_1, \hat{F}_2, \hat{F}_3)$, where $\hat{F}_i : X_A \times X_Z \to \underline{2}$. In terms of (6.28), *then:*

- **Nature** *then requires that Λ is given by (6.22) (so that $F_i : X \to \underline{2}$).*
- **Freedom** *states that A and Z are* independent *in the sense that the function*

$$\begin{aligned} A \times Z : X &\to X_A \times X_Z \\ x &\mapsto (A(x), Z(x)) \end{aligned} \tag{6.31}$$

is surjective; in other words, for each $(a, z) \in X_A \times X_Z$ there is an $x \in X$ for which $A(x) = a$ and $Z(x) = z$ (making a and z free variables).

- **Non-contextuality** *(cf. Lemma 6.6) finally stipulates that $\hat{F}$ take the form*

$$\hat{F}((\mathbf{u}_1, \mathbf{u}_2, \mathbf{u}_3), z) = (\tilde{F}(\mathbf{u}_1, z), \tilde{F}(\mathbf{u}_2, z), \tilde{F}(\mathbf{u}_3, z)), \tag{6.32}$$

for a single function $\tilde{F} : S^2 \times X_Z \to \underline{2}$ that also satisfies

$$\tilde{F}(-\mathbf{u}, z) = \tilde{F}(\mathbf{u}, z). \tag{6.33}$$

"Nature" may be taken to be either an experimental result or an uncontroversial prediction of (some corner of) quantum mechanics. The function Z (including its domain X_Z) describes anything relevant to the experiment (such as the behaviour of the particle) *except* the variables determining the settings (which do form part of X). The goal of the freedom assumption is to remove any potential dependencies between the variables (a, z), and hence between the physical system Alice perform her measurements *on*, and the devices she performs her measurements *with*.

Corollary 6.12. *Determinism, Nature, Freedom, and Non-contextuality are contradictory.*

Proof. For each $z \in X_Z$, define a function $\tilde{V}_z : S^2 \to \underline{2}$ by $\tilde{V}_z(\mathbf{u}) = \tilde{F}(\mathbf{u}, z)$. The assumptions combine to give $\tilde{V}_z$ the same properties as $\tilde{V}$ in Lemma 6.6 (where z "goes along for a free ride"). According to Corollary 6.8 (which applies because by *Freedom* one can freely vary a for any given z), the function $\tilde{V}_z$ cannot exist. □

This "mini FWT" is a good exercise for the Free Will Theorem in the next section. For example, let us note, as a warning, that if Determinism is seen as the culprit (and hence falls), then the other assumptions in the (min) FWT are no longer defined. This blocks a direct inference from Freedom to Indeterminism à la Conway & Kochen.

6.2 The Free Will Theorem

The Free Will Theorem is similar in spirit to Corollary 6.12, with the difference that the experiment now has two wings and the *non-contextuality* assumption is replaced by a certain *locality* condition. This condition relates to the setting introduced by Einstein, Podolsky, and Rosen in 1935 and further studied by Bohm, Bell, and others, in which (in current jargon) two physicists, called Alice and Bob, are far apart whilst performing simultaneous experiments on some correlated two-particle state (technically speaking, their measurements need to be *spacelike separated*). In the situation considered by EPR each particle had a spatial degree of freedom and hence required the infinite-dimensional Hilbert space $L^2(\mathbb{R}^3)$ for its description, but, as recognized by Bohm, the thrust of the argument comes out more clearly if each particle merely has an internal degree of freedom (and is "frozen" otherwise).

Bell (1964) considered a pair of spin $\frac{1}{2}$ particles (cf. §6.5), each of which has Hilbert space $\mathbb{C}^2$ (although the famous experiments of Aspect testing the violation of Bell's inequalities used photons, which have the "same" Hilbert space), but because of its reliance on the Kochen–Specker Theorem (which fails for $\mathbb{C}^2$) the Free Will Theorem requires one dimension more, i.e., $H = \mathbb{C}^3$. As before, we see this as the state space of a massive spin-1 particle. The price of this extra dimension is that the pertinent experiment whose outcome provides the *Nature* input for the Free Will Theorem has not actually been performed, but, as in the Bell case, the predictions of quantum mechanics are uncontroversial and will serve as input instead.

These predictions are as follows. Alice and Bob measure on the correlated state

$$\psi_0 = (\mathbf{e}_1 \otimes \mathbf{e}_1 + \mathbf{e}_2 \otimes \mathbf{e}_2 + \mathbf{e}_3 \otimes \mathbf{e}_3)/\sqrt{3}, \tag{6.34}$$

where we recall that $(\mathbf{e}_1, \mathbf{e}_2, \mathbf{e}_3)$ is the standard basis of $\mathbb{R}^3$, now seen as a basis of $\mathbb{C}^3$. This state is rotation-invariant, which means that nonzero angular momentum in one particle must be compensated for in the other, creating the desired correlations.

As before, we denote Alice's setting by $A = a$, which remains the choice of some basis of $\mathbb{R}^3$, but this time also Bob picks some basis b, so that we write $B = b$ for his choice. Similar to Alice's outcome $F = \lambda$ we denote Bob's by $G = \gamma$, and quantum mechanics provides all (Born) probabilities

$$P_{\psi_0}(F = \lambda, G = \gamma | A = a, B = b) \equiv p_{J^2_{\mathbf{u}_1}, J^2_{\mathbf{u}_2}, J^2_{\mathbf{u}_3}, J^2_{\mathbf{v}_1}, J^2_{\mathbf{v}_2}, J^2_{\mathbf{v}_3}}(\lambda_1, \lambda_2, \lambda_3, \gamma_1, \gamma_2, \gamma_3),$$

which are well defined because Alice's squared angular momentum operators $J^2_{\mathbf{u}_1}$ commute with Bob's $J^2_{\mathbf{v}_1}$ as a consequence of Einstein locality (stating that spacelike separated observables commute). Note that similarly to $a = (\mathbf{u}_1, \mathbf{u}_2, \mathbf{u}_3)$ for Alice's basis, we write $b = (\mathbf{v}_1, \mathbf{v}_2, \mathbf{v}_3)$ for Bob's. If Alice merely measures F_i whilst Bob measures G_j, then, as in the previous section, it does not matter which other (commuting) operators are measured and/or whether Alice and Bob know about this, cf. (6.25). Thus we may write either $(A = a, B = b)$ or $A_i = \mathbf{u}_i, B_i = \mathbf{v}_i$ for the settings, and simple calculations show that the Born probabilities are given by:

$$P_{\psi_0}(F_i = 1, G_j = 1 | A = a, B = b) = \tfrac{1}{3}(1 + \langle \mathbf{u}_i, \mathbf{v}_j \rangle^2); \quad (6.35)$$
$$P_{\psi_0}(F_i = 0, G_j = 0 | A = a, B = b) = \tfrac{1}{3}\langle \mathbf{u}_i, \mathbf{v}_j \rangle^2; \quad (6.36)$$
$$P_{\psi_0}(F_i = 1, G_j = 0 | A = a, B = b) = \tfrac{1}{3}(1 - \langle \mathbf{u}_i, \mathbf{v}_j \rangle^2); \quad (6.37)$$
$$P_{\psi_0}(F_i = 0, G_j = 1 | A = a, B = b) = \tfrac{1}{3}(1 - \langle \mathbf{u}_i, \mathbf{v}_j \rangle^2), \quad (6.38)$$

where $\langle \mathbf{u}_i, \mathbf{v}_j \rangle^2 = |\langle \mathbf{u}_i, \mathbf{v}_j \rangle|^2$, etc., since the vectors are real, In terms of the notation

$$P_{\psi_0}(F_i = G_j | \cdot) = P_{\psi_0}(F_i = 0, G_j = 0 | \cdot) + P_{\psi_0}(F_i = 1, G_j = 1 | \cdot); \quad (6.39)$$
$$P_{\psi_0}(F_i \neq G_j | \cdot) = P_{\psi_0}(F_i = 0, G_j = 1 | \cdot) + P_{\psi_0}(F_i = 1, G_j = 0 | \cdot), \quad (6.40)$$

this yields

$$P_{\psi_0}(F_i = G_j | A = a, B = b) = \tfrac{1}{3}(1 + 2\langle \mathbf{u}_i, \mathbf{v}_j \rangle^2); \quad (6.41)$$
$$P_{\psi_0}(F_i \neq G_j | A = a, B = b) = \tfrac{2}{3}(1 - \langle \mathbf{u}_i, \mathbf{v}_j \rangle^2). \quad (6.42)$$

The crucial point for the Free Will Theorem is that this implies ***perfect correlation***:

$$P_{\psi_0}(F_i = G_j | A_i = B_j) = 1, \quad (6.43)$$

in agreement with the intuition about angular momentum expressed earlier.

We now move to a (possibly counterfactual) deterministic description of this experiment along the lines of the previous section. It is straightforward to adapt all of Definition 6.11 except Non contextuality (which after all is the assumption we would like to get rid of!). With the obvious changes, we obtain:

- ***Determinism*** again *first* claims there is a state space X with associated functions

$$A : X \to X_A; \quad (6.44)$$
$$B : X \to X_B; \quad (6.45)$$
$$F : X \to \Lambda; \quad (6.46)$$
$$G : X \to \Lambda, \quad (6.47)$$

  where $X_A = X_B$ is the set of all bases in $\mathbb{R}^3$, and Λ is some set of possible outcomes, which completely describe the experiment in the sense that each state $x \in X$ determines both its settings $(a = A(x), b = B(x))$ and its outcome $(\lambda = F(x), \gamma = G(x))$. Here $A = (A_1, A_2, A_3)$ and $B = (B_1, B_2, B_3)$ where the functions $A_i : X \to S^2$ (where S^2 is seen as the space of unit vectors in $\mathbb{R}^3$) combine to define a basis (similarly for $B_j : X \to S^2$), and $F = (F_1, F_2, F_3)$. *Secondly*, there exists some set X_Z and an additional function $Z : X \to X_Z$ such that

$$F = F(A, B, Z); \quad (6.48)$$
$$G = G(A, B, Z), \quad (6.49)$$

  in that for each $x \in X$ one has the functional relationships

$$F(x) = \hat{F}(A(x), B(x), Z(x)); \tag{6.50}$$
$$G(x) = \hat{G}(A(x), B(x), Z(x)), \tag{6.51}$$

for certain functions $\hat{F} : X_A \times X_B \times X_Z \to \Lambda$ and $\hat{G} : X_A \times X_B \times X_Z \to \Lambda$, each of which is a triple $\hat{F} = (\hat{F}_1, \hat{F}_2, \hat{F}_3)$ with $\hat{F}_i : X_A \times X_B \times X_Z \to \mathbb{R}$, etc. The value $z = Z(x)$ is just the traditional "hidden variable" (which is often denoted by λ).

- ***Freedom*** then states that A, B, and Z are *independent* in that for each $(a, b, z) \in X_A \times X_B \times X_Z$ there is an $x \in X$ for which $A(x) = a$, $B(x) = b$, and $Z(x) = z$.
- ***Nature*** requires that:
  - Λ is given by (6.22), i.e. F_i and G_j, and hence $\hat{F}_i$ and $\hat{G}_j$ take values in $\{0, 1\}$;
  - The experiment measures *squares* of angular momenta, so that

$$\hat{F}(a', b', z) = \hat{F}(a, b, z); \tag{6.52}$$
$$\hat{G}(a', b', z) = \hat{G}(a, b, z), \tag{6.53}$$

  whenever (a', b') differ from (a, b) by changing the sign of any basis vector;
  - *Perfect correlation* obtains, cf. (6.43), i.e., writing $a = (\mathbf{u}_1, \mathbf{u}_2, \mathbf{u}_3)$ for Alice's basis and $b = (\mathbf{v}_1, \mathbf{v}_2, \mathbf{v}_3)$ for Bob's, one has

$$\mathbf{u}_i = \mathbf{v}_j \Rightarrow \hat{F}_i(a, b, z) = \hat{G}_j(a, b, z). \tag{6.54}$$

We now come to the locality condition that is to replace *Non-contextuality*. This condition was first clearly stated by Bell (1964, p. 196), who attributes it to Einstein:

> 'The vital assumption is that the result G for particle 2 does not depend on the setting a of the magnet for particle 1, nor F on b.'

Noting various other notions of locality (such as *Einstein locality* in local quantum physics, which requires spacelike separated operators to commute, or *Bell locality*, discussed below), the above idea might be called *Context locality*, but we will simply refer to it as *Locality*. In our deterministic setting, a precise formulation is this:

- ***Locality*** means that $F(A, B, Z)$ is independent of B and $G(A, B, Z)$ is independent of A. In other words, we have $F = F(A, Z)$ and $G = G(B, Z)$, so that (with slight abuse of notation) $\hat{F} : X_A \times X_Z \to \Lambda$ and $\hat{G} : X_B \times X_Z \to \Lambda$, or, then again, $F(x) = \hat{F}(A(x), Z(x))$ and $G(x) = \hat{G}(B(x), Z(x))$, for each $x \in X$.

This finally brings us to (our reformulation of) the ***Free Will Theorem***:

Theorem 6.13. *Determinism, Freedom, Nature, and Locality are contradictory.*

Proof. The *Freedom* assumption allows us to treat (a, b, z) as free variables, a fact that will tacitly be used all the time. First, taking $i = j$ in (6.54) shows that $\hat{F}_i(\mathbf{u}_1, \mathbf{u}_2, \mathbf{u}_3, z)$ only depends on $(\mathbf{u}_i, z)$, whilst $\hat{G}_j(\mathbf{v}_1, \mathbf{v}_2, \mathbf{v}_3, z)$ only depends on $(\mathbf{v}_j, z)$. Hence we write $\hat{F}_i(a, z) = \tilde{F}_i(\mathbf{u}_i, z)$, etc. Next, taking $i \neq j$ in (6.54) shows that $\tilde{F}_1(\mathbf{u}, z) = \tilde{F}_2(\mathbf{u}, z) = \tilde{F}_3(\mathbf{u}, z)$. Consequently, the function $\hat{F} : X_A \times X_Z \to X_F$ is given by (6.32). We are now back to the proof of Corollary 6.12, concluding that such a function does not exist by Corollary 6.8. □

6.3 Philosophical intermezzo: Free will in the Free Will Theorem

> 'The determinism-free will controversy has all of the earmarks of a dead problem. The positions are well staked out and the opponents manning them stare at each other in mutual incomprehension.' (Earman, 1986, p. 235)

The question arises which specific notion of free will is among the assumptions of the FWT (in the reformulation just given). To put this question in perspective, let us briefly recall the main point of the debate about free will. This concept has two poles. One is the "will" itself, requiring a sense of ***agency***, deliberation, and control. This pole seems to require some form of determinism. A powerful expressions is:

> 'Fürst! Was Sie sind, sind Sie durch Zufall und Geburt. Was ich bin, bin ich durch mich.'[1] (Beethoven, to his benefactor (!) Prince Lichnowsky)

The other pole of free will is the adjective "free", i.e., ***the ability to do otherwise***, which at first sight requires indeterminism. *The problem of free will is that these poles seem contradictory*. Many authors conflate free will with moral responsibility:

> 'free will can be defined as the unique ability of persons to exercise control over their conduct in the manner necessary for moral responsibility.' (McKenna & Coates, 2015)

This aspect is irrelevant to our discussion, concerned as it is with the question what it would mean for Alice and Bob to choose their settings "freely" if determinism is assumed (it would have been different if one setting launched a nuclear missile).

Even in our narrow context, the traditional philosophical stances are relevant:

- ***Compatibilism*** denies the contradiction, claiming that free will and determinism coexist. This position may be defended in many ways, among which one finds:
  - Reconceptualizing "the ability to do otherwise" in a deterministic world. This will be our focus in what follows, especially in a version inspired by Lewis.
  - Belittling the relevance of "the ability to do otherwise", as e.g. by Dennett:

    > 'So if anyone at all is interested in the question of whether one could have done otherwise in *exactly* the same circumstances (and internal state) this will have to be a particularly pure metaphysical curiosity—that is to say, a curiosity so pure as to be utterly lacking in any ulterior motive, since the answer could not conceivably make any noticeable difference to the way the world went.' (Dennett, 1984, p. 559).

- ***Incompatibilism*** accepts the contradiction, once again branching off into:
  - *Libertarianism*, arguing that free will requires an indeterministic world.
  - *Hard determinism*, claiming determinism (which is assumed) blocks free will:

    > 'Ein Mensch kann zwar tun was er will, aber nicht wollen was er will.'[2] (Schopenhauer)
  - *Hard incompatibilism*, asserting that 'every way you look at it you lose': free will makes no sense in either a deterministic or an indeterministic world.

---

[1] 'Lord! What you are, you are through chance and birth. What I am, I am because of myself.'

[2] 'One can admittedly do what one wants, but one cannot want what one wants.'

Although hard incompatibilism has our sympathy, our opening question concerning the notion of free will in the FWT drives us into the compatibilist direction, since determinism is among the assumptions shown to be contradictory by Theorem 6.13. Within compatibilism, we will be close to the well-known 'local miracle' variant thereof proposed by the philosopher David Lewis. Like other compatibilists before him (starting at least with G.E. Moore), Lewis attempts to make sense of the intuition that even in a deterministic world one in principle has the ability to act differently from the way one actually does, despite the fact that the latter was predetermined. A simple example is Alice's choosing setting a by moving her hand in a certain way, although she was able to choose a'. On the other hand, she could not have moved her hand with a speed greater than that of light, so her ability remains constrained by the laws of nature. Lewis asks us to distinguish between:

- 'I am able to do something such that, if I did it, a law would be broken.'
- 'I am able to break a law.'

The latter is impossible, but the former is not on Lewis's own theory of counterfactuals, according to which the phrase 'if I did it' leads us to consider the possible world in which doing 'something' is actually true, whilst in the possible worlds under consideration as many other features as possible are kept the same as in the actual world (the precise underlying measure of similarity is not important here). Thus the phrase 'a law would be broken' refers to the laws of the actual world (in which the alternative action is not realized). It seems to be of great importance to Lewis that in the first case it is not the agent who would break a law; instead, it is the breaking of some law of our actual world at an earlier time that enables the subject to do in an alternative possible world what she could not do in our actual world, .

By making this distinction, Lewis claims that he invalidates the seemingly lethal ***Consequence Argument*** against compatibilist free will, of which a simple version reads (assuming determinism, on which compatibilist free will is predicated):

1. Alice's actions are a necessary consequence of the laws of nature plus the state of the universe (or the relevant part thereof) at any earlier time;
2. Alice is unable to render both (laws and earlier states) false;
3. Alice is unable to render the consequences of laws and earlier states false;
4. *Ergo*: Alice is unable to do otherwise than what she actually does.

Lewis claims that statement 3 is ambiguous, in that it fails to distinguish between the two senses in his two bullet points above. The Consequence Argument requires the latter (which is false), whereas this argument itself is unsound on the former (which is true). This disambiguation of assumption 3 in the Consequence Argument, then, is supposed to save (compatibilist) free will. However, a considerable philosophical literature suggests that the tension between Lewis's denying the second bullet point whilst accepting the first is pretty uncomfortable, reflecting the corresponding tension between the conjunction of determinism and freedom in general; indeed, this is what the FWT makes precise! Let us first point out that, at least in his terminology Lewis fails to make a clear distinction between *laws of nature* and *initial states*; from the point of view of modern physics, this distinction is absolutely fundamental (although it may dispappear in post-modern physics based in e.g. quantum gravity).

Lewis's examples of law-breaking events in our actual world typically refer to violations of some law of nature (like exceeding the speed of light), whereas the (alleged) law-breaking in his counterfactuals, such as choosing a' (where in fact Alice did not do so) amounts to a change in some earlier state. Thus it might have been more appropriate if the paper in which Lewis laid out his version of compatibalism had been entitled *Are we free to change the states?* instead of *Are we free to break the laws?*. On this revision, his distinction of the two cases takes the following form:

- I am able to do something such that, if I did it, the state of the actual world at some earlier time would have been different.
- I am able to change the actual state of the world.

The latter remains impossible, while it is the former that enables free will. Applied to Alice, the former should mean (still in the compatibilist spirit of Lewis):

- A slight alteration in the state of the actual world (which would have made it a different but very similar world according to Lewis) would have led Alice to do something (such as choosing a') that she did not do in the actual world (because according to determinism its actual state at any earlier time—as opposed to the counterfactual alternative state in the discussion—led her to choose a).

We now make this revised version of Lewis's local miracle compatibilism mathematically precise, in a way that has the additional advantage of involving not only "the ability do do otherwise", but also the other component free will, i.e. agency. Here the intuition is that free will involves a separation between the agent, Alice, (who is to exercise it) and the rest of the world, under whose influence she acts. Namely, as in the FWT, let X be the state space of the Universe, and let

$$a = A(x) \tag{6.55}$$

again be Alice's setting, where $A : X \to X_A$, as before. We now assume that a is determined by her "inner state" I as well as the "outer state" O of the rest of the world, under whose influence she acts. These, in turn, are determined by the state $x \in X$ of the world. That is, $A = A(O,I)$, which expresses the existence of functions

$$O : X \to X_O; \tag{6.56}$$

$$I : X \to X_I; \tag{6.57}$$

$$\hat{A} : X_O \times X_I \to X_A, \tag{6.58}$$

where X_O and X_I are certain sets, such that for each $x \in X$ one has

$$A(x) = \hat{A}(O(x), I(x)). \tag{6.59}$$

In other words, for some given state x of the world we have

$$o = O(x); \tag{6.60}$$

$$i = I(x); \tag{6.61}$$

$$a = \hat{A}(o,i). \tag{6.62}$$

Note that, in the spirit of Conway and Kochen, in the above analysis Alice (whose free choice they after all believe to be ultimately a consequence of the free choice of elementary particles) now plays the role of the spin-1 particles in the bipartite experiment. Thus the analogy is between the triples:

$$(a,z,\lambda) \in X_A \times Z \times \Lambda; \tag{6.63}$$

$$(o,i,a) \in X_O \times X_I \times X_A. \tag{6.64}$$

- The first triple is defined in the experimental context of the FWT, where a is the setting of Alice's wing of the experiment (which from the perspective of the spin-1 particle plays the role of the outer state of the world), z is the inner state of the particle, and λ is the outcome of Alice's measurement.
- The second pertains to the analysis of Alice's "free" choice of the setting of her experiment, where o is the outer state of the world, i is her inner state, and a is her actual setting, given $x \in X$ and hence $(o,i) = (O(x), I(x))$.

Beyond ***Determinism***, which is expressed by the above framework, our fundamental assumption underpinning compatibilist free will is ***Freedom***, defined exactly as in the FWT: O and I are *independent* in that the following function is surjective:

$$\begin{aligned} O \times I : X &\to X_O \times X_I \\ x &\mapsto (O(x), I(x)), \end{aligned} \tag{6.65}$$

i.e., for each pair $(o,i) \in X_I \times X_O$ there is $x \in X$ for which (6.60) and (6.61) hold.

Rephrasing our earlier analysis in this elementary mathematical language, Lewis wants to make sense of the idea that although Alice's choice (6.62) at some fixed time t was determined by the state x of the Universe at that time through (6.60) - (6.61), or, equivalently, through (6.59), and hence—and this is the whole point of the Consequence Argument Lewis challenges—by any earlier state x_p of the Universe at time t_p, *nonetheless* Alice was "able to act otherwise" at time t, e.g. in choosing

$$a' = \hat{A}(o', i'), \tag{6.66}$$

but did not do so, since choosing a' would illegally have changed the state x to x' (both at time t), and, equivalently (given determinism), would have changed x_p to x'_p. On our reading of Lewis's theory of counterfactuals, Alice's ability to choose a' simply means that there exists a state x' of the world close to x in the sense that

$$O(x') = O(x) = o, \tag{6.67}$$

making the environment in which Alice acts the same as in the actual world, but

$$i' = I(x') \neq I(x) = i, \tag{6.68}$$

where i' should be close to i in some appropriate sense (such as a slight change in the state of Alice's brain), such that (6.66) holds, with $o' = o$ as required by (6.67).

The point, then, is that according to our *Freedom* assumption, there indeed *is* such a nearby state x', for any given i' and (o,i). Thus the freedom Alice has is precisely what we have formalized as *Freedom*: even *given* the state o of the causal influences on her behaviour (and possibly even the entire state of the rest of the world), there is a different admissible state x' of the world such that, had this state been actual, she would have chosen a' (although she in fact, necessarily, picked a).

It should be clear now that at least in the context of the Free Will Theorem, our precise technical formulation of all assumptions implies that the freedom Alice and Bob have in choosing their settings is an instance of the local miracle compatibilist form of free will proposed by Lewis (1981), at least if one accepts our reformulation thereof. The theorem then establishes a contradiction between:

- the physics assumptions, i.e., *Nature*, and *Locality*;
- the compatibilist free will assumption, i.e., *Determinism* and *Freedom*.

Accepting the former, the latter must fall. Making this choice, one should realize that the physics assumptions on the one hand just form a small corner of modern physics (from which point of view they are weak), but on the other hand have singled out the corner in which the two fundamental theories of quantum mechanics and special relativity meet and are brought to a head (from which perspective they are strong).

The challenge their theorem puts to compatibalism was recognized by Conway & Kochen (2009), who write:

> 'The tension between human free will and physical determinism has a long history. Long ago, Lucretius made his otherwise deterministic particles swerve unpredictably to allow for free will. It was largely the great success of deterministic classical physics that led to the adoption of determinism by so many philosophers and scientists, particularly those in fields remote from current physics. (This remark also applies to "compatibilism", a now unnecessary attempt to allow for human free will in a deterministic world.)'

This quotation does not use a precise version of compatibilism, but, as Conway explains elsewhere, what they mean is that compatibilism in whatever form was a desperate pre-twentieth-century attempt to save the notion of free will for e.g. Christianity in the face of the physics of the time, which assumed that the universe was a mechanical clockwork. Such attempts, then, would no longer be necessary if the world is, in fact, indeterministic (as Conway and Kochen claim to have at last proved). Our reformulation of their theorem (which removes the threat of circularity) gives a more subtle picture: the FWT uses modern physics to challenge one particular version of *compatibilist free will*. As such, it only provides indirect support for *libertarian free will*, namely by weakening one of its competitors.

To close this philosophical intermezzo, let us note that determinism is seen as a property of *theories*. Since it is the job of a deterministic theory to predict the outcome of any experiment, whether or not it is performed, this obviates the need for assumptions like counterfactuality in the sense that 'unperformed experiments have results' (which was famously denied by Asher Peres). Such controversial notions of counterfactuality have effectively been replaced by the considerably more refined modal counterfactuality of Lewis (at least in our slight reformulation thereof).

6.4 Technical intermezzo: The GHZ-Theorem

The essence of the proof of the Free Will Theorem lies in the argument that perfect correlation together with context-locality implies non-contextuality. Remarkably, context-locality is at the same time a special case of non-contextuality, as the following example illustrates. We take $H = \mathbb{C}^2 \otimes \mathbb{C}^2$, equipped with the ***Bell basis***

$$\upsilon_0 = (|01\rangle - |10\rangle)/\sqrt{2}; \tag{6.69}$$
$$\upsilon_1 = (|01\rangle + |10\rangle)/\sqrt{2}; \tag{6.70}$$
$$\upsilon_2 = (|00\rangle - |11\rangle)/\sqrt{2}; \tag{6.71}$$
$$\upsilon_3 = (|00\rangle + |11\rangle)/\sqrt{2}, \tag{6.72}$$

where we use the physicists' notation

$$|1\rangle = (1,0); \tag{6.73}$$
$$|0\rangle = (0,1); \tag{6.74}$$
$$|ij\rangle = |i\rangle \otimes |j\rangle. \tag{6.75}$$

Of course, $\mathbb{C}^2 \otimes \mathbb{C}^2 \cong \mathbb{C}^4$ contains the spin-1 Hilbert space $\mathbb{C}^3$ of the Kochen–Specker Theorem as the subspace orthogonal to the vector υ_0. Thus we identify $\mathbb{C}^3$ with the subspace $\tilde{\mathbb{C}}^3$ of $\mathbb{C}^4$ spanned by the basis vectors $\upsilon_1, \upsilon_2, \upsilon_3$. The operators

$$\tilde{J}_{\mathbf{u}} = \tfrac{1}{2}(\sigma_{\mathbf{u}} \otimes 1_2 + 1_2 \otimes \sigma_{\mathbf{u}}), \tag{6.76}$$

where $\mathbf{u} \in \mathbb{R}^3$ is a unit vector as before, and

$$\sigma_{\mathbf{u}} = \sum_{i=1}^{3} \sigma^i u_i \tag{6.77}$$

in terms of the Pauli matrices σ^i, map υ_1 to zero and leave its orthogonal complement $\tilde{\mathbb{C}}^3$ stable. Elementary group theory or direct calculation then shows that the operator $J_{\mathbf{u}}$ on $\mathbb{C}^3$ in (6.11) is (unitarily) equivalent to the operator $\tilde{J}_{\mathbf{u}}$ on $\tilde{\mathbb{C}}^3$. Since

$$\tilde{J}_{\mathbf{u}}^2 = \tfrac{1}{2}(\sigma_{\mathbf{u}} \otimes \sigma_{\mathbf{u}} + 1_2 \otimes 1_2), \tag{6.78}$$

the Kochen–Specker argument can be rephrased in terms of the operators $\sigma_{\mathbf{u}} \otimes \sigma_{\mathbf{u}}$. In particular, for each frame $a = (\mathbf{u}_1, \mathbf{u}_2, \mathbf{u}_3)$, the three operators

$$(\sigma_{\mathbf{u}_1} \otimes \sigma_{\mathbf{u}_1}, \sigma_{\mathbf{u}_2} \otimes \sigma_{\mathbf{u}_2}, \sigma_{\mathbf{u}_3} \otimes \sigma_{\mathbf{u}_3}) \tag{6.79}$$

commute, they each square to one, and their joint eigenvalues are one of the triples:

$$(-1,-1,-1), (-1,1,1), (1,-1,1), (1,1,-1).$$

The eigenvector corresponding to the first one is υ_0, and hence the others must lie in $\tilde{\mathbb{C}}^3$. Hence by Lemma 6.4 any quasi-linear non-contextual hidden variable must also assign these values, which by Lemma 6.7 is impossible for arbitrary bases.

The key mathematical property of the three operators (6.79) is that they commute, and together with the unit $1_2 \otimes 1_2$ form a maximal set of commuting self-adjoint matrices on $\mathbb{C}^4$. But other such sets could have been chosen by Alice (under whose sole control the situation so far has been assumed to be), such as a triple of the kind

$$(\sigma_{\mathbf{u}} \otimes 1_2, 1_2 \otimes \sigma_{\mathbf{v}}, \sigma_{\mathbf{u}} \otimes \sigma_{\mathbf{v}}),$$

where $\mathbf{u}$ and $\mathbf{v}$ are arbitrary unit vectors in $\mathbb{R}^3$. Since the third operator is the product of the first two, the joint eigenvalues of this triple, and hence also the assignments by a quasi-linear non-contextual hidden variable, must be one of the four triples

$$(1,1,1),(-1,1,-1),(1,-1,-1),(-1,-1,1).$$

The non-contextuality assumption would then dictate that the outcome of Alice's measurement of $\sigma_{\mathbf{u}} \otimes 1_2$ be independent of her choice of the setting $\mathbf{v}$ in a possible simultaneous measurement of $1_2 \otimes \sigma_{\mathbf{v}}$, and *vice versa*. Therefore, in a (non-local) bipartite setting where Alice is only able to measure operators of the type $a \otimes 1_2$, whilst Bob can measure $1_2 \otimes b$, on the above choice of (commuting) operators, *non-contextuality in the situation where Alice controls everything is mathematically equivalent to (context) locality in the bipartite Alice & Bob setting.*

Further constraints then arise if the system is prepared in a correlated state like ψ_0, which is an eigenstate of $\sigma_{\mathbf{u}} \otimes \sigma_{\mathbf{v}}$ with eigenvalue -1 whenever $\mathbf{u} = \mathbf{v}$. So in that case the values of $(\sigma_{\mathbf{u}} \otimes 1_2, 1_2 \otimes \sigma_{\mathbf{v}})$ can only be $(1,-1)$ or $(-1,1)$, yielding perfect anti-correlation. This is not enough, however, to derive a Free Will Theorem; to do so with the small single-site Hilbert space $\mathbb{C}^2$, one needs a third (non-local) party.

Indeed, the well-known tripartite GHZ-argument may be rephrased as a Free Will Theorem, as follows. The underlying Hilbert space is

$$H = \mathbb{C}^2 \otimes \mathbb{C}^2 \otimes \mathbb{C}^2 \cong \mathbb{C}^8, \tag{6.80}$$

and hence as a warm-up we first (re)prove Theorem 6.5 for $n = 8$. Suppose we have a map $V : H_8(\mathbb{C}) \to \mathbb{R}$ as in Definition 6.1. Write

$$\lambda_1^{(a)} = V(\sigma_a \otimes 1_2 \otimes 1_2), \lambda_2^{(b)} = V(1_2 \otimes \sigma_b \otimes 1_2), \lambda_3^{(c)} = V(1_2 \otimes 1_2 \otimes \sigma_c),$$

where a,b,c can be $1,2,3$. From Lemma 6.4 we then have

$$V(\sigma_1 \otimes \sigma_2 \otimes \sigma_2) = \lambda_1^{(1)} \lambda_2^{(2)} \lambda_3^{(2)}; \tag{6.81}$$

$$V(\sigma_2 \otimes \sigma_1 \otimes \sigma_2) = \lambda_1^{(2)} \lambda_2^{(1)} \lambda_3^{(2)}; \tag{6.82}$$

$$V(\sigma_2 \otimes \sigma_2 \otimes \sigma_1) = \lambda_1^{(2)} \lambda_2^{(2)} \lambda_3^{(1)}; \tag{6.83}$$

$$V(\sigma_1 \otimes \sigma_1 \otimes \sigma_1) = \lambda_1^{(1)} \lambda_2^{(1)} \lambda_3^{(1)}. \tag{6.84}$$

Furthermore, the four operators on the left-hand side commute and turn out to satisfy

$$\sigma_1 \otimes \sigma_2 \otimes \sigma_2 \cdot \sigma_2 \otimes \sigma_1 \otimes \sigma_2 \cdot \sigma_2 \otimes \sigma_2 \otimes \sigma_1 = -\sigma_1 \otimes \sigma_1 \otimes \sigma_1, \tag{6.85}$$

so that again by Lemma 6.4,

$$\lambda_1^{(1)}\lambda_2^{(2)}\lambda_3^{(2)} \cdot \lambda_1^{(2)}\lambda_2^{(1)}\lambda_3^{(2)} \cdot \lambda_1^{(2)}\lambda_2^{(2)}\lambda_3^{(1)} = -\lambda_1^{(1)}\lambda_2^{(1)}\lambda_3^{(1)}, \tag{6.86}$$

i.e. $(\lambda_1^{(1)}\lambda_2^{(2)}\lambda_3^{(2)})^2 = -1$. Since $\lambda_j^{(i)} = \pm 1$, this is impossible, so that V cannot exist.

Now, using the notation in the preceding discussion, consider the unit vector

$$\psi_{GHZ} = (|111\rangle - |000\rangle)/\sqrt{2}, \tag{6.87}$$

which is a joint eigenstate of each of the four operators on the left-hand side of (6.81) - (6.84), with eigenvalue $+1$ for the first three, and hence eigenvalue -1 for the fourth, i.e., $\sigma_1 \otimes \sigma_1 \otimes \sigma_1$. So if setting $A = a$ for Alice (where $a \in \{1,2\}$) means that she measures $F = \sigma_a \otimes 1_2 \otimes 1_2$ with outcome $\lambda_1^{(a)} = \pm 1$, and similarly $B = b$ for Bob and $C = c$ for Cindy mean that they measure $G = 1_2 \otimes \sigma_b \otimes 1_2$ and $H = 1_2 \otimes 1_2 \otimes \sigma_c$ with outcomes $\lambda_2^{(b)} = \pm 1$ and $\lambda_3^{(c)} = \pm 1$, respectively, then in the state ψ_{GHZ} each of the settings gives the correlation

$$\text{settings } (a,b,c) = (1,2,2),(2,1,2),(2,2,1) \Rightarrow \lambda_1^{(a)}\lambda_2^{(b)}\lambda_3^{(c)} = 1; \tag{6.88}$$

$$\text{setting } (a,b,c) = (1,1,1) \Rightarrow \lambda_1^{(a)}\lambda_2^{(b)}\lambda_3^{(c)} = -1. \tag{6.89}$$

Theorem 6.14. *The conjunction of the following assumptions is contradictory:*

- **Determinism**: *there is a state space X with associated functions*

$$A,B,C : X \to \{1,2\}, F,G,H : X \to \Lambda,$$

which completely describes the experiment, in that $x \in X$ determines both settings (a,b,c) and outcomes $(\lambda_1,\lambda_2,\lambda_3) \in \Lambda^3$ through $a = A(x)$, $\lambda_1 = F(x)$, etc.
- **Nature:** *the experiment (performed in the state ψ_{GHZ}) has possible outcomes in $\Lambda = \{-1,1\}$, subject to the correlations* (6.88) - (6.89)*;*
- **Freedom**: *there is a further function $Z : X \to X_Z$, in terms of which*

$$F = F(A,B,C,Z),\ G = G(A,B,C,Z),\ H = H(A,B,C,Z),$$

and F, G, H, Z are independent, i.e. for each (a,b,c,z) there is $x \in X$ such that

$$A(x) = a,\ B(x) = b,\ C(x) = c,\ Z(x) = z.$$

- **Locality**: *$F = F(A,Z)$, $G = G(B,Z)$, and $H = H(C,Z)$.*

Proof. Using notation as in the proof of Theorem 6.13, for fixed $z \in Z$ we obtain $\hat{F}(a,z) = \lambda_1^{(a)}$ etc. *Nature* then leads to the contradiction derived after (6.86). □

6.5 Bell's theorems

Two different results are known as "Bell's Theorem": the first, from his paper in 1964, is Theorem 6.15 below, and the second, dating from 1976, is Theorem 6.18. The first is similar to the Free Will Theorem in both its assumptions and its conclusion, and to make this similarity more obvious we first state it for $\mathbb{C}^3$ instead of $\mathbb{C}^2$. The difference lies in the probabilistic flavour of Bell's Theorem, whose empirical input is not given by the only non-probabilistic consequence to be drawn from the quantum-mechanical formulae (6.35) - (6.38), viz. the certainty (6.43) of perfect correlation on identical settings, but rather by the probabilistic formula (6.40), i.e.,

$$P_{\psi_0}(F_i \neq G_j | A_i = \mathbf{u}_i, B_j = \mathbf{v}_j) = \tfrac{2}{3}\sin^2\theta_{\mathbf{u}_i,\mathbf{v}_j} \ \ (i,j = 1,2,3), \tag{6.90}$$

where $\theta_{\mathbf{u},\mathbf{v}}$ is the angle between two unit vectors $\mathbf{u}$ and $\mathbf{v}$. Furthermore, the state space X must be upgraded to a probability space (X,Σ,μ), carrying functions A and B (for the settings, which unlike Bell himself—who treated them as labels—we include among the random variables), F and G (for the outcomes) and finally Z (for the hidden variable traditionally called λ) as random variables, i.e., measurable functions. This also implies that the target spaces X_A to X_Z (which is traditionally called Λ) must be equipped with some σ-algebra of measurable subsets. But this is not a big deal, since $X_A = X_B$ carries a natural Borel structure and $X_F = X_G$ is finite. The probability measure μ is assumed independent of (A,B,F,G), and *vice versa*.

The measure μ, which gives the "hidden state" of the system that allegedly underlies its quantum-mechanical description, is chosen in such a way that empirical probabilities (typically obtained from long runs of repeated measurements) are recovered as joint conditional probabilities defined by μ and the random variables, i.e., assuming the settings (a,b) are possible in that $P(A=a,B=b) > 0$, we put

$$P(F=\lambda, G=\gamma | A=a, B=b) = \frac{P(F=\lambda, G=\gamma, A=a, B=b)}{P(A=a, B=b)}, \tag{6.91}$$

where the joint probabilities on the right-hand side are given by

$$P(A=a,B=b) = \mu(A=a,B=b\}; \tag{6.92}$$

$$P(F=\lambda,G=\gamma,A=a,B=b) = \mu(F=\lambda,G=\gamma,A=a,B=b\}, \tag{6.93}$$

where $\mu(A=a,B=b)$ is shorthand for $\mu(x \in X \mid A(x)=a, B(x)=b\}$, etc. This implies that μ depends on (but may not be determined by) the quantum state ψ_0.

On this understanding, the assumptions of ***Determinism*** and ***Locality*** are the same as for the Free Will Theorem (except that equations like $F(x) = \hat{F}(A(x),Z(x))$ are merely supposed to hold almost everywhere with respect to μ). ***Freedom*** is now taken to mean that (A,B,Z) are *probabilistically independent* relative to μ. By definition, this also means that the pairs (A,B), (A,Z), and (B,Z) are independent, so that for any $\mathsf{A} \subset X_A$, $\mathsf{B} \subset X_B$, and (measurable) $\mathsf{Z} \subset X_Z$, defining

$$P(A \in \mathsf{A}, B \in \mathsf{B}, Z \in \mathsf{Z}) = \mu(x \in X \mid A(x) \in \mathsf{A}, B(x) \in \mathsf{B}, Z(x) \in \mathsf{Z}), \tag{6.94}$$

and analogous expressions for $P(A \in \mathsf{A})$ and $P(A \in \mathsf{A}, B \in \mathsf{B})$, etc., we have

$$P(A \in \mathsf{A}, B \in \mathsf{B}) = P(A \in \mathsf{A})P(B \in \mathsf{B}); \quad (6.95)$$
$$P(A \in \mathsf{A}, Z \in \mathsf{Z}) = P(A \in \mathsf{A})P(Z \in \mathsf{Z}); \quad (6.96)$$
$$P(B \in \mathsf{B}, Z \in \mathsf{Z}) = P(B \in \mathsf{B})P(Z \in \mathsf{Z}); \quad (6.97)$$
$$P(A \in \mathsf{A}, B \in \mathsf{B}, Z \in \mathsf{Z}) = P(A \in \mathsf{A})P(B \in \mathsf{B})P(Z \in \mathsf{Z}). \quad (6.98)$$

If we finally define ***Nature*** as the claim that $\hat{F}$ and $\hat{G}$ are $\underline{2}$-valued and that

$$P(F_i \neq G_j | A_i = \mathbf{u}_i, B_j = \mathbf{v}_j) = \tfrac{2}{3}\sin^2\theta_{\mathbf{u}_i,\mathbf{v}_j} \quad (i,j = 1,2,3), \quad (6.99)$$

where the left-hand side is the *conditional* probability defined by μ and the random variables in question (whereas the left-hand side of (6.90) is the *empirical* probability for the experiment in question, or, equivalently, the quantum-mechanical prediction thereof), then we obtain the following spin-1 version of ***Bell's first theorem***:

Theorem 6.15. *Determinism, Freedom, Nature, and Locality are contradictory.*

This formulation is literally the same as Theorem 6.13, but the terms have acquired a different technical meaning now, especially *Freedom* and *Nature*. Moreover, purists would add *Probability Theory* as an assumption in Bell's Theorem, as its formalism is decidedly non-tautological and its interpretation is far from obvious, even in a classical setting. In any case, the proof is practically the same as in the more familiar optical version of the EPR-experiment, to which we now turn.

In the classical (sic) form of the experiment, Alice and Bob perform measurements on incoming photons by letting them pass through a polaroid glass whose axis of polarization makes angle a (Alice) or b (Bob) with (say) the horizontal axis in the plane orthogonal to the direction of propagation of the photons. Considered in the light of the previous experiment on spin-1 particles, such a choice of settings may also be seen as a choice of basis for $\mathbb{R}^3$, with the proviso that, assuming (by convention) the photons move along the y-axis, one basis element $\mathbf{u}_2 = (0,1,0)$ is fixed so that the remaining two vectors $(\mathbf{u}_1, \mathbf{u}_3)$ must lie in the x-z plane (in which, on a naive picture, the photons may "vibrate"). This constraint gives rise to bases

$$\mathbf{u}_1 = (\cos a, 0, \sin a), \mathbf{u}_2 = (0,1,0), \mathbf{u}_3 = (-\sin a, 0, \cos a), \quad (6.100)$$

the first of which (say) gives the actual direction of the axis of polarization. In any case, Alice writes down $F = 1$ if her photon passes her glass at angle a, and $F = 0$ if it does not; similarly Bob writes $G = 1$ (pass) or $G = 0$ (fail) at setting b.

In a quantum-mechanical description of the experiment, the Hilbert space of the photon pair is $\mathbb{C}^2 \otimes \mathbb{C}^2$, and the correlated photon state is taken to be

$$\psi_0 = (\mathbf{e}_1 \otimes \mathbf{e}_1 + \mathbf{e}_2 \otimes \mathbf{e}_2)/\sqrt{2}, \quad (6.101)$$

where $\mathbf{e}_1 = (1,0)$ and $\mathbf{e}_2 = (0,1)$ form the standard basis of $\mathbb{C}^2$. The probabilities (6.35) - (6.38) as predicted by quantum mechanics are now replaced by

$$P_{\psi_0}(F=1,G=1|A=a,B=b) = \tfrac{1}{2}\cos^2(a-b); \tag{6.102}$$
$$P_{\psi_0}(F=0,G=0|A=a,B=b) = \tfrac{1}{2}\cos^2(a-b); \tag{6.103}$$
$$P_{\psi_0}(F=1,G=0|A=a,B=b) = \tfrac{1}{2}\sin^2(a-b), \tag{6.104}$$
$$P_{\psi_0}((F=0,G=1|A=a,B=b) = \tfrac{1}{2}\sin^2(a-b), \tag{6.105}$$

which are also the experimentally measured ones. Instead of (6.90) we then obtain

$$P_{\psi_0}(F\neq G|A=a,B=b) = \sin^2(a-b); \tag{6.106}$$
$$P_{\psi_0}(F=G|A=a,B=b) = \cos^2(a-b). \tag{6.107}$$

In particular, if their settings are the same (i.e., $a=b$), then Alice and Bob will always find the same outcome (***perfect correlation***), whereas in case they are orthogonal (i.e., $a=b\pm\pi/2$), they obtain ***perfect anti-correlation***, in that Alice's photon passes whenever Bob's is blocked, and *vice versa*. However, this will not be used. Although it should be obvious from the previous case what the assumptions in Theorem 6.15 mean for this particular experiment, we make them explicit:

- ***Determinism*** means that there is a probability space (X,Σ,μ) with associated (measurable) functions

$$A:X\to[0,\pi], B:X\to[0,\pi], F:X\to\{0,1\}, G:X\to\{0,1\}, \tag{6.108}$$

  which completely describe the experiment in the sense that $x\in X$ determines *both* its settings $a=A(x), b=B(x)$ *and* its outcomes $\lambda=F(x), \gamma=G(x)$.
- ***Freedom*** stipulates that there is a (measurable) function $Z:X\to X_Z$ such that:
  - $F=F(A,B,Z)$ and $G=G(A,B,Z)$;
  - (A,B,Z) are probabilistically independent relative to μ.
- ***Locality*** means that $F(A,B,Z)=F(A,Z)$ and $G(A,B,Z)=G(B,Z)$.
- ***Nature*** states that the empirical as well as theoretical probabilities (6.106) for the experiment are reproduced as conditional joint probabilities given by μ through

$$P(F\neq G|A=a,B=b) = \sin^2(a-b). \tag{6.109}$$

Theorem 6.15 then holds *verbatim* for this situation, with the following proof.

Proof. Determinism and *Freedom* imply

$$P(F=\lambda,G=\gamma|A=a,B=b) = P_{ABZ}(\hat{F}=\lambda,\hat{G}=\gamma|\hat{A}=a,\hat{B}=b), \tag{6.110}$$

where we use the notation (6.50) - (6.51), the function $\hat{A}:X_A\times X_B\times X_Z\to X_A$ is projection on the first coordinate, likewise the function $\hat{B}:X_A\times X_B\times X_Z\to X_B$ is projection on the second, and P_{ABZ} is the joint probability on $X_A\times X_B\times X_Z$ induced by the triple (A,B,Z) and the probability measure μ; by independence, P_{ABZ} is a product measure on $X_A\times X_B\times X_Z$. According to *Locality*, $\hat{F}(a,b,z)$ does not depend on b, whilst $\hat{G}(a,b,z)$ does not depend on a.

For fixed settings (a,b), we may therefore define the following functions on X_Z:

$$\hat{F}_a(z) = \hat{F}(a,z); \tag{6.111}$$
$$\hat{G}_b(z) = \hat{G}(b,z). \tag{6.112}$$

A brief computation then yields

$$P_{ABZ}(\hat{F}=\lambda,\hat{G}=\gamma|\hat{A}=a,\hat{B}=b) = P_Z(\hat{F}_a=\lambda,\hat{G}_b=\gamma), \tag{6.113}$$

where P_Z is the joint probability on X_Z defined by Z and μ. Therefore, from (6.110),

$$P(F=\lambda,G=\gamma|A=a,B=b) = P_Z(\hat{F}_a=\lambda,\hat{G}_b=\gamma). \tag{6.114}$$

Nature then gives the crucial result

$$P_Z(\hat{F}_a \neq \hat{G}_b) = \sin^2(a-b). \tag{6.115}$$

Lemma 6.16. *Any four* $\{0,1\}$*-valued random variables* (F_1,F_2,G_1,G_2) *satisfy*

$$P(F_1 \neq G_1) \leq P(F_1 \neq G_2) + P(F_2 \neq G_1) + P(F_2 \neq G_2). \tag{6.116}$$

This lemma (said to go back to Boole) is very easy to prove directly, but for completeness's sake we mention that it also follows from Proposition 6.17 below.

Taking $F_1=\hat{F}_{a_1}$, $F_2=\hat{F}_{a_2}$, $G_1=\hat{G}_{b_1}$, $G_2=\hat{G}_{b_2}$, and $P=P_Z$, for suitable values of (a_1,a_2,b_1,b_2) this inequality is violated by (6.115). Take, for example, $a_2=b_2=3x$, $a_1=0$, and $b_1=x$. The inequality (6.116) then assumes the form $f(x)\geq 0$ for

$$f(x) = \sin^2(3x)+\sin^2(2x)-\sin^2(x).$$

But in fact, $f(x)<0$ for continuously many values of $x\in[0,2\pi]$, see plot. □

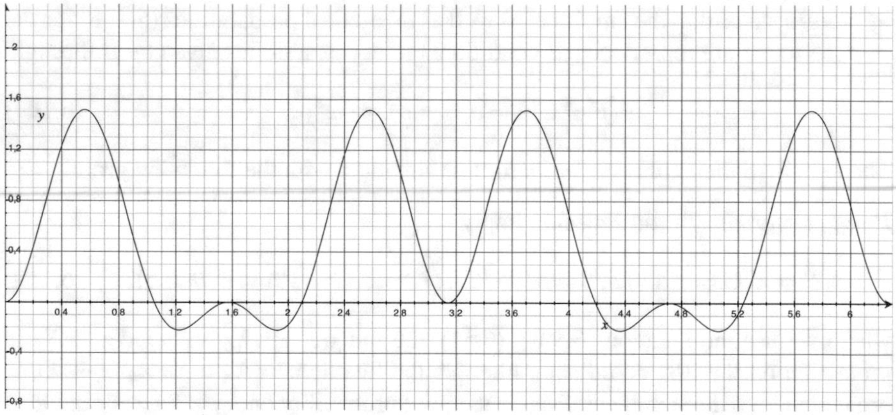

Graph of $x\mapsto \sin^2(3x)+\sin^2(2x)-\sin^2(x)$, *showing (in the region where it is negative) that quantum mechanics violates the Bell inequality* (6.116).

Lemma 6.16 is a special case of a more general result.

Proposition 6.17. *Let $F_i : X \to [-1,1]$ and $G_j : X \to [-1,1]$, where (X,Σ,μ) is some probability space, be two parametrized random variables, $i,j = 1,2$. Then the* **two-point function** $\langle F_i G_j \rangle = \int_X d\mu\, F_i G_j$ *satisfies the* CHSH-**inequality**

$$|\langle F_1 G_1 \rangle + \langle F_1 G_2 \rangle + \langle F_2 G_1 \rangle - \langle F_2 G_2 \rangle| \leq 2. \tag{6.117}$$

If F_i and G_j just take the values ± 1, then (6.116) is a special case of (6.117).

Proof. In terms of the function $\Phi = F_1 \cdot (G_1 + G_2) + F_2 \cdot (G_1 - G_2)$, we may write

$$\langle F_1 G_1 \rangle + \langle F_1 G_2 \rangle + \langle F_2 G_1 \rangle - \langle F_2 G_2 \rangle = \int_X d\mu\, \Phi. \tag{6.118}$$

Since $|F_i(x)| \leq 1$ and $|G_j(x)| \leq 1$ by assumption, we have $|\Phi(x)| \leq 2$ and hence

$$\left| \int_X d\mu(x)\, \Phi(x) \right| \leq \int_X d\mu(x)\, |\Phi(x)| \leq 2, \tag{6.119}$$

since μ is a probability measure. To prove the the last claim, we just note that

$$P(F_i = G_j) - P(F_i \neq G_j) = \langle F_i G_j \rangle;$$
$$P(F_i = G_j) + P(F_i \neq G_j) = 1.$$

□

In Bell's second (1976) theorem on ***stochastic hidden variables***, the assumption of *Determinism* is dropped, and all we have is a theory stating conditional probabilities $P(F = \lambda, G = \gamma | A = a, B = b, x)$ for the outcomes of the above bipartite experiment given some hidden variable x, as well as the single-wing versions $P(F = \lambda | A = a)$ and $P(G = \gamma | B = b, x)$. Here F, G, A, B are just notational devices to record such outcomes, *which are no longer (necessarily) represented as random variables*. On this new understanding of the notation, the ***Nature*** assumption is formulated just as before, cf. (6.109). We do assume the existence of a probability space (X, Σ, μ) and of conditional probabilities

$$P(F = \lambda, G = \gamma | A = a, B = b, x), \;\; P(F = \lambda | A = a, x), \;\; P(G = \gamma | B = b, x),$$

defined μ-a.e. in x, in which the state of the world is specified as being $x \in X$. In terms of this space, the ***Freedom*** assumption means that

$$P(F = \lambda, G = \gamma | A = a, B = b) = \int_X d\mu(x)\, P(F = \lambda, G = \gamma | A = a, B = b, x), \tag{6.120}$$

for any settings (a,b), of which μ is independent (as the notation already indicated).
The crucial assumption replacing *Determinism* is ***Bell locality***, which reads

$$P(F = \lambda, G = \gamma | A = a, B = b, x) = P(F = \lambda | A = a, x) \cdot P(G = \gamma | B = b, x). \tag{6.121}$$

Bell's second theorem for stochastic hidden variable theories reads as follows.

Theorem 6.18. *Nature, Freedom, and Bell locality are contradictory.*

Proof. The idea of the proof is to introduce an artificial probability space in order to recover the framework of Theorem 6.15. To this end, we take

$$\tilde{X} = [0,1] \times [0,1] \times X; \tag{6.122}$$
$$d\tilde{\mu}(s,t,x) = ds \cdot dt \cdot d\mu(x). \tag{6.123}$$

where we denoted the elements of $\tilde{X}$ by (s,t,x). On $\tilde{X}$, define random variables

$$\tilde{F}_a(s,t,x) = 1_{[0,P(F=1|A=a,x)]}(s); \tag{6.124}$$
$$\tilde{G}_b(s,t,x) = 1_{[0,P(G=1|B=b,x)]}(t), \tag{6.125}$$

where 1_Δ is the indicator function for $\Delta \subseteq [0,1]$. Writing, as usual,

$$\tilde{P}(\tilde{F}_a = \lambda, \tilde{G}_b = \gamma) = \int_{\tilde{X}} d\tilde{\mu}(s,t,x)\,\{(s,t,x) \in \tilde{X} \mid \tilde{F}_a(s,t,x) = \lambda, \tilde{G}_b(s,t,x) = \gamma\},$$

we obtain (first for $\lambda = \gamma = 1$, from which the other cases follow):

$$\tilde{P}(\tilde{F}_a = \lambda, \tilde{G}_b = \gamma) = \int_X d\mu(x)\, P(F = \lambda|A = a,x) \cdot P(G = \gamma|B = b,x). \tag{6.126}$$

With *Freedom* and *Bell locality*, this yields

$$P(F = \lambda, G = \gamma|A = a, B = b) = \tilde{P}(\tilde{F}_a = \lambda, \tilde{G}_b = \gamma), \tag{6.127}$$

as in (6.114), so that the proof may be completed as for Theorem 6.15. □

Let us note that since in Bell's second theorem the settings (a,b) are treated as free parameters to begin with, the difference between X and Z evaporates, so that in the above formulae one might as well have replaced (X,μ) by the space (X_Z,μ_Z) that describes all relevant degrees of freedom *except the settings* (i.e., the experimentalist, in either human or machine form). Either way, Bell's locality condition may be disentangled into the following conditions (introduced by Jarrett and Shimony):

1. ***Parameter Independence*** (PI):

$$P(\lambda|a,b,x) = P(\lambda|a,x); \tag{6.128}$$
$$P(\gamma|a,b,x) = P(\gamma|b,x); \tag{6.129}$$

2. ***Outcome Independence*** (OI):

$$P(\lambda|a,b,\gamma,x) = P(\lambda|a,b,x); \tag{6.130}$$
$$P(\gamma|a,b,\lambda,x) = P(\gamma|a,b,x), \tag{6.131}$$

where we have abbreviated $P(F = \lambda|A = a, B = b,x)$ by $P(\lambda|a,b,x)$, etc., and have used the following notation (which states identities in case one has (6.91) - (6.93)):

$$P(\lambda|a,b,x) \equiv \sum_{\gamma} P(\lambda,\gamma|a,b,x); \tag{6.132}$$

$$P(\gamma|a,b,x) \equiv \sum_{\lambda} P(\lambda,\gamma|a,b,x); \tag{6.133}$$

$$P(\lambda|a,b,\gamma,x) \equiv \frac{P(\lambda,\gamma|a,b,x)}{P(\gamma|a,b,x)}; \tag{6.134}$$

$$P(\gamma|a,b,\lambda,x) \equiv \frac{P(\lambda,\gamma|a,b,x)}{P(\lambda|a,b,x)}, \tag{6.135}$$

It is easy to see that *Bell locality is equivalent to the conjunction of* PI *and* OI.

Note that the former (PI), akin to *Locality*, is a hidden or 'subsurface' version of the ***no signaling*** property of the 'surface' probabilities, which states that

$$P(\lambda|a,b) \equiv \sum_{\gamma} P(\lambda,\gamma|a,b)$$

is independent of b (and *vice versa*). But a violation of PI only leads to signaling if x can be operationally controlled, similar to the way in which experimental physicists prepare quantum states ψ. Hence it is reassuring that quantum mechanics satisfies PI if we see the quantum state ψ as a hidden variable: assuming

$$P(\lambda,\gamma|a,b,x) = P_{\psi_0}(F=\lambda,G=\gamma|A=a,B=b), \tag{6.136}$$

as computed in (6.102) - (6.105), PI is valid but OI is not. First, for $\lambda = 0$ or $\lambda = 1$,

$$P(\lambda|a,b,x) = \sum_{\gamma=0,1} P_{\psi_0}(F=\lambda,G=\gamma|a,b) = \tfrac{1}{2}\cos^2(a-b) + \tfrac{1}{2}\sin^2(a-b) = \tfrac{1}{2}, \tag{6.137}$$

which is independent of b, and likewise $P(\gamma|a,b,x) = \frac{1}{2}$, independently of a. This yields PI, which a similar computation shows to be true for any quantum state. On the other hand, given this result, OI would require

$$P_{\psi_0}(F=\lambda,G=\gamma|A=a,B=b) = P_{\psi_0}(F=\lambda|A=a)\cdot P_{\psi_0}(G=\gamma|B=b),$$

which is false, since by (6.102) - (6.105), Alice's and Bob's outcomes are correlated.

Hence Bell locality is violated by quantum mechanics, but this does not imply that "quantum mechanics is nonlocal" (as some say). Bell's is a very specific locality condition invented as a constraint on hidden variable theories. In another important sense, viz. *Einstein locality*, quantum mechanics *is* local, in that observables with spacelike separated localization regions commute (this is the case in quantum field theory, but also in any bipartite experiment of the type considered here, where Alice's operators commute with Bob's just by definition of the tensor product).

On the other hand, deterministic theories, which in the present context are defined as those for which all conditional probabilities like $P(\lambda,\gamma|a,b,x)$ are either zero or one (in which case one may introduce random variables reproducing these probabilities), violate PI but satisfy OI, at least if they reproduce the Born probabilities (such as Bohmian mechanics). Hence such theories violate Bell locality.

Finally, Bell-type inequalities like (6.117) also give information about quantum mechanics itself, particularly about the degree of entanglement of states. Let H_1 and H_2 be Hilbert spaces, with tensor product $H_1 \otimes H_2$. A unit vector $\psi \in H_1 \otimes H_2$ is called ***uncorrelated*** if it is of the form $\psi = \varphi_1 \otimes \varphi_2$, where $\varphi_k \in H_k$ are unit vectors, $k = 1,2$, and ***correlated*** otherwise. Clearly, the vectors (6.34) and (6.101) used in the experiments so far are correlated. The simplest result is then as follows.

Theorem 6.19. *Let a_1 and a_2 be self-adjoint operators on H_1, and let b_1 and b_2 be self-adjoint operators on H_2, each with spectrum contained in $[-1,1]$ (equivalently $\|X_a\| \leq 1$, etc.). Let ψ be a unit vector in $H_1 \otimes H_2$, and define two-point functions*

$$\langle F_i G_j \rangle = \langle \psi, a_i \otimes b_j \psi \rangle. \tag{6.138}$$

If ψ is uncorrelated, then the Bell inequality (6.117) *holds.*

Proof. This follows from the factorization property

$$\langle F_i G_j \rangle = \langle \varphi_1 \otimes \varphi_2, a_i \otimes b_j \varphi_1 \otimes \varphi_2 \rangle = \langle \varphi_1, a_i \varphi_1 \rangle \cdot \langle \varphi_2, b_j \varphi_2 \rangle = \langle F_i \rangle \cdot \langle G_j \rangle, \tag{6.139}$$

where $\langle F_i \rangle = \langle \varphi_1, a_i \varphi_1 \rangle$ and $\langle G_j \rangle = \langle \varphi_2, b_j \varphi_2 \rangle$. For either sign, this property yields

$$\langle F_2(G_1 - G_2) \rangle = \langle F_2 \rangle \langle G_1 \rangle (1 \pm \langle F_1 \rangle \langle G_2 \rangle) - \langle F_2 \rangle \langle G_2 \rangle (1 \pm \langle F_1 \rangle \langle G_1 \rangle). \tag{6.140}$$

The spectral assumption implies that $|\langle F_i \rangle| \leq 1$ and $|\langle G_j \rangle| \leq 1$, which will be used directly below, as well as its consequence $|1 \pm \langle F_1 \rangle \langle G_2 \rangle| = 1 \pm \langle F_i \rangle \langle G_j \rangle$. Hence

$$\begin{aligned} |\langle F_2(G_1 - G_2) \rangle| &\leq |1 \pm \langle F_1 \rangle \langle G_2 \rangle| + |1 \pm \langle F_1 \rangle \langle G_1 \rangle| \\ &= 1 \pm \langle F_1 \rangle \langle G_2 \rangle + 1 \pm \langle F_1 \rangle \langle G_1 \rangle \\ &= 2 \pm \langle F_1 (G_1 + G_2) \rangle. \end{aligned} \tag{6.141}$$

Similarly,

$$|\langle F_1(G_1 + G_2) \rangle| \leq 2 \pm \langle F_2(G_1 - G_2) \rangle, \tag{6.142}$$

so that, writing $\Phi = \langle F_1 G_1 \rangle + \langle F_1 G_2 \rangle + \langle F_2 G_1 \rangle - \langle F_2 G_2 \rangle$, for either sign $\pm$ we have

$$|\Phi| \leq |\langle F_1(G_1 + G_2) \rangle| + |\langle F_2(G_1 - G_2) \rangle| \leq 4 \pm \Phi \tag{6.143}$$

If $\Phi \geq 0$ we choose the minus sign, whereas for $\Phi < 0$ we take the plus sign. Either way, we obtain $|\Phi| \leq 2$, which is the inequality (6.117). □

This result is actually much more general (as hinted at by the way that the proof only uses the uncorrelated vector state $\psi = \varphi_1 \otimes \varphi_2$). The simplest generalization is to replace pure states by mixed states, where we say that a density operator ρ on $H_1 \otimes H_2$ is *uncorrelated* if it is of the form $\rho = \sum_i p_i \rho_1 \otimes \rho_2$, where the p_i are probabilities and ρ_k is a density matrix on H_k, $k = 1,2$. Then all uncorrelated density matrices satisfy the inequality (6.117). Even more generally, uncorrelated states on C*-algebras or von Neumann algebras $A \otimes B$ satisfy (6.117), see Notes.

6.6 The Colbeck–Renner Theorem

One may try to strengthen Bell's second theorem by weakening its assumptions. A remarkable result in this direction states that, roughly speaking, any probabilistic hidden variable theory that satisfies *Freedom* and *Parameter Independence* and is compatible with quantum mechanics adds nothing to quantum mechanics. In other words, it appears that quantum mechanics "cannot be extended", or "is complete".

In fact, the result turns out to be more modest than this summary suggests, since the reasoning required to prove the claim hinges on certain assumptions which are satisfied by quantum mechanics itself, but might seem unnatural for a hidden variable theory. In any case, we have to state our notation and assumptions very clearly.

Definition 6.20. *A* **hidden variable theory** $\mathscr{T}$ *underlying quantum mechanics consists of a measurable space* (X,Σ) *whose points x label conditional probabilities*

$$P(a_1=\lambda_1,\ldots,a_n=\lambda_n|x)\equiv P(\mathbf{a}=\lambda|x)$$

for the possible outcomes $\lambda=(\lambda_1,\ldots,\lambda_n)$ *of a measurement of any family* $\mathbf{a}=(a_1,\ldots,a_n)$ *of n commuting self-adjoint operators on any Hilbert space H.*

These formal conditional probabilities are a priori *only supposed to satisfy*

$$0\leq P(\mathbf{a}=\lambda|x)\leq 1; \tag{6.144}$$
$$\sum_{\lambda} P(\mathbf{a}=\lambda|x)=1. \tag{6.145}$$

Apart from these probabilities, for each Hilbert space H and any pure state $e\in\mathscr{P}_1(H)$, *the theory* $\mathscr{T}$ *yields a classical state* μ_e, *i.e., a probability measure on X.*

As the notation indicates, μ_e depends on e only and hence is independent of a and λ. From the point of view of $\mathscr{T}$, a quantum state *is* a probability measure on X! In what follows we assume for simplicity that H is finite-dimensional, so that $e=e_\psi$ for some unit vector $\psi\in H$. With slight abuse of notation we then write μ_ψ for μ_{e_ψ}.

An important special case will be the bipartite setting $H=H_1\otimes H_2$, where Alice and Bob measure self-adjoint operators X and Y on H_1 and H_2, respectively, so that

$$n=2,\ a_1=X\otimes 1_{H_2},\ a_2=1_{H_1}\otimes Y.$$

We then introduce settings $c=(a,b)$, as in the previous sections, so that we typically look at expressions like $P(X_a=\lambda_1,Y_b=\lambda_2|x)$. The other case of interest will simply be $n=1$ with $a_1\equiv a$, $\lambda_1\equiv\lambda$; indeed, this will be the case in the statement of the theorem (the bipartite case playing a role only in the proof, though a crucial one!).

The following notation will be quite important to the argument. An equality

$$P_\psi(\mathbf{a}=\lambda|x)=\alpha(x), \tag{6.146}$$

where $\alpha: X\to[0,1]$ is measurable (often even constant), abbreviates:

$$P(\mathbf{a}=\lambda|x)=\alpha(x) \text{ for almost every } x \text{ with respect to the measure } \mu_\psi.$$

That is, there is a subset $X' \subset X$ such that $\mu_\psi(X') = 0$ and $P_\psi(\mathbf{a} = \lambda|x) = \alpha(x)$ holds for any $x \in X\backslash X'$. If X is finite, this simply means that the equality holds for any x for which $\mu_\psi(\{x\}) > 0$. Since this notation may render equalities like

$$P_\psi(\mathbf{a} = \lambda|x) = P_\varphi(\mathbf{a}' = \lambda'|x), \tag{6.147}$$

ambiguous, we explicitly define (6.147) as the double implication

$$P_\psi(\mathbf{a} = \lambda|x) = \alpha(x) \Leftrightarrow P_\varphi(\mathbf{a}' = \lambda'|x) = \alpha(x).$$

Furthermore, for $\varepsilon \to 0$ we write

$$P_\psi(\mathbf{a} = \lambda|x) \overset{\varepsilon}{\approx} P_\varphi(\mathbf{a}' = \lambda'|x) \Leftrightarrow P_\psi(\mathbf{a} = \lambda|x) = P_\varphi(\mathbf{a}' = \lambda'|x) + O(\sqrt{\varepsilon}), \tag{6.148}$$

as well as

$$\psi \overset{\varepsilon}{\approx} \varphi \Leftrightarrow (1-\varepsilon) \leq |\langle\psi, \varphi\rangle| \leq 1. \tag{6.149}$$

We are now ready to state our assumptions for the Colbeck–Renner Theorem:

- ***Compatibility with Quantum Mechanics (CQ):*** for any unit vector $\psi \in H$,

$$\int_X d\mu_\psi(x)\, P(\mathbf{a} = \lambda|x) = p_\psi(\mathbf{a} = \lambda), \tag{6.150}$$

  where the quantum-mechanical prediction $p_\psi(\mathbf{a} = \lambda)$ is given by the Born rule

$$p_\psi(\mathbf{a} = \lambda) = \langle\psi, e^{(1)}_{\lambda_1} \cdots e^{(n)}_{\lambda_n} \psi\rangle, \tag{6.151}$$

  cf. (2.21), where $e^{(i)}_{\lambda_i}$ is the spectral projection on the eigenspace $H_{\lambda_i} \subset H$ of a_i.
- ***Unitary Invariance (UI):*** for any unit vector $\psi \in H$ and unitary u on H,

$$P_{u\psi}(\mathbf{a} = \lambda|x) = P_\psi(u^{-1}\mathbf{a}u = \lambda|x). \tag{6.152}$$

- ***Continuity of Probabilities (CP:*** If $\psi \overset{\varepsilon}{\approx} \varphi$, then $P_\psi(\mathbf{a} = \lambda|x) \overset{\varepsilon}{\approx} P_\varphi(\mathbf{a} = \lambda|x)$.

In the remaining axioms, $H = H_1 \otimes H_2$, and a and b are self-adjoint operators on H_1 and H_2, respectively (duly identified with operators $a \otimes 1_{H_2}$ and $1_{H_1} \otimes b$ on H).

- ***Parameter Independence (PI):***

$$\sum_{\gamma\in\sigma(b)} P(a = \lambda, b = \gamma|x) = P(a = \lambda|x); \tag{6.153}$$

$$\sum_{\lambda\in\sigma(a)} P(a = \lambda, b = \gamma|x) = P(b = \gamma|x). \tag{6.154}$$

- ***Product Extension (PE):*** for any pair of states $\psi_1 \in H_1$, $\psi_2 \in H_2$,

$$P_{\psi_1}(a = \lambda|x) = P_{\psi_1\otimes\psi_2}(a = \lambda|x). \tag{6.155}$$

- ***Schmidt Extension (SE):*** if $\upsilon_i \in H_1$ $(i = 1, \ldots, \dim(H))$ are eigenstates of a, then for arbitrary orthogonal states $u_i \in H_2$ and coefficients $c_i > 0$ with $\sum_i c_i^2 = 1$,

$$P_{\sum_i c_i \cdot \upsilon_i}(a = x|x) = P_{\sum_i c_i \cdot \upsilon_i \otimes u_i}(a = x|x). \tag{6.156}$$

Note that **PI** makes sense, because (6.151) and (6.150) imply that for $p_\psi(\mathbf{a} = \lambda)$ to be nonzero we must have $\lambda_i \in \sigma(a_i)$ for each i. All assumptions are satisfied by quantum mechanics itself (seen as a hidden variable theory with ψ as the "hidden" variable x). In the context of hidden variable theories, though, one might doubt the plausibility of ***UI***, ***CP***, and ***SE***. But we need all these assumptions to prove:

Theorem 6.21. *If $\mathscr{T}$ satisfies* **CQ**, **UI**, **CP**, **PI**, **PE**, *and* **SE**, *then for any (finite-dimensional) Hilbert space H, unit vector $\psi \in H$, and operator $a \in B(H)_{\text{sa}}$,*

$$P_\psi(a = \lambda|x) = p_\psi(a = \lambda). \tag{6.157}$$

In other words, the hidden variable x is even more hidden than expected, since knowing its value has no effect on the probabilities for the outcomes of experiments.

Proof. We first assume (without loss of generality) that a is nondegenerate as a self-adjoint matrix, in that it has distinct eigenvalues $(\lambda_1, \ldots, \lambda_{\dim(H)})$; this assumption will be removed at the end of the proof. The proof consists of three steps.

1. The theorem holds for $H = \mathbb{C}^2$ and any pair (a, ψ) for which

$$p_\psi(a = \lambda_1) = p_\psi(a = \lambda_2) = 1/2, \tag{6.158}$$

   This only requires assumptions ***CQ***, ***PI***, and ***SE***.
2. The theorem holds for $H = \mathbb{C}^l$, $l < \infty$ arbitrary, and any pair (a, ψ) for which

$$p_\psi(a = \lambda_1) = \cdots = p_\psi(a = \lambda_l) = 1/l. \tag{6.159}$$

   This is just a slight extension of step 1 and uses the same three assumptions.
3. The theorem holds in general. This requires all assumptions (as well as step 2).

Proof of step 1. Let $H = \mathbb{C}^2$, with basis (υ_1, υ_2) of eigenvectors of a, so that $\psi \in \mathbb{C}^2$ may be written as

$$\psi = (\upsilon_1 + \upsilon_2)/\sqrt{2}. \tag{6.160}$$

Without loss of generality, we may assume that $\lambda_1 = 1$ and $\lambda_2 = -1$. We now relabel a as a_0 and extend it to a family of operators $(a_k)_{k=0,1,\ldots,2N-1}$ by fixing an integer $N > 1$, putting $\theta_k = k\pi/2N$, and defining

$$c_k = e_{\theta_{k+\pi}} - e_{\theta_k}, \tag{6.161}$$

where, for any angle $\theta \in [0, 2\pi]$, the operator $e_\theta = |\theta\rangle\langle\theta|$ is the orthogonal projection onto the one-dimensional subspace spanned by the unit vector

$$|\theta\rangle = \sin(\theta/2) \cdot \upsilon_1 + \cos(\theta/2) \cdot \upsilon_2. \tag{6.162}$$

In the bipartite setting, we have operators $a_k = c_k \otimes 1_2$ and $b_k = 1_2 \otimes c_k$ on $\mathbb{C}^2 \otimes \mathbb{C}^2$, as well as a maximally correlated (Bell) state $\psi_{AB} \in \mathbb{C}^2 \otimes \mathbb{C}^2$, given by

$$\psi_{AB} = \frac{1}{\sqrt{2}}(\upsilon_1 \otimes \upsilon_1 + \upsilon_2 \otimes \upsilon_2). \tag{6.163}$$

Using assumptions ***PI*** and ***SE***, we then have, for $i = 1,2$ $\lambda_1 = 1$, and $\lambda_2 = -1$,

$$P_\psi(a = \lambda_i | x) = P_{\psi_{AB}}(a_0 = \lambda_i | x). \tag{6.164}$$

The quantum-mechanical prediction is

$$p_{\psi_{AB}}(a_0 = 1) = p_{\psi_{AB}}(a_0 = -1) = \tfrac{1}{2}. \tag{6.165}$$

Our goal is to show that also for each $x \in X$, knowing x is irrelevant in that

$$P_{\psi_{AB}}(a_0 = 1|x) = P_{\psi_{AB}}(a_0 = -1|x) = \tfrac{1}{2}. \tag{6.166}$$

To this effect we introduce the combination of probabilities

$$I^{(N)}(x) = P(a_0 = b_{2N-1}|x) + \sum_{k \in K_N, l \in L_N, |k-l|=1} P(a_k \neq b_l|x), \tag{6.167}$$

where $K_N = \{0, 2, \ldots, 2N-2\}$ and $L_N = \{1, 3, \ldots, 2N-1\}$. Our first inequality is

$$\begin{aligned}
|P(a_k = \lambda_i|x) - P(b_l = \lambda_i|x)| &= |P(a_k = \lambda_i, b_l = \lambda_i|x) + P(a_k = \lambda_i, b_l \neq \lambda_i|x) \\
&\quad - P(a_k = \lambda_i, b_l = \lambda_i|x) + P(a_k \neq \lambda_i, b_l = \lambda_i|x)| \\
&= |P(a_k = \lambda_i, b_l \neq \lambda_i|x) - P(a_k \neq \lambda_i, b_l = \lambda_i|x)| \\
&\leq P(a_k = \lambda_i, b_l \neq \lambda_i|x) + P(a_k \neq \lambda_i, b_l = \lambda_i|x) \\
&= P(a_k \neq b_l|x),
\end{aligned} \tag{6.168}$$

where $i = 1, 2$, and we used ***PI***. This implies a second inequality: since $a_{2N} = -a_0$,

$$\begin{aligned}
|P(a_0 = 1|x) - P(a_0 = -1|x)| &= |P(a_0 = 1|x) - P(a_{2N} = 1|x)| \\
&\leq \sum_{k,l,|k-l|=1} |P(a_k = 1|x) - P(b_l = 1|x)| \\
&\leq \sum_{k,l,|k-l|=1} P(a_k \neq b_l|x) \leq I^{(N)}(x).
\end{aligned}$$

Integrating this with respect to the measure $\mu_{\psi_{AB}}$ and using ***CQ*** gives

$$\int_X d\mu_{\psi_{AB}}(x)\, |P(a_0 = 1|x) - P(a_0 = -1|x)| \leq \int_X d\mu_{\psi_{AB}}(x)\, I^{(N)}(x) = I^{(N)}_{\psi_{AB}}. \tag{6.169}$$

We wish to invoke the corresponding quantum-mechanical expression, defined by

$$I^{(N)}_{\psi_{AB}} = p_{\psi_{AB}}(a_0 = b_{2N-1}) + \sum_{k \in K_N, l \in L_N, |k-l|=1} p_{\psi_{AB}}(a_k \neq b_l). \tag{6.170}$$

A straightforward calculation shows that this expression is equal to

$$I^{(N)}_{\psi_{AB}} = 2N \sin^2(\pi/4N). \tag{6.171}$$

Since $\lim_{N\to\infty} I^{(N)}_{\psi_{AB}} = 0$, letting $N \to \infty$ in (6.169) therefore yields (6.166). From (6.164) we then obtain (6.158).

Proof of step 2. Let $H = \mathbb{C}^l$ and let $(\upsilon_i)_{i=1}^l$ be an orthonormal basis of eigenvectors of a, with corresponding eigenvalues λ_i, and phase factors for the eigenvectors υ_i such that $c_i > 0$ (and of course, $\sum_i c_i^2 = 1$) in the expansion

$$\psi = \sum_i c_i \upsilon_i. \tag{6.172}$$

The case of interest will be $c_1 = \cdots = c_l = 1/l$, but first we merely assume that $c_1 = c_2$ (the same reasoning applies to any other pair), with $\lambda_1 = 1$ and $\lambda_2 = -1$ (which involves no loss of generality either and just simplifies the notation). The other positive coefficients c_i are arbitrary. Generalizing (6.166), we will show that

$$P_\psi(a = 1|x) = P_\psi(a = -1|x). \tag{6.173}$$

This shows that if two Born probabilities defined by some quantum state e_ψ are equal, then the underlying hidden variable probabilities must be equal μ_ψ-a.e., too. Eq. (6.159) immediately follows from this result by taking all c_i to be equal.

As in step 1, we pass to the bipartite setting, introducing two copies of $H = \mathbb{C}^l$ denoted by $H_A = H_B = \mathbb{C}^l$, and define the correlated state

$$\psi_{AB} = \sum_i c_i \cdot \upsilon_i \otimes \upsilon_i \tag{6.174}$$

in $H_A \otimes H_B$. Eq. (6.164) again follows from assumptions ***PI*** and ***SE***. Throughout the argument of step 1, we now replace each probability $P(a_k = \lambda_i, b_l = \gamma_j|x)$ by an adapted probability $P^{(1)}(a_k = \lambda_i, b_l = \gamma_j|x)$, defined as the conditional probability

$$\begin{aligned} P^{(1)}(a_k = \lambda_i, b_l = \gamma_2|x) &= P(a_k = \lambda_i, b_l = \gamma_2 ||\lambda_i| = |\gamma_2| = 1, x) \\ &= \frac{P(a_k = \lambda_i, b_l = \gamma_2, |\lambda_i| = |\gamma_2| = 1|x)}{P(|\lambda_i| = |\gamma_2| = 1|x)}, \end{aligned} \tag{6.175}$$

for all x for which $P(|\lambda_i| = |\gamma_2| = 1|x) > 0$, whereas

$$P^{(1)}(a_k = \lambda_i, b_l = \gamma_2|x) = 0 \tag{6.176}$$

whenever $P(|\lambda_i| = |\gamma_2| = 1|x) = 0$. The same argument then yields (6.169), with P replaced by $P^{(1)}$ but with the same right-hand side. As in step 1,

$$P^{(1)}_{\psi_{AB}}(a_0 = 1|x) = P^{(1)}_{\psi_{AB}}(a_0 = -1|x), \tag{6.177}$$

which implies that

$$P_{\psi_{AB}}(a_0 = 1|x) = P_{\psi_{AB}}(a_0 = -1|x), \tag{6.178}$$

either because both sides vanish (if $P(|\lambda_i| = |\gamma_2| = 1|x) = 0$), or because (in the opposite case) the denominator $P(|\lambda_i| = |\gamma_2| = 1|x)$ cancels from both sides of (6.177).

Combined with (6.164), eq. (6.178) proves (6.173) and hence establishes step 2.

Proof of step 3. This is the most difficult step in the proof, relying on a technique wittily called ***embezzlement*** (which we only need for maximally entangled states). We will deal with three Hilbert spaces, namely $H = \mathbb{C}^l$, $H' = \mathbb{C}^m$, and $H'' = \mathbb{C}^n$ (where $n = m^N$ for some large N, see below), each with some fixed orthonormal basis $(\upsilon_i)_{i=1}^l$, $(\upsilon'_j)_{j=1}^m$, and $(\upsilon''_k)_{k=1}^n$, respectively. Given a further number $m_i \leq m$, we now list the nm basis vectors $\upsilon''_k \otimes \upsilon'_j$ of $H'' \otimes H'$ in two different orders:

1. $\upsilon''_1 \otimes \upsilon'_1, \ldots, \upsilon''_n \otimes \upsilon'_1, \upsilon''_1 \otimes \upsilon'_2, \ldots, \upsilon''_n \otimes \upsilon'_2, \ldots, \upsilon''_1 \otimes \upsilon'_m, \ldots, \upsilon''_n \otimes \upsilon'_m$;
2. $\upsilon''_1 \otimes \upsilon'_1, \ldots, \upsilon''_1 \otimes \upsilon'_{m_i}, \upsilon''_2 \otimes \upsilon'_1, \ldots, \upsilon''_2 \otimes \upsilon'_{m_i}, \ldots, \upsilon''_n \otimes \upsilon'_1, \ldots, \upsilon''_n \otimes \upsilon'_{m_i}, \ldots$,

where the remaining vectors (i.e., those of the form $\upsilon''_k \otimes \upsilon'_j$ for $1 \leq k \leq n$ and $j > m_i$) are listed in some arbitrary order.

Define

$$u^{(m_i)} : H'' \otimes H' \to H'' \otimes H' \tag{6.179}$$

as the unitary operator that maps the first list on the second. We will need the explicit expression

$$u^{(m_i)}(\upsilon''_k \otimes \upsilon'_1) = \upsilon''_{s^i_k} \otimes \upsilon'_{j^i_k}, \tag{6.180}$$

where for given $k = 1, \ldots, n$ the numbers $s^i_k = 1, \ldots, n_i$ (where n_i is the smallest integer such that $n_i m_i \geq n$) and $j^i_k = 1, \ldots, n_i$ are uniquely determined by

$$k = (s^i_k - 1)m_i + j^i_k. \tag{6.181}$$

We will actually work with two copies of $H'' \otimes H'$, called $H''_A \otimes H'_A$ and $H''_B \otimes H'_B$, with ensuing copies of $u_A^{(m_i)}$ and $u_B^{(m_i)}$ of $u^{(m_i)}$, and hence, leaving the isomorphism

$$H''_A \otimes H'_A \otimes H''_B \otimes H'_B \cong H''_A \otimes H''_B \otimes H'_A \otimes H'_B \tag{6.182}$$

implicit, we obtain a unitary operator

$$u_A^{(m_i)} \otimes u_B^{(m_i)} : H''_A \otimes H''_B \otimes H'_A \otimes H'_B \to H''_A \otimes H''_B \otimes H'_A \otimes H'_B. \tag{6.183}$$

The point of all this is that the unit vector

$$\kappa_n \in H''_A \otimes H''_A; \tag{6.184}$$

$$\kappa_n = \frac{1}{\sqrt{C(n)}} \sum_{k=1}^n \upsilon''_k \otimes \upsilon''_k, \tag{6.185}$$

where $C(n) = \sum_{k=1}^{n} 1/k$, acts as a "catalyst" in producing the maximally entangled state

$$\varphi \in H'_A \otimes H'_B; \tag{6.186}$$

$$\varphi = \frac{1}{\sqrt{m_i}} \sum_{j=1}^{m_i} \upsilon'_j \otimes \upsilon'_j, \tag{6.187}$$

from the uncorrelated state $\upsilon'_1 \otimes \upsilon'_1 \in H'_A \otimes H'_B$, in that for any $m_i \leq m$,

$$u_A^{(m_i)} \otimes u_B^{(m_i)} (\kappa_n \otimes \upsilon'_1 \otimes \upsilon'_1) \overset{\varepsilon/2}{\approx} \kappa_n \otimes \varphi. \tag{6.188}$$

Here $\varepsilon = 1/N$ if $n = m^{2N}$. This follows straightforwardly from (6.183) - (6.187).

After this preparation we are ready for the proof of step 3, continuing to use the notation established at the beginning of step 2, especially (6.172). As in step 1, we introduce two copies $H_A = H_B = \mathbb{C}^l$ of H, as well as two states

$$\psi_{AB} = \sum_i c_i \cdot \upsilon_i \otimes \upsilon_i \in H_A \otimes H_B; \tag{6.189}$$

$$\psi'''_{AB} = \kappa_n \otimes \upsilon'_1 \otimes \upsilon'_1 \otimes \psi_{AB} \in H'''_A \otimes H'''_B, \tag{6.190}$$

where κ_n is given by (6.185), we put

$$H''' = H'' \otimes H' \otimes H, \tag{6.191}$$

and in our notation we have ignored the obvious permutations of factors in the tensor product. For any $\varepsilon > 0$, pick $c'_i \in \mathbb{R}^+$ such that $(c'_i)^2 \in \mathbb{Q}^+$ and

$$|c'_i - c_i| < \varepsilon/\dim(H), \tag{6.192}$$

which implies that, in the sense of (6.149), we have

$$\sum_i c'_i \upsilon_i \overset{\varepsilon/2}{\approx} \sum_i c_i \upsilon_i. \tag{6.193}$$

Suppose

$$c'_i = \sqrt{p_i/q_i}, \tag{6.194}$$

with $p_i, q_i \in \mathbb{N}$ and $\gcd(p_i, q_i) = 1$, and define

$$m_i = p_i \prod_{i' \neq i} q_{i'}. \tag{6.195}$$

Consequently, writing

$$q = 1/\sqrt{\sum_{i'} m_{i'}}, \tag{6.196}$$

the following quotient is independent of i:

$$\frac{c_i'}{\sqrt{m_i}} = q. \tag{6.197}$$

Given the integers m_i thus obtained, we define a unitary operator

$$u : H''' \to H'''; \tag{6.198}$$

$$u = \sum_{i=1}^{l} u^{(m_i)} \otimes |\upsilon_i\rangle\langle\upsilon_i|, \tag{6.199}$$

where $u^{(m_i)}$ is defined in (6.180). From this definition, with additional labels to denote the copies $u_A : H_A''' \to H_A'''$ and $u_B : H_B''' \to H_B'''$, and (6.188), and writing

$$\xi^{ij} = \upsilon_i \otimes \upsilon_j' \in H \otimes H', \tag{6.200}$$

with corresponding copies

$$\xi_{AA'}^{ij_i} \in H_A \otimes H_A'; \tag{6.201}$$

$$\xi_{BB'}^{ij_i} \in H_B \otimes H_B', \tag{6.202}$$

we then obtain the important relations

$$1_{H_A'''} \otimes 1_{H_B'''}(\psi_{AB}''') = \kappa_n \otimes \sum_{i=1}^{l} c_i \cdot \xi_{AA'}^{i1} \otimes \xi_{BB'}^{i1}; \tag{6.203}$$

$$u_A \otimes 1_{H_B'''}(\psi_{AB}''') = \frac{1}{\sqrt{C(n)}} \sum_{i=1}^{l} \sum_{k=1}^{n} \frac{c_i}{\sqrt{k}} \cdot \upsilon_{s_k}'' \otimes \upsilon_k'' \otimes \xi_{AA'}^{ij_k^i} \otimes \xi_{BB'}^{i1}; \tag{6.204}$$

$$1_{H_A'''} \otimes u_B(\psi_{AB}''') = \frac{1}{\sqrt{C(n)}} \sum_{i=1}^{l} \sum_{k=1}^{n} \frac{c_i}{\sqrt{k}} \cdot \upsilon_k'' \otimes \upsilon_{s_k}'' \otimes \xi_{AA'}^{i1} \otimes \xi_{BB'}^{ij_k^i}; \tag{6.205}$$

$$u_A \otimes u_B(\psi_{AB}''') \overset{\varepsilon}{\approx} q \cdot \kappa_n \otimes \sum_{i=1}^{l} \sum_{j_i=1}^{m_i} \xi_{AA'}^{ij_i} \otimes \xi_{BB'}^{ij_i}. \tag{6.206}$$

Here the right-hand sides of (6.203) - (6.206) have been arranged so as to obtain vectors in the six-fold tensor product

$$H_A'' \otimes H_B'' \otimes H_A \otimes H_A' \otimes H_B \otimes H_B'.$$

We will repeatedly invoke the following lemma, whose proof just unfolds the notation (on the appropriate identification of a with $a \otimes 1_{H_2}$ and of b with $1_{H_1} \otimes b$).

Lemma 6.22. *Assume* **PI** *and* **UI**. *For any pair of unitary operators u_1 on H_1 and u_2 on H_2, and any unit vector $\psi \in H_1 \otimes H_2$, one has*

$$P_{(u_1 \otimes 1_{H_2})\psi}(b = \gamma|x) = P_\psi(b = \gamma|x); \tag{6.207}$$

$$P_{(1_{H_1} \otimes u_2)\psi}(a = \lambda|x) = P_\psi(\lambda = x|x). \tag{6.208}$$

Since we assume that a is nondegenerate, there is a bijective correspondence between its eigenvalues $a = \lambda_i$ and its eigenvectors υ_i. Instead of $P(a = \lambda_i)$ dressed with whatever parameters x or ψ, we may then write $P(\upsilon_i)$, where a is understood, and analogously for the more complicated operators on tensor products of Hilbert space appearing below. Repeatedly using Lemma 6.22, we proceed as follows.

- From Step 2, using the notation explained below (6.172),

$$P_{q\cdot\sum_{i=1}^{l}\sum_{j_i=1}^{m_i}\xi_{BB'}^{ij_i}}(\xi_{BB'}^{ij}|x) = q^2. \tag{6.209}$$

- From (6.156) in ***PE*** and (6.209),

$$P_{q\cdot\sum_{i,j_i}\xi_{AA'}^{ij_i}\otimes\xi_{BB'}^{ij_i}}(\xi_{BB'}^{ij}|x) = q^2. \tag{6.210}$$

- From (6.155) in ***SE*** and (6.210),

$$P_{q\cdot\kappa_n\otimes\sum_{i,j_i}\xi_{AA'}^{ij_i}\otimes\xi_{BB'}^{ij_i}}(\xi_{BB'}^{ij}|x) = q^2. \tag{6.211}$$

- From (6.211), ***CP*** (whose notation we use), and (6.206),

$$P_{(u_A\otimes u_B)\psi_{AB}'''}(\xi_{BB'}^{ij}|x) \overset{\varepsilon}{\approx} q^2. \tag{6.212}$$

- Recall the number m (satisfying $m \geq m_i$ for all i). From (6.212) and Lemma 6.22,

$$\begin{aligned} P_{(1_{H_A'''}\otimes u_B)\psi_{AB}'''}(\xi_{BB'}^{ij_i}|x) &\overset{\varepsilon}{\approx} q^2 \ (j_i = 1,\ldots,m_i); \\ P_{(1_{H_A'''}\otimes u_B)\psi_{AB}'''}(\xi_{BB'}^{ij_i}|x) &\overset{\varepsilon}{\approx} 0 \ (j_i = m_i+1,\ldots,m). \end{aligned} \tag{6.213}$$

We now start a different line of argument, to be combined with (6.213) in due course.

- From ***PE***, ***SE***, and (6.172), with $\upsilon_A^i \in H_A$ denoting $\upsilon_i \in H$, we have

$$P_\psi(a = \lambda_i|x) \equiv P_\psi(\upsilon_i|x) = P_{\kappa_n\otimes\sum_i c_i\cdot\xi_{AA'}^{i1}\otimes\xi_{BB'}^{i1}}(\upsilon_A^i|x). \tag{6.214}$$

- Using Lemma 6.22, (6.203), and (6.204),

$$P_{\kappa_n\otimes\sum_i c_i\cdot\xi_{AA'}^{i1}\otimes\xi_{BB'}^{i1}}(\upsilon_A^i|x) = P_{(1_{H_A'''}\otimes u_B)\psi_{AB}'''}(\upsilon_A^i|x), \tag{6.215}$$

  and hence

$$P_\psi(a = \lambda_i|x) = P_{(1_{H_A'''}\otimes u_B)\psi_{AB}'''}(\upsilon_A^i|x). \tag{6.216}$$

- From quantum mechanics, notably (6.151), and (6.205), for any $i' \neq i$ we have

$$P_{(1_{H_A'''}\otimes u_B)\psi_{AB}'''}(\upsilon_A^{i'}\otimes\xi_{BB'}^{ij_i}) = 0. \tag{6.217}$$

- From ***CQ*** and (6.217), for any $i' \neq i$,

$$P_{(1_{H_A'''}\otimes u_B)\psi_{AB}'''}(\upsilon_A^{i'},\xi_{BB'}^{ij_i}|x)=0. \tag{6.218}$$

- From *PI*,

$$P(\upsilon_A^{i'}|x)=\sum_{i,j_i}P(\upsilon_A^{i'},\xi_{BB'}^{ij_i}|x); \tag{6.219}$$

$$P(\xi_{BB'}^{ij_i}|x)=\sum_{i'}P(\upsilon_A^{i'},\xi_{BB'}^{ij_i}|x). \tag{6.220}$$

- From (6.218), (6.219), and (6.220),

$$P_{(1_{H_A'''}\otimes u_B)\psi_{AB}'''}(\upsilon_A^{i}|x)=\sum_{j_i}P_{(1_{H_A'''}\otimes u_B)\psi_{AB}'''}(\xi_{BB'}^{ij_i}|x). \tag{6.221}$$

Finally, from (6.214), (6.221), (6.213), and (6.197) we obtain

$$P_\psi(a=\lambda_i|x)\overset{\varepsilon}{\approx}\sum_{j_i}^{m_i}q^2=m_i\cdot q^2=c_i^2. \tag{6.222}$$

Since $c_i>0$ we have $c_i^2=|c_i|^2$; using (6.192) and letting $\varepsilon\to 0$ then proves step 3:

$$P_\psi(a=\lambda_i|x)=|c_i|^2=p_\psi(a=\lambda_i). \tag{6.223}$$

Finally, we remove our standing assumption that the spectrum of a be nondegenerate. In the degenerate case one has

$$p_\psi(a=\lambda_i)=\sum_{j_i}p_\psi(\upsilon_{j_i}), \tag{6.224}$$

where the sum is over any orthonormal basis $(\upsilon_{j_i})_{j_i}$ of the eigenspace of λ_i. Similarly, since each vector υ_{j_i} gives $a=\lambda_i$, probability theory gives for all x,

$$P(a=\lambda_i|x)=\sum_{j_i}P(\upsilon_{j_i}|x). \tag{6.225}$$

The nondegenerate case of the theorem (which distinguishes the states υ_{j_i}) yields

$$P_\psi(\upsilon_{j_i}|x)=p_\psi(\upsilon_{j_i}), \tag{6.226}$$

from which (6.157) follows once again:

$$P_\psi(a=\lambda_i|x)=\sum_{j_i}P_\psi(\upsilon_{j_i}|x)=\sum_{j_i}p_\psi(\upsilon_{j_i})=p_\psi(a=\lambda_i).$$

Our proof of the Colbeck–Renner Theorem is now complete. □

Under less stringent assumptions this theorem might have been regarded as the conclusion of von Neumann's program to disprove the possibility of completing quantum mechanics by adding hidden variables, but as yet this seems unwarranted.

Notes

§6.1. **From von Neumann to Kochen–Specker**

> 'For decades nobody spoke up against von Neumann's arguments, and his conclusions were quoted by some as the gospel'. (Belinfante, 1973, pp. 24)

Theorem 6.2 is due to non Neumann (1932, §IV.2); it was the first result to impose useful constraints on hidden variable theories, anticipating all later literature on the subject. Unfortunately (as part of their general anti-Copenhagen rhetoric), Bell and his followers left the realm of decent academic discourse by calling von Neumann's arguments against hidden variables 'silly' and 'foolish', through which they merely displayed the depth of their own misunderstanding of von Neumann's reasoning; see Caruana (1995), Bub (2011a), and especially Dieks (2016b). In fact, von Neumann (1932, p. 172) carefully qualifies his Theorem 6.2 by stating that it follows '*im Rahmen unserer Bedingungen*' (i.e. '*given our assumptions*'), of which he earlier (on p. 164) admits that linearity is physically reasonable only for *commuting* operators, but nonetheless justifies this assumption through an ensemble argument (now outdated, but by no means 'silly'). Though couched in agreeable academic parlance, the earlier critique by Hermann (1935) was misguided, too (Dieks, 2016b).

The Kochen–Specker Theorem is due to Kochen & Specker (1967); the authors were originally logicians. A similar but less precise statement had appeared earlier in Bell (1966), who was not cited by Kochen and Specker; some authors refer to the ***Bell–Kochen–Specker Theorem***. The *Nature* assumption has been experimentally verified, cf. Huang et al (2003). The proof of the fundamental Lemma 6.7 we present is essentially due to Kochen and Specker, as simplified by Peres (1995). Our independent proof for $\mathbb{C}^4$ is taken from Cabello et al (1996). Surveys of various proofs are given by Brown (1992) and Gould (2009); see also Waegell & Aravind (2012) and references therein, as well as Bub (1997) for another proof. From the Netherlands, we cannot fail to mention the short proof by Gill & Keane (1996). For geometric aspects (and even a link with M.C. Escher) see Zimba & Penrose (1993).

One finds two opposite directions of research around the Kochen–Specker Theorem. A computational one, which seems hardly relevant to conceptual issues in physics (the goal rather being *The Guinness Book of Records*), consists of attempts to find a *minimal* set of vectors that *cannot* be coloured. See, for example, Pavicic et al (2005) for arbitrary dimension and Arends (2009) and Uijlen & Westerbaan (2015) for $\mathbb{R}^3$, the latter paper showing that at least 22 vectors are needed.

The other, which is of significant conceptual importance and hence is worth some more extensive discussion, consists of attempts to find a *maximal* set of vectors that *can* be coloured. That is, one looks for large (preferably dense and measurable) subsets S^2_c of S^2 for which there exists a function $\tilde{V} : S^2_c \to \{0,1\}$ that satisfies:

- $\tilde{V}(-\mathbf{u}) = \tilde{V}(\mathbf{u})$ for each $\mathbf{u} \in S^2_c$;
- $\tilde{V}(\mathbf{u}_1) + \tilde{V}(\mathbf{u}_2) + \tilde{V}(\mathbf{u}_3) = 2$, for each (orthonormal) basis $(\mathbf{u}_1, \mathbf{u}_2, \mathbf{u}_3)$ of $\mathbb{R}^3$ whose elements lie in S^2_c.

The first result in this direction was obtained by Meyer (1999) and Havlicek et al (2001), who showed that one may take $S^2_c = S^2 \cap \mathbb{Q}^3$; this choice was motivated by invoking finite precision arguments to circumvent the Kochen–Specker Theorem, see below. To write down a suitable function $\tilde{V} : S^2 \cap \mathbb{Q}^3 \to \{0,1\}$, we first define an auxiliary function $S : S^2 \cap \mathbb{Q}^3 \to \mathbb{Z}$ by

$$S\left(\frac{n_1}{m_1}, \frac{n_2}{m_2}, \frac{n_3}{m_3}\right) = \frac{n_3}{m_3} \cdot \frac{\mathrm{lcm}(m_1,m_2,m_3)}{\gcd(n_1,n_2,n_3)}, \tag{6.227}$$

where lcm is the *least common multiple* and gcd is the *greatest common divisor* of the argument. This function is obviously well defined. Then the following works:

$$\tilde{V}(x,y,z) = 0 \text{ if } S(x,y,z) \text{ is odd;} \tag{6.228}$$
$$\tilde{V}(x,y,z) = 1 \text{ if } S(x,y,z) \text{ is even.} \tag{6.229}$$

More generally, for an arbitrary n-dimensional) Hilbert space H, with $n < \infty$, Clifton & Kent (2000) proved the existence of a countable dense colorable subset $\mathscr{P}_1(H)_c$ of $\mathscr{P}_1(H)$ (cf. Definition 6.9), with the additional property that different resolutions of the identity drawn from $\mathscr{P}_1(H)_c$ never share a projection (so that the key strategy proof of Lemma 6.7, which is based on the existence of overlapping bases, falls apart). Given some enumeration $(e_i^{(1)}), (e_i^{(2)}), \ldots$ of the countable set of all resolutions of the identity drawn from $\mathscr{P}_1(H)_c$, so that each $(e_1^{(k)}, \ldots, e_n^{(k)})$ is a basis of H, $k \in \mathbb{N}$, each possible coloring $W = W_f$ bijectively corresponds to some function $f : \mathbb{N} \to \{1, \ldots, n\}$ through

$$W_f(e) = 1 \text{ if } e = e^{(k)}_{f(k)}; \tag{6.230}$$
$$W_f(e) = 0 \text{ otherwise.} \tag{6.231}$$

Note that because of the total incompatibility of the projections, each $e \in \mathscr{P}_1(H)_c$ belongs to a unique resolution $(e_i^{(k)})$, so that W_f is well defined. The statistical predictions of quantum mechanics may then be recovered as follows. For each density operator $\rho \in \mathscr{D}(H)$ we may define a probability measure μ_ρ on the set $\underline{n}^{\mathbb{N}}$ of all functions $f : \mathbb{N} \to \{1, \ldots, n\}$ by imposing the conditions

$$\mu_\rho\left(\{f \in \underline{n}^{\mathbb{N}} \mid W_f(e_i^{(k)}) = \lambda_i^{(k)} \forall i = 1, \ldots, n, k \in K\}\right) = \prod_{k \in K} \mathrm{Tr}\left(\rho \prod_{i=1}^{n} [e_i^{(k)} = \lambda_i^{(k)}]\right), \tag{6.232}$$

where $\lambda_i^{(k)} \in \{0,1\}$, $K \subset \mathbb{N}$ is finite, and $[e_i^{(k)} = \lambda_i^{(k)}]$ is the projection onto the corresponding eigenspace $H_{\lambda_i^{(k)}}$ of the projection $e_i^{(k)}$ (more generally, for $a \in B(H)_{\mathrm{sa}}$ we write $[a = \lambda]$ for the spectral projection e_λ defined by a and $\lambda \in \sigma(a)$). The subset of $\underline{n}^{\mathbb{N}}$ in the argument of μ_ρ is hereby declared measurable; existence and uniqueness of the measure μ_ρ on a suitable σ-algebra follow from the Kolmogorov extension theorem of measure theory, which applies because the marginals (6.232) satisfy the appropriate consistency conditions, cf. Hermens (2009) for details.

This formula guarantees that the left-hand side vanishes if $\lambda_i^{(k)} = 0$ for each i, and also if $\lambda_i^{(k)} = 1$ for more than one value of i. If $K = \{k_0\}$ is a singleton and $\lambda = (\lambda_1, \ldots, \lambda_n)$, then the right-hand side (and hence the left-hand side) is the Born probability for the outcome $e_i^{(k_0)} = \lambda_i$ for each i, i.e.,

$$\mu_\rho\left(\{f \in \underline{n}^{\mathbb{N}} \mid W_f(e_i^{(k_0)}) = \lambda_i \forall i = 1, \ldots, n\}\right) = \mathrm{Tr}\left(\rho \prod_{i=1}^{n} [e_i^{(k_0)} = \lambda_i]\right). \quad (6.233)$$

Consequently, it is true by construction that for any admissible measurement in quantum mechanics (in that all observables commute), i.e., for each $k_0 \in \mathbb{N}$, averaging over the 'hidden variable' $f \in \underline{n}^{\mathbb{N}}$ reproduces the statistical predictions of quantum mechanics. This success is achieved at a high cost, however:

- Two random variables $e_i^{(k)}$ and $e_{i'}^{(k')}$ are statistically independent (with respect to μ_ρ) whenever $k \neq k'$, even though $\|e_i^{(k)} - e_{i'}^{(k')}\|$ may be arbitrarily small.
- For each $f \in \underline{n}^{\mathbb{N}}$ the associated coloring W_f is maximally discontinuous, in that for each $\mathbf{u} \in \mathscr{P}_1(H)_c$ and each $\varepsilon > 0$ there is $\mathbf{u}' \in \mathscr{P}_1(H)_c$ such that although $\|e_{\mathbf{u}} - e_{\mathbf{u}'}\| < \varepsilon$ one has $W_f(e_{\mathbf{u}}) \neq W_f(e_{\mathbf{u}'})$, so that in fact $|W_f(e_{\mathbf{u}}) - W_f(e_{\mathbf{u}'})| = 1$.

These facts were noted by Clifton & Kent themselves, and Appleby (2005) proved that they are a necessary feature of all constructions that involve sufficiently large subsets of $\mathscr{P}_1(H)$ that can be colored.

Without challenging their mathematical significance, these discontinuities undermine any potential physical relevance such models might have, and this in turn challenges the reason such models were introduced in the first place (Meyer, 1999), namely the (alleged) *finite precision loophole* of the Kochen–Specker Theorem.

The thrust of this loophole is that it would be an illusion for an experimentalist like Alice to claim that she measures some observable a with infinite accuracy; in fact, given $\varepsilon > 0$ she might equally well measure some a' with $\|a - a'\| < \varepsilon$. Consequently, finding a dense colorable subset $\mathscr{P}_1(H)_c \subset \mathscr{P}_1(H)$ should suffice for a hidden variable interpretation of quantum mechanics, since if Alice believes she measures some projection e, the model assigns a value $W(e')$ to the projection $e' \in \mathscr{P}_1(H)_c$ she actually measures (where e' is selected by some algorithm that is part of the theory itself, cf. Clifton & Kent (2000)), and presents that value to Alice as the outcome of her measurement. However, owing to the discontinuities just mentioned, this value is as arbitrary as the identification of e'.

As emphasized by Barrett & Kent (2004), this arbitrariness, although perhaps undesirable, does not by itself affect the ability of the Clifton–Kent model to reproduce the statistical predictions of quantum mechanics. On the other hand, it would be pretty awkward to have a theory whose individual value attributions are completely arbitrary, especially since the finite precision argument is predicated on the idea that observables close to the one Alice believes herself to measure (i.e., e) should have approximately the same value as the one she actually does measure (namely, e'). If this is not the case, her measurements are pointless and the hidden variable W_f would be empirically inaccessible and hence truly "hidden" (Appleby, 2005).

See also Hermens (2009, 2016). This last point applies to Corollary 6.12, which would no longer be true if the set X_A of all bases of $\mathbb{R}^3$ in Definition 6.11 would be replaced by some subset $X_A^c \subset X_A$ drawn from a colorable subset S_c^2 of S^2. Each $z \in X_Z$ would then correspond to some coloring $\mathbf{u} \mapsto \tilde{F}(\mathbf{u}, z)$ of S_c^2, which, by the above discussion, would be maximally discontinuous and hence empirically inaccessible. Nonetheless, such a theory does exist in principle.

The aim of maximizing colorable sets was pursued in a different direction by Bub & Clifton (1996); see also Bub (1997). Given a "preferred" observable $a \in B(H)_{\mathrm{sa}}$ and a pure state $e \in \mathscr{P}_1(H)$, these authors look for a maximal sublattice $\mathscr{P}(e,a)$ of $\mathscr{P}(H)$ that contains all spectral projections of a (but, despite the notation $\mathscr{P}(e,a)$, does not necessarily contain e!), admits sufficiently many lattice homomorphism $h : \mathscr{P}(e,a) \to \{0,1\}$ (i.e., binary valuations) such that the Born measure μ_e on $\sigma(a)$, i.e., $\mu_e(\Delta) = \mathrm{Tr}(ee_\Delta)$, $\Delta \subseteq \sigma(a)$, can be reproduced by averaging over these homomorphisms, and finally is invariant under all unitary isomorphisms of $\mathscr{P}(H)$ that commute with both e and a. Equivalently, one wants a maximal C*-subalgebra $A(a,e)$ of $B(H)$ that contains a, admits sufficiently many dispersion-free states so as to reproduce the Born probabilities defined by a in the given state e, and is invariant in the said way (a fourth condition used by Bub and Clifton is superfluous; see Bub, 1997, p. 128). Asuming for simplicity that $n = \dim(H) < \infty$, the answer is

$$A(a,e) = C^*(e_\lambda e e_\lambda, \lambda \in \sigma(a))' \tag{6.234}$$

where, as always, e_λ is the projection into the eigenspace H_λ for $\lambda \in \sigma(a)$, and the prime denotes the commutant (one might as well take the commutant of the *set* of all $e_\lambda e e_\lambda$). Equivalently, putting $e = e_\psi = |\psi\rangle\langle\psi|$, eq. (6.234) is the C*-algebra generated by all projections f_λ onto the nonzero components $e_\lambda \psi$ of ψ in each H_λ and all one-dimensional projections that are orthogonal to all f_λ (given that $\dim(H) < \infty$, this is the same as the linear span of these projections). Thus $A(a,e)$ always contains $C^*(a)$, since it contains each e_λ, $\lambda \in \sigma(a)$), but note that $A(a,e)$ need not be commutative. In comparison, if the requirement had been the reproduction of all Born probabilities for *arbitrary* pure states e rather than for some *given* e, the answer would have been any maximal abelian C*-algebra in $B(H)$ that contains $C^*(a)$; if a has non-degenerate spectrum, this is just $C^*(a)$ itself. The simplest possibility is

$$A(1_H, e) = C^*(e)' = \{e\}', \tag{6.235}$$

which is the linear span of all projections $f \in \mathscr{P}(H)$ for which either $e \leq f$ or $e \leq 1_H - f$ (i.e., if $e = e_\psi$, then either $\psi \in fH$ or $\psi \in (fH)^\perp$). In other words, we have $a \in A(1_H, e)$ iff ψ is an eigenvector of a (i.e. the eigenvector-eigenvalue link).

Each dispersion-free state on $A(a,e)$, or, equivalently, each homomorphism h_λ : $\mathscr{P}(e,a) \to \{0,1\}$, corresponds to one of the projections f_λ through $h_\lambda(f_\lambda) = 1$ and $h_\lambda(f) = 0$ for all other one-dimensional projections f in $\mathscr{P}(e,a)$. The Born probabilities from e are then recovered by assigning (Born) measure $\mathrm{Tr}(ef_\lambda)$ to h_λ.

Though interesting, this result mainly supports so-called modal interpretations of quantum mechanics, which we reject, since they tell us nothing physical about the measurement process and address the measurement problem only philosophically.

§6.2. **The Free Will Theorem**

The Free Will Theorem was published in two versions by Conway & Kochen (2006, 2009). Analogous results had previously been published by Heywood & Redhead (1983), Stairs (1983), Brown & Svetlichny (1990), and Clifton (1993), of which only the first paper was cited by Conway and Kochen. Moreover, the close relationship to Bell's (1964) Theorem might well be insisted on as a topic that should have been discussed in the original papers. Other critical literature (making the points listed in the preamble to this chapter) includes Bassi & Ghirardi (2007), 't Hooft (2007), Goldstein et al (2010), Wüthrich (2011), Hemmick & Shakur (2012), Cator & Landsman (2014), Hermens (2014, 2015), and Walleczek (2016).

The original (Strong) Free Will Theorem (FWT) states that three assumptions, called SPIN, TWIN, and MIN, imply that the response of a spin-one particle to the bipartite experiment with spin-one particles described above 'is not a function of properties of that part of the universe that is earlier than this response (...).' Here SPIN and TWIN are the first and second half of our *Nature* axiom, whilst MIN expresses a form of context-locality as well as the loose assumption that Alice and Bob may 'freely choose' their settings a and b, respectively. Accordingly, in our notation, Conway and Kochen only use the parameter space Z, rather than the full space X we need in order to consistently axiomatize determinism. Their formulation contains an implicit assumption of determinism, whose precise nature only becomes clear from their proof, and which is akin to our formulation, except for the crucial difference that the function they allude to only acts on the particle variables and not on the settings of the experiment (of which, as already noted, Conway and Kochen just say that the experimenters can 'freely choose' them).

Conway and Kochen paraphrase their theorem as follows:

> 'if indeed we humans have free will, then elementary particles already have their own small share of this valuable commodity. More precisely, if the experimenter can freely choose the directions in which to orient his apparatus in a certain measurement, then the particles response (to be pedantic—the universe's response near the particle) is not determined by the entire previous history of the universe. (...) our theorem asserts that if experimenters have a certain freedom, then particles have exactly the same kind of freedom. Indeed, it is natural to suppose that this latter freedom is the ultimate explanation of our own. (...) Granted our three axioms [i.e., the physical ones and freedom of choice], the Free Will Theorem shows that nature itself is nondeterministic.'

However, such far-reaching conclusions seem unwarranted by the actual technical content of the theorem. Indeed, though it is also assumed in Bell's first theorem (see §6.5 below), the conjunction of *Determinism* and *Freedom* is *a priori* is uncomfortable, especially since the main novelty of the FWT lies in the emphasis Conway and Kochen (unlike Bell) put on free will. The authors acknowledge at least this point already on the first page of their first paper (Conway & Kochen, 2006), in which they anticipate criticism of the kind:

> '"I saw you put the fish in!" said a simpleton to an angler who had used a minnow to catch a bass.'

Indeed, also after more serious philosophical analysis, it has been concluded that:

'Their [Conway & Kochen's] case against determinism thus has all the virtues of theft over honest toil. It is truly indeterminism in, indeterminism out.' (Wüthrich, 2011)

Our formulation of the FWT, in which the original allusion to undefined free will in allowing arbitrary settings of the experiment has been replaced by complete determinism including the settings, avoids this criticism.

To derive (6.35) - (6.38), we use (6.21) to write down the formulae

$$\begin{aligned} P_{\psi_0}(F_i=1,G_j=1|A=a,B=b) &= \langle\psi_0,(1_3-|\mathbf{u}_i\rangle\langle\mathbf{u}_i|)\otimes(1_3-|\mathbf{v}_j\rangle\langle\mathbf{v}_j|)\psi_0\rangle;\\ P_{\psi_0}(F_i=0,G_j=0|A=a,B=b) &= \langle\psi_0,|\mathbf{u}_i\rangle\langle\mathbf{u}_i|\otimes|\mathbf{v}_j\rangle\langle\mathbf{v}_j|\psi_0\rangle;\\ P_{\psi_0}(F_i=1,G_j=0|A=a,B=b) &= \langle\psi_0,(1_3-|\mathbf{u}_i\rangle\langle\mathbf{u}_i|)\otimes|\mathbf{v}_j\rangle\langle\mathbf{v}_j|\psi_0\rangle;\\ P_{\psi_0}(F_i=0,G_j=1|A=a,B=b) &= \langle\psi_0,|\mathbf{u}_i\rangle\langle\mathbf{u}_i|\otimes(1_3-|\mathbf{v}_j\rangle\langle\mathbf{v}_j|)\psi_0\rangle. \end{aligned}$$

For example, for any pair of unit vectors $\mathbf{u},\mathbf{v}$ we have

$$\begin{aligned} &\langle\psi_0,|\mathbf{u}\rangle\langle\mathbf{u}|\otimes|\mathbf{v}\rangle\langle\mathbf{v}|\psi_0\rangle =\\ &\tfrac{1}{3}\langle\mathbf{e}_1\otimes\mathbf{e}_1+\mathbf{e}_2\otimes\mathbf{e}_2+\mathbf{e}_3\otimes\mathbf{e}_3,\mathbf{u}|\otimes|\mathbf{v}\rangle\langle\mathbf{v}|(\mathbf{e}_1\otimes\mathbf{e}_1+\mathbf{e}_2\otimes\mathbf{e}_2+\mathbf{e}_3\otimes\mathbf{e}_3)\rangle =\\ &\tfrac{1}{3}\langle\mathbf{e}_1\otimes\mathbf{e}_1+\mathbf{e}_2\otimes\mathbf{e}_2+\mathbf{e}_3\otimes\mathbf{e}_3,\langle\mathbf{u},\mathbf{v}\rangle\mathbf{u}\otimes\mathbf{v}\rangle\\ &=\tfrac{1}{3}\langle\mathbf{u},\mathbf{v}\rangle^2, \end{aligned}$$

which gives (6.36). The other cases are similar.

The implications of the finite precision loophole of the Kochen–Specker Theorem for the Free Will Theorem were analyzed by Hermens (2014), who concluded that this loophole does not apply. We give a more precise argument to this effect.

We have dense colorable subsets $X_A^c \subset X_A$ and $X_B^c \subset X_B = X_A$, where X_A^c may or may not coincide with X_B^c. If not, the perfect correlation condition (6.54) in the *Nature* assumption cannot even be stated, but even if $X_A^c = X_B^c$, since finite precision of experiment has been declared to be an issue it would be quite out of character to impose (6.54). Instead, one needs a probabilistic version of this condition, of which it will turn out that it cannot be satisfied. As in the notes to the previous section, for each density matrix ρ one needs a probability measure μ_ρ on Z that reproduces the statistical quantum-mechanical predictions for the associated quantum state. Compared to the notes to the previous section, the role of W is now played by z, in that for given F and G one might write

$$W(a,b) = (\hat{F}(a,z),\hat{G}(b,z). \tag{6.236}$$

This measure may be constructed analogously to (6.232), i.e., for any sequence $(a^{(k)})$ of bases drawn from X_A^c, any sequence $(b^{(k)})$ of bases drawn from X_B^c, and any sequences $(\lambda^{(k)})$ and $(\gamma^{(k)})$ in Λ, cf. (6.22), where $k\in K\subset\mathbb{N}$ is arbitrary, we define

$$\begin{aligned} &\mu_\rho(\{z\in Z\mid \hat{F}(a^{(k)},z)=\lambda^{(k)},\hat{G}(b^{(k)},z)=\gamma^{(k)},k\in K\} =\\ &\prod_{k\in K}\mathrm{Tr}\left(\rho\prod_{i,j=1}^{3}[J_{\mathbf{u}_i}^2=\lambda_i^{(k)}]\cdot[J_{\mathbf{v}_j}^2=\gamma_j^{(k)}]\right), \end{aligned} \tag{6.237}$$

where, as in the main text,

$$a = (\mathbf{u}_1, \mathbf{u}_2, \mathbf{u}_3); \tag{6.238}$$

$$b = (\mathbf{v}_1, \mathbf{v}_2, \mathbf{v}_3). \tag{6.239}$$

Note that $J^2_{\mathbf{u}_i}$ acts on Alice's Hilbert space $\mathbb{C}^3$ whilst $J^2_{\mathbf{v}_j}$ acts on Bob's. In particular, for fixed $k_0 \in K$ and $\lambda, \gamma \in \Lambda$, we have the special case of (6.237) for compatible measurements, viz.

$$\mu_\rho(\{z \in Z \mid \hat{F}(a^{(k_0)}, z) = \lambda, \hat{G}(b^{(k_0)}, z) = \gamma\} = \mathrm{Tr}\left(\rho \prod_{i,j=1}^{3} [J^2_{\mathbf{u}_i} = \lambda_i] \cdot [J^2_{\mathbf{v}_j} = \lambda_j]\right),$$

where in the main text we would have written $P_\rho(F = \lambda, G = \mu | A = a, B = b)$ for the right-hand side. Hence for the correlated state $\rho = |\psi_0\rangle\langle\psi_0|$ we obtain from (6.42):

$$\mu_{\psi_0}(\{z \in Z \mid \hat{F}_i(a,z) \neq \hat{G}_j(b,z)\}) = \tfrac{2}{3}(1 - \langle \mathbf{u}_i, \mathbf{v}_j\rangle^2), \tag{6.240}$$

which of course vanishes if $\mathbf{u}_i = \mathbf{v}_j$. If the expression $1 - \langle \mathbf{u}_i, \mathbf{v}_j\rangle^2$ appearing here is small, then the projections $e_{\mathbf{u}_i}$ and $e_{\mathbf{v}_j}$ are close (in norm), since

$$\|e_{\mathbf{u}_i} - e_{\mathbf{v}_j}\|^2 \leq 2(1 - \langle \mathbf{u}_i, \mathbf{v}_j\rangle^2). \tag{6.241}$$

Eq. (6.240) therefore allows us to make rigorous sense of Hermens' (2014) heuristic idea that the assumption (6.54) in the FWT should be modified as follows:

'if $\|e_{\mathbf{u}_i} - e_{\mathbf{v}_j}\|$ is small, then in most of the cases $\hat{F}_i(a,z) = \hat{G}_j(b,z)$.'

Namely, we replace (6.54) by the following approximate correlation condition:

- For every $\varepsilon > 0$ there is $\delta > 0$ such that if $1 - \langle \mathbf{u}_i, \mathbf{v}_j\rangle^2 < \delta$, then

$$\mu_{\psi_0}(\{z \in Z \mid \hat{F}_i(a,z) \neq \hat{G}_j(b,z)\}) < \varepsilon. \tag{6.242}$$

Indeed, if the theory existed, on could simply take $\delta = \varepsilon$. However, a theory satisfying (6.242) does not exist, as can be proved by contradiction: if $\hat{F}_i(a,z) = \hat{G}_j(b,z)$ for all pairs $(\mathbf{u}_i, \mathbf{v}_j)$ such that $1 - \langle \mathbf{u}_i, \mathbf{v}_j\rangle^2 < \varepsilon$, then the proof of Theorem 6.13 shows not only that (6.32) still holds on the modified *Nature* assumption (so that $\tilde{F}(\cdot, z)$ again defines a coloring of S^2), but that *in addition* we have

$$1 - \langle \mathbf{u}, \mathbf{u}'\rangle^2 < \delta \Rightarrow \tilde{F}(\mathbf{u}, z) = \tilde{F}(\mathbf{u}', z). \tag{6.243}$$

In particular, the apparently weaker correlation condition ending with (6.242) is actually *stronger* than its exact counterpart (6.54).

Thus Theorem 6.13 still holds on this revised *Nature* assumption, so that unlike the Kochen–Specker Theorem, the Free Will Theorem is immune to the finite precision loophole. The price for this immunity is that, quite against the spirit of the FWT, some probabilistic reasoning had to be invoked, so that the difference between the FWT and Bell's first theorem has blurred even further.

§6.3. **Philosophical intermezzo: Free will in the Free Will Theorem**

The literature on free will is immense. Introductory accounts include Walter (2001), which focuses on the connection with neuroscience, Doyle (2011), and Beebee (2013), the second of which remains largely philosophical, the third even completely. A very sophisticated recent defense of compatibilism is Ismael (2016). Lewis's 'local miracle compatibilism' was proposed in Lewis (1981). What's more:

> '[Lewis's paper is] the finest essay that has ever been written in defense of compatibilism—possibly the finest essay that has ever been written about any aspect of the free will problem.' (van Inwagen, 2008).

Saunders (1968) already made a point similar to Lewis's; see also Moore (1912, Ch. 6). For Lewis's theory of counterfactuals see Lewis (1973, 1979, 2000), as well as Menzies (2014). See also Fischer (1994), Beebee (2003, 2013), and Vihvelin (2013).

Although Lewis's position is called *local miracle compatbilism*, a miracle takes place neither in the actual world where Alice's hand is at rest nor in the possible world where she raises it, i.e., a law is broken neither in the former nor in the latter:

> 'This is what Lewis means by a 'miracle': an event M is a miracle if and only if M occurs at *possible world* w, and M is contrary to some *actual* law (or combination of laws) L. The point here is that while M is a miracle in Lewis's sense, it is not contrary to any of w's laws of nature. At w, L simply isn't a law in the first place. So, as things *actually* happened—in the *actual* world—L is a law, and m does not occur, so there is no miracle in the usual sense of 'miracle'. m is only a 'miracle' in Lewis's special sense of 'miracle': something (m) happens in w that is contrary to the laws of nature in the *actual* world.'
> (Beebee, 2013, p. 62)

Unfortunately, confusion may arise if the quotation in the main text 'if I did it, a law would be broken' from Lewis (1981) is subjected to the following explanation:

> 'On Lewis's account of counterfactuals, the *truth conditions* for counterfactuals—what makes them true—are as follows. Suppose we have the counterfactual 'if A had been the case, B would have been the case' (so if A is 'I miss the bus' and B is 'I'm late', this counterfactual just says, 'if I'd missed the bus, I would have been late'). This counterfactual will be true if and only if, *at the closest possible world to the actual world* at which A is true, B is also true. So, our sample counterfactual, 'if I'd missed the bus, I would have been late', is true if and only if: *at the closest possible world to the actual world* at which I miss the bus, I'm late.' (Beebee, 2013, p. 60).

Removing any possible remaining doubt, on p. 62 she mentions that the closest possible world where I miss the bus is the world w. According to this explanation, then, Lewis's sentence 'if I did it, a law would be broken', would mean that *at the closest possible world to the actual world* in which I did it, a law *is* broken, i.e., in w. But according to Beebee's definition quoted in the main text of what Lewis means by a miracle, apparently this is not the right reading (and indeed it would, in our view, be nonsensical). Moreover, Lewis (1981) emphasizes that in the first bullet point in the main text above—which he defends—it is not the agent who would break a law, whereas in the second bullet point —rejected by Lewis—it is; in the first it is the breaking of some law at an earlier time that enables the agent to do what she, in our actual world, did not do. Thus Lewis's phrasing seems awkward.

Our development of Lewis's argument is indebted to Vihvelin (2013, pp. 164–165), who (re)states Lewis's first bullet point as the following conjunction:

1. **Slightly Different Past**: If I had raised my hand, the past would still have been exactly the same until shortly before the time of my decision.
2. **Slightly Different Laws:** If I had raised my hand, the laws would have been ever so slightly different in a way that permitted a divergence from the lawful course of actual history shortly before the time of my decision.

A second way in which Alice could (counterfactually) have raised here hand is through an instant (counterfactual) modification of the state of the world, as in Bennett (1984). This has been explicated by Vihvelin (2013, p. 165), too:

1. **Same Laws:** If I had raised my hand, the laws would still have been the same.
2. **Completely Different Past:** If I had raised my hand, past history would have been different all the way back to the Big Bang.

Here we prefer to write **Different Past**, since even though in this scenario the state indeed (by determinism) would have been different all the way back to the Big Bang, the entire trajectory of the world may or may not be close to the actual one. In this scenario, the two cases Lewis distinguishes take the form in the main text.

Since the main novelty of their papers lies in the emphasis on free will, the reader might wonder what Conway & Kochen themselves have to say about the subject. As we can read in the delightful biography of Conway by Roberts (2015), or watch in his video lectures on the Free Will Theorem (Conway, 2009), free will is indeed of great importance to at least the first author of the theorem. Unfortunately, his interest in free will seems unaccompanied by any philosophical sophistication, e.g.:

> 'Compatibilism in my view is silly. Sorry, I shouldn't just say straight off that it is silly. Compatibilism is an old viewpoint from previous centuries when philosophers were talking about free will. The were accustomed to physical theory being deterministic. And then there's the question: How can we have free will in this deterministic universe? Well, they sat and thought for ages and ages and ages and read books on philosophy and God knows what and they came up with compatibilism, which was a tremendous wrenching effect to reconcile 2 things which seemed incompatible. And they said they were compatible after all. But nobody would *ever* have come up with compatibilism if they thought, as turns out to be the case, that science wasn't deterministic. The whole business of compatibilism was to reconcile what science told you at the time, centuries ago down to 1 century ago: Science appeared to be totally deterministic, and how can we reconcile that with free will, which is not deterministic? So compatibilism, I see it as out of date, really. It's doing something that doesn't need to be done. However, compatibilism hasn't gone out of date, certainly, as far as the philosophers are concerned. Lots of them are still very keen on it. How can I say it? If you do anything that seems impossible, you're quite proud when you appear to have succeeded. And so really the philosophers don't want to give up this notion of compatibilism because it seems to damned clever. But my view is it's really nonsense. And it's not necessary. So whether it actually is nonsense or not doesn't matter.'
>
> (Conway, quoted in Roberts, 2015, pp. 361–362).

Finally, our version of van Inwagen's (1975) Consequence Argument is due to Beebee (2003), and the novel parts of this section are based on Landsman (2016c). For interesting philosophical criticism of this approach, see De Mola (2016).

§6.4. **Technical intermezzo: The GHZ-Theorem**

The GHZ Theorem appeared in Greenberger et al (1990) See also Clifton, Redhead, & Butterfield (1991) and Bub (1997). Innumerable variations on and generalizations of such arguments may be given, leading to equally many Free Will Theorems. All of these have their roots in algebraic properties of matrices, which hidden variable theories (in vain) try to reproduce.

§6.5. **Bell's theorems**

The original contributions to the theme of this section are Bell (1964, 1976), of which the first is one of the most famous papers of 20th century theoretical physics. Since there are more than 10,000 papers citing Bell (1964) alone, it is impossible to discuss all literature relevant to Bell's work. What we call his first theorem originates with Bell (1964), which incidentally was written after Bell (1966), but our treatment of the settings (taken from Cator & Landsman, 2014) is different. Though originally motivated as an attempt to make the Free Will Theorem look less of a *petitio principii*, it also addresses a problem Bell faced even according to some of his staunchest supporters (Norsen, 2009; Seevinck & Uffink, 2011), namely the tension between the idea that the hidden variables (in the pertinent causal past) should on the one hand include all ontological information relevant to the experiment, but on the other hand should leave Alice and Bob free to choose any settings they like.

His second theorem comes from Bell (1976), followed by Bell (1990a). Apart from his own papers, which are reprinted in Bell, Gottfried & Veltman (2001), treatments of Bell's Theorems we regard as sound include Fine (1982), Jarrett (1984), Pitowsky (1989), van Fraassen (1991), Butterfield (1992a,b), Bub (1997), Werner, & Wolf (2001), Liang, Spekkens, & Wiseman (2011), Shimony (2013), Wiseman (2014), and Brown & Timpson (2015). Recent and mathematically innovative approaches include Abramsky & Brandenburger (2011), Acín et al (2015), and Fritz (2016). For history, see Gilder (2008) and Kaiser (2010).

Unfortunately, we have not been able to come to grips with (and hence do not cite) literature claiming that Bell's theorems are false, or have nothing to do with hidden variables, or prove that quantum mechanics (if not nature itself!) is nonlocal *per se*, or that he never changed his mind and only has one theorem saying it all.

The verification of (6.102) - (6.105) is analogous to the above computations deriving (6.35) - (6.38). In terms of the unit vector

$$v_a = \begin{pmatrix} \cos a \\ \sin a \end{pmatrix}, \tag{6.244}$$

the observable F Alice measures on setting $A = a$ is the projection $e_a = |v_a\rangle\langle v_a|$, and similarly for Bob. Hence the corresponding Born probabilities are given by

$$\begin{aligned}
P_{\psi_0}(F=1, G=1|A=a, B=b) &= \langle \psi_0, e_a \otimes e_b \psi_0 \rangle; \\
P_{\psi_0}(F=0, G=0|A=a, B=b) &= \langle \psi_0, (1_2 - e_a) \otimes (1_2 - e_b) \psi_0 \rangle; \\
P_{\psi_0}(F=1, G=0|A=a, B=b) &= \langle \psi_0, e_a \otimes (1_2 - e_b) \psi_0 \rangle; \\
P_{\psi_0}((F=0, G=1|A=a, B=b) &= \langle \psi_0, (1_2 - e_a) \otimes e_b \psi_0 \rangle.
\end{aligned}$$

For example, we have

$$\begin{aligned}\langle\psi_0, e_a\otimes e_b\psi_0\rangle &= \tfrac{1}{2}\langle \mathbf{e}_1\otimes\mathbf{e}_1+\mathbf{e}_2\otimes\mathbf{e}_2, |v_a\rangle\langle v_a|\otimes|v_b\rangle\langle v_b|(\mathbf{e}_1\otimes\mathbf{e}_1+\mathbf{e}_2\otimes\mathbf{e}_2)\rangle\\ &= \tfrac{1}{2}\langle \mathbf{e}_1\otimes\mathbf{e}_1+\mathbf{e}_2\otimes\mathbf{e}_2, (\cos a\cos b+\sin a\sin b)v_a\otimes v_b\rangle\\ &= \tfrac{1}{2}(\cos a\cos b+\sin a\sin b)^2\\ &= \tfrac{1}{2}\cos^2(a-b).\end{aligned}$$

The CHSH-inequality (6.117) is due to Clauser, Horne, Shimony, & Holt (1969). The definitive (i.e., loophole-free) experimental verification of its violation in nature is Henson et al. (2015). A direct proof starts of (6.117) from the simpler inequality

$$P(F\neq H)\leq P(F\neq G)+P(G\neq H), \tag{6.245}$$

for three $\{0,1\}$-valued random variables F,G,H, which implies (6.117). To prove (6.245), one just writes

$$\begin{aligned}P(F\neq H) &= P(F=1,G=1,H=0)+P(F=1,G=0,H=0)\\ &\quad + P(F=0,G=1,H=1)+P(F=0,G=0,H=1),\end{aligned}$$

etc., and notes that each term on the left-hand side of (6.245) also occurs on the right-hand side. Since each term lies in $[0,1]$ and hence is positive, this implies (6.245). Our proof of Proposition 6.17 follows Werner & Wolf (2001), as does our proof of Theorem 6.18 (though not our formulation thereof, which once again derives from Cator & Landsman (2014). This proof shows that, as first noted by Fine (1982) and analyzed more deeply in Butterfield (1992b), there is no real distinction between the possibility of reproducing given (empirical) probabilities $P(F=\lambda, G=\gamma|A=a, B=b)$ *that satisfy Bell locality* by a *local deterministic hidden variable theory* or by a *local stochastic hidden variable theory*. Most current research in this direction, sparked by Popescu & Rohlich (1994), is therefore concerned with theories defined by formal joint conditional probabilities that satisfy a no signaling condition like OI instead of Bell locality, cf. Bub (2011b) and Brunner et al (2014) for reviews.

Formal conditional probabilities of the kind that Bell's second theorem uses have been axiomatized by e.g. Popper (1938) and Rényi (1955); the following axioms are theorems if conditional probabilities are defined à la Kolmogorov by (1.1). Let Σ be some σ-algebra and let $\mathscr{F}\subset\Sigma\backslash\{\emptyset\}$ be an ideal in Σ in the sense that if $B\in\Sigma$ and $C\in\mathscr{F}$, then $B\cap C\in\mathscr{F}$. A ***conditional probability*** on $(\Sigma,\mathscr{F})$ is a map

$$P:\Sigma\times\mathscr{F}\to[0,1]; \tag{6.246}$$

$$(A,C)\mapsto P(A|C), \tag{6.247}$$

such that:

1. For each $C\in\mathscr{F}$ the map $A\mapsto P(A|C)$ is a probability measure on Σ;
2. $P(A\cap B|C)=P(A|B\cap C)\cdot P(B|C)$, for each $A,B\in\Sigma$ and $C\in\mathscr{F}$.

Van Fraassen (1991) noted that if (6.121) holds, then the variable x is a ***common cause*** in the sense of Reichenbach for Alice's and Bob's outcomes (see Hofer-Szabó (2015) for a recent paper in this direction). To explain this observation, suppose two random processes F and G (like Alice's and Bob's measurements) are correlated, i.e., $P(F=\lambda, G=\gamma) \neq P(F=\lambda)P(G=\gamma)$. What might cause the correlation?

1. *Chance*. If Alice and Bob independently throw dice but always get the same result, there is a computable nonzero probability for this to happen without any reason. But this probability decreases as the number of occurrences grows.
2. *Causation*. One outcome influences or even determines the other. Maybe Bob, whose experiment is genuinely random, is able to manipulate Alice's experiment once he has seen his outcome. But according to relativity theory or other basic notions of causality in space-time, this should be impossible if Alice and Bob perform their measurements simultaneously and far from each other.
3. *Ur-determinism*. The initial conditions at the Big Bang plus deterministic Laws of Nature imply the correlation. However, physics becomes pointless if we endorse this option. The notion of *explanation* as the purpose of science is defeated and there is little difference between this argument and Divine Predestination.
4. *Identity*. The motions of my mirror image are strongly correlated with me, but that is because this image is really the same as me (at least in so far as motion is concerned, as opposed to e.g. thoughts). This example might also be explained using causation. Another example consists of Alice and Bob filming the same random process (which may also be explained using the following concept).
5. *Common Cause* A random process X is said to be a ***common cause*** for two correlated random processes if it precedes both and satisfies

$$P(F=\lambda, G=\gamma|X=x) = P(F=\lambda|X=x)P(G=\gamma|X=x). \tag{6.248}$$

   Another way to write this is $P(F=\lambda|G=\gamma, X=x) = P(F=\lambda|X=x)$, which shows that a common cause X screens off the dependence of F on G. Often the common cause is hidden and has to be inferred from the observed correlation (having excluded other explanations, like the ones above). A nice example of this is the inference of a manuscript called Q in New Testament studies. It is clear that the Gospels of Matthew and Luke both draw on Mark, but they also contain strikingly similar or even identical non-Markan passages. For various reasons it is unlikely that either one copied these from the other, so that the main hypothesis is that they both rely on Q, which is now lost. See e.g. Mack (1993).

From this perspective, the amazing fact is that the correlations in the Alice and Bob experiment with either spin-1 particle or photons cannot be explained by a common cause, since its existence (in the form of x) would imply the Bell inequality. However, of the four other explanations described above, no. 1 is ridiculous given the statistics of the relevant experiments, no. 2 is at odds with relativity, and no. 4 seems inapplicable. This leaves no. 3, which seems only supported by 't Hooft (2016), who denies the independence assumptions (i.e. between the settings and the state of the pair of particles undergoing measurement) lying at the basis of both the Free Will Theorem and Bell's theorems. Every way you look at it you lose!

Generalizations of Theorem 6.19 to operator algebras were given e.g. by Baez (1987), Raggio (1988), Werner (1989), and Bacciagaluppi (1993), as follows. Let A and B be unital C*-algebras, with projective tensor product $A\hat{\otimes}B$ (i.e., the completion of the algebraic tensor product $A\otimes B$ in the maximal C*-cross-norm), cf. §C.13; the choice of the projective tensor product guarantees that each state on $A\otimes B$ extends to a state on $A\hat{\otimes}B$ by continuity; conversely, since $A\otimes B$ is dense in $A\hat{\otimes}B$, each state on the latter is uniquely determined by its values on the former. In particular, product states $\rho\otimes\sigma$ and mixtures $\omega=\sum_i p_i\rho_i\otimes\sigma_i$ thereof are well defined on $A\hat{\otimes}B$. If $A\subset B(H_1)$ and $B\subset B(H_2)$ are von Neumann algebras, and all states considered are normal, it is easier to work with the ***spatial*** tensor product $A\overline{\otimes}B$, defined as the double commutant (or weak completion) of $A\otimes B$ in $B(H_1\otimes H_2)$. Any *normal* state on $A\otimes B$ extends to a normal state on $A\overline{\otimes}B$ by continuity. Below we use $\hat{\otimes}$, but the results also work for $\overline{\otimes}$. In what follows, A and B are *unital* C*-algebras.

Definition 6.23. *Let ω be a state on $A\hat{\otimes}B$.*

1. *A* **product state** *is a state of the form $\omega=\rho\otimes\sigma$, i.e., ω is defined by linear (and continuous) extension of $\omega(a\otimes b)=\rho(a)\sigma(b)$.*
2. *A state ω is* **uncorrelated** *when it is in the w*-closure of the convex hull of the product states on $A\hat{\otimes}B$. In particular, states $\omega=\sum_i p_i\rho_i\otimes\sigma_i$, where $p_i>0$ and $\sum_i p_i=1$, are uncorrelated (w*-convergent infinite sums are allowed here).*
3. *A state is* **correlated** *when it is* not *uncorrelated.*

An uncorrelated state ω is pure precisely when it is a product of pure states. This has the important consequence that both its restrictions $\omega_{|A}$ and $\omega_{|B}$ to A and B, respectively, are pure as well (the restriction $\omega_{|A}$ of a state ω on $A\hat{\otimes}B$ to, say, A is given by $\omega_{|A}(a)=\omega(a\otimes 1_B)$, where 1_B is the unit element of B, etc.). A correlated *pure* state has the property that its restriction to A or B is *mixed*.

Proposition 6.24. *The following conditions are equivalent:*

- *Each state on $A\hat{\otimes}B$ is uncorrelated;*
- *Each pure state on $A\hat{\otimes}B$ is a product state;*
- *At least one of the C*-algebras A and B is commutative.*

For the proof see Takesaki (2002), Theorem 4.14.

Corollary 6.25. *Correlated states exist iff A and B are both noncommutative.*

As one might expect, this result is closely related to the Bell inequalities:

Proposition 6.26. *For any $\omega\in S(A\hat{\otimes}B)$, the following conditions are equivalent:*

- *ω is uncorrelated.*
- *For all self-adjoint operators $a_1,a_2\in A$ and $b_1,b_2\in B$ of norm ≤ 1 we have*

$$|\omega(a_1(b_1+b_2)+a_2(b_1-b_2))|\leq 2. \tag{6.249}$$

See Baez (1987), Raggio (1988), Bacciagaluppi (1993), and Landsman (2006a).

Corollary 6.27. *If A or B is commutative, then* (6.249) *holds for all states ω.*

An elegant geometric approach to the Bell inequalities was developed by Pitowsky (1989, 1994), which we now summarize (also cf. Werner & Wolf, 2001).

Suppose we have a bipartite experiment with m different settings $A = a_1, \ldots a_m$ and $B = b_1, \ldots, b_m$ on each wing, and binary outcomes, i.e., in $\{0, 1\}$. We now denote the probability $P(F = 1|A = a_i)$ that $F(a_i)$ (i.e. the particular property measured by experiment F at setting a_i) is true by p_i $(i = 1, \ldots, m)$, and likewise we write p_{j+m} for $P(G|B = b_j)$, i.e., the probability that $G(b_j)$ is true, once again for $j = 1, \ldots, m$. Furthermore, we abbreviate the probability that $F(a_i)$ and $G(b_j)$ are both true by

$$p_{i,j+m} \equiv P(F = 1, G = 1|A = a_i, B = b_j) \;\; (i, j = 1, \ldots, m). \tag{6.250}$$

The $2m + m^2$ "surface probabilities" $\mathbf{p} = (p_1, \ldots, p_{2m}, p_{1,m+1}, \ldots, p_{m,2m})$ form a vector in $\mathbb{R}^{2m+m^2}$, which we wish to constrain by the following assumption: there is a fact of the matter underlying each experiment according to which the pair $(F(a_i), G(b_j))$ already had a truth value for each possible setting (a_i, b_j), independently of any measurement being carried out or not ("***local realism***"). Thus the probabilities $\mathbf{p}$ (which now arguably have an ignorance interpretation) must lie in the convex polytope in $\mathbb{R}^{|2m+m^2|}$ defined as the convex hull C_m of the following set of (extreme) points: for each $2m$-tuple $\lambda = (\lambda_1, \ldots, \lambda_{2m})$, where $\lambda_i \in \{0, 1\}$, define

$$\mathbf{x}_\lambda = (\lambda_1, \ldots, \lambda_{2m}, \lambda_1 \cdot \lambda_{m+1}, \ldots, \lambda_m \cdot \lambda_{2m}) \in \mathbb{R}^{2m+m^2}, \tag{6.251}$$

i.e., the entry at place k is λ_k $(k = 1, \ldots, 2m)$ and the entry at place (i, j) is $\lambda_i \cdot \lambda_{m+j}$, where $i, j = 1, \ldots, m$. The interpretation of this is that $\mathbf{x}_\lambda$ represents the particular fact of the matter where $F(a_i)$ has truth value λ_i and $G(b_j)$ has truth value λ_{m+j}, so that their conjunction $(F(a_i), G(b_j))$ has truth value $\lambda_i \cdot \lambda_{m+j}$. In this state the probability of the said configuration is one and all other states have probability zero; arbitrary probability assignments then lie in C_m. The point, then, is to characterize the convex polytope $C_m \subset \mathbb{R}^{2m+m^2}$ through a finite set of inequalities, which turn out to be generalized Bell inequalities. Seeing this result requires some background.

Let V be a real topological vector space with (continuous) dual V^*; if $V = \mathbb{R}^n$ we may also put $V^* = \mathbb{R}^n$ and write $\varphi(v)$ as an inner product $\langle \varphi, v \rangle$ in what follows.

1. Any (not necessarily convex) subset $S \subset V$ has a ***polar*** $S^o \subset V^*$ defined by

$$S^o = \{\varphi \in V^* \mid \varphi(v) \leq 1 \forall v \in S\}, \tag{6.252}$$

which is a closed convex subset of V^*. If $S = K$ is a compact convex set, we have

$$K^o = \{\varphi \in V^* \mid \varphi(v) \leq 1 \forall v \in \partial_e K\}. \tag{6.253}$$

2. The ***bipolar theorem*** (cf. e.g. Simon (2011, Theorem 5.5) states that

$$S^{oo} = \mathrm{co}(S \cup \{0\}). \tag{6.254}$$

In particular, if K a closed convex set containing the origin, then

$$K^{oo} = K, \tag{6.255}$$

and hence, if K^o is a compact convex set, we may reconstruct K from K^o as

$$K = \{v \in V \mid \varphi(v) \leq 1 \forall \varphi \in \partial_e K^o\}. \tag{6.256}$$

3. In particular, if K is a convex polytope in a finite-dimensional vector space containing the origin, then so is K^o. In that case, $\partial_e K^o$ is a finite set and so points in K are characterized by a *finite* set of *linear* inequalities (6.256), which describe the faces of the polytope. In this case, the associated (dual) description of K is called the ***Minkowski–Weyl Theorem***, see e.g. Paffenholz (2010) for applications.

For example, among the five Platonic solids (i.e. in $\mathbb{R}^3$) the cube and the octahedron are dual to each other, as are the dodecahedron and the icosahedron, whereas the terahedron is self-dual. *A propos*, the latter arises as the convex polytope C_1 for $m = 1$ in the above story: clearly $2m + m^2 = 3$, and for the vertices of C_1 one takes the four points $\mathbf{x}_\lambda$ ensuing from the four possibilities $\lambda = (0,0),(1,0),(0,1),(1,1)$, i.e., $\mathbf{x}_\lambda = (0,0,0),(1,0,0),(0,1,0),(1,1,1)$. Then the inequalities in (6.256) are

$$p_{1,2} \geq 0, \; p_1 \geq p_{1,2}, \; p_2 \geq p_{1,2}, \; p_1 + p_2 - p_{1,2} \leq 1. \tag{6.257}$$

For $m = 2$ the ensuing convex polytope $C_2 \subseteq \mathbb{R}^8$ is the convex hull of 16 extreme points, whose inequalities may be found in Pitowsky (1989, p. 27); these imply the CHSH inequality, whose violation in quantum mechanics therefore shows that the probabilities in question have no local realistic model.

More generally, suppose we have n yes-no experiments $(E_1,\ldots,E_n)$ and some subset S_n of the set $\{(i,k) \mid 1 \leq l < k \leq n\}$ (above we had $n = 2m$, $E_i = F(a_i)$ for $i = 1,\ldots,m$, $E_{m+j} = G(b_j)$ for $j = 1,\ldots,m$, and $S_n = \{(i,m+j) \mid 1 \leq i,j \leq m\}$). This gives surface probabilities $(p_1,\ldots,p_n,p_{i,k})$, where $(i,k) \in S_n$), which form a vector $\mathbf{p}$ in $\mathbb{R}^{n+|S_n|}$. As in (6.251), each truth assignment $\lambda = (\lambda_1,\ldots,\lambda_n)$, $\lambda \in \{0,1\}$, then defines a point $\mathbf{x}_\lambda \in \mathbb{R}^{n+|S_n|}$ with coordinates $(\lambda_1,\ldots,\lambda_n,\lambda_i \cdot \lambda_k)$, where once again $(i,k) \in S_n$. This set of 2^n points in turn spans a convex polytope C_{S_n} characterized by inequalities following from the dual characterization (6.256). Classical thinking would constrain the $\mathbf{p}$ so as to lie in C_{S_n}, and indeed we have $\mathbf{p} \in C_{S_n}$ iff there is a probability space (X, G, μ) such that $p_i = \mu(A_i)$ and $p_{i,k} = \mu(A_i \cap A_k)$ for certain events $A_i \in \Sigma$, cf. Theorem 2.3 in Pitowsky (1989), which is based on Fine (1982).

Some authors claim on this basis that Bell-type inequalities have nothing to do with physics, but surely the point is that some physical assumptions (notably local realism) have to be made in order to justify the "classical thinking" behind C_{S_n}.

§6.6. **The Colbeck–Renner Theorem**

This section is based on Colbeck & Renner (2011, 2012a, 2012b), where the main idea originates (alas with unclear assumptions and at best heuristic "proofs"), Braunstein & Caves (1990), who provided steps 1 and 2 of the proof, and Landsman (2015), whom we follow closely. See also Leegwater (2016) for a technically different approach (by a far more complicated argument, Leegwater seems to manage to do without our **CP** assumption, i.e., continuity of probabilities).

Chapter 7
Limits: Small $\hbar$

Limits are essential to the asymptotic Bohrification program. It was recognized at an early stage in the development of quantum mechanics that the limit $\hbar \to 0$ of Planck's constant going to zero should play a role in the derivation of classical physics from quantum theory, and later on also the thermodynamic limit (which often means "$\lim_{N\to\infty}$", where N is the number of particles in the system) became a subject of interest in quantum statistical mechanics. The conceptual status of these limits will be discussed in Chapter 10; in the present one we mainly explain the underlying mathematics. However, one question needs to be addressed immediately, since it is a source of much confusion. Varying N seems a realistic thing to do in the lab or on paper, whereas $\hbar$ is a *constant*, so how can it be varied? The answer is that $\hbar$ is a *dimensionful* constant, from which one forms dimensionless combinations of $\hbar$ and other parameters; this combination then re-enters the theory as if it were a dimensionless version of $\hbar$ that can indeed be varied. The oldest example is Planck's radiation formula $E_\nu/N_\nu = h\nu/(e^{h\nu/kT}-1)$, with temperature T as the pertinent variable. Indeed, the observation of Einstein and Planck that in the limit $\hbar\nu/kT \to 0$ this formula converges to the classical equipartition law $E_\nu/N_\nu = kT$ may well be the first use of the $\hbar \to 0$ limit of quantum theory; note that Einstein put $\hbar\nu/kT \to 0$ by letting $\nu \to 0$ at fixed T and $\hbar$, whereas Planck took $T \to \infty$ at fixed ν and $\hbar$!

Another example is the Hamiltonian $h = -\frac{\hbar^2}{2m}\Delta + V(x)$ in the Schrödinger equation of non-relativistic quantum mechanics, where m is the mass of the pertinent particle. Here one may pass to dimensionless parameters by introducing an energy scale ε typical of H, like $\varepsilon = \sup_x |V(x)|$, as well as a typical length scale ℓ, such as $\ell = \varepsilon/\sup_x |\nabla V(x)|$ (if these quantities are finite). In terms of the dimensionless variable $\tilde{x} = x/\ell$, the rescaled Hamiltonian $\tilde{h} = h/\varepsilon$ is then dimensionless and equal to $\tilde{h} = -\tilde{\hbar}^2\tilde{\Delta} + \tilde{V}(\tilde{x})$, where $\tilde{\hbar} = \hbar/\ell\sqrt{2m\varepsilon}$, the operator $\tilde{\Delta}$ is the Laplacian for $\tilde{x}$, and $\tilde{V}(\tilde{x}) = V(\ell\tilde{x})/\varepsilon$. Here $\tilde{\hbar}$ is dimensionless, and one might study the regime where it is small. Similarly, it is often realistic to rescale the potential V by a positive number λ, in which case $h_\lambda = -\frac{\hbar^2}{2m}\Delta + \lambda V(x)$ can be rescaled to $h_\lambda/\lambda = -\frac{\tilde{\hbar}^2}{2m}\Delta + V(x)$, with $\tilde{\hbar} = \hbar/\sqrt{\lambda}$, so that the "large V limit" $\lambda \to \infty$ comes down to $\tilde{\hbar} \to 0$.

K. Landsman, *Foundations of Quantum Theory*,
Fundamental Theories of Physics 188, DOI 10.1007/978-3-319-51777-3_7

In (older) textbooks on quantum mechanics the limit $\hbar \to 0$ is typically studied using the so-called WKB-approximation. This may be justified on historical grounds, but in fact this approximation is rarely applicable, and is extremely delicate even when it applies. Fortunately, a much more satisfactory and almost universally applicable framework has become available since the 1990s, namely ***(strict) deformation quantization***, where the word "strict" (which we will henceforth omit) refers to the fact that in this approach $\hbar$ is a real number that can "really" (!) be varied and hence can be made small (as opposed to *formal* deformation quantization, where $\hbar$ is a formal parameter having no actual value). Also, "strict" sometimes refers to the use of C*-algebras and the high mathematical standards this brings. In the formalism that follows, (deformation) quantization and the classical limit of quantum mechanics are seen as two sides of the same coin, as the axioms of quantization are predicated on recovering the correct classical limit, while conversely the classical limit only makes sense in the context of some correct notion of quantization.

The starting point of deformation quantization is a phase space X, mathematically described as a Poisson manifold, i.e., a manifold equipped with a Poisson bracket $\{\cdot,\cdot\}$ on its algebra of smooth functions $C^{\infty}(X)$, see §3.2. We recall that a Poisson bracket is a Lie bracket on $C^{\infty}(X)$ with the additional property that for each $h \in C^{\infty}(X)$, the map $\delta_h(f) = \{h, f\}$ is a derivation of $C^{\infty}(X)$ with respect to its structure as a commutative algebra under pointwise multiplication, i.e.,

$$\delta_h(fg) = f\delta_h(g) + \delta_h(f)g. \tag{7.1}$$

Furthermore, like pointwise multiplication, the Poisson bracket preserves real-valuedness, i.e., if $f \in C^{\infty}(X,\mathbb{R})$ and $g \in C^{\infty}(X,\mathbb{R})$, then also $\{f,g\} \in C^{\infty}(X,\mathbb{R})$.

As early as 1925, Dirac noted the formal analogy between Poisson brackets of functions on phase space and commutators of operators on Hilbert space (i.e., $[a,b] = ab - ba$). Indeed, if A is any C*-algebra, the commutator is a Lie bracket on A, and if we use $[a,b]' = i[ab - ba]$, then also self-adjointness is preserved (in that $a^* = a$ and $b^* = b$ implies that also $[a,b]'$ is self-adjoint, which fails to be the case for the commutator itself unless it vanishes). Thus $[-,-]'$ is a Lie bracket on A_{sa}. Moreover, if for fixed $a \in A$ we define $\delta_a(b) = [a,b]'$, then we have the product rule

$$\delta_a(bc) = \delta_a(b)c + b\delta_a(c), \tag{7.2}$$

which makes $\delta_a : A \to A$ a derivation. A problem arises if one wishes to restrict δ_a to A_{sa}, since this subspace is not stable under multiplication. This may be remedied by passing to the Jordan product (5.14), i.e., $a \circ b = \frac{1}{2}(ab + ba)$, which is defined on A_{sa}. If $a^* = a$, then $\delta_a : A_{\mathrm{sa}} \to A_{\mathrm{sa}}$ satisfies the rule (7.2) also with respect to $\circ$.

All this remains true if $[-,-]'$ is rescaled by a nonzero real number. Which number this should be was suggested by Schrödinger's construction of momentum and position operators on the Hilbert space $H = L^2(\mathbb{R})$ through the substitutions

$$p \rightsquigarrow \hat{p} = \frac{\hbar}{i}\frac{d}{dx}; \tag{7.3}$$

$$q \rightsquigarrow \hat{q} = x, \tag{7.4}$$

where "x" is the multiplication operator m_{id} (with $\mathrm{id}(x) = x$), i.e., $\hat{q}\psi(q) = x\psi(x)$; for the moment we will not be bothered by the fact that these operators are unbounded; let us say they are both defined on the domain $C_c^\infty(\mathbb{R}) \subset L^2(\mathbb{R})$.

This yields the ***canonical commutation relations*** (which formally hold on $C_c^\infty(\mathbb{R})$):

$$\frac{i}{\hbar}[\hat{p}, \hat{q}] = 1_H, \tag{7.5}$$

Noting the Poisson brackets (in which p, q are the coordinate functions on $X = \mathbb{R}^2$)

$$\{p, q\} = 1_X, \tag{7.6}$$

it it clear that analogy should be between $\{-,-\}$ and $(i/\hbar)[-,-]$. Thus Dirac wrote:

> 'The strong analogy between the quantum P.B. defined by [$(i/\hbar)$ times the commutator] and the classical P.B. (...) leads us to make the assumption that the quantum P.B.'s, or at any rate the simpler ones of them, have the same values as the corresponding classical P.B.'s.'

Combined with Heisenberg's decisive idea that quantum mechanics should be an *Umdeutung* (i.e., reinterpretation) of classical mechanics, one is led to the idea that "quantization" should be given by a linear map

$$f \mapsto Q_\hbar(f), \tag{7.7}$$

where f is some (smooth) function on phase space X and $Q_\hbar(f)$ is some operator on some "corresponding" Hilbert space, whose identification or construction is a separate problem (but for $X = \mathbb{R}^2$ it should apparently be $L^2(\mathbb{R})$), such that

$$\frac{i}{\hbar}[Q_\hbar(f), Q_\hbar(g)] = Q_\hbar(\{f, g\}), \tag{7.8}$$

at least for functions $f, g \in C^\infty(X)$ with 'the simpler' Poisson brackets. If only to do justice to Schrödinger's example (7.3) - (7.4) with (7.5), one should also require

$$Q_\hbar(1_X) = 1_H. \tag{7.9}$$

The act of quantization should also preserve the adjoint, i.e., writing $f^*(x) = \overline{f(x)}$,

$$Q_\hbar(f^*) = Q_\hbar(f)^*. \tag{7.10}$$

Putting $\hbar$ on the right-hand side of eqs. (7.5) and (7.8), Dirac (and similarly the *Dreimännerarbeit* Born–Heisenberg–Jordan) concluded from these equations that:

> '*classical mechanics may be regarded as the limiting case of quantum mechanics when $\hbar$ tends to zero.*'

In the remainder of this chapter we try to do justice to this fabulous insight of Dirac's (and also of Born, Heisenberg, and Jordan, or even Planck, Einstein, and Bohr, none of whom seem to have quite appreciated the stupendous complexity of the claim).

7.1 Deformation quantization

Recall Definition C.121 of a continuous bundle of C*-algebras over some space I, which below is taken to be a subset of the unit interval $[0,1]$ that contains 0 as an accumulation point (so one may have e.g. $I = [0,1]$ itself, or $I = (1/\mathbb{N}) \cup \{0\}$).

Definition 7.1. *A* **deformation quantization** *of a Poisson manifold X consists of a continuous bundle of C*-algebras $(A, \{\varphi_\hbar : A \to A_\hbar\}_{\hbar \in I})$ over I, along with maps*

$$Q_\hbar : \tilde{A}_0 \to A_\hbar \ (\hbar \in I), \tag{7.11}$$

where $\tilde{A}_0$ is a dense subspace of $A_0 = C_0(X)$, such that:

1. *Q_0 is the inclusion map $\tilde{A}_0 \hookrightarrow A_0$;*
2. *Each map $Q_\hbar$ is linear and satisfies* (7.10)*;*
3. *For each $f \in \tilde{A}_0$ the following map is a continuous section of the bundle:*

$$0 \mapsto f; \tag{7.12}$$
$$\hbar \mapsto Q_\hbar(f) \ (\hbar > 0); \tag{7.13}$$

4. *For all $f,g \in \tilde{A}_0$ one has the* **Dirac–Groenewold–Rieffel condition**

$$\lim_{\hbar \to 0} \left\| \frac{i}{\hbar}[Q_\hbar(f), Q_\hbar(g)] - Q_\hbar(\{f,g\}) \right\|_\hbar = 0. \tag{7.14}$$

It follows from the definition of a continuous bundle that continuity properties like

$$\lim_{\hbar \to 0} \|Q_\hbar(f)\| = \|f\|_\infty; \tag{7.15}$$
$$\lim_{\hbar \to 0} \|Q_\hbar(f)Q_\hbar(g) - Q_\hbar(fg)\| = 0, \tag{7.16}$$

are automatically satisfied. Let us note that condition (7.9) is absent from this definition, because $1_X \notin C_0(X)$ whenever X is not compact, in which case typically also the C*-algebras $A_\hbar$ have no unit (see below). However, the given conditions turn out to be sufficiently powerful to produce the "right" examples. We give one of the main such examples without proof (the underlying analysis is quite forbidding). We put

$$A_0 = C_0(T^*\mathbb{R}^n); \tag{7.17}$$
$$A_\hbar = B_0(L^2(\mathbb{R}^n)) \ (\hbar > 0), \tag{7.18}$$

where $T^*\mathbb{R}^n \cong \mathbb{R}^{2n}$ carries the canonical Poisson structure (3.34), and $A_\hbar$ is the C*-algebra of compact operators on the familiar Hilbert space $L^2(\mathbb{R}^n)$ of wave-functions on $\mathbb{R}^n$. For the sake of completeness we also mention that

$$A = C_r^*((\mathbb{R}^n \times \mathbb{R}^n)^T) \tag{7.19}$$

is the (reduced) C*-algebra of the tangent groupoid $(\mathbb{R}^n \times \mathbb{R}^n)^T$ to the pair groupoid $\mathbb{R}^n \times \mathbb{R}^n$ on $\mathbb{R}^n$, see §§C.16,C.19, where one may also find the maps $\varphi_\hbar$.

Let us summarize the situation. Continuity of the limit $\hbar \to 0$ is hard to envisage if one merely has the classical phase space $X = T^*\mathbb{R}^n$ and the quantum Hilbert space $L^2(\mathbb{R}^n)$ in mind. However, the move to either: the underlying Lie groupoids $T\mathbb{R}^n$ and $\mathbb{R}^n \times \mathbb{R}^n$, which jointly comprise the smooth tangent groupoid $\mathbb{R}^n \times \mathbb{R}^n)^T$, or: the corresponding canonically defined C*-algebras $C_0(T^*\mathbb{R}^n)$ and $B_0(L^2(\mathbb{R}^n))$, which are glued together as a continuous bundle (7.17) - (7.19), does give rise to a satisfactory structure that makes the limit $\hbar \to 0$ "continuous".

In this example, various possibilities for the quantization maps $Q_\hbar$ arise. As explained in §C.19, the groupoid structure underlying (7.17) - (7.18) suggests Weyl's prescription (C.549), which for convenience we reproduce:

$$Q_\hbar^W(f)\psi(x) = \int_{T^*\mathbb{R}^n} \frac{d^npd^ny}{(2\pi\hbar)^n} e^{ip(x-y)/\hbar}\psi(y)f(\tfrac{1}{2}(x+y),p), \tag{7.20}$$

where f lies in the image of $C_c^\infty(T\mathbb{R}^n)$ under the fiberwise Fourier transform (C.547). This image, then, is the space $\tilde{A}_0$ in Definition 7.1. We may rewrite (7.20) as

$$Q_\hbar^W(f) = \int_{T^*\mathbb{R}^n} \frac{d^npd^nq}{(2\pi\hbar)^n} f(q,p)\Omega_\hbar^W(q,p), \tag{7.21}$$

where the operators in the integrand are given by

$$\Omega_\hbar^W(q,p)\psi(x) = 2^n e^{2ip(x-q)/\hbar}\psi(2q-x). \tag{7.22}$$

The purpose of (7.21) is that for each $\psi \in L^2(\mathbb{R}^n)$ we then obviously have

$$\langle\psi, Q_\hbar^W(f)\psi\rangle = \int_{T^*\mathbb{R}^n} \frac{d^npd^nq}{(2\pi\hbar)^n} f(q,p)W_\hbar^\psi(p,q), \tag{7.23}$$

where $W_\hbar^\psi : T^*\mathbb{R}^n \to \mathbb{R}$ is the ***Wigner function***, given by

$$W_\hbar^\psi(p,q) = \hbar^{-n}\langle\psi, \Omega_\hbar^W(q,p)\psi\rangle \tag{7.24}$$

$$= \int_{\mathbb{R}^n} d^nv e^{ipv}\overline{\psi(q+\tfrac{1}{2}\hbar v)}\psi(q-\tfrac{1}{2}\hbar v). \tag{7.25}$$

If $\|\psi\| = 1$, then $W_\hbar^\psi$ gives a "phase space portrait" of the corresponding pure state e_ψ on $B_0(L^2(\mathbb{R}))$. However, this portrait cannot be interpreted as a probability density on $T^*\mathbb{R}^n$, since the Wigner function is not necessarily positive. This reflects a problem with Weyl's quantization map $Q_\hbar^W$ itself (at fixed $\hbar > 0$). We say that $Q_\hbar$ as introduced in (7.11) is ***positive*** if, for each $f \in \tilde{A}_0 \subset A_0$ (seen as a C*-algebra),

$$f \geq 0 \Rightarrow Q_\hbar(f) \geq 0, \tag{7.26}$$

where positivity of $Q_\hbar(f)$ is defined in the C*-algebra $A_\hbar$ (which in the case at hand is $B_0(L^2(\mathbb{R}^n))$). This is not the case for $Q_\hbar^W$. Moreover, $Q_\hbar^W$ fails to be continuous, and for this reason it cannot be extended to A_0 (at least not in the obvious way, viz. by continuity). Fortunately, both problems can be resolved by a change in $Q_\hbar$.

A strict deformation quantization of $\mathbb{R}^2$ that *is* positive exists under the name of ***Berezin quantization***, denoted by $Q^B_\hbar$. However, the fundamental idea of the underlying coherent states goes back to Schrödinger. For each $(p,q) \in \mathbb{R}^2$ and $\hbar > 0$, define a unit vector $\phi_\hbar^{(p,q)} \in L^2(\mathbb{R})$, called a ***coherent state***, by

$$\phi_\hbar^{(p,q)}(x) = (\pi\hbar)^{-n/4} e^{-ipq/2\hbar} e^{ipx/\hbar} e^{-(x-q)^2/2\hbar}. \tag{7.27}$$

Writing $z = p + iq$, the transition probability between two coherent states is

$$|\langle \phi_\hbar^{(z)}, \phi_\hbar^{(z')} \rangle|^2 = e^{-|z-z'|^2/2\hbar}. \tag{7.28}$$

In terms of these coherent states, we define $Q^B_\hbar : C_0(T^*\mathbb{R}^n) \to B_0(L^2(\mathbb{R}^n))$ by

$$Q^B_\hbar(f) = \int_{T^*\mathbb{R}^n} \frac{d^n p d^n q}{2\pi\hbar} f(p,q) |\phi_\hbar^{(p,q)}\rangle\langle\phi_\hbar^{(p,q)}|, \tag{7.29}$$

where the integral is meant in the sense that for each $\psi, \varphi \in L^2(\mathbb{R}^n)$ we have

$$\langle \varphi, Q_\hbar(f)\psi \rangle = \int_{\mathbb{R}^{2n}} \frac{d^n p d^n q}{2\pi\hbar} f(p,q) \langle \varphi, \phi_\hbar^{(p,q)} \rangle \langle \phi_\hbar^{(p,q)}, \psi \rangle. \tag{7.30}$$

In particular, for each unit vector $\psi \in L^2(\mathbb{R}^n)$ we may write

$$\langle \psi, Q_\hbar(f)\psi \rangle = \int_{T^*\mathbb{R}^n} d\mu_\psi f, \tag{7.31}$$

where μ_ψ is the probability measure on $T^*\mathbb{R}^n$ with density

$$B^\psi_\hbar(p,q) = |\langle \phi_\hbar^{(p,q)}, \psi \rangle|^2, \tag{7.32}$$

called the ***Husimi function*** of $\psi \in L^2(\mathbb{R}^n)$; in other words, μ_ψ is given by

$$d\mu_\psi(p,q) = \frac{d^n p d^n q}{2\pi\hbar} B^\psi_\hbar(p,q). \tag{7.33}$$

Weyl and Berezin quantization are related in many ways, for example, by

$$Q^B_\hbar(f) = Q^W_\hbar(e^{\frac{\hbar}{4}\Delta_{2n}} f), \tag{7.34}$$

where $\Delta_{2n} = \sum_{j=1}^n (\partial^2/\partial p_j^2 + \partial^2/\partial (q^j)^2)$, from which it follows that Weyl and Berezin quantization are *asymptotically equal* in the sense that for any $f \in \tilde{A}_0$,

$$\lim_{\hbar \to 0} \|Q^B_\hbar(f) - Q^W_\hbar(f)\| = 0. \tag{7.35}$$

Indeed, this provides one way (among various others) of proving that $Q^B_\hbar$ satisfies Definition 7.1, where we note that even though $Q^B_\hbar$ is defined on all of $C_0(T^*\mathbb{R}^n)$, eq. (7.14) only holds on a suitable dense subspace thereof, such as $C_0^\infty(T^*\mathbb{R}^n)$.

7.2 Quantization and internal symmetry

In the presence of symmetries, Dirac's condition (7.8) can often be met by suitable functions f and g related to the symmetries in question, though such functions may be unbounded. This blasts the C*-algebraic framework, but it does so in a controlled way. We start with internal symmetries, like spin, which will be coupled to motion in the next step. Let G be a Lie group with Lie algebra $\mathfrak{g}$, to which we associate:

- The "classical" ***Lie–Poisson manifold*** $\mathfrak{g}^*$, see (3.98), whose Poisson bracket we now preface with a minus sign, so that instead of (3.98) and (3.99) we now have

$$\{f,g\}_-(\theta) = -C^c_{ab}\theta_c \frac{\partial f(\theta)}{\partial \theta_a}\frac{\partial g(\theta)}{\partial \theta_b}; \tag{7.36}$$

$$\{\hat{A},\hat{B}\}_- = -\widehat{[A,B]}. \tag{7.37}$$

  We write $\mathfrak{g}^*_-$ for this Poisson manifold.
- The "quantum-mechanical" reduced ***group(oid) C*-algebra*** $C^*_r(G)$, cf. §C.18, defined as the norm-closure of $\pi(C^\infty_c(G))$ within $B(L^2(G))$, where

$$\pi(\check{f})\psi = \check{f} * \psi; \tag{7.38}$$

$$\check{f} * \psi(x) = \int_G dy\, \check{f}(xy)\psi(y^{-1}), \tag{7.39}$$

  where $\check{f} \in C^\infty_c(G)$ and $\psi \in L^2(G)$, cf. (C.481), and dy is Haar measure on G (which also provides the measure defining the Hilbert space $L^2(G)$).

We then obtain a continuous bundle of C*-algebras, with fibers and total C*-algebra

$$A_0 = C^*_r(\mathfrak{g}); \tag{7.40}$$

$$A_\hbar = C^*_r(G)\ (\hbar > 0); \tag{7.41}$$

$$A = C^*_r(G^T), \tag{7.42}$$

where $\mathfrak{g}$ is seen as an abelian Lie group under addition, cf. Theorem C.123. We have

$$C^*_r(\mathfrak{g}) \cong C_0(\mathfrak{g}^*_-), \tag{7.43}$$

which isomorphism (i.e. of C*-algebras) is given by the Fourier transform

$$f(\theta) = \int_{\mathfrak{g}} d^nA\, e^{-i\theta(A)}\check{f}(A); \tag{7.44}$$

$$\check{f}(A) = \int_{\mathfrak{g}^*} \frac{d^n\theta}{(2\pi)^n}\, e^{i\theta(A)}\, f(\theta), \tag{7.45}$$

where initially $\check{f} \in C^\infty_c(G)$, and the map $\check{f} \mapsto f$ is subsequently extended to $C^*_r(G)$ by continuity. Here the normalization of Lebesgue measure d^nA on $\mathfrak{g}$ is arbitrary, but the normalization of $d^n\theta$ is thereby fixed. In what follows, we take a (left-invariant)

Haar measure dx on G and fix the normalization of d^nA by the condition

$$J(0) = 1 \tag{7.46}$$

in the definition of the Jacobian under the exponential map $\exp : \mathfrak{g} \to G$, i.e.,

$$J(A) = \frac{d(\exp(A))}{d^nA}. \tag{7.47}$$

With $\tilde{A}_0 = C_c^\infty(\mathfrak{g})$, the quantization map $Q_\hbar : C_c^\infty(\mathfrak{g}) \to C_r^*(G)$ is then given by

$$Q_\hbar(\check{f})(e^A) = \hbar^{-n}\check{f}(A/\hbar), \tag{7.48}$$

where $n = \dim(G)$ and we assume that $\hbar > 0$ is small enough that $\hbar$ times the support of $\check{f} \in C_c^\infty(\mathfrak{g})$ is contained in an open neighbourhood U of $0 \in \mathfrak{g}$ where the exponential map is a diffeomorphism onto some open neighbourhood U' of $e \in G$; otherwise a cutoff function should be included. Equivalently, defining $\tilde{A}_0 \subset C_0(\mathfrak{g}^*_-)$ as the image of $C_c^\infty(\mathfrak{g})$ under the Fourier transform $\check{f} \mapsto f$ (which consists of the so-called Paley–Wiener functions on $\mathfrak{g}^*$), the map $Q_\hbar : \tilde{A}_0 \to C_r^*(G)$ is given by

$$Q_\hbar(f)(e^A) = \int_{\mathfrak{g}^*} \frac{d^n\theta}{(2\pi\hbar)^n} e^{i\theta(A)/\hbar} f(\theta). \tag{7.49}$$

Although these maps satisfy (7.14), if G is non-abelian there are no natural functions on $\mathfrak{g}^*$ whose quantizations satisfy the exact Dirac condition (7.8). This is a limitation of the C*-algebraic framework, since candidate functions like

$$\hat{A} : \mathfrak{g}^* \to \mathbb{R}; \tag{7.50}$$
$$\hat{A}(\theta) = \theta(A), \tag{7.51}$$

whose Poisson brackets (3.99) are promising, are unbounded. However, this is easily remedied by regarding $C_r^*(G)$ as an algebra of bounded operators on the Hilbert space $L^2(G)$—which indeed is the way it was originally defined—rather than abstractly. This "spatial" context allows the passage to the Lie algebra, as reviewed in §5.6, see especially (5.156) - (5.161). First note that (7.38) - (7.39) is a special case of (5.172), where $H = L^2(G)$ and $u = u_L$, i.e., the ***left-regular representation***

$$u_L(y)\psi(x) = \psi(y^{-1}x). \tag{7.52}$$

In this representation, the construction (5.156) then realizes $\mathfrak{g}$ as right-invariant differential operators on the Gårding domain$D_G \subset C^\infty(G)$. By definition of $C_r^*(G)$, seen as an operator on $L^2(G)$ the function $Q_\hbar(f)$ is given in coordinates by

$$Q_\hbar(f) = \int_{\mathfrak{g}} d^nX\, J(X) \int_{\mathfrak{g}^*} \frac{d^n\theta}{(2\pi\hbar)^n} e^{i\theta(X)/\hbar} f(\theta) u_L\left(\exp\left(\sum_j X_j T_j\right)\right). \tag{7.53}$$

Here $(X_1,\dots,X_n)$ in (7.53) are coordinates on $\mathfrak{g}$ defined by a basis choice $(T_1,\dots,T_n)$, i.e., $A=\sum_i X_i T_i$. The function $\hat{T}_j$ on $\mathfrak{g}^*$ is then simply given by the coordinate function $\hat{T}_j(\theta)=\theta_j$. Now take $A\in\mathfrak{g}$ and assume that $f=\hat{A}$. This function is unbounded, but the following formal calculation is rigorously correct on the Gårding domain and may be justified by some distribution theory. For simplicity we assume that G is unimodular, in which case $J(X)=1+O(X^2)$ as $X\to 0$, so that all first derivatives of J vanish at $X=0$. Taking $f=\hat{T}_j$ in (7.53) then gives

$$
\begin{aligned}
Q_\hbar(\hat{T}_j) &= \int_{\mathfrak{g}} d^nX\, J(X) \int_{\mathfrak{g}^*} \frac{d^n\theta}{(2\pi\hbar)^n} e^{i\theta(X)/\hbar}\, \theta_j u_L\left(\exp\left(\sum_j X_j T_j\right)\right) \\
&= -i \int_{\mathfrak{g}} d^nX\, J(\hbar X) u_L\left(\exp\left(\hbar\sum_j X_j T_j\right)\right) \frac{\partial}{\partial X_j}\delta(X) \\
&= i\hbar u_L'(X_j),
\end{aligned}
\tag{7.54}
$$

from which we obtain

$$Q_\hbar(\hat{A}) = i\hbar u_L'(A) = \pi_L(A). \tag{7.55}$$

This explains the need for *minus* the Lie–Poisson bracket, since instead of (3.99) we now have (7.37), so that (5.160) gives the exact result (7.8) for $f=\hat{A}$ and $g=\hat{B}$:

$$\frac{i}{\hbar}[Q_\hbar(\hat{A}), Q_\hbar(\hat{B})] = Q_\hbar(\{\hat{A},\hat{B}\}_-). \tag{7.56}$$

The minus sign in the Lie–Poisson bracket could have been avoided by writing $\check{f}(-A/\hbar)$ in (7.48), whose minus sign would have propagated into (5.159) and hence in the commutation relations (5.160), but the latter are so engrained in the physics literature that we see the minus sign on the bracket in (7.56) as the lesser evil.

Any continuous unitary representation u_λ of G (where λ is some label) induces a representation $u_\lambda^\int$ of $C_c^\infty(G)$ by (5.173), which may be extended to a representation of $C^*(G)$ by continuity (the same is true for $C_r^*(G)$ provided u_λ is weakly contained in $L^2(G)$, cf. §C.18). This gives operators $u^\int(Q_\hbar(f))$ which, by the same formal computation as for the case $u=u_L$ above, for $A\in\mathfrak{g}$ rigorously give rise to operators

$$\pi_\lambda(A) = i\hbar u_\lambda'(A), \tag{7.57}$$

satisfying the like of (5.160) for fixed values of $\hbar$ (but without control over the limit $\hbar\to 0$). Many commutation relations in quantum mechanics take this form, where both irreducible and reducible representations u give rise to interesting examples. The reducible case typically comes from group actions and is best studied using the formalism of action groupoids reviewed in the next section, where we will see that further operators start playing a role. The irreducible case, on the other hand, gives rise to intriguing new examples of continuous bundles of C*-algebras, where $\hbar$ (now related the label λ) takes values in a discrete set and may be sent to zero, cf. §8.1.

7.3 Quantization and external symmetry

We now generalize the setting of the preceding section from groups taken by themselves to group actions. Let a Lie group G act smoothly on some manifold Q; for example, we may have $Q = \mathbb{R}^3$ with either $G = SO(3)$ acting by rotations, or $G = \mathbb{R}^3$ action by translations. We now take $X = \mathfrak{g}^* \times Q$. Recalling the notation (3.71) and writing $\delta_a \equiv \delta_{T_a}$, we define the ***action Poisson bracket***

$$\{f,g\} = -C_{ab}^c \theta_c \frac{\partial f}{\partial \theta_a}\frac{\partial g}{\partial \theta_b} + \xi_a f \frac{\partial g}{\partial \theta_a} - \frac{\partial f}{\partial \theta_a}\xi_a g. \tag{7.58}$$

Interesting special cases arise if we take $A \in \mathfrak{g}$ and define $\hat{A} \in C^\infty(\mathfrak{g}^*)$ as before, i.e., $\hat{A}(\theta) = \theta(A)$, now regarded as a function on $\mathfrak{g}^* \times Q$ (ignoring the *second* argument q). Similarly, if $\tilde{f} \in C^\infty(Q)$ we write $\hat{f}$ for the corresponding function on $\mathfrak{g}^* \times Q$ (ignoring the *first* argument θ). This gives the coordinate-independent expressions

$$\{\hat{A},\hat{B}\} = -\widehat{[A,B]}; \tag{7.59}$$
$$\{\hat{A},\hat{f}\} = -\delta_A f; \tag{7.60}$$
$$\{\hat{f},\hat{g}\} = 0. \tag{7.61}$$

Clearly, if Q is a point (with trivial G-action) we recover (minus) the Lie–Poisson structure on $\mathfrak{g}^*$. If, on the other hand, $Q = \mathbb{R}^3$ and $G = \mathbb{R}^3$ acts on Q by translation, i.e., $\mathbf{a} \cdot \mathbf{x} = \mathbf{x} + \mathbf{a}$, we recover the canonical Poisson bracket (3.34), where the momenta p_a ($a = 1, \ldots, n$) are identified with the coordinates θ_a on the dual of the Lie algebra of $\mathbb{R}^3$, which is just $\mathbb{R}^3$ itself (with the usual basis (e_1, e_2, e_3)). Therefore, the Poisson bracket (3.34) on $\mathbb{R}^{2n}$ may be generalized in two ways:

1. By passing to arbitrary cotangent bundles T^*M, whose canonical Poisson bracket is still given in local coordinates by (3.34), which emphasizes the role of momenta as fiber coordinates on T^*M.
2. By passing to the setting discussed here, which emphasizes the role of momenta as generators of global translations of the base space $\mathbb{R}^3$ (a property that breaks the p-q symmetry and cannot be generalized to arbitrary cotangent bundles).

A richer structure emerges if we keep $Q = \mathbb{R}^3$ but now take $G = E(3)$, i.e.,

$$E(3) = SO(3) \ltimes \mathbb{R}^3, \tag{7.62}$$

known as the ***Euclidean group***. To explain its group structure, let some group L act on a vectors space V, seen as an abelian group under addition. Then the operations

$$(\lambda, v) \cdot (\lambda', v') = (\lambda\lambda', v + \lambda \cdot v'); \tag{7.63}$$
$$(\lambda, v)^{-1} = (\lambda^{-1}, -\lambda^{-1} \cdot v), \tag{7.64}$$

turn $G = L \ltimes V$ into a group, called the ***semi-direct product*** of L and V.

Then $E(3)$ acts on $\mathbb{R}^3$ in the obvious way, giving rise to the Poisson manifold $\mathfrak{g}^* \times Q = \mathbb{R}^3 \times \mathbb{R}^3 \times \mathbb{R}^3$ (since $\mathfrak{so}(3) \cong \mathbb{R}^3$). We now also have generators (J_1, J_2, J_3) of the Lie algebra of $SO(3)$, with corresponding functions $\hat{J}_i$, as well as standard coordinate functions (q_1, q_2, q_3) on $Q = R^3$, giving rise to the Poisson brackets

$$\{\hat{J}_i, \hat{J}_j\} = -\varepsilon_{ijk}\hat{J}_k; \ \{\hat{J}_i, p_j\} = -\varepsilon_{ijk}p_k; \ \{p_i, p_j\} = 0; \tag{7.65}$$
$$\{\hat{J}_i, q_j\} = -\varepsilon_{ijk}q_k; \ \{p_i, q_j\} = \delta_{ij}; \ \{q_i, q_j\} = 0. \tag{7.66}$$

The appropriate target C*-algebra $C_r^*(G,Q)$ for quantization is a generalization of $C_r^*(G)$, constructed in a similar way, as explained in §C.18. For the moment it is enough to know that $C_r^*(G,Q)$ is the completion of the function space $C_c^\infty(G \times Q)$, seen as a *-algebra in the operations (C.526) - (C.527), in a suitable norm, namely

$$\|f\|_r = \|\tilde{\rho}(f)\|, \tag{7.67}$$

where the representation $\tilde{\rho} : C_c^\infty(G \times Q) \to B(L^2(G \times Q))$ is given by (C.530). In case that Q has a G-invariant measure ν (still with support Q), the operator

$$w : L^2(G \times Q) \to L^2(G \times Q); \tag{7.68}$$
$$w\psi(x,q) = \psi(x, x^{-1}q), \tag{7.69}$$

is unitary, and in terms of the notation

$$\tilde{u}(y) = wu(y)w^*, \ \tilde{\pi}(\tilde{f}) = w\pi(\tilde{f})w^*, \ \tilde{\rho}(f) = w\rho(f)w^*, \tag{7.70}$$

the formulae (C.528) - (C.530) take the slightly more appealing form

$$\tilde{u}(y)\psi(x,q) = \psi(y^{-1}x, y^{-1}q); \tag{7.71}$$
$$\tilde{\pi}(\tilde{f})\psi(x,q) = \tilde{f}(q)\psi(x,q); \tag{7.72}$$
$$\tilde{\rho}(f)\psi(x,q) = \int_G dy\, f(y,q)\psi(y^{-1}x, y^{-1}q). \tag{7.73}$$

The simplification thus gained especially concerns the position functions (7.72).

Analogously to (7.49), the quanitzation maps are given by

$$Q_\hbar : C_0(\mathfrak{g}^* \times Q) \to C_r^*(G,Q); \tag{7.74}$$
$$Q_\hbar(f)(e^A, q) = \int_{\mathfrak{g}^*} \frac{d^n\theta}{(2\pi\hbar)^n} e^{i\theta(A)/\hbar} f(\theta, e^{-\frac{1}{2}A} \cdot q), \tag{7.75}$$

where, as in the pure group case, strictly speaking f must lie in the dense subspace of $C_0(\mathfrak{g}^* \times Q)$ consisting of Paley–Wiener functions (in A) that are the Fourier transform (in the first argument) of functions that lie in $C_c^\infty(\mathfrak{g} \times Q)$.

Computations similar to (7.54) then establish, for $A \in \mathfrak{g}$ and $\tilde{f} \in C^\infty(Q)$ as before,

$$Q_\hbar(\hat{A}) = i\hbar\tilde{u}'(A); \tag{7.76}$$
$$Q_\hbar(\hat{f}) = \tilde{\pi}(\tilde{f}). \tag{7.77}$$

Form these formulae and (7.59) - (7.60), it is easy to verify that Dirac's exact condition (7.8) holds in the following special cases:

$$\frac{i}{\hbar}[Q_\hbar(\hat{A}), Q_\hbar(\hat{B})] = Q_\hbar(\{\hat{A}, \hat{B}\}); \quad (7.78)$$

$$\frac{i}{\hbar}[Q_\hbar(\hat{A}), Q_\hbar(\hat{f})] = Q_\hbar(\{\hat{A}, \hat{f}\}); \quad (7.79)$$

$$\frac{i}{\hbar}[Q_\hbar(\hat{f}), Q_\hbar(\hat{g})] = Q_\hbar(\{\hat{f}, \hat{g}\}) = 0. \quad (7.80)$$

These might be regarded as infinitesimal versions of the covariance condition (C.514), specialized to the case at hand. We formalize this special case as follows.

Definition 7.2. *Let G be a locally compact group and let Q be a space equipped with some continuous G-action. A* **system of imprimitivity** $(u(G), \pi(C_0(Q)))$ *for the given group action $G \circlearrowright Q$ is a combination of a strongly continuous unitary representation u of G and a nondegenerate representation π of $C_0(Q)$, both defined on the same Hilbert space, that for each $x \in G$ and $\tilde{f} \in C_0(Q)$ satisfies*

$$u(x)\pi(\tilde{f})u(x)^* = \pi(\widetilde{L_x f}). \quad (7.81)$$

Here $\widetilde{L_x f}(q) = \tilde{f}(x^{-1}q)$, as usual. We recall from §C.18 that such systems of imprimitivity bijectively correspond to degenerate representations $\rho \equiv \pi \rtimes u^{\int}$ of $C^*(G,Q)$ through (C.515), which in the special case (C.524) - (C.525) comes down to

$$\rho(f) = \int_G dx\, \pi(f(x,\cdot))u(x). \quad (7.82)$$

The formulae (7.71) - (7.73) define such a system of imprimitivity on the Hilbert space $H = L^2(G \times Q)$. However, this cannot be the end result of quantization, since this space is typically reducible under the pair $(u(G), \pi(C_0(Q)))$, or, equivalently, under $\rho(C^*(G,Q))$. For example, this is the case for $G = \mathbb{R}^3$ or $G = E(3)$ acting on $Q = \mathbb{R}^3$ in the natural way discussed above, for which we obtain $H = L^2(\mathbb{R}^3 \times \mathbb{R}^3)$ or even $H = L^2(E(3) \times \mathbb{R}^3)$. In the former case we do obtain the correct position operators q^i, but for the momentum operators we find the curious expression $-i\hbar(\partial/\partial x^i + \partial/\partial q^i)$—to their credit, these do satisfy the canonical commutation relations (7.5), since these follow from (7.78) - (7.80), which in turn follow from the covariance condition (7.81) defining a system of imprimitivity.

Instead, we would prefer the Hilbert space $H = L^2(\mathbb{R}^3)$ expected from elementary quantum mechanics (without spin), equipped with the system of imprimitivity

$$u(y)\psi(q) = \psi(y^{-1}q); \quad (7.83)$$

$$\pi(\tilde{f})\psi(q) = \tilde{f}(q)\psi(q). \quad (7.84)$$

The answer lies in the search for *irreducible* systems of imprimitivity $(u(G), \pi(C_0(Q)))$, or, equivalently, *irreducible* representations of $\rho(C^*(G,Q))$; see §7.5.

7.4 Intermezzo: The Big Picture

First, however, we summarize and generalize the results in this chapter so far into what we call ***The Big Picture***. This arose in the 1990s from efforts to relate Mackey's quantization theory based on systems of imprimitivity (which Mackey himself saw as the natural implementation of what he called ***Weyl's Program***, i.e. the construction of the basic operators of quantum mechanics from group-theoretical considerations) to deformation quantization (and hence to the tradition started by Dirac, as continued by Groenewold, Moyal, Berezin, Flato, Rieffel, and others).

The Big Picture is technically based on the theory of ***Lie groupoids*** (already alluded to in the preceding sections) and ***Lie algebroids***. For a precise definition of the former we refer to Definition C.115; briefly, a ***groupoid*** G is an object like a group, where however *multiplication* is defined only partially (although the *inverse* is defined for each element). To see which elements can be multiplied, one has maps $s,t : G_1 \to G_0$ from the *total space* G_1 of the groupoid to its *base space* G_0, such that the product $xy \in G_1$ of $x,y \in G_1$ is defined whenever $s(x) = t(y)$, and satisfies $s(xy) = s(y)$, $t(xy) = t(x)$, and $s(x^{-1}) = t(x)$. Four relevant examples are:

- ***Spaces***, where $G_1 = G_0 = Q$ for some set Q, with $s(x) = t(x) = x$ for all $x \in G_1$, and hence xy is defined iff $y = x$, with result $xx = x$; furthermore, $x^{-1} = x$.
- ***Groups***, where $G_1 = G$ and $G_0 = \{e\}$, with $s(x) = t(x) = e$ for all x, so that all elements can be multiplied and the notion of a groupoid reduces to a group.
- ***Pair groupoids*** over a set Q have base space $G_0 = Q$, total space $G_1 = Q \times Q$, and projections $s(q,q') = q'$ and $t(q,q') = q$, so that $(q,q')(r,r')$ is defined iff $q' = r$, resulting in $(q,q')(q',r') = (q,r')$. The inverse is given by $(q,q')^{-1} = (q',q)$.
- ***Action groupoids*** (also called ***semi direct product groupoids***) are important in what follows. These originate in some group action we denote by $G \circlearrowright Q$, where G is a group and Q is a set. The ensuing groupoid is called $\Gamma = G \ltimes Q$, where

$$\Gamma_1 = G \times Q,\ \Gamma_0 = Q,\ s(x,q) = x^{-1}q,\ t(x,q) = q, \tag{7.85}$$

so that products $(x,q)(y,q')$ are defined iff $q' = x^{-1}q$, with result

$$(x,q)(y,x^{-1}q) = (xy,q). \tag{7.86}$$

Finally, the inverse is (necessarily) given by

$$(x,q)^{-1} = (x^{-1},x^{-1}q). \tag{7.87}$$

A ***Lie groupoid*** is a groupoid G where G_1 and G_0 are manifolds and all operations are smooth. In all examples just given this requires Q to be a manifold, and in the last one G should be a Lie group, and the given action $G \times Q \to Q$ must be smooth.

Generalizing the construction of a Lie algebra $\mathfrak{g}$ from a given Lie group G, a Lie groupoid comes with an associated linearized (or "infinitesimal") structure, called a *Lie algebroid*. As in the group case, this differential-geometric notion can also be defined independently of its origin in the theory of Lie groupoids, as follows:

Definition 7.3. *A* **Lie algebroid** *E over a manifold Q is a vector bundle $E \stackrel{\pi}{\to} Q$ with a vector bundle map $E \stackrel{\alpha}{\to} TQ$ (called the* **anchor**), *as well as with a Lie bracket $[\,,]$ on the space $C^\infty(Q,E)$ of smooth cross-sections of E, satisfying the Leibniz rule*

$$[\sigma_1, f\cdot\sigma_2] = f\cdot[\sigma_1,\sigma_2] + (\alpha\circ\sigma_1 f)\cdot\sigma_2 \tag{7.88}$$

for all $\sigma_1,\sigma_2 \in C^\infty(Q,E)$ and $f \in C^\infty(Q)$ (here $\alpha\circ\sigma_1$ is a vector field on Q).

It follows that the map $\sigma \mapsto \alpha\circ\sigma : C^\infty(Q,E) \to C^\infty(Q,TQ)$ induced by the anchor is a homomorphism of Lie algebras, where the latter is equipped with the usual commutator of vector fields (this homomorphism property used to be part of the definition of a Lie algebroid, but in fact it follows from the stated definition).

Lie algebroids generalize (finite-dimensional) Lie algebras as well as tangent bundles, and the (infinite-dimensional) Lie algebra $C^\infty(Q,E)$ could be said to be of geometric origin in the sense that it derives from an underlying finite-dimensional geometrical object. Similar to the above list of examples of Lie groupoids, one has the following basic classes of Lie algebroids.

- ***Manifolds***, where $E = Q$, seen as the zero-dimensional vector bundle over Q, evidently with identically vanishing Lie bracket and anchor.
- ***Lie algebras***, where $E = \mathfrak{g}$ and Q is a point (which may be identified with the identity element of any Lie group with Lie algebra g) and anchor $\alpha = 0$.
- ***Tangent bundles*** over a manifold Q, where $E = TQ$ and $\alpha = \text{id} : TQ \to TQ$, with the Lie bracket given by the usual commutator of vector fields (or derivations).
- ***Action algebroids*** (or ***semi-direct product algebroids***) are defined by a $\mathfrak{g}$-action on a manifold Q, i.e. a Lie algebra homomorphism $\mathfrak{g} \to C^\infty(Q,TQ)$, $A \mapsto \delta_A$, where we identify vector fields on Q with derivations on $C^\infty(Q)$—these are often, but not necessarily, obtained from a G-action on Q via see (3.71). We write $E = \mathfrak{g} \ltimes Q$, which is $E = \mathfrak{g}\times Q$ as a trivial bundle (with π the projection on the second space), and $\alpha(A,q) = -\delta_A(q) \in T_qQ$, where $A \in \mathfrak{g}$. The Lie bracket is given by

$$[\sigma_1,\sigma_2](q) = [\sigma_1(q),\sigma_2(q)]_{\mathfrak{g}} + \delta_{\sigma_2}\sigma_1(q) - \delta_{\sigma_1}\sigma_2(q). \tag{7.89}$$

These examples may also be recovered as special cases of the following construction that canonically associates a Lie algebroid $\text{Lie}(G)$ to a Lie groupoid G: as a vector bundle, $\text{Lie}(G)$ is the restriction of $\ker(t_*)$ to G_0 (where $t_* : TG_1 \to TG_0$ is the derivative map of the source projection $t : G_1 \to G_0$), and the anchor is $\alpha = s_*$ (one may alternatively define $\text{Lie}(G)$ as the normal bundle to the object inclusion map $i : G_0 \hookrightarrow G_1$, cf. Definition C.115, but this makes the definition of the anchor a bit more complicated). As in the Lie group case, one may identify sections of $\text{Lie}(G)$ with left-invariant vector fields on G, and under this identification the Lie bracket on $C^\infty(G_0, \text{Lie}(G))$ is by definition given by the commutator of vector fields.

Conversely, one may ask whether a given Lie algebroid E is ***integrable***, in that $E \cong \text{Lie}(G)$ for some Lie groupoid G (where the isomorphism sign $\cong$ means that a pertinent vector bundle isomorphism $E \cong \ker(t_*)_{|G_0}$ should preserve all relevant structure). Unlike the special case of Lie groups (where Lie's Third Theorem 5.41 settles this in the positive), this is not necessarily the case, but that is of no concern.

We now state a crucial connection between Lie algebroids and Poisson geometry.

Proposition 7.4. *The dual vector bundle E^* of a Lie algebroid E is a Poisson manifold, whose Poisson bracket on $C^\infty(E^*)$ is defined by the following special cases:*

$$\{f,g\} = 0 \ (f,g \in C^\infty(Q)); \tag{7.90}$$

$$\{\tilde{\sigma}, f\} = -\alpha \circ \sigma f \ (\sigma \in C^\infty(Q,E), f \in C^\infty(Q)); \tag{7.91}$$

$$\{\tilde{\sigma}_1, \tilde{\sigma}_2\} = -\widetilde{[\sigma_1, \sigma_2]}, \tag{7.92}$$

where $\tilde{\sigma} \in C^\infty(E^*)$ is defined by a given section σ of E through the obvious pairing.

Conversely, if the dual F^* to a given vector bundle $F \to Q$ is a Poisson manifold such that the Poisson bracket of two linear functions is linear, then $F \cong E$ for some Lie algebroid E over Q, with the above Poisson structure on E^*.

Following our earlier lists, the main examples are:

- A manifold Q, seen as the dual to the zero-dimensional vector bundle $Q \to Q$, carries the zero Poisson structure.
- The dual $\mathfrak{g}^*$ of a Lie algebra $\mathfrak{g}$ acquires (minus) the Lie–Poisson structure (3.98).
- A cotangent bundle T^*Q acquires (minus) the Poisson structure defined by its standard symlectic structure, cf. (3.34).
- The dual $\mathfrak{g}^* \ltimes Q$ of an action algebroid acquires the Poisson bracket (7.58).

The following theorem displays a rich and physically relevant class of examples of Definition 7.1 of deformation quantization. The key point is that a Lie groupoid G defines both classical and quantum data, namely the (reduced) Lie groupoid C*-algebra $C^*_{(r)}(G)$ (cf. §C.17) and the Poisson manifold $\mathrm{Lie}(G)^*$ (cf. Proposition 7.4), and these are continuously (even smoothly) related through the tangent groupoid G^T (cf. Proposition C.117) and its associated Lie groupoid C*-algebra $C^*_{(r)}(G^T)$.

Theorem 7.5. *For any Lie groupoid G, the bundle of C*-algebras given by*

$$A_0 = C_0(\mathrm{Lie}(G)^*) \ \ (\hbar = 0); \tag{7.93}$$

$$A_\hbar = C^*(G) \qquad (0 < \hbar \leq 1); \tag{7.94}$$

$$A = C^*(G^T), \tag{7.95}$$

defines a deformation quantization of the Poisson manifold $\mathrm{Lie}(G)^*$ *over $I = [0,1]$. The same statement holds for the corresponding* reduced *groupoid C*-algebras.*

The key lemma for this theorem is Theorem C.123, which provides the continuity of the given bundle of C*-algebras. A lengthy computation shows that also the Dirac–Groenewold–Rieffel condition (7.14) is met. In this light, the quantization of the phase space $T^*\mathbb{R}^n$ in §7.1 then corresponds to the pair groupoid $G = \mathbb{R}^n \times \mathbb{R}^n$ on $\mathbb{R}^n$, the one in §7.2 follows from the special case where the Lie groupoid G is "simply" a Lie group, and the case of §7.3, which puts Mackey's quantization theory in a deformation framework, is obviously given by the action groupoid $G \ltimes Q$. Finally, the space groupoid $G_0 = G_1 = Q$ gives a trivial continuous bundle of C*-algebras, where $A_\hbar = C_0(Q)$ for all $\hbar \in [0,1]$, and Q carries the zero Poisson bracket.

7.5 Induced representations and the imprimitivity theorem

Returning to §7.3, we recall the bijective correspondence between systems of imprimitivity $(u(G), \pi(C_0(Q)))$ and non-degenerate representations of the C*-algebra $C^*(G,Q)$ of the action groupoid defined by the given action $G \circlearrowright Q$. This correspondence preserves irreducibility, and our task is to find irreducible representations.

It was recognized at least 50 years ago that this task can be carried out if the group action satisfies a certain regularity condition, and is hopeless otherwise. This is sometimes called the ***Mackey–Glimm dichotomy***. The condition in question may be stated in a number of equivalent ways (whose equivalence is not at all obvious).

First, we recall some terminology from topology. Let X be a space. One calls $Y \subset Y' \subseteq X$ ***relatively open*** in Y' if there is an open set $U \subset X$ such that $Y = Y' \cap U$. A subset $Y \subset X$ is ***locally closed*** if each $y \in Y$ has an open neighbourhood U in X such that $U \cap Y$ is closed, and finally "X is T_0" if for any two distinct points there is an open set that contains exactly one of them. Furthermore, each $q \in Q$ defines a G-orbit through q denoted by $G \cdot q$, as well as a stabilizer (or "little group")

$$G_q = \{x \in G \mid x \cdot q = q\}. \tag{7.96}$$

For any subgroup $H \subset G$, we denote the equivalence class of x in G/H by $[x]$.

Definition 7.6. *A smooth action of a Lie group G on a manifold Q is called* **regular** *if one and hence each of the following equivalent conditions is satisfied:*

1. *Each G-orbit in Q is relatively open in its closure;*
2. *Each G-orbit in Q is locally closed;*
3. *The quotient space Q/G of G-orbits in Q is T_0;*
4. *Each map $[x] \mapsto xq$ is a homeomorphism from G/G_q to the orbit $G \cdot q$ ($q \in Q$).*

Probably the simplest example of a non-regular action is the action $\mathbb{Z} \circlearrowright \mathbb{T}$ given by

$$n : z \mapsto e^{2\pi i n\theta} z, \tag{7.97}$$

where $\theta \in \mathbb{R}\backslash\mathbb{Q}$ (here $\mathbb{Z}$ may be seen as a zero-dimensional Lie group with infinitely many components—in fact, Definition 7.6 more generally applies to second countable locally compact groups and spaces that are "almost Hausdorff"). Indeed, each orbit is dense in $\mathbb{T}$ (but not open), and the orbit space $\mathbb{T}/\mathbb{Z}$ has no proper open sets.

Theorem 7.7. *Let a group action $G \circlearrowright Q$ be regular. Then the irreducible representations of the associated action groupoid C*-algebra $C^*(G,Q)$—and hence also the irreducible systems of imprimitivity $(u(G), \pi(C_0(Q)))$—are classified up to unitary equivalence by pairs $(\mathcal{O}, u_\chi)$, where $\mathcal{O}$ is a G-orbit in Q and u_χ is an irreducible representation of the stabilizer G_q of an arbitrary point $q \in \mathcal{O}$, with an explicit construction of the corresponding representation $\rho_{(\mathcal{O},u_\chi)}(C^*(G,Q))$. Two such representations $\rho_{(\mathcal{O},u_\chi)}$ and $\rho_{(\mathcal{O}',u'_\chi)}$ are equivalent iff $\mathcal{O} = \mathcal{O}'$ and, given that $q' = xq$ and hence $G_{q'} = xG_qx^{-1}$ for some $x \in G$, u'_χ is unitarily equivalent to $u_\chi \circ \mathrm{Ad}(x)$. Finally, any irreducible representation ρ is unitarily equivalent to some $\rho_{(\mathcal{O},u_\chi)}$.*

In the simplest case, Q is equal to a point, so that $C^*(G,Q) = C^*(G)$, and we find that irreducible representations of $C^*(G)$ (which are necessarily non-degenerate) bijectively correspond to unitary irreducible representations of G. In the next easiest case, G acts nontrivially but still transitively on Q, in which case the action is clearly regular and $Q \cong G/H$ through the G-equivariant map in no. 4 of the above definition (read in the opposite direction), i.e., we pick some $q_0 \in Q$, define $H = G_{q_0}$, and finally map Q to G/H by $q \mapsto [x]$, where $q = xq_0$ (this map is well defined); in that case, we might as well have assumed that $Q = G/H$ to begin with. The following important corollary of Theorem 7.7 is called the ***Imprimitivity Theorem***.

Corollary 7.8. *Up to unitary equivalence, irreducible representations of* $C^*(G,G/H)$ *(or, equivalently, of pairs* $(\pi(C_0(G/H)),u(G))$ *satisfying the covariance condition* (7.81)*) bijectively correspond to unitary irreducible representations of* H.

In preparation for the general case stated in Theorem 7.7, and also as a goal in itself, we first give an explicit construction of the irreducible representation ρ^χ of $C^*(G,G/H)$ corresponding to a given unitary irreducible representation $u_\chi(H)$, where we label the unitary irreducible representations of H (up to unitary equivalence) by $\chi \in \hat{H}$ (where $\hat{H}$ is the set of unitary equivalence classes of unitary irreducible representations of H, cf. §C.15 for the abelian case), and let the corresponding representation $\rho^\chi(C^*(G,G/H))$—or the pair $\pi^\chi(C_0(G/H))$ and $u^\chi(G)$—inherit this label (in raised form, in order to prevent confusion between $u_\chi(H)$ and $u^\chi(G)_{|H}$).

The construction of $\rho^\chi(C^*(G,G/H))$—or, equivalently, of a system of imprimitivity $(\pi^\chi(C_0(G/H)),u^\chi(G))$—from $u_\chi(H)$ proceeds by the technique of ***induced representations*** (which physicists may be familiar with from the representation theory of the Poincaré group, see Theorem 7.9 below). We start from a specific realization of $u_\chi(H)$ on a Hilbert space H_χ (which is finite-dimensional if H is compact or abelian). From this, we construct a new Hilbert space H^χ, whose realization depends on the choice of a ***quasi-invariant measure*** ν on G/H, i.e., a (non-zero) measure whose null-sets are G-invariant in the sense that if $\nu(A) = 0$ for some (Borel) measurable $A \subset G/H$, then also $\nu(x \cdot A) = 0$ for each $x \in G$. This will surely be the case if ν is ***invariant***, i.e., if $\nu(x \cdot A) = \nu(A)$ for each measurable A, but invariant measures on G/H may not exist, whereas quasi-invariant measures always do.

We now consider (measurable) functions $\psi : G \to H_\chi$ that satisfy

$$\psi(xh) = u_\chi(h^{-1})\psi(x), \tag{7.98}$$

for every $x \in G$ and $h \in H$; equivalently, we may say that

$$u_\chi(h) \circ R_h \psi = \psi, \tag{7.99}$$

for each $h \in H$, where $R_h\psi(x) = \psi(xh)$. Now if ψ and φ both satisfy (7.98), then, by unitarity of u_χ, their inner product $\langle \varphi(x), \psi(x) \rangle_{H_\chi}$ in H_χ is H-invariant, in that

$$\langle \varphi(xh), \psi(xh) \rangle_{H_\chi} = \langle \varphi(x), \psi(x) \rangle_{H_\chi}. \tag{7.100}$$

Hence the function $x \mapsto \langle \varphi(x), \psi(x) \rangle_{H_\chi}$, *a priori* defined from G to $\mathbb{C}$, induces a function $[x] \mapsto \langle \varphi(x), \psi(x) \rangle_{H_\chi}$ from G/H to $\mathbb{C}$. We write the latter function as $\langle \varphi, \psi \rangle_{H_\chi}[x]$; in particular, taking $\varphi = \psi$, we write $\|\psi\|^2_{H_\chi}[x] = \langle \psi(x), \psi(x) \rangle_{H_\chi}$. We may then define a new Hilbert space H^χ that consists of all measurable functions $\psi : G \to H_\chi$ that for each $h \in H$ satisfy (7.98), and are square-integrable on G/H:

$$\int_{G/H} d\nu([x]) \, \|\psi\|^2_{H_\chi}[x] < \infty. \tag{7.101}$$

This space turns out to be complete in the natural inner product

$$\langle \varphi, \psi \rangle = \int_{G/H} d\nu([x]) \, \langle \varphi, \psi \rangle_{H_\chi}[x] \tag{7.102}$$

It also carries a system of imprimitivity: in case that ν is G-invariant we simply have

$$u^\chi(y)\psi(x) = \psi(y^{-1}x) \ \ (x, y \in G); \tag{7.103}$$

$$\pi^\chi(\tilde{f})\psi(x) = \tilde{f}([x])\psi(x) \ \ (\tilde{f} \in C_0(G/H)), \tag{7.104}$$

where we note that $u^\chi(y)\psi$ satisfies (7.98) if ψ does. Unitarity of u^χ as well as the covariance condition (7.81) are easily checked. In general, we replace (7.103) by

$$u^\chi(y)\psi(x) = \sqrt{\frac{d\nu([y^{-1}x])}{d\nu([x])}} \, \psi(y^{-1}x), \tag{7.105}$$

where $d\nu([y^{-1}\cdot])/d\nu([\cdot])$ is the Radon–Nikodym derivative of the translated measure $L_y^*\nu$ with respect to ν, cf. (B.137), which is well defined because by the assumption of quasi-invariance, $L_y^*\nu$ is absolutely continuous with respect to ν (indeed, on this assumption they are even equivalent). Here $L_y^*\nu(A) = \nu(L_y^{-1}(A))$, $A \subset G/H$.

Physicists do not like the Hilbert space H^χ, preferring a different realization

$$\tilde{H}^\chi = L^2(G/H) \otimes H_\chi, \tag{7.106}$$

in which the wave-function ψ is not constrained and one has a clean separation between the (typically) spatial degree of freedom $Q = G/H$ and the internal degree of freedom H_χ. One half of the system of imprimitivity will then be given nicely by

$$\tilde{\pi}^\chi(\tilde{f})\tilde{\psi} = \tilde{f}\tilde{\psi} \ \ (\tilde{f} \in C_0(G/H)), \tag{7.107}$$

but this cleanliness comes at the cost of a more complicated formula for $\tilde{u}^\chi(y)$, as follows. Pick a (measurable) cross-section $s : G/H \to G$, i.e., a *right* inverse to the projection $p : G \to G/H$, $p(x) = [x]$, in other words, we have

$$p \circ s = \mathrm{id}_{G/H}. \tag{7.108}$$

It may not be possible to make s continuous, and, crucially, s is not a *left* inverse to p; instead, there exists a unique function $h_s : G \to H$ such that $s \circ p(x) = xh_s(x)$, i.e.,

$$h_s(x) = x^{-1}s([x]). \tag{7.109}$$

Such a cross-section s gives rise to a unitary isomorphism

$$w_s : H^\chi \to \tilde{H}^\chi; \tag{7.110}$$
$$w_s\psi(q) = \psi(s(q)); \tag{7.111}$$
$$w_s^{-1}\tilde{\psi}(x) = u_\chi(h_s(x))\tilde{\psi}([x]), \tag{7.112}$$

which enables us to move the system of imprimitivity (u^χ, π^χ) to $\tilde{H}^\chi$ by defining

$$\tilde{u}^\chi(y) = w_s u^\chi(y) w_s^* \quad (y \in G); \tag{7.113}$$
$$\tilde{\pi}^\chi(\tilde{f}) = w_s \pi^\chi(\tilde{f}) w_s^* \quad (\tilde{f} \in C_0(G/H)). \tag{7.114}$$

This duly leads to (7.107), but instead of (7.105), we obtain the more cumbersome

$$\tilde{u}^\chi(y)\tilde{\psi}(q) = \sqrt{\frac{d\nu(y^{-1}q)}{d\nu(q)}}\,u_\chi(s(q)^{-1}ys(y^{-1}q))\,\tilde{\psi}(y^{-1}q), \tag{7.115}$$

where of course the square root may be omitted if ν is G-invariant, as in (7.103). The argument $h = s(q)^{-1}ys(y^{-1}q)$ of u_χ appearing here is called the ***Wigner cocycle*** (after the physicist who first introduced it in his classification of the irreducible representations of the Poincaré group). One may verify that $h \in H$ by applying p, which by construction is G-equivariant (i.e., $p(xy) = xp(y)$), which gives

$$p(h) = p(s(q)^{-1}ys(y^{-1}q)) = s(q)^{-1}yp(s(y^{-1}q)) = s(q)^{-1}yy^{-1}q = s(q)^{-1}q,$$

where in the third step we used (7.108). For any $x \in G$ we have $x^{-1}[x] = [x^{-1}x] = [e]$, so taking $x = s(q)$ in this computation we find $p(h) = [e]$, which is true iff $h \in H$.

Given an irreducible system of imprimitivity $(\tilde{u}^\chi, \tilde{\pi}^\chi)$, we obtain generalized momentum operators by passing to the associated representation of the Lie algebra $\mathfrak{g}$ of G through (5.156) and (7.57), i.e.,

$$\tilde{\pi}^\chi(A) = i\hbar(\tilde{u}^\chi)'(A), \tag{7.116}$$

where $A \in \mathfrak{g}$, so that, cf. (7.78) - (7.80), we obtain from (5.160) and (7.81):

$$[\tilde{\pi}^\chi(A), \tilde{\pi}^\chi(B)] = i\hbar\tilde{\pi}^\chi([A,B]); \tag{7.117}$$
$$[\tilde{\pi}^\chi(A), \tilde{\pi}^\chi(\tilde{f})] = i\hbar\tilde{\pi}^\chi(\delta_A\tilde{f}); \tag{7.118}$$
$$[\tilde{\pi}^\chi(\tilde{f}), \tilde{\pi}^\chi(\tilde{g})] = 0, \tag{7.119}$$

where $A, B \in \mathfrak{g}$ and $\tilde{f}, \tilde{g} \in C_0(Q)$ (in fact, these formulae—defined on the right domain—work also for many unbounded functions on Q, see below), and δ_A is defined in (3.71). Let us take a look at a few illustrative special cases:

- If $H = G$, then Q is a point, so that $C^*(G,Q) = G^*(G)$, and systems of imprimitivity are just irreducible representations of G. We have $H^\chi \cong H_\chi$ through the map $w : H^\chi \to H_\chi$ defined by $\psi \mapsto \psi(e) \equiv \psi' \in H_\chi$, with inverse $\psi(x) = u_\chi(x^{-1})\psi'$. This gives $wu^\chi(y)w^{-1} = u_\chi(y)$. Similarly, in (7.115) we take $s = e$, which gives $\tilde{u}^\chi(y) = u_\chi(y)$ on $\tilde{H}^\chi = H_\chi$.
- If $H = \{e\}$ we have $Q = G$ and $C^*(G,G) \cong B_0(L^2(G))$, which quantizes the underlying classical phase space $\mathfrak{g}^* \times G \cong T^*G$. We now have $H = L^2(G)$ carrying the left-regular representation of G.
- Let $G = E(3)$ act canonically on $Q = \mathbb{R}^3$. Taking $q_0 = 0$ gives $H = SO(3)$, so irreducible systems of imprimitivity are classified by $j = 0, 1, \ldots$, with corresponding irreducible representations $D_j(SO(3))$ on $H_j = \mathbb{C}^{2j+1}$, cf. §5.8. Hence

$$\tilde{H}^j = L^2(\mathbb{R}^3) \otimes H_j, \tag{7.120}$$

and using the cross-section $s(q) = (1_3, q)$ from $\mathbb{R}^3$ to $E(3)$ we obtain, from (7.115) with (7.63) - (7.64) and (7.107), the expressions

$$\tilde{u}^j(R,a))\tilde{\psi}(q) = D_j(R)\tilde{\psi}(R^{-1}(q-a)); \tag{7.121}$$
$$\tilde{\pi}^j(\tilde{f}))\tilde{\psi}(q) = \tilde{f}(q)\tilde{\psi}(q). \tag{7.122}$$

For $j = 0$ this gives the usual quantum theory of a spinless particle:

1. The Hilbert space is $\tilde{H}^0 = L^2(\mathbb{R}^3)$.
2. For the generators of $\mathbb{R}^3 \subset E(3)$ we duly obtain the momentum operators

$$P_i = -i\hbar\frac{\partial}{\partial q^i}, \tag{7.123}$$

where $P_i = \tilde{\pi}^0(e_i)$ is defined in terms of the standard basis (e_1, e_2, e_3) of $\mathbb{R}^3$, now seen as the Lie algebra of $\mathbb{R}^3$.
3. Using the basis (3.66) of the Lie algebra of $SO(3) \subset E(3)$, we obtain the orbital angular momentum operators (which pick up extra terms for $j > 0$):

$$\tilde{\pi}^0(J_1) = i\hbar\left(q^3\frac{\partial}{\partial q^2} - q^2\frac{\partial}{\partial q^3}\right); \tag{7.124}$$
$$\tilde{\pi}^0(J_2) = i\hbar\left(q^1\frac{\partial}{\partial q^3} - q^3\frac{\partial}{\partial q^1}\right); \tag{7.125}$$
$$\tilde{\pi}^0(J_3) = i\hbar\left(q^2\frac{\partial}{\partial q^1} - q^1\frac{\partial}{\partial q^2}\right). \tag{7.126}$$

4. The coordinate functions $\tilde{f}(q) = q^i$ yield the position operators $Q_i = \tilde{\pi}^0(q^i)$:

$$Q_i\tilde{\psi}(q) = q^i\tilde{\psi}(q). \tag{7.127}$$

5. Thus we obtain all the familiar commutation relations like $[Q_i, P_j] = i\hbar\delta_{ij}$, $[\tilde{\pi}^0(J_1), \tilde{\pi}^0(J_2)] = i\hbar\tilde{\pi}^0(J_3)$, etc., cf. (7.65) - (7.66).

- Let $G = \mathbb{R}$ act on $Q = \mathbb{T}$, which we parametrize by $z = \exp(2\pi i q)$, $q \in [0,1)$, by

$$a : \exp(2\pi i q) \mapsto \exp(2\pi i (q+a)), \tag{7.128}$$

so that $H = \mathbb{Z}$, with $\hat{H} = \mathbb{T}$ under $u_z(n) = z^n$, $z \in \mathbb{T}$, $n \in \mathbb{Z}$, cf. (C.349). We parametrize $\hat{H}$ by $z = \exp(i\theta)$, $\theta \in [0,2\pi)$, so that (with slight abuse of notation) $u_\theta(n) = e^{in\theta}$. In the second description (i.e. the one of the physicists) we have

$$\tilde{H}^\theta = L^2(\mathbb{T}) = L^2(0,1), \tag{7.129}$$

where topology of Q is lost for the moment. Using the cross-section

$$s\left(e^{2\pi i q}\right) = q, \tag{7.130}$$

where $q \in [0,1)$, we obtain

$$\tilde{u}^\theta(a)\tilde{\psi}(q) = e^{in(a,q)\theta}\,\tilde{\psi}(q-a+n(a,q)), \tag{7.131}$$

where $n(a,q) \in \mathbb{Z}$ is the unique integer such that $q - a + n(a,q) \in [0,1)$. The corresponding momentum operator is formally given by the usual expression $P = -i\hbar\partial/\partial q$, cf. (7.123), which appears to be independent of θ (since for any $q \in (0,1)$ and a small enough we have $n(a,q) = 0$), but in fact the θ-dependence is in its domain, which can be shown to consist of the subspace of the Sobolev space $H^1(0,1)$—i.e. the closure of $C^\infty([0,1])$ in the inner product (5.318) adapted to $L^2(0,1)$, which implies $H^1(0,1) \subset C([0,1])$—whose elements satisfy

$$\psi(1) = e^{-i\theta}\psi(0). \tag{7.132}$$

To see this, we recall that

$$P\tilde{\psi} = i\hbar \lim_{\varepsilon\to 0}\left(\frac{\tilde{u}^\theta(\varepsilon)\tilde{\psi} - \tilde{\psi}}{\varepsilon}\right), \tag{7.133}$$

where the limit is taken in the L^2-norm, so that we need existence of

$$\lim_{\varepsilon\to 0}\varepsilon^{-2}\int_0^1 dq\,|e^{in(a,q)\theta}\,\tilde{\psi}(q-\varepsilon+n(\varepsilon,q)) - \tilde{\psi}(q)|^2.$$

For $0 < q < \varepsilon$ we have $n(\varepsilon,q) = 1$, whereas for $\varepsilon < q < 1$ we have $n(\varepsilon,q) = 0$, so it is convenient to split the integral as a sum of $\int_0^\varepsilon$ and $\int_\varepsilon^1$. The second term enforces the existence of derivatives in the L^2-sense (which in turn makes $\tilde{\psi}$ continuous on $[0,1]$) and is unproblematic, but the first requires the existence of

$$\lim_{\varepsilon\to 0}\varepsilon^{-2}\int_0^\varepsilon dq\,|e^{i\theta}\,\tilde{\psi}(q-\varepsilon+1) - \tilde{\psi}(q)|^2.$$

This strange expression, then, enforces the boundary condition (7.132). In this case there is no single position operator, but the algebra $C(\mathbb{T})$ plays its role.

7.6 Representations of semi-direct products

The case $Q = G/H$ also provides the key for the general case, as long as the G-action on Q is *regular*, cf. Theorem 7.7. In that case, the construction of the irreducible system of imprimitivity $(u(G), \pi(C_0(Q)))$ corresponding to a pair $(\mathcal{O}, u_\chi(H))$, where $\mathcal{O}$ is a G-orbit in Q, requires no new ideas: we have $\mathcal{O} \cong G/H$, and hence $u = u^\chi$ and $\pi = \pi^\chi$ as described in §7.5 (where the function $\tilde{f}$ in formulae like (7.104) or (7.114), which in these expression was defined on G/H, should be seen as the restriction of $\tilde{f} \in C_0(Q)$ to $\mathcal{O} \subset Q$). An important application of this construction is the representation theory of ***regular semi-direct products*** $L \ltimes V$ (cf. §7.3), where regularity means that the *dual* L-action on V^* is regular; this action is given by

$$\lambda \cdot \theta(v) = \theta(\lambda^{-1} \cdot v) \; (\lambda \in L, \theta \in V^*, v \in V). \tag{7.134}$$

Theorem 7.9. *Up to unitary equivalence, the irreducible unitary representations of a regular semi-direct product $G = L \ltimes V$ are classified by pairs $(\mathcal{O}, \sigma)$, where $\mathcal{O}$ is an L-orbit in V^* and σ is an element of the unitary dual of the stabilizer $L_0 \subset L$ of an arbitrary point $\theta_0 \in \mathcal{O}$. The corresponding representation $\tilde{u}^{(\mathcal{O},\sigma)}(G)$ may be realized from an irreducible representation u_σ of L_0 on a Hilbert space H_σ combined with a cross-section $s : L/L_0 \to L$ of the canonical projection $p : L \to L/L_0$, namely through*

$$\tilde{H}^{(\mathcal{O},\sigma)} = L^2(L/L_0) \otimes H_\sigma; \tag{7.135}$$

$$\tilde{u}^{(\mathcal{O},\sigma)}(\lambda, v)\tilde{\psi}(\theta) = e^{i\theta(v)} u_\sigma(s(\theta)^{-1}\lambda s(\lambda^{-1}\theta))\tilde{\psi}(\lambda^{-1}\theta). \tag{7.136}$$

Proof. Let u be a unitary representation of G. This implies

$$u(\lambda)u(v)u(\lambda^{-1}) = u(\lambda \cdot v), \tag{7.137}$$

in which $\lambda \equiv (\lambda, 0)$ and $v \equiv (e, v)$. Since $V \subset G$ is abelian, we have $C^*(V) \cong C_0(V^*)$ by the Fourier transform (cf. Theorem C.109 in §C.15), which here is given by (7.44) - (7.45), with $A \rightsquigarrow v$. Hence the representation $u^{\int}(C^*(V))$ defined by $u(V)$ via (5.172), seen as a representation of $C_0(V^*)$ via the Fourier transform, is given by

$$u^{\int}(f) = (2\pi)^{-n} \int_{V \times V^*} d^n v d^n \theta \, e^{i\theta(v)} f(\theta) u(v). \tag{7.138}$$

Using invariance of the measure $d^n v d^n \theta$ under the joint transformation $(v, \theta) \rightsquigarrow (\lambda \cdot v, \lambda \cdot \theta)$, from (7.137) we obtain, for $f \in C_0(V^*)$ in the image of $\check{f} \in C_c^\infty(V)$,

$$\begin{aligned} u(\lambda)u^{\int}(f)u(\lambda)^* &= (2\pi)^{-n} \int_{V \times V^*} d^n v d^n \theta \, e^{i\theta(v)} f(\theta) u(\lambda \cdot v) \\ &= (2\pi)^{-n} \int_{V \times V^*} d^n v d^n \theta \, e^{i(\lambda \cdot \theta)(\lambda \cdot v)} f(\lambda^{-1} \cdot \lambda \cdot \theta) u(\lambda \cdot v) \\ &= (2\pi)^{-n} \int_{V \times V^*} d^n v d^n \theta \, e^{i\theta(v)} f(\lambda^{-1} \cdot \theta) u(v) \\ &= u^{\int}(L_\lambda f). \end{aligned} \tag{7.139}$$

Consequently, a unitary representation $u(L \ltimes V)$ defines a system of imprimitivity $(u(L), u^{\int}(C_0(V^*)))$, and *vice versa*, since any pair of representations $(u(L), u(V))$ that satisfies (7.137) gives rise to a representation $u(G)$ by $u(\lambda, v) = u(v)u(\lambda)$.

Now apply Theorem 7.7 with $G \rightsquigarrow L$ and $Q \rightsquigarrow V^*$. All we need in order to obtain (7.135) - (7.136) from (7.106) and (7.107) - (7.115) is to find the representation $u(V)$ that induces the representation $u^{\int}(C_0(V^*))$ given by (7.107), namely

$$u(v)\tilde{\psi}(\theta) = e^{-i\theta(v)}\tilde{\psi}(\theta), \tag{7.140}$$

as is easily checked from (7.138). □

In view of this, we have a remarkable group–groupoid C*-algebra isomorphism

$$C^*(L \ltimes V) \cong C^*(L \ltimes V^*), \tag{7.141}$$

where the left-hand side is just the C*-algebra of the *group* $L \ltimes V$, whereas the right-hand side is the C*-algebra of the action group*oid* $L \ltimes V^*$ relative to (7.134). Also, a computation shows that the same formulae (7.135) - (7.136) are obtained if, given $\theta_0 \in V^*$ and hence given L_0 as its stabilizer, we define a subgroup $H \subset G$ by

$$H = L_0 \ltimes V, \tag{7.142}$$

and induce from the representation $u_{(\theta_0,\sigma)}$ of H defined by

$$u_{(\theta_0,\sigma)}(\lambda, v) = e^{i\theta_0(v)}u_\sigma(\lambda). \tag{7.143}$$

We briefly discuss four basic examples from physics, each of which is easily seen to be regular. We write a instead of v in $(\lambda, v) \in G$ so as to emphasize the "spatial" character of V, whereas V^* is labeled by a dual "momentum" variable p.

- $G = E(2) = SO(2) \ltimes \mathbb{R}^2$, defined like $E(3)$, i.e., with respect to the usual action of $SO(2)$ on $\mathbb{R}^2$ (this group will play a role in the representation theory of the Poincaré-group). We find the same action of $SO(2)$ on $(\mathbb{R}^2)^* = \mathbb{R}^2$, so that the orbits are $\mathcal{O}_0 = \{0\}$ with $G_0 = SO(2)$ and $\mathcal{O}_r = \{(x,y) \in \mathbb{R}^2 \mid x^2 + y^2 = r^2\}$ for $r > 0$, with $G_r = \{e\}$. Thus the Hilbert spaces and representations are given by

$$\tilde{H}^{(0,n)} = \mathbb{C}; \tag{7.144}$$
$$\tilde{u}^{(0,n)}(\lambda, a) = e^{2\pi i n\lambda}; \tag{7.145}$$
$$\tilde{H}^r = L^2(0,1); \tag{7.146}$$
$$\tilde{u}^r(\lambda, a)\tilde{\psi}(p) = e^{ir(a_1 \cos p' + a_2 \sin p')}\psi(p - \lambda|\text{mod}\,1), \tag{7.147}$$

  where $n \in \mathbb{Z}$, $\lambda \in [0,1)$, $p \in (0,1)$, and $p' = 2\pi p$. In the first case $\mathbb{R}^2 \subset E(2)$ is represented trivially, whereas in the second the r-dependence of the representation lies entirely in $\mathbb{R}^2$ (since $\tilde{H}^r$ and $\tilde{u}^r(\lambda, 0)$ are evidently independent of r). The projective representations of G are of considerable interest, too, cf. §5.10.

Lemma 7.10. *If* $G = SO(p,q) \ltimes \mathbb{R}^{p+q}$ $(p > 0, q \geq 0)$*, then* $H^2(\mathfrak{g}, \mathbb{R}) = 0$.

Here $SO(p,q)$ is the subgroup of $SL_{p+q}(\mathbb{R}^{p+q})$ whose elements leave the form

$$x^2 = x_1^2 + \cdots + x_p^2 - (x_{p+1}^2 + \cdots + x_{p+q}^2)$$

invariant; the best-known example is the (proper) Lorentz group $SO(3,1)$, see below. This lemma may be proved by a straightforward but lengthy computation. By Theorem 5.59, the projective unitary representations of G then correspond to the ordinary unitary representations of the universal covering

$$\tilde{G} = \mathbb{R} \ltimes \mathbb{R}^2, \tag{7.148}$$

where $\mathbb{R}$ acts on $\mathbb{R}^2$ through the covering projection $\tilde{\pi} : \mathbb{R} \to SO(2) = \mathbb{R}/\mathbb{Z}$, cf. Theorem 5.41 (with $D \rightsquigarrow \mathbb{Z}$). This changes the expressions (7.144) - (7.147) into

$$\tilde{H}^{(0,s)} = \mathbb{C}; \tag{7.149}$$

$$\tilde{u}^{(0,s)}(\lambda, a) = e^{is\lambda}; \tag{7.150}$$

$$\tilde{H}^{(r,\theta)} = L^2(0,1); \tag{7.151}$$

$$\tilde{u}^{(r,\theta)}(\lambda, a)\tilde{\psi}(p) = e^{ir(a_1 \cos p' + a_2 \sin p')} e^{in(\lambda,p)\theta}\, \tilde{\psi}(p - \lambda + n(\lambda, p)), \tag{7.152}$$

where $\lambda \in \mathbb{R}$, $s \in \mathbb{R}$, $\theta \in [0, 2\pi)$, $p \in (0,1)$, and $n(\lambda, p)$ is defined as in (7.131).

- $G = E(3) = SO(3) \ltimes \mathbb{R}^3$, as before with the defining action of $SO(3)$. The $SO(3)$-orbits in $(\mathbb{R}^3)^* = \mathbb{R}^3$ are spheres $S_r^2 \cong SO(3)/SO(2)$ with radius $r > 0$, as well as the origin ($r = 0$) with stabilizer $SO(3)$, so that for the Hilbert spaces we obtain

$$\tilde{H}^{(0,j)} = \mathbb{C}^{2j+1}; \tag{7.153}$$

$$\tilde{H}^{(r,n)} = L^2(S^2); \tag{7.154}$$

where $j = 0, 1, \ldots$ labels the unitary irreducible representations of $SO(3)$ on $H_j = \mathbb{C}^{2j+1}$, whereas $n \in \mathbb{Z}$ labels the irreducible representations of $SO(2)$ on $\mathbb{C}$ (we write $S^2 \equiv S_1^2$). In the second case, the representation $u^{(r,n)}$ of $SO(3) \subset E(3)$ depends explicitly on n through the Wigner cocycle; for $n = 0$ we simply obtain

$$\tilde{u}^{(r,0)}(R, a)\tilde{\psi}(p) = e^{irp\cdot a}\tilde{\psi}(R^{-1}p). \tag{7.155}$$

For $n \neq 0$ we just give a formula for $\tilde{u}^{(r,n)}(R, a)$ in case that R is a rotation around the z-axis and $a = 0$; this is enough to make the point. To this end we parametrize $SO(3)$ by the well-known Euler angles, i.e., in terms of the matrices J_i, cf. (3.66),

$$R(\phi, \theta, \alpha) = e^{\phi J_3} e^{\theta J_2} e^{\alpha J_3}, \tag{7.156}$$

and write $q \in S^2$ as $q = (\phi, \theta) = R(\phi, \theta, 0)e_3$ with $e_3 = (0,0,1)$ (the spherical coordinates of q are $(\phi - \frac{1}{2}\pi, \theta)$). This also provides S^2 with an $SO(3)$-invariant measure $d\nu(\phi, \theta) = d\phi d\theta \sin\theta$. A convenient choice of $s : S^2 \to SO(3)$ is

$$s(\phi, \theta) = R(\phi, \theta, -\phi), \tag{7.157}$$

in which case we simply obtain, writing $R_z(\alpha) = R(\alpha,0,0)$,

$$\tilde{u}^{(r,n)}(R_z(\alpha),0)\tilde{\psi}(\phi,\theta) = e^{in\alpha}\tilde{\psi}(\phi-\alpha,\theta). \tag{7.158}$$

The universal covering group of $E(3)$ is

$$\widetilde{E(3)} = SU(2) \ltimes \mathbb{R}^3, \tag{7.159}$$

where $SU(2) = \widetilde{SO(3)}$ acts on $\mathbb{R}^3$ through its covering projection $\tilde{\pi}$ onto $SO(3)$, as in the previous case. By Theorem 5.59 and Lemma 7.10, the projective unitary irreducible representations of $E(3)$ are given by the unitary irreducible representations of $SU(2) \ltimes \mathbb{R}^3$. This obviously leads to additional half-integral values for j in (7.153), since this number now labels the unitary irreducible representations of $SU(2)$. As to n in (7.154), the subgroup $H \subset SU(2)$ that stabilizes $(0,0,r) \in S^2_r$ consists of all matrices $u_z = \mathrm{diag}(z,\overline{z})$, where $z \in \mathbb{T}$, so $H \cong \mathbb{T}$ and hence $\hat{H} = \mathbb{Z}$ under $u_z \mapsto z^m$, $m \in \mathbb{Z}$. We now recall from the proof of Proposition 5.5 that

$$u = \cos(\theta/2)\cdot 1_2 + i\sin(\theta/2)\mathbf{u}\cdot\sigma \in SU(2), \tag{7.160}$$

where $\mathbf{u}$ is a unit vector in $\mathbb{R}^3$, projects to $\tilde{\pi}(u) = R_\theta(\mathbf{u}) \in SO(3)$, i.e., the rotation around $\mathbf{u}$ by an angle θ. Parametrizing $z = \cos(\alpha/2) + i\sin(\alpha/2)$, $\alpha \in [0,4\pi)$, therefore gives $\tilde{\pi}(u_z) = \exp(\alpha J_3)$. Besides (7.157), we now also need a cross-section $s : S^2_r \to SU(2)$, for which the above analysis suggests we take

$$s(\phi,\theta) = u^{(3)}(\phi)u^{(2)}(\theta)u^{(3)}(-\phi); \tag{7.161}$$

$$u^{(2)}(\theta) = \cos(\tfrac{1}{2}\theta)\cdot 1_2 + i\sin(\tfrac{1}{2}\theta)\cdot\sigma_2, \tag{7.162}$$

$$u^{(3)}(\phi) = \cos(\phi/2)\cdot 1_2 + i\sin(\phi/2)\cdot\sigma_3; \tag{7.163}$$

note that $u_z = u^{(3)}(\alpha)$. A calculation similar to the one leading to (7.158) gives

$$\tilde{u}^{(r,m)}(u_z,0)\tilde{\psi}(\phi,\theta) = e^{im\alpha/2}\tilde{\psi}(\phi-\alpha,\theta). \tag{7.164}$$

Comparing (7.158) and (7.164), we see that if m is even, then $n = m/2$ (of course, by convention we may replace $m/2$ in (7.164) by n on the understanding that n may now be half-integral). If m is odd, choosing $\alpha = 2\pi$ we famously obtain

$$\tilde{u}^{(r,m)}(-1_2,0)\tilde{\psi} = -\tilde{\psi}. \tag{7.165}$$

More generally, if we take a closed path $t \mapsto R_{2\pi t}(\mathbf{u})$, $t \in [0,1]$ in $SO(3)$, which starts and ends at 1_3, and lift it (with respect to the covering projection $\tilde{\pi} : SU(2) \to SO(3)$) to a path $t \mapsto u(t) \equiv \cos(\pi t) + i\sin(\pi t)\mathbf{u}\cdot\sigma$ in $SU(2)$, which now starts at 1_2 and ends at -1_2, then the corresponding representation $\tilde{u}^{(r,m)}(u(t),0)$ takes the wave-function $\tilde{\psi}$ to itself if m is even, whereas it takes $\tilde{\psi}$ to $-\tilde{\psi}$ whenever m is odd (this is an embryonic version of the connection between spin and statistics, fully realized only in quantum field theory).

- $G = L \ltimes \mathbb{R}^{3+1}$, the ***Poincaré group***, where the ***Lorentz group*** $L = O(3,1)$ consists of all real 4×4 matrices that leave the indefinite quadratic form

$$x^2 = x_0^2 - x_1^2 - x_2^2 - x_3^2 \tag{7.166}$$

invariant; in this context the standard coordinates on $\mathbb{R}^4$ are labeled as (x_0,x_1,x_2,x_3). The Lorentz group has four connected components, which may be identified by the (independent) conditions $\det(\lambda) = \pm 1$ and $\pm\lambda_{00} \geq 1$. For simplicity we restrict ourselves to the connected component $L_+^\uparrow$ of the identity, in which $\det(\lambda) = 1$ and $\lambda_{00} \geq 1$. This group is called the ***proper orthochronous Lorentz group***, which in turn defines the ***proper orthochronous Poincaré group*** $P_+^\uparrow = L_+^\uparrow \ltimes \mathbb{R}^4$. Writing $p^2 = p_0^2 - p_1^2 - p_2^2 - p_3^2$, the $L_+^\uparrow$-orbits in $(\mathbb{R}^4)^* = \mathbb{R}^4$ are seen to be:

1. $\mathscr{O}_0 = \{(0,0,0,0)\}$, with stabilizer $(L_+^\uparrow)_0 = L_+^\uparrow$;
2. $\mathscr{O}_m^\pm = \{p \in \mathbb{R}^4 \mid p^2 = m^2, \pm p_0 \geq 0\}$, $m > 0$, with $(L_+^\uparrow)_0 = SO(3)$;
3. $\mathscr{O}_0^\pm = \{p \in \mathbb{R}^4 \mid p^2 = 0, \pm p_0 \geq 0\}$, with $(L_+^\uparrow)_0 = E(2)$;
4. $\mathscr{O}_{im} = \{p \in \mathbb{R}^4 \mid p^2 = -m^2, \pm p_0 \geq 0\}$, $m > 0$, with $(L_+^\uparrow)_0 = SO(2,1)$.

Here the stabilizers L_0 are found by taking the reference points $(\pm m,0,0,0)$ in case 2, $(\pm 1,0,0,-1)$ in case 3, and $(0,0,0,m)$ in case 4. The physically relevant cases are probably $\mathscr{O}_{m^2}^+$ and $\mathscr{O}_0^+$. We pass straight to the universal covering group

$$\tilde{P}_+^\uparrow = SL(2,\mathbb{C}) \ltimes \mathbb{R}^4, \tag{7.167}$$

where the covering projection $\tilde{\pi} : SL(2,\mathbb{C}) \to L_+^\uparrow$ is given analogously to the case (5.46). We again start from the four matrices $(\sigma_0,\sigma_1,\sigma_2,\sigma_3)$ in (5.42), and note:

- These form a basis for the (real) vector space of all self-adjoint 2×2 matrices;
- For any $x \in \mathbb{R}^4$ we have $\det(\sum_{\mu=0}^3 x_\mu \sigma_\mu) = x^2$ as defined in (7.166);
- For any $\tilde{\lambda} \in SL(2,\mathbb{C})$ and $a \in M_2(\mathbb{C})$ we have $\det(\tilde{\lambda} a \tilde{\lambda}^*) = \det(a)$;
- For any $\tilde{\lambda} \in SL(2,\mathbb{C})$ and self-adjoint $a \in M_2(\mathbb{C})$, $\tilde{\lambda} a \tilde{\lambda}^*$ is again self-adjoint.

Taking $a = \sum_\mu x_\mu \sigma_\mu$, it follows that for $\tilde{\lambda} \in SL(2,\mathbb{C})$ and $x \in \mathbb{R}^4$ there must be $\lambda \in O(3,1)$ such that $\tilde{\lambda} \sum_\mu x_\mu \sigma_\mu \tilde{\lambda}^* = \sum_\mu (\lambda \cdot x)_\mu \sigma_\mu$. By continuity and the fact that $SL(2,\mathbb{C})$ is connected it follows that in fact $\lambda \in L_+^\uparrow$, so we put $\tilde{\pi}(\lambda) = \lambda$. As for (5.46), the kernel is $\ker(\tilde{\pi}) = \mathbb{Z}_2 = \{\pm 1_2\}$. This enlarges the stabilizers:

1. For $\mathscr{O}_{m^2}^+$ we now obtain $(\tilde{L}_+^\uparrow)_0 = SU(2)$, leading to a family of unitary irreducible representations $u^{m,j}$ labeled by ***mass*** $m > 0$ and ***spin*** $j = 0, \frac{1}{2}, 1, \ldots$.
2. For $\mathscr{O}_0^+$ the stabilizer $(\tilde{L}_+^\uparrow)_0$ of $(1,0,0,1)$ is a double cover $E(2)'$ of $E(2)$, whose unitary irreducible representations are labeled by either $(0,n)$ with $n \in \mathbb{Z}/2$ (called ***helicity***) or by $r > 0$. The latter case does not occur in nature.

On the one hand, this classification is a triumph of mathematical physics, but on the other hand, it fails to single out which cases actually occur in nature: as far as we know, these are spin $j = 0$ and $j = \frac{1}{2}$ and helicity $n = \pm 1$ and $n = \pm 2$.

- $G = E(3) \ltimes \mathbb{R}^4$, the Galilei group, defined via the following $E(3)$-action on $\mathbb{R}^4$:

$$(R, \mathbf{v}) : (a_0, \mathbf{a}) \mapsto (a_0, R\mathbf{a} + a_0\mathbf{v}). \tag{7.168}$$

Note that $\mathbf{v}$ is physically interpreted as a velocity, whereas earlier $\mathbf{a} \in \mathbb{R}^3 \subset E(3)$ was a position variable. This is clear from the defining G-action on $\mathbb{R}^4$, given by

$$(R, \mathbf{v}, a_0, \mathbf{a}) : (t, \mathbf{x}) \mapsto (t + a_0, R\mathbf{x} + \mathbf{a} + t\mathbf{v}), \tag{7.169}$$

which in fact determines the action (7.168). Either way, we obtain the group law

$$(R, \mathbf{v}, a_0, \mathbf{a}) \cdot (R', \mathbf{v}', a_0', \mathbf{a}') = (RR', \mathbf{v} + R\mathbf{v}', a_0 + a_0', \mathbf{a} + R\mathbf{a}' + a_0'\mathbf{v}). \tag{7.170}$$

We therefore see that the role of the Lorentz group $SO(3,1)$ is now played by the Euclidean group $E(3)$. Since from (7.170) the inverse is found to be

$$(R, \mathbf{v}, a_0, \mathbf{a})^{-1} = (R^{-1}, -R^{-1}\mathbf{v}, -a_0, -R^{-1}(\mathbf{a} - a_0\mathbf{v})), \tag{7.171}$$

the dual $E(3)$-action on $(\mathbb{R}^4)^* \cong \mathbb{R}^4$ is given (in non-relativistic notation) by

$$(R, \mathbf{v}) : (E, \mathbf{p}) \mapsto (E - \langle \mathbf{v}, R\mathbf{p} \rangle, R\mathbf{p}). \tag{7.172}$$

Hence the dual $E(3)$-orbits in $\mathbb{R}^4$ are labeled by $E \in \mathbb{R}$ and $r > 0$, as follows:

$$\mathcal{O}_E = \{(E, \mathbf{0})\}, \tag{7.173}$$
$$\mathcal{O}_{(r)} = \{(E, \mathbf{p}), E \in \mathbb{R}, \|\mathbf{p}\| = r\}. \tag{7.174}$$

The representations of G corresponding to the first type are basically the representations of $E(3)$, whereas in the second case the stability group of say $(0,0,0,r)$ is isomorphic to $E(2)$. None of the ensuing induced representations of G reproduces some recognizable version of non-relativistic quantum mechanics, for which we need to pass to projective representations of G. These may be found from Theorem 5.62, which here applies in full glory, since $H^2(\mathfrak{g}, \mathbb{R}) \neq 0$. A (lengthy) computation shows that $H^2(\mathfrak{g}, \mathbb{R})$ has a single generator

$$\varphi((M, \mathbf{v}, a_0, \mathbf{a}), (M', \mathbf{v}', a_0', \mathbf{a}')) = \langle \mathbf{v}, \mathbf{a}' \rangle - \langle \mathbf{v}', \mathbf{a} \rangle, \tag{7.175}$$

where $M \in \mathfrak{so}(3)$, and $(\mathbf{v}, a_0, \mathbf{a}) \in \mathbb{R}^3 \times \mathbb{R}^4 \subset \mathfrak{g} = \mathfrak{so}(3) \oplus \mathbb{R}^3 \oplus \mathbb{R}^4$ are identified with the corresponding Lie group elements. Following the procedure culminating in Theorem 5.62, the central extension $\check{G}$ is found to be (cf. (7.159) and (5.46))

$$\check{G} = \widetilde{E(3)} \ltimes \mathbb{R}^5, \tag{7.176}$$

where, writing $\tilde{\pi}(u) \equiv R(u)$, the covering group $\widetilde{E(3)}$ acts on $\mathbb{R}^5$ through

$$(u, \mathbf{v}) : (a_0, \mathbf{a}, c) \mapsto (a_0, R(u)\mathbf{a} + a_0\mathbf{v}, c + \tfrac{1}{2}a_0\|\mathbf{v}\|^2 + \langle \mathbf{v}, R(u)\mathbf{a} \rangle). \tag{7.177}$$

Consequently, writing $\tilde{x} = (R, \mathbf{v}, a_0, \mathbf{a})$, for the group law in $\check{G}$ we obtain

$$(\tilde{x}, c) \cdot (\tilde{x}', c') = (\tilde{x} \cdot \tilde{x}', c + c' + \langle \mathbf{v}, R(u)\mathbf{a}' \rangle + \tfrac{1}{2} a_0' \|\mathbf{v}\|^2). \tag{7.178}$$

Eq. (7.177) implies the following dual $\widetilde{E(3)}$-action on $(\mathbb{R}^5)^* = \mathbb{R}^5$:

$$(u, \mathbf{v}) : (E, \mathbf{p}, m) \mapsto (E - \langle \mathbf{v}, R(u)\mathbf{p} \rangle + \tfrac{1}{2} m \|\mathbf{v}\|^2, R(u)\mathbf{p} - m\mathbf{v}, m). \tag{7.179}$$

This time, the $\widetilde{E(3)}$-orbits in $\mathbb{R}^5$ are:

1. $\mathscr{O}_E = \{(E, \mathbf{0}, 0)\}$ $(E \in \mathbb{R})$, with stabilizer $\widetilde{E(3)}$;
2. $\mathscr{O}_{(r,0)} = \{(E, \mathbf{p}, 0) \mid E \in \mathbb{R}, \|\mathbf{p}\| = r\}$ $(r > 0)$, with stabilizer $E(2)'$;
3. $\mathscr{O}_{U,m} = \{(E, \mathbf{p}, m) \mid E - E_{\mathbf{p}} = U\}$ $(m \in \mathbb{R}\backslash\{0\}, U \in \mathbb{R})$, with stabilizer $SU(2)$.

Here $E(2)' \subset \widetilde{E(3)}$ is a double cover of $E(2)$, like the subgroup of $SL(2,\mathbb{C})$ stabilizing the point $(1,0,0,1) \in \mathbb{R}^4$ in the theory of the Poincaré-group. This time we take any point $(E, 0, 0, r, 0) \in \mathbb{R}^5$, which is stabilized by pairs $(u, \mathbf{v}) \in \widetilde{E(3)}$ for which $R(u)$ is a rotation around the z-axis and $\mathbf{v} = (v_1, v_2, 0)$; the image of these pairs in $E(3)$ is $E(2) = SO(2) \ltimes \mathbb{R}^2$, where $SO(2) \subset SO(3)$ consists of rotations around the z-axis and $\mathbb{R}^2$ is the x-y plane. In the third case we write $E_{\mathbf{p}} = \|\mathbf{p}\|^2/2m$ and take $(U, \mathbf{0}, m)$, whose stabilizer in $E(3)$ is evidently $SO(3)$.

Thus we have massless as well as massive particles both in relativistic and in non-relativistic quantum physics. The simplest case of all is formed by massive non-relativistic particles, which correspond to the orbits $\mathscr{O}_{U,m}$ above, supplemented with a spin j labelling the underlying irreducible representation D_j of $SU(2)$. Such orbits are diffeomorphic to $\mathbb{R}^3$ under the identification $(U + E_{\mathbf{p}}, \mathbf{p}, m) \leftrightarrow \mathbf{p}$, and a convenient choice of the cross-section $s : \mathscr{O}_{U,m} \to \widetilde{E(3)}$ is $s(\mathbf{p}) = (1_2, -\mathbf{p}/m)$, since in that case the Wigner cocycle simply becomes $s(\mathbf{p})^{-1}(u,\mathbf{v})s((u,\mathbf{v})^{-1}\mathbf{p}) = u$. Since different values of U turn out to give equivalent representations of $\check{G}$ (in the sense explained at the end of §5.10), we take $U = 0$, and eqs. (7.135) - (7.136) become

$$\tilde{H}^{m,j} = L^2(\mathbb{R}^3) \otimes H_j; \tag{7.180}$$
$$\tilde{u}^{m,j}(u, \mathbf{v}, a_0, \mathbf{a})\tilde{\psi}(\mathbf{p}) = e^{i(a_0 E_{\mathbf{p}} + \langle \mathbf{a}, \mathbf{p} \rangle)} D_j(u) \tilde{\psi}(R(u)^{-1}(\mathbf{p} + m\mathbf{v})). \tag{7.181}$$

Here $L^2(\mathbb{R}^3)$ simply carries Lebesgue measure $d^3\mathbf{p}$, which is $\widetilde{E(3)}$-invariant.

The massive relativistic case is slightly more involved: we again have $\mathscr{O}_m^+ \cong \mathbb{R}^3$ under $(\omega_{\mathbf{p}}, \mathbf{p}) \leftrightarrow \mathbf{p}$, where $\omega_{\mathbf{p}} = \sqrt{\|\mathbf{p}\|^2 + m^2}$, but the Lorentz-invariant measure on $\mathscr{O}_m^+$ is $d^3\mathbf{p}/\omega_{\mathbf{p}}$. For each $\mathbf{p} \in \mathbb{R}^3$ there is a unique boost $b_{\mathbf{p}} \in L_+^\uparrow$ that maps $(m, 0, 0, 0)$ to $(\omega_{\mathbf{p}}, \mathbf{p})$, with pre-image $\tilde{b}_{\mathbf{p}}$ in $SL(2,\mathbb{C})$, so we take $s(\mathbf{p}) = \tilde{b}_{\mathbf{p}}$. The Hilbert space is (*mutatis mutandis*) still given by (7.180), but instead of (7.181) we now obtain

$$\tilde{u}^{m,j}(\tilde{\lambda}), \mathbf{a})\tilde{\psi}(\mathbf{p}) = e^{i(a_0 \omega_{\mathbf{p}} - \langle \mathbf{a}, \mathbf{p} \rangle)} D_j(\tilde{b}_{\mathbf{p}}^{-1} \tilde{\lambda} \tilde{b}_{\lambda^{-1}\mathbf{p}}) \tilde{\psi}(\lambda^{-1}\mathbf{p}), \tag{7.182}$$

where $a = (a_0, \mathbf{a})$, $\tilde{\lambda} \in SL(2,\mathbb{C})$, and $\lambda \in L_+^\uparrow$ the image of $\tilde{\lambda}$ under the covering projection. We leave the corresponding formulae for the massless case to the reader.

7.7 Quantization and permutation symmetry

Another interesting application of the quantization theory developed in this chapter is to ***indistinghuishable particles***. Since all elementary particles come in families of indistinghuishable sorts (such as electrons, photons, ...), this topic is obviously of fundamental importance to physics. It is also puzzling, since (as we shall see) mathematically one expects more possibilities than those realized in Nature (namely bosons and fermions). This topic is also interesting philosophically, because it appears to be a testing ground for Leibniz's ***Principle of the Identity of Indiscernibles*** (PII), which states that two different objects cannot have exactly the same properties (in other words, two objects that have exactly the same properties must be identical).

After a period of confusion but growing insight, involving some of the greatest physicists such as Planck, Einstein, Ehrenfest, Fermi, and especially Heisenberg, the modern point of view on quantum statistics was introduced by Dirac.

Using modern notation, and abstracting from his specific example (which involved electronic wave-functions), Dirac's argument is as follows. Let H be the Hilbert space of a single quantum system, called a *particle* in what follows. The two-fold tensor product $H^2 \equiv H \otimes H$ then describes two distinguishable copies of this particle. The permutation group $\mathfrak{S}_2$ on two objects, with nontrivial element (12), acts on the state space H^2 by linear extension of $u(12)\psi_1 \otimes \psi_2 = \psi_2 \otimes \psi_1$. Praising Heisenberg's emphasis on defining everything in terms of observable quantities only, Dirac then declares the two particles to be indistinguishable if $u(12)au(12)^* = a$ for any two-particle observable a; by unitarity, this is to say that a commutes with $u(12)$. Dirac notes that such operators map symmetrized vectors (i.e. those $\psi \in H \otimes H$ for which $u(12)\psi = \psi$) into symmetrized vectors, and likewise map anti-symmetrized vectors (i.e. those $\psi \in H \otimes H$ for which $u(12)\psi = -\psi$) into anti-symmetrized vectors, and these are the only possibilities; we would now say that under the action of the $\mathfrak{S}_2$-invariant (bounded) operators one has

$$H^2 \cong H^2_+ \oplus H^2_-; \tag{7.183}$$

$$H^2_+ = \{\psi \in H^2 \mid u(12)\psi = \psi\}; \tag{7.184}$$

$$H^2_- = \{\psi \in H^2 \mid u(12)\psi = -\psi\}. \tag{7.185}$$

Arguing that in order to avoid double counting (in that ψ and $u(12)\psi$ should not both occur as independent states) one has to pick one of these two possibilities, Dirac concludes that state vectors of a system of two indistinguishable particles must be either symmetric or anti-symmetric. He then generalizes this to N identical particles: if (ij) is the element of the permutation group $\mathfrak{S}_N$ on N objects that permutes i and j $(i, j = 1,\ldots,N)$, then according to Dirac, $\psi \in H^N \equiv H^{\otimes N}$ should satisfy either $u(ij)\psi = \psi$, in which case $\psi \in H^2_+$, or $u(ij)\psi = -\psi$, in which case $\psi \in H^2_-$, where u is the natural unitary representation of $\mathfrak{S}_N$ on H^N, given, on $p \in \mathfrak{S}_N$, by linear (and if necessary continuous) extension of

$$u(p)\psi_1 \otimes \cdots \otimes \psi_N = \psi_{p(1)} \otimes \cdots \otimes \psi_{p(N)}. \tag{7.186}$$

Equivalently, $\psi \in H^2_+$ if it is invariant under all permutations, and $\psi \in H^2_-$ if it is invariant under even permutations and picks up a minus sign under odd permutations.

A slightly more sophisticated version of this argument often finds runs as follows:

> 'Since, in the case of indistinguishable particles, $\psi \in H^N$ and $u(p)\psi$ must represent the same state for any $p \in \mathfrak{S}_N$, and since two unit vectors represent the same state iff they differ by a phase vector, by unitarity it must be that $u(p)\psi = c(p)\psi$, for some $c(p) \in \mathbb{C}$ satisfying $|c(p)| = 1$. The group property $u(pp') = u(p)u(p')$ then implies that $c(p) = 1$ for even permutations and $c(p) = \pm 1$ for odd permutations. The choice $+1$ in the latter leads to bosons, whereas -1 leads to fermions, so these are the only possibilities.'

Alas, where Dirac's argument is incomplete, this one is even inconsistent: the claim that two unit vectors represent the same state iff they differ by a phase vector, presumes that the particles are distinguishable! Indeed, the only physical argument to the effect that two unit vectors ψ and ψ' are equivalent iff $\psi' = z\psi$ with $|z| = 1$, is that it guarantees that expectation values coincide, i.e., that

$$\langle \psi, a\psi \rangle = \langle \psi', a\psi' \rangle, \tag{7.187}$$

for *all* (bounded) operators a, i.e., not merely for the permutation-invariant operators (in which case (7.187) does not follow). But, following Heisenberg and Dirac, the whole point of having indistinguishable particles is that an operator a represents a physical observable iff it is invariant under all permutations (acting by conjugation)!

Although the above arguments therefore seem feeble at best, their conclusion that only bosons and fermions can exist seems validated by Nature, despite the mathematical fact that the orthogonal complement of $H^2_+ \oplus H^2_-$ in H_N (describing particles with ***parastatistics***) is non-zero as soon as $N > 2$. This should be a source of concern, and indeed, much research on indistinguishable particles (in $d > 2$) has had the goal of *explaining away parastatistics*. Distinguished by the different actions of $\mathfrak{S}_N$ they depart from, these explanations have traditionally been based on:

- ***Quantum observables.*** $\mathfrak{S}_N$ acts on the C*-algebra $B(H^N)$ of bounded operators on H^N by conjugation of the unitary representation $u(\mathfrak{S}_N)$ on H^N, cf. (7.186). One implements permutation invariance by postulating that the physical observables of the N-particle system under consideration be the $\mathfrak{S}_N$-invariant operators: with u given by (7.186), the algebra of observables is therefore taken to be

$$M_N = B(H^N)^{\mathfrak{S}_N} \equiv \{a \in B(H^N) \mid [a, u(p)] = 0\, (p \in \mathfrak{S}_N)\}. \tag{7.188}$$

- ***Quantum states.*** By restriction, $\mathfrak{S}_N$ then also acts on the (normal) state space

$$\mathscr{S}_n(H^N) \cong \mathscr{D}(H^N) \subset B(H^N), \tag{7.189}$$

  from which it is postulated that the physical state space is $\mathscr{D}(H^N)^{\mathfrak{S}_N}$.
- ***Classical states.*** $\mathfrak{S}_N$ acts on M^N, the N-fold cartesian product of the classical one-particle phase space M, by permutation. If $M = T^*Q$ for some configuration space Q, we might as well start from the natural action of $\mathfrak{S}_N$ on Q^N (pulled back to M^N), and this is indeed what we shall do, often further simplifying to $Q = \mathbb{R}^d$.

Unsurprisingly, the first two approaches equivalent. Define a linear map

$$E_N : B(H^N) \to B(H^N)^{\mathfrak{S}_N}; \quad (7.190)$$

$$a \mapsto \frac{1}{n!} \sum_{p \in \mathfrak{S}_N} u(p)au(p)^*; \quad (7.191)$$

this is a (normal) ***conditional expectation*** from the von Neumann algebra $B(H^N)$ to the von Neumann algebra $B(H^N)^{\mathfrak{S}_N}$, i.e., $E_N(a^*) = E_N(a)^*$ for all $a \in B(H^N)$, $E_N^2 = E_N$, and $\|E_N\| = 1$. Moreover, E_N preserves positivity as well as the trace, so that it also maps the state space $\mathscr{D}(H^N)$ onto the invariant states $\mathscr{D}(H^N) \subset B(H^N)$. Simple computations also establish the properties

$$\mathrm{Tr}(\rho a) = \mathrm{Tr}(E_N(\rho)a) \;\; (\rho \in \mathscr{D}(H^N), a \in B(H^N)^{\mathfrak{S}_N}); \quad (7.192)$$

$$\mathrm{Tr}(\rho a) = \mathrm{Tr}(\rho E_N(a)) \;\; (\rho \in \mathscr{D}(H^N)^{\mathfrak{S}_N}, a \in B(H^N)). \quad (7.193)$$

Finally, the reduction of H^N under $u(\mathfrak{S}_N)$ described below may equally well be described in terms of the state space, since a subspace $eH^N \subset H^N$ (where $e \in \mathscr{P}(H^N)$ is a projection) is stable under u iff $e \in \mathscr{P}(H^N)^{\mathfrak{S}_N}$, in which case it may be described in terms of the associated density operator $\rho = e/\mathrm{Tr}(e) \in \mathscr{D}(H^N)^{\mathfrak{S}_N}$. With some more effort, in can be even be shown that $\rho \in \partial_e(\mathscr{D}(H^N)^{\mathfrak{S}_N})$ iff eH is irreducible.

We may therefore focus on the first and the third approaches, starting with the first, based on (7.188). Note that the C*-algebra of invariant *compact* operators, i.e.,

$$A_N = B_0(H^N)^{\mathfrak{S}_N} = \{a \in B_0(H^N) \mid [a, u(p)] = 0 \, (p \in \mathfrak{S}_N)\}, \quad (7.194)$$

induces the same decomposition of H^N as M_N does (since $M = A_N''$), so if H is infinite-dimensional one may use A_N rather than M_N as the algebra of quantum observables; this is convenient for comparison with the classical state space approach.

As long as $\dim(H) > 1$ and $N > 1$, the algebras M_N and A_N act reducibly on H^N. The reduction of H^N under M_N (and hence of A_N and of $u(H)^N$) is traditionally carried out by ***Schur duality***. This rests on the following concepts.

Definition 7.11. • *A* **partition** λ *of N is a way of writing*

$$N = n_1 + \cdots + n_k, \; n_1 \geq \cdots \geq n_k > 0, \; k = 1, \ldots, N. \quad (7.195)$$

- *The corresponding* **frame** *(or* **Young diagram***)* F_λ *is a picture of N boxes with* n_i *boxes in the i'th row,* $i = 1, \ldots, k$.
- *For each frame* F_λ*, one has* $N!$ *possible* **Young tableaux** T*, each of which is a particular way of writing all of the numbers 1 to N into the boxes of* F_λ.
- *A Young tableau is* **standard** *if the entries in each row increase from left to right and the entries in each column increase from top to bottom. The set of all (standard) Young tableaux on* F_λ *is called* $\mathscr{T}_\lambda$ $(\mathscr{T}_\lambda^S)$.
- *To each* $T \in \mathscr{T}_\lambda$ *we associate the subgroup* $\mathrm{Row}(T) \subset \mathfrak{S}_N$ *of all permutations* $p \in \mathfrak{S}_N$ *that preserve each row (i.e., each row of T is permuted within itself); likewise* $\mathrm{Col}(T) \subset \mathfrak{S}_N$ *consists of all* $p \in \mathfrak{S}_N$ *that preserve each column.*

The set Par(N) of all partitions λ of N parametrizes the conjugacy classes of $\mathfrak{S}_N$ and hence also the (unitary) dual of $\mathfrak{S}_N$; in other words, up to (unitary) equivalence each (unitary) irreducible representation u_λ of $\mathfrak{S}_N$ bijectively corresponds to some partition λ of N; the dimension of any vector space V_λ carrying u_λ is $N_\lambda = |\mathscr{T}_\lambda^S|$, that is, the number of different standard Young tableaux on the frame F_λ.

Returning to (7.186), to each $\lambda \in \text{Par}(N)$ and each Young tableau $T \in \mathscr{T}_\lambda$ we associate an operator e_T on H^N by the formula

$$e_T = \frac{N_\lambda}{N!} \sum_{p \in \text{Col}(T)} \text{sgn}(p)u(p) \sum_{p' \in \text{Row}(T)} u(p'), \tag{7.196}$$

which happens to be a projection. Its image $e_T H^N \subset H^N$ is denoted by H_T^N, and the restriction of M_N to H_T^N is called $M_N(T)$. One may now write the decomposition of H^N under the action of M_N (up to unitary equivalence) as

$$H^N \cong \bigoplus_{\lambda \in \text{Par}(N)} H_{T_\lambda}^N \otimes V_\lambda, \tag{7.197}$$

$$M_N \cong \bigoplus_{\lambda \in \text{Par}(N)} M_N(T_\lambda) \otimes 1_{V_\lambda}, \tag{7.198}$$

$$u(\mathfrak{S}_N) \cong \bigoplus_{\lambda \in \text{Par}(N)} 1_{H_{T_\lambda}^N} \otimes u_\lambda, \tag{7.199}$$

where the labeling is by the partitions λ of N, the multiplicity spaces V_λ are irreducible $\mathfrak{S}_N$-modules, and T_λ is an arbitrary choice of a Young tableau defined on F_λ. For simplicity we here assume that $\dim(H) \geq N$; if $\dim(H) < N$, then only partitions (7.195) with $k \leq \dim(H)$ occur. For example, the partitions (7.195) of $N = 2$ are $2 = 2$ and $2 = 1 + 1$, each of which admits only one standard Young tableau, which we denote by S and A, respectively. With $N_2 = N_{1+1} = 1$ and hence $V_1 \cong V_{1+1} \cong \mathbb{C}$ as vector spaces, this recovers (7.183); the corresponding projections e_+ and e_-, respectively, are given by $e_+ = \frac{1}{2}(1 + u(12))$ and $e_- = \frac{1}{2}(1 - u(12))$. The bosonic states ψ_+, i.e., the solutions of $\psi_+ \in H_+^2$, or $e_+\psi_+ = \psi_+$, are just the symmetric vectors, whereas the fermionic states $\psi_- \in H_-^2$ are the antisymmetric ones. These sectors exist for all $N > 1$ and they always occur with multiplicity one.

However, and this is the bite of the topic, for $N \geq 3$ additional irreducible representations of M_N appear, always with multiplicity greater than one; states in such sectors are said to describe ***paraparticles*** and/or are said to have ***parastatistics***. For example, for $N = 3$ one new partition $3 = 2 + 1$ occurs, with $N_{2+1} = 2$, and hence

$$H^3 \cong H_+^3 \oplus H_-^3 \oplus H_P^3 \oplus H_{P'}^3, \tag{7.200}$$

where H_P^3 and $H_{P'}^3$ are the images of the projections $e_P = \frac{1}{3}(1 - u(13))(1 + u(12))$ and $e_{P'} = \frac{1}{3}(1 - u(12))(1 + u(13))$, respectively. The corresponding two classes of ***parastates*** (i.e. states carrying parastatistics) ψ_P and $\psi_{P'}$ then by definition satisfy $e_P\psi_P = \psi_P$ and $e_{P'}\psi_{P'} = \psi_{P'}$, respectively. In other words, the Hilbert spaces carrying each of the four sectors are the following closed linear spans:

$$H_+^3 = \text{span}^-\{\psi_{123} + \psi_{213} + \psi_{321} + \psi_{312} + \psi_{132} + \psi_{231}\}; \quad (7.201)$$
$$H_-^3 = \text{span}^-\{\psi_{123} - \psi_{213} - \psi_{321} + \psi_{312} - \psi_{132} + \psi_{231}\}; \quad (7.202)$$
$$H_P^3 = \text{span}^-\{\psi_{123} + \psi_{213} - \psi_{321} - \psi_{312}\}; \quad (7.203)$$
$$H_{P'}^3 = \text{span}^-\{\psi_{123} + \psi_{321} - \psi_{213} - \psi_{231}\}, \quad (7.204)$$

where $\psi_{ijk} \equiv \psi_i \otimes \psi_j \otimes \psi_k$ and the ψ_i vary over H (and span$^-$ is closed linear span).

For any $N > 2$, let us note that instead of the decomposition (7.197) - (7.198), which is defined up to unitary equivalence, one may alternatively decompose H^N as

$$H^N = \bigoplus_{T \in \mathscr{T}_\lambda^S, \lambda \in \text{Par}(N)} H_T^N; \quad (7.205)$$

$$M_N = \bigoplus_{T \in \mathscr{T}_\lambda^S, \lambda \in \text{Par}(N)} M_N(T), \quad (7.206)$$

which has the advantage over (7.197) - (7.198) that the H_T^N are subspaces of H^N. The disadvantage is that $M_N(T)$ is unitarily equivalent to $M_N(T')$ iff T and T' both lie in $\mathscr{T}_\lambda^S$ (i.e., for the same λ), so that unlike (7.197) - (7.198), the decomposition (7.205) - (7.206) is non-unique (for example, Young tableaux different from standard ones might have been chosen in the parametrization). The analogue of the third line (7.199) in the earlier decomposition would therefore be a mess. Indeed, although $\mathfrak{S}_N$ maps each of the subspaces H_+ and H_- into itself (the former is even pointwise invariant under $\mathfrak{S}_N$, whereas elements of the latter at most pick up a minus sign), this is no longer the case for parastatistics. For example, for $N = 3$ some permutations map H_P^3 into $H_{P'}^3$, and *vice versa*. This is clear from (7.205) - (7.206): for $\lambda = P$, one has $\dim(V_P) = 2$, and choosing a basis (υ_1, υ_2) of V_P one may identify $H_P^{\otimes 3}$ and $H_{P'}^{\otimes 3}$ in (7.205) with (say) $H_P^{\otimes 3} \otimes \upsilon_1$ and $H_P^{\otimes 3} \otimes \upsilon_2$ in (7.197), respectively. And analogously for $N > 3$, where $\dim(V_\lambda) > 1$ for all $\lambda \neq S, A$.

A (or perhaps *the*) competing approach to permutation invariance in quantum mechanics starts from classical (rather than quantal) data. Let Q be the classical single-particle configuration space, e.g., $Q = \mathbb{R}^d$; to avoid irrelevant complications, we assume that Q is a connected and simply connected manifold. The associated configuration space of N identical but distinguishable particles is Q^N. Depending on the assumption of (in)penetrability of the particles, we may define one of

$$\breve{Q}_N = Q^N / \mathfrak{S}_N; \quad (7.207)$$
$$Q_N = (Q^N \backslash \Delta_N) / \mathfrak{S}_N, \quad (7.208)$$

as the configuration space of N indistinguishable particles, where Δ_N is the extended diagonal in Q^N, i.e., the set of points $(q_1, \ldots, q_N) \in Q^N$ where $q_i = q_j$ for at least one pair (i, j), $i \neq j$ (so that for $Q = \mathbb{R}$ and $N = 2$ this is the usual diagonal in $\mathbb{R}^2$). At first sight, these two choices should lead to exactly the same quantum theory, based on the Hilbert space $L^2(\breve{Q}_N) = L^2(Q_N)$, since Δ_N is a subset of measure zero for any measure used to define L^2 that is locally equivalent to Lebesgue measure.

However, the effect of Δ_N is noticeable as soon as one represents physical observables as operators on L^2 through any serious quantization procedure, which should be sensitive to both the topological and the smooth structure of the underlying configuration space. In the case at hand, Q_N is multiply connected as a topological space, but as a manifold it is smooth and has no singularities. In contrast, $\check{Q}_N$ is simply connected as a topological space, but in the smooth setting it is a so-called *orbifold*. This leads to interesting complications, but following tradition (i.e., in the configuration space approach to indistinguishable particle) we continue with Q_N.

To quantize Q_N we use the language of Lie groupoids and their C*-algebras, cf. §§C.16–C.17. Let Q be any (possibly) multiply connected manifold, with universal covering space $\tilde{Q}$. In particular, the first homotopy group $\pi_1(Q)$ acts (say from the right) on $\tilde{Q}$ in such a way that $Q = \tilde{Q}/\pi_1(Q)$. We denote the canonical projection by $\pi : \tilde{Q} \to Q$. One may have the example $Q = \mathbb{T}$, $\tilde{Q} = \mathbb{R}$, $\pi_1(Q) = \mathbb{Z}$ in mind here.

As a variation on the pair groupoid $G = Q \times Q$, we now consider the Lie groupoid

$$\tilde{G}_Q = \tilde{Q} \times_{\pi_1(Q)} \tilde{Q}, \tag{7.209}$$

whose elements are equivalence classes $[\tilde{q}_1, \tilde{q}_2]$ in $\tilde{Q} \times \tilde{Q}$ under the equivalence relation $\sim$ defined by $(\tilde{q}_1, \tilde{q}_2) \sim (\tilde{q}'_1, \tilde{q}'_2)$ iff $\tilde{q}_1 = \tilde{q}'_1 x$ and $\tilde{q}_2 = \tilde{q}'_2 x$ for some $x \in \pi_1(Q)$; the source and target projections are $s([\tilde{q}_1, \tilde{q}_2]) = \pi(\tilde{q}_2)$ and $t([\tilde{q}_1, \tilde{q}_2]) = \pi(\tilde{q}_1)$, respectively, the inverse is $[\tilde{q}_1, \tilde{q}_2]^{-1} = [\tilde{q}_2, \tilde{q}_1]$, and multiplication is the obvious one borrowed from the pair groupoid $\tilde{Q} \times \tilde{Q}$ over $\tilde{Q}$ (which is well defined on $\tilde{G}_Q$). The tangent groupoid $\tilde{G}_Q^T$ of $\tilde{G}_Q$ (cf. Proposition C.117) has the following fiber at $\hbar = 0$:

$$(\tilde{G}_Q)_0^T = TQ, \tag{7.210}$$

to be contrasted with the corresponding fiber $G_0^T = T\tilde{Q}$ of the pair groupoid on the covering space $\tilde{Q}$. In particular, for our configuration space $Q = Q_N$ we have

$$\tilde{G}_{Q_N} = \tilde{Q}_N \times_{\pi_1(Q_N)} \tilde{Q}_N; \tag{7.211}$$

$$(\tilde{G}_{Q_N})_0^T = TQ_N, \tag{7.212}$$

which gives the fibers of the corresponding continuous bundle of C*-algebras as

$$A_0 = C_0(T^*Q_N) \ (\hbar = 0); \tag{7.213}$$

$$A_\hbar = C^*(\tilde{G}_Q) \quad (0 < \hbar \leq 1), \tag{7.214}$$

cf. §C.19. This gives a generalization of the fibers (7.17) - (7.18) for $Q = \mathbb{R}^n$, and also now we have an example of Definition 7.1: the fibers (7.213) - (7.214) combine to form a continuous bundle of C*-algebras with total C*-algebra $A = C^*(\tilde{G}_Q^T)$, yielding a deformation quantization of the Poisson manifold T^*Q_N (i.e., the usual phase space defined by the configuration space Q_N). We now define the ***inequivalent quantizations*** of Q_N as the inequivalent irreducible representations of the corresponding C*-algebra of quantum observables $C^*(\tilde{G}_{Q_N})$, as follows.

Theorem 7.12. *1. Let Q be multiply connected. The inequivalent irreducible representations π^λ of the C^*-algebra $C^*(\tilde{G}_Q)$ bijectively correspond to the inequivalent irreducible unitary representations u_λ of the first homotopy group $\pi_1(Q)$.*
2. Each representation π^λ has a natural realization on the Hilbert space

$$H^\lambda = L^2(Q) \otimes H_\lambda, \tag{7.215}$$

where H_λ is a specific carrier space for the representation u_λ. More fancifully, one may use the Hilbert space $L^2(Q, E_\lambda)$ of L^2-sections of the vector bundle

$$E_\lambda = \tilde{Q} \times_{\pi_1(Q)} H_\lambda \tag{7.216}$$

associated to the principal bundle $\pi : \tilde{Q} \to Q$ by the representation u_λ.

Provided one accepts (7.208), this theorem in principle gives a complete solution to the problem of quantizing multiply connected configuration spaces, and hence, taking $Q = Q_N$, of the problem of quantizing systems of indistinguishable particles.

Proof. We just prove Theorem 7.12 in the case we need, where $\pi_1(Q)$ is finite. Then

$$C^*(\tilde{Q} \times_{\pi_1(Q)} \tilde{Q}) \cong B_0(L^2(\tilde{Q}))^{\pi_1(Q)}; \tag{7.217}$$

$$B_0(L^2(\tilde{Q}))^{\pi_1(Q)} \cong B_0(L^2(Q)) \otimes C^*(\pi_1(Q)), \tag{7.218}$$

where (in our usual notation) $B_0(L^2(\tilde{Q}))^{\pi_1(Q)}$ is the C^*-algebra of $\pi_1(Q)$-invariant compact operators on $L^2(\tilde{Q})$, and $C^*(\pi_1(Q))$ is the group C^*-algebra of $\pi_1(Q)$ (which is finite-dimensional and hence nuclear, given the assumption that $\pi_1(Q)$ is finite, so that the choice of the C^*-algebraic tensor product does not matter).

To prove (7.217), we first exploit finiteness of $\pi_1(Q)$ in order to identify functions $\tilde{a} \in C_c^\infty(\tilde{G}_Q)$ with constrained C_c^∞ functions a on $\tilde{Q} \times \tilde{Q}$ that satisfy

$$a(\tilde{q}h, \tilde{q}'h) = a(\tilde{q}, \tilde{q}') \;\; (h \in \pi_1(Q)). \tag{7.219}$$

This identification is explicitly given by

$$a(\tilde{q}, \tilde{q}') = \tilde{a}([\tilde{q}, \tilde{q}']), \tag{7.220}$$

where $[\tilde{q}, \tilde{q}']$ denotes the equivalence class of $(\tilde{q}, \tilde{q}') \in \tilde{Q} \times \tilde{Q}$ under the diagonal action of $\pi_1(Q)$. This makes the space $C_c^\infty(\tilde{G}_Q)$ a dense subset of $C^*(\tilde{G}_Q)$. We write $a \in C_c^\infty(\tilde{Q} \times \tilde{Q})^{\pi_1(Q)}$; for (7.208) this just means that a is a permutation-invariant kernel. Second, we equip $\tilde{Q}$ with some measure $d\tilde{q}$ that is locally equivalent to the Lebesgue measure, and in addition is $\pi_1(Q)$-invariant under the regular action R of $\pi_1(Q)$ on functions on $\tilde{Q}$, given, as usual, by $R_h \tilde{\psi}(\tilde{q}) = \tilde{\psi}(\tilde{q}h)$. In that case, one also has a measure dq on Q that is locally equivalent to the Lebesgue measure, so that the measures $d\tilde{q}$ and dq on $\tilde{Q}$ and Q, respectively, are related by

$$\int_{\tilde{Q}} d\tilde{q}\, f(\tilde{q}) = \frac{1}{|\pi_1(Q)|} \sum_{h \in \pi_1(Q)} \int_Q dq\, f(s(q)h). \tag{7.221}$$

Here $f \in C_c(\tilde{Q})$, $|\pi_1(Q)|$ is the number of elements of $\pi_1(Q)$, and $s : Q \to \tilde{Q}$ is any (measurable) cross-section of $\tau : \tilde{Q} \to Q$. We may then define a Hilbert space $L^2(\tilde{Q})$ with respect to $d\tilde{q}$, on which elements a of $C_c^\infty(\tilde{Q} \times \tilde{Q})^{\pi_1(Q)}$ act faithfully by

$$a\tilde{\psi}(\tilde{q}) = \int_{\tilde{Q}} d\tilde{q}'\, a(\tilde{q},\tilde{q}')\tilde{\psi}(\tilde{q}'). \tag{7.222}$$

The product of two such operators is given by the multiplication of the kernels on $\tilde{Q}$, and involution is defined as expected, too, namely by hermitian conjugation:

$$a^*(\tilde{q},\tilde{q}') = \overline{a(\tilde{q}',\tilde{q})}. \tag{7.223}$$

The norm-closure of $C_c^\infty(\tilde{Q} \times \tilde{Q})^{\pi_1(Q)}$, represented as operators on $L^2(\tilde{Q})$ by (7.222), is then given by $B_0(L^2(\tilde{Q}))^{\pi_1(Q)}$. This proves (7.217).

Eq. (7.218) is a special case of the following: let X be a manifold carrying a *free* action of a *compact* group G. If $L^2(X)$ is defined by some G-invariant "locally Lebesgue" measure on X, as in the construction above, then one has an isomorphism

$$B_0(L^2(X))^G \cong B_0(L^2(X/G) \otimes C^*(G). \tag{7.224}$$

This is proved in a similar way, realizing $B_0(H)$ as the norm-completion of the Hilbert–Schmidt operators $B_2(H)$ (for general H), and, in the L^2-case at hand, identifying $B_2(L^2(X))$ with the algebra of operators with kernels in $L^2(X \times X)$.

Part 2 of the theorem now follows from the fact that for any Hilbert space H the C*-algebra $B_0(H)$ of compact operators on H has exactly one irreducible representation (up to unitary equivalence), i.e. the defining one (this can be proved in many ways, e.g. from Rieffel's theory of Morita equivalence of C*-algebras), combined with the bijective correspondence between continuous unitary representations u of any locally compact group G and non-degenerate representations of its associated group C*-algebra $C^*(G)$; see §C.18, Definition C.119 etc. □

As mentioned in Theorem 7.12, there are two ways of realizing the Hilbert space H^λ, where λ labels some irreducible representation of $\pi_1(Q)$. This is very similar to the discussion in §7.5, so we will be relatively brief here. The first realization corresponds to having constrained wave-functions defined on the covering space $\tilde{Q}$; for example, the usual description of bosonic or fermonic wave-functions is of this sort. The second realization uses unconstrained wave-functions on the actual configuration space Q (*bad hombres* confusingly call such functions "multi-valued").

1. The space $C^\infty(Q, E_\lambda)$ of *smooth* cross-sections of E_λ may be given by the smooth maps $\tilde{\psi} : \tilde{Q} \to H_\lambda$ satisfying the equivariance condition ("constraint")

$$\tilde{\psi}(\tilde{q}h) = u_\lambda(h^{-1})\tilde{\psi}(\tilde{q}), \tag{7.225}$$

for all $h \in \pi_1(Q)$, $\tilde{q} \in \tilde{Q}$. The Hilbert space

$$H^\lambda = L^2(\tilde{Q}, H_\lambda)^{\pi_1(Q)}, \tag{7.226}$$

then, is defined as the usual L^2-completion of the space of all $\tilde{\psi} \in \Gamma(Q, E_\lambda)$ for which $\langle\tilde{\psi}, \tilde{\psi}\rangle < \infty$. The irreducible representation $\pi^\lambda(C^*(G_Q))$ is then given on elements $\tilde{a}$ of the dense subspace $C_c^\infty(G_Q)$ of $C^*(G_Q)$ by the expression

$$\pi^\lambda(\tilde{a})\psi(\tilde{q}) = \int_{\tilde{Q}} d\tilde{q}'\, \tilde{a}([\tilde{q}, \tilde{q}'])\psi(\tilde{q}'); \tag{7.227}$$

any $\pi_1(Q)$-invariant operator on $L^2(\tilde{Q})$ acts on H^λ in this way (ignoring H_λ). If $\pi_1(Q)$ is finite, then two simplifications occur. Firstly, H_λ is finite-dimensional, and secondly each Hilbert space H^λ may be regarded as a subspace of $L^2(\tilde{Q})$; the above action of $C^*(G_Q)$ on H^λ is then simply given by restriction of its action on $L^2(\tilde{Q})$. In that case one may equivalently realize this irreducible representation in terms of the right-hand side of (7.217), in which case the action of $\pi^\lambda(a)$ on H^λ as defined in (7.226) is given by

$$\pi^\lambda(a)\psi(\tilde{q}) = \int_{\tilde{Q}} d\tilde{q}'\, a(\tilde{q}, \tilde{q}')\psi(\tilde{q}'). \tag{7.228}$$

This is true as it stands if $a \in C_c^\infty(\tilde{Q} \times \tilde{Q})^{\pi_1(Q)}$, cf. (7.219), and may be extended to general $\pi_1(Q)$-invariant compact operators $a \in B_0(L^2(\tilde{Q}))^{\pi_1(Q)}$ by norm continuity, and, furthermore, even to $B(L^2(\tilde{Q}))^{\pi_1(Q)}$ by strong or weak continuity.

2. Elements of the Hilbert space $L^2(\tilde{Q}, H_\lambda)^{\pi_1(Q)}$ are typically (equivalence classes of) *discontinuous* cross-sections of E_λ. Possibly discontinuous cross-sections may simply be given directly as functions $\psi : Q \to H_\lambda$, with inner product

$$\langle\psi, \varphi\rangle = \int_Q dq\, \langle\psi(q), \varphi(q)\rangle_{H_\lambda}. \tag{7.229}$$

This specific realization of $L^2(Q, E_\lambda)$ will be denoted by $L^2(Q) \otimes H_\lambda$. If $H_\lambda = \mathbb{C}$,

$$L^2(Q) \otimes H_\lambda \cong L^2(Q). \tag{7.230}$$

These equivalent descriptions of π^λ may be related once a (typically discontinuous) cross-section $\sigma : Q \to \tilde{Q}$ of the projection $\tau : \tilde{Q} \to Q$ has been chosen (i.e., $\tau \circ \sigma = \mathrm{id}_Q$), in which case $\psi(q) = \tilde{\psi}(\sigma(q))$. We formalize this in terms of a unitary

$$u : L^2(\tilde{Q}, H_\lambda)^{\pi_1(Q)} \to L^2(Q) \otimes H_\lambda \tag{7.231}$$

$$u\tilde{\psi}(q) = \tilde{\psi}(\sigma(q)); \tag{7.232}$$

$$u^{-1}\psi(\tilde{q}) = u_\lambda(h)\psi(q), \tag{7.233}$$

where $q = \tau(\tilde{q})$, and h is the unique element of $\pi_1(Q)$ for which $\tilde{q}h = \sigma(q)$. The action $\pi_\sigma^\lambda(a) = u\pi^\lambda(a)u^{-1}$ on $L^2(Q) \otimes H_\lambda$ now follows from (7.228) - (7.233): If a is a $\pi_1(Q)$-invariant kernel on $L^2(\tilde{Q})$, then using (7.221) we obtain

$$\pi_\sigma^\lambda(a)\psi(q) = \sum_{h \in \pi_1(Q)} \int_Q dq'\, a(\sigma(q), \sigma(q')h)u_\lambda(h)\psi(q'). \tag{7.234}$$

We now apply this formalism to N indistinguishable particles moving on the (single-particle) configuration space $\mathbb{R}^3$. Eq. (7.208) then gives the N-particle space

$$Q_N = ((\mathbb{R}^3)^N - \Delta_N)/\mathfrak{S}_N. \tag{7.235}$$

The universal covering space of this multiply connected space is

$$\tilde{Q}_N = \mathring{\mathbb{R}}^{3N} \equiv (\mathbb{R}^3)^N - \Delta_N, \tag{7.236}$$

which (unlike its counterpart in $d = 2$) is connected and simply connected, so that

$$\pi_1(Q_N) = \mathfrak{S}_N. \tag{7.237}$$

It follows from (7.217) and (7.237) that the algebra of observables is given by

$$C^*(\tilde{G}_{Q_N}) = B_0(L^2(\mathbb{R}^3)^{\otimes N})^{\mathfrak{S}_N}. \tag{7.238}$$

Comparing (7.238) with (7.194), we obtain a complete equivalence between the "quantum observables" approach and the deformation quantization approach based on Theorem 7.12, in that the configuration space approach through the representation theory of the groupoid C*-algebra $C^*(\tilde{G}_{Q_N})$ leads to the same classification as the "quantum observables" approach based in (7.188) above, cf. (7.197) - (7.199).

We discuss a few interesting special cases.

N = 1. Here $\tilde{Q}_1 = Q_1 = \mathbb{R}^3$ and $\pi_1(Q_1) = \{e\}$, so the algebra of observables is

$$C^*(\tilde{G}_{Q_1}) = B_0(L^2(\mathbb{R}^3)), \tag{7.239}$$

which has a unique irreducible representation on $L^2(\mathbb{R}^3)$.

N = 2. This time, the pertinent homotopy group is

$$\pi_1(Q_2) = \mathfrak{S}_2 = \mathbb{Z}_2 = \{e, (12)\}, \tag{7.240}$$

which has two irreducible representations: firstly, $u_B(p) = 1$ for both $p \in \mathfrak{S}_2$, and secondly, $u_F(e) = 1$, $u_F(12) = -1$, each realized on $H_\lambda = \mathbb{C}$. Hence with $q = (x, y, z) \in \mathbb{R}^3$, eq. (7.225) yields

$$H_B^2 = \{\psi \in L^2(\mathbb{R}^3)^2 \mid \psi(q_2, q_1) = \psi(q_1, q_2)\}; \tag{7.241}$$
$$H_F^2 = \{\psi \in L^2(\mathbb{R}^3)^2 \mid \psi(q_2, q_1) = -\psi(q_1, q_2)\}. \tag{7.242}$$

Here $L^2(\mathbb{R}^3)^2 \equiv L^2(\mathbb{R}^3) \otimes L^2(\mathbb{R}^3) \cong L^2(\mathbb{R}^6)$. The C*-algebra

$$C^*(\tilde{G}_{Q_2}) = B_0(L^2(\mathbb{R}^3) \otimes L^2(\mathbb{R}^3))^{\mathfrak{S}_2} \cong B_0(L^2(\mathbb{R}^3 \times \mathbb{R}^3))^{\mathfrak{S}_2} \tag{7.243}$$

consists of all $\mathfrak{S}_2$-invariant compact operators on $L^2(\mathbb{R}^3 \times \mathbb{R}^3)$, acting on H_B^2 or H_F^2 in the same way as they do on $L^2(\mathbb{R}^6)$; cf. (7.228), noting that the constraints in (7.241) and (7.242) are preserved due to the $\mathfrak{S}_2$-invariance of $A \in C^*(\tilde{G}_{Q_2})$. This recovers Dirac's description of statistics given earlier in this section.

N = 3. Here we have a non-abelian homotopy group

$$\pi_1(Q_3) = \mathfrak{S}_3, \tag{7.244}$$

which, besides the irreducible boson and fermion representations on $\mathbb{C}$, has an irreducible ***parafermionic*** representation u_P on $H_P = \mathbb{C}^2$. This representation is most easily obtained explicitly by reducing the natural action of $\mathfrak{S}_3$ on $\mathbb{C}^3$. Define an orthonormal basis of the latter by

$$e_0 = \frac{1}{\sqrt{3}}\begin{pmatrix}1\\1\\1\end{pmatrix}; \; e_1 = \frac{1}{\sqrt{2}}\begin{pmatrix}0\\1\\-1\end{pmatrix}; \; e_2 = \frac{1}{\sqrt{6}}\begin{pmatrix}-2\\1\\1\end{pmatrix}. \tag{7.245}$$

It follows that $\mathbb{C}\cdot e_0$ carries the trivial representation of $\mathfrak{S}_3$, whereas the linear span of e_1 and e_2 carries a two-dimensional irreducible representation u_P, given on the generators (12), (13), and (23) of $\mathfrak{S}_3$ by

$$u_P(12) = \tfrac{1}{2}\begin{pmatrix}1 & -\sqrt{3}\\ -\sqrt{3} & -1\end{pmatrix}; \; u_P(13) = \tfrac{1}{2}\begin{pmatrix}1 & \sqrt{3}\\ \sqrt{3} & -1\end{pmatrix}; \; u_P(23) = \begin{pmatrix}-1 & 0\\ 0 & 1\end{pmatrix}. \tag{7.246}$$

We already gave realizations of the Hilbert space H_P^3 of three parafermions in (7.203) and (7.204),where it emerged as a subspace of $L^2(\mathbb{R}^3)\otimes L^2(\mathbb{R}^3)\otimes L^2(\mathbb{R}^3) \cong L^2(\mathbb{R}^3\times\mathbb{R}^3\times\mathbb{R}^3)$. An equivalent realization $H^P \equiv \tilde{H}_P^3$ may be given on the basis of (7.225), according to which H^P is the subspace of $L^2(\mathbb{R}^3)^3\otimes\mathbb{C}^2 \cong L^2(\mathbb{R}^9)\otimes\mathbb{C}^2$ that consists of doublet wave-functions ψ_i $(i = 1,2)$ that satisfy

$$\psi_i(q_{p(1)}, q_{p(2)}, q_{p(3)}) = \sum_{j=1}^{2} u_{ij}(p)\psi_j(q_1,q_2,q_3), \tag{7.247}$$

for any permutation $p \in \mathfrak{S}_3$, where $u \equiv u_P$, cf. (7.246). I.e., the parafermionic wave-functions in this realization of H_P^3 are constrained by the conditions

$$\psi_1(q_2,q_1,q_3) = \tfrac{1}{2}\psi_1(q_1,q_2,q_3) - \tfrac{1}{2}\sqrt{3}\,\psi_2(q_1,q_2,q_3); \tag{7.248}$$
$$\psi_2(q_2,q_1,q_3) = -\tfrac{1}{2}\sqrt{3}\,\psi_1(q_1,q_2,q_3) - \tfrac{1}{2}\psi_2(q_1,q_2,q_3); \tag{7.249}$$
$$\psi_1(q_3,q_2,q_1) = \tfrac{1}{2}\psi_1(q_1,q_2,q_3) + \tfrac{1}{2}\sqrt{3}\,\psi_2(q_1,q_2,q_3); \tag{7.250}$$
$$\psi_2(q_3,q_2,q_1) = \tfrac{1}{2}\sqrt{3}\,\psi_1(q_1,q_2,q_3) - \tfrac{1}{2}\psi_2(q_1,q_2,q_3); \tag{7.251}$$
$$\psi_1(q_1,q_3,q_2) = -\psi_1(q_1,q_2,q_3); \tag{7.252}$$
$$\psi_2(q_1,q_3,q_2) = \psi_2(q_1,q_2,q_3). \tag{7.253}$$

The algebra of observables $C^*(\tilde{G}_{Q_3})$ of three indistinguishable particles without internal degrees of freedom, i.e., then acts on $H^P \subset L^2(\mathbb{R}^3)^3\otimes\mathbb{C}^2$ as in (7.234), identifying $a \in C^*(\tilde{G}_{Q_3})$ with $a\otimes 1_2$ (so that a ignores the internal degree of freedom $\mathbb{C}^2$). This representation π^P is irreducible by Theorem 7.12.

N > 3. The above construction may be generalized to any $N > 3$. There will now be many parafermionic representations u_λ of $\mathfrak{S}_N$ (given by Young tableaus), each of which induces an irreducible representation of the C*-algebra (7.238).

The question now arises whether parastatistics is to be found in Nature—or, indeed, if this question is even well defined! As a warm-up to the case $N = 3$, where the question first plays a role, let us give an alternative realization of $\pi^F(C^*(\tilde{G}_{Q_2}))$, cf. Theorem 7.12. Take two isospin doublet bosons (which by definition transform under the defining spin-$\frac{1}{2}$ representation $D_{1/2}$ of $SU(2)$ on $\mathbb{C}^2$). With

$$H^{(2)} = (L^2(\mathbb{R}^3) \otimes \mathbb{C}^2)^{\otimes 2}, \tag{7.254}$$

and using indices $a_1, a_2 = 1, 2$, the Hilbert space of these bosons is

$$H_B^{(2)} = \{\psi \in H^{(2)} \mid (\psi_{a_2 a_1}(q_2, q_1) = \psi_{a_1 a_2}(q_1, q_2)\}, \tag{7.255}$$

with corresponding projection $e_B^{(2)} : H^{(2)} \to H_B^{(2)}$ given by

$$e_B^{(2)} \psi_{a_1 a_2}(q_1, q_2) = \tfrac{1}{2}(\psi_{a_2 a_1}(q_2, q_1) + \psi_{a_1 a_2}(q_1, q_2)). \tag{7.256}$$

Subsequently, define a partial isometry $w : H^{(2)} \to L^2(\mathbb{R}^3)^{\otimes 2}$ by

$$w\psi(q_1, q_2) \equiv \psi_0(q_1, q_2) = \frac{1}{\sqrt{2}}(\psi_{12}(q_1, q_2) - \psi_{21}(q_1, q_2)). \tag{7.257}$$

Physically, this singles out an isospin singlet Hilbert subspace $H^{(0)} = e_0 H^{(2)}$ within $H^{(2)}$, where $e_0 = w^* w$ (which is a projection). This singlet subspace may be constrained to the bosonic sector by passing to

$$H_B^{(0)} = e_0 e_B^{(2)} H^{(2)}; \tag{7.258}$$

note that e_0 and $e_B^{(2)}$ commute. Now extend the defining representation of $C^*(\tilde{G}_{Q_2})$ on $L^2(\mathbb{R}^3)^{\otimes 2}$ to $H^{(2)}$ by ignoring the indices a_1, a_2 (i.e., isospin is deemed unobservable). This extended representation commutes with e_0 and with $e_B^{(2)}$, and hence is well defined on $H_B^{(0)} \subset H^{(2)}$. Let us denote this representation of $\tilde{G}_{Q_2}$ by $\pi_B^{(0)}$. It is then immediate from the property $\psi_0(q_2, q_1) = -\psi_0(q_1, q_2)$ that:

Proposition 7.13. *The representations $\pi_B^{(0)}(C^*(\tilde{G}_{Q_2}))$ on $H_B^{(0)}$ and $\pi^F(C^*(\tilde{G}_{Q_2}))$ on H^F are unitarily equivalent.*

In other words, two *fermions* without internal degrees of freedom are equivalent to the singlet state of two *bosons* with an isospin degrees of freedom, at least if the observables are isospin-blind. Similarly, two *bosons* without internal degrees of freedom are equivalent to the singlet state of two *fermions* with isospin, and two fermions without internal degrees of freedom are equivalent to the isospin triplet state of two fermions (this corresponds to the Schur decomposition of $(\mathbb{C}^2)^{\otimes 2}$ under the commuting actions of $\mathfrak{S}_2$ and $SU(2)$).

For $N = 3$ we may carry out a similar trick as for $N = 2$, and replace parafermions without (further) degrees of freedom by either bosons or fermions. We discuss the former and leave the explicit description of the various alternative descriptions to the reader. We proceed as for $N = 2$, *mutatis mutandis*. We have a Hilbert space

$$H^{(3)} = (L^2(\mathbb{R}^3) \otimes \mathbb{C}^2)^{\otimes 3}, \tag{7.259}$$

of three distinguishable isospin doublets, containing the Hilbert space $H_B^{(3)}$ of three bosonic isospin doublets as a subspace, that is,

$$H_B^{(3)} = \{\psi \in H^{(3)} \mid \psi_{a_{p(1)}a_{p(2)}a_{p(3)}}(q_{p(1)}, q_{p(2)}, q_{p(3)}) = \psi_{a_1a_2a_3}(q_1, q_2, q_3)\,(p \in \mathfrak{S}_3)\}. \tag{7.260}$$

The corresponding projection, denoted by $e_B^{(3)} : H^{(3)} \to H_B^{(3)}$, will not be written down explicitly. Define an $SU(2)$ doublet (ψ_1, ψ_2) within the space $H^{(3)}$ through a partial isometry

$$w : H^{(3)} \to L^2(\mathbb{R}^3)^{\otimes 3} \otimes \mathbb{C}^2; \tag{7.261}$$

$$w\psi_1(q_1, q_2, q_3) = \frac{1}{\sqrt{2}}(\psi_{121}(q_1, q_2, q_3) - \psi_{112}(q_1, q_2, q_3)); \tag{7.262}$$

$$w\psi_2(q_1, q_2, q_3) = \frac{1}{\sqrt{6}}(-2\psi_{211}(q_1, q_2, q_3) + \psi_{121}(q_1, q_2, q_3) + \psi_{112}(q_1, q_2, q_3)). \tag{7.263}$$

Defining a projection $e_2 = w^* w$ on $H^{(3)}$, the Hilbert space $H^{(3)}$ contains a closed subspace $H_B^{(2)} = e_2 e_B^{(3)} H^{(3)}$, which is stable under the natural representation of $C^*(\tilde{G}_{Q_3})$ (since e_2 and $e_B^{(3)}$ commute). We call this representation $\pi_B^{(2)}$. An easy calculation then establishes:

Proposition 7.14. *The representations $\pi_B^{(2)}(C^*(\tilde{G}_{Q_3}))$ on $H_B^{(2)}$ and $\pi^P(C^*(\tilde{G}_{Q_3}))$ on H^P (as defined by Theorem 7.12) are unitarily equivalent.*

In other words, three *parafermions* without internal degrees of freedom are quivalent to an isospin doublet formed by three identical *bosonic* isospin doublets (corresponding to the Schur decomposition of $(\mathbb{C}^2)^{\otimes 3}$ under the commuting actions of $\mathfrak{S}_3$ and $SU(2)$; in this decomposition, the spin 3/2 representation of $SU(2)$ couples to the bosonic representation of $\mathfrak{S}_3$, whilst the spin-$\frac{1}{2}$ representation of $SU(2)$ couples to the parafermionic representation of $\mathfrak{S}_3$), at least if the observables of the latter are isospin-blind. Many other realizations of parafermions in terms of fermions or bosons with an internal degree of freedom can be constructed in a similar way.

For $N > 3$ we similarly find that the representation of the C*-algebra (7.238) induced by some parafermionic representations u_χ of $\mathfrak{S}_N$ is unitarily equivalent to a representation on some $SU(n)$ multiplet of bosons with an internal degree of freedom; the appropriate multiplet is the one coupled to u_χ in the Schur reduction of $(\mathbb{C}^n)^{\otimes N}$ with respect to the natural and commuting actions of $\mathfrak{S}_N$ and $SU(n)$.

The moral of this story is that one cannot tell from glancing at some Hilbert space whether the world consists of fermions or bosons or parafermions; what matters is the Hilbert space *as a carrier of some (irreducible) representation of the algebra of observables*. From that perspective we already see for $N = 2$ that being bosonic or fermionic is not an invariant property of such representations, since one may freely choose between fermions/bosons *without* internal degrees of freedom and bosons/fermions *with* internal degrees of freedom. In a more systematic discussion using superselection theory one may impose some physical selection criterion in order to restrict attention to "physically interesting" sectors. Such criteria (which, for example, would have the goal of excluding parastatistics) should be formulated with reference to some algebra of observables. Such issues cannot be settled at the level of quantum mechanics and instead require quantum field theory, where parastatistics can always be removed in terms of either bose- or fermi-statistics, in somewhat similar vein to our discussion. For (nonlocal) charges in gauge theories there are no rigorous results, but historically a similar goal played a role in the road to quantum chromodynamics (QCD), which is one of the ingredients of the Standard Model.

A different argument against parastatistics arises from the state space approach based on the compact convex set $\mathscr{D}(H^N)^{\mathfrak{S}_N}$ studied at the beginning of this section. The extreme boundary $\partial_e\left(\mathscr{D}(H^N)^{\mathfrak{S}_N}\right)$ consists of one part that is contained in $\partial_e\mathscr{D}(H^N) = \mathscr{P}_1(H^N)$, and one that is not. The first part consists of those one-dimensional invariant projections $e \in \mathscr{P}_1(H^N)^{\mathfrak{S}_N}$ whose image eH^N belongs to either the bosonic subspace H^N_+ (in which case $u(p)e = e$ for each $p \in \mathfrak{S}_N$) or the fermionic subspace H^N_- of H^N (in which case $u(p)e = \text{sgn}(p)e$ for each $p \in \mathfrak{S}_N$); in other words, pure bosonic on fermionic states on $B(H^N)^{\mathfrak{S}_N}$ are also pure on $B(H^N)$. The second part, then, consists of parastatistical pure states on $B(H^N)^{\mathfrak{S}_N}$, which are therefore mixed on $B(H^N)$. Furthermore, pure bosonic or fermionic states on $B(H^N)^{\mathfrak{S}_N}$ both extend and restrict to pure bosonic or fermionic states on $B(H^{N+1})^{\mathfrak{S}_{N+1}}$ and $B(H^{N-1})^{\mathfrak{S}_{N-1}}$, respectively, whereas parastatistical pure states turn out to have neither property and hence are "isolated" at the given value of N.

Finally, in $d = 2$ the equivalence between the operator and configuration space approaches breaks down, because $\mathfrak{S}_N \neq \pi_1(Q_N) = B_N$, i.e., the braid group on N strings. Even defining the operator quantum theory on $H_N = L^2(\tilde{Q}_N)$, with algebra of observables $M_N = B(L^2(\tilde{Q}_N))^{B_N}$, fails to rescue the equivalence, because the decomposition of H_N under M_N by no means contains all irreducible representations of B_N. In this case deformation quantization gives many more sectors than the improved operator approach (which already gave more sectors than the approach using 'multi-valued' scalar wave-functions).

Notes

The quotations in the preamble are from Dirac (1947), p. 87. Similarly, the *Dreimännerarbeit* (Born, Heisenberg, & Jordan, 1926) bluntly states (in Ch. 1, §1) that:

> 'one can see from eq. (5) [i.e., $pq - qp = -i\hbar \cdot 1_H$, cf. our eq. (7.5)] that in the limit $\hbar = 0$, the new theory would converge to the classical theory, as is physically required.'

§7.1. **Deformation quantization**

In the wake of Dirac's famous insight on the analogy between the Poisson bracket and the commutator in quantum mechanics, the idea of deformation quantization (in the form of what we now call *star products*) may be traced back to Groenewold (1946) and Moyal (1949). The mathematical (physics) literature on the subject started with Berezin (1975) and Bayen et al (1978), who introduced what we now call ***formal deformation quantization***, in which $\hbar$ is not a real number but a formal parameter occurring in formal power series. The C*-algebraic setting for deformation quantization we use was introduced by Rieffel (1989, 1994); see also Landsman (1998a), Chapter 2, for a detailed treatment.

§7.2. **Quantization and internal symmetry**

This section is based on Rieffel (1990) and Landsman (1998a), Chapter 3.

§7.3. **Quantization and external symmetry**

§7.4. **Intermezzo: The Big Picture**

§7.5. **Induced representations and the imprimitivity theorem**

§7.6. **Representations of semi-direct products**

The action Poisson bracket (7.58) was introduced by Krishnaprasad & Marsden (1987); see also Marsden & Ratiu (1994).

Systems of imprimitivity and their applications to representation theory, semi-direct products, and quantum mechanics are due to Mackey (1958, 1968), who was inspired by Weyl (1927, 1928), von Neumann (1932), and Wigner (1939). As Mackey (1978, 1992) describes, he saw his work as the development of what he calls ***Weyl's Program***. Weyl (1927) posed two questions in quantum mechanics:

1. 'How to construct the matrix of Hermitian form[1] that represents some quantity given in the context of a known physical system?'[2]
2. 'Given this Hermitian form, what is their physical meaning, and which physical statements can we make about it?'[3]

Weyl considered the second question to have been resolved by von Neumann's recent work, and so he concentrated on the first, which he tried to answer using group theory. The main achievement of Weyl (1927), elaborated in his subsequent

[1] Like Hilbert himself, Weyl at the time still thought of operators in terms of matrices or Hermitian forms, rather than abstractly, like von Neumann. Also cf. our Introduction.

[2] 'Wie komme ich zu der Matrix, der Hermiteschen Form, welche eine gegebene Größe in einem seiner Konstitution nach bekannten physikalischen System repräsentiert?' (Weyl, 1927, p. 1)

[3] 'Wenn einmal die Hermitesche Form gewonnen ist, was ist ihre physikalische Bedeutung, was für physikalische Aussagen kann ich ihr entnehmen?' (*ibid.*)

book Weyl (1928), was a reformulation of the canonical commutation relations $i[p,q] = \hbar \cdot 1_H$ in terms of projective unitary representations of the additive group $\mathbb{R}^2$ (or, equivalently, of unitary representations of the associated Heisenberg group). He also introduced the formula (7.21) in an equivalent form where the (classical) Fourier expansion of f, i.e.,

$$f(p,q) = \int_{\mathbb{R}^2} da\,db\, e^{iap+ibq} \hat{f}(a,b), \tag{7.264}$$

is "quantized" by the operator in which $\exp(iap + ibq)$ in the above formula is replaced by the (projective) unitary representative $u(a,b)$ of $(a,b) \in \mathbb{R}^2$ just mentioned, i.e., the real numbers p and q are replaced by the corresponding operators $\hat{p}$ and $\hat{q}$, as in (7.3) - (7.4). In particular, Weyl treated p and q symmetrically.

In his development of Weyl's Program, Mackey broke the symmetry between p and q, in that he saw the momentum operator $\hat{p}$ as the ("infinitesimal") generator of a unitary representation of the additive group $\mathbb{R}$, whereas the position operator $\hat{q}$ was replaced by a projection-valued measure on the real line; this is equivalent to a nondegenerate representation of the commutative C*-algebra $C_0(Q)$, as in our discussion in §7.3. This way of tearing p and q apart was the key to the general case of quantizing group actions on configuration space discussed in §7.3.

In their independent elaboration of Weyl's ideas, Groenewold (1946) and Moyal (1949) emphasized the deformation aspect of quantization (including the classical limit) rather than its group-theoretical underpinning; the former aspect is completely absent in Mackey's work. "The Big Picture" (Landsman, 1998a, Ch. 3; Landsman & Ramazan, 2001; Landsman, 2007) is an attempt to have the best of both worlds, in that the role of Lie groupoids delivers the symmetry aspect of quantization, whereas our (i.e. Rieffel's) very definition of quantization puts the deformation aspect in the front seat. The underlying theory of Lie groupoids and Lie algebroids may be found in Moerdijk & Mrčun (2003) or Mackenzie (2005); see also Landsman (1998a).

A comprehensive study of the Mackey–Glimm dichotomy may be found in Williams (2007), which contains a wealth of information on crossed product C*-algebras and induced representations in general.

The representation theory of the Poincaré-group was first studied (using somewhat heuristic methods) by Wigner (1939) using induced representations. The entire subject was subsequently taken up and finished by Mackey. For treatments in the spirit of (mathematical) physics see e.g. Simms (1968), Niederer & O'Raifeartaigh (1974), and Barut & Raçka (1977). Lemma 7.10 is proved by Bargmann (1954).

Among the known elementary particles, the case $j = 0$ (and $m > 0$) corresponds to the Higgs boson, whereas $j = \frac{1}{2}$ gives all known fermionic particles (i.e., electrons, quarks, neutrino's, and their antiparticles). If one counts the gauge bososn $W_\pm$ and Z_0 as massive, they provide the case $j = 1$, but in the fundamental Lagrangian they are massless and correspond to helicity $n = \pm 1$, like the photon. Helicity ± 2 gives the graviton. We discard particles predicted by supersymmetry, which evidently does not exist in nature (this evidence seems lost on string theorists).

§7.7. **Quantization and permutation symmetry**

This section is based on Landsman (2016a). The literature on indistinguishable particles is enormous, initiated by Heisenberg (1926) and Dirac (1926). What we call the "quantum observables" approach goes back to Messiah & Greenberg (1964); see also Drühl, Haag, & Roberts (1970). Key papers in the configuration space approach are Souriau (1967), Laidlaw & DeWitt-Morette (1971) and Leinaas & Myrheim (1977). More generally, for the quantization of multiply connected space see Dowker (1972), Schulman (1981), Isham (1984), Horvathy, Morandi, & Sudarshan (1989), Morchio & Strocchi (2007), and Morandi (1992). The state space approach to indistinguishable particles was proposed by Bach (1997), who proves (7.192) - (7.193), as well as the claim following these equations to the effect that $\rho \in \partial_e \mathscr{D}(H^N)^{\mathfrak{S}_N}$ iff eH is irreducible. The state space arguments against parastatistics given near the end of this section are also due to Bach (1997).

The representation theory used in this section may be found in many books, such as Weyl (1928), Fulton (1997), or Goodman & Wallach (2000).

The groupoid (7.209) is a special case of the so-called ***gauge groupoid*** defined by a principal H-bundle $P \xrightarrow{\pi} Q$, where $G_1 = P \times_H P$ (which stands for $(P \times P)/H$ with respect to the diagonal H-action on $P \times P$), $G_0 = Q$, and the operations are

$$s([p,q]) = \pi(q), \;\; t([p,q]) = \pi(p), \;\; [x,y]^{-1} = [y,x], \;\; [p,q][q,r] = [p,r];$$

here $[p,q][q',r]$ is defined whenever $\pi(q) = \pi(q')$, but to write down the product one picks some element $q \in \pi^{-1}(q')$.

Recent philosophical literature on indistinguishable particles includes French & Krause (2006), Earman (2010), Caulton & Butterfield (2012), Saunders (2013), and Baker, Halvorson, & Swanson (2015). This philosophical literature stills needs to be integrated with the mathematical approach launched in this section, and it was indeed the goal of Earman, Halvorson, & Landsman (2013ish) to do so. Alas!

Chapter 8
Limits: large N

Beside the limit $\hbar \to 0$, we consider the limit $N \to \infty$, where N could be the principal quantum number labeling orbits in atomic physics (as in Bohr's Correspondence Principle), or the number of particles or lattice sites, or the number of identical experiments in a long run measuring the relative frequencies of possible outcomes.

The case of large quantum numbers will be dealt with first: as our toy model of an classical orbit we take a *coadjoint orbit* in the dual $\mathfrak{g}^*$ of the Lie algebra $\mathfrak{g}$ of a compact connected Lie group G, see §5.9; for $G = SU(2)$ or $SO(3)$ these are simply two-spheres S^2_r. The corresponding quantum theories are indexed by their spin $j = \frac{1}{2}n$, where $n \in \mathbb{N}$, which we send to infinity in order to recover the classical orbit. This can be done more generally by rescaling the highest weight λ of some fixed irreducible representation of G to $n\lambda$ and again letting $n \to \infty$.

The second case, where the limit $N \to \infty$ is typically the thermodynamic limit (namely if the density N/V is kept fixed, where V is the volume of the system sent to infinity, too), has been rigorously studied using operator algebras since the 1960s. In such work the system constructed *at* the limit $N = \infty$ is typically quantum statistical mechanics in infinite volume, whose existence (followed by the establishment of e.g. phase transitions) was a major achievement of mathematical physics.

However, our goal in taking the limit $N \to \infty$ is quite different, in that—in the spirit of Bohrification—our limiting system will be *classical*; from the traditional point of view we look at the macroscopic rather than the quasi-local observables. Nonetheless, for each finite value of $N \in \mathbb{N}$ our (quantum) system will be the same as in the usual theory! Like the first case, in which increasingly large matrix algebras converge to an algebra of continuous functions on some compact space, this apparent miracle is described by the theory of continuous bundles of C*-algebras, as outlined in §C.19. As in the case $\hbar \to 0$ studied in the previous chapter, this theory provides a convenient mathematical machinery for studying the limit $N \to \infty$ also.

We then apply the the limit $N \to \infty$ to N repeated experiments, and, applying the doctrine of classical concepts, rederive the Born rule (avoiding the conceptual and mathematical pitfalls of various previous attempts to do so).

Bridging the gap to the next two chapters, we close with an introduction to quantum spin systems (as a later playing ground for spontaneous symmetry breaking).

K. Landsman, *Foundations of Quantum Theory*,
Fundamental Theories of Physics 188, DOI 10.1007/978-3-319-51777-3_8

8.1 Large quantum numbers

As in §5.9, let G be a compact connected Lie group with Lie algebra $\mathfrak{g}$ and dual $\mathfrak{g}^*$, and let $T \subset G$ be a maximal torus with Lie algebra $\mathfrak{t}$ and dual $\mathfrak{t}^*$. Let $\mathscr{O}_\lambda$ be a ***regular integral coadjoint orbit*** in $\mathfrak{g}^*$, labeled by a dominant weight $\lambda \in \Lambda_d$. This means that there is a point $\theta \in \mathscr{O}_\lambda$ whose stabilizer G_θ is T, and $\lambda = \theta_{|\mathfrak{t}}$; conversely, $\lambda \in \mathfrak{t}^*$ determines $\theta \in \mathfrak{g}^*$, which vanishes on each generator E_α of $\mathfrak{g}_{\mathbb{C}}$ $(\alpha \in \Delta)$.

Following Theorems 5.49 and 5.51, we associate a unitary irreducible representation $u_\lambda : G \to U(H_\lambda)$ to $\mathscr{O}_\lambda$ (or rather to λ), whose underlying Hilbert space H_λ contains a unique highest weight vector υ_λ. We then have (5.228). We abbreviate

$$d_\lambda = \dim(H_\lambda). \tag{8.1}$$

For $SU(2)$ we have $\lambda \in \mathbb{N}_0/2 = \{0, \frac{1}{2}, 1, \ldots\}$, usually called j, and the (regular) coadjoint orbits in $\mathfrak{g}^* \cong \mathbb{R}^3$ are the spheres S^2_j with radius j (with $j \neq 0$). The corresponding highest weight representation u_j is carried by H_j with $d_j = 2j+1$, whose highest weight vector υ_j is an eigenvector of $L_3 = iu'(S_3)$ with eigenvalue j.

We are going to define a continuous bundle of C*-algebras over the base space

$$I = (1/\mathbb{N}) \cup \{0\} \equiv 1/\dot{\mathbb{N}}, \tag{8.2}$$

where $\mathbb{N} = \{1, 2, \ldots\}$ and $\dot{N} = \mathbb{N} \cup \{\infty\}$; as required, I contains 0 as an accumulation point. One may think of elements of I as "quantized" values of Planck's constant $\hbar = 1/N$, upon which the limit $N \to \infty$ is formally the same as the limit $\hbar \to 0$.

If $\lambda \in \Lambda_d$, then $n\lambda \in \Lambda_d$ for all $n \in \mathbb{N}$. We may therefore define the C*-algebras

$$A_0 = C(\mathscr{O}_\lambda); \tag{8.3}$$
$$A_{1/n} = B(H_{n\lambda}). \tag{8.4}$$

For each $f \in C(\mathscr{O}_\lambda)$ we define $f_\lambda = \pi^* f$ under the canonical projection $\pi : G \to G/G_\theta \cong \mathscr{O}_\lambda$ (i.e., $f_\lambda(x) = f(\pi(x))$), which enables us to define the operators

$$Q_{1/n}(f) = d_{n\lambda} \int_G dx\, f_\lambda(x) |u_{n\lambda}(x)\upsilon_{n\lambda}\rangle\langle u_{n\lambda}(x)\upsilon_{n\lambda}| \in A_{1/n}. \tag{8.5}$$

In fact, the entire integrand in (8.5) is a function on $\mathscr{O}_\lambda$, because for $z \in T$ we have

$$u_{n\lambda}(xz)\upsilon_{n\lambda} = u_{n\lambda}(x)u_{n\lambda}(z)\upsilon_{n\lambda} = \chi_{n\lambda}(z)u_{n\lambda}(x)\upsilon_{n\lambda},$$

and $\chi_{n\lambda}(z) \in \mathbb{T}$ cancels the factor $\overline{\chi_{n\lambda}(z)}$ from the last term in (8.5). Note that

$$Q_{1/n}(1_{\mathscr{O}_\lambda}) = 1_{H_{n\lambda}}, \tag{8.6}$$

as follows by taking $\psi_2 = \psi_3 = \upsilon_{n\lambda}$ in Schur's well-known orthogonality relations

$$d_{n\lambda} \int_G dx \langle \psi_1, u_{n\lambda}(x)\psi_2\rangle\langle u_{n\lambda}(x)\psi_3, \psi_4\rangle = \langle \psi_1, \psi_4\rangle\langle \psi_3, \psi_2\rangle \;\; (\psi_i \in H_{n\lambda}). \tag{8.7}$$

Other properties of the maps $Q_{1/n}: C(\mathscr{O}_\lambda) \to B(H_{n\lambda})$ (between C*-algebras) are:

- ***Self-adjointness***, i.e., $Q_{1/n}(f)^* = Q_{1/n}(f^*)$.
- ***Positivity***, i.e., $Q_{1/n}(f) \geq 0$ whenever $f \geq 0$.
- ***Equivariance***, i.e., writing $L_y f(x) = f(y^{-1}x)$ as usual, for any $y \in G$ we have

$$Q_{1/n}(L_y f) = u_{n\lambda}(y) Q_{1/n}(f) u_{n\lambda}(y)^*. \tag{8.8}$$

Positivity does not follows from self-adjointness, as $Q_{1/n}$ is not a homomorphism.

Theorem 8.1. *There exists a continuous bundle of C*-algebras A over I as defined in* (8.2)*, with fibers* (8.3) - (8.4)*, whose continuous sections are given by all sequences* $(a_{1/n})_{n\in\dot{\mathbb{N}}} \in \prod_{n\in\dot{\mathbb{N}}} A_{1/n}$ *for which* $a_0 \in C(\mathscr{O}_\lambda)$ *and* $a_{1/n} \in B(H_{n\lambda})$*, and the sequence* $(a_{1/n})_{n\in\mathbb{N}}$ *is asymptotically equivalent to* $(Q_{1/n}(a_0))_{n\in\mathbb{N}}$*, in the sense that*

$$\lim_{n\to\infty} \|a_{1/n} - Q_{1/n}(a_0)\| = 0. \tag{8.9}$$

In particular, if $f \in C(\mathscr{O}_\lambda)$, then the cross-section of $\prod_{n\in\dot{\mathbb{N}}} A_{1/n}$ defined by

$$a_0 = f; \tag{8.10}$$
$$a_{1/n} = Q_{1/n}(f), \tag{8.11}$$

is continuous. In fact, we have a deformation quantization of $\mathscr{O}_\lambda$ in the sense of Definition 7.1, where the Poisson structure of $\mathscr{O}_\lambda$ is inherited from (minus) the canonical one on the Poisson manifold $\mathfrak{g}^*$, but we shall merely prove the claim of the theorem.

Proof. This will follow from Proposition C.124, in whose notation $\tilde{A}$ (which will actually coincide with A) consists of all $\tilde{a} = (\tilde{a}_\hbar)_{\hbar\in I}$ where f runs through $C(\mathscr{O}_\lambda)$ in

$$\tilde{a}_0 = f; \tag{8.12}$$
$$\tilde{a}_{1/n} = Q_{1/n}(f). \tag{8.13}$$

To verify the conditions for Proposition C.124 we start with the property that the set $\{\tilde{a}_\hbar \mid \tilde{a} \in \tilde{A}\}$ be dense in $A_\hbar$; we will show that it even coincides with $A_\hbar$. At $\hbar = 0$ this is true by construction. At $\hbar = 1/n$, the required property

$$Q_{1/n}(C(\mathscr{O}_\lambda)) = B(H_{n\lambda}) \tag{8.14}$$

can be proved in two steps. For simplicity we set $n = 1$; the proof is the same for any $n \in \mathbb{N}$. The first step is to define a function L_a on G for each $a \in B(H_\lambda)$ by

$$L_a(x) = \mathrm{Tr}\,(a|u_\lambda(x)\upsilon_\lambda\rangle\langle u_\lambda(x)\upsilon_\lambda|) = \langle \upsilon_\lambda, u_\lambda(x)^* a u_\lambda(x)\upsilon_\lambda\rangle. \tag{8.15}$$

This function is continuous and is right-invariant under T, so that L_a is really an element of $C(\mathscr{O}_\lambda)$. Thus we have a map $L: B(H_\lambda) \to C(\mathscr{O}_\lambda)$, $a \mapsto L_a$. Furthermore,

$$\langle a, Q_1(f)\rangle_{HS} = \langle L_a, f\rangle_2, \tag{8.16}$$

where the Hilbert–Schmidt inner product on left-hand side is $\langle a,b\rangle_{HS} = \mathrm{Tr}(a^*b)$, cf. (B.495)—which is well defined since H_λ is finite-dimensional—and the right-hand side is the inner product on $L^2(\mathscr{O}_\lambda)$ with respect to the measure induced by the subspace of $L^2(G, d_\lambda \cdot dx)$ consisting of T-invariant functions. Now $Q_{1/n}(C(\mathscr{O}_\lambda))$ is a (necessarily closed) linear subspace of $B(H_\lambda)$, which coincides with $B(H_\lambda)$ iff its orthogonal complement in the Hilbert–Schmidt inner product is zero.

Hence (8.14) is equivalent to the implication: $a \in (Q_{1/n}(C(\mathscr{O}_\lambda)))^\perp \Rightarrow a = 0$. By (8.16), the antecedent holds iff $\langle L_a, f\rangle_2 = 0$ for each $f \in C(\mathscr{O}_\lambda)$, which, because $C(\mathscr{O}_\lambda)$ is dense in $L^2(\mathscr{O}_\lambda)$, holds iff $L_a = 0$. Hence the the above implication is equivalent to: $L_a = 0 \Rightarrow a = 0$, i.e., $\ker L = \{0\}$. We must therefore prove the latter.

If $L_a(x) = 0$ for all $x \in G$, then, taking $x = \exp(t_1A_1)\cdots\exp(t_nA_n)$, where each $A_i \in \mathfrak{g}$, and applying (5.156) for each t_i to the right-hand side of (8.15), we obtain

$$\langle \upsilon_\lambda, [u'_\lambda(A_n), \cdots [u'_\lambda(A_2), [u'_\lambda(A_1), a]] \cdots] \upsilon_\lambda\rangle = 0. \tag{8.17}$$

This equality extends to $\mathfrak{g}_\mathbb{C}$, so we may take $A_i = E_{\alpha_i}$ for some positive root $\alpha_i \in \Delta^+$. Since $u'_\lambda(E_\alpha)\upsilon_\lambda = 0$ for $\alpha \in \Delta^+$, of each commutator $[u'_\lambda(E_{\alpha_i}), a]$ only the term $u'_\lambda(E_{\alpha_i})a$ contributes. Moving the $u'_\lambda(E_{\alpha_i})$ to act as $u'_\lambda(E_{\alpha_i})^* = u'_\lambda(E_{-\alpha_i})$ on the vector on the left in the inner product in (8.17) gives all other eigenvectors of $\mathfrak{t}$, so that (8.17) implies $\langle \psi, a\upsilon_\lambda\rangle = 0$ for each $\psi \in H_\lambda$, and hence $a\upsilon_\lambda = 0$. Now it is clear from (8.15) that $L_{u_\lambda(y)^*au_\lambda(y)}(x) = L_a(yx)$, so if $L_a(x) = 0$ for all $x \in G$, then also $L_{u_\lambda(y)^*au_\lambda(y)}(x) = 0$ for all $x \in G$. Hence we may replace a by $u_\lambda(y)^*au_\lambda(y)$ in the above argument, finding $u_\lambda(y)^*au_\lambda(y)\upsilon_\lambda = 0$ and hence $au_\lambda(y)\upsilon_\lambda = 0$ for each $y \in G$. Since u_λ is irreducible, this implies $a\psi = 0$ for any $\psi \in H_\lambda$, and hence $a = 0$.

This completes the proof of (8.14). Proposition C.124 furthermore requires

$$\lim_{n\to\infty} \|Q_{1/n}(f)\| = \|f\|_\infty, \tag{8.18}$$

This follows from the following key property (to be proved at the end):

$$\lim_{n\to\infty} \langle u_{n\lambda}(y)\upsilon_{n\lambda}, Q_{1/n}(f)u_{n\lambda}(y)\upsilon_{n\lambda}\rangle = f_\lambda(y), \tag{8.19}$$

for any $y \in G$ and $f \in C(\mathscr{O}_\lambda)$. Indeed, for any $y \in G$ we obviously have

$$\|Q_{1/n}(f)\| \geq \langle u_{n\lambda}(y)\upsilon_{n\lambda}, Q_{1/n}(f)u_{n\lambda}(y)\upsilon_{n\lambda}\rangle. \tag{8.20}$$

Since G and hence $\mathscr{O}_\lambda$ is compact, by Weierstrass's Theorem there is an $y \in G$ such that $|f_\lambda(y)| = \|f\|_\infty$. Using this y in (8.20) and (8.19), the two of these imply

$$\liminf_{n\to\infty} \|Q_{1/n}(f)\| \geq \|f\|_\infty. \tag{8.21}$$

Conversely, for any unit vector $\psi \in H_{n\lambda}$, eqs. (8.5) and (8.7) imply

$$\langle \psi, Q_{1/n}(f)\psi\rangle = |\langle \psi, Q_{1/n}(f)\psi\rangle| \leq \|f\|_\infty. \tag{8.22}$$

If f is real-valued, then $Q_{1/n}(f)^* = Q_{1/n}(f^*) = Q_{1/n}(f)$. In that case, (8.22) implies

$$\|Q_{1/n}(f)\| \leq \|f\|_{\infty}. \tag{8.23}$$

By the C*-identity $\|a^*a\| = \|a\|^2$, this is true for any $f \in C(\mathscr{O}_\lambda)$. Therefore,

$$\limsup_{n\to\infty} \|Q_{1/n}(f)\| \leq \|f\|_{\infty}. \tag{8.24}$$

Eqs. (8.21) and (8.24) yield (8.18). It remains to prove (8.19), i.e.,

$$\lim_{n\to\infty} d_{n\lambda} \int_G dx\, f_\lambda(x) |\langle u_{n\lambda}(y) \upsilon_{n\lambda}, u_{n\lambda}(x) \upsilon_{n\lambda}\rangle|^2 = f_\lambda(y). \tag{8.25}$$

The key to the proof is the fact that if λ and μ are dominant weights, with associated highest weight representations u_λ and u_μ, respectively, for any $x \in G$ one has

$$\langle \upsilon_\lambda, u_\lambda(x)\upsilon_\lambda\rangle \cdot \langle \upsilon_\mu, u_\mu(x)\upsilon_\mu\rangle = \langle \upsilon_{\lambda+\mu}, u_{\lambda+\mu}(x)\upsilon_{\lambda+\mu}\rangle. \tag{8.26}$$

Namely, because the exponential map is surjective for compact connected Lie groups, eq. (8.26) is equivalent to the property

$$\langle \upsilon_\lambda, u'_\lambda(A)\upsilon_\lambda\rangle + \langle \upsilon_\mu, u'_\mu(A)\upsilon_\mu\rangle = \langle \upsilon_{\lambda+\mu}, u'_{\lambda+\mu}(A)\upsilon_{\lambda+\mu}\rangle, \tag{8.27}$$

for any $A \in \mathfrak{g}$. For $A \in \mathfrak{t}$ this amounts to $\lambda + \mu = \lambda + \mu$, cf. (5.228), whereas for $A = E_\alpha$ for some root $\alpha \in \Delta$ we have $0 = 0$, so that (8.27) is true for all $A \in \mathfrak{g}$. This also proves (8.26), of which we need the special (and iterated) case

$$\langle \upsilon_{n\lambda}, u_{n\lambda}(x)\upsilon_{n\lambda}\rangle = \langle \upsilon_\lambda, u_\lambda(x)\upsilon_\lambda\rangle^n. \tag{8.28}$$

This motivates us to introduce a sequence (μ_n) of probability measures on G by

$$d\mu_n(x) = d_{n\lambda} \cdot dx\, |\langle \upsilon_\lambda, u_\lambda(x)\upsilon_\lambda\rangle|^{2n}, \tag{8.29}$$

so that, after a change $x \mapsto yx$ of the integration variable, eq. (8.25) reads

$$\lim_{n\to\infty} d_{n\lambda} \int_G d\mu_n(x)\, f_\lambda(yx) = f_\lambda(y), \tag{8.30}$$

for any $f \in C(\mathscr{O}_\lambda)$. Now $F(x) = |\langle \upsilon_\lambda, u_\lambda(x)\upsilon_\lambda\rangle|$ takes values in $(0,1]$ and hence the measure (8.29) is $d\mu_n(x) \sim \exp(-nS(x))$ for $S(x) = -\ln(F(x))$, with $S \geq 0$ and $S(x) = 0$ iff $x \in G_{\theta_\lambda} = T$ (using regularity of the orbit). In that case, i.e., if $z \in T$, then $f_\lambda(yz) = f(\pi(yz)) = f(\pi(y)) = f_\lambda(y)$. The method of steepest descent shows that any part of G (of positive Haar measure) where $S(x) > 0$ makes no contribution as $n \to \infty$, so that we may replace $f_\lambda(yx)$ in (8.30) by $f_\lambda(y)$, obtaining

$$\lim_{n\to\infty} \int_G d\mu_n(x)\, f_\lambda(yx) = f_\lambda(y) \lim_{n\to\infty} \int_G d\mu_n(x) = f_\lambda(y) \lim_{n\to\infty} 1 = f_\lambda(y). \tag{8.31}$$

We have now verified conditions 1 and 2 in Proposition C.124, and no. 3 is trivially satisfied since in condition 1 we have equality with $A_\hbar$, as shown above. □

8.2 Large systems

We now move from large quantum numbers within a single system to large quantum systems that consist of N identical sites, where we eventually study what happens as $N \to \infty$ (as is customary in quantum statistical mechanics we change notation from $n \in \mathbb{N}$ to $N \in \mathbb{N}$). This limit gives rise to two different continuous bundles $A^{(q)}$ and $A^{(c)}$ of C*-algebras over I as given by (8.2), which have exactly the same fibers at $1/N$ but, amazingly, differ dramatically at $N = \infty$, i.e., $1/N = 0$. This difference reflects two choices one may make for the N-particle observables that have a limit as $N \to \infty$, namely *local* ones, giving rise to a highly *non-commutative* limit algebra $A_0^{(q)}$ (which is the one usually studied in quantum statistical mechanics of infinite systems), and *macroscopic* ones, which generate a *commutative* algebra $A^{(c)}$ of observables of an infinite quantum system (describing classical thermodynamics as a limit of quantum statistical mechanics). It is the latter that we need for Bohrification.

Let B be a fixed *unital* C*-algebra, describing a single quantum system. The case of a two-level system, where $B = M_2(\mathbb{C})$, is already fascinating, and many other interesting examples are described by finite-dimensional C*-algebras. Though irrelevant in finite dimension, we note that the constructions below are generally valid if (for technical reasons to be found in Proposition C.97) we use the *projective* tensor product $\hat{\otimes}_{\max}$ between C*-algebras; see §C.13. For any $N \in \mathbb{N}$ we put

$$A_{1/N}^{(c)} = A_{1/N}^{(q)} = B^N, \tag{8.32}$$

i.e., the N-fold (projective) tensor product $\hat{\otimes}_{\max}^N B$ of B with itself. Furthermore,

$$A_0^{(c)} = C(S(B)); \tag{8.33}$$

$$A_0^{(q)} = B^\infty, \tag{8.34}$$

where $S(B)$ is the state space of B, seen as a compact convex set in the weak*-topology, as usual, and B^∞ is the infinite (projective) tensor product of B with itself as described in §C.14; see especially (C.318) with $C_i = B$ for each i. For example, the state space of $B = M_2(\mathbb{C})$ is affinely homeomorphic to the unit ball in $\mathbb{R}^3$, whose boundary is the familiar Bloch sphere of qubits; see Proposition 2.9.

We now explain how (8.32) and (8.33) - (8.34) give rise to continuous bundles $A^{(c)}$ and $A^{(q)}$ of C*-algebras, starting with the former. First, for each $N \in \mathbb{N}$, let $\mathfrak{S}_N$ be the permutation group (i.e. symmetric group) on N objects, acting on B^N in the obvious way, i.e., by linear and continuous extension of

$$\alpha_p^{(N)}(b_1 \otimes \cdots \otimes b_N) = b_{p(1)} \otimes \cdots \otimes b_{p(N)}, \tag{8.35}$$

where $b_i \in B$. This yields a ***symmetrization operator*** $S_N : B^N \to B^N$ defined by

$$S_N = \frac{1}{N!} \sum_{p \in \mathfrak{S}_N} \alpha_p^{(N)}. \tag{8.36}$$

If B is infinite-dimensional, these maps can be extended by continuity to the completion $B^\infty = \hat{\otimes}^\infty_{\max} B$ of the *algebraic* tensor product $\otimes^\infty B$; indeed, passing to any faithful representation of B it is easy to see that S^N is even continuous with respect to the minimal cross-norm (cf. §C.13). For $N \geq M$ we then define

$$S_{M,N} : B^M \to B^N \tag{8.37}$$

by linear (and if necessary continuous) extension of

$$S_{M,N}(a_{1/M}) = S_N(a_{1/M} \otimes 1_B \otimes \cdots \otimes 1_B) \ (a_{1/M} \in B^M), \tag{8.38}$$

with $N - M$ copies of the unit $1_B \in B$ so as to obtain an element of B^N. Clearly, $S_{N,N} = S_N$. In particular, $S_{1,N} : B \to B^N$ gives the average of b over N copies of B:

$$S_{1,N}(b) = \frac{1}{N} \sum_{k=1}^{N} 1_B \otimes \cdots \otimes b_{(k)} \otimes 1_B \cdots \otimes 1_B,. \tag{8.39}$$

For example, take $B = M_n(\mathbb{C})$ for simplicity, and pick some $a = a^* \in B$ and $\lambda \in \sigma(a)$, with associated spectral projection e_λ. Putting $b = e_\lambda$ in (8.39) gives

$$f_N^{(\lambda)} = S_{1,N}(e_\lambda). \tag{8.40}$$

This is a ***frequency operator***: applied to states of the kind $\upsilon_1 \otimes \cdots \otimes \upsilon_N \in (\mathbb{C}^n)^N$, where each υ_i is an eigenstate of a, so that $a\upsilon_i = \lambda_i \upsilon_i$ for some $\lambda_i \in \sigma(a)$, the corresponding operator counts the relative frequency of λ in the list $(\lambda_1, \ldots, \lambda_N)$. The commutative case $B = C(X)$ provides a classical analogue. Eq. (C.271) gives

$$B^N = C(X)^N \cong C(X^N), \tag{8.41}$$

so that, identifying elements of B^N with functions on X^N, for $f \in C(X)$ we have

$$S_{1,N}(f)(x_1, \ldots, x_N) = \frac{1}{N} \sum_{k=1}^{N} (f(x_1) + \cdots f(x_N)). \tag{8.42}$$

We return to the construction of a continuous bundle of C*-algebras with fibers (8.32) and (8.33). As in §8.1, we construct this bundle by specifying a preliminary family of continuous cross-sections and then using Proposition C.124 to finish.

Definition 8.2. *We say that a sequence $(a_{1/N})_{N\in\mathbb{N}}$, with $a_{1/N} \in B^N$, is* **symmetric** *when there exist $M \in \mathbb{N}$ and $a_{1/M} \in B^M$ such that for each $N \geq M$ one has*

$$a_{1/N} = S_{M,N}(a_{1/M}). \tag{8.43}$$

This implies $a_{1/M} = S_M(a_{1/M})$. Symmetric sequences can start in any finite way they like, but their infinite tails consist of averaged observables. Hence *symmetric sequences asymptotically commute*: if $(a_{1/N})$ and $(b_{1/N})$ are symmetric, then

$$\lim_{N\to\infty}\|a_{1/N}b_{1/N}-b_{1/N}a_{1/N}\|_{B^N}=0, \tag{8.44}$$

simply because the commutators of single-site operators are nonvanishing only at finitely many positions, upon which the factor $1/N$ in (8.39) guarantees (8.44).

For example, if $B=M_2(\mathbb{C})$, and (σ_i) are the Pauli matrices, we have

$$[S_{1,N}(\tfrac{1}{2}\hbar\sigma_1),S_{1,N}(\tfrac{1}{2}\hbar\sigma_2)]=i\frac{\hbar}{N}S_{1,N}(\tfrac{1}{2}\hbar\sigma_3), \tag{8.45}$$

et cetera, showing that the averaged spin-$\frac{1}{2}$ operators effectively rescale $\hbar$ by $\hbar/N$.

In view of this, it is reasonable to expect that we may be able to assemble the algebra B^N into a continuous bundle whose limit algebra at $N=\infty$ is commutative.

For each symmetric sequence $(a_{1/N})$ we define a function $a_0:S(B)\to\mathbb{C}$ by

$$a_0(\omega)=\lim_{N\to\infty}\omega^N(a_{1/N}), \tag{8.46}$$

where $\omega\in S(B)$, and $\omega^N\in S(B^N)$ is defined by linear (and continuous) extension of

$$\omega^N(b_1\otimes\cdots\otimes b_N)=\omega(b_1)\cdots\omega(b_N); \tag{8.47}$$

continuity of ω^N on the algebraic tensor product $\otimes^N B$ (and hence extendibility to $A_{1/N}$) is guaranteed by Proposition C.98, although this is not really needed here because a_0 only requires the values of ω^N on $\otimes^N B$ itself. In any case, the limit exists by definition of a symmetric sequence, from which we also see that $a_0\in C(S(B))$, because it is a finite sum of finite products of the type $\omega(b_1)\cdots\omega(b_M)$, each of which is continuous in ω by definition of the w^*-topology on $S(B)$.

For example, the frequency operators (8.40) define a symmetric sequence $(f_N^\lambda)_{N\in\mathbb{N}}$, whose the limit function $f_0^\lambda:S(B)\to\mathbb{C}$ in the sense of (C.560) or (8.46) is

$$f_0^\lambda(\omega)=\omega(e_\lambda). \tag{8.48}$$

Thus (8.46) gives the Born probability for the outcome $a=\lambda$ in the state ω; see §8.4. Classically, identifying elements of $S(C(X))$ with probability measures μ on X, the limit of the sequence $a_{1/N}=S_{1,N}(f)$ for fixed $f\in C(X)$, cf. (8.42), is

$$a_0(\mu)=\int_X d\mu\, f. \tag{8.49}$$

This convergence is an example of the strong law of large numbers, see §8.3.

We return to the general case.

Definition 8.3. *A sequence $(a_{1/N})_{N\in\mathbb{N}}$ as above is* **quasi-symmetric** *if for each $N\in\mathbb{N}$ one has $a_{1/N}=S_N(a_{1/N})$ and for any $\varepsilon>0$ there is a symmetric sequence $(\tilde{a}_{1/N})$ and some $M\in\mathbb{N}$ such that $\|a_{1/N}-\tilde{a}_{1/N}\|<\varepsilon$ for all $N>M$.*

For example, if $\lim_{N\to\infty}\|a_{1/N}-\tilde{a}_{1/N}\|=0$ for some fixed symmetric sequence $(\tilde{a}_{1/N})$, then $(a_{1/N})_{N\in\mathbb{N}}$ is obviously quasi-symmetric.

Theorem 8.4. *For any unital C*-algebra B, the C*-algebras* (8.32) *and* (8.33), *i.e.,*

$$A_0^{(c)} = C(S(B)); \tag{8.50}$$

$$A_{1/N}^{(c)} = B^N, \tag{8.51}$$

where B^N is N-fold projective tensor power $\hat{\otimes}_{\max}^N B$, are the fibers of a continuous bundle $A^{(c)}$ of C-algebras over $I = (1/\mathbb{N}) \cup \{0\} \equiv 1/\dot{\mathbb{N}}$ whose continuous cross-sections are the quasi-symmetric sequences $(a_{1/N})$ with limit a_0 given by* (8.46).

As in Theorem 8.1, also here we have a deformation quantization of $S(B)$ in the sense of Definition 7.1, where the Poisson bracket on $S(B)$ may be defined by specifying its value on linear function $\hat{b} \in C(S(B))$, where $b \in B$ and $\hat{b}(\omega) = \omega(b)$, by

$$\{\hat{a}, \hat{b}\} = i\widehat{[a,b]}. \tag{8.52}$$

Unfortunately, this involves the theory of infinite-dimensional Poisson manifolds, which we prefer to omit. Thus we shall only prove Theorem 8.4 as stated.

The proof relies on Størmer's ***quantum De Finetti Theorem*** 8.6 below.

Definition 8.5. *Let B be a unital C*-algebra. A state ρ on B^N is called:*

- **permutation-invariant** *if $\rho \circ \alpha_p^{(N)} = \rho$ for any $p \in \mathfrak{S}_N$.*
- ***K*-exchangeable** *($K \in \mathbb{N}$) if it is permutation-invariant and in addition ρ is the restriction to B^N of some permutation-invariant state on B^{N+K}.*
- **Infinitely exchangeable** *if it is K-exchangeable for all $K \in \mathbb{N}$.*

The set of all permutation-invariant states / K-exchangeable states / infinitely exchangeable states on B^N is denoted by $S^{\mathfrak{S}_N}(B^N)$ / $S_K^{\mathfrak{S}_N}(B^N)$ / $S_\infty^{\mathfrak{S}_N}(B^N)$.

Theorem 8.6. *Let B be a unital C*-algebra. For any $N \in \mathbb{N}$ the correspondence $\omega^N \leftrightarrow \omega$, where $\omega \in S(B)$ and $\omega^N \in S(B^N)$, cf.* (8.47), *gives a bijection*

$$\partial_e S_\infty^{\mathfrak{S}_N}(B^N) \cong S(B). \tag{8.53}$$

This theorem was originally stated (in the language of infinite tensor products) as Theorem 8.9 in §8.3, where it (and hence Theorem 8.6) will also be proved.

We also need a formula for the norm of any self-adjoint element a of any C*-algebra A in terms of the state space A and the pure state space $P(A)$, viz.

$$\|a\| = \sup\{|\omega(a)| : \omega \in S(A)\} = \sup\{|\omega(a)|, \omega \in P(A)\}. \tag{8.54}$$

This follows from Proposition C.15, the spectral radius formula (B.254), and compactness of $\sigma(a)$, implying that the supremum in (B.254) is reached on $\sigma(a)$.

Proof. The proof of Theorem 8.4 is quite similar to the proof of Theorem 8.1, in that we once again rely on Proposition C.124, where the symmetric sequences are going to play the role of $\tilde{A}$. To apply Proposition C.124, we should prove that:

1. The set $\tilde{A}_0$ (consisting of all $\tilde{a}_0 \in A_0 = C(S(B))$ as defined by (8.46), where $(\tilde{a}_{1/N})$ runs through all symmetric sequences) is a *-algebra which is dense in A_0.
2. For any symmetric sequence $(\tilde{a}_{1/N})$ with limit $\tilde{a}_0$ as given by (8.46), one has

$$\lim_{N\to\infty} \|\tilde{a}_{1/N}\| = \|\tilde{a}_0\|_\infty. \tag{8.55}$$

To prove the first claim, we first note that $\tilde{a}_0$ is the linear span of all finite products $\omega(b_1)\cdot\omega(b_N)$, where $N \in \mathbb{N}$ and $b_1,\ldots,b_N \in B$. Since $\overline{\omega(b)} = \omega(b^*)$ this is obviously a *-algebra. The monomials $\hat{b}(\omega) = \omega(b)$ already separate points of $S(B) \subset B^*$, since if $\omega' \neq \omega$ then clearly is there some $b \in B$ for which $(\omega - \omega')(b) \neq 0$. Hence claim no. 1 follows from the Stone–Weierstrass Theorem B.51.

For the second, let $(\tilde{a}_{1/N})$ be a symmetric sequence. Since there are $M \in \mathbb{N}$ and $\tilde{a}_{1/M} \in B^M$ with $\tilde{a}_{1/M} = S_M(\tilde{a}_{1/M})$ and $\tilde{a}_{1/(M+K)} = S_{M,M+K}(\tilde{a}_{1/M})$ for all $K \in \mathbb{N}$,

$$\begin{aligned}
\|\tilde{a}_{1/M}\| &= \sup\{|\rho(\tilde{a}_{1/M})| : \rho \in S(B^M)\} = \sup\{|\rho(\tilde{a}_{1/M})| : \rho \in S^{\mathfrak{S}_M}(B^M)\};\\
\|\tilde{a}_{1/(M+K)}\| &= \sup\{|\rho(S_{M,M+K}(\tilde{a}_M))| : \rho \in S^{\mathfrak{S}_{M+K}}(B^{M+K})\}\\
&= \sup\{|\rho(\tilde{a}_{1/M})| : \rho \in S^{\mathfrak{S}_M}_K(B^M)\},
\end{aligned}$$

where we used (8.54) and (8.43). Theorem 8.6 and (8.46) then yield (8.55):

$$\begin{aligned}
\lim_{N\to\infty} \|\tilde{a}_{1/N}\| &= \lim_{K\to\infty} \|\tilde{a}_{1/(M+K)}\|\\
&= \sup\{|\rho(\tilde{a}_{1/M})| : \rho \in S^{\mathfrak{S}_M}_\infty(B^M)\}\\
&= \sup\{|\rho(\tilde{a}_{1/M})| : \rho \in \partial_e S^{\mathfrak{S}_M}_\infty(B^M)\} = \sup\{|\omega^M(\tilde{a}_{1/M})| : \omega \in S(B)\}\\
&= \sup\{|\lim_{N\to\infty} \omega^N(\tilde{a}_{1/N})| : \omega \in S(B)\} = \sup\{|\tilde{a}_0(\omega)| : \omega \in S(B)\}\\
&= \|\tilde{a}_0\|_\infty
\end{aligned}$$

The proof that the sequences $(a_{1/N})$ for which condition (C.552) in Proposition C.124 holds are precisely the approximately symmetric sequences is the same as the proof of the equivalence of the two conditions in Lemma C.125, taking $\hbar_0 = 0$.

Finally, it is easy to show that the limit (8.46) exists also for quasi-symmetric observables a: take $\varepsilon > 0$ and find $\tilde{a}$ and M as in Definition 8.3. For this $\tilde{a}$, let M_0 be such that (8.43) holds (with $M \rightsquigarrow M_0$). For all N, N' greater than both M and M_0,

$$\begin{aligned}
|\omega^N(a_{1/N}) - \omega^{N'}(a_{1/N'})| &\leq |\omega^N(a_{1/N} - \tilde{a}_{1/N}) - \omega^{N'}(a_{1/N'} - \tilde{a}_{1/N'})|\\
&\quad + |\omega^N(\tilde{a}_{1/N}) - \omega^{N'}(\tilde{a}_{1/N'})|\\
&\leq \|a_{1/N} - \tilde{a}_{1/N}\| + \|a_{1/N'} - \tilde{a}_{1/N'})\| + 0\\
&< 2\varepsilon,
\end{aligned} \tag{8.56}$$

since $\|\omega^N\| = 1$. Hence $(\omega^N(a_{1/N}))$ is a Cauchy sequence (in $\mathbb{C}$). □

Our second continuous bundle of C*-algebras of interest is described by the following changes in Definitions 8.2 and 8.3.

Definition 8.7. *Let B be a unital C*-algebra and let $a_{1/N} \in B^N$ for each $N \in \mathbb{N}$.*

- *A sequence $(a_{1/N})_{N\in\mathbb{N}}$ is called* **local** *when there exist $M \in \mathbb{N}$ and $a_{1/M} \in B^M$ such that for each $N \geq M$ one has*

$$a_{1/N} = a_{1/M} \otimes 1_B \otimes \cdots \otimes 1_B, \tag{8.57}$$

with $N - M$ copies of the unit $1_B \in B$ (so that indeed $a_{1/N} \in B^N$).
- *A sequence $(a_{1/N})_{N\in\mathbb{N}}$ is* **quasi-local** *if for any $\varepsilon > 0$ there is a local sequence $(\tilde{a}_{1/N})$ and some $M \in \mathbb{N}$ such that $\|a_{1/N} - \tilde{a}_{1/N}\| < \varepsilon$ for all $N > M$.*

For the right analogue of Theorem 8.4 we recall the description of the infinite tensor product B^∞; cf. §C.14, especially the explanation preceding (C.315). Accordingly, a dense subspace of B^∞ is given by equivalence classes of local sequences $(a_{1/N})_{N\in\mathbb{N}}$ under the equivalence relation $a \sim a'$ iff $\lim_{N\to\infty} \|a_{1/N} - a'_{1/N}\| = 0$; the C*-algebraic operations in B^∞ are inherited from the B^N, and if we denote the equivalence class of $(a_{1/N})_N$ by $[a_{1/N}]_N$, the norm in B^∞ is given by

$$\|[a_{1/N}]_N\| = \lim_{N\to\infty} \|a_{1/N}\|. \tag{8.58}$$

By construction, this number is independent of the representative $(a_{1/N})_N$ in the class $[a_{1/N}]_N$. By definition, B^∞ is the completion of the space of these equivalence classes in the norm (8.58). As explained after (C.315), for each $M \in \mathbb{N}$ we have an injective (and hence isometric) homomorphism $\varphi_M : B^M \to B^\infty$ that maps $a_{1/M} \in B^M$ to the equivalence class $[a_{1/N}]_N$ of the sequence $(a_{1/N})_N$ defined by

$$a_{1/N} = 0, \;\; (N < M); \tag{8.59}$$
$$a_{1/N} = a_{1/M}, \;\; (N = M); \tag{8.60}$$
$$a_{1/(M+K)} = a_{1/M} \otimes 1_B \otimes \cdots \otimes 1_B, \;\; (K > 0), \tag{8.61}$$

with K copies of 1_B. It is easy to verify that one might as well have started from quasi-local sequences and their equivalence classes, for which the limit (8.58) exists by an argument similar to (8.56). In that case the ensuing C*-algebra is already complete, which leads to a direct description of the elements of B^∞ as equivalence classes of quasi-local sequences. This fact also follows from the following analogue of Theorem 8.4, which may be proved in the same way, i.e., from Proposition C.124, where this time the elements of $\tilde{A}$ are local sequences rather than symmetric ones (in fact, the proof is much easier, since this time we obtain (C.552) for free):

Theorem 8.8. *For any unital C*-algebra B, the C*-algebras* (8.32) *and* (8.34), *i.e.,*

$$A_0^{(q)} = B^\infty; \tag{8.62}$$
$$A_{1/N}^{(q)} = B^N, \tag{8.63}$$

are the fibers of a continuous bundle $A^{(q)}$ of C-algebras over $I = 1/\dot{\mathbb{N}}$ whose continuous cross-sections are the quasi-local sequences $(a_{1/N})$ with limit $a_0 = [a_{1/N}]_N$.*

8.3 Quantum de Finetti Theorem

As an initial step in exploring the connection between the bundles $A^{(c)}$ and $A^{(q)}$ we prove Theorem 8.6, which we first restate in an equivalent form. Let $\mathfrak{S}_\infty$ be the group of bijections of $\mathbb{N}$ that differ from the identity only on a finite set. Each such finite permutation $p \in \mathfrak{S}_\infty$ defines a map $\alpha_p : B^\infty \to B^\infty$, as follows. Let $S \subset \mathbb{N}$ be the finite subset of $\mathbb{N}$ on which p acts nontrivially (if $S = \emptyset$ we have $p = \mathrm{id}_{\mathbb{N}}$, in which case also $\alpha_p = \mathrm{id}_{B^\infty}$, see below). Take a local sequence $(a_{1/N})_N$, so that (8.57) holds, in which we may assume $M \geq \max S$; we also redefine $a_{1/N} = 0$ for each $N < M$. For each $N \geq M \geq \max S$, the map p may be regarded as an element p_N of $\mathfrak{S}_N$ by restriction to $\{1,\ldots,N\} \subset \mathbb{N}$ and hence p acts on B^N by permuting the entries in elementary tensor products of operators, cf. (8.35). For each $p \in \mathfrak{S}_\infty$, define a map

$$\alpha_p : B^\infty \to B^\infty; \tag{8.64}$$

$$\alpha_p([a_{1/N}]_N) = [\alpha_p^{(N)}(a_{1/N})]_N. \tag{8.65}$$

This uses a specific representative of the equivalence class $[a_{1/N}]_N \in B^\infty$, but nonetheless the map α_p is well defined. Furthermore, since each $\alpha_p^{(N)} : B^N \to B^N$ is an automorphism (i.e., an invertible homomorphism), it is an isometry, so that also α_p is an isometry on its domain and hence extends to an automorphism of B^∞. The ensuing map $p \mapsto \alpha_p$ from $\mathfrak{S}_\infty$ to the group $\mathrm{Aut}(B^\infty)$ of all automorphisms of B^∞ is a homomorphism of groups, and we say that $\mathfrak{S}_\infty$ is an *automorphism group* of B^∞.

Writing $S^{\mathfrak{S}_\infty}(B^\infty)$ for the set of all $\mathfrak{S}_\infty$-invariant states on B^∞, i.e., $\rho \in S^{\mathfrak{S}_\infty}(B^\infty)$ iff $\rho \circ \alpha_p = \rho$ for each $p \in \mathfrak{S}_\infty$, we may now rephrase Theorem 8.6 as follows:

Theorem 8.9. *Let B be a unital C^*-algebra. There is a bijection*

$$\partial_e S^{\mathfrak{S}_\infty}(B^\infty) \cong S(B), \tag{8.66}$$

given by $\omega^\infty \leftrightarrow \omega$, where $\omega \in S(B)$, and $\omega^\infty \in S(B^\infty)$ is defined by, cf. (8.47),

$$\omega^\infty([a_{1/N}]_N) = \lim_{N\to\infty} \omega^N(a_{1/N}). \tag{8.67}$$

This is essentially the same as Theorem 8.6: for any $M \in \mathbb{N}$, a state on B^M is infinitely exchangeable iff it is the restriction of an element of $S^{\mathfrak{S}_\infty}(B^\infty)$ to $B^M \subset B^\infty$, where the inclusion is given by the map φ_M defined below (8.58).

Proof. Let $S(B) \subset S^{\mathfrak{S}_\infty}(B^\infty)$ under the map $\omega \mapsto \omega^\infty$. We first show the inclusion

$$\partial_e S^{\mathfrak{S}_\infty}(B^\infty) \subseteq S(B) \tag{8.68}$$

contrapositively, i.e., if $\rho \in S^{\mathfrak{S}_\infty}(B^\infty)$ does not lie in $S(B)$, then ρ has a nontrivial convex decomposition in $S^{\mathfrak{S}_\infty}(B^\infty)$. We identify B^N with $\varphi_N(B^N) \subset B^\infty$ and denote the restriction of ρ to B^N by ρ_N. If $\rho = \omega^\infty$ for some $\omega \in S(B)$, then

$$\rho_{M+K}(a'_{1/M} \otimes a'_{1/K}) = \rho_M(a'_{1/M})\rho_K(a'_{1/K}), \tag{8.69}$$

for each $a'_{1/M} \in B^M$ and $a'_{1/K} \in B^K$. If (8.69) holds whenever $0 \leq a'_{1/M} \leq 1_{B^M}$, then by Lemma C.53 and (C.8) it always holds. Adding suitable multiples of the unit and rescaling, it follows that if (8.69) holds whenever

$$\tfrac{1}{3} \cdot 1_{B^M} \leq a'_{1/M} \leq \tfrac{2}{3} \cdot 1_{B^M}; \tag{8.70}$$

then it always holds. Therefore, if (8.69) fails, then it fails for some $a'_{1/M}$ satisfying (8.70) and some and $a'_{1/K}$, in which case $\frac{1}{3} \leq \rho_M(a'_{1/M}) \leq \frac{2}{3}$. However, such a failure implies the existence of a nontrivial convex decomposition

$$\rho = t\rho' + (1-t)\rho'', \tag{8.71}$$

with $t = \rho_M(a'_{1/M})$, and the functionals ρ' and ρ'' on B^∞ are defined by

$$\rho'([a_{1/N}]_N) = \lim_{N\to\infty} \rho_{M+N}(a'_{1/M} \otimes a_{1/N})/\rho_M(a'_{1/M}); \tag{8.72}$$

$$\rho''([a_{1/N}]_N) = \lim_{N\to\infty} \rho_{M+N}((1_{B^N} - a'_{1/M}) \otimes a_{1/N})/\rho_M(1_{B^M} - a'_{1/M}). \tag{8.73}$$

These limits exist on symmetric sequences (where they stabilize), and hence they exists in general. Furthermore, since $\rho_M(1_{B^M} - a'_{1/M}) = 1-t$, the property (8.71) is obvious. Both ρ' and ρ'' belong to $S^{\mathfrak{S}_\infty}(B^\infty)$, since each functional ρ_{M+N} is an element of $S^{\mathfrak{S}_{M+N}}(B^{M_N})$. Finally, (8.71) is nontrivial, since if $\rho' = \rho''$, then $\rho'_K = \rho''_K$, and hence (8.69) would hold (whose violation we assumed). This proves (8.68).

Though it is always true, for simplicity we prove the converse inclusion

$$S(B) \subseteq \partial_e S^{\mathfrak{S}_\infty}(B^\infty) \tag{8.74}$$

just for the case where B is generated by projections, as in the case $B = M_n(\mathbb{C})$, $B = B(H)$, or B a von Neumann algebra, or more generally an AW*-algebra (see §C.24). In that case also each B^N is generated by its projections.

For each $\rho \in S^{\mathfrak{S}_\infty}(B^\infty)$, each $N \in \mathbb{N}$, and each projection $e \in B^N$, we have

$$\rho_N(e)^2 \leq \rho_{2N}(e \otimes e), \tag{8.75}$$

see below. Assuming (8.75), suppose $\omega \in S(B)$ and $\omega^\infty = t\rho' + (1-t)\rho''$ for some $t \in (0,1)$ and $\rho', \rho'' \in S^{\mathfrak{S}_\infty}(B^\infty)$. Since $\omega^\infty_N = \omega^N$, we then have

$$\begin{aligned}
\omega^N(e)^2 &= (t\rho'_N(e) + (1-t)\rho''_N(e))^2 = \left\langle \begin{pmatrix} \sqrt{t} \\ \sqrt{1-t} \end{pmatrix}, \begin{pmatrix} \rho'_N(e)\sqrt{t} \\ \rho''_N(e)\sqrt{1-t} \end{pmatrix} \right\rangle^2 \\
&\leq \left\langle \begin{pmatrix} \sqrt{t} \\ \sqrt{1-t} \end{pmatrix}, \begin{pmatrix} \sqrt{t} \\ \sqrt{1-t} \end{pmatrix} \right\rangle \cdot \left\langle \begin{pmatrix} \rho'_N(e)\sqrt{t} \\ \rho''_N(e)\sqrt{1-t} \end{pmatrix}, \begin{pmatrix} \rho'_N(e)\sqrt{t} \\ \rho''_N(e)\sqrt{1-t} \end{pmatrix} \right\rangle \\
&= t\rho'_N(e)^2 + (1-t)\rho''_N(e)^2 \\
&\leq t\rho'_{2N}(e \otimes e) + (1-t)\rho''_{2N}(e \otimes e) \\
&= \omega^{2N}(e \otimes e) = \omega^N(e)^2,
\end{aligned}$$

where the inner product in the first line is the usual one in $\mathbb{R}^2$, and, noting it is positive, we have used the Cauchy–Schwarz inequality for this inner product, as well as (8.75). Hence both inequalities must be equalities, and for the first one this implies $\rho'_N(e) = \rho''_N(e)$. Since this is true for all N and all projections in B^N, this implies $\rho' = \rho'' = \omega^\infty$, so that $\omega^\infty \in \partial_e S^{\mathfrak{S}_\infty}(B^\infty)$, and (8.74) has been established, up to the proof of (8.75). To this effect, note for each $M \in \mathbb{N}$ and $t \in \mathbb{R}$ we have

$$\rho_{MN}((1_{B^N} \otimes \cdots \otimes 1_{B^N} \otimes e + \cdots + e \otimes 1_{B^N} \otimes \cdots \otimes 1_{B^N} + t \cdot 1_{B^{MN}})^2) \quad (8.76)$$
$$= M(M-1)\rho_{2N}(e \otimes e) + M\rho_N(e) + 2tM\rho_N(e) + t^2, \quad (8.77)$$

with $M-1$ copies of 1_{B^N} and e moving from right to left in the first line, leaving M terms before the final one $t \cdot 1_{B^{MN}}$ in (8.76). In working out the square in (8.76) and moving to the second line we used $e^2 = e$ as wel as permutation invariance of the state ρ_{MN}. The point is that (8.76) is positive, so that (8.77) must be positive, too, for all $M \in \mathbb{N}$ and $t \in \mathbb{R}$. Now a function $f(t) = t^2 + 2bt + c = (t+b)^2 - b^2 + c$ obviously satisfies $f(t) \geq 0$ for each t iff $b^2 \leq c$, so that (8.76) is positive for all t iff

$$M^2 \rho_N(e)^2 \leq M(M-1)\rho_{2N}(e \otimes e) + M\rho_N(e).$$

Letting $M \to \infty$ gives (8.75). □

Taking $B = C(X)$ for some compact Hausdorff space X, in view of (8.41) the situation may be transferred to the Cartesian product X^N, equipped with the product topology (which is generated by products $A_1 \times \cdots \times A_N \subset X^N$ with each $A_i \subset X$ open) and the ensuing Borel σ-algebra (generated by the above products with each A_i Borel). If $\mu_1, \ldots, \mu_N$ are (probability) measures on X (in which case we write $\mu_i \in \Pr(X)$), then there is a unique (probability) measure $\mu_1 \times \cdots \times \mu_N$ whose value on a product as above is equal to $\mu_1(A_1) \cdots \mu_N(A_N)$. In particular, any probability measure $\mu \in \Pr(X)$ on X defines a probability measure μ^N on X^N.

The symmetric group $\mathfrak{S}_N$ acts on X^N in the obvious way, and hence its acts on the power set $\mathscr{P}(X^N)$. We call the latter action $\sigma^{(N)}$, so that for $p \in \mathfrak{S}_N$ we have

$$\sigma_p^{(N)}(A_1 \times \cdots \times A_N) = A_{p(1)} \times \cdots \times A_{p(N)}. \quad (8.78)$$

The Cartesian product $X^\infty \equiv X^{\mathbb{N}}$ is well defined both topologically and measure-theoretically (the topology is generated by all products $\prod_i A_i$ with finitely many A_i open and different from X, and likewise for the Borel structure), and the infinite symmetric group $\mathfrak{S}_\infty = \cup_N \mathfrak{S}_N$ acts on it in the obvious way, in that $p \in \mathfrak{S}_N \subset \mathfrak{S}_\infty$ permutes the first N coordinates. Specializing Definition 8.5 to $B = C(X)$, we obtain:

Definition 8.10. *A probability measure ν_N on X^N is called:*

- **permutation-invariant** *if $\nu_N \circ \sigma_p^{(N)} = \nu_N$ for any $p \in \mathfrak{S}_N$.*
- ***K*-exchangeable** *($K \in \mathbb{N}$) if it is permutation-invariant and in addition ν_N is the restriction to B^N of some permutation-invariant probability measure on X^{N+K}.*
- **exchangeable** *if it is K-exchangeable for all $K \in \mathbb{N}$.*

A probability measure ν_∞ on X^∞ is called **permutation-invariant** *if $\nu_\infty \circ \sigma_p^{(N)} = \nu_\infty$ for any $p \in \mathfrak{S}_N$ and $N \in \mathbb{N}$, where $\sigma_p^{(N)}$ acts on $\prod_i A_i$ by (8.78) on the first N factors $A_1, \ldots, A_N$ whilst acting trivally on all remaining A_i's.*

The connection between the two parts of this definition is that ν_N is exchangeable iff it is the restriction to X^N of some permutation-invariant measure ν_∞ on X^∞.

From Theorems 8.6 and 8.3 we obtain the ***Hewitt–Savage Theorem***:

Corollary 8.11. *Let X be a compact Hausdorff space. For any $N \in \mathbb{N}$, any infinitely exchangeable probability measure ν_N on X^N takes the form*

$$\nu_N = \int_{\Pr(X)} dP(\mu)\, \mu^N \tag{8.79}$$

for some probability measure P on $\Pr(X)$ that is uniquely determined by ν_N, and similarly for $N = \infty$, where ν_∞ is a permutation-invariant probability measure.

The two claims in the theorem are equivalent by the remark after Definition 8.10.

The probability measure $P \in \Pr(\Pr(X))$ has the following interpretation. For $N \in \mathbb{N}$ and $(x_1, \ldots, x_N) \in X^N$, define the so-called ***empirical measure*** $E_N^{(x_1,\ldots,x_N)}$ on X as

$$E_N^{(x_1,\ldots,x_N)} = \frac{1}{N}\sum_{i=1}^{N} \delta_{x_i}, \tag{8.80}$$

where δ_x is the Dirac measure on X. Seen as a map on $C(X)$, this is the same as

$$\int_X dE_N^{(x_1,\ldots,x_N)}\, f = \frac{1}{N}\sum_{i=1}^{N} f(x_i). \tag{8.81}$$

Given a probability measure ν_N on X^N, these formulae give a random probability measure on X depending on a drawing from X^N, i.e., a map

$$E_N : X^N \to \Pr(X); \tag{8.82}$$
$$(x_1, \ldots, x_N) \mapsto E_N^{(x_1,\ldots,x_N)}. \tag{8.83}$$

Proposition 8.12. *The probability measure P in Corollary 8.11 is given by*

$$\lim_{N\to\infty} \int_{\Pr(X)} dP_N\, F = \int_{\Pr(X)} dP\, F, \tag{8.84}$$

for each $F \in C(\Pr(X))$ (that is, $P = \lim_{N\to\infty} P_N$ weakly), where $P_N \in \Pr(\Pr(X))$ is the probability measure on $\Pr(X)$ defined by $\nu_N \in \Pr(X^N)$ and (8.82) - (8.83), i.e.,

$$P_N(A) = \nu_N(E_N^{-1}(A)) \;\; (A \subset \Pr(X)). \tag{8.85}$$

Proof. By the Stone–Weierstrass Theorem it suffices to prove (8.84) for linear combinations of monomials like $F(\mu) = \mu(f_1)\cdots\mu(f_K)$, where $f_1, \ldots, f_K \in C(X)$ are arbitrary and $\mu(f) = \int_X d\mu\, f$. This is a simple computation: using (8.85), we have

$$\begin{aligned}\int_{\Pr(X)} dP_N F &= \int_{X^N} d\nu_N(x_1,\ldots,x_N)\, F(E_N^{(x_1,\ldots,x_N)})\\ &= \int_{X^N} d\nu_N(x_1,\ldots,x_N) \prod_{j=1}^{K}\left(\frac{1}{N}\sum_{i=1}^{N} f_j(x_i)\right)\\ &= \int_{\Pr(X)} dP(\mu) \int_{X^N} d\mu^N(x_1,\ldots,x_N) \prod_{j=1}^{K}\left(\frac{1}{N}\sum_{i=1}^{N} f_j(x_i)\right),\end{aligned}$$

where in the third step we used (8.79). The result follows, since clearly

$$\lim_{N\to\infty} \int_{\Pr(X)} dP(\mu) \int_{X^N} d\mu^N(x_1,\ldots,x_N) \prod_{j=1}^{K}\left(\frac{1}{N}\sum_{i=1}^{N} f_j(x_i)\right) =$$

$$\int_{\Pr(X)} dP(\mu) \int_X d\mu(x_1)\, f_1(x_1) \cdots \int_X d\mu(x_K)\, f_k(x_K) = \int_{\Pr(X)} dP F. \quad \square$$

We can also say more about the limit of the sum (8.81), So far, we have been dealing with the Borel σ-algebras $\mathscr{B}_N \subset \mathscr{P}(X^N)$ and $\mathscr{B}_\infty \subset \mathscr{P}(X^\infty)$ generated by the topology (i.e., by the open sets). On top of this, consider $\mathscr{S}_N \subset \mathscr{B}_N$, defined as the σ-algebra generated by the permutation-invariant Borel subsets of X^N, or, equivalently, as the smallest σ-algebra for which the permutation-invariant Borel measurable functions on X^N are measurable. Likewise, $\mathscr{S}_\infty \subset \mathscr{B}_\infty$; regarding $A \subset X^N$ as a subset $A \times \prod_{K>N} X$ of X^∞, we have $\mathscr{S}_\infty = \cap_{N\in\mathbb{N}} \mathscr{S}_N$. For any permutation-invariant probability measure ν_N on X^N, the Hilbert space $L^2(X, \mathscr{S}_N, \nu_N)$ is a closed subspace of $L^2(X^N, \mathscr{B}_N, \nu_N)$, and the associated conditional expectation

$$E_{(\mathscr{S}_N,\nu_N)} : L^2(X^N, \mathscr{B}_N, \nu_N) \to L^2(X, \mathscr{S}_N, \nu_N) \tag{8.86}$$

is defined as the corresponding orthogonal projection. Since $C(X^N) \subset L^2(X^N)$, this map restricts to $C(X^N)$. Similarly for $N = \infty$. For each $N \in \mathbb{N}$, and also for $N = \infty$, we may regard $f \in C(X)$ as a function f_K on X^N through

$$f_K(x_1,\ldots,x_N) = f(x_K) \;\; K = 1,\ldots,N. \tag{8.87}$$

Proposition 8.13. *Let ν_∞ be a permutation-invariant probability measure on X^∞, with restriction ν_N to X^N. Recall* (8.42). *For any $f \in C(X)$ we have* pointwise:

$$S_{1,N}(f) = E_{(\mathscr{S}_N,\nu_N)}(f_1), \;\; \nu_N\text{-almost surely}; \tag{8.88}$$
$$\lim_{N\to\infty} S_{1,N}(f) = E_{(\mathscr{S}_\infty,\nu_\infty)}(f_1), \;\; \nu_\infty\text{-almost surely}, \tag{8.89}$$

where the left-hand sides of (8.88) *and* (8.89) *are functions on X^N and X^∞, respectively. Furthermore, if $\nu_\infty = \mu^\infty$ for some $\mu \in \Pr(X)$, then pointwise on X^∞,*

$$\lim_{N\to\infty} S_{1,N}(f) = \int_X d\mu\, f, \;\; \mu^\infty\text{-almost surely } (f \in C(X)). \tag{8.90}$$

Equivalently, if $L_\mu \subset X^\infty$ is the set of infinite sequences $(x_1, x_2, \ldots)$ in X^∞ for which the limit in (8.90) exists for each $f \in C(X)$ and equals $\int_X d\mu\, f$, then

$$\mu^\infty(L_\mu) = 1. \tag{8.91}$$

Proof. Eq. (8.88) is almost trivial, since $S_{1,N}(f)$ is permutation invariant and hence already lies in $L^2(X, \mathscr{S}_N, \nu_N)$, so that the equality just expresses the projection property $E^2_{(\mathscr{S}_N,\nu_N)} = E_{(\mathscr{S}_N,\nu_N)}$. Eq. (8.89) follows from the ergodic theorem, applied to the probability space $(X^\infty, \mathscr{B}_\infty, \nu_\infty)$, the ***unilateral shift***

$$T : (x_1, x_2, \ldots) \mapsto (x_2, x_3, \ldots),$$

and the random variable f_1 defined by $f \in C(X)$ via (8.87). Since ν_∞ is permutation invariant, it is also T-invariant (in the sense that $\nu_\infty(T^{-1}(A)) = \nu_\infty(A)$ for any $A \subset B_\infty$). This follows either directly, where one has to realize firstly that

$$T^{-1}(A_1 \times A_2 \times \cdots A_n \times \cdots) = X \times A_1 \times A_2 \times \cdots \times \cdots A_n \times \cdots,$$

and secondly that $\mathscr{B}_\infty$ is generated by products $\prod_i A_i$ with finitely many A_i different from X, or, more easily, from Corollary 8.11. The (pointwise) ergodic theorem gives

$$\lim_{N\to\infty} S_{1,N}(f) = E_{(\mathscr{B}_T,\nu_\infty)}(f_1), \ \ \nu_\infty\text{-almost surely } (f \in C(X)), \tag{8.92}$$

where $\mathscr{B}_T$ is the σ-algebra within $\mathscr{B}_\infty$ by the T-invariant sets, and $f_1 \in C(X^\infty)$ is still defined by (8.87). Since $\mathscr{S}_\infty \subset \mathscr{B}_S$ and the left-hand side of (8.89) is $\mathscr{S}_\infty$-measurable (provided it exists, as we have just shown), eq. (8.89) follows from (8.92).

If $\nu_\infty = \mu^\infty$, then the unilateral shift on X^∞ is ergodic by Kolmogorov's 0–1 law, and hence the ergodic theorem gives (8.90). Alternatively, if $\nu_\infty = \mu^\infty$, then the random variables (f_N), defined by (8.87) with $N = \infty$, are i.i.d. (i.e., independent and identically distributed) and (8.90) follows from the strong law of large numbers (which, coherently, in turn may be derived from the ergodic theorem!). □

Note that (8.92) has been proved for $f \in C(X)$, but it holds for many other functions, including $f = 1_A$, where $A \in \mathscr{B}$. This gives ***Borel's law of large numbers***

$$\lim_{N\to\infty} S_{1,N}(1_A) = \mu(A), \ \ \mu^\infty\text{-almost surely.} \tag{8.93}$$

For example, take $X = \{0,1\}$ (e.g., a coin toss with outcomes $1 =$ heads and $0 =$ tails). With $f(x) = x$ in (8.90) or $A = \{1\}$ in (8.93), writing $p = \mu(\{1\})$, we obtain

$$\lim_{N\to\infty} \frac{1}{N}\sum_{i=1}^N x_i = p, \ \ \mu^\infty\text{-almost surely on } \underline{2}^{\mathbb{N}}. \tag{8.94}$$

Equivalently, if $L_p \subset \underline{2}^{\mathbb{N}}$ is the set of infinite binary sequences $x_1 x_2 \cdots$ for which the limit in (8.94) exists and equals p, then $\mu^\infty(L_p) = 1$, cf. (8.91).

8.4 Frequency interpretation of probability and Born rule

Results like (8.90), (8.93), and (8.94) give a relationship between the single-case probabilities $\mu(A)$ or p and the limits of long series of trials on samples drawn according to μ or p. Despite the seemingly comforting appearance of $N < \infty$ on the left-hand side, this relationship depends in an essential way on the infinite idealization X^∞, which is strictly necessary in order to be able to say that the limit (8.94) holds almost surely relative to the measure μ^∞. This violates Earman's Principle (cf. the Introduction), which is the reason why we prefer the limit (8.49) over (8.93).

Although these results are mathematically equivalent, both formalizing the idea that if $(x_1,\ldots,x_N)$ are sampled from X according to some probability measure μ, then $(1/N)\sum_{i=1}^N f(x_i)$ converges to $\int_X d\mu\, f$ as $N \to \infty$, in (8.49) we never need to work with the "actual infinity" $N = \infty$ and (8.49) holds everywhere on $\Pr(X)$ rather than almost everywhere on X^∞. One reason for this is that in (8.93) etc. the choice of the sampling measure μ has to be made at the beginning, whereas in (8.49) it only comes in at the very end. But it has to made either way, and similarly for any other serious effort to relate probability to frequencies in long runs of measurements.

The extreme delicacy of such efforts is clear from the fact that limiting results like (8.90), (8.93), and (8.94) are insensitive to any finite part of the sum, whereas any practical use of probability only involves finite trials. As Lord Keynes once said:

> 'In the long run we are all dead.'

The founder of the mathematical theory of probability expressed himself likewise:

> 'The frequency concept based on the notion of limiting frequency as the number of trials increased to infinity, does not contribute anything to substantiate the applicability of the results of probability theory to real practical problems where we have always to deal with a finite number of trials.' (Kolmogorov).

Moreover, a *definition* of probability based on e.g. (8.93) is well known to be circular: although superficially the "almost sure" terminology in the statement of the result might instill confidence in the reader, in fact it is an exceptionally strong constraint on the sequences $(x_n) \in X^\infty$ in question that the limit should exist *and* has the right value $\mu(A)$, i.e., that $(x) \in L_\mu$, cf. (8.91), and we see that this constraint can only be formulated if the single-case probability μ was already defined in the first place. This shows that the link between probability and frequencies of outcomes of long runs of trials only exists and makes sense if single-case probabilities are prior.

On the other hand, if single-case probabilities are "objective", as those provided by the Born measure in quantum mechanics ought to be at least in remotely realistic interpretations of the theory (as opposed to "personal" or "subjective" probabilities construed as "degrees of belief" or "rationality constraints" or whatever other decision-theoretic concept in human psychology), then it is hard to say what they really mean, since it is precisely about single cases that they do not seem to say anything. This brings us to what we propose to call the ***Paradox of Probability***:

Although single-case probabilities must be logically prior to probabilities construed as frequencies, the numerical values of the former have no bearing on single trials and can only be validated through their predictions about (finite) frequencies.

This paradox imposes the following consistency requirement (which philosophers may want to compare with Lewis's "Principal Principle" that regulates credences):

The assumption that a single-case probability measure be μ must imply that the probabilities for the various outcomes of long runs of repetitions of identical experiments (provided these are possible) are distributed according to μ.

This describes the relationship between theoretical and experimental physics quite well, but still leaves us in the dark as to the meaning of single-case probabilities!

We are now ready to revisit the Born rule, which we already discussed from a purely mathematical point of view in §§§2.1, 2.5, and 4.1. To repeat the main point, if $a = a^* \in B(H)$ is a bounded self-adjoint operator on a Hilbert space, with spectrum $\sigma(a)$, then any state ω on $B(H)$ defines a unique probability measure μ_ω on $\sigma(a) \subset \mathbb{R}$, called the ***Born measure***, such that

$$\omega(f(a)) = \int_{\sigma(a)} d\mu_\omega f, \; f \in C(\sigma(a)), \tag{8.95}$$

where $f(a) \in C^*(a) \subset B(H)$ is defined through the continuous functional calculus (Theorem 4.3). For example, for $f = \mathrm{id}_{\sigma(a)}$, i.e., the function $x \mapsto x$, eq. (8.95) yields

$$\omega(a) = \int_{\sigma(a)} d\mu_\omega(\lambda)\lambda. \tag{8.96}$$

The point of this construction of the Born measure is that it is obtained by simply restricting the state ω, initially defined on $B(H)$, to its commutative C^*-subalgebra $C^*(a)$. If, in the spirit of (exact) Bohrification, such commutative algebras are identified with corners of classical physics within quantum theory, one may argue that Heisenberg gave the right picture of the origin of probability in quantum mechanics:

> 'One may call these uncertainties objective, in that they are simply a consequence of the fact that we describe the experiment in terms of classical physics; they do not depend in detail on the observer. One may call them subjective, in that they reflect our incomplete knowledge of the world.' (Heisenberg, 1958, pp. 53–54)

See, however, §11.1. In any case, there are extensions of this construction to unbounded self-adjoint operators as well as to families of commuting self-adjoint operators, to which the following discussion applies, too, *mutatis mutandis*.

The ***Born rule*** relates the Born measure for a to measurements of a and as such is responsible for most predictions of quantum physics, especially in quantum field theory, where the connection between theory and experiment mainly involves the measurement of cross-sections computed from the Born measure via Feynman rules. The Born rule and the Heisenberg uncertainty relations are often seen as a turning point where indeterminism entered fundamental physics. Nonetheless, it is hard to say what this Born rule actually states! We made a first attempt in §4.1:

If an observable a is measured in a state ω, then the probability $P_\omega(a \in A)$ that the outcome lies in some measurable subset $A \subseteq \sigma(a) \subset \mathbb{R}$ is given by

$$P_\omega(a \in A) = \mu_\omega(A). \tag{8.97}$$

Two questions immediately arise:

1. What is meant by a "measurement" of a (and by its "outcome")?
2. What does the "probability" $P_\omega(a \in A)$ mean?

Perhaps these are even the main questions in the foundations of quantum mechanics. The first will be taken up in Chapter 11; for now, we simply assume that measurements of quantum-mechanical observables a are defined and have outcomes in $\sigma(a)$. The second has just been answered (or some might say evaded): through the Born measure, the formalism of quantum mechanics provides numerical values of $\mu_\omega(A)$, whose *mathematical* meaning seems unquestionable, and whose *operational* meaning is given by the predictions they give for outcomes of long runs of repetitions of identical experiments. Therefore, all that remains to be done is derive these predictions by analogy with the results in §8.3 for the commutative C*-algebra $C(X)$.

One such attempt is—in its strengths and its weaknesses—quite analogous to the Borel's law of large numbers (8.93). Although we will soon move to $B = B(H)$, the following result is valid for any unital C*-algebra B, with infinite tensor product B^∞ as defined in §C.14 and recalled at the end of §8.2, including the map $\varphi_M : B^M \to B^\infty$.

Proposition 8.14. *If $\omega \in S(B)$, there is a unique state ω^∞ on B^∞ such that*

$$\omega^\infty(\varphi_M(b_1 \otimes \cdots \otimes b_M)) = \prod_{n=1}^{M} \omega(b_n), \; M \in \mathbb{N}, b_1, \ldots, b_M \in B. \tag{8.98}$$

Moreover, ω^∞ is pure iff ω is pure.

This is a special case of Proposition C.105, with $C_i = B$ and $\omega_i = \omega$ for all $i \in \mathbb{N}$.

We now take $B = B(H)$ for some separable Hilbert space H, some observable $a = a^* \in B(H)$ with spectrum $\sigma(a) \subset \mathbb{R}$, and some unit vector $\upsilon \in H$, with associated (normal) pure state ω_υ in $B(H)$ defined by $\omega_\upsilon(b) = \langle \upsilon, b\upsilon \rangle$, and Born measure $\mu_{\omega_\upsilon} \equiv \mu_\upsilon$ on $\sigma(a)$. Now take the corresponding pure state ω_υ^∞ on $B(H)^\infty$ and construct the associated GNS-representation $\pi_{\omega_\upsilon^\infty}(B(H)^\infty)$. The Hilbert space $H_{\omega_\upsilon^\infty}$ carrying this representation is an example of an ***infinite tensor product of Hilbert spaces*** in the sense of von Neumann, which may also be defined directly, as follows.

Take sequences $(\psi_n) \equiv (\psi_1, \psi_2, \ldots)$ with $\psi_n \in H$ satisfying the condition

$$\sum_n |\|\psi_n\| - 1| < \infty; \tag{8.99}$$

the rationale behind this condition is that for any sequence (z_n) of complex numbers, the product $\prod_n z_n$ converges *and* has a nonzero limit iff $\sum_n |z_n - 1| < \infty$, so (8.99) is equivalent to the requirement that $\prod_n \|\psi_n\|$ converges to some nonzero value. Following von Neumann, we now introduce the convention that if, for some sequence (z_n) of complex numbers, $\prod_n |z_n|$ converges but $\prod_n z_n$ does not, we define the latter to be zero. On this convention, linear and continuous extension of the expression

$$\langle (\psi_n), (\psi'_n) \rangle = \prod_n \langle \psi_n, \psi'_n \rangle_H, \tag{8.100}$$

defines an inner product on the finite linear span H_0^∞ of all sequences (ψ_n) satisfying (8.99); the ***complete tensor product*** H^∞ is defined as the closure of H_0^∞ in the ensuing norm. However, this is not the Hilbert space of interest, since it is far too large (e.g., it is not separable even if H is). To define interesting separable subspaces of H^∞, we call sequences (ψ_n) and (ψ'_n) that both satisfy (8.99) ***equivalent*** if

$$\sum_n |\langle \psi_n, \psi'_n\rangle - 1| < \infty; \tag{8.101}$$

this turns out to be a *bona fide* equivalence relation. In particular, if (ψ_n) and (ψ'_n) are *in*equivalent, then $\langle(\psi_n),(\psi'_n)\rangle = 0$. For any unit vector $\upsilon \in H$, we now define the ***incomplete tensor product*** H_υ^∞ as the closure of the linear span of all sequences (ψ_n) that satisfy (8.99) and are equivalent to υ^∞ (i.e., the sequence (ψ'_n) with $\psi'_n = \upsilon$ for each n), with inner product borrowed from H^∞ (note that von Neumann's terminology "incomplete" is somewhat confusing, since H_υ^∞ is complete as a normed vector space and in particular it is a Hilbert space). By construction, $\upsilon^\infty \in H_\upsilon^\infty$, and it is easy to show that H_υ^∞ is the closed linear span of all sequences (ψ_n) that differ from $\upsilon \in H$ in at most finitely many places. We often write $\otimes_n \psi_n$ or $\psi_1\otimes\psi_2\otimes\cdots$ for (ψ_n). Furthermore, for any $M \in \mathbb{N}$, any $b \in B(H)$ defines a bounded operator $b_\upsilon^{(M)}$ on H_υ^∞ by continuous linear extension of

$$b_\upsilon^{(M)}(\psi_1\otimes\psi_2\otimes\cdots\otimes\psi_M\otimes\cdots) = \psi_1\otimes\psi_2\otimes\cdots\otimes b\psi_M\otimes\cdots. \tag{8.102}$$

This extends to a representation π_υ^∞ of B^∞ on H_υ^∞, as follows. Define $b^{(M)} \in B^\infty$ by

$$b^{(M)} = \varphi_M(1_H\otimes\cdots\otimes 1_H\otimes b), \tag{8.103}$$

in which $1_H\otimes\cdots\otimes 1_H\otimes b \in B^M$, and $\varphi_M : B^M \to B^\infty$ was defined after (8.58). In other words, for $b \in B(H)$, the operator $b^{(M)}$ is the element of B^∞ given by the equivalence class $[a_{1/N}]_N$ of the sequence $(a_{1/N})_N$ with 1_B in every place except $a_{1/M} = b$. We then define $\pi_\upsilon^\infty(B^\infty)$ by linear and continuous extension of

$$\pi_\upsilon^\infty(b_1^{(M_1)}\cdots b_N^{(M_N)}) = b_{1\upsilon}^{(M_1)}\cdots b_{N\upsilon}^{(M_N)}. \tag{8.104}$$

Proposition 8.15. *For any unit vector $\upsilon \in H$, the* GNS*-representation $\pi_{\omega_\upsilon^\infty}(B^\infty)$ on $H_{\omega_\upsilon^\infty}$ is unitarily equivalent with $\pi_\upsilon^\infty(B^\infty)$ on H_υ^∞, under which equivalence the cyclic vector $\Omega_{\omega_\upsilon^\infty} \in H_{\omega_\upsilon^\infty}$ corresponds with $\upsilon^\infty \in H_\upsilon^\infty$.*

Proof. This is a simple consequence of Proposition C.91 and the equality

$$\omega_\upsilon^\infty(a) = \langle \upsilon^\infty, \pi_\upsilon^\infty(a)\upsilon^\infty\rangle_{H_\upsilon^\infty}, \tag{8.105}$$

initially for $a = b^{(M)}$, subsequently for $a = b_1^{(M_1)}\cdots b_N^{(M_N)}$, and finally, by linearity and continuity, for any $a \in B^\infty$. □

In view of this, we will henceforth identify the two Hilbert spaces etc., so that:

$$H_{\omega_\upsilon^\infty} = H_\upsilon^\infty; \tag{8.106}$$
$$\pi_{\omega_\upsilon^\infty}(b^{(M)}) = b_\upsilon^{(M)}; \tag{8.107}$$
$$\Omega_{\omega_\upsilon^\infty} = \upsilon^\infty. \tag{8.108}$$

Recall that $\mathscr{P}(H)$ is the set of all projections on H, seen as a lattice ordered by $e \leq f$ iff $ef = e$, which is equivalent to $eH \subseteq fH$, and coincides with the order in $B(H)_{\mathrm{sa}}$, cf. Proposition C.170. Also, $\mathscr{B}$ is the Boolean lattice of Borel subsets of $\sigma(a)$, ordered by inclusion. For each Borel set $A \subset \sigma(A)$ we have an associated spectral projection $e_A \in \mathscr{P}(H)$, and the map $A \mapsto e_A$ defined by the Borel functional calculus, i.e., Theorem B.102, is a lattice homomorphism from $\mathscr{B}$ to $\mathscr{P}(H)$. This follows because from the perspective of the Borel functional calculus the map $A \mapsto e_A$ is really the map $1_A \mapsto e_A$, which is the restriction of a homomorphism between C*-algebras and hence preserves positivity. Let $\mathscr{B}^\infty$ be the Boolean lattice of Borel sets $\mathscr{B}^\infty$ in $\sigma(a)^\infty$. As above, take some unit vector $\upsilon \in H$, with corresponding vector state ω_υ on $B(H)$ and associated state ω_υ^∞ on $B(H)^\infty$ as defined in Proposition 8.14, which in turn defines the GNS-representation $\pi_{\omega_\upsilon^\infty}$ of $B(H)^\infty$ on the Hilbert space $H_{\omega_\upsilon^\infty}$. The lattice homomorphism $A \mapsto e_A$ then extends to a homomorphism

$$e^\infty : \mathscr{B}^\infty \to \mathscr{P}(H_{\omega_\upsilon^\infty}); \tag{8.109}$$
$$A_1 \times \cdots \times A_M \times \prod_{M+1}^{\infty} \sigma(a) \mapsto \pi_{\omega_\upsilon^\infty}(e_{A_1}^{(1)} \cdots e_{A_M}^{(M)}); \tag{8.110}$$

this defines e^∞ on the basis Borel sets in $\sigma(a)^\infty$ and extends to all of $\mathscr{B}^\infty$. Realizing $H_{\omega_\upsilon^\infty}$ as the infinite tensor product H_υ^∞, cf. (8.106) - (8.108), we rewrite this as

$$e^\infty\left(A_1 \times \cdots \times A_M \times \prod_{M+1}^{\infty} \sigma(a)\right) = e_{A_1\upsilon}^{(1)} \cdots e_{A_M\upsilon}^{(M)}. \tag{8.111}$$

Theorem 8.16. *Let $a = a^* \in B(H)$, let μ_υ be the Born measure on $\sigma(a)$ defined by some unit vector $\upsilon \in H$, and define e^∞ by* (8.111). *Let $\sigma(a)_\upsilon^\infty$ be the set of all points in $\sigma(a)^\infty$ for which* (8.92), *or, equivalently,* (8.93) *holds (with $\mu \rightsquigarrow \mu_\upsilon$). Then*

$$e^\infty(\sigma(a)_\upsilon^\infty) = 1_{H_{\omega_\upsilon^\infty}}. \tag{8.112}$$

Furthermore, if $A \subseteq \sigma(a)$ is Borel measurable, then, using the notation (8.39),

$$\lim_{N\to\infty} S_{1,N}(e_A) = \mu_\upsilon(A) \cdot 1_{H_{\omega_\upsilon^\infty}}, \tag{8.113}$$

in the strong operator topology (i.e., applied to each fixed vector in $H_{\omega_\upsilon^\infty}$).

This is the ***quantum-mechanical law of strong numbers***, plus its Borel version. In comparison, the strong law of large numbers or Borel's law of large numbers gives

$$\mu_\upsilon^\infty(\sigma(a)_\upsilon^\infty) = 1. \tag{8.114}$$

Proof. For any probability measure μ on any σ-finite compact space X, the corresponding probability measure μ^∞ on X^∞ is characterized by the property

$$\mu^\infty\left(A_1\times\cdots\times A_M\times A\times\prod_{M+2}^{\infty}\sigma(a)\right)=\mu(A)\mu^\infty\left(A_1\times\cdots\times A_M\times\prod_{M+1}^{\infty}\sigma(a)\right),$$

for any $M\in\mathbb{N}$ and Borel sets $A_i\subseteq X$. The measure ν on $\sigma(a)^\infty$ defined by

$$\nu\left(A_1\times\cdots\times A_M\times\prod_{M+1}^{\infty}\sigma(a)\right)=\omega_\upsilon^\infty\left(e^{(1)}_{A_1}\cdots e^{(M)}_{A_M}\right) \tag{8.115}$$

satisfies the above property for $\mu=\mu_\upsilon$ and hence coincides with μ_υ. In view of this, eqs. (C.196) and (8.114) give

$$\langle\Omega_{\omega_\upsilon^\infty},e^\infty(\sigma(a)^\infty_\upsilon)\Omega_{\omega_\upsilon^\infty}\rangle=1. \tag{8.116}$$

For any projection e' and any unit vector $\psi'\in H'$ in any Hilbert space H', the properties $\langle\psi',e'\psi'\rangle=1$, $\|e'\psi'\|=1$, and $e'\psi'=\psi'$ are equivalent. Therefore,

$$e^\infty(\sigma(a)^\infty_\upsilon)\Omega_{\omega_\upsilon^\infty}=\Omega_{\omega_\upsilon^\infty}. \tag{8.117}$$

Consider a vector $\otimes_n\psi_n\in H^\infty_\upsilon$, where only $\psi_1,\ldots,\psi_K$ possibly differ from υ ($K<\infty$). Noting that by (8.106) - (8.107) the right-hand side of (8.115) may be written as

$$\begin{aligned}\omega_\upsilon^\infty\left(e^{(1)}_{A_1}\cdots e^{(M)}_{A_M}\right)&=\langle\Omega_{\omega_\upsilon^\infty},\pi_{\omega_\upsilon^\infty}\left(e^{(1)}_{A_1}\cdots e^{(M)}_{A_M}\right)\Omega_{\omega_\upsilon^\infty}\rangle\\&=\langle\upsilon^\infty,(e^{(1)}_{A_1\upsilon}\otimes\cdots\otimes e^{(M)}_{A_M\upsilon})\upsilon^\infty\rangle,\end{aligned} \tag{8.118}$$

we modify (8.115) so as to define a new measure ν' on $\sigma(a)^\infty$ by

$$\nu'\left(A_1\times\cdots\times A_M\times\prod_{M+1}^{\infty}\sigma(a)\right)=\langle\otimes_n\psi_n,(e^{(1)}_{A_1\upsilon}\otimes\cdots\otimes e^{(M)}_{A_M\upsilon})\otimes_n\psi_n\rangle.$$

Generalizing the above case of μ^∞, the measure $\nu''=\mu_{\psi_1}\times\cdots\times\mu_{\psi_K}\times\prod_{K+1}^{\infty}\mu_\upsilon$ on σ^∞ is characterized by the following two properties:

$$\nu''\left(A_1\times\cdots\times A_K\times\prod_{K+1}^{\infty}\sigma(a)\right)=\mu_{\psi_1}(A_1)\cdots\mu_{\psi_K}(A_K); \tag{8.119}$$

$$\nu''\left(A_1\times\cdots\times A_M\times A\times\prod_{M+2}^{\infty}\sigma(a)\right)=\mu_\upsilon(A)\nu''\left(A_1\times\cdots\times A_M\times\prod_{M+1}^{\infty}\sigma(a)\right),\quad (M>K), \tag{8.120}$$

and hence $\nu'=\nu''$. Therefore, even though $\nu'\neq\mu_\upsilon^\infty$, we have $\nu'(\sigma(a)^\infty_\upsilon)=1$, since membership of $\sigma(a)^\infty_\upsilon$ is entirely defined by the tail of the event. Hence we obtain

$$e^{\infty}(\sigma(a)_{\upsilon}^{\infty}) \otimes_n \psi_n = \otimes_n \psi_n, \tag{8.121}$$

by the same reasoning as for $\upsilon^{\infty} \equiv \Omega_{\omega_{\upsilon}^{\infty}}$. Since the linear span of such vectors is dense in $H_{\upsilon}^{\infty} \equiv H_{\omega_{\upsilon}^{\infty}}$ and the projection $e^{\infty}(\sigma(a)_{\upsilon}^{\infty})$ is bounded, we obtain (8.112).

To derive (8.113), we use the definition of the Born measure μ_{υ} to find

$$\|(S_{1,N}(e_A) - \mu_{\upsilon}(A))\upsilon^{\infty}\| = \frac{1}{N}(\mu_{\upsilon}(A) - 2\mu_{\upsilon}(A)^2), \tag{8.122}$$

which vanishes as $N \to \infty$, so that (8.113) holds on υ^{∞}. A similar computation proves (8.113) on vectors $\otimes_n \psi_n$ as above, since the initial K terms where possibly $\psi_n \neq \upsilon$ drop out in the limit $N \to \infty$. Thus we have (8.113) on a dense subspace of $H_{\omega_{\upsilon}^{\infty}}$. Since the strong limit operator $\mu_{\upsilon}(A) \cdot 1_{H_{\omega_{\upsilon}^{\infty}}}$ is bounded, this proves (8.113). □

An alternative argument shows the mere existence of the limit on the left-hand side of (8.113) on the same dense set, upon which the limit operator is seen to commute with all local and hence (by norm-continuity) with all quasi-local operators. Since ω_{υ} is pure, so is $\omega_{\upsilon}^{\infty}$, and hence $\pi_{\omega_{\upsilon}^{\infty}}$ is irreducible. Thus the limit is a multiple of the unit, and the coefficient $\mu_{\upsilon}(A)$ then follows from the computation

$$\lim_{N \to \infty} \langle \upsilon^{\infty}, S_{1,N}(e_A)\upsilon^{\infty} \rangle = \mu_{\upsilon}(A). \tag{8.123}$$

To reduce the level of abstraction and since it is an important case, we now specialize Theorem 8.16 to a two-level system, i.e., $B = M_2(\mathbb{C})$. In other words, we take $H = \mathbb{C}^2$, and pick a simple observable $a = \mathrm{diag}(1,0)$ with non-degenerate spectrum $\sigma(a) = \underline{2} = \{0,1\}$, so that measurements outcomes are just strings of zero's and one's. Furthermore, we take a unit vector $\upsilon = c_0|0\rangle + c_1|1\rangle$, where $|0\rangle = (1,0)$ and $|1\rangle = (0,1)$ form the standard basis of $\mathbb{C}^2$, and $|c_0|^2 + |c_1|^2 = 1$. We write $p = |c_1|^2$. The Born measure μ_{υ} on $\sigma(a) = \{0,1\}$ is then given by $\mu_{\upsilon}(\{1\}) = p$ and $\mu_{\upsilon}(\{0\}) = 1 - p$; cf. (2.10) - (2.11). Taking $A = \{1\}$, we have $e_A = |1\rangle\langle 1|$. The Hilbert space $(\mathbb{C}^2)_{\upsilon}^{\infty}$ is the closure of the finite linear span of vectors of the kind $\psi_1 \otimes \psi_2 \cdots$ with $\psi_n \in \mathbb{C}^2$ and only finitely many ψ_n possibly different from υ. For $M \in \mathbb{N}$, the operator$|1\rangle\langle 1|^{(M)})$ sends such a vector to $\psi_1 \otimes \psi_2 \cdots \otimes (|1\rangle\langle 1|\psi_M) \otimes \cdots$, with all ψ_n unaffected except for $n = M$. Eqs. (8.112) - (8.113) then simply read

$$e^{\infty}(\underline{2}_p^{\infty}) = 1_{(\mathbb{C}^2)_{\upsilon}^{\infty}}; \tag{8.124}$$

$$\lim_{N \to \infty} \frac{1}{N} \sum_{M=1}^{N} (|1\rangle\langle 1|^{(M)}) = p \cdot 1_{(\mathbb{C}^2)_{\upsilon}^{\infty}}, \tag{8.125}$$

where $\underline{2}_p^{\infty}$ denotes the set of all infinite binary strings $x_1 x_2 \cdots$ for which $x_i \in \underline{2}$ and

$$\lim_{N \to \infty} \frac{1}{N} \sum_{i=1}^{N} x_1 = p, \tag{8.126}$$

and once again the limit in (8.125) is meant strongly, i.e., the expression on the left-hand side must be applied to a fixed vector in $(\mathbb{C}^2)_{\upsilon}^{\infty}$.

Theorem 8.16 forms the (mathematical) culmination of attempts that started in 1960s to derive the Born rule from other postulates of quantum mechanics, notably the so-called ***eigenvalue-eigenvector link***, according to which a quantum-mechanical observable has a definite value if and only if the current quantum state is an eigenvector of the associated operator. This link is applied to the state υ^{∞} (or to any other state with approximately the same tail) and the operators $e^{\infty}(\sigma(a)^{\infty}_{\upsilon})$ and $\lim_{N\to\infty} S_{1,N}(e_A)$. The idea, then, is that according to (8.112), the property expressed by the projection $e^{\infty}(\sigma(a)^{\infty}_{\upsilon})$ is certain in the state υ^{∞} (for qubits this means that any possible infinite string of binary measurement outcomes has average value p). This is reinforced by (8.113), which states that the frequency operator for the outcome A has a sharp limit equal to $\mu(A)$ (for qubits, with $A = \{1\}$ this limit is p).

However, although the mathematics is suggestive, apart from the fact that the eigenvalue-eigenvector link itself falls prey to Earman's Principle (in that sharp eigenvalues and eigenvectors are an idealization in a world full of continuous spectra), this particular application of the link makes sense only *at* $N = \infty$. In this respect, eq. (8.124) has the same drawback as the strong law of large numbers (on which its derivation indeed relies), including the fact that attempts to define probabilities through (8.113) or its special case (8.125) are inherently circular. Moreover, υ^{∞} fails to be an eigenvector of any finite-N approximant to (8.125), and by the same token, the limit operator defined by (8.125) can only be measured via its individual contributions $|1\rangle\langle 1|^{(M)}$, none of which has υ^{∞} as an eigenvector; in fact, it can be shown that any joint eigenvector of all projections $|1\rangle\langle 1|^{(M)}$ is orthogonal to the entire space $(\mathbb{C}^2)^{\infty}_{\upsilon}$ with the complete infinite tensor product $(\mathbb{C}^2)^{\infty}$.

Problems with Earman's Principle are avoided if we use Theorem 8.4 (applied to $B = B(H)$) rather than Theorem 8.16: the sequence of operators $S_{1,N}(e_A)$ forms a continuous section of the continuous bundle of C*-algebras with fibers (8.50) - (8.51), whose limit at $N = \infty$, in the sense of (8.46) or (C.560), is given by

$$S_{1,\infty}(e_A) : \omega \mapsto \omega(A); \tag{8.127}$$

recall that $S_{1,\infty}(e_A) \in C(S(B(H)))$. In particular, for pure states $\omega = \omega_{\upsilon}$ we obtain the Born probability $\mu_{\upsilon}(A)$. As we have also seen in the commutative case, this limit avoids infinite idealizations and other problems with the law of large numbers.

From the point of view of (asymptotic) Bohrification, $C(S(B(H)))$ provides a classical description of a long run of identical experiments, which becomes increasingly accurate as $N \to \infty$; this is the whole point of the limits (8.46) and (C.560). In particular, the unsound eigenvalue-eigenvector link has been replaced by the role of points $\omega \in S(B(H))$ as truthmakers, which is uncontroversial in classical physics. If the quantum state in each identical experiment on the given (single) system is ω, then the above derivation shows that in the limit $N \to \infty$, this state acquires a classical meaning (which according to Bohr would even be the *only* meaning it has), namely as the point in the "classical phase space" $S(B(H))$ that gives the relative frequencies of outcomes of the given long runs of identical experiments. Short of deriving the Born rule, this at least provides the reasoning that links the Born measure (which is canonically given by the theory) to experiment.

8.5 Quantum spin systems: Quasi-local C*-algebras

Beside the Born rule, our second application of the previous formalism is to ***quantum spin systems***, especially to spontaneous symmetry breaking (SSB), see Chapter 10. Postponing a conceptual discussion of infinite systems in their role of idealizations of finite systems to the preamble of that chapter, for the moment we just describe infinite quantum spin systems mathematically. As in §C.14, we take a Hilbert space H, here assumed *finite-dimensional*, i.e., $H \cong \mathbb{C}^n$, and use the standard lattice $\mathbb{Z}^d \subset \mathbb{R}^d$ in dimension d. For any *finite* subset $\Lambda \subset \mathbb{Z}^d$, i.e., $\Lambda \in \mathscr{P}_f(\mathbb{Z}^d)$, we put

$$H_\Lambda = \otimes_{x\in\Lambda} H_x; \tag{8.128}$$
$$A_\Lambda = B(H_\Lambda) \cong \otimes_{x\in\Lambda} B(H_x), \tag{8.129}$$

where $H_x = H$ for each $x \in \Lambda$, cf. (C.297) and (C.303). The symbolic notations

$$A = \otimes_{x\in\mathbb{Z}^d} B(H) = \underline{\lim}_\Lambda A_\Lambda = \overline{\bigcup_{\Lambda\in\mathscr{P}_f(\mathbb{Z}^d)} A_\Lambda}^{\|\cdot\|}, \tag{8.130}$$

all come down to the same thing—see §C.14, notably (C.323) and (C.317)—and define a ***quasi-local C*-algebra***. Elements of each $A_\Lambda \subset A$ are called ***local observables***, those in the closure of their union are referred to as ***quasi-local observables***.

Eq. (8.129) defines a map $\Lambda \mapsto A_\Lambda$, which has three important properties:

$$A_{\Lambda^{(1)}} \subseteq A_{\Lambda^{(2)}} \text{ if } \Lambda^{(1)} \subseteq \Lambda^{(2)} \textit{ (Isotony)}; \tag{8.131}$$
$$[A_{\Lambda^{(1)}}, A_{\Lambda^{(2)}}] = 0 \text{ if } \Lambda^{(1)} \cap \Lambda^{(2)} = \emptyset \textit{ (Einstein locality)}; \tag{8.132}$$
$$A'_\Lambda = A_{\Lambda'} \textit{ (Haag duality)}, \tag{8.133}$$

where A'_Λ in (8.133) is the commutant of A_Λ within A, and, in cute notation, we put $\Lambda' = \mathbb{Z}^d\backslash\Lambda$ (which is infinite), so that the right-hand side of (8.133) denotes

$$A_{\Lambda'} = \otimes_{x\in\Lambda'} B(H) = \overline{\bigcup_{\Lambda^{(1)}\in\mathscr{P}_f(\mathbb{Z}^d\backslash\Lambda)} A_{\Lambda^{(1)}}}^{\|\cdot\|}, \tag{8.134}$$

which is a C*-subalgebra of A. Since $\Lambda^{(2)} \subset \mathbb{Z}^d\backslash\Lambda^{(1)}$ whenever $\Lambda^{(1)} \cap \Lambda^{(2)} = \emptyset$, Haag duality implies Einstein locality (and sharpens it), but it is still worth mentioning these properties separately: although in quantum spin systems (8.133)—and hence (8.132)—holds, Einstein locality is a more fundamental property (e.g. it is also valid in algebraic quantum field theory, where Haag duality may well fail).

We now discuss some C*-algebraic concepts that will be needed for the analysis of SSB. Through the associated GNS-representation $\pi_\omega : A \to B(H_\omega)$, any state ω on A defines two interesting subalgebras of $B(H_\omega)$, which *a priori* may be different:

- The ***center*** $A^c_\omega = \pi_\omega(A)'' \cap \pi_\omega(A)'$;
- The ***algebra at infinity*** $A^\infty_\omega = \bigcap_{\Lambda\in\mathscr{P}_f(\mathbb{Z}^d)} \pi_\omega(A_{\Lambda'})''$.

Recall that the center of a von Neumann algebra $M \subset B(H)$ is $M \cap M'$, and that M is called a factor if $M \cap M = \mathbb{C} \cdot 1$ (cf. §C.21), so A^c_ω is the center of the von Neumann algebra $\pi_\omega(A)''$. It is easy to show from Einstein locality that $A^\infty_\omega \subseteq A^c_\omega$. If each local algebra A_Λ is simple, Haag duality yields the opposite inclusion, so in that case,

$$A^\infty_\omega = A^c_\omega. \tag{8.135}$$

Given (8.129), this applies as long as $\dim(H) < \infty$, in which case also A is simple.

The algebra at infinity provides a new perspective on the macroscopic observables in §8.2. Averages like $|\Lambda|^{-1} \sum_{x \in \Lambda} b(x)$, where $b \in B(H)$, do not have a limit in A as $\Lambda \uparrow \mathbb{Z}^d$, but (depending on ω) their representatives $|\Lambda|^{-1} \sum_{x \in \Lambda} \pi_\omega(b(x))$ may have a weak limit in $B(H_\omega)$. If they do, Einstein locality implies that the limit operator lies in algebra at infinity A^∞_ω (and hence, assuming (8.135), in A^c_ω). If the algebra of infinity is trivial (i.e. $\mathbb{C} \cdot 1_{H_\omega}$), macroscopic observables are therefore "c-numbers", i.e., multiples of the unit operator. In particular, they do not fluctuate, which is among the defining properties of *pure* thermodynamic phases. Formally, this idea is captured by the following generalization of the notion of a pure state:

Definition 8.17. *A representation $\pi(A)$ is* **primary** *if $\pi(A)'' \cap \pi(A)'$ is trivial.*
A state $\omega \in S(A)$ is **primary** *if the* GNS*-representation π_ω is primary.*

For compact groups G (or rather their group C*-algebras $C^*(G)$), all representations are completely reducible, and a representation is primary iff it is a (possibly infinite) multiple of some irreducible representation. However, this is not the right picture for general groups or C*-algebras, which requires some discussion. In preparation, we call some representation $\pi'(A)$ on a Hilbert space $H' \subset H$ a ***subrepresentation*** of a representation $\pi(A)$ on H, written $\pi' \subset \pi$, if $\pi' = \pi_{|H'}$. Subrepresentations π' of π correspond to projections $e \in \pi(A)'$, such that $\pi'(a) = e\pi(a)$. It follows that $\pi_1(A)$ and $\pi_2(A)$ have equivalent subrepresentations iff there exists a nonzero partial isometry $w : H_1 \to H_2$ such that $w\pi_1(a) = \pi_2(a)w$ for all $a \in A$.

Definition 8.18. *Two representations π_1 and π_2 of a C*-algebra A are called:*

1. **equivalent** *if there is a unitary $u : H_1 \to H_2$ such that $u\pi_1(a)u^* = \pi_2(a)$ $(a \in A)$;*
2. **quasi-equivalent** *if every subrepresentation of π_1 has a subrepresentation that is equivalent to some subrepresentation of π_2, and* vice versa*;*
3. **disjoint** *if they do not have any equivalent subrepresentations.*

We say that two states ω_1 and ω_2 on A equivalent, disjoint, or quasi-equivalent if the corresponding GNS*-representations π_{ω_1} and π_{ω_2} have the said property.*

In other words, π_1 and π_2 are quasi-equivalent iff π_1 has no subrepresentations disjoint from π_2, and *vice versa*. This, in turn, is equivalent to the property that the set of ***π_i-normal states*** on A, i.e. states of the form $a \mapsto \mathrm{Tr}(\rho\pi_i(a))$ with $\rho \in \mathscr{D}(H_i)$, is the same for $i = 1$ as it is for $i = 2$. Contrapositively, π_1 and π_2 are disjoint iff no state exists that is both π_1-normal and π_2-normal. For example, taking $A = C(X)$, in which case states are probability measures μ on X, equivalence and disjointness of states recovers the usual notions of equivalence and disjointness of measures, respectively (i.e., having the same null sets and having disjoint supports).

Proposition 8.19. *For any state ω, if $\omega = t\omega_1 + (1-t)\omega_2$ for some $t \in (0,1)$, then ω_1 and ω_2 are disjoint iff there is a projection $e \in A^c_\omega = \pi_\omega(A)' \cap \pi_\omega(A)''$ such that*

$$\pi_\omega(A)_{|eH_\omega} \cong \pi_{\omega_1}(A); \tag{8.136}$$
$$\pi_\omega(A)_{|e^\perp H_\omega} \cong \pi_{\omega_2}(A). \tag{8.137}$$

Since subrepresentations of $\pi_\omega(A)$ always correspond to projections $e \in \pi_\omega(A)'$; the key assumption being made here is that e also lies in the weak closure $\pi_\omega(A)''$.

Proof. One direction is easy: if (8.136) - (8.137) hold, then (arguing by contradiction) equivalent subrepresentations $\pi_1(A)$ of $\pi_{\omega_1}(A)$ and $\pi_2(A)$ of $\pi_{\omega_2}(A)$ are given by projections $e_1 \leq e$ and $e_2 \leq e^\perp = 1_{H_\omega} - e$, respectively, through

$$\pi_i(a) = \pi_\omega(a)_{|e_i H_\omega}, \;\; (i = 1,2, a \in A), \tag{8.138}$$

and the partial isometry w on H_ω whose restriction to $e_1 H_\omega$ implements a (unitary) equivalence between $\pi_1(A)$ and $\pi_2(A)$ by definition satisfies $w^*w = e_1$, $ww^* = e_2$. Moreover, $e_1 \leq e$ implies $we = w$ and $e_2 \leq e^\perp$ implies $e^\perp w = w$, which together give $e^\perp we = w$. Furthermore, again by definition, $w \in \pi_\omega(A)'$. If now $e \in \pi_\omega(A)''$, then $we = ew$. Combining these equalities gives $w = 0$, which is the desired contradiction.

Lemma 8.20. *For any functional $\omega' \in A^*$ such that $0 \leq \omega' \leq \omega$, where $\omega \in S(A)$, there is an operator $c \in \pi_\omega(A)'$ on H_ω such that $0 \leq c \leq 1_H$ and*

$$\omega'(a) = \langle \Omega_\omega, c\pi_\omega(a)\Omega_\omega \rangle \;\; (a \in A). \tag{8.139}$$

In particular, there is a vector $\xi \in H_\omega$ such that

$$\omega'(a) = \langle \xi, \pi_\omega(a)\xi \rangle_{H_\omega}. \tag{8.140}$$

Proof. Cauchy–Schwarz for the positive semidefinite form $\langle a,b \rangle' = \omega'(a^*b)$ gives

$$|\omega'(a^*b)|^2 \leq \omega'(a^*a)\omega'(b^*b) \leq \omega(a^*a)\omega(b^*b) = \|\pi_{\omega_i}(a)\Omega_{\omega_i}\|^2 \|\pi_{\omega_i}(b)\Omega_{\omega_i}\|^2.$$

Hence we obtain a well-defined positive quadratic form B on H_ω, initially defined on the dense domain $\pi_\omega(A)\Omega_\omega \times \pi_\omega(A)\Omega_\omega$ by the formula

$$B(\pi_\omega(a)\Omega_\omega, \pi_\omega(b)\Omega_\omega) = \omega'(a^*b), \tag{8.141}$$

and extended to $H_\omega \times H_\omega$ by continuity; the above inequality immediately gives $|B(\varphi,\psi)| \leq \|\varphi\|\|\psi\|$, and hence Proposition B.79 yields an operator $0 \leq c \leq 1_H$ such that $B(\varphi,\psi) = \langle \varphi, c\psi \rangle$. With (8.141), this gives (8.139). We now compute

$$\begin{aligned}\omega'(a^*b^*d) &= B(\pi_\omega(ba)\Omega_\omega, \pi_\omega(d)\Omega_\omega) = \langle \pi_\omega(a)\Omega_\omega, \pi_\omega(b^*)c\pi_\omega(d)\Omega_\omega \rangle \\ &= B(\pi_\omega(a)\Omega_\omega, \pi_\omega(b^*d)\Omega_\omega = \langle \pi_\omega(a)\Omega_\omega, c\pi_\omega(b^*)\pi_\omega(d)\Omega_\omega \rangle,\end{aligned}$$

so that $[c, \pi_\omega(b^*)] = 0$ for each $b \in A$, i.e., $c \in \pi_\omega(A)'$. Writing $c = c_1^2$ with $c_1^* = c_1$, and then $\xi = c_1\Omega_\omega$, completes the proof. □

We continue the proof of Proposition 8.19 in the converse direction. Assume

$$\omega = t\omega_1 + (1-t)\omega_2 = \omega_1' + \omega_2', \tag{8.142}$$

with $\omega_1' = t\omega_1$ and $\omega_2' = (1-t)\omega_2$, so that $0 \leq \omega_1' \leq \omega$ and $0 \leq \omega_2' \leq \omega$. It follows from the first claim in Lemma 8.20 that there is $c \in B(H_\omega)$ as stated such that

$$\omega_1'(a) = \langle \Omega_\omega, c\pi_\omega(a)\Omega_\omega \rangle; \tag{8.143}$$
$$\omega_2'(a) = \langle \Omega_\omega, (1_{H_\omega} - c)\pi_\omega(a)\Omega_\omega \rangle, \tag{8.144}$$

where (8.144) follows from (8.143), (C.196), and $\omega = \omega_1' + \omega_2'$. Define $\omega' \in A^*$ by

$$\omega'(a) = \langle \Omega_\omega, c(1_{H_\omega} - c)\pi_\omega(a)\Omega_\omega \rangle. \tag{8.145}$$

We have $0 \leq \omega' \leq \omega_1'$ (since $c(1_{H_\omega} - c) \leq c$) as well as $0 \leq \omega' \leq \omega_2'$ (since also $c(1_{H_\omega} - c) \leq 1_{H_\omega} - c$). Now assume that ω_1 and ω_2 are disjoint. Applying (8.140) with $\omega \rightsquigarrow \omega_i$ shows that ω' is π_1-normal as well as π_2-normal, so that it follows from the remarks following Definition 8.18 that $\omega' = 0$. Since Ω_ω is cyclic for $\pi_\omega(A)$ by the GNS-construction, this implies $c(1_{H_\omega} - c) = 0$, and hence $c^2 = c$. Since $c \geq 0$, which implies $c^* = c$, it follows that c is a projection, henceforth called e. Therefore,

$$\omega_1(a) = \langle \Omega_\omega, e\pi_\omega(a)\Omega_\omega \rangle / \|e\Omega_\omega\|^2; \tag{8.146}$$
$$\omega_2(a) = \langle \Omega_\omega, e^\perp \pi_\omega(a)\Omega_\omega \rangle / \|e^\perp \Omega_\omega\|^2, \tag{8.147}$$

where $t = \|e\Omega_\omega\|^2$. We see from these formulae and Proposition C.91 that π_{ω_1} and π_{ω_2} are equivalent to the restrictions of π_ω to eH_ω and $e^\perp H_\omega$, respectively; under this equivalence, the cyclic vectors Ω_{ω_1} and Ω_{ω_2} correspond with $e\Omega_\omega/\|e\Omega_\omega\|$ and $e^\perp\Omega_\omega/\|e^\perp\Omega_\omega\|$, respectively. Since $e \in \pi_\omega(A)'$ by Lemma 8.20, it only remains to be shown that $e \in \pi_\omega(A)''$. To this effect, for any $b \in \pi_\omega(A)'$ and $\psi \in H_\omega$, define

$$\omega'' \in A^*;$$
$$\omega''(a) = \langle e^\perp be\psi, \pi_\omega(a)e^\perp be\psi \rangle. \tag{8.148}$$

Then ω'' is positive, as well as π_{ω_2}-normal, the latter because of the presence of the projection $e^\perp$ and (8.147). But for $a \in A^+$ we have the inequalities

$$0 \leq \omega''(a) \leq \|e^\perp b\|^2 \langle e\psi, \pi_\omega(a)e\psi \rangle, \tag{8.149}$$

so that $0 \leq \omega'' \leq \omega_1''$ for the state (assuming $e\psi$ is a unit vector)

$$\omega_1''(a) = \langle \psi, e\pi_\omega(a)e\psi \rangle. \tag{8.150}$$

Since $e\psi \in eH_\omega$, the latter state is π_{ω_1}-normal, so that ω_1'' is itself π_{ω_1}-normal by Lemma 8.20 (which argument by now should sound familiar). Again invoking disjointness of ω_1 and ω_2, it follows that $\omega'' = 0$, which, since ψ was arbitrary, in turn yields $e^\perp be = 0$ for any $b \in \pi_\omega(A)'$. This forces $e \in \pi_\omega(A)''$. □

The first of the following corollaries to Proposition 8.19 is ***Hepp's Lemma:***

Lemma 8.21. *Let $\pi : A \to B(H)$ be a representation of A, and let ψ_1, ψ_2 be unit vectors in H. Then the vector states $\omega_i(a) = \langle\psi_i, \pi(a)\psi_i\rangle$ $(i = 1,2)$ are disjoint iff*

$$\langle\psi_1, \pi(a)\psi_2\rangle = 0 \ (a \in A). \tag{8.151}$$

Proof. Take, for example, $\omega = \frac{1}{2}(\omega_1 + \omega_2)$ in Proposition 8.19. □

Corollary 8.22. *1. Two primary states are either disjoint or quasi-equivalent.*
2. A state is primary iff it has no convex decomposition into disjoint *states.*

Recall that a state is pure if it has no nontrivial convex decomposition *whatsoever*. The analogy between pure states and primary states may be completed as follows:

- ω pure $\leftrightarrow \pi_\omega(A)' = \mathbb{C} \cdot 1$ (cf. Theorem C.90);
- ω primary $\leftrightarrow \pi_\omega(A)' \cap \pi_\omega(A)'' = \mathbb{C} \cdot 1$ (cf. Definition 8.17).

A physical property of primary states is that the corresponding correlation functions have a clustering property of a kind that may even be experimentally accessible:

Theorem 8.23. *A state ω on a quasi-local C*-algebra A* (8.130) *has trivial algebra at infinity, i.e., $A_\omega^\infty = \mathbb{C} \cdot 1$, iff it is* **clustering**, *in the following sense: for each $a \in A$ and $\varepsilon > 0$ there is a finite $\Lambda \subset \mathbb{Z}^d$ such that for all $b \in A_{\Lambda'}$ with $\|b\| = 1$ one has*

$$|\omega(ab) - \omega(a)\omega(b)| \leq \varepsilon. \tag{8.152}$$

In particular, if ω is primary, then it is clustering and hence (8.152) *holds.*

Proof. The complete proof is quite technical, but the main idea is as follows. Choose finite regions Λ_n moving to infinity (i.e., eventually avoiding any given Λ), and pick elements $c_n \in A_{\Lambda_n}$), $\|c_n\| = 1$. The sequence $(\pi_\omega(c_n))$ in $B(H_\omega)$ has a weakly convergent subsequence with limit $c \in B(H_\omega)$. This follows from the Banach–Alaoglu Theorem B.48, applied to $B(H_\omega)$ seen as the dual space of $B_1(H_\omega)$): on the unit ball, the corresponding weak*-topology on $B(H_\omega)$ coincides with the weak operator topology, so that the unit ball in $B(H_\omega)$ is weakly compact and the theorem applies.

- By von Neumann's Bicommutant Theorem C.127 we have $c \in \pi_\omega(A)''$.
- By Einstein locality (8.132) and the delocalization of the Λ_n, also $c \in \pi_\omega(A)'$.

Hence $c \in A_\omega^c$, and by a more refined argument (which is unnecessary if if $A_\omega^\infty = A_\omega^c$), even $c \in A_\omega^\infty$. So if $A_\omega^\infty = \mathbb{C} \cdot 1$ we have $c = (\Omega_\omega, c\Omega_\omega) \cdot 1$. On the other hand,

$$\langle\Omega_\omega, c\Omega_\omega\rangle = \lim_n \langle\Omega_\omega, \pi_\omega(c_n)\Omega_\omega\rangle = \lim_n \omega(c_n),$$

so that we may compute:

$$\lim_n \omega(ac_n) = \lim_n \langle\Omega_\omega, \pi_\omega(a)\pi_\omega(c_n)\Omega_\omega\rangle = \langle\Omega_\omega, \pi_\omega(a)c\Omega_\omega\rangle = \omega(a) \lim_n \omega(c_n).$$

Thus for any $\varepsilon > 0$ there is an N such that $|\omega(ac_n) - \omega(a)\omega(c_n)| \leq \varepsilon$ for all $n > N$. To derive (8.152) from this, an easy *reductio ad absurdum* argument suffices.

The converse direction follows from Kaplansky's Density Theorem C.131. □

8.6 Quantum spin systems: Bundles of C*-algebras

In this section we reformulate the theory of quantum spin systems in the continuous C*-bundle language of §8.2. First, for each $N \in \mathbb{N}$ we define $\Lambda_N \in \mathscr{P}_f(\mathbb{Z}^d)$ by

$$\Lambda_N = \{x \in \mathbb{Z}^d \mid \|x\| \leq N\}. \tag{8.153}$$

We then have the following analogue of the continuous bundle of C*-algebras $A^{(q)}$ of C*-algebras of Theorem 8.8. The base space remains $I = 1/\dot{\mathbb{N}} \subset [0,1]$, where $\dot{\mathbb{N}} = \{1, 2, \ldots, \infty\}$ (seen as possible values of $1/\hbar$), and the fibers are given by

$$A_0 = A = \underline{\lim}_N A_{\Lambda_N} = \overline{\bigcup_{N \in \mathbb{N}} A_{\Lambda_N}}^{\|\cdot\|}; \tag{8.154}$$

$$A_{1/N} = A_{\Lambda_N} = B(H_{\Lambda_N}) \ (N \in \mathbb{N}), \tag{8.155}$$

cf. (8.128) - (8.130), still assuming $\dim(H) < \infty$. As before, the topology of this bundle is defined through its continuous cross-sections $(a_{1/N})_{N \in \dot{\mathbb{N}}}$, which are the analogues of the quasi-local sequences of Definition 8.7. Given (8.154) - (8.155), each fiber algebra $A_{1/N}$ is a subalgebra of A_0, and some sequence $(a_{1/N})_{N \in \dot{\mathbb{N}}}$ simply defines a continuous cross-section of the bundle iff within A (i.e. in norm) we have

$$\lim_{N \to \infty} a_{1/N} = a_0. \tag{8.156}$$

In other words, a sequence $(a_{1/N})_{N \in \mathbb{N}}$ with $a_{1/N} \in A_{1/N} \subset A$ is quasi-local in the sense of Definition 8.7 iff it converges in A (i.e., iff it is Cauchy in the norm of A).

The continuous bundle of Theorem 8.4 makes equally good sense for quantum spin systems. First, with $B = B(H) \cong M_n(\mathbb{C})$, the fibers are obviously given by

$$A_0^{(c)} = C(S(B(H))); \tag{8.157}$$

$$A_{1/N}^{(c)} = B(H_{\Lambda_N}). \tag{8.158}$$

Second, the continuous sections are once again specified via symmetrization maps

$$S_{M,N} : B(H_{\Lambda_M}) \to B(H_{\Lambda_N}), \tag{8.159}$$

defined similarly to (8.39), namely via canonical symmetrizers

$$S_N : B(H_{\Lambda_N}) \to B(H_{\Lambda_N}) \tag{8.160}$$

that are defined à la (8.35) - (8.36), where this time the tensor product and ensuing permutation in (8.35) are over all sites $x \in \Lambda_N$. Regarding $a_{1/M} \in B(H_{\Lambda_M})$ as an element $a'_{1/M}$ of $B(H_{\Lambda_N})$ via the embedding $A_{\Lambda_M} \hookrightarrow A_{\Lambda_N}$, we finally define $S_{M,N}$ by

$$S_{M,N}(a_{1/M}) = S_N(a'_{1/M}). \tag{8.161}$$

Symmetric and quasi-symmetric sequences may then be defined exactly as in Definitions 8.2 and 8.3; each quasi-symmetric sequence $(a_{1/N})_{N\in\mathbb{N}}$ duly has a limit $a_0 \in A_0^{(c)}$ given by (8.46), where ω^N is defined as in (8.47), once again with a tensor product over all sites $x \in \Lambda_N$. By definition, the continuous sections of the bundle (8.157) - (8.158) are then given by the quasi-symmetric sequences.

Although the fibers A in (8.154) and $C(S(B(H)))$ in (8.157) are as wide apart as they could possibly be, they stunningly arise as limit algebras at $\hbar = 0$ (i.e., $N = \infty$ or $\Lambda = \mathbb{Z}^d$) for the same fiber algebras (8.155) and (8.158) at $\hbar > 0$ (i.e., $N < \infty$ or $\Lambda \in \mathscr{P}_f(\mathbb{Z}^d)$). As in §8.2, the difference lies in the choice of the topology on the bundle, defined via the continuous sections, which in the first case are the quasi-local sequences, and in the second are the quasi-symmetric (i.e., macroscopic) ones.

An interesting connection between these bundles can be obtained via the following concept, which in a way justifies the introduction of the bundles themselves.

Definition 8.24. *A* **continuous field of states** *on a continuous bundle of C*-algebras with fibers $(A_{1/N})_{N\in\dot{\mathbb{N}}}$ is a family $(\omega_{1/N})_{N\in\dot{\mathbb{N}}}$ where*

$$\omega_{1/N} \in S(A_{1/N}); \tag{8.162}$$
$$\lim_{N\to\infty} \omega_{1/N}(a_{1/N}) = \omega_0(a_0), \tag{8.163}$$

for each continuous cross-sections $(a_{1/N})$. In that case, we write

$$\omega_0 = \lim_{N\to\infty} \omega_{1/N}, \tag{8.164}$$

despite the fact that all states in question may be defined on different C-algebras.*

For example, any state ω on $A_0 = A$ as in (8.154) defines a continuous field:

Proposition 8.25. *For any state $\omega \in S(A)$, the set $(\omega_{1/N})_{N\in\dot{\mathbb{N}}}$ of states defined by*

$$\omega_0 = \omega; \tag{8.165}$$
$$\omega_{1/N} = \omega_{|A_{1/N}}, \tag{8.166}$$

is a continuous field of states on the bundle with fibers (8.154) - (8.155).

Proof. We use the notation of Definition 8.7. For local sequences (8.57) we have

$$\omega_{1/N}(a_{1/N}) = \omega(a_{1/N}) = \omega(a_{1/M}),$$

for all $N \geq M$. Since $a_0 = a_{1/M}$, this equals $\omega_0(a_0)$. For quasi-local sequences, a_0 is the limit of the sequence $(a_{1/N})$ in the norm of A, so that $\omega(a_{1/N}) \to \omega(a_0)$. □

Definition 8.26. *A state $\omega \in S(A)$ is* **macroscopic** *if $\lim_{N\to\infty} \omega(a_{1/N})$ exists for any (quasi-) symmetric sequence $(a_{1/N})$.*

It does not matter whether we put "symmetric" or "quasi-symmetric" here, since existence of the limit for symmetric sequences implies its existence on quasi-symmetric sequences. Indeed, using the fact that $\|\omega\| = 1$, we may estimate

$$|\omega(a_{1/N}) - \omega(a_{1/M})| \leq |\omega(\tilde{a}_{1/N}) - \omega(\tilde{a}_{1/N})| \\ + |\|a_{1/N} - \tilde{a}_{1/N}\| + \|a_{1/M} - \tilde{a}_{1/M}\|, \quad (8.167)$$

for any sequence $(\tilde{a}_{1/M})$. Using Definition 8.3, and hence taking $(\tilde{a}_{1/M})$ symmetric, we see that if $(\omega(\tilde{a}_{1/N}))$ is a Cauchy sequence, then so is $(\omega(a_{1/N}))$.

Proposition 8.27. *A macroscopic state ω determines a state $\omega_0^{(c)}$ on $C(S(B))$ by*

$$\omega_0^{(c)}(a_0) = \lim_{N\to\infty} \omega(a_{1/N}), \quad (8.168)$$

where $(a_{1/N})$ is any quasi-symmetric sequence with limit $a_0 \in C(S(B))$, cf. (8.46).

Proof. First, note that $\omega_0^{(c)}$ is independent of the choice of the approximating sequence $(a_{1/N})$, since by the same argument as in the proof of Proposition C.126, if $a_{1/N} \to a_0$ as well as $a'_{1/N} \to a_0$, we have

$$\lim_{N\to\infty} \|a_{1/N} - a'_{1/N}\| = \|a_0 - a_0\| = 0, \quad (8.169)$$

and because $\|\omega\| = 1$ for any state ω, we also have

$$|\omega(a_{1/N} - a'_{1/N})| \leq \|a_{1/N} - a'_{1/N}\|. \quad (8.170)$$

Eqs. (8.169) - (8.170) obviously imply

$$\lim_{N\to\infty} \omega(a_{1/N}) = \lim_{N\to\infty} \omega(a'_{1/N}). \quad (8.171)$$

We next show that if $a_{1/N} \to a_0$ and $b_{1/N} \to b_0$ in the sense of (C.560), then

$$a_{1/N}b_{1/N} \to a_0 b_0.$$

If $(a_{1/N})$ is a symmetric sequence à la (8.43), and likewise $(b_{1/N})$, where we may assume without loss of generality that M is the same for both, then

$$a_0(\rho) = \rho^M(a_{1/M}), \quad (8.172)$$

where $\rho \in S(B)$, and likewise for b_0. Using (8.38), we obtain

$$\lim_{N\to\infty} \rho^N(a_{1/N}b_{1/N}) = \rho^M(a_{1/M})\rho^M(b_{1/M}) = a_0(\rho)b_0(\rho) = (a_0b_0)(\rho). \quad (8.173)$$

In particular, if $a_{1/N} \to a_0$, then $a^*_{1/N}a_{1/N} \to a^*_0 a_0$. Since ω is a state, it follows that $\omega_0^{(c)}(a^*_0 a_0) \geq 0$, and since also $\omega_0^{(c)}(1_{S(B)}) = 1$ (because the sequence with $a_{1/N} = 1_{H_{\Lambda_N}}$ converges to $1_{S(B(H))}$), the claim follows for symmetric sequences. For quasi-symmetric sequences $(a_{1/N})$ the result follows by approximating $(a_{1/N})$ with symmetric sequences (cf. Definition 8.3). □

Each state $\omega_0^{(c)} \in S(A_0^{(c)})$ is represented by a probability measure μ on the state space $S(B(H))$ of $B(H)$. We compute this measure if $\omega \in S(A)$ is ***permutation-invariant*** in that each restriction $\omega_{1/N} = \omega_{|B(H_{\Lambda_N})}$ is invariant under the natural action of the permutation group $\mathfrak{S}_{|\Lambda_N|}$ on $B(H_{\Lambda_N}) \cong \otimes_{x\in\Lambda_N} B(H)$, where $N \in \mathbb{N}$ and $|\Lambda_N|$ is the number of points in Λ_N (as in the case of B^∞ in §8.2). It follows from the Quantum De Finetti Theorem 8.9 (and the fact that that the set $S^{\mathfrak{S}_\infty}(A)$ of permutation-invariant states on A is a so-called *Bauer simplex*) that each permutation-invariant state $\omega \in S^{\mathfrak{S}_\infty}(A)$ takes the form

$$\omega = \int_{S(B(H))} d\mu(\rho)\, \rho^\infty, \tag{8.174}$$

where μ is some probability measure on $S(B(H))$, and $\rho \in S(B(H))$; the associated state ρ^∞ on A is defined by its values on each $A_{\Lambda_N} \subset A$ via the isomorphism

$$A_{\Lambda_N} \cong \otimes_{x\in\Lambda_N} B(H). \tag{8.175}$$

Furthermore, the integral in (8.174) is defined weakly, i.e., for any $a \in A$ the number $\omega(a)$ is obtained by integrating the function $\rho \mapsto \rho^\infty(a)$ on $S(B(H))$ with respect to μ. In particular, $\omega \in \partial_e S^{\mathfrak{S}_\infty}(A)$ iff μ is a Dirac measure on $S(B(H))$.

Proposition 8.28. *Each permutation-invariant state $\omega \in S^{\mathfrak{S}_\infty}(A)$ is macroscopic (cf. Definition 8.26), and the probability measure μ on $S(B(H))$ defined by $\omega_0^{(c)}$ via* (8.168) *coincides with the one appearing in* (8.174).

Proof. Let $(a_{1/N})$ be a symmetric sequence (the quasi-symmetric case follows from this), so that $a_{1/N} = S_{M,N}(a_{1/M})$ for some M whenever $N > M$, cf. (8.43). The limit $a_0 \in C(S(B(H)))$ is given by (8.172), so that state $\omega_0^{(c)}$ on $C(S(B(H)))$ defined by

$$\omega_0^{(c)}(f) = \int_{S(B(H))} d\mu(\rho)\, f(\rho) \tag{8.176}$$

satisfies the required condition

$$\lim_{N\to\infty} \omega_{1/N}(a_{1/N}) = \omega_{1/M}(a_{1/M}) = \int_{S(B(H))} d\mu(\rho)\, \rho^M(a_{1/M}) = \omega_0^{(c)}(a_0). \qquad \square$$

To proceed we make the following technical assumption on $\omega \in S(A)$ (which is satisfied in typical physical models): if $\pi_\omega(a_{1/N}) \to 0$ weakly in $B(H_\omega)$, for some sequence $(a_{1/N})$ where $a_{1/N} \in A_{1/N}$, then $\pi_\omega(a_{1/N})\Omega_\omega \to 0$ in $B(H_\omega)$ (in norm).

Theorem 8.29. *Assume that the state ω in part 1 below (and likewise the states ω_1 and ω_2 in part 2) satisfies the above technical condition. Then:*

1. *If ω is a* primary *macroscopic state on A, then the corresponding state $\omega_0^{(c)}$ is* pure, *i.e., the probability measure μ on $S(B(H))$ is a Dirac measure.*
2. *If ω_1 and ω_2 are quasi-equivalent primary macroscopic state on A, then $\mu_1 = \mu_2$ (and hence if $\mu_1 \neq \mu_2$, then ω_1 and ω_2 are disjoint).*

The techniques in the proof below can be used to show that our additional assumption is equivalent to: if (8.178) below holds weakly in $B(H_\omega)$, then it also holds strongly. Thus we could have redefined a macroscopic state ω as one for which the strong limit $\lim_{N\to\infty}\pi_\omega(a_{1/N})$ exists in $B(H_\omega)$ (and some authors indeed do so).

Proof. We first show that if ω is a primary macroscopic state on A, and $(a_{1/N})$ is symmetric (from which the quasi-symmetric case duly follows) such that

$$\lim_{N\to\infty}\omega(a_{1/N})=\alpha, \tag{8.177}$$

then, in the weak operator topology on the GNS-representation space $B(H_\omega)$,

$$\lim_{N\to\infty}\pi_\omega(a_{1/N})=\alpha\cdot 1_{H_\omega}. \tag{8.178}$$

To this end, we first note that $\|a_{1/N}\|$ is uniformly bounded in N: if $(a_{1/N})$ is symmetric, as in (8.43), then obviously $\|a_{1/N}\|=\|a_{1/M}\|$ for all $N>M$, so that if $(a_{1/N})$ is merely quasi-symmetric we have $\|a_{1/N}\|\leq\|a_{1/M}\|+\varepsilon$ for all $N>M$, where ε and M are the quantities appearing in Definition 8.3. Hence it is enough to establish the weak limit (8.178) between states in a dense set, viz. $\pi_\omega(b)\Omega_\omega$, where $b\in A$, or even in $\cup_N A_{1/N}$. Furthermore, using the polarization identity (A.5) and (C.8) - (C.9), it is enough to prove that for each $K\in\mathbb{N}$ and $b\in A_{1/K}$, we have

$$\lim_{N\to\infty}\omega(b^*a_{1/N}b)=\alpha\omega(b^*b), \tag{8.179}$$

since by the GNS-construction we obviously have

$$\langle\pi_\omega(b)\Omega_\omega,\pi_\omega(a_{1/N})\pi_\omega(b)\Omega_\omega\rangle=\omega(b^*a_{1/N}b). \tag{8.180}$$

Theorem 8.23 implies (or even states) that if ω is primary, for each $b\in A$ and $\varepsilon>0$ there is $M\in\mathbb{N}$ such that for all $a\in A'_{\Lambda_M}$ with $\|a\|=1$, we have

$$|\omega(b^*ba)-\omega(b^*b)\omega(a)|\leq\varepsilon. \tag{8.181}$$

Assuming $b\in A_{1/K}$, we first note that $\lim_{N\to\infty}[a_{1/N},b]=0$ in norm (even though $\lim_{N\to\infty}a_{1/N}$ does not exist in norm), and secondly that, for any given $M\in\mathbb{N}$, if $\tilde{a}_{1/N}$ is the same as $a_{1/N}$ except that in any term $b_1\otimes\cdots\otimes b_{|\Lambda_N|}$ that contributes to $a_{1/N}$ we replace $b_i\rightsquigarrow 1_H$ whenever $b_i\in A_{1/M}$, then

$$\lim_{N\to\infty}\|\tilde{a}_{1/N}-a_{1/N}\|=0. \tag{8.182}$$

Given (8.177), these facts with (8.181) immediately give (8.179) and hence (8.178).

According to (8.177) and (8.178), the state $\omega_0^{(c)}\in S(C(S(B(H))))$ is given by

$$\omega_0^{(c)}(a_0)=\lim_{N\to\infty}\langle\Omega_\omega,\pi_\omega(a_{1/N})\Omega_\omega\rangle, \tag{8.183}$$

where $a_{1/N}$ is some symmetric sequence converging to $\to a_0$ in the sense of (C.560); as in the proof of Proposition 8.27, the left-hand side is independent of the particular choice of this sequence. The proof of Proposition 8.27 also showed that if $a_{1/N} \to a_0$ and $b_{1/N} \to b_0$, then $a_{1/N}b_{1/N} \to a_0 b_0$, so that

$$\begin{aligned} \omega_0^{(c)}(a_0 b_0) &= \lim_{N\to\infty} \langle \Omega_\omega, \pi_\omega(a_{1/N}b_{1/N})\Omega_\omega\rangle \\ &= \lim_{N\to\infty} \langle \Omega_\omega, \pi_\omega(a_{1/N}) - \alpha\cdot 1_{H_\omega})\pi_\omega(b_{1/N})\Omega_\omega\rangle + \alpha\beta, \end{aligned}$$

where α is defined by (8.177), and likewise β. At this point that we need our additional assumption, which, together with uniform boundedness of $\|\pi_\omega(a_{1/N})\|$ and hence of $\|\pi_\omega(a_{1/N})\Omega_\omega\|$ in N yields that the first term in the second line is zero. Therefore, $\omega_0^{(c)}$ is multiplicative and hence pure (cf. Proposition C.14).

To prove the second claim, first suppose ω_1 and ω_2 are quasi-equivalent. In that case, up to unitary equivalence, either π_{ω_1} is a subrepresentation of π_{ω_2}, or *vice versa*; assume the former. We then have a projection $e \in \pi_{\omega_2}(A)'$ such that

$$\pi_{\omega_1}(a) = e\pi_{\omega_2}(a), \tag{8.184}$$

for each $a \in A$, and since $e = 1_{H_{\omega_1}}$ by construction, eq. (8.178) gives

$$\lim_{N\to\infty} \pi_{\omega_1}(a_{1/N}) = \alpha_1 \cdot e; \tag{8.185}$$

$$\lim_{N\to\infty} \pi_{\omega_2}(a_{1/N}) = \alpha_2 \cdot 1_{H_{\omega_2}}. \tag{8.186}$$

Multiplying both sides of (8.186) with e gives $\alpha_1 = \alpha_2$. □

Corollary 8.30. *A permutation-invariant state $\omega \in S^{\mathfrak{S}_\infty}(A)$ is primary iff the corresponding measure μ in* (8.174) *is a Dirac measure, and it is pure iff the latter is supported by a pure state on $B(H)$.*

Proof. In the first claim, the inference from "primary" to "Dirac" obviously follows from Theorem 8.29. The converse direction is a consequence of the commutation theorem (C.329) for von Neumann algebras, combined with the fact that each representation of $B(H)$ for finite-dimensional H is primary (which in turn follows from the fact, not proved in this book, that $B(H)$ has just one irreducible representation, up to equivalence). The second claim follows from Proposition C.105. □

Finally, *one* macroscopic state generates many others. A ***folium*** in the state space $S(A)$ of a C*-algebra A is a convex, norm-closed subspace $\mathscr{F}$ of $S(A)$ with the property that if $\omega \in \mathscr{F}$ and $b \in A$ such that $\omega(b^*b) > 0$, then the "reduced" state $\omega_b : a \mapsto \omega(b^*ab)/\omega(b^*b)$ must be in $\mathscr{F}$. For example, if π is a representation of A on a Hilbert space H, then the set of all density matrices on H (i.e. the π-normal states on A) comprises a folium $\mathscr{F}_\pi$. In particular, each state ω on A defines a folium $\mathscr{F}_\omega \equiv \mathscr{F}_{\pi_\omega}$ through its GNS-representation π_ω. It then follows from cyclicity of the GNS-representation that each state in the folium $\mathscr{F}_\omega$ of a macroscopic state $\omega \in S(A)$ is automatically macroscopic and even has the same limit state $\omega^{(c)}$ as ω.

Notes

§8.1. **Large quantum numbers**

Theorem 8.1 has been adapted from Landsman (1998b); the proof relies on Simon (1980), who, generalizing the case of $SU(2)$ treated by Lieb (1973), in turn uses the coherent states for Lie groups introduced by Perelomov (1972, 1986). Duffield (1999) gives the details of the method of steepest descent used in proving (8.30). Although this material was inspired by Bohr's Correspondence Principle, at the end of the day the relationship may seem remote.

§8.2. **Large systems**

The theory in this section, which elaborates on Landsman (2007), is a reformulation in terms of continuous bundles of C*-algebras of the formal parts of a series of papers on quantum mean-field systems by Raggio & Werner (1989, 1991), Duffield & Werner (1992a,b,c), and Duffield, Roos, & Werner (1992). These models have their origin in the treatment of the BCS theory of superconductivity due to Bogoliubov (1958) and Haag (1962); for further references see the notes to §10.8.

§8.3. **Quantum de Finetti Theorem**

Theorem 8.9 is due to Størmer (1969), whose proof was based on the fact that the $\mathfrak{S}_\infty$-action on B^∞ is ***asymptotically abelian***, in that for any $a, a' \in B^\infty$ one has

$$\inf\{\|[\alpha_p(a), a']\|, p \in \mathfrak{S}_\infty\} = 0.$$

This implies that $S^{\mathfrak{S}_\infty}(B^\infty)$ is a Choquet simplex, which quickly leads to (8.66). Our proof is taken from Hudson & Moody (1975). See also Caves, Fuchs, & Schack (2002a). Finite-size corrections to Theorem 8.9 are studied e.g. in König & Mitchison (2009). Corollary 8.11 is due to Hewitt & Savage (1955), who credit Jules Haag (rather than De Finetti) for the binary case (i.e., $X = \{0,1\}$). See Kallenberg (2005) for an exhaustive account of such results (in classical probability theory).

Proposition 8.12 is taken from Diaconis & Freedman (1980), who also give finite-size corrections to Corollary 8.11, as follows. Let a permutation-invariant probability measure ν_N on X^N be K-exchangeable, so that there is a permutation-invariant probability measure ν_{N+K} on X^{N+K} whose restriction to X^N is ν_N. Let P_{N+K} be the probability measure on $\Pr(X)$ defined by ν_{N+K} as in (8.85), i.e., $P_{N+K}(A) = \nu_{N+K}(E_{N+K}^{-1}(A))$, and finally define

$$\nu'_{N+K} = \int_{\Pr(X)} dP_{N+K}(\mu)\, \mu^{N+K},$$

as in (8.79). Then, in terms of the usual norm on the Banach dual $C(X^N)^*$,

$$\|\nu_N - \nu'_N\| \leq \frac{K(K-1)}{N}.$$

Proposition 8.13 is stated without proof in Kingman (1978). See Mackey (1974) or Gray (2009) for ergodic theory in connection with probability theory.

Of course, there are numerous results in probability theory that do not share the problems of the law of large numbers. For example, in the situation (8.94), for any $\varepsilon > 0$ one has the ***Chernoff–Hoeffding bound***

$$\mu^N\left(|\frac{1}{N}\sum_{i=1}^{N} x_i - p| \geq \varepsilon|\right) \leq e^{-2N\varepsilon^2},$$

which is superior to the weak law of large numbers, i.e., for every $\varepsilon > 0$,

$$\lim_{N\to\infty} \mu^N\left(|\frac{1}{N}\sum_{i=1}^{N} x_i - p| \geq \varepsilon|\right) = 0,$$

which from the point of view of Earman's Principle is already a marked conceptual improvement over the strong law (but which is mathematically weaker).

§8.4. **Frequency interpretation of probability and Born rule**

The Kolmogorov quote is from Fine (1973, p. 94), which even 40 years later is still to be recommended as one of the best (technical) book on the foundations of probability theory. See also Hájek & Hitchcock (2016) for a comprehensive recent survey of the philosophy of probability. The Keynes quote is from Hacking (2001, p. 149), which is a very elementary introduction to the foundations of probability At a more advanced level see also Gillies (2000), whilst Howson (1995) is a useful brief survey.

The original version of the *Principal Principle* (Lewis, 1980) equated probability (or chance) as subjective degree of belief (i.e. credence) with objective chance (though in the single case as opposed to relative frequency. Our own version in the main text is meant to clarify the relationship between singe-case probabilities and long run frequencies, both seen as objective.

Attempts to derive the Born rule started with Finkelstein (1965) and were continued e.g. by Hartle (1968), Farhi, Goldstone, & Gutmann (1989), Van Wesep (2006), Aguirre & Tegmark (2011), Moulay (2014), and others, partly based on indubitable mathematical arguments in the spirit of the strong law of large numbers supplied by e.g. Ochs (1977, 1980), Bugajski & Motyka (1981), Pulmannová & Stehlková (1986). Such attempts (typically presented as claims) provoked valid critiques of the kind mentioned in the main text from e.g. Cassinelli & Sánchez-Gómez (1996) and Caves & Schack (2005). For a balanced account see also Cassinelli & Lahti (1989). Infinite tensor products of Hilbert spaces were introduced by von Neumann (1938).

Our approach, which is sympathetic to both sides of the dispute, is a vast expansion of Landsman (2008). The existence of $e\infty$ as in (8.109) - (8.110) is based on the same extension argument that proves the Kolmogorov existence theorem for infinite product probabilities, see e.g. Dudley (1989), proof of Theorem 8.2.2, and Van Wesep (2006), who carries out the proof for $X = \{0,1\}$.

There is also a large (and inconclusive) literature on alleged derivations of the Born rule in the context of the Many-Worlds (i.e. Everettian) Interpretation of quantum mechanics, which may be traced back from Wallace (2012), who supports such derivations, and Dawid & Thébault (2015), who criticize them.

§8.5. Quantum spin systems: Quasi-local C*-algebras

Basic references are Ruelle (1969), Israel (1979), Bratteli & Robinson (1987, 1997), and Simon (1993); for macroscopic states see Hepp (1972) and Sewell (2002). Naaijkens (2013) is a useful brief introduction to quantum spin systems.

The proof that Haag duality holds for quantum spin systems is far from trivial: see Simon (1993), Prop. IV.1.6. In the proof of (8.135), simplicity of A given simplicity of each A_Λ is easily inferred from the fact that if $I \subset A$ is an ideal, then $I_\Lambda = I \cap A_\Lambda$ is an ideal in $A_\Lambda = B(H_\Lambda)$, which must be either zero or A_Λ, both of which contradict non-triviality of I. Theorem 8.23 is a famous result due to Lanford & Ruelle (1969), partly anticipated by Powers (1967). For a complete proof see also Simon (1993), Theorem IV.1.4.

§8.5. Quantum spin systems: Bundles of C*-algebras

This section was inspired by Landsman (2007), §6, and Gerisch (1993).

Folia of states (in the sense meant here) were introduced by Haag, Kadison, & Kastler (1970), but note that the name "folium" is poorly chosen, since $S(A)$ is by no means foliated by its folia (for example, a folium may contain subfolia).

Chapter 9
Symmetry in algebraic quantum theory

In §3.9 we defined symmetries of classical physics as symmetries of either Poisson manifolds or Poisson algebras; these notions are equivalent. At the bare level of the underlying phase space X, merely seen as a locally compact space (rather than a Poisson manifold), the key result establishing this equivalence is this:

Theorem 9.1. *Let X and Y be locally compact Hausdorff spaces. Each isomorphism $\alpha : C_0(Y) \to C_0(X)$ is induced by a homeomorphism $\varphi : X \to Y$ via $\alpha = \varphi^*$ (and so each automorphism of $C_0(X)$ is induced by a homeomorphism of X).*

More generally, if A and B are commutative C-algebras, then each isomorphism $\alpha : A \to B$ is induced by a homeomorphism $\varphi : \Sigma(B) \to \Sigma(A)$ of the corresponding Gelfand spectra via $\alpha = G_B^{-1} \circ \varphi^* \circ G_A$, where $G_A : A \to C_0(\Sigma(A))$ is the Gelfand ismomorphism, cf.* (C.79), *and similarly for B (and so each automorphism of A is induced by a homeomorphism of its Gelfand spectrum $\Sigma(A)$).*

This immediately follows from Theorems C.8 and C.45, and Corollary C.48.

In Chapter 5 we saw that even in elementary quantum mechanics, where $A = B(H)$ for some Hilbert space H, the concept of a symmetry is more diverse, as least apparently, since a non-commutative C*-algebra like $B(H)$ gives rise to numerous "quantum structures". The ones we looked at were listed after Proposition 5.3, viz.

1. The ***normal pure state space*** $\mathscr{P}_1(H)$, dressed with a transition probability (2.44).
2. The ***normal (total) state space*** $\mathscr{D}(H)$, seen as a convex set; see Theorem 2.8.
3. The ***self-adjoint operators*** $B(H)_{\text{sa}}$ on H, seen as a Jordan algebra.
4. The ***effects*** $\mathscr{E}(H) = [0,1]_{B(H)}$ on H, seen as a convex poset.
5. The ***projections*** $\mathscr{P}(H)$ on H, seen as an orthocomplemented lattice.
6. The ***unital commutative C*-subalgebras*** $\mathscr{C}(B(H))$ ***of*** $B(H)$, seen as a poset.

Each structure comes with its own notion of a symmetry, see Definition 5.1. This raises two questions, which for $B(H)$ were completely answered in Chapter 5:

- The possible equivalence of the various notions of quantum symmetry;
- Unitary implementability of symmetries.

Indeed, it was found that if $\dim(H) > 2$, then all these notions of symmetry are equivalent, as well as unitarily implementable à la Wigner; see Theorem 5.4.

K. Landsman, *Foundations of Quantum Theory*,
Fundamental Theories of Physics 188, DOI 10.1007/978-3-319-51777-3_9

9.1 Symmetries of C*-algebras and Hamhalter's Theorem

In this chapter we generalize this analysis from $A = B(H)$ to arbitrary C*-algebras A, which for simplicity we assume to have a unit 1_A. See §C.25 for terminology.

Definition 9.2. *Let A be a unital C*-algebra.*

1. *The* **pure state space** $P(A) = \partial_e S(A)$ *is the extreme boundary of the state space* $S(A)$*, seen as a uniform space equipped with a transition probability*

$$\tau(\omega,\omega') = \inf\{\omega(a) \mid a \in A, 0 \leq a \leq 1_A, \omega'(a) = 1\}. \tag{9.1}$$

A **Wigner symmetry** *of A is a uniformly continuous bijection* $\mathsf{W} : P(A) \to P(A)$ *with uniformly continuous inverse that preserves transition probabilities, i.e.,*

$$\tau(\mathsf{W}(\omega)\mathsf{W}(\omega')) = \tau(\omega,\omega'), \; \omega,\omega' \in P(A). \tag{9.2}$$

If $A = B(H)$, Proposition C.177 guarantees that the above expression reproduces the standard quantum-mechanical transition probabilities (2.44), but compared to this special case, one novel aspect of $P(A)$ is that all pure states are now taken into account (as opposed to merely the normal ones, which notion is undefined for general C*-algebras anyway). Another is that in order to obtain the desired equivalence with other structures, the set $P(A)$ should carry a uniform structure, namely the w^*-uniformity inherited from A^*.

2. *The* **state space** $S(A)$ *is the set of all states on A, seen as a compact convex set in the w^*-topology inherited from the embedding* $S(A) \subset A^*$*. A* **Kadison symmetry** *of A is an affine homeomorphism* $\mathsf{K} : S(A) \to S(A)$.
Compared to $A = B(H)$, firstly *all* states are now taken into account (instead of all *normal* states), and secondly we have added a continuity condition on K.

3. *Any C*-algebra A defines an associated* **Jordan algebra** *(more precisely, a JB-algebra), namely* $A_{\rm sa}$ *equipped with the commutative product* $a \circ b = \frac{1}{2}(ab+ba)$. *A* **Jordan symmetry** J *of A is a Jordan isomorphism of* $(A_{\rm sa}, \circ)$ *(or, equivalently, an invertible unital linear isometry of* $(A_{\rm sa}, \|\cdot\|)$*, which in turn is the same as a unital linear order isomorphism of* $(A_{\rm sa}, \leq)$*, cf. Lemma C.173). A* **weak Jordan symmetry** *of A is an invertible map* $\mathsf{J} : A_{\rm sa} \to A_{\rm sa}$ *whose restriction to each subspace* $C_{\rm sa}$ *of* $A_{\rm sa}$*, where* $C \in \mathscr{C}(A)$*, is linear and preserves the Jordan product.*

4. *The* **effects** *in A comprise the order unit interval* $\mathscr{E}(A) = [0, 1_A]$*, i.e., the set of all* $a \in A_{\rm sa}$ *such that* $0 \leq a \leq 1_A$*, seen as a convex poset in the obvious way. A* **Ludwig symmetry** *of A is an affine order isomorphism* $\mathsf{L} : \mathscr{E}(A) \to \mathscr{E}(A)$.

5. *The projections* $\mathscr{P}(A)$ *in A form an orthomodular poset (cf. Definition D.1) with* $e \leq f$ *iff* $ef = e$ *and* $e^{\perp} = 1_A - e$*; if A is a von Neumann algebra (cf. Proposition C.136), or more generally an AW*-algebra or a Rickart C*-algebra (see §C.24),* $\mathscr{P}(A)$ *is even an orthomodular lattice. A* **von Neumann symmetry** *of A is an isomorphism* $\mathsf{N} : \mathscr{P}(A) \to \mathscr{P}(A)$ *of orthomodular posets.*

6. *The poset* $\mathscr{C}(A)$ *(lying at the heart of exact Bohrification) consists of all* commutative *C*-subalgebras of A that contain the unit* 1_A*, partially ordered by inclusion. A* **Bohr symmetry** *of A, then, is an order isomorphism* $\mathsf{B} : \mathscr{C}(A) \to \mathscr{C}(A)$.

The structures 1, 2, 3 (with Jordan symmetries), and 4 are equivalent; see Theorem C.179 for $1 \leftrightarrow 2$ and Theorem C.172 for $2 \leftrightarrow 3$; the equivalence $3 \leftrightarrow 4$ is proved in exactly the same way as in Proposition 5.21, with Lemma 5.20 for the special case $A = B(H)$ replaced by Lemma C.173 (which has the same proof). From 1–4 we pick the Jordan algebra structure of A, since it gives the most straightforward results.

Henceforth, A and B are unital C*-algebras, and we define a ***weak Jordan isomorphism*** of A and B as an invertible map $\mathsf{J} : A_{\mathrm{sa}} \to B_{\mathrm{sa}}$ whose restriction to each subspace C_{sa} of A_{sa}, where $C \in \mathscr{C}(A)$, is linear and preserves the Jordan product $\circ$ (so that a Jordan symmetry of A alone is a weak Jordan automorphism of of A). Such a map complexifies to a map $\mathsf{J}_{\mathbb{C}} : A \to B$ in the usual way, i.e. writing $a \in A$ as $a = b + ic$, with $b^* = b$ and $c^* = c$, cf. (C.9), and put $\mathsf{J}_{\mathbb{C}}(a) = \mathsf{J}(b) + i\mathsf{J}(c)$). If no confusion arises, we just write J for $\mathsf{J}_{\mathbb{C}}$. We first turn to Bohr symmetries.

Proposition 9.3. *Given a weak Jordan isomorphism* $\mathsf{J} : A_{\mathrm{sa}} \to B_{\mathrm{sa}}$, *the ensuing map* $\mathsf{B} : \mathscr{C}(A) \to \mathscr{C}(B)$ *defined by* $\mathsf{B}(C) = \mathsf{J}_{\mathbb{C}}(C) \equiv \mathsf{J}(C)$ *is an order isomorphism.*

Note that as an argument of B the symbol C is a point in the poset $\mathscr{C}(A)$, whereas as an argument of $\mathsf{J}_{\mathbb{C}}$ it is a subset of A, so that $\mathsf{J}_{\mathbb{C}}(C)$ stands for $\{\mathsf{J}_{\mathbb{C}}(c) \mid c \in C\}$.

Proof. The restriction $\mathsf{J}_{|C} : C \to B$ is a homomorphism of C*-algebras on each *commutative* C*-algebra $C \subset A$ (although $\mathsf{J} : A \to B$ may not be). Since $\mathsf{J}_{|C}$ is injective on C_{sa} (where it coincides with J), it is also injective on C. Hence $\mathsf{J}_{|C}$ is isometric by Theorem C.62.3, so that its range is closed and therefore $\mathsf{J}(C)$ is a commutative C*-algebra in B, which is unital if C is. Trivially, if $C \subseteq D$ in A (so that $C \leq D$ in $\mathscr{C}(A)$), then $\mathsf{J}(C) \subseteq \mathsf{J}(D)$ in B (so that $\mathsf{J}(C) \leq \mathsf{J}(D)$ in $\mathscr{C}(B)$). □

The converse, however, is a deep result, which we call ***Hamhalter's Theorem***:

Theorem 9.4. *Let A and B be unital C*-algebras and let* $\mathsf{B} : \mathscr{C}(A) \to \mathscr{C}(B)$ *be an order isomorphism. Then there is a weak Jordan isomorphism* $\mathsf{J} : A_{\mathrm{sa}} \to B_{\mathrm{sa}}$ *such that* $\mathsf{B} = \mathsf{J}_{\mathbb{C}}$. *Moreover, if A is isomorphic to neither* $\mathbb{C}^2$ *nor* $M_2(\mathbb{C})$, *then* J *is uniquely determined by* B, *so in that case there is a bijective correspondence* $\mathsf{J} \leftrightarrow \mathsf{B}$ *between weak Jordan symmetries* J *of A and Bohr symmetries* B *of A.*

Before proving this, let us explain why $\mathbb{C}^2$ and $M_2(\mathbb{C})$ are exceptional. In the first case, $\mathscr{C}(\mathbb{C}^2) \cong \{0,1\}$ (with $0 \equiv \mathbb{C} \cdot 1_2$ and $1 \equiv \mathbb{C}^2$), which admits just one order isomorphism (viz. the identity map), which is induced by both the map $(a,b) \mapsto (b,a)$ and by the identity map on $\mathbb{C}^2$ (each of which is a weak Jordan automorphism).

In the second case, the poset $\mathscr{C}(M_2(\mathbb{C}))$ has a bottom element $0 \equiv \mathbb{C} \cdot 1_2$, as before, but no top element; each element $C \neq \mathbb{C} \cdot 1_2$ of $\mathscr{C}(M_2(\mathbb{C})$ is a unitary conjugate of the diagonal subalgebra $D_2(\mathbb{C})$, with $0 \leq C$ but no other orderings. Furthermore, $C \cap C' = \mathbb{C} \cdot 1_2$ whenever $C \neq C'$. Hence any order isomorphism of $\mathscr{C}(M_2(\mathbb{C}))$ maps $\mathbb{C} \cdot 1_2$ to itself and permutes the C's. Thus each map $\mathsf{J} : M_2(\mathbb{C})_{\mathrm{sa}} \to M_2(\mathbb{C})_{\mathrm{sa}}$ whose complexification $\mathsf{J}_{\mathbb{C}} : M_2(\mathbb{C}) \to M_2(\mathbb{C})$ shuffles the C's isomorphically (as C*-algebras) gives a weak Jordan automorphism. For example, take $(a,b) \mapsto (b,a)$ on $D_2(\mathbb{C})$ and the identity on each $C \neq D_2(\mathbb{C})$); this induces the identity map on $\mathscr{C}(M_2(\mathbb{C})$. It follows that there are vastly more weak Jordan automorphisms of $M_2(\mathbb{C})$ than there are order isomorphisms of $\mathscr{C}(M_2(\mathbb{C}))$.

Proof. The key to the proof lies in the commutative case, which can be reduced to topology. If $A = C(X)$, any $C \in \mathscr{C}(A)$ induces an equivalence relation $\sim_C$ on X by

$$x \sim_C y \text{ iff } f(x) = f(y) \, \forall f \in C. \tag{9.3}$$

This, in turn, defines a partition $X = \bigsqcup_\lambda K_\lambda$ of X (henceforth called π), whose blocks $K_\lambda \subset X$ are the equivalence classes of $\sim_C$. To study a possible inverse of this procedure, for any closed subset $K \subset X$ we define the ideal

$$I_K = C(X;K) = \{f \in C(X) \mid f(x) = 0 \forall x \in K\}, \tag{9.4}$$

in $C(X)$, and its unitization $\dot{I}_K = I_K \oplus \mathbb{C} \cdot 1_X$, which evidently consists of all continuous functions on X that are constant on K. If X is finite (and discrete), each partition π of X defines some unital C*-algebra $C \subseteq C(X)$ through

$$C = \bigcap_{K_\lambda \in \pi} \dot{I}_{K_\lambda}, \tag{9.5}$$

which consists of all $f \in C(X)$ that are constant on each block K_λ of the given partition π. In that case, the correspondence $C \leftrightarrow \pi$, where π is defined by the equivalence relation $\sim_C$ in (9.3), gives a bijection between $\mathscr{C}(C(X))$ and the set $\mathfrak{P}(X)$ of all partitions of X. For example, the subalgebra $C = \dot{I}_K$ corresponds to the partition consisting of K and all singletons not lying in K. Given the already defined partial order on $\mathscr{C}(C(X))$ (i.e., $C \leq D$ iff $C \subseteq D$), we may promote this bijection to an order isomorphism of posets if we define the partial order $\leq'$ on $\mathfrak{P}(X)$ to be the *opposite* of the natural one $\leq$ in which $\pi \leq \pi'$ (where π and π' consist of blocks $\{K_\lambda\}$ and $\{K'_{\lambda'}\}$, respectively) iff each K_λ is contained in some $K'_{\lambda'}$ (i.e., π is finer than π'). The partial ordering $\leq'$ makes $\mathfrak{P}(X)$ a complete lattice, whose top element consists of all singletons on X and whose bottom element just consists of X itself: the former corresponds to $C(X)$, which is the top element of $\mathscr{C}(C(X))$, whilst the latter corresponds to $\mathbb{C} \cdot 1_X$, which is the bottom element of $\mathscr{C}(C(X))$.

For general compact Hausdorff spaces X, since $C(X)$ is sensitive to the topology of X the equivalence relation (9.3) does not induce arbitrary partitions of X. It turns out that each $C \in \mathscr{C}(C(X))$ induces an ***upper semicontinuous partition*** (abbreviated by *u.s.c. decomposition*) of X, i.e.,

- Each block K_λ of the partition π is closed;
- For each block K_λ of π, if $K_\lambda \subseteq U$ for some open $U \in \mathscr{O}(X)$, then there is $V \in \mathscr{O}(X)$ such that $K_\lambda \subseteq V \subseteq U$ and V is a union of blocks of π (in other words, if K is such a block, then $V \cap K = \emptyset$ implies $K = \emptyset$).

This can be seen as follows. Firstly, if we equip π with the quotient topology with respect to the the natural map $q : X \to \pi$, $x \mapsto K_\lambda$ if $x \in K_\lambda$, then π is compact, for X is compact. Moreover, π is Hausdorff. To see this, let K_λ and K_μ be two distinct *points* in π. Recall that $x, y \in K_\lambda$ if and only if $f(x) = f(y)$ for each $f \in C$. Since $K_\lambda \neq K_\mu$, there is some $x \in K_\lambda$, some $y \in K_\mu$ and some $f \in C$ such that $f(x) \neq f(y)$, whence there are open disjoint $U, V \subseteq \mathbb{C}$ such that $f(x) \in U$ and $f(y) \in V$.

Define $\hat{f} : \pi \to \mathbb{C}$ by $\hat{f}(K_\lambda) = f(x)$ for some $x \in K_\lambda$. By definition of K_λ, this is independent of the choice of $x \in K_\lambda$, hence $\hat{f}$ is well defined. Again by definition, we have $f = \hat{f} \circ q$, hence $q^{-1}(\hat{f}^{-1})[U] = f^{-1}[U]$, which is open in X since f is continuous. Since π is equipped with the quotient topology, it follows that $\hat{f}^{-1}[U]$ is open in π, and similarly $\hat{f}^{-1}[V]$ is open. Moreover, we have $\hat{f}(K_\lambda) = f(x)$ and $f(x) \in U$, hence $K_\lambda \in \hat{f}^{-1}[U]$, and similarly, $K_\mu \in \hat{f}^{-1}[V]$. We conclude that π is also Hausdorff. Since q is a continuous map between compact Hausdorff spaces, it follows that q is closed. It is a standard result in topology that q is closed iff π is a u.s.c. decomposition, so we have now proved the latter.

Consequently, by the same maps (9.3) and (9.5), the poset $\mathscr{C}(C(X))$ is anti-isomorphic to the poset $\mathfrak{F}(X)$ of all u.s.c. decompositions of X in the natural ordering $\leq$ (which proves that $\mathfrak{F}(X)$ is a complete lattice, since $\mathscr{C}(C(X))$ is). This is still a complicated poset; assuming X to be larger than a singleton, the next step is to identify the simpler poset $\mathscr{F}_2(X)$ of all closed subsets of X containing at least two elements within $\mathfrak{F}(X)$, where (as above) we identify a closed $K \subseteq X$ with the (u.s.c.) partition π_K of X whose blocks are K and all singletons not lying in K (note that the poset $\mathscr{F}(X)$ of all closed subsets of X is less useful, since any singleton in $\mathscr{F}(X)$ gives rise to the bottom element of $\mathfrak{F}(X)$). To do so, we first recall that β is said to *cover* α in some poset if $\alpha < \beta$, and $\alpha \leq \gamma < \beta$ implies $\alpha = \gamma$. If the poset has a bottom element, then its covers are precisely its *atoms*. Furthermore, note that since the bottom element 0 of $\mathfrak{F}(X)$ consists of singletons, the atoms in $\mathfrak{F}(X)$ are the partitions of the form $\pi_{\{x_1,x_2\}}$ (where $x_1 \neq x_2$). It follows that some partition $\pi \subset \mathfrak{F}(X)$ lies in $\mathscr{F}_2(X) \subset \mathfrak{F}(X)$ iff exactly one of the following conditions holds:

- π is an atom in $\mathfrak{F}(X)$, i.e., $\pi = \pi_{\{x_1,x_2\}}$ for some $x_1, x_2 \in X$, $x_1 \neq x_2$;
- π covers three (distinct) atoms in $\mathfrak{F}(X)$, in which case $\pi = \pi_{\{x_1,x_2,x_3\}}$ where all x_i are different, which covers the atoms $\pi_{\{x_1,x_2\}}$, $\pi_{\{x_1,x_3\}}$, and $\pi_{\{x_2,x_3\}}$;
- If $\alpha \neq \beta$ are atoms in $\mathfrak{F}(X)$ such that $\alpha \leq \pi$ and $\beta \leq \pi$, there is an atom $\gamma \leq \pi$ such that there are three (distinct) atoms covered by $\alpha \vee \gamma$ and three (distinct) atoms covered by $\beta \vee \gamma$. In that case, $\pi = \pi_K$ where K has more than three elements: if $\alpha = \pi_{\{x_1,x_2\}}$ and $\beta = \pi_{\{x_3,x_4\}}$, then due to the assumption $\alpha \neq \beta$, the set $\{x_1,x_2,x_3,x_4\}$ (which lies in K) has at least three distinct elements, say $\{x_1,x_2,x_3\}$. Hence we may take $\gamma = \pi_{\{x_2,x_3\}}$, in which case $\alpha \vee \gamma = \pi_{\{x_1,x_2,x_3\}}$, which covers the atoms α, γ, and $\pi_{\{x_1,x_3\}}$. Likewise, we have $\beta \vee \gamma = \pi_{\{x_2,x_3,x_4\}}$, which covers three atoms β, γ, and $\pi_{\{x_2,x_4\}}$.

In order to see that π satisfying the third condition must be of the form π_K, assume the converse. So π contains two blocks K_λ and K_μ consisting of two or more elements. Say $\{x_1,x_2\} \subseteq K_\lambda$ and $\{x_3,x_4\} \subseteq K_\mu$. Then $\alpha = \pi_{\{x_1,x_2\}}$ and $\beta_{\{x_3,x_4\}}$ are atoms such that $\alpha, \beta < \pi$, and there is an atom $\gamma = \pi_{\{x_5,x_6\}} \leq \pi$ such that there are three atoms covered by $\alpha \vee \gamma$, and there are three atoms covered by $\beta \vee \gamma$. It follows from the second condition that $\alpha \vee \gamma = \pi_L$ with L a three-point set. This implies that $\{x_1,x_2\} \cap \{x_5,x_6\}$ is not empty, from which it follows that $\alpha \vee \gamma = \pi_{\{x_1,x_2,x_5,x_6\}}$. Similarly, we find $\beta \vee \gamma = \pi_{\{x_3,x_4,x_5,x_6\}}$. Since $\{x_1,x_2,x_5,x_6\}$ and $\{x_3,x_4,x_5,x_6\}$ overlap, we obtain $\alpha \vee \beta \vee \gamma = \pi_{\{x_1,x_2,x_3,x_4,x_5,x_6\}}$. Moreover, $\alpha, \beta, \gamma \leq \pi$, so $\alpha \vee \beta \vee \gamma \leq \pi$. However, since $x_1, x_2 \in K_\lambda$, we must have $\{x_1,x_2,x_3,x_4,x_5,x_6\} \subseteq K_\lambda$ by definition of the order on $\mathfrak{F}(X)$. But since $x_3, x_4 \in K_\mu$, we must also have $\{x_1, x_2, x_3, x_4, x_5, x_6\} \subseteq K_\mu$, which is not possible, since K_λ and K_μ are distinct blocks, hence disjoint. We conclude that π can have only one block K of two or more elements, hence $\pi = \pi_K$.

Thus $\mathscr{F}_2(X) \subset \mathfrak{F}(X)$ has been characterized order-theoretically. Moreover,

$$\pi = \vee_{x \in X} \pi_{K(x)}, \tag{9.6}$$

where $K(x)$ is the unique block of X that contains x. Hence $\mathscr{F}_2(X)$ determines $\mathfrak{F}(X)$.

Let X and Y be compact Hausdorff spaces of cardinality at least two (so that the empty set and singletons are excluded). By the previous analysis, an order isomorphism $\mathsf{B} : \mathscr{C}(C(X)) \to \mathscr{C}(C(Y))$ is equivalent to an order isomorphism $\mathfrak{F}(X) \to \mathfrak{F}(Y)$, which in turn restricts to an order isomorphism $\mathscr{F}_2(X) \to \mathscr{F}_2(Y)$.

Lemma 9.5. *If X and Y are compact Hausdorff spaces of cardinality at least two, then any order isomorphism $\mathsf{F} : \mathscr{F}_2(X) \to \mathscr{F}_2(Y)$ is induced by a homeomorphism $\varphi : X \to Y$ via $\mathsf{F}(F) = \varphi(F)$, i.e., $\mathsf{F}(F) = \cup_{x \in F}\{\varphi(x)\}$. Moreover, if X and Y have cardinality at least three, then φ is uniquely determined by F.*

To see the idea, we first prove this for finite X, where $\mathscr{F}_2(X)$ simply consists of all subsets of X having at least two elements, etc. It is easy to see that X and Y must have the same cardinality $|X| = |Y| = n$. If $n = 2$, then $\mathscr{F}_2(X) = X$ etc., so there is only one map F, which is induced by each of the two possible maps $\varphi : X \to Y$, so that φ exists but fails to be unique. If $n > 2$, then F must map each subset of X with $n-1$ elements to some subset of Y with $n-1$ elements, so that taking complements we obtain a unique bijection $\varphi : X \to Y$. To show that φ induces F, note that the meet $\wedge$ in $\mathscr{F}_2(X)$ is simply intersection $\cap$, and also that for any $F \in \mathscr{F}_2(X)$,

$$F = \cup_{x \in F}\{x\} = \cap_{x \notin F}\{x\}^c = (\cup_{x \notin F}\{x\})^c, \tag{9.7}$$

where $A^c = X \backslash A$. Since F is an order isomorphism, it preserves $\wedge = \cap$, so that

$$\mathsf{F}(F) = \cap_{x \notin F}\mathsf{F}(\{x\}^c) = \cap_{x \notin F} X \backslash \{\varphi(x)\} = (\cup_{x \notin F}\{\varphi(x)\})^c = \cup_{x \in F}\{\varphi(x)\}. \tag{9.8}$$

Now assume that X is infinite. Let $x \in X$. If x is not isolated, we define $\varphi(x)$ as follows. Let $\mathscr{O}(x)$ denote the set of all open neighborhoods of x. Since x is not isolated, each $O \in \mathscr{O}(x)$ contains at least another element, so $\overline{O} \in \mathscr{F}_2(X)$. Moreover, finite intersections of elements of $\{\overline{O} : O \in \mathscr{O}(x)\}$ are still in $\mathscr{F}_2(X)$. Indeed, if $O_1, \ldots, O_n \in \mathscr{O}(x)$, then $O_1 \cap \ldots \cap O_n$ is an open set containing x, and since $\overline{O_1 \cap \ldots \cap O_n} \subseteq \overline{O_1} \cap \ldots \cap \overline{O_n}$, it follows that $\overline{O_1} \cap \ldots \cap \overline{O_n} \in \mathscr{F}_2(X)$. Since F is an order isomorphism, we find that finite intersections of $\{\mathsf{F}(\overline{O}) : O \in \mathscr{O}(x)\}$ are contained in $\mathscr{F}_2(Y)$. This implies that $\{\mathsf{F}(\overline{O}) : O \in \mathscr{O}(x)\}$ satisfies the finite intersection property. As Y is compact, it follows that $I_x = \bigcap_{O \in \mathscr{O}(x)} \mathsf{F}(\overline{O})$ is nonempty. We can say more: it turns out that I_x contains exactly one element. Indeed, assume that there are two different points $y_1, y_2 \in I_x$. Then $\{y_1, y_2\} \in \mathscr{F}_2(Y)$, so $\mathsf{F}^{-1}(\{y_1, y_2\}) \in \mathscr{F}_2(X)$. Since $\{y_1, y_2\} \in \mathsf{F}(\overline{O})$ for each $O \in \mathscr{O}(x)$, we also find that $\mathsf{F}^{-1}(\{y_1, y_2\}) \subseteq \overline{O}$ for each $O \in \mathscr{O}(x)$. This implies that

$$\mathsf{F}^{-1}(\{y_1,y_2\}) \subseteq \bigcap_{O\in\mathscr{O}(x)} \overline{O} = \{x\}, \tag{9.9}$$

where the last equality holds by normality of X. But this is a contradiction with $\mathsf{F}: \mathscr{F}_2(X) \to \mathscr{F}_2(Y)$ being a bijection. So I_x contains exactly one point. We define $\varphi(x)$ such that $\{\varphi(x)\} = I_x$. Notice that $\varphi(x)$ cannot be isolated in Y, since if we assume otherwise, then $Y \setminus \{\varphi(x)\}$ must be a co-atom in $\mathscr{F}_2(Y)$, whence $\mathsf{F}^{-1}(Y \setminus \{\varphi(x)\})$ is a co-atom in $\mathscr{F}_2(X)$, which must be of the form $X \setminus \{z\}$ for some isolated $z \in X$. Since x is not isolated, we cannot have $x = z$, so $X \setminus \{z\}$ is an open neighborhood of x, which is even clopen since z is isolated. By definition of $\varphi(x)$, we must have $\varphi(x) \in \mathsf{F}(X \setminus \{z\})$, but $\mathsf{F}(X \setminus \{z\}) = Y \setminus \{\varphi(x)\}$. We found a contradiction, hence $\varphi(x)$ cannot be isolated. Now assume that x is an isolated point. Then $X \setminus \{x\}$ is a co-atom in $\mathscr{F}_2(X)$, so $\mathsf{F}(X \setminus \{x\})$ is a co-atom in $\mathscr{F}_2(Y)$, too. Clearly this implies that $\mathsf{F}(X \setminus \{x\}) = Y \setminus \{y\}$ for some unique $y \in Y$, which must be isolated, since $Y \setminus \{y\}$ is closed. We define $\varphi(x) = y$.

In an analogous way, F^{-1} induces a map $\psi : Y \to X$. We shall show that φ and ψ are each other's inverses. Let $x \in X$ be isolated. We have seen that $\varphi(x)$ must be isolated as well, and that $\varphi(x)$ is defined by the equation $\mathsf{F}(X \setminus \{x\}) = Y \setminus \{\varphi(x)\}$. Since F is an order isomorphism, we have $X \setminus \{x\} = \mathsf{F}^{-1}(Y \setminus \{\varphi(x)\})$. Since $\varphi(x)$ is isolated, we find by definition of ψ that $\psi(\varphi(x)) = x$. In a similar way we find that $\varphi(\psi(y)) = y$ for each isolated $y \in Y$. Now assume that x is not isolated and let $F \in \mathscr{F}_2(X)$ such that $x \in F$. Then

$$\begin{aligned} \{\varphi(x)\} &= \bigcap_{O\in\mathscr{O}(x)} \mathsf{F}(\overline{O}) \subseteq \bigcap\{\mathsf{F}(\overline{O}) : O \text{ open}, F \subseteq O\} \\ &= \mathsf{F}\left(\bigcap\{\overline{O} : O \text{ open}, F \subseteq O\}\right) = \mathsf{F}(F), \end{aligned} \tag{9.10}$$

where the last equality follows by completely regularity of X. The penultimate equality follows from the following facts. Firstly, the set $\bigcap\{\overline{O} : O \text{ open}, F \subset O\}$ is closed since it is the intersection of closed sets. Moreover, the intersection contains more than one point, since F contains two or more points and $F \subseteq \overline{O}$ for each O. Hence $\bigcap\{\overline{O} : O \text{ open}, F \subseteq O\} \in \mathscr{F}_2(X)$, and since F is an order isomorphism, it preserves infima, which justifies the penultimate equality. Hence $\varphi(x) \in \mathsf{F}(F)$ for each $F \in \mathscr{F}_2(X)$ containing x. Since x is not isolated, $\varphi(x)$ is not isolated either. Hence in a similar way, we find that $\psi(\varphi(x)) \in \mathsf{F}^{-1}(G)$ for each $G \in \mathscr{F}_2(Y)$ containing $\varphi(x)$. Let $z = \psi(\varphi(x))$. Combining both statements, we find that $z \in F$ for each $F \in \mathscr{F}_2(X)$ such that $x \in F$. In other words, $z \in \bigcap\{F \in \mathscr{F}_2(X) : x \in F\}$. Since x is not isolated, we each $O \in \mathscr{O}(x)$ contains at least two points. Hence

$$\bigcap\{F \in \mathscr{F}_2(X) : x \in F\} \subseteq \bigcap\{\overline{O} : O \in \mathscr{O}(x)\} = \{x\}, \tag{9.11}$$

where we used complete regularity of X in the last equality. We conclude that $z = x$, so $\psi(\varphi(x)) = x$. In a similar way, we find that $\varphi(\psi(y)) = y$ for each non-isolated $y \in Y$. We conclude that φ is a bijection with inverse $\varphi^{-1} = \psi$.

Continuing the proof of Lemma 9.5, we have to show that if $F \in \mathscr{F}_2(X)$, then $\varphi[F] = \mathsf{F}(F)$. Let $x \in F$. In the proof that φ is a bijection we already noticed that $\varphi(x) \in \mathsf{F}(F)$ if x is not isolated. If x is isolated in X, then we first assume that F has at least three points. Since $\{x\}$ is open, $G = F \setminus \{x\}$ is closed. Since F contains at least three points, $G \in \mathscr{F}_2(X)$. So G is covered by F in $\mathscr{F}_2(X)$, so $\mathsf{F}(F)$ covers $\mathsf{F}(G)$. It follows that there must be an element $y_G \in Y \setminus \mathsf{F}(G)$ such that

$$\mathsf{F}(F) = \mathsf{F}(G \cup \{x\}) = \mathsf{F}(G) \cup \{y_G\}. \tag{9.12}$$

Both $G \cup \{x\}$ and $X \setminus \{x\}$ are elements of $\mathscr{F}_2(X)$, so

$$\begin{aligned} \mathsf{F}(G) &= \mathsf{F}(G \cup \{x\} \cap X \setminus \{x\}) = \mathsf{F}(G \cup \{x\}) \cap \mathsf{F}(X \setminus \{x\}) \\ &= (\mathsf{F}(G) \cup \{y_G\}) \cap (Y \setminus \{\varphi(x)\}), \end{aligned} \tag{9.13}$$

where $\mathsf{F}(X \setminus \{x\}) = Y \setminus \{\varphi(x)\}$ by definition of values of φ at isolated points. Since $x \notin G$ and F preserves inclusions, this latter equation also implies $\mathsf{F}(G) \subseteq Y \setminus \{\varphi(x)\}$. Hence we find

$$\mathsf{F}(G) = (\mathsf{F}(G) \cup \{y_G\}) \cap (Y \setminus \{\varphi(x)\}) = \mathsf{F}(G) \cup (\{y_G\} \cap Y \setminus \{\varphi(x)\}). \tag{9.14}$$

Thus we obtain $\{y_G\} \cap Y \setminus \{\varphi(x)\} \subseteq \mathsf{F}(G)$, but since $y_G \notin \mathsf{F}(G)$, we must have $\varphi(x) = y_G$. As a consequence, we obtain $\mathsf{F}(F) = \mathsf{F}(G) \cup \{\varphi(x)\}$, so $\varphi(x) \in \mathsf{F}(F)$.

Summarizing, if F has at least three points, then $\varphi(x) \in \mathsf{F}(F)$ for $x \in F$, regardless whether x is isolated or not. So $\varphi[F] \subseteq \mathsf{F}(F)$ for each $F \in \mathscr{F}_2(X)$ such that F has at least three points. Let $F \in \mathscr{F}_2(X)$ have exactly two points. Then there are $F_1, F_2 \in \mathscr{F}_2(X)$ with exactly three points such that $F = F_1 \cap F_2$. Then since φ is a bijection and F as an order isomorphism both preserve intersections in $\mathscr{F}_2(X)$, we find

$$\varphi[F] = \varphi[F_1 \cap F_2] = \varphi[F_1] \cap \varphi[F_2] \subseteq \mathsf{F}(F_1) \cap \mathsf{F}(F_2) = \mathsf{F}(F_1 \cap F_2) = \mathsf{F}(F). \tag{9.15}$$

So $\varphi[F] \subseteq \mathsf{F}(F)$ for each $F \in \mathscr{F}_2(X)$. In a similar way, we find $\varphi^{-1}[G] \subseteq \mathsf{F}^{-1}[G]$ for each $G \in \mathscr{F}_2(Y)$. So if we substitute $G = \mathsf{F}(F)$, we obtain $\varphi^{-1}[\mathsf{F}(F)] \subseteq F$. Since φ is a bijection, it follows that $\mathsf{F}(F) = \varphi[F]$ for each $F \in \mathscr{F}_2(X)$. As a consequence, φ induces a one-one correspondence between closed subsets of X and closed subsets of Y. Hence φ is a homeomorphism. This proves Lemma 9.5. □

The special case of Theorem 9.4 where A and B are commutative now follows if we combine all steps so far:

1. The Gelfand isomorphism allows us to assume $A = C(X)$ and $B = C(Y)$, as above.
2. The order isomorphism $\mathsf{B} : \mathscr{C}(A) \to \mathscr{C}(B)$ determines an order isomorphism $F : \mathfrak{F}(X) \to \mathfrak{F}(Y)$ of the underlying lattices of u.s.c. decompositions, and *vice versa*.
3. Because of (9.6), the order isomorphism F in turn determines and is determined by an order isomorphism $\mathsf{F} : \mathscr{F}_2(X) \to \mathscr{F}_2(Y)$.
4. Lemma 9.5 yields a homeomorphism $\varphi : X \to Y$ inducing $\mathsf{F} : \mathscr{F}_2(X) \to \mathscr{F}_2(Y)$.
5. The inverse pullback $(\varphi^{-1})^* : C(X) \to C(Y)$ is an isomorphism of C*-algebras, which (running backwards) reproduces the initial map $\mathsf{B} : \mathscr{C}(C(X)) \to \mathscr{C}(C(Y))$.

Therefore, in the commutative case we apparently obtain rather more than a weak Jordan isomorphism $\mathsf{J} : A_{\mathrm{sa}} \to B_{\mathrm{sa}}$; we even found an isomorphism $\mathsf{J} : A \to B$ of C*-algebras. However, if A and B are commutative, the condition of linearity on each commutative C*-subalgebra C of A includes $C = A$, so that (after complexification) weak Jordan isomorphisms are the same as isomorphisms of C*-algebras.

We now turn to the general case, in which A and B are both noncommutative (the case where one, say A, is commutative but the other is not, cannot occur, since $\mathscr{C}(A)$ would be a complete lattice but $\mathscr{C}(B)$ would not). Let D and E be maximal abelian C*-subalgebras of A, so that the corresponding elements of $\mathscr{C}(A)$ are maximal in the order-theoretic sense. Given an order isomorphism $\mathsf{B} : \mathscr{C}(A) \to \mathscr{C}(B)$, we restrict the map B to the down-set $\downarrow D = \mathscr{C}(D)$ in $\mathscr{C}(A)$ so as to obtain an order homomorphism $\mathsf{B}_{|D} : \mathscr{C}(D) \to \mathscr{C}(B)$. The image of $\mathscr{C}(D)$ under B must have a maximal element (since B is an order isomorphism), and so there is a maximal commutative C*-subalgebra $\tilde{D}$ of B such that $\mathsf{B}_{|D} : \mathscr{C}(D) \to \mathscr{C}(\tilde{D})$ is an order isomorphism. Applying the previous result, we obtain an isomorphism $\mathsf{J}_D : D \to \tilde{D}$ of commutative C*-algebras that induces $\mathsf{B}_{|D}$. The same applies to E, so we also have an isomorphism $\mathsf{J}_E : E \to \tilde{E}$ of commutative C*-algebras that induces $\mathsf{B}_{|E}$. Let $C = D \cap E$, which lies in $\mathscr{C}(A)$. We now show that J_D and J_E coincide on C. There are three cases.

1. $\dim(C) = 1$. In that case $C = \mathbb{C} \cdot 1_A$ is the bottom element of $\mathscr{C}(A)$, so it must be sent to the bottom element $\tilde{C} = \mathbb{C} \cdot 1_B$ of $\mathscr{C}(B)$, whence the claim.
2. $\dim(C) = 2$. This the hard case dealt with below.
3. $\dim(C) > 2$. This case is settled by the uniqueness claim in Lemma 9.5.

So assume $\dim(C) = 2$. In that case, $C = C^*(e)$ for some proper projection $e \in \mathscr{P}(A)$, which is equivalent to C being an atom in $\mathscr{C}(A)$. Recall that all our C*-algebras are unital, and that by assumption C*-subalgebras C share the unit of the ambient C*-algebra A, hence $C^*(e)$ contains the unit of A. Hence $\tilde{C} \equiv \mathsf{B}(C) = \mathsf{B}_{|D}(C) = \mathsf{B}_{|E}(C)$ is an atom in $\mathscr{C}(B)$, which implies that $\tilde{C} = C^*(\tilde{e})$ for some projection $\tilde{e} \in \mathscr{P}(B)$. If $\mathsf{J}_D(e) = \mathsf{J}_E(e)$ we are ready, so we must exclude the case $\mathsf{J}_D(e) = \tilde{e}$, $\mathsf{J}_E(e) = 1_B - \tilde{e}$. This exclusion again requires a case distinction:

$$\dim(eAe) = \dim(e^{\perp}Ae^{\perp}) = 1; \tag{9.16}$$
$$\dim(eAe) = 1, \ \dim(e^{\perp}Ae^{\perp}) > 1; \tag{9.17}$$
$$\dim(eAe) > 1, \ \dim(e^{\perp}Ae^{\perp}) > 1, \tag{9.18}$$

where $e^{\perp} = 1_A - e$. Each of these cases is nontrivial, and we need another lemma.

Lemma 9.6. *Let $C \in \mathscr{C}(A)$ be maximal (i.e., $C \subset A$ is maximal abelian).*

1. *For each projection $e \in \mathscr{P}(C)$ we have* $\dim(eCe) = 1$ *iff* $\dim(eAe) = 1$.
2. *We have* $\dim(C) = 2$ *iff either $A \cong \mathbb{C}^2$ or $A \cong M_2(\mathbb{C})$.*

Proof. For the first claim $\dim(eAe) = 1$ clearly implies $\dim(eCe) = 1$. For the converse implication, assume *ad absurdum* that $\dim(eAe) > 1$, so that there is an $a \in A$ for which $eae \neq \lambda \cdot e$ for any $\lambda \in \mathbb{C}$. If also $\dim(eCe) = 1$, then any $c \in C$ takes the form $c = \mu \cdot e + e^{\perp}ce^{\perp}$ for some $\mu \in \mathbb{C}$. Indeed, since $c, e, e^{\perp}$ commute within C,

$$c = ce + ce^{\perp} = ce^2 + c(e^{\perp})^2 = ece + e^{\perp}ce^{\perp} = \mu e + e^{\perp}ce^{\perp}, \tag{9.19}$$

where the last equality follows since $ece \in eCe$, which is spanned by e. This implies that $eae \in C'$ (where C' is the commutant of C within A), and since C is maximal abelian, we have $C = C'$, whence $eae \in C$. Now $eae = e(eae)e$, hence $eae \in eCe$, whence $eae = \lambda \cdot e$ for some $\lambda \in \mathbb{C}$. Contradiction. According to Theorem C.169.1, the assumption $\dim(C) = 2$ implies that A is finite-dimensional, upon which Theorem C.163 and (C.641) yield the second claim. □

Having proved Lemma 9.6, we move on the analyze the cases (9.16) - (9.18).

- Eq. (9.16) implies that C is maximal, as follows. Any element $a \in A$ is a sum of eae, $e^{\perp}ae^{\perp}$, $eae^{\perp}$, and $e^{\perp}ae$; nonzero elements of $C' = \{e\}'$ can only be of the first two types. If (9.16) holds, then $\dim(C') = 2$, but since C is abelian we have $C \subseteq C'$ and since $\dim(C) = 2$ we obtain $C' = C$. Lemma 9.6.2 then implies that either $A \cong \mathbb{C}^2$ or $A \cong M_2(\mathbb{C})$. These C*-algebras have been analyzed after the statement of Theorem 9.4, and since those two A's conversely imply (9.16), we may exclude them in dealing with (9.17) - (9.18). By Lemma 9.6.2 (applied to D and E instead of C), in what follows we may assume that $\dim(D) > 2$ and $\dim(E) > 2$ (as D and E are maximal).
- Eq. (9.17) implies $\dim(eD) = 1$. Assuming $\mathrm{J}_D(e) = \tilde{e}$, this implies $\dim(\tilde{e}\tilde{D}) = 1$ (since J_D is an isomorphism). Applying Lemma 9.6.1 to B gives $\dim(\tilde{e}B\tilde{e}) = 1$ (since $\tilde{D}$ is maximal). If also $\dim((1_B - \tilde{e})B(1_B - \tilde{e})) = 1$, then $\dim(\tilde{D}) = 2$, whence $\dim(D) = 2$, which we excluded. Hence
$$\dim((1_B - \tilde{e})B(1_B - \tilde{e})) > 1. \tag{9.20}$$
Applied to J_E this gives $\mathrm{J}_E(e) = \tilde{e}$, and hence J_D and J_E coincide on $C = C^*(e)$.
- Eq. (9.18) implies that $\dim(eDe) > 1$ as well as $\dim(e^{\perp}Ee^{\perp}) > 1$ (apply Lemma 9.6.1 to D and E). Since $\dim(eDe) > 1$, there is some $a \in D$ such that e and $a' = eae \in D$ are linearly independent, and similarly there is some $b \in E$ such that $b' = e^{\perp}be^{\perp}$ is linearly independent of $e^{\perp}$. Then a', b', e commute (in fact, $a'b' = b'a' = 0$), so that we may form the abelian C*-algebras $C_1 = C^*(e, a') \subseteq D$ and $C_2 = C^*(e, b') \subseteq E$, which (also containing the unit 1_A) both have dimension at least three. We also form $C_3 = C^*(e, a', b')$, which contains C_1 and C_2 and hence is at least three-dimensional, too. Because D and E are maximal abelian, C_3 must lie in both D and E. Applying the abelian case of the theorem already proved to D and E, as before, but replacing C used so far by C_3, we find that J_D and J_E coincide on C_3 (as its dimension is > 2). In particular, $\mathrm{J}_D(e) = \mathrm{J}_E(e)$.

To finish the proof, we first note that Theorem 9.4 holds for $A = B = \mathbb{C}$ by inspection, whereas the cases $A \cong B \cong \mathbb{C}^2$ or $\cong M_2(\mathbb{C})$ have already been discussed.

In all other cases we define $\mathrm{J} : A_{\mathrm{sa}} \to B_{\mathrm{sa}}$ by putting $\mathrm{J}(a) = \mathrm{J}_D(a)$ for any maximal abelian unital C*-subalgebra D containing $C = C^*(a)$ and hence a; as we just saw, this is independent of the choice of D. Since each J_D is an isomorphism of commutative C*-algebras, J is a weak Jordan isomorphism. Finally, uniqueness of J (under the stated restriction on A) follows from Lemma 9.5. □

Theorem 9.4 begs the question if we can strengthen weak Jordan isomorphisms to Jordan isomorphism (i.e. invertible linear maps that preserve the Jordan product, cf. Appendix C.25). This hinges on the extendibility of weak Jordan isomorphisms to linear maps (which of course continue to preserve the Jordan product and hence are automatically Jordan isomorphisms). A general result in this direction is:

Theorem 9.7. *Let A and B be unital AW*-algebras, where A contains no summand of type* I_2. *Then there is a bijective correspondence between order isomorphisms* $\mathsf{B} : \mathscr{C}(A) \to \mathscr{C}(B)$ *and Jordan isomorphisms* $\mathsf{J} : A_{sa} \to B_{sa}$.

This follows from Gleason's Theorem for AW*-algebras, which we will neither state nor prove. If $A = B = B(H)$, then the ordinary Gleason Theorem suffices to yield the crucial lemma for Wigner's Theorem for Bohr symmetries (i.e. Theorem 5.4.6):

Lemma 9.8. *Let H be a Hilbert space of dimension greater than two. Then any Bohr symmetry of* $\mathscr{C}(B(H))$ *is induced by a Jordan symmetry of* $B(H)_{sa}$.

Proof. This follows from Theorem 9.4 and Corollary 5.22, which for the case at hand turns weak Jordan isomorphisms into Jordan isomorphisms. □

We finally turn to symmetries of projection lattices. Theorem C.174 shows that for von Neumann algebras (and more generally for AW*-algebras) A (without summand of type I_2) and B, any isomorphism $\mathsf{N} : \mathscr{P}(A) \to \mathscr{P}(B)$ of the corresponding orthocomplemented projection lattices (which automatically preserves arbitrary suprema) is the restriction of a unique Jordan isomorphism $\mathsf{J} : A_{sa} \to B_{sa}$.

This completes the argument to the effect that for many C*-algebras of observables A (including $B(H)$ for $\dim(H) > 1$ as far as nos. 1–4 are concerned, and having $\dim(H) > 2$ if we also include nos. 5–6) our six seemingly different notions of symmetry of a quantum system described by a C*-algebra are equivalent. In particular, they are equivalent to Jordan isomorphisms, which are also the easiest ones to use, as they involve a readily identifiable part A_{sa} of A, and (by complexification, as explained above) may even be defined on A itself (namely as those complex-linear isomorphisms that preserve the involution $*$ as well as the Jordan product $\circ$).

Putting $B = A$ and assuming (without loss of generality) that $A \subseteq B(H)$, Theorem C.175 then yields a separation of Jordan automorphisms into three disjoint classes:

Corollary 9.9. *If* J *is a Jordan symmetry of a unital C*-algebra* $A \subseteq B(H)$, *then there are three mutually orthogonal projections* e_1, e_2, e_3 *in* $A' \cap A''$ *such that:*

1. $e_1 + e_2 + e_3 = 1_H$;
2. *The map* $a \mapsto \mathsf{J}(a)e_1$ *from A to* $B(e_1H)$ *is a homomorphism (of C*-algebras);*
3. *The map* $a \mapsto \mathsf{J}(a)e_2$ *from A to* $B(e_2H)$ *is an anti-homomorphism* (ibid.)*;*
4. *The map* $a \mapsto \mathsf{J}(a)e_3$ *from A to* $B(e_3H)$ *is both a homomorphism and an anti-homomorphism of C*-algebras (so that the "corner"* $\mathsf{J}(A)e_3$ *is commutative).*

If in addition $a \mapsto \mathsf{J}(a)e_1$ *is not an anti-homomorphism and* $a \mapsto \mathsf{J}(a)e_2$ *is not a homomorphism, then* e_1, e_2, *and* e_3 *are uniquely determined by these conditions.*

As we shall now see, if the symmetries form a (Lie) group, then this result often justifies restricting our attention simply to homomorphisms of C*-algebras.

9.2 Unitary implementability of symmetries

There are good reasons for the dichotomy (or even trichotomy) between homomorphisms and anti-homomorphisms of C*-algebras left by Corollary 9.9, since in physics certain discrete symmetries of quantum theory indeed give rise to anti-homomorphisms: the best-known examples are time inversion T and charge conjugation C combined with space inversion (i.e. parity) P, giving CP (there are also other examples in condensed matter physics, like quantum spin flip). However, for the kind of problems mainly addressed in this book it is sufficient to restrict our attention to homomorphisms. One reason is that even if we use discrete symmetries (where the simplest non-trivial group $\mathbb{Z}_2$ often suffices to make our point), the models we treat simply realize these symmetries as homomorphisms. Another reason is that if symmetries join to form a *connected* topological group G (typically a Lie group) and the maps $x \mapsto \mathrm{J}_x$ sending $x \in G$ to some Jordan symmetry J_x of the given C*-algebra A of observables form a (strongly) continuous homomorphism (see below), then the identity $e \in G$ must be mapped to the identity id_A, which of course is a homomorphism of A. Continuity then implies that all J_x must be homomorphisms.

In what follows we therefore assume that G is a (topological) group and that we are given a (continuous) homomorphism $x \mapsto \alpha_x$from G into the group $\mathrm{Aut}(A)$ of all automorphisms of A; note that, given our restriction to homomorphisms, we switch notation from J to the customary symbol α. Continuity here always means ***strong continuity***, in that for each $a \in A$ the map $x \mapsto \alpha_x(a)$ from G to A is continuous (so that the map $G \times A \to A$ given by $(x,a) \mapsto \alpha_x(a)$ is continuous, as usually required for group actions in a topological setting, cf. Proposition 5.35).

It follows from Theorem 5.4 (technically, from part 4 of that theorem, but "morally" from all of it, including the equivalences between all kinds of symmetries) that if $A = B(H)$, then a homomorphism $\alpha : G \to \mathrm{Aut}(B(H))$ is always implemented by a family $u(x)$ of unitary operators on H, in that

$$\alpha_x(a) = u(x)au(x)^* \quad (x \in G). \tag{9.21}$$

The group representation property $\alpha_x\alpha_y = \alpha_{xy}$ does not enforce $u(x)u(y) = u_{xy}$: indeed, as we saw in detail in §5.10 one may have a projective unitary representation $g \mapsto u(x)$ of G on H. However, by Theorem 5.62 one may usually pass to a central extension $\check{G}$ of G for which this problem does not arise (e.g., $\check{SO}(3) = SU(2)$). In Corollary 9.12 below (unbroken symmetry), even such a passage is not necessary.

For general C*-algebras A—especially those modeling either classical systems (in which case A is commutative) or infinite quantum systems (where A is typically an infinite tensor product), one rarely has $\alpha(a) = uau^*$ for some $u \in A$ even for single automorphisms α, let alone for a whole group of them. Instead, we settle for a weaker notion of unitary implementability, where the unitary u need not be in A.

Definition 9.10. *Let $\pi : A \to B(H)$ be a representation of A. An automorphism $\alpha \in$* $\mathrm{Aut}(A)$ *is* implemented *in H if there exists a unitary operator $u : H \to H$ such that*

$$\pi(\alpha(a)) = u\pi(a)u^* \quad (a \in A). \tag{9.22}$$

The fundamental criterion for implementability uses the pullback $\alpha^* : S(A) \to S(A)$ of $\alpha : A \to A$ to the state space $S(A)$, defined by $\alpha^*\omega = \omega \circ \alpha^{-1}$; cf. §C.25.

Theorem 9.11. *An automorphism $\alpha : A \to A$ can be implemented in the* GNS*-representation π_ω defined by a state ω on A iff $\pi_{\alpha^*\omega}$ and π_ω are unitarily equivalent.*

Proof. Whether or not $\pi_{\alpha^*\omega}$ and π_ω are unitarily equivalent, we may define

$$w : H_\omega \to H_{\alpha^*\omega}; \tag{9.23}$$
$$w\pi_\omega(a)\Omega_\omega = \pi_{\alpha^*\omega}(\alpha(a))\Omega_{\alpha^*\omega}. \tag{9.24}$$

This operator is well defined and unitary, and satisfies $w\Omega_\omega = \Omega_{\alpha^*\omega}$ as well as $w\pi_\omega(a)w^* = \pi_{\alpha^*\omega}(\alpha(a))$; these properties even characterize w. If $\pi_{\alpha^*\omega} \simeq \pi_\omega$, there exists a unitary $v : H_\omega \to H_{\alpha^*\omega}$ satisfying $v\pi_\omega(a)v^* = \pi_{\alpha^*\omega}(a)$, $a \in A$. Then $u = v^*w$ satisfies (9.22) for $\pi = \pi_\omega$. The converse is similar. □

An important special case arise if ω is invariant under α.

Corollary 9.12. *If $\alpha^*\omega = \omega$ (that is, $\omega(\alpha(a)) = \omega(a)$ for all $a \in A$), then α is implemented by a unitary operator $u_\omega : H_\omega \to H_\omega$ satisfying $u_\omega\Omega_\omega = \Omega_\omega$. In particular, given a continuous homomorphism $\alpha : G \to \mathrm{Aut}(A)$ such that $\alpha_x^*\omega = \omega$ for each $x \in G$, one has a family of unitaries $u_\omega(x) : H_\omega \to H_\omega$ that for all $x \in G$ satisfy*

$$u_\omega(x)\Omega_\omega = \Omega_\omega; \tag{9.25}$$
$$\pi_\omega(\alpha_x(a)) = u_\omega(x)\pi_\omega(a)u_\omega(x)^*, \tag{9.26}$$

and form a continuous unitary representation of G on H_ω.

Proof. One easily checks that the following operators do the job:

$$u_\omega(x)\pi_\omega(a)\Omega_\omega = \pi_\omega(\alpha_x(a))\Omega_\omega. \qquad \square$$

Given some $\alpha \in \mathrm{Aut}(A)$, a weak form of ***spontaneous symmetry breaking*** (SSB) is that some state ω—it is always a *state* that breaks a symmetry—satisfies $\alpha^*\omega \neq \omega$; a stronger one states that the two equivalent conditions in Theorem 9.11 are violated, i.e., that α cannot be implemented in the GNS-representation $\pi_\omega(A)$ (cf. Definition 9.10). In order to be physically relevant, the weaker notion has to be supplemented with additional structure, which also guarantees that generically the weak form implies the strong one. Part of this structure involves the identification of suitable classes of states within which we define SSB; these classes are predicated on a time-evolution on A. We also need a symmetry *group* instead of a single automorphism α (which implicitly uses the group $\mathbb{Z}_p = \mathbb{Z}/p\cdot\mathbb{Z}$, where p is the smallest integer such that $\alpha^p = \mathrm{id}_A$; if no such p exists the group is just $\mathbb{Z}$). Thus we need:

- A C*-algebra A with ***time-evolution***, i.e., a homomorphism $\alpha : \mathbb{R} \to \mathrm{Aut}(A)$;
- A preferred class of states defines via α, viz. ***ground states*** or ***equilibrium states***;
- A ***symmetry group*** G acting on A via a homomorphism$\gamma : G \to \mathrm{Aut}(A)$ satisfying

$$\alpha_t\gamma_g = \gamma_g\alpha_t \quad (t \in \mathbb{R}, g \in G). \tag{9.27}$$

9.3 Motion in space and in time

The C*-algebras A we are going to use are the ***quasi-local*** ones introduced in §8.5 for quantum spin systems; especially recall (8.130). Also, the C*-algebra $A = B^{\infty}$ in §8.2 is a case in point, but this would require some changes in what follows. The last expression in (8.130) is convenient for introducing ***spatial translation symmetry***

$$\tau : \mathbb{Z}^d \to \mathrm{Aut}(A) \tag{9.28}$$

of $\mathbb{Z}^d$, as follows: for $x \in \mathbb{Z}^d$, define $\tau_x : A_\Lambda \to A_{x+\Lambda}$ initially by

$$\tau_x(b(y)) = b(x+y), \tag{9.29}$$

where, for given $b \in B(H)$ and $y \in \Lambda$, the operator $b(y) \in A_\Lambda$ is the element $\otimes_{z\in\Lambda} a_z$ with $a_y = b$ and $a_z = 1_H$ whenever $z \neq y$. Since arbitrary elements of A_Λ are (norm-limits of) finite linear combinations of products of such operators $b(y)$, the automorphic (and hence isometric) property of τ_x defines its action on all of A_Λ (if necessary by continuous extension). Note that for $a \in A_\Lambda$ the operator $\tau_x(a)$ thus defined is independent of the (typically non-unique) realization of a in terms of the $b(y)$, because τ_x is an isometry. The group homomorphism property of the map (9.28) thus constructed is guaranteed by (9.29), whilst continuity is no issue since $\mathbb{Z}^d$ is discrete.

Since $A_\Lambda = \otimes_{y\in\Lambda} A_y$ with $A_y = B(H)$, an equivalent way to define τ_x is to use identifications $\mathrm{id}_{yz} : A_y \to A_z$ (since $A_y = A_z = B(H)$), which, taking tensor products, yield isomorphisms $\mathrm{id}_{\Lambda,\Lambda'} : A_\Lambda \to A_{\Lambda'}$ whenever some bijection $\Lambda \cong \Lambda'$ is given. In terms of those, we simply have $(\tau_x)_{|A_\Lambda} = \mathrm{id}_{\Lambda,x+\Lambda}$. Either way, the maps $(\tau_x)_{|A_\Lambda}$ extend to $\tau_x : A \to A$ by continuity. The following property then holds:

Proposition 9.13. *An automorphic action τ of $\mathbb{Z}^d$ on a quasi-local C*-algebra A is* **asymptotically abelian** *in the sense that $\lim_{x\to\infty}[a, \tau_x(b)] = 0$ for all $a, b \in A$.*

Here $x \to \infty$ means that any sequence (x_n) with $|x_n| \to \infty$ with respect to the Euclidean norm on $\mathbb{Z}^d$ has a subsequence (x'_n) for which the stated result holds.

Proof. For a and b local, i.e., $a \in A_{\Lambda^{(1)}}$ and $b \in A_{\Lambda^{(2)}}$ this follows from Einstein locality. The general case follows by approximating a and b by local elements. □

Thus quasi-local C*-algebras A satisfy the assumptions in the following theorem, which will be important in linking the various notions of SSB discussed earlier.

Theorem 9.14. *Let A be a C*-algebra A equipped with an asymptotically abelian action τ of $\mathbb{Z}^d$, and let ω be a translation-invariant primary state on A (i.e., $\tau_x^* \omega = \omega$ for all $x \in \mathbb{Z}^d$). Then Ω_ω is the only translation-invariant vector in H_ω. Moreover,*

$$\lim_{x\to\infty} \omega(a\tau_x(b)) = \omega(a)\omega(b) \ \ (a,b \in A); \tag{9.30}$$

$$\lim_{x\to\infty} \pi_\omega(\tau_x(b)) = \omega(b)\cdot 1_{H_\omega} \ \ (b \in A); \tag{9.31}$$

$$\lim_{\Lambda\uparrow\mathbb{Z}^d} |\Lambda|^{-1} \sum_{x\in\Lambda} \pi_\omega(\tau_x(b)) = \omega(b)\cdot 1_{H_\omega} \ \ (b \in A). \tag{9.32}$$

Here (9.31) and (9.32) hold in the weak operator topology on $B(H_\omega)$, and the limit $\Lambda \uparrow \mathbb{Z}^d$ in is taken along the hypercubes Λ_N in (8.153) as $N \to \infty$.

Proof. If ω is primary, Theorem 8.23 (or its proof) yields

$$\lim_{x\to\infty} |\omega(a\tau_x(b)) - \omega(a)\omega(\tau_x(b))| = 0. \tag{9.33}$$

Translation-invariance of ω then yields (9.30), which also is a lemma for (9.31) - (9.32). Towards (9.31) we compute $\omega(a\tau_x(b))$ in terms of the projection

$$e_0 = \lim_{\Lambda\uparrow\mathbb{Z}^d} |\Lambda|^{-1} \sum_{x\in\Lambda} u(x) \tag{9.34}$$

onto the translation-invariant subspace of H_ω, where u is the unitary representation of $\mathbb{Z}^d$ on H_ω from Corollary 9.12 (with $G = \mathbb{Z}^d$), and the limit is taken in the strong operator topology. Eq. (9.34) is a special case of von Neumann's L^2 ergodic theorem (which generalizes the Peter–Weyl–Schur relation $e_0 = \int_G dx\, u(x)$ for compact groups G to amenable groups like $\mathbb{Z}^d$ or $\mathbb{R}^d$). Since $e_0\Omega_\omega = \Omega_\omega$, we have

$$\begin{aligned}
\omega(a\tau_x(b)) &= \langle \Omega_\omega, \pi_\omega(a)\pi_\omega(\tau_x(b))\Omega_\omega\rangle & (9.35)\\
&= \langle \Omega_\omega, \pi_\omega(a)([\pi_\omega(\tau_x(b)), e_0] + e_0\pi_\omega(b))\Omega_\omega\rangle. & (9.36)
\end{aligned}$$

We now let $x \to \infty$. The commutator then vanishes, because the weak limit of $\pi_\omega(\tau_x(b))$ lies in the center of $\pi_\omega(A)''$, which is trivial since ω is primary. The remaining term matches with (9.30) iff e_0 is one-dimensional, so that Ω_ω is the only translation-invariant vector in H_ω, and $e_0 = |\Omega_\omega\rangle\langle\Omega_\omega|$. A similar trick then yields

$$\pi_\omega(\tau_x(b))\pi_\omega(a)\Omega_\omega = ([\pi_\omega(\tau_x(b)), \pi_\omega(a)] + \pi_\omega(a)([\pi_\omega(\tau_x(b)), e_0] + \omega(b)))\Omega_\omega.$$

Both commutators vanish (weakly) as $x \to \infty$, proving (9.31). Similarly, write

$$\pi_\omega(\tau_x(b))\pi_\omega(a)\Omega_\omega = ([\pi_\omega(\tau_x(b)), \pi_\omega(a)] + \pi_\omega(a)u(x)\pi_\omega(b))\Omega_\omega, \tag{9.37}$$

and use (9.34) and the previous formula for e_0 to prove (9.32). □

In the C*-algebraic formalism, ***dynamics*** is described by a continuous homomorphism $\alpha : \mathbb{R} \to \mathrm{Aut}(A)$, $t \mapsto \alpha_t$. For $A = B(H')$, where H' is some Hilbert space (not to be confused with our earlier H in the quasi-local setting), Theorem 5.4 yields

$$\alpha_t(a) = u_t a u_t^* \tag{9.38}$$

for some family of unitaries $u_t \equiv u(t)$, $t \in \mathbb{R}$. Eq. (5.268) and Proposition 5.53 then imply that the family u_t may be redefined so as to make the map $t \mapsto u_t$ a continuous unitary representation of $\mathbb{R}$ on H'. Stone's Theorem 5.73 finally gives the familiar expression for time evolution in the so-called Heisenberg picture in terms of the ***Hamiltonian*** h, which is a (possibly unbounded) self-adjoint operator on H', i.e.,

$$\alpha_t(a) = e^{ith} a e^{-ith}. \tag{9.39}$$

For arbitrary (unital) C*-algebras A one has no counterpart of Theorem 5.4, and one cannot rely on Theorem 9.11 either because there are no preferred states to begin with; such states typically require a time-evolution for their definition (see below). For quantum spin systems (still with $H=\mathbb{C}^n$ and hence $B(H)\cong M_n(\mathbb{C})$), one tries to construct the map $t\mapsto\alpha_t$ from local approximations: with A_Λ given by (8.129) with (8.128), we pick local Hamiltonians $h_\Lambda\in B(H_\Lambda)$ and define maps $t\mapsto \mathrm{Aut}(A_\Lambda)$ by

$$\alpha_t^\Lambda(a)=e^{ith_\Lambda}ae^{-ith_\Lambda}, \tag{9.40}$$

where $a\in A_\Lambda$. Letting $\Lambda\nearrow\mathbb{Z}^d$, we would then like to assemble the family α^Λ into a single automorphism group $\alpha:\mathbb{R}\to\mathrm{Aut}(A)$, which describes the dynamics of the corresponding infinite quantum system. Towards this aim, we start from a ***potential*** (also called an ***interaction***) $\Phi(X)\in B(H_X)$, which is defined for any finite sublattice X of $\mathbb{Z}^d$, in terms of which the local Hamiltonians h_Λ take the form

$$h_\Lambda=\sum_{X\subseteq\Lambda}\Phi(X), \tag{9.41}$$

where the sum is over all sublattices X of Λ. For ***nearest-neighbour interactions***, $\Phi(X)$ is nonzero iff $X=\{x,y\}$ is a pair of neighbours, and in the presence of an external magnetic field one also has terms proportional to $\Phi(\{x\})$. For example, the ***quantum Ising model*** is defined by $H=\mathbb{C}^2$ and $\Phi(\{x,y\})=-J\sigma_3(x)\sigma_3(y)$ for nearest neighbours and $\Phi(\{x\})=-B\sigma_1(x)$ for all x, where $J>0$ and $B\in\mathbb{R}$. The local Hamiltonians are therefore given by

$$h_\Lambda=-J\sum_{\langle xy\rangle\in\Lambda}\sigma_3(x)\sigma_3(y)-B\sum_{x\in\Lambda}\sigma_1(x), \tag{9.42}$$

where the sum over $\langle xy\rangle\in\Lambda$ denotes summing over nearest neighbours in Λ. The expression (9.42) implicitly has so-called ***free boundary conditions***, in that only neighbours inside Λ take part in h_Λ. Alternatively, one could use ***periodic boundary conditions***, which in $d=1$ define the ***quantum Ising chain***

$$h_N=-J\left(\sum_{x=1}^{N-1}(\sigma_3(x)\sigma_3(x+1)+\sigma_3(N)\sigma_3(1)\right)-B\sum_{x=1}^{N}\sigma_1(x). \tag{9.43}$$

In (9.42) - (9.43) the operators $\sigma_i(x)$ in A_Λ is defined as explained after (9.29). We are going to study the quantum Ising chain in detail in connection with SSB; for the moment, we just mention another popular spin model, namely the ***Heisenberg model*** for magnetism. This also has $H=\mathbb{C}^2$, but the local Hamiltonians are

$$h_\Lambda=J\sum_{\langle xy\in\Lambda\rangle}\sum_{i=1}^{3}\sigma_i(y)\sigma_i(y), \tag{9.44}$$

with free boundary conditions, where $J<0$ ($J>0$) yields (anti) ferromagnetism.

Although we do not have (9.38) for any $u_t\in A$, we may construct α_t as follows.

Theorem 9.15. *Let Φ be a* **short-range potential** *in that there is $r \in \mathbb{N}$ such that $\Phi(X) \neq 0$ only if $|x-y| \leq r$ for all $x, y \in X$, and define local Hamiltonians h_Λ by (9.41). For fixed finite $\Lambda \subset \mathbb{Z}^d$ and $a \in A_\Lambda$, the following (norm) limit exists and defines an automorphism α_t of $\cup_{\Lambda \subset \mathbb{Z}^d} A_\Lambda$ and hence by continuity also of A:*

$$\alpha_t(a) = \lim_{N \to \infty} e^{ith_{\Lambda_N}} a e^{-ith_{\Lambda_N}}, \tag{9.45}$$

Proof. Note that for large enough N, the hypercube Λ_N contains any $\Lambda \in \mathscr{P}_f(\mathbb{Z}^d)$. Take $a \in A_\Lambda$, take $\Lambda_{N_2} \supset \Lambda_{N_1} \supset \Lambda$, and use (9.40) and (9.41) to compute

$$\begin{aligned}
\|\alpha_t^{(\Lambda_{N_2})}(a) - \alpha_t^{(\Lambda_{N_1})}(a)\| &= \left\| \int_0^t ds \frac{d}{ds}(\alpha_s^{(\Lambda_{N_2})} \circ \alpha_{t-s}^{(\Lambda_{N_1})}(a)) \right\| \\
&= \left\| \int_0^t ds \left([h_{\Lambda_{N_2}}, \alpha_s^{(\Lambda_{N_1})} \circ \alpha_{t-s}^{(\Lambda_{N_1})}(a)] - \alpha_s^{(\Lambda_{N_2})}([h_{\Lambda_{N_1}}, \alpha_{t-s}^{(\Lambda_{N_1})}(a)]) \right) \right\| \\
&= \left\| \int_0^t ds\, \alpha_s^{(\Lambda_{N_2})}([h_{\Lambda_{N_2}} - h_{\Lambda_{N_1}}, \alpha_{t-s}^{(\Lambda_{N_1})}(a)]) \right\| \\
&\leq \int_0^t ds\, \|\alpha_s^{(\Lambda_{N_2})}([h_{\Lambda_{N_2}} - h_{\Lambda_{N_1}}, \alpha_{t-s}^{(\Lambda_{N_1})}(a)])\| \\
&\leq \int_0^t ds\, \|[h_{\Lambda_{N_2}} - h_{\Lambda_{N_1}}, \alpha_{t-s}^{(\Lambda_{N_1})}(a)]\| \\
&= \int_0^t ds \left\| \sum_{x \in \Lambda_{N_2} \backslash \Lambda_{N_1}} \sum_{X \ni x} [\Phi(X), \alpha_{t-s}^{(\Lambda_{N_1})}(a)] \right\| \\
&\leq \sum_{x \in \Lambda_{N_2} \backslash \Lambda_{N_1}} \sum_{X \ni x} \int_0^t ds\, \|[\Phi(X), \alpha_{t-s}^{(\Lambda_{N_1})}(a)]\|.
\end{aligned} \tag{9.46}$$

We now show that the left-hand side of the first line is a Cauchy sequence. Since

$$\alpha_{t-s}^{(\Lambda_{N_1})}(a) = e^{i(t-s)\sum_{Y \subseteq \Lambda_{N_1}} \Phi(Y)} a e^{-i(t-s)\sum_{Y \subseteq \Lambda_{N_1}} \Phi(Y)} \in B(H_{\Lambda_{N_1}}), \tag{9.47}$$

which is finite-dimensional (as Λ_{N_1} is finite), we have a norm-convergent expansion

$$\alpha_t^{(\Lambda_{N_1})}(a) = a + it \sum_{Y_1 \subseteq \Lambda_{N_1}} [\Phi(Y_1), a] + \frac{(it)^2}{2!} \sum_{Y_1, Y_2 \subseteq \Lambda_{N_1}} [\Phi(Y_2), [\Phi(Y_1), a]] + \cdots \tag{9.48}$$

Let $\Lambda(r)$ consist of all $y \in \mathbb{Z}^d$ for which there is some $x \in \Lambda$ for which $|x-y| \leq r$. Then the zeroth term a in (9.48) is in A_Λ, the first is in $A_{\Lambda(r)}$, ..., the n'th is in $A_{\Lambda(nr)}$. Therefore, we can find $n = n(N_1, N_2, 3)$ such that the only terms in (9.48) that contribute to the commutator in (9.46) are the n'th and beyond. Taking Λ_{N_1} and Λ_{N_2} large enough, this tail can be made arbitrarily small, so that $(\alpha_t^{(\Lambda_N)}(a))_N$ is a Cauchy sequence in A. This gives convergence of (9.45) for $a \in A_\Lambda$, where Λ is arbitrary (but finite), yielding an automorphism α_t in $\cup_\Lambda A_\Lambda$. Being an automorphism, α_t is isometric, so that it extends to A by continuity. □

9.4 Ground states of quantum systems

A ground state of a finite system $A_\Lambda = B(H_\Lambda)$ is an eigenstate of the local Hamiltonian h_Λ with the lowest eigenvalue; because $\dim(H_\Lambda) < \infty$, the spectrum of h_Λ is discrete and hence local ground states exist. For infinite systems, no Hamiltonian is yet defined, so we need to define ground states in terms of the dynamics α_t.

Definition 9.16. *Let A be a C*-algebra with time evolution, i.e., a continuous homomorphism $\alpha : \mathbb{R} \to \mathrm{Aut}(A)$ (which gives the dynamics of the underlying physical system). A* **ground state** *of (A, α) is a state ω on A such that:*

1. *ω is time-independent, i.e. $\alpha_t^* \omega = \omega$ (or $\omega(\alpha_t(a)) = \omega(a)$ for all $a \in A$) $\forall t \in \mathbb{R}$;*
2. *The generator h_ω of the ensuing continuous unitary representation*

$$t \mapsto u_t = e^{ith_\omega} \tag{9.49}$$

of $\mathbb{R}$ on H_ω has positive spectrum, i.e., $\sigma(h_\omega) \subseteq \mathbb{R}^+$, or, equivalently,

$$\langle \psi, h_\omega \psi \rangle \geq 0 \ (\psi \in D(h_\omega)). \tag{9.50}$$

Note that the existence of the operator h_ω is guaranteed by Corollary 9.12 and the arguments after (9.38). Since Corollary 9.12 yields

$$h_\omega \Omega_\omega = 0; \tag{9.51}$$

$$\pi_\omega(\alpha_t(a)) = e^{ith_\omega} \pi_\omega(a) e^{-ith_\omega}, \tag{9.52}$$

it follows that h_ω is a Hamiltonian in the usual sense, implementing the Heisenberg-picture time evolution (albeit in the representation $\pi_\omega(A)$ rather than in A itself). Moreover, in view of (9.51) and the assumed positivity of $\sigma(h_\omega)$, the unit vector Ω_ω of the GNS-representation π_ω induced by a ground state ω is a ground state for the Hamiltonian h_ω in the usual sense. If ω is pure (see below for a discussion of this desirable possibility), then obviously $\exp(ith_\omega) \in \pi_\omega(A)''$, since the latter equals $B(H_\omega)$. A deep result states that this is always the case (***Borchers Theorem***):

Theorem 9.17. *If ω is a ground state on A, then $\exp(ith_\omega) \in \pi_\omega(A)''$ for all $t \in \mathbb{R}$.*

As we shall see, this contrasts with equilibrium states. The Heisenberg equation of motion for operators $a(t)$ has a counterpart in the C*-algebraic formalism, which requires a concept already encountered in §3.1, but repeated here for convenience:

Definition 9.18. *A* **derivation** *on a C*-algebra A is a linear map $\delta : A \to A$ with*

$$\delta(ab) = \delta(a)b + a\delta(b), \ (a, b \in A) \ \textbf{(Leibniz rule)}. \tag{9.53}$$

An **unbounded derivation** *is a linear map $\delta : \mathrm{Dom}(\delta) \to A$, where the domain $\mathrm{Dom}(\delta) \subset A$ of δ is a dense linear subspace of A, that satisfies the Leibniz rule.*

An (unbounded) derivation δ is **symmetric** *when $\delta(a^*) = \delta(a)^*$ for all a (in $\mathrm{Dom}(\delta)$, which must be self-adjoint in that $a \in \mathrm{Dom}(\delta)$ iff $a^* \in \mathrm{Dom}(\delta)$).*

Bounded derivations are rare in classical physics; nonzero derivations of $A = C_0(\mathbb{R}^d)$ do not even exist, but it has plenty of *unbounded* derivations, viz. $\delta(f) = \xi f$ for some vector field ξ on $\mathbb{R}^d$. In quantum mechanics, $A = B(H')$ does have derivations, all given by $\delta(a) = i[h,a]$ for some bounded (self-adjoint) operator h on H'.

Proposition 9.19. *Any continuous homomorphism $\alpha : \mathbb{R} \to \mathrm{Aut}(A)$ on any C^*-algebra A defines an unbounded symmetric derivation δ on A by the norm limit*

$$\delta(a) = \frac{d}{dt}\alpha_t(a)_{|t=0} \equiv \lim_{t\to 0}\frac{\alpha_t(a)-a}{t}, \tag{9.54}$$

where $\mathrm{Dom}(\delta)$ *consists of all $a \in A$ for which this limit exists. Moreover, this domain is stable under α_t in that if $a \in \mathrm{Dom}(\delta)$, then $\alpha_t(a) \subset \mathrm{Dom}(\delta)$ ($t \in \mathbb{R}$).*

The proof is an elementary verification (cf. Theorem 5.73). On H_ω we then have

$$\pi_\omega(\delta(a)) = i[h_\omega, \pi_\omega(a)], \tag{9.55}$$

which, then, is "Heisenberg's equation of motion revisited." One may also reformulate Definition 9.16 in terms of the derivation δ associated to α by (9.54):

Proposition 9.20. *A state $\omega \in S(A)$ is a ground state for given dynamics α iff*

$$-i\omega(a^*\delta(a)) \geq 0 \ \ (a \in \mathrm{Dom}(\delta)). \tag{9.56}$$

Proof. If ω is a ground state according to Definition 9.16, we may use (9.55), (C.196), (9.51), and finally (9.50) to compute

$$\begin{aligned}-i\omega(a^*\delta(a)) &= -i\langle\Omega_\omega, \pi_\omega(a^*\delta(a))\Omega_\omega\rangle = \langle\Omega_\omega, \pi_\omega(a)^*[h_\omega, \pi_\omega(a)]\Omega_\omega\rangle \\ &= \langle\pi_\omega(a)\Omega_\omega, h_\omega\pi_\omega(a)\Omega_\omega\rangle \geq 0. \end{aligned}\tag{9.57}$$

Conversely, we first show that if ω satisfies (9.56), then it is α_t-invariant. We initially assume $a = a^*$, so that $\delta(a)^* = \delta(a^*) = \delta(a)$, as δ is symmetric by construction. Since ω is a state, one has $\omega(b^*) = \overline{\omega(b)}$ for any $b \in A$, so taking $b = \delta(a)a$, using (9.56) just in that $\omega(a^*\delta(a)) \in i\mathbb{R}$, we obtain $\omega(\delta(a)a) = -\omega(a\delta(a))$. Hence

$$\omega(\delta(a^2)) = 0, \tag{9.58}$$

by (9.53), so also $\omega(\delta(\alpha_s(a)^2)) = 0$, $s \in \mathbb{R}$. With (9.54), we find

$$\begin{aligned}0 &= \int_0^u ds\,\omega(\delta(\alpha_s(a)^2)) = \int_0^u ds\,\omega\left(\frac{d}{dt}\alpha_t(\alpha_s(a)^2)_{|t=0}\right) \\ &= \int_0^u ds\frac{d}{dt}\omega(\alpha_{t+s}(a)^2))_{|t=0} = \int_0^u ds\frac{d}{ds}\omega(\alpha_s(a)^2)) = \omega(\alpha_u(a^2)) - \omega(a^2).\end{aligned}$$

Hence $\omega(\alpha_u(a^2)) = \omega(a^2)$ for each $u > 0$ (and analogously for each $u < 0$), whenever $a^* = a$, i.e., $\omega(\alpha_u(b) = \omega(b)$ for each $b \geq 0$. But any $b \in A$ may be written as a sum of at most four positive elements, so $\omega \circ \alpha_u = \omega$ for all $u \subset \mathbb{R}$. We therefore have a Hamiltonian h_ω, whose positivity follows from (9.57), ran backwards. □

9.5 Ground states and equilibrium states of classical spin systems

Thermal equilibrium states are arguably physically more relevant than ground states, as the latter rely on the idealization of temperature zero. Since in statistical mechanics infinite systems are used to approximate very large ones, it will be of particular interest to define equilibrium states in infinite volume. If only to highlight contrasts with quantum theory, we take a long run and start with the classical case.

Classical spin systems on a lattice are defined by a single-site configuration space $\underline{n} \cong \{0,1,\ldots,n\}$, where $m \in \underline{n}$ may either be interpreted as some spin-like degree of freedom (as in the Ising model, where $n = 2$) or as the number of (structureless) particles occupying a given site (in which case one has a ***lattice gas***). As in (C.310), for any finite sublattice $\Lambda \subset \mathbb{Z}^d$, the local algebra of observables is given by

$$A_\Lambda^{(c)} = C(\underline{n}^\Lambda), \tag{9.59}$$

where $\underline{n}^\Lambda = C(\Lambda, \underline{n})$ consists of all functions $s : \Lambda \to \underline{n}$. For finite Λ this is a finite set (of cardinality $n^{|\Lambda|}$), so that all functions in question are continuous and hence $C(\underline{n}^\Lambda)$ just stands for the commutative C*-algebra of *all* functions from $\underline{n}^\Lambda$ to $\mathbb{C}$. If $\Lambda_1 \subseteq \Lambda^{(2)}$, we have maps $\iota^{(c)}_{\Lambda_1 \Lambda^{(2)}} : A^{(c)}_{\Lambda_1} \hookrightarrow A^{(c)}_{\Lambda^{(2)}}$, written $f_1 \mapsto f_2$, which are given by

$$f_2(s) = f_1(s_{|\Lambda_1}), \tag{9.60}$$

where $s : \Lambda^{(2)} \to \underline{n}$. As these maps are injective, the ensuing inductive limit is simply

$$A^{(c)} = \cup_{\Lambda \subset \mathbb{Z}^d} A_\Lambda^{(c)} \cong C\left(\underline{n}^{\mathbb{Z}^d}\right), \tag{9.61}$$

where $\underline{n}^{\mathbb{Z}^d} = \prod_{x \in \mathbb{Z}^d} \underline{n}$ is endowed with the product topology and hence (by Tychonoff's theorem) is compact (for $n = 2, d = 1$ this is a model of the Cantor set).

As in the quantum case, local Hamiltonians are defined via an ***interaction*** Φ, which now is an assignment $X \mapsto \Phi(X)$, where $X \subset \mathbb{Z}^d$ is finite and $\Phi(X) \in A_X^{(c)}$. If $X \subset Y$, we regard $\Phi(X)$ an an element in $A_Y^{(c)}$ through the inclusion $A_X^{(c)} \subset A_Y^{(c)}$, indicating this explicitly by writing $\Phi(X)_Y \in A_Y^{(c)}$. We then define $h_\Lambda \in A_\Lambda^{(c)}$ by

$$h_\Lambda = \sum_{X \subset \Lambda} \Phi(X)_\Lambda, \tag{9.62}$$

where the the sum is over all subsets X of Λ. For example, the Ising Hamiltonian

$$h_\Lambda(s) = -J \sum_{\langle ij \rangle_\Lambda} s_i s_j - B \sum_{i \in \Lambda} s_i, \tag{9.63}$$

where the sum is over nearest neighbours in Λ, and we assume $\underline{2} = \{-1, 1\}$ (rather than the usual c-bit $\{0,1\}$), comes from the following potential:

- $\Phi(X) = 0$ if either $|X| > 2$ or, if $|X| = 2$, its elements are not nearest neighbours;
- $\Phi(\{i\}) : s \mapsto -Bs_i$, and $\Phi(\{i,j\}) : s \mapsto -Js_i s_j$ if i and j are nearest neighbours.

As in (9.41), the prescription (9.62) has free boundary conditions, in that it only involves spins inside Λ. Another possibility is to fix a "boundary" spin configuration $b \in \underline{n}^{\mathbb{Z}^d}$, and define $h^b_\Lambda \in A^{(c)}_\Lambda$ by

$$h^b_\Lambda = \sum_{X \subset \mathbb{Z}^d, |X|<\infty, X\cap\Lambda \neq \emptyset} \Phi(X)^b_\Lambda. \tag{9.64}$$

This involves some new notation $\Phi(X)^b_\Lambda$, which means the following. In principle, $\Phi(X) \in A^{(c)}_X$ is a function on $\underline{n}^X$. We now turn $\Phi(X)$ into a function $\Phi(X)^b_\Lambda$ on $\underline{n}^\Lambda$ (so that h^b_Λ is a function on $\underline{n}^\Lambda$ as required): for given $s : \Lambda \to \underline{n}$ and given $b : \mathbb{Z}^d \to \underline{n}$ we define $s' : X \to \underline{n}$ by putting $s' = s$ on $X \cap \Lambda$ and $s' = b$ on the remainder of X (which is $X \cap \Lambda^c$, with $\Lambda^c = \mathbb{Z}^d \backslash \Lambda$). Then

$$\Phi(X)^b_\Lambda(s) = \Phi(X)(s'). \tag{9.65}$$

Physically, this simply means that those spins outside Λ that interact with spins inside Λ are set at a fixed value determined by the boundary condition b. For example, consider the Ising model in $d = 1$. If we take $\Lambda = \{2,3\}$, then from (9.62) we obtain $h_\Lambda = -Js_2s_3 - B(s_2 + s_3)$; spins outside Λ do not contribute. From (9.64), on the other hand, we obtain $h^b_\Lambda = h_\Lambda - J(b_1s_2 + s_3b_4)$. Although the boundary condition b is arbitrary, one may think of simple choices like $b_i = 1$ or -1 for each i.

We may actually rewrite (9.64) as a difference between Hamiltonians with free boundary conditions. To do so, for given finite Λ we pick some finite $\Lambda' \supset \Lambda$ large enough that it contains all spins outside Λ that interact with spins inside Λ (provided this is possible). With the conventional notation $h_\Lambda(s|b) \equiv h^b_\Lambda(s)$, this yields

$$h_\Lambda(s|b) = h_{\Lambda'}(s,b) - h_{\Lambda'\backslash\Lambda}(b) = \sum_{X' \subset \Lambda'} \Phi(X')_{\Lambda'}(s,b) - \sum_{Y \subset \Lambda'\backslash\Lambda} \Phi(Y)_{\Lambda'\backslash\Lambda}(b).$$

Analogous to (9.65), the notation $\Phi(X')_{\Lambda'}(s,b)$ here means $\Phi(X')_{\Lambda'}(s')$, for the function $s' : \Lambda' \to \underline{n}$ that on $\Lambda \subset \Lambda'$ coincides with $s : \Lambda \to \underline{n}$, whilst on $(\Lambda'\backslash\Lambda) \subset \Lambda'$ it coincides with the restriction of b to $\Lambda'\backslash\Lambda$. Thus we may also write

$$h_\Lambda(s|b) = \lim_{\Lambda' \uparrow \mathbb{Z}^d} (h_{\Lambda'}(s,b) - h_{\Lambda'\backslash\Lambda}(b)), \tag{9.66}$$

although neither $h_{\mathbb{Z}^d}(s,b)$ nor $h_{\mathbb{Z}^d\backslash\Lambda}(b)$ makes sense by itself. ***Periodic*** boundary conditions for local Hamiltonians may be defined for arbitrary interactions Φ and special lattices. For example, the Ising chain in $d = 1$ has local Hamiltonians

$$h^{pbc}_{\{1,2,\ldots,n\}}(s) = J\left(s_1s_n + \sum_{i=1}^{n-1} s_is_{i+1}\right) - B\sum_{i=1}^{n} s_i. \tag{9.67}$$

Naively, a ***ground state*** of a *finite* classical spin system, i.e., a system of the above kind defined on a *fixed* finite lattice $\Lambda \subset \mathbb{Z}^d$, is a spin configuration $s_0 \in \underline{n}^\Lambda$ that minimizes the local Hamiltonian h_Λ (9.62), or its counterpart (9.64), that is,

$$h_\Lambda(s_0) \leq h_\Lambda(s), \tag{9.68}$$

for all $s \in \underline{n}^\Lambda$. For example, if Λ is a hypercube Λ_N, then the Ising model (9.63) has a unique ground state for $B > 0$, namely $s_0(x) = 1$ for all $x \in \Lambda$, whereas it has two ground states $s_0^\pm$ for $B = 0$, given by $s_0^\pm(x) = \pm 1$ for all x. Ground states of finite classical systems always exist (since the space on which h_Λ is finite), but they are not necessarily unique; we just gave a counterexample! The same is true for quantum theory, since for $B = 0$ also the quantum Ising model (9.42) has two degenerate symmetry-breaking ground states. Nonetheless, this case is special, since for nonzero small values of B the ground state of the quantum Ising model is unique for finite Λ, whereas on the infinite lattice $\mathbb{Z}^d$ it is degenerate (cf. §10.7).

The definition of ground states of *infinite* classical spin systems is just slightly more involved: for local Hamiltonians h_Λ with free boundary conditions defined by an interaction Φ à la (9.62), a ground state is a point $s_0 \in \underline{n}^{\mathbb{Z}^d}$ for which

$$h_\Lambda(s_{0|\Lambda}) \leq h_\Lambda(s_{|\Lambda}), \tag{9.69}$$

for any finite $\Lambda \subset \mathbb{Z}^d$ and any spin configuration $s \in \underline{n}^{\mathbb{Z}^d}$. Alternatively, one may ask

$$h_\Lambda^{s_0}(s_0) \leq h_\Lambda^{s_0}(s), \tag{9.70}$$

for all finite $\Lambda \subset \mathbb{Z}^d$ and all spin configurations $s \in \underline{n}^{\mathbb{Z}^d}$ *that coincide with s_0 outside* Λ, where $h_\Lambda^{s_0}$ stands for (9.64) with $b = s_0$. In other words, s_0 provides a boundary condition b, which is fixed for all s that compete with s_0 in minimizing the local Hamiltonian $h_\Lambda^{b=s_0}$. Both definitions give the usual two ground states for the Ising model with $B = 0$ (in which all spins are either "up" or "down"), but the second one also opens the possibility of *domain walls*, where infinite chains of "spin up" alternate with infinite chains of "spin down", and similarly in higher d.

If different ground states in the above ("pure") sense exist, we may reinterpret such states s_0 as Dirac measures δ_{s_0} on the space $\underline{n}^\Lambda$ of all spin configurations on Λ, and may also allow convex combinations of ground states as ground states. This, as well as the analogy with Definition 9.16 (in which no purity condition is imposed) inspires a more liberal definition of a ground state, which is predicated on Boltzmann's idea that a state of a classical system of the kind we consider is a probability measure μ_Λ^0 on $\underline{n}^\Lambda$, and likewise for $\underline{n}^{\mathbb{Z}^d}$. In the C*-algebraic formalism we use, this follows from (9.61) and the identification of states on $C(X)$ with completely regular probability measures on X (assumed to be a compact Hausdorff space, cf. §B.5). A state μ on $C(\underline{n}^{\mathbb{Z}^d})$, i.e., a probability measure on $\underline{n}^{\mathbb{Z}^d}$, induces a state on each local algebra $C(\underline{n}^\Lambda)$, i.e., a probability measure μ_Λ on $\underline{n}^\Lambda$ simply by restriction, since

$$C(\underline{n}^\Lambda) \subset C(\underline{n}^{\mathbb{Z}^d}) \tag{9.71}$$

through the injection (9.60), according to which $f_\Lambda \in C(\underline{n}^\Lambda)$ has image $f \in C(\underline{n}^{\mathbb{Z}^d})$ defined by $f(s) = f_\Lambda(s_{|\Lambda})$. The measure μ_Λ, then, is given in terms of μ by

$$\mu_\Lambda(f_\Lambda) = \mu(f); \tag{9.72}$$

the corresponding probability distribution p_Λ (i.e., $p_\Lambda(s) = \mu_\Lambda(\{s\})$) is given by

$$p_\Lambda(s) = \mu\left(\{s' \in \underline{n}^{\mathbb{Z}^d} \mid s'_{|\Lambda} = s\}\right), \quad s \in \underline{n}^\Lambda. \tag{9.73}$$

The family of probability measures (μ_Λ) defined by μ is ***consistent*** in that if $\Lambda^{(1)} \subset \Lambda^{(2)}$ and $f_1 \in C(\underline{n}^{\Lambda^{(1)}})$ and $f_2 \in C(\underline{n}^{\Lambda^{(2)}})$ are related as in (9.60), then

$$\mu_{\Lambda^{(1)}}(f_1) = \mu_{\Lambda^{(2)}}(f_2). \tag{9.74}$$

Conversely, a consistent family of probability measures (μ_Λ) defines a unique probability measure μ on $\underline{n}^{\mathbb{Z}^d}$ which induces the given family through (9.72).

Definition 9.21. *For given finite $\Lambda \subset \mathbb{Z}^d$, a probability measure μ_Λ^0 on $\underline{n}^\Lambda$ is a* **ground state** *of a local Hamiltonian h_Λ (with free boundary conditions) if, in terms of the probabilities $p_\Lambda^0(s) = \mu_\Lambda^0(\{s\})$, for any probability measure μ_Λ on $\underline{n}^\Lambda$,*

$$\sum_{s \in \underline{n}^\Lambda} p_\Lambda^0(s) h_\Lambda \leq \sum_{s \in \underline{n}^\Lambda} p_\Lambda(s) h_\Lambda. \tag{9.75}$$

A probability measure μ_0 on $\underline{n}^{\mathbb{Z}^d}$ is a **ground state** *for some interaction Φ if* (9.75) *holds for any probability measure μ on $\underline{n}^{\mathbb{Z}^d}$ and any finite subset $\Lambda \subset \mathbb{Z}^d$, where this time p_Λ^0 (and analogously p_Λ) is defined by* (9.73).

In particular, convex sums of pure ground states are ground states in this more general sense, so that, if all pure ground states break some symmetry (as is the case for the $\mathbb{Z}_2$-symmetry $s \mapsto -s$ of the Ising model at $B = 0$), symmetric convex sums will restore the symmetry. The set of all ground states of a given interaction Φ is a convex set, whose extreme points are the pure ground states (at least, under suitable hypotheses on Φ). This leads to a discussion of SSB similar to the quantum case.

In the following discussion of equilibrium states, we use the notation

$$\Pr(X) \cong S(C(X)) \tag{9.76}$$

for the compact convex set of all completely regular probability measures on X, which as above will either be the finite set $\underline{n}^\Lambda$ (with discrete topology)—on which of course any probability measure is completely regular—or the compact space $\underline{n}^{\mathbb{Z}^d}$. In the first case we may as well use probability *distributions* p_Λ (instead of probability measures) on $\underline{n}^\Lambda$. In the second, we could also use Baire measures.

Given an interaction Φ and the ensuing family (9.62) of local Hamiltonians h_Λ, we define the local ***energy*** for each finite $\Lambda \subset \mathbb{Z}^d$ as a function $\mathscr{E}_\Lambda : \Pr(\underline{n}^\Lambda) \to \mathbb{R}$ by

$$\mathscr{E}_\Lambda(p_\Lambda) = \sum_{s \in \underline{n}^\Lambda} p_\Lambda(s) h_\Lambda(s). \tag{9.77}$$

Of course, this is just the expectation value of the Hamiltonian in the state p_Λ. The local ***entropy*** $S_\Lambda : \mathrm{Pr}(\underline{n}^\Lambda) \to \mathbb{R}$ is a more subtle concept; rather than the expectation value of some (local) observable, it specifies a property of the probability distribution itself. With Boltzmann's constant k_B, we have

$$S_\Lambda(p_\Lambda) = -k_B \sum_{s \in \underline{n}^\Lambda} p_\Lambda(s) \ln(p_\Lambda(s)). \tag{9.78}$$

Note that $S_\Lambda(p_\Lambda) \geq 0$, with equality iff p_Λ is a pure state (i.e., p_Λ is supported at a single spin configuration). The local ***free energy*** $\mathscr{F}_\Lambda^\beta : \mathrm{Pr}(\underline{n}^\Lambda) \to \mathbb{R}$ is defined as

$$\mathscr{F}_\Lambda^\beta = \mathscr{E}_\Lambda - TS_\Lambda, \tag{9.79}$$

where $\beta = 1/k_B T$. A ***local equilibrium state***, then, is a probability distribution p_Λ^β that minimizes the free energy (for fixed temperature T).

Theorem 9.22. *For each $T > 0$, there is a* unique *local equilibrium state, given by the* **Boltzmann distribution** *(and associated* **partition function***)*

$$p_\Lambda^\beta(s) = (Z_\Lambda^\beta)^{-1} e^{-\beta h_\Lambda(s)}; \tag{9.80}$$

$$Z_\Lambda^\beta = \sum_{s' \in \underline{n}^\Lambda} e^{-\beta h_\Lambda(s')}. \tag{9.81}$$

The associated ***free energy in equilibrium*** is then given by

$$F_\Lambda^\beta = \mathscr{F}_\Lambda^\beta(p_\Lambda^\beta) = -\beta^{-1} \ln Z_\Lambda^\beta. \tag{9.82}$$

Proof. The claim follows from the fact that any $p_\Lambda \in \mathrm{Pr}(\underline{n}^\Lambda)$ satisfies the inequality

$$\mathscr{F}_\Lambda^\beta(p_\Lambda) \geq -\beta^{-1} \ln Z_\Lambda^\beta, \tag{9.83}$$

with equality iff $p = p_\Lambda^\beta$, i.e., using (9.79), (9.77), and (9.78), we need to show that

$$\sum_{s \in E^\Lambda} p(s)(h_\Lambda(s) + \beta^{-1} \ln p(s)) + \beta^{-1} \ln Z_\Lambda^\beta \geq 0. \tag{9.84}$$

Using (9.80), for each $s \in E^\Lambda$ we obtain

$$-\beta h_\Lambda(s) = \ln Z_\Lambda^\beta + \ln p_\Lambda^\beta(s). \tag{9.85}$$

Substituting this in (9.84), using $\sum_s p(s) = 1$, omitting the ensuing prefactor β^{-1}, and noting that $p_\Lambda^\beta(s) > 0$ for all s, the inequality (9.84) to be proved becomes

$$\sum_{s \in E^\Lambda} p(s) \ln\left(\frac{p(s)}{p_\Lambda^\beta(s)}\right) \geq 0. \tag{9.86}$$

Hence we need to prove the inequality

$$\sum_{s\in E^{\Lambda}} p^{\beta}_{\Lambda}(s)\cdot\left(\frac{p(s)}{p^{\beta}_{\Lambda}(s)}\right)\ln\left(\frac{p(s)}{p^{\beta}_{\Lambda}(s)}\right)\geq 0, \tag{9.87}$$

with equality iff $p(s)=p^{\beta}_{\Lambda}(s)$ for all s. Let us note that the function $f(x)=x\ln x$ is strictly convex for all $x\geq 0$, that is, for any finite set of numbers $p'(s)\in(0,1)$ with $\sum_s p'(s)=1$ and any set of positive real numbers $(x_s)_s\geq 0$, we have

$$\sum_s p'(s)f(x_s)\geq f\left(\sum_s p'(s)x_s\right), \tag{9.88}$$

with equality iff all numbers x_s are the same. Applying this with $p'(s)=p^{\beta}_{\Lambda}(s)$ and $x_s=p(s)/p^{\beta}_{\Lambda}(s)$, so that $p'(s)x_s=p(s)$ and hence $\sum_s p'(s)x_s=\sum_s p(s)=1$, which makes the right-hand side of (9.88) vanish since $\ln(1)=0$, finally leads to (9.87). Equality arises iff $p(s)/p^{\beta}_{\Lambda}(s)$ equals the same numer c for all s; summing over all s forces $c=1$, so that one has equality iff $p(s)=p^{\beta}_{\Lambda}(s)$ for all s, as desired. □

Neither the local Hamiltonians (9.62) nor the local partition functions (9.81) have a limit as $\Lambda\uparrow\mathbb{Z}^d$. A precise definition equilibrium states of infinite classical systems was given in 1968 by Dobrushin and by Lanford and Ruelle (DLR).

Definition 9.23. *For fixed inverse temperature $\beta\in(0,\infty)$ and fixed interaction Φ, a* **Gibbs measure** *μ^{β} is a (Baire = regular Borel) probability measure on $\underline{n}^{\mathbb{Z}^d}$ such that for each finite $\Lambda\subset\mathbb{Z}^d$ and each pair (s,b) of a spin configuration $s:\Lambda\to\underline{n}$ plus boundary condition $b:\Lambda^c\to\underline{n}$, the conditional probability $\mu^{\beta}(\mathsf{s}|\mathsf{b})$ for the events*

$$\mathsf{s}=\{s'\in\underline{n}^{\mathbb{Z}^d}\mid s'_{|\Lambda}=s\}\subset\underline{n}^{\mathbb{Z}^d}; \tag{9.89}$$

$$\mathsf{b}=\{s''\in\underline{n}^{\mathbb{Z}^d}\mid s''_{|\Lambda^c}=b\}\subset\underline{n}^{\mathbb{Z}^d}, \tag{9.90}$$

is given in terms of the local Hamiltonian $h_{\Lambda}(s|b)$ as defined by (9.66) *by*

$$\mu^{\beta}(\mathsf{s}|\mathsf{b})=(Z^{\beta}_{\Lambda}(b))^{-1}e^{-\beta h_{\Lambda}(s|b)}, \tag{9.91}$$

$$Z^{\beta}_{\Lambda}(b)=\sum_{s\in\underline{n}^{\Lambda}}e^{-\beta h_{\Lambda}(s|b)}. \tag{9.92}$$

Recall that $\mu^{\beta}(\mathsf{s}|\mathsf{b})=\mu^{\beta}(\mathsf{s}\cap\mathsf{b})/\mu^{\beta}(\mathsf{b})$, where $\mathsf{s}\cap\mathsf{b}=\{s_b\}$ consists of the single spin configuration $s_b:\mathbb{Z}^d\to\underline{n}$ that coincides with s on Λ and coincides with b on Λ^c. Thus we may write $\mu^{\beta}(\mathsf{s}|\mathsf{b})=p^{\beta}(s_b)/\mu^{\beta}(\mathsf{b})$, where $p^{\beta}(s)=\mu^{\beta}(\{s\})$ as usual.

It was initially unclear how to generalize this highly fruitful definition of equilibrium states in classical statistical mechanics to the quantum case, where conditional probabilities are not well defined (this was eventually resolved, however, through Definition 10.9 below). Thus a different (equally fruitful) approach to equilibrium states of (infinite) quantum systems was developed, to which we now turn.

9.6 Equilibrium (KMS) states of quantum systems

For *finite* quantum spin systems we have expressions for the energy $\hat{\mathscr{E}}^{\beta}_{\Lambda}$, the entropy $\hat{S}_{\Lambda}$, and the free energy $\hat{\mathscr{F}}_{\Lambda}$ that are analogous to their classical counterparts (9.77), (9.78), and (9.79). In particular, these quantities are functions on the state space $S(A_{\Lambda})$. Since $A_{\Lambda} = B(H_{\Lambda})$, where we assume that H and hence H_{Λ} is finite-dimensional, each state $\omega_{\Lambda} \in S(A_{\Lambda})$ is given by a density operator ρ_{Λ}, so that

$$\hat{\mathscr{E}}_{\Lambda}(\omega_{\Lambda}) = \omega_{\Lambda}(h_{\Lambda}) = \mathrm{Tr}\,(\rho_{\Lambda} h_{\Lambda}); \tag{9.93}$$

$$\hat{S}_{\Lambda}(\omega_{\Lambda}) = -k_B \mathrm{Tr}\,(\rho_{\Lambda} \ln \rho_{\Lambda}); \tag{9.94}$$

$$\hat{\mathscr{F}}^{\beta}_{\Lambda} = \hat{\mathscr{E}}_{\Lambda} - T\hat{S}_{\Lambda}. \tag{9.95}$$

Defining a local equilibrium state as a density matrix ρ^{β}_{Λ} that minimizes the free energy (for fixed T), we have the following quantum analogue of Theorem 9.22:

Theorem 9.24. *For each $T > 0$, there is a* unique *local equilibrium state ω^{β}_{Λ}, viz.*

$$\omega^{\beta}_{\Lambda}(a) = \mathrm{Tr}\left(\rho^{\beta}_{\Lambda} a\right); \tag{9.96}$$

$$\rho^{\beta}_{\Lambda} = (\hat{Z}^{\beta}_{\Lambda})^{-1} e^{-\beta h_{\Lambda}}; \tag{9.97}$$

$$\hat{Z}^{\beta}_{\Lambda} = \mathrm{Tr}\left(e^{-\beta h_{\Lambda}}\right). \tag{9.98}$$

Accordingly, the free energy F^{β}_{Λ} in equilibrium is given by

$$F^{\beta}_{\Lambda} = \hat{\mathscr{F}}^{\beta}_{\Lambda}(\rho^{\beta}_{\Lambda}) = -\beta^{-1} \ln \hat{Z}^{\beta}_{\Lambda}. \tag{9.99}$$

Proof. One proof is analogous to the classical case, in that for all $\rho_{\Lambda} \in \mathscr{D}(B(H_{\Lambda}))$,

$$\hat{\mathscr{F}}^{\beta}_{\Lambda}(\rho_{\Lambda}) \geq -\beta^{-1} \ln \hat{Z}^{\beta}_{\Lambda}, \tag{9.100}$$

with equality iff $\rho_{\Lambda} = \rho^{\beta}_{\Lambda}$. This, in turn, follows from the inequality

$$\mathrm{Tr}\,(a(\ln b - \ln a)) \leq \mathrm{Tr}\,(b - a), \tag{9.101}$$

with equality iff $b = a$, which is valid for matrices a, b for which $a \geq 0$ (in the usual sense that $\lambda \geq 0$ for each $\lambda \in \sigma(a)$) and $b > 0$ in that $\lambda > 0$ for each $\lambda \in \sigma(b)$. The case $a = \rho_{\Lambda}$ and $b = \rho^{\beta}_{\Lambda}$ immediately gives the claim. □

What remains to be done, however, is to define equilibrium states for infinite systems. This is achieved through the so-called KMS-***condition***, which is based on the observation that for any $a, b \in A_{\Lambda}$, in terms of (9.40) the state (9.96) satisfies

$$\omega^{\beta}_{\Lambda}(\alpha^{(\Lambda)}_t(a)b) = \omega^{\beta}_{\Lambda}(b\alpha^{(\Lambda)}_{t+i\beta}(a)) \;\; (t \in \mathbb{R}). \tag{9.102}$$

Moreover, in finite systems this condition (even at $t=0$) fully characterizes ω_Λ^β:

Proposition 9.25. *Let h be a self-adjoint operator on a finite-dimensional Hilbert space H', with associated density operator ρ and (complex) time-evolution given by*

$$\rho = \frac{e^{-h}}{\mathrm{Tr}\,(e^{-h})}; \tag{9.103}$$

$$\alpha_z(a) = e^{izh} a e^{-izh}, \; z\in\mathbb{C}, a\in B(H'), \tag{9.104}$$

respectively (the exponentials being defined by a norm-convergent power series).
Then the associated two-point functions defined by $\omega(a) = \mathrm{Tr}\,(\rho a)$ satisfy

$$\omega(ab) = \omega(b\alpha_i(a)) \; (a,b\in B(H)). \tag{9.105}$$

Conversely, any state for which (9.105) *holds for given h and α_z is given by* (9.103).

Proof. Eq. (9.105) follows from (9.103) - (9.104) and cyclicity of the trace, i.e., (A.78). Similarly, given non-degeneracy of the Hilbert-Schmidt inner product (B.495) on $B(H)$, eq. (9.105) is equivalent to the condition

$$\rho a = e^{-h} a e^{h} \rho, \tag{9.106}$$

for each $a\in B(H')$. Multiplying with $\exp(h)$ shows that $\exp(h)\rho$ commutes with every $a\in B(H')$. Since $B(H')' = \mathbb{C}\cdot 1_{H'}$, we obatin $\exp(h)\rho = \lambda\cdot 1_H$. Since $\exp(h)$ is invertible with inverse $\exp(-h)$, we obtain $\rho = \lambda\cdot\exp(-h)$, upon which the normalization condition $\mathrm{Tr}\,(\rho) = 1$ yields (9.103). □

For arbitrary C*-algebras A with time-evolution $t\mapsto\alpha_t$, expressions like $\alpha_{t+i\beta}(a)$ may not be defined, so one has to proceed more carefully, but the idea is the same.

Definition 9.26. *Let A be a C*-algebra with an automorphism group $\mathbb{R}$. A* **KMS state** *at "inverse temperature" $\beta\in\mathbb{R}$ is a state ω on A with the following property:*

1. *For any $a,b\in A$, the function $F_{a,b} : t\mapsto\omega(b\alpha_t(a))$ from $\mathbb{R}$ to $\mathbb{C}$ has an analytic continuation to the strip*

$$\mathscr{S}_\beta = \{z\in\mathbb{C}\mid 0\leq \mathrm{Im}(z)\leq\beta\}, \tag{9.107}$$

where it is holomorphic in the interior and continuous on the boundary

$$\partial\mathscr{S}_\beta = \mathbb{R}\cup(\mathbb{R}+i\beta). \tag{9.108}$$

2. *The boundary values of $F_{a,b}$ are related, for all $t\in\mathbb{R}$, by*

$$F_{a,b}(t) = \omega(b\alpha_t(a)); \tag{9.109}$$

$$F_{a,b}(t+i\beta) = \omega(\alpha_t(a)b). \tag{9.110}$$

If this is the case, ω satisfies the KMS-**condition** *at (inverse temperature) β.*

It is easy to show that A has a dense subset A_α such that for any $a \in A_\alpha$ the function $t \mapsto \alpha_t(a)$ from $\mathbb{R}$ to A extends to an entire A-valued analytic function, written $z \mapsto \alpha_z(a)$ (i.e., for each $\varphi \in A^*$ the function $z \mapsto \varphi(\alpha_z(a))$ from $\mathbb{C}$ to $\mathbb{C}$ is entire analytic). Namely, for any $a \in A$ and $\varepsilon > 0$, define

$$a_\varepsilon = \int_{-\infty}^{\infty} \frac{dt}{\sqrt{2\pi\varepsilon}} e^{-t^2/2\varepsilon} \alpha_t(a), \tag{9.111}$$

which satisfies $a_\varepsilon \in A_\alpha$ and $\lim_{\varepsilon \downarrow 0} a_\varepsilon = a$. If $A = B(H')$ with $\dim(H') < \infty$, we even have $B(H')_\alpha = B(H')$, since (9.104) is entire analytic in z for any $a \in B(H')$. For any A, the KMS-condition on ω is then equivalent to the simpler requirement

$$\omega(ab) = \omega(b\alpha_{i\beta}(a)) \quad (a \in A_\alpha, b \in A). \tag{9.112}$$

Corollary 9.27. *If $A = B(H')$ with $\dim(H') < \infty$, then KMS states (at fixed β) are necessarily given by the equilibrium states of Theorem 9.24 and hence are unique.*

Although initially the characterization of equilibrium states of infinite systems by the KMS condition was tentative, in the 1970s and '80s it became clear that it was spot on, being equivalent to local and global thermodynamic stability (against perturbations of the dynamics), the (local) maximum entropy principle, etc. Also:

Proposition 9.28. *A KMS state at $\beta \in \mathbb{R}\backslash\{0\}$ is time-independent.*

Proof. We just sketch the proof if A is unital. Taking $b = 1_A$, for fixed $a \in A_\alpha$ the function $F_{a,1_A} \equiv F$ defined by $F(z) = \omega(\alpha_z(a))$ is entire analytic on $\mathbb{C}$. Writing $z = t + is$ (with $s, t \in \mathbb{R}$), we have $\alpha_z = \alpha_t \circ \alpha_{is}$ and hence (since each α_t is an automorphism and hence an isometry), $|F(t+is)| \leq \|\alpha_{is}(a)\|$. Also, (9.112) yields $F(t + i(s+\beta)) = F(t+is)$. Hence $F(t+is)$ is bounded in t and periodic in s; by the latter property its supremum on $\mathbb{C}$ may be computed by its supremum on the strip $\mathscr{S}_\beta$, and by the former property this supremum is finite. Therefore, F is bounded, and so by Liouville's Theorem it must be constant, especially if $z = t \in \mathbb{R}$. Hence $\alpha_t^*\omega(a) = \omega(a)$ for each $a \in A_\alpha$, and since this is a dense set, $\alpha_t^*\omega = \omega$. □

By the argument for ground states following Definition 9.16, the automorphism group $t \mapsto \alpha_t$ is unitarily implemented in the GNS-representation π_ω induced by a KMS state ω, such that (9.51) - (9.52) hold. However, the operator h_ω in this construction should not be confused with the Hamiltonian of the system. For example suppose $A = B(H')$ for some (not necessarily finite-dimensional) Hilbert space H', so that (9.39) holds for some (not necessarily bounded) Hamiltonian h with discrete spectrum, such that $\exp(-\beta h) \in B_1(H')$. If we now define the density operator

$$\rho = \frac{e^{-\beta h}}{\mathrm{Tr}\left(e^{-\beta h}\right)}, \tag{9.113}$$

then the corresponding state ω satisfies the KMS-condition at β. Generalizing the computations around (2.66) in §2.4, we then find (up to unitary equivalence):

$$H_\omega = B_2(H'); \tag{9.114}$$
$$\pi_\omega(a)b = ab; \tag{9.115}$$
$$\Omega_\omega = \rho^{1/2}; \tag{9.116}$$
$$e^{ith_\omega} = \pi_\omega\left(e^{ith}\right)\pi'_\omega\left(e^{-ith}\right), \tag{9.117}$$

where for any $a \in B(H')$, the operator $\pi'_\omega(a)$ on $B_2(H')$ is defined by

$$\pi'_\omega(a)b = ba. \tag{9.118}$$

Note that (9.115) is well defined, since $\rho \geq 0$ and $\rho \in B_1(H')$, whence $\rho^{1/2} \in B_2(H')$, and hence also $ab \in B_2(H')$ and $ba \subset B_2(H')$, since $B_2(H')$ is a two-sided ideal in $B(H')$. If h happens to be bounded, we may therefore write

$$h_\omega = \pi_\omega(h) - \pi'_\omega(h). \tag{9.119}$$

Note that the π'_ω term in (9.117) is not needed for (9.52), since $[\pi_\omega(a), \pi'_\omega(b)] = 0$ for any $a, b \subset B(H')$, but it *is* necessary to secure (9.51). Another feature of this example is that the vector Ω_ω is not only *cyclic* for $\pi_\omega(B(H'))$, which it has to be by virtue of the GNS-construction, but also *separating*, i.e., $\pi_\omega(a)\Omega_\omega = 0$ implies $\pi_\omega(a) = 0$. In other words, one has $\omega(a^*a) = 0$ iff $a = 0$ (which is by no means the case for ground states). If $\dim(H') < \infty$, this is obvious, because $\pi_\omega(a)\Omega_\omega = a\rho^{1/2}$ and $\rho^{1/2}$ is invertible. In general, for arbitrary C*-algebras A we have:

Proposition 9.29. *Let ω be a* KMS *state on A at $\beta \in \mathbb{R}$. Then Ω_ω is both cyclic and separating for $\pi_\omega(A)$ and hence also for $\pi_\omega(A)''$ (as well as for $\pi_\omega(A)'$).*

Proof. Since $\omega(a^*a) = \|\pi_\omega(a)\Omega_\omega\|^2$, we have $\omega(a^*a) = 0$ iff $\pi_\omega(a)\Omega_\omega = 0$, so that

$$\omega(a^*\alpha_t(a)) = \langle \pi_\omega(a)\Omega_\omega, \pi_\omega(\alpha_t(a))\Omega_\omega\rangle = 0 \ (t \in \mathbb{R})$$

if $\omega(a^*a) = 0$, and hence $F_{a^*,a}(t) = 0$, cf. (9.109). The "edge of the wegde" theorem then gives $F_{a^*,a}(z) = 0$ for all $z \in \mathscr{S}_\beta$, upon which the KMS-condition gives

$$\omega(aa^*) = F_{a^*,a}(i\beta) = 0.$$

This means that $\omega(a^*a) = 0$ iff $\omega(aa^*) = 0$, or $\pi_\omega(a)\Omega_\omega = 0$ iff $\pi_\omega(a)^*\Omega_\omega = 0$, and hence $\pi_\omega(b^*)\pi_\omega(a)\Omega_\omega = 0$ iff $\pi_\omega(a^*)\pi_\omega(b)\Omega_\omega = 0$. Since Ω_ω is cyclic for $\pi_\omega(A)$, the assumption $\pi_\omega(a)\Omega_\omega = 0$ therefore implies that the bounded operator $\pi_\omega(a^*)$ vanishes on a dense domain in H_ω and hence vanishes. Since $\pi_\omega(a) = (\pi_\omega(a^*))^*$, it follows that $\pi_\omega(a) = 0$. The extension to $\pi_\omega(A)''$ (and $\pi_\omega(A)'$) is obvious. □

Corollary 9.30. *If ω is a* KMS *state on a quasi-local algebra A, i.e., given by* (8.130) *with* $\dim(H) < \infty$, *then $\omega(a^*a) = 0$ iff $a = 0$ and hence the* GNS*-representation $\pi_\omega : A \to B(H_\omega)$ is injective.*

Proof. By the previous proof, the closed left-ideal (C.204) is actually a two-sided ideal, which must be zero, since A is simple (as is easily shown from the simplicity of $B(H)$ for finite-dimensional H, cf. §8.5). □

Proposition 9.29 shows that the von Neumann algebra $\pi_\omega(A)''$ is in standard form (see Definition C.158), so that the KMS condition bring us into the realm of the Tomita–Takesaki theory. In particular, Theorem C.159 provides us with another time-evolution, namely the one given by the modular group. In the situation of Theorem C.159, we take $a \in M_\alpha$ and $b \in M$, and compute

$$\begin{aligned} \langle \Omega, b\alpha_{-i}(a)\Omega \rangle &= \langle \Omega, b\Delta a \Delta^{-1}\Omega \rangle = \langle \Omega, b\Delta a\Omega \rangle \\ &= \langle \Delta^{1/2} b^* \Omega, \Delta^{1/2} a\Omega \rangle = \langle J\Delta^{1/2} a\Omega, J\Delta^{1/2} b^*\Omega \rangle \\ &= \langle Sa\Omega, Sb^*\Omega \rangle = \langle a^*\Omega, b\Omega \rangle \qquad (9.120) \\ &= \langle \Omega, ab\Omega \rangle, \qquad (9.121) \end{aligned}$$

where we used the property $\Delta^{1/2}\Omega = \Omega$ as well as anti-unitarity of J, which implies $\langle J\psi, J\varphi \rangle = \langle \varphi, \psi \rangle$; these facts follow from the definitions of Δ and J via S. Therefore, the state ω on M defined by $\omega(a) = \langle \Omega, a\Omega \rangle$ $(a \in M)$ satisfies the KMS-condition for the modular group at $\beta = -1$. If, on the other hand, we start with a β-KMS state ω on a C*-algebra A with respect to some given time-evolution α_t, and take $H = H_\omega, M = \pi_\omega(A)''$, and $\Omega = \Omega_\omega$, the normal extension of ω to $\pi_\omega(A)''$ given by $\langle \Omega_\omega, \cdot \Omega_\omega \rangle$ still satisfies the KMS condition with respect to the time-evolution on $\pi_\omega(A)''$ given by conjugation with $\exp(ith_\omega)$, as in (9.52). Comparing the latter with the time-evolution on M defined by conjugation with Δ^{it} (cf. Theorem C.159) gives

$$e^{ith_\omega} = \Delta^{-it/\beta}, \qquad (9.122)$$

since both one-parameter groups of unitary operators satisfy the KMS-condition at β, and some time-evolution α_t that satisfies the KMS-condition relative to a given state ω and inverse temperature β is unique. To see this (barring technicalities about unbounded operators that are easily dealt with), take $\beta = -1$ for simplicity, assume α_t is conjugation by $\Delta^{it} = \exp(ith)$ (i.e., $\Delta = \exp(h)$), and rewrite (9.112) as

$$\omega(ab) = \langle b^*\Omega, \Delta a\Omega \rangle. \qquad (9.123)$$

This determines $\langle \varphi, \Delta\psi \rangle$ between a dense set of vectors φ, ψ, and hence fixes Δ.

The operators J and Δ from the Tomita–Takesaki theory can explicitly be computed in the example (9.113); the antilinear operator $J : B_2(H') \to B_2(H')$ reads

$$Jb = b^*, \qquad (9.124)$$

so that the isomorphism $a \mapsto JaJ$ between $\pi_\omega(A)'' = B(H')$ (where $B(H')$ acts on $B_2(H')$ by left multiplication) and its commutant $\pi_\omega(A)' = B(H')$ (which copy of $B(H')$ now acts on $B_2(H')$ by right multiplication) is given by $JaJb = ba$. Furthermore, the (generally unbounded) linear operator $\Delta : B_2(H') \to B_2(H')$ is given by

$$\Delta b = \rho b \rho^{-1}, \qquad (9.125)$$

which strictly speaking is defined as the closure of the expression (9.125) on the domain of all $b \in B_2(H')$ for which $b\rho^{-1/2} \in B(H')$.

Theorem 9.31. *For given unital C*-algebra A, dynamics* $\alpha : \mathbb{R} \to \mathrm{Aut}(\mathbb{R})$*, and inverse temperature* $\beta \in \mathbb{R}$*, let* $S_\beta(A)$ *be the compact convex set of* KMS *states. Then*

$$\partial_e S_\beta(A) = S_\beta(A) \cap S_p(A), \tag{9.126}$$

where $S_p(A)$ *is the set of primary states on A (cf. Definition 8.17). Consequently, extreme* KMS *states at fixed inverse temperature* β *are either equal or disjoint.*

This suggests that extreme KMS states define ***pure thermodynamics phases***.

Proof. We enlarge $S_\beta(A)$ to the set $\hat{K}_\beta(A) \subset A^*$ of all continuous linear functionals on A that satisfy the β-KMS condition (so that $S_\beta(A)$ consists of all positive elements in $\hat{K}_\beta(A)$ of unit norm). The key to the proof is a bijection between the set $S(\omega)$ of functionals $\rho \in \hat{K}_\beta(A)$ for which $0 \leq \rho \leq \omega$, where $\omega \in S_\beta(A)$ is fixed, and the set $T(\omega)$ of operators $c \in \pi_\omega(A)' \cap \pi_\omega(A)''$ such that $0 \leq c \leq 1_{H_\omega}$, given by

$$\rho(a) = \langle \Omega_\omega, c\pi_\omega(a)\Omega_\omega \rangle. \tag{9.127}$$

This implies the claim, since $\omega \in \partial_e S_\beta$ iff any $\rho \in S(\omega)$ takes the form $\rho = t\omega$ for some $t \in [0,1]$ (cf. Lemma C.17), which in turn is the case iff $c = t \cdot 1_{H_\omega}$.

First, for any state $\omega \in S(A)$ there is a bijection between the set of linear functionals $\rho \in A^*$ for which $0 \leq \rho \leq \omega$ and the set of operators $c \in \pi_\omega(A)'$ such that $0 \leq c \leq 1_{H_\omega}$, given by (9.127). Indeed, in one direction, given $a = b^*b \geq 0$, we have

$$(\omega - \rho)(a) = \langle \pi_\omega(b)\Omega_\omega, (1_{H_\omega} - c)\pi_\omega(b)\Omega_\omega \rangle \geq 0, \tag{9.128}$$

for if $0 \leq c \leq 1_{H_\omega}$, then $0 \leq (1_{H_\omega} - c) \leq 1_{H_\omega}$. Hence $\rho \leq \omega$, whilst from (9.127) we similarly find $\rho \geq 0$. Conversely, ρ induces a quadratic form R on H_ω, defined initially on the dense domain $\pi_\omega(A)H_\omega$ by the formula

$$R(\pi_\omega(a)\Omega_\omega, \pi_\omega(b)\Omega_\omega) = \rho(a^*b), \tag{9.129}$$

which is easily seen to be well defined, positive, and bounded, and so Proposition B.79 supplies the operator c, which a simple computation shows to be in $\pi_\omega(A)'$.

For the bijection $S(\omega) \cong T(\omega)$, where ω is a β-KMS state as above, we therefore need the additional property $c \in \pi_\omega(A)''$. Putting $\beta = -1$ for convenience and using the notation of Theorem C.159, we first show that $\Delta^{-it}c\Delta^{it} = c$ for any $t \in \mathbb{R}$: indeed, since ρ satisfies the KMS condition, it is time-translation invariant, so that

$$\begin{aligned}
\langle \pi_\omega(a^*)\Omega_\omega, \Delta^{-it}c\Delta^{it}\pi_\omega(b)\Omega_\omega \rangle &= \langle \Omega_\omega, c\Delta^{it}\pi_\omega(a)\Delta^{-it}\Delta^{-it}\pi_\omega(b)\Delta^{-it}\Omega_\omega \rangle \\
&= \langle \Omega_\omega, c\pi_\omega(\alpha_t(ab))\Omega_\omega \rangle \\
&= \rho(\alpha_t(ab)) = \rho(ab) \\
&= \langle \pi_\omega(a^*)\Omega_\omega, c\pi_\omega(b)\Omega_\omega \rangle,
\end{aligned}$$

so that $\Delta^{-it}c\Delta^{it} = c$ between a dense set of states, and hence this is valid as an operator equation. This also implies that c commutes with any power of Δ. Define $c' = JcJ$, which by Theorem C.159 is an element of $\pi_\omega(A)''$, and compute

$$\begin{aligned}\langle\Omega_\omega,\pi_\omega(a)c'\Omega_\omega\rangle &= \langle\Omega_\omega,\pi_\omega(a)Jc\Delta^{1/2}\Omega_\omega\rangle = \langle\Omega_\omega,\pi_\omega(a)J\Delta^{1/2}c\Omega_\omega\rangle\\ &= \langle\Omega_\omega,\pi_\omega(a)Sc\Omega_\omega\rangle = \langle\Omega_\omega,\pi_\omega(a)c^*\Omega_\omega\rangle\\ &= \langle\Omega_\omega,\pi_\omega(a)c\Omega_\omega\rangle\\ &= \rho(a), \end{aligned} \tag{9.130}$$

where we used the properties $J\Omega_\omega = \Omega_\omega$, $\Delta^{1/2}\Omega_\omega = \Omega_\omega$, $c\Delta^{1/2} = \Delta^{1/2}c$ as just mentioned, $S = J\Delta^{1/2}$, and $c^* = c$ (since $c \geq 0$). Finally, it follows from the KMS condition (applied to the normal extension of the state ω to $\pi_\omega(A)''$ given by $\langle\Omega_\omega,\cdot\Omega_\omega\rangle$ as well as to the normal extension of ρ to $\pi_\omega(A)''$ given by $\langle\Omega_\omega,\cdot c'\Omega_\omega\rangle$ just computed) that $c' \in \pi_\omega(A)'$, since for arbitrary $a,b,d \in A_\alpha$ we have

$$\begin{aligned}\omega(ac'bd) &= \omega(\alpha_i(bd)ac') = \rho(\alpha_i(bd)a) = \rho(\alpha_i(b)\alpha_i(d)a)\\ &= \rho(\alpha_i(d)ab) = \omega(\alpha_i(d)abc') = \omega(abc'd).\end{aligned}$$

In other words, for any $a,b,d \in A$ we have

$$\langle\pi_\omega(a^*)\Omega_\omega, c'\pi_\omega(b)\pi_\omega(d)\Omega_\omega\rangle = \langle\pi_\omega(a^*)\Omega_\omega,\pi_\omega(b)\pi_\omega(d)c'\Omega_\omega\rangle, \tag{9.131}$$

so that $c'\pi_\omega(b) = \pi_\omega(b)c'$ between vectors in a dense domain, so that this is an operator equality. Hence $c' \in \pi_\omega(A)'$, and in view of this we may rewrite (9.130) as $\rho(a) = \langle\Omega_\omega, c'\pi_\omega(a)\Omega_\omega\rangle$. Since the operator $c' \in \pi_\omega(A)'$ in (9.127) is uniquely determined by ρ, this shows that $c' = c$. Since we already had $c' \in \pi_\omega(A)''$, it follows that $c \in \pi_\omega(A)' \cap \pi_\omega(A)''$. □

It can also be shown that $S_\beta(A)$ is a (Choquet) *simplex*, which is a property rather more typical of the state space of a *commutative* unital C*-algebra; this makes it especially remarkable for the set of β-KMS states on a highly *non-commutative* C*-algebra like the infinite tensor product of $B = M_n(\mathbb{C})$. In the physically relevant case where $S_\beta(A)$ is metrizable, this implies that for any given KMS state $\omega \in S_\beta(A)$ there is a *unique* probability measure μ on $\partial_e S_\beta(A)$, such that for each $a \in A$,

$$\omega(a) = \int_{\partial S_\beta(A)} d\mu(\omega')\,\omega'(a). \tag{9.132}$$

Conversely, any probability measure μ on $\partial_e S_\beta(A)$ *defines* a β-KMS state by reading this equality from right to left. Towards the next chapter, suppose for example that there is a G-action on A, i.e., a continuous homomorphism $\gamma : G \to \mathrm{Aut}(A)$ (where G is a locally compact group). Then G also acts on $S(A)$ via the dual maps $\gamma_g^*(\omega) = \omega \circ \gamma_{g^{-1}}$, and if G is a symmetry of the dynamics in that $\alpha_t \circ \gamma_g = \gamma_g \circ \alpha_t$ for each $t \in \mathbb{R}$ and $g \in G$, then this dual action maps both $S_\beta(A)$ and $\partial_e S_\beta(A)$ into themselves. If G is compact with normalized Haar measure μ, then for any fixed extremal KMS state $\omega_0 \in \partial_e S_\beta(A)$, by (left) invariance of μ one obtains a G-invariant state by

$$\omega = \int_G d\mu(g)\,\gamma_g^*\omega_0. \tag{9.133}$$

Notes

§9.1. **Symmetries of C*-algebras and Hamhalter's Theorem**

Theorem 9.4 is due to Hamhalter (2011). Our proof, taken almost *verbatim* from Landsman & Lindenhovius (2016) roughly follow his, but adds various details and also takes some different turns. The main differences with the original proof by Hamhalter are the following. Firstly, we give an order-theoretic characterization of u.s.c. decompositions of the form π_K (and hence of the commutative algebras in $\mathscr{C}(C(X))$ that are the unitization of some ideal) by the three axioms stated in Lemma 3.1.1 in Firby (1973), whereas Hamhalter uses Proposition 7 in Mendivil (1999), which gives a different characterization of unitizations of ideals. Furthermore, Hamhalter only treats Lemma 9.5 in full generality, whereas in our opinion it is very instructive to take the case of finite sets first, where many of the key ideas already appear in a setting where they are not overshadowed by topological complications. Finally, our proof of Lemma 9.6.2 differs from Hamhalter's proof. The topology of partitions may be found in Willard (1970), especially Theorem 9.9.

Theorem 9.7 is due to Hamhalter (2015). Corollary 9.9 has a long history, starting with Jacobson & Rickart (1950) and ending with Thomsen (1982).

§9.2. **Unitary implementability of symmetries**

See Bratteli & Robinson (1987), §4.3.

§9.3. **Motion in space and in time**

For a far more detailed study of asymptotic abelianness see Bratteli & Robinson (1987), §4.3.2 and Bratteli & Robinson (1997), §5.4.1. Results like Theorem 9.14 may also be found in Sewell (2002). Theorem 9.14 is also valid for ***ergodic states*** with respect to the given $\mathbb{Z}^d$-action, where we say that a state on a C*-algebra A with G-action is ergodic if it is an element of $\partial_e(S(A)^G)$, i.e., extreme in the convex set of G-invariant states on A. Also Theorem 9.15 holds (with a more complicated proof, of course) under weaker conditions on Φ, typically exponential decay in X.

Theorem 9.15 is the simplest result in this direction; for similar results under weaker assumptions on the interaction Φ, see Bratteli & Robinson (1997), §6.2.1.

§9.4. **Ground states of quantum systems**

The idea of a ground state of a quantum system may be attributed to Bohr (1913), who postulated that an atom has a state of lowest energy (which he called a "permanent state"). See e.g. Pais (1986), p. 199. In this section, which merely present some key points treated in far more detail in Bratteli & Robinson (1997), §5.3.3. and §6.2.7, we have just scratched the surface of the topic, which is basic to physics.

§9.5. **Ground states and equilibrium states of classical spin systems**

Basic references for the mathematical physics of classical spin systems on a lattice are Israel (1979), Simon (1993), van Enter, Fernandez, & Sokal (1993), and Georgii (2011). One may now define pure thermodynamics phases as extreme elements of the compact convex set of all Gibbs measures (or of the set of all translation-invariant Gibbs measures, as in Simon, 1993, §III.5), but there is no identification between pure thermodynamics phases with primary equilibrium states (as in the quantum case), because a state on a commutative C*-algebra like $C(\underline{n}^{\mathbb{Z}^d})$ is primary iff it is pure. Fortunately, the specific measure-theoretic setting of classical statistical mechanics provides its own resources. For any $\Lambda \subset \mathbb{Z}^d$, let Σ_Λ be the smallest σ-algebra (within the Borel σ-algebra for $\underline{n}^{\mathbb{Z}^d}$) for which each $f \in C(\underline{n}^\Lambda)$ is measurable, and let

$$\Sigma_\infty = \bigcap_\Lambda \Sigma_\Lambda, \tag{9.134}$$

where each Λ is *finite*, be the ***σ-algebra at infinity***, with associated commutative C*-algebra $\mathscr{B}_\infty(\underline{n}^{\mathbb{Z}^d})$ of all bounded measurable functions on $\underline{n}^{\mathbb{Z}^d}$ that are Σ_∞-measurable. This is the home of the macroscopic observables, defined as averages analogously to the quantum case. The role of primary states (or rather of states whose algebra of observables is trivial at infinity, as in Theorem 8.23) is now played by ***states that are trivial at infinity***, that is, probability measures μ on $\underline{n}^{\mathbb{Z}^d}$ for which either $\mu(X) = 0$ or $\mu(X) = 1$ for $X \in \Sigma_\infty$ (cf. the Kolmogorov 0-1 law of probability theory). Indeed there is a classical version of Theorem 8.23, making exactly the same claim *mutatis mutandis*, see Theorem III.1.6 in Simon (1993). The main result (cf. Theorem 7.7 in Georgii, 2011), is that a state is extreme in the compact convex set of all Gibbs measures (at fixed temperature and potential, of course) iff it is a Gibbs measure that is trivial at infinity. It follows that two distinct extreme Gibbs measures are mutually singular on Σ_∞ (which is the pertinent classical version of disjointness of primary states).

§9.6. **Equilibrium (KMS) states of quantum systems**

The KMS condition was introduced by Haag, Hugenholtz, and Winnink (1967), in the following equivalent form:

$$\int_{-\infty}^{\infty} dt\, f(t - i\beta)\omega(a\alpha_t(b)) = \int_{-\infty}^{\infty} dt\, f(t)\omega(\alpha_t(b)a), \tag{9.135}$$

for each $a, b \in A$ and each Schwartz function $f \in \mathscr{D}(\mathbb{R})$. The name KMS derives from the earlier observation (9.102) of Kubo (1957) and independently Martin & Schwinger (1957). See also Haag (1992), Simon (1993), Borchers (2000), Sewell (2002), Thirring (2002), Emch (2007), and perhaps also, at a heuristic level, Landsman & van Weert (1987), especially for applications of the KMS condition to quantum field theory at finite temperature and the quark-gluon plasma (this, incidentally, was the MSc thesis as well as the first major published paper by the author).

The KMS condition also plays a major role in operator algebras and noncommutative geometry; see Connes (1994) and Connes & Marcolli (2008).

For a proof of (9.101) see Bratteli & Robinson (1997, Lemma 6.2.21); this book is the bible about the KMS condition and its application to quantum spin systems.

The proof of Proposition 9.25 is taken from Simon (1993), Lemma IV.4.1 and Proposition IV.4.2. The terminology of pure thermodynamical phases for primary KMS states (introduced after Theorem 9.31) is not completely standard; also ergodic states are sometimes called 'pure phases'.

Chapter 10
Spontaneous Symmetry Breaking

As we shall see, the undeniable natural phenomenon of ***spontaneous symmetry breaking*** (SSB) seems to indicate a serious mismatch between theory and reality. This mismatch is well expressed by what is sometimes called ***Earman's Principle***:

> 'While idealizations are useful and, perhaps, even essential to progress in physics, a sound principle of interpretation would seem to be that no effect can be counted as a genuine physical effect if it disappears when the idealizations are removed.' (Earman, 2004, p. 191)

To describe the various examples apparently violating Earman's Principle (and hence the link between theory and reality) in a general way (so general even that it will encapsulate the measurement problem), it is convenient to install a definition:

Definition 10.1. Asymptotic emergence *is the conjunction of three conditions:*

1. *A* **higher-level theory** H *(which is often called a* **phenomenological theory** *or a* **reduced theory***) is a limiting case of some* **theory!lower-level** L *(often called* **fundamental theory** *or a* **reducing theory***).*
2. *Theory* H *is well defined and understood* by itself *(typically predating* L*).*
3. *Theory* H *has features that cannot be explained by* L*, e.g. because* L *does not have any property inducing those feature(s) in the pertinent limit to* H*.*

In connection with SSB (as item 3.) we will look at the following pairs (H,L):

- – H is classical mechanics (notably of a particle on the real line $\mathbb{R}$);
  – L is quantum mechanics (on the pertinent Hilbert space $L^2(\mathbb{R})$);
  – The limiting relationship between the two theories is as described in §7.1 (notably by the continuous bundle of C*-algebras (7.17) - (7.19) for $n = 1$).
- – H is classical thermodynamics of a spin system;
  – L is statistical mechanics of a quantum spin system on a *finite* lattice;
  – Their limiting relationship is as described in §8.6 (cf. Theorem 8.4).
- – H is statistical mechanics of an *infinite* quantum spin system;
  – L is statistical mechanics of a quantum spin system on a *finite* lattice;
  – The limiting relationship between H and L is given in §8.6 (cf. Theorem 8.8).

K. Landsman, *Foundations of Quantum Theory*,
Fundamental Theories of Physics 188, DOI 10.1007/978-3-319-51777-3_10

Of course, there are many other interesting example of (apparent) asymptotic emergence not treated in this book, such as geometric optics (as H) versus wave optics (as L), where the new feature of H would be the *absence* of interference of light rays—foreshadowing the measurement problem of quantum mechanics!— or hydrodynamics (as H) versus molecular dynamics (as L), where the new feature is irreversibility. Perhaps space-time asymptotically emerges from quantum gravity.

The "unexplained" features of H mentioned in the third part of Definition 10.1 are often called ***emergent***, although this term has to be used with great care. Its meaning here reflects the original use of the term by the so-called "British Emergentists" (whose pioneer was J.S. Mill), as expressed in 1925 by C.D. Broad:

> 'The characteristic behaviour of the whole *could* not, even in theory, be deduced from the most complete knowledge of the behaviour of its components, taken separately or in other combinations, and of their proportions and arrangements in this whole. This is what I understand by the 'Theory of Emergence'. I cannot give a conclusive example of it, since it is a matter of controversy whether it actually applies to anything.' (Broad, 1925, p. 59)

In quotations like these, the notion "emergence" is meant to be the very opposite of the idea of "reduction" (or "mechanicism", as Broad called it); in fact, for many authors this opposition seems to be the principal attraction of emergence. In principle, two rather different notions of reduction then lead (contrapositively) to two different kinds of emergence, which are sometimes mixed up but should be distinguished:

1. The reduction of a whole (i.e., a composite system) to its parts;
2. The reduction of a theory H to a theory L.

In older literature concerned with the reduction of biology to chemistry (challenged by Mill) and of chemistry to physics (still contested by Broad), the first notion also referred to wholes consisting of a *small* number of particles. That notion of emergence seems a lost cause, since, as noted by Hempel,

> 'the properties of hydrogen include that of forming, if suitably combined with oxygen, a compound which is liquid, transparent, etc.' (Hempel, 1965, p. 260)

A similar comment applies to e.g. the tertiary structure of proteins, but also to cases of emergence such as ant hills, slime mold, and even large cities (Johnson, 2001), all of which are actually fascinating success stories for *reductionism*.

More recently, the apparent possibility that *very large* assemblies of parts might give rise to emergent properties of the corresponding wholes has become increasingly popular, both in physics and in the philosophy of mind (where consciousness has been proposed as an emergent property of the brain). In physics, the modern discussion on emergence in physics was initiated by P.W. Anderson, who in a famous essay from 1972 called 'More is different' emphasized the possibility of emergence in very large systems (surprisingly, Anderson actually avoids the term 'emergence', instead speaking of 'new laws' and 'a whole new conceptual structure'). In particular, Anderson claimed SSB to be an example (if not *the* example) of emergence, duly adding that one really had to take the $N \to \infty$ limit. Thus at least in physics, the interesting case for emergence in the first (i.e. whole-part) sense arises if the 'whole' is strictly infinite, as in the thermodynamic limit of quantum statistical mechanics.

This example confirms that 1. and 2. often go together, but they do not always do: the classical limit of quantum mechanics is a case of pure theory reduction.

A clear description of emergence has also been given by Jaegwon Kim:

> 1. *Emergence of higher-level properties:* All properties of higher-level entities arise out of the properties and relations that characterize their constituent parts. Some properties of these higher, complex systems are "emergent", and the rest merely "resultant". Instead of the expression "arise out of", such expressions as "supervene on" and "are consequential upon" could have been used. In any case, the idea is that when appropriate lower-level conditions are realized in a higher-level system (that is, the parts that constitute the system come to be configured in a certain relational structure), the system will necessarily exhibit certain higher-level properties, and, moreover, that no higher-level property will appear unless an appropriate set of lower-level conditions is realized. Thus, "arise" and "supervene" are neutral with respect to the emergent/resultant distinction: both emergent and resultant properties of a whole supervene on, or arise out of, its microstructural, or micro-based, properties. The distinction between properties that are emergent and those that are merely resultant is a central component of emergentism. As we have already seen, it is standard to characterize this distinction in terms of predictability and explainability.
> 2. *The unpredictability of emergent properties:* Emergent properties are not predictable from exhaustive information concerning their "basal conditions". In contrast, resultant properties are predictable from lower-level information.
> 3. *The unexplainability/irreducibility of emergent properties:* Emergent properties, unlike those that are merely resultant, are neither explainable nor reducible in terms of their basal conditions.' (Kim, 1999, p. 21, italics added)

Similarly, Silberstein (2002) states (paraphrased) that a higher-level theory H:

> 'bears predictive/explanatory emergence with respect to some lower-level theory L if L cannot *replace* H, if H cannot be *derived* from L [i.e., L cannot reductively explain H], or if L cannot be shown to be *isomorphic* to H.'

A key point here is Kim's no. 1: not even "emergentists" deny that the whole consists of its parts, or, in asymptotic emergence, that the higher-level theory H in fact originates from the lower-level theory L. The essence of emergence, then, would be that H nonetheless has "acquired" properties *not* reducible to L. One possibility for this to happen could be that the (allegedly) emergent property of H refers to some concept that does not even make sense in L, such as the experience of pain, which is hard to make sense of at a neural level, but another possibility, which is indeed the one relevant to physics and especially to SSB, is that some particular concept possessed by H (such as SSB) is admittedly *defined* within L, but *banned*.

In describing the relationship between H and L we have to be clear about the difference between *approximations* and *idealizations*. Following Norton (2012):

- An ***approximation*** is an inexact description of a target system.
- An ***idealization*** is a fictitious system, distinct from the target system, some of whose properties provide an inexact description of aspects of the target system.

Thus idealizations also provide approximations, but as systems they stand on their own and are defined independently of the target system. In our cases, the target system is a real physical system such as a ferromagnet or a quantum particle, which is supposed to be described exactly by theory L, i.e., the lower-level theory. In fact, L is a family of theories parametrized by $1/N$ ($N \in \mathbb{N}$) or $\hbar \in (0,1]$, and our real material relates to some very small value of this parameter (which may also be seen as a certain regime of L, seen as a single, unparametrized theory).

The pertinent theory H is an idealization in the above sense, through which one approximates very large systems by infinite ones and highly semi-classical ones (where $\hbar$ is very small) by classical ones (where $\hbar = 0$). It is in this setting that asymptotic emergence would violate Earman's Principle and hence would blast the relationship between theory and reality: the abstract point (made concrete for SSB earlier on) is that if some real property of a real system is described by H but is not approximated in any sense by L in any regime (as is the threat with SSB), although H is supposed to be a limit of L, then the latter theory L fails to describe the real system it is supposed to describe, whereas this systems *is* described by the theory H, which portrays fictitious systems. This marks a difference with other cases of emergence, where H (including some "whole") is not an idealization but a real system itself (as might be the case with consciousness and other examples from neuroscience and the philosophy of mind). Thus our discussion does not apply to such cases.

The tension between SSB and Earman's Principle has not quite gone unnoticed in the philosophy of physics literature. For example, Liu and Emch (2005) first write that it is a mistake to regard idealizations as acts of '*neglecting the negligible*' (p. 155, which already appears to deny Earman's Principle), and continue by:

> "The broken symmetry in question is *not reducible* to the configurations of the microscopic parts of any *finite* systems; but it should *supervene* on them in the sense that for any two systems that have the exactly (sic) duplicates of parts and configurations, both will have the same spontaneous symmetry breaking in them because both will behave identically in the limit. In other words, the result of the macroscopic limit is determined by the non-relational properties of parts of the finite system in question." (Liu & Emch, 2005, p. 156)

It is not easy to make sense of this, but the authors genuinely seem to believe in asymptotic emergence and hence they (again) appear to deny Earman's Principle. Another suggestion, made by Ruetsche, is to modify Earman's Principle to:

> 'No effect predicted by a non-final theory can be counted as a genuine physical effect if it disappears from that theory's successors.' (Ruetsche, 2011, p. 336)

For example, the theory L explaining SSB should not be quantum statistical mechanics but quantum field theory (which has an infinite number of ultraviolet degrees of freedom even in finite volume, and hence in principle allows SSB). This does make sense within physics, but, as Ruetsche herself notices, her principle 'has the pragmatic shortcoming that we can't apply it until we know what (all) successors to our present theories are.' With due respect, we will describe a rather different way out, based on unexpectedly implementing ***Butterfield's Principle***, which is a corollary to Earman's Principle that removes the reduction-emergence opposition:

> 'there is a weaker, yet still vivid, novel and robust behaviour that occurs before we get to the limit, i.e. for finite N. And it is this weaker behaviour which is physically real.' (Butterfield, 2011, p. 1065)

To do so, we now turn our attention to specific (classes of) models of SSB.

10.1 Spontaneous symmetry breaking: The double well

The simplest example of SSB is undoubtedly the equation $x^2 = 1$ (where $x \in \mathbb{C}$), which is invariant under a $\mathbb{Z}_2$ symmetry given by $x \mapsto -x$. Its solutions $x = \pm 1$, then, do not share this symmetry; instead $\mathbb{Z}_2$ acts nontrivially on the solution space.

Another example that is simple at least compared to quantum spin systems is provided by elementary quantum mechanics. Thus we are now in the context of the first of the three pairs (H, L) listed in the preamble to this chapter, where, in detail:

- H is classical mechanics of a particle moving on the real line, with associated phase space $\mathbb{R}^2 = \{(p,q)\}$ and ensuing C*-algebra of observables $A_0 = C_0(\mathbb{R}^2)$;
- L is the corresponding quantum theory, with a C*-algebra of observables $A_\hbar$ ($\hbar > 0$) taken to be the compact operators $B_0(L^2(\mathbb{R}))$ on the Hilbert space $L^2(\mathbb{R})$;
- The relationship between H and L is given by the continuous bundle of C*-algebras (7.17) - (7.19), for $n = 1$, notably in the classical limit $\hbar \to 0$.

At the level of states, the passage to the classical limit $\hbar \to 0$ of any $\hbar$-dependent wave-function $\psi_\hbar \in L^2(\mathbb{R})$, if it exists, is described via the associated probability measure $\mu_{\psi_\hbar}$ on $\mathbb{R}^2$, which is defined by (7.31); in other words,

$$\mu_{\psi_\hbar}(\Delta) = \int_\Delta \frac{d^n p d^n q}{2\pi\hbar} |\langle \phi_\hbar^{(p,q)}, \psi_\hbar \rangle|^2 \quad (\Delta \subset \mathbb{R}^{2n}), \tag{10.1}$$

where the (Schrödinger) ***coherent states*** $\phi_\hbar^{(p,q)} \subset L^2(\mathbb{R})$ are given by (7.27), i.e.,

$$\phi_\hbar^{(p,q)}(x) = (\pi\hbar)^{-n/4} e^{-ipq/2\hbar} e^{ipx/\hbar} e^{-(x-q)^2/2\hbar}. \tag{10.2}$$

In terms of the associated vector states $\omega_{\psi_\hbar}$ on the C*-algebra $B_0(L^2(\mathbb{R}))$, one has

$$\omega_{\psi_\hbar}(Q_\hbar^B(f)) = \langle \psi_\hbar, Q_\hbar^B(f) \psi_\hbar \rangle = \int_{\mathbb{R}^{2n}} d\mu_\psi(p,q) f(p,q), \tag{10.3}$$

where $f \in C_0(\mathbb{R}^2)$. We then say that the wave-functions $\psi_\hbar$ have a classical limit if

$$\lim_{\hbar \to 0} \int_{\mathbb{R}^{2n}} d\mu_\psi f = \int_{\mathbb{R}^{2n}} d\mu_0 f, \tag{10.4}$$

for any $f \in C_0(\mathbb{R}^2)$, where μ_0 is some probability measure on $\mathbb{R}^2$. Seen as a state ω_0 on the classical C*-algebra of observables $C_0(\mathbb{R}^2)$, the probability measure μ_0 is regarded as the classical limit of the family $\omega_{\psi_\hbar}$ of states on the C*-algebra $B_0(L^2(\mathbb{R}))$ of quantum-mechanical observables. This family is continuous in the sense that the function $\hbar \mapsto \omega_{\psi_\hbar}(\sigma(\hbar))$ from $[0,1]$ to $\mathbb{C}$ is continuous for every continuous cross-section σ of the given bundle of C*-algebras. An example of such a continuous cross-section is $\sigma(0) = f$ and $\sigma(\hbar) = Q_\hbar^B(f)$, for any $f \in C_0(\mathbb{R}^2)$), cf. (C.550) - (C.551), and indeed this example reproduces (10.4), which after all is just

$$\lim_{\hbar \to 0} \omega_{\psi_\hbar}(Q_\hbar^B(f)) = \omega_0(f) \quad (f \in C_0(\mathbb{R}^2)). \tag{10.5}$$

First, let us illustrate this formalism for the ground state of the one-dimensional harmonic oscillator. Taking $m = 1/2$ and $V(x) = \frac{1}{2}\omega^2 x^2$ in the usual Hamiltonian

$$h_\hbar = -\hbar^2 \frac{d^2}{dx^2} + V(x), \tag{10.6}$$

it is well known that the ground state is unique and that its wave-function, i.e.,

$$\psi_\hbar(x) = \left(\frac{\omega}{2\pi\hbar}\right)^{1/4} e^{-\omega x^2/4\hbar}, \tag{10.7}$$

is a Gaussian, peaked above $x = 0$. As $\hbar \to 0$, this ground state has a classical limit, namely the Dirac measure μ_0 concentrated at the origin $(p = 0, q = 0)$, i.e.,

$$\lim_{\hbar \to 0} \int_{\mathbb{R}^{2n}} d\mu_{\psi_\hbar} f = f(0,0) \;\; (f \in C_0(\mathbb{R}^2)). \tag{10.8}$$

This is just the unique ground state of the corresponding classical Hamiltonian

$$h_0(p,q) = p^2 + V(q), \tag{10.9}$$

seen as a point in the phase space $\mathbb{R}^2$ minimizing h_0, reinterpreted as a probability measure on phase space as explained in the context of Theorem 3.3. Note that we kept the mass fixed at $m = 1/2$, but instead we could have kept $\hbar$ fixed and take the limit $m \to \infty$ instead of $\hbar \to 0$; cf. the preamble to Chapter 7.

The same features hold for the *an*harmonic oscillator (with small $\lambda > 0$), i.e.,

$$V(x) = \tfrac{1}{2}\omega^2 x^2 + \tfrac{1}{4}\lambda x^4. \tag{10.10}$$

However, a new situation arises for the symmetric double-well potential

$$V(x) = -\tfrac{1}{2}\omega^2 x^2 + \tfrac{1}{4}\lambda x^4 + \tfrac{1}{4}\omega^4/\lambda = \tfrac{1}{4}\lambda(x^2 - a^2)^2, \tag{10.11}$$

where $a = \omega/\sqrt{\lambda} > 0$ (assuming $\omega > 0$ as well as $\lambda > 0$). This time, the ground state of the classical Hamiltonian is doubly degenerate, being given by the points $(p = 0, q = \pm a) \in \mathbb{R}^2$, with ensuing Dirac measures $\mu_0^\pm$ given by

$$\int_{\mathbb{R}^{2n}} d\mu_0^\pm f = f(0, \pm a). \tag{10.12}$$

But it is a deep and counterintuitive fact of quantum theory that the corresponding quantum Hamiltonian (10.6) with (10.11) has a unique ground state. Indeed:

Theorem 10.2. *Let* $V \in L^2_{\mathrm{loc}}(\mathbb{R}^m)$ *be positive and suppose that* $\lim_{|x| \to \infty} V(x) = \infty$. *Then* $-\Delta + V$ *has a nondegenerate (and strictly positive) ground state.*

Roughly speaking, the proof is based on an infinite-dimensional version of the Perron–Frobenius Theorem in linear algebra (applied to $\exp(-th_\hbar)$ rather than to the Hamiltonian $h_\hbar$ itself, so that the largest eigenvalue of the former corresponds to the smallest eigenvalue of the latter, i.e., the energy of the ground state).

And yet there are two quantum-mechanical shadows of the classical degeneracy:

- The wave-function $\psi_\hbar^{(0)}$ of the ground state (which by a suitable choice of phase may be taken to be real) is positive definite and has two peaks, above $x = \pm a$, with exponential decay $|\psi_\hbar^{(0)}(x)| \sim \exp(-1/\hbar)$ in the classically forbidden region.
- Energy eigenfunctions (and the associated eigenvalues) come in pairs.

In what follows, we will be especially interested in the first excited state $\psi_\hbar^{(1)}$, which like $\psi_\hbar^{(0)}$ is real, but has one peak *above* $x = a$ and another peak *below* $x = -a$. See Figure 10.1. The eigenvalue splitting (or "gap") vanishes exponentially in $-1/\hbar$ like

$$\Delta_\hbar = E_\hbar^{(1)} - E_\hbar^{(0)} \sim (\hbar\omega/\sqrt{\tfrac{1}{2}e\pi}) \cdot e^{-d_V/\hbar} \; (\hbar \to 0), \tag{10.13}$$

where the typical WKB-factor is given by

$$d_V = \int_{-a}^{a} dx \sqrt{V(x)}. \tag{10.14}$$

Also, the probability density of each of the wave-functions $\psi_\hbar^{(0)}$ or $\psi_\hbar^{(1)}$ contains approximate δ-function peaks above *both* classical minima $\pm a$. See Figure 10.2, displayed just for $\psi_\hbar^{(0)}$, the other being analogous. We can make the correspondence between the *nondegenerate* pair $(\psi_\hbar^{(0)}, \psi_\hbar^{(1)})$ of low lying quantum-mechanical wave-functions and the pair (μ_0^+, μ_0^-) of *degenerate* classical ground states more transparent by invoking the above notion of a classical limit of states. Indeed, in terms of the corresponding algebraic states $\omega_{\psi_\hbar^{(0)}}$ and $\omega_{\psi_\hbar^{(1)}}$, one has

$$\lim_{\hbar\to 0} \psi_\hbar^{(0)} = \lim_{\hbar\to 0} \psi_\hbar^{(1)} = \mu_0^{(0)}, \tag{10.15}$$

$$\mu_0^{(0)} \equiv \tfrac{1}{2}(\mu_0^+ + \mu_0^-), \tag{10.16}$$

where $\mu_0^\pm$ are the pure classical ground states (10.12) of the double-well Hamiltonian. To see this, one may consider numerically computed Husimi functions, as shown in Figure 10.3 (just for $\psi_\hbar^{(0)}$, as before). From this, it is clear that the *pure* (algebraic) quantum ground state $\psi_\hbar^{(0)}$ converges to the *mixed* classical state (10.16).

In contrast, the localized (but now time-dependent) wave-functions

$$\psi_\hbar^\pm = \frac{\psi_\hbar^{(0)} \pm \psi_\hbar^{(1)}}{\sqrt{2}}, \tag{10.17}$$

which of course define pure states as well, converge to *pure* classical states, i.e.,

$$\lim_{\hbar\to 0} \psi_\hbar^\pm = \mu_0^\pm. \tag{10.18}$$

In conclusion, one has SSB in H, but at first sight the underlying theory L seems to forbid it. Yet we will now show that (10.17) - (10.18), will save Earman's Principle.

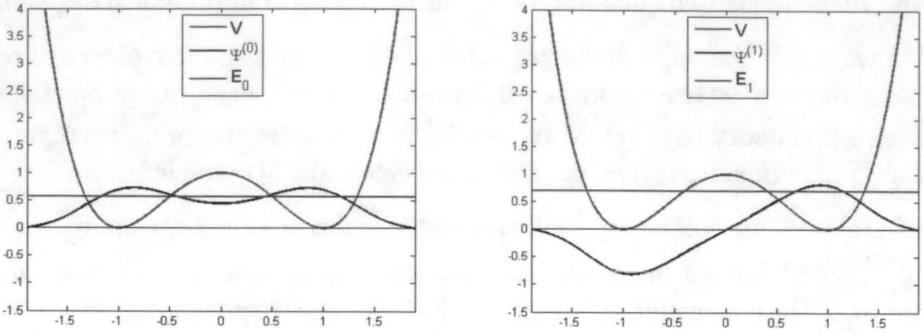

Fig. 10.1 Double-well potential with ground state $\psi^{(0)}_{\hbar=0.5}$ and first excited state $\psi^{(1)}_{\hbar=0.5}$.

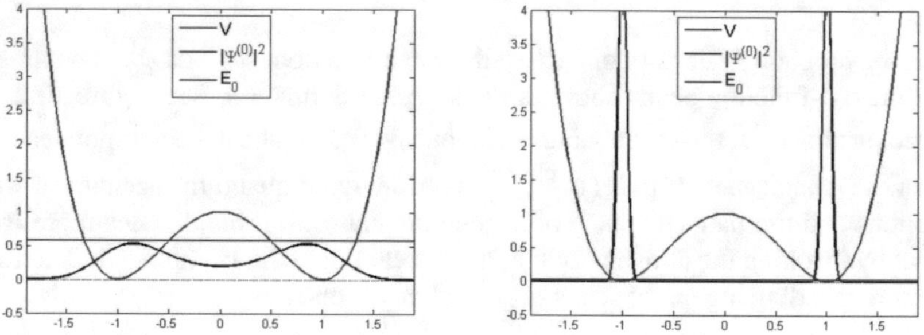

Fig. 10.2 Probability densities for $\psi^{(0)}_{\hbar=0.5}$ (left) and $\psi^{(0)}_{\hbar=0.01}$ (right).

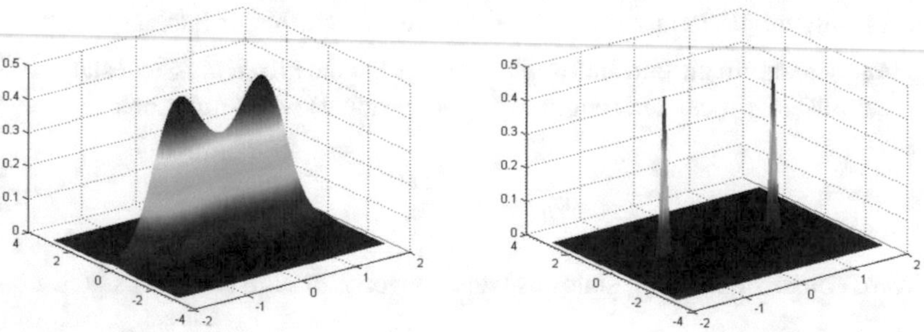

Fig. 10.3 Husimi functions for $\psi^{(0)}_{\hbar=0.5}$ (left) and $\psi^{(0)}_{\hbar=0.01}$ (right).

10.2 Spontaneous symmetry breaking: The flea

Regarding the doubly-peaked ground state $\psi_\hbar^{(0)}$ of the symmetric double well as the quantum-mechanical counterpart of a hung parliament, the analogue of a small party that decides which coalition is formed is a tiny *asymmetric* perturbation δV of the potential. Indeed, the following spectacular phenomenon in the theory of Schrödinger operators was discovered in 1981 by Jona-Lasinio, Martinelli and Scoppola. In view of the extensive (and very complicated) ensuing mathematical literature, we just take it as our goal to explain the main idea in a heuristic way.

Replace V in (10.6) by $V + \delta V$, where δV (i.e., the "flea") is assumed to:

1. Be real-valued with fixed sign, and C_c^∞ (hence bounded) with connected support not including the minima $x = a$ or $x = -a$;
2. Satisfy $|\delta V| >> e^{-d_V/\hbar}$ for sufficiently small $\hbar$ (e.g., by being independent of $\hbar$);
3. Be localized not too far from at least one the minima, in the following sense. First, for $y, z \in \mathbb{R}$ and $A \subset \mathbb{R}$, we extend the notation (10.14) to

$$d_V(y,z) = \left| \int_y^z dx\, \sqrt{V(x)} \right|; \tag{10.19}$$

$$d_V(y,A) = \inf\{d_V(y,z), z \in A\}. \tag{10.20}$$

Second, we introduce the symbols

$$d_V' = 2 \cdot \min\{d_V(-a, \mathrm{supp}\ \delta V), d_V(a, \mathrm{supp}\ \delta V)\}; \tag{10.21}$$

$$d_V'' = 2 \cdot \max\{d_V(-a, \mathrm{supp}\ \delta V), d_V(a, \mathrm{supp}\ \delta V)\}. \tag{10.22}$$

The localization assumption on δV is that one of the following conditions holds:

$$d_V' < d_V < d_V''; \tag{10.23}$$

$$d_V' < d_V'' < d_V. \tag{10.24}$$

In the first case, the perturbation is typically localized either on the left or on the right edge of the double well, whereas in the second it resides on the middle bump (symmetric perturbations are excluded by 3, as these would satisfy $d_V' = d_V''$).

Under these assumptions, the ground state wave-function $\psi_\hbar^{(\delta)}$ of the perturbed Hamiltonian (which had two peaks for $\delta V = 0$!) localizes as $\hbar \to 0$, in a direction which *given that localization happens* may be understood from energetic considerations. For example, if δV is positive and is localized to the right, then the relative energy in the left-hand part of the double well is lowered, so that localization will be to the left. See Figures 10.4 - 10.6. Eqs. (10.17) - (10.18) then yield Butterfield's Principle (with $N \rightsquigarrow 1/\hbar$), so that also Earman's Principle is saved: the essence of the argument is that (at least in the presence of a flea-perturbation) SSB is already foreshadowed in quantum mechanics *for small yet positive* $\hbar$, if only approximately.

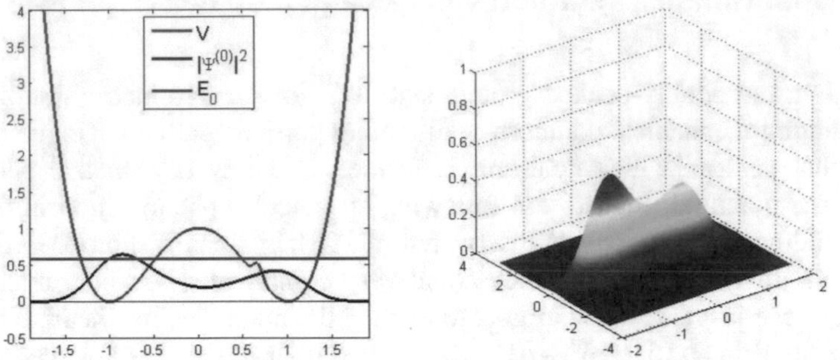

Fig. 10.4 Flea perturbation of ground state $\psi^{(\delta)}_{\hbar=0.5}$ with corresponding Husimi function. For such relative large values of $\hbar$, little (but some) localization takes place.

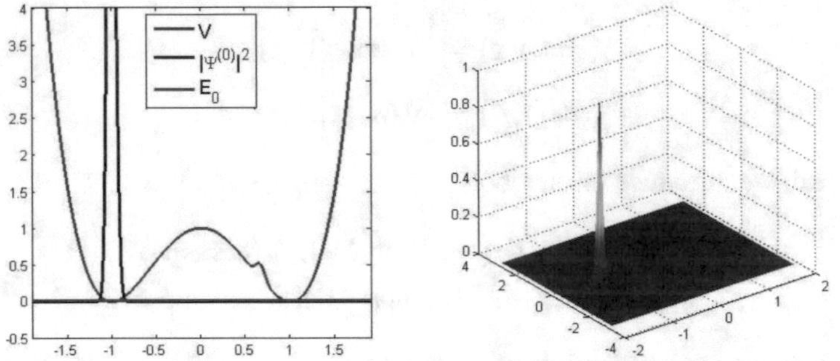

Fig. 10.5 Same at $\hbar = 0.01$. For such small values of $\hbar$, localization is almost total.

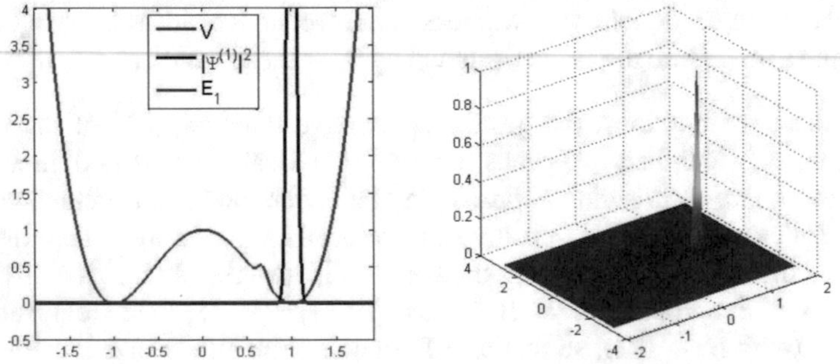

Fig. 10.6 First excited state for $\hbar = 0.01$. Note the opposite localization area.

In more detail, for the perturbed ground state we have (subject to assumptions 1–3):

$$\frac{\psi_\hbar^{(\delta)}(a)}{\psi_\hbar^{(\delta)}(-a)} \sim e^{\mp d_V/\hbar} \ (\pm\delta V > 0, \ \mathrm{supp}(V) \subset \mathbb{R}^+); \tag{10.25}$$

$$\frac{\psi_\hbar^{(\delta)}(a)}{\psi_\hbar^{(\delta)}(-a)} \sim e^{\pm d_V/\hbar} \ (\pm\delta V > 0, \ \mathrm{supp}(V) \subset \mathbb{R}^-), \tag{10.26}$$

with the opposite localization for the perturbed first excited state (so as to remain orthogonal to the ground state). A more precise version of the energetics used above is as follows. The ground state tries to minimize its energy according to the rules:

- The cost of localization (if $\delta V = 0$) is $\mathcal{O}(e^{-d_V/\hbar})$.
- The cost of turning on δV is $\mathcal{O}(e^{-d'_V/\hbar})$ when the wave-function is delocalized.
- The cost of turning on δV is $\mathcal{O}(e^{-d''_V/\hbar})$ when the wave-function is localized in the well around $x_0 = \pm a$ for which $d_V(x_0, \mathrm{supp}\ \delta V) = d''_V$.

In any case, these results only depend on the support of δV, but not on its size: this means that the tiniest of perturbations may cause collapse in the classical limit.

Although the collapse of the perturbed ground state for small $\hbar$ is a mathematical theorem, it remains enigmatic. Indeed, despite the fact that in quantum theory the localizing effect of the flea is enhanced for small $\hbar$, the corresponding classical system has no analogue of it. Trivially, a classical particle residing at one of the two minima of the double well at zero (or small) velocity, i.e., in one of its degenerate ground states, will not even notice the flea; the ground states are unchanged. But even under a stochastic perturbation, which leads to a nonzero probability for the particle to be driven from one ground state to the other in finite time (as some form of classical "tunneling", where in this case the necessary fluctuations come from Brownian motion), the flea plays a negligible role. For example, in the case at hand the standard ***Eyring–Kramers formula*** for the mean transition time reads

$$\langle\tau\rangle \simeq \frac{2\pi}{\sqrt{V''(a)V''(0)}} e^{V(0)/\varepsilon}, \tag{10.27}$$

where ε is the parameter in the Langevin equation $dx_t = -\nabla V(x_t)dt + \sqrt{2\varepsilon}dW_t$, in which W_t is standard Brownian motion. Clearly, this expression only contains the height of the potential at its maximum and its curvature at its critical points; most perturbations satisfying assumptions 1–3 above do not affect these quantities.

The instability of the ground state of the double-well potential under "flea" perturbations as $\hbar \to 0$ is easy to understand (at least heuristically) if one truncates the infinite-dimensional Hilbert space $L^2(\mathbb{R})$ to a two-level system. This simplification is accomplished by keeping only the lowest energy states $\psi_\hbar^{(0)}$ and $\psi_\hbar^{(1)}$, in which case the full Hamiltonian (10.6) with (10.11) is reduced to the 2×2 matrix

$$H_0 = \tfrac{1}{2}\begin{pmatrix} 0 & -\Delta \\ -\Delta & 0 \end{pmatrix}, \tag{10.28}$$

with $\Delta > 0$ given by (10.13). Dropping $\hbar$, the eigenstates of H_0 are given by

$$\varphi_0^{(0)} = \frac{1}{\sqrt{2}}\begin{pmatrix}1\\1\end{pmatrix},\ \varphi_0^{(1)} = \frac{1}{\sqrt{2}}\begin{pmatrix}1\\-1\end{pmatrix}, \tag{10.29}$$

with energies $E_0 = -\frac{1}{2}\Delta$ and $E_1 = \frac{1}{2}\Delta$, respectively; in particular, $E_1 - E_0 = \Delta$. If

$$\varphi_0^{\pm} = \frac{\varphi_0^{(0)} \pm \varphi_0^{(1)}}{\sqrt{2}}, \tag{10.30}$$

as in (10.17), then

$$\varphi_0^{+} = \begin{pmatrix}0\\1\end{pmatrix},\ \varphi_0^{-} = \begin{pmatrix}1\\0\end{pmatrix}. \tag{10.31}$$

Hence in this approximation φ_0^+ and φ_0^- play the role of wave-functions (10.17) localized above the classical minima $x = +a$ and $x = -a$, respectively, with classical limits $\mu_0^{\pm}$. The "flea" is introduced as follows. If its support is in $\mathbb{R}^+$, we put

$$\delta_+ V = \begin{pmatrix}0 & 0\\0 & \delta\end{pmatrix}, \tag{10.32}$$

where $\delta \in \mathbb{R}$ is a constant. A perturbation with support in $\mathbb{R}^-$ is approximated by

$$\delta_- V = \begin{pmatrix}\delta & 0\\0 & 0\end{pmatrix}. \tag{10.33}$$

Without loss of generality, take the latter (a change of sign of δ leads to the former). The eigenvalues of $H^{(\delta)} = H_0 + \delta_- V$ are $E_0 = E_-$ and $E_1 = E_+$, with energies

$$E_{\pm} = \tfrac{1}{2}(\delta \pm \sqrt{\delta^2 + \Delta^2}), \tag{10.34}$$

and normalized eigenvectors

$$\varphi_\delta^{(0)} = \frac{1}{\sqrt{2}}\left(\delta^2 + \Delta^2 + \delta\sqrt{\delta^2+\Delta^2}\right)^{-1/2}\begin{pmatrix}\Delta\\ \delta + \sqrt{\delta^2+\Delta^2}\end{pmatrix}; \tag{10.35}$$

$$\varphi_\delta^{(1)} = \frac{1}{\sqrt{2}}\left(\delta^2 + \Delta^2 - \delta\sqrt{\delta^2+\Delta^2}\right)^{-1/2}\begin{pmatrix}\Delta\\ \delta - \sqrt{\delta^2+\Delta^2}\end{pmatrix}. \tag{10.36}$$

Note that $\lim_{\delta\to 0}\varphi_\delta^{(i)} = \varphi_0^{(i)}$ for $i = 0, 1$. Now, if $\hbar \to 0$, then $|\delta| >> \Delta$, in which case $\varphi_\delta^{(0)} \to \varphi_0^{\pm}$ for $\pm\delta > 0$ (and starting from (10.32) instead of (10.33) would have given the opposite case, i.e., $\varphi_\delta^{(0)} \to \varphi_0^{\mp}$ for $\pm\delta > 0$). Thus the ground state localizes as $\hbar \to 0$, which resembles the situation (10.25) - (10.26) for the full double-well.

In conclusion, in the (practically unavoidable) presence of asymmetric "flea" perturbations, *explicit* (rather than spontaneous) symmetry breaking already takes place for positive $\hbar$, so that Butterfield's Principle holds, and hence also Earman's.

10.3 Spontaneous symmetry breaking in quantum spin systems

Before discussing SSB in quantum spin systems, we return to ground states and KMS states as discussed in the generality of §§9.4–9.6. Starting with the former, it is natural to ask whether ground states are pure, as would be expected on physical grounds; indeed, this question goes to the heart of SSB. Proposition 9.20 implies that ground states (for given dynamics) form a compact convex subset $S(A)_{\infty}$ of the total state space $S(A)$; the notation $S_{\infty}(A)$ (rather than e.g. $S_0(A)$) will be motivated shortly by the analogy with equilibrium states. It would be desirable that

$$\partial_e S_{\infty}(A) = S_{\infty}(A) \cap \partial_e S(A), \tag{10.37}$$

in which case extreme ground states are necessarily pure. This will indeed be the case in the simple models we study in this book, but it is provably the case in general only under additional assumptions, such as ***weak asymptotic abeliannnes*** of the dynamics, i.e., $\lim_{t\to\infty} \omega([\alpha_t(a), b]) = 0$ for all $a, b \in A$. A weaker sufficient condition for (10.37) is that $\pi_{\omega}(A)'$ be commutative (which is the case if ω is pure).

We are now in a position to define SSB, at least in the context of ground states.

Definition 10.3. *Suppose we have a (topological) group G and a (continuous) homomorphism $\gamma : G \to \mathrm{Aut}(A)$, which is a symmetry of the dynamics in that*

$$\alpha_t \circ \gamma_g = \gamma_g \circ \alpha_t \quad (g \in G, t \in \mathbb{R}). \tag{10.38}$$

The G-symmetry is said to be **spontaneously broken** *(at temperature $T = 0$) if*

$$(\partial_e S_{\infty}(A))^G = \emptyset, \tag{10.39}$$

and **weakly broken** *if $(\partial_e S_{\infty}(A))^G \neq \partial_e S_{\infty}(A)$, i.e., there is* at least one $\omega \in \partial_e S_{\infty}(A)$ *that fails to be G-invariant (although invariant extreme ground states may exist).*

Here $\mathscr{S}^G = \{\omega \in \mathscr{S} \mid \omega \circ \gamma_g = \omega \forall g \in G\}$, defined for any subset $\mathscr{S} \subset S(A)$, is the set of G-invariant states in $\mathscr{S}$. Assuming (10.37), eq. (10.39) means that there are no *pure* G-invariant ground states. This by no means implies that there are no G-invariant ground states at all, quite to the contrary: for compact, or, more generally, amenable groups G, one can always construct G-invariant ground states by averaging over G, exploiting the fact that if G is a symmetry of the dynamics, then each affine homeomorphism γ_g^* of $S(A)$ (defined by $\gamma_g^*(\omega) = \omega \circ \gamma_g$) maps $S_{\infty}(A)$ to itself. Definition 10.3 therefore implies that if SSB occurs, then one has a dichotomy:

- *Pure ground states are not invariant, whilst invariant ground states are not pure.*

Definition 10.4. *We call a G-symmetry* **spontaneously broken** *at inverse temperature $\beta \in (0, \infty)$ if there are no G-invariant extreme β-KMS states, i.e.,*

$$(\partial_e S_{\beta}(A))^G = \emptyset, \tag{10.40}$$

and **weakly broken** *if there is* at least one *non-G-invariant extreme* KMS *state.*

By Theorem 9.31 we may replace *extreme* β-KMS states by *primary* β-KMS states, so that, similarly to ground states, SSB at nonzero temperature means that:

- *Primary* KMS *states are invariant, whilst invariant* KMS *states are not primary.*

For the next result, please recall Definition 9.10 and Theorem 9.11.

Proposition 10.5. *Let A be a quasi-local C*-algebra of the kind* (8.130) *and suppose the given G-action* γ *commutes not only with time translations* α_t *but also with space translation* τ_x. *If* $\gamma_g^*\omega \neq \omega$ *for some* $\omega \in \partial_e S_\beta(A)$ *and* $g \in G$, *then the automorphism* γ_g *cannot be unitarily implemented in the* GNS*-representation* π_ω.

This is true also at $\beta = \infty$, i.e., for ground states.

Proof. This is an obvious corollary of Proposition 9.13 and Theorems 9.14 and 9.31: if γ_g were implementable by a unitary u_g, then $u_g\Omega_\omega \neq \Omega_\omega$ (not even up to a phase), since $\gamma_g^*\omega \neq \omega$. But in that case, since $\tau_x \circ \gamma_g = \gamma_g \circ \tau_x$ for each $x \in \mathbb{Z}^d$, we would have $u_x u_g = u_g u_x$ and hence $u_x(u_g\Omega_\omega) = u_g\Omega_\omega$. Thus $u_g\Omega_\omega$ would be another translation-invariant ground state, contradicting Theorem 9.14. □

This result is worth mentioning, since some authors *define* SSB through the conclusion of this proposition, that is, they call a symmetry γ_g (spontaneously) broken by some state ω iff γ_g cannot be unitarily implemented in π_ω. This definition seems physically dubious, however, because quantum spin systems may have ground states ω that are not G-invariant but in which nonetheless all of G is unitarily implementable (in such states translation invariance has to be broken, of course). For example, the Ising model in $d = 1$ with ferromagnetic nearest-neighbour interaction and vanishing external magnetic field (where $G = \mathbb{Z}_2$) has an infinite number of such ground states, in which a "domain wall" separates infinitely many "spins up" to the left from infinitely many "spins down" to the right. Although this model has a unique KMS state at any nonzero temperature, such ground states (and perhaps analogous states at $\beta \neq \infty$ in different models, so far understood only heuristically) seem far from pathological and play a major role in modern condensed matter physics. Hence we trust this alternative definition only if the states it singles out also satisfy Definition 10.3 or 10.4, for which Proposition 10.5 gives a sufficient condition: for translation-invariant states and symmetries on quasi-local algebras, our definition of SSB through (10.40) is compatible with the one based on unitary implementability.

This is fortunate, since the physicist's notion of an *order parameter*, through which at least *weak* SSB may be detected, is tailored to translation-invariant states:

Definition 10.6. *Let A be a quasi-local C*-algebra A as in* (8.130)*, with symmetry group G. A* **(strong) order parameter** *in A is an n-tuple* $\phi = (\phi_1, \ldots, \phi_n) \in A^n$ *for which* $\omega(\phi) = 0$ *if (and only if)* ω *is G-invariant, for any* $\mathbb{Z}^d$-invariant *state* ω *on A.*

An order parameter defines an accompanying vector field $x \mapsto \phi(x)$ by $\phi_i(x) = \tau_x(\phi)$. Since ω is translation-invariant, $\omega(\phi) = 0$ is equivalent to $\omega(\phi(x)) = 0$ for all x. In the Ising model, with $G = \mathbb{Z}_2$, $\sigma_3(0)$ is an order parameter, which can be extended to a strong one $\phi = (\sigma_2(0), \sigma_3(0))$. In the Heisenberg model, where $G = SO(3)$, the triple $(\sigma_1(0), \sigma_2(0), \sigma_3(0))$ provides a strong order parameter.

Theorem 10.7. *Suppose that ϕ is a (strong) order parameter, as in Definition 10.6. Then a G-invariant and translation-invariant* KMS *state $\omega \in S_\beta(A)^G$ (including $\beta = \infty$, i.e., a ground state) displays weak* SSB*—in the sense that at least one of the components in its extremal decomposition fails to be G-invariant—if (and only if) the associated two-point function exhibits* **long-range order**, *in that*

$$\lim_{x\to\infty} \omega\left(\sum_{i=1}^{n} \phi_i(0)^* \phi_i(x)\right) > 0. \tag{10.41}$$

Proof. The "if" part of the theorem is equivalent to the vanishing of the limit in question in the *absence* of SSB. Let (9.132) be the extremal decomposition of ω. If (almost) each extreme state φ is invariant, then $\omega'(\phi_i(x)) = 0$ for all i by definition of an order parameter, and similarly $\omega'(\phi_i(x)^*) = \overline{\omega'(\phi_i(x))} = 0$. Interchanging $\lim_{x\to\infty}$ with the integral over $\partial S_\beta(A)$ (which is allowed because μ is a probability measure), and using (9.30) then shows that the left-hand side of (10.41) vanishes.

To avoid difficult measure-theoretic aspects of the extremal decomposition theory, and also for pedagogical purposes, we prove the "only if" part only in the case

$$\omega = \int_G dg\, \omega'_g, \tag{10.42}$$

weakly, where $\omega' \in \partial S_\beta(A)$ and $\omega'_g = \gamma_g^* \omega'$. Since the expression

$$\omega'_g(\sum_{i=1}^{n} \phi_i(0)^* \phi_i(x))$$

is independent of $g \in G$ (by definition of an order parameter), we may replace ω'_g by ω' in the expression for ω; the term $\int_G dg$ then factors out and is equal to unity. Thus we may replace ω in (10.41) by ω'. Since ω' is a primary state, we may now use (9.30) once again, so that the left-hand side of (10.41) becomes $\sum_{i=1}^n |\omega'(\phi_i)|^2$. By assumption, ω' is not G-invariant, so that (by definition of a strong order parameter) at least one of the terms $|\omega'(\phi_i)|$ is nonzero. □

If G is compact, for any C*-algebra A, invariant KMS states (including ground states) can always be constructed via (9.133), provided, of course, KMS states (or ground states) exist in the first place. Fortunately, existence can be shown in the following way. Let A be a quasi-local C*-algebra à la (8.130), in which:

1. $\dim(H) < \infty$ (and hence also $\dim(H_\Lambda) < \infty$ for any finite $\Lambda \subset \mathbb{Z}^d$);
2. Dynamics is defined locally on each algebra $A_\Lambda = B(H_\Lambda)$ via (9.40) and (9.41), i.e., with free boundary conditions, having a global limit α as in Theorem 9.15.

In that case, by Corollary 9.27 each C*-algebra A_Λ has a unique β-KMS state ω_Λ^β, given by the local Gibbs state (9.96). However, if $\Lambda^{(1)} \subset \Lambda^{(2)}$, then the restriction of the β-KMS state $\omega^\beta_{\Lambda^{(2)}}$ to $A_{\Lambda^{(1)}} \subset A_{\Lambda^{(2)}}$ is not given as naively expected, namely by the β-KMS state $\omega^\beta_{\Lambda^{(1)}}$, because the former involves boundary terms.

Fortunately, this complication may be overcome, since at least *for models with short-range forces* (cf. Theorem 9.15) one may put

$$\omega_G^{\beta}(a) = \lim_{N\to\infty} \omega_{\Lambda_N}^{\beta}(a), \tag{10.43}$$

where Λ_N is defined in (8.153). This limit exists for $a \in \cup_\Lambda A_\Lambda$, from which ω^β extends by continuity to all of A, on which it is a β-KMS state (cf. Theorem 10.10).

Alternatively, by the Hahn–Banach Theorem (in the form of Corollary B.41) combined with Lemma C.4 (which guarantees that any Hahn–Banach extension of a state remains a state), each local Gibbs state ω_Λ^β on $A_\Lambda \subset A$ extends, in a non-unique way, to a state $\hat{\omega}_\Lambda^\beta$ on A. This gives a net of states $(\hat{\omega}_\Lambda^\beta)$ on A indexed by the finite subsets Λ of $\mathbb{Z}^d$; one may also work with sequences $(\hat{\omega}_{\Lambda_N}^\beta)$. Since A has a unit, its state space $S(A)$ is a *compact* convex set, so the above net (or sequence) has at least one limit point, or, equivalently, has at least one convergent subnet (or subsequence), which—despite its potential lack of uniqueness in two respects, i.e. the choice of the extensions $\hat{\omega}_\Lambda^\beta$ and the choice of a limit point—one might write as

$$\hat{\omega}^\beta = \lim_{\Lambda\nearrow\mathbb{Z}^d} \hat{\omega}_\Lambda^\beta. \tag{10.44}$$

Without proof, we quote the relevant technical result (assuming 1–2 above):

Proposition 10.8. *Each limit state $\hat{\omega}^\beta$ is a β-KMS state (i.e. for the dynamics α).*

Anticipating the existence of SSB in models, one should now feel a little uneasy:

- It follows from Corollary 9.27 that (at fixed β) there is a unique KMS state on each local algebra A_Λ for the given local dynamics $\alpha_t^{(\Lambda)}$, namely the local Gibbs state ω_Λ^β on A_Λ. If—as is the case in all our examples—the globally broken G-symmetry is induced by local automorphisms $\gamma_g^{(\Lambda)} : A_\Lambda \to A_\Lambda$ that commute with the local dynamics $\alpha_t^{(\Lambda)}$, then each local Gibbs state is G-invariant: this follows explicitly from G-invariance of the local Hamiltonian h_Λ and the formulae (9.96) - (9.98), or, more abstractly, from the fact if ω_Λ^β were not invariant under all $\gamma_g^{(\Lambda)}$, it would not be unique (as its translate $\omega_\Lambda^\beta \circ \gamma_g^{(\Lambda)}$ would be another KMS state).
- And yet (in case of SSB) there exist non-invariant (and hence non-unique) KMS states on A, which are even limits in the sense of (10.44) of the above invariant (and hence unique) local KMS states on A_Λ!
- Real samples are finite and hence are described by the local algebras A_Λ, with their *unique* invariant equilibrium states ω_Λ^β. Yet finite samples do display SSB, e.g., ferromagnetism (broken $\mathbb{Z}_2$-symmetry), superconductivity (broken $U(1)$).
- Therefore, the theory that *should* describe SSB in real materials, namely the finite theory A_Λ, apparently fails to do so (as it seems to forbid SSB), whereas the idealized theory A, which describes strictly *infinite* systems and in those systems allows SSB, in fact turns out to describe key properties of *finite* samples.

10.4 Spontaneous symmetry breaking for short-range forces

We continue our discussion of SSB in quantum spin systems, especially of the construction of global KMS states in the previous section, see (10.44) and preceding text. Recall that each finite system A_Λ has a unique β-KMS state ω_Λ^β, namely the local Gibbs state (9.96), but that these states are incompatible for different Λ's, in that, if $\Lambda^{(1)} \subset \Lambda^{(2)}$, then the restriction of $\omega^\beta_{\Lambda^{(2)}}$ to $A_{\Lambda^{(1)}} \subset A_{\Lambda^{(2)}}$ is not given by $\omega^\beta_{\Lambda^{(1)}}$ because of boundary terms. To correct for this, one introduces the ***surface energy***

$$b_{\Lambda^{(1)},\Lambda^{(2)}} = \sum_{X\subseteq \Lambda^{(2)}: X\cap\Lambda^{(1)}\neq\emptyset, X\cap\Lambda_1^c\neq\emptyset} \Phi(X), \tag{10.45}$$

with ensuing ***interaction energy***

$$b_\Lambda = \lim_{\Lambda^{(2)}\nearrow\mathbb{Z}^d} b_{\Lambda,\Lambda^{(2)}} = \sum_{X\cap\Lambda\neq\emptyset, X\cap\Lambda^c\neq\emptyset} \Phi(X), \tag{10.46}$$

provided this limit exists (which it does for short-range forces). Now perturb $\omega^\beta_{\Lambda^{(2)}}$ by replacing $h_{\Lambda^{(2)}}$ in (9.96) - (9.98) (with $\Lambda \rightsquigarrow \Lambda^{(2)}$) by $h_{\Lambda^{(2)}} - b_{\Lambda^{(1)},\Lambda^{(2)}}$. Denoting this modification of $\omega^\beta_{\Lambda^{(2)}}$ by $\omega^\beta_{\Lambda^{(1)},\Lambda^{(2)}}$, we obtain (10.47), which implies (10.48):

$$\omega^\beta_{\Lambda^{(1)},\Lambda^{(2)}} = \omega^\beta_{\Lambda^{(1)}} \otimes \omega^\beta_{\Lambda\backslash\Lambda^{(1)}}; \tag{10.47}$$

$$(\omega^\beta_{\Lambda^{(1)},\Lambda^{(2)}})_{|A_{\Lambda^{(1)}}} = \omega^\beta_{\Lambda^{(1)}}. \tag{10.48}$$

If (10.46) exists, we may likewise perturb any ι-invariant state ω on A to $\tilde{\omega}_\Lambda$, i.e.,

$$\tilde{\omega}_\Lambda(a) = \frac{\langle e^{-\beta(h_\omega - \pi_\omega(b_\Lambda))/2}\Omega_\omega, \pi_\omega(a) e^{-\beta(h_\omega-\pi_\omega(b_\Lambda))/2}\Omega_\omega\rangle}{\|e^{-\beta(h_\omega-\pi_\omega(b_\Lambda))/2}\Omega_\omega\rangle\|^2}, \tag{10.49}$$

where $\Lambda \subset \mathbb{Z}^d$ is finite, h_ω is defined as in (9.51) - (9.52), and Ω_ω is in the domain of the unbounded operator $\exp(-\beta(h_\omega - \pi_\omega(b_\Lambda))/2)$; the reason is that $\pi_\omega(b_\Lambda)$ is bounded, whereas $\exp(-\beta h_\omega/2)\Omega_\omega = \Omega_\omega$ (since $h_\omega\Omega_\omega = 0$). For example,

$$(\tilde{\omega}^\beta_{\Lambda^{(2)}})_{\Lambda^{(1)}} = \omega^\beta_{\Lambda^{(1)},\Lambda^{(2)}}, \tag{10.50}$$

where $\omega = \omega^\beta_{\Lambda^{(2)}}$ is a Gibbs state on $A = A_{\Lambda^{(2)}}$, as in Theorem 9.24 (with $\Lambda \rightsquigarrow \Lambda^{(2)}$). Indeed, using (9.114) - (9.117) and the relation $h_\omega = h_{\Lambda^{(2)}} - Jh_{\Lambda^{(2)}}J$, where the operator J is defined in (9.124), we compute the numerator in (10.49) as

$$\begin{aligned} &\mathrm{Tr}\Big(\Big(\Big(e^{-\beta(h_{\Lambda^{(2)}}-Jh_{\Lambda^{(2)}}J-b_\Lambda)/2}e^{-\beta h_{\Lambda^{(2)}}/2}\Big)^* a e^{-\beta(h_{\Lambda^{(2)}}-Jh_{\Lambda^{(2)}}J-b_\Lambda)/2}e^{-\beta h_{\Lambda^{(2)}}/2}\Big) \\ &= \mathrm{Tr}\left(e^{-\beta(h_{\Lambda^{(2)}}-b_\Lambda)}a\right), \end{aligned} \tag{10.51}$$

since $Jh_{\Lambda(2)}J$ commutes with $h_{\Lambda(2)} - b_{\Lambda}$. This subsequently gives

$$e^{-\beta(h_{\Lambda(2)} - Jh_{\Lambda(2)}J - b_{\Lambda})/2} = e^{-\beta(h_{\Lambda(2)} - b_{\Lambda})/2} e^{\beta Jh_{\Lambda(2)}J};$$
$$e^{\beta Jh_{\Lambda(2)}J} e^{-\beta h_{\Lambda(2)}/2} = e^{-\beta h_{\Lambda(2)}/2} e^{\beta h_{\Lambda(2)}/2} = 1_H. \tag{10.52}$$

Likewise, the denominator in (10.49) equals $\mathrm{Tr}\,(\exp(-\beta(h_{\Lambda(2)} - b_{\Lambda})))$.

Eqs. (10.50) and (10.48) suggest that if $\omega = \omega^{\beta}$ is a β-KMS state, then although ω^{β} itself does not localizes to a Gibbs state ω^{β}_{Λ} on A_{Λ}, its perturbed version $\tilde{\omega}^{\beta}_{\Lambda}$ does. Under assumptions 1–2 stated in §10.3, i.e., in the situation of Theorem 9.15 with $\dim(H) < \infty$, this motivates the following quantum analogue of the DLR approach to classical equilibrium states, i.e., of Definition 9.23:

Definition 10.9. *For fixed inverse temperature $\beta \in \mathbb{R}\backslash\{0\}$ and fixed interaction Φ, a* **Gibbs state** *ω^{β} on a quasi-local algebra A with dynamics given by some potential Φ is an α_t-independent state such that for each finite region $\Lambda \subset \mathbb{Z}^d$ one has*

$$\tilde{\omega}^{\beta}_{\Lambda} = \omega^{\beta}_{\Lambda} \otimes \omega'_{\Lambda^c}, \tag{10.53}$$

where ω^{β}_{Λ} is the local Gibbs state (9.96) on A_{Λ} and ω'_{Λ^c} is some state on A_{Λ^c}.

Theorem 10.10. *Under assumptions 1–2 in §10.3, and if in addition the subspace $D = \cup_{\Lambda} A_{\Lambda} \subset A$ is a core for the derivation (9.54) (i.e., the closure of δ defined on D is δ as defined in Proposition 9.19), then Gibbs states coincide with KMS states.*

The proof is rather technical and so we omit it. It follows that if $\omega^{\beta} \in S_{\beta}(A)$, then

$$(\tilde{\omega}^{\beta}_{\Lambda})_{|A_{\Lambda}} = \omega^{\beta}_{\Lambda}. \tag{10.54}$$

Even so, we still need to define in precisely which sense the net $((\tilde{\omega}^{\beta}_{\Lambda})_{|A_{\Lambda}})_{\Lambda}$ converges to ω_{Λ} (or when perhaps even the net $(\omega^{\beta}_{\Lambda})$ converges to ω_{Λ}); for simplicity we take $\Lambda = \Lambda_N$ as in (8.153), and just consider sequences indexed by N (rather than nets). To this end, let $(\omega_{1/N})_N$ be a sequence of states with $\omega_{1/N} \in S(A_{\Lambda_N})$. As in Definition 8.24, given some $\omega_0 \in S(A)$ (if it exists), we say that

$$\lim_{N\to\infty} \omega_{1/N} = \omega_0 \tag{10.55}$$

iff for any sequence $(a_{1/N})_N$ in A with $a_{1/N} \in A_{\Lambda_N} \subset A$ that converges to $a \in A$ one has

$$\lim_{N\to\infty} \omega_{1/N}(a_{1/N}) = \omega_0(a). \tag{10.56}$$

For example, if we take $\omega_0 \in S(A)$ and define $\omega_{1/N} = \omega_{0|A_{\Lambda_N}}$, then (10.55) holds by continuity of ω_0 (as $\|\omega_0\| = 1$), which implies that $\lim_{N\to\infty} \omega_0(a_{1/N}) = \omega_0(a)$.

It follows from the comments preceding Definition 8.24 that the above notion (10.55) - (10.56) of convergence is the same as the one given by (8.164), so that it is similar to the convergence of states we defined for the other two classes of examples of listed earlier, viz. classical mechanics (cf. §10.1) and thermodynamics.

We denote the restriction of some global KMS state ω^β (defined on A) to $A_{\Lambda_N} \subset A$ by $\omega^\beta_{1/N}$, whereas as usual we write $\omega^\beta_{\Lambda_N}$ for the unique local Gibbs state on A_{Λ_N}. Keeping Definition 8.24 and Proposition 8.25 in mind, the situation is as follows:

1. Any KMS state ω^β equals the limit ω^β_0 of its restrictions $\omega^\beta_{1/N}$ (i.e. to A_{Λ_N}).
2. Each state $\omega^\beta_{1/N}$ differs from the local Gibbs state $\omega^\beta_{\Lambda_N}$ (even if ω^β is unique).
3. The local Gibbs states $\omega^\beta_{\Lambda_N}$ typically converge to a KMS state ω^β_G, cf. (10.43).
4. In models with symmetry, this global Gibbs state ω^β_G is invariant (like the $\omega^\beta_{\Lambda_N}$).

The first claim follows from the argument given after (10.55). The second is the contrapositive to (10.54) and has been explained in §10.3: although the states $\omega^\beta_{1/N}$ and $\omega^\beta_{\Lambda_N}$ are both of local Gibbs type, their Hamiltonians differ from h_{Λ_N} by the boundary term b_Λ. The third claim cannot be proved in general, but in models with short-range forces it holds in both forms (10.43) and (10.55) - (10.56). In such models the G-symmetry is local, i.e., G acts on each A_Λ through unitaries

$$u_g^{(\Lambda)} = \otimes_{x\in\Lambda} u_g(x); \tag{10.57}$$

$$\gamma_g^{(\Lambda)}(a_\Lambda) = u_g^{(\Lambda)} a (u_g^{(\Lambda)})^* \ (a_\Lambda \in A_\Lambda, g \in G), \tag{10.58}$$

where $u_g(x) \in B(H_x)$, leaving each local Hamiltonian h_Λ and hence each local Gibbs state $\omega^\beta_{\Lambda_N}$ invariant. If $a \in A$ is local, i.e., $a \in \cup_\Lambda A_\Lambda$, then

$$\gamma_g(a) = \lim_{N\to\infty} \gamma_g^{(\Lambda_N)}(a_N), \tag{10.59}$$

followed by continuous extension to $a \in A$, so that, assuming (10.55),

$$\omega_0(\gamma_g(a)) = \lim_{N\to\infty} \omega_{1/N}(\gamma_g(a_N)) = \lim_{N\to\infty} \omega_{1/N}(\gamma_g^{(\Lambda_N)}(a_N)) = \lim_{N\to\infty} \omega_{1/N}(a_N) = \omega_0(a),$$

since $\omega_{1/N} \circ \gamma_g^{(\Lambda_N)} = \omega_{1/N}$ by assumption. Thus the global Gibbs state ω^β_G inherits the G-invariance of its local approximants $\omega^\beta_{\Lambda_N}$. In case of SSB, the restrictions $\omega^\beta_{1/N}$ of some non-invariant extreme KMS state ω^β determine ω^β, so that in principle SSB is detectable through the local states $\omega^\beta_{1/N}$. It would be question-begging to construct the latter from the global states ω^β, though, so Butterfield's Principle (and hence in its wake Earman's Principle) holds only if we can show how and why the states of sufficiently large yet *finite* systems A_{Λ_N} tend to $\omega^\beta_{1/N}$ rather than to $\omega^\beta_{\Lambda_N}$.

Unfortunately, showing any of this in specific models at finite (inverse) temperature $0 < \beta < \infty$ is pretty complicated. For example, in the quantum Ising model (9.42) in $d = 1$, KMS states are unique for any B, so that for SSB one must go to $d \geq 2$. In that case, it can be shown from Theorem 10.7 that for $B = 0$, below some critical temperature (i.e. for $\beta > \beta_c$) the $\mathbb{Z}_2$ symmetry defined in (10.68) below is broken, but this takes considerable effort and is beyond the scope of this book.

10.5 Ground state(s) of the quantum Ising chain

It is much simpler to put $\beta = \infty$ and hence turn to the *ground state(s)* of the quantum Ising model (9.42) in $d = 1$, which is manageable. The interesting case is $B > 0$, with $J = 1$ and free boundary conditions, so that for $\Lambda = \Lambda_N$ (with N even), we have

$$h_N = -\sum_{x\in\Lambda_N} (\sigma_3(x)\sigma_3(x+1) + B\sigma_1(x)); \tag{10.60}$$

$$\Lambda_N = \{-\tfrac{1}{2}N, \ldots, \tfrac{1}{2}N - 1\}; \tag{10.61}$$

$$H_{\Lambda_N} = H_N = \otimes_{x\in\Lambda_N} H_x; \tag{10.62}$$

$$H_x = \mathbb{C}^2 \ (x \in \Lambda_N), \tag{10.63}$$

where the operator $\sigma_i(x)$ acts as the Pauli matrix σ_i on H_x and as the unit matrix 1_2 elsewhere. This model describes a chain of N immobile spin-$\frac{1}{2}$ particles with ferromagnetic coupling in a transverse magnetic field (it is a special case of the so-called XY-model, to which similar conclusions apply). The local Hamiltonians h_N define time evolution on the local algebras

$$A_{\Lambda_N} \equiv A_N = B(H_N) \tag{10.64}$$

by (9.40), i.e.,

$$\alpha_t^{(N)}(a_N) = e^{ith_N} a_N e^{-ith_N} \ (a \in A_N), \tag{10.65}$$

which by Theorem 9.15 defines a time evolution on the quasi-local C*-algebra

$$A = \overline{\bigcup_{N\in\mathbb{N}} A_N}^{\|\cdot\|} = \bigotimes_{x\in\mathbb{Z}} B(H_x), \tag{10.66}$$

namely by regarding the unitaries $\exp(ith_N) \in A_N \subset A$ as elements of A and putting

$$\alpha_t(a) = \lim_{N\to\infty} e^{ith_N} a e^{-ith_N} \ (a \in A), \tag{10.67}$$

which exists (although the sequence $(\exp(ith_N))_N$ in A does not converge in A).

For any $B \in \mathbb{R}$, the quantum Ising chain has a $\mathbb{Z}_2$-symmetry given by a 180 degree rotation around the x-axis, locally implemented by the unitary operator $u(x) = \sigma_1(x)$, which at each $x \in \Lambda_N$ yields $(\sigma_1, \sigma_2, \sigma_3) \mapsto (\sigma_1, -\sigma_2, -\sigma_3)$, since $\sigma_i\sigma_j\sigma_i^* = -\sigma_j$ for $i \neq j$. Thus $u(x)$ sends each $\sigma_3(x)$ to $-\sigma_3(x)$ but maps each $\sigma_1(x)$ to itself. As in (10.57), this symmetry is implemented by the unitary operator

$$u^{(N)} = \otimes_{x\in\Lambda_N} \sigma_1(x) \tag{10.68}$$

on H_N, which satisfies $[h_N, u^{(N)}] = 0$, or, equivalently,

$$u^{(N)} h_N (u^{(N)})^* = h_N. \tag{10.69}$$

The ensuing $\mathbb{Z}_2$-symmetry is given by the automorphism $\gamma^{(N)}$ of A_N defined by

$$\gamma^{(N)}(a) = u^{(N)}a(u^{(N)})^* \ (a \in A_N), \tag{10.70}$$

which induces a global automorphism $\gamma \in \mathrm{Aut}(A)$ as in (10.59), i.e.,

$$\gamma(a) = \lim_{N\to\infty} u^{(N)}a(u^{(N)})^* \ (a \in A), \tag{10.71}$$

which limit once again exists despite the fact that the sequence $u^{(N)}$ has no limit in A. Thus $\mathbb{Z}_2$-invariance of the model follows from the local property

$$\alpha_t^{(N)} \circ \gamma^{(N)} = \gamma^{(N)} \circ \alpha_t^{(N)}, \tag{10.72}$$

which in the limit $N \to \infty$ gives

$$\alpha_t \circ \gamma = \gamma \circ \alpha_t \ (t \in \mathbb{R}). \tag{10.73}$$

Since $\gamma^2 = \mathrm{id}_A$, we have an action of the group $\mathbb{Z}_2 = \{-1,1\}$ on A, where the nontrivial element (i.e., $g = -1$) is sent to γ. By (10.72) this group acts on the set $S_\infty(A_N)$ of ground states of A_N relative to the dynamics $\alpha^{(N)}$, and by (10.73) the same is true for the set $S_\infty(A)$ of ground states of the corresponding infinite system for α (and analogously for β-KMS states). These sets may be described as follows.

Theorem 10.11. *1. For any $N < \infty$ and $B = 0$ the ground state of the quantum Ising model* (10.60) *is doubly degenerate and breaks the $\mathbb{Z}_2$ symmetry of the model.*
2. For $N < \infty$ and any $B > 0$ the ground state $\omega^{(0)}_{1/N}$ is unique and hence $\mathbb{Z}_2$-invariant.
3. At $N = \infty$ with magnetic field $0 \leq B < 1$, the model has a doubly degenerate translation-invariant ground state $\omega_0^{\pm}$, which again breaks the $\mathbb{Z}_2$ symmetry.
4. At $N = \infty$ and $B \geq 1$ the ground state is unique (and hence $\mathbb{Z}_2$-invariant).
5. Recall Definition 8.24. For $0 \leq B < 1$ the states $(\omega^{(0)}_{1/N})_{N\in\mathbb{N}}$ (as in no. 2) with

$$\omega_0^{(0)} = \tfrac{1}{2}(\omega_0^+ + \omega_0^-) \tag{10.74}$$

form a continuous field of states on the continuous bundle $A^{(q)}$; in particular,

$$\lim_{N\to\infty} \omega^{(0)}_{1/N} = \omega_0^{(0)}. \tag{10.75}$$

The two ground states in no. 1 and no. 3 are tensor products of $|\uparrow\rangle$ and $|\downarrow\rangle$, respectively (where $\sigma_3|\uparrow\rangle = |\uparrow\rangle$ and $\sigma_3|\downarrow\rangle = -|\downarrow\rangle$), so that $\sigma_3(0)$ is an order parameter in the sense of Definition 10.6. In no. 4, on the other hand, each spin aligns with the magnetic field in the x-direction, so that the ground state is an infinite tensor product of states $|\rightarrow\rangle$, where $\sigma_1|\rightarrow\rangle = |\rightarrow\rangle$, and this time $\sigma_1(0)$ is an order parameter.

Case no. 2 becomes more transparent if we realize the Hilbert space H_N as $\ell^2(S_N)$, where S_N is the set of all spin configurations s on N sites, that is,

$$s : \{-\tfrac{1}{2}N, -\tfrac{1}{2}N+1, \ldots, \tfrac{1}{2}N-1\} \to \{-1,1\}.$$

In terms of the eigenvectors $|1\rangle \equiv |\uparrow\rangle$ and $|-1\rangle \equiv |\downarrow\rangle$ of σ_3, and the orthonormal basis $(\delta_s)_{s\in S_N}$ of $\ell^2(S_N)$ (where $\delta_s(t) = \delta_{st}$), a suitable unitary equivalence

$$v_N : \ell^2(S_N) \to H_N \tag{10.76}$$

is given by linear extension of

$$v_N\delta_s = |s(-\tfrac{1}{2}N)\cdots s(\tfrac{1}{2}N-1)\rangle,\ s,t \in S_N. \tag{10.77}$$

For example, the state $|1\cdots 1\rangle$ corresponds to $\delta_{s_\uparrow}$, where $s_\uparrow(x) = 1$ for all x, and analogously $s_\downarrow(x) = -1$ for the state $|-1\cdots -1\rangle$. Using $\ell^2(S_N)$, we may talk of localization of states in spin configuration space (similar to localization of wave-functions in $L^2(\mathbb{R}^n)$), in the sense that some $\psi \in \ell^2(S_N)$ may be peaked on just a few spins configurations. Provided $0 < B < 1$ this is indeed the case for the unique ground state in case no. 2, which is similar to the ground state of the double-well potential discussed in §§10.1–10.2, replacing $\mathbb{R}$ by S_N (and $\hbar > 0$ by $1/N$).

Theorem 10.11 and related results used below, such as eq. (10.82), follow from the exact solution of the model for both $N < \infty$ and $N = \infty$, to be discussed in §§10.6–10.7. This solution is rather involved, but a rough picture of the various ground states may already be obtained from a classical approximation in the spirit of §8.1. This approximation assumes that the spin-1/2 operators $\frac{1}{2}\sigma_i$ are replaced by their counterparts for spin $n\cdot\frac{1}{2}$, upon which one takes the limit $n \to \infty$. In this limit, the spin operators are turned into the corresponding coordinate functions on the coadjoint orbit $\mathscr{O}_{1/2} \subset \mathbb{R}^3$ for $SU(2)$, which is the two-sphere $S^2_{1/2}$ with radius $r = 1/2$. In principle, this should be done for each of the N spins separately, yielding a classical Hamiltonian h_c that is a function on the N-fold cartesian product of $S^2_{1/2}$ with itself. However, if we *a priori* assume translation invariance of the classical ground state, only one such copy remains. Using spherical coordinates

$$(x = \tfrac{1}{2}\sin\theta\cos\phi, y = \tfrac{1}{2}\sin\theta\sin\phi, z = \tfrac{1}{2}\cos\theta), \tag{10.78}$$

the ensuing trial Hamiltonian becomes just a function on $\mathscr{O}_{1/2}$, given by

$$h(\theta,\phi) \approx -(\tfrac{1}{2}\cos^2\theta + B\sin\theta\cos\phi). \tag{10.79}$$

Minimizing gives $\cos\phi = 1$ and hence $y = 0$ for any B, upon which

$$h(\theta) \approx -(\tfrac{1}{2}\cos^2\theta + B\sin\theta) \tag{10.80}$$

yields the phase portrait of Theorem 10.11 for $N = \infty$, as follows. For $0 \leq B < 1$, the global minimum is reached at the two different solutions $\theta_\pm$ of $\cos\theta_\pm = B$, with ensuing spin vectors

$$\mathbf{x}_\pm(B) = (\tfrac{1}{2}B, 0, \pm\tfrac{1}{2}\sqrt{1-B^2}), \tag{10.81}$$

starting at $\mathbf{x}_\pm(0) = (0,0,\pm\frac{1}{2})$ and merging at $B = 1$ to $\mathbf{x}_+(1) = \mathbf{x}_-(1) = (\frac{1}{2},0,0)$. This remains the unique ground state for $B \geq 1$, where all spins align with the field.

In the regime $0 < B < 1$ with large but finite N, one finds a far-reaching analogy between the double-well potential and and the quantum Ising chain, namely:

- The ground state of (10.60) is doubly peaked in spin configuration space, similar to its counterpart for the double-well potential in real configuration space.
- One has convergence to localized ground states (10.15) - (10.16) for the quantum Ising chain and (10.74) - (10.75) for the double well.
- For the energy difference $\Delta_N = E_N^{(1)} - E_N^{(0)}$ between the first excited state and the ground state one has (10.17) - (10.18) for the double well, and

$$\Delta_N \approx (1-B^2)B^N \ (N \to \infty), \tag{10.82}$$

  for the quantum Ising chain. Thus both models show exponential decay, i.e. of (10.82) in N as $N \to \infty$, and of (10.13) in $1/\hbar$ as $\hbar \to 0$.

It should be mentioned that *exponential* decay of the energy gap seems a low-dimensional luxury, which is not really needed for SSB. All that counts is that $\lim_{N\to\infty} \Delta_N = 0$, which guarantees that the first excited state is asymptotically degenerate with the ground state, so that appropriate linear combinations like $\omega_0^{\pm}$ can be formed that converge to the degenerate symmetry-breaking *pure* (and hence physical) ground states (or *extreme* and hence physical KMS states) of the limit system, which are localized and stable (as is clear from the double well). The fact that in the two models at hand only *one* excited state participates in this mechanism is due to the simple $\mathbb{Z}_2$ symmetry that is being broken; SSB of continuous symmetries requires a large number of low-lying states that are asymptotically degenerate with the ground state and hence also with each other—one speaks of a *thin* energy spectrum).

The existence of low-lying excited states may be proved abstractly (i.e., in a model-independent way), as follows. For $N < \infty$, let $\psi_N^{(0)}$ be the ground state (assumed unique) of some model defined on $\Lambda_N \subset \mathbb{Z}^d$, and let ϕ be an order parameter (cf. Theorem 10.7) with accompanying vector field $\Phi_N = \sum_{x\in\Lambda_N} \phi(x)$; in the quantum Ising chain, we take $\phi = \sigma_1$. Then the key assumptions are expressed by

$$\langle \psi_N^{(0)}, \Phi_N \psi_N^{(0)} \rangle = 0; \tag{10.83}$$

$$\langle \Phi_N \psi_N^{(0)}, \Phi_N \psi_N^{(0)} \rangle \geq C_1 \cdot N^2 \ (N \to \infty, C_1 > 0); \tag{10.84}$$

$$\| [[\Phi_N, h_N], \Phi_N] \| \leq C_2 \cdot N \ (N \to \infty, C_2 > 0). \tag{10.85}$$

The first states that the ground state is symmetric, the second enforces long-range order, as in (10.41), and the third follows from having short-range forces. A simple computation then shows that the unit vector $\tilde{\psi}_N^{(1)} = \Phi_N \psi_N^{(0)} / \|\Phi_N \psi_N^{(0)}\|$ satisfies

$$\langle \tilde{\psi}_N^{(1)}, h_N \tilde{\psi}_N^{(1)} \rangle - \langle \psi_N^{(0)}, h_N \psi_N^{(0)} \rangle \leq C_2/(C_1 N) \ (N \to \infty). \tag{10.86}$$

Since $\tilde{\psi}_N^{(1)}$ is orthogonal to $\psi_N^{(0)}$ by (10.83), the variational principle for eigenvalues (note that h_N has discrete spectrum, as $\dim(H_{\Lambda_N}) < \infty$) then gives $\Delta_N \leq C_2/(C_1 N)$, so that Δ_N vanishes as $N \to \infty$, though perhaps not as quickly as (10.82) indicates.

10.6 Exact solution of the quantum Ising chain: $N < \infty$

The solution of the quantum Ising chain is based on a transformation to fermionic variables. Let H be a Hilbert space and let $F_-(H)$ be its ***fermionic Fock space***, i.e.,

$$F_-(H) = \oplus_{k=0}^{\infty} H_-^k, \tag{10.87}$$

where $H^0 = \mathbb{C}$, and for $k > 0$ the Hilbert space $H_-^k = e_-^{(k)} H^k$ is the totally antisymmetrized k-fold tensor product of H with itself, see also §7.7. Here the projection $e_-^{(k)} : H^k \to H^k$ is defined by linear extension of

$$e_-^{(k)} f_1 \otimes \cdots \otimes f_k = \frac{1}{k!} \sum_{p \in \mathfrak{S}_k} \text{sgn}(p) f_{p(1)} \otimes \cdots \otimes f_{p(k)}, \tag{10.88}$$

where $\mathfrak{S}_k$ is the permutation group on k objects, and $\text{sgn}(p)$ is $+1/-1$ if p is an even/odd permutation. With the (total) Fock space $F(H) = \oplus_{k=0}^{\infty} H^k$ we have $F_-(H) = e_- F(H)$, where $e = \sum_k e_-^{(k)}$ (strongly) is a projection. For $f \in H$ we define the (unbounded) ***annihilation operator*** $a(f)$ on $F(H)$ by (finite) linear extension of

$$a(f) f_1 \otimes \cdots \otimes f_k = \sqrt{k} \langle f, f_1 \rangle_H \otimes \cdots \otimes f_k, \tag{10.89}$$

for $k > 0$, with $a(f)z = 0$ on $H^0 = \mathbb{C}$. This gives the adjoint $a(f)^* \equiv a^*(f)$ as

$$a^*(f) f_1 \otimes \cdots \otimes f_k = \sqrt{k+1} f \otimes f_1 \otimes \cdots \otimes f_k. \tag{10.90}$$

For each $f \in H$, we then define the following operators on $F_-(H)$:

$$c(f) = e_- a(f) e_-; \tag{10.91}$$

$$c^*(f) = e_- a^*(f) e_-. \tag{10.92}$$

Note that the map $f \mapsto c(f)$ is *antilinear* in f, whereas $f \mapsto a^*(f)$ is *linear* in f. It follows that $c^*(f) = c(f)^*$, that each operator $c(f)$ and $c(f)$ on $F_-(H)$ is bounded with $\|c(f)\| = \|c^*(f)\| = \|f\|$, and the ***canonical anticommutation relations*** hold:

$$[c(f), c^*(g)]_+ = \langle f, g \rangle_H \cdot 1_{F_-(H)}; \tag{10.93}$$

$$[c(f), c(g)]_+ = [c^*(f), c^*(g)]_+ = 0. \tag{10.94}$$

Thus we may define $\text{CAR}(H)$ as the C*-algebra within $B(F_-(H))$ generated by all $c(f)$, where $f \in H$. This is called the C*-algebra of ***canonical anticommutation relations*** over H, which have constructed in its defining representation on $F_-(H)$. Choosing an orthonormal basis (e_i) of H and writing $c(e_i) = c_i$ etc. clearly yields

$$[c_i, c_j^*]_+ = \delta_{ij} \cdot 1_{F_-(H)}; \tag{10.95}$$

$$[c_i, c_j]_+ = [c_i^*, c_j^*]_+ = 0. \tag{10.96}$$

If $\dim(H) = N < \infty$, then $\mathrm{CAR}(H) = B(F_-(H))$. First, a dimension count yields

$$F_-(\mathbb{C}^N) = \oplus_{k=0}^N H_-^k \cong \mathbb{C}^{2^N} \cong \otimes^N \mathbb{C}^2. \tag{10.97}$$

By Theorem C.90, the C*-algebra $\mathrm{CAR}(H)$ acts irreducibly on $F_-(H)$, so that

$$\mathrm{CAR}(\mathbb{C}^N) \cong M_{2^N}(\mathbb{C}). \tag{10.98}$$

This is already nontrivial for $N = 1$. In that case, $F_-(\mathbb{C}) = \mathbb{C} \oplus \mathbb{C} = \mathbb{C}^2$, and

$$c = \sigma_- = \begin{pmatrix} 0 & 0 \\ 1 & 0 \end{pmatrix}; \tag{10.99}$$

$$c^* = \sigma_+ = \begin{pmatrix} 0 & 1 \\ 0 & 0 \end{pmatrix}, \tag{10.100}$$

where $\sigma_\pm = \frac{1}{2}(\sigma_1 \pm i\sigma_2)$. This realization explicitly shows that

$$\mathrm{CAR}(\mathbb{C}) = M_2(\mathbb{C}). \tag{10.101}$$

To generalize this to $N > 1$, we introduce a lattice (or chain) $\underline{N} = \{1, \ldots, N\}$, and for each $x \in \underline{N}$ we define operators c_x, c_x^* by the ***Jordan–Wigner transformation***

$$c_x = e^{\pi i \sum_{y=1}^{x-1} \sigma_+(y)\sigma_-(y)} \sigma_-(x) = \left(\prod_{y=1}^{x-1} (-\sigma_3)(y) \right) \cdot \sigma_-(x), \tag{10.102}$$

$$c_x^* = e^{-\pi i \sum_{y=1}^{x-1} \sigma_+(y)\sigma_-(y)} \sigma_+(x) = \left(\prod_{y=1}^{x-1} (-\sigma_3)(y) \right) \cdot \sigma_+(x), \tag{10.103}$$

where $x > 1$, and $c_1 = \sigma_1^-$ and $c_1^* = \sigma_1^+$ (here $\sigma_\pm(x) = \frac{1}{2}(\sigma_1(x) \pm i\sigma_2(x))$ etc.). These operators satisfy (10.95) - (10.96); the second expression on each line follows because the operators $\sigma_+(y)\sigma_-(y)$ commute for different sites y, and

$$e^{\pi i \sigma^+ \sigma^-} = -\sigma_3. \tag{10.104}$$

Furthermore, since

$$c_x^* c_x = \sigma_+(x)\sigma_-(x) = \begin{pmatrix} 1 & 0 \\ 0 & 0 \end{pmatrix}(x); \tag{10.105}$$

$$c_x c_x^* = \sigma_-(x)\sigma_+(x)) = \begin{pmatrix} 0 & 0 \\ 0 & 1 \end{pmatrix}(x), \tag{10.106}$$

the inverse of the Jordan–Wigner transformation is given by

$$\sigma_-(x) = e^{-\pi i \sum_{y=1}^{x-1} c_y^* c_y} c_x; \tag{10.107}$$

$$\sigma_+(x) = c_x^* e^{\pi i \sum_{y=1}^{x-1} c_y^* c_y}. \tag{10.108}$$

We return to the quantum Ising model (10.60) with free boundary conditions, where we relabel the sites as $\{1,\ldots,N\}$, as above, and change to the Hamiltonian

$$h_N^{\rm QI} = -\tfrac{1}{2}\left(\sum_{x=1}^{N-1}\sigma_1(x)\sigma_1(x+1)+\lambda\sum_{x=1}^{N}\sigma_3(x)\right), \tag{10.109}$$

where, in order to avoid notational confusion with the operator B in (10.111) below, we henceforth replace $B \rightsquigarrow \lambda$. In terms of the unitary operator $u=\sqrt{1/2}(1_2+i\sigma_2)$ on $\mathbb{C}^2$ and hence $u^{(N)}=\otimes_{x=1}^N u(x)$ on $\otimes^N\mathbb{C}^2$, we have $u^{(N)}h_N(u^{(N)})^* = h'_N$.

Using (10.102) - (10.103), up to an additive constant $\lambda N\cdot 1_N$ we omit, we find

$$h_N^{\rm QI} = -\sum_{x=1}^{N}(\lambda c_x^*c_x+\tfrac{1}{2}(c_x^*-c_x)(c_{x+1}^*+c_{x+1})), \tag{10.110}$$

so we now show how to diagonalize quadratic fermionic Hamiltonians of the type

$$h_N = -\sum_{x,y=1}^{N}\left(A_{xy}c_x^*c_y+\tfrac{1}{2}B_{xy}(c_x^*c_y^*-c_xc_y)\right), \tag{10.111}$$

where A and B are real $N\times N$ matrices, with $A^*=A$ and $B^*=-B$. Indeed, taking

$$A = \tfrac{1}{2}(S+S^*)+\lambda\cdot 1_N; \tag{10.112}$$
$$B = \tfrac{1}{2}(S-S^*), \tag{10.113}$$

recovers (10.110), where $S:\mathbb{C}^N\to\mathbb{C}^N$ is the ***shift operator***, defined by

$$Sf(x) = f(x+1); \tag{10.114}$$
$$S^*f(x) = f(x-1). \tag{10.115}$$

By convention, $f(N+1)=f(0)=0$ (i.e., $Sf(N)=S^*f(0)=0$ for any $f\in\mathbb{C}^N$); in terms of the standard basis (υ_x) of $\mathbb{C}^N$ we have $S\upsilon_1=0$ and $S\upsilon_x=\upsilon_{x-1}$ for $x=\{2,\ldots,N\}$, and likewise $S^*\upsilon_N=0$ and $S\upsilon_x=\upsilon_{x+1}$ for $x=\{1,\ldots,N-1\}$.

The smart thing to do now turns out to be diagonalizing the $2N\times 2N$-matrix

$$M=\begin{pmatrix} A & B\\ -B & -A\end{pmatrix}, \tag{10.116}$$

which by a unitary transformation may be brought into the simpler form

$$M'=\begin{pmatrix}\sqrt{1/2} & -\sqrt{1/2}\\ \sqrt{1/2} & \sqrt{1/2}\end{pmatrix}\begin{pmatrix} A & B\\ -B & -A\end{pmatrix}\begin{pmatrix}\sqrt{1/2} & \sqrt{1/2}\\ -\sqrt{1/2} & \sqrt{1/2}\end{pmatrix}=\begin{pmatrix}0 & C\\ C^* & 0\end{pmatrix}, \tag{10.117}$$

where $C=A+B$. For example, for the model (10.111) we simply have

$$C=S+\lambda\cdot 1_N. \tag{10.118}$$

The equations for the eigenvalues ε_k and eigenvectors of M', i.e.,

$$M' \begin{pmatrix} \varphi_k \\ \psi_k \end{pmatrix} = \varepsilon_k \begin{pmatrix} \varphi_k \\ \psi_k \end{pmatrix} \tag{10.119}$$

where $\varphi_k, \psi_k \in \mathbb{C}^N$, are equivalent to both the coupled system of equations

$$C\psi_k = \varepsilon_k \varphi_k; \tag{10.120}$$
$$C^* \varphi_k = \varepsilon_k \psi_k; \tag{10.121}$$
$$C = A + B, \tag{10.122}$$

where the eigenvalues ε_k are real (since $M^* = M$), and to the uncoupled version

$$CC^* \varphi_k = \varepsilon_k^2 \varphi_k; \tag{10.123}$$
$$C^* C\psi_k = \varepsilon_k^2 \psi_k; \tag{10.124}$$
$$CC^* = A^2 - B^2 - [A,B]; \tag{10.125}$$
$$C^* C = A^2 - B^2 + [A,B]. \tag{10.126}$$

Without loss of generality we may (and will) assume that the φ_k, ψ_k are unit vectors in $\mathbb{C}^N$, so that the corresponding unit vector in $\mathbb{C}^{2N}$ is $(\varphi_k, \psi_k)/\sqrt{2}$). Furthermore, since C (or M) is a matrix with real entries and the ε_k are real, by a suitable choice of phase we may (and will) also arrange that φ_k, ψ_k have real components. Finally, it follows from (10.120) - (10.120) that $(-\varphi_k, \psi_k)$ is an eigenvector of C with eigenvalue $-\varepsilon_k$, so that the unitary transformation U' that diagonalizes M', i.e.,

$$(U')^{-1} M' U' = \begin{pmatrix} -E & 0 \\ 0 & E \end{pmatrix}, \tag{10.127}$$

where $E = \mathrm{diag}(\varepsilon_1, \ldots, \varepsilon_N)$, takes the form

$$U' = \frac{1}{\sqrt{2}} \begin{pmatrix} \varphi & -\varphi \\ \psi & \psi \end{pmatrix}, \tag{10.128}$$

where φ is the $N \times N$ matrix $(\varphi_1, \ldots, \varphi_N)$, seeing each vector φ_i as a column, etc. Combined with (10.117), we obtain

$$U^{-1} M U = \begin{pmatrix} -E & 0 \\ 0 & E \end{pmatrix}; \tag{10.129}$$

$$U = \tfrac{1}{2} \begin{pmatrix} 1 & 1 \\ -1 & 1 \end{pmatrix} \cdot \begin{pmatrix} \varphi & -\varphi \\ \psi & \psi \end{pmatrix} = \tfrac{1}{2} \begin{pmatrix} \psi + \varphi & \psi - \varphi \\ \psi - \varphi & \psi + \varphi \end{pmatrix} \equiv \begin{pmatrix} u & v \\ v & u \end{pmatrix}, \tag{10.130}$$

where we introduced $N \times N$ matrices

$$u = \tfrac{1}{2}(\psi + \varphi); \tag{10.131}$$
$$v = \tfrac{1}{2}(\psi - \varphi). \tag{10.132}$$

Using orthonormality and completeness of both the (φ_k) and the (ψ_k), one obtains

$$u^*u + v^*v = 1_H; \tag{10.133}$$
$$u^*v + v^*u = 0; \tag{10.134}$$
$$uu^* + vv^* = 1_H; \tag{10.135}$$
$$uv^* + vu^* = 0. \tag{10.136}$$

Of course, u and v are far from unique, as they depend on both the ordering and the phases of the vectors φ_k and ψ_k. In partial remedy of the former ambiguity we assume that $0 \leq \varepsilon_0 \leq \varepsilon_1 \leq \cdots \leq \varepsilon_N$ (which can be arranged by a suitable ordering as well as choice of sign of the eigenvectors φ_k). Towards the latter, we already agreed that both the φ_k and ψ_k are real, so that also our matrices u and v have real entries.

We now explain the purpose of diagonalizing M in (10.116) using u and v.

Proposition 10.12. *Let* u *and* v *be operators on a Hilbert space* H*, where* u *is* linear *and* v *is* anti-linear. *Let* $c(f)$ *and* $c^*(f)$ *be the operators* (10.91) - (10.92), *satisfying the CAR* (10.93) - (10.94). *Define the* **Bogoliubov transformation**

$$\eta(f) = c(\mathsf{u}f) + c^*(\mathsf{v}f); \tag{10.137}$$
$$\eta^*(f) = c^*(\mathsf{u}f) + c(\mathsf{v}f), \tag{10.138}$$

which extends to a linear map $\alpha : \mathrm{CAR}(H) \to \mathrm{CAR}(H)$, *where* $\eta(f) = \alpha(c(f))$ *etc. Then* α *is a homomorphism of C*-algebras, or, equivalently, one has the CAR*

$$[\eta(f), \eta^*(g)]_+ = \langle f, g\rangle_H \cdot 1_H; \tag{10.139}$$
$$[\eta(f), \eta(g)]_+ = [\eta^*(f), \eta^*(g)]_+ = 0, \tag{10.140}$$

iff u *and* v *satisfy* (10.133) - (10.134), *with* $u \rightsquigarrow \mathsf{u}, v \rightsquigarrow \mathsf{v}$. *Moreover,* α *is invertible (and hence defines an automorphism of* $\mathrm{CAR}(H)$*) iff in addition* (10.135) - (10.136) *are valid (again with with* $u \rightsquigarrow \mathsf{u}, v \rightsquigarrow \mathsf{v}$*), in which case the inverse is*

$$c(f) = \eta(\mathsf{u}^*f) + \eta^*(\mathsf{v}^*f); \tag{10.141}$$
$$c^*(f) = \eta^*(\mathsf{u}^*f) + \eta(\mathsf{v}^*f). \tag{10.142}$$

Note that anti-linearity of v is needed to make $f \mapsto \eta(f)$ anti-linear, like $f \mapsto c(f)$. With respect to a base (e_i) of H, the transformations (10.137) - (10.142) reads

$$\eta_i = \sum_j (\overline{u}_{ji} c_j + v_{ji} c_j^*); \tag{10.143}$$
$$\eta_j^* = \sum_j (u_{ji} c_j^* + \overline{v}_{ji} c_j); \tag{10.144}$$
$$c_i = \sum_j (u_{ij} \eta_j + \overline{v}_{ij} \eta_j^*); \tag{10.145}$$
$$c_i^* = \sum_j (\overline{u}_{ij} \eta_j^* + v_{ij} \eta_j). \tag{10.146}$$

Proof. The proof is a straightforward computation. □

In comparison with the preceding diagonalization process, where $H = \mathbb{C}^N$, we notice that in this process u and v were both linear, whereas in Proposition 10.12 u is linear whereas v is antilinear. This difference is easily overcome by taking $\mathsf{u} = u$ and $\mathsf{v} = Jv$, where $J : \mathbb{C}^N \to \mathbb{C}^N$ is the anti-linear map $Jf(x) = \overline{f(x)}$, so that J is a ***conjugation*** in being an anti-linear map that satisfies $J^* = J^{-1} = J$.

Returning to our generic Hamiltonian (10.111), a straightforward computation using (10.145) - (10.146), (10.116), (10.129), and (10.133) - (10.136) yields

$$h_N = \sum_{k=1}^{N} \varepsilon_k \eta_k^* \eta_k, \tag{10.147}$$

up to a (computable) constant, where we recall that $\varepsilon_k \geq 0$ $(k = 1, \ldots, N)$. Note that h_N is still defined on the fermionic Fock space $F_-(\mathbb{C}^N)$, as h_N is a (complicated) *quadratic* expression in the operators c_i and c_i^* on $F_-(\mathbb{C}^N)$. The point is that (as a consequence of Proposition 10.12) the η_k and η_k^* also satisfy the CAR, i.e.,

$$[\eta_i, \eta_j^*]_+ = \delta_{ij} \cdot 1_{F_-(H)}; \tag{10.148}$$

$$[\eta_i, \eta_j]_+ = [\eta_i^*, \eta_j^*]_+ = 0. \tag{10.149}$$

Theorem 10.13. *Let $A = \mathrm{CAR}(\mathbb{C}^N)$ be the CAR algebra over $H = \mathbb{C}^N$ with dynamics $\alpha_t(a) = e^{ith_N} a e^{-ith_N}$ given by (10.111) and hence by(10.147). Then α has a unique (and hence pure and symmetric) ground state ω_0, specified by the property*

$$\pi_{\omega_0}(\eta(f))\Omega_{\omega_0} = 0 \;\; (f \in \mathbb{C}^N). \tag{10.150}$$

Proof. Recall that α defines a derivation $\delta : \mathrm{CAR}(\mathbb{C}^N) \to \mathrm{CAR}(\mathbb{C}^N)$ defined by (9.54), which in the case at hand is simply by $\delta(a) = i[h_N, a]$ (since A is finite-dimensional, δ is bounded and hence defined everywhere). Using the identity

$$[ab, c] = a[b, c]_+ - [c, a]_+ b, \tag{10.151}$$

as well as the relations (10.148) - (10.149), we obtain $\delta(\eta_k) = -i\varepsilon_k \eta_k$, and hence

$$-i\omega_0(\eta_k^* \delta(\eta_k)) = -\omega_0(\eta_k^* \eta_k). \tag{10.152}$$

The condition $-i\omega_0(a^*\delta(a)) \geq 0$, i.e., eq. (9.56) from Proposition (9.20), therefore implies that $\omega_0(\eta_k^* \eta_k) \leq 0$, and hence $\omega_0(\eta_k^* \eta_k) = 0$ by positivity of ω_0. Since $F_0(H)$ is finite-dimensional and $A \cong B(F_0(H))$, cf. (10.98), we may assume ground state(s) to be pure and normal, i.e., there is some unit vector $\psi_0 \in F_-(H)$ with $\omega(a) = \langle \psi_0, a\psi_0 \rangle$ for each $a \in A$. Hence $\langle \psi_0, \eta_k^* \eta_k \psi_0 \rangle = 0$, which enforces

$$\eta_k \psi_0 = 0 \;\; (k = 1, \ldots, N). \tag{10.153}$$

This property makes ψ_0 unique up to a phase. Indeed, together with (10.148) - (10.149), eq. (10.153) implies the values of all one- and two-point functions, i.e.,

$$\omega_0(\eta(f)) = \omega_0(\eta^*(f)) = 0; \tag{10.154}$$
$$\omega_0(\eta^*(f)\eta(g)) = \omega_0(\eta^*(f)\eta^*(g)) = \omega_0(\eta(f)\eta(g)) = 0; \tag{10.155}$$
$$\omega_0(\eta(f)\eta^*(g)) = \langle f,g\rangle_H. \tag{10.156}$$

Furthermore, the value of ω_0 on any product of an odd number of $\eta(f)$ and $\eta^*(g)$ vanishes; for an even number the value $\omega_0(\prod_{i=1}^n \eta(f_i)\prod_{j=1}^n \eta^*(g_j))$ it is given by

$$\sum_{p=1}^{n}(-1)^{n-p}\omega_0(\eta(f_1)\eta^*(g_p)\omega_0\left(\prod_{i=2}^{n}\eta(f_i)\prod_{j=1,j\neq p}^{n}\eta^*(g_j)\right).$$

Hence (10.153) gives ω_0 on all of $\mathrm{CAR}(\mathbb{C}^N)$. Since $\mathrm{CAR}(H) = B(F_-(H))$, this fixes ψ_0 up to a phase. Eq. (10.150) is just a fancy way of rewriting (10.153). □

By construction, the ground state energy of (10.147) is zero. In connection with our approach to SSB via Butterfield's Principle it is of interest to compute the energy ε_1 of the first excited state. This may be done from (10.120) - (10.121) with (10.122) and the specific expression (10.118) for the quantum Ising chain. Thus we solve

$$\lambda\psi_k(x) + \psi_k(x+1) = \varepsilon_k\varphi_k(x) \ \ (x = 1,\ldots,N, \psi_k(N+1) = 0); \tag{10.157}$$
$$\lambda\varphi_k(x) + \varphi_k(x-1) = \varepsilon_k\psi_k(x) \ \ (x = 1,\ldots,N, \varphi_k(0) = 0). \tag{10.158}$$

A solution of this system (with real wave-functions and positive energy) is given by

$$\varphi_k(x) = C(-1)^k\sin(q_k(x-N-1)); \tag{10.159}$$
$$\psi_k(x) = -C\sin(q_k x); \tag{10.160}$$
$$\varepsilon_k = \sqrt{1+\lambda^2+2\lambda\cos(q_k)}, \tag{10.161}$$

where $C > 0$ is a normalization constant, and q_k should be solved from

$$(N+1)q_k = (k-1)\pi + \arctan\left(\frac{\sin q_k}{\lambda+\cos q_k}\right). \tag{10.162}$$

For example, for $\lambda = 0$ (i.e. no transverse magnetic field) we obtain $q_k = k\pi/N$, where $k = 1,\ldots,N$. For $\lambda > 1$ there is a unique real solution q_k for each k, too, and even as $N \to \infty$ there is an energy gap $\varepsilon_k > 0$ for each k. For $0 < \lambda < 1$, however, there is a *complex* solution $q_1 = \pi + i\rho$, where $\rho \in \mathbb{R}$ is a solution to

$$\tanh((N+1)\rho) = \frac{\sinh\rho}{\cosh\rho - \lambda}. \tag{10.163}$$

As $N \to \infty$, we find $\rho = -\ln(\lambda) - (1-\lambda^2)\lambda^{2(N-1)}$. Eq. (10.161) then gives

$$\varepsilon(q_1) \approx (1-\lambda^2)\lambda^N \ \ (N \to \infty), \tag{10.164}$$

which, recalling that $E_N^{(1)} = \varepsilon_1$ and $E_N^{(0)} = 0$ and hence $\Delta_N = \varepsilon_1$, confirms (10.82).

10.7 Exact solution of the quantum Ising chain: $N = \infty$

The (two-sided) infinite quantum Ising chain is described by the C*-algebra

$$F = \mathrm{CAR}(\ell^2(\mathbb{Z})); \tag{10.165}$$

one may also consider a one-sided chain, but it lacks translation symmetry. By the construction at the beginning of the previous section, F is isomorphic to the infinite tensor product $A = M_2(\mathbb{C})^{\infty}$. We consider F to be generated by the operators $c_x^{\pm}$ ($x \in \mathbb{Z}$), where $c_x^- \equiv c_x$ and $c_x^+ \equiv c_x^*$. In this notation, the CAR (10.95) - (10.96) read

$$[c_x^{\pm}, c_y^{\mp}]_+ = \delta_{xy}; \tag{10.166}$$

$$[c_x^{\pm}, c_y^{\pm}]_+ = 0. \tag{10.167}$$

Although the local Hamiltonians (10.111) do not have a limit as $N \to \infty$, as explained in §10.5 they do generate a time-evolution on F in the sense of a continuous homomorphism $\alpha : \mathbb{R} \to \mathrm{Aut}(F)$ via (10.65) and (10.67); see also Theorem 9.15.

Let us first extend the approach in the previous section to $N = \infty$, in which case $\mathbb{C}^N$ is replaced by $H = \ell^2(\mathbb{Z})$, assuming the theory has already been brought into fermionic form with local Hamiltonians (10.111) (as we will see, it is this step, i.e., the Jordan–Wigner transformation, that marks the difference between $N < \infty$ and $N = \infty$). Thus we define operators $A : \ell^2(\mathbb{Z}) \to \ell^2(\mathbb{Z})$ and $B : \ell^2(\mathbb{Z}) \to \ell^2(\mathbb{Z})$ as the obvious extensions of the $N \times N$ matrices A and B to operators on $\ell^2(\mathbb{Z})$, and similarly $S : \ell^2(\mathbb{Z}) \to \ell^2(\mathbb{Z})$ is the "full" shift operator, defined by $(Sf)(x) = f(x+1)$. Instead of the somewhat clumsy explicit solution procedure sketched in the previous section for $N < \infty$, we may now simply rely on the Fourier transformation

$$\mathscr{F} : \ell^2(\mathbb{Z}) \to L^2([-\pi, \pi]); \tag{10.168}$$

$$(\mathscr{F} f)(k) \equiv \hat{f}(k) = \sum_{x \in \mathbb{Z}} e^{-ikx} f_j; \tag{10.169}$$

$$(\mathscr{F}^{-1} \hat{f})(x) \equiv f(x) = \int_{-\pi}^{\pi} \frac{dk}{2\pi} e^{ikx} \hat{f}(k), \tag{10.170}$$

which diagonalizes A and B to operators $\hat{A}, \hat{B} : L^2([-\pi, \pi]) \to L^2([-\pi, \pi])$. For the quantum Ising Hamiltonian (10.110) these are given by the multiplication operators

$$\hat{A}\hat{\psi}(k) = -(\cos k + \lambda)\hat{\psi}(k); \tag{10.171}$$

$$\hat{B}\hat{\psi}(k) = -i \sin k \, \hat{\psi}(k). \tag{10.172}$$

For fixed k, the eigenvalues and eigenvectors of the 2×2 matrix

$$M_k = \begin{pmatrix} -(\cos k + \lambda) & -i \sin k \\ i \sin k & \cos k + \lambda \end{pmatrix}, \tag{10.173}$$

are $\pm\varepsilon_k$, given by (10.161) with $q_k \rightsquigarrow k$. It is then routine to find a unitary 2×2 matrix $U_k = \begin{pmatrix} u_k & v_k \\ v_k & u_k \end{pmatrix}$ that diagonalizes M_k in the sense that $U_k^{-1} M_k U_k = \begin{pmatrix} -\varepsilon_k & 0 \\ 0 & \varepsilon_k \end{pmatrix}$. Fourier transforming these multiplication operators back to $\ell^2(\mathbb{Z})$ then yields an operator U on $\ell^2(\mathbb{Z}) \oplus \ell^2(\mathbb{Z})$ that satisfies (10.129). This yields a unique ground state ω_0 characterized by a property like (10.150) or (10.153), where

$$\eta(\hat{f}) = \int_{-\pi}^{\pi} \frac{dk}{2\pi} \hat{f}(k)(u_k \hat{c}_k + v_k \hat{c}^*_{-k}); \tag{10.174}$$

$$\hat{c}_k = \sum_{j \in \mathbb{Z}} e^{-ijk} c_j; \tag{10.175}$$

$$\hat{c}^*_k = \sum_{j \in \mathbb{Z}} e^{ijk} c^*_j. \tag{10.176}$$

In summary, one-dimensional fermionic models with quadratic Hamiltonians like (10.111) have a unique ground state even at $N = \infty$. Thus one wonders where SSB in the quantum Ising chain could possibly come from. We will answer this question.

Almost every argument to follow relies on $\mathbb{Z}_2$-symmetry. In general, a $\mathbb{Z}_2$-action on a C*-algebra A corresponds to an automorphism $\theta : A \to A$ such that $\theta^2 = \mathrm{id}_A$, i.e. θ represents the nontrivial element of $\mathbb{Z}_2$. For example, define $\theta : F \to F$ by

$$\theta(c_x^{\pm}) = -c_x^{\pm} \ (j \in \mathbb{Z}), \tag{10.177}$$

which is an example of a Bogoliubov transformation (cf. Proposition 10.12) and hence extends to an automorphism of F (which implies that $\theta(1_F) = 1_F$). Clearly, $\theta^2 = \mathrm{id}_F$, and in addition each local Hamiltonian (10.111) is invariant under θ; by implication, so is the dynamics α, i.e., $\alpha_t \circ \theta = \theta \circ \alpha_t$ for all $t \in \mathbb{R}$.

A C*-algebra A carrying a $\mathbb{Z}_2$-action decomposes as

$$A = A_+ \oplus A_-; \tag{10.178}$$

$$A_\pm = \{a \in A \mid \theta(a) = \pm a\}, \tag{10.179}$$

where the ***even*** part A_+ is a subalgebra of A, whereas the ***odd*** part A_- is not: one has $ab \in A_+$ for a, b both in either A_+ or A_-, and $ab \in A_-$ if one is in A_+ and the other in A_-. For example, if $A = B(H)$ for some Hilbert space H and $w : H \to H$ is a untitary operator satisfying $w^2 = 1$ (and hence $w^* = w$), then

$$\theta(a) = waw^* \ (= waw) \tag{10.180}$$

defines a $\mathbb{Z}_2$-action on A. In that case, A_+ and A_- consist of all $a \in A$ that commute and anticommute with w, respectively, that is,

$$A_\pm = \{a \in A \mid aw \mp wa = 0\}. \tag{10.181}$$

In case of (10.165) with (10.177), the subspace F_+ (F_-) is just the linear span of all products of an even (odd) number of $c_j^{\pm}$'s.

Let us move to Theorem C.90 and reconsider the proof of the claim that if $\pi_\omega(A)' \neq \mathbb{C} \cdot 1$, then ω is mixed. If the commutant $\pi_\omega(A)'$ is nontrivial, then it contains a nontrivial projection $e_+ \in \pi_\omega(A)'$. It then follows that $e_+\Omega_\omega \neq 0$: for if $e_+\Omega_\omega = 0$, then $ae_+\Omega_\omega = e_+a\Omega_\omega = 0$ for all $a \in A$, so that $e_+ = 0$, since π_ω is cyclic. Similarly, $e_-\Omega_\omega \neq 0$ with $e_- = 1_H - e_+$, so we may define the unit vectors

$$\Omega_\pm = e_\pm\Omega_\omega / \|e_\pm\Omega_\omega\|, \tag{10.182}$$

and the associated states $\omega_\pm(a) = \langle\Omega_\pm, \pi_\omega(a)\Omega_\pm\rangle$on A. This yields a convex decomposition $\omega = \lambda\omega_+ + (1-\lambda)\omega_-$, with $\lambda = \|\Omega_-\|^2$. Since $\lambda \neq 0, 1$ and $\omega_+ \neq \omega_-$, it follows that ω is mixed. The associated reduction is effected by writing

$$H = H_+ \oplus H_-; \tag{10.183}$$

$$H_\pm = e_\pm H, \tag{10.184}$$

in that A (more precisely, $\pi_\omega(A)$) maps each subspace $H_\pm$ into itself. Now pass from the projections $e_\pm$ to the operator $w = e_+ - e_-$, which by construction satisfies

$$w^* = w^{-1} = w. \tag{10.185}$$

In particular, w is unitary. Conversely, if some unitary w satisfies $w^2 = 1_H$, then

$$e_\pm = \tfrac{1}{2}(1_H \pm w) \tag{10.186}$$

are projections satisfying $e_+ + e_- = 1_H$, giving rise to the decomposition (10.184). Group theoretically, this means that one has a unitary $\mathbb{Z}_2$-action on $H = H_\omega$, in which the nontrivial element of $\mathbb{Z}_2 = \{-1, 1\}$ is represented by w. The decomposition (10.184) then simply means that $\mathbb{Z}_2$ acts trivially on H_+ (in that both group elements are represented by the unit operator) and acts nontrivially on H_- (in that the nontrivial element is represented by *minus* the unit operator). In conclusion, one has a $\mathbb{Z}_2$ perspective on the reduction of H_ω, and instead of a projection $e \in \pi_\omega(A)'$ one may equivalently look for an operator $w \in \pi_\omega(A)'$ that satisfies (10.185).

Proposition 10.14. *Suppose A carries a $\mathbb{Z}_2$-action θ and consider a state $\omega : A \to \mathbb{C}$ that is $\mathbb{Z}_2$-invariant in the sense that $\omega(\theta(a)) = \omega(a)$ for all $a \in A$. We write this as $\theta^*\omega = \omega$, with $\theta^*\omega = \omega \circ \theta$. Then there is a unitary operator $w : H_\omega \to H_\omega$ satisfying $w^2 = 1_H$, $w\Omega = \Omega$, and and $w\pi_\omega(a)w^* = \pi_\omega(\theta(a))$ for each $a \in A$.*

Cf. Corollary 9.12. In this situation, we obtain a decomposition of $H \equiv H_\omega$ according to (10.183), where the projections $e_\pm$ are given by (10.186), so that, equivalently,

$$H_\pm = \{\psi \in H \mid w\psi = \pm\psi\} = A_\pm\Omega_-. \tag{10.187}$$

In terms of the decomposition (10.178), it is easily seen that each subspace $H_\pm$ is stable under A_+, whereas A_- maps $H_\pm$ into $H_\mp$. We denote the restriction of $\pi_\omega(A_+)$ to $H_\pm$ by $\pi_\pm$, so that a $\mathbb{Z}_2$-invariant state θ on A not just gives rise to the GNS-representation π_ω of A on H_ω, but also induces two representations $\pi_\pm$ of the even part A_+ on $H_\pm$. This leads to a refinement of Theorem C.90:

Theorem 10.15. *Suppose A carries a $\mathbb{Z}_2$-action θ, and let $\omega : A \to \mathbb{C}$ be a $\mathbb{Z}_2$-invariant state. With the above notation, suppose the representation $\pi_+(A_+)$ on H_+ is irreducible. Then also the representation $\pi_-(A_+)$ on H_- is irreducible, and there are the following two possibilities for the representation $\pi_\omega(A)$ on $H = H_+ \oplus H_-$:*

1. *$\pi_\omega(A)$ is irreducible (and ω is pure) iff $\pi_+(A_+)$ and $\pi_-(A_+)$ are inequivalent.*
2. *$\pi_\omega(A)$ is reducible (and ω is mixed) iff $\pi_+(A_+)$ and $\pi_-(A_+)$ are equivalent.*

Proof. The proof of this theorem is much more difficult than one would expect (given its simple statement), so we restrict ourselves to the easy steps, as well as to two examples illustrating each of the two possibilities. To start with the latter:

1. $A = M_2(\mathbb{C})$, with $\theta(a) = \sigma_3 a \sigma_3$; note that $\sigma_3^2 = 1$ and $\sigma_3^* = \sigma_3$. Then

$$A_+ = \left\{ \begin{pmatrix} z_+ & 0 \\ 0 & z_- \end{pmatrix}, z_\pm \in \mathbb{C} \right\} \equiv D_2(\mathbb{C}); \tag{10.188}$$

$$A_- = \left\{ \begin{pmatrix} 0 & z_1 \\ z_2 & 0 \end{pmatrix}, z_1, z_2 \in \mathbb{C} \right\}, \tag{10.189}$$

where $D_n(\mathbb{C})$ denotes the C*-algebra of diagonal $n \times n$ matrices. Take $\Omega = (1,0)$, with associated state

$$\omega(a) = \langle \Omega, a\Omega \rangle, \tag{10.190}$$

where $a \in M_2(\mathbb{C})$. It follows from §2.4 that the associated GNS-representation $\pi_\omega(A)$ is just (equivalent to) the defining representation of $M_2(\mathbb{C})$ on $H_\omega = \mathbb{C}^2$, in which the cyclic vector Ω_ω of the GNS-construction is Ω itself. Since $\sigma_3 \Omega = \Omega$, the state defined by (10.190) is $\mathbb{Z}_2$-invariant, and the unitary operator w in Proposition 10.14 is simply $w = \sigma_3$. Hence the decomposition (10.183) of $H = \mathbb{C}^2$ is simply $\mathbb{C}^2 = \mathbb{C} \oplus \mathbb{C}$, i.e.,

$$H_+ = \{(z,0), z \in \mathbb{C}\}; \tag{10.191}$$

$$H_- = \{(0,z), z \in \mathbb{C}\}. \tag{10.192}$$

Of course, we then have $H_\pm = A_\pm \Omega$. Identifying $H_\pm \cong \mathbb{C}$, this gives the one-dimensional representations $\pi_\pm(D_2(\mathbb{C}))$ as

$$\pi_\pm \begin{pmatrix} z_+ & 0 \\ 0 & z_- \end{pmatrix} = z_\pm, \tag{10.193}$$

which are trivially inequivalent. Hence by Theorem 10.15 the defining representation of $M_2(\mathbb{C})$ on $\mathbb{C}^2$ is irreducible, as it should be.

2. $A = D_2(\mathbb{C})$, with

$$\theta(\mathrm{diag}(z_+, z_-)) = \mathrm{diag}(z_-, z_+), \tag{10.194}$$

where we have denoted the matrix in (10.188) by $\mathrm{diag}(z_+, z_-)$. This time,

$$A_\pm = \{\mathrm{diag}(z, \pm z), z \in \mathbb{C}\}. \tag{10.195}$$

We once again define a $\mathbb{Z}_2$-invariant state ω by (10.190), but this time we take

$$\Omega = \frac{1}{\sqrt{2}} \begin{pmatrix} 1 \\ 1 \end{pmatrix}. \tag{10.196}$$

Hence

$$H_{\pm} = \{(z, \pm z), z \in \mathbb{C}\}.. \tag{10.197}$$

We may now identify each $A_{\pm}$ with $\mathbb{C}$ under the map $\mathrm{diag}(z, \pm z) \mapsto z$ from $A_{\pm}$ to $\mathbb{C}$. Similarly, we identify each each subspace $H_{\pm}$ with $\mathbb{C}$ under the map $H_{\pm} \to \mathbb{C}$ defined by $(z, \pm z) \mapsto z$. Under these identifications, we have two one-dimensional representations $\pi_{\pm}$ of the C*-algebra $\mathbb{C}$ on the Hilbert space $\mathbb{C}$, given by $\pi_{\pm}(z) = z$. Clearly, these are equivalent: they are even identical. Hence by Theorem 10.15 the defining representation of $D_2(\mathbb{C})$ on $\mathbb{C}^2$ is reducible, as it should be: the explicit decomposition of $\mathbb{C}^2$ in $D_2(\mathbb{C})$-invariant subspaces is just the one (10.191) - (10.192) of the previous example.

The first-numbered claim of Theorem 10.15 is relatively easy to prove from Theorem C.90. Suppose $\pi_{\pm}(A_+)$ are inequivalent and take $b \in \pi_{\omega}(A)'$: we want to show that $b = \lambda \cdot 1$ for some $\lambda \in \mathbb{C}$. Relative to $H = H_+ \oplus H_-$, we write

$$b = \begin{pmatrix} b_{++} & b_{+-} \\ b_{-+} & b_{--} \end{pmatrix}, \tag{10.198}$$

where the four operators in this matrix act as follows:

$$b_{++} : H_+ \to H_+, b_{+-} : H_- \to H_+, b_{-+} : H_+ \to H_-, b_{--} : H_- \to H_-. \tag{10.199}$$

Since $A_+ \subset A$, we also have $b \in \pi_{\omega}(A_+)'$. The condition $[b, a] = 0$ for each $a \in A_+$ is equivalent to the four conditions

$$[b_{++}, \pi_+(a)] = 0; \tag{10.200}$$
$$[b_{--}, \pi_-(a)] = 0; \tag{10.201}$$
$$\pi_+(a) b_{+-} = b_{+-} \pi_-(a); \tag{10.202}$$
$$\pi_-(a) b_{-+} = b_{-+} \pi_+(a). \tag{10.203}$$

We now use the fact (which we state without proof) that, as in group theory, the irreducibility and inequivalence of $\pi_{\pm}(A_+)$ implies that there can be no nonzero operator $c : H_+ \to H_-$ such that $c\pi_+(a) = \pi_-(a)c$ for all $a \in A_+$, and vice versa. Hence $b_{+-} = 0$ as well as $b_{-+} = 0$. In addition, the irreducibility of $\pi_{\pm}(A_+)$ implies that $b_{++} = \lambda_+ \cdot 1_{H_+}$ and $b_{--} = \lambda_- \cdot 1_{H_-}$. Finally, the property $[b, a] = 0$ for each $a \in A_-$ implies $\lambda_+ = \lambda_-$. Hence $b = \lambda \cdot 1$, and $\pi_{\omega}(A)$ is irreducible.

To prove the second-numbered claim of Theorem 10.15, let $\pi_+(A_+) \cong \pi_-(A_+)$, so by definition (of equivalence) there is a unitary operator $v : H_- \to H_+$ such that

$$v\pi_-(a) = \pi_+(a)v, \forall a \in A_+. \tag{10.204}$$

Extend v to an operator $w : H \to H$ by

$$w = \begin{pmatrix} 0 & v \\ v^* & 0 \end{pmatrix}. \tag{10.205}$$

It is easy to verify from (10.204) that $[w, \pi(a)] = 0$ for each $a \in A_+$. To check that the same is true for each $a \in A_-$, one needs the difficult analytical fact that w is a (weak) limit of operators of the kind $\pi(a_n)$, where $a_n \in A_-$, which also implies that $w^*\pi(a) \in \pi(A_+)''$. Since $\pi(A_+)''' = \pi(A_+)'$ and $w \in \pi(A_+)'$, we obtain $[w^*\pi(a), w] = 0$ for each $a \in A_-$. But for unitary operators w this is the same as $[w, \pi(a)] = 0$. So $w \in \pi(A)'$, and hence $\pi(A)$ is reducible by Theorem C.90. □

In determining the ground state(s) of the quantum Ising chain, we will apply Theorem 10.15 to the C*-algebra (10.87). This application relies on the representation theory of F. For the moment we leave the Hilbert space H general, equipped though with a conjugation $J : H \to H$. It turns out to be convenient to use the ***self-dual formulation of the CAR***, which treats c and c^* on an equal footing. Define

$$K = H \dot{\oplus} H, \tag{10.206}$$

whose elements are written as $h = (f, g)$ or $h = f \dot{+} g$, with inner product

$$\langle h_1, h_2 \rangle_K = \langle f_1, f_2 \rangle_H + \langle g_1, g_2 \rangle_H. \tag{10.207}$$

We then introduce a new operator in CAR(H), namely the ***field***

$$\Phi(h) = c^*(f) + c(Jg), \tag{10.208}$$

which is *linear* in $h = f \dot{+} g$, because the antilinearity of $c(f)$ in f is canceled by the antilinearity of J. This yields the anti-commutation relations

$$[\Phi^*(h_1), \Phi(h_2)]_+ = \langle h_1, h_2 \rangle_K, \tag{10.209}$$

but be aware that generally $[\Phi^*(h_1), \Phi^*(h_2)]_+$ and $[\Phi(h_1), \Phi(h_2)]_+$ do not vanish. Indeed, in terms of the antilinear operator $\Gamma : K \to K$, defined by

$$\Gamma = \begin{pmatrix} 0 & J \\ J & 0 \end{pmatrix} \tag{10.210}$$

we have the following expression for the adjoint $\Phi(h)^* \equiv \Phi^*(h)$:

$$\Phi^*(h) = \Phi(\Gamma h). \tag{10.211}$$

If we identify $f \in H$ with $f \dot{+} 0 \in K$, we may reconstruct c and c^* from Φ through

$$c^*(f) = \Phi(f); \tag{10.212}$$

$$c(f) = \Phi(\Gamma f). \tag{10.213}$$

Bogoliubov transformations now take an extremely elegant form. For any unitary operator S on K that satisfies $[S, \Gamma] = 0$, we define the transform Φ_S of Φ by

$$\Phi_S(h) = \Phi(Sh), \tag{10.214}$$

with associated creation- and annihilation operators (where $H \ni f \equiv f \dotplus 0$, as above)

$$c_S^*(f) = \Phi_S(f); \tag{10.215}$$
$$c_S(f) = \Phi_S^*(f). \tag{10.216}$$

To see the equivalence with the original formulation of the Bogoliubov transformation, note that for unitary S, the condition $[S, \Gamma] = 0$ is equivalent to the structure

$$S = \begin{pmatrix} u & vJ \\ Jv & JuJ \end{pmatrix}, \tag{10.217}$$

where $u : H \to H$ is linear, $v : H \to H$ is antilinear, and u and v satisfy (10.133) - (10.134). Moreover, from (10.137) - (10.138) we obtain

$$c_S(f) = \eta(f); \tag{10.218}$$
$$c_S^*(f) = \eta^*(f). \tag{10.219}$$

An interesting class of pure states on $\mathrm{CAR}(H)$ arises as follows.

Theorem 10.16. *There is a bijective correspondence between:*

- *Projections $e : K \to K$ that (apart form the properties $e^2 = e^* = e$) satisfy*

$$\Gamma e \Gamma = 1_K - e; \tag{10.220}$$

- *States ω_e on F that satisfy*

$$\omega_e(\Phi(h)^*\Phi(h)) = \langle h, eh \rangle \; \forall h \in K. \tag{10.221}$$

Such a state ω_e is automatically pure (so that the corresponding GNS*-representation π_e is irreducible), and is explicitly given by*

$$\omega_e(\Phi(h_1) \cdots \Phi(h_{2n+1})) = 0; \tag{10.222}$$
$$\omega_e(\Phi(h_1) \cdots \Phi(h_{2n})) = \sum_{p \in \mathfrak{S}_{2n}}{}' \operatorname{sgn}(p) \prod_{j=1}^{n} \langle e h_{\operatorname{sgn}(2j)}, \Gamma h_{\operatorname{sgn}(2j-1)} \rangle, \tag{10.223}$$

the sum Σ' is over all permutations p of $1, \ldots, 2n$ such that

$$p(2j-1) < p(2j); \tag{10.224}$$
$$p(1) < p(3) < \cdots < p(2n-1). \tag{10.225}$$

We omit the proof. Note that (10.221) is a special case of (10.223), because of (10.211). States like ω_e, which are determined by their two-point functions, are called ***quasi-free***; the ground state ω_0 on $\mathrm{CAR}(\mathbb{C}^N)$ constructed in the previous section is an example (one also has mixed quasi-free states, e.g. certain KMS states).

As a warm-up, we reconstruct the ground state of the free fermionic Hamiltonian on F using the above formalism. That is, we assume that h_N in (10.111) reads

$$h_N = \sum_{x=-N/2}^{N/2-1} \varepsilon_x c_x^* c_x, \tag{10.226}$$

initially defining dynamics on $F_N = \mathrm{CAR}(\mathbb{C}^N)$. In that case, the projection e_0 onto the second copy of $H = \mathbb{C}^N$ in K, i.e.

$$e_0 = \begin{pmatrix} 0 & 0 \\ 0 & 1 \end{pmatrix}, \tag{10.227}$$

reproduces the ground state $\omega_0(a) = \langle 0|a|0\rangle$, where $|0\rangle$ is the vector $1 \in \mathbb{C}$ in $F_-(H)$, such that $c(f)|0\rangle = 0$ for all $f \in H$. This also works for $N = \infty$, i.e., we construct dynamics on $\mathrm{CAR}(\ell^2(\mathbb{Z}))$ from the local Hamiltonians (10.226) as indicated at the beginning of this section, and use the same formula for e_0, this time with $H = \ell^2(\mathbb{Z})$.

In the more general case (10.111), we replace e_0 in (10.227) by

$$e_0^{(S)} = S e_0 S^{-1}, \tag{10.228}$$

where S is given by (10.217), in which for $N < \infty$ the operators u and v were constructed in (10.131) - (10.132). This time, the associated state $\omega_{e_0^{(S)}} \equiv \omega_S$ is the state called ω_0 in Theorem 10.13. As explained at the beginning of this section, this procedure even works for $N = \infty$ and hence $H = \ell^2(\mathbb{Z})$.

Having understood fermionic models with quadratic Hamiltonians, what remains to be done now is to reformulate the original quantum Ising chain, defined in terms of the local spin matrices $\sigma_i(x)$, in terms of the fermionic variables c_x and c_x^*. For finite N this was done through the Jordan–Wigner transformation (10.102) - (10.103). This time we need a similar isomorphism between A and F, where

$$A = \otimes_{j\in\mathbb{Z}} M_2(\mathbb{C}); \tag{10.229}$$
$$F = \mathrm{CAR}(\ell^2(\mathbb{Z})), \tag{10.230}$$

and hence we would need to start the sums in the right-hand side of (10.102) - (10.103) at $j = -\infty$. At first sight this appears to be impossible, though, because operators like $\exp(\pi i \sum_{y=-\infty}^{x-1} \sigma_+(y)\sigma_-(y))$ do not lie in A (whose elements have infinite tails of 2×2 unit matrices). Fortunately, this problem can be solved by adding a formal operator T to A, which plays the role of the "tail"

$$\text{“}T = e^{\pi i \sum_{y=-\infty}^{0} \sigma_+(y)\sigma_-(y))}\text{”}. \tag{10.231}$$

This formal expression (to be used only heuristically) suggests the relations:

$$T^2 = 1; \tag{10.232}$$
$$T^* = T; \tag{10.233}$$
$$TaT = \theta_-(a), \tag{10.234}$$

where $\theta_- : A \to A$ is a $\mathbb{Z}_2$-action defined by (algebraic) extension of

$$\theta_-(\sigma_\pm(y)) = -\sigma_\pm(y) \ (y \leq 0); \tag{10.235}$$
$$\theta_-(\sigma_\pm(y)) = \sigma_\pm(y) \ (y > 0); \tag{10.236}$$
$$\theta_-(\sigma_3(y)) = \sigma_3(y) \ (y \in \mathbb{Z}); \tag{10.237}$$
$$\theta_-(\sigma_0(y) = \sigma_0(y) \ (y \in \mathbb{Z}), \tag{10.238}$$

where $\sigma_0 = 1_2$. Formally, define an algebra extension

$$\hat{A} = A \oplus A \cdot T, \tag{10.239}$$

with elements of the type $a + bT$, $a, b \in A$, and algebraic relations given by (10.232) - (10.233). That is, we have

$$(a + bT)^* = a^* + \theta_-(b^*)T; \tag{10.240}$$
$$(a + bT) \cdot (a' + b'T) = aa' + b\theta_-(b') + (ab' + b\theta_-(a'))T. \tag{10.241}$$

Within $\hat{A}$, the correct version of (10.102) - (10.103) may now be written down as

$$c_x^\pm = Te^{\mp\pi i \sum_{y=x}^{0} \sigma_+(y)\sigma_-(y)} \sigma_x^\pm \ (x < 1); \tag{10.242}$$
$$c_x^\pm = T\sigma_1^\pm; \tag{10.243}$$
$$c_x^\pm = Te^{\mp\pi i \sum_{y=1}^{x-1} \sigma_+(y)\sigma_-(y)} \sigma_x^\pm \ (x > 1), \tag{10.244}$$

with formal inverse transformation given by

$$\sigma_\pm(x) = Te^{\pm\pi i \sum_{y=x}^{0} c_y^+ c_y^-} c_x^\pm \ (x < 1); \tag{10.245}$$
$$\sigma_\pm(x) = Tc_1^\pm; \tag{10.246}$$
$$\sigma_\pm(x) = Te^{\pm\pi i \sum_{y=1}^{x-1} \sigma_+(y)\sigma_-(y)} \sigma_\pm(x) \ (x > 1), \tag{10.247}$$

where this time we regard T as an element of the extended fermionic algebra

$$\hat{F} = F \oplus F \cdot T, \tag{10.248}$$

satisfying the same rules (10.232) - (10.234), but now in terms of a "fermionic" $\mathbb{Z}_2$-action $\theta_y : F \to F$ given by extending the following action on elementary operators:

$$\theta_-(c_y^\pm) = -c_y^\pm \ (y \leq 0); \tag{10.249}$$
$$\theta_-(c_y^\pm) = c_y^\pm \ (y > 0). \tag{10.250}$$
$$\tag{10.251}$$

Because of T, the Jordan–Wigner transformation does not give an isomorphism $A \cong F$, but it does give an isomorphism $\hat{A} \cong \hat{F}$. More importantly, if, having already defined the $\mathbb{Z}_2$-action θ on F by (10.177), we define a similar $\mathbb{Z}_2$-action on A by

$$\theta(\sigma_{\pm}(y)) = -\sigma_{\pm}(y) \;\; (y \in \mathbb{Z}); \tag{10.252}$$

$$\theta(\sigma_3(y)) = \sigma_3(y) \;\; (y \in \mathbb{Z}); \tag{10.253}$$

$$\theta(\sigma_0(y)) = \sigma_0(y) \;\; (y \in \mathbb{Z}), \tag{10.254}$$

and decompose $A = A_+ \oplus A_-$ and $F = F_+ \oplus F_-$, according to this action, cf. (10.178), we have isomorphisms

$$A_+ \cong F_+; \tag{10.255}$$

$$A_- \cong F_- T; \tag{10.256}$$

$$A \cong F_+ \oplus F_- T. \tag{10.257}$$

For given dynamics (10.111), suppose ω_0^A is a $\mathbb{Z}_2$-*invariant* ground state on A. Then ω_0^A also defines a $\mathbb{Z}_2$-invariant ground state ω_0^F on F by (10.255) and $\omega_0^F(f) = 0$ for all $f \in F_-$. Conversely, a $\mathbb{Z}_2$-invariant ground state ω_0^F on F defines a state ω_0^A on A by (10.255) and $\omega_0^A(a) = 0$ for all $a \in A_-$. But F has a unique ground state, so:

- Either ω_0 is pure on A, in which case it is the unique ground state on A;
- Or ω_0 is mixed on A, in which case $\omega_0 = \frac{1}{2}(\omega_0^+ + \omega_0^-)$, where $\omega_0^{\pm}$ are pure but transform under the above $\mathbb{Z}_2$-action θ as $\omega_0^{\pm} \circ \theta = \omega_0^{\mp}$.

Theorem 10.15 gives a representation-theoretical criterion deciding between these possibilities, but to apply it we need some information on the restriction of $\mathbb{Z}_2$-invariant quasi-free pure states on F to its even part F_+. The abstract setting involves a $\mathbb{Z}_2$-action W on K that commutes with Γ (so that W is unitary, $W^2 = 1$, and $[\Gamma, W] = 0$), which induces a $\mathbb{Z}_2$-action θ on F by linear and algebraic extension of $\theta(\Phi(h)) = \Phi(Wh)$. A quasi-free state ω_e, defined according to Theorem 10.16 by a projection $e : K \to K$ that satisfies (10.220), is then $\mathbb{Z}_2$-invariant iff $[W, e] = 0$.

In our case, this simplifies to $\theta(\Phi(h)) = -\Phi(h)$, so that $W = -1$, and every projection commutes with W. In any case, with considerable effort one can prove:

Lemma 10.17. *Given some $\mathbb{Z}_2$-action W on K, as well as a projection $e : K \to K$ satisfying* (10.220), *such that $[W, \Gamma] = [W, e] = 0$:*

1. *The quasi-free state ω_e of Theorem 10.16 is $\mathbb{Z}_2$-invariant (i.e., $\omega_e \circ \theta = \omega_e$);*
2. *The corresponding* GNS*-representation space $H_e \equiv H_{\omega_e}$ for $F = F_+ \oplus F_-$ decomposes as $H_e = H_e^+ \oplus H_e^-$, with $H_e^{\pm} = \overline{F_{\pm}\Omega_e}$. Each subspace $H_e^{\pm}$ is stable under $\pi_e(F_+)$, and the restriction $\pi_e^{\pm}$ of $\pi(F_+)$ to $H_e^{\pm}$ is irreducible.*

Theorem 10.15 then leads to a lemma, which also summarizes the discussion so far.

Lemma 10.18. *1. For given $\mathbb{Z}_2$-invariant dynamics, let ω_0^F be the (unique, $\mathbb{Z}_2$-invariant) ground state on $F = F_+ \oplus F_-$. Under $F_+ \subset F$ the associated* GNS*-representation space H_0 decomposes as $H_0 = H_0^+ \oplus H_0^-$, with $H_0^{\pm} = \overline{F_{\pm}\Omega_0}$, and we denote the restriction of $\pi_0(F_+)$ to $H_0^{\pm}$ by $\pi_0^{\pm}$. Then $\pi_0^{\pm}(F_+)$ are irreducible.*

2. *Regard ω_0^F also as a state ω_0^T on $F_+ \oplus F_-T$ by putting $\omega_0^T(a) = 0$ for all $a \in F_-T$, and similarly as a state ω_0^A on A by invoking* (10.255) *and putting $\omega_0^A(a) = 0$ for all $a \in A_-$. Let $H_0^T = H_+^T \oplus H_-^T$ be the* GNS*-representation space of $F_+ \oplus F_-T$ defined by ω_0^T, where $H_+^T = \overline{F_+\Omega}$ and $H_-^T = \overline{F_-T\Omega}$. Here H_+^T and H_-^T are stable under F_+; we denote the restriction of F_+ to $H_\pm^T$ by $\pi_\pm^T$, so that $\pi_+^T \cong \pi_0^+$.*
    a. *Then ω_0^A is a ground state on A. Any $\mathbb{Z}_2$-invariant ground state on A arises in this way (via F), so that there is a unique $\mathbb{Z}_2$-invariant ground state on A.*
    b. *The state ω_0^A is pure on A iff the irreducible representations $\pi_+^T(F_+)$ (or $\pi_0^+(F_+)$) and $\pi_-^T(F_+)$ are inequivalent.*

It turns out to be difficult to directly check the (in)equivalence of $\pi_\pm^T(F_+)$. Fortunately, we can circumvent this problem by passing to yet another (irreducible) representation of F_+. We first enlarge F to a new algebra

$$\hat{F} = F \oplus FT = F_+ \oplus F_- \oplus F_+T \oplus F_-T, \tag{10.258}$$

and extend the state ω_0^F on F to a state $\hat{\omega}_0$ on $\hat{F}$ by putting $\hat{\omega}_0(FT) = 0$, so that $\hat{\omega}_0$ is nonzero only on $F_+ \subset \hat{F}$. Let $\hat{\pi}_0$ be the associated GNS-representation of $\hat{F}$ on the Hilbert space $\hat{H}_0 = \overline{\hat{F}\hat{\Omega}}$. Under $\hat{\pi}(F_+)$ this space decomposes as

$$\hat{H}_0 = \overline{F_+\hat{\Omega}_0} \oplus \overline{F_-\hat{\Omega}_0} \oplus \overline{F_+T\hat{\Omega}_0} \oplus \overline{F_-T\hat{\Omega}_0}, \tag{10.259}$$

with corresponding restrictions $\hat{\pi}_\pm(F_+)$ and $\hat{\pi}_\pm^T(F_+)$; more precisely, $\hat{\pi}_\pm$ is the restriction of $\hat{\pi}(F_+)$ to $\overline{F_\pm\hat{\Omega}_0}$, whilst $\hat{\pi}_\pm^T$ is is the restriction of $\hat{\pi}(F_+)$ to $\overline{F_\pm T\hat{\Omega}_0}$. Clearly, $\hat{\pi}_\pm(F_+)$ is the same as $\pi_0^\pm(F_+)$, and $\hat{\pi}_-^T(F_+)$ is just our earlier $\pi_-^T(F_+)$, but $\hat{\pi}_+^T(F_+)$ is new. To understand the latter, we rewrite (10.259) as

$$\hat{H}_0 = H_0 \oplus \hat{H}_0^T; \tag{10.260}$$

$$H_0 = \overline{F_+\hat{\Omega}_0} \oplus \overline{F_-\hat{\Omega}_0} \cong \overline{F_+\Omega_0} \oplus \overline{F_-\Omega_0}; \tag{10.261}$$

$$\hat{H}_0^T = \overline{F_+T\hat{\Omega}_0} \oplus \overline{F_-T\hat{\Omega}_0}, \tag{10.262}$$

the point being that $\hat{\pi}(F)$ evidently restricts to both H_0 and $\hat{H}_0^T$. We know the action of $\hat{\pi}(F)$ on H_0 quite well: it is the representation induced by the ground state ω_0. As to $\hat{H}_0^T$, we define a state $\hat{\omega}_0^T$ on F by

$$\hat{\omega}_0^T(a) = \langle \hat{\pi}(T)\hat{\Omega}_0, \hat{\pi}(a)\hat{\pi}(T)\hat{\Omega}_0 \rangle_{\hat{H}_0} = \langle \hat{\Omega}_0, \hat{\pi}(\theta_-(a))\hat{\Omega}_0 \rangle_{\hat{H}_0}, \tag{10.263}$$

where the second equality follows from (10.234). Comparing H_0 and $\hat{H}_0$, for all $b \in F$ (and hence especially for $b = \theta_-(a)$) we simply have

$$\langle \hat{\Omega}_0, \hat{\pi}(b)\hat{\Omega}_0 \rangle_{\hat{H}_0} = \hat{\omega}_0(b) = \omega_0^F(b), \tag{10.264}$$

so that $\hat{\omega}_0^T = \omega_0^F \circ \theta_- \equiv \theta_-^* \omega_0^F$. Decomposing the GNS-representation space $H_{\theta_-^*\omega_0^F}$ of $\pi_{\theta_-^*\omega_0^F}(F)$ as $H_{\theta_-^*\omega_0^F} = H^+_{\theta_-^*\omega_0^F} \oplus H^-_{\theta_-^*\omega_0^F}$, it follows that $\hat{\pi}_+^T(F_+)$ is the restriction of $\pi_{\theta_-^*\omega_0^F}(F_+)$ to $H^+_{\theta_-^*\omega_0^F}$. Therefore, the representation $\hat{\pi}(F)$ restricted to $\hat{H}_0^T$ is the GNS-representation $\pi_{\theta_-^*\omega_0^F}(F)$, so that in turn $\hat{\pi}_+^T(F_+)$ is $\pi_{\theta_-^*\omega_0^F}(F_+)$, restricted to $H^+_{\theta_-^*\omega_0}$. Hence, further to (10.260) - (10.262), we obtain the decomposition

$$\hat{\pi}(F) \cong \pi_{\omega_0^F}(F) \oplus \pi_{\theta_-^*\omega_0^F}(F). \tag{10.265}$$

The point is that for the quantum Ising chain Hamiltonian (10.110), we have:

Lemma 10.19. *1. For each $\lambda \neq \pm 1$, we have $\pi_{\omega_0^F}(F) \cong \pi_{\theta_-^*\omega_0^F}(F)$.*

2. If this holds, then the representations $\pi_0^+(F_+) \equiv \pi^+_{\omega_0^F}(F_+)$ and $\pi_-^T(F_+)$ are inequivalent iff the representations $\pi^+_{\omega_0^F}(F_+)$ and $\pi^+_{\theta_-^*\omega_0^F}(F_+)$ are equivalent.

3. For each $\lambda \neq \pm 1$, the ground state ω_0^A is pure on A iff the representations $\pi_{\omega_0^F}(F_+)$ and $\pi_{\theta_-^*\omega_0^F}(F_+)$ are equivalent.

The first claim follows from Theorem 10.20 below. The third follows from Lemma 10.18 and the previous claims. The second claim is proved by repeatedly applying Theorem 10.15 to $\hat{\pi}(\hat{F})$. Given this lemma, the real issue now lies in comparing $\pi_{\omega_0^F}$ and $\pi_{\theta_-^*\omega_0^F}$, both as representations of F (as they are defined) and as representations of $F_+ \subset F$. This can be settled in great generality by first looking at Theorem 10.16, and thence, recalling the positive-energy projection (10.228), realizing that

$$\pi_{\omega_0^F} = \pi_{e_0^{(S)}}; \tag{10.266}$$

$$\pi_{\theta_-^*\omega_0^F} = \pi_{W_- e_0^{(S)} W_-}. \tag{10.267}$$

Here $W_- : K \to K$ is the $\mathbb{Z}_2$-action on K defining the $\mathbb{Z}_2$-action θ_- on F as explained above Lemma 10.17; specifically, W_- is the direct sum of two copies of $w_- : \ell^2(\mathbb{Z}) \to \ell^2(\mathbb{Z})$, defined by $w_-(f_j) = f_j$ $(j > 0)$ and $w_-(f_j) = -f_j$ $(j \leq 0)$.

Subsequently, without proof we invoke a basic result on the CAR-algebra:

Theorem 10.20. *Let e and e' be projections on K that satisfy* (10.220). *Then:*

1. $\pi_e(F) \cong \pi_{e'}(F)$ iff $e - e' \in B_2(K)$;

2. $\pi_e^+(F_+) \cong \pi_{e'}^+(F_+)$ iff $e - e' \in B_2(K)$ and $\dim(eK \cap (1-e')K)$ is even.

If the first condition is satisfied, the dimension in the second part is finite, so that one may indeed say it is even or odd. From Lemmas 10.18 and 10.19 and Theorem 10.20, we finally obtain the phase structure of the infinite quantum Ising chain:

Theorem 10.21. *The unique $\mathbb{Z}_2$-invariant ground state ω_0 of the Hamiltonian* (10.110) *is pure (and hence forms the unique ground state) iff both of the following hold:*

$$e_0^{(S)} - W_- e_0^{(S)} W_- \in B_2(K); \tag{10.268}$$

$$\dim(e_0^{(S)} K \cap (1 - W_- e_0^{(S)} W_-)K) \text{ is even.} \tag{10.269}$$

This is true for all λ with $|\lambda| \geq 1$. If $|\lambda| < 1$, then $\omega_0 = \frac{1}{2}(\omega_0^+ + \omega_0^-)$, where $\omega_0^\pm$ are pure and transform under the $\mathbb{Z}_2$-action θ as $\omega_0^\pm \circ \theta = \omega_0^\mp$.

10.8 Spontaneous symmetry breaking in mean-field theories

We are now going to study SSB in so-called ***mean-field theories***: these are quantum spin systems with Hamiltonians like the ***Curie–Weiss-model*** for ferromagnetism:

$$h_{\Lambda}^{\mathrm{CW}} = -\frac{J}{2|\Lambda|}\sum_{x,y\in\Lambda}\sigma_3(x)\sigma_3(y) - B\sum_{x\in\Lambda}\sigma_1(x), \tag{10.270}$$

where $J > 0$ scales the spin-spin coupling, and B is an external magnetic field. Similar to the quantum Ising model, (10.270) has a $\mathbb{Z}_2$-symmetry $(\sigma_1,\sigma_2,\sigma_3)\mapsto(\sigma_1,-\sigma_2,-\sigma_3)$, which at each site x is implemented by $u(x)=\sigma_1(x)$. This model differs from its short-range counterpart (9.42), i.e, the quantum Ising model, or the Heisenberg model (9.44), in that every spin now interacts with every other spin. It falls into the class of ***homogeneous mean-field theories***, which are defined by a single-site Hilbert space $H_x = H = \mathbb{C}^n$ and local Hamiltonians of the type

$$h_{\Lambda} = |\Lambda|\tilde{h}(T_0^{(\Lambda)}, T_1^{(\Lambda)}, \ldots, T_{n^2-1}^{(\Lambda)}). \tag{10.271}$$

Here $T_0 = 1_n$, and the matrices $(T_i)_{i=1}^{n^2-1}$ in $M_n(\mathbb{C})$ form a basis of the real vector space of traceless self-adjoint $n\times n$ matrices; the latter may be identified with i times the Lie algebra $\mathfrak{su}(n)$ of $SU(n)$, so that $(T_0, T_1, \ldots, T_{n^2-1})$ is a basis of i times the Lie algebra $\mathfrak{u}(n)$ of the unitary group $U(n)$ on $\mathbb{C}^n$. In those terms, we define

$$T_i^{(\Lambda)} = \frac{1}{|\Lambda|}\sum_{x\in\Lambda} T_i(x), \tag{10.272}$$

Finally, $\tilde{h}$ is a polynomial (which is sensitive to operator ordering). For example, to cast (10.270) (with $J=1$) in the form (10.271), take $n=2$, $T_i = \frac{1}{2}\sigma_i$ $(=1,2,3)$, and

$$\tilde{h}^{\mathrm{CW}}(T_1,T_2,T_3) = -2(T_3^2 + BT_1). \tag{10.273}$$

The assumptions of Theorem 9.15 do not hold now, and indeed the local dynamics (9.40) fails to converge to global dynamics on the quasi-local C*-algebra A defined by (8.130). Fortunately, it does converge to a global dynamics on the C*-algebra $C(S(B))$, where $B = M_n(\mathbb{C})$ is the single-site algebra. In order to describe the limiting dynamics of (homogeneous) mean-field models as $\Lambda \nearrow \mathbb{Z}^d$, we equip the state space $S(B)$ with the Poisson structure (8.52), which we now elucidate.

For unital C*-algebras B, we may regard $S(B)$ as a w^*-compact subspace of either the complex vector space B^* or the real vector space B^*_{sa}; in the latter case we regard states as linear maps $\omega : B^*_{\mathrm{sa}} \to \mathbb{R}$ that satisfy $\omega(1_B)=1$ and $\omega(a^2)\geq 0$ for each $a\in B_{\mathrm{sa}}$. If $B = M_n(\mathbb{C})$, which is all we need, we may furthermore identify B^*_{sa} with $i\mathfrak{u}(n)^*$, and since the value of each state $\omega\in S(M_n(\mathbb{C}))$ is fixed on $T_0 = 1_B \in i\mathfrak{u}(n)$, it follows that $S(M_n(\mathbb{C}))$ is a compact convex subset of $i\mathfrak{su}(n)^*$. In that case, the Poisson bracket (8.52) on $S(M_n(\mathbb{C}))$ is none other than the restriction of (minus) the canonical Lie-Poisson bracket on $\mathfrak{su}(n)^* \cong i\mathfrak{su}(n)^*$ to $S(M_n(\mathbb{C}))$, cf. (3.98) - (3.99).

For example, for $n = 2$ we have $S(M_2(\mathbb{C})) \cong B^3 \subset \mathbb{R}^3$ by Proposition 2.9, i.e.,

$$\omega_{(x,y,z)}(a) = \mathrm{Tr}(\rho(x,y,x)a) \;\; ((x,y,z) \in B^3, a \in M_2(\mathbb{C})); \tag{10.274}$$

$$\rho(x,y,z) = \tfrac{1}{2}\begin{pmatrix} 1+z & x-iy \\ x+iy & 1-z \end{pmatrix}. \tag{10.275}$$

We also have $\mathfrak{su}(2)^* \cong \mathbb{R}^3$ upon the choice of the basis $(T_i = \frac{1}{2}\sigma_i)$, $i = 1,2,3$, of $i\mathfrak{su}(2)$, which means that $\theta_{(x,y,z)} \in i\mathfrak{su}(2)^*$ maps (T_1, T_2, T_3) to (x,y,z) (where this time $(x,y,z) \in \mathbb{R}^3$), cf. §5.8). If we now regard the matrices T_i as functions $\hat{T}_i$ on B^3 by $\hat{T}_i(\omega) = \omega(T_i)$, we find that the corresponding functions on B^3 are given by

$$\hat{T}_1(x,y,z) = \tfrac{1}{2}x, \;\; \hat{T}_2(x,y,z) = \tfrac{1}{2}y, \;\; \hat{T}_3(x,y,z) = \tfrac{1}{2}z. \tag{10.276}$$

The corresponding Poisson brackets (8.52) are $\{T_1, T_2\} = -2T_3$ etc., i.e., $\{x,y\} = -2z$ etc.; this is -2 times the bracket defined in (3.43) or (3.97) - (3.98). This factor 2 could have been avoided by moving to the three-ball with radius $r = 1/2$ instead of $r = 1$, whose boundary is the coadjoint orbit $\mathcal{O}_{1/2}$ naturally associated to spin-$\frac{1}{2}$.

We now return to our continuous bundle of C*-algebras $A^{(c)}$ of Theorem 8.4, of course in the slightly adapted form appropriate to quantum spin systems, see §8.6. In particular, we recall that $A_0^{(c)} = C(S(B))$ and $A_{1/N}^{(c)} = B(H_{\Lambda_N})$, cf. (8.157) - (8.158), and hence we see the limit $N \to \infty$ as a specific way of taking the limit $\Lambda \nearrow \mathbb{Z}^d$ along the hypercubes Λ_N. Symmetric and quasi-symmetric sequences $(a_{1/N})_{N\in\mathbb{N}}$ are defined as explained after (8.161). The following observation is fundamental.

Theorem 10.22. *Let $B = M_n(\mathbb{C})$. If $(a_{1/N})_{N\in\mathbb{N}}$ and $(b_{1/N})_{N\in\mathbb{N}}$ are symmetric sequences with limits a_0 and b_0 as defined by* (8.46), *respectively (so that $(a_{1/N})_{N\in\dot{\mathbb{N}}}$ and $(b_{1/N})_{N\in\dot{\mathbb{N}}}$ are continuous sections of the continuous bundle $A^{(c)}$), then the sequence*

$$(\{a_0, b_0\}, i[a_1, b_1], \ldots, i|\Lambda_N|[a_{1/N}, b_{1/N}], \cdots) \tag{10.277}$$

defines a continuous section of $A^{(c)}$. In particular, for each $\omega \in S(B)$ we have

$$i \lim_{N\to\infty} \omega^{|\Lambda_N|}(|\Lambda_N|[a_{1/N}, b_{1/N}]) = \{a_0, b_0\}(\omega). \tag{10.278}$$

Proof. The proof is a straightforward combinatorial exercise, and we just mention the simplest case where $d = 1$ and $a_{1/N} = S_{1,N}(a_1)$ and $b_{1/N} = S_{1,N}(b_1)$, where $a_1 \in B$ and $b_1 \in B$, cf. (8.39). Then $a_0 = \hat{a}_1$, $b_0 = \hat{b}_1$, and similarly to (8.45) we find

$$[S_{1,N}(a_1), S_{1,N}(b_1)] = \frac{1}{N} S_{1,N}([a_1, b_1]), \tag{10.279}$$

Using (8.52), we find that (10.277) is equal to $(i\widehat{[a_1, b_1]}, \ldots, S_{1,N}([a_1, b_1]), \ldots)$. Since $\omega^N(S_{1,N}([a_1, b_1])) = \omega([a_1, b_1])$, the left-hand side of (10.278) is therefore equal to $i\omega([a_1, b_1])$, which by (8.52) equals the right-hand side. □

In other words, although the sequence of commutators $[a_{1/N}, b_{1/N}]$ converges to zero (which is why $A_0^{(c)}$ has to be commutative!), the rescaled commutators $iN[a_{1/N}, b_{1/N}]$ converge to the macroscopic observable $\{a_0, b_0\} \in C(S(B))$. This reconfirms the analogy between the limit $N \to \infty$ and the limit $\hbar \to 0$ of Chapter 7, see especially Definitions 7.1 and 8.2. With $B = M_n(\mathbb{C})$, Theorem 10.22 implies the central result about the macroscopic (and hence classical!) dynamics of mean-field theories:

Corollary 10.23. *Let $(h_{1/N})_{N\in\dot{\mathbb{N}}}$ be a continuous section of $A^{(c)}$ defined by a symmetric sequence, and let $(a_{1/N})_{N\in\dot{\mathbb{N}}}$ be an arbitrary continuous section of $A^{(c)}$ (i.e. a quasi-symmetric sequence). Then, writing $h_{1/N} = h_{\Lambda_N}$ for clarity, the sequence*

$$\left(a_0(t), e^{ih_{\Lambda_1}t}a_1e^{-ih_{\Lambda_1}t}, \cdots e^{ih_{\Lambda_N}t}a_{1/N}e^{-ih_{\Lambda_N}t}, \cdots\right), \tag{10.280}$$

where $a_0(t)$ is the solution of the equations of motion on $S(M_n(\mathbb{C}))$ with classical Hamiltonian h_0 and Poisson bracket (8.52)*, defines a continuous section of $A^{(c)}$.*

In other words, the Heisenberg dynamics on $A_{\Lambda_N} = B(H_{\Lambda_N})$ defined by the quantum Hamiltonians h_{Λ_N} converges to the classical dynamics on the Poisson manifold $S(M_n(\mathbb{C}))$ that is generated by their classical limit, viz. the Hamiltonian h_0.

For example, since the operators $T_i^{(\Lambda)}$ form symmetric sequences, so do Hamiltonians of the type (10.271). The limit $h_0 \in C(S(M_n(\mathbb{C})))$ of the family (h_Λ) in (10.271) is simply obtained by replacing the operators $T_i^{(\Lambda)}$ in the function $\tilde{h}$ by the functions $\hat{T}_i$ on $S(M_n(\mathbb{C}))$. Equivalently, one may replace the $T_i^{(\Lambda)}$ by the canonical coordinates (θ_i) of $i\mathfrak{su}(n)^*$ dual to the basis $(T_1, \ldots, T_{n^2-1})$ of $i\mathfrak{su}(n)^*$, i.e., $\theta_i(T_j) = \delta_{ij}$, and restricting the ensuing function on $i\mathfrak{su}(n)^*$ to $S(M_n(\mathbb{C})) \subset i\mathfrak{su}(n)^*$.

Using (10.276), for the Curie–Weiss model (10.270) with $J = 1$ this gives

$$h_0^{\mathrm{CW}}(x,y,z) = -\tfrac{1}{2}z^2 - Bx. \tag{10.281}$$

The ground states of this Hamiltonian are simply its minima, viz.

$$\mathbf{x}_{\pm} = (B, 0, \pm\sqrt{1-B^2}) \;\; (0 \leq B < 1); \tag{10.282}$$
$$\mathbf{x} = (1,0,0)) \;\; (B \geq 1), \tag{10.283}$$

all of which lie on the boundary S^2 of B^3. Note that the points $\mathbf{x}_\pm$ coalesce as $B \to 1$, where they form a saddle point. Modulo our use of radius $r = 1$ instead of $r = 1/2$, this result coincides with (10.81) for classical limit of the quantum Ising model.

We now turn to symmetry and its possible breakdown. Suppose there is some subgroup of $U(n)$, typically the image of a unitary representation $g \mapsto u_g$ of a compact group G on $\mathbb{C}^n$, under which $\tilde{h}(T_0, T_1, \ldots, T_{n^2-1})$ in (10.271) satisfies

$$\tilde{h}(T_0, u_gT_1u_g^*, \ldots, u_gT_{n^2-1}u_g^*) = \tilde{h}(T_0, T_1, \ldots, T_{n^2-1}) \;\; (g \in G). \tag{10.284}$$

For example, in the Curie–Weiss model one has $G = \mathbb{Z}_2$, whose nontrivial element is represented by σ_1. For (10.271) itself this implies $u^{(N)}h_N(u^{(N)})^* = h_N$, cf. (10.69).

Hence also in homogeneous mean-field models we obtain the structure (10.57), (10.58), and (10.59) familiar from the case of short-range forces. For the limit theory this implies that the classical Hamiltonian h_0 on $S(M_n(\mathbb{C}))$ is invariant under the coadjoint action of $G \subset U(n)$ on $i\mathfrak{su}(\mathfrak{n})^*$, restricted to $S(M_n(\mathbb{C})) \subset i\mathfrak{su}(\mathfrak{n})^*$: in the Curie–Weiss model this "classical shadow" of the $\mathbb{Z}_2$ symmetry of the quantum theory is simply the map $(x,y,z) \mapsto (x,-y,-z)$ on B^3.

In the regime $0 < B < 1$, the degenerate ground states of this model break this symmetry. In contrast, it can be shown from the Perron–Frobenius Theorem (which applies since both σ_3 and σ_1 are real matrices) that for $B > 0$ each quantum-mechanical Hamiltonian (10.270) has a unique ground state $\psi_N^{(0)}$. Being unique, this vector must share the invariance of h_N under the permutation group $\mathfrak{S}_N$, so that

$$\psi_N^{(0)} = \sum_{n_+=0}^{N} c(n_+/N)|n_+, n_-\rangle, \tag{10.285}$$

where $|n_+, n_-\rangle$ is the totally symmetrized unit vector in $\otimes^N \mathbb{C}^2$ with n_+ spins up and $n_- = N - n_+$ spins down, and $c : \{0, 1/N, 2/N, \ldots, (N-1)/N, 1\} \to [0,1]$ is some function such that $\sum_{n_+} c(n_+/N)^2 = 1$ (we may assume $c \geq 0$ by the Perron–Frobenius Theorem). The asymptotic behaviour of c as $N \to \infty$ has been studied, and as expected, c to converges pointwise to $c(0) = c(1) = \sqrt{1/2}$ and $c(x) = 0$, and zero elsewhere (at $B = 0$ one of course has either $c(0) = 1$ or $c(1) = 1$ for all N).

Thus we encounter a familiar headache: the "higher-level" theory $C(S(M_n(\mathbb{C})))$ at $N = \infty$ breaks the $\mathbb{Z}_2$ symmetry, whereas the "lower-level" quantum theories $B(H_{\Lambda_N})$ $(N < \infty)$ do not, although the former should be a limiting case of the latter. Indeed, the situation for the Curie–Weiss model in the regime $0 < B < 1$ is exactly analogous to the double-well potential as well as to the quantum Ising model in the same regime: if the two degenerate ground states $\mathbf{x}_\pm \in B^3$ of h_0^{CW} are reinterpreted as Dirac measures $\delta_\pm$ on B^3, which in turn are seen as (pure) states $\omega_\pm$ on the classical algebra of observables $C(S(M_2(\mathbb{C})))$, then (10.74) holds, *mutatis mutandis*.

The resolution of this problem through the restoration of Butterfield's Principle should also be the same as for the previous two cases: there is a first excited state $\psi_N^{(1)}$ such that as $N \to \infty$, the energy difference with the ground state approaches zero and one has approximate symmetry breaking as in (10.75)). Alas, for the Curie–Weiss model so far only numerical evidence is available supporting this scenario.

Equilibrium states of homogeneous mean-field models at any inverse temperature $0 < \beta < \infty$ exist, despite the fact that in such models time-evolution α_t on the infinite system A (and hence the KMS condition characterizing equilibrium states) is ill-defined (unless one passes to certain representations of A, which would be question-begging). Instead, one invokes the quasi-local C*-algebra A, cf. (8.130), and *in lieu* of KMS states looks for limit points $\hat{\omega}^\beta \in S(A)$ of the local Gibbs states $\omega_{\Lambda_N}^\beta$ defined by (9.96) as $N \to \infty$; see (10.44) and surrounding discussion. Proposition 10.8 does not apply now, but Theorem 8.9 does: since each local Hamiltonian h_{Λ_N} is permutation-invariant (because each $T_i^{(\Lambda_N)}$ is), so is each local Gibbs state $\omega_{\Lambda_N}^\beta$, and accordingly, each w^*-limit point of this sequence must share this property.

As in (8.174), from the quantum De Finetti Theorem 8.9 we therefore have:

$$\hat{\omega}^{\beta} = \int_{S(M_n(\mathbb{C}))} d\mu_{\beta}(\theta) \left(\omega_{\theta}^{\beta}\right)^{\infty}, \tag{10.286}$$

for some probability measure μ_β on the single-spin state space $S(M_n(\mathbb{C}))$. By Proposition 8.28, this measure may also be regarded as a limit of the local Gibbs states, but now regarded as a state on the limit algebra $A_0^{(c)} = C(S(M_n(\mathbb{C})))$ rather than as a state on $A_0^{(q)} = A$. By the same token, each state ω_θ^β in the decomposition (10.286) is a pure state on $A_0^{(c)}$ (though seen as a state on $M_n(\mathbb{C})$ it will be mixed!). The states ω_θ^β are computed as follows. Given a classical Hamiltonian h_0 computed from (10.271) as explained after Corollary 10.23, for each point $\theta = (\theta_0, \ldots, \theta_{n^2-1}) \in i\mathfrak{u}(n)^*$ we define a new self-adjoint operator $\hat{h}_\theta \in M_n(\mathbb{C})$ by

$$\hat{h}_{\theta} = h_0(\theta) \cdot 1_n + \sum_{i=0}^{n^2-1} \frac{\partial h_0}{\partial \theta_i}(\theta) \cdot T_i. \tag{10.287}$$

For example, in the Curie–Weiss model, from (10.273) we have

$$h_0^{\mathrm{CW}}(\theta) = -2(\theta_3^2 + B\theta_1); \tag{10.288}$$
$$\hat{h}_{\theta}^{\mathrm{CW}} = h_0^{\mathrm{CW}}(\theta) - 2\theta_3\sigma_3 - B\sigma_1. \tag{10.289}$$

Eq. (10.287) has the following origin. Let ω be any state on A for which the strong limit $T_i^{(\omega)}$ of each operator $\pi_\omega(T_i^{(\Lambda_N)})$ on H_ω exists as $N \to \infty$ (for example, as in the proof of Theorem 8.16 one may show that this is the case when ω is a permutation-invariant state of A). It easily follows that $T_i^{(\omega)}$ lies in the algebra at infinity for π_ω, and hence in the center of $\pi_\omega(A)''$, cf. §8.5. If, in addition, ω is primary, then

$$T_i^{(\omega)} = \theta_i \cdot 1_{H_\omega}; \tag{10.290}$$
$$\theta_i = \lim_{N\to\infty} \omega(T_i^{(\Lambda_N)}). \tag{10.291}$$

Under these assumptions, we compute the commutator

$$[\pi_\omega(h_{\Lambda_N}), \pi_\omega(a)] = \sum_i \frac{\partial h_0}{\partial \theta_i}\left(T_0^{(\Lambda)}, \ldots, T_{n^2-1}^{(\Lambda)}\right) \cdot \sum_{x\in\Lambda_N} [\pi_\omega(T_i(x)), \pi_\omega(a)] + O\left(\frac{1}{|\Lambda_N|}\right),$$

where $a \in \cup_\Lambda A_\Lambda$, and $O(1/|\Lambda_N|)$ denotes a finite sum of (multiple) commutators between some power of $T_i^{(\Lambda)}$ and operators that are (norm-) bounded in N. For example, for the Curie–Weiss model the $O(1/|\Lambda_N|)$ term is a multiple of

$$\sum_{x\in\Lambda_N} [[\pi_\omega(\sigma_3(x)), \pi_\omega(a)], \sigma_3^{(\Lambda_N)}]. \tag{10.292}$$

Since a is local, all commutators $\sum_{x\in\Lambda_N}[\pi_\omega(T_i(x)),\pi_\omega(a)]$ are in $\pi_\omega(A)$, so that further commutators à la (10.292) vanish as $N\to\infty$. Also, in this limit the terms $T_i^{(\Lambda)}$ in the argument of $\sum_i \partial h_0/\partial\theta_i$ assume their c-number values θ_i, so that

$$\lim_{N\to\infty}[\pi_\omega(h_{\Lambda_N}),\pi_\omega(a)] = [h_\omega,\pi_\omega(a)], \tag{10.293}$$

where formally (i.e. on a suitable domain) we have an ω-dependent Hamiltonian

$$h_\omega = \sum_{x\in\mathbb{Z}^d}\pi(\hat{h}_\theta(x)), \tag{10.294}$$

where the θ_i depend on ω via (10.291). Also, for each $a\in A$ one has strong limits

$$\lim_{N\to\infty}\pi_\omega\left(e^{ih_{\Lambda_N}t}ae^{-ih_{\Lambda_N}t}\right) = e^{ih_\omega t}\pi(a)e^{-ih_\omega t}. \tag{10.295}$$

Hence in the limit $N=\infty$ (provided it makes sense, which it does under the stated assumptions), the original mean-field Hamiltonian (10.271) with its homogeneous long-range forces converges to a sum of single-body Hamiltonians, in which the original forces between the spins have been incorporated into the parameters θ_i.

Returning to (10.286), for any $\beta = T^{-1}$, we now determine ω_θ^β from the *Ansatz*

$$\omega_\theta^\beta(a) = \frac{\mathrm{Tr}\,(e^{-\beta\hat{h}_\theta}\,a)}{\mathrm{Tr}\,(e^{-\beta\hat{h}_\theta})}, \tag{10.296}$$

where θ is found by by solving the ***self-consistency equation***

$$\omega_\theta^\beta = \theta. \tag{10.297}$$

As explained after Corollary 10.23, here $\omega_\theta^\beta : M_n(\mathbb{C})_{\mathrm{sa}}\to\mathbb{R}$ is defined by its values on $i\mathfrak{su}(\mathfrak{n})$ and hence should be seen as a map $i\mathfrak{su}(\mathfrak{n})\to\mathbb{R}$, like $\theta\in\mathfrak{su}(\mathfrak{n})^*$, so that (10.297) consists of n^2-1 equations $\omega_\theta^\beta(T_i)=\theta_i$ $(i=1,\dots,n^2-1)$. Alternatively, one may extend θ from $i\mathfrak{su}(\mathfrak{n})$ to $i\mathfrak{u}(\mathfrak{n})$ by prescribing $\theta(1_n)=1$, and subsequently extend it further to $M_n(\mathbb{C})$ by complex linearity. Clearly, the constant $h_0(\theta)$ in (10.287) drops out of (5.152) and may be ignored in solving (10.297).

For example, if we take (10.289) with $B=0$, then (10.297) forces $\theta_1=\theta_2=0$, whereas the magnetization $2\theta_3\equiv m=\omega_\theta^\beta(\sigma_3)$ satisfies the famous ***gap equation***

$$\tanh(\beta m) = m. \tag{10.298}$$

For any β this has a solution $m=0$, i.e., $\theta=0$ in B^3, which corresponds to the tracial state $\omega(a)=\frac{1}{2}\mathrm{Tr}\,(a)$ normally associated with infinite temperature (i.e., $\beta=0$). This state is evidently $\mathbb{Z}_2$-invariant. For $T\geq T_c=1/4$ (i.e. $\beta\leq 4$) this is the only solution. For $T<T_c$ (or $\beta>4$), two additional solutions $\pm m_\beta$ (with $m_\beta>0$) appear, which break the $\mathbb{Z}_2$ symmetry. For $B>0$ computations become tedious, but for $\beta\to\infty$, where ω_θ^β converges to the ground state of $\hat{h}_\theta$, one recovers our earlier conclusions.

Proposition 10.24. *The self-consistency equation* (10.297) *has at least one solution.*

Proof. This follows from Brouwer's Fixed Point Theorem (stating that any continuous map f from a compact compact set $K \subset \mathbb{R}^k$ to itself has a fixed point), applied to $K = S(M_n(\mathbb{C}))$ and $f(\theta) = \omega_\theta^\beta$, where $\theta \in S(M_n(\mathbb{C}))$, as just explained. □

The key result on equilibrium states of homogeneous mean-field theories, then, is:

Theorem 10.25. *Let h_Λ in* (10.271) *define a homogeneous mean-field theory with compact symmetry group G. The sequence $(\omega_{\Lambda_N}^\beta)$ of local Gibbs states defined by* (9.96) *and* (10.271) *has a unique G-invariant limit point $\hat{\omega}^\beta$, whose decomposition into primary states is given by* (10.286)*. The G-invariant probability measure μ_β is concentrated on some G-orbit in $S(M_n(\mathbb{C}))$, and the states ω_θ^β on $M_n(\mathbb{C})$ are given by* (10.296)*, with Hamiltonians $\hat{h}_\theta$ defined by* (10.287)*, where θ satisfies* (10.297).

Proof. We just sketch the proof, which is based on the Quantum De Finetti Theorem 8.9. Each operator $T_i^{(\Lambda_N)}$ is permutation-invariant, which property is transferred first to each local Hamiltonian h_{Λ_N}, thence to each local Gibbs state $\omega_{\Lambda_N}^\beta$ defined by h_{Λ_N}, and finally to each limit point of this sequence. As already noted, Theorem 8.9 then gives the decomposition (10.286), which by Theorem 8.29 (whose assumption holds in mean-field models) also gives the primary decomposition of $\hat{\omega}^\beta$ (i.e., each state $(\omega_\theta^\beta)^\infty$ is primary on the quasi-local algebra A). By our earlier argument centered on (10.294) - (10.295), time-evolution is implemented in the GNS-representation induced by such a state. An important step in the proof—which we omit because it requires various reformulations of the KMS condition we have not discussed—is that $(\omega_\theta^\beta)^\infty$ satisfies the KMS condition with respect to the dynamics (10.295). This, in turn, implies (10.296), which, by definition of θ through (10.290) - (10.291), gives the self-consistency condition (10.297). The proof is completed by a tricky argument (which again uses alternatives to the KMS condition) to the effect that if some ω_θ^β breaks the G-symmetry, the probability measure μ_β on the G-orbit in $S(M_n(\mathbb{C}))$ through ω_θ^β induced by the normalized Haar measure on G, defines the only possible limit point of the local Gibbs states, and hence must be unique. □

Thus SSB can be detected by solving (10.297) and checking if the ensuing state(s) ω_θ^β on $M_n(\mathbb{C})$ is (are) G-invariant. As we have seen, in the Curie–Weiss model this is the case for $\beta \leq 4$, whereas for $\beta > 4$ the measure μ_β in (10.286) is given by

$$\mu_\beta = \tfrac{1}{2}(\delta_{(0,0,m_\beta/2)} + \delta_{(0,0,-m_\beta/2)}), \tag{10.299}$$

where $\delta_\theta(f) = f(\theta)$. In such cases, since each local Gibbs state is invariant, one faces the (by now) familiar threat to Earman's Principle. In response, we expect Butterfield's Principle to be restored through the introduction of asymmetric flea-type perturbations to h_Λ that are localized in spin configuration space, although at nonzero temperature all excited states (rather than just the first) will start to play a role, and the precise details of the "flea" scenario remain to be settled.

10.9 The Goldstone Theorem

So far, we have only discussed the simplest of all symmetry groups, namely $G = \mathbb{Z}_2$, which is both finite and abelian. Although it will not change our picture of SSB, for the sake of completeness (and interest to foundations) we also present a brief introduction to continuous symmetries, culminating in the Goldstone Theorem and the Higgs mechanism (which at first sight contradict each other and hence require a very careful treatment). The former results when the broken symmetry group G is a Lie group, whereas the latter arises when it is an infinite-dimensional gauge group.

Let us start with the simple case $G = SO(2)$, acting on $\mathbb{R}^2$ by rotation. This induces the obvious action on the classical phase space $T^*\mathbb{R}^2$, i.e.,

$$R(p,q) = (Rp, Rq), \tag{10.300}$$

cf. (3.94), as well as on the quantum Hilbert space $H = L^2(\mathbb{R}^2)$, that is,

$$u_R\psi(x) = \psi(R^{-1}x). \tag{10.301}$$

Let us see what changes with respect to the action of $\mathbb{Z}_2$ on $\mathbb{R}$ considered in §10.1. We now regard the double-well potential V in (10.11) as an $SO(2)$-invariant function on $\mathbb{R}^2$ through the reinterpretation of x^2 as $x_1^2 + x_2^2$. This is the ***Mexican hat potential***. Thus the classical Hamiltonian $h(p,q) = p^2/2m + V(q)$, similarly with $p^2 = p_1^2 + p_2^2$, is $SO(2)$-invariant, and the set of classical ground states

$$\mathscr{E}_0 = \{(p,q) \in T^*\mathbb{R}^2 \mid p = 0, q^2 = a^2\} \tag{10.302}$$

is the SO(2)-orbit through e.g. the point $(p_1 = p_2 = 0, q_1 = a, q_2 = 0)$. Unlike the one-dimensional case, the set of ground states is now connected and forms a circle in phase space, on which the symmetry group $SO(2)$ acts. The intuition behind the Goldstone Theorem is that a particle can freely move in this circle at no cost of energy. If we look at mass as inertia, such motion is "massless", as there is no obstruction. However, this intuition is only realized in quantum field theory. In quantum mechanics, the ground state of the Hamiltonian (10.6) (now acting on $L^2(\mathbb{R}^2)$) remains unique, as in the one-dimensional case. In polar coordinates (r,ϕ) we have

$$h_\hbar = -\frac{\hbar^2}{2m}\left(\frac{\partial^2}{\partial r^2} + \frac{1}{r}\frac{\partial}{\partial r} + \frac{1}{r^2}\frac{\partial^2}{\partial \phi^2}\right) + V(r), \tag{10.303}$$

with $V(r) = \frac{1}{4}\lambda(r^2 - a^2)^2$. With

$$L^2(\mathbb{R}^2) \cong L^2(\mathbb{R}^+) \otimes \ell^2(\mathbb{Z}) \tag{10.304}$$

under Fourier transformation in the angle variable, this becomes

$$h_\hbar\psi(r,n) = \left(-\frac{\hbar^2}{2m}\left(\frac{\partial^2}{\partial r^2} + \frac{1}{r}\frac{\partial}{\partial r} - \frac{n^2}{r^2}\right) + V(r)\right)\psi(r,n). \tag{10.305}$$

Since $\hbar^2 n^2/2mr^2$ is positive, the ground state $\psi_\hbar^{(0)}$ has $\psi_\hbar^{(0)}(r,n) = 0$ for all $n \neq 0$, and hence it is $SO(2)$-invariant, since the $SO(2)$-action on $L^2(\mathbb{R}^2)$ becomes

$$u_\theta \psi(r,n) = \exp(in\theta)\psi(r,n), \tag{10.306}$$

after a Fourier-transform. Indeed, from a group-theoretical point of view, the unitary isomorphism (10.304) is nothing but the decomposition

$$L^2(\mathbb{R}^2) \cong \bigoplus_{n\in\mathbb{Z}} H_n, \tag{10.307}$$

where $H_n = L^2(\mathbb{R}^+)$ for all n, but with $\phi_n \in H_n$ transforming under $SO(2)$ as

$$u_\theta \phi_n(r) = \exp(in\theta)\phi_n(r) \ \ (\theta \in [0,2\pi]). \tag{10.308}$$

The $SO(2)$-invariant subspace of $L^2(\mathbb{R}^2)$, then, is precisely the space H_0 in which $\psi_\hbar^{(0)}$ lies. This is analogous to the situation occurring in one dimension higher (i.e. $\mathbb{R}^3$) with e.g. the hydrogen atom: in that case, the symmetry group is $SO(3)$, and $L^2(\mathbb{R}^3)$ decomposes accordingly as

$$L^2(\mathbb{R}^3) \cong \bigoplus_{j\in\mathbb{N}} H_j; \tag{10.309}$$

$$H_j = L^2(\mathbb{R}^+) \otimes \mathbb{C}^{2j+1}. \tag{10.310}$$

The ground state for a spherically symmetric potential, then, lies in H_0 and is $SO(3)$-invariant. For our purposes the relevant comparison is with the one-dimensional case: the decomposition of $L^2(\mathbb{R})$ under the natural $\mathbb{Z}_2$-action $u_{-1}\psi(x) = \psi(-x)$ is

$$L^2(\mathbb{R}) = H_0 \oplus H_1 \tag{10.311}$$

$$H_i = \{\psi \in L^2(\mathbb{R}) \mid \psi(x) = (-1)^i \psi(-x)\}, \ i = 0,1. \tag{10.312}$$

This time, H_+ is the $\mathbb{Z}_2$-invariant subspace containing the ground state $\psi_\hbar^{(0)}$. Being $\mathbb{Z}_2$-invariant, $\psi_\hbar^{(0)}$ is has peaks above both classical minima $\pm a$; in fact, $\psi_\hbar^{(0)}$ is real-valued and strictly positive. The ground state of the corresponding two-dimensional system, seen as an element of $L^2(\mathbb{R}^2)$, is just this wave-function $\psi_\hbar^{(0)}$ extended from $\mathbb{R}$ to $\mathbb{R}^2$ by rotational invariance. Hence the ground state remains real-valued and strictly positive, with peaks about the circle of classical minima in $\mathbb{R}^2$.

Let us recall the situation for $d = 1$ (cf. §10.1). The first excited state $\psi_\hbar^{(1)}$ lies in H_1; it is real-valued, like $\psi_\hbar^{(0)}$, but since it has to satisfy $\psi_\hbar^{(1)}(-x) = -\psi_\hbar(x)$, it cannot be positive. Indeed, with a suitable choice of phase, $\psi_\hbar^{(1)}$ has one positive peak above a and the same peak but now negative below $-a$. Then the wave-function

$$\psi_\hbar^\pm = (\psi_\hbar^{(0)} \pm \psi_\hbar^{(1)})\sqrt{2}, \tag{10.313}$$

is peaked above $\pm a$ alone (i.e., the negative peak of $\pm\psi_\hbar^{(1)}$ below $\mp a$ exactly cancels the corresponding peak of $\psi_\hbar^{(0)}$). The classical limit of $\psi_\hbar^{(0)}$ comes out as the mixed state $\frac{1}{2}(\omega_0^+ + \omega_0^-)$, where $\omega_0^\pm = (p = 0, \pm a)$, but each state $\psi_\hbar^\pm$ has the pure state $\omega_0^\pm$ as its classical limit. The latter are ground states, and hence in particular they are time-independent, because the energy difference $E^{(1)} - E^{(0)}$ between $\psi_\hbar^{(1)}$ and $\psi_\hbar^{(0)}$ vanishes (even exponentially fast) as $\hbar \to 0$.

A similar but more complicated situation arises in $d = 2$. The role of the pair

$$\left(\psi_\hbar^{(0)} \in H_0, \psi_\hbar^{(1)} \in H_1\right)$$

is now played by an infinite tower of unit vectors

$$\left(\psi_\hbar^{(n)} \in H_n, n \in \mathbb{Z}\right),$$

where $\psi_\hbar^{(n)}$ is the lowest energy eigenstate (for $h_\hbar$ in (10.305)) in $H_n \subset L^2(\mathbb{R}^2)$. The analogue of the states $\psi_\hbar^\pm$ for $d = 1$ involves a limit which heuristically is like

$$\lim_{N\to\infty} \psi_\hbar^{(N,\theta)} = \frac{1}{\sqrt{2N+1}} \sum_{n=-N}^{N} u_\theta \psi_\hbar^{(n)}, \tag{10.314}$$

but this limit does not exist in $L^2(\mathbb{R}^2)$. As in §10.1, we instead rely on the technique explained around (10.4), which makes the unit vectors $\psi_\hbar^{(N,\theta)}$ converge to some probability measure $\mu_\hbar^\theta$ on $\mathbb{R}^2$ as $N \to \infty$. In the subsequent limit $\hbar \to 0$, one obtains a probability measure μ_0^θ concentrated on a suitable point in the orbit of classical ground states (10.302). Similarly, in the same sense the ground state $\psi_\hbar^{(0)}$ converges to a probability measure supported by all of $\mathscr{E}_0$.

To the extent that there is a Goldstone Theorem in classical mechanics, it would state that motion in the orbit $\mathscr{E}_0$ is free. That is, at fixed $(r = a, p_r = 0)$, where p_r is the radial component of momentum, one has an effective Hamiltonian

$$h_a(p_\phi, \phi) = \frac{p_\phi^2}{2ma^2}, \tag{10.315}$$

whose time-independent states $(p_\phi = 0, \phi_0)$ for arbitrary $\phi_0 \in [0, 2\pi)$ yield the ground states of the system, and whose "excited states"

$$(p_\phi(t), \phi(t)) = \left(p_\phi(0), \phi(0)) + \frac{p_\phi(0)t}{ma^2}\right) \tag{10.316}$$

give motion along the orbit $\mathscr{E}_0$ with effective mass ma^2, whose energy converges to zero as $p_\phi \to 0$. However, since massless particles (whose existence is the main conclusion of the usual Goldstone Theorem) are not defined in classical mechanics, we now turn to relativistic field theory (with which we assume some familiarity).

We now illustrate SSB in classical field theory through a simple example, where the symmetry group is $G = SO(N)$, but whenever write things down in such a way that the generalization to arbitrary scalar field theories is obvious. Suppose we have N real scalar fields $\varphi \equiv (\varphi_1, \dots, \varphi_N)$, on which $SO(N)$ acts in the defining representation on $\mathbb{R}^N$. Following the physics literature, from now on we sum over repeated indices like i and μ (***Einstein summation convention***). Let the Lagrangian

$$\mathscr{L} = \tfrac{1}{2}\partial_\mu \varphi_i \partial^\mu \varphi_i - V(\varphi), \tag{10.317}$$

contain an $SO(N)$-invariant potential V, typically of the form (with $\varphi^2 \equiv \sum_{i=1}^N \varphi_i^2$)

$$V(\varphi) = -\frac{m^2}{2}\varphi^2 + \frac{\lambda}{4}\varphi^4, \tag{10.318}$$

where $\lambda > 0$, but m^2 may have either sign. If $m^2 < 0$, the minimum of V lies at $\varphi = 0$, but if $m^2 > 0$ the minima form the $SO(N)$-orbit through

$$\varphi^c = (v, 0, \cdots, 0); \tag{10.319}$$

$$v \equiv m/\sqrt{\lambda} = \|\varphi^c\|. \tag{10.320}$$

The idea is that the physical fields are excitations of the "vacuum state" φ^c, so that, instead of φ, as the appropriate "small oscillation" field one should use

$$\chi(x) = \varphi(x) - \varphi^c. \tag{10.321}$$

Consequently, the potential is expanded in a Taylor series for small χ as

$$V(\varphi) = V(\varphi^c) + \tfrac{1}{2}V''_{ij}\chi_i\chi_j + O(\chi^3); \tag{10.322}$$

$$V''_{ij} \equiv \frac{\partial^2 V}{\partial \varphi_i \partial \varphi_j}(\varphi^c). \tag{10.323}$$

Note that the linear term vanishes because $V'(\varphi^c) = 0$. We now use the $SO(N)$-invariance of V, i.e., $V(g\varphi) = V(\varphi)$ for all $g \in SO(N)$. For $T_a \in \mathfrak{g}$ (i.e. the Lie algebra of G, realized by anti-symmetric traceless $N \times N$ matrices) this yields

$$\frac{d}{dt}V(e^{tT_a}\varphi)_{t=0} = 0 \Leftrightarrow \frac{\partial V(\varphi)}{\partial \varphi_i}(T_a)_{ij}\varphi_j = 0. \tag{10.324}$$

Differentiation with respect to φ_k and putting $\varphi = \varphi^c$ then gives

$$V''_{ik}(T_a)_{ij}\varphi^c_j = 0. \tag{10.325}$$

In general, let $H \subset G$ be the stabilizer of φ^c, i.e., $g \in H$ iff $g\varphi^c = \varphi^c$. In our example (10.318) - (10.319), we evidently have $H = SO(N-1)$. Then $T_a\varphi^c = 0$ for all generators T_a of the Lie algebra $\mathfrak{h}$ of H, so that there are

$$M \equiv \dim(G) - \dim(H) = \dim(G/H) = \dim(G \cdot \varphi^c) \tag{10.326}$$

linearly independent null eigenvectors of V'' (seen as an $N \times N$ matrix). This number equals the dimension of the submanifold of $\mathbb{R}^N$ where V assumes its minimum. In our example we have $M = N - 1$, since $\dim(SO(N)) = \frac{1}{2}N(N-1)$. We now perform an affine field redefinition, based on an affine coordinate transformation in $\mathbb{R}^N$ that diagonalizes the matrix V''. The original (real) fields were $\varphi = (\varphi_1, \ldots, \varphi_N)$, and the new (real) fields are $(\chi_1, \theta_2, \cdots, \theta_N)$, with

$$\chi_1 = \varphi_1 - v, \tag{10.327}$$

as in (10.321), and the ***Goldstone fields*** are defined, also in general, by

$$\theta_a = \frac{1}{v}\langle T_a \varphi^c, \varphi \rangle = \frac{1}{v}(T_a)_{ij}\varphi_j^c \varphi_i. \tag{10.328}$$

Here $\langle \cdot, \cdot \rangle$ denotes the inner product in $\mathbb{R}^N$, and we have chosen a basis of $\mathfrak{g}$ in which the elements $(T_1, \ldots, T_{\dim(H)})$ form a basis of $\mathfrak{h}$, completed by M further elements $(T_{\dim(H)+1}, \ldots T_{\dim(G)+1})$, so as to have basis of $\mathfrak{g}$. The index a in (10.328), then, runs from $\dim(H) + 1$ to $\dim(G)$, so that there are M Goldstone fields, cf. (10.326). In our running example, this number was shown to be $M = N - 1$, and in view of (10.319), the field $\theta_a = (T_a)_{i1}\varphi_i$ is a linear combination of the φ_2 till φ_N.

The simplest example is $N = 2$, with potential (10.318) and $m^2 > 0$. With the single generator $T = -i\sigma_2$, we obtain $\theta = \varphi_2$. Since $V'' = \mathrm{diag}(2m^2, 0)$, we see that the mass term $-\frac{1}{2}m^2\varphi_1^2$ in (10.318) (with $\varphi^2 = \varphi_1^2 + \varphi_2^2$) changes from the "wrong" sign $-m^2$ to the 'right' sign $+2m^2$ in (10.322), whilst $-\frac{1}{2}m^2\varphi_2^2$ in (10.318) disappears, so that the field θ comes out to be massless. Indeed, this is the point of the introduction of the Goldstone fields: in view of (10.325) and (10.328), the Goldstone fields do not occur in the quadratic term in (10.322) and hence they are massless, in satisfying a field equation of the form $\partial_\mu \partial^\mu \theta_a = \cdots$, where $\cdots$ does not contain any term linear in any field. This proves the ***classical Goldstone Theorem***:

Theorem 10.26. *Suppose that a compact Lie group $G \subset SO(N)$ acts on N real scalar fields $\varphi = (\varphi_1, \ldots, \varphi_N)$, leaving the potential V in the Lagrangian* (10.317) *invariant. If G is spontaneously broken to an unbroken subgroup $H \subset G$ (in the sense that the stability group of some point φ^c in the G-orbit minimizing V is H), then there are at least* $\dim(G/H)$ *massless fields, i.e., there is a field transformation*

$$(\varphi_1, \ldots, \varphi_N) \mapsto (\chi_1, \ldots, \chi_{N-M}, \theta_1, \ldots, \theta_M) \ (M = \dim(G) - \dim(H)), \tag{10.329}$$

that is invertible in a neighborhood of $\varphi = \varphi^c$, such that the potential $V(\varphi)$, re-expressed in the fields χ and θ, has no quadratic terms in θ.

The local invertibility of the field redefinition around $\varphi^c \neq 0$ is crucial; in our example, where $\chi \equiv \chi_1 = \varphi_1 - v$ and $\theta_a = T_{i1}^a \varphi_i$, this may be checked explicitly.

An alternative proof of Theorem 10.26 uses nonlinear Goldstone fields, viz.

$$\varphi(x) = e^{\frac{1}{v}\theta_a(x)T_a}(\varphi^c + \chi(x)), \tag{10.330}$$

where the sum over a (implicit in the Einstein summation convention) ranges from 1 to M, $v = \|\varphi^c\|$, and the fields $\chi = (\chi_1, \dots, \chi_{N-M})$ are chosen orthogonal (in $\mathbb{R}^N$) to each $T_a\varphi_c$, $a = 1, \dots, M$, and hence to the θ_a. Provided that the generators of $SO(N)$ (and hence of $G \subset SO(N)$) have been chosen such that

$$\langle T_a\varphi^c, T_b\varphi^c\rangle = v^2\delta^{ab}, \tag{10.331}$$

the fields θ^a defined by (10.330) coincide with the fields in (10.328) up to quadratic terms in χ and θ; to see this, expand the exponential and also use the fact that both $\langle T_a\varphi^c, \varphi^c\rangle$ and $\langle T_a\varphi^c, \chi\rangle$ vanish. This transformation is only well defined if $v \neq 0$, i..e., if SSB from G to H occurs, and its existence implies the Goldstone Theorem 10.26, for by (10.330) and G-invariance, $V(\varphi)$ is independent of θ.

The Goldstone Theorem can be derived in quantum field theory, but in the spirit of this chapter we will discuss it rigorously for quantum spin systems. Far from considering the most general case, we merely treat the simplest setting. We assume that A is a quasi-local C*-algebra given by (8.130), with $H = \mathbb{C}^n$. Furthermore:

1. The group of space translations $\mathbb{Z}^d$ acts on A by automorphisms τ_x, and so does the group $\mathbb{R}$ of time translations by automorphisms α_t commuting with the τ_x (cf. §9.3); we often write $\alpha_{(x,t)}$ for $\alpha_t \circ \tau_x$ as well as $a(x,t)$ for $\alpha_t \circ \tau_x(a)$.
2. A compact Lie group G acts on $H = \mathbb{C}^n$ through a unitary representation u and hence acts on on A by automorphisms γ_g as in (10.58) - (10.59), such that
$$\gamma_g \circ \alpha_{(x,t)} = \alpha_{(x,t)} \circ \gamma_g \;\; ((x,t) \in \mathbb{Z}^d \times \mathbb{R}, g \in G). \tag{10.332}$$
3. There exists a pure translation-invariant ground state ω.
4. One has SSB in that $\omega \circ \gamma_g \neq \omega$ for all $g \in G_a \subset G$, where
$$G_a = \{\exp(sT_a), s \in \mathbb{R}, T_a \in \mathfrak{g}\}. \tag{10.333}$$
5. There is an n-tuple $\varphi = (\varphi_1, \dots, \varphi_n)$ of local operators $\varphi_\alpha \in M_n(\mathbb{C})$ that transforms under G by $\varphi \mapsto u_g\varphi u_g^* = \gamma_g(\varphi)$, and defines an order parameter ϕ_a by
$$\phi_a = \delta_a\varphi \equiv \frac{d}{ds}\left(\gamma_{\exp(sT_a)}(\varphi)\right)_{|s=0}, \tag{10.334}$$
at least for SSB of G_a (as above) in that, cf. Definition 10.6,
$$\omega(\delta_a\varphi) \neq 0. \tag{10.335}$$
6. Writing $j_a^0 = iu'(T_a) \in M_n(\mathbb{C})$, it follows that $\delta_a\varphi = -i[j_a^0, \varphi]$, and hence that
$$\delta_a\varphi(x) = -i \lim_{\Lambda \nearrow \mathbb{Z}^d} \sum_{y \in \Lambda} [j_a^0(y), \varphi(x)] \;\; (x \in \mathbb{Z}^d), \tag{10.336}$$
since by (8.132) (i.e., Einstein locality) only the term $y = x$ will contribute. Physicists then wish to define a charge by $Q_a = \sum_{y \in \mathbb{Z}^d} j_a^0(y)$ and write (10.336) as $\delta_a\varphi(x) = -i[Q_a, \varphi(x)]$, but Q_a does not exist precisely in the case of SSB!

Eq. (10.336) motivates the crucial assumption for the Goldstone Theorem, viz.

$$\omega(\delta_a\varphi(x,t)) = -i \lim_{\Lambda\nearrow\mathbb{Z}^d} \sum_{y\in\Lambda} \omega([j_a^0(y),\varphi(x,t)]) \ (x\in\mathbb{Z}^d, t\in\mathbb{R}), \tag{10.337}$$

which incorporates the condition that the sum over y converge absolutely. Although (10.337) at first sight *softens* (10.336) in turning an operator equation into a numerical one, in fact (10.337) decisively *sharpens* (10.336) by involving the time-dependence of φ, whose propagation speed should be sufficiently small for enabling the limit in (10.336) to catch up with the limit in (10.337). As such, eq. (10.337) is satisfied with short-range forces, but the Meissner effect in superconductivity and the closely related Higgs mechanism in gauge theories (both of which circumvents the Goldstone Theorem) are possible precisely because in those cases (10.337) fails (at least in physical gauges, see also §10.10).

7. Finally, we make two assumptions just for convenience, namely

$$\varphi_\alpha(x)^* = \varphi_\alpha(x); \tag{10.338}$$
$$\omega(\varphi_\alpha(x)) = 0. \tag{10.339}$$

If these are not the case, one could simply take real and imaginary components of φ_α and/or redefine φ_α as $\tilde{\varphi}_\alpha = \varphi_\alpha - \omega(\varphi_\alpha)\cdot 1_A$, so that $\omega(\tilde{\varphi}_\alpha(x)) = 0$.

The Goldstone Theorem provides information about the joint-energy momentum spectrum of the theory at hand. To define this notion, we exploit the fact that from assumption no. (3) and Corollary 9.12 we obtain a unitary representation u_ω of the (locally compact) abelian space-time translation group $A = \mathbb{Z}^d\times\mathbb{R}$ on the GNS-representation space H_ω induced by ω. The SNAG-Theorem C.114 applied to A, with dual $\hat{A} = \mathbb{T}^d\times\mathbb{R}$ (cf. Proposition C.108), then yields a projection-valued measure

$$e_\omega : \mathscr{B}(\mathbb{R}\times\mathbb{T}^d) \to \mathscr{P}(H_\omega), \tag{10.340}$$

as a map from the Borel sets in $\mathbb{R}\times\mathbb{T}^d$ to the projection lattice in $B(H_\omega)$, such that

$$1_{H_\omega} = \int_{\mathbb{T}^d}\int_0^\infty de(E,k); \tag{10.341}$$
$$u_\omega(y,t) = \int_{\mathbb{T}^d}\int_0^\infty de(E,k)\, e^{i(Et-y\cdot k)} \ (y\in\mathbb{Z}^d, t\in\mathbb{R}). \tag{10.342}$$

Here $k = (k_1,\ldots,l_d)$, $y\cdot k = \sum_{i=1}^d y_i k_i$, and we have reduced the integration range over E (which *a priori* would be $\mathbb{R}$) to $\mathbb{R}^+$. Indeed, by Stone's Theorem we have $u_\omega(t) = \exp(ith_\omega)$, where $\sigma(h_\omega)\subset[0,\infty)$ because ω is a ground state by assumption, and the support of e is evidently contained in $\mathbb{Z}^d\times\sigma(h_\omega)$ (cf. Definition A.16).

Definition 10.27. *The* **joint energy-momentum spectrum** $\sigma(h_\omega, p_\omega)$ *of a space-time invariant state* ω *(i.e.,* $\omega\circ\alpha_{(x,t)} = \omega$, $(x,t)\in\mathbb{Z}^d\times\mathbb{R}$*) is the support of the projection-valued measure* e_ω *associated to the* GNS*-representation* π_ω*, i.e., the smallest closed set* $\sigma(h_\omega,p_\omega)\subset\mathbb{T}^d\times\mathbb{R}$ *such that* $e((\mathbb{T}^d\times\mathbb{R})\backslash\sigma(h_\omega,p_\omega)) = 0$.

The notation $\sigma(h_\omega, p_\omega)$ is purely symbolic here, since (as opposed to the continuum case) the group $\mathbb{Z}^d$ of spatial translations is discrete and hence has no generators p_ω.

Since $u_\omega(x,t)\Omega_\omega = \Omega_\omega$, the origin $(0,0)$ certainly lies in $\sigma(h_\omega, p_\omega)$, with

$$e_\omega(0,0) = |\Omega_\omega\rangle\langle\Omega_\omega|, \tag{10.343}$$

which by Theorem 9.14 is the unique $\mathbb{T}^d \times \mathbb{R}$-invariant state in H_ω. Denoting this contribution to e_ω by $e_\omega^{(0)}$, in many physical theories one has $e_\omega = e_\omega^{(0)} + e_\omega^{(1)} + \cdots$, where $e_\omega^{(1)}$ is supported on the graph of some continuous function $k \mapsto \varepsilon_k \geq 0$, i.e.,

$$\{(k, \varepsilon_k), k \in \mathbb{T}^d\} \subset \sigma(h_\omega, p_\omega) \subset \mathbb{T}^d \times \mathbb{R}. \tag{10.344}$$

The joint energy-momentum spectrum may be studied in part by considering

$$\begin{aligned} f(\varepsilon, p) &= \sum_{y\in\mathbb{Z}^d} \int_{-\infty}^{\infty} dt\, e^{-i\varepsilon t + ip\cdot(x-y)}\omega([j_a^0(y), \varphi(x,t)]) \\ &= 2i \sum_{y\in\mathbb{Z}^d} \int_{-\infty}^{\infty} dt\, e^{-i\varepsilon t + ip\cdot(x-y)} \mathrm{Im}\langle\Omega_\omega, \pi_\omega(j_a^0(0))e^{ith_\omega}u_\omega(y)\pi_\omega(\varphi_\alpha(0))\Omega_\omega\rangle \\ &= \int_{\mathbb{T}^d}\int_0^{\infty} (\langle\Omega_\omega, \pi_\omega(j_a^0(0))de_\omega(E,k)\,\pi_\omega(\varphi_\alpha(0))\Omega_\omega\rangle\delta(\varepsilon - E)\delta(p-k) \\ &- \langle\Omega_\omega\pi_\omega(\varphi_\alpha(0))de_\omega(E,k)\,\pi_\omega(j_a^0(0))\Omega_\omega\rangle\delta(\varepsilon+E)\delta(p+k)), \end{aligned} \tag{10.345}$$

i.e., the Fourier transform of the two-point function defined by j_a^0 and φ, which is a distribution on the dual group $\mathbb{T}^d \times \mathbb{R}$; for the third equality we used a distributional version of the Fourier inversion formula (C.382). For example, if we replace $e_\omega(E,k)$ by $e_\omega^{(1)}(E,k)$, then, since $e_\omega^{(1)}$ is absolutely continuous with respect to Haar measure d^dk on $\mathbb{T}^d$, we see that $f(\varepsilon, p)$ is proportional to $\delta(\varepsilon - \varepsilon_p)$.

Theorem 10.28. *Under assumptions 1–7 (notably* (10.337) *and* SSB *of some continuous symmetry), the Hamiltonian h_ω has continuous spectrum starting at zero and hence has no gap. If there is an excitation spectrum $e_\omega^{(1)}$ as explained above, with*

$$\int \langle\Omega_\omega, \pi_\omega(j_a^0(0))de_\omega^{(1)}(E,k)\,\pi_\omega(\varphi_\alpha(0))\Omega_\omega\rangle \neq 0, \tag{10.346}$$

then the continuous function $k \mapsto \varepsilon_k$ defining the spectrum satisfies $\varepsilon_0 = 0$.

Proof. Since the sum in (10.337) converges absolutely, the Fourier transform $\check{f}(t,p)$ of $y \mapsto \omega([j_a^0(y), \varphi(x,t)])$ in y alone is continuous in p, and by (10.337) we have

$$i\omega(\delta_a\varphi(x,t)) = \check{f}(t,0). \tag{10.347}$$

By (10.332), the left-hand side is independent of x and t, hence the Fourier transform $f(\varepsilon,0)$ of the right-hand side in t is proportional to $\delta(\varepsilon)$. Since (10.343) does not contribute to f by (10.339), the calculation (10.345) shows that $f(\varepsilon,0) = 0$ if $\sigma(h_\omega)$ has a gap. But $f(\varepsilon,0) \neq 0$ by (10.335), and so $\sigma(h_\omega)$ has no gap. Similarly, for the final claim note that $f(\varepsilon,0) \sim \delta(\varepsilon - \varepsilon_0)$ as well as $f(\varepsilon,0) \sim \delta(\varepsilon)$. □

10.10 The Higgs mechanism

We proceed to a discussion of SSB in gauge theories, especially with an eye on the Higgs Mechanism, which plays a central role in the Standard Model of high-energy physics (whose empirical confirmation was more or less finished with the discovery of the Higgs boson at CERN, announced on July 4, 2012).

We look at the ***Abelian Higgs Model***, given by the Lagrangian

$$\mathscr{L} = -\tfrac{1}{4}F_A^2 + \tfrac{1}{2}\langle D_\mu^A\varphi, D_\mu^A\varphi\rangle - V(\varphi), \tag{10.348}$$

where $\varphi = (\varphi_1, \varphi_2)$ is a scalar doublet, the usual electromagnetic field strength is

$$F_{\mu\nu} = \partial_\mu A_\nu - \partial_\nu A_\mu, \tag{10.349}$$

in terms of which $F_A^2 = F_{\mu\nu}F^{\mu\nu}$, and the covariant derivative is

$$D_\mu^A \equiv \partial_\mu - eA_\mu \cdot T = \partial_\mu \cdot 1_2 + ieA_\mu \cdot \sigma_2. \tag{10.350}$$

Here e is some coupling constant, identified with the unit of electrical charge. We still assume that V only depends on $\|\varphi\|^2 = \langle\varphi,\varphi\rangle$ and hence is $SO(2)$-invariant.

The novel situation compared to (10.317) and the like is that, whereas (10.317) is invariant under *global* $SO(2)$ transformations, the Lagrangian (10.348) is invariant under *local* $SO(2)$ ***gauge transformations*** that depend on x, namely

$$\varphi(x) \mapsto e^{\alpha(x)\cdot T}\varphi(x) = \begin{pmatrix} \cos\alpha(x) & -\sin\alpha(x) \\ \sin\alpha(x) & \cos\alpha(x) \end{pmatrix} \cdot \begin{pmatrix} \varphi_1(x) \\ \varphi_2(x) \end{pmatrix}; \tag{10.351}$$

$$A_\mu(x) \mapsto A_\mu(x) + \frac{1}{e}\partial_\mu\alpha(x). \tag{10.352}$$

We say that the ***local gauge group*** $\mathscr{G} = C^\infty(\mathbb{R}^d, U(1))$ acts on the space of fields (A,φ) by (10.351) - (10.352). Now suppose V has a minimum at some constant value $\varphi^c \neq 0$. In that case, any field configuration

$$\varphi(x) = \exp(\alpha(x)\cdot T)\varphi^c; \tag{10.353}$$

$$A_\mu(x) = (1/e)\partial_\mu\alpha(x))\ (\alpha \in \mathscr{G}), \tag{10.354}$$

minimizes the action. Hence the possible "vacua" of the model comprise the (infinite-dimensional) orbit $\mathscr{V}$ of the gauge group through $(A = 0, \varphi = \varphi^c)$. Note that $D_\mu^A\varphi = 0$ for $(A,\varphi) \in \mathscr{V}$, i.e., φ is *covariantly* constant along the vacuum orbit (whereas for global symmetries it is constant full stop). Relative to the (arbitrary) choice $(0,\varphi^c) \in \mathscr{V}$, we then introduce real fields χ and θ, called the ***Higgs field*** and the ***would-be Goldstone boson***, respectively, by (10.330), which now simply reads

$$\begin{pmatrix} \varphi_1(x) \\ \varphi_2(x) \end{pmatrix} = e^{\frac{1}{v}\theta(x)\cdot T} \cdot \begin{pmatrix} v + \chi(x) \\ 0 \end{pmatrix}. \tag{10.355}$$

After this redefinition of the scalar fields, the Lagrangian (10.348) becomes

$$\mathscr{L} = -\tfrac{1}{4}F_B^2 + \tfrac{1}{2}\partial_\mu \chi \partial^\mu \chi + \tfrac{1}{2}e^2(v+\chi)^2 B_\mu B^\mu - V(v+\chi, 0), \tag{10.356}$$

where $B_\mu = A_\mu - (1/ev)\partial_\mu \theta$, and $F_B^2 = F_{\mu\nu}F^{\mu\nu}$ for $F_{\mu\nu} = \partial_\mu B_\nu - \partial_\nu B_\mu$. This describes a vector boson B with mass term $\frac{1}{2}m_B^2 B_\mu B^\mu$, with $m_B^2 = \frac{1}{2}e^2 v^2 > 0$ (as opposed to the massless vector field A), and a scalar field χ with mass term $\frac{1}{2}m_\chi^2 \chi^2$, with $m_\chi^2 = (\partial^2 V/\partial \phi_1^2)_{|(v,0)} > 0$ (since V supposedly has a minimum at $\varphi^c = (v,0)$).

This is the ***Higgs mechanism***: the gauge field becomes massive, whilst the massless ("would-be") Goldstone boson disappears from the theory: it is (allegedly) "eaten" by the gauge field. Thus the scalar degree of freedom θ that seems lost is recovered as the longitudinal component of the massive vector field (which for a gauge field would have been an unphysical gauge degree of freedom, see below).

In the description just given, the Higgs mechanism in classical field theory is seen as a consequence of SSB. Remarkably, there is an alternative account of the Higgs mechanism, according to which it has nothing to do with SSB! Namely, we now perform a field redefinition analogous to (10.355) etc. straight away, viz.

$$\begin{pmatrix} \varphi_1(x) \\ \varphi_2(x) \end{pmatrix} = e^{\theta(x)\cdot T} \cdot \begin{pmatrix} \rho(x) \\ 0 \end{pmatrix}; \tag{10.357}$$

$$A_\mu = B_\mu + (1/e)\partial_\mu \theta. \tag{10.358}$$

This transformation is defined and invertible in a neighbourhood of any point $(\rho_0, \theta_0, B_0$, where $\rho_0 > 0$, $\theta_0 \in (-\pi, \pi)$, and B_0 is arbitrary. Each of these new fields is gauge-invariant: for the gauge transformation (10.351) becomes

$$\theta(x) \mapsto \theta(x) + \alpha(x); \tag{10.359}$$

$$\rho(x) \mapsto \rho(x), \tag{10.360}$$

and in view of (10.352), B does not transform at all. The Lagrangian becomes

$$\mathscr{L} = -\tfrac{1}{4}F_B^2 + \tfrac{1}{2}\partial_\mu \rho \partial^\mu \rho + \tfrac{1}{2}e^2 \rho^2 B_\mu B^\mu - V(\rho), \tag{10.361}$$

with $V(\rho) \equiv V(\rho, 0)$. This is a Lagrangian without any internal symmetries at all (not even $\mathbb{Z}_2$, since $\rho > 0$), but of course one can still look for classical vacua that minimize the energy and hence the potential $V(\rho)$. If $\rho = 0$ is the absolute minimum, then the above field redefinition is *a fortiori* invalidated, but if $V'(v) = 0$ for some $v > 0$, we proceed as before, introducing a Higgs field $\chi(x) = \rho(x) - v$, and recovering the Lagrangian (10.356). This once again leads to the Higgs mechanism.

This can be generalized to the nonabelian case; since it suffices to explain the idea, we just discuss the $SU(2)$ case. In (10.348), the scalar field $\varphi = (\varphi_1, \varphi_2)$ is now complex, forming an $SU(2)$ doublet, the brackets $\langle \cdot, \cdot \rangle$ now denote the inner product in $\mathbb{C}^2$, the nonabelian gauge field is $A = A^a \sigma_a$ (where the Pauli matrices σ_a, $a = 1,2,3$, form a self-adjoint basis of the Lie algebra of $SU(2)$), with associated field strength $F_{\mu\nu} = \partial_\mu A_\nu - \partial_\nu A_\mu + g[A_\mu, A_\nu]$ and covariant derivative $D_\mu^A = \partial_\mu + igA_\mu$.

With $F_A^2 = F_{\mu\nu}^a F_a^{\mu\nu}$, the Lagrangian (10.348) is invariant under the transformations

$$\varphi(x) \mapsto e^{i\alpha_a(x)\sigma_a(x)}\varphi(x); \tag{10.362}$$
$$A_\mu(x) \mapsto e^{i\alpha_a(x)\sigma_a(x)}(A_\mu(x) - (i/g)\partial_\mu)e^{-i\alpha_a(x)\sigma_a(x)}. \tag{10.363}$$

The definition of the gauge-invariant fields B and ρ à la (10.357) - (10.358) is now

$$\begin{pmatrix} \varphi_1(x) \\ \varphi_2(x) \end{pmatrix} = e^{i\theta_a(x)\cdot\sigma_a} \cdot \begin{pmatrix} \rho(x) \\ 0 \end{pmatrix}; \tag{10.364}$$
$$A_\mu(x) = e^{i\theta_a(x)\sigma_a(x)}(B_\mu(x) - (i/g)\partial_\mu)e^{-i\theta_a(x)\sigma_a(x)}, \tag{10.365}$$

which leads, *mutatis mutandis*, to the very same Lagrangian (10.361).

As a compromise between these two derivations of the Higgs mechanism, it is also possible to fix the gauge by picking the representative (φ, A) in each $\mathscr{G}$-orbit for which $\varphi_2(x) = 0$ and $\varphi_1(x) > 0$; note that this so-called ***unitary gauge*** is ill-defined if $\varphi_1(x) = 0$. Calling this unique representative (ρ, B), we are again led to (10.361).

Gauge field theories are ***constrained systems***, in which the *apparent* degrees of freedom in the Lagrangian are not the *physical* ones. For free electromagnetism, the Lagrangian is $\mathscr{L}(A) = -\frac{1}{4}F_{\mu\nu}F^{\mu\nu}$, with $F_{\mu\nu} = \partial_\mu A_\nu - \partial_\nu A_\mu$. In terms of the gauge-invariant fields $E_i = F_{i0} = \partial_i A_0 - \partial_0 A_i$ and $\mathbf{B} = \nabla \times \mathbf{A}$, Maxwell's equations

$$\nabla \cdot \mathbf{E} = 0; \tag{10.366}$$
$$\partial \mathbf{E}/\partial t = \nabla \times \mathbf{B}; \tag{10.367}$$
$$\frac{\partial \mathbf{B}}{\partial t} = -\nabla \times \mathbf{E}; \tag{10.368}$$
$$\nabla \cdot \mathbf{B} = 0, \tag{10.369}$$

then arise as follows: eqs. (10.366) and (10.367) correspond to the Euler–Lagrange equation for A_0 and A_i, respectively, whereas (10.368) and (10.369) immediately follow from the definitions of $\mathbf{B}$ and $\mathbf{E}$ in terms of A. The Maxwell equations are in Hamiltonian form, with canonical momenta $\Pi_\mu = \partial\mathscr{L}/\partial\dot{A}_\mu$; this yields $\Pi_i = -E_i$, as well as the ***primary constraint*** $\Pi_0 = 0$. Nonetheless, the canonical Hamiltonian

$$h = \int d^3x\, (\Pi_\mu(x)\dot{A}_\mu(x) - \mathscr{L}(x)) = \int d^3x\, (\tfrac{1}{2}\mathbf{E}^2(x) + \tfrac{1}{2}\mathbf{B}^2(x) - A_0(x)\nabla\cdot\mathbf{E}(x))$$

is well defined. In the Hamiltonian formalism, Gauss' Law resurfaces as the ***secondary constraint*** stating that the primary constraint be preserved in time, viz.

$$\dot{\Pi}_0(x) = -\frac{\delta h}{\delta A_0(x)} = \nabla \cdot \mathbf{E}(x) \equiv 0. \tag{10.370}$$

Since

$$\frac{d}{dt}\nabla \cdot \mathbf{E}(x) = -\partial_i(\delta h/\delta A_i(x)) = -\partial_i(\Delta A_i - \partial_i \nabla \cdot \mathbf{A}) = 0, \tag{10.371}$$

there are no "tertiary" constraints. Thus we have canonical phase space variables $(\mathbf{E},\mathbf{A})$ and (Π_0,A_0), subject to (10.366) and to $\Pi_0(x)=0$ for each $x\in\mathbb{R}^3$, i.e.,

$$\Pi_0(\lambda_0)\equiv\int d^3x\,\Pi_0(x)\lambda_0(x)=0; \tag{10.372}$$

$$\Pi(\lambda)\equiv\int d^3x\,\nabla\cdot\mathbf{E}(x)\lambda(x)=0, \tag{10.373}$$

for all (reasonable) functions λ_0 and λ on $\mathbb{R}^3$. The constraints (10.372) - (10.373) are ***first class*** in the sense of Dirac, which means that their Poisson brackets are equal to existing constraints (or zero). In the Hamiltonian formalism, the role of the space-time dependent gauge transformations of the Lagrangian theory is played by the canonical transformations generated by the first class constraints, i.e.,

$$\delta_{\lambda_0}A_0(x)=\{\Pi_0(\lambda_0),A_0(x)\}=\lambda_0(x); \tag{10.374}$$

$$\delta_{\lambda_0}A_i(x)=\delta_{\lambda_0}E_i(x)=0; \tag{10.375}$$

$$\delta_{\lambda}\mathbf{A}(x)=\nabla\lambda(x); \tag{10.376}$$

$$\delta_{\lambda}\mathbf{E}(x)=0; \tag{10.377}$$

$$\delta_{\lambda}A_0(x)=0. \tag{10.378}$$

The holy grail of the Hamiltonian formalism is to find variables that are both *gauge invariant* and *unconstrained*. In our case, $A_\mu=(A_0,\mathbf{A})$ are unconstrained but gauge variant, whilst $\Pi_\mu=(\Pi_0,-\mathbf{E})$ are gauge invariant but constrained! Now write some vector field $\mathbf{V}$ as $\mathbf{V}=\mathbf{V}^L+\mathbf{V}^T$, where $\mathbf{V}^L=\Delta^{-1}\nabla(\nabla\cdot\mathbf{V})$ is the longitudinal component, so that $V_i^T=(\delta_{ij}-\Delta^{-1}\partial_i\partial_j)V_j$ is the transverse part. Then the physical variables of free electromagnetism are $\mathbf{A}^T$ and $\mathbf{E}^T$. The physical Hamiltonian

$$h=\tfrac{1}{2}\int d^3x(\mathbf{E}^T\cdot\mathbf{E}^T-\mathbf{A}^T\cdot\Delta\mathbf{A}^T), \tag{10.379}$$

then, is well defined on the physical (or reduced) phase space, which is the subset of all (A_μ,Π_μ) where the constraints (10.373) hold, modulo gauge equivalence.

After this preparation, we now revisit the abelian Higgs model as a constrained Hamiltonian system. It is convenient to combine the two real scalar fields φ_1 and φ_2 into a single complex scalar field $\varphi=(\varphi_1+i\varphi_2)/\sqrt{2}$, and treat φ and its complex conjugate $\overline{\varphi}$ as independent variables. The Lagrangian (10.348) then becomes

$$\mathscr{L}=-\tfrac{1}{4}F_A^2+\overline{D_\mu^A\varphi}\cdot D_\mu^A\varphi-V(\varphi,\overline{\varphi}), \tag{10.380}$$

with $D_\mu^A\varphi=(\partial_\mu-ieA_\mu)\varphi$, etc. The conjugate momenta Π_μ to A_μ are the same as for free electromagnetism, i.e., $\Pi_0=0$ and $\Pi_i=-E_i$, and for φ we obtain

$$\pi=\partial\mathscr{L}/\partial\dot{\varphi}=\overline{D_0^A\varphi}; \tag{10.381}$$

$$\overline{\pi}=\partial\mathscr{L}/\partial\dot{\overline{\varphi}}=D_0^A\varphi. \tag{10.382}$$

The associated Hamiltonian h is equal to

$$\int d^3x\left(\tfrac{1}{2}\mathbf{E}^2+\tfrac{1}{2}\mathbf{B}^2-A_0(\nabla\cdot\mathbf{E}-j_0)+\overline{\pi}\pi+\overline{D_i^A\varphi}\cdot D_i^A\varphi+V(\varphi,\overline{\varphi})\right), \quad (10.383)$$

where $j_0 = ie(\pi\varphi - \overline{\pi\varphi})$ is the zero'th component of the Noether current. Hence the primary constraint remains $\Pi_0 = 0$, but the secondary constraint picks up an additional term and becomes $\nabla\cdot\mathbf{E} = j_0$ (which remains Gauss' law!). The physical (i.e., gauge invariant and unconstrained) variables can be computed as

$$\varphi_A = e^{ie\Delta^{-1}\nabla\cdot\mathbf{A}}\varphi,\ \overline{\varphi}_A = e^{-ie\Delta^{-1}\nabla\cdot\mathbf{A}}\overline{\varphi}; \quad (10.384)$$

$$\pi_A = e^{-ie\Delta^{-1}\nabla\cdot\mathbf{A}}\varphi,\ \overline{\pi}_A = e^{ie\Delta^{-1}\nabla\cdot\mathbf{A}}\overline{\pi}, \quad (10.385)$$

plus the same transverse fields $\mathbf{A}^T$ and $\mathbf{E}^T$, as in free electromagnetism. In terms of the transverse covariant derivative $D_i^T = \partial_i - ieA_i^T$, the physical Hamiltonian h is

$$\int d^3x\left(\tfrac{1}{2}(\mathbf{E}^T\cdot\mathbf{E}^T-\mathbf{A}^T\cdot\Delta\mathbf{A}^T-j_0^A\Delta^{-1}j_0^A)+\overline{\pi}_A\pi_A+\overline{D_i^T\varphi_A}\cdot D_i^T\varphi_A+V(\varphi_A,\overline{\varphi}_A)\right). \quad (10.386)$$

The third term in (10.386) is the Coulomb energy, in which the charge density

$$j_0^A = ie(\pi_A\varphi_A - \overline{\pi}_A\overline{\varphi}_A) \quad (10.387)$$

is the same as j_0 (since the latter is gauge invariant). Remarkably, the physical field variables carry a residual *global* $U(1)$-symmetry, viz.

$$\varphi_A \mapsto \exp(i\alpha)\varphi_A; \quad (10.388)$$

$$\pi_A \mapsto \exp(-i\alpha)\pi_A; \quad (10.389)$$

$$\overline{\varphi}_A \mapsto \exp(-i\alpha)\overline{\varphi}_A; \quad (10.390)$$

$$\overline{\pi}_A \mapsto \exp(i\alpha)\overline{\pi}_A, \quad (10.391)$$

and no change for $\mathbf{A}^T$ and $\mathbf{E}^T$, under which the Hamiltonian (10.386) is invariant.

If V has a minimum at $\varphi = \overline{\varphi} = v$, we recover the Higgs mechanism: redefining

$$\varphi_A = \exp(i\theta/v)(v+\chi), \quad (10.392)$$

and complex conjugate, and the reintroduction of the longitudinal components

$$A_i^L = -(1/ev)\partial_i\theta;\ E_i^L = -ev\Delta^{-1}\partial_i\pi_\theta, \quad (10.393)$$

of the gauge field and its conjugate momentum, the Hamiltonian (10.386) becomes

$$\tfrac{1}{2}\int d^3x\left(\mathbf{E}^2+\mathbf{B}^2+\pi_\chi^2+\partial_i\chi\partial_i\chi+\frac{(\nabla\cdot\mathbf{E})^2}{e^2v^2}+e^2v^2\mathbf{A}^2+V(v+\chi)\right), \quad (10.394)$$

where $\mathbf{A} = \mathbf{A}^T + \mathbf{A}^L$ and $\mathbf{E} = \mathbf{E}^T + \mathbf{E}^L$. This describes a massive vector field, and the would-be Goldstone boson θ has disappeared, as befits the Higgs mechanism!

It is fair to say that the Higgs mechanism in quantum field theory—and more generally, the notion of SSB in gauge theories—is poorly understood. Indeed, the entire quantization of gauge theories is not well understood, except at the perturbative level or on a lattice. The problems already come out in the abelian case with $d = 3$. The main culprit is Gauss' Law $\nabla \cdot \mathbf{E} = j_0$. One would naively expect this constraint to remain valid in quantum field theory as an operator equation, and this is indeed the case in so-called physical gauges like the Coulomb gauge (i.e. $\partial_i A_i = 0$). If we now look at condition (10.337) in §10.9, which for $G = U(1)$ and for example $\delta\varphi_1 = \varphi_2$ and $\delta\varphi_2 = -\varphi_1$ for a charged field $\varphi = (\varphi_1 + i\varphi_2)/\sqrt{2}$, or $\delta\varphi = i\varphi$, reads

$$\lim_{\Lambda \nearrow \mathbb{R}^3} \int_\Lambda d^3y\, \omega([j_0(y,0), \varphi_\alpha(x,t)]) = -i\omega(\delta\varphi_\alpha(x,t)), \tag{10.395}$$

then it is clear that (10.395) can only hold if charged fields are nonlocal. For by Gauss' Law the commutator $[j_0(y,0), \varphi_\alpha(x,t)]$ equals $[\nabla \cdot \mathbf{E}(0,y), \varphi_\alpha(x,t)]$, and by Gauss'(!) Theorem in vector calculus, all contributions to the left-hand side of (10.395) come from terms $[E_i(0,y), \varphi_\alpha(x,t)]$, with $y \in \partial\Lambda$ (i.e., the boundary of Λ). These must remain nonzero if $\Lambda \nearrow \mathbb{R}^3$, at least if (10.395) holds. On the other hand, such nonlocality must be enforced by massless fields, which idea leads to one of the very few rigorous result about the Higgs mechanism (in the continuum):

Theorem 10.29. *In the Coulomb gauge the following conditions are equivalent:*

- *The electromagnetic field* $\mathbf{A}$ *is massless;*
- *Eq.* (10.395) *holds for any field* φ_α*;*
- *The charge operator* $Q = \lim_{\Lambda \uparrow \mathbb{R}^3} \int_\Lambda d^3y\, j_0(y,0)$ *exists (on some suitable domain in* H_ω *containing* Ω_ω*) and satisfies* $Q\Omega_\omega = 0$.

Hence (contrapositively), SSB *of* $U(1)$ *by the state* ω *is only possible if* $\mathbf{A}$ *is massive. In that case, the Fourier transform of the two-point function* $\langle 0|\varphi_\alpha(x,x_0) j_0^a(y,y_0)|0\rangle$ *(cf. the proof of the Goldstone Theorem 10.28 in §10.9) has a pole at the mass of* $\mathbf{A}$.

This theorem indeed yields the Higgs mechanism for say the abelian Higgs model in a specific physical gauge: note that the idea that the would-be Goldstone boson is eaten by the gauge field is already suggested by Gauss' Law, through which (minus) the canonical momentum $\mathbf{E}$ to $\mathbf{A}$ acquires j_0 as its longitudinal component; that is, the very same field that creates the Goldstone boson from the ground state.

In covariant gauges, all fields remain local, but (10.395) is rescued by the gauge-fixing term added to the Lagrangian. For example, adding $\mathscr{L}_{gf} = -(1/2\xi)(\partial_\mu A^\mu)^2$ to (10.348) leads to an equation of motion $\partial_\mu F^\mu_\nu = j_\nu - \partial_\nu \partial_\mu A^\mu$, so that (discarding all surface terms by locality), one obtains

$$-i\omega(\delta\varphi_\alpha(x,t)) = \int_{\mathbb{R}^3} d^3y\, \omega([\partial_0^2 A_0(y,0), \varphi_\alpha(x,t)]). \tag{10.396}$$

In the proof of the Goldstone Theorem, the massless Goldstone bosons do emerge, but they turn out to lie in some "unphysical subspace" of H_ω (which, for local gauges, is not a Hilbert space but has zero- and negative norm states).

Notes

In a philosophical context, the notion of ***emergence*** is usually traced to J.S. Mill (1843), who drew attention to 'a distinction so radical, and of so much importance, as to require a chapter to itself', namely the one between what Mill calls the principle of the 'Composition of Causes', according to which the joint effect of several causes is identical with the sum of their separate effects, and the negation of this principle. For example, in the context of his overall materialism, Mill believed that although all 'organised bodies' are composed of material parts,

> 'the phenomena of life, which result from the juxtaposition of those parts in a certain manner, bear no analogy to any of the effects which would be produced by the action of the component substances considered as mere physical agents. To whatever degree we might imagine our knowledge of the properties of the several ingredients of a living body to be extended and perfected, it is certain that no mere summing up of the separate actions of those elements will ever amount to the action of the living body itself.'
> Mill (1952 [1843], p. 243)

Mill launched what is now called ***British Emergentism*** (Stephan, 1992; McLaughlin, 2008; O'Connor & Wong, 2012), a school of thought which seems to have ended with C.D. Broad, who has our sympathy over Mill because of the doubt he expresses in our quotation in the preamble. Among the British Emergentists, the most modern views seem to have been those of S. Alexander, who, as paraphrased in O'Connor & Wong (2012), was committed to a view of emergence as

> 'the appearance of novel qualities and associated, high-level causal patterns which cannot be directly expressed in terms of the more fundamental entities and principles. But these patterns do not supplement, much less supersede, the fundamental interactions. Rather, they are macroscopic patterns running through those very microscopic interactions. Emergent qualities are something truly new (...), but the world's fundamental dynamics remain unchanged.'

Alexander's idea that emergent qualities 'admit no explanation' and had 'to be accepted with the "natural piety" of the investigator foreshadowed the later notion of ***explanatory emergence***. Indeed, philosophers distinguish between *ontological* and *epistemological* reduction or emergence, but ontological emergence seems a relic from the days of vitalism and other immature understandings of physics and (bio)chemistry (including the formation of chemical compounds, which Broad and some of his contemporaries still saw as an example of emergence in the strongest possible sense, i.e., falling outside the scope of the laws of physics). Recent literature, including the present chapter, is concerned with epistemological emergence, of which explanatory emergence is a branch. For example, Hempel wrote:

> 'The concept of *emergence* has been used to characterize certain phenomena as 'novel', and this not merely in the psychological sense of being unexpected, but in the theoretical sense of being unexplainable, or unpredictable, on the basis of information concerning the spatial parts or other constituents of the systems in which the phenomena occur, and which in this context are often referred to as "wholes".' (Hempel, 1965, p. 62)

See also Batterman (2002), Bedau & Humpreys (2008), Norton (2012), Silberstein (2002), Wayne & Arciszewski (2009), and many other surveys of emergence.

§10.1. **Spontaneous symmetry breaking: The double well**

The facts we use about the double-well Hamiltonian may be found in Garg (2000) or Landau & Lifshitz (1977) at a heuristic level (but with correct conclusions), or, rigorously, in Reed & Simon (1978), Simon (1985), Helffer (1988), and Hislop & Sigal (1996). Theorem 10.2 is Theorem XIII.47 in Reed & Simon (1978).

§10.2. **Spontaneous symmetry breaking: The flea**

The flea perturbation and its effect on the ground state were first described in Jona-Lasinio, Martinelli, & Scoppola (1981a,b), who used methods from stochastic mechanics. See also Claverie & Jona-Lasinio (1986). Using more conventional methods, their results were reconfirmed and analyzed further by e.g. Combes, Duclos, & Seiler (1983), Graffi, Grecchi, & Jona-Lasinio (1984), Helffer & Sjöstrand (1985), Simon (1985), Helffer (1988), and Cesi (1989). The "Flea on the Elephant" terminology used by Simon (1985) motivated the title of Landsman & Reuvers (2013), who, as will be explained in the next chapter, identified the proper host animal as a cat. All pictures in this section are taken from the latter paper (and were prepared by the second author). For the Eyring–Kramers formula see Berglund (2011) for mathematicians or Hänggi, Talkner, & Borkovec (1990) for physicists.

§10.3. **Spontaneous symmetry breaking in quantum spin systems**

The translation-non-invariant ground states mentioned after Proposition 10.5 are discussed e.g. in Example 6.2.56 in Bratteli & Robinson (1997). See also Liu & Emch (2005), which was in important source for this section, and Ruetsche (2011) for a discussion of the definition of SSB through non-implementability. For order parameters see e.g. Sewell (2002), §3.3. A proof of Proposition 10.8 may be found in Bratteli & Robinson (1997), Proposition 6.2.15.

§10.4. **Spontaneous symmetry breaking for short-range forces**

The idea of SSB goes back to Heisenberg(1928). The C*-algebraic approach in quantum spin systems with short-range forces is reviewed in Bratteli & Robinson (1997); see also Nachtergaele (2007). Theorem 10.10 is due to Araki (1974); see also Simon (1993), Theorem IV.5.6, and Bratteli & Robinson (1997), Theorem 6.2.18. In Definition 10.9, Araki required Ω_ω to be separating for $\pi_\omega(A)''$ instead of ω to be α_t-invariant, but in the presence of (10.53) and hence (10.53) these conditions are equivalent. The fact that (for short-range forces) global Gibbs states defined by (10.43) satisfy the KMS condition follows from Theorem 10.10, but this was the starting point of Haag, Hugenholtz, & Winnink (1967); see Winnink (1972).

Uniqueness of KMS states for one-dimensional quantum spin systems with short-range forces at any positive temperature (which also holds for the classical case, e.g. the one-dimensional Ising model) has been proved by Araki (1975). See also Mattis (1965) and Altland & Simons (2010) for some of the underlying physical intuition.

§10.5. **Ground state(s) of the quantum Ising chain**

Theorem 10.11.1 was first established in Pfeuty (1970) by explicit calculation, based on Lieb, Schultz, & Mattis (1961). For more information on the quantum Ising model (also in higher dimension) see e.g. Karevski (2006), Sachdev (2011), Suzuki et al (2013), and Dutta et al (2015). Uniqueness of the ground state of the quantum Ising model with $B \neq 0$ holds in any dimension d, as first shown by Campanino,

Klein & Perez (1991) on the basis of Perron–Frobenius type arguments similar to those for Schrödinger operators. The singular case $B = 0$ leads to a violation of the strict positivity conditions necessary to apply the Perron–Frobenius Theorem, and this case indeed features a degenerate ground state even when $N < \infty$.

The overall picture of SSB described in this section arose from the work of Horsch & von der Linden (1988), Kaplan, Horsch, & von der Linden (1989), Kaplan, von der Linden, & Horsch (1990), and especially Koma & Tasaki (1993, 1994). See also van Wezel (2007, 2008), van Wezel & van den Brink (2007), and Fraser (2016).

The analogy between the quantum Ising chain and the double-well potential may not be surprising physically, since the latter was originally derived from the former: in potassium dihydrogen phosphate, i.e. KH_2PO_4, each proton of the hydrogen bond would reside in one of the two minima of an effective double-well potential originating in the oxygen atoms, if it were not for tunneling, parametrized by the field B, which at small values yields a symmetric ground state (De Gennes, 1963).

§10.6. **Exact solution of the quantum Ising chain:** $N < \infty$

The general set-up to this solution is due to Lieb, Schultz, & Mattis (1961), and was adapted to the quantum Ising by Pfeuty (1970), with further details by Karevski (2006). The complex solution q_0 was already noted by Lieb et al. The energy splitting in higher dimensions does not seem to be known, but Koma & Tasaki (1994, eq. (1.5)) expect similar behaviour as in $d = 1$.

§10.7. **Exact solution of the quantum Ising chain:** $N = \infty$

The solution described in this section is due to Araki & Matsui (1985), where further details may be found; this is a highlight of modern mathematical physics! Theorem 10.20 is due to Araki (1987), although such results have a long history going back to Shale & Stinespring (1964, 1965). For a very clear exposition see Ruijsenaars (1987). See also Evans & Kawahigashi (1998), Chapter 6.

The reason the one-sided chain $\Lambda = \mathbb{N}$ is problematic is that although the bosonic algebra $\otimes_{j\in\mathbb{N}} M_2(\mathbb{C})$ and its fermionic counterpart $CAR(\ell^2(\mathbb{N}))$ are well defined, and are isomorphic through the Jordan–Wigner transformation (10.102) - (10.103), the limiting dynamics has no simple form on either A or F, because the Fourier transform of $\ell^2(\mathbb{N})$ is the Hardy space $H^2(-\pi,\pi)$ of L^2-functions with positive Fourier coefficents, instead of the usual $L^2(-\pi,\pi)$. Unlike on L^2, The energies sgn_k of the fermionic quasiparticles do not define a multiplication operator on H^2.

§10.8. **Spontaneous symmetry breaking in mean-field theories**

The Poisson structure on $S(B)$ was introduced by Bona (1988) and more generally by Duffield & Werner (1992a); see also Bona (2000). Theorem 10.22 and Corollary 10.23 are due to Duffield & Werner (1992a). The symplectic leaves of the given Poisson structure on $S(B)$ (for which notion see e.g. Marsden & Ratiu (1994) or Landsman (1998a)) were determined by Duffield & Werner (1992a): Two states ρ and σ lie in the same symplectic leaf of $\mathscr{S}(B)$ iff $\rho(a) = \sigma(uau^*)$ for some unitary $u \in B$. If ρ and σ are pure, this is the case iff the GNS-representations $\pi_\rho(B)$ and $\pi_\sigma(B)$ are unitarily equivalent, cf. Thm. 10.2.6 in Kadison & Ringrose (1986). In general the implication holds only in one direction: if ρ and σ lie in the same leaf, then they have unitarily equivalent GNS-representations.

Our survey of equilibrium states of homogeneous mean-field models is based on Fannes, Spohn, & Verbeure (1980) and Bona (1989). For rigorous results on the Curie–Weiss model see Chayes et al (2008) and Ioffe & Levit (2013). Numerical evidence for the restoration of Butterfield's Principle may be found in Botet, Julien & Pfeuty (1982) and Botet & Julien (1982), which are up to $N \sim 150$, and Vidal et al (2004), which reaches $N = 1000$. Note that experimental samples have $N < 10$.

In the context of the BCS model of superconductivity in the strong coupling limit), the Hamiltonian, $\hat{h}_\theta$ in (10.287) or h_ω in (10.294) is called the ***Bogoliubov–Haag Hamiltonian***, after Bogoliubov (1958) and Haag (1962). Further contributions to mean-field theories include Thirring & Wehrl (1967), Thirring (1968), Hepp (1972), Hepp & Lieb (1973), van Hemmen (1978), Rieckers (1984), Morchio & Strocchi (1987), Duffner & Rieckers (1988), Bona (1988, 1989, 2000), Unnerstall (1990a, 1990b), and Sewell (2002). For a nice proof of Theorem 10.25, which originates in Fannes, Spohn, &Verbeure (1980) and Bona (1989), see Gerisch (1993).

Even in the absence of a global KMS condition for $\hat{\omega}^\beta$, one is justified in interpreting the primary states $(\omega_\theta^\beta)^\infty$ as pure thermodynamic phases of the given infinite quantum system, whose thermodynamics is described by the "phase space" $S(M_n(\mathbb{C}))$. Though somewhat against the spirit of Bohrification (according to which the commutative C*-algebra $C(M_n(\mathbb{C}))$ is the right one to look at), the argument can be strengthened by enlarging A to $A \otimes C(M_n(\mathbb{C}))$ (where the choice of the tensor product does not matter, since $C(M_n(\mathbb{C}))$ is commutative and hence nuclear, see §C.13). This larger C*-algebra was introduced by Bona (1990), who proved:

Theorem 10.30. *1. There is a unique time-evolution α on $A \otimes C(M_n(\mathbb{C}))$ such that for any primary permutation-invariant state ω on A and $a \in A$ one (strongly) has*

$$\lim_{N\to\infty} \pi_\omega\left(e^{ith_{\Lambda_N}} a e^{-ith_{\Lambda_N}}\right) = \pi_\omega(\alpha_t(a)). \qquad (10.397)$$

2. The states $\hat{\omega}^\beta$ and ω_θ^β in (10.286), which are defined on A, extend to the tensor product $A \otimes C(M_n(\mathbb{C}))$ as $\hat{\omega}^\beta \otimes \mu_\beta$ and $\omega_\theta^\beta \otimes \delta_\theta$, respectively, and as such satisfy the KMS condition at inverse temperature β with respect to the dynamics α.

§10.9. **The Goldstone Theorem**

There is a large amount of literature on the Goldstone Theorem, both heuristic and rigorous. The former started with Goldstone, Salam, & Weinberg (1962), whereas the latter originates in Kastler, Robinson, & Swieca (1966); see also Buchholz et al (1992). For a survey, see Strocchi (2008, 2012), whose approach (based on Morchio & Strocchi, 1987) we follow. See also Berzi (1979, 1981), Landau, Perez, & Wreszinski (1981), Fannes, Pule, & Verbeure (1982), and Wreszinski (1987).

§10.10. **The Higgs mechanism**

The original reference is Higgs (1964ab). Our discussion is based on Lusanna & Valtancoli (1996ab) and Struyve (2011), both of whom derive the physical variables in the abelian Higgs model. See also Rubakov (2002), Strocchi (2008), where Theorem 10.29 may be found, and Stöltzner (2014) for some history and sociology.

Chapter 11
The measurement problem

The measurement problem of quantum mechanics was probably born in 1926:

> 'Thus Schrödinger's quantum mechanics gives a very definite answer to the question of the outcome of a collision; however, this does not involve any causal relationship. One obtains *no* answer to the question "what is the state after the collision," but only to the question "how probable is a specific outcome of the collision" (in which the quantum-mechanical law of [conservation of] energy must of course be satisfied). This raises the entire problem of determinism. From the standpoint of our quantum mechanics, there is no quantity that could causally establish the outcome of a collision in each individual case; however, so far we are not aware of any experimental clue to the effect that there are internal properties of atoms that enforce some particular outcome. Should we hope to discover such properties that determine individual outcomes later (perhaps phases of the internal atomic motions)? Or should we believe that the agreement between theory and experiment concerning our in ability to give conditions for a causal course of events is some pre-established harmony that is based on the non-existence of such conditions? I myself tend to relinquish determinism in the atomic world. But this is [also] a philosophical question, for which physical arguments alone are not decisive.' (Born, 1926a, p. 866; translation by the author)

In other words, quantum mechanics stipulates that the state after some collision (or measurement) is $\psi = \sum_n c_n \psi_n$, whereas experiment demonstrates that in fact the final state is just one of the ψ_n, with (Born) probability $|c_n|^2$. Quantum mechanics, then, seems unable to account for single outcomes of experiments and has to satisfy physicists with merely probabilistic predictions. This, in a nutshell, is the measurement problem—although very substantial analysis is needed to flesh it out.

Giving up determinism was soon incorporated in the Copenhagen Interpretation of Bohr and Heisenberg (cf. the Introduction) and more broadly became part of what might be called "***orthodoxy***", which represents the apparent (but not actual) consensus among Bohr, Heisenberg, Pauli, Born, Jordan, Dirac, von Neumann, and many others, which they supposedly reached around 1930 after the formal completion of quantum mechanics. This "orthodoxy", which later gave rise to the unfortunate "shut up and calculate" attitude most physicists seem to have (especially towards the measurement problem), should be distinguished from the Copenhagen Interpretation. For example, von Neumann never endorsed the doctrine of classical concepts, which in the above attitude has been replaced by the different and far more superficial idea that it is the entire goal of physics to explain experiments.

K. Landsman, *Foundations of Quantum Theory*,
Fundamental Theories of Physics 188, DOI 10.1007/978-3-319-51777-3_11

11.1 The rise of orthodoxy

Even within the strict Copenhagen Interpretation, there were sharp differences between Bohr and Heisenberg, beyond the one concerning classical concepts reviewed in the Introduction. However, it seems that they agreed about the following point made by Bohr in his Como lecture concerning measurement:

> 'According to the quantum theory, just the impossibility of neglecting the interaction with the agency of measurement means that every observation introduces a new uncontrollable element.' (Bohr, 1928, p. 584)

This placed measurement squarely outside quantum mechanics for the second time: the first time was in the insistence that the measurement device ("if it is to serve its purpose") had to be described classically (cf. the Introduction), and now we also learn that the interaction between the quantum object undergoing measurement and the apparatus in question is "uncontrollable", *despite the fact that Bohr and Heisenberg regarded quantum mechanics as a complete theory*: their argument was apparently that precisely the classical nature of the apparatus makes the interaction uncontrollable. This in turn justified the classical description of the device, in that registration of a measurement result ought to be "objective", so that reading it out by performing a measurement on the apparatus, so to speak, should not introduce any further disturbance and hence uncontrollability (or so the argument goes).

Consistent with Bohr's point, a more detailed conceptual analysis of the measurement process was given by Heisenberg (1958, pp. 46–47, 54–55), who consistently refers to the quantum state or wave-function as the "probability function":

> 'Therefore, the theoretical interpretation of an experiment requires three distinct steps:
>
> 1. the translation of the initial experimental situation into a probability function;
> 2. the following up of this function in the course of time;
> 3. the statement of a new measurement to be made of the system, the result of which can then be calculated from the probability function.
>
> (...) After [the] interaction [with the measuring device] has taken place, the probability function contains the objective element of tendency and the subjective element of incomplete knowledge, even if it has been a "pure case" before [i.e., it has become a mixture]. It is for this reason that the result of the observation cannot generally be predicted with certainty; what can be predicted is the probability of a certain result of the observation, and this statement about the probability can be checked by repeating the experiment many times. (...) The observation itself [i.e., the act of registration of the result by the mind of the observer] changes the probability function discontinuously; it selects of all possible events the actual one that has taken place. Since through the observation our knowledge of the system has changed discontinuously, its mathematical representation also has undergone the discontinuous change and we speak of a "quantum jump."'

Here we find the typical Copenhagen view of measurement as a two-step process:

1. Measurement turns an initial pure state (of the measured object) into a mixture;
2. One term in this mixture is singled out (by Nature and thence by the observer).

Note that Heisenberg's last comment puts him squarely into the camp of what is now called "QBism" (i.e., ***Quantum Bayesianism***, see §11.2 below)!

Von Neumann (1932, §VI.1) gave a more formal (and highly influential) presentation of the (alleged) two stages of the measurement process:

'In the discussion so far we have treated the relation of quantum mechanics to the various causal and statistical methods of describing nature. In the course of this we found a peculiar dual nature of the quantum mechanical procedure which could not be satisfactorily explained. Namely, we found that on the one hand a state ϕ is transformed into the state ϕ' under the action of an energy operator H in the time interval $0 \leq \tau \leq t$:

$$\frac{\partial}{\partial \tau}\phi_\tau = -\frac{2\pi i}{h}\mathsf{H}\phi_\tau \ : \ 0 \leq \tau \leq t$$

so if we write $\phi_0 = \phi$, $\phi_t = \phi'$ then $\phi' = e^{-\frac{2\pi i}{h}t\mathsf{H}}\phi$, which is purely causal. A mixture U is correspondingly transformed into

$$\mathsf{U}' = e^{-\frac{2\pi i}{h}t\mathsf{H}}\mathsf{U}e^{+\frac{2\pi i}{h}t\mathsf{H}}$$

Therefore, as a consequence of the causal change of ϕ into ϕ' the [pure] states $\mathsf{U} = \mathsf{P}_{[\phi]}$ [$=|\phi\rangle\langle\phi|$] go over into the [pure] states $\mathsf{U}' = \mathsf{P}_{[\phi']}$ (process **2** in V.1.). On the other hand, the state ϕ—which may measure a quantity with discrete spectrum, distinct eigenvalues and eigenfunctions $\phi_1, \phi_2, \ldots$—undergoes in a measurement a non-causal change in which each of the states $\phi_1, \phi_2, \ldots$ can result, and in fact does result with the respective probabilities $|\langle\phi,\phi_1\rangle|^2, |\langle\phi,\phi_2\rangle|^2, \ldots$. That is, the mixture

$$\mathsf{U}' = \sum_{n=1}^{\infty} |\langle\phi,\phi_n\rangle|^2 \mathsf{P}_{[\phi']}$$

obtains (...) (process **1** in V.1.). Since the [pure] states [i.e. $\mathsf{P}_{[\phi]}$] go over into mixtures, the process is not causal. The difference between these two processes $\mathsf{U} \mapsto \mathsf{U}'$ is a very fundamental one: aside from their different behaviors in regard to the principle of causality, they are also different in that the former is (thermodynamically) reversible, while the latter is not.' (pp. 417–418 in von Neumann (1955); translation: R.T. Beyer)

All this concerns merely the first stage of the measurement, in which a pure state is transformed into a mixed one. The second stage, in which a single outcome is obtained, is already alluded to above (though clouded by von Neumann's ensemble language), but is described (in prose) later on through what is now called a ***von Neumann chain***: one redefines system plus apparatus as the system, and couples it to a new apparatus, etc. This chain supposedly ends with the "ego" of the "individual" whose "intellectual inner life" is finally responsible for a single outcome.

It is very remarkable that von Neumann nowhere seems to use the central Copenhagen dogma that the apparatus be described classically (cf. the Introduction), especially since the mathematics of operator algebras he was inventing at almost exactly the same time is tailor-made for incorporating this dogma (which fact indeed forms the motivation for the present book). One clue for his lack of enthusiasm may come from the very end of his book (i.e., §VI.3), where he challenges 'an explanation often proposed to account for the statistical character of the process 1', namely the idea that (the non-unitary) process 1 might have its origin in an initial mixed state of the apparatus. Indeed, even if the apparatus as a quantum-mechanical system is in a pure state (as any system should be ontologically), its description as a classical system generally renders its state mixed—and the same conclusion may be drawn on epistemic grounds, arguing that the state of macroscopic or otherwise complicated systems cannot be known exactly. Many writings by the Copenhagen school, then, suggest that the alleged unanalyzable nature of the measurement and the randomness of its outcome should be attributed to the classical description of the apparatus and its ensuing mixed state, including our earlier quotation (cf. §8.4) from Heisenberg (1958) on the origin of probabilities in quantum mechanics:

> 'these uncertainties (...) are simply a consequence of the fact that we describe the experiment in terms of classical physics' (Heisenberg, 1958, p. 53)

To counter this argument, von Neumann argues that physics requires the (Born) probabilities for the various outcomes to depend only on the initial state ϕ of the quantum system undergoing measurement (as opposed to the state of the apparatus, be it classical or quantum), whereas any "process 2" (i.e. unitary) time evolution would merely push the coefficients w_n in the (alleged) mixed apparatus state into the role of probabilities for the possible outcomes. However, 'the w_n are characteristic of the observer alone (and therefore independent of ϕ)', and hence

> 'the non-causal nature of the process 1. is not produced by any incomplete knowledge of the state of the observer.' (von Neumann, 1955, p. 439).

Von Neumann's argument became the mother of all "insolubility theorems" for the measurement problem, some of which will be reviewed in §11.3 below.

Pauli (1933, §9) also includes some comments on measurement and the interpretation of quantum mechanics in general. These display a bizarre hybrid between the ideas of Bohr and von Neumann, somehow mediated by Heisenberg. Thus Pauli endorses (even starts with) some notion of Complementarity, but he relates this to the mathematical formalism rather than to the doctrine of classical concepts (which he nowhere invokes). Similarly, his treatment of measurement on the one hand follows the disturbance ideology of Bohr and Heisenberg (but without grounding this in the classical description of the apparatus), whilst technically he quotes and follows von Neumann, claiming that measurement leads to mixtures which subsequently reduce to one term through '*ein besonderer, naturgesetzlich nicht im Voraus determinierter Akt*' (i.e., special process that does not follow deterministic laws of nature). A rather more systematic review of early measurement theory was written by London & Bauer (1939), whose opening is highly promising and almost poetic:

> 'The majority of introductions to quantum mechanics follow a rather dogmatic path from the moment that they reach the statistical interpretation of the theory. In general they are content to show, by more or less intuitive considerations, how the actual measuring devices always introduce an element of indeterminism, as this interpretation demands. However, care is rarely taken to verify explicitly that the formalism of the theory, applied to that special process which constitutes the measurement, truly implies a transition of the system under study to a state of affairs less fully determined than before. A certain uneasiness arises. One does not see exactly with what right and up to what point one may, in spite of this loss of determinism, attribute to the system an appropriate state of its own. Physicists are to some extent sleepwalkers, who try to avoid such issues and try to concentrate on concrete problems. But it is exactly these questions of principle which nevertheless interest nonphysicists and all who wish to understand what modern physics says about the analysis of the act of observation itself.' (London & Bauer, 1939, pp. 218-219)

Yet the authors mainly repeat von Neumann's analysis (confirming its lofty status):

> 'The interaction with the apparatus does not put the object into a new pure state. Alone, it does not confer to the object a new wave function. On the contrary, it actually gives nothing but a statistical mixture: It leads to one mixture for the object and one mixture for the apparatus. For either system regarded individually there results uncertainty, incomplete knowledge. Yet nothing prevents our reducing this uncertainty by further observation.
>
> And this is our opportunity. So far we have only coupled one apparatus with one object. But a coupling even with a measuring device is not yet a measurement. A measurement is achieved only when the position of the pointer has been observed. It is precisely the increase of knowledge, acquired by the observation, that gives the observer the right to choose among the different components of the mixture predicted by the theory, to reject those which are not observed, and to attribute thenceforth to the object a new wave function, that of the pure case which he has found. We note the essential role played by the consciousness of the observer in this transition from the mixture to the pure state. Without his effective intervention, one would never obtain a new ψ function.' (*ibid.*, p. 251)

Accordingly, at the end of the golden era of quantum mechanics, the view of measurement as a two-stage process in which a pure state is first transformed into a mixture in a more or less scientific way, upon which unanalyzable and possibly mental phenomena bring about a single outcome, was firmly established, although—the point deserves to be repeated—in their formal treatments neither von Neumann nor London & Bauer incorporated the key claim Bohr and Heisenberg made about measurement, namely that the corresponding apparatus *must* be described classically.

Opponents of the Copenhagen Interpretation (the most prominent among whom were Einstein and Schrödinger) were well aware of this tension between formalism and ideology, which in the form of ***Schrödinger's Cat*** even reached immortality (!):

> 'One may also construct highly burlesque cases. A cat is confined in a box of steel together with the following hellish machine (which one should secure against a direct attack by the cat): A Geiger counter contains a tiny amount of radioactive material, *so* little that during one hour *possibly* one of its atoms decays, but equally likely also none does; if it does, then the counter is triggered and activates, via a relais, a little hammer which breaks a small container of hydrocyanic acid. Having left this system to itself for one hour, one will say that the cat is still alive *if* meanwhile no atom has decayed. The first decay of an atom would have poisoned her. The ψ-function of the entire system would express this in such a way that in it the living and the dead cat would be mixed or spread out on equal terms. What is typical about these cases is that an uncertainty which is originally limited to the atomic domain has been transformed into a coarse-grained uncertainty, which may then be *decided* by direct observation. This prevents us from regarding a "faded model" as an image of reality in such a naive way. As such [this model] contains nothing that is unclear or contradictory. There is a difference between a moved or poorly focused photograph and a record of clouds and fog banks.' (Schrödinger, 1935, p. 812; translation by the author)

The last sentence is particularly powerful, contrasting Schrödinger's (as well as Einstein's) view that physics should describe some sharply defined reality (of which quantum mechanics at best produces blurred pictures) with the Copenhagen view, according to which reality itself lacks focus (with quantum mechanics providing the best possible picture of it). This contrast confirms our idea that Schrödinger's Cat metaphor specifically draws attention to the problems that arise from the Copenhagen "duality postulate" that macroscopic systems (such as measurement devices and cats) admit both a classical and a quantum-mechanical description.

11.2 The rise of modernity: Swiss approach and Decoherence

Despite Schrödinger's Cat, the measurement problem was not an active field of research until Wigner (1963) rekindled interest in the topic. Even so, his paper mainly reiterated von Neumann's views—which already had been repeated by London and Bauer—including his omission of the doctrine of classical concepts. In particular, it continued to promulgate the suggestion that measurement is a two-step process for which the clarification of the first step (i.e. of turning a pure state into a mixture) would already be a major part of the solution of the measurement problem.

Wigner's paper inspired for example the "'Swiss" approach to the measurement problem, which was remarkable in being the first serious mathematical attempt to take into account the Bohr–Heisenberg dogma that the apparatus be described classically, whilst also paying tribute to von Neumann in insisting on mathematical rigour. Indeed, the Swiss approach relies on the formalism of operator algebras, which also marks a conceptual break with all earlier—and indeed most later—approaches in taking the observables rather than the states as a starting point. The aim of the Swiss approach is to show that relative to a suitable class of observables, the pure state

$$\rho = |\psi\rangle\langle\psi|, \ \psi = \sum c_n \psi_n,$$

coincides with the corresponding mixture without the off-diagonal terms, i.e.,

$$\rho' = \sum_n |c_n|^2 |\psi_n\rangle\langle\psi_n|.$$

Thus the ambition of this approach is limited, in that no attempt is made to explain (at least the appearance of) single outcomes, except by appealing to the ignorance interpretation of probability (in vain, see below). The alleged equivalence between pure states and mixtures can typically be achieved if the apparatus is infinite and the measurement time is infinite, too. The infinite character of the apparatus (here seen as an idealization of a macroscopic device, as is standard in quantum statistical mechanics), is no guarantee for its classicality, but it is certainly a step in the right direction (cf. Chapter 8). Thus two closely related problems must be overcome:

1. In its reliance on superselection sectors (technically, on disjoint states on a suitable algebra of observables of the apparatus, see Definition 8.18), the program only works in the limit of infinite apparatus and infinite measurement time. Indeed, any approximation ruins the equivalence between pure states and mixtures; and hence even this limited solution to the problem violates Earman's Principle.
2. In so far as the subsequent problem of obtaining single outcomes to measurement is recognized in the Swiss approach at all, it seems to be addressed by an appeal to the ignorance interpretation of probability. Despite the fact that the mathematical situation in this respect is better than in ordinary quantum mechanics (where the ignorance interpretation of the formal probability distribution given by the coefficients in a diagonal density operator is nonsensical, if only because the state space is not a simplex), there is still no valid argument for this move.

To explain the last point, we quote Leggett (though somewhat out of context):

> 'Now, following Schrödinger, let us consider a thought experiment in which the quantum-mechanical description of the final state, as obtained by appropriate solution of the time dependent Schrdinger equation, contains simultaneously nonzero probability amplitudes for two or more states of the universe that are, by some reasonable criterion, macroscopically distinct (in Schrödingers example, this would be "cat alive" and "cat dead"). Of course, just about everyone, including me, would accept that because of, inter alia, the effects of decoherence, it is likely to be impossible, at least for the foreseeable future, to experimentally demonstrate the interference of such states. (On the other hand, as the late John Bell was fond of pointing out, the foreseeable future is not a very well-defined concept. In fact, as late as 1999, not a few people were confidently arguing that because of the inevitable effects of decoherence, the projected experiments to demonstrate interference at the level of flux qubits would never work. In this case, the foreseeable future lasted approximately one year. As Bell used to emphasize, the answers to fundamental interpretive questions should not depend on the accident of what is or is not currently technologically feasible.) But the crucial point is that the formalism of quantum mechanics itself has changed not one whit between the microscopic and macroscopic levels. Are we then entitled to embrace, at the macrolevel, an interpretation that was forbidden at the microlevel, simply because the evidence against it is no longer available? I would argue very strongly that we are not, and would therefore draw the conclusion: also at the macrolevel, when the quantum-mechanical description assigns simultaneously nonzero [probabilities] to two or more macroscopically distinct possibilities, then it is not the case that each system of the relevant ensemble realizes either one possibility or the other.' (Leggett, in Schlosshauer, 2011, p. 155)

This argument of Leggett's (which is a special case of Earman's Principle) was originally targeted at decoherence, but it also applies *verbatim* to the Swiss approach (which is closely related to decoherence, as both heavily rely on limits and superselection rules—which are absolute in the former and dynamically induced in the latter). In an even earlier hunch of Earman's Principle, Bell— this time aiming directly at the Swiss approach—in fact made a related point about its reliance on the $t \to \infty$ limit (in that even at extremely large but finite time the state remains pure).

Jumping to the modern era, a striking point of continuity with the 1920s and 1930s is the idea that the measurement procedure (and hence the measurement problem) consists of two stages; only the terminology and the scope have changed:

> 'There are two distinct measurement problems in quantum mechanics: what Pitowsky has called a "big" measurement problem and a "small" measurement problem. The "big" measurement problem is the problem of explaining how measurements can have definite outcomes, given the unitary dynamics of the theory: it is the problem of explaining how individual measurement outcomes come about dynamically. The "small" measurement problem is the problem of accounting for our familiar experience of a classical, or Boolean, macroworld, given the non-Boolean character of the underlying quantum event space: it is the problem of explaining the dynamical emergence of an effectively classical probability.' (Bub, in Schlosshauer, 2011, pp. 145–146)

Clearly, the "small" measurement problem is modern parlance for the problem how to turn a superposition into a mixture, upon which the "big" problem—if it is noticed at all—still concerns the old issue of selecting *one* term from this mixture.

Furthermore, the measurement problem seems to have acquired increased scope and importance, as exemplified by the following quotations:

'One of the most ancient philosophical questions (Heidegger thought is was *the* question) is this: why is there something rather than nothing? In terms of events rather than substances, the question would be: how come anything happens at all? That question is the measurement problem.' (Fine, in Schlosshauer, 2011, p. 146)

'The measurement problem has been called "the reality problem" by Philip Pearle. This is a better name for it. We perceive objects in the world as being in definite states. A door is either open or shut, a given ball either is in a given box or it is not. The wave function, however, can have superpositions of these things, suggesting that thc door can be simultaneously open and shut at the same time, and that the ball can be both in the box and not in the box at the same time. The reality problem is that there is a discrepancy between the version of reality we perceive, and the version presented to us by the most obvious interpretation of the wave function.' (Hardy, in Schlosshauer, 2011, p. 153)

'Fundamentally, the measurement problem is the problem of connecting probability with truth in the quantum world, that is to say, it is the problem of how to relate quantum probabilities to the objective occurrence and non-occurrence of events. The problem arises because there appears to be a difficulty in reconciling the objectivity of a particular measurement outcome with the entangled state at the end of a measurement.' (Bub, *ibid.*, p. 145)

More technically, the measurement problem has come to be seen as a special case of the problem of explaining at least the *appearance* of the classical world from quantum theory. If the measurement problem is seen from the Copenhagen perspective this is eminently reasonable, as both problems involve the dual description of either the apparatus or the world around us as both classical and quantum (and its possible failure). In this context, an alleged solution to the "small" problem, such as Decoherence, is often also seen as this explanation (as if there were no issue about the derivation of the laws of classical physics, including the dynamical ones).

A propos, another characteristic feature of the modern era is undoubtedly the dominance of ***Decoherence*** (if only over the Swiss approach), for example:

'I think the whole discussion about whether measurements in quantum mechanics are indeed problematic somewhat misses the point. Measurement interactions are only one of many examples of quantum interactions that lead to superpositions of macroscopically distinct states. Nature has been producing macroscopic superpositions for millions of years, well before any quantum physicist cared to artificially engineer such a situation. The key concept here is decoherence. Environmental interactions tend to produce superpositions of classically distinct states. This raises the issue of how one could describe a classical regime in quantum mechanics, quite irrespective of the existence of measuring apparatuses. (...)

If decoherence and its applications had been developed early in the history of quantum theory, then the idea that measurements play a special role in the theory might not have risen to such prominence, and the foundations of quantum mechanics would have focused instead on the problem of how to derive a classical regime within the theory.'
(Bacciagaluppi, in Schlosshauer, 2011, p. 143)

Mathematically, decoherence boils down to the idea of adding one more link to the von Neumann chain (see §11.1) beyond $S+A$ (i.e. the system and the apparatus). Conceptually, however, there is a fundamental conceptual as well as technical difference between Decoherence and older approaches that took such a step: whereas previously (e.g., in the hands of von Neumann, London & Bauer, and Wigner) the chain *converged towards the observer*, in Decoherence it *diverges away from the observer*. Namely, the third and final link is now taken to be the ***environment***.

This notion is often taken in a fairly literal sense in agreement with the intuitive meaning of the word, but it may also (we would even say: preferably) refer to internal degrees of freedom of the apparatus, as in the Spehner–Haake model in §11.4. Either way, the "environment" is usually treated as an infinite system (necessitating a limit like $N \to \infty$), which (in simple models where the pointer has discrete spectrum) has the consequence that the post-measurement state $\sum_n c_n \psi_n \otimes \phi_n \otimes \chi_n$ (in which the χ_n are mutually orthogonal) is only reached not only in the limit $N \to \infty$ of infinitely many degrees of freedom but also in the limit $t \to \infty$ of infinite time. In that case, the restriction of the above state to $S+A$ (i.e. the trace of the corresponding density operator over the degrees of freedom of the environment) is mixed, which means that the quantum-mechanical interference between the states $\psi_n \otimes \phi_n$ for different values of n has become "delocalized" to the environment, and accordingly is deemed irrelevant if the latter is not observed (i.e. omitted from the description).

Unfortunately, in so far as it claims to provide a solution to the measurement problem, Decoherence is an unmitigated disaster:

1. Decoherence actually *aggravates* the measurement problem: where previously this problem was believed to be man-made and relevant only to rather unusual laboratory situations, it has now become clear that "measurement" of a quantum system *by the environment* (instead of by an experimental physicist) happens everywhere and all the time: hence it remains even more miraculous than before that there is a single outcome after each such measurement.
2. Even the need for *one* of the two limits $N \to \infty$ or $t \to \infty$ makes Decoherence vulnerable to Earman's Principle; see Bell's and Leggett's critiques above.
3. Like the Swiss approach, Decoherence suffers from the difficulty that even if it were able to reach its goal of reducing pure states to mixtures (about which ability one may have doubts), there is no sound follow-up step to solve the next problem of selecting one term from the mixture produced in the previous step. The ignorance interpretation seems blocked by Leggett's argument quoted above (i.e. his continuity argument to the effect that Decoherence just removes the *evidence* for a given Schrödinger's cat state to be a superposition, elsewhere charging those claiming that Decoherence solves the measurement problem of committing the logical fallacy that removal of the evidence for a crime would undo the crime).

Thus Decoherence is parasitic on some interpretation of quantum mechanics that solves the measurement problem, which in turn is typically strengthened by it. In this context, the most popular of these has been the Everett (i.e., Many-Worlds) Interpretation, which, after decades of obscurity or even derision, suddenly started to be greeted with a flourish of trumpets in the wake of the popularity of Decoherence. However, even if such extravagant interpretations are coherent, these should in our opinion be a very last resort, acceptable only if truly everything else has failed.

On the positive side, Decoherence has led to the important idea of ***einselection*** (for *environment-induced superselection*), where a pure state ψ of some system (possibly plus apparatus) is "einselected" if it remains pure after coupling to the environment and subsequent restriction. The hope (or rather program), then, is to show that classical states are classical precisely because they are robust in this way.

Finally, it may be appropriate to close this historical introduction to the measurement problem by mentioning another modern approach, namely outright *denial*:

> 'I remember giving a talk at a meeting at the London School of Economics seven or so years ago. In the audience was an Oxford philosophy professor, and I suppose he didn't much like my brash cowboy dismissal of a good bit of his life's work. When the question session came around, he took me to task with the most proper and polite scorn I had ever heard (I guess that's what they do). "Excuse me. You seem to have made an important point in your talk, and I want to make sure that I have not misunderstood anything. Are you saying that you have solved the measurement problem? This problem that has plagued quantum mechanics for seventy-five years? The message of your talk is that, using quantum information theory, you have finally solved it?" (Funny the way the words could be put together as a question, but have no intended usage but as a statement.) I don't know that I did anything but turn the screw on him a bit further, but I remember my answer. "No, not me; I havent done anything. What I am saying is that a "measurement problem" never existed in the first place. (...)
>
> The "measurement problem" is purely an artefact of a wrong-headed view of what quantum states and/or quantum probabilities ought to be. (...) quantum states are not real things from a Quantum Bayesian view (...) but a personal judgment, a quantified degree of belief. A quantum state is a set of numbers an agent uses to guide the gambles he might take on the consequences of his potential interactions with a quantum system. It has no more substantiality than that. Aren't epistemic states real things? Well ... yes, in a way. They are as real as the people who hold them. But no one would consider a person to be a property of the quantum system he happens to be contemplating. And one shouldn't think of a quantum state in that way either—one shouldnt think of it as a property of the quantum system to which it is assigned. Take the source of the paradox away, we say, and the paradox itself will go away.' (Fuchs, in Schlosshauer, 2011, pp. 146–147)

These words have been quoted at some length, because the view that "physics is information" and its alleged corollary that all foundational problems are solved by Bayesian reasoning (perhaps with a quantum flavour) is becoming increasingly popular. Physicist are now seen as punters (or, in academic parlance, "agents") who in smoky offices bet on the outcomes of experiments, and hence use (quantum) Dutch Book arguments to justify some sort of strictly epistemic (quantum) probability calculus. However, the ideology of "***QBism***" thus expressed appears to have adopted precisely the weakest ingredients of the Copenhagen Interpretation—viz. the idea that the wave-function is just a catalogue of the probabilities for possible outcomes of measurements whose details are supposedly beyond our grasp, cf. the Introduction—at the expense of its one strong component, namely the doctrine of classical concepts. Although there may have been pragmatic reasons for this attitude in the 1920s, (mathematical) physics has moved forward since then, enabling much more detailed analysis and hence justifying considerably greater ambition in understanding the measurement process than Bohr and Heisenberg *cum suis* had.

In any case, the fact that one competent author regards the measurement problem as the key to reality whilst another flatly denies even its very existence should give pause for thought. As in the Bohr–Einstein debate, different perspectives on reality and on the task of physics seem to play a role here, culminating in contrasting views of quantum-mechanical states: the more "reality" one attributes to states, the more serious the measurement problem is. Or, contrapositively, the more operationalist one's attitude, the further the problem disappears behind the horizon.

11.3 Insolubility theorems

Since in §11.4 we will "propose the impossible", namely miraculously solving the measurement problem within unitary quantum mechanics, it is helpful to review the arguments why this is generally felt to be impossible. Such arguments take the form of so-called ***insolubility theorems***. As already mentioned, such theorems ultimately go back to von Neumann: especially those that prove the impossibility of explaining his process 1 (i.e. the transition from a pure state to a mixture) from process 2 (unitary time evolution according to the Schrödinger equation). Another kind of insolubility theorem shows that single outcomes are impossible from process 2.

It might be argued that both kinds of theorem add little to the basic mathematical intuition behind the measurement problem, which is as follows (it goes without saying that we disagree with this traditional description of measurement, see below). Let $s \in B(H_S)$ be the observable being measured (where H_S is some Hilbert space associated to a quantum object S undergoing measurement) and let $a \in B(H_A)$ be a "pointer observable" correlated to S (where H_A is a second Hilbert space). In particular, the measurement apparatus A is described quantum mechanically. For the moment we assume both Hilbert spaces to be finite-dimensional and both operators to be non-degenerate, even having the same spectrum $\{\lambda_1, \ldots, \lambda_n\}$; this of course implies that $\dim(H_S) = \dim(H_A) = n$. Thus H_S has a basis $(\upsilon_i^{(s)})$ of eigenvectors of s and likewise H_A has a basis $(\upsilon_i^{(a)})$ of eigenvectors of a, with $s\upsilon_i^{(s)} = \lambda_i \upsilon_i^{(s)}$ and $a\upsilon_i^{(a)} = \lambda_i \upsilon_i^{(a)}$ $(i = 1, \ldots, n)$. The (erroneous) argument, then, is as follows:

1. Measurement should establish a correlation between values of s of S and values of a of A, which with the above labeling implies that for each i the initial system state $\upsilon_i^{(s)}$ should push the pointer from some initial state $\psi_0^{(A)}$ into a final (post-measurement) state $\upsilon_i^{(a)}$. Hence the dynamics, described by some unitary operator $u \in B(H_S \otimes H_A)$, should be such that
$$u(\upsilon_i^{(s)} \otimes \psi_0) = \upsilon_i^{(s)} \otimes \upsilon_i^{(a)} \equiv \varphi_i. \tag{11.1}$$
2. If the initial system state is $\psi_0^{(S)} = \sum_i c_i \upsilon_i^{(s)}$ (with $\sum_i |c_i|^2 = 1$), then, by linearity of u, the final state is $\varphi = \sum_i c_i \varphi_i$. But if A is sufficiently macroscopic this conflicts with observation, which always shows one of the terms in the sum. In other words, in theory, a—more precisely, $1_{H_S} \otimes a$—has no value in this state, whereas in practice it does, since in the real world measurements do have outcomes.
3. Hence the final state should be the *mixed* density operator $\sum_i |c_i|^2 |\varphi_i\rangle\langle\varphi_i|$ (rather than the *pure* one $|\varphi\rangle\langle\varphi|$), whose ignorance interpretation (allegedly) yields one of the states φ_i with probability $|c_i|^2$. But it is impossible to transform the initial pure state $|\varphi_0\rangle\langle\varphi_0|$ into the above mixture by any unitary operator, let alone by the u defined by (11.1), which by construction yields
$$u|\psi_0^{(S)}\rangle\langle\psi_0^{(S)}|u^* = |\varphi\rangle\langle\varphi| \neq \sum_i |c_i|^2 |\varphi_i\rangle\langle\varphi_i|. \tag{11.2}$$

As we already discussed, for some authors the measurement problem is the clash between nos. 1 and 3 (this is the "small" problem), whereas for others it is the conflict between nos. 1 and 2 (i.e. the "big" one). Either way, the goal of insolubility theorems is to show that the problem is not a consequence of idealizations in primitive arguments like the one just given, but remains even under very general assumptions. In particular, both the purity of the initial system as well as apparatus states (and hence of their tensor product), and the exact system-apparatus correlation assumed (including the premise of point spectra and finite-dimensional Hilbert spaces), can be considerably relaxed. To illustrate the kind of discussion, we present one example of an insolubility proof along the former lines and one along the latter. These proofs even remain valid if the notion of an observable itself is relaxed, too, namely from a self-adjoint operator to a POVM (see (2.178)), but we will not discuss this utmost generality (if only because it would not circumvent our critique below). It should be noted that insolubility theorems tacitly assume that the mathematical objects in the quantum-mechanical formalism describe all there is physically.

In the first direction, we have Theorem 11.2 below, which we may summarize as the ***problem of statistics***: there is a contradiction between the following postulates:

1. *System and apparatus are both described quantum-mechanically.*
2. *The wave-function of the system is complete.*
3. *The wave-function always evolves linearly (e.g., by the Schrödinger equation).*
4. *Measurements with identical initial wave-functions may have different outcomes, and the probability of each possible outcome is given by the Born rule.*

Here the second and third postulates may be consequences of the first, but even so it is useful to list them separately, since denying or circumventing nos. 1, 2, and 3 is typically done in completely different ways (see the end of this section).

Formally, let $s = s^* \in B(H_S)$ be an arbitrary self-adjoint operator on an arbitrary (separable) Hilbert space H_S, with associated spectral projections $e^{(s)}_\Delta \in \mathscr{P}(H_S)$, $\Delta \subset \sigma(s)$, and likewise $a \in B(H_A)$. It is convenient (and entails no genuine loss of generality) to still assume that $\sigma(s) = \sigma(a)$. Recall that the Born measure $\mu^{(s)}_{\rho_S}$ on the spectrum $\sigma(s)$ induced by some density operator $\rho_S \in \mathscr{D}(H_S)$ is given by

$$\mu^{(s)}_{\rho_S}(\Delta) = \mathrm{Tr}\left(\rho_S e^{(s)}_\Delta\right) = \omega_S\left(e^{(s)}_\Delta\right) = \mu^{(s)}_{\omega_S}(\Delta), \tag{11.3}$$

cf. (4.9), where ω_S is the state associated to ρ_S by (2.33), and no notational confusion between $\mu^{(s)}_{\rho_S}$ and $\mu^{(s)}_{\omega_S}$ should arise (they are the same thing). Likewise for a.

Definition 11.1. *1. Let H be a Hilbert space and let $b \in B(H)_{\mathrm{sa}}$. Two (normal) states ω, ω' on $B(H)$ are called* **b-distinguishable** *if $\mu^{(b)}_\omega \neq \mu^{(b)}_{\omega'}$; in other words, there is some $\Delta \subset \sigma(b)$ such that $\mu^{(b)}_\omega(\Delta) \neq \mu^{(b)}_{\omega'}(\Delta)$. Similarly for $\rho, \rho' \in \mathscr{D}(H)$.*
2. In the situation described before (11.3)*, a pair (ρ_A, u), where ρ_A is a density operator on $B(H_A)$ and u is a unitary operator on $H_S \otimes H_A$, is a* **measurement scheme** *for s if s-distinguishability of two density operators ρ_S, ρ'_S on H_S implies $1_{H_S} \otimes a$ -distinguishability of the two states $u(\rho_S \otimes \rho_A)u^*$ and $u(\rho'_S \otimes \rho_A)u^*$.*

3. *A measurement scheme* (ρ_A, u) *for* s **preserves probabilities** *if for any density operator* $\rho_S \in \mathscr{D}(H_S)$ *the probability measure on* $\sigma(a) = \sigma(1_{H_S} \otimes a)$ *induced by* $u(\rho_S \otimes \rho_A)u^*$ *equals the Born measure* $\mu^{(s)}_{\rho_S}$ *on* $\sigma(s) = \sigma(a)$ *induced by* ρ_S.
4. *A density operator* $\rho \in \mathscr{D}(H_S \otimes H_A)$ **objectifies** *the pointer observable a relative to some countable partition* $\sigma(a) = \bigsqcup_i \Delta_i$ *of its spectrum if* $\rho = \sum_i p_i e_{\upsilon_i}$, *where each unit vector* $\upsilon_i \in H_S \otimes H_A$ *is an eigenvector of* $1_{H_S} \otimes e^{(a)}_{\Delta_i}$ $(p_i \geq 0, \sum_i p_i = 1)$.

For example, in case of a discrete spectrumf or simplicity, if $\lambda_1 \neq \lambda_2$ in $\sigma(b)$, then any two unit eigenvectors $\upsilon^{(b)}_i$ $(i = 1,2)$ give rise to b-distinguishable vector states $\rho_i = |\upsilon^{(b)}_i\rangle\langle\upsilon^{(b)}_i|$. If $\psi = c_1\upsilon^{(b)}_1 + c_2\upsilon^{(b)}_2$ with $|c_1|^2 + |c_2|^2 = 1$ and $c_1 \neq 0,1$, then also the trio (ρ_1, ρ_2, e_ψ) is pairwise b-distinguishable. If, the other hand, $\lambda \in \sigma(b)$ is degenerate, then e_ψ and $e_{\psi'}$ fail to b-distinguishable whenever $\psi, \psi' \in H_\lambda$.

Clause 2 of Definition 11.1—which incorporates a vast number of at least theoretical scenario's—is a considerable weakening of the scheme (11.1), while clause 3 sharpens the second, implying that measurement transfers all Born probabilities for the object to the apparatus, probabilistically making the latter a mirror image of the former. Clause 4 firstly takes care of continuous spectra; if $\sigma(a)$ is discrete, one may simply partition it by its points (a partition of $\sigma(a)$ is sometimes called a ***reading scale***). The "objectification" terminology is questionable (if not outright misleading), as it is motivated by the ignorance interpretation of mixtures (see below), but we follow the literature in using it. In what follows, we exclude the trivial cases where $\sigma(s)$ consist of a single point, and/or $\sigma(a)$ is partitioned by itself.

Theorem 11.2. *For any nontrivial object observable* s *and partitioning of* $\sigma(a)$, *there exists no measurement scheme* (ρ_A, u) *for* s *whose final state* $u(\rho_S \otimes \rho_A)u^*$ *objectifies* a *for any initial system state* ρ_S *(let alone one that preserves probabilities).*

Proof. Since we will not use this theorem (except for pointing out that it attacks a straw man), we just prove it in the special case where $\sigma(a)$ is discrete and partitioned by its points, and also the spectral decomposition $\rho_A = \sum_n p_n e_n$ of the initial apparatus state is unique, cf. (B.490). For any unit vector in $\upsilon^{(s)} \in H_S$ we then have

$$u(e_{\upsilon^{(s)}} \otimes \rho_A)u^* = \sum_n p_n u(e_{\upsilon^{(s)}} \otimes e_n)u^*. \tag{11.4}$$

Take $\lambda_1 \neq \lambda_2$ in $\sigma(s)$, with associated eigenvectors $\upsilon^{(s)}_1$ and $\upsilon^{(s)}_2$. If $e_n = |\alpha_n\rangle\langle\alpha_n|$, for unit vectors $\alpha_n \in H_A$, then objectification of a requires that each of the vectors

$$u(\upsilon^{(s)}_1 \otimes \alpha_n), \;\; u(\upsilon^{(s)}_2 \otimes \alpha_n), \;\; u((c_1\upsilon^{(s)}_1 + c_2)\upsilon^{(s)}_2 \otimes \alpha_n),$$

with $|c_1|^2 + |c_2|^2 = 1$ and $c_1 \neq 0,1$, must be an eigenvector of $1_{H_S} \otimes a$. This is only possible if the first two vectors (and hence the third) lie in the same eigenspace for $1_{H_S} \otimes a$, but in that case condition no. 2 in Definition 11.1 is violated, since the three given initial system states are pairwise s-distinguishable whereas the corresponding outcomes states just listed evidently fail to be $1_{H_S} \otimes a$-distinguishable. □

Insolubility theorems of the second kind describe the ***problem of outcomes***, according to which clauses 1., 2., and 3. of the problem of statistics also contradict:

4'. Measurements have determinate outcomes.

Technical statements to this effect are even more straightforward than those formalizing the problem of statistics. We keep H_S and $s \in B(H_S)$ as they were, but this time, H_A may refer to the rest of the Universe outside the quantum object described by H_S (which includes the pointer, of course). Here is the key assumption.

Definition 11.3. *Let $s \in B(H_S)_{\mathrm{sa}}$ be an object observable with partition $\sigma(s) = \bigsqcup_{i \in I} \Delta_i$ of its spectrum (if $\sigma(s) = \{\lambda_1, \ldots\}$ is discrete, one may take $\Delta_i = \{\lambda_i\}$), and let H_A be a second Hilbert space. A* **sound measurement scheme** *consists of:*

- *A collection $(S_i)_{i \in I}$ of* **outcome spaces**, *i.e. subsets of the (normal) state space,*

$$S_i \subset S_n(H_S \otimes H_A) \cong \mathscr{D}(H_S \otimes H_A), \tag{11.5}$$

for which there is $0 \leq \eta < 1/2$ such that for $i \neq j$, one has

$$2\sqrt{1-\eta} \leq \|\omega_i - \omega_j\| \leq 2 \;\; (\omega_i \in S_i, \omega_j \in S_j). \tag{11.6}$$

- *A pair (ρ_A, u), where ρ_A is a density operator on $B(H_A)$ and u is a unitary on $H_S \otimes H_A$, such that for each $i \in I$ and each unit vector $\upsilon_i^{(s)} \in H_{\Delta_i}$ (i.e., $e_{\Delta_i} \upsilon_i^{(s)} = \upsilon_i^{(s)}$), the state $u(e_{\upsilon_i^{(s)}} \otimes \rho_A)u^*$ (i.e. the outcome of the measurement) lies in S_i.*

In (11.6) the first bound (which for small η is $\approx (2-\eta) \leq \cdots$) is the key one, as the last one ≤ 2 is always satisfied and has been included for clarity. In particular,

$$\|\omega_i - \omega_j\| > \sqrt{2}. \tag{11.7}$$

Note that (11.6) implies that the S_i must be disjoint, since assuming $\omega \in S_i$ gives $\|\omega - \omega_j\| \geq 2\sqrt{1-\eta}$ for all $\omega_j \in S_j$, whereas $\omega \in S_j$ allows one to take $\omega_j = \omega$ in this inequality, leading to the contradiction $0 \geq 2\sqrt{1-\eta}$. Note that in terms of density operators we have

$$\|\omega_i - \omega_j\| = \|\rho_i - \rho_j\|_1, \tag{11.8}$$

where $\omega_i(a) = \mathrm{Tr}\,(\rho_i a)$, cf. (B.481) and Theorem B.146. If ω_i and ω_j are pure, induced by unit vectors ψ_i and ψ_j in $H_S \otimes H_A$, then by (C.637), eq. (11.6) comes down to

$$0 \leq |\langle \psi_i, \psi_j \rangle|^2 \leq \eta. \tag{11.9}$$

For example, in the von Neumann measurement scheme (11.1), the subspace S_i just consist of the vector state defined by $\upsilon_i^{(s)} \otimes \upsilon_i^{(a)}$, hence (11.6) holds with $\eta = 0$.

Theorem 11.4. *For any nontrivial object observable s and partitioning of $\sigma(s)$, any sound measurement scheme $((S_i), \eta, \rho_A, u)$ admits initial states $\upsilon \in H_S$ such that $u(e_\upsilon \otimes \rho_A)u^*$ (i.e. the post-measurement state) does not lie in any outcome space S_i.*

Proof. Let $\upsilon = (\upsilon_i + \upsilon_j)/\sqrt{2}$, where $i \neq j$ and for the moment υ_i and υ_j are merely orthonormal vectors in H_S. For each $i = 1,2$ we then compute:

$$\begin{aligned}\|u(e_\upsilon \otimes \rho_A)u^* - u(e_{\upsilon_i} \otimes \rho_A)u^*\|_1^{(H_S \otimes H_A)} &= \|e_\upsilon \otimes \rho_A - e_{\upsilon_i} \otimes \rho_A\|_1^{(H_S \otimes H_A)} \\ &= \|e_\upsilon - e_{\upsilon_i}\|_1^{(H_S)} \\ &= \|\omega_\upsilon - \omega_{\upsilon_i}\| \\ &= 2\sqrt{1 - |\langle \upsilon, \upsilon_i \rangle|^2} \\ &= \sqrt{2}, \end{aligned} \tag{11.10}$$

where $\|\cdot\|_1^{(H)}$ denotes the trace norm relative to H. Now take $\upsilon_i = \upsilon_i^{(s)}$ as in Definition 11.3. Since $\omega_i \equiv u(e_{\upsilon_i^{(s)}} \otimes \rho_A)u^* \in S_i$ by definition of a sound measurement, it follows from (11.7) and (11.10) that $\omega \equiv u(e_\upsilon \otimes \rho_A)u^*$ cannot lie in any subspace S_k, since that would require $\|\omega - \omega_l\| > \sqrt{2}$ for all $l \neq k$, whereas (11.10) shows that this inequality fails for at least two values of l, viz. $l = i$ and $l = j \neq i$. □

In order to circumvent Theorems 11.2 and 11.4, one should deny at least one of their explicit premises. Moreover, we note that postulate no. 3 (i.e. linearity of time-evolution) is always implicitly used in the form of the following counterfactual:

> *If* ψ_n *were* the initial state, then *for each n* it *would* evolve (linearly) according to the Schrödinger equation with *given* Hamiltonian h. *If* the initial state *were* $\sum_n c_n \psi_n$, also then it *would* evolve according to the *same* Hamiltonian h.

This counterfactual should be added as a tacit assumption to all insolubility proofs (and also to informal statements of the measurement problem). As such, it may reasonably be denied (see §11.4), and such a denial puts assumption no. 4 in the *problem of statistics* in perspective, namely by denying the possibility that identical initial states can always be prepared in such a way that they evolve through exactly the same Hamiltonian. This leaves room for the following denials of some premise:

¬ 1. The apparatus is not described quantum-mechanically;
¬ 2. The wave-function of the system is not complete;
¬ 3. The wave-function does not always evolve by the Schrödinger equation;
¬ 4. Identical initial wave-functions always yield identical outcomes;
¬ 4'. Measurements do not have determinate outcomes.

Current programs for solving the measurement problem neatly fall into this scheme:

¬ 1. Copenhagen Interpretation and Swiss Approach;
¬ 2. Hidden-variable theories, most prominently Bohmian mechanics;
¬ 3. Dynamical collapse theories (such as GRW);
¬ 4. Instability approaches, e.g., the Flea on Schrödinger's Cat (which keeps 3);
¬ 4'. Many-Worlds Interpretation, i.e., Everettian quantum mechanics.

Leaving most of these to the literature, we now turn to the instability approach (¬4).

11.4 The Flea on Schrödinger's Cat

The conclusion of this lengthy historical and technical introduction is that there are (at least) two different formulations of the measurement problem, whose insolubility is expressed by Theorems 11.2 and 11.4, respectively (leaving apart lavish opportunities for disagreement about the precise formulation of the underlying assumptions, and not even speaking about the outright dismissal of the whole issue as a *Scheinproblem*). Thus the problem in question is evidently of a different kind from say the famous open conjectures in mathematics (like the Riemann hypothesis), where it is clear what the theorem is that needs to be proved. Nonetheless, despite its undeniable philosophical aspects, we see the measurement problem as a genuine physics problem concerned with the discrepancy between (quantum) theory and experiment, to be addressed by mathematical, physical, and philosophical analysis.

Well aware that different people typically draw different lessons from history, we will now, in the interest of motivating our approach to follow, draw our own (necessarily subjective) conclusions from the history of the measurement problem.

1. Though grounded in genius and tradition (Heisenberg, von Neumann, Wigner), the two-step way of looking at the measurement process (i.e. in terms of firstly a reduction of the wave-function by some non-unitary "process 1" and secondly a registration of a single outcome), with ensuing separation of the measurement problem into a "small" and a "big" problem, is fruitless and should be abandoned. It has no basis whatsoever in experimental physics (where the alleged mixed post-measurement states are conspicuously absent), it reflects obsolete ensemble thinking, and it is unsound also theoretically, as shown both by the first kind of insolubility results (à la von Neumann and Theorem 11.2), as well as by the failure of programs addressing just the "small" problem (like the Swiss approach and Decoherence). These approaches are unable to deal with the "big" problem (except perhaps through desperate remedies like Many Worlds) and hence, even if they work, they deliver Pyrrhic victories at best. The problem of obtaining single outcomes should be solved directly, before it is too late. Since such a solution would leave nothing to interfere, the "small" problem automatically disappears. This does not mean that it is sufficient to obtain definite outcomes alone; among all remaining challenges, deriving the Born rule stands out in particular.
2. Too much formal analysis has been done on the measurement problem (including the insolubility theorems just reviewed) without taking the special nature of measurement devices into account; alas, this negligence has its roots in the work of von Neumann. These devices are typically treated as ordinary quantum systems, as a consequence of which the notion of an "outcome" has to be defined within quantum mechanics and hence has to be identified e.g. with an eigenstate of some operator describing the apparatus (as in Theorem 11.2) or with some subspace of the quantum-mechanical state space (as in Theorem 11.4). Such identifications are purely formal and have little basis in experimental physics: as long as one defines outcomes of measurements within quantum mechanics, there is no measurement problem (but at worst some unease concerning value indefiniteness)!

Fig. 11.1 *The waves crashed between the towering cliff of Scylla and the jagged rocks of Charybdis*. Colour litograph by Gino D'Antonio. Reprinted with permission from Look and Learn Ltd.

On the other hand, both the Copenhagen Interpretation and the Swiss approach seem to have gone too far in the opposite direction: the former because it simply assumed (without providing any justification) that measurements have outcomes as soon as the apparatus is described classically, the latter in treating apparatuses as strictly infinite, and hence falling victim to Earman's Principle. The right approach, then, must be to define measurement as in the Copenhagen Interpretation, i.e. using a classical description of the apparatus whilst realizing it is ontologically a quantum system, and thusly navigate between Scylla (who treats measurement devices as arbitrary *quantum* systems) and Charybdis (who is too enthusiastic in taking infinite limits and hence in using a *classical* description).

3. Some kind of reality has to be attributed to the state of the system (though this reality cannot be "absolute", as in classical physics). In the algebraic approach to quantum theory adopted throughout the present book, the starting point is provided by the observables, relative to which states are defined. Since the doctrine of classical concepts drives us to switch between quantum-mechanical and classical descriptions, the reality of the quantum state is therefore *perspectival*. However, their perspectival nature does not make states less real; they say everything there is to say (at least by quantum theory) about some given level of description (which may be said to be chosen by the observer, and hence is intersubjective).

Thus the measurement problem arises in the way Schrödinger (rather than von Neumann) described it, although a precise framework has to be added to his poetry.

A framework that is precise both conceptually and mathematically is offered by ***asymptotic emergence***, which we already encountered in our discussion of SSB in the previous chapter (see especially its preamble). To repeat the main points, we speak of asymptotic emergence if the following three conditions are all satisfied:

1. A “higher-level theory” H (which in the context of the measurement problem is either classical mechanics or classical thermodynamics, depending on the measurement setup) is a limiting case of some “lower-level theory” L (viz. quantum mechanics, including quantum statistical mechanics of a finite system).
2. Theory H is well defined and understood by itself (typically predating L).
3. Theory H has “emergent” features that cannot be explained by L, e.g. because L does not have any property inducing those feature(s) in the limit pertinent to H.

The root of the measurement problem (and hence the relevance of asymptotic emergence), then, lies in Bohr’s requirement that the outcomes of measurements on systems defined within L be recorded in (at the least the language of) H, so that, crucially, *measurement according to* L *is a notion external to* L (if only partly), in particular involving the relationship between L and H. None of the insolubility proofs of the measurement problem take this into account (although due to Butterfield’s Principle these proofs remain relevant in a secondary way). The typical feature of H that would be emergent in the above sense if the measurement problem were unresolved is that every physical system subject to the theory H is ontologically in a pure state; in Schrödinger’s words quoted in §11.1: in H, sharply focused photographs of states are always possible (and hence any uncertainty or chance is due to ignorance, as in classical physics). Now, whatever the ontological nature of states in L, the states they induce in H should be real in the above sense, i.e., pure. But this is precisely what does *not* seem to be the case in typical measurement situations (e.g., Schrödinger’s Cat), where the post-measurement state on L induces a *mixed* state on H. Just as in the case of SSB, this violates Butterfield’s Principle, which in the case at hand states that since H is an idealization of L, any physical effect in H must be foreshadowed in L: *as* L *approaches* H, sharp measurement outcomes (defined as pure states in H) must arise from at least approximate single measurement outcomes (i.e. “singly-peaked wave-functions”) in *the relevant asymptotic regime of* L (since only these wave-functions gives rise to pure classical states on H).

As noted before in the setting of SSB: violating Butterfield’s Principle means violating Earman’s Principle, which in turn leads to a violation of the link between theory and reality. It is worth spelling this out for the measurement problem:

- Reality is described by quantum mechanics (even in the Copenhagen Interpretation, classical mechanics is an idealization of quantum mechanics);
- Real phenomena—in this case, sharp measurement outcomes— are correctly described by classical mechanics *although this is an idealization*;
- Quantum mechanics (allegedly) cannot possibly induce these phenomena in its limit towards classical mechanics *although it is the theory that should apply*;
- Hence quantum mechanics contradicts reality. Classical mechanics does not contradict the reality of sharp measurement outcomes, but it is not the appropriate theory to explain them; this explanation should come from quantum mechanics.

It may now seem that invoking Butterfield’s Principle has reduced the measurement problem to the usual one(s) described in the preceding sections. But look at the small print: in the Copenhagen Interpretation, single measurement outcomes only appear in some limiting “classical” regime of quantum mechanics.

"Deep inside" quantum mechanics, there is no need at all for the typical superposition $\sum_n c_n \psi_n$ to collapse into one of the states ψ_n (unless one conflates the physical measurement problem with the philosophical problem of value indefiniteness). The external and asymptotic nature of measurement outcomes *causes* the measurement problem, but, as we shall see, at the same time it provides the key for its *solution*, since the collapse mechanism we propose is only effective asymptotically (so that it operates where it should and does not act where it should not). More precisely, by taking into account perturbations of the Hamiltonian that are tiny and ineffective in the quantum regime, but become hugely destabilizing in the classical regime (even before the actual limit), the wave-function of the apparatus will collapse.

Summarizing the preceding discussion, "our" measurement problem states that:

- *Certain* pure *post-measurement states of an (ontologically quantum-mechanical!) apparatus coupled to a microscopic quantum object induce* mixed *states on the apparatus (and on the composite)* **once the apparatus is described classically**.

This is a precise version of Schrödinger's Cat problem (rather than von Neumann's purely quantum-mechanical measurement problem), making it clear that at heart the problem does not lie with the (dis)appearance of interference terms (which is a red herring) but with the inability of quantum mechanics to predict single outcomes.

We now show by means of a simple example what it means to describe an ontologically quantum-mechanical apparatus classically, and outline the scenario we envisage for the solution of the measurement problem on the basis of this example. The ***Spehner–Haake model*** of the apparatus described below is too simple to be realistic, but nonetheless it may serve its purpose (as Bohr would say). The model involves a double-well potential like (10.11), modified however by a little basin in the middle, as shown below (including ground states for one large and one small value of $\hbar$). Also here, SSB will play a crucial role, so please recall §10.1.

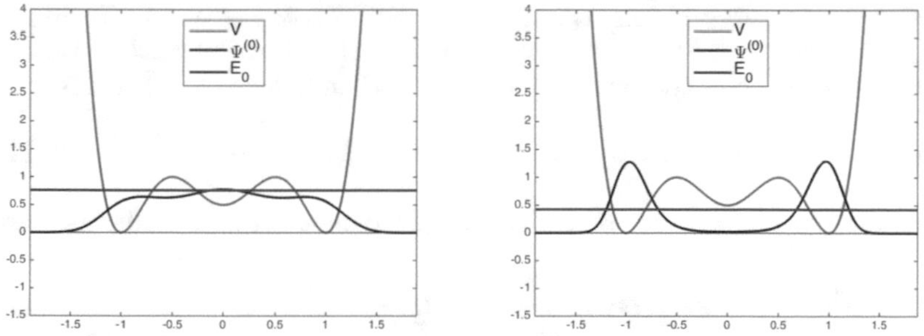

Fig. 11.2 Double-well potential with basin; ground state $\psi^{(0)}_{\hbar=0.5}$ and $\psi^{(0)}_{\hbar=0.01}$.

Consider $N' \equiv N+1$ non-interacting particles, each with mass m, moving on the real line under the influence of a one-particle potential V (note that although the zero'th particle with be handled lightly differently from the others, it is not the pointer!). In terms of the canonical coordinates $(\mathbf{p}',\mathbf{q}') = (p_0,\dots,p_N,q_0,\dots,q_N) \in \mathbb{R}^{2N'}$ on the phase space $X = T^*\mathbb{R}^{N'}$ the classical Hamiltonian is

$$h(\mathbf{p}',\mathbf{q}') = \sum_{n'=0}^{N}\left(\frac{p_{n'}^2}{2m} + V(q_{n'})\right). \tag{11.11}$$

Now perform a canonical transformation to center of mass and relative coordinates

$$P = \sum_{n'=0}^{N} p_{n'} \qquad Q = \frac{1}{N'}\sum_{n'=0}^{N} q_{n'}; \tag{11.12}$$

$$\pi_n = \sqrt{N'}p_n - \frac{1}{\sqrt{N'}}\sum_{n'=0}^{N} p_{n'} \qquad \rho_n = \frac{1}{\sqrt{N'}}(q_n - q_0)\ (n=1,\dots,N); \tag{11.13}$$

the center of mass (P,Q) will be the pointer. The inverse transformation is given by

$$p_0 = \frac{P}{N'} - \frac{1}{\sqrt{N'}}\sum_{n=1}^{N}\pi_n; \tag{11.14}$$

$$p_n = \frac{P}{N'} + \frac{1}{\sqrt{N'}}\pi_n; \tag{11.15}$$

$$q_0 = Q - \frac{1}{\sqrt{N'}}\sum_{n=1}^{N}\rho_n; \tag{11.16}$$

$$q_n = Q + \sqrt{N'}\rho_n - \frac{1}{\sqrt{N'}}\sum_{k=1}^{N}\rho_k. \tag{11.17}$$

Granted that $\{p_{n'},q_{k'}\} = \delta_{n'k'}$, $\{p_{n'},p_{k'}\} = 0$, and $\{q_{n'},q_{k'}\} = 0$, we then duly have $\{P,Q\} = 1$ and $\{\pi_n,\rho_k\} = \delta_{nk}$, with all other elementary Poisson brackets vanishing.

In terms of the new coordinates, the classical Hamiltonian (11.11) reads

$$h(P,Q,\pi,\rho) = h_A(P,Q) + h_{AE}(Q,\rho) + h_E(\pi), \tag{11.18}$$

where $\pi = (\pi_1,\dots,\pi_N)$, $\rho = (\rho_1,\dots,\rho_N)$, and the three partial Hamiltonians are

$$h_A(P,Q) = \frac{P^2}{2M} + N'V(Q); \tag{11.19}$$

$$h_E(\pi) = \frac{1}{2M}\left(\sum_{n=1}^{N}\pi_n^2 + \left(\sum_{n=1}^{N}\pi_n\right)^2\right); \tag{11.20}$$

$$h_{AE}(Q,\rho) = \sum_{k=1}^{\infty}\frac{1}{k!}f_k(\rho)V^{(k)}(Q), \tag{11.21}$$

where $M = Nm$ is the total mass of the system, for simplicity we assumed V to be analytic (it will even be taken to be polynomial), and we abbreviated

$$f_k(\rho) = \left(-\frac{1}{\sqrt{N'}}\sum_{l=1}^{N}\rho_l\right)^k + \sum_{n=1}^{N}\left(\sqrt{N'}\rho_n - \frac{1}{\sqrt{N'}}\sum_{l=1}^{N}\rho_l\right)^k. \tag{11.22}$$

Note that $f_1(\rho) = 0$, so that to lowest order (i.e. $k = 2$) we have

$$h_{AE}(Q,\rho) = \left(\tfrac{1}{2}N\sum_{n=1}^{N}\rho_n^2 - \sum_{k\neq l}^{N}\rho_k\rho_l\right)V''(Q) + \cdots \tag{11.23}$$

We pass to the corresponding quantum-mechanical Hamiltonians in the usual way, and couple a two-level quantum system to the apparatus through the Hamiltonian

$$h_{SA} = \mu \cdot \sigma_3 \otimes P, \tag{11.24}$$

where the object observable $s = \sigma_3$, acting on $H_S = \mathbb{C}^2$, is to be measured. The idea is that h_A is the Hamiltonian of a pointer that registers outcomes by localization on the real line, h_E is the (free) Hamiltonian of the "environment", realized as the internal degrees of the freedom of the total apparatus that are not used in recording the outcome of the measurement, and h_{AE} describes the pointer-environment interaction. The classical description of the apparatus then involves two approximations:

- Ignoring all degrees of freedom except those of A, which classically are (P,Q);
- Taking the classical limit of h_A, here realized as $N \to \infty$ (*in lieu of* $\hbar \to 0$).

The measurement of s is now expected to unfold according to the following scenario:

1. The apparatus is initially in a metastable state (this is a very common assumption), whose wave-function is e.g. a Gaussian centered at the origin.
2. If the object state is "spin up", i.e., $\psi_S = (1,0)$, then it kicks the pointer to the right, where it comes to a standstill at the bottom of the double well. If spin is down, likewise to the left. If $\psi_S = (1,1)/\sqrt{2}$, the pointer moves to a superposition of these, which is close to the ground state of V displayed in Figure 11.2.
3. In the last case, the Flea mechanism of §10.2 comes into play: tiny asymmetric perturbations irrelevant for small N localize the ground state as $N \to \infty$.
4. Mere localization of the ground state of the perturbed (apparatus) Hamiltonian in the classical regime is not enough: there should be a *dynamical* transition from the ground state of the original (unperturbed) Hamiltonian (which has become metastable upon perturbation) to the ground state of the perturbed one. This dynamical mechanism in question should also recover the Born rule.

Thus the classical description of the apparatus is at the same time the root of the measurement problem and the key to its solution: it creates the problem because at first sight a Schrödinger Cat state has the wrong classical limit (namely a mixture), but it also solves it, because precisely in the classical limit Cat states are destabilized even by the tiniest (asymmetric) perturbations and collapse to the "right" states.

The "flea" perturbation might itself be a genuine random process, perhaps ultimately of quantum origin. In that case, the measurement merely amplifies the randomness that was already inherent in the flea by transferring it to the apparatus.

Alternatively, the flea might be fundamentally deterministic (though it may nonetheless be modeled stochastically for pragmatic reasons). In principle, this would open the door to a restoration of determinism: for the flea now transfers its *determinism* (rather than its *randomness*) to the apparatus. The mistaken impression that quantum theory implies the irreducible randomness of nature then arises because although measurement outcomes are determined, they are unpredictable "for all practical purposes", even in a way that (because of the exponential sensitivity to the flea in $1/\hbar$ or N) dwarfs the unpredictability of classical chaotic systems.

Either way, the flea perturbation would naturally be different at each different run of an experiment under otherwise identical initial conditions, which motivates our critique of the counterfactual discussed after the proof of Theorem 11.4.

The location of the flea plays a similar role to the position variable in Bohmian mechanics, i.e., it is essentially a hidden variable. Recall the notions of *Outcome Independence* (OI) and *Parameter Independence* (PI), reviewed in §6.5. Briefly, the conjunction of OI and PI is equivalent to Bell's locality condition, and if the latter is satisfied, then the Bell inequalities hold. Since these are violated by quantum mechanics, any hidden variable theory compatible with quantum mechanics must violate OI or PI. Deterministic hidden variable theories necessarily satisfy OI, in which case Bell's Theorem or the Free Will Theorem shows that they must violate PI in order to be compatible with quantum mechanics. A violation of PI leads to possible superluminal signaling only if the hidden variable z can be controlled. If the wave-function ψ is regarded as the hidden variable, then quantum theory itself satisfies PI but violates OI (since ψ *can* be prepared, the other way round would be disastrous). *Qua* deterministic hidden variable theory, Bohmian mechanics satisfies OI, and hence it violates PI; for the GRW collpase theory it is the other way round.

The fate of the flea therefore depends on the nature of the perturbation: if it is deterministic, the theory behaves like Bohmian mechanics in this respect and hence violates PI, whereas stochastic perturbations typically violate OI (and possibly also PI). Either way, no conflict with the said theorems arises. Moreover, in the Colbeck–Renner Theorem, assumption **CP** fails for the flea scenario—assuming, in view of its limitation to finite-dimensional Hilbert spaces, the theorem is applicable at all!

Besides such issues, others remain to be resolved, of which we just mention two:

1. Collapse of the wave-function has become a tunneling process, whose static effects are exponentially enhanced as $N \to \infty$ (or $\hbar \to 0$, as in §10.2). However, tunneling times *increase* in the same way, so that the environment is needed not only to provide the perturbation, but also to speed up the dynamics of collapse.
2. The flea not only destabilizes the Schrödinger Cat state (as desired), but also destabilizes the intended outcome states (like those in S_i, cf. Theorem 11.4). Also here the environment should play a decisive role in (re)stabilizing the latter but not the former, possibly through the mechanism of *einselection*, cf. §11.2.

Notes

§11.1. **The rise of orthodoxy**

The literature on the measurement problem is vast. Apart from the annotated reprint volume Wheeler & Zurek (1983), relatively recent surveys of and books include Bell (1990b), Maudlin (1995), Busch, Lahti, & Mittelstaedt (1996), Bassi & Ghirardi (2003), Mittelstaedt (2004), Wallace (2012), Allahverdyan, Balian, & Nieuwenhuizen (2013), and Busch, Lahti, Pellonpää, & Ylinen, (2016). In modal interpretations of quantum mechanics, the measurement problem is (dubiously) conflated with the far milder problem of value indefiniteness, see e.g. Bub (1997).

§11.2. **The rise of modernity: Swiss approach and Decoherence**

The Swiss approach to the measurement problem was initiated by Jauch (1964), to be continued by e.g. Hepp (1972), Emch & Whitten-Wolfe (1976), and recently also by Hepp's former student Fröhlich; see e.g. Fröhlich & Schubnel (2013) and Blanchard, Fröhlich & Schubnel (2016). In addition, see Landsman (1991, 1995)—now seen as naive—, Breuer, Amann & Landsman (1993), and Sewell (2005).

Key early papers on decoherence were Zeh (1970), Zurek (1981), and Joos & Zeh (1985), and standard reviews are Zurek (2003), Joos et al (2003), and Schlosshauer (2007). Penetrating critiques include Janssen (2008) and Tanona (2013). See also Camilleri (2009a) and Freire (2009) for some history.

A defence of QBism may be found in Caves, Fuchs, & Schack (2002b).

§11.3. **Insolubility theorems**

Insolubility theorems of the first kind kind go back to von Neumann (1932) and, in his wake, Wigner (1963) and Fine (1970). Theorem 11.2 is (in even more general form) due to Busch & Shimony (1996); with slightly different assumptions, the special case proved in the main text is due to Brown (1986). The monographs by Busch, Lahti, & Mittelstaedt (1996) and Mittelstaedt (2004) contain detailed discussions of theorems of this kind. See also Bacciagaluppi (2014).

The formulation of the problem of statistics and the problem of outcomes is taken from Maudlin (1995). Theorem 11.4 is due to Bassi & Ghirardi (2003), although here it is presented in a form inspired by Grübl (2003).

For Bohmian mechanics see e.g. Goldstein (2013) and Bricmont (2016). A recent review of the GRW program and related dynamical collapse theories is Bassi et al (2013). Nowadays, the *locus classicus* for Many Worlds is Wallace (2012).

The time-evolution counterfactual discussed in the main text was inspired by the problem of free will, see the quotation of Dennett at the beginning of §6.3.

S11.4. **The Flea on Schrödinger's Cat**

The approach to the measurement problem discussed here has its roots in Landsman & Reuvers (2013) and Landsman (2013), whose model at the time only involved the apparatus. This was criticized in van Heugten & Wolters (2016), many of whose points may be addressed by turning to the Spehner–Haake model, introduced by Spehner & Haake (2008). The ABN-model of Allahverdyan, Balian, & Nieuwenhuizen (2013) gives a similar picture; for a comparison see Spehner (2009).

Chapter 12
Topos theory and quantum logic

The topos-theoretic approach to quantum mechanics (also known as ***quantum toposophy***) has the same origin as the quantum logic programme initiated by Birkhoff and von Neumann, namely the feeling that classical logic is inappropriate for quantum theory and needs to be replaced by something else. For example, Schrödinger's Cat serves as an "intuition pump" for this feeling (at least in the naive view—dispensed with in Chapter 11—that it is neither alive nor dead). However, we feel that the quantum logic proposed by Birkhoff and von Neumann is:

- *too radical* in giving up distributivity (rendering it problematic to interpret the logical operations $\wedge$ and $\vee$ as conjunction and disjunction, respectively);
- *not radical enough* in keeping the law of excluded middle, which is precisely what intuition pumps like Schrödinger's cat and the like challenge.

Thus it would be preferable to have a quantum logic with exactly the opposite features, i.e., one that is distributive but drops the law of excluded middle: this suggest the use of *intuitionistic logic*. It is interesting to note that Birkhoff and von Neumann (who had earlier corresponded with Brouwer about possible intuitionistic aspects of game theory, notably chess) actually considered intuitionistic logic, but rejected it:

> 'The models for propositional calculi which have been considered in the preceding sections are also interesting from the standpoint of pure logic. Their nature is determined by quasi-physical and technical reasoning, different from the introspective and philosophical considerations which have had to guide logicians hitherto. Hence it is interesting to compare the modifications which they introduce into Boolean algebra, with those which logicians on "intuitionist" and related grounds have tried introducing. The main difference seems to be that whereas logicians have usually assumed that properties L71–L73 [i.e. $(a')' = a$, $a \cap a' = \perp$, $a \cup a' = \top$, and $a \subset b$ implies $a' \supset b'$] of negation were the ones least able to withstand a critical analysis, the study of mechanics points to the *distributive identities*as the weakest link in the algebra of logic. (...) Our conclusion agrees perhaps more with those critiques of logic, which find most objectionable the assumption that $a' \cup b = \top$ implies $a \subset b$ (or, dually, the assumption that $a \cap b' = \perp$ implies $b \supset a$—the assumption that to deduce an absurdity from the conjunction of a and not b, justifies one in inferring that a implies b).' (Birkhoff & von Neumann, 1936, p. 837).

As already made clear, then, our view is exactly the opposite. It is perhaps more striking that our position on (quantum) logic also differs from Bohr's:

K. Landsman, *Foundations of Quantum Theory*,
Fundamental Theories of Physics 188, DOI 10.1007/978-3-319-51777-3_12

> 'All departures from common language and ordinary logic are entirely avoided by reserving the word "phenomenon" solely for reference to unambiguously communicable information, in the account of which the word "measurement" is used in its plain meaning of standardized comparison.' (Bohr, 1996, p. 393)

Rather than *postulate* the logical structure of quantum mechanics, our goal is to *derive* it from our Bohrification ideology, more specifically, from the poset $\mathscr{C}(A)$ of all unital commutative C*-subalgebras of a unital C*-algebra A, ordered by inclusion. One may think of this poset as a mathematical home for Bohr's notion of ***Complementarity***, in that each $C \in \mathscr{C}(A)$ represents some classical or experimental context, which has been decoupled from the others, *except for the inclusion relations, which relate* compatible *experiments* (in general there seem to be no preferred *pairs* of complementary subalgebras $C, C' \in \mathscr{C}(A)$ that jointly generate A, although Bohr typically seems to have had such pairs in mind, e.g. position and momentum).

Quantum toposophy also accommodates the feeling that quantum mechanics is so radical that not just the actors of classical mechanics, but its whole stage must be replaced. This need is well expressed by the following quotation from Grothendieck, who created topos theory (but never witnessed its application to quantum theory):

> 'Passer de la mécanique de Newton à celle d'Einstein doit être un peu, pour le mathématicien, comme de passer du bon vieux dialecte provençal à l'argot parisien dernier cri. Par contre, passer à la mécanique quantique, j'imagine, c'est passer du français au chinois.' (Grothendieck, 1986, p. 61).[1]

Indeed, topos theory replaces even set theory, seen as the stage of classical mathematics and physics, by some other stage: each topos provides a "universe of discourse" in which to do mathematics. One major difference with set theory, then, is that logic in most toposes (including the ones we will use) is . . . intuitionistic!

This chapter presupposes familiarity with §C.11 on the logical side of the Gelfand isomorphism for commutative C*-algebras, Appendix D on lattice theory and logic, and Appendix E on topos theory. Since this material is off the beaten track, as in Chapter 6 it may be helpful to provide a very brief guided tour through this chapter.

In §12.1 we first define the "quantum mechanical" topos $\mathsf{T}(A)$ that will act as the mathematical stage for the remainder of the chapter; it depends some given (unital) C*-algebra A only via the poset $\mathscr{C}(A)$. We then define C*-algebras internal to any topos T (in which the natural numbers and hence the rationals can be defined), which notion we then apply to $\mathsf{T} = \mathsf{T}(A)$, so as to define an internal C*-algebra $\underline{A}$, which turns out to be *commutative*. Following an interlude on constructive Gelfand spectra in §12.2, in §12.3 we then compute the internal Gelfand spectrum of $\underline{A}$ for $A = M_n(\mathbb{C})$, and derive our intuitionistic logic of quantum mechanics from this, given by eqs. (12.95) - (12.96) and (12.103) - (12.107). We also discuss its (Kripke) semantics. In §12.4 we generalize these computations to arbitrary (unital) C*-algebras A, culminating in Corollary 12.22. Finally, in §12.5 we relate this material to both the Kochen–Specker Theorem (which provided the original motivation for quantum toposophy), as well as to an attempt at ontology called "Daseinisation."

---

[1] 'For a mathematician, switching from Newton's mechanics to Einstein's must to some extent be like switching from a good old provincial dialect to Paris slang. In contrast, I imagine that switching to quantum mechanics amounts to switching to Chinese.' Translation by the author.

12.1 C*-algebras in a topos

Let A be a unital C*-algebra (in Sets), with associated poset $\mathscr{C}(A)$ of all unital commutative C*-subalgebras $C \subset A$ ordered by inclusion. Regarding $\mathscr{C}(A)$ as a (posetal) category, in which there is a unique arrow $C \to D$ iff $C \subseteq D$ and there are no other arrows, we obtain the topos $\mathsf{T}(A)$ of functors $\underline{F} : \mathscr{C}(A) \to \mathsf{Sets}$ (F underlined!), i.e.,

$$\mathsf{T}(A) = [\mathscr{C}(A), \mathsf{Sets}]. \tag{12.1}$$

Since for any poset X we have an isomorphism of categories $[X, \mathsf{Sets}] \simeq \mathsf{Sh}(X)$, where X is endowed with the Alexandrov topology, see (E.84), we may alternatively write

$$\mathsf{T}(A) \simeq \mathsf{Sh}(\mathscr{C}(A)). \tag{12.2}$$

This alternative description will turn out to be very useful in computing the Gelfand spectrum of the internal commutative C*-algebra $\underline{A}$ to be defined shortly. Since we occasionally switch between $\mathsf{T}(A)$ and the topos Sets, we underline objects (i.e., functors $\underline{F} : \mathscr{C}(A) \to \mathsf{Sets}$) of the former. In order to do some kind of Analysis in $\mathsf{T}(A)$, we need real numbers. In many toposes this is a tricky concept, but:

Proposition 12.1. *In* $\mathsf{T}(A)$, *the* **Dedekind reals** *are given by the constant functor*

$$\underline{\mathbb{R}}_0 : C \mapsto \mathbb{R}, \tag{12.3}$$

where $C \in \mathscr{C}(A)$, *with associated frame given by the functor*

$$\underline{\mathscr{O}(\mathbb{R})}_0 : C \mapsto \mathscr{O}((\uparrow C) \times \mathbb{R}). \tag{12.4}$$

Similarly, we have complex numbers $\underline{\mathbb{C}}$ and their frame $\mathscr{O}(\underline{\mathbb{C}})$ in $\mathsf{T}(A)$.

Proof. In a general sheaf topos $\mathsf{Sh}(X)$, the Dedekind real numbers object is the sheaf (E.150), with frame (E.149). The point now is that each continuous function $f \in C(\mathscr{C}(A), \mathbb{R})$ on $X = \mathscr{C}(A)$ with the Alexandrov topology is locally constant.

To see this, suppose $C \leq D$ in U, and take $V \subseteq \mathbb{R}$ open with $f(C) \in V$. Then $C \in f^{-1}(V)$ and $f^{-1}(V)$ is open by continuity of f. But the smallest open set containing C is $\uparrow C$, which contains D, so that $f(D) \in V$. Taking $V = (f(C) - \varepsilon, \infty)$ gives the inequality $f(D) > f(C) - \varepsilon$ for all $\varepsilon > 0$, whence $f(D) \geqslant f(C)$, whereas $V = (-\infty, f(C) + \varepsilon)$ yields $f(D) \leq f(C)$. Hence $f(C) = f(D)$.

Thus we obtain (12.3) - (12.4) as special cases of (E.150) - (E.149). □

Other objects of interest in $\mathsf{T}(A)$ that we will steadily use are:

- The ***terminal object*** $\underline{1}$, i.e., the constant functor $C \mapsto *$, where $*$ is a singleton.
- The ***truth object*** $\underline{\Omega}$, which according to (E.86) - (E.87) is given by

$$\underline{\Omega}_0(C) = \mathrm{Upper}(C); \tag{12.5}$$
$$\underline{\Omega}_1(C \subseteq D) = (-) \cap (\uparrow D), \tag{12.6}$$

where Upper(C) is the set of all upper sets above C (i.e., $S \in \text{Upper}(C)$ iff $S \subset \mathscr{C}(A)$ such that: (i) $C \subseteq D$ for each $D \in S$, and (ii) $D \in S$ and $D \subseteq E$ imply $E \in S$).

- The ***subobject classifier*** $t : \underline{1} \to \underline{\Omega}$, which is a natural transformation whose components t_C are given, according to (E.88), as

$$t_C(*) = \uparrow C, \tag{12.7}$$

i.e., the set of all $D \supseteq C$ in $\mathscr{C}(A)$; this is the maximal element of Upper(C).

Furthermore, exponentials in $\mathsf{T}(A)$ have the following straightforward description:

$$\underline{F}_0^{\underline{G}}(C) = \text{Nat}(\underline{G}_{\uparrow C}, \underline{F}_{\uparrow C}) \quad (C \in \mathscr{C}(A)), \tag{12.8}$$

where $\underline{F}_{\uparrow C}$ is the restriction of the functor $\underline{F} : \mathscr{C}(A) \to \mathsf{Sets}$ to $\uparrow C \subseteq \mathscr{C}(A)$, and Nat$(-,-)$ denotes the set of natural transformations between the functors in question. In particular, since $\mathbb{C} \cdot 1$ is the bottom element of the poset $\mathscr{C}(A)$, one has

$$\underline{F}^{\underline{G}}(\mathbb{C} \cdot 1) = \text{Nat}(\underline{G}, \underline{F}). \tag{12.9}$$

One way to derive (12.8) is to start from general sheaf toposes $\mathsf{Sh}(X)$, where

$$F_0^G(U) = \text{Nat}(G_{|U}, F_{|U}), \tag{12.10}$$

both restricted to $\mathscr{O}(U)$ (i.e. defined on each open $V \subseteq U$ instead of all $V \in \mathscr{O}(X)$), and use (E.84). Combining these observations, one has

$$\underline{\Omega}^{\underline{F}}(C) \cong \text{Sub}(\underline{F}_{\uparrow C}), \tag{12.11}$$

i.e., the set of subfunctors of $\underline{F}_{\uparrow C}$. In particular, like in (12.9), we find

$$\underline{\Omega}^{\underline{F}}(\mathbb{C} \cdot 1) \cong \text{Hom}(\underline{F}, \underline{\Omega}) \cong \text{Sub}(\underline{F}), \tag{12.12}$$

the set of subfunctors of $\underline{F}$ itself. Recall that, as explained after Lemma E.16, a subfunctor $\underline{Z} \in \text{Sub}(\underline{F})$ is a functor $\underline{Z} : \uparrow\mathscr{C}(A) \to \mathsf{Sets}$ for which $\underline{Z}_0(C) \subseteq \underline{F}_0(C)$ for all $C \in \mathscr{C}(A)$ and $\underline{Z}_1$ is the restriction of $\underline{F}_1$. If $C \subseteq D$, then the set-theoretic map $\underline{\Omega}^{\underline{F}}(C) \to \underline{\Omega}^{\underline{F}}(D)$ defined by $\underline{\Omega}^{\underline{F}}$, identified with a map $\text{Sub}(\underline{F}_{\uparrow C}) \to \text{Sub}(\underline{F}_{\uparrow D})$, is simply given by restricting a given subfunctor of $\underline{F}_{\uparrow C}$ to $\uparrow D$.

Using either the internal language of a topos (see §E.5) or direct object-arrow constructions, one can copy standard definitions in set theory so as to define mathematical objects "internal" to any given topos, *as long as these definitions make sense in first-order intuitionistic logic* (which roughly speaking means that they are "constructive", in not using the axiom of choice or the law of the excluded middle).

As a case in point, let us now define ***internal C*-algebras*** in $\mathsf{T}(A)$ (this may be done even more generally in any topos T in which at least the natural numbers $\mathbb{N}$, and hence the rationals $\mathbb{Q}$, are defined). Vector spaces (over $\underline{\mathbb{R}}$ or $\underline{\mathbb{C}}$) and (commutative) *-algebras may be defined in $\mathsf{T}(A)$ through straightforward object-arrow translations of the usual constructions in Sets, i.e., one has an object $\underline{A}$ and arrows:

$$\cdot : \underline{\mathbb{C}} \times \underline{A} \to \underline{A} \quad \text{(scalar multiplication);} \tag{12.13}$$
$$+ : \underline{A} \times \underline{A} \to \underline{A} \quad \text{(addition);} \tag{12.14}$$
$$\times : \underline{A} \times \underline{A} \to \underline{A} \quad \text{(multiplication);} \tag{12.15}$$
$$* : \underline{A} \to \underline{A} \quad \text{(involution),} \tag{12.16}$$

subject to the usual axioms. Syntactically, a ***unit (internal)*** in $\underline{A}$ is a constant

$$1_{\underline{A}} : \underline{\mathbf{1}} \to \underline{A},$$

with $\underline{\mathbf{1}}$ the terminal object in $\mathsf{T}(A)$, such that

$$\left(\underline{A} \xrightarrow{\cong} \mathbf{1} \times \underline{A} \xrightarrow{(1_{\underline{A}}, \mathrm{id}_{\underline{A}})} \underline{A} \times \underline{A} \xrightarrow{\times} \underline{A}\right) = \left(\underline{A} \xrightarrow{\mathrm{id}_{\underline{A}}} \underline{A}\right). \tag{12.17}$$

The notions of norm and completeness are less easily defined internally, and hence one starts reinterpreting the notion of a ***seminorm*** in Sets as a subset

$$N \subset A \times \mathbb{Q}^+, \tag{12.18}$$

for which

$$(a,q) \in \underline{N} \text{ iff } \|a\| < q. \tag{12.19}$$

In our topos $\mathsf{T}(A)$, we interpret $\underline{N} \subset \underline{A} \times \underline{\mathbb{Q}}^+$ as a subfunctor $\underline{N} \to \underline{A} \times \underline{\mathbb{Q}}^+$ (or, equivalently by λ-conversion (F.153), as an arrow $\mathbf{1} \to \underline{\Omega}^{\underline{A} \times \underline{\mathbb{Q}}^+}$), subject to the axioms:

$$\forall_p\, p > 0 \to (0,p) \in \underline{N}; \tag{12.20}$$
$$\exists_q\, q > 0 \wedge (a,q) \in \underline{N}; \tag{12.21}$$
$$\forall_a \forall_p\, (a,p) \in \underline{N} \to (a^*,p) \in \underline{N}; \tag{12.22}$$
$$\forall_a \forall_q\, ((a,q) \in \underline{N} \leftrightarrow \exists_p p < q \wedge (a,p) \in \underline{N}); \tag{12.23}$$
$$\forall_a \forall_p\, ((a,p) \in \underline{N} \wedge (b,q) \in \underline{N} \to (a+b, p+q) \in \underline{N}); \tag{12.24}$$
$$\forall_a \forall_p\, ((a,p) \in \underline{N} \wedge (b,q) \in \underline{N} \to (a \cdot b, p \cdot q) \in \underline{N}); \tag{12.25}$$
$$\forall_a \forall_p \forall_z ((a,p) \in \underline{N} \wedge (|z| < q) \to (z \cdot a, p \cdot q) \in \underline{N}). \tag{12.26}$$

Here a,b are variables of type $\underline{A}$, p and q are variables of type $\underline{\mathbb{Q}}$, z is a variable of type $\underline{\mathbb{C}}$, 0 is the zero constant in $\underline{A}$, etc. For a unital *-algebra (whose internal definition we leave to the reader), with unit denoted by 1_A as usual, we also require

$$\Vdash \forall_a \forall_p p > 1 \to (1_A, p) \in \underline{N}. \tag{12.27}$$

If the seminorm relation furthermore satisfies

$$(a^* \cdot a, q^2) \in \underline{N} \leftrightarrow (a,q) \in \underline{N} \tag{12.28}$$

for all $a \in A$ and $q \in \mathbb{Q}^+$, then A is said to be a ***pre-semi-C*-algebra***.

To proceed to a C*-algebra, one requires $a = 0$ whenever $(a,q) \in N$ for all q in $\mathbb{Q}^+$, making the seminorm into a norm, and subsequently this normed space should be complete. The latter condition is quite complicated, since in a topos one has no Cauchy sequences in the usual sense, because $\underline{A}$ may not have global elements (in the sense of arrows $\mathbf{1} \to \underline{A}$). Indeed, our algebra $\underline{A}$ defined below only has trivial global elements, namely multiples of the the unit operator.

Hence one needs a generalization of Cauchy sequences in the general spirit of topos theory, where global elements are replaced by general elements.

Definition 12.2. *With $\underline{\mathbb{N}}$ the natural numbers object in $\mathsf{T}(A)$ (which is simply the constant functor $C \mapsto \mathbb{N}$), a* **Cauchy approximation** *in $\underline{A}$ is an arrow $s : \underline{\mathbb{N}} \to \underline{\Omega}^{\underline{A}}$ (or, equivalently, by λ-conversion* (E.153)*, an arrow $\chi : \underline{\mathbb{N}} \times \underline{A} \to \underline{\Omega}$, which in turn is the same as a subobject $\underline{S}$ of $\underline{\mathbb{N}} \times \underline{A}$) such that:*

$$\forall_n \exists_a \, a \in s_n; \tag{12.29}$$

$$\forall_k \exists_m \forall_n \forall_{n'} (n > m, n' > m, a \in s_n, a' \in s_{n'}) \to (a - a', 1/k) \in \underline{N}. \tag{12.30}$$

Here (for brevity) the first three comma's (but not the last!) stand for $\wedge$, and $a \in s_n$ denotes $(n,a) \in \underline{S}$, where $\underline{S}$ is the above subobject of $\underline{\mathbb{N}} \times \underline{A}$ classified by χ (we use the notation explained in item 9 at the end of §E.5, where the variable $x : X$ is now the pair (n,a) of type $\underline{\mathbb{N}} \times \underline{A}$). Moreover, a Cauchy approximation **converges** *to b if:*

$$\forall_k \exists_m \forall_n \, (n > m, a \in s_n) \to (a - b, 1/k) \in \underline{N}, \tag{12.31}$$

and we call A **complete** *if each Cauchy approximation in A converges.*

Finally, a **C*-algebra in** $\mathsf{T}(A)$ *(and similarly in any topos with natural numbers) is a complete pre-semi-C*-algebra in which the semi-norm is a norm.*

Homomorphisms and isomorphisms between such (internal) C*-algebras may be defined in the usual way, bijections in set theory being replaced by isomorphisms of objects. We only consider internal C*-algebras with unit, so that we may define internal categories $\underline{\mathsf{CA}}_1$ (and $\underline{\mathsf{CCA}}_1$) of (commutative) unital C*-algebras in $\mathsf{T}(A)$ in the obvious way (where the homomorphisms are required to preserve the unit).

We now come to the basic construction that underlies "quantum toposophy".

Theorem 12.3. *Let A be a unital C*-algebra. Define a functor $\underline{A} \in \mathsf{T}(A)$ by*

$$\underline{A} : \mathscr{C}(A) \to \mathsf{Sets}; \tag{12.32}$$

$$\underline{A}_0(C) = C; \tag{12.33}$$

$$\underline{A}_1(C \subseteq D) = (C \hookrightarrow D). \tag{12.34}$$

Then $\underline{A}$ is an internal unital commutative *C*-algebra under pointwise operations.*

Here A is meant to be an "ordinary" unital C*-algebra, i.e., defined in Sets. Note that the symbol C in (12.33) changes character from left to right: on the left-hand side it is a *point* in $\mathscr{C}(A)$, whereas on the right-hand side it is a *subset* of A. Nonetheless, one might describe $\underline{A}$ as the ***tautological functor*** in $[\mathscr{C}(A), \mathsf{Sets}]$.

The pointwise operations in $\underline{A}$ are the obvious natural transformations that are ultimately defined by the corresponding operations in each commutative C*-algebra C. For exampe, addition $+ : \underline{A} \times \underline{A} \to \underline{A}$ is a natural transformation with components $+_C : C \times C \to C$ defined in C, etc. Commutativity of $\underline{A}$ then trivially follows from commutativity of each *commutative* C*-subalgebra C.

As already mentioned, the unit $1_{\underline{A}}$ is syntactically a constant $1_{\underline{A}} : \mathbf{1} \to \underline{A}$, whose components $(1_{\underline{A}})_C : * \to C$ are just the units 1_C in each C (recall that elements of our poset $\mathscr{C}(A)$ were defined as *unital* commutative C*-subalgebras of A!).

Finally, we regard the (semi) norm $\underline{N}$ as a subobject of $\underline{A} \times \underline{\mathbb{R}}^+$ (or $\underline{A} \times \underline{\mathbb{Q}}^+$), hence as a natural transformation, with components $\underline{N}_C \subset C \times \mathbb{R}^+$ defined by

$$(c,q) \in \underline{N}_C \text{ iff } \|c\| < q, \tag{12.35}$$

where $\|\cdot\|$ is the norm in C (which of course is inherited from A).

Proof. The proof is a straightforward verification, expect perhaps for completeness. First, the above subobject $\underline{S}$ of $\underline{\mathbb{N}} \times \underline{A}$, realized as a subfunctor as usual, looks as follows: for each $C \in \mathscr{C}(A)$ we have a subset $\underline{S}_C \subset \mathbb{N} \times C$, regarded as a sequence (C_n) of subsets of C through the identification $(n,c) \in \underline{S}_C$ iff $c \in C_n$, such that $C_n \subset D_n$ whenever $C \subset D$. Unfolding axiom (12.29) using the Kripke–Joyal semantics rules listed at the end of §E.5, we find that this axiom holds iff:

$$\forall_{C \in \mathscr{C}(A)} \forall_{n \in \mathbb{N}} \exists_{c \in C} \forall_{D \supseteq C}\, c \in D_n, \tag{12.36}$$

which is satisfied iff each of the above subsets $C_n \subseteq C$ is non-empty. By a similar analysis, axiom (12.30) is satisfied iff for each $\varepsilon > 0$ there is $m \in \mathbb{N}$ such that for all $n, n, > m$ and all $c \in C_n$, $c' \in C_{n'}$ one has $\|c - c'\| < \varepsilon$ in C. This simply means that any choice (c_n) where $c_n \in C_n$ is a Cauchy sequence in C. Accordingly, $\underline{A}$ is complete provided each such sequence converges, i.e., iff each $C \in \mathscr{C}(A)$ is complete. Since these C's are C*-subalgebras of C, this is simply true by construction. □

In a similar way, one easily proves the following generalization of Theorem 12.3:

Theorem 12.4. *Let* C *be a small category. Any internal C*-algebra in the associated presheaf topos* $[\mathsf{C}^{\mathrm{op}}, \mathsf{Sets}]$ *is given by a contravariant functor* $\underline{A} : \mathsf{C} \to \mathsf{CA}$, *where* CA *is the category that has C*-algebras as objects and homomorphims as arrows. Moreover,* $\underline{A}$ *is unital/commutative iff each C*-algebra* $\underline{A}(C)$ *is unital/commutative.*

It should be mentioned that internal C*-algebras on sheaf toposes $\mathsf{T} = \mathsf{Sh}(X)$ are not covered by this theorem (except in the somewhat degenerate case we use, namely $X = \mathscr{C}(A)$ with the Alexandrov topology). As a case in point, we just mention the beautiful fact that internal C*-algebras in $\mathsf{Sh}(X)$ correspond to continuous bundles of C*-algebras over X (in Sets).

12.2 The Gelfand spectrum in constructive mathematics

In this chapter we rely on a particular construction of the frame $\mathcal{O}(\Sigma(A))$ (cf. §C.11) that can be generalized to topos theory (in which the Gelfand spectrum $\Sigma(A)$ of an internal commutative C*-algebra A is a locale). We start with some lattice lore.

Definition 12.5. *Let L be a* distributive *lattice with top $\top$ and bottom $\bot$.*

1. *A* **lower set** *in L is a subset $S \subseteq L$ such that if $x \in S$ and $y \leq x$, then $y \in S$. We denote the poset of all lower subsets of L, ordered by inclusion, by $\mathrm{D}(L)$.*
2. *An* **ideal** *in a lattice L is a lower set I in L such that $x, y \in I$ implies $x \vee y \in I$. The poset all ideals in a lattice L, ordered by inclusion, is denoted by $\mathrm{Idl}(L)$.*
3. *We say that $x \ll y$ (in words: "x* **is well inside** *y" or "x* **is rather below** *y") iff there exists z such that $x \wedge z = \bot$ and $y \vee z = \top$. Note that $x \ll y$ implies $x \leq y$, as*

$$x = x \wedge (y \vee z) = (x \wedge y) \vee (x \wedge z) = x \wedge y \leq y. \tag{12.37}$$

4. *An ideal $I \in \mathrm{Idl}(L)$ is* **regular** *if the condition $I \supseteq \{y \in L \mid y \ll x\}$ implies $x \in I$. The poset of regular ideals in L, ordered by inclusion, is called $\mathrm{RIdl}(L)$, i.e.,*

$$\mathrm{RIdl}(L) = \{I \in \mathrm{Idl}(L) \mid (\forall_{y \in L} y \ll x \Rightarrow y \in I) \Rightarrow x \in I\}. \tag{12.38}$$

The posets $\mathrm{D}(L)$, $\mathrm{Idl}(L)$ and $\mathrm{RIdl}(L)$ are easily seen to be frames. Any ideal $I \in \mathrm{Idl}(L)$ can be *regularized*, i.e., turned into a regular ideal $\mathscr{A}(I)$, by means of the restriction to $\mathrm{Idl}(L) \subset \mathrm{D}(L)$ of the "closure" map $\mathscr{A} : \mathrm{D}(L) \to \mathrm{D}(L)$ defined by

$$\mathscr{A}(I) = \{x \in L \mid \forall_{y \in L} y \ll x \Rightarrow y \in I\}. \tag{12.39}$$

In terms of $\mathscr{A}$, the canonical map $x \mapsto \downarrow x$ from L to $\mathrm{Idl}(L)$ "regularizes" to a map

$$f : L \to \mathrm{RIdl}(L); \tag{12.40}$$

$$x \mapsto \mathscr{A}(\downarrow x). \tag{12.41}$$

For $I \in \mathrm{RIdl}(L)$ we obviously have $\mathscr{A}(I) = I$, and hence we may write

$$\mathrm{RIdl}(L) = \{I \in \mathrm{Idl}(L) \mid \mathscr{A}(I) = I\}. \tag{12.42}$$

Definition 12.6. *1. A frame $\mathcal{O}(X)$ with top element $\top$ is called* **compact** *if every subset $S \subset \mathcal{O}(X)$ with $\bigvee S = \top$ has a finite subset $F \subset S$ with $\bigvee F = \top$.*
2. A frame $\mathcal{O}(X)$ is called **regular** *if each $V \in \mathcal{O}(X)$ satisfies*

$$V = \bigvee\{U \in \mathcal{O}(X) \mid U \ll V\}. \tag{12.43}$$

When $\mathcal{O}(X)$ is the topology of some space X, the frame $\mathcal{O}(X)$ is compact (regular) iff X is compact (regular) as a space. Furthermore, X is compact and Hausdorff iff it is compact and regular, and hence the Gelfand spectrum $\Sigma(A)$ of a commutative unital C*-algebra A will be a compact and regular frame; see Theorem 12.8 below.

Recall that the self-adjoint part A_{sa} of any C*-algebra A is partially ordered by putting $a \leq b$ iff $b - a \in A^+$, cf. §C.7. This partial order is, of course, inherited by the positive cone $A^+ \subset A_{\text{sa}}$. If A is commutative, this partial ordering makes A_{sa} a lattice; for example, if $A = C(X)$ the lattice operations are $a \vee b = \max\{a,b\}$ and $a \wedge b = \min\{a,b\}$ (taken pointwise). In general, one may then compute $\vee$ and $\wedge$ from the Gelfand isomorphism $A \cong C(X)$, but they are intrinsically defined via $\leq$.

Let A be a commutative unital C*-algebra. For $a,b \in A^+$, define $a \preccurlyeq b$ iff there exists $n \in \mathbb{N}$ such that $a \leq nb$. Define $a \approx b$ iff $a \preccurlyeq b$ and $b \preccurlyeq a$. This is an equivalence relation. Moreover, $\approx$ is a ***congruence***, that is, an equivalence relation $\sim$ on a lattice L that is compatible with $\wedge$ and $\vee$ in the sense that $x \sim y$ and $x' \sim y'$ imply $x \wedge x' \sim y \wedge y'$ and $x \vee x' \sim y \vee y'$. Given some congruence $\sim$ on L, one may define $\wedge$ and $\vee$ on $L/\sim$ by $[x] \wedge [y] = [x \wedge y]$ and $[x] \vee [y] = [x \vee y]$, respectively, so that the set-theoretic quotient $L/\sim$ inherits the lattice structure of L and hence is a lattice in its own right.

This quotient construction by a congruence preserves distributivity, so that

$$L_A = A^+/\approx . \tag{12.44}$$

is a distributive lattice. We will use the elements $\mathsf{D}_a \equiv [a^+]$ of L_A (indexed by $a \in A_{\text{sa}}$), where $[a^+]$ is the equivalence class in L_A of the positive part a^+ in the canonical decomposition $a = a^+ - a^-$, with $a^\pm \geq 0$ and $a_+ a_- = 0$; lattice theoretically, one has $a_+ = a \vee 0$ and $a_- = a \wedge 0$. This gives a lattice homomorphism $A_{\text{sa}} \to L_A$, $a \mapsto \mathsf{D}_a$, whose restriction to A^+ is just the canonical projection $A^+ \to L_A$. These D_a satisfy:

$$\mathsf{D}_1 = \top; \tag{12.45}$$

$$\mathsf{D}_a \wedge \mathsf{D}_{-a} = \bot; \tag{12.46}$$

$$\mathsf{D}_a = \bot \ (a \leq 0); \tag{12.47}$$

$$\mathsf{D}_{a+b} \leqslant \mathsf{D}_a \vee \mathsf{D}_b; \tag{12.48}$$

$$\mathsf{D}_a \wedge \mathsf{D}_b \leqslant \mathsf{D}_{ab}; \tag{12.49}$$

$$\mathsf{D}_{ab} \leqslant \mathsf{D}_a \vee \mathsf{D}_{-b}, \tag{12.50}$$

where the inequalities may also be written as equalities, since $x \leq y$ iff $x = x \wedge y$. These relations are easy to check for $A = C(X)$, and hence they are true for any A. The elements D_a obviously exhaust A^+, and eqs. (12.45) - (12.50) imply:

$$a \leq b \Rightarrow \mathsf{D}_a \leqslant \mathsf{D}_b; \tag{12.51}$$

$$\mathsf{D}_a = \mathsf{D}_{a^+}; \tag{12.52}$$

$$\mathsf{D}_{na} = \mathsf{D}_a \ (n \in \mathbb{N}); \tag{12.53}$$

$$\mathsf{D}_{ab} = (\mathsf{D}_a \wedge \mathsf{D}_b) \vee (\mathsf{D}_a \wedge \mathsf{D}_{-b}; \tag{12.54}$$

$$\mathsf{D}_a \wedge \mathsf{D}_b = \mathsf{D}_{a \wedge b}. \tag{12.55}$$

For the Gelfand spectrum we need the frame $\text{RIdl}(L_A)$, and hence the relation $\ll$.

Lemma 12.7. *For all $\mathsf{D}_a, \mathsf{D}_b \in L_A$, we have (with both $q \in \mathbb{Q}^+$ and $q \in \mathbb{R}^+$):*

$$\mathsf{D}_b \ll \mathsf{D}_a \textit{ iff } \exists_{q>0} \mathsf{D}_b \leqslant \mathsf{D}_{a-q}. \tag{12.56}$$

Proof. From right to left, just choose $\mathsf{D}_c = \mathsf{D}_{q-a}$. Conversely, if $A = C(X)$, it is easy to see that if there exists $\mathsf{D}_c \in L_A$ such that $\mathsf{D}_c \vee \mathsf{D}_a = \top$ and $\mathsf{D}_c \wedge \mathsf{D}_b = \bot$, then there exists $q > 0$ such that $\mathsf{D}_{c-q} \vee \mathsf{D}_{a-q} = \top$. Hence $\mathsf{D}_c \vee \mathsf{D}_{a-q} = \top$, so that

$$\mathsf{D}_b = \mathsf{D}_b \wedge (\mathsf{D}_c \vee \mathsf{D}_{a-q}) = \mathsf{D}_b \wedge \mathsf{D}_{a-q} \leqslant \mathsf{D}_{a-q}.$$ □

Note that by construction the map f in (12.40) is given by

$$f(\mathsf{D}_a) = \{\mathsf{D}_c \in L_A \mid \forall_{\mathsf{D}_b \in L_A}\, \mathsf{D}_b \ll \mathsf{D}_c \Rightarrow \mathsf{D}_b \leqslant \mathsf{D}_a\}, \tag{12.57}$$

and, by Lemma 12.7, satisfies

$$f(\mathsf{D}_a) \leqslant \bigvee \{f(\mathsf{D}_{a-q}) \mid q > 0\}. \tag{12.58}$$

For later use, also note that (12.57) implies

$$f(\mathsf{D}_a) = \top \Leftrightarrow \mathsf{D}_a = \top. \tag{12.59}$$

Theorem 12.8. *The topology $\mathcal{O}(\Sigma(A))$ of the Gelfand spectrum $\Sigma(A)$ of a commutative unital C*-algebra A is isomorphic to the frame of all regular ideals of L_A:*

$$\mathcal{O}(\Sigma(A)) \cong \mathrm{RIdl}(L_A); \tag{12.60}$$
$$\{\omega \in \Sigma(A) \mid \omega(a) > 0\} \leftrightarrow \mathsf{D}_a, \tag{12.61}$$

or, equivalently, for the opens $(r,s) \in \mathcal{O}(\mathbb{R})$ with ensuing opens $\hat{a}^{-1}(r,s)$ in $\mathcal{O}(\Sigma(A))$,

$$\hat{a}^{-1}(r,s) \equiv \{\omega \in \Sigma(A) \mid \omega(a) \in (r,s)\} \leftrightarrow f(\mathsf{D}_{s-a} \wedge \mathsf{D}_{a-r}) \ (r < s). \tag{12.62}$$

Moreover, on this isomorphism, $\mathcal{O}(\Sigma(A))$ is a compact regular frame.

The proof of this theorem is unfortunately beyond our reach; instead, we now give an alternative descriptions of the frame $\mathrm{RIdl}(L_A)$, which will be useful for computational purposes in topos theory. This again requires some more background in lattice theory. Let $(L, \leqslant)$ be a meet semilattice (i.e., a poset in which any pair of elements has an infimum; in most of our applications $(L, \leqslant)$ is actually a distributive lattice).

Definition 12.9. *A* covering relation *on L is a relation $\lhd \subseteq L \times \mathcal{P}(L)$—equivalently, a function $L \to \mathcal{P}(\mathcal{P}(L))$—written $x \lhd U$ when $(x,U) \in \lhd$, such that:*

1. *If $x \in U$ then $x \lhd U$.*
2. *If $x \lhd U$ and $U \lhd V$ (i.e., $y \lhd V$ for all $y \in U$) then $x \lhd V$.*
3. *If $x \lhd U$ then $x \wedge y \lhd U$.*
4. *If $x \in U$ and $x \in V$, then $x \lhd U \wedge V$ (where $U \wedge V = \{x \wedge y \mid x \in U, y \in V\}$).*

For example, if $(L, \leqslant) = (\mathcal{O}(X), \subseteq)$ one may take $x \lhd U$ iff $x \leqslant \bigvee U$, i.e., iff U covers x. Also here we have a closure operation $\mathscr{A} : \mathrm{D}(L) \to \mathrm{D}(L)$, given by

$$\mathscr{A}U = \{x \in L \mid x \lhd U\}. \tag{12.63}$$

This operation has the following properties:

$$\downarrow U \subseteq \mathscr{A}U; \tag{12.64}$$
$$U \subseteq \mathscr{A}V \Rightarrow \mathscr{A}U \subseteq \mathscr{A}V; \tag{12.65}$$
$$\mathscr{A}U \cap \mathscr{A}V \subseteq \mathscr{A}(\downarrow U \cap \downarrow V). \tag{12.66}$$

The frame $\mathscr{F}(L, \lhd)$ generated by such a structure is then defined by

$$\mathscr{F}(L, \lhd) = \{U \in \mathrm{D}(L) \mid \mathscr{A}U = U\} = \{U \in \mathscr{P}(L) \mid x \lhd U \Rightarrow x \in U\}; \tag{12.67}$$

the second equality follows because firstly the property $\mathscr{A}U = U$ guarantees that $U \in \mathrm{D}(L)$, and secondly one has $\mathscr{A}U = U$ iff $x \lhd U$ implies $x \in U$. Defining

$$U \sim V \text{ iff } U \lhd V \text{ and } V \lhd U, \tag{12.68}$$

an equivalent description of the frame $\mathscr{F}(L, \lhd)$ that is occasionally useful is

$$\mathscr{F}(L, \lhd) \cong \mathscr{P}(L)/\sim. \tag{12.69}$$

Indeed, the map $U \mapsto [U]$ from $\mathscr{F}(L, \lhd)$ (as defined in (12.67)) to $\mathscr{P}(L)/\sim$ is a frame map with inverse $[U] \mapsto \mathscr{A}U$. The idea behind the isomorphism (12.69) is that the map $\mathscr{A}$ picks a unique representative in the equivalence class $[U]$, namely $\mathscr{A}U$. As in (12.40) - (12.71), also here we have a canonical map

$$f : L \to \mathscr{F}(L, \lhd); \tag{12.70}$$
$$x \mapsto \mathscr{A}(\downarrow x), \tag{12.71}$$

which satisfies $f(x) \leqslant \bigvee f(U)$ if $x \lhd U$. In fact, f is universal with this property, in that any homomorphism $g : L \to \mathscr{G}$ of meet semilattices into a frame $\mathscr{G}$ such that $g(x) \leqslant \bigvee g(U)$ whenever $x \lhd U$ has a factorisation $g = \varphi \circ f$ for some unique frame map $\varphi : \mathscr{F}(L, C) \to \mathscr{G}$. This may suggest the following result:

Proposition 12.10. *Suppose one has a frame $\mathscr{F}$ and a meet semilattice L with a map $f : L \to \mathscr{F}$ of meet semilattices that generates $\mathscr{F}$ in the sense that for each $U \in \mathscr{F}$ one has $U = \bigvee\{f(x) \mid x \in L, f(x) \leq U\}$. Define a cover relation $\lhd$ on L by*

$$x \lhd U \textit{ iff } f(x) \leqslant \bigvee f(U). \tag{12.72}$$

Then one has a frame isomorphism $\mathscr{F} \cong \mathscr{F}(L, \lhd)$.

We now turn to maps between frames, from the point of view of coverings.

Definition 12.11. *Let $(L, \lhd)$ and $(M, \blacktriangleleft)$ be meet semilattices with covering relation as above, and let $f^* : L \to \mathscr{P}(M)$ be such that:*

1. *$f^*(L) = M$;*
2. *$f^*(x) \wedge f^*(y) \blacktriangleleft f^*(x \wedge y)$;*
3. *$x \lhd U \Rightarrow f^*(x) \blacktriangleleft f^*(U)$ (where $f^*(U) = \bigcup_{u \in U} f(U)$).*

If L and M have top elements $\top_L$ *and* $\top_M$*, respectively, then the first condition may be replaced by* $f^*(\top_L) = \top_M$*. Define two such maps* f_1^*, f_2^* *to be equivalent if* $f_1^*(x) \sim f_2^*(x)$ *(i.e.,* $f_1^*(x) \blacktriangleleft f_2^*(x)$ *and* $f_2^*(x) \blacktriangleleft f_1^*(x)$*) for all* $x \in L$*. A continuous map* $f : (M, \blacktriangleleft) \to (L, \lhd)$ *is an equivalence class of such maps* $f^* : L \to \mathscr{P}(M)$*.*

Our main interest in continuous maps lies in the following result:

Proposition 12.12. *Each continuous map* $f : (M, \blacktriangleleft) \to (L, \lhd)$ *is equivalent to a frame map* $\mathscr{F}(f) : \mathscr{F}(L, \lhd) \to \mathscr{F}(M, \blacktriangleleft)$*, given by*

$$\mathscr{F}(f) : U \mapsto \mathscr{A} f^*(U). \tag{12.73}$$

We may now equip L_A with the covering relation defined by (12.72), given (12.60) and the ensuing map (12.57). Consequently, by Proposition 12.10 one has

$$\mathcal{O}(\Sigma) \cong \mathscr{F}(L_A, \lhd), \tag{12.74}$$

which yields the following expression for the constructive Gelfand spectrum:

$$\mathcal{O}(\Sigma) \cong \{U \in \mathrm{D}(L_A) \mid x \lhd U \Rightarrow x \in U\}. \tag{12.75}$$

This lattice becomes computable through a lemma that is crucial for what follows:

Lemma 12.13. *In any topos, the covering relation* $\lhd$ *on* L_A *defined by* (12.72) *with* (12.60) *and* (12.57)*, is given by* $\mathsf{D}_a \lhd U$ *iff for all* $q > 0$ *there exists a (Kuratowski) finite* $U_0 \subseteq U$ *such that* $\mathsf{D}_{a-q} \leqslant \bigvee U_0$*. If U is directed, this means that there exists* $\mathsf{D}_b \in U$ *such that* $\mathsf{D}_{a-q} \leqslant \mathsf{D}_b$*.*

Proof. The easy part is the "$\Leftarrow$" direction: from (12.58) and the assumption we have $f(\mathsf{D}_a) \leqslant \bigvee f(U)$ and hence $\mathsf{D}_a \lhd U$ by definition of the covering relation.

In the opposite direction, assume $\mathsf{D}_a \lhd U$ and take some $q > 0$. From (the proof of) Lemma 12.7, $\mathsf{D}_a \vee \mathsf{D}_{q-a} = \top$, hence $\bigvee f(U) \vee f(\mathsf{D}_{q-a}) = \top$. Since $\mathcal{O}(\Sigma)$ is compact, there is a finite $U_0 \subset U$ for which $\bigvee f(U_0) \vee f(\mathsf{D}_{q-a}) = \top$, so that by (12.59) we have $\mathsf{D}_b \vee \mathsf{D}_{q-a} = \top$, with $\mathsf{D}_b = \bigvee U_0$. By (12.46) we have

$$\mathsf{D}_{a-q} \wedge \mathsf{D}_{q-a} = \perp, \tag{12.76}$$

and hence

$$\mathsf{D}_{a-q} = \mathsf{D}_{a-q} \wedge \top = \mathsf{D}_{a-q} \wedge (\mathsf{D}_b \vee \mathsf{D}_{q-a}) = \mathsf{D}_{a-q} \wedge \mathsf{D}_b \leqslant \mathsf{D}_b = \bigvee U_0. \qquad \square$$

If A is finite-dimensional, L_A is a finite lattice. In that case, since $\mathsf{D}_{a-q} = \mathsf{D}_a$ for small enough q, one simply has $x \lhd U$ iff $x \leq \bigvee U$, and the condition $x \lhd U \Rightarrow x \in U$ in (12.75) holds iff U is a (principal) down set, i.e. $U = \downarrow x$ for some $x \in L_A$ (not the same x as the placeholder x in (12.75)). Hence for finite-dimensional A we obtain

$$\mathcal{O}(\Sigma(A)) \cong \mathrm{Idl}(L_A) = \{\downarrow x \mid x \in L_A\}. \tag{12.77}$$

12.3 Internal Gelfand spectrum and intuitionistic quantum logic

We are now going to combine the (*a priori* independent) material in the previous two sections. The point of the above description of the topology $\mathcal{O}(\Sigma(A))$ of the Gelfand spectrum $\Sigma(A)$ of a unital commutative C*-algebra A is that it may be "internalized" to any topos (with natural number object, i.e., in which C*-algebras may be defined internally in the first place). The key to the ensuing generalization of Gelfand duality is that in topos theory (and more generally in constructive mathematics) the *space* $\Sigma(A)$ in set theory needs to be replaced by the corresponding *frame* $\mathcal{O}(\Sigma(A))$, or preferably by its associated *locale*, which confusingly is denoted by $\Sigma(A)$, even though it is the same thing as $\mathcal{O}(\Sigma(A))$ and neither may be spatial (in being the topology of some space); see §C.11 and §E.4 for this bizarre notation. Similarly, we write $f : X \to Y$ for a map between locales, which is essentially the same as the frame map $f^{-1} : \mathcal{O}(Y) \to \mathcal{O}(X)$, but seen as a map in the opposite direction (where once again nothing is assumed about possible spatiality of the frames in question).

Using this notation, the ***constructive Gelfand isomorphism*** (which is valid in any topos T in which commutative C*-algebras make sense) states:

Theorem 12.14. *For each (internal) commutative unital C*-algebra A in* T *there exists a compact regular locale* $\Sigma(A)$ *such that one has a Gelfand isomorphism*

$$A \cong C(\Sigma(A), \mathbb{C}). \tag{12.78}$$

Furthermore, the locale $\Sigma(A)$ *is uniquely determined by A up to isomorphism and its corresponding frame is given by Theorem 12.8 (or, more explicitly, by* (12.75) *in conjunction with Lemma 12.13, all of which makes sense internally).*

Here $\cong$ denotes (internal) isomorphism of (commutative) C*-algebras, and the notation $C(\Sigma(A), \mathbb{C})$ stands for the object of all frame maps from $\mathcal{O}(\mathbb{C})$ to $\mathcal{O}(\Sigma(A))$ (which object turns out to be a commutative C*-algebra in any case). As usual, we denote the Gelfand transform $A \to C(\Sigma(A), \mathbb{C})$ by $a \mapsto \hat{a}$, where, as explained above, the locale map $\hat{a} : \Sigma(A) \to \mathbb{C}$ is really the reverse reading of the frame map

$$\hat{a}^{-1} : \mathcal{O}(\mathbb{C}) \to \mathcal{O}(\Sigma(A)). \tag{12.79}$$

Note that in Sets, the latter is given by its literal meaning, given $\hat{a} : \omega \mapsto \omega(a)$.

We will shortly apply this formalism to our internal C*-algebra $\underline{A}$ in the topos $\mathsf{T}(A)$, but since these computations are a bit involved, as a warm-up we first apply our machinery to a very simple case, namely $A = \mathbb{C}^n$ in Sets. Recall (12.44) etc.

For $A = \mathbb{C}^n$ we have $A^+ = (\mathbb{R}^n)^+$, in which $(r_1, \ldots, r_n) \approx (s_1, \ldots, s_n)$ just in case $r_i = 0$ iff $s_i = 0$ for all $i = 1, \ldots n$. Hence each equivalence class under $\approx$ has a unique representative of the form $[k_1, \ldots, k_n]$ with $k_i = 0$ or $k_i = 1$; the pre-images of such an element of L_A in A^+ under the natural projection $A^+ \to A^+/\approx$ are the diagonal matrices whose i'th entry is zero if $k_i = 0$ and any nonzero positive number if $k_i = 1$. The partial order in L_A is pointwise, i.e. $[k_1, \ldots, k_n] \leq [l_1, \ldots, l_n]$ iff $k_i \leq l_i$ for all i. Hence $L_{\mathbb{C}^n}$ is isomorphic as a distributive lattice to the lattice $\mathcal{P}(D_n(\mathbb{C})) \equiv \mathcal{P}(\mathbb{C}^n)$ of projections in $D_n(\mathbb{C})$, i.e. the lattice of diagonal projections in $M_n(\mathbb{C})$.

Under this isomorphism, $[k_1,\ldots,k_n]$ corresponds to the matrix $\mathrm{diag}(k_1,\ldots,k_n)$. If we equip $\mathscr{P}(\mathbb{C}^n)$ with the usual partial ordering of projections on the Hilbert space $\mathbb{C}^n$, viz. $e \leq f$ whenever $e\mathbb{C}^n \subseteq f\mathbb{C}^n$ (which coincides with their ordering as element of positive cone of the C*-algebra $M_n(\mathbb{C})$), then this is even a lattice isomorphism. Hence by (12.77), the frame $\mathscr{O}(\Sigma(\mathbb{C}^n))$ consists of all sets of the form $\downarrow e$, $e \in \mathscr{P}(\mathbb{C}^n)$, partially ordered by inclusion. This means that

$$\mathscr{O}(\Sigma(\mathbb{C}^n)) \cong \mathscr{P}(\mathbb{C}^n), \tag{12.80}$$

under the further identification of $\downarrow p \subset \mathscr{P}(\mathbb{C}^n)$ with $p \in \mathscr{P}(\mathbb{C}^n)$. This starts out just as an isomorphism of posets, and turns out to be one of frames (which in the case at hand happen to be Boolean). To draw the connection with the usual spectrum $\hat{\mathbb{C}}^n = \{1,2,\ldots,n\}$ of $\mathbb{C}^n$, we note that the right-hand side of (12.80) is isomorphic to the discrete topology $\mathscr{O}(\hat{\mathbb{C}}^n)$ of $\hat{\mathbb{C}}^n$ (i.e. its power set) under the frame isomorphism

$$\begin{aligned} \mathscr{P}(\mathbb{C}^n) &\stackrel{\cong}{\to} \mathscr{O}(\hat{\mathbb{C}}^n); \\ \mathrm{diag}(k_1,\ldots,k_n) &\mapsto \{i \in \{1,2,\ldots,n\} \mid k_i = 1\}. \end{aligned} \tag{12.81}$$

We now describe the Gelfand transform (12.78) - (12.79) for self-adjoint a, so that one has a (locale) map $A_{\mathrm{sa}} \to C(\Sigma(A),\mathbb{R})$. Let $a = (a_1,\ldots,a_n) \in \mathbb{C}^n_{\mathrm{sa}} = \mathbb{R}^n$. With $\Sigma(\mathbb{C}^n)$ realized as $\hat{\mathbb{C}}^n$, this just reads $\hat{a}(i) = a_i$, for $\hat{a} : \hat{\mathbb{C}}^n \to \mathbb{C}$. The induced frame map $\hat{a}^{-1} : \mathscr{O}(\mathbb{C}) \to \mathscr{O}(\hat{\mathbb{C}}^n)$ is given by $U \mapsto \{i \in \{1,2,\ldots,n\} \mid a_i \in U\}$, and by (12.81), this is equivalent to

$$\begin{aligned} \hat{a}^{-1} : \mathscr{O}(\mathbb{R}) &\to \mathscr{P}(\mathbb{C}^n); \\ U &\mapsto 1_U(a), \end{aligned} \tag{12.82}$$

where $U \in \mathscr{O}(\mathbb{R})$, and the right-hand side denotes the spectral projection $1_U(a)$ defined by the self-adjoint operator a on the Hilbert space $\mathbb{C}^n$.

After this warm-up, we now compute the Gelfand spectrum $\mathscr{O}(\underline{\Sigma}(\underline{A}))$ in our topos $\mathsf{T}(A)$, for the special case $A = M_n(\mathbb{C})$ (which is still an exercise for the general case). For simplicity we write $\underline{L}$ for the lattice $L_{\underline{A}}$ in $\mathsf{T}(A)$; similarly, $\underline{\Sigma}$ stands for $\underline{\Sigma}(\underline{A})$.

First, for arbitrary A, the lattice functor $\underline{L}$ can be computed "locally", in the sense that $\underline{L}_0(C) = L_C$, see Proposition 12.17 in §12.4 below, so that by (12.44) one has

$$\underline{L}_0(C) = C^+/\approx. \tag{12.83}$$

Let $\mathscr{P}(C)$ be the (Boolean) lattice of projections in C, and consider the functor

$$\underline{\mathscr{P}}_0(C) = \mathscr{P}(C); \tag{12.84}$$
$$\underline{\mathscr{P}}_1(C \subseteq D) = (\mathscr{P}(C) \hookrightarrow \mathscr{P}(D)). \tag{12.85}$$

As in the case $A = \mathbb{C}^n$ just discussed, it follows that we may identify $\underline{L}_0(C)$ with $\mathscr{P}(C)$ and hence we may and will identify the functor $\underline{L}$ with the functor $\underline{\mathscr{P}}$.

Second, whereas in Sets eq. (12.77) makes $\mathscr{O}(\Sigma)$ a subset of L, in the topos $\mathsf{T}(A)$ the frame $\mathscr{O}(\underline{\Sigma})$ is a subobject $\mathscr{O}(\underline{\Sigma}) \rightarrowtail \underline{\Omega}^{\underline{L}}$. It then follows from (12.11) that $\mathscr{O}(\underline{\Sigma})(C)$ is a subset of $\mathrm{Sub}(\underline{\mathscr{P}}_{\uparrow C})$, the set of subfunctors of the functor $\underline{\mathscr{P}} : \mathscr{C}(A) \to$ Sets restricted to $\uparrow C \subset \mathscr{C}(A)$. To see which subset, define

$$\mathrm{Sub}_d(\underline{\mathscr{P}}_{\uparrow C}) = \{\tilde{S} \in \mathrm{Sub}(\underline{\mathscr{P}}_{\uparrow C}) \mid \forall D \supseteq C\, \exists x_D \in \mathscr{P}(D) : \tilde{S}(D) = \downarrow x_D\}. \quad (12.86)$$

Thus $\mathrm{Sub}_d(\underline{\mathscr{P}}_{\uparrow C})$ consists of subfunctors S of $\underline{\mathscr{P}}_{\uparrow C}$ that are locally down-sets. It then follows from (12.77) and the local interpretation of the relation $\lhd$ in $\mathsf{T}(A)$ (see Lemma 12.18 in §12.4 below) that the subobject $\mathscr{O}(\underline{\Sigma}) \rightarrowtail \underline{\Omega}^{\underline{L}}$ in $\mathsf{T}(A)$ is the functor

$$\mathscr{O}(\underline{\Sigma})_0(C) = \mathrm{Sub}_d(\underline{\mathscr{P}}_{\uparrow C}); \quad (12.87)$$
$$\mathscr{O}(\underline{\Sigma})_1(C \subseteq D) = (\mathscr{O}(\underline{\Sigma})(C) \hookrightarrow \mathscr{O}(\underline{\Sigma})(D)), \quad (12.88)$$

where $\mathscr{O}(\underline{\Sigma})_1$ is inherited from $\underline{\Omega}^{\underline{L}}$ (of which $\mathscr{O}(\underline{\Sigma})$ is a subobject), and hence is just given by restricting an element of $\mathscr{O}(\underline{\Sigma})(C)$ to $\uparrow D$. Writing

$$\mathrm{Sub}_d(\underline{\mathscr{P}}) = \{\tilde{S} \in \mathrm{Sub}(\underline{\mathscr{P}}) \mid \forall D \in \mathscr{C}(A)\, \exists x_D \in \mathscr{P}(D) : \tilde{S}(D) = \downarrow x_D\}, \quad (12.89)$$

it is convenient to embed $\mathrm{Sub}_d(\underline{\mathscr{P}}_{\uparrow C}) \subseteq \mathrm{Sub}_d(\underline{\mathscr{P}})$ by requiring elements of the left-hand side to vanish whenever D does not contain C. We also note that if $\tilde{S}$ is to be a subfunctor of $\underline{\mathscr{P}}_{\uparrow C}$, one must have $\tilde{S}(D) \subseteq \tilde{S}(E)$ whenever $D \subseteq E$, and that $\downarrow x_D \subseteq \downarrow x_E$ iff $x_D \leq x_E$ in $\mathscr{P}(E)$. Thus one may simply describe elements of $\mathscr{O}(\underline{\Sigma})(C)$ via maps $S : \mathscr{C}(A) \to \mathscr{P}(A)$ such that:

$$S(D) \in \mathscr{P}(D); \quad (12.90)$$
$$S(D) = 0 \text{ if } D \notin \uparrow C \text{ (i.e. } C \nsubseteq D); \quad (12.91)$$
$$S(D) \leq S(E) \text{ if } C \subseteq D \subseteq E. \quad (12.92)$$

The corresponding element $\tilde{S}$ of $\mathscr{O}(\underline{\Sigma})(C)$ is then given by

$$\tilde{S}(D) = \downarrow S(D), \quad (12.93)$$

seen as a subset of $\mathscr{P}(D)$. Hence it is convenient to introduce the notation

$$\mathscr{O}(\Sigma)_{\uparrow C} = \{S : \uparrow C \to \mathscr{P}(A) \mid S(D) \in \mathscr{P}(D), S(D) \leq S(E) \text{ if } D \subseteq E\}, \quad (12.94)$$

of which we single out the case $C = \mathbb{C} \cdot 1_A$, which will be of great importance:

$$\mathscr{O}(\Sigma) = \{S : \mathscr{C}(A) \to \mathscr{P}(A) \mid S(C) \in \mathscr{P}(C), S(C) \leq S(D) \text{ if } C \subseteq D\}. \quad (12.95)$$

Both are posets and even frames in the pointwise partial order with respect to the usual ordering of projections (which algebraically means $e \leq f$ iff $ef = e$), i.e.,

$$S \leq T \Leftrightarrow S(C) \leq T(C) \text{ for all } C \in \mathscr{C}(A). \quad (12.96)$$

In terms of (12.94) - (12.95), we then have isomorphisms

$$\mathcal{O}(\underline{\Sigma})_0(\mathbb{C}\cdot 1) \cong \mathcal{O}(\Sigma); \tag{12.97}$$
$$\mathcal{O}(\underline{\Sigma})(C)_0 \cong \mathcal{O}(\Sigma)_{\uparrow C}. \tag{12.98}$$

More importantly, the frame $\mathcal{O}(\Sigma)$ in Sets is the key to the *external description* of the *internal frame* $\mathcal{O}(\underline{\Sigma})$ in $\mathsf{T}(A)$; see the end of §E.4. Since $\mathscr{C}(A)$ carries the Alexandrov topology, by (E.84) this description is given by the frame map

$$\pi_{\Sigma}^{-1} : \mathcal{O}(\mathscr{C}(A)) \to \mathcal{O}(\Sigma), \tag{12.99}$$

given on the basic opens $\uparrow D \in \mathcal{O}(\mathscr{C}(A))$ by

$$\pi_{\Sigma}^{-1}(\uparrow D) = \chi_{\uparrow D} : E \mapsto 1 \ (E \supseteq D);$$
$$E \mapsto 0 \ (E \not\supseteq D). \tag{12.100}$$

As explained before, even in Sets, in principle $\mathcal{O}(\Sigma)$ is just a notation for a frame, without suggesting that there exists an underlying space Σ whose topology it is. In this case, however, there is such a space (as we shall show in the next section), and also (12.99) is in fact the inverse image map to a genuine map $\pi_{\Sigma} : \Sigma \to \mathscr{C}(A)$ between spaces (as opposed to the formal notation used for a locale map).

We now state the Heyting algebra structure of $\mathcal{O}(\Sigma)$. First, top and bottom are

$$\top(C) = 1 \text{ for all } C; \tag{12.101}$$
$$\bot(C) = 0 \text{ for all } C. \tag{12.102}$$

The logical operations on $\mathcal{O}(\Sigma)$ may be computed from the partial order as

$$(S \wedge T)(C) = S(C) \wedge T(C); \tag{12.103}$$
$$(S \vee T)(C) = S(C) \vee T(C); \tag{12.104}$$
$$(S \dashrightarrow T)(C) = \bigwedge_{D \supseteq C}^{\mathscr{P}(C)} S(D)^{\perp} \vee T(D); \tag{12.105}$$
$$(\neg S)(C) = \bigwedge_{D \supseteq C}^{\mathscr{P}(C)} S(D)^{\perp}; \tag{12.106}$$
$$(\neg\neg S)(C) = \bigwedge_{D \supseteq C}^{\mathscr{P}(C)} \bigvee_{E \supseteq D}^{\mathscr{P}(D)} S(E), \tag{12.107}$$

where the right-hand side of (12.105) (and similarly (12.106) - (12.107)) is short for

$$\bigwedge_{D \supseteq C}^{\mathscr{P}(C)} S(D)^{\perp} \vee T(D) \equiv \bigvee \{e \in \mathscr{P}(C) \mid e \leq S(D)^{\perp} \vee T(D) \, \forall D \supseteq C\}. \tag{12.108}$$

Recall that a Heyting algebra is Boolean iff $\neg\neg S = S$ for each S. One sees from (12.107) that (at least if $n > 1$) the property $\neg\neg S = S$ only holds iff S is either $\top$ or $\bot$, so that the Heyting algebra $\mathcal{O}(\Sigma) \equiv CO(\Sigma(A))$ is properly intuitionistic.

Since from both a physical and a logical point of view the Heyting algebra $\mathcal{O}(\Sigma(A))$ has vast advantages over the projection lattice $\mathscr{P}(A)$ of Birkhoff and von Neumann, we propose it as a candidate for a new quantum logic. Let us explain why.

Physically, in von Neumann's approach each projection $e \in \mathscr{P}(A)$ defines an elementary proposition, whereas in Bohr's (where the classical context C is crucial) an elementary proposition is a *pair* (C,e), where $e \in \mathscr{P}(C)$ is a proposition à la von Neumann (who lost sight of the context C). If for each such pair (C,e) we define

$$S_{(C,e)} : \mathscr{C}(A) \to \mathscr{P}(A); \tag{12.109}$$

$$D \mapsto e \ (C \subseteq D); \tag{12.110}$$

$$D \mapsto \bot \ \text{otherwise}, \tag{12.111}$$

we see that each pair (C,e) injectively defines an element of $\mathcal{O}(\Sigma)$. Furthermore, each element S of $\mathcal{O}(\Sigma)$ is a disjunction over such elementary propositions, since

$$S = \bigvee_{C \in \mathscr{C}(A)} S_{(C,S(e))}. \tag{12.112}$$

In contrast to traditional quantum logic, both logical connectives $\wedge$ and $\vee$ on $\mathcal{O}(\Sigma)$ are physically meaningful, as they only involve *local* conjunctions $S(C) \wedge T(C)$ and disjunctions $S(C) \vee T(C)$, for which $S(C) \in \mathscr{P}(C)$ and $T(C) \in \mathscr{P}(C)$ commute.

Logically, the absence of an implication arrow in quantum logic has always been worrying; this has now been put straight in $\mathcal{O}(\Sigma)$, where $\dashrightarrow$ belongs to the defining structure and behaves well logically. Truth attribution in quantum logic is equally suspicious: for any state ω on A one declares a proposition $e \in \mathscr{P}(A)$ *true* iff $\omega(e) = 1$, and *false* iff $\omega(e) = 0$, with no verdict otherwise (except probabilistically).

We, however, define a natural ***Kripke semantics*** (cf. §D.3) on $P = \mathscr{C}(A)$ by

$$V_\omega : \mathcal{O}(\Sigma) \to \text{Upper}(\mathscr{C}(A)) = \mathcal{O}(\mathscr{C}(A)); \tag{12.113}$$

$$V_\omega(S) = \{C \in \mathscr{C}(A) \mid \omega(S(C)) = 1\}, \tag{12.114}$$

where $\mathscr{C}(A)$ carries the Alexandrov topology as usual. Note that $V_\omega(S)$ indeed defines an upper set in $\mathscr{C}(A)$, for if $C \subseteq D$ then $S(C) \leq S(D)$, so that $\omega(S(C)) \leq \omega(S(D))$ by positivity of states, and hence $\omega(S(D)) = 1$ whenever $\omega(S(C)) = 1$ (given that $\omega(S(D)) \leq 1$, which is true since $0 \leq \omega(e) \leq 1$ for any projection e).

As explained in §D.3, a proposition $S \in \mathcal{O}(\Sigma)$ is ***true*** in a state ω if $V_\omega(S) = \mathscr{C}(A)$, i.e. the top element of the frame $\mathcal{O}(\mathscr{C}(A))$; we also declare it ***false*** if $V_\omega(S) = \emptyset$, i.e. the bottom element of $\mathcal{O}(\mathscr{C}(A))$. Then $\neg S$ is true iff S is false, and $S \vee T$ is true iff either S or T is true (since $V_\omega(S) = \mathscr{C}(A)$ iff $S(\mathbb{C} \cdot 1) = 1$, which forces $S(C) = 1$ for all C). Consequently, (12.114) simply lists the contexts C in which $S(C)$ is true.

12.4 Internal Gelfand spectrum for arbitrary C*-algebras

In this section we compute the internal Gelfand spectrum $\underline{\Sigma}(\underline{A}) \equiv \underline{\Sigma}$ in $\mathsf{T}(A)$ for an arbitrary unital C*-algebra A. Recall Definition D.6 (in §D.1) of a free lattice $\mathscr{L}_S$ on a set S, and its refinement in quotienting by a congruence on $\mathscr{L}_S$ explained after that definition. According to Definition E.21, lattices can be defined in any topos. The following "locality lemma" shows that the construction of a free lattice on some object makes sense in functor toposes, and so does its refinement just mentioned, at least as long as the congruence in question is defined through equalities.

Lemma 12.15. *Let* $\mathsf{T} = [\mathsf{C}, \mathsf{Sets}]$ *be any functor topos (where* C *is some category).*

1. *There exists a free distributive lattice* $\underline{\mathscr{L}_S} \in \mathsf{T}$ *on any object* $\underline{S} \in \mathsf{T}$*, which can be computed locally: the object part of* $\underline{\mathscr{L}_S}$ *is given by*
$$(\underline{\mathscr{L}_S})_0(C) = \mathscr{L}_{\underline{S}_0(C)}, \tag{12.115}$$
where $\mathscr{L}_{\underline{S}_0(C)}$ *defined in* Sets*, and the arrow part is defined as follows. If* $f : C \to D$*, then* $(\underline{\mathscr{L}_S})_1(f)$ *is the unique arrow making the following diagram commute:*
$$\begin{array}{ccc} \underline{S}_0(C) & \xrightarrow{\underline{S}_1(f)} & \underline{S}_0(D) \\ \mathscr{L}\downarrow & & \downarrow\mathscr{L} \\ \mathscr{L}_{\underline{S}_0(C)} & \xrightarrow{(\underline{\mathscr{L}_S})_1(f)} & \mathscr{L}_{\underline{S}_0(D)} \end{array} \tag{12.116}$$
2. *The same is true if* $\underline{\mathscr{L}_S}$ *is subject to relations defined by* equalities *among elements of* $\underline{\mathscr{L}_S}$ *(as long as these equalities generate a congruence).*

Proof. The proof is an elaborate verification, which may be summarized as follows.

1. Existence and uniqueness of the arrow $(\underline{\mathscr{L}_S})_1(f)$ in (12.116) follows from the universal property of the free distributive lattice $\mathscr{L}_{\underline{S}_0(C)}$ in Sets; just consider the function $\mathscr{L} \circ \underline{S}_1(f) : \underline{S}_0(C) \to \mathscr{L}_{\underline{S}_0(D)}$. The claim follows from the fact that $\underline{\mathscr{L}_S}$ (defined locally) has the required universal property (as can be established locally, from the corresponding property of each $(\underline{\mathscr{L}_S})_0(C)$) and hence is unique.
2. This is proved in a similar way, since also a free distributive lattice $\mathscr{L}_S/\sim$ on generators S with relations given by equalities has a universal property, cf. the final part of §D.1. This works locally in a functor topos by rule no. 7 of Kripke–Joyal semantics, cf. §E.5 (which states that equalities are enforced locally). □

We will apply this lemma to $\mathsf{T} = \mathsf{T}(A)$, as in (12.1), with $\mathsf{C} = \mathscr{C}(A)$. This hinges on a lemma of independent interest, which we first state for Sets, i.e., for "ordinary" commutative unital C*-algebras A, to be subsequently internalized to our topos $\mathsf{T}(A)$.

Lemma 12.16. *The lattice* L_A *in* (12.44) *is (constructively) isomorphic to the lattice* L'_A *freely generated by the symbols* D_a*,* $a \in A_{\mathrm{sa}}$ *and the relations* (12.45) - (12.50).

Proof. The point is that the map $a \mapsto \mathsf{D}_a$ from A_{sa} to L'_A is surjective; this follows from the relations (12.45) - (12.50) through their consequences (12.51) - (12.55). The pertinent isomorphism $L'_A \cong L_A$ is then given by mapping $\mathsf{D}_a \leftrightarrow [a^+]$ on generators (note that in the original discussion of L_A following (12.44) this map was the *definition* of D_a; this time, these play an independent role as generators of the lattice L'_A, and in the present proof they are *related* to the elements $[a^+] \in L_A$). □

Now let A be a (not necessarily commutative) unital C*-algebra (in Sets), with ensuing internal commutative C*-algebra $\underline{A}$ in the functor topos $\mathsf{T}(A)$, cf. Theorem 12.3. Our goal is to apply the constructive definition of the Gelfand spectrum $\Sigma(A)$, or rather of its topology $\mathcal{O}(\Sigma(A))$ (seen as a frame, so that $\Sigma(A)$ is seen as a locale) in §12.2 to $\underline{A}$. The first step concerns the lattice L_A, which in $\mathsf{T}(A)$ is denoted by $\underline{L_A}$. Here and in what follows, we try to avoid notational confusion by writing D_a for the formal variable indexed by a (which is a variable of type $\underline{A}$ in $\mathsf{T}(A)$), whilst writing D_c for the actual element $[c^+]$ of L_C if we apply (12.44) etc. to $C \in \mathscr{C}(A)$.

Proposition 12.17. *For each $C \in \mathscr{C}(A)$ one has*

$$\underline{L_A}(C) = L_C, \tag{12.117}$$

where L_C is defined in Sets *through* (12.44) *(with $A \rightsquigarrow C$), where it may be computed through Lemma 12.16. Furthermore, if $C \subseteq D$, then the map $\underline{L_A}(C) \to \underline{L_A}(D)$ given by the functoriality of $\underline{L_A}$, i.e., $L_C \to L_D$, maps each generator D_c in L_C (where $c \in C_{\mathrm{sa}}$) to the same generator in L_D. This is well defined, because $c \in D_{\mathrm{sa}}$, and this inclusion preserves the relations* (12.45) - (12.50). *We write this as $L_C \hookrightarrow L_D$.*

Proof. Internalizing Lemma Lemma 12.16 to our functor topos $\mathsf{T}(A)$, it follows that the internal lattice $\underline{L_A}$ in $\mathsf{T}(A)$ is isomorphic to a distributive lattice freely generated by generators and relations given by equalities. Hence Lemma 12.15 applies to it. □

The next step is to move from $\underline{L_A}$ to the corresponding frame of regular ideals, cf. Theorem 12.8. Abbreviating $\mathcal{O}(\Sigma(A)) \equiv \mathcal{O}(\Sigma)$, we first rewrite (12.60) as

$$\mathcal{O}(\Sigma) \cong \{U \in \mathrm{Idl}(L_A) \mid \forall_{q>0}\, \mathsf{D}_{a-q} \in U \Rightarrow \mathsf{D}_a \in U\}. \tag{12.118}$$

To apply this to our functor topos $\mathsf{T}(A)$, we apply Kripke–Joyal semantics for the internal language of the topos $\mathsf{T}(A)$ (which is reviewed §E.5) to the formula $\mathsf{D}_a \lhd U$. This is a formula φ with two free variables, namely D_a of type L_A, and U of type

$$\mathscr{P}(L_A) \equiv \Omega^{L_A}. \tag{12.119}$$

Hence in the forcing statement $C \Vdash \varphi(\alpha)$ in $\mathsf{T}(A)$, we have to insert

$$\alpha \in (\underline{L_A} \times \underline{\Omega}^{\underline{L_A}})(C) \cong L_C \times \mathrm{Sub}(\underline{L_A}_{|\uparrow C}),$$

where $\underline{L_A}_{|\uparrow C}$ is the restriction of the functor

$$\underline{L_A} : \mathscr{C}(A) \to \mathsf{Sets} \tag{12.120}$$

to $\uparrow C \subset \mathscr{C}(A)$. Here we have used (12.117), as well as the isomorphism (12.11). Consequently, we have

$$\alpha = (D_c, \underline{U}), \tag{12.121}$$

where $D_c \in L_C$ for some $c \in C_{\text{sa}}$, and $\underline{U} : \uparrow C \to \mathbf{Sets}$ is a subfunctor of $\underline{L_A}_{|\uparrow C}$. In particular, $\underline{U}(D) \subseteq L_D$ is defined whenever $D \supseteq C$, and the subfunctor condition on $\underline{U}$ simply boils down to $\underline{U}(D) \subseteq \underline{U}(E)$ whenever $C \subseteq D \subseteq E$.

Lemma 12.18. *In the topos* $\mathsf{T}(A)$, *the cover* $\lhd$ *of Lemma 12.13 may be computed locally, in the sense that for any* $C \in \mathscr{C}(A)$, $D_c \in L_C$ *and* $\underline{U} \in \mathrm{Sub}(\underline{L_A}_{|\uparrow C})$, *one has*

$$C \Vdash \mathsf{D}_a \lhd U(D_c, \underline{U}) \textit{ iff } D_c \lhd_C \underline{U}(C),$$

in that for all $q > 0$ *there exists a finite* $U_0 \subseteq \underline{U}(C)$ *such that* $D_{c-q} \leqslant \bigvee U_0$.

Proof. We assume that $\bigvee U_0 \in U$, so that we may replace U_0 by $\mathsf{D}_b = \bigvee U_0$; the general case is analogous. We then have to inductively analyze the formula $\mathsf{D}_a \lhd U$, which, under the stated assumption, in view of Lemma 12.13 may be taken to mean

$$\forall_{q>0} \exists_{\mathsf{D}_b \in L_A} (\mathsf{D}_b \in U \wedge \mathsf{D}_{a-q} \leqslant \mathsf{D}_b). \tag{12.122}$$

We now infer from the rules for Kripke–Joyal semantics in a functor topos that

$$C \Vdash (\mathsf{D}_a \in U)(D_c, \underline{U}) \tag{12.123}$$

iff for all $D \supseteq C$ one has $D_c \in \underline{U}(D)$; since $\underline{U}(C) \subseteq \underline{U}(D)$, this happens to be the case iff $D_c \in \underline{U}(C)$. Furthermore,

$$C \Vdash (\mathsf{D}_b \leqslant \mathsf{D}_a)(D_{c'}, D_c) \tag{12.124}$$

iff $D_{c'} \leqslant D_c$ in L_C. Also,

$$C \Vdash (\exists_{\mathsf{D}_b \in L_A} \mathsf{D}_b \in U \wedge \mathsf{D}_{a-q} \leqslant \mathsf{D}_b)(D_c, \underline{U}) \tag{12.125}$$

iff there is $D_{c'} \in \underline{U}(C)$ such that $D_{c-q} \leqslant D_{c'}$. Finally,

$$C \Vdash (\forall_{q>0} \exists_{\mathsf{D}_b \in L_A} \mathsf{D}_b \in U \wedge \mathsf{D}_{a-q} \leqslant \mathsf{D}_b)(D_c, \underline{U}) \tag{12.126}$$

iff for all $D \supseteq C$ and all $q > 0$ there is $D_d \in \underline{U}(D)$ such that $D_{c-q} \leqslant D_d$, where $D_c \in L_C$ is seen as an element of L_D through the injection $L_C \hookrightarrow L_D$ of Proposition 12.17, and $\underline{U} \in \mathrm{Sub}(\underline{L_A}_{|\uparrow C})$ is seen as an element of $\mathrm{Sub}(\underline{L_A}_{|\uparrow D})$ by restriction. This, however, is true at all $D \supseteq C$ iff it is true at C, because $\underline{U}(C) \subseteq \underline{U}(D)$ and hence one can take $D_d = D_{c'}$ for the $D_{c'} \in L_C$ that makes the condition true at C. □

Lemma 12.19. *The spectrum* $\mathscr{O}(\underline{\Sigma})$ *of* $\underline{A}$ *in* $\mathsf{T}(A)$ *may be computed as follows:*

1. *At* $C \in \mathscr{C}(A)$, *the set* $\mathscr{O}(\underline{\Sigma})(C)$ *consists of those subfunctors* $\underline{U} \in \mathrm{Sub}(\underline{L_A}_{|\uparrow C})$ *such that for all* $D \supseteq C$ *and all* $D_d \in L_D$ *one has:*

$$D_d \lhd_D \underline{U}(D) \Rightarrow D_d \in \underline{U}(D).$$

2. *At $\mathbb{C}\cdot 1$, the set $\mathscr{O}(\underline{\Sigma})(\mathbb{C}\cdot 1)$ consists of those subfunctors $\underline{U} \in \mathrm{Sub}(\underline{L_A})$ such that for all $C \in \mathscr{C}(A)$ and all $D_c \in L_C$ one has:*

$$D_c \lhd_C \underline{U}(C) \Rightarrow D_c \in \underline{U}(C).$$

3. *The condition that $\underline{U} = \{\underline{U}(C) \subseteq L_C\}_{C\in\mathscr{C}(A)}$ be a subfunctor of $\underline{L_A}$ comes down to the requirement that:*

$$C \subseteq D \Rightarrow \underline{U}(C) \subseteq \underline{U}(D).$$

4. *The map $\mathscr{O}(\underline{\Sigma})(C) \to \mathscr{O}(\underline{\Sigma})(D)$ given by the functoriality of $\mathscr{O}(\underline{\Sigma})$ whenever $C \subseteq D$ is given by truncating an element $\underline{U} :\uparrow C \to \mathsf{Sets}$ of $\mathscr{O}(\underline{\Sigma})(C)$ to $\uparrow D$.*
5. *The external description of $\mathscr{O}(\underline{\Sigma})$ is the frame map*

$$\pi_{\Sigma}^{*} : \mathscr{O}(\mathscr{C}(A)) \to \mathscr{O}(\underline{\Sigma})(\mathbb{C}\cdot 1), \tag{12.127}$$

given on the basic opens $\uparrow D \in \mathscr{O}(\mathscr{C}(A))$ by

$$\begin{aligned} \pi_{\Sigma}^{*}(\uparrow D) = \chi_{\uparrow D} : E &\mapsto \top \ (E \supseteq D); \\ E &\mapsto \bot \ (E \not\supseteq D), \end{aligned} \tag{12.128}$$

where the top and bottom $\top, \bot$ at E are given by $\{L_E\}$ and $\emptyset$, respectively.

Proof. By (12.75), $\mathscr{O}(\underline{\Sigma})$ is the subobject of $\underline{\Omega}^{\underline{L_A}}$ defined by the formula φ given by

$$\forall_{\mathsf{D}_a \in L_A} \mathsf{D}_a \lhd U \Rightarrow \mathsf{D}_a \in U, \tag{12.129}$$

whose interpretation in $\mathsf{T}(A)$ is an arrow from $\underline{\Omega}^{\underline{L_A}}$ to $\underline{\Omega}$. In view of (12.11), we may identify an element $\underline{U} \in \mathscr{O}(\underline{\Sigma})(C)$ with a subfunctor of $\underline{L_A}_{|\uparrow C}$, and by (12.129) and Kripke–Joyal semantics in functor topoi, we have $\underline{U} \in \mathscr{O}(\underline{\Sigma})(C)$ iff $C \Vdash \varphi(\underline{U})$, with φ given by (12.129). Unfolding this using Kripke–Joyal semantics and using Lemma 12.18 (including part 1 of its proof), we find that $\underline{U} \in \mathscr{O}(\underline{\Sigma})(C)$ iff

$$\forall_{D\supseteq C} \forall_{D_d \in L_D} \forall_{E \supseteq D} D_d \lhd_E \underline{U}(E) \Rightarrow D_d \in \underline{U}(E), \tag{12.130}$$

where D_d is regarded as an element of L_E. This condition, however, is equivalent to the apparently weaker condition

$$\forall_{D\supseteq C} \forall_{D_d \in L_D} D_d \lhd_D \underline{U}(D) \Rightarrow D_d \in \underline{U}(D); \tag{12.131}$$

indeed, condition (12.130) clearly implies (12.131), but the latter applied at $D = E$ actually implies the first, since $D_d \in L_D$ also lies in L_E.

Clauses 2 to 4 should now be obvious. Clause 5 follows by the explicit prescription for the external description of frames (which has been recalled in the previous section, after its initial description the end of §E.4). Note that each $\mathscr{O}(\underline{\Sigma})(C)$ is a frame in Sets, inheriting the frame structure of the ambient frame $\mathrm{Sub}(\underline{L_A}_{|\uparrow C})$. □

We now present the computation of $\mathscr{O}(\underline{\Sigma}) \equiv \mathscr{O}(\underline{\Sigma}(\underline{A}))$ for general unital C*-algebras A. To explain the final formula, topologize the disjoint union

$$\Sigma^A = \bigsqcup_{C \in \mathscr{C}(A)} \Sigma(C), \tag{12.132}$$

where $\Sigma(C)$ is the Gelfand spectrum of $C \in \mathscr{C}(A)$, as follows, abbreviating

$$\mathscr{U}_C \equiv \mathscr{U} \cap \Sigma(C). \tag{12.133}$$

One has $\mathscr{U} \in \mathscr{O}(\Sigma^A)$ iff the following two conditions are satisfied for all $C \in \mathscr{C}(A)$:

1. $\mathscr{U}_C \in \mathscr{O}(\Sigma(C))$.
2. For all $D \supseteq C$, if $\lambda \in \mathscr{U}_C$ and $\lambda' \in \Sigma(D)$ such that $\lambda'_{|C} = \lambda$, then $\lambda' \in \mathscr{U}_D$.

In fact, $\mathscr{O}(\Sigma^A)$ is simply the weakest topology making the canonical projection

$$\pi : \Sigma^A \to \mathscr{C}(A); \tag{12.134}$$

$$\pi(\sigma) = C \;\; (\sigma \in \Sigma(C) \subset \Sigma^A), \tag{12.135}$$

continuous with respect to the Alexandrov topology on $\mathscr{C}(A)$. For $U \in \mathscr{O}(\mathscr{C}(A))$,

$$\Sigma^A_U = \bigsqcup_{C \in U} \Sigma(C) \tag{12.136}$$

is a subset of Σ^A, with relative topology inherited from Σ^A. In particular, for the basic opens $U = \uparrow C$ of the Alexandrov topology on $\mathscr{C}(A)$ we have

$$\Sigma^A_{\uparrow C} = \bigsqcup_{D \supseteq C} \Sigma(D). \tag{12.137}$$

Theorem 12.20. *Let A be a unital C*-algebra A. The internal Gelfand spectrum $\mathscr{O}(\underline{\Sigma}(\underline{A}))$ of our internal commutative C*-algebra $\underline{A}$ in the topos $\mathsf{T}(A)$ is the functor*

$$\mathscr{O}(\underline{\Sigma}(\underline{A}))_0 : C \mapsto \mathscr{O}(\Sigma^A_{\uparrow C}), \tag{12.138}$$

i.e., the frame (in Sets*) of the open sets of $\Sigma^A_{\uparrow C}$ in the topology defined after* (12.132)*; if $C \subseteq D$, the arrow-part of the functor in question is given by*

$$\mathscr{O}(\underline{\Sigma}(\underline{A}))_1 : \mathscr{O}(\Sigma^A_{\uparrow C}) \to \mathscr{O}(\Sigma^A_{\uparrow D}); \tag{12.139}$$

$$\mathscr{U} \mapsto \mathscr{U} \cap \uparrow D. \tag{12.140}$$

Similarly, in the description of $\mathsf{T}(A)$ as the category of sheaves $\mathsf{Sh}(\mathscr{C}(A))$, cf. (E.84)*, the Gelfand spectrum is given by the sheaf (where $U \subseteq V$ in* (12.142)*):*

$$\mathscr{O}(\underline{\Sigma}(\underline{A}))_0 : U \mapsto \mathscr{O}(\Sigma^A_U) \;\; (U \in \mathscr{O}(\mathscr{C}(A))); \tag{12.141}$$

$$\mathscr{O}(\underline{\Sigma}(\underline{A}))_1 : \mathscr{U} \mapsto \mathscr{U} \cap \Sigma^A_U \;\; (\mathscr{U} \in \mathscr{O}(\Sigma^A_V)). \tag{12.142}$$

Proof. The proof is based on Lemma 12.19, which implies that the internal frame $\underline{\mathrm{RIdl}}(\underline{L_A})$ in $\mathsf{T}(A)$ is given by the functor

$$\underline{\mathrm{RIdl}}(\underline{L_A}) : C \mapsto \{\underline{F} \in \mathrm{Sub}(\underline{L_A}_{|\uparrow C}) \mid \underline{F}(D) \in \mathrm{RIdl}(L_D) \text{ for all } D \supseteq C\}. \quad (12.143)$$

Here, since D is a commutative unital C*-algebra in Sets, according to (12.60) the set $\mathrm{RIdl}(L_D)$ may be identified with the topology $\mathcal{O}(\Sigma(D))$, where $\Sigma(D)$ is the Gelfand spectrum of D in the usual sense. We will make this identification in the following step, which is the last step of the proof of Theorem 12.20.

Lemma 12.21. *The transformation $\theta : \mathrm{RIdl}(\underline{L_A}) \to \mathcal{O}(\underline{\Sigma}(\underline{A}))$ with components*

$$\begin{aligned} \theta_C : \{\underline{F} \in \mathrm{Sub}(\underline{L_A}_{|\uparrow C}) \mid \underline{F}(D) \in \mathcal{O}(\Sigma(D)) \text{ for all } D \supseteq C\} &\to \mathcal{O}(\Sigma^A_{\uparrow C}); \\ \underline{F} \mapsto \bigsqcup_{D \supseteq C} \underline{F}(D), & \quad (12.144) \end{aligned}$$

is a natural isomorphism of functors—i.e., an isomorphism of objects in $\mathsf{T}(A)$.

Since $\underline{\mathrm{RIdl}}(\underline{L_A})$ and $\mathcal{O}(\underline{\Sigma})$ are internal frames in $\mathsf{T}(A)$, it suffices to prove that each θ_C is an isomorphism of frames in Sets. Unfortunately, even this proof is a very lengthy (though straightforward) affair, for which we refer to the literature. □

Corollary 12.22. *The external description (in* Sets*) of the internal locale $\underline{\Sigma}(\underline{A})$ (in* $\mathsf{T}(A)$*) is given by the canonical projection* (12.134).

Note that both Σ^A and $\mathscr{C}(A)$ are topological spaces, so that (12.134) is a *bona fide* continuous map between spaces. This is worth stressing, since in general, an external description of an internal locale in a sheaf topos, though defined in Sets, is a map between locales (or, equivalently, between frames) that are not necessarily topological spaces. But in the case (12.134) at hand they are, so at least this time there is no confusion between $\mathcal{O}(X)$ as both formal notation for a frame (not necessarily coming from a topology) and notation for the topology of a space X; see §C.11.

Note that (12.95) is a special case of Theorem 12.20 or Corollary 12.22, for

$$A = M_n(\mathbb{C}). \quad (12.145)$$

To see this, we identify $\mathscr{U} = \bigsqcup_{C \in \mathscr{C}(A)} \mathscr{U}_C$ as an element of $\mathcal{O}(\Sigma^A)$ with

$$S : \mathscr{C}(A) \to \mathscr{P}(A)$$

on the right-hand side of (12.95), where $S(C) \in \mathscr{P}(C)$ is the image of $\mathscr{U}_C \in \mathcal{O}(\Sigma(C))$ under the isomorphism $\mathcal{O}(\Sigma(C)) \to \mathscr{P}(C)$ between the (discrete) topology of the (finite) Gelfand spectrum of C and the (Boolean) projection lattice of C derived earlier, see (12.80). Similarly, for $U \in \mathcal{O}(\mathscr{C}(A))$, the frame $\mathcal{O}(\Sigma^A_U)$ may be identified with maps

$$S : U \to \mathscr{P}(A)$$

satisfying the conditions in (12.95). Of course, the special case (12.145) leading to (12.95) is very appealing, and was well worth treating in its own right!

Theorem 12.20 and Corollary 12.22 also give an explicit description of the general internal Gelfand isomorphism (12.78), whose real part in $\mathsf{T}(A)$ reads

$$\underline{A}_{\text{sa}} \cong C(\underline{\Sigma}, \underline{\mathbb{R}}) \equiv \text{Frm}(\mathscr{O}(\underline{\mathbb{R}}), \mathscr{O}(\underline{\Sigma})), \tag{12.146}$$

where the right-hand side, which denotes the object of frame homomrphisms from $\mathscr{O}(\underline{\mathbb{R}})$ to $\mathscr{O}(\underline{\Sigma})$ within $\mathsf{T}(A)$, is the *definition* of the middle term (which is just a notation). To understand the situation in $\mathsf{T}(A)$, one has to distinguish between:

1. The *object* $\text{Frm}(\mathscr{O}(\underline{\mathbb{R}}), \mathscr{O}(\underline{\Sigma}))$ in $\mathsf{T}(A)$, defined as the subobject of the exponential $\mathscr{O}(\underline{\Sigma})^{\mathscr{O}(\underline{\mathbb{R}})}$ consisting of (internal) frame maps from $\mathscr{O}(\underline{\mathbb{R}})$ to $\mathscr{O}(\underline{\Sigma})$.
2. The *set* $\text{Hom}_{\text{Frm}}(\mathscr{O}(\underline{\mathbb{R}}), \mathscr{O}(\underline{\Sigma}))$ of internal frame maps from the frame $\mathscr{O}(\underline{\mathbb{R}})$ of (Dedekind) real numbers in $\mathsf{T}(A)$ to the frame $\mathscr{O}(\underline{\Sigma})$ (i.e., the set of those arrows from $\mathscr{O}(\underline{\mathbb{R}})$ to $\mathscr{O}(\underline{\Sigma})$ that happen to be frame maps as seen from within $\mathsf{T}(A)$).

The connection between 1. and 2. is given by λ-conversion, i.e., the bijective correspondence between $C \to B^A$ and $A \times C \to B$, cf. (E.153). Taking $C = \mathbf{\underline{1}}$ (i.e. the terminal object in $\mathsf{T}(A)$), we see that an *element* of the set $\text{Hom}(A, B)$ corresponds to an *arrow* $1 \to B^A$. Eq. (12.8) yields

$$\text{Frm}(\mathscr{O}(\underline{\mathbb{R}}), \mathscr{O}(\underline{\Sigma}))(C) = \text{Nat}_{\text{Frm}}(\mathscr{O}(\underline{\mathbb{R}})_{\uparrow C}, \mathscr{O}(\underline{\Sigma})_{\uparrow C}), \tag{12.147}$$

the set of all natural transformations between the functors $\mathscr{O}(\underline{\mathbb{R}})$ and $\mathscr{O}(\underline{\Sigma})$, both restricted to $\uparrow C \subset \mathscr{C}(A)$, that are frame maps. This set can be computed from the external description of frames and frame maps in §E.4. Recall (12.4) etc. The frame $\mathscr{O}(\underline{\mathbb{R}})_{\uparrow C}$ has external description

$$\pi_{\mathbb{R}}^{-1} : \mathscr{O}(\uparrow C) \to \mathscr{O}(\uparrow C \times \mathbb{R}), \tag{12.148}$$

where $\pi_{\mathbb{R}} : \uparrow C \times \mathbb{R} \to \uparrow C$ is projection on the first component. The special case $C = \mathbb{C} \cdot 1$ yields the external description of $\mathscr{O}(\underline{\mathbb{R}})$ itself, namely

$$\pi_{\mathbb{R}}^{-1} : \mathscr{O}(\mathscr{C}(A)) \to \mathscr{O}(\mathscr{C}(A) \times \mathbb{R}), \tag{12.149}$$

where this time (with abuse of notation) the projection is $\pi_{\mathbb{R}} : \mathscr{C}(A) \times \mathbb{R} \to \mathscr{C}(A)$. By Corollary 12.22, the Gelfand frame $\mathscr{O}(\underline{\Sigma})_{\uparrow C}$ has external description

$$\pi_{\Sigma}^{-1} : \mathscr{O}(\uparrow C) \to \mathscr{O}(\Sigma)_{\uparrow C}, \tag{12.150}$$

given by (12.128), with the understanding that $D \supseteq C$ (the special case $C = \mathbb{C} \cdot 1$ then recovers the external description (12.99) of $\mathscr{O}(\underline{\Sigma})$ itself). It follows that there is a bijective correspondence between two classes of frame maps:

$$\underline{\varphi}_C^{-1} : \mathscr{O}(\underline{\mathbb{R}})_{\uparrow C} \to \mathscr{O}(\underline{\Sigma})_{\uparrow C} \ \ (\text{in } \mathsf{T}(A)); \tag{12.151}$$

$$\varphi_C^{-1} : \mathscr{O}(\uparrow C \times \mathbb{R}) \to \mathscr{O}(\Sigma)_{\uparrow C} \ \ (\text{in Sets}), \tag{12.152}$$

where φ_C must satisfy the condition that for any $D \supseteq C$,

$$\varphi_C^{-1}(\uparrow D \times \mathbb{R}) = \chi_{\uparrow D}. \tag{12.153}$$

Indeed, such a map φ_C^{-1} defines an element $\underline{\varphi}_C^{-1}$ of $\mathrm{Nat}(\mathscr{O}(\underline{\mathbb{R}})_{\uparrow C}, \mathscr{O}(\underline{\Sigma})_{\uparrow C})$ in the obvious way: for $D \in \uparrow C$, the components

$$\underline{\varphi}_C^{-1}(D) : \mathscr{O}(\underline{\mathbb{R}})(D) \to \mathscr{O}(\underline{\Sigma})(D) \tag{12.154}$$

of the natural transformation $\underline{\varphi}_C^{-1}$, i.e.

$$\underline{\varphi}_C^{-1}(D) : \mathscr{O}(\uparrow D \times \mathbb{R}) \to \mathscr{O}(\Sigma)_{\uparrow D}, \tag{12.155}$$

are simply given by the restriction of φ_C^{-1} to $\mathscr{O}(\uparrow D \times \mathbb{R}) \subset \mathscr{O}(\uparrow C \times \mathbb{R})$; cf. (E.147). This is consistent, because (12.153) implies that for any $U \in \mathscr{O}(\mathbb{R})$ and $C \subseteq D \subseteq E$,

$$\varphi_C^{-1}(\uparrow E \times U)(F) \leq \varphi_C^{-1}(\uparrow D \times \mathbb{R})(F), \tag{12.156}$$

which by (12.153) vanishes whenever $F \not\supseteq D$. Consequently,

$$\varphi_C^{-1}(\uparrow E \times U)(F) = 0 \text{ if } F \not\supseteq D, \tag{12.157}$$

so that $\underline{\varphi}_C^{-1}(D)$ actually takes values in $\mathscr{O}(\Sigma)_{\uparrow D}$ (rather than in $\mathscr{O}(\Sigma)_{\uparrow C}$, as might be expected). Denoting the set of frame maps (12.152) that satisfy (12.153) by $\mathrm{Frm}'(\mathscr{O}(\uparrow C \times \mathbb{R}), \mathscr{O}(\Sigma)_{\uparrow C})$, we obtain a functor

$$\mathrm{Frm}'(\mathscr{O}(\uparrow(-) \times \mathbb{R}), \mathscr{O}(\Sigma)_{\uparrow -}) : \mathscr{C}(A) \to \mathsf{Sets}, \tag{12.158}$$

with the stipulation that for $C \subseteq D$ the induced map

$$\mathrm{Frm}'(\mathscr{O}(\uparrow C \times \mathbb{R}), \mathscr{O}(\Sigma)_{\uparrow C}) \to \mathrm{Frm}'(\mathscr{O}(\uparrow D \times \mathbb{R}), \mathscr{O}(\Sigma)_{\uparrow D})$$

is given by restricting an element of the left-hand side to $\mathscr{O}(\uparrow D \times \mathbb{R}) \subset \mathscr{O}(\uparrow C \times \mathbb{R})$; this is consistent by the same argument (12.157).

The Gelfand isomorphism (12.78) is therefore a natural transformation

$$\underline{A} \xrightarrow{\cong} \mathrm{Frm}'(\mathscr{O}(\uparrow - \times \mathbb{R}), \mathscr{O}(\Sigma)_{\uparrow -}), \tag{12.159}$$

which means that one has a compatible (i.e. natural) family of isomorphisms

$$\begin{aligned} C &\xrightarrow{\cong} \mathrm{Frm}'(\mathscr{O}(\uparrow C \times \mathbb{R}), \mathscr{O}(\Sigma)_{\uparrow C}); \\ a &\mapsto \hat{a}^{-1} : \mathscr{O}(\uparrow C \times \mathbb{R}) \to \mathscr{O}(\Sigma)_{\uparrow C}. \end{aligned} \tag{12.160}$$

On basic opens $\uparrow D \times U \in \mathscr{O}(\uparrow C \times \mathbb{R})$, with $D \supseteq C$, we obtain

$$\begin{aligned} \hat{a}^{-1}(\uparrow D \times U) : E &\mapsto 1_U(a) \text{ if } E \supseteq D; \\ E &\mapsto 0 \text{ if } E \not\supseteq D. \end{aligned} \tag{12.161}$$

Here $1_U(a)$ is the spectral projection of a in U, cf. (12.82); as it lies in $\mathscr{P}(C)$ and $C \subseteq D \subseteq E$, the projection $1_U(a)$ certainly lies in $\mathscr{P}(E)$, as required. Furthermore, we need to extend $\hat{a}^{-1}$ to general opens in $\uparrow C \times \mathbb{R}$ by the frame map property, and note that (12.153) for $\varphi_C^{-1} = \hat{a}^{-1}$ is satisfied.

This analysis also holds in the topos $\mathsf{Sh}(\mathscr{C}(A))$ of sheaves in $\mathscr{C}(A)$ (as always, equipped with the Alexandrov topology, cf. (E.84). It then follows from (12.159) and (12.141) that as a sheaf,

$$C(\underline{\Sigma}, \underline{\mathbb{C}}) : U \mapsto C(\Sigma_U^A, \mathbb{C}), \tag{12.162}$$

where Σ_U^A is given by (12.136); if $U \subseteq V$, the map $C(\Sigma_V^A, \mathbb{C}) \to C(\Sigma_U^A, \mathbb{C})$ is given by the pullback of the inclusion $\Sigma_U^A \hookrightarrow \Sigma_V^A$ (that is, by restriction). It then follows from (12.162) that the isomorphism (12.146) is given by its components

$$\underline{A}(U) \cong C(\Sigma_U^A, \mathbb{C}). \tag{12.163}$$

In particular, the component of the natural isomorphism in (12.146) at $U = \uparrow C$ is

$$C \cong C(\Sigma_{\uparrow C}^A, \mathbb{C}). \tag{12.164}$$

A glance at the topology of Σ^A shows that the so-called ***Hausdorffication***, which for a general compact space may be defined either directly, or C*-algebraically by $X^H = \Sigma(C(X))$, and coincides with the left adjoint of the forgetful functor from the category of compact Hausdorff spaces (and continuous maps) to the category of compact spaces (and continuous maps), is given by $(\Sigma_{\uparrow C}^A)^H \cong \Sigma(C)$, so that

$$C(\Sigma_{\uparrow C}^A, \mathbb{C}) \cong C(\Sigma(C), \mathbb{C}), \tag{12.165}$$

where the isomorphism is given by restricting $f \in C(\Sigma_{\uparrow C}^A, \mathbb{C})$ to $\Sigma(C) \subset \Sigma_{\uparrow C}^A$.

Corollary 12.23. *The internal Gelfand isomorphism*

$$\underline{A} \xrightarrow{\cong} C(\underline{\Sigma}, \underline{\mathbb{C}}), \tag{12.166}$$

which is a natural isomorphism between functors $\mathscr{C}(A) \to$ Sets, is given at each $C \in \mathscr{C}(A)$ by the usual Gelfand isomorphism for the commutative C-algebra C:*

$$\underline{A}_0(C) = C \xrightarrow{\cong} C(\Sigma(C), \mathbb{C}) \cong C(\underline{\Sigma}, \underline{\mathbb{C}})_0(C). \tag{12.167}$$

At the end of the day, the Gelfand isomorphism (12.146) therefore turns out to simply assemble all isomorphisms (12.167) for the commutative C*-subalgebras C of A into a single sheaf-theoretic construction. Incidentally, taking $C = \mathbb{C} \cdot 1$ in (12.164) shows that $(\Sigma^A)^H$ is a point, which is also obvious from the fact that any open set containing the point $\Sigma(\mathbb{C} \cdot 1)$ of Σ^A must be all of Σ^A.

12.5 “Daseinisation” and Kochen–Specker Theorem

The internal Gelfand transform (12.166) constructed in the previous section acts on each commutative subalgebra $A \in \mathscr{C}(A)$. What about A itself? There is a more subtle transform, inspired by the remarkable “Daseinisation” construction of Döring and Isham (whose name has unfortunately been inspired by the controversial German philosopher Heidegger), which turns self-adjoint elements a of A into continuous functions $\delta(a)$ on the topos-theoretical phase space $\underline{\Sigma}^A$, whose range is the so-called *interval domain* $\mathbb{IR}$ (which is a fuzzy version of $\mathbb{R}$). Hence we will define a map

$$\delta : A_{\text{sa}} \to C(\underline{\Sigma}^A, \mathbb{IR}), \tag{12.168}$$

which, alas, is defined only if A is a von Neumann algebra; we shall therefore assume this throughout this section. Similarly, the notation $\mathscr{C}(A)$ will now stand for the poset of abelian *von Neumann* subalgebras of A (as opposed to abelian C^*-subalgebras of A, as in the remainder of this book).

“Daseinisation” requires two slightly unusual concepts, the first of which is the said ***interval domain*** $\mathbb{IR}$. To motivate its definition, consider Brouwer’s approximation of real numbers by nested intervals with endpoints in $\mathbb{Q}$. For example, the real number π can be described by specifying the sequence

$$[3,4],[3.1,3.2],[3.14,3.15],[3.141,3.142],\ldots$$

This description of the reals is formalized by $\mathbb{IR}$, defined as the poset whose elements are *compact* intervals $[a,b]$ in $\mathbb{R}$ (including singletons $[a,a] = \{a\}$), ordered by *reverse* inclusion (for a smaller interval means that we have more information about the real number that the ever smaller intervals converge to). This poset is a so-called ***dcpo*** (for ***directed complete partial order***); directed suprema are simply intersections. As such, it carries the ***Scott topology***, whose open sets are upper subsets U of $\mathbb{IR}$ with the additional property that for every directed set D with $\bigvee D \in U$ the intersection $D \cap U$ is nonempty. This means that each open interval (p,q) in $\mathbb{R}$ (with $p = -\infty$ and $q = +\infty$ allowed) corresponds to a Scott open

$$U_{(p,q)} = \{[a,b] \mid p < a \le b < q\}. \tag{12.169}$$

Indeed, these opens form a basis of the Scott topology $\mathscr{O}_{\text{Scott}}(\mathbb{IR}) \equiv \mathscr{O}(\mathbb{IR})$ of $\mathbb{IR}$. This topology is, of course, a frame, so far defined in Sets. However, this frame is easily internalized to any (pre)sheaf topos, similar to the Dedekind reals (12.3) - (E.149); in particular, in $\mathsf{T}(A)$ we have

$$\mathscr{O}(\underline{\mathbb{IR}})_0 : C \mapsto \mathscr{O}((\uparrow C) \times \mathbb{IR}), \tag{12.170}$$

with external description as a locale (see §E.4) given by the canonical projection

$$\pi_1 : \mathscr{C}(A) \times \mathbb{IR} \to \mathscr{C}(A). \tag{12.171}$$

The second ingredient of "Daseinisation" is the *spectral order* on $A_{\rm sa}$. The partial order $\leq$ defined in §C.7 (in which $a \leq b$ iff $\omega(a) \leq \omega(b)$ for all states ω on A) has good linearity properties in that it makes A_+ a convex cone in the real vector space $A_{\rm sa}$ (cf. Definition C.50), but it is terrible from a lattice point of view (unless A is abelian): for example, for $A = B(H)$, suprema $a \vee b$ and infima $a \wedge b$ exist iff either $a \leq b$ or $b \leq a$ (and indeed $A_{\rm sa}$ is a lattice with respect to $\leq$ iff A is abelian). However, there is a different order on $A_{\rm sa}$ that turns it into a ***conditionally*** (or ***boundedly***)***complete lattice***, i.e., a poset X with the property that if some subset $S \subseteq X$ has an upper bound (i.e., there is $x \in X$ such that $s \leq x$ for each $s \in S$), then it has a *lowest* upper bound (i.e., $\bigvee S$ exists), and similarly for (greatest) lower bounds.

Definition 12.24. *For $a, b \in A_{\rm sa}$ we say that $a \leq_s b$ (i.e., a is less or equal than b in the* **spectral order***) iff $a^n \leq b^n$ for each $n \in \mathbb{N}$.*

It can be shown that $a \leq_s b$ iff $e^{(b)}_{(\lambda)} \leq e^{(a)}_{(\lambda)}$ for each $\lambda \in \mathbb{R}$ (note the change of order), where $e^{(a)}_{(\lambda)}$ is the spectral projection $1_{(-\infty,\lambda]\cap\sigma(a)}(a)$, etc. This, in turn, implies, that

$$a \leq_s b \text{ iff } \mu_\omega(a \leq \lambda) \geq \mu_\omega(b \leq \lambda), \tag{12.172}$$

for each (normal) state ω on A and each $\lambda \in \mathbb{R}$, where

$$\mu_\omega(a \leq \lambda) = \omega(1_{(-\infty,\lambda]\cap\sigma(a)}(a)) \tag{12.173}$$

is the Born probability for the outcome $a \leq \lambda$ in state ω (and similarly for b). Furthermore, if a and b commute, or if a and b are both projections, the $a \leq_s b$ iff $a \leq b$, i.e., $\leq_s$ coincides with the usual partial order $\leq$ iff A is abelian, and $\leq_s$ restricts to $\leq$ on the projection lattice $\mathscr{P}(A)$ of A. For each $a \in A_{\rm sa}$ and $C \in \mathscr{C}(A)$, we define

$$\delta^i_C(a) = \bigvee\{b \in C_{\rm sa} \mid b \leq_s a\}; \tag{12.174}$$

$$\delta^o_C(a) = \bigwedge\{b \in C_{\rm sa} \mid a \leq_s b\}, \tag{12.175}$$

called the ***inner*** and ***outer Daseinisation*** of a with respect to C, respectively; those objecting to Heidegger might prefer to simply call these the inner and outer ***localizations*** of a with respect to C. For projections, these expressions simplify to

$$\delta^i_C(e) = \bigvee\{f \in \mathscr{P}(C) \mid f \leq_s e\}; \tag{12.176}$$

$$\delta^o_C(e) = \bigwedge\{f \in \mathscr{P}(C) \mid e \leq_s f\}, \tag{12.177}$$

and in fact one has a very nice categorical description, in that $\delta^i_C : \mathscr{P}(A) \to \mathscr{P}(C)$ and $\delta^o_C : \mathscr{P}(A) \to \mathscr{P}(C)$ are the right and left adjoint, respectively, of the inclusion functor $\mathscr{P}(C) \hookrightarrow \mathscr{P}(A)$ in the category of complete orthomodular lattices.

We are now in a position to define the map (12.168): for $a \in A_{\rm sa}$ we put

$$\delta(a) : (C, \omega) \mapsto [\omega(\delta^i_C(a)), \omega(\delta^o_C(a))], \tag{12.178}$$

where (as the notation indicates) the point $(C, \omega) \in \Sigma(C) \subset \Sigma^A$ is just $\omega \in \Sigma(C)$.

It is easily checked that the right-hand side of (12.178) makes sense, since positivity of states and (12.174) - (12.175) obviously imply $\omega(\delta^i_C(a)) \leq \omega(\delta^o_C(a))$. Also, $\delta(a)$ is continuous, so that δ is well defined. If we define a closely related map

$$\hat{\delta}(a) : \Sigma^A \to \mathscr{C}(A) \times \mathbb{IR}; \tag{12.179}$$

$$\hat{\delta}(a)(C,\omega) = (C, \delta(a)(C,\omega)), \tag{12.180}$$

then $\hat{\delta}(a)$ is the external description of an internal locale map

$$\underline{\delta}(a) : \underline{\Sigma}(\underline{A}) \to \underline{\mathbb{IR}}. \tag{12.181}$$

In view of this, we may regard (12.168) as a hybrid (i.e. "category mistake") map

$$\underline{\delta} : A_{\text{sa}} \to C(\underline{\Sigma}(\underline{A}), \underline{\mathbb{IR}}); \tag{12.182}$$

see the text below (12.146), with $\mathbb{R} \rightsquigarrow \mathbb{IR}$, for the meaning of the right-hand side.

The relationship between δ and the Gelfand transform (12.166) is as follows. For $a \in A_{\text{sa}}$, let $W^*(a)$ be the unital commutative von Neumann algebra generated by $a = a^*$ and 1_A within A. Using (12.164), we then have a Gelfandish isomorphism

$$W^*(a)_{\text{sa}} \xrightarrow{\cong} C(\Sigma^A_{\uparrow W^*(a)}, \mathbb{R}); \tag{12.183}$$

$$c \mapsto \hat{c}. \tag{12.184}$$

In particular, since $a \in W^*(a)$, we obtain a continuous function

$$\hat{a} : \Sigma^A_{\uparrow W^*(a)} \to \mathbb{R}. \tag{12.185}$$

Furthermore, we have an inclusion

$$\iota : \mathbb{R} \hookrightarrow \mathbb{IR}; \tag{12.186}$$

$$x \mapsto [x,x], \tag{12.187}$$

which is continuous, and hence induces a map $C(\Sigma^A, \mathbb{R}) \to C(\Sigma^A, \mathbb{IR})$, as well as maps $C(\Sigma^A_{\uparrow W^*(a)}, \mathbb{R}) \to C(\Sigma^A_{\uparrow W^*(a)}, \mathbb{IR})$. Then the following diagram commutes:

$$\begin{array}{ccc} \Sigma^A_{\uparrow W^*(a)} & \xrightarrow{\delta(a)} & \mathbb{IR} \\ & \searrow^{\hat{a}} & \uparrow \iota \\ & & \mathbb{R} \end{array} \tag{12.188}$$

In words, the restriction of the "Daseinisation" $\delta(a) : \Sigma^A \to \mathbb{IR}$ of a to the open subset $\Sigma^A_{\uparrow W^*(a)} \subset \Sigma^A$ takes values in $\mathbb{R} \subset \mathbb{IR}$, and as such coincides with the Gelfand transform $\hat{a}$ of a, seen as a map (12.185). Hence, as might be expected in quantum mechanics, any fuzziness of $\delta(a)$ is only noticeable outside its own context $W^*(a)$.

The "Daseinisation" construction enables one to interpret propositions $a \in (p,q)$ as open subsets of the "phase space" Σ^A, as in classical physics, where $a : X \to \mathbb{R}$ would be a continuous function on a phase space X, and one would say that

$$[[a \in (p,q)]]_{\mathrm{CM}} = a^{-1}(p,q) \in \mathcal{O}(X). \tag{12.189}$$

In quantum mechanics, one would interpret $a \in (p,q)$ as the spectral projection

$$[[a \in (p,q)]]_{\mathrm{QM}} = e^{(a)}_{(p,q)} \equiv 1_{(p,q)\cap\sigma(a)}(a), \tag{12.190}$$

or, equivalently, with the corresponding closed subset of the ambient Hilbert space. In our quantum toposophy setting, however, we may adapt (12.189) as

$$[[a \in (p,q)]]_{\mathrm{QT}} = \delta(a)^{-1}(U_{(p,q)}) \in \mathcal{O}(\Sigma^A). \tag{12.191}$$

Similarly, one may interpret $a \in (p,q)$ as an internal open subset of the internal Gelfand spectrum $\underline{\Sigma}(\underline{A})$, as follows. For any locale Y in a topos T, an internal open in $\mathcal{O}(Y)$ is defined as an arrow $\mathbf{1} \to \mathcal{O}(Y)$, where as usual $\mathbf{1}$ is the terminal object in T. In the case at hand we have $Y = \underline{\Sigma}(\underline{A})$, and use the composition

$$\underline{\mathbf{1}} \xrightarrow{\underline{(p,q)}} \mathcal{O}(\underline{\mathbb{IR}}) \xrightarrow{\underline{\delta}(a)^{-1}} \mathcal{O}(\underline{\Sigma}(\underline{A})), \tag{12.192}$$

where the natural transformation $\underline{(p,q)}$ has components

$$\underline{(p,q)}_C(*) = \uparrow C \times U_{(p,q)}, \tag{12.193}$$

cf. (12.170), and $\underline{\delta}(a)^{-1} : \mathcal{O}(\underline{\mathbb{IR}}) \to \mathcal{O}(\underline{\Sigma}(\underline{A}))$ is the frame version of the locale map (12.181), whose component at C, i.e.,

$$\underline{\delta}(a)_C^{-1} : \mathcal{O}((\uparrow C) \times \mathbb{IR}) \to \mathcal{O}(\Sigma^A_{\uparrow C}), \tag{12.194}$$

is given on basic opens in $(\uparrow C) \times \mathbb{IR}$, with $D \supseteq C$ and $p < q$, by

$$\underline{\delta}(a)_C^{-1}(\uparrow D \times U_{(p,q)}) = \delta(a)^{-1}(U_{(p,q)}) \cap \Sigma^A_{\uparrow D}. \tag{12.195}$$

We therefore obtain the quantum-toposophical interpretation of $a \in (p,q)$ as:

$$\underline{[[a \in (p,q)]]}_{\mathrm{QT}} : \underline{\mathbf{1}} \to \mathcal{O}(\underline{\Sigma}(\underline{A})); \tag{12.196}$$

$$\underline{[[a \in (p,q)]]}_{\mathrm{QT}} = \underline{\delta}(a)^{-1} \circ \underline{(p,q)}. \tag{12.197}$$

We are now going to combine this expression with a construction relating states $\omega \in S(A)$ to arrows from $\mathcal{O}(\underline{\Sigma}(\underline{A}))$ to the truth object $\underline{\Omega}$ in $\mathsf{T}(A)$. This construction generalizes the fundamental bijective correspondence between states on commutative (unital) C*-algebras A and probability measures on its Gelfand spectrum $\Sigma(A)$ (cf. Theorem B.24) to the non-commutative case.

To this end, we first need to replace probability measures on *spaces* by probability measures on *locales*. This, in turn, requires the ***lower real numbers*** $\mathbb{R}_l$, which may be identified with proper subsets $x_l \subset \mathbb{Q}$ with the following two properties:

1. If $p \in x_l$, then there exists $q \in x_l$ with $p < q$.
2. If $p < q \in x_l$, then $p \in x_l$ (i.e., x_l is a lower subset of $\mathbb{Q}$).

In Sets, the lower reals may be identified with $\mathbb{R}$ (in Hilbert's definition) by identifying x_l with its supremum $x = \sup x_l$, but in arbitrary toposes (that admit internal natural and hence rational numbers) they drift apart. Similarly, one defines the ***upper real numbers*** $\mathbb{R}_u$ as proper upper subsets $x_u \subset \mathbb{Q}$ such that $p \in x_u$ implies that there exists $q \in x_u$ with $p > q$; once again, in Sets, $\mathbb{R}_u$ may be identified with Hilbert's $\mathbb{R}$ by taking $x = \inf x_u$. The ***Dedekind real numbers*** $\mathbb{R}_d$, then, are pairs (x_l, x_u) where $x_l \in \mathbb{R}_l$ and $x_u \in \mathbb{R}_u$ are such that $x_l \cap x_u = \emptyset$ and for each $p, q \in \mathbb{Q}$ with $p < q$, either $p \in x_l$ or $q \in x_u$. In Sets these may be identified with $\sup x_l = \inf x_u = x$, so that $\mathbb{R}_d \cong \mathbb{R}$, but in many toposes $\mathbb{R}_l$, $\mathbb{R}_u$, and $\mathbb{R}_d$ are all different. For example, we have already seen that in sheaf toposes $\mathsf{Sh}(X)$, the Dedekind reals are given by the sheaf (E.150), but the lower reals turn out to be defined by

$$(\mathbb{R}_l)_0 : U \mapsto L(U, \mathbb{R}), \tag{12.198}$$

where $U \in \mathcal{O}(X)$ and $L(U, \mathbb{R})$ is the set of all lower semicontinuous functions from U to $\mathbb{R}$ that are locally bounded from above (and similarly for $\mathbb{R}_u$, *mutatis mutandis*). In particular, in $\mathsf{T}(A)$ we have the functor

$$(\underline{\mathbb{R}}_l)_0 : C \mapsto L(\uparrow C, \mathbb{R}), \tag{12.199}$$

which is quite different from (12.3) (and similarly for $\underline{\mathbb{R}}_u$).

Definition 12.25. *A* **probability measure on a locale** X *is a monotone map*

$$\mu : \mathcal{O}(X) \to [0,1]_l, \tag{12.200}$$

where $[0,1]_l$ *is the collection of lower reals between 0 and 1 (defined by replacing* $\mathbb{Q}$ *in the definition of* $\mathbb{R}_l$ *by the set of all rationals* $0 \leq q \leq 1$*), that satisfies*

$$\mu(\top) = 1; \tag{12.201}$$

$$\mu(U) + \mu(V) = \mu(U \wedge V) + \mu(U \vee V); \tag{12.202}$$

$$\mu\left(\bigvee_\lambda U_\lambda\right) = \bigvee_\lambda \mu(U_\lambda), \tag{12.203}$$

for any directed family (U_λ) *in* $\mathcal{O}(X)$.

Compared with (probability) measures on σ-algebras, we see that (probability) measures on locales are merely defined on *open* sets (as opposed to *measurable* sets, which include opens), but this weakening is compensated for by the much stronger (i.e. uncountable) additivity axiom (12.203). Indeed, in Sets, if X is a compact Hausdorff space, one even has a bijective correspondence between *regular* probability measures μ' on X as a space and probability measures μ on X as a locale.

This definition makes sense in constructive mathematics, and hence it may be internalized to $\mathsf{T}(A)$. Doing so, probability measures on the internal Gelfand spectrum $\underline{\Sigma}(\underline{A})$ turn out to correspond to the following notion (cf. Definition 2.26).

Definition 12.26. *A* **quasi-state** *on a unital C*-algebra A is a map $\omega : A \to \mathbb{C}$ that is positive and normalized ($\omega(1_A) = 1$), satisfies $\omega(b+ic) = \omega(b) + i\omega(c)$ for $b^* = b$ and $c^* = c$, and is linear on each commutative unital C*-algebra in A.*

Theorem 12.27. *There is a bijective correspondence between quasi-states ω on A and probability measures $\underline{\mu}_{\omega}$ on the internal Gelfand spectrum $\underline{\Sigma}(\underline{A})$.*

The proof uses the fact that given the (Alexandrov) topology on $\mathscr{C}(A)$, a function $\uparrow C \to [0,1]$ is lower semicontinuous iff it is order-preserving (i.e., monotone); since $[0,1]$ is bounded, the condition of local boundedness is trivially satisfied and hence $L(\uparrow C, [0,1])$ consists of all order-preserving functions from $\uparrow C \subset \mathscr{C}(A)$ to $[0,1]$.

Proof. Any probability measure on $\underline{\Sigma}(\underline{A})$ is a natural transformation

$$\underline{\mu} : \underline{\Sigma}(\underline{A}) \to \underline{[0,1]}_l, \tag{12.204}$$

whose component at $C \in \mathscr{C}(A)$, according to (12.138) and (12.199), is a map

$$\underline{\mu}_C : \mathscr{O}(\Sigma^A_{\uparrow C}) \to L(\uparrow C, [0,1]), \tag{12.205}$$

satisfying properties dictated by Definition 12.25. In particular, if C is maximal abelian in A, then by the comment preceding the proof, $\underline{\mu}_C$ is simply a function $\mathscr{O}(\Sigma(C)) \to [0,1]$ that satisfies (12.201) - (12.203) and hence is a (regular) probability measure μ_C on $\Sigma(C)$. Thus by Riesz–Markov one obtains a state ω_C on each maximal abelian C. From the topology on Σ^A and (12.137) we see that if D is not maximal, $\underline{\mu}_D$ is determined by $\underline{\mu}_C$ for any $C \supset D$, so that we also obtain a probability measure μ_D on $\Sigma(D)$, or, equivalently, a state ω_D, by restriction of ω_C to D. One might fear that μ_D and ω_D could depend on the chosen embedding $D \subset C$, but naturality of $\underline{\mu}$ implies that if $D \subset C$ as well as $D \subset C'$, where both C and C' are maximal, then the ensuing measures μ_D are the same. This implies the same property for the corresponding states ω_D, which in turn shows that the collection of all μ_D and μ_C thus obtained organizes itself into a single quasi-state ω on A.

The converse follows by running this argument backwards. □

Combining (12.196) with Theorem 12.27, we obtain a state-proposition pairing that is no longer probabilistic, as in ordinary quantum mechanics, but defines a proposition in the internal language of $\mathsf{T}(A)$ and as such may or may not be true at each stage $C \in \mathscr{C}(A)$. The final ingredient for this is an arrow

$$\underline{1} : \underline{\Sigma}(\underline{A}) \to \underline{[0,1]}_l, \tag{12.206}$$

defined by its components $\underline{1}_C : \mathscr{O}(\Sigma^A_{\uparrow C}) \to L(\uparrow C, [0,1])$ that map each open subset of $\Sigma^A_{\uparrow C}$ to the constant function on $\uparrow C$ taking the value $1 \in [0,1]$. The internal language of $\mathsf{T}(A)$ (cf. §E.5) turns this into a formula $\underline{\mu}_{\omega} = \underline{1}$ with the following interpretation:

$$[[\underline{\mu}_{\omega} = \underline{1}]] : \underline{\Sigma}(\underline{A}) \to \underline{\Omega}. \tag{12.207}$$

We combine this with (12.196) so as to obtain an internal state-proposition pairing

$$[[\underline{\mu}_{\omega}(\underline{a \in (p,q)}) = \underline{1}]]_{\mathrm{QT}} : \mathbf{1} \to \underline{\Omega}, \tag{12.208}$$

where we have abbreviated

$$[[\underline{\mu}_{\omega}(\underline{a \in (p,q)}) = \underline{1}]]_{\mathrm{QT}} = [[\underline{\mu}_{\omega} = \underline{1}]] \circ [[\underline{a \in (p,q)}]]_{\mathrm{QT}}. \tag{12.209}$$

The truth of the proposition (12.208) at stage C may be determined from Kripke–Joyal semantics; a straightforward computation for $A = B(H)$ shows that

$$C \Vdash \underline{\mu}_{\omega}(\underline{a \in (p,q)}) = \underline{1} \tag{12.210}$$

iff there exists a projection $e \in \mathscr{P}(C)$ with $e \leq e^{(a)}_{(p,q)}$ and $\omega(e) = 1$. Assuming ω is a vector state $\omega(a) = \langle \psi, a\psi \rangle$ for some unit vector $\psi \in H$, this means that (12.210) holds iff $\psi \in eH \subseteq e^{(a)}_{(p,q)}H$ for some $e \in \mathscr{P}(C)$, i.e., if the proposition $a \in (p,q)$ has (Born) probability one in state ψ *and* there is a yes-no measurement in context C verifying this probability. In comparison, in classical mechanics a pure state $x \in X$ makes $a \in (p,q)$ true iff $a(x) \in (p,q)$, where $a \in C(X,\mathbb{R})$ as before.

We close this chapter with a topos-theoretical (or, one might say, topological) reinterpretation of the Kochen–Specker Theorem, which to some extent explains why the previous construction had to use the fuzzy interval domain $\mathbb{IR}$ rather than the sharp reals $\mathbb{R}$. To this end, we first generalize the notion of a quasi-linear non-contextual hidden variable (cf. Definitions 6.1 and 6.3) to any (unital) C*-algebra:

Definition 12.28. *1. A* **valuation** *on a unital C*-algebra A is a unital map*

$$V : A_{\mathrm{sa}} \to \mathbb{R} \tag{12.211}$$

that is dispersion-free (i.e. multiplicative) and linear on commuting operators.
2. A **point** *in a frame $\mathscr{O}(X)$ in some topos T is defined as a frame homomorphism*

$$p : \mathscr{O}(X) \to \Omega, \tag{12.212}$$

where Ω is the truth object in T.

If A is commutative, the Gelfand spectrum $\Sigma(A)$ consists of the valuations on A. The second part generalizes the notion of a point of a frame in set theory (cf. §C.11).

Theorem 12.29. *For any unital C*-algebra A, there are canonical bijective correspondences between:*

- *Valuations on A.*
- *Points of $\underline{\Sigma}(\underline{A})$ in $\mathsf{Sh}(\mathscr{C}(A))$.*
- *Continuous cross-sections $\sigma : \mathscr{C}(A) \to \Sigma^A$ of the bundle $\pi : \Sigma^A \to \mathscr{C}(A)$.*

Proof. We first give the external description of points of a locale $\underline{Y}$ in a sheaf topos $\mathsf{Sh}(X)$ (cf. §E.4). The subobject classifier in $\mathsf{Sh}(X)$ is the sheaf $\underline{\Omega} : U \mapsto \mathcal{O}(U)$, in terms of which a point of $\underline{Y}$ is a frame map $\mathcal{O}(\underline{Y}) \to \underline{\Omega}$. Externally, the point-free space defined by the frame $\underline{\Omega}$ is given by the identity map $\mathrm{id}_X : X \to X$, so that a point of $\underline{Y}$ externally correspond to a continuous cross-section $\sigma : X \to Y$ of the bundle $\pi : Y \to X$ (i.e., $\pi \circ \sigma = \mathrm{id}_X$). In principle, π and σ are by definition frame maps in the opposite direction, but in the case at hand, namely $X = \mathscr{C}(A)$ and $Y = \Sigma^A$, the map $\sigma : \mathscr{C}(A) \to \Sigma^A$ may be interpreted as a continuous cross-section of the projection (12.134) in the usual sense. Being a cross-section simply means that $\sigma(C) \in \Sigma(C)$. As to continuity, by definition of the Alexandrov topology, σ is continuous iff the following condition is satisfied:

$$\text{For all } \mathscr{U} \in \mathcal{O}(\Sigma^A) \text{ and all } C \subseteq D, \text{ if } \sigma(C) \in \mathscr{U}, \text{ then } \sigma(D) \in \mathscr{U}.$$

Hence, given the definition of $\mathcal{O}(\Sigma^A)$, the following condition is sufficient for continuity: if $C \subseteq D$, then $\sigma(D)_{|C} = \sigma(C)$. However, this condition is also necessary. To explain this, let $\rho_{DC} : \Sigma(D) \to \Sigma(C)$ again be the restriction map. This map is continuous and open. Suppose $\rho_{DC}(\sigma(D)) \neq \sigma(C)$. Since $\Sigma(D)$ is Hausdorff, there is an open neighbourhood $\mathscr{U}_D$ of $\rho_{DC}^{-1}(\sigma(C))$ not containing $\sigma(D)$. Let $\mathscr{U}_C = \rho_{DC}(\mathscr{U}_D)$ and take any $\mathscr{U} \in \mathcal{O}(\Sigma^A)$ such that $\mathscr{U} \cap \mathcal{O}(\Sigma(C)) = \mathscr{U}_C$ and $\mathscr{U} \cap \mathcal{O}(\Sigma(D)) = \mathscr{U}_D$. This is possible, since $\mathscr{U}_C$ and $\mathscr{U}_D$ satisfy both conditions in the definition of $\mathcal{O}(\Sigma^A)$. By construction, $\sigma(C) \in \mathscr{U}$ but $\sigma(D) \notin \mathscr{U}$, so that σ is not continuous. Hence σ is a continuous cross-section of π iff

$$\sigma(D)_{|C} = \sigma(C) \text{ for all } C \subseteq D. \tag{12.213}$$

Now define a map $V : A_{\mathrm{sa}} \to \mathbb{C}$ by $V(a) = \sigma(C^*(a))(a)$, where $C^*(a)$ is the commutative unital C*-algebra generated by a. If $b^* = b$ and $[a,b] = 0$, then $V(a+b) = V(a) + V(b)$ by (12.213), applied to $C^*(a) \subset C^*(a,b)$ as well as to $C^*(b) \subset C^*(a,b)$. Furthermore, since $\sigma(C) \in \Sigma(C)$, the map V is dispersion-free.

Conversely, a valuation V defines a cross-section σ by complex linear extension of $\sigma(C)(a) = V(a)$, where $a \in C_{\mathrm{sa}}$. By the criterion (12.213) this cross-section is continuous, since the value $V(a)$ is independent of the choice of C containing a. □

Corollary 12.30. *The bundle $\pi : \Sigma^A \to \mathscr{C}(A)$ (cf. Corollary 12.22) admits no continuous cross-sections as soon as A has no valuations (e.g. if $A = M_n(\mathbb{C})$, $n > 2$).*

The contrast between the pointlessness of the internal spectrum $\underline{\Sigma}$ and the spatiality of the external spectrum Σ^A is striking, but easily explained: a point of Σ^A (in the usual sense, but also in the frame-theoretic sense if Σ^A is sober) necessarily lies in some $\Sigma(C) \subset \Sigma^A$, and hence is defined (and dispersion-free) only in the context C. For example, for $A = M_n(\mathbb{C})$, a point $V \in \Sigma(C)$ corresponds to a map

$$V^* : \mathcal{O}(\Sigma^A) \to \{0,1\},\ S \mapsto V(S(C)), \tag{12.214}$$

where $\mathcal{O}(\Sigma^A)$ is given by (12.95). Thus V^* is only sensitive to the value of S at C.

Notes

Previous advocates of intuitionistic logic for quantum mechanics include Popper (1968) and Coecke (2002). The earliest use of topos theory in quantum mechanics was probably by Adelman & Corbett (1995), but the founding papers of the topos approach to quantum mechanics as further developed in this chapter are Isham & Butterfield (1998), Butterfield & Isham (1999, 2002), and Hamilton, Isham & Butterfield (2000). This series of papers was predated by Isham (1997) and was followed by Döring & Isham (2008abcd, 2010); see also Flori (2013) for an introduction. Wolters (2013ab) gives a detailed comparison between the "contravariant" Butterfield–Döring–Isham approach and the "covariant" approach in this chapter.

The original motivation behind our approach to "quantum toposophy" was the ***Principle of General Tovariance*** (Heunen, Landsman, & Spitters, 2008), which was a pun on Einstein's *Principle of General Covariance* underlying General Relativity (Norton, 1993, 1995). Einstein based his theory of gravity and space-time on the mathematical postulate that all equations of physics be invariant under arbitrary coordinate transformation, and similarly we proposed that all physical theories should be invariant under so-called *geometric morphisms* between toposes and hence should be formulated in terms of what (confusingly) is called *geometric logic* (cf. Mac Lane & Moerdijk, 1992; Johnstone, 2002). Since in fact some of our constructions turned out be non-geometric in this sense, we subsequently dropped this principle and stopped even referring to the above paper. However, as Raynaud (2014) and, more generally, Henry (2015) show, our theory can actually be made geometric (in the topos-theoretical sense) provided one puts the entire theory of (internal) C*-algebras on a localic (i.e., pointfree) basis, as in Henry (2014ab). Other recent developments of the program (which are not discussed here) may be found in e.g. van den Berg & Heunen (2012, 2014), Spitters, Vickers, & Wolters (2014), Heunen (2014ab), and Heunen & Lindenhovius (2015).

§12.1. **C*-algebras in a topos**

C*-algebras in a topos, including a constructive version of Gelfand duality for commutative unital C*-algebras that is valid in arbitrary Grothendieck toposes, were first studied by Banaschewski & Mulvey (2000ab, 2006). The topos $\mathsf{T}(A)$ and the internal commutative C*-algebra $\underline{A}$ were introduced by Heunen, Landsman, & Spitters (2009). All these papers rely crucially on the theory of internal locales in toposes, which owes much to Johnstone (1982) and Joyal & Tierney (1984). See also Johnstone (1983) and Vickers (2007). It is possible to realize $\mathsf{T}(A)$ as the topos of sheaves on the locale $\mathrm{Idl}(\mathscr{C}(A))$, which is the ideal completion of the "mere" poset $\mathscr{C}(A)$, but we will not use this description (Raynaud, 2014).

§12.2. **The Gelfand spectrum in constructive mathematics**

This section is based on Coquand (2005) and Coquand & Spitters (2005, 2009), where also the missing details may be found. All necessary background on lattice theory is provided by Johnstone (1982), except the ingredients for the proof that the constructive Gelfand spectrum is compact and regular, which is due to Cederquist & Coquand (2000). Proposition 12.10 may be found in Aczel (2006).

§12.3. **Internal Gelfand spectrum and intuitionistic quantum logic**

This section is based on Caspers, Heunen, Landsman, & Spitters (2009), except for the final part on Kripke semantics, which is taken from Heunen, Landsman, & Spitters (2012). An interesting philosophical analysis of the intuitionistic logic emerging from this program may be found in Hermens (2016), to whom the interpretation elements of the frame $\mathcal{O}(\Sigma^A)$ as disjunctions is due.

§12.4. **Internal Gelfand spectrum for arbitrary C*-algebras**

This section is based on Caspers (2008), Caspers, Heunen, Landsman, & Spitters (2009), and Heunen, Landsman, & Spitters (2009). Complete proofs of Lemma 12.15 and Lemma 12.16 may be found in Caspers (2008), §5.2. For different proofs of these lemmas see Heunen, Landsman, & Spitters (2009) and Coquand (2005), respectively. A proof of Lemma 12.21 may be found in Wolters (2013b), Theorem 2.17, also available as `http://arxiv.org/pdf/1010.2031v2.pdf`.

§12.5. **"Daseinisation" and Kochen–Specker Theorem**

The spectral order was introduced by Olson (1971) and was rediscovered by De Groote (2011). For a devastating critique of Heidegger's philosophy see Philipse (1999). The first construction of a "Daseinisation" map was given by Döring & Isham (2008b). The version presented here is an improvement, due to Wolters (2013ab), of a previous adaptation of the Döring–Isham appraoch to the topos $\mathsf{T}(A)$ in Heunen, Landsman, & Spitters (2009). Similarly, Theorem 12.29, first published in Heunen, Landsman, Spitters, & Wolters (2012), is an improvement due to Wolters (2013a) of an earlier result in this direction in Heunen, Landsman, & Spitters (2009).

The work of Isham & Butterfield (1998), which, as already mentioned, started the entire quantum toposophy program, was actually motivated by an topos-theoretica reformulation of the Kochen–Specker Theorem. Isham and Butterfield started from the following observation. Let $\mathscr{C}(B(H))$ be the poset of commutative *von Neumann* subalgebras of $B(H)$, partially ordered by set-theoretic inclusion, seen as a category in the usual way. Consider the presheaf topos $[\mathscr{C}(H)^{\mathrm{op}}, \mathrm{Set}]$ of *contravariant* functors $\underline{F} : \mathscr{C}(H) \to \mathrm{Set}$, where Set is the category of sets. The ***spectral presheaf*** is the contravariant functor $\underline{\Sigma}$ defined on objects by $\underline{\Sigma}_0(C) = \Sigma(C)$, and by the natural map on arrows, that is, $\underline{\Sigma}_1(C \subset D)$ maps $\omega \in \Sigma(D)$ (which is a map $D \to \mathbb{C}$) to its restriction to C, i.e., to $\omega_{|C} \in \Sigma(C)$. A ***point*** of some object $\underline{F}$ in $[\mathscr{C}(B(H))^{\mathrm{op}}, \mathrm{Set}]$ is defined as a natural transformation $\mathbf{1} \to \underline{F}$, where $\mathbf{1}$ is the terminal object, i.e., the presheaf that maps everything into the singleton set $*$.

The Kochen–Specker Theorem à la Butterfield & Isham, then, states that if $\dim(H) > 2$ as usual, *the spectral presheaf has no points*.

Appendix A
Finite-dimensional Hilbert spaces

Although we assume the reader to be familiar with linear algebra, some of the points below may not be emphasized at that level and hence need to be recalled.

Unless explicitly stated otherwise, all vector spaces (and hence also all algebras) are defined over the *complex numbers* $\mathbb{C}$. Moreover, from §A.2 until the end of this appendix, V will be ***finite-dimensional***; the infinite-dimensional case will be treated in the next appendix on functional analysis and general Hilbert spaces.

A.1 Basic definitions

Definition A.1. *Let V be a vector space (not necessarily finite-dimensional).*

1. *A* **sesquilinear form** *on V is a map $V \times V \to \mathbb{C}$, written $(v,w) \mapsto \langle v,w \rangle$ (or, occasionally, to distinguish it from an inner product, as $(v,w) \mapsto B(v,w)$) that is real-bilinear and satisfies $\langle iv,w \rangle = -i\langle v,w \rangle$ and $\langle v,iw \rangle = i\langle v,w \rangle$ for all $v,w,x \in V$.*
2. *A* **hermitian** *form on V is a sesquilinear form that satisfies $\langle w,v \rangle = \overline{\langle v,w \rangle}$.*
3. *A* **pre-inner product** *on V is a* **positive** *hermitian form, i.e., $\langle v,v \rangle \geq 0$.*
4. *An* **inner product** *on V is, in addition,* **positive definite***: $\langle v,v \rangle = 0$ iff $v = 0$.*
5. *A* **norm** *on V is a function $\|\cdot\| : V \to \mathbb{R}^+$ satisfying, for all $v,w,h \in V$ and $\lambda \in \mathbb{C}$:*

   a. *$\|v+w\| \leq \|v\| + \|w\|$ (***triangle inequality***);*
   b. *$\|\lambda v\| = |\lambda|\|v\|$ (***homogeneity***);*
   c. *$\|v\| = 0$ iff $v = 0$ (***positive definiteness***).*

Many analytical arguments in functional analysis are based on the fundamental ***Cauchy–Schwarz inequality***, which is satisfied by any (pre-) inner product:

$$|\langle v,w \rangle|^2 \leq \langle v,v \rangle \langle w,w \rangle. \tag{A.1}$$

Proposition A.2. *An inner product on V defines a norm on V by means of*

$$\|v\| = \sqrt{\langle v,v \rangle}. \tag{A.2}$$

K. Landsman, *Foundations of Quantum Theory*,
Fundamental Theories of Physics 188, DOI 10.1007/978-3-319-51777-3

The Cauchy–Schwarz inequality (A.1) *then reads*

$$|\langle v,w\rangle| \leq \|v\|\,\|w\|, \tag{A.3}$$

with equality iff v and w are linearly dependent.

The question arises when a norm comes from an inner product via (A.2).

Theorem A.3. *A norm* $\|\cdot\|$ *comes from an inner product through* (A.2) *iff*

$$\|v+w\|^2+\|v-w\|^2 = 2(\|v\|^2+\|w\|^2). \tag{A.4}$$

In that case, one has the **polarization identity**

$$\langle v,w\rangle = \tfrac{1}{4}(\|v+w\|^2-\|v-w\|^2+i\|v-iw\|^2-i\|v+iw\|^2). \tag{A.5}$$

Proof. Easy computations show that (A.2) holds, that $\langle w,v\rangle = \overline{\langle v,w\rangle}$, and, with a bit more effort, that $\langle v,w_1+w_2\rangle = \langle v,w_1\rangle + \langle v,w_2\rangle$. Now suppose we know that

$$\langle w,sv\rangle = s\langle w,v\rangle \tag{A.6}$$

for certain $s\in\mathbb{R}$. Then this property clearly also holds for s^{-1} instead of s. Furthermore, having (A.6) for s as well as $t\in\mathbb{R}$ implies the same property also for $s+t$ and st. Starting with $s=t=1$, this generates (A.6) for each $s\in\mathbb{Q}$. Now if $s_n\to s$ for $s_n\in\mathbb{Q}$ and $s\in\mathbb{R}$, then by continuity and homogeneity of the norm, $\langle w,s_nv\rangle\to\langle w,sv\rangle$. Consequently, (A.6) holds for each $s\in\mathbb{R}$. Finally, from (A.5) we also find $\langle w,iv\rangle = i\langle w,v\rangle$, and hence (A.6) holds for each $s\in\mathbb{C}$. □

There is an analogous result for continuous hermitian forms, with practically the same proof (where continuity is once again needed to pass from $\mathbb{Q}$ to $\mathbb{R}$). Let V be a vector space with inner product, and let $B: V\times V\to\mathbb{C}$ be a hermitian form. The associated ***quadratic form*** $Q: V\to\mathbb{R}$, defined by

$$Q(v) = B(v,v), \tag{A.7}$$

then satisfies

$$Q(zv) = |z|^2Q(v)\ \ (z\in\mathbb{C}); \tag{A.8}$$

$$Q(v+w)+Q(v-w) = 2(Q(v)+Q(w)). \tag{A.9}$$

Proposition A.4. *Let* V *be a vector space with inner product. A map* $Q: V\to\mathbb{R}$ *that is continuous in the associated norm* (A.2) *is derived from a hermitian form* $B: H\times H\to\mathbb{C}$ *through* (A.7) *iff* Q *satisfies* (A.8) - (A.9), *in which case*

$$B(v,w) = \tfrac{1}{4}(Q(v+w)-Q(v-w)+iQ(v-iw)-iQ(v+iw)). \tag{A.10}$$

A.2 Functionals and the adjoint

In the remainder of this appendix, V is a ***finite-dimensional*** *complex vector space with inner product.* Since this is automatically a (finite-dimensional) ***Hilbert space*** (as defined in the next appendix), we rename it as H. The archetypal example is $H = \mathbb{C}^n$, with elements $z = (z_1, \ldots, z_n)$, $z_i \in \mathbb{C}^n$, and standard inner product

$$\langle z, w \rangle = \sum_{i=1}^{n} \overline{z}_i w_i. \tag{A.11}$$

In that case, we hardly make a difference between a linear map $a : H \to H$ and the corresponding matrix (a_{ij}), where $(az)_i = \sum_j a_{ij} z_j$, or, equivalently,

$$a_{ij} = \langle \upsilon_i, a\upsilon_j \rangle, \tag{A.12}$$

where $(\upsilon_1 = (1, 0, \ldots, 0), \ldots \upsilon_n = (0, \ldots, 0, 1))$ is the standard basis of $\mathbb{C}^n$. More generally, we will only consider ***orthonormal bases*** of Hilbert spaces H, i.e., bases (υ_i) for which $\langle \upsilon_i, \upsilon_j \rangle = \delta_{ij}$. In fact, in the present (finite-dimensional) case, any orthonormal set of $n = \dim(H)$ vectors is automatically a basis. *Throughout this book, the word "basis" will be synonymous with* orthonormal *basis.*

Let H^* be the vector space of linear maps $f : H \to \mathbb{C}$, also called (linear) ***functionals*** (on H). Since the inner product is positive definite, it is also non-degenerate:

Proposition A.5. *The map $\psi \mapsto f_\psi$, where*

$$f_\psi(\varphi) = \langle \psi, \varphi \rangle, \tag{A.13}$$

is an anti-linear isomorphism $H \to H^*$ (i.e., one has $\lambda\psi \mapsto \overline{\lambda} f_\psi$ for any $\lambda \in \mathbb{C}$).

Proof. Injectivity is obvious. For surjectivity, note that coker(f) (i.e., the orthogonal complement of the kernel ker(f) of f) is one-dimensional (assuming f is nonzero), and take a unit vector $\tilde{\psi} \in \text{coker}(f)$. Then $\psi = \overline{f(\tilde{\psi})}\tilde{\psi}$ does the job: by linearity of f, we have $f(\varphi)\tilde{\psi} - f(\tilde{\psi})\varphi \in \ker(f)$ for any $\varphi \in H$ (and even any $\tilde{\psi} \in H$), so that $\langle \tilde{\psi}, f(\varphi)\tilde{\psi} - f(\tilde{\psi})\varphi \rangle = 0$. Since $\langle \tilde{\psi}, \tilde{\psi} \rangle = \|\tilde{\psi}\|^2 = 1$, this yields $f = f_\psi$. □

A linear map $a : H \to H$ is also called an ***operator***; we denote the algebra of all operators on H by $B(H)$. For example, we have $B(\mathbb{C}^n) \cong M_n(\mathbb{C})$. Two arbitrary vectors $\psi, \varphi \in H$ define an operator $|\psi\rangle\langle\varphi|$ through Dirac's "bra-ket" notation

$$|\psi\rangle\langle\varphi|\chi = \langle \varphi, \chi \rangle \psi. \tag{A.14}$$

The ***adjoint*** a^* of an operator a is defined by the property

$$\langle a^*\psi, \varphi \rangle = \langle \psi, a\varphi \rangle, \quad (\psi, \varphi \in H). \tag{A.15}$$

Indeed, for given χ (and a), define a functional $f_{a,\chi} : H \to \mathbb{C}$ by $f_{a,\chi}(\varphi) = \langle \chi, a\varphi \rangle$. Then, as we just saw, $f_{a,\chi} = f_\psi$ for some unique $\psi \in H$; define a^* by $a^*\chi = \psi$. This map is linear by construction.

Clearly, one has

$$a^{**} = a. \tag{A.16}$$

For $H = \mathbb{C}^n$, the matrix corresponding to the adjoint a^* is given by the well-known formula $a^*_{ij} = \overline{a_{ji}}$. A more abstract example of an adjoint is given by

$$|\psi\rangle\langle\varphi|^* = |\varphi\rangle\langle\psi|. \tag{A.17}$$

The ***(operator) norm*** of $a : H \to H$ is defined by

$$\|a\| = \sup\{\|a\psi\|, \psi \in H_1\}. \tag{A.18}$$

where the ***unit sphere*** $H_1 \subset H$ is defined by

$$H_1 = \{\psi \in H, \|\psi\| = 1\}. \tag{A.19}$$

Proposition A.6. *One has $\|a\| < \infty$ for any linear map $a : H \to H$.*

Proof. Recall that $\dim(H) = n < \infty$! Map H to $\mathbb{C}^n$ by the choice of some basis (υ_i). Thus $\psi \in H$ is mapped to $\tilde{\psi} = (\psi_1, \dots, \psi_n) \in \mathbb{C}^n$, with $\psi_i = \langle\upsilon_i, \psi\rangle$, and we have $\|\tilde{\psi}\|_2 = \|\psi\|$, where $\|z\|_2^2 = \sum_i |z_i|^2$ is the usual norm on $\mathbb{C}^n$, which is given by (A.2) with (A.11). This also transfers the operator $a : H \to H$ to a linear map $\tilde{a} : \mathbb{C}^n \to \mathbb{C}^n$ defined by the matrix (A.12). Then $\|a\| = \|\tilde{a}\| = \sup\{\|\tilde{a}z\|_2, z \in \mathbb{C}^n_1\}$, where $\mathbb{C}^n_1 = \{z \in \mathbb{C}^n, \|z\|_2 = 1\}$. Now $\tilde{a}$ is continuous because it is linear, and hence it maps $\mathbb{C}^n_1$ (which is compact by Heine–Borel) to some compact set $\tilde{a}(\mathbb{C}^n_1)$ in $\mathbb{C}^n$. It is easy to see that the norm $\|\cdot\|_2 : \mathbb{C}^n \to \mathbb{R}^+$ is continuous, and according to Weierstrass the norm therefore assumes a finite maximum (as well as a minimum) on any compact set K. Taking $K = \tilde{a}(\mathbb{C}^n_1)$ proves the claim. □

Proposition A.7. *Let $a, b : H \to H$ be linear maps, and let $\psi \in H$. Then:*

$$\|a\psi\| \leq \|a\|\|\psi\|; \tag{A.20}$$
$$\|ab\| \leq \|a\|\|b\|; \tag{A.21}$$
$$\|a^*\| = \|a\|; \tag{A.22}$$
$$\|a^*a\| = \|a\|^2. \tag{A.23}$$

Proof. The first two inequalities are immediate from (A.18). Next, if $\|\psi\| = 1$, by (A.3), (A.15), and (A.20) we have

$$\|a^*\psi\|^2 = \langle a^*\psi, a^*\psi\rangle = \langle\psi, aa^*\psi\rangle \leq \|\psi\|\|aa^*\psi\| \leq \|a\|\|a^*\psi\|, \tag{A.24}$$

so $\|a^*\psi\| \leq \|a\|$, and hence from (A.18), $\|a^*\| \leq \|a\|$. But (A.16) gives the opposite inequality, whence (A.22). Finally, (A.21) and (A.22) yield $\|a^*a\| \leq \|a^*\|\|a\| = \|a\|^2$. From (A.3) and (A.20), on the other hand, we obtain

$$\|a\psi\|^2 = \langle a\psi, a\psi\rangle = \langle\psi a^*a\psi\rangle \leq \|a^*a\|, \tag{A.25}$$

so $\|a\|^2 \leq \|a^*a\|$ by (A.18), and hence (A.23) is proved. □

A.3 Projections

The most important examples (and also, as will see shortly, building blocks) of self-adjoint operators are ***projections*** $e : H \to H$, defined by the property

$$e^2 = e^* = e. \tag{A.26}$$

Proposition A.8. *There is a bijective correspondence $e \leftrightarrow L$ between:*

- *projections e on H;*
- *linear subspaces L of H,*

given by

$$L = eH; \tag{A.27}$$
$$e = \sum_i |\upsilon_i\rangle\langle\upsilon_i|, \tag{A.28}$$

where $eH = \{e\psi, \psi \in H\}$ is the image of e, and (υ_i) is an arbitrary basis of L.

The proof is routine, including the fact that (A.28) is independent of the basis. Whenever convenient, we write (A.28) as e_L. For example, the "sub"space $L = H$ corresponds to $e_H = 1_H$, whereas $L = \{0\}$ corresponds to $e_{\{0\}} = 0$.

Define the ***orthogonal complement!of subset of Hilbert space*** $L^{\perp}$ of *any* subset $L \subset H$ by

$$L^{\perp} = \{\psi \in H \mid \langle\psi, \varphi\rangle = 0 \forall \varphi \in L\}. \tag{A.29}$$

In particular, if L is a *linear* subspace of H, one easily checks that

$$e_{L^{\perp}} = 1 - e_L. \tag{A.30}$$

Corollary A.9. *For each linear subspace $L \subset H$ one has*

$$H = L \oplus L^{\perp}, \tag{A.31}$$

in the sense that $L \cap L^{\perp} = \{0\}$, and each vector $\psi \in H$ has a unique *decomposition*

$$\psi = \psi^{\parallel} + \psi^{\perp}, \tag{A.32}$$

where $\psi^{\parallel} \in L$ and $\psi^{\perp} \in L^{\perp}$.

Proof. Existence of the decomposition is given by

$$\psi^{\parallel} = e_L \psi; \tag{A.33}$$
$$\psi^{\perp} = (1 - e_L)\psi. \tag{A.34}$$

Uniqueness follows by assuming $\psi = \chi^{\parallel} + \chi^{\perp}$ with $\chi^{\parallel} \in L$ and $\chi^{\perp} \in L^{\perp}$: one then has $\psi^{\parallel} - \chi^{\parallel} = \psi^{\perp} - \chi^{\perp}$, but since the left-hand side is in L and the right-hand side is in $L^{\perp}$, both sides lie in $L \cap L^{\perp} = 0$. □

A.4 Spectral theory

An ***eigenvector*** of an operator a is a nonzero element $\psi \in H$ such that

$$a\psi = \lambda\psi \tag{A.35}$$

for some $\lambda \in \mathbb{C}$, called an ***eigenvalue*** of a. We also define the ***eigenspace*** H_λ by

$$H_\lambda = \{\psi \in H \mid a\psi = \lambda\psi\}, \tag{A.36}$$

with associated projection e_λ (in that $H_\lambda = e_\lambda H$, cf. Proposition A.8). In case that $\dim(H_\lambda) = 1$ the eigenvalue λ is called ***non-degenerate*** (or ***simple***). Otherwise it is said to be ***degenerate***, with ***multiplicity*** $m_\lambda = \dim(H_\lambda)$. In linear algebra, the set of all eigenvalues of a is called the ***spectrum*** of a, denoted by $\sigma(a)$ (for infinite-dimensional H, this turns out to be the wrong definition of the spectrum, see §B.14).

We now give two formulations of the ***spectral theorem for self-adjoint operators***.

Theorem A.10. *Let a be a self-adjoint operator on H. Then $\sigma(a) \subset \mathbb{R}$, eigenspaces for different eigenvectors $\lambda \neq \mu$ are orthogonal (i.e., $e_\lambda e_\mu = \delta_{\lambda\mu} e_\lambda$), and*

$$a = \sum_{\lambda \in \sigma(a)} \lambda \cdot e_\lambda; \tag{A.37}$$

$$1_H = \sum_{\lambda \in \sigma(a)} e_\lambda. \tag{A.38}$$

Equivalently, we may reformulate the above **spectral resolution** *of a in terms of the existence of a basis (υ_i) of H consisting of eigenvectors of a. In that case, we have*

$$a = \sum_{i=1}^{\dim(H)} \lambda_i |\upsilon_i\rangle\langle\upsilon_i|; \tag{A.39}$$

$$1_H = \sum_{i=1}^{\dim(H)} |\upsilon_i\rangle\langle\upsilon_i|, \tag{A.40}$$

where λ_i is the eigenvalue corresponding to the eigenvector υ_i (i.e., $a\upsilon_i = \lambda_i \upsilon_i$).

Note that the eigenvalues λ occurring in (A.37) are all different, whereas the λ_i in (A.40) need not be: the number of times an eigenvalue $\lambda_i \in \sigma(a)$ occurs is given by its multiplicity. This also implies that the spectral resolution (A.37) - (A.38) is canonical (i.e. free of any choices), whereas (A.39) - (A.40) depends on arbitrary choices of bases in all subspaces H_λ with dimension greater than one. Nonetheless, it is easier to prove (A.39) - (A.40), which obviously imply (A.37) - (A.38): just collect all λ_i that are equal to λ and realize that, as in (A.28), one has

$$e_\lambda = \sum_{i|\lambda_i=\lambda} |\upsilon_i\rangle\langle\upsilon_i|. \tag{A.41}$$

More generally, for some (at the moment) arbitrary (but later: measurable) subset $\Delta \subset \mathbb{R}$ it turns out to be convenient to introduce the ***spectral projection*** e_Δ on H and the associated ***spectral subspace*** $H_\Delta \subseteq H$: if $\Delta \cap \sigma(a) = \emptyset$ we put $e_\Delta = 0$ and $H_\Delta = \{0\}$, and otherwise,

$$e_\Delta = \sum_{\lambda \in \Delta \cap \sigma(a)} e_\lambda; \tag{A.42}$$

$$H_\Delta = e_\Delta H. \tag{A.43}$$

We now prepare for the proof of Theorem A.10. First, note from (A.15) that

$$2i \cdot \mathrm{Im}(\langle \psi, a\psi \rangle) = \langle \psi, a\psi \rangle - \overline{\langle \psi, a\psi \rangle} = \langle \psi, a\psi \rangle - \langle \psi, a^* \psi \rangle. \tag{A.44}$$

If $a^* = a$, from (A.35) and (A.44) one obtains $\mathrm{Im}(\lambda) = 0$ and hence $\sigma(a) \subset \mathbb{R}$.

Lemma A.11. *A self-adjoint operator a has an eigenvalue λ for which $|\lambda| = \|a\|$.*

Proof. As in the previous proof, the norm $\|\cdot\|$ assumes a maximum on the compact set aH_1, where $H_1 = \{\psi \in H, \|\psi\| = 1\}$. Suppose this happens at $a\psi_1$, where by construction $\|\psi_1\| = 1$. By definition of the norm, this maximum must be $\|a\|$, so that $\|a\| = \|a\psi_1\|$. Hence, using $a^* = a$, (A.3), and (A.23), we may estimate

$$\|a\|^2 = \|a\psi_1\|^2 = \langle a\psi_1, a\psi_1 \rangle = \langle \psi_1, a^2\psi_1 \rangle \leq \|a^2\psi_1\| \leq \|a^2\| = \|a\|^2. \tag{A.45}$$

Hence we need equality at the $\leq$ sign in (A.45), which according to the remark below (A.3) can only be the case if $a^2\psi_1 = \|a\|^2\psi_1$. Define $\chi_1 = a\psi_1 - \|a\|\psi_1$. There are two possibilities: if $\chi_1 = 0$, then $a\psi_1 = \|a\|\psi_1$, and $\chi_1 \neq 0$, then

$$a\chi_1 = a^2\psi_1 - \|a\|a\psi_1 = \|a\|^2\psi_1 - \|a\|a\psi_1 = -\|a\|\chi_1. \tag{A.46}$$

Hence either $a\psi_1 = \|a\|\psi_1$ or $a\chi_1 = -\|a\|\chi_1$, which proves the claim. □

We are now in a position to prove Theorem A.10.

Proof. By Lemma A.11, we already found one eigenvector υ_1 of a, viz. either $\upsilon_1 = \psi_1$ or $\upsilon_1 = \chi_1$. Furthermore, it is easy to show that if a self-adjoint operator a leaves a linear subspace $L \subset H$ stable (in that $a\varphi \in L$ whenever $\varphi \in L$), then it also leaves $L^\perp$ stable, and remains self-adjoint as an operator $a : L^\perp \to L^\perp$. First use this with $L_1 = \mathbb{C} \cdot \upsilon_1$. Lemma A.11, now applied to $a : L^\perp \to L^\perp$, gives a second eigenvector υ_2. Now take L_2 to be the linear span of υ_1 and υ_2, and restrict a to $L_2^\perp$, etc. Since H is finite-dimensional, this procedure ends after $\dim(H)$ steps.

This leaves us with a basis (υ_i) of H that by construction entirely consists of eigenvectors. The mutual orthogonality of these eigenvectors (and hence of the spectral projections e_λ) follows from a simple calculation. □

Corollary A.12. *The norm of a self-adjoint operator a is given by*

$$\|a\| = \sup\{|\lambda|, \lambda \in \sigma(a)\}. \tag{A.47}$$

Proof. This rapidly follows from Theorem A.10 by expanding ψ in (A.18) with respect to the basis of H given in (A.39) - (A.40). □

Corollary A.13. *A self-adjoint operator a is a projection iff:*

- $\sigma(a) = \{0\}$, *in which case* $a = 0$;
- $\sigma(a) = \{1\}$, *in which case* $a = 1$;
- $\sigma(a) = \{0, 1\}$, *in which case a is called a* **proper** *projection.*

In particular, if e is a nonzero projection, then

$$\|e\| = 1. \tag{A.48}$$

Proof. Only the third case is nontrivial. If $a = e$ is a proper projection, then by Corollary A.9 its eigenvectors can only lie in $L = eH$ (with eigenvalue $\lambda = 1$) or in $L^{\perp} = (1-e)H$ (with eigenvalue $\lambda = 0$). The converse implication follows from Theorem A.10, notably from (A.37). Eq. (A.48) then follows from (A.47). □

A less elementary but more powerful approach to the spectral theorem is as follows. For the notion of a C*-algebra see Definition C.1 in Appendix C.

Definition A.14. *Let* $a \in B(H)$. *Then* $C^*(a)$ *is the C*-algebra generated by a and* 1_H *(i.e., the algebra of all polynomials in a).*

Theorem A.15. *If a is self-adjoint, then* $C^*(a)$ *is commutative, and:*

1. *There is an isomorphism of (commutative) C*-algebras*

$$C(\sigma(a)) \cong C^*(a), \tag{A.49}$$

written $f \mapsto f(a)$, *which is unique if it is subject to the following conditions:*

- *the unit function* $1_{\sigma(a)} : \lambda \mapsto 1$ *corresponds to the unit operator* 1_H;
- *the identity function* $\mathrm{id}_{\sigma(a)} : \lambda \mapsto \lambda$ *is mapped to the given operator a.*

2. *In terms of the spectral projections* e_λ *of the operator a we have*

$$C^*(a) = C^*(e_\lambda, \lambda \in \sigma(a)) = \mathrm{span}(e_\lambda, \lambda \in \sigma(a)), \tag{A.50}$$

where the middle term is the C-algebra generated by the projections* e_λ.

3. *Under the isomorphism* (A.49),

$$e_\lambda = \delta_\lambda(a), \tag{A.51}$$

where the delta-function $\delta_{\lambda'}$ *on* $\sigma(a)$ *is defined by* $\delta_{\lambda'} : \lambda \mapsto \delta_{\lambda\lambda'}$.

Proof. For any complex (finite) polynomial $p(x) = \sum_n c_n x^n$ on $\mathbb{R}$, define an operator

$$p(a) = \sum_n c_n a^n. \tag{A.52}$$

Simple computations then show that, for arbitrary polynomials p, and $t \in \mathbb{C}$,

$$(tp+q)(a) = tp(a)+q(a); \tag{A.53}$$
$$(pq)(a) = p(a)q(a); \tag{A.54}$$
$$p(a)^* = \overline{p}(a). \tag{A.55}$$

Hence the space $P^*(a)$ of all such polynomials in a forms a *-algebra of $B(H)$. As a linear subspace of the finite-dimensional vector space $B(H)$, $P^*(a)$ must itself be finite-dimensional, hence it is C*-algebra. Moreover, $P^*(a)$ clearly contains 1_H (take $p(x) = 1$) as well as a (take $p(x) = x$), and since $P^*(a) \subseteq C^*(a)$ by definition of the latter, we must have $P^*(a) = C^*(a)$. Since $pq = qp$ and hence $p(a)q(a) = q(a)p(a)$ by the above computations, it follows that $P^*(a)$ and hence $C^*(a)$ is commutative. This proves the first claim.

To establish the isomorphism (A.49), we are going to define a map

$$C(\sigma(a)) \ni f \mapsto f(a) \in C^*(a). \tag{A.56}$$

We initially do this for polynomials $f = p$, so that $f(a) = p(a)$ is defined by (A.52). Since $C^*(a) = P^*(a)$ consists of polynomials in a, the map (A.56) is evidently surjective. It is also injective, for suppose $p(a) = q(a)$. Applying this to an eigenvector $v_\lambda \in H_\lambda$ yields $p(\lambda) = q(\lambda)$, for each $\lambda \in \sigma(a)$, and hence $p = q$ as functions on $\sigma(a)$. Hence $f \mapsto f(a)$ is, at least, a bijection of sets. Moreover, the properties (A.53) - (A.55) turn it into an isomorphism of C*-algebras, evidently with the properties stated after (A.49). Finally, for any given function $f : \sigma(a) \to \mathbb{C}$ there exists some polynomial p that coincides with f on the finite set $\sigma(a) \subset \mathbb{R}$, so that we may define $f(a)$ in (A.56) by $p(a)$, as in (A.52); by the above proof of injectivity, the ensuing operator $f(a)$ is independent of the choice of p.

We prove the last two claims, using the orthogonality property $e_\lambda e_\mu = \delta_{\lambda\mu} e_\lambda$ of *spectral* projections and the defining properties $e_\lambda^2 = e_\lambda = e_\lambda$ of *general* projections, see (A.26). From eq. (A.37) in Theorem A.10 we obtain (for polynomials f):

$$f(a) = \sum_{\lambda \in \sigma(a)} f(\lambda) \cdot e_\lambda. \tag{A.57}$$

If we now define $C^*(a)'$ as the linear span of the spectral projections e_λ and 1_H (which is a unital commutative C*-algebra by the properties of the e_λ just mentioned), then (A.57) shows that $C^*(a) \subseteq C^*(a)'$. Conversely, (A.57) gives (A.51), which shows that $C^*(a)' \subseteq C^*(a)$, and hence $C^*(a) = C^*(a)$. □

A second approach to the final claims of Theorem A.15 is more ambitious, as it includes a derivation of Theorem A.10 (instead of assuming it, as we just did). We now use (A.51) to *define* the spectral projections e_λ; from (A.54) - (A.55) we have

$$e_\lambda^2 = \delta_\lambda(a)^2 = \delta_\lambda^2(a) = \delta_\lambda(a) = e_\lambda;$$
$$e_\lambda^* = \delta_\lambda(a)^* = \overline{\delta_\lambda}(a) = \delta_\lambda(a) = e_\lambda,$$

showing that e_λ is indeed a projection. Also note the following identities in $C(\sigma(a))$:

$$\mathrm{id}_{\sigma(a)} = \sum_{\lambda\in\sigma(a)} \lambda\cdot\delta_\lambda; \tag{A.58}$$

$$1_{\sigma(a)} = \sum_{\lambda\in\sigma(a)} \delta_\lambda. \tag{A.59}$$

Transferring these from $C(\sigma(a))$ to $C^*(a)$ via the isomorphism (A.49) then yields (A.37) - (A.38). To analyse the projections e_λ defined by (A.51), we first compute

$$e_\lambda e_\mu = \delta_\lambda(a)\delta_\mu(a) = (\delta_\lambda\delta_\mu)(a) = \delta_{\lambda\mu}\delta_\lambda(a) = \delta_{\lambda\mu}e_\lambda, \tag{A.60}$$

which shows that the e_λ are mutually orthogonal. Second, we compute

$$ae_\lambda\psi = a\delta_\lambda(a)\psi = \mathrm{id}_{\sigma(a)}(a)\delta_\lambda(a)\psi = (\mathrm{id}_{\sigma(a)}\cdot\delta_\lambda)(a)\psi = \lambda\cdot\delta_\lambda(a)\psi = \lambda e_\lambda\psi,$$

which shows that $e_\lambda H \subseteq H_\lambda$. Third, (A.60) and (A.59) give $\oplus_{\lambda\in\sigma(a)}e_\lambda H = H$, which together with the second step gives $e_\lambda H = H_\lambda$. Hence the e_λ are indeed the spectral projections of a. Since we have already proved (A.37) - (A.38), we conclude that Theorem A.10 follows from the first part of Theorem A.15. By the argument in the main proof above, this first part then also yields the second part.

The generalization of Theorem A.15 to a family $\underline{a} = (a_1,\dots,a_n)$ of commuting self-adjoint operators is as follows.

Definition A.16. *Let $\underline{a} = (a_1,\dots,a_n)$ be commuting self-adjoint operators.*

1. *A* **joint eigenvector** *of $\underline{a}$ is a nonzero vector $\psi\in H$ such that $\underline{a}\psi = \underline{\lambda}\psi$, where $\underline{\lambda} = (\lambda_1,\dots,\lambda_n)$ with $\lambda_i\in\mathbb{C}$, i.e., for each $i = 1,\dots,n$, one has $a_i\psi = \lambda_i\psi$. We call $\underline{\lambda}$ a* **joint eigenvalue** *of $\underline{a}$.*
2. *The* **joint spectrum** *$\sigma(a_1,\dots,a_n)\equiv\sigma(\underline{a})$ consists of all joint eigenvalues of $\underline{a}$.*
3. *$C^*(\underline{a})$ is the smallest unital C*-subalgebra of $B(H)$ that contains each a_i.*

Clearly, we have

$$\sigma(\underline{a}) \subseteq \sigma(a_1)\times\cdots\times\sigma(a_n)\subset\mathbb{R}^n. \tag{A.61}$$

Furthermore, since $\dim(H) < \infty$, once again $C^*(\underline{a})$ is just the algebra of complex polynomials in all operators a_i.

Theorem A.17. *Let $\underline{a} = (a_1,\dots,a_n)$ be commuting self-adjoint operators on H. Then the C*-algebra $C^*(\underline{a})$ generated by these operators is commutative, and:*

1. *There is a unique isomorphism of C*-algebras*

$$C(\sigma(\underline{a}))\cong C^*(\underline{a}), \tag{A.62}$$

written $f\mapsto(f(\underline{a})$, subject to the following conditions:

- *the unit function $1_{\sigma(\underline{a})}:\underline{\lambda}\mapsto 1$ corresponds to the unit operator 1_H;*
- *the coordinate function $\pi_i:\underline{\lambda}\mapsto\lambda_i$ is mapped to a_i, for each $i = 1,\dots,n$.*

2. *In terms of the spectral projections $e^{(a_i)}_{\lambda_i}$ of the operators a_i, we have*

$$C^*(\underline{a}) = C^*(e^{(a_i)}_{\lambda_i}, i = 1,\dots,n, \lambda_i\in\sigma(a_i)). \tag{A.63}$$

3. If for each $\underline{\lambda} \in \sigma(\underline{a})$ we define the operator

$$e_{\underline{\lambda}} = e^{(a_1)}_{\lambda_1} \cdots e^{(a_n)}_{\lambda_n}, \tag{A.64}$$

then $e_{\underline{\lambda}}$ is a projection, in terms of which the joint spectrum may be rewritten as

$$\sigma(\underline{a}) = \{\underline{\lambda} \in \sigma(a_1) \times \cdots \times \sigma(a_n) \mid e_{\underline{\lambda}} \neq 0\}. \tag{A.65}$$

4. Finally, we have

$$C^*(\underline{a}) = C^*(e_{\underline{\lambda}}, \underline{\lambda} \in \sigma(\underline{a})) = \text{span}(e_{\underline{\lambda}}, \underline{\lambda} \in \sigma(\underline{a})). \tag{A.66}$$

We will not prove this in any detail, as the reasoning is quite analogous to the proof of Theorem A.15; for example, in (A.56) one just has to replace a by $\underline{a}$. The only nontrivial point is that since all a_i commute, so do all their spectral projections $e^{(i)}_{\lambda_i}$; this follows from (A.51), which makes these operators elements of the *commutative* C*-algebra $C^*(\underline{a})$ (which by definition contains each $C^*(a_i)$ and, in fact, is just the smallest C*-algebra in $B(H)$ with this property). Using (A.38) for each a_i and multiplying the n versions of the unit 1_H thus obtained with each other, yields

$$H = \bigoplus_{\underline{\lambda} \in \sigma(\underline{a})} H_{\underline{\lambda}}. \tag{A.67}$$

Since $p(\underline{a})v_{\underline{\lambda}} = p(\underline{\lambda})v_{\underline{\lambda}}$ for each joint eigenvector $v_{\underline{\lambda}} \subset H_{\underline{\lambda}}$, eq. (A.67) gives injectivity of the map (A.56) (*mutatis mutandis*) by the same argument as for $n = 1$.

This leads to a multi-spectral theorem for the commuting family $\underline{a}$, which is most conveniently stated in the following form. First, for any polynomial

$$p(x_1, \ldots, x_n) = \sum_{k_1, \ldots, k_n} x_1^{k_1} \cdots x_n^{k_n}, \tag{A.68}$$

in n real variables, we generalize (A.52) to

$$p(\underline{a}) = \sum_{k_1, \ldots, k_n} a_1^{k_1} \cdots a_n^{k_n}. \tag{A.69}$$

Theorem A.18. *Let $\underline{a} = (a_1, \ldots, a_n)$ be commuting self-adjoint operators on H. Then for any polynomial p in n real variables, with associated operator* (A.69),

$$p(\underline{a}) = \sum_{\underline{\lambda} \in \sigma(\underline{a})} p(\underline{\lambda}) \cdot e_{\underline{\lambda}}, \tag{A.70}$$

where the spectral projections $e_{\underline{\lambda}}$ are given by (A.64).

The special case $p(x_1, \ldots, x_n)$ then recovers (A.67). As for $n = 1$, eq. (A.70) may be generalized to arbitrary continuous functions $f(x_1, \ldots, x_n)$, either by replacing f by a polynomial that coincides with f on the joint spectrum $\sigma(\underline{a})$, or by approximating f by polynomials on some compact set K containing $\sigma(\underline{a})$.

Proposition A.19. *Let $\underline{a} = (a_1, \ldots, a_n)$ be a family of commuting self-adjoint operators on H. Then there is a self-adjoint operator $a \in B(H)$ such that $C^*(\underline{a}) = C^*(a)$.*

Proof. Take $a = \sum_{\underline{\lambda} \in \sigma(\underline{a})} c_{\underline{\lambda}} e_{\underline{\lambda}}$, with all $c_{\underline{\lambda}}$ different from each other. Then

$$C^*(a) = C^*(e_{\underline{\lambda}}, \underline{\lambda} \in \sigma(\underline{a})), \tag{A.71}$$

by (A.50), and hence the claim follows from (A.66). □

Corollary A.20. *Every (unital) commutative C*-algebra C in $B(H)$ is generated by a single self-adjoint operator a (and the unit 1_H), i.e., $C = C^*(a)$.*

Proof. Just take a basis (c_k) of C as a vector space and decompose $c_k = a_k + ia'_k$ with a_k and a'_k self-adjoint (namely, $a_k = \frac{1}{2}(c_k + c_k^*)$ and $a'_k = -\frac{1}{2}i(c_k - c_k^*)$). If C is to be commutative, each c_k must be normal, i.e., $c_k^* c_k = c_k c_k^*$, which is equivalent to commutativity of a_k and a'_k, and all c_k must commute, i.e., all a_k and a'_k must commute for different k. Hence $C = C^*(a_k, a'_k)$, which is of the form $C^*(\underline{a})$ for an appropriate family $\underline{a}$, and so by Proposition A.19 it takes the form $C^*(a)$. □

We say that a unital commutative C*-algebra $C \subset B(H)$ is ***maximal*** if it is not contained in some bigger unital commutative C*-algebra in $B(H)$. Also, we call a self-adjoint operator *a* ***maximal*** iff $\sigma(a)$ has cardinality $\dim(H)$, or, in other words, if each eigenvalue of a is nondegenerate. In finite dimension it is easy to classify maximal unital commutative C*-algebras in $B(H)$ *up to unitary equivalence*.

Here we say (as usual) that a linear map $u : H \to H'$ is ***unitary*** when it is invertible and satisfies $\langle u\varphi, u\psi \rangle' = \langle \varphi, \psi \rangle$ for each $\varphi, \psi \in H$ (note that the inverse u^{-1} is automatically linear). Two *-algebras $C \subset B(H)$ and $C' \subset B(H')$ are called ***unitarily equivalent***, then, if there is a unitary map $u : H \to H'$ such that $C' = uCu^{-1}$.

Theorem A.21. *A unital commutative C*-algebra $C \subset B(H)$ is maximal iff it is unitarily equivalent to the algebra $D_n(\mathbb{C})$ of all diagonal matrices on $H' = \mathbb{C}^n$.*

Proof. First, $D_n(\mathbb{C})$ is indeed maximal abelian in $M_n(\mathbb{C})$; any extension of $D_n(\mathbb{C})$ would have to contain some additional matrix $b \in M_n(\mathbb{C})$ that commutes with all $a \in D_n(\mathbb{C})$, but by elementary linear algebra this very property implies $b \in D_n(\mathbb{C})$.

By Corollary A.20, we have $C = C^*(a)$, where $a^* = a$. Then C is maximal iff a is maximal. For if not, some eigenvalue $\lambda' \in \sigma(a)$ would have multiplicity $m_{\lambda'} > 1$, and hence the corresponding spectral projection $e_{\lambda'}$ could be decomposed as $e_{\lambda'} = e_{\lambda'}^{(1)} + e_{\lambda'}^{(2)}$, where both terms are orthogonal and hence commute. We could then extend $C^*(a)$, as in (A.50), to $C^*(e_\lambda, e_{\lambda'}^{(1)}, e_{\lambda'}^{(2)}, \lambda \in \sigma(a), \lambda \neq \lambda')$, which remains commutative, and we have a contradiction with the alleged maximality of $C^*(a)$.

Thus a is maximal, in which case we list the spectrum as $\sigma(a) = \{\lambda_1, \ldots, \lambda_n\}$, with corresponding eigenvectors $\{\upsilon_{\lambda_1}, \ldots, \upsilon_{\lambda_n}\}$. This gives rise to a unitary map $u : H \to \mathbb{C}^n$ defined by $u\upsilon_{\lambda_i} = \upsilon_i$, where $(\upsilon_1, \ldots, \upsilon_n)$ is the standard basis of $\mathbb{C}^n$, and clearly $uau^{-1} = \mathrm{diag}(\lambda_1, \ldots, \lambda_n)$. If (as is the case) all entries $\lambda_i \in \mathbb{R}$ are different, any $(z_1, \ldots, z_n) \in \mathbb{C}^n$ may be written as $z_i = p(\lambda_i)$, $i = 1, \ldots, n$, where p is some complex polynomial $p(x) = \sum_i c_i x^n$, $x \in \mathbb{R}$, $c_i \in \mathbb{C}$. Hence $uC^*(a)u^{-1} = D_n(\mathbb{C})$. □

A.5 Positive operators and the trace

Operators $a : H \to H$ satisfying one (and hence all) of the conditions in the next proposition are called ***positive***, written $a \geq 0$ or $0 \leq a$. More generally, we write $a \leq b$ iff $b - a \geq 0$. Positive operators play a very important role in quantum mechanics.

Proposition A.22. *The following conditions on an operator a are equivalent:*

1. *$\langle \psi, a\psi \rangle \geq 0$ for arbitrary $\psi \in H$.*
2. *$a^* = a$ and $\sigma(a) \subset \mathbb{R}^+$.*
3. *$a = c^2$ for some self-adjoint operator c.*
4. *$a = b^*b$ for some operator b.*

Proof. $1 \to 2$: Putting $\langle \psi, a\psi \rangle \geq 0$ in (A.44) gives $\langle \psi, a\psi \rangle = \langle \psi, a^*\psi \rangle$ for all ψ. But for any operator b and vectors $\chi, \varphi \in H$, as in (A.10) we have the identity

$$\begin{aligned} 4\langle \chi, b\varphi \rangle &= \langle \chi + \varphi, b(\chi + \varphi) \rangle - \langle \chi - \varphi, b(\chi - \varphi) \rangle \\ &\quad + i\langle \chi - i\varphi, b(\chi - i\varphi) \rangle - i\langle \chi + i\varphi, b(\chi + i\varphi) \rangle. \end{aligned} \tag{A.72}$$

So $b = 0$ iff $\langle \psi, b\psi \rangle = 0$ for all $\psi \in H$, and hence condition 1 implies $a^* = a$. We therefore know that $\sigma(a) \subset \mathbb{R}$, and since an eigenvalue $\lambda < 0$ would contradict the first condition 1, the second condition follows.

$2 \to 3$: define $c = \sqrt{a}$, where (since $\lambda_i \geq 0$) the square root is (well) defined by

$$\sqrt{a} = \sum_{i=1}^{\dim(H)} \sqrt{\lambda_i} |\upsilon_i\rangle\langle\upsilon_i|. \tag{A.73}$$

$3 \to 4$ is trivial (take $b = c$), as is $4 \to 1$, since $\langle \psi, a\psi \rangle = \|b\psi\|^2$. □

Combining this with Proposition A.5, we obtain the following result.

Proposition A.23. *The relationship $\langle \varphi, \psi \rangle' = \langle \varphi, a\psi \rangle$ gives a bijective correspondence between (hermitian/positive) sesquilinear forms $\langle \cdot, \cdot \rangle'$ on H and (hermitian/positive) operators a on H.*

Proof. One direction is trivial. For the other, fix $\chi \in H$ and define a functional $f(\psi) = \langle \chi, \psi \rangle'$. By Proposition A.5, $f = f_\varphi$ for some unique $\varphi \in H$. Define an operator $b : H \to H$ by $b\chi = \psi$ and put $a = b^*$. □

Proposition A.24. *Any self-adjoint operator $a \in B(H)$ has a decomposition*

$$a = a_+ - a_-, \tag{A.74}$$

where $a_\pm \geq 0$. These are unique if they also satisfy $a_+a_- = a_-a_+ = 0$.

Proof. Using Theorem A.10, we may take

$$a_\pm = \pm \sum_{\lambda \in \sigma(a) \cap \mathbb{R}^\pm} \lambda \cdot e_\lambda. \tag{A.75}$$

Equivalently, we may use Theorem A.15 to rewrite (A.75) as

$$a_{\pm} = (|\mathrm{id}_{\sigma(a)}| \cdot 1_{\mathbb{R}^{\pm}})(a) \equiv f_{\pm}(a), \tag{A.76}$$

where $|\mathrm{id}_{\sigma(a)}|$ is the function $\lambda \mapsto |\lambda|$, $\mathbb{R}^+ = [0,\infty)$ and $\mathbb{R}^- = (-\infty,0)$. To prove uniqueness, we note that since $\sigma(a) \subset \mathbb{R}$ is finite, there is a polynomial p such that $f_+ = p$, and hence $a_+ = p(a)$. If $a = a'_+ - a'_-$ with $a'_\pm \geq 0$ and $a'_+ a'_- = a'_- a'_+ = 0$, then for any polynomial p we have $p(a) = p(a'_+) + p(-a'_-)$. For the one just taken, this gives $p(a) = a'_+$ by positivity of the $a'_\pm$, and hence $a'_+ = a_+$, etc. □

We now introduce a construction of great significance to quantum mechanics.

Lemma A.25. *If (υ_i) and (υ'_i) are bases of H, then for any operator $a : H \to H$,*

$$\sum_i \langle \upsilon_i, a\upsilon_i \rangle = \sum_i \langle \upsilon'_i, a\upsilon'_i \rangle.$$

Proof. A simple computational proof uses the identity (A.40) for any basis (υ_i) (i.e., the υ_i need not be eigenvectors of a, as in (A.39)). Then, as in physics books,

$$\sum_i \langle \upsilon'_i, a\upsilon'_i \rangle = \sum_{i,j,k} \langle \upsilon_k, \upsilon'_i \rangle \langle \upsilon'_i, \upsilon_j \rangle \langle \upsilon_j, a\upsilon_k \rangle = \sum_{j,k} \langle \upsilon_k, \upsilon_j \rangle \langle \upsilon_j, a\upsilon_k \rangle = \sum_i \langle \upsilon_i, a\upsilon_i \rangle. \square$$

This lemma allows us to define the ***trace*** of a by

$$\mathrm{Tr}\,(a) = \sum_i \langle \upsilon_i, a\upsilon_i \rangle, \tag{A.77}$$

where (υ_i) is any basis of H. By almost the same proof as Lemma A.25 we obtain

$$\mathrm{Tr}\,(ab) = \sum_{i,j} \langle \upsilon_i, a\upsilon_j \rangle \langle \upsilon_j, b\upsilon_i \rangle = \sum_{i,j} \langle \upsilon_i, b\upsilon_j \rangle \langle \upsilon_j, a\upsilon_i \rangle = \mathrm{Tr}\,(ba). \tag{A.78}$$

If u is ***unitary*** (in that $uu^* = u^*u = 1$,) then from either Lemma A.25 or eq. (A.78),

$$\mathrm{Tr}\,(uau^*) = \mathrm{Tr}\,(a). \tag{A.79}$$

Finally, if $a^* = a$, then (A.37) and taking the trace over the basis in (A.39) yields

$$\mathrm{Tr}\,(a) = \sum_{\lambda \in \sigma(a)} m_\lambda \cdot \lambda. \tag{A.80}$$

Definition A.26. *A* **density operator** *is a positive operator ρ on H such that*

$$\mathrm{Tr}\,(\rho) = 1. \tag{A.81}$$

The analysis of density operators hinges on the introduction of a second operator norm, beside the canonical one (A.18). In finite dimension these norms are equivalent, but in general they are not, and it makes sense to introduce both already here.

For any $a \in B(H)$, the operator a^*a is positive and hence self-adjoint, so that

$$a^*a = \sum_{\mu\in\sigma(a^*a)} \mu e_\mu = \sum_{i=1}^{n} \mu_i |\upsilon_i\rangle\langle\upsilon_i| \tag{A.82}$$

for certain eigenvalues $\mu_i \geq 0$ (including possible multiplicities) or $\mu \in \sigma(a^*a)$ (excluding multiplicities), all necessarily non-negative by positivity of a^*a, and some normalized eigenvectors υ_i or spectral projections e_μ; cf. (A.37) - (A.39). Then put

$$\|a\|_1 = \sum_{\mu\subset\sigma(a^*a)} \sqrt{\mu}m_\mu = \sum_{i=1}^{n} \sqrt{\mu_i}. \tag{A.83}$$

It is not immediately clear that $\|\cdot\|_1$ *is* a norm on $B(H)$, but we will shortly prove that it is; we provisionally refer to $B(H)$, equipped with the norm (A.83), as $B_1(H)$.

Another way to defined this ***trace-norm*** is to first introduce the ***absolute value***

$$|a| = \sqrt{a^*a} \tag{A.84}$$

of any operator $a \in B(H)$, where the square root is simply defined as

$$\sqrt{a^*a} = \sum_{\mu\in\sigma(a^*a)} \sqrt{\mu}e_\mu = \sum_{i=1}^{n} \sqrt{\mu_i}|\upsilon_i\rangle\langle\upsilon_i|, \tag{A.85}$$

which coincides with $f(a^*a)$ for $f(x) = \sqrt{x}$ as defined in Theorem A.15, see (A.57). If a is positive, then $|a| = a$. Some other useful properties of the absolute value are

$$\ker |a| = \ker a = (\mathrm{ran}\, |a|)^\perp; \tag{A.86}$$

$$\| |a|\psi\| = \|a\psi\|, \ \psi \in H. \tag{A.87}$$

For the first equality in (A.86),

$$a\psi = 0 \Rightarrow a^*a\psi = 0 \Leftrightarrow \sqrt{a^*a}\psi = 0 \Leftrightarrow |a|\psi = 0,$$

but also $a^*a\psi = 0 \Rightarrow \langle\psi, a^*a\psi\rangle = 0 \Leftrightarrow \|a\psi\|^2 = 0 \Leftrightarrow a\psi = 0$. For the second,

$$\ker a = (\mathrm{ran}\, a^*)^\perp, \tag{A.88}$$

which in turn is immediate from the definition of the adjoint. Eq. (A.87) is similar.

Though once again lacking transparency as a norm, by construction we now have

$$\|a\|_1 = \mathrm{Tr}\,(|a|), \tag{A.89}$$

so if (λ_i) are the (positive) eigenvalues of $|a|$, including multiplicities, then

$$\|a\|_1 = \sum_{i=1}^{n} \lambda_1. \tag{A.90}$$

To obtain suitable estimates for the trace norm we need some further techniques.

Definition A.27. *Let H be a finite-dimensional Hilbert space.*

1. *A* **partial isometry** *is an operator $u \in B(H)$ for which $u^*u = e$ is a projection.*
2. *A* **unitary** *is an invertible partial isometry.*

For immediate and later reference, we collect some properties of such operators.

Lemma A.28. *Let H be a Hilbert space with a partial isometry $u \in B(H)$.*

- *Also u^* is a partial isometry, or, equivalently, $uu^* = f$ is a projection.*
- *The kernel of u is $(pH)^\perp$, and its range is fH.*
- *The given partial isometry u is unitary from eH to fH.*
- *Conversely, an operator v on H for which there is a (closed) subspace $L \subset H$ on which v is isometric, whilst it is identically zero on $L^\perp$, is a partial isometry.*
- *If $u \neq 0$, then $\|u\| = 1$.*
- *An partial isometry u is unitary iff $u^*u = uu^* = 1_H$ (i.e., $e = f = 1_H$).*

The proof is an easy verification. In the infinite-dimensional case, a distinction arises between isometries (i.e, *injective* partial isometries, so that $u^*u = 1_H$) and unitaries, but if $\dim(H) < \infty$, injectivity implies subjectivity and hence bijectivity.

We now come to von Neumann's highly convenient ***polar decomposition*** of an operator, which mimics the polar decomposition $z = r\exp(i\varphi)$ of $z \in \mathbb{C}$.

Proposition A.29. *For $a \in B(H)$, assumed nonzero, the operator u given by*

$$u|a|\psi = a\psi, \quad (|a|\psi \in \mathrm{ran}\,|a|); \tag{A.91}$$
$$u\psi = 0, \quad (\psi \in (\mathrm{ran}\,|a|)^\perp = \ker|a|): \tag{A.92}$$

1. *Is well defined;*
2. *Is a partial isometry (and hence has norm $\|u\| = 1$);*
3. *Is unitary from $\mathrm{ran}\,|a|$ to $\mathrm{ran}\,a$ (if $\dim(H) = \infty$, take closures $(\mathrm{ran}\,|a|)^-$, $(\mathrm{ran}\,a)^-$);*
4. *Satisfies*

$$\||a|\psi\| = \|a\psi\|; \tag{A.93}$$
$$u^*u|a| = |a| = |a|u^*u. \tag{A.94}$$

Given that u is a partial isometry, it is characterized by the two properties:

$$\ker u = \ker a; \tag{A.95}$$
$$a = u|a|. \tag{A.96}$$

Furthermore, if $a \neq 0$, then a is invertible iff u is unitary.

Proof. This follows from (A.86) - (A.87), except the claim that (A.95) - (A.96) uniquely define u, which we will not use and whose proof we therefore omit. □

Recall from the easy Theorem 2.7 that there is a bijective correspondence between linear maps $\omega : B(H) \to \mathbb{C}$ and operators $\rho \in B_1(H)$, given by (2.33), i.e.,

$$\omega(a) = \mathrm{Tr}\,(\rho a). \tag{A.97}$$

Proposition A.30. *If H is finite-dimensional, the map $\omega \mapsto \rho$ from $B(H)^*$ to $B_1(H)$, defined by* (A.97) *gives an isometric isomorphism of Banach spaces*

$$B(H)^* \cong B_1(H); \tag{A.98}$$

in particular, one has

$$\|\omega\| = \|\rho\|_1. \tag{A.99}$$

Proof. Bijectivity being known already, the basic estimate towards (A.99) is

$$|\mathrm{Tr}\,(\rho a)| \leq \|\rho\|_1 \|a\|. \tag{A.100}$$

This follows from the polar decomposition $\rho = u|\rho|$ and the spectral decomposition

$$|\rho| = \sum_{i=1}^{m \leq n} p_i |\upsilon_i\rangle\langle\upsilon_i|, \tag{A.101}$$

where $p_i > 0$ (but not necessarily $\sum_i p_i = 1$). Assuming $\rho \neq 0$, using (A.101), (A.78), Cauchy–Schwarz, (A.20), (A.21), $\|u\| = \|\upsilon_i\| = 1$, and (A.90), we indeed have

$$|\mathrm{Tr}\,(\rho a)| = |\mathrm{Tr}\,(u|\rho|a)| = |\mathrm{Tr}\,(|\rho|au)| = |\sum_i p_i \langle\upsilon_i, au\upsilon_i\rangle| \tag{A.102}$$

$$\leq \sum_i p_i |\langle\upsilon_i, au\upsilon_i\rangle| \leq \sum_i p_i \|a\|\|u\|\|\upsilon_i\| = \|\rho\|_1 \|a\|. \tag{A.103}$$

To prove saturation of this bound, take $a = u^*$, which is isometric on the space $\mathrm{ran}|\rho| = \mathrm{span}(\upsilon_1, \ldots, \upsilon_m)$ and hence satisfies $\|a\| = 1$ as well as $\langle\upsilon_i, au\upsilon_i\rangle$=1. Consequently, from (A.102) we find $|\mathrm{Tr}\,(\rho a)| = \sum_i p_i$. By (A.90) for ρ instead of a, i.e., $\|\rho\|_1 = \mathrm{Tr}\,(|\rho|) = \sum_i p_i$, we obtain $|\mathrm{Tr}\,(\rho a)| = \|\rho\|_1$, which yields (A.99). □

Corollary A.31. *The trace-norm $\|\cdot\|_1$ is (indeed) a norm on $B_1(H)$.*

As explained in more detail in §B.9, for any vector space V with norm, with double dual V^{**}, we have a canonical map $V \to V^{**}$ given by $v \mapsto \hat{v}$, where

$$\hat{v}(\theta) = \theta(v), \tag{A.104}$$

where $v \in V$, $\hat{v} \in V^{**}$, and $\theta \in V^*$. By the general theory, this map is always isometric (and hence injective), and if V is finite-dimensional, it is also surjective and hence an isomorphism. Therefore, taking $V = B(H)$, we infer from (A.98) that

$$B_1(H)^* \cong B(H), \tag{A.105}$$

where $a \in B(H)$ corresponds to $\hat{a} \in B_1(H)^*$ by means of

$$\hat{a}(\rho) = \mathrm{Tr}\,(\rho a). \tag{A.106}$$

This new role of $B(H)$ as the dual of $B_1(H)$ also equips it with a new topology (besides the norm topology it already has), viz. the accompanying w^*-topology.

This topology is defined by saying that $a_n \to a$ iff $\hat{a}_n(\rho) \to \hat{a}(\rho)$ for each $\rho \in B_1(H)$. For historical reasons this is called the ***σ-weak*** topology on $B(H)$, so we say that $a_n \to a$ ***σ-weakly*** in $B(H)$ iff $\mathrm{Tr}\,(\rho a_n) \to \mathrm{Tr}\,(\rho a)$ for each $\rho \in B_1(H)$.

To close, it is interesting to ut the trace-norm into a classical perspective. As explained in Chapter 1, at least on finite-dimensional Hilbert spaces, density operators are the quantum counterparts of probability measures (or distributions). If X is a *finite set*, the associated function space $C(X)$ carries the ***supremum-norm***

$$\|f\|_\infty = \sup\{|f(x)|, x \in X\}, \tag{A.107}$$

cf. (1.24). We equip the space $C(X)^*$ of all linear maps $\omega : C(X) \to \mathbb{C}$ with the norm

$$\|\omega\| = \sup\{|\omega(f)|, f \in C(X), \|f\|_\infty = 1\}. \tag{A.108}$$

Let $L^1(X)$ be the vector space of all functions $\rho : X \to \mathbb{C}$, equipped with the norm

$$\|\rho\|_1 = \sum_{x \in X} |\rho(x)|. \tag{A.109}$$

As in the quantum case just discussed, even for finite X it is not immediate that this expression indeed defines a norm; this follows from the next proposition.

Each $\rho \in L^1(X)$ defines a linear map $\omega : C(X) \to \mathbb{C}$ by

$$\omega(f) = \sum_{x \in X} \rho(x) f(x). \tag{A.110}$$

Conversely, each $\omega \in C(X)^*$ defines an element $\rho \in L^1(X)$ by

$$\rho(x) = \omega(\delta_x), \tag{A.111}$$

with $\delta_x \in C(X)$ defined by $\delta_x(y) = \delta_{xy}$ as usual.

Proposition A.32. *If X is finite, the map $\omega \mapsto \rho$ from $C(X)^*$ to $L^1(X)$, defined by* (A.111), *has inverse* (A.110) *and gives an isometric isomorphism*

$$C(X)^* \cong L^1(X) \tag{A.112}$$

of Banach spaces; in particular, one has

$$\|\omega\| = \|\rho\|_1. \tag{A.113}$$

Proof. The vector space isomorphism in question can be checked effortlessly. To verify (A.113), note that trivially $|\omega(f)| \leq \|\rho\|_1 \|f\|_\infty$, whence $\|\omega\| \leq \|\rho\|_1$. To show saturation of this bound, given $\rho \in L^1(X)$ take $f(x) = |\rho(x)|/\rho(x)$ if $\rho(x) \neq 0$ and $f(x) = 0$ elsewhere; if $\rho \neq 0$ this gives $\|f\|_\infty = 1$ and $|\omega(f)| = \|\rho\|_1$. □

Notes

The material in this appendix has been collected from numerous functional analysis books (some of which are mentioned in the Notes to the next appendix), adapted to the finite-dimensional case. Though not used in preparing this text, Halmos (1958, 1970) are classics. Theorem A.3 is due to Jordan & von Neumann (1935); Amir (1986) contains many other characterizations of inner product spaces.

Appendix B
Basic functional analysis

This appendix contains all technical information on general Hilbert spaces (as opposed to the finite-dimensional ones of the previous appendix) and, more generally, infinite-dimensional Banach spaces, that is either directly needed in the main text, or forms necessary preparation for the next appendix on operator algebras (which in turn play a central role in this book). Since most interesting examples of both Hilbert spaces and more general Banach spaces require some measure theory, which at the same time provides the mathematical foundation of probability theory, we include a brief introductory overview to this area as well (restricted, though, to the case we need, viz. measures and integrals on locally compact spaces).

Functional analysis has its roots in both mathematics and physics. In particular, the general area of ***spectral theory***, which emerged during the period 1900-1930 in the hands of Hilbert and his school, largely owes its existence to mathematical physics, as well as to Hilbert's genius in finding the right combination of examples and abstract theory (including his innovative definition of the spectrum). Hilbert's school culminated in the books *Methoden der mathematischen Physik* by Courant and Hilbert (1924), *Gruppentheorie und Quantenmechanik* by Weyl (1928), and *Mathematische Grundlagen der Quantenmechanik* by von Neumann (1932), all of whom were at Göttingen at the time (as were such giants in the history of quantum mechanics like Born, Heisenberg, and Jordan). Whereas Courant & Hilbert at least *thought* they described classical physics (although it soon turned out that their discussion of eigenvalue problems paved the way for the Schrödinger equation discovered two years later), von Neumann explicitly developed the Hilbert space formalism in order to describe quantum physics (for example, the modern abstract definition of a Hilbert space was his), as did Weyl (in connection with group theory).

What seems to have come from pure mathematics, though, is the idea, central to functional analysis, of looking at functions as points in some (infinite-dimensional) vector space. This emerged from the French school of Hadamard and his student Fréchet, requiring considerable interaction between the (then) new fields of linear algebra and topology. Eventually, this also led to the fundamental work of Banach.

We hope that the combination of logical setup, examples, theorems, and proofs in this appendix helps convince the reader of the sober elegance of functional analysis.

K. Landsman, *Foundations of Quantum Theory*,
Fundamental Theories of Physics 188, DOI 10.1007/978-3-319-51777-3

B.1 Completeness

A notable difference between finite-dimensional vector spaces with norm and infinite-dimensional ones is that the former are always *complete* in a sense to be defined now, whereas the latter may or may not be. This distinction has major consequences, especially where idealizations (and hence limits) are concerned.

As before, all vector spaces are defined over $\mathbb{C}$ (unless stated otherwise).

Definition B.1. *Let V be a vector space (or, more generally, a set).*
A **metric** *on V is a function $d : V \times V \to \mathbb{R}^+$ satisfying, for all $f,g,h \in V$:*

1. *$d(f,g) \leq d(f,h) + d(h,g)$* (**triangle inequality**)*;*
2. *$d(f,g) = d(g,f)$ for all $f,g \in V$* (**symmetry**)*;*
3. *$d(f,g) = 0$ iff $f = g$* (**positive definiteness**)*.*

Our main example is a vector space V with norm $\|\cdot\|$, which, as an easy exercise shows, gives rise to a metric on V via

$$d(f,g) = \|f - g\|. \tag{B.1}$$

In particular, an inner product on V induces a metric on V through (A.2) and (B.1).

The reader should have some experience with metric spaces from an undergraduate Analysis course, but for convenience we repeat the definition of completeness.

Definition B.2. *1. Let $(v_n) = \{v_n\}_{n\in\mathbb{N}}$ be a sequence in a metric space (V,d). We say that $v_n \to v$ for some $v \in V$ when $\lim_{n\to\infty} d(v_n,v) = 0$, or, more precisely: for any $\varepsilon > 0$ there is $N \in \mathbb{N}$ such that $d(v_n,v) < \varepsilon$ for all $n > N$. In a normed space, this means that $v_n \to v$ iff $\lim_{n\to\infty} \|v_n - v\| = 0$.*

2. *A sequence (v_n) in (V,d) is called a* **Cauchy sequence** *when $d(v_n,v_m) \to 0$ when $n,m \to \infty$, or, more precisely: for any $\varepsilon > 0$ there is $N \in \mathbb{N}$ such that $d(v_n,v_m) < \varepsilon$ for all $n,m > N$. In a normed space, this means that (v_n) is Cauchy when $\|v_n - v_m\| \to 0$ for $n,m \to \infty$, in other words, when $\lim_{n,m\to\infty} \|v_n - v_m\| = 0$.*
3. *A metric space (V,d) is called* **complete** *when every Cauchy sequence in V converges (i.e., to an element of V).*

A convergent sequence is Cauchy: from the triangle inequality and symmetry one has $d(v_n,v_m) \leq d(v_n,v) + d(v_m,v)$, so for given $\varepsilon > 0$ there is $N \in \mathbb{N}$ such that $d(v_n,v) < \varepsilon/2$, et cetera. However, the converse statement does not hold in general: for example, take the vector space $\ell_c(\mathbb{N})$ of all functions $f : \mathbb{N} \to \mathbb{C}$ that are zero expect at finitely many places (with the obvious pointwise operations), or, equivalently, the vector space $\mathbb{C}^\infty$ of all sequences (x_n) with finitely many nonzero entries. This vector space is incomplete in any conceivable norm, like the sup-norm

$$\|f\|_\infty = \sup\{|f(x)|, x \in \mathbb{N}\}. \tag{B.2}$$

Indeed, the sequence (f_n), where $f_n(x) = 1/x$ for $x = 1,\ldots,n$ and $f(x) = 0$ for $x > n$, which corresponds to the sequence $(1,1/2,1/3,\ldots,1/n,0,0,\ldots)$ in $\mathbb{C}^\infty$ is Cauchy, but its obvious limit $f(x) = 1/x$ for *each* $x \in \mathbb{N}$, or $x_n = 1/n$, does not lie in $\ell_c(\mathbb{N})$.

Definition B.3. • *A* **Banach space** *is a vector space with norm that is complete in the associated metric* (B.1).

- *A* **Hilbert space** *is vector space with inner product that is complete in the associated metric* (B.1)*, in which the norm is defined by* (A.2)*. Equivalently, a Hilbert space is a Banach space whose norm comes from an inner product via* (A.2).

As we have seen, $\ell_c(\mathbb{N})$ fails to be a Banach space in the sup-norm, but (its completion) $\ell^\infty(\mathbb{N})$, which consists of all bounded functions $f : \mathbb{N} \to \mathbb{C}$, is (see §B.2).

Definition B.4. *Two norms* $\|\cdot\|$ *and* $\|\cdot\|'$ *on the same vector space* V *are* **equivalent** *if there are constants* $M > 0$ *and* $m > 0$ *such that for any* $v \in V$,

$$m\|v\|' \leq \|v\| \leq M\|v\|'. \tag{B.3}$$

In that case, the two metric topologies on X defined by these norms coincide, so that in particular completeness and convergence in $\|\cdot\|$ and $\|\cdot\|'$ are the same.

Proposition B.5. *Let* V *be a* finite-dimensional *vector space. All norms on* V *are equivalent, and hence* V *is complete in any norm.*

Proof. We derive this from a basic fact of Analysis, namely that $\mathbb{C}^n$ is complete in the (Euclidean) norm $\|\cdot\|_2$ derived from the standard inner product (A.11), that is,

$$\|z\|_2^2 = \sum_{i=1}^{n} |z_i|^2. \tag{B.4}$$

So the first step is to transfer the problem from V to $\mathbb{C}^n$, where $n = \dim(V)$, by choosing a basis (υ_i) of V, and mapping υ_i to the standard basis vector u_i of $\mathbb{C}^n$. Linear extension then maps $v = \sum_i z_i \upsilon_i \in V$ to $z = (z_1, \ldots, z_n) \in \mathbb{C}^n$, which gives an isomorphism $V \to \mathbb{C}^n$. This maps endows $\mathbb{C}^n$ with a new norm $\|z\| = \|v\|$ (i.e. the given norm on V), which we now prove to be equivalent to $\|\cdot\|_2 \equiv \|\cdot\|'$. The second inequality in (B.3) easily follows from Cauchy–Schwarz, viz.

$$\|z\| = \|\sum_i z_i u_i\| \leq \sum_i |z_i|\|u_i\| \leq \sqrt{\sum_i \|u_i\|^2}\sqrt{\sum_j |z_i|^2} \equiv M\|z\|_2.$$

This inequality, together with the elementary but extremely useful estimate

$$|\|v\| - \|w\|| \leq \|v - w\|, \tag{B.5}$$

which is valid for any norm in any dimension, implies that the function $\|\cdot\| : \mathbb{C}^n \to \mathbb{R}$ is continuous with respect to the Euclidean metric on $\mathbb{C}^n$. Now the unit ball $\mathbb{C}^n_1 = \{x \in \mathbb{C}^n \mid \|x\|_2 = 1\}$ in $\mathbb{C}^n$ is compact, so according to Weierstrass, the norm $\|\cdot\|$ assumes a minimum on $\mathbb{C}^n_1$. Hence there exists $\mu \in \mathbb{C}^n_1$ such that $\|\mu\| \leq \|z\|$ for all $z \in \mathbb{C}^n_1$. For arbitrary nonzero $z \in \mathbb{C}^n$, the rescaled vector $z' = z/\|z\|_2$ lies in $\mathbb{C}^n_1$, so $\|\mu\| \leq \|z'\|$, which is nothing but the first inequality in (B.3) with $m = \|\mu\|$. □

B.2 ℓ^p spaces

The simplest examples of infinite-dimensional Banach spaces are the ℓ^p-spaces, where $1 \leq p \leq \infty$ (for $p < 1$ the Minkowski inequality (B.14) below goes in the wrong direction, so that, by failure of the triangle inequality, eq. (B.8) below fails to define a norm). Such spaces are defined on some set X, hence we write $\ell^p(X)$.

If $X = \{x_1, \ldots, x_n\}$ is finite, with cardinality $n = |X|$, then $\ell^p(X)$ consist of all function $f : X \to \mathbb{C}$ with pointwise operations, so that $\ell^p(X) \cong \mathbb{C}^n$ as vector spaces through the map $f \mapsto (f(x_1), \ldots, f(x_n))$, where $\mathbb{C}^n$ is equipped with a specific (and, for $p \neq 2$, unusual) norm. However, by Proposition (B.5) we may as well take $p = 2$ and nothing has been gained compared with the linear algebra of Appendix A.

Therefore, life starts with infinite sets X, and we begin with the simplest of those, viz. $X = \mathbb{N}$ (but to avoid unnecessary duplication with regard to later generalization, although for the moment we assume $X = \mathbb{N}$, we still write X for the underlying set). We define $\ell^p \equiv \ell^p(X)$ as the set of functions $f : X \to \mathbb{C}$ that satisfy

$$\sum_{x \in X} |f(x)|^p < \infty \; (1 \leq p < \infty); \tag{B.6}$$

$$\sup_{x \in X} |f(x)| < \infty \; (p = \infty). \tag{B.7}$$

As will be shown in far greater generality (cf. Theorem B.9), the point is that for any $1 \leq p \leq \infty$, the set $\ell^p(X)$ thus defined is not merely a vector space (under pointwise operations); it is even a Banach space in the norm

$$\|f\|_p = \left(\sum_{x \in X} |f(x)|^p \right)^{1/p} \quad (1 \leq p < \infty); \tag{B.8}$$

$$\|f\|_\infty = \sup\{|f(x)|, x \in X\} = \inf\{C > 0 \mid |f(x)| \leq C \, \forall x \in X\}. \tag{B.9}$$

The case $p = 2$ is unique in that $\ell^2(X)$ is also a Hilbert space in the inner product

$$\langle f, g \rangle = \sum_{x \in X} \overline{f(x)} g(x). \tag{B.10}$$

As we now outline, these expressions may be generalized to any set, to which end we should define the meaning of (possibly uncountable) sums $\sum_{x \in X}$. Although the generality below will only be used in §B.12, it is convenient (at little extra cost) to cover more general codomains for f than just the complex numbers $\mathbb{C}$.

Definition B.6. *Let X be a set, V a normed vector space, $f : X \to V$ some function, and $v \in V$. The sentence $\sum_{x \in X} f(x) = v$ means that for each $\varepsilon > 0$ there is a finite subset $F \subset X$ such that for each finite subset $G \subset X$ with $F \subseteq G$, we have*

$$\|\sum_{x \in G} f(x) - v\| < \varepsilon.$$

In terms of nets, this means that the net $s = (s_F)_{F\in\mathscr{P}_f(X)}$ in V indexed by finite subsets $F \subset X$ (ordered by inclusion), where $s_F(x) = \sum_{x\in F} f(x)$, converges to v.

For $X = \mathbb{N}$ and $V = \mathbb{C}$ we may take F to be $\{1,\ldots,N\}$ and G to be $\{1,\ldots,n\}$, where $n \geq N$, in which case we recover the usual notion of convergence of sums (i.e. $\forall\varepsilon > 0 \exists N \in \mathbb{N} \forall n \geq N : |\sum_{x=1}^{n} f(x) - v| < \varepsilon$). However, since also more general F and G are allowed, Definition B.6 is in fact equivalent to absolute convergence:

Lemma B.7. *Let X be a set and let $f : X \to \mathbb{C}$ be some function.*

1. *There exists $z \in \mathbb{C}$ such that $\sum_{x\in X} f(x) = z$ iff $\sum_{x\in X} |f(x)| < \infty$.*
2. *If $f(x) \geq 0$ for each $x \in X$, then, in the sense of Definition B.6,*

$$\sum_{x\in X} f(x) = \sup\left\{\sum_{x\in F} f(x), F \subset X \text{ finite}\right\}, \tag{B.11}$$

which is true even if the supremum on the right-hand side is infinite (in which case the left-hand side simply does not converge).

Therefore, for $f : X \to \mathbb{C}$, one may use (B.11) to check if $\sum_{x\in X} |f(x)| < \infty$, in which case it makes sense to try and find the value v of $\sum_{x\in X} f(x)$ as in Definition B.6.

Proof. 1. We write $f = f_1 + if_2$, with $f_i : X \to \mathbb{R}$, and for given $G \subset X$, write $G_{i\pm} = \{x \in G \mid \pm f_i(x) > 0\}$ (the ambiguity at those x where $f(x) = 0$ is irrelevant). Then

$$\begin{aligned} |\sum_{x\in G} f(x)| &\leq \sum_{x\in G} |f(x)| \leq \sum_{x\in G} |f_1(x)| + \sum_{x\in G} |f_2(x)| \\ &= \sum_{x\in G_{1+}} f_1(x) - \sum_{x\in G_{1-}} f_1(x) + \sum_{x\in G_{2+}} f_2(x) - \sum_{x\in G_{2-}} f_2(x) \\ &\leq 4\sup\left\{|\sum_{x\in G_\alpha} f(x)|, \alpha \in \{1+,1-,2+,2-\}\right\}. \end{aligned} \tag{B.12}$$

Using Proposition B.8 below, the first inequality in (B.12) shows that absolute convergence implies convergence in the sense of Cauchy, whereas the last inequality (i.e., $\sum_{x\in G} |f(x)| \leq 4\sup\cdots$) shows the converse.

2. We pick $\varepsilon > 0$ and abbreviate the right-hand side of (B.11) as σ. By definition of the supremum (which we assume finite) there is a finite $F \subset X$ for which $\sigma \geq \sum_{x\in F} f(x) \geq \sigma - \varepsilon$. Since the terms are positive, for any finite $G \supseteq F$ we have $\sum_{x\in G} f(x) \geq \sum_{x\in F} f(x)$ and hence also $\sigma \geq \sum_{x\in G} f(x) \geq \sigma - \varepsilon$, from which $|\sum_{x\in G} f(x) - \sigma| < \varepsilon$. Hence $\sum_{x\in X} f(x) = \sigma$ by Definition B.6.
The same argument works if $\sigma = \infty$, in which case for any $0 < M < \infty$ there is a finite $F \subset X$ for which $\sum_{x\in F} f(x) > M$, and hence certainly $\sum_{x\in G} f(x) > M$. □

Leaving its proof to the reader, we state the Cauchy condition for convergence:

Proposition B.8. *We have $\sum_{x\in X} f(x) = v$ for some (necessarily unique) $v \in V$, in the sense of Definition B.6, iff for each $\varepsilon > 0$ there is a finite subset $F \subset X$ such that for each finite subset $G' \subset X\backslash F$ we have $\|\sum_{x\in G'} f(x)\| < \varepsilon$.*

For uncountable set X, Definition B.6 is not as bad as it may sound, since whenever $\sum_{x\in X}|f(x)|<\infty$, only a *countable* number of terms can be nonzero (proof by contradiction: if not, there must be an $n\in\mathbb{N}$ for which infinitely many x satisfy $|f(x)|>1/n$ (nested proof by contradiction: if not, then for all n, only finitely many x satisfy $|f(x)|>1/n$, and hence, a countable union of finite sets remaining countable, only a countable number of x can have $f(x)\neq 0$), so the sum of $|f(x)|$ over those x alone already diverges). In particular, for $X=\mathbb{N}$ the sum in (B.6) has its usual meaning. However, even for $X=\mathbb{N}$, the sums just defined *only* have their usual meaning if the series in question is absolutely convergent (the standard counterexample of a real series $\sum_n x_n$ that is convergent but not absolutely convergent is given by $x_n=(-1)^n/n$; in the above light, taking $G=F\cup E$, where E is a large but finite set of even numbers, then makes $|\sum_{i\in G}x_n-x|$ as big as you do not like).

Using the triangle inequality for the norm and the Cauchy criterion for convergence, it is easy to show that if V is a Banach space and $\sum_{x\in X}\|f(x)\|<\infty$, then the sum $\sum_{x\in X}f(x)$ exists in V (i.e., it equals some $v\in V$ in the sense of Definition B.6). The implication is one-sided, though: the latter sum may exist even if the former does not. For example, take $V=\ell^2(\mathbb{N})$, pick some $\tilde{f}\in\ell^2(\mathbb{N})$, and define $f:\mathbb{N}\to\ell^2(\mathbb{N})$ by $f(x)=\tilde{f}(x)\delta_x$, where $\delta_x(y)=\delta_{xy}$ (and hence $\|\delta_x\|_2=1$). Then

$$\sum_{x\in\mathbb{N}}\|f(x)\|_2=\sum_{x\in\mathbb{N}}|\tilde{f}(x)|=\|\tilde{f}\|_1.$$

Now $\sum_{x\in\mathbb{N}}f(x)=\tilde{f}$ exists *per* assumption that $\tilde{f}\in\ell^2(\mathbb{N})$ and hence $\|\tilde{f}\|_2<\infty$, which is implied by, but is not equivalent to $\|\tilde{f}\|_1<\infty$. See also §B.12 below.

In any case, the meaning of the possibly uncountable sums in (B.6) and (B.8) should be clear now, as only finite sums (B.11) are involved; for (B.10), by Hölder's inequality (B.15) below for $p=q=2$, the sum in question is absolutely convergent, and hence it falls within the scope of Definition B.6 and Lemma B.7.

Theorem B.9. *For any $1\leq p\leq\infty$, the set $\ell^p(X)$ is a vector space under pointwise operations. Moreover, $\ell^p(X)$ is a Banach space in the norm* (B.8) - (B.9).

Proof. 1. *ℓ^p is a vector space.* The case $p=\infty$ is obvious. For $1\leq p<\infty$, use the convexity of the function $t\mapsto t^p$ for $t\in[0,\infty)$. For convex functions one has $f(\frac{1}{2}(t_1+t_2))\leq\frac{1}{2}(f(t_1)+f(t_2))$, so that $(\frac{1}{2}(t_1+t_2))^p\leq\frac{1}{2}(t_1^p+t_2^p)$. Combined with monotonicity of the function $t\mapsto t^p$ on $[0,\infty)$, i.e. $s\leq t\Rightarrow s^p\leq t^p$, this gives

$$|f(x)+g(x)|^p\leq(|f(x)|+|g(x)|)^p\leq 2^{p-1}(|f(x)|^p+|g(x)|^p), \tag{B.13}$$

so that summing over x gives $\|f+g\|_p^p\leq 2^{p-1}(\|f\|_p^p+\|g\|_p^p)<\infty$.
Hence if $f\in\ell^p$ and $g\in\ell^p$, then $f+g\in\ell^p$.

2. *$\|\cdot\|_p$ is a norm on ℓ^p.* The case $p=\infty$ is, once again, obvious. For $1\leq p<\infty$, the only nontrivial part is the triangle inequality

$$\|f+g\|_p\leq\|f\|_p+\|g\|_p, \tag{B.14}$$

called the ***Minkowski inequality***. This follows from ***Hölder's inequality***:

$$\|fg\|_1 \leq \|f\|_p \|g\|_q, \tag{B.15}$$

which is valid for $f \in \ell^p$ and $g \in \ell^q$, where $1 \leq p \leq \infty$ and $1 \leq q \leq \infty$ satisfy

$$\frac{1}{p} + \frac{1}{q} = 1. \tag{B.16}$$

Thus one has $q = p/(p-1)$ for $1 < p < \infty$, or $q = \infty$ for $p = 1$, or $q = 1$ for $p = \infty$. One calls p and q ***conjugate exponents*** (so that $p = 2$ is self-conjugate).

3. *ℓ^p is complete in the norm $\|\cdot\|_p$.* We must prove that some Cauchy sequence (f_k) in ℓ^p converges. This takes three steps, which we first prove for $1 \leq p < \infty$.

   a. *Find a candidate f for the limit.* Since (f_k) is Cauchy, for each $\varepsilon > 0$ there exists $K \in \mathbb{N}$ such that $\|f_k - f_l\|_p < \varepsilon$ for all $k, l > K$, or

   $$\|f_k - f_l\|_p^p = \sum_{x \in X} |f_k(x) - f_l(x)|^p < \varepsilon^p. \tag{B.17}$$

   Hence $|f_k(x) - f_l(x)|^p < \varepsilon^p$ for all x, so $(f_k(x))_k$ is a Cauchy sequence in $\mathbb{C}$. Since $\mathbb{C}$ is complete, $(f_k(x))_k$ converges, hence we may define $f : X \to \mathbb{C}$ by

   $$f(x) = \lim_{k \to \infty} f_k(x). \tag{B.18}$$

   b. *Show that $f \in \ell^p$.* Note that

   $$\|g\|_p^p = \sup_{F \subset X} \sum_{x \in F} |g(x)|^p, \tag{B.19}$$

   where the supremum is over all finite subsets $F \subset X$. For fixed F we have

   $$\sum_{x \in F} |f_k(x) - f_l(x)|^p < \varepsilon^p.$$

   Since the sum is finite, we may take $\lim_{k \to \infty}$, giving $\sum_{x \in F} |f(x) - f_l(x)|^p < \varepsilon^p$. By (B.19), the sup over all finite F yields: $\forall \varepsilon > 0 \exists K \in \mathbb{N}$ such that $\forall l > K$, we have $\|f - f_l\|_p^p < \varepsilon^p$. For fixed ε and l, this says that $f - f_l \in \ell^p$, so $f \in \ell^p$, because $f = (f - f_l) + f_l$ with $f_l \in \ell^p$, and we know that ℓ^p is a vector space.

   c. *Show that $f_k \to f$ in ℓ^p.* This is contained in the previous step, since we had

   $$\forall \varepsilon > 0 \exists K \in \mathbb{N} \forall_{l > K} : \|f - f_l\|_p < \varepsilon. \tag{B.20}$$

   But this is the same as $\lim_{l \to \infty} \|f - f_l\|_p = 0$, or $f_l \to f$ in ℓ^p.

   The proof for $p = \infty$ is virtually the same, with (B.19) replaced by

   $$\|g\|_\infty = \sup_{F \subset X} \sup_{x \in F} \{|g(x)|\}. \tag{B.21}$$

   Within the finite supremum $\sup_{x \in F} |f_k(x) - f_l(x)| < \varepsilon$, we may take the limit $k \to \infty$ once again, followed by a supremum over $F \subset X$. □

B.3 Banach spaces of continuous functions

Further Banach spaces that can be defined without measure theory come from topology, notably from the class of ***locally compact*** spaces X (like $\mathbb{N}$, or $\mathbb{R}^n$, etc.).

For any $f: X \to \mathbb{C}$, define the ***support*** of f as the closure of the set where $f \neq 0$.

Definition B.10. *Let X be a locally compact space. Then:*

- *$C(X)$ is the set of all continuous functions $f: X \to \mathbb{C}$;*
- *$C_c(X)$ is the set of all continuous functions $f: X \to \mathbb{C}$* **with compact support***;*
- *$C_0(X)$ is the set of all continuous functions $f: X \to \mathbb{C}$ that* **vanish at infinity***, i.e., for any $\varepsilon > 0$ the set $\{x \in X \mid |f(x)| \geq \varepsilon\}$ is compact, or, equivalently, for any $\varepsilon > 0$ there is a compact set $K \subset X$ such that $|f(x)| < \varepsilon$ for all $x \notin K$;*
- *$C_b(X)$ is the set of all continuous functions $f: X \to \mathbb{C}$ that are* **bounded***, i.e., there is a constant $C > 0$ (which depends on f) such that $|f(x)| \leq C$ for all $x \in X$.*

In general, one has the obvious inclusions

$$C_c(X) \subseteq C_0(X) \subseteq C_b(X) \subseteq C(X), \tag{B.22}$$

with strict inclusions iff X is non-compact, and equalities iff X is compact.

For example, if $X = \mathbb{R}$, then $f(x) = \exp(-x^2)$ lies in C_0, whereas $f(x) = 1$ is in C_b. If X is discrete, the space $\ell_c(X)$ and $\ell^\infty(X)$ of the previous section are the same as $C_c(X)$ and $C_b(X)$, respectively, and we may also write $\ell_0(X) \equiv C_0(X)$.

Theorem B.11. *The sets $C_c(X)$, $C_0(X)$, $C_b(X)$, and $C(X)$ are vector spaces under pointwise operations, and $C_0(X)$ and $C_b(X)$ are Banach spaces in the* **sup-norm**

$$\|f\|_\infty = \sup_{x \in X}\{|f(x)|\}. \tag{B.23}$$

In particular, if X is compact, then $C(X)$ is a Banach space in the norm (1.24)

Proof. Only completeness in the sup-norm (B.23) is nontrivial. We use the fact from elementary analysis that sup-norm (i.e., uniform) limits f of sequences (f_n) of continuous functions exist (they are given by the pointwise limit $f(x) = \lim_n f(x)$) and are continuous. Therefore, concerning $C_0(X)$ we just need to show that the limit f of some sequence (f_n) in $C_0(X)$ vanishes at infinity. Indeed, for given $\varepsilon > 0$, since $f_n \to f$ uniformly, we can find N such that $|f(x) - f_n(x)| < \varepsilon/2$ for all x and all $n > N$. Since $f_n \in C_0(X)$, we can also find some compact $K \subset X$ such that $|f_n(x)| < \varepsilon/2$ for all $x \notin K$ and all n. Hence for $x \notin K$ and $n > N$,

$$|f(x)| \leq |f(x) - f_n(x)| + |f_n(x)| < \varepsilon/2 + \varepsilon/2 = \varepsilon. \tag{B.24}$$

To show that the limit f of a sequence (f_n) in C_b is again bounded, note that for $\varepsilon > 0$ we have $|f(x) - f_n(x)| < \varepsilon$ for $n > N$ and $|f_n(x)| < C_n$, both for all x, whence

$$|f(x)| \leq |f(x) - f_n(x)| + |f_n(x)| < \varepsilon + C_n < \infty, \tag{B.25}$$

so f is bounded and hence lies in $C_b(X)$. □

B.4 Basic measure theory

Measure theory studies ***measure spaces*** (X, Σ, μ), where X is a set, and:

- $\Sigma \subseteq \mathscr{P}(X)$ is a so-called ***σ-algebra*** of subsets of X, which means that:
  1. $X \in \Sigma$;
  2. If $A \in \Sigma$, then $A^c \in \Sigma$ (where $A^c \equiv X \backslash A$ is the complement of A);
  3. If $A_n \in \Sigma$ for $n \in \mathbb{N}$, then $\cup_n A_n \in \Sigma$ (i.e., Σ *is closed under countable unions*).

  It follows that $\emptyset \in \Sigma$, and that Σ is closed under countable *intersections*, too.
- $\mu : \Sigma \to [0, \infty]$, called a (positive) ***measure***, is ***countably additive***, i.e.,

$$\mu(\cup_n A_n) = \sum_n \mu(A_n), \tag{B.26}$$

  whenever $A_n \in \Sigma$, $n \in \mathbb{N}$, $A_i \cap A_j = \emptyset$ for all $i \neq j$. The obvious convention here is that $t + \infty = \infty$ for any $t \in \mathbb{R}^+$, as well as $\infty + \infty = \infty$. Countable additivity is indispensable in almost every limit argument in measure theory.

A ***probability space*** is a measure space (X, Σ, μ) for which $\mu(X) = 1$. More generally, a measure space is called ***finite*** if $\mu(X) < \infty$, which evidently implies $\mu(A) < \infty$ for any $A \in \Sigma$, and ***σ-finite*** if X is a countable union $X = \cup_n A_n$ with $\mu(A_n) < \infty$ for each n. For example, $X = \mathbb{R}$ is σ-finite, whilst $X = [0, 1]$ with Lebesgue measure is finite. The non-σ-finite case is pathological and hardly occurs in practice.

This definition of a σ-algebra marks a difference with a topology on X, which is a collection $\mathscr{O}(X)$ of *open* subsets (containing X and the empty set $\emptyset$) that is closed under *arbitrary* unions and *finite* intersections (but *not* under complementation!).

Nonetheless, topology and measure theory are closely related:

1. Any topological space X gives rise to a σ-algebra $\mathscr{B}(X)$, viz. the smallest σ-algebra in $\mathscr{P}(X)$ that contains $\mathscr{O}(X)$ (this exists and equals the intersection of all σ-algebra that contain $\mathscr{O}(X)$, where one notes that the intersection of any family of σ-algebras is again a σ-algebra). Elements of $\mathscr{B}(X)$ are called ***Borel sets***.
2. The definition of a ***continuous*** function $f : X \to Y$ between topological spaces X and Y as a function for which $f^{-1}(V) \in \mathscr{O}(X)$ for each $V \in \mathscr{O}(Y)$, is copied by saying that $f : X \to Y$ is ***measurable*** with respect to given σ-algebras Σ_X (on X) and Σ_Y (on Y) if $f^{-1}(B) \in \Sigma_X$ for any $B \in \Sigma_Y$.
3. If X and Y are topological spaces and $\Sigma_X = \mathscr{B}(X)$, $\Sigma_Y = \mathscr{B}(Y)$, then it it easy to show that f is (Borel) measurable iff $f^{-1}(B) \in \Sigma_X$ merely for any $B \in \mathscr{O}(Y)$, from which it follows that each continuous function is measurable. For $f : X \to \mathbb{R}$ to be measurable it is even sufficient that $f^{-1}((t, \infty)) \in \Sigma_X$ for each $t \in \mathbb{R}$.
4. The above condition of σ-finiteness is often used just in case the A_i are *compact*.

An important goal of measure theory is to provide a rigorous theory of ***integration***; here the key idea (due to Lebesgue) is that in defining the integral of some *measurable* function $f : X \to \mathbb{R}$, one should partition the *range* $\mathbb{R}$ rather than the *domain* X, as had been done in the Calculus since Newton (where typically $X \subseteq \mathbb{R}^n$). This, in turn, suggests that f should first be approximated by ***simple*** functions.

These are *measurable* functions $s : X \to \mathbb{R}^+$ with finite range, or, equivalently,

$$s = \sum_i \lambda_i 1_{A_i}, \tag{B.27}$$

where $\lambda_i \geq 0$, $A_i \in \Sigma$, and $n < \infty$. Such a representation is unique if we require that the sets A_i are mutually disjoint and the coefficients λ_i are distinct; namely, if $\{x_1, \ldots, x_n\}$ are the distinct values of s, one takes $A_i = s^{-1}(x_i)$ and $\lambda_i = x_i$. Given some measure μ, we further restrict the class of simple functions to those for which $\mu(A_i) < \infty$. One then first defines the integral of a simple function s, as in (B.27), by

$$\int_X d\mu\, s = \sum_i \lambda_i \mu(A_i); \tag{B.28}$$

a nontrivial argument shows that the right-hand side is independent of the particular representation (B.27) of s used on the left. Granting this, linearity of the integral on simple functions is immediate. Subsequently, for *positive* measurable functions $f \geq 0$, writing $s \leq f$ iff $s(x) \leq f(x)$ for each $x \in X$, one defines the integral by

$$\int_X d\mu\, f = \sup \left\{ \int_X d\mu\, s \mid 0 \leq s \leq f, s \text{ simple} \right\}. \tag{B.29}$$

For measurable functions $f : X \to \mathbb{C}$, one first decomposes f as

$$f = \sum_{k=0}^{3} i^k f_k, \; f_k \geq 0, \tag{B.30}$$

where, writing $f = \mathrm{Re}(f) + i\,\mathrm{Im}(f) \equiv f' + i f''$, $f_0 \equiv f'_+$, $f_2 \equiv f'_-$, $f_1 \equiv f''_+$, and $f_3 \equiv f''_-$, so that $f^\bullet = f^\bullet_+ - f^\bullet_-$ for $^\bullet = {}'{,}''$ one may take $f^\bullet_\pm = \frac{1}{2}(|f^\bullet| - f^\bullet)$.

On this basis, one then defines the integral by linear extension of (B.29), that is,

$$\int_X d\mu\, f = \sum_{k=0}^{3} i^k \int_X d\mu\, f_k. \tag{B.31}$$

We call f ***integrable*** with respect to μ, writing $f \in \mathscr{L}^1(X, \Sigma, \mu)$, if

$$\int_X d\mu\, |f| < \infty; \tag{B.32}$$

this implies that each positive part f_k, and hence also f itself, is integrable, i.e.,

$$\int_X d\mu\, f < \infty. \tag{B.33}$$

However, (B.33) does not imply (B.32). From (B.32) one has the useful estimates

$$\left| \int_X d\mu\, f \right| \leq \int_X d\mu\, |f| \leq \|f\|^{\mathrm{ess}}_\infty \mu(X), \tag{B.34}$$

where the ***essential supremum*** of f (with respect to μ) is defined by

$$\|f\|_\infty^{\text{ess}} = \inf\{t \in [0,\infty] \mid |f| \leq t \ \mu\text{-almost everywhere}\}, \tag{B.35}$$

where $|f| \leq t$ μ-a.e. means that $\mu(\{x \in X \mid |f(x)| > t\}) = 0\}$. In (B.34), the expressions $\|f\|_\infty^{\text{ess}}$ and/or $\mu(X)$ may well be infinite (in which case the second estimate still holds, of course!). However, if X is a locally compact space (see the next section), μ is finite, and $f \in C_0(X)$ or even $f \in C_b(X)$, then all of (B.34) is finite.

Linearity of the integral is far from trivial: the proof relies on linearity for simple functions, as well as on a fundamental approximation lemma:

Lemma B.12. *If $f \geq 0$ is measurable, there is a monotone increasing sequence of simple functions s_n, i.e., such that $0 \leq s_1 \leq s_2 \leq \cdots \leq s_n \leq s_{n+1} \leq \cdots \leq f$ pointwise, for which $s_n \to f$ pointwise (i.e., $\lim_{n\to\infty} s_n(s) = f(x)$ for each $x \in X$).*

Furthermore, one needs one of the two great convergence theorems of measure theory named after Lebesgue, both of which (for future use) we now state. In these theorems (as well as in many others), we say that a measurable functions $f : X \to \mathbb{C}$ has some property ***μ-almost everywhere (μ-a.e.)*** if the set where f does *not* have the said property has measure zero. For example $f = 0$ μ-a.e. means that $f(x) = 0$ for each $x \notin N$, for some measurable set N with $\mu(N) = 0$ (as they say, "morally", the behaviour of measurable functions on subsets of measure zero should not matter).

Theorem B.13. *Let (f_n) be a sequence of (complex-valued) measurable functions.*

1. **Dominated Convergence**: *if (f_n) converges pointwise μ-a.e. to some function f and $|f_n(x)| \leq g(x)$ μ-a.e. for some $g \in \mathscr{L}^1(X,\Sigma,\mu)$, then $f \in \mathscr{L}^1(X,\Sigma,\mu)$, and*

$$\lim_{n\to\infty} \int_X d\mu\, f_n = \int_X d\mu\, f. \tag{B.36}$$

2. **Monotone Convergence**: *if $f_n \geq 0$ and (f_n) is monotone increasing μ-a.e., and*

$$\sup_n \left\{ \int_X d\mu\, f_n \right\} < \infty, \tag{B.37}$$

then $\lim_{n\to\infty} f_n(x) \equiv f(x)$ exists μ-a.e., $f \in \mathscr{L}^1(X,\Sigma,\mu)$, and (B.36) *holds.*

Note that the first *conclusion* of the monotone convergence theorem is an *assumption* in the dominated one! Either way, the fact that the pointwise limit function f is integrable, being implicit in the notation $f \in \mathscr{L}^1(X,\Sigma,\mu)$, is part of the *result*.

Corollary B.14. *Integration is linear, i.e., if f_1, f_2 are integrable and $\lambda_1, \lambda_2 \in \mathbb{C}$,*

$$\int_X d\mu\, (\lambda_1 f + \lambda_2 f_2) = \lambda_1 \int_X d\mu\, f_1 + \lambda_2 \int_X d\mu\, f_2. \tag{B.38}$$

Proof. If $f_1 \geq 0, f_2 \geq 0$, let $s_n^{(1)} \to f_1$ and $s_n^{(2)} \to f_2$, as in Lemma B.12. Then the conditions of the monotone convergence theorem hold, because integration is itself a monotone operation (i.e., if $f \leq g$, then $\int_X d\mu\, f \leq \int_X d\mu\, g$). Combined with linearity on simple functions (as already established above), this yields the claim. □

B.5 Measure theory on locally compact Hausdorff spaces

For us it suffices to deal with ***locally compact Hausdorff spaces*** X. Our main goal is Corollary B.21. We say that a map $\varphi : C(X) \to \mathbb{C}$ is ***positive*** if $\varphi(f) \geq 0$ whenever $f \geq 0$ (pointwise). We also write $\mathscr{O}(X)$ for the set of *open* subsets of X, whilst $\mathscr{K}(X)$ denotes the set of all *compact* subsets of X. We first assume that X is compact. Any finite measure $\mu : \mathscr{B}(X) \to [0, \infty)$ gives rise to a positive linear map $\varphi : C(X) \to \mathbb{C}$,

$$\varphi(f) = \int_X d\mu\, f,\ f \in C(X). \tag{B.39}$$

Conversely, any such map canonically defines a finite measure μ at least on opens $U \in \mathscr{O}(X)$ and on compacta $K \in \mathscr{K}(X)$ (which are key examples of Borel sets) by

$$\mu(U) = \sup\{\varphi(f) \mid f \in C_c(U), 0 \leq f \leq 1_X\}; \tag{B.40}$$
$$\mu(K) = \inf\{\varphi(f) \mid f \in C_c(X), 0 \leq f \leq 1_X, f_{|K} = 1_K\}. \tag{B.41}$$

Subsequently, this preliminary measure is (hopefully!) to be extended to at least all of $\mathscr{B}(X)$, i.e., to all Borel sets, in such a way that μ recovers φ via (B.39).

This works, and one even obtains a bijective correspondence between finite measure spaces (X, Σ, μ) and positive linear maps $\varphi : C(X) \to \mathbb{C}$ if the former are subjected to two additional conditions, predicated on having $\mathscr{B}(X) \subset \Sigma$, namely:

- ***completeness***, in that $\mu(B) = 0$ and $A \subset B$ for $A \in \mathscr{P}(X)$, $B \in \Sigma$ imply $A \in \Sigma$;
- ***regularity***, i.e., for a given measure $\mu : \Sigma \to [0, \infty]$, for any $A \in \Sigma$, one has

$$\mu^*(A) = \mu_*(A) = \mu(A), \tag{B.42}$$

where the ***outer measure*** μ^* and ***inner measure*** μ_* are defined by

$$\mu^*(A) = \inf\{\mu(U) \mid U \supseteq A, U \in \mathscr{O}(X)\}; \tag{B.43}$$
$$\mu_*(A) = \sup\{\mu(K) \mid K \subseteq A, K \in \mathscr{K}(X)\}, \tag{B.44}$$

respectively. These expressions apparently make sense for *all* subsets $A \subset X$, but lovers of the Banach–Tarski Paradox may be reassured that μ^* and μ_* typically fail to be countable additive if they are seen as maps from $\mathscr{P}(X)$ to $[0, \infty]$.
For future reference we also define (X, Σ, μ) to be ***inner regular*** if (merely) $\mu_*(A) = \mu(A)$ for $A \in \Sigma$, and ***outer regular*** if (merely) $\mu^*(A) = \mu(A)$, $A \in \Sigma$.
So *a regular measure is both inner and outer regular*. We are now in a position to state the ***Riesz Representation Theorem*** (often attributed also to ***Radon***).

Theorem B.15. *Let X be a compact Hausdorff space. There is a bijective correspondence between* complete regular finite measure spaces (X, Σ, μ) *and* positive linear maps $\varphi : C(X) \to \mathbb{C}$*, explicitly given as follows:*

- *The measure space (X, Σ, μ) defines φ through* (B.39)*, assuming* (B.29) - (B.31)*;*
- *The map φ defines the pair (Σ, μ) in three steps:*

1. *μ is given on opens U and on compacta K by* (B.40) *and* (B.41), *respectively;*
2. *Σ is defined as the collection of all sets $A \in \mathscr{P}(X)$ for which $\mu^*(A) = \mu_*(A)$;*
3. *μ is given on all of Σ by $\mu(A) = \mu^*(A)$, using* (B.43), *or, equivalently (given the previous point), by $\mu(A) = \mu_*(A)$, based on* (B.44).

We omit the lengthy proof, expect by announcing that Theorem B.15 may be seen as a special case of the more advanced Choquet theory reviewed in §B.11. For now, just note that expressions like (B.40) and (B.41) are really desperate attempts to define "$\mu(A) = \varphi(1_A)$", which is OK for finite X, but in general is ill defined because even for Borel sets A, the characteristic function 1_A is rarely continuous on X.

We note that μ has to be finite, since obviously $\mu(X) = \varphi(1_X)$. One can say a little more about this. A linear map $\varphi : C(X) \to \mathbb{C}$ is ***bounded*** if, for some $0 < C < \infty$,

$$|\varphi(f)| \leq C\|f\|_\infty \ (f \in C(X). \tag{B.45}$$

In that case, the following expression, called the ***norm*** of φ, is $\leq C$, hence finite:

$$\|\varphi\| = \sup\{|\varphi(f)|, f \in C(X), \|f\|_\infty = 1\}. \tag{B.46}$$

Proposition B.16. *Let X be a compact Hausdorff space. If a linear map $\varphi : C(X) \to \mathbb{C}$ is positive, then it is bounded, with norm*

$$\|\varphi\| = \varphi(1_X). \tag{B.47}$$

Proof. Positivity makes $\langle f, g \rangle = \varphi(f^* g)$ a pre-inner product on $C(X)$, so by (A.1) with $v = 1_X$ and $w = f$, we find $|\varphi(f)|^2 \leq \varphi(|f|^2)\varphi(1_X)$ for any f. If $\|f\|_\infty = 1$, then pointwise $0 \leq |f|^2 \leq 1_X$, so by positivity, $\varphi(|f|^2) \leq \varphi(1_X)$. Hence $|\varphi(f)| \leq \varphi(1_X)$, so that $\|\varphi\| \leq \varphi(1_X)$. Finally, taking $f = 1_X$ in (B.46) gives equality. □

A ***state*** on $C(X)$ is a positive linear functional $\omega : C(X) \to \mathbb{C}$ with $\omega(1_X) = 1$.

Corollary B.17. *If X is a compact Hausdorff space, there is a bijective correspondence between states on $C(X)$ and complete regular probability measures on X.*

We now move to the next case in difficulty, where X is assumed to be ***σ-compact***, in being a countable union of compact sets, i.e., $X = \cup_n K_n$, where $K_n \in \mathscr{K}(X)$. Using a little topology, this is actually equivalent to X being a perhaps more appealing union $X = \cup_n U_n$, where each U_n is open with compact closure U_n^-, and $U_n^- \subseteq U_{n+1}$. This, in turn, implies that $X = \cup_n K_n'$ with $K_n' \subseteq K_{n+1}'$ all compact. If (X, μ, σ) is a measure space where X is σ-compact topologically, $\mathscr{B}(X) \subseteq \Sigma$, and

$$\mu(K) < \infty, \ (K \in \mathscr{K}(X)), \tag{B.48}$$

then X is also σ-finite measure-theoretically. Since these are the only σ-finite measure spaces we will consider, with a slight change in terminology we call a locally compact measure space (X, Σ, μ) ***σ-finite*** if it is also σ-compact and (B.48) holds.

The new point compared to the compact case is that functionals like the above φ should now be defined on the space $C_c(X)$ of continuous functions on X *with compact support*. Otherwise, Theorem B.15 may be repeated almost *verbatim*:

Theorem B.18. *Let X be a σ-compact Hausdorff space. There is a bijective correspondence between* complete regular σ-finite measure spaces (X,Σ,μ) *and* positive linear maps $\varphi : C_c(X) \to \mathbb{C}$*, explicitly given as in Theorem B.15.*

For the sake of completeness we also state Theorem B.18 in the case where X is not even assumed to be σ-compact. In that case, inner regularity may be lost:

Theorem B.19. *Let X be a locally compact Hausdorff space. There is a bijective correspondence between* complete outer regular measure spaces (X,Σ,μ) *satisfying* (B.48)*, and* positive linear maps $\varphi : C_c(X) \to \mathbb{C}$*, explicitly given as in Theorem B.15, except for the fact that Σ now consists of all $A \in \mathscr{P}(X)$ for which $\mu(A\cap K) < \infty$ and $\mu^*(A\cap K) = \mu(A\cap K)$ for any $K \in \mathscr{K}(X)$. In that case, μ is* defined *by*

$$\mu(A) = \mu^*(A), \ A \in \Sigma. \tag{B.49}$$

However, this generality will not really be needed for our purposes, which will only require *finite* measures, in which case outer regularity implies regularity.

In order to generalize Corollary B.17 to the σ-compact case, or even to the locally compact case, we must involve the Banach spaces $C_c(X)$ and $C_0(X)$ of the previous section. Also for linear maps $\varphi : C_c(X) \to \mathbb{C}$ or $\varphi : C_0(X) \to \mathbb{C}$ we use the notation (B.46), where now the supremum is taken over $f \in C_c(X)$ and $f \in C_0(X)$, respectively. For example, in the latter case, provided (B.45) holds, we have

$$\|\varphi\| = \sup\{|\varphi(f)|, f \in C_0(X), \|f\|_\infty = 1\}. \tag{B.50}$$

Lemma B.20. *Let X be a locally compact Hausdorff space.*

1. *$C_c(X)$ is a dense subspace of $C_0(X)$ with respect to the norm* (B.23)*.*
2. *For a* positive *linear map $\varphi : C_c(X) \to \mathbb{C}$, the following are equivalent:*
   a. *φ is bounded, as in* (B.45)*;*
   b. *φ can be extended to a* positive *linear map $\varphi : C_0(X) \to \mathbb{C}$.*

In particular, a positive linear map $\varphi : C_0(X) \to \mathbb{C}$ is automatically bounded.

Proof. 1. The first claim means either of the following two equivalent properties:

- For any $f \in C_0(X)$ there is a sequence (f_n) in $C_c(X)$ converging to f;
- For any $f \in C_0(X)$ and $\varepsilon > 0$ there is $g \in C_c(X)$ with $\|f-g\| < \varepsilon$.

We prove both. For some given $f \in C_0$ and $\varepsilon > 0$, find the usual compact K such that $|f(x)| < \varepsilon$ outside K. Urysohn's Lemma gives $h \in C_c(X)$ with $0 \le h(x) \le 1$ for all $x \in X$ and $h(x) = 1$ for all $x \in K$. Take $g = fh \in C_c(X)$, so that $\|f-g\|_\infty < \varepsilon$. For $\varepsilon = 1/n$, rename the g thus constructed as f_n. Then $\|f-f_n\|_\infty \to 0$.

2. To go from 2.a to 2.b, using the previous item, let $f_n \to f$ uniformly (i.e., in the sup-norm), and define the extension $\varphi : C_0(X) \to \mathbb{C}$ by $\varphi(f) = \lim_n \varphi(f_n)$. This limit exists, since $|\varphi(f_m) - \varphi(f_n)| \le C\|f_m - f_n\|_\infty$, so that, (f_n) being convergent and hence Cauchy in $C_0(X)$, the sequence $(\varphi(f_n))$ is Cauchy in $\mathbb{C}$. The value $\varphi(f)$ is easily verified to be independent of the approximating sequence (f_n).

Finally, the approximation in 2.a preserves positivity, i.e., if $f \geq 0$ then $f_n \geq 0$, so also $\varphi(f) \geq 0$, as it has been defined as the limit of a positive sequence.

By definition, the converse implication 2.b $\to$ 2.a is equivalent to the claim that

$$\sup\{|\varphi(f)|, f \in C_0(X), \|f\|_\infty \leq 1\} < \infty, \tag{B.51}$$

which in turn is equivalent to the apparently weaker claim to the effect that

$$\sup\{|\varphi(f_n)|, n \in \mathbb{N}\} < \infty, \tag{B.52}$$

for any sequence (f_n) with $\|f_n\| \leq 1$. Indeed, if the first supremum were infinite, then for each $n \in \mathbb{N}$ there is f_n such that $|\varphi(f_n)| > n$, and (B.52) could not possibly hold. Furthermore, (B.52) need only hold for *non-negative* functions $f_n \geq 0$ (still with $\|f_n\| \leq 1$, of course) cf. (B.31), since $|\varphi(f_k)| = \varphi(f_k) < C$ for each $k = 0, \ldots, 3$ implies $|\varphi(f)| < 4C$. And this, finally, reduces to the claim that

$$\sum_{n=1}^{\infty} g(n)\varphi(f_n) < \infty, \ \forall g \in \ell^1(\mathbb{N}), g(n) \geq 0. \tag{B.53}$$

Namely, if the sequence $(\varphi(f_n))$ where unbounded, it would be trivial to find such a summable function g for which the sum in (B.53) diverges (for example, take a subsequence for which $\varphi(f_{n_m}) > m$ and take g such that $g_{n_m} = 1/m^2$).

To prove (B.53), then, given that $f_n \geq 0$ and hence $\varphi(f_n) \geq 0$, with $\|f_n\| \leq 1$, first note that $\sum_n g(n) f_n$ converges in $C_0(X)$ (since it is obviously absolutely convergent, and any absolutely convergent series in a Banach space converges). Calling the sum h, for any $N < \infty$ we have $\sum_{n=1}^N g(n) f_n \leq h$ and hence, by positivity of φ, also $\sum_{n=1}^N g(n)\varphi(f_n) \leq \varphi(h) < \infty$. Letting $N \to \infty$ gives (B.53). □

We now define a ***state*** on $C_0(X)$ as a positive (and hence bounded) linear functional $\omega : C_0(X) \to \mathbb{C}$ with $\|\omega\| = 1$; this is consistent with the terminology for the compact case because of (B.47), as well as with the terminology for C*-algebras.

Corollary B.21. *Let X be a locally compact Hausdorff space. There is a bijective correspondence between positive linear functionals on $C_0(X)$ and complete regular finite measures on X, explicitly given as in the bullet points of Theorem B.15.*

In particular, states *on $C_0(X)$ correspond to regular* probability measures *on X.*

Proof. All that remains to be shown is that, under (B.39), we have

$$\|\varphi\| = \mu(X), \tag{B.54}$$

so that, in particular, the case $\|\varphi\| = 1$ corresponds to $\mu(X) = 1$. For compact X, eq. (B.54) is immediate from (B.47). For locally compact X, we immediately see from (B.39) and (B.50) that $\|\varphi\| \leq \mu(X)$. To saturate this inequality, we use inner regularity of the measure μ corresponding to φ, cf. Theorem B.19 and subsequent comment. From (B.42) and (B.44), for any $\varepsilon > 0$ we can find $K \in \mathscr{K}(X)$ with $\mu(X) - \mu(K) < \varepsilon$. Now use Urysohn's Lemma to find $f \in C_c(X)$ such that $0 \leq f \leq 1$ and $f_{|K} = 1$. Then $\varphi(f) \geq \mu(K)$, and, letting $\varepsilon \to 0$, eq. (B.54) follows. □

Finally, we extend the above corollaries to the entire (Banach) dual $C_0(X)^*$, i.e., the space of *all* (i.e. not necessarily positive) bounded linear maps $\varphi : C_0(X) \to \mathbb{C}$, equipped with the norm (B.50). As we shall see more generally in §B.9, this is a vector space (under pointwise operations) and even a Banach space in its own right.

From the point of view of measure theory, the relevant concept is that of a ***complex measure***. This is a map $\mu : \Sigma \to \mathbb{C}$ satisfying the countable additivity condition (B.26), as in the positive case. In the complex case this condition implies that μ is finite. One then (trivially) has a decomposition $\mu = \mu' + i\mu''$, where μ' and μ'' are countably additive maps $\Sigma \to \mathbb{R}$ (just take $\mu' = \frac{1}{2}(\mu + \mu^*)$ and $\mu'' = -\frac{1}{2}i(\mu - \mu^*)$, where $\mu^*(A) = \overline{\mu(A)}$), and (nontrivially) has the ***(Hahn)–Jordan decomposition***:

Theorem B.22. *Let Σ be a σ-algebra on a set X and let μ be a (finite)* **signed measure**, *i.e.,a countably additive map $\Sigma \to \mathbb{R}$. Then there is a unique decomposition*

$$\mu = \mu_+ - \mu_-, \tag{B.55}$$

where the measures $\mu_\pm : \Sigma \to \mathbb{R}^+$ are given by:

$$\mu_+(A) = \sup\{\mu(B) \mid B \subseteq A, B \in \Sigma\}; \tag{B.56}$$
$$\mu_-(A) = -\inf\{\mu(B) \mid B \subseteq A, B \in \Sigma\}, \tag{B.57}$$

and μ_+ and μ_- are **mutually singular** *in that there is a set $N \in \Sigma$ such that*

$$\mu_+(N) = \mu_-(X \backslash N) = 0. \tag{B.58}$$

We will not prove this, just noting that in terms of the ***total variation*** $|\mu|$ of μ, i.e.,

$$|\mu|(A) = \sup\left\{\sum_{n\in\mathbb{N}} |\mu(A_n)|\right\}, \tag{B.59}$$

where the supremum is taken over all measurable partitions $A = \cup_n A_n$, one has

$$\mu_\pm = \tfrac{1}{2}(|\mu| \pm \mu). \tag{B.60}$$

Fom the point of view of C*-algebras it is more natural to start from bounded linear functionals on $C_0(X)$. First, we call a map $\varphi : C_0(X) \to \mathbb{C}$ ***hermitian*** if

$$\varphi(f^*) = \overline{\varphi(f)} \;\; (f^*(x) \equiv \overline{f(x)}). \tag{B.61}$$

Theorem B.23. *1. Any functional $\varphi \in C_0(X)^*$ has a unique decomposition*

$$\varphi = \varphi' + i\varphi'', \tag{B.62}$$

where the functionals $\varphi' \in C_0(X)^*$ and $\varphi'' \in C_0(X)^*$ are hermitian.
2. Any hermitian *functional $\varphi \in C_0(X)^*$ has a decomposition*

$$\varphi = \varphi_+ - \varphi_-, \tag{B.63}$$

where the functionals $\varphi_\pm \in C_0(X)^*$ are positive, and are given on $f \geq 0$ by

$$\varphi_+(f) = \sup\{\varphi(g), g \in C_0(X), 0 \leq g \leq f\}; \tag{B.64}$$
$$\varphi_-(f) = -\inf\{\varphi(h), h \in C_0(X), 0 \leq h \leq f\}. \tag{B.65}$$

3. *These expressions satisfy*

$$\|\varphi\| = \|\varphi_+\| + \|\varphi_-\|, \tag{B.66}$$

and any positive functionals $\varphi_\pm \in C_0(X)^*$ that satisfy (B.63) *as well as* (B.66) *are necessarily given by* (B.64) - (B.65).

4. *Any functional $\varphi \in C_0(X)^*$ is a linear combination of at most four states.*

Proof. 1. Take $\varphi' = \frac{1}{2}(\varphi + \varphi^*)$ and $\varphi'' = -\frac{1}{2}i(\varphi - \varphi^*)$, where $\varphi^*(f) = \overline{\varphi(f^*)}$.

2. The range $h : 0 \leq h \leq f$ is the same as the range $h : 0 \leq f - h \leq f$, so that (B.64) - (B.65) gives (B.63). Positivity of φ_+ follows because the value $\varphi(0) = 0$ is included in the supremum in (B.64), which therefore can only be ≥ 0, and likewise $-\varphi_-$ is negative (and hence φ_- is positive) because $\varphi(0) = 0$ is included in the infimum in (B.65), which therefore can only be ≤ 0.

3. We first prove (B.66) for compact X, so that $1_X \in C_0(X) = C(X)$. From (B.47),

$$\|\varphi\| \leq \|\varphi_+\| + \|\varphi_-\| = \varphi_+(1_X) + \varphi_-(1_X) \tag{B.67}$$
$$= \sup\{\varphi(g), 0 \leq g \leq 1_X\} - \inf\{\varphi(h), 0 \leq h \leq 1\}. \tag{B.68}$$

For any $\varepsilon > 0$, there is g such that $\varphi(g)$ is close to the supremum in (B.68) by $\frac{1}{2}\varepsilon$, and likewise there is h such that $\varphi(h)$ is close to the infimum in (B.68) by the same amount, so that

$$|\varphi_+(1_X) + \varphi_-(1_X) - \varphi(g - h)| < \varepsilon. \tag{B.69}$$

Since $0 \leq g \leq 1_X$ and $0 \leq h \leq 1$, we have $\|g - h\| \leq 1$, and thereore

$$\varphi(g - h) \leq \|\varphi\| \|g - h\| \leq \|\varphi\|. \tag{B.70}$$

Hence (B.67) gives

$$\|\varphi\| \leq \|\varphi_+\| + \|\varphi_-\| \leq \|\varphi\| + \varepsilon, \tag{B.71}$$

so letting $\varepsilon \to 0$ yields (B.66).

For locally compact X, we reduce the proof to the compact case by forming the ***one-point compactification*** $\dot{X}$ of X, cf. §C.6. As a set, this is $\dot{X} = X \cup \{\infty\}$, where ∞ is a singleton. As a space, the open sets in $\dot{X}$ are the open sets in X plus those subsets of $\dot{X}$ whose complement is compact in X. The obvious injection $i : X \hookrightarrow \dot{X}$ is continuous, and any $f \in C_0(X)$ extends uniquely to a function $f \in C(\dot{X})$ that vanishes at the compactification point, i.e., $f(\infty) = 0$. This yields an *isometric* embedding $C_0(X) \hookrightarrow C(\dot{X})$. Furthermore, as vector spaces one has

$$C(\dot{X}) = C_0(X) \oplus \mathbb{C} \cdot 1_{\dot{X}}. \tag{B.72}$$

Any linear map φ on $C_0(X)$ may then be extended to a linear map $\dot{\varphi}$ on $C(\dot{X})$ via

$$\dot{\varphi}(f + \lambda 1_{\dot{X}}) = \varphi(f) + \lambda \|\varphi\|, \; f \in C_0(X), \lambda \in \mathbb{C}. \tag{B.73}$$

From the point of view of (B.39), this extension may alternatively be described as follows: extend the measure μ on X that underlies φ to a measure $\dot{\mu}$ on $\dot{X}$ by $\dot{\mu}(A \cup \{\infty\}) = \mu(A)$, $A \in \Sigma$. This shows that $\dot{\varphi}$ remains positive when φ is, and using (B.54) and the analogue of (B.47) for $\dot{X}$ instead of X, we also obtain

$$\|\dot{\varphi}\| = \dot{\varphi}(1_{\dot{X}}) = \dot{\mu}(\dot{X}) = \mu(X) = \|\varphi\|. \tag{B.74}$$

One may then repeat the proof of the compact case, using $\dot{\varphi}$ instead of φ.
We just prove uniqueness for the compact case (in general, add dots as in the previous proof). Suppose $\varphi = \varphi'_+ - \varphi'_-$. For $f \geq 0$, using (B.64) and $\varphi'_-(g) \geq 0$,

$$\begin{aligned}\varphi_+(f) &= \sup\{\varphi'_+(g) - \varphi'_-(g), 0 \leq g \leq f\} \\ &\leq \sup\{\varphi'_+(g), 0 \leq g \leq f\} \leq \varphi'_+(f),\end{aligned}$$

so $\psi \equiv \varphi'_+ - \varphi_+ \geq 0$. With $\varphi'_\pm = \varphi_\pm + \psi$, imposing $\|\varphi\| = |\varphi'_+\| + \|\varphi'_-\|$ and repeatedly using (B.47), we find $\|\psi\| = 0$, and hence $\psi = 0$.
4. This is trivial from parts 1–2, noting that any nonzero positive functional $\varphi = t\omega$ is a multiple of a state $\omega = \varphi/\|\varphi\|$, with $t = \|\varphi\|$, since obviously $\|\omega\| = 1$.

Combining this proposition with Corollaries B.17 and B.21, we finally obtain:

Theorem B.24. *Let X be a locally compact Hausdorff space. The Banach dual $C_0(X)^*$ of all bounded linear maps $\varphi : C_0(X) \to \mathbb{C}$ is isometrically isomorphic with the space $M(X)$ of all complete regular complex measures μ on X, with norm*

$$\|\mu\| = |\mu|(X). \tag{B.75}$$

In particular, if μ is real (i.e., hermitian as a functional on $C(X)$), then (cf. (B.55)*)*

$$\|\mu\| = \mu_+(X) + \mu_-(X). \tag{B.76}$$

This implies Corollary B.21, including its crucial final claim to the effect that states on $C_0(X)$ correspond to regular probability measures on X.

We briefly sketch an analogous result for *finitely additive measures*. Instead of a σ-algebra of subsets of some set X, we now start from a so-called *semiring*:

Definition B.25. *A* **semiring** *of subsets of X is a family $\mathscr{R} \subseteq \mathscr{P}(X)$ such that:*

1. *$\emptyset \in \mathscr{R}$;*
2. *if $A, B \in \mathscr{R}$, then $A \cap B \in \mathscr{R}$;*
3. *if $A, B \in \mathscr{R}$ and $B \subset A$, then for the complement of B in A we have $A \backslash B = \cup_{i=1}^n B_i$, where $n < \infty$, each $B_i \in \mathscr{R}$, and the B_i are pairwise disjoint.*

In fact, in all our examples a stronger version of axiom 3 holds: if $A, B \in \mathscr{R}$ and $B \subset A$, then $A \backslash B \in \mathscr{R}$. Indeed, we will typically have $X = \mathbb{N}$ and either $\mathscr{R} = \mathscr{P}(\mathbb{N})$ or $\mathscr{R} = \mathscr{P}_f(\mathbb{N})$ (i.e. the collection of *finite* subsets of $\mathbb{N}$).

Using the ***fundamental lemma for semirings***, which states that if $A_1, \ldots, A_n \in \mathscr{R}$, there are finitely many pairwise disjoint $B_1, \ldots, B_m$ in $\mathscr{R}$ such that $\cup_n A_n = \cup_m B_m$, it can be shown that the complex linear span $\mathrm{Step}(X, \mathscr{R})$ of the characteristic functions 1_A $(A \in \mathscr{R})$ is a commutative algebra under obvious pointwise operations. Since functions on $\mathrm{Step}(X, \mathscr{R})$ are bounded, we may form the closure of $\mathrm{Step}(X, \mathscr{R})$ in the supremum-norm; adding pointwise complex conjugation this yields a commutative C*-algebra called $\ell^\infty(X, \mathscr{R})$ (which has a unit iff $X = \cup \mathscr{R}$). For example, we have

$$\ell^\infty(\mathbb{N}, \mathscr{P}(\mathbb{N})) = \ell^\infty(\mathbb{N}) \equiv \ell^\infty; \tag{B.77}$$

$$\ell^\infty(\mathbb{N}, \mathscr{P}_f(\mathbb{N})) = \ell_0(\mathbb{N}) \equiv c_0. \tag{B.78}$$

Definition B.26. *A* **finitely additive measure** *on $(X, \mathscr{R})$ is a map $\mu : \mathscr{R} \to [0, \infty]$ such that $\mu(A \cup B) = \mu(A) + \mu(B)$ whenever $A, B \in \mathscr{R}$, $A \cup B \in \mathscr{R}$, and $A \cap B = \emptyset$.*

Similarly, we have finitely additive ***signed*** measures taking values in $\mathbb{R}$, which admit a Jordan–Hahn decomposition (B.55) with (B.56) - (B.57), just as in the σ-additive case. We say that a finitely additive signed measure μ is ***finite*** if $|\mu(A)| < \infty$ for each $A \in \mathscr{R}$, and ***bounded*** if $\sup\{|\mu(A)|, A \in \mathscr{R}\} < \infty$. With $|\mu| = \mu_+ + \mu_-$, the bounded finitely additive signed measures form a real Banach space $\mathrm{ba}(X, \mathscr{R})$ in the norm

$$\|\mu\| = \sup\{|\mu|(A), A \in \mathscr{R}\}. \tag{B.79}$$

Within this space, the ***probability measures*** stand out as those measures μ that take values in $[0, 1]$ (so that $\mu = \mu_+$) and satisfy $\|\mu\| = 1$.

Functions in $\mathrm{Step}(X, \mathscr{R})$ may be integrated against measures in $\mathrm{ba}(X, \mathscr{R})$ in the obvious way, cf. (B.27) - (B.28). This is well defined, and one easily infers that

$$|\int_X d\mu\, s| \leq \|\mu\| \|s\|_\infty, \tag{B.80}$$

for any $s \in \mathrm{Step}(X, \mathscr{R})$. Hence we may extend the integral to any $f \in \ell^\infty(X, \mathscr{R})$ by

$$\int_X d\mu\, f = \lim_{n \to \infty} \int_X d\mu\, s_n, \tag{B.81}$$

where (s_n) is any sequence in $\mathrm{Step}(X, \mathscr{R})$ converging to f in the sup-norm $\| \cdot \|_\infty$. This is well defined by the usual arguments. At the end of the day, we obtain:

Theorem B.27. *Let X be a set equipped with some semiring $\mathscr{R} \subseteq \mathscr{P}(X)$.*

- *There is a bijective correspondence between finitely additive probability measures μ on $(X, \mathscr{R})$ and states φ on $\ell^\infty(X, \mathscr{R})$, given by* (B.39) *and* (B.81).
- *This correspondence extends to an isometric isomorphism between* $\mathrm{ba}(X, \mathscr{R})$ *and the real Banach space of bounded hermitian functionals on $\ell^\infty(X, \mathscr{R})$.*
- *This isomorphism of real Banach spaces extends (i.e. complexifies) to an isomorphism between the complexification* $\mathrm{ba}(X, \mathscr{R})_{\mathbb{C}}$ *and the (Banach) dual $\ell^\infty(X, \mathscr{R})^*$.*

B.6 L^p spaces

We return to the usual, countably additive setting for measure theory. In the previous section, the notion of a measure space (X,Σ,μ) has mainly been used to provide an integration theory for *continuous* functions on X, though (B.29) suggested greater generality. In what follows, we keep the restriction to locally compact spaces X (although the theory is more general), but we expand the class of functions that can be integrated over X "against the measure μ". This, then, leads to an important class of Banach spaces, called $L^p(X) \equiv L^p(X,\Sigma,\mu)$; some authors write $L^p(X,\Sigma)$, others $L^p(\mu)$. One may have examples like $X = \Omega \subset \mathbb{R}^n$ in mind, with Ω measurable (typically open or closed, like $X = \mathbb{R}^n$ or $X = [0,1]$), and μ being Lebesgue measure. On the other hand, one may think of X as a discrete space with counting measure (i.e., $\mu(\{x\}) = 1$ for each $x \in X$), in which case the space $L^p(X)$ will reduces to the space $\ell^p(X)$ we already know; the typical case will be $X = \mathbb{N}$.

Definition B.28. *Given a measure space (X,Σ,μ) and a real number $1 \leq p \leq \infty$:*

- *For $1 \leq p < \infty$, the set $\mathscr{L}^p(X) \equiv \mathscr{L}^p(X,\Sigma,\mu)$ consists of all of measurable functions $f : X \to \mathbb{C}$ that are* **essentially bounded** *(with respect to μ), i.e.,*

$$\int_X d\mu\, |f|^p < \infty. \tag{B.82}$$

- *$\mathscr{L}^\infty(X) \equiv \mathscr{L}^\infty(X,\Sigma,\mu)$ is the set of measurable functions $f : X \to \mathbb{C}$ for which*

$$\inf\{t \in [0,\infty] : |f| \leq t \, (\mu\text{-almost everywhere})\} < \infty. \tag{B.83}$$

- *$\mathscr{N}_\mu$ is the set of all measurable functions $f : X \to \mathbb{C}$ that vanish μ-a.e., that is,*

$$\mu(\{x \in X \mid f(x) \neq 0\}) = 0. \tag{B.84}$$

- *Noting that $\mathscr{N}_\mu \subset \mathscr{L}^p(X)$ for all $1 \leq p \leq \infty$, we put*

$$L^p(X,\Sigma,\mu) \equiv L^p(X) = \mathscr{L}^p(X)/\mathscr{N}_\mu. \tag{B.85}$$

To appreciate the perhaps somewhat mysterious condition (B.83), we write

$$\inf\{t \in [0,\infty] : |f| \leq t \, \mu - \text{a.e.}\} = \inf\{t \in [0,\infty] : \mu(\{x \in X, |f(x)| > t\}) = 0\}.$$

Compare this with the expressions (defined for any function $f : X \to \mathbb{C}$):

$$\begin{aligned} \sup\{|f(x)| \mid x \in X\} &= \inf\{t \in [0,\infty] : |f(x)| \leq t \, \forall x \in X\} \\ &= \inf\{t \in [0,\infty] : \{x \in X, |f(x)| > t\} = \emptyset\} < \infty, \end{aligned} \tag{B.86}$$

which state the condition that f be bounded. Consequently, the stipulation that f be *essentially* bounded is the same as the condition that it is bounded, expect that the empty set in (B.86) has been replaced by a measure-zero set.

Theorem B.29. *For $1 \leq p < \infty$, the set $L^p(X)$ is a vector space under pointwise operations, as well as a Banach space, in the norm*

$$\|f\|_p = \left(\int_X d^n x |f(x)|^p \right)^{1/p}. \tag{B.87}$$

Likewise, $L^\infty(X)$ is a Banach space in the norm

$$\|f\|_\infty^{\text{ess}} = \inf\{t \in [0,\infty] : \mu(\{x \in X, |f(x)| > t\}) = 0\}. \tag{B.88}$$

Strictly speaking, elements of L^p are therefore equivalence classes of functions rather than functions, the pertinent equivalence relation $\sim_\mu$ being

$$f \sim_\mu g \text{ iff } \mu(\{x \in X \mid f(x) \neq g(x)\}) = 0, \tag{B.89}$$

but whenever no confusion can arise, we write $f \in L^p$ instead of $f \in \mathscr{L}^p$ or $[f] \in L^p$, as we have already done, for example, in (B.87) and (B.88); that is, the left-hand sides of these equations should officially be written as $\|[f]\|_p$ for $1 \leq p \leq \infty$. Note in this respect that in (B.87) - (B.88) the function f on the right-hand side could be any representative of its equivalence class $[f]$. However, one cannot replace the right-hand side of (B.88) by $\|f\|_\infty$, because (B.86) *does* depend on the representative f. Those who dislike (B.88) may, equivalently, write

$$\|f\|_\infty^{\text{ess}} = \inf\{\|g\|_\infty, g \sim_\mu f\}. \tag{B.90}$$

One should be aware of the need to pass to the quotient (B.85) in the first place: the natural expressions (B.87) and (B.88) fail to define norms on $\mathscr{L}^p$ and $\mathscr{L}^\infty$, respectively, because the positive definiteness axiom in Definition A.1.5c might fail. Indeed, although any f that is nonzero just on some null set is nonzero as an element of the vector space $\mathscr{L}^p$, one has $\|f\|_p = 0$. This problem is solved by passing to L^p.

The proof of Theorem B.29 uses both parts of Theorem B.13, which is concerned with a sequence (f_n) of functions in $\mathscr{L}^1(X)$, where (X, Σ, μ) is an arbitrary measure space. Note that on our definition of L^p spaces, these pointwise limits themselves might not lie in $\mathscr{L}^1$, but it is part of the conclusion of the convergence theorems that they do so up to some null set, and hence do define elements of L^1. For this reason, at this point one must distinguish between $f \in \mathscr{L}^1$ and $[f] \in L^1$. Let us mention in this context that L^p spaces are often constructed from measurable functions $f : X \to \overline{\mathbb{C}}$, whose positive real parts f_k (cf. (B.31)) by definition take values in $[0,\infty]$. This also leads to slightly more general versions of the Lebesgue convergence theorems, in which the f_n are allowed to be infinite on null sets. However, if $f \in L^p$, then $|f| < \infty$ μ-a.e., so little is lost by starting from functions $f : X \to \mathbb{C}$ or $f : X \to \mathbb{R}$.

Proof. We first prove Theorem B.29 for $1 \leq p < \infty$. Minkowski's Inequality (B.14) holds for $L^p \equiv L^p(X)$ just as it does for ℓ^p, as does Hölder's Inequality (B.15), so it remains to prove completeness. To this effect, let (f_n) a Cauchy sequence in L^p. Then (f_n) has a subsequence $(f_{n_k})_k$ such that

$$\|f_{n_{k+1}} - f_{n_k}\|_p < 2^{-k} \tag{B.91}$$

for each $k \in \mathbb{N}$ (indeed, for given $\varepsilon = 2^{-k}$, take n_k to be the famously existing N for which $\|f_n - f_m\|_p < \varepsilon$ for all $n, m > N$, etc.), and if $\lim_{k\to\infty} \|f_{n_k} - f\|_p = 0$ for some f, then $\lim_{n\to\infty} \|f_n - f\|_p = 0$ (this is a standard feature of Cauchy subsequences).

We now rewrite $f_{n_{k+1}}$ using a little trick, and introduce an auxiliary function g by

$$f_{n_k} = f_{n_1} + \sum_{l=1}^{k-1} (f_{n_{l+1}} - f_{n_l}); \tag{B.92}$$

$$g_{n_k} = |f_{n_1}| + \sum_{l=1}^{k-1} |f_{n_{l+1}} - f_{n_l}|. \tag{B.93}$$

Using (B.91), we estimate $\|g_{n_k}\|_p \leq \|f_{n_1}\|_p + \sum_{l=1}^{k-1} 2^{-l}$, which converges as $k \to \infty$. Hence $\sup_k \|g_{n_k}^p\|_1 < \infty$, so by the *Monotone* Convergence Theorem, $\lim_{k\to\infty} g_{n_k}^p \equiv h$ exists pointwise μ-a.e., with $h \in L^1$. Since $g_{n_k} \geq 0$, we have $h \geq 0$ at least μ-a.e., and with $g = h^{1/p}$, by continuity of $x \mapsto x^{1/p}$, we have $g_{n_k} \to g$ pointwise μ-a.e., with $g \in L^p$. Thus the series (B.92) converges (absolutely pointwise μ-a.e.) to some f. Since $|f| \leq g$, we also have $f \in L^p$. To prove that $f_{n_k} \to f$ in L^p (and not just pointwise μ-a.e.), we estimate

$$\begin{aligned} |f(x) - f_{n_k}(x)|^p &\leq (2\max\{|f(x)|, |f_{n_k}(x)|\})^p \\ &\leq 2^p(|f(x)| + |f_{n_k}(x)|)^p \leq 2^{p+1} g(x)^p, \end{aligned}$$

so, already knowing that $g^p \in L^1$, we may use (B.36) in the *Dominated* Convergence Theorem (with f_n replaced by $f - f_{n_k}$, and hence f replaced by the zero function) to conclude that $\lim_{k\to\infty} \int_X d\mu |f(x) - f_{n_k}(x)|^p = 0$, i.e., $\|f - f_{n_k}\|_p \to 0$.

We continue for $p = \infty$. For any fixed measurable subset $E \subset X$ we define

$$\|f\|_\infty^{(E)} = \sup\{|f(x)| \mid x \in E\} = \inf\{t \in [0,\infty] \mid |f(x)| \leq t \, \forall x \in E\}. \tag{B.94}$$

If $X \backslash E$ has measure zero, as we assume in what follows, then

$$\|f\|_\infty^{\text{ess}} \leq \|f\|_\infty^{(E)}, \tag{B.95}$$

since E might be expanded to a larger set of measure zero, which might decrease the infimum in (B.88). It follows that convergence with respect to the norm $\|\cdot\|_\infty^{(E)}$ implies convergence in $\|\cdot\|_\infty^{\text{ess}}$. We use this insight to prove the completeness of L^∞ by reducing this to a limiting problem with respect to the norm $\|\cdot\|_\infty^{(E)}$, for a suitable choice of $E \subset X$. Namely, let (f_n) be a Cauchy sequence in L^∞. This means

$$\forall \varepsilon > 0 \, \exists n \, \forall_{j,k>n} \, \|f_j - f_k\|_\infty^{\text{ess}} < \varepsilon.$$

Parametrizing $\varepsilon = 1/m$ for large $m \in \mathbb{N}$, and using (B.88), this implies:

$$\forall m \exists n \forall_{j,k>n} \exists N_{(j,k,m)} : \mu(N_{(j,k,m)}) = 0 \text{ and } \forall x \in X \backslash N_{(j,k,m)} : |f_j(x) - f_k(x)| < 1/m.$$

Now define $N = \cup_{j,k,m\in\mathbb{N}} N_{(j,k,m)}$. Since measures are countably additive by definition and N is a countable union of the measure zero sets, N has measure zero. With $E = X\backslash N$, so that $X\backslash E = N$ has measure zero, as above, we then have

$$\forall m \exists n \forall_{j,k>n} \forall x \in E\, |f_j(x) - f_k(x)| < 1/m.$$

Thus (f_n) (strictly speaking, the corresponding sequence of restrictions of each f_n to E) is a Cauchy sequence of bounded functions on E in the supremum norm (B.94), so that we are back in the $\ell^\infty(X)$ case with $X = E$, with the three-step proof we gave: the pointwise limits $f(x) = \lim_{n\to\infty} f_n(x)$ exist, the function f thus defined on E is bounded, i.e., $\|f\|_\infty^{(E)} < \infty$, and $f_n \to f$ not just pointwise but also in the norm $\|\cdot\|_\infty^{(E)}$. Extending f from E to X in an arbitrary way (the ensuing equivalence class in L^∞ does not depend on the behaviour of f on the null set $X\backslash E$), we first conclude from (B.95) that $\|f\|_\infty^{\text{ess}} < \infty$, and secondly infer that $f_n \to f$ also in $\|\cdot\|_\infty^{\text{ess}}$. □

Without proof, we state some useful results about the place of continuous functions in L^p-spaces. For simplicity, we assume that μ is regular and has support X (in that X has no open subset U with $\mu(U) = 0$). In that case, $C_b(X)$ and its subspaces $C_0(X)$ and $C_c(X)$ may be seen as subspaces of $L^\infty(X)$, on which the norm (B.88) or (B.90) simply reduces to the ordinary sup-norm (1.24).

Theorem B.30. • *If $1 \leq p < \infty$, then $C_c(X)$ is dense in $L^p(X)$ (in the L^p norm).*
• *If $p = \infty$, one has an inclusion of Banach spaces (all carrying the L^∞-norm)*

$$C_0(X) \subset C_b(X) \subset L^\infty(X). \tag{B.96}$$

Compare (B.22). Since the closure $C_c(X)$ is $C_0(X)$, it follows that $C_c(X)$ is dense in $L^\infty(X)$ only in the exceptional case where $L^\infty(X) = C_0(X)$ (e.g., for finite X). So in this respect, the values $1 \leq p < \infty$ behave quite differently from $p = \infty$.

The first claim is based on two facts, of which the first is true for all $1 \leq p \leq \infty$, whereas the second is valid only for $1 \leq p < \infty$ (i.e. it fails for $p = \infty$):

1. The set $S(X)$ of simple functions $s = \sum_i \lambda_i 1_{A_i}$, where $\mu(A_i) < \infty$ for each i, is dense in $L^p(X)$;
2. For each measurable subset $A \subset X$ with $\mu(A) < \infty$, and each $\varepsilon > 0$ there is a function $g \in C_c(X)$ such that $\|1_A - g\|_p < \varepsilon$.

Similarly to Theorems B.27 and B.24, we know the state space of $L^\infty(X, \nu)$:

Theorem B.31. *Let (X, Σ, ν) be a measure space. There is a bijective correspondence between states on $L^\infty(X, \nu)$ and finitely additive probability measures μ on (X, Σ) that are absolutely continuous with respect to ν (i.e., $\nu(A) = 0$ implies $\mu(A) = 0$), given by* (B.39) *and* (B.81).

In this case, the role of the semiring $\mathscr{R}$ is of course played by Σ, so that $\text{Step}(X, \Sigma)$ is simply the complex linear span of the simple functions on (X, Σ), and (B.28) duly applies. Since it may once again be shown that $\text{Step}(X, \Sigma)$ is dense in $L^\infty(X, \nu)$, the definition (B.81) of integration "by continuity" makes sense in this situation, too.

B.7 Morphisms and isomorphisms of Banach spaces

We often want to say that two Banach spaces are *isomorphic*. For example, in the next section the dual of a given Banach space is typically identified with some known Banach space; such identifications even belong to the nicest results in functional analysis. Of course, this issue is predicated on the correct definition of (not necessarily invertible) maps between Banach spaces in the first place.

Definition B.32. *A* **morphism** $a : V \to W$ *between Banach spaces* V,W *(or, more generally, normed spaces) is a* **bounded linear map**, *i.e., a linear map for which there is a constant* $C > 0$ *such that for each* $v \in V$,

$$\|av\|_W \leq C\|v\|_V, \tag{B.97}$$

or, equivalently,

$$\sup\{\|av\|_W, v \in V, \|v\|_V \leq 1\} < \infty. \tag{B.98}$$

It is extremely important (yet easy to show) that bounded maps are automatically continuous (and even uniformly continuous); conversely, a continuous linear map between vector spaces with norm is bounded. We note two important special cases:

- If $W = V$, a morphism $a : V \to V$ is called a (bounded) ***operator*** on V.
- If $W = \mathbb{C}$, a morphism $\varphi : V \to \mathbb{C}$ is called a (bounded linear) ***functional*** on V.

Theorem B.33. *Let* V *be a normed vector space and* W *a Banach space. The space* $B(V,W)$ *of all morphisms (i.e., bounded linear maps)* $a : V \to W$ *is a Banach space with respect to pointwise operations (e.g.,* $(\lambda a + b)v = \lambda av + bv$*), and the norm*

$$\|a\| = \sup\{\|av\|_W, v \in V, \|v\|_V \leq 1\}. \tag{B.99}$$

Proof. Only completeness is nontrivial; the idea is that if (a_n) is a Cauchy sequence in $B(V,W)$, we define $a : V \to W$ by $av = \lim_n a_n v$. This limit exists, since we have $\|a_n v - a_m v\|_W \leq \|a_n - a_m\|\|v\|_V$. Furthermore, it is easy to show (e.g., by contradiction) that a Cauchy sequence must be bounded, say $\|a_n\| \leq K$, and that, if $a_n v \to w$, then also $\|a_n v\|_W \to \|w\|_W$. Hence $\|av\|_W = \lim_n \|a_n v\|_W \leq K\|v\|_V$, so $a \in B(V,W)$. Finally, $a_n \to a$, since for $\|v\|_V \leq 1$ and, given $\varepsilon > 0$, the usual N for which $\|a_n - a_m\| < \varepsilon/2$ for all $n,m > N$ and $\|av - a_m v\|_W < \varepsilon/2$ for all $m > N$,

$$\|av - a_n v\|_W \leq \|av - a_m v\|_W + \|a_m v - a_n v\|_W < \tfrac{1}{2}\varepsilon + \tfrac{1}{2}\varepsilon = \varepsilon. \tag{B.100}$$

Since this holds for any $v \in V$ with $\|v\|_V \leq 1$, eq. (B.99) gives $\|a - a_n\| < \varepsilon$. □

Clearly, if $a \in B(V,W)$, then one has the useful estimate, cf. (A.20),

$$\|av\|_W \leq \|a\|\|v\|_V, \tag{B.101}$$

and if $W = V$ and $a,b \in B(V) \equiv B(V,V)$, we also have (cf. (A.21))

$$\|ab\| \leq \|a\|\|b\|. \tag{B.102}$$

Indeed, $B(V)$ is a ***Banach algebra***, which is just to say that it is a Banach space as well as an algebra, in which (B.102) holds (a C*-algebra will be a special case).

Returning to our opening theme, the level of discourse now suddenly becomes quite advanced. We start with Banach's famous ***Open Mapping Theorem***.

Theorem B.34. *if V and W are Banach spaces and $a \in B(V,W)$ is surjective, then a is open (in mapping open sets to open sets).*

Proof. For fixed $u \in V$ we write $V_r(u) = \{v \in V : \|u - v\| < r\}$ for the open r-ball around u, with $V_r \equiv V_r(0)$ and hence $V_r(u) = u + V_r$. Furthermore, the closure of $U \subset V$ is denoted by U^-. Likewise for W. The theorem follows if $aV_1 \equiv a(V_1) \subset W$ contains an open ball W_s, for some $s > 0$ (in which case, by linearity, aV_r contains an open ball W_{rs} for any $r > 0$). By the theory of metric spaces, some subset $U \subset V$ is open iff for any $u \in U$ there is $r > 0$ such that $V_r(u) \subset U$. Then aU contains the open set $W_{rs}(au)$, and since $au \in aU$ is arbitrary, aU is open by the same criterion.

To prove that aV_1 contains an open ball, first note that since $a : V \to W$ is surjective, $W = \cup_n aV_n$, so that by the Baire Category Theorem (which applies because Banach spaces are complete metric spaces by definition) some $(aV_n)^-$ contains an open set, and hence an open ball. Since a is linear this must then be true for all n; let us take $n = 1$, so that $W_\varepsilon(w_0) \subset (aV_1)^-$ for some $w_0 \in (aV_1)^-$. Since any point in the closure of some $U \subset W$ can be approximated by points in U, there is $w_1 \in aV_1$ such that $\|w_1 - w_0\| < \frac{1}{2}\varepsilon$. Hence for any $w \in W_{\varepsilon/2}$ we have

$$\|(w_1 - w) - w_0\| \leq \|w_1 - w_0\| + \|w\| < \tfrac{1}{2}\varepsilon + \tfrac{1}{2}\varepsilon = \varepsilon, \tag{B.103}$$

so $w_1 - w \in W_\varepsilon(w_0)$ and hence $w_1 - w \in (aV_1)^-$. Similarly, $w_1 + w \in (aV_1)^-$. Since $w = \frac{1}{2}(w_1 + w) - \frac{1}{2}(w_1 - w)$, we obtain $w \in (aV_1)^-$, for if $x, y \in (aV_1)^-$, then we have $\frac{1}{2}(x \pm y) \in (aV_1)^-$. Since $w \in W_{\varepsilon/2}$ was arbitrary, it follows that $W_{\varepsilon/2} \subset (aV_1)^-$.

To produce an open ball in aV_1 rather than in its closure, let $w'_0 \in W_{\varepsilon/4}$, so that $2w'_0 \in W_{\varepsilon/2}$. Hence there exists $w'_1 \in aV_1$ such that $\|2w'_0 - w'_1\| < \varepsilon/4$. And because $2(2w'_0 - w'_1) \in W_{\varepsilon/2}$, there exists $w'_2 \in aV_1$ such that $\|2(2w'_0 - w'_1) - w'_2\| < \varepsilon/4$, *et cetera*. Because $2(2(2w'_0 - w'_1) - w'_2) \in W_{\varepsilon/2}$, there exists $w'_3 \in aV_1, \ldots$

Repeating this N times, we obtain a sequence (w'_n) in aV_1 such that for any $N \in \mathbb{N}$,

$$\|2^N w'_0 - 2^{N-1} w'_1 - \cdots - 2^1 w'_{N-1} - 2^0 w'_N\| < \varepsilon/4, \tag{B.104}$$

i.e., $\|w'_0 - \sum_{n=1}^N 2^{-n} w'_n\| < 2^{-N-2}\varepsilon$. Letting $N \to \infty$ then gives $w'_0 = \sum_{n=1}^\infty 2^{-n} w'_n$.

Since $w'_n \in aV_1$, there is a corresponding sequence (v'_n) in V_1 such that $av'_n = w'_n$, with $\|v'_n\| < 1$ for each n. Hence we may estimate $\sum_{n=1}^\infty \|2^{-n} v'_n\| < \sum_{n=1}^\infty 2^{-n} = 1$, so the series $\sum_n 2^{-n} v'_n$ in V is absolutely convergent and hence convergent. Since V is assumed complete, it has a limit $v' = \sum_{n=1}^\infty 2^{-n} v'_n$. Since

$$\left\| \sum_{n=1}^N 2^{-n} v'_n \right\| \leq \sum_{n=1}^N \|2^{-n} v'_n\| < \sum_{n=1}^\infty \|2^{-n} v'_n\| < 1,$$

letting $N \to \infty$ gives $\|v'\| \leq 1$, or $v' \in V_1^-$. Since a is bounded and hence continuous,

$$av' = a\left(\sum_{n=1}^{\infty} 2^{-n} v'_n\right) = \sum_{n=1}^{\infty} 2^{-n} a v'_n = \sum_{n=1}^{\infty} 2^{-n} w'_n = w'_0. \tag{B.105}$$

We now recall that $w'_0 \in W_{\varepsilon/4}$ was arbitrary, so we have shown that $W_{\varepsilon/4} \subset a(V_1^-)$. By linearity of a, it follows that $W_s \subset aV_1$ for any $s < \varepsilon/4$. □

Corollary B.35. *Let V and W be Banach spaces. The (set-theoretic) inverse a^{-1} of a* bijective *morphism $a \in B(V,W)$ is automatically linear and bounded.*

In other words, a^{-1} lies in $B(W,V)$. Corollary B.35 suggests defining two Banach spaces V and W to be isomorphic if there exists a bijective morphism $a \in B(V,W)$ (in which case they would be isomorphic as objects in the category of Banach spaces with bounded linear maps). However, we often prefer to use a sharper notion.

Definition B.36. *Let V and W be normed spaces.*

1. An **isometry** *from V to W is a linear map $u : V \to W$ satisfying*

$$\|av\|_W = \|v\|_V, \; v \in V. \tag{B.106}$$

2. An **isometric isomorphism** *from V to W is a surjective isometry $u : V \to W$.*

Since an isometry is clearly bounded as well as injective, by Corollary B.35 a surjective isometry has a bounded linear inverse, which is easily seen to be isometric, too. In practice, it is the conditions in Definition B.36 that one typically checks.

Nonetheless, the non-isometric case is also quite important. As a case in point, we prove a classical result of functional analysis, called the *Closed Graph Theorem*. In preparation, note that two normed spaces V, W define a third one, called their ***direct sum*** $V \oplus W$, which as a set is $V \times W$, turned into a vector space by the operations $(v_1, w_1) + (v_2, w_2) = (v_1 + v_2, w_1 + w_2)$ and $\lambda(v,w) = (\lambda v, \lambda w)$, etc., with norm

$$\|(v,w)\| = \|v\|_V + \|w\|_W. \tag{B.107}$$

It is easily shown that if V and W are Banach spaces, then so is $V \oplus W$.

Furthermore, if $a : V \to W$ is any linear map, the ***graph*** of a is the vector space

$$G(a) = \{(v, av), v \in V\} \subset V \oplus W. \tag{B.108}$$

If a is bounded, then $G(a)$ is closed (i.e. in the norm inherited from the Banach space $V \oplus W$). The converse, then, is the ***Closed Graph Theorem***:

Theorem B.37. *Let V and W be Banach spaces and let $a : V \to W$ be a linear map. If the graph $G(a)$ is closed (in the norm inherited from $V \oplus W$), then a is bounded.*

Proof. Let $b : G(a) \to V$ be the linear map $(v, av) \mapsto v$, which is clearly a bijection, with inverse $b^{-1} : V \to G(a)$, $b^{-1}(v) = (v, av)$. Furthermore, $\|b(v,av)\| = \|v\|_V \leq \|v\|_V + \|av\|_W = \|(v,av)\|$, so b is bounded. Hence Corollary B.35 makes b^{-1} bounded as well, i.e., $\|b^{-1}(v)\| \leq C\|v\|_V$ for some $C > 0$. Hence $\|(v,av)\| = \|v\|_V + \|av\|_W \leq C\|v\|_V$. So $\|av\|_W \leq (C-1)\|v\|_V$, and hence a is bounded. □

B.8 The Hahn–Banach Theorem

In this section we present another traditional pillar of functional analysis.

Definition B.38. *A* **sublinear functional** *on a* real *vector space V is a map $p : V \to \mathbb{R}$ that for each $v, w \in V$ and scalars $t \geq 0$ satisfies*

$$p(v+w) \leq p(v) + p(w); \tag{B.109}$$
$$p(tv) = tp(v). \tag{B.110}$$

We will deal with two examples of such functionals. One is simply a norm (even on a complex vector space, which in particular is a real vector space). For the other, recall that a subset K of a real vector space V is called ***convex*** if whenever $v, w \in K$ and $t \in (0,1)$, one has $tv + (1-t)w \in K$. Even without a topology on V, we can define an ***interior point*** of K (or indeed of any subset of V) as a point $v \in K$ such that for each $v' \in V$ there is $\varepsilon > 0$ such that $v + tv' \in K$ for any $0 < t < \varepsilon$. We denote the set of interior points of K by int(K). For example, if V is normed (with associated topology), or is the dual of a normed space equipped with the w^*-topology (or, even more generally, if V is a ***topological vector space***, i.e., a vector space carrying a Hausdorff topology in which addition and scalar multiplication are continuous), then each point of an open set U is interior in the above sense, so that $U = \text{int}(U)$.

Let $K \subset V$ be convex and suppose it contains 0 as an interior point. Then the indexfunctional!Minkowski***Minkowski functional*** (also called ***gauge***) $p : V \to \mathbb{R}^+$ of K is defined by

$$p(v) = \inf\{a > 0 \mid v/a \in K\}. \tag{B.111}$$

Note that $p(v) < \infty$, because $0 \in K$ is interior, so that there is $\varepsilon > 0$ such that $\varepsilon v \in K$, and hence $a = 1/\varepsilon$ lies in the set in (B.111). It is clear that if $v \in K$, then $a = 1$ lies in the set in (B.111), so that $p(v) \leq 1$. As a simple example, for the (open or closed) unit ball B in a normed space (both of which are convex), we have $p(v) = \|v\|$.

Proposition B.39. *Let $K \subset V$ be convex and let $0 \in K$ be an interior point of K. Then the Minkowski functional p of K satisfies* (B.109) - (B.110). *Furthermore, we may recover the set* int(K) *of interior points of K through*

$$\text{int}(K) = \{v \in V \mid p(v) < 1\}. \tag{B.112}$$

Conversely, if some function $p : V \to \mathbb{R}^+$ satisfies (B.109) - (B.110), *then the set*

$$K = \{v \in V \mid p(v) \leq 1\} \tag{B.113}$$

is convex, with interior given by (B.112).

For example, if K is open (in a topological vector space), then (B.112) equals K.

Proof. Given (B.111), eq. (B.110) is obvious. To prove (B.109), find $a > 0$ and $b > 0$ such that $v/a \in K$ and $w/b \in K$; cf. the comment after (B.111). Since K is convex, with $t = a/(a+b)$ and hence $1 - t = b/(a+b)$ we have $t \cdot v/a + (1-t) \cdot w/b \in K$.

Hence $p(t\cdot v/a+(1-t)\cdot w/b)\leq 1$, which, using (B.110), reads $p(v+w)\leq a+b$. Taking the infimum over a and b constrained by $v/a\in K$, $w/b\in K$ then turns the right-hand side into $p(v)+p(w)$, so that we have proved (B.109).

The proof of the converse claims is almost trivial, except perhaps for the last claim. To prove that $p(v)<1$ implies $v\in \text{int}(K)$, we note that for any $v'\in V$ and $\varepsilon>0$, from (B.109) - (B.110) we have $p(v+\varepsilon v')\leq p(v)+\varepsilon p(v')$. If $p(v')=0$, this gives $p(v+\varepsilon v')\leq p(v)<1$, so that $v+\varepsilon v'\in K$. If not, assume $p(v)=1-\delta$ for some $\delta\in(0,1]$, and we find that $p(v+\varepsilon v')<1$ for any $0<\varepsilon<\delta/p(v')$. □

Having motivated Definition B.38, we now state the ***Hahn–Banach Theorem***:

Theorem B.40. *Let V be a real vector space equipped with a sublinear functional p, and let $W\subset V$ be a linear subspace carrying a linear map $\varphi_W:W\to\mathbb{R}$ that is dominated by p in the sense that for each $w\in V$ we have $\varphi_W(w)\leq p(w)$.*

Then φ_W has a linear extension $\varphi:V\to\mathbb{R}$ that for each $v\in V$ satisfies

$$\varphi(v)\leq p(v). \tag{B.114}$$

Proof. Take $v_1\in V$, $v_1\notin W$, and extend φ_W to $W\oplus\mathbb{R}\cdot v_1$ by

$$\varphi(w+tv_1)=\varphi_W(w)+t\varphi(v_1), \tag{B.115}$$

with $t\in\mathbb{R}$ and $\varphi(v_1)$ to be determined. In order to satisfy (B.114), we need

$$\varphi(w+tv_1)\leq p(w+tv_1), \tag{B.116}$$

for each $w\in W$ and $t\in\mathbb{R}$. Using (B.110), this is true iff it is true for $t\pm 1$, which yields two conditions (in two variables $w,w'\in W$), which may jointly be written as

$$\varphi(w')-p(w'-v_1)\leq\varphi(v_1)\leq p(w+v_1)-\varphi(w). \tag{B.117}$$

Since φ is linear, this can obviously be satisfied by some $\varphi(v_1)\in\mathbb{R}$ iff

$$\varphi(w+w')\leq p(w+v_1)+p(w'-v_1), \tag{B.118}$$

which is indeed the case: for by assumption we have $\varphi(w+w')\leq p(w+w')$, whence

$$\varphi(w+w')\leq p(w+v_1+w'-v_1)\leq p(w+v_1)+p(w'-v_1), \tag{B.119}$$

where we used (B.109). Hence any choice of $\varphi(v_1)$ that satisfies (B.117) provides an extension (B.115) of φ to $W\oplus\mathbb{R}\cdot v_1$, which by construction satisfies (B.114).

Lovers of Zorn's Lemma may now complete the proof as follows. Let F be the set of all pairs (φ,X), where $X\subseteq V$ is a linear subspace and $\varphi:X\to\mathbb{R}$ is a linear extension of φ_W that satisfies (B.114). We partially order F by

$$(\varphi_1,X_1)\leqslant(\varphi_2,X_2) \text{ iff } X_1\subseteq X_2 \text{ and } \varphi_1(v)=\varphi_2(v)\,\forall v\in X_1. \tag{B.120}$$

Then F is clearly nonempty, and every totally ordered subset $\{(X_\lambda,\varphi_\lambda)\}$ of F has an upper bound (φ,X), where $X=\cup_\lambda X_\lambda$ and $\varphi(v)=\varphi_\lambda(v)$ whenever $v\in X_\lambda$.

Thus Zorn's Lemma applies, "giving" a maximal element (φ, Z). If $Z \neq V$, one may extend Z by the first step of the proof (applied to $W \rightsquigarrow Z$), contradicting maximality of (φ, Z). Hence $Z = V$, and φ is the desired functional. □

If V is finite-dimensional, then Zorn's Lemma is unnecessary, and a constructive proof may be given by repeating the first step of the proof a finite number of times.

Corollary B.41. *Let V be a normed vector space, with dual V^*, and let $W \subset V$ be a linear subspace (inheriting the norm from V, with associated dual W^*).*

Then each $\varphi_W \in W^*$ has an extension $\varphi \in V^*$ to V with the same norm.

Proof. We take $p(v) = \|\varphi_W\|\|v\|$, which clearly satisfies (B.109) - (B.110). If V is real, Theorem B.40 gives $\varphi : V \to \mathbb{R}$ satisfying $|\varphi(v)| \leq \|\varphi_W\|\|v\|$ for each $v \in V$, and hence $\|\varphi\| \leq \|\varphi_W\|$. But $\|\varphi_W\| \leq \|\varphi\|$ since $W \subset V$, hence $\|\varphi\| = \|\varphi_W\|$,

If V is complex, we first regard it as a real vector space, take the real part φ'_W of φ_W, and isometrically extend φ'_W to a linear functional $\varphi' : X \to \mathbb{R}$ as above, so that $\|\varphi'\| = \|\varphi'_W\|$. Then define $\varphi : X \to \mathbb{C}$ by

$$\varphi(v) = \varphi'(v) - i\varphi'(iv). \tag{B.121}$$

One checks that $\varphi((s+it)v) = (s+it)\varphi(v)$. Since $\varphi'(v)$ is the real part of $\varphi(v)$, with $|\varphi(v)|^2 = |\varphi'(v)|^2 + |\varphi'(iv)|^2$, we have $|\varphi'(v)| \leq |\varphi(v)|$ and hence $\|\varphi'\| \leq \|\varphi\|$. Conversely, for any v with $\varphi(v) \neq 0$, take $z = |\varphi(v)|/\varphi(v)$, so that $|\varphi(v)| = \varphi(zv)$. Hence $\varphi(zv)$ is real and therefore it is equal to its real part, so that, since $|z| = 1$,

$$\varphi(zv) = \varphi'(zv) \leq \|\varphi'\|\|zv\| = \|\varphi'\|\|v\|.$$

Therefore, $\|\varphi\| \leq \|\varphi'\|$, and hence $\|\varphi\| = \|\varphi'\|$. The same computation applies to φ_W, yielding $\|\varphi_W\| = \|\varphi'_W\|$, so that finally $\|\varphi\| = \|\varphi'\| = \|\varphi'_W\| = \|\varphi_W\|$. □

In fact, this trick to pass from the real to the complex case was overlooked by Hanh and Banach themselves, whose arguments were much more involved.

As to Zorn's Lemma, if V is infinite-dimensional but still separable, using (countable) induction one may construct a sequence (v_n) of linearly independent unit vectors in $V \backslash W$, such that V is the closed linear span of W and the v_n. The above procedure then gives φ in the real algebraic linear span of W and the v_n, which is bounded by construction and may be extended to all of V by continuity. However, the construction of (v_n) still requires a weaker form of the Axiom of Choice (which is equivalent to Zorn's Lemma), namely the so-called *Axiom of Dependent Choice*.

In the situation of Corollary B.41, the extension φ is unique iff the normed space V is ***strictly convex***, which by definition means that its unit sphere is strictly convex, i.e., if $\|v\| = \|w\|$ for $v \neq w$ and $t \in (0,1)$, then $\|tv + (1-t)w\| < 1$. Equivalently, if $\|v\| = \|w\| = \frac{1}{2}\|v+w\|$, then $v = w$. This is the case, for example, in Hilbert spaces H, as easily follows from the comment after (A.3). Indeed, anticipating Theorem B.66, if $W \subset H$ is closed (as we may assume, since φ_W is continuous), we may identify $\varphi_W : W \to \mathbb{C}$ with some vector $\varphi_W \in W$, and if we do, the unique extension $\varphi : H \to \mathbb{C}$ corresponds to the same vector φ_W, now regarded as an element of H.

Corollary B.42. *Let V be a normed vector space, with dual V^*, and fix some nonzero vector $v_0 \in V$. There exists a functional $\varphi \in V^*$ such that*

$$\varphi(v_0) = \|v_0\|; \tag{B.122}$$
$$\|\varphi\| = 1. \tag{B.123}$$

Proof. Take $W = \mathbb{C} \cdot v_0$ in Corollary B.41, so that $\|\varphi_W\| = 1$ by construction. □

We now turn to an application of Theorem B.40 to convexity theory, which we will need for the Krein–Milman Theorem (and hence eventually for the existence of pure states on C*-algebras). Although we will apply the lemma below to the dual of a normed vector space in its w^*-topology, the setting is more general; all we need is a few easily established facts for topological vector spaces V, namely that if $U \subset V$ is open, then so is every translate $U + v$ of U, and so is εU, for any $\varepsilon > 0$, and hence also $(-\varepsilon U) \cap (\varepsilon U)$. Furthermore, a linear map $\varphi : V \to \mathbb{R}$ is continuous iff it is continuous at 0. These elementary facts will be used in the proof below.

Theorem B.43. *Let V be a real topological vector space and let A and B be disjoint nonempty convex subsets of V, with A open. Then there is a continuous linear functional $\varphi : V \to \mathbb{R}$ and some $t \in \mathbb{R}$ such that $\varphi(a) < t \leq \varphi(b)$ for all $a \in A, b \in B$.*

Proof. From $C = A - B = \{a - b \mid a \in A, b \in B\}$, which is convex and open (as it is a union of open sets $A + b$ over $b \in B$). Then move C so that it contains 0, by taking any $a_0 \in A$ and $b_0 \in B$ and defining $K = C + v_0$, with $v_0 = b_0 - a_0$. Thus K has its associated Minkowski functional p_K, cf. (B.111). Noting that $v_0 \notin K$ (since $A \cap B = \emptyset$), we have $p_K(v_0) \geq 1$. With $W = \mathbb{R} \cdot v_0$, define a functional $\varphi_W : W \to \mathbb{R}$ by $\varphi_W(sv_0) = s$ for $s \in \mathbb{R}$. This implies $\varphi_W(v) \leq p_K(v)$ for $v \in \mathbb{R} \cdot v_0$: if $v = sv_0$ with $s \geq 0$, this is obvious from (B.110) and $\varphi_W(v_0) = 1$, and if $s < 0$, then $\varphi_W(v) < 0$ whereas $p_K(v) \geq 0$. We now use Theorem B.40 to extend φ_W to a functional $\varphi : V \to \mathbb{R}$ satisfying (B.114), which implies $\varphi(v) \leq p_K(v) < 1$ for any $v \in K$. Taking $v = a - b + v_0$ gives $\varphi(a) < \varphi(b)$ for any $a \in A, b \in B$. Taking $t = \inf\{\varphi(b) \mid b \in B\}$, the last claim of the lemma follows. Finally, since $\varphi(v) < 1$ for each $v \in K$, we have $\varphi^{-1}(-\varepsilon, \varepsilon) \subset (-\varepsilon K) \cap (\varepsilon K)$, which is open, so that φ is continuous. □

This is the precise result we will need, but variations abound. If A *and* B are open, in which case $\varphi(B)$ is open, we have $\varphi(a) < t < \varphi(b)$. If V is ***locally convex***, in that its topology has a basis consisting of convex sets, then if A is closed and B is compact, there are disjoint open convex sets A' and B' containing A and B, respectively, so that also in this case we obtain the strict inequalities just mentioned.

Finally, even if V has no topology, we can still show that $\varphi(a) \leq t \leq \varphi(b)$ on the mere assumption that A has an interior point (φ then lacks continuity, of course).

Result like this are often called ***separation theorems***. Namely, a plane H in $\mathbb{R}^3$ always takes the form $x_0 + \ker\varphi = \varphi^{-1}(c)$, where $x_0 \in \mathbb{R}^3$ and $\varphi : \mathbb{R}^3 \to \mathbb{R}$ is a (nonzero) linear map. Equivalently, $H = \varphi^{-1}(c)$, where $c = \varphi(x_0)$. More generally, a ***hyperplane*** in a vector space V is a (nonempty) subspace of the form $H = \varphi^{-1}(c)$, where φ is a linear functional on V; clearly, H has codimension one and if V is a topological vector space and φ is continuous, then H is closed. So Theorem B.43 shows that A *and* B *are separated by the closed hyperplane* $H = \varphi^{-1}(t)$.

B.9 Duality

We now turn to duality theory. For any normed (but not necessarily complete) vector space V, Theorem B.33 shows that the space V^* of all morphisms $\varphi : V \to \mathbb{C}$ is a Banach space, called the ***dual*** of V. By (B.99), the norm of $\varphi \in V^*$ is given by

$$\|\varphi\| = \sup\{|\varphi(v)|, v \in V, \|v\|_V \leq 1\}. \tag{B.124}$$

Any morphism $a \in B(V,W)$ induces a ***dual morphism*** $a^* \in B(W^*,V^*)$ by

$$(a^*\varphi)(v) = \varphi(av), \;\; \varphi \in W^*. \tag{B.125}$$

By definition of the various norms involved here, we find

$$\|a^*\| = \sup\{|\varphi(av)|, \varphi \in W^*, v \in V, \|\varphi\| = \|v\| = 1\}. \tag{B.126}$$

Since $|\varphi(av) \leq \|\varphi\|\|av\| \leq \|a\|$, this immediately yields

$$\|a^*\| \leq \|a\|. \tag{B.127}$$

In fact, one even has

$$\|a^*\| = \|a\|, \tag{B.128}$$

but unexpectedly heavy machinery (namely the Hahn–Banach Theorem) is required to prove this. By Corollary B.42 (applied to W), for any $v \in V$, there exists $\varphi \in W^*$ with $\|\varphi\| = 1$ and $\varphi(av) = \|av\|$, so from (B.126) we have $\|a^*\| \geq \|av\|$ for any $v \in V$ with $\|v\| = 1$. Taking the supremum over such v and using (B.99) gives $\|a^*\| \geq \|a\|$. With our earlier (B.127), this gives (B.128).

Another application of Corollary B.42 lies in the ***double dual*** $V^{**} = (V^*)^*$.

Proposition B.44. *For any normed space V, the map $v \mapsto \hat{v}$ from V to V^{**}, given by*

$$\hat{v}(\varphi) = \varphi(v), \;\; \varphi \in V^*, \tag{B.129}$$

is isometric (and hence injective), mapping V onto a closed subspace $\hat{V} \subseteq V^{**}$.

This will follow from part 1 of the following consequence of Corollary B.42:

Corollary B.45. *Let V be a normed vector space, with dual V^**

1. For any $v \in V$, one has

$$\|v\| = \sup\{|\varphi(v)|, \varphi \in V^*, \|\varphi\| = 1\}. \tag{B.130}$$

2. For any $w \neq v$, there exists $\varphi \in V^*$ with $\varphi(w) \neq \varphi(v)$.
3. For any $a \in B(V,W)$, we have

$$\|a\| = \sup\{|\tau(av)|, v \in V, \tau \in W^*, \|v\| = \|\tau\| = 1\}. \tag{B.131}$$

Proof. This is the proof of Corollary B.45.

1. If $\|\varphi\| = 1$, then $|\varphi(v)| \leq \|v\|$, so the supremum is $\leq \|v\|$. But according to Corollary B.42 the supremum is $\geq \|v\|$.
2. Take $v_0 = v - w$ in Corollary B.42 and use the previous item.
3. Apply part 1 in W to $\|av\|$ and use (B.99). □

Proof. And this is the proof of Proposition B.44. Note that $\|\hat{v}\| \leq \|v\|$, since

$$\|\hat{v}\| = \sup\{|\varphi(v)\|, \varphi \in V^*, \|\varphi\| = 1\}, \tag{B.132}$$

and $|\varphi(v)\| \leq \|\varphi\|\|v\| = \|v\|$. Corollary B.42 shows this bound is saturated. □

If V is finite-dimensional, Proposition B.44 gives a *natural* isomorphism $V^{**} \cong V$, in contrast with the "unnatural" isomorphisms $V^* \cong V$ that require the choice of a basis (this terminology is made precise in category theory, see Appendix E).

In addition to their (metric) topology coming from the norm, both V and V^* naturally carry another topology (which will be of great importance in operator algebras and hence in quantum theory), defined in an almost identical way:

- The ***weak topology*** on V is the weakest topology that makes all functions $\varphi : V \to \mathbb{C}$ continuous, $\varphi \in V^*$. Equivalently, one has convergence $v_n \to v$ (of sequences, or, more generally, of nets) iff $\varphi(v_n) \to \varphi(v)$ for each $\varphi \in V^*$.
- The ***weak* topology*** (or ***w*-topology***) on V^* is the weakest topology that makes all functions $\hat{v} : V^* \to \mathbb{C}$ continuous, $v \in V$. Equivalently, it is the *topology of pointwise convergence*, in that $\varphi_n \to \varphi$ iff $\varphi(v_n) \to \varphi(v)$ for each $v \in V$ (etc.).

As their names suggest, these topologies are weaker than the norm topologies (except when V is finite-dimensional): indeed, if $\|v_n - v\| \to 0$ and $\varphi \in V^*$, then certainly $|\varphi(v_n) - \varphi(v)| \leq \|\varphi\|\|v_n - v\| \to 0$, and similarly for V^*. Consequently, a functional $\varphi : V \to \mathbb{C}$ is norm-continuous if it is weakly continuous, but the converse may be false. Nonetheless, the weak dual of V *coincides* with its norm dual, and we combine this with a contrasting result for the weak* continuous functionals V^*, which *en passant* locates the image $\hat{V}$ of V in V^{**} under (B.129):

Proposition B.46. • *Any functional $\varphi \in V^*$ is weakly continuous.*
• *A functional $\theta \in V^{**}$ is weak* continuous iff $\theta \in \hat{V}$.*

We just mention that, because of Corollary B.45.2, this proposition is a special case of a very general result on topological vector spaces. Namely, let V and W two vector spaces in ***separating duality***, that is, there is a bilinear form

$$\langle -, - \rangle : V \times W \to \mathbb{C}$$

such that for each $v, v' \in V$ there is $w \in W$ with $\langle v, w \rangle \neq \langle v', w \rangle$, and for each $w, w' \in W$ there is $v \in V$ with $\langle v, w \rangle \neq \langle v, w' \rangle$. Then V can be given the so-called $\sigma(V,W)$***-topology***, which is the weakest topology making each map $v \mapsto \langle v, w \rangle$ continuous ($w \in W$), and W likewise carries the $\sigma(W,V)$ topology (sometimes also called the $\sigma(V,W)$-topology). In particular, the weak topology on V is just the $\sigma(V,V^*)$-topology, whereas the weak* topology on V^* is the $\sigma(V^*,V)$-topology.

Theorem B.47. *Let V and W be vector spaces in separating duality. The space of $\sigma(V,W)$-continuous linear functionals on V coincides with W, and likewise, the space of $\sigma(W,V)$-continuous linear functionals on W coincides with V.*

This follows from elementary topology, and hence omit the proof. From this point of view, the apparent difference between the two parts of Proposition B.46 originates in the fact that the weak* topology on V^* is defined by its separating duality with V (or, equivalently, with $\overline{V}$), rather than its separating duality with V^{**}.

Next, the ***Banach–Alaoglu Theorem*** shows an unexpected but important property of the weak* topology (a least when V is infinite-dimensional). For example, in quantum theory this theorem implies w^*-compactness of the state space, and this, in turn (through the Krein–Milman Theorem), leads to an abundance of pure states.

Theorem B.48. *If V is a normed vector space, any d-ball*

$$V_d^* = \{\varphi \in V^*, \|\varphi\| \leq d\} \tag{B.133}$$

is compact in the weak topology. More generally, if U is any neighborhood of 0 in V, the set $V_U^* = \{\varphi \in V^*, |\varphi(x)| \leq d \forall v \in U\}$ is w^*-compact.*

Clearly, $U = V_1$ yields (B.133). Omitting the proof, we just note that the first claim is based on the fact that V_U^* is a closed subset of the space

$$\prod_{v\in V}\{z \in \mathbb{C} \mid |z| \leq d\|v\|\},$$

which is compact by Tychonoff's Theorem in topology (such reliance on awful non-constructive results is unfortunately typical of traditional functional analysis).

After this abstract theory, it is high time to turn to some examples; see Table B.1.

No.	V	V^*	*V^*-V-pairing*	*comment*
1.	$C_0(X)$	$M(X)$	$\langle\mu,f\rangle = \int_X d\mu\, f$	X locally compact Hausdorff space
2.	$C_b(X)$	$M(\beta X)$	$\langle\mu,f\rangle = \int_X d\mu\, f$	βX Čech–Stone compactification of X
3.	$\ell_0(X)$	$\ell^1(X)$	$\langle f,g\rangle = \sum_{x\in X} f(x)g(x)$	X countable set, $\ell_0(\mathbb{N})$ often called c_0
4.	$\ell^1(X)$	$\ell^\infty(X)$	$\langle f,g\rangle = \sum_{x\in X} f(x)g(x)$	X countable set
5.	$\ell^\infty(X)$	$\mathrm{ba}(X,\mathscr{P}(X))_{\mathbb{C}}$	$\langle\mu,g\rangle = \int_X d\mu\, g$	bounded finitely additive signed measures on X
6.	$\ell^p(X)$	$\ell^q(X)$	$\langle f,g\rangle = \sum_{x\in X} f(x)g(x)$	$\frac{1}{p}+\frac{1}{q}=1$, $p,q\neq 1,\infty$, X countable
7.	$\ell^2(X)$	$\overline{\ell^2(X)}$	$\langle f,g\rangle = \sum_{x\in X} \overline{f(x)}g(x)$	ℓ^2 treated as a Hilbert space
8.	H	$\overline{H}$	$\langle f,g\rangle = \langle f,g\rangle_H$	H general Hilbert space
9.	$L^2(X)$	$\overline{L^2(X)}$	$\langle f,g\rangle = \int_X d\mu\, \overline{f}g$	L^2 treated as a Hilbert space
10.	$L^1(X)$	$L^\infty(X)$	$\langle f,g\rangle = \int_X d\mu\, fg$	(X,Σ,μ) σ-finite measure space
11.	$L^p(X)$	$L^q(X)$	$\langle f,g\rangle = \int_X d\mu\, fg$	$\frac{1}{p}+\frac{1}{q}=1$, $p,q\neq 1,\infty$
12.	$B_0(H)$	$B_1(H)$	$\langle\rho,a\rangle = \mathrm{Tr}(\rho a)$	$B_0(H)$ compact operators, $B_1(H)$ trace class
13.	$B_1(H)$	$B(H)$	$\langle a,\rho\rangle = \mathrm{Tr}(\rho a)$	$B(H)$ bounded operators on H
14	M_*	M	$\langle a,\varphi\rangle = \varphi(a)$	M_* predual of von Neumann algebra M

Table B.1 Some Banach spaces and their duals, up to isometric isomorphism

1. The first entry is Theorem B.24.
2. This one is true by definition if we define the Čech–Stone compactification βX of a locally compact (Hausdorff) space as the Gelfand spectrum of $C_b(X)$ as a commutative C*-algebra, or, equivalently, by

$$C_b(X) \cong C(\beta X); \tag{B.134}$$

The compact Hausdorff space βX then has the feature that each $f \in C_b(X)$ has a unique continuous extension to βX. More generally, let X be a topological space. Provided it exists, "the" ***Čech–Stone compactification*** of X, denoted by βX, is a compact Hausdorff space together with a continuous map $\beta_X : X \to \beta X$ such that for each compact Hausdorff space K and each continuous function $f : X \to K$ there is a *unique* continuous function $\beta f : \beta X \to K$ such that the following diagram commutes:

$$\begin{array}{ccc} X & \xrightarrow{\beta_X} & \beta X \\ & \searrow^{f} & \downarrow{\scriptstyle \exists! \beta f} \\ & & K \end{array} \tag{B.135}$$

This universal property makes βX unique up to homeomorphism (if it exists). If X is locally compact Hausdorff, then βX exists and β_X is injective, making $\beta_X(X) \cong X$ a dense subspace of βX. The above diagram then implies (B.134) through $f \mapsto \beta f$; just take $K = \mathrm{Ran}(f)^-$, which is compact since f is bounded. Specializing this case to arbitrary sets X seen as discrete topological spaces, we can give an explicit description of βX as the set of all ultrafilters on X.

Definition B.49. *Let X be any set (seen as a discrete topological space).*

- *A* **filter** *on X is a non-empty collection F of subsets of X such that $A \in F$ and $B \in F$ implies $A \cap B \in F$, $A \in F$ and $A \subset B$ implies $B \in F$, and finally $\emptyset \notin F$.*
- *An* **ultrafilter** *is a filter that is maximal in the set of all proper filters F (i.e. $F \neq \mathscr{P}(X)\backslash\emptyset$), ordered by inclusion. It is straightforward to show that a filter F is maximal iff one and hence all of the following equivalent conditions hold:*
  a. for any $A \subset X$ we have either $A \in F$ or $A^c \in F$;
  b. if $A \cup B \in F$, then $A \in F$ or $B \in F$ (i.e., F is **prime***);*
  c. if $A \cap B \neq \emptyset$ for all $B \in F$, then $A \in F$.
- *For any $x \in X$, the set U_x of all subsets of X that contain x forms an ultrafilter, called* **principal***; any ultrafilter not of this kind is called* **free** *(if $|X| = \infty$, the existence of free ultrafilters on X follows from Zorn's Lemma).*

For discrete X, the set of all ultrafilters on X, endowed with the topology generated by all sets of the form $U_A = \{U \in \beta X \mid A \in U\}$, where $A \subset X$, is a realization of the Čech–Stone compactification of X, and may therefore be denoted by βX. Note that each U_A is clopen in βX. The embedding β_X maps $x \in X$ to the principal ultrafilter U_x, and the continuous extension βf of $f : X \to K$ is given by

$$\beta f(U) \equiv \lim_U f = \bigcap_{A \in U} f(A)^-, \tag{B.136}$$

Theorem 4.24 then explains the pairing in no. 2 of Table B.1 (see also no. 5).

3. • This is a special case of no. 1, since $\ell_0(X) = C_0(X)$, given that X is discrete (as a topological space). We then use the (Lebesgue–) ***Radon–Nikodym Theorem*** of measure theory: if (X, Σ, μ) is a σ-finite measure space and ν is a complex measure on Σ that is *absolutely continuous* with respect to μ (i.e., $\mu(A) = 0$ implies $\nu(A) = 0$, $A \in \Sigma$), then there is a function $d\nu/d\mu \in L^1(X)$ such that

$$\int_X d\nu\, f = \int_X d\mu \frac{d\nu}{d\mu} f, \; f \in L^\infty(X). \tag{B.137}$$

In the case at hand, X is countable and μ is the counting measure, with respect to which any measure is absolutely continuous. This yields $M(X) \cong \ell^1(X)$.

• Secondly, this duality is also a special case of Theorem B.27: as in (B.78),

$$\ell_0(X) = \ell^\infty(X, \mathscr{P}_f(X)), \tag{B.138}$$

so that bounded hermitian functionals $\varphi : \ell_0(X) \to \mathbb{C}$ (which in this case correspond to bounded real-linear functionals $\ell_0(X, \mathbb{R}) \to \mathbb{R}$) are given by

$$\varphi(g) = \lim_{n \to \infty} \int_A d\mu\, s_n,$$

where $g \in \ell_0(X)$, (s_n) is a sequence in $\mathrm{Step}(X, \mathscr{P}_f(X))$, which simply consists of functions on X with finite support, and μ is a finitely additive bounded signed measure on $\mathscr{P}_f(X)$, which is given by its values on any singleton $x \in X$ and hence is just a real-valued function

$$f(x) = \mu(\{x\}); \tag{B.139}$$

boundedness of μ gives $f \in \ell^1(X)$. Writing $X = \cup_n X_n$, where the X_n are finite and $X_n \subset X_{n+1}$ (e.g., for $X = \mathbb{N}$ one may take $X_n = \{1, \ldots, n\}$, so that $\sum_{x \in X_n} = \sum_{x=1}^n$), we may use $s_n = f_{|X_n}$ on X_n and $s_n(x) = 0$ outside X_n, which gives

$$\varphi(g) = \lim_{n \to \infty} \sum_{x \in X_n} f(x) g(x) = \sum_{x \in X} f(x) g(x). \tag{B.140}$$

One easily verifies that indeed $\|f\|_1 = \|\mu\|$, since (B.56) - (B.57) yield

$$\|\mu\| = \sup\{\mu_+(A) + \mu_-(A) \mid A \in \mathscr{P}_f(X)\} = \sup\left\{\sum_{x \in A} |f(x)|, A \in \mathscr{P}_f(X)\right\},$$

whose right-hand side in turn is equal to $\|f\|_1$.

• As a third approach, we give a direct proof of the desired duality

$$\ell_0(X)^* \cong \ell^1(X). \tag{B.141}$$

To start, for $f \in \ell^1(X)$ and $g \in \ell_0(X)$, we define an expression $\varphi_f(g)$ by

$$\varphi_f(g) = \langle f, g\rangle \equiv \sum_{x\in X} f(x)g(x). \tag{B.142}$$

By the obvious estimate

$$|\varphi_f(g)| \leq \|f\|_1 \|g\|_\infty, \tag{B.143}$$

which is Hölder's inequality for $p = 1$ and $q = \infty$, the sum (B.142) is absolutely convergent, and hence defines a linear map $\varphi_f : \ell_0(X) \to \mathbb{C}$, which satisfies $\|\varphi_f\| \leq \|f\|_1$. Thus the map $f \mapsto \varphi_f$ is well defined from $\ell^1(X)$ to $\ell_0(X)^*$. To prove surjectivity of this map, for given $\varphi \in \ell_0(X)^*$, define $f : X \to \mathbb{C}$ by

$$f(x) = \varphi(\delta_x). \tag{B.144}$$

It follows from continuity of φ that $\varphi = \varphi_f$, cf. (B.140), but it remains to be shown that $f \in \ell^1(X)$. To do so, for each $n \in \mathbb{N}$ we define $\varphi_n : \ell_0(X) \to \mathbb{C}$ by

$$\varphi_n(g) = \sum_{x\in X_n} f(x)g(x). \tag{B.145}$$

This operator is bounded, with

$$\|\varphi_n\| = \|s_n\|_1, \tag{B.146}$$

where s_n was defined prior to (B.140). To see this, we have

$$\|\varphi_n\| \leq \|s_n\|_1, \tag{B.147}$$

from (B.143), whereas the opposite inequality follows from a trick: define

$$g_n(x) = \overline{f(x)}/|f(x)| \ (x \in X_n, f(x) \neq 0); \tag{B.148}$$
$$g_n(x) = 0 \text{ (otherwise)}, \tag{B.149}$$

so that, assuming $\varphi \neq 0$, we have $\|g_n\|_\infty = 1$ and $\varphi_n(g_n) = \|s_n\|_1$, and hence

$$\|\varphi_n\| \geq \|s_n\|_1. \tag{B.150}$$

Since $\varphi(g) = \varphi_f(g)$ is finite by assumption, as in (B.140) $\lim_{n\to\infty} \varphi_n(g)$ exists for each $g \in \ell_0(X)$. Hence $\lim_{n\to\infty} \|\varphi_n(g)\|$ exists, so $\sup_n\{\|\varphi_n(g)\|\} < \infty$. The Principle of Uniform Boundedness (cf. Theorem B.78 below) then gives $\sup_n\{\|\varphi_n\|\} < \infty$, and this supremum equals $\sup_n\{\|s_n\|\} = \|f\|_1$.

Comparing the first two approaches, we see that bounded *finitely additive* measures on $\mathscr{P}_f(X)$ bijectively correspond to bounded σ*-additive* measures on $\mathscr{P}(X)$, both of which in turn are given by positive functions $f \in \ell^1(X)$.

4. This is similar to the third proof of the previous case. For $f \in \ell^\infty(X)$ and $g \in \ell^1(X)$, we define $\varphi_f(g)$ by (B.142), and instead of (B.143) we now obtain

$$|\varphi_f(g)| \leq \|f\|_\infty \|g\|_1. \tag{B.151}$$

Thus we have a map $f \mapsto \varphi_f$ from $\ell^\infty(X)$ to $\ell_1(X)^*$, satisfying $\|\varphi\| \leq \|f\|_\infty$. To prove surjectivity, for some $\varphi \in \ell^1(X)^*$ we once again define $f : X \to \mathbb{C}$ by (B.144), so that $\varphi = \varphi_f$ by continuity. Then for any $x \in X$,we obtain $|f(x)| \leq \|\varphi\| \|\delta_x\|_1 = \|\varphi\|$, so $\|f\|_\infty \leq \|\varphi\|$ and hence $\|\varphi\| = \|f\|_\infty$. In particular, $f \in \ell^\infty(X)$ and the bijection $\varphi_f \leftrightarrow f$ gives an isometric isomorphism à la (B.141):

$$\ell^1(X)^* \cong \ell^\infty(X). \tag{B.152}$$

5. Similar to no. 3, this is a special case of two more general dualities, namely

$$\ell^\infty(X)^* \cong M(\beta X); \tag{B.153}$$
$$\ell^\infty(X,\mathbb{R})^* \cong \mathrm{ba}(X,\mathscr{P}(X)), \tag{B.154}$$

cf. no. 2, and Theorem B.27, respectively. Thus bounded *finitely additive* measures μ on X (with underlying semiring $\mathscr{R} = \mathscr{P}(X)$) bijectively correspond to bounded σ-*additive* measures μ_β on βX (equipped with the Borel σ-algebra) by

$$\int_X d\mu\, f = \int_{\beta X} d\mu_\beta\, \beta f, \tag{B.155}$$

for any $f \in \ell^\infty(X)$. This is not as surprising as it seems, because there is a bijective correspondence between ultrafilters U on X and finitely additive probability measures μ on X that take values in $\{0,1\}$. This correspondence is given by:

$$U = \{A \subset X \mid \mu(A) = 1\}; \tag{B.156}$$
$$\mu(A) = 1 \text{ iff } A \in U. \tag{B.157}$$

Principal ultrafilters U_x thereby correspond to Dirac measures δ_x on X, whereas free ultrafilters U correspond to (finitely additive) measures μ_U on X that vanish on any finite subset of X. For general ultrafilters $U \in \beta X$ we have, for $f \in \ell^\infty(X)$,

$$\int_X d\mu\, f = \bigcap_{A \in U} f(A)^-, \tag{B.158}$$

where $f(A) = \{f(x) \mid x \in A\}$ as usual, and $f(A)^-$ is the closure of this set in $\mathbb{C}$. Thus (B.158) is equal to the unique $z \in \mathbb{C}$ with the property that for each $\varepsilon > 0$ the set $\{x \in \mathbb{N} : |f(x) - z| < \varepsilon\}$ lies in U; for $U = U_x$, this recovers $z = f(x)$.
6. This is similar to nos. 3 and 4, but is slightly more involved. For $f \in \ell^q(X)$ and $g \in \ell^p(X)$, with (B.16) and $p, q \neq 1, \infty$, we again define $\varphi_f(g)$ by (B.142), upon which Hölder's inequality yields $\|\varphi_f\| \leq \|f\|_q$. Conversely, for $\varphi \in \ell^p(X)^*$, once again define f by (B.144), so that $\varphi = \varphi_f$. We now show that $\|f\|_q \leq \|\varphi\|$.

Pick $X_n \subset X$ as defined below (B.139), and define $f_n : X \to \mathbb{C}$ by $f_n(x) = f(x)$ if $x \in X_n$ and $f_n(x) = 0$ if $x \notin X_n$. If $\|f\|_q < \infty$, then $\|f\|_q = \sup_n \|f_n\|_q$. Now define

$$g_n(x) = |f_n(x)|^q / f_n(x) \ (f_n(x) \neq 0); \tag{B.159}$$

$$g_n(x) = 0 \ (f_n(x) = 0). \tag{B.160}$$

Using (B.142), we obtain

$$\|f_n\|_q \|f_n\|_q^{q-1} = \|f_n\|_q^q = \langle f_n, g_n \rangle = \varphi(g_n) \leq \|\varphi\| \|g_n\|_p = \|\varphi\| \|f_n\|_q^{q-1}, \tag{B.161}$$

whence $\|f_n\|_q \leq \|\varphi\|$. Taking $\sup_n$ gives $\|f\|_q \leq \|\varphi\|$, and hence

$$\ell^p(X)^* \cong \ell^q(X). \tag{B.162}$$

7. $p = q = 2$ stands out as a special, self-dual case. As the next item explains, this is because $\ell^2(X)$ is a Hilbert space with inner product (B.10). This differs from the pairing (B.142) by the complex conjugation of the first term, making it appropriate to redefine the pairing between $\ell^2(X)^*$ and $\ell^2(X)$ in terms of the inner product. This leads to an *antilinear* isometric isomorphism $\ell^2(X)^* \cong \ell^2(X)$, as opposed to the *linear* isometric isomorphisms for all other values of p, q.
8. Proposition A.5 generalizes to infinite-dimensional Hilbert spaces (in which case it is often named after Riesz and Fréchet), with the following additions to the proof. First, boundedness of f guarantees that $\ker(f)$ is a closed subspace of H, so that (if $f \neq 0$) the orthogonal complement $\ker(f)^\perp$ is not empty by Proposition B.57 below. Second, uniqueness of the representing vector ψ in (A.13) now needs to be shown. This is easy: if $\langle \psi, \varphi \rangle = \langle \psi', \varphi \rangle$ for all $\varphi \in H$, then, taking $\varphi = \psi - \psi'$, it follows that $\langle \psi - \psi', \psi - \psi' \rangle = \|\psi - \psi'\|^2 = 0$, hence $\psi' = \psi$.

No. 9 follows from no. 8, whilst 10 and 11 are similar to 4 and 6, except for some tricky measure-theoretic details. We only sketch the main idea (where for simplicity we assume μ is finite; using an approximation procedure the result is valid also for the σ-finite case, but not beyond!). Namely, the function f representing the functional $\varphi \in L^p(X)^*$ is constructed by first defining a complex measure ν on Σ by $\nu(A) = \varphi(1_A)$, $A \in \Sigma$. Using (B.85), we see that ν is absolutely continuous with respect to μ, and we put

$$f = d\nu/d\mu. \tag{B.163}$$

Using definition (B.29) of integration, this yields

$$\varphi(g) = \langle f, g \rangle = \int_X d\mu \, fg, \tag{B.164}$$

and similar arguments as in the discrete case show that $f \in L^q(X)$.

Nos. 12–13 follow from Theorem B.146 below, and no. 14, which is forward-looking, too, is true by definition of the predual of a von Neumann algebra (whose *existence* is highly nontrivial); see Theorem C.132 in Appendix §C.

B.10 The Krein–Milman Theorem

Returning to the abstract theory, we now apply the Hahn–Banach Theorem and duality theory to prove one of the most beautiful results in functional analysis.

The ***boundary*** $\partial_e K$ of a convex set K consists of all $v \in K$ satisfying:

$$\textit{if } v = tw + (1-t)x \textit{ for certain } w, x \in K \textit{ and } t \in (0,1), \textit{ then } v = w = x.$$

Hence Caratheodory's Theorem 1.12, which, we recall, states that if K is a nonempty *compact* convex subset of $\mathbb{R}^n$, then $\partial_e K \neq \emptyset$, and each point of K is a convex sum of at most $n+1$ points in $\partial_e K$, implies, in particular, that $\partial_e K$ is not empty. This is readily visualized: the simplest example is $K = [0,1]$, where $\partial_e K = \{0,1\}$. One also has triangles in the plane, whose boundaries consist of their vertices (rather than their sides, which are among their *faces*, see below). Furthermore, the *closed* (unit) three-ball B^3 in $\mathbb{R}^3$ is convex, with boundary $\partial_e B^3 = S^2$, cf. Proposition 2.9. In these examples the interior of K, which is still convex, would have an empty boundary, so that the assumption of compactness in Theorem 1.12 is absolutely essential.

Caratheodory's Theorem follows from a straightforward induction argument in the dimension of K, and the following ***Krein–Milman Theorem***. The ***convex hull*** $\text{co}(X)$ of a subset X of a vector space is defined as the set of all convex sums $tx + (1-t)y$, where $t \in (0,1)$ and $x, y \in X$; this is the smallest convex set containing X.

Theorem B.50. *Let V be a real normed vector space with dual V^*, and let K be a convex subset of V^* that is compact in the w^*-topology. Then $\partial_e K \neq \emptyset$, and each point of K lies in the w^*-closure of the convex hull of $\partial_e K$. In other words,*

$$K = (\text{co}(\partial_e K))^-. \tag{B.165}$$

Zorn's Lemma will be used twice in the proof: both directly and through Theorem B.43, which relies on the Hahn–Banach Theorem B.40, whose proof uses Zorn. Furthermore, a ***face*** of a convex set K is a nonempty convex subset $F \subseteq K$ such that:

$$\textit{If } z = tx + (1-t)y \textit{ for } z \in F \textit{ with } t \in (0,1) \textit{ and } x, y \in K, \textit{ then } x, y \in F.$$

In particular, each extreme point $x \in \partial_e K$ is a face in its own right; conversely, a face consisting of a single point lies in $\partial_e K$ (as should be clear from the definitions).

Proof. 1. Let $\mathscr{F}(K)$ be the set of all *closed* faces in K, partially ordered by *inverse* inclusion, i.e., $F_1 \leqslant F_2$ iff $F_2 \subseteq F_1$. The intersection of any finite subset of a totally ordered subset $\{F_\lambda\}$ of $\mathscr{F}(K)$ is obviously nonempty, so that, by compactness of K, we also have $\cap_\lambda F_\lambda \neq \emptyset$. (Proof by contradiction: if $\cap_\lambda F_\lambda = \emptyset$, then $\cup_\lambda F_\lambda^c = (\cap_\lambda F_\lambda)^c = \emptyset^c = K$, so that $\{F_\lambda^c\}$ is an open cover of K, which definition of compactness has a finite subcover $\{F_{\lambda'}^c\}$. By the same argument, $\cap_{\lambda'} F_{\lambda'} = \emptyset$.) Hence $\cap_\lambda F_\lambda$ is an upper bound of $\{F_\lambda\}$, so that Zorn gives us a (not necessarily unique) maximal element F_0 in $\mathscr{F}(K)$ (which set-theoretically is *minimal* because of the reverse ordering, i.e., F_0 contains no strictly smaller closed face).

2. We now show that F_0 must be a singleton (and hence an extreme point of K). For any $v \in V$, the function $\hat{v} : V^* \to \mathbb{R}$ defined by $\hat{v}(\varphi) = \varphi(v)$ is w^*-continuous, see Propositions B.44 and B.46. Since $F_0 \subset K$ is compact, $\hat{v}$ assumes a minimum on F_0, say m. The set
$$F_m = \{\varphi \in F_0 \mid \hat{v}(\varphi) = m\} \tag{B.166}$$
is not only closed (by continuity of $\hat{v}$), and hence compact (since F is), but it is again a face in K: first, if $\varphi \in F_m$ takes the form
$$\varphi = t\varphi_1 + (1-t)\varphi_2, \tag{B.167}$$
with $\varphi_1, \varphi_2 \in F_0$, then
$$\hat{v}(\varphi) = m = t\hat{v}(\varphi_1) + (1-t)\hat{v}(\varphi_2), \tag{B.168}$$
which, given that $\hat{v}(\varphi_i) \geq m$, is only possible if $\hat{v}(\varphi_1) = \hat{v}(\varphi_2) = m$, so that $\varphi_i \in F_m$. Hence F_m is a face in F_0, but this implies that it is equally well a face in K. Namely, if (B.167) holds for $\varphi \in F_m$ and $\varphi_i \in K$, then regarding φ as an element of F_0 gives $\varphi_i \in F_0$, because F_0 is a face in K, upon which the previous step, where we regard φ as an element of F_m, gives $\varphi_i \in F_m$.
Since F_0 is maximal, we must have $F_m = F_0$, so that each functional $\hat{v}$ is constant on F_0. Now we know (even without the Hahn–Banach Theorem) that the functionals $\hat{v}$ separate points in V^*, since the very statement that $\varphi_1 \neq \varphi_2$ means that there is some $v \in V$ such that $\varphi_1(v) \neq \varphi_2(v)$ and hence $\hat{v}(\varphi_1) \neq \hat{v}(\varphi_2)$. So if F_0 contains more than one point, there must be a functional $\hat{v}$ that is not constant on F_0. Hence F_0 is a singleton, and therefore an element of $\partial_e K$. That is, $\partial_e K \neq \emptyset$.
3. The same argument applies to any closed face F in K, showing that each $F \in \mathscr{K}(K)$ contains at least one point in $\partial_e F$. But such a point is a face in F and hence in K, and being a one-point face in K, it must lie in $\partial_e K$. So we may strengthen the previous point by concluding that $F \cap \partial_e K \neq \emptyset$ for any closed face $F \subseteq K$.
4. To prove (B.165) by *reductio ad absurdum*, define
$$B = (\text{co}(\partial_e K))^-, \tag{B.169}$$
and assume $B \neq K$. First note that $\text{co}(\partial_e K)$ is convex by construction, and that its closure B remains convex (because the vector space operations, and *a fortiori* the convex sums, are continuous). Its complement in V^* is open, and hence any point $\alpha \in K \backslash B$ has an open convex neighbourhood $A \subset V^* \backslash B$ (see below), which is therefore disjoint from B. Hence Theorem B.43 applies (with $V \rightsquigarrow V^*$ and $\varphi \rightsquigarrow \hat{v}$), giving us $v \in V$ and $t \in \mathbb{R}$ for which $\hat{v}(\alpha) < t \leq \hat{v}(\beta)$ for any $\beta \in B$.
Now define $s = \min\{\hat{v}(\varphi) \mid \varphi \in K\}$, which exists since K is w^*-compact and $\hat{v}$ is w^* continuous. Since $\alpha \in K \backslash B \subset K$ and $\hat{v}(\alpha) < t$, we have $s < t$. Subsequently, define $F_s = \{\varphi \in K \mid \hat{v}(\varphi) = s\}$. As in step 2 above, it follows that F_s is a closed face in K. According to step 3, there is a point $\omega \in F_s \cap \partial_e K$, so that $\hat{v}(\omega) = s$. This contradicts $s < t \leq \hat{v}(\beta)$ for any $\beta \in B$, as $\omega \in \partial_e K \subset B$. □

The existence of A in step 4 above arises from the fact that open sets of the form

$$\mathscr{O}^{(\varepsilon)}_{v_1,\dots,v_n} = \{\varphi \in V^*, |\varphi(v_i)| < \varepsilon\, (i = 1,\dots,n)\}, \tag{B.170}$$

where $\varepsilon > 0$ and all $v_i \in V$, form a basis of w^*-neighbourhoods of $0 \in V^*$, and hence its translates $\omega + \mathscr{O}^{(\varepsilon)}_{v_1,\dots,v_n}$ form such a basis for any $\omega \in V^*$; the point is that such sets are convex, because if $|\varphi_i(v)| < \varepsilon$ for $i = 1,2$ and $t \in (0,1)$, then

$$|(t\varphi_1 + (1-t)\varphi_2)(v)| \leq t|\varphi_1(v)| + (1-t)|\varphi_2(v)| < (t+1-t)\varepsilon = \varepsilon. \tag{B.171}$$

Although the Krein–Milman Theorem is of considerable interest and beauty in itself, our main use of it lies in a few corollaries. Among those is Choquet's Theorem in the next section, but we first turn to the ***Stone–Weierstrass Theorem***:

Theorem B.51. *Let X be a compact Hausdorff space. Let B be an involutive subalgebra of $C(X)$ (regarded as a commutative C*-algebra) that separates points on X (i.e., if $x \neq y$ there is $f \in B$ such that $f(x) \neq f(y)$) and contains the unit function 1_X. Then B is dense in $C(X)$ in the sup-norm. In particular, if B is closed, then $B = C(X)$.*

In other words, B is a linear subspace of $C(X)$ such that if $f, g \in B$, then $fg \in B$, and if $f \in B$, then $f^* \in B$, where $f^*(x) = \overline{f(x)}$. Furthermore, $C(X)$ and hence B are equipped with the sup-norm. The assumptions could even be weakened: instead of asking that $1_X \in B$ and that B separate points, for the proof we just need that for each $x, y \in X$ and $s, t \in \mathbb{R}$ there is $f \in B$ such that $f(x) = s$ and $f(y) = t$.

We are going to derive Theorem B.51 from Theorem B.50 and the following:

Lemma B.52. *Let B be a linear subspace of some Banach space V. Then B is dense in V iff the only element $\varphi \in V^*$ that satisfies $\varphi(v) = 0$ for all $v \in B$ is $\varphi = 0$.*

Proof. The "$\Rightarrow$" direction (which will not be needed) is immediate from the fact that $\varphi \in V^*$ is bounded and therefore, if $v = \lim v_\lambda$ for (v_λ) in B, then $\varphi(v) = \lim \varphi(v_\lambda)$, so that $\varphi(v) = 0$ for all $v \in B$ implies $\varphi(v) = 0$ for all $v \in V$ and hence $\varphi = 0$.

Conversely, if $B^- \neq V$, we will exhibit some nonzero $\varphi \in V^*$ with $\varphi_{|B} = 0$. Take some $w \notin B^-$ and define $W \subset V$ by $W = \mathbb{C} \cdot w + B^-$, along with a map $\varphi_W : W \to \mathbb{C}$ given by $\varphi_W(\lambda w + v) = \lambda$ for any $\lambda \in \mathbb{C}$ and $v \in B^-$. This map is trivially linear, as well as bounded: since $w \notin B^-$ we have $\|w - v\| \geq d$ for some $d > 0$, for each $v \in B^-$; since then also $-v \in B^-$, we have $\|\lambda w + v\| \geq |\lambda| d$, and therefore

$$|\varphi_W(\lambda w + v)| = |\lambda| \leq d^{-1} \|\lambda w + v\|.$$

By Corollary B.41, our φ_W extends to some $\varphi \in V^*$, with $\varphi_{|B} = \varphi_{W|B} = 0$. □

Proof. We now prove Theorem B.51. We define a subspace $B^0 \subset M(X)$ by

$$B^0 = \{\mu \in M(X) \mid \mu(f^*) = \overline{\mu(f)}, \|\mu\| \leq 1, \mu(f) = 0 \forall f \in B\}, \tag{B.172}$$

where $f^*(x) = \overline{f(x)}$ as usual. Our aim is to show that

$$B^0 = \{0\}. \tag{B.173}$$

Since any $\varphi \in M(X)$ is a multiple of some μ in the unit ball $\|\mu\| \leq 1$, eq. (B.173) gives the antecedent of the "$\Leftarrow$" part of Lemma B.52, which gives Theorem B.51.

Noting that the w^*-topology in $M(X)$ is just the topology in which $\mu_\lambda \to \mu$ iff $\mu_\lambda(f) \to \mu(f)$ for each $f \in C(X)$, we see that B^0 is closed in the unit ball of $M(X)$, so that it is w^*-compact by the Banach–Alaoglu Theorem. Furthermore, B^0 is convex, so the Krein–Milman Theorem gives $\partial_e B^0 \neq \emptyset$. Any $\mu \in \partial_e B^0$ has either $\|\mu\| = 0$, in which case (B.173) holds and we are ready, or, as we assume in what follows,

$$\|\mu\| = 1. \tag{B.174}$$

Indeed, if $0 < \|\mu\| < 1$, then

$$\mu = t\mu_1 + (1-t)\mu_2, \tag{B.175}$$

with $t = \|\mu\|$, $\mu_1 = \mu/\|\mu\|$, and $\mu_2 = 0$ would give a nontrivial decomposition of μ.

For $g \in C(X)$, define

$$L_g : M(X) \to M(X); \tag{B.176}$$
$$L_g\mu(f) = \mu(gf), \tag{B.177}$$

or "$L_g d\mu = g \cdot d\mu$". It follows from the assumptions on B in Theorem B.51 that if $0 < g < 1_X$ and $g \in B$ (as we will now assume), then L_g maps B^0 into itself, and also $0 < 1_X - g < 1_X$. Hence $L_{1_X - g}$ maps B^0 into itself. Given (B.174), we then have

$$\|L_{1_X-g}\mu\| = 1 - \|L_g\mu\|. \tag{B.178}$$

This follows from (B.76): the Hahn-Jordan decomposition (B.55) of μ also gives $(L_g\mu)_\pm = L_g\mu_\pm$ and $(L_{1_X-g}\mu)_\pm = L_{1_X-g}\mu_\pm$ (since $g > 0$ and $1_X - g > 0$), so that

$$\|L_{1_X-g}\mu\| = L_{1_X-g}\mu_+(X) + L_{1_X-g}\mu_-(X) \tag{B.179}$$
$$= \mu_+(X) + \mu_-(X) - L_g\mu_+(X) - L_g\mu_+(X) = \|\mu\| - \|L_g\mu\|. \tag{B.180}$$

Because of (B.178), we obtain a convex decomposition (B.175) with $t = \|L_g\mu\|$, $\mu_1 = L_g\mu/\|L_g\mu\|$, and $\mu_2 = L_{1_X-g}\mu/\|L_{1_X-g}\mu\|$, which are well defined because of (B.174), which guarantees that the two denominators are nonzero. Since μ is extreme by assumption (i.e., it lies in $\partial_e B^0$), it must be that

$$\frac{L_g\mu}{\|L_g\mu\|} = \frac{L_{1_X-g}\mu}{\|L1_X - g\mu\|} = \mu. \tag{B.181}$$

Hence $g(x) = \|L_g\mu\|$ almost everywhere with respect to μ; in particular, this must hold for each $x \in \text{supp}(\mu)$. Suppose there are at least two different points $x, y \in \text{supp}(\mu)$. Since B separates points and contains 1_X, we can easily find $0 < g < 1_X$ such that $g(x) \neq g(y)$, contradicting constancy of g on $\text{supp}(\mu)$. So $\text{supp}(\mu) = \{x\}$, which, given (B.174), implies that $\mu = \pm\delta_x$, so that $\mu(1_X) = \pm 1$. Since $1_X \in B$, this contradicts (B.172). Hence (B.174) leads to a contradiction, and we are left with the other possibility $\|\mu\| = 0$. This gives $\mu = 0$, that is, (B.173). □

B.11 Choquet's Theorem

Choquet's Theorem B.53 beautifully follows up on the Krein–Milman Theorem. To state it, we need the ***support*** $\mathrm{supp}(\mu)$ of a measure μ on a space X, defined as the *smallest* closed set F such that $\mu(X\backslash F)=0$, or, equivalently, as the *largest* closed set F such that each open neighbourhood U of each $x\in F$ has strictly positive measure $\mu(U)>0$, provided such a set exists. This is the case, for example, if X is locally compact Hausdorff and μ is (inner) regular. To see this, let $\{U_\lambda\}$ be set of all open $U_\lambda\in\mathscr{O}(X)$ such that $\mu(U_\lambda)=0$, and let $U=\cup_\lambda U_\lambda$. By inner regularity, $\mu(U)=\sup\{\mu(K)\mid K\subset U, K\in\mathscr{K}(X)\}$. Since each such K is compact, $K\subset\cup_{i=1}^n U_{\lambda_i}$, whence $\mu(K)\leq\sum_i\mu(U_{\lambda_i})=0$. Hence $\mu(U)=0$, and $\mathrm{supp}(\mu)=X\backslash U$.

Theorem B.53. *In the notation of Theorem B.50, for each $\varphi\in K$ there is a probability measure μ on K whose support is contained in $\partial_e K^-$ such that for each $v\in V$,*

$$\varphi(v)=\int_{\partial_e K^-} d\mu(\omega)\,\omega(v). \tag{B.182}$$

Moreover, if K is metrizable, then the support of μ may be restricted to $\partial_e K$.

Here $\partial_e K^-\equiv(\partial_e K)^-$ is the closure of $\partial_e K$; in many examples (e.g., state spaces of C*-algebras of infinite quantum systems), $\partial_e K$ is not closed or even Borel.

Reading (B.182) from right to left, the point $\varphi\in K$ is called the ***barycenter*** of μ. Preparing for the proof, we note that if X is a compact Hausdorff space, the dual $C(X)^*$ of $C(X)$ as a Banach space (in the sup-norm) is the space $M(X)$ of all complete regular complex measures μ on X; cf. Theorem B.24. The set $M_1^+(X)$ of all complete regular probability measures on X is a closed subset of the unit ball of $M(X)$, since $\|\mu\|=\mu(X)=1$ if $\mu\in M_1^+(X)$, cf. (B.54), and hence $M_1^+(X)$ is w^*-compact by the Banach–Alaoglu Theorem. We will use these facts with $X=\partial_e K^-$.

We also recall that a (not necessarily continuous) function $f:K\to\mathbb{R}$ is ***affine*** if

$$f(t\varphi_1+(1-t)\varphi_2)=tf(\varphi_1)+(1-t)f(\varphi_2), \tag{B.183}$$

for $t\in(0,1)$ and $\varphi_1\neq\varphi_2\in K$, ***concave*** if one has $\geq$ instead of $=$ in (B.183), ***convex*** with $\leq$ instead of $=$, and ***strictly convex*** if (B.183) holds with $=\rightsquigarrow<$.

For example, $f(x)=x^2$ is strictly convex on $[-1,1]$. The assumption of metrizability will only be used to prove the existence of a strictly convex continuous function on K, so this existence could have been assumed instead of metrizability. Finally, we denote the space of real-valued continuous affine functions on K by $A(K)$.

Proof. By Theorem B.50, $\varphi=\lim\varphi_\lambda$, where (φ_λ) is some net in $\mathrm{co}(\partial_e K)$, so that $\varphi_\lambda=\sum_i p_i^{(\lambda)}\omega_i^{(\lambda)}$, where the sum is finite, $p_i^{(\lambda)}\geq 0$, and $\sum_i p_i^{(\lambda)}=1$. Then $\mu_\lambda=\sum_i p_i^{(\lambda)}\delta_{\omega_i^{(\lambda)}}$ is a probability measure on $\partial_e K$ and hence also on its (compact) closure $\partial_e K^-$. Since $M_1^+(\partial_e K^-)$ is w^*-compact, the previous net has a subnet that w^*-converges to some $\mu\in M_1^+(\partial_e K^-)$. Noting that $\varphi(v)=\hat{\varphi}(v)$, where $\hat{v}\in V^{**}$ is w^*-continuous by Proposition B.46, this μ by construction satisfies (B.182).

We now prove the last claim. If K is metrizable, then $C(K)$ is separable, so that its subspace $A(K)$ is separable, too. Thus we can find some countable dense subset $(f_n)_{n>0}$ of $A(K)$, in terms of which we define a function $f_0 : K \to \mathbb{R}$ by

$$f_0(\varphi) = \sum_{n=1}^{\infty} 2^{-n}(\|f_n\|_\infty + 1)^{-2}|f_n(\varphi)|^2. \tag{B.184}$$

First, continuity of f_0 follows from uniform convergence of this series and continuity of each f_n; recall that $A(K) \subset C(K,\mathbb{R})$. Second, the x^2 example just given implies that if $f \in A(K)$, then f^2 is convex, and it is even strictly convex provided there is at least one $n > 0$ for which $f_n(\varphi_1) \neq f_n(\varphi_2)$. To show that this is the case, we note that since $V \subset V^{**}$ separates points in V^* and each $\hat{v} \in V^{**}$ defines an element of $A(K)$ by restriction, $A(K)$ separates points in K. Therefore, by density of the family (f_n), the claim follows, and f_0 is strictly convex. This will be crucial.

For each real-valued $f \in C(K,\mathbb{R})$, define the ***concave envelope*** $\hat{f}$ by

$$\hat{f}(\varphi) = \inf\{g(\varphi) \mid g \in A(K), g \geq f\}. \tag{B.185}$$

The terminology comes from the fact that $f \leq \hat{f}$ for any $f \in C(K)$, with equality if f is concave; this is because for any continuous concave function f we may write

$$f(\varphi) = \inf\{g(\varphi) \mid h \in A(K), g \geq f\}. \tag{B.186}$$

In terms of this, for any fixed element $\varphi_0 \in K$ we define $p : C(K,\mathbb{R}) \to \mathbb{R}$ by

$$p(f) = \hat{f}(\varphi_0). \tag{B.187}$$

Since $\widehat{f+g} \leq \hat{f} + \hat{g}$ and $\widehat{tf} = t\hat{f}$ for $t \geq 0$, as is easily verified, it follows that p is sublinear (cf. Definition B.38). We define a linear subspace $W \subset C(K,\mathbb{R})$ by

$$W = A(K) + \mathbb{R} \cdot f_0, \tag{B.188}$$

endowed with the 'hatted' evaluation map $\widehat{\mathrm{ev}}_{\varphi_0} : W \to \mathbb{R}$ defined by

$$\widehat{\mathrm{ev}}_{\varphi_0}(g + sf_0) = g(\varphi_0) + s\hat{f}_0(\varphi_0); \tag{B.189}$$

since $g = \hat{g}$ for any $g \in A(K)$, for $s \geq 0$ we have $\widehat{\mathrm{ev}}_{\varphi_0}(g + sf_0) = \mathrm{ev}_{\varphi_0}(\hat{g} + \widehat{sf_0})$.

It is easy to show that p dominates $\widehat{\mathrm{ev}}_{\varphi_0}$, so that the Hahn–Banach Theorem B.40 yields an extension $\widehat{\mathrm{ev}}'_{\varphi_0}$ of $\widehat{\mathrm{ev}}_{\varphi_0}$ to $C(K,\mathbb{R})$ that satisfies $\widehat{\mathrm{ev}}'_{\varphi_0}(f) \leq \hat{f}(\varphi_0)$. This implies that $\widehat{\mathrm{ev}}'_{\varphi_0}$ is positive; to see this, take $f \leq 0$. Since the zero function is in $A(K)$ we have $\hat{f} \leq 0$ also, so that $\widehat{\mathrm{ev}}'_{\varphi_0}(f) \leq 0$. Passing to $-f$, we find that $\widehat{\mathrm{ev}}'_{\varphi_0}(f) \geq 0$ whenever $f \geq 0$. Furthermore, since $1_K \in A(K) \subset W$, we have

$$\widehat{\mathrm{ev}}'_{\varphi_0}(1_K) = \widehat{\mathrm{ev}}_{\varphi_0}(1_K) = 1_K(\varphi_0) = 1.$$

Therefore, $\widehat{\text{ev}}'_{\varphi_0}$ is a state on $C(K)$. Corollary B.17 then turns $\widehat{\text{ev}}'_{\varphi_0}$ into a probability measure μ on K. Taking $f = \hat{v}$ for some $v \in V$, we have $f \in A(K) \subset W$, so that

$$\int_K d\mu(\omega)\,\omega(v) \equiv \int_K d\mu\,\hat{v} \equiv \mu(\hat{v}) = \widehat{\text{ev}}_{\varphi_0}(\hat{v}) = \hat{v}(\varphi_0) = \varphi_0(v). \tag{B.190}$$

This is almost (B.182) with $\varphi \rightsquigarrow \varphi_0$; what we still need to prove is the property

$$\text{supp}(\mu) \subseteq \partial_e K. \tag{B.191}$$

This will be proved in two steps. For any $f \in C(K)$, we define $K(f) \subset K$ by

$$K(f) = \{\varphi \in K \mid f(\varphi) = \hat{f}(\varphi)\}. \tag{B.192}$$

We will separately show that

$$\text{supp}(\mu) \subseteq K(f_0); \tag{B.193}$$
$$K(f_0) \subseteq \partial_e K. \tag{B.194}$$

Towards (B.193) we start showing that

$$\mu(f_0) = \mu(\hat{f}_0), \tag{B.195}$$

which is a conjunction of $\mu(f_0) \leq \mu(\hat{f}_0)$ and $\mu(f_0) \geq \mu(\hat{f}_0)$. The first is true for any $f \in C(K)$, since μ is positive and $f \leq \hat{f}$ (pointwise). The second is specific to f_0:

$$\begin{aligned}\mu(f_0) &= \widehat{\text{ev}}'_{\varphi_0}(f_0) = \widehat{\text{ev}}_{\varphi_0}(f_0) = \hat{f}_0(\varphi_0)\\ &= \inf\{g(\varphi_0) \mid g \in A(K), g \geq f_0\}\\ &= \inf\{\mu(g) \mid g \in A(K), g \geq f_0\},\end{aligned} \tag{B.196}$$

since for $g \in A(K)$ we have $g(\varphi_0) = \mu(g)$ because $A(K) \subset W$. If in addition $g \geq f_0$, we have $g \geq \hat{f}_0$, which implies $\mu(g) \geq \mu(\hat{f}_0)$. This inequality survives the infimum in (B.196), so that we finally obtain $\mu(f_0) \geq \mu(\hat{f}_0)$, and hence (B.195).

We now prove (B.193) from (B.195). Since $f_0 \leq \hat{f}_0$, for each $n > 0$ we may define

$$K_n = \{\varphi \in K \mid \hat{f}_0(\varphi) - f_0(\varphi) \geq 1/n\}. \tag{B.197}$$

Then $0 \leq \mu(K_n) \leq n \cdot \int_K d\mu\,(\hat{f}_0 - f_0)$, which vanishes by (B.195). Hence $\mu(K_n) = 0$ for each n, and therefore $\mu(\cup_n K_n) = 0$. But $\cup_n K_n = K(f_0)^c$, so (B.193) follows.

Eq. (B.194) is equivalent to the inclusion $(\partial_e K)^c \subseteq K(f_0)^c$, i.e., the implication:

if $\varphi = t\varphi_1 + (1-t)\varphi_2$ for some $t \in (0,1)$ and $\varphi_1 \neq \varphi_2$, then $\hat{f}_0(\varphi) \neq f_0(\varphi)$.

Indeed, strict convexity of f_0 (used at last!) and the familiar property $f_0 \leq \hat{f}_0$ give

$$\begin{aligned}\hat{f}_0(\varphi) &= \inf\{tg(\varphi_1) + (1-t)g(\varphi_2) \mid g \in A(K), g \geq f_0\}\\ &\geq t\inf\{g(\varphi_1) \mid g \in A(K), g \geq f_0\} + (1-t)\inf\{g(\varphi_2) \mid g \in A(K), g \geq f_0\}\\ &= t\hat{f}_0(\varphi_1) + (1-t)\hat{f}_0(\varphi_2) \geq tf_0(\varphi_1) + (1-t)f_0(\varphi_2) > f_0(\varphi).\end{aligned}$$

□

In turn, the existence of some measure μ in (B.182) representing an arbitrary point $\varphi \in K$ implies the Krein–Milman Theorem. We rewrite (B.182) as

$$\hat{v}(\varphi) = \int_{\partial_e K^-} d\mu\, \hat{v}, \tag{B.198}$$

where $\varphi \in K$ is arbitrary and $\hat{v} \in C(K)$ is the (affine) continuous function on $K \subset V^*$ induced by the functional $\hat{v} \in V^{**}$ on V^* defined by $v \in V$ under the canonical injection $V \hookrightarrow V^{**}$, $v \mapsto \hat{v}$, see Proposition B.44. From (B.198) and (B.34) we obtain

$$|\hat{v}(\varphi)| \leq \|\hat{v}\|_\infty^{(\partial_e K^-)},$$

which, because $\partial_e K^- \subset (\text{co}(\partial_e K))^-$, also gives the inequality

$$|\hat{v}(\varphi)| \leq \|\hat{v}\|_\infty^{((\text{co}(\partial_e K))^-)}.$$

This forces $\varphi \in (\text{co}(\partial_e K))^-$, for if $\varphi \notin (\text{co}(\partial_e K))^-$ we would obtain a contradiction with Theorem B.43 (which is a version of the Hanh–Banach Theorem), or more precisely, with the alternative version thereof stated after its proof, with $A = \{\varphi\}$ closed and $B = (\text{co}(\partial_e K))^-$ compact and convex (and, of course, $\varphi \rightsquigarrow -\varphi$). Therefore, $K \subseteq (\text{co}(\partial_e K))^-$, which implies (B.165).

If only to illustrate Choquet's Theorem, we note that existence of the probability measure μ in the Riesz Representation Theorem B.15 follows from it. To see this, fix some compact Hausdorff space X, and take $V = C(X,\mathbb{R})$ (as a real Banach space in the supremum-norm) and $K = S(C(X,\mathbb{R})) \subset V^*$, i.e., the set of positive linear functionals $\varphi : C(X,\mathbb{R}) \to \mathbb{R}$ that satisfy $\varphi(1_X) = 1$. By the argument following Definition 1.14, K coincides with the state space $S(C(X))$ of the commutative C*-algebra $C(X)$, which is a *complex* Banach space (cf. Appendix C), in that each $\varphi \in K$ extends uniquely to a state $\varphi : C(X) \to \mathbb{C}$ by complex linearity, which extension remains positive in the sense of Definition C.3. From Propositions C.14 and C.19, the map $X \to V^*$ given by $x \mapsto \text{ev}_x$, where $\text{ev}_x(f) = f(x)$ is the evaluation map at x, takes values in $\partial_e K$ and yields a homeomorphism

$$\partial_e K \cong X. \tag{B.199}$$

In particular, $\partial_e K$ is closed in V^* (and in K), so (B.182) comes down to (B.39).

The part of Theorem B.15 that does not follow from Theorem B.53 is the possible uniqueness of the measure μ on $\partial_e K^-$ that represents the point $\varphi \in K$. Uniqueness of the measure in Choquet's Theorem is settled by the following notion.

Definition B.54. *A* **(Choquet) simplex** *is a compact convex set $K \subset V^*$ whose associated convex cone $\tilde{K} = \mathbb{R}^+ \cdot K \equiv \{t\omega \mid t \geq 0, \omega \in K\}$ (cf. Definition C.50) is a lattice in the partial ordering $\leq$ defined by $\rho \leq \sigma$ iff $\sigma - \rho \in \tilde{K}$*

Here we assume that for any $\rho \in \tilde{K}$ there is a unique $t \in \mathbb{R}^+$ and $\omega \in K$ such that $t\omega = \rho$; this is the case if $K = \tilde{K} \cap H$ for some closed hyperplane H in V^* that does not contain the origin. For example, if $K = S(A)$ is the state space of some unital C*-algebra A, then $H = \{\varphi \in A^* \mid \varphi(1_A) = 1\}$ and $\tilde{K} = \{\varphi \in A^* \mid \varphi \geq 0\}$).

In finite dimension, Choquet simplices are special convex polytopes called ***simplices***. Recall that the so-called ***regular polyhedra*** were classified (up to affine isomorphism) by Schläfli in 1852, who showed that the only possibilities are:

- The *simplices* $\Delta_n = \{x \in \mathbb{R}^{n+1} \mid x_i \geq 0, \sum_i x_i = 1\}$, $n \geq 1$;
- The *cubes* $Q_n = \{x \in \mathbb{R}^n \mid -1 \leq x_i \leq 1\}$, $n > 1$;
- The *cross-polytopes* $O_n = \{x \in \mathbb{R}^n \mid \sum_i |x_i| \leq 1\}$, $n > 1$;
- The countably many *regular polygons* in $\mathbb{R}^2$ (which include Q_2, O_2, Δ_2);
- The five *Platonic solids* in $\mathbb{R}^3$ (which include Q_3, O_3, Δ_3);
- The six *regular polychora* in $\mathbb{R}^4$ (which include Q_4, O_4, Δ_4).

An n-dimensional simplex is affinely homeomorphic to the convex hull of $n+1$ *linearly independent* points (or, equivalently, $|\partial_e K| = n+1$). In particular, the simplex Δ_n is the set $\Pr(\underline{n+1})$ of all probability distributions on a set $X = \underline{n+1}$ of cardinality $n+1$, cf. Definition 1.9. Generalizing this idea, if X is a compact Hausdorff space, then the state space $S(C(X))$ of the associated commutative C*-algebra $C(X)$, which as we know consists of all probability measures on X, is a Choquet simplex.

In the notation of Theorem B.53, the simplest result (again due to Choquet) is:

Theorem B.55. *Suppose K is metrizable, and assume* $\mathrm{supp}(\mu) \subseteq \partial_e K$ *in* (B.182). *Then μ is uniquely determined by its barycenter φ iff K is a Choquet simplex.*

However, we note that without any assumption on K, conversely the barycenter φ for which (B.182) holds for all $v \in V$ is uniquely determined by μ. This observation gives rise to a map B from the compact convex set $M(K)_1^+$ of all probability measures on K to K itself, such that $B(\mu)$ is the unique point in K such that (B.198) with $\varphi = B(\mu)$ holds for all $v \in V$. This map B is, in fact, affine as well as continuous.

Theorem B.55 covers finite phase spaces in classical mechanics as well as, negatively, finite-dimensional Hilbert spaces in quantum mechanics: in the former case, any state admits a unique decomposition into pure states (cf. Proposition 1.13), whereas in the latter this fails. For example, for $H = \mathbb{C}^2$, the state space $S(B(H)) \cong B^3$ (see Proposition 2.9) is not a simplex. See also Proposition 2.14.

To explain the general (i.e., non-metrizable) case, we first define the ***Choquet ordering*** $\prec$ on the set of probability measures on K by $\mu \prec \nu$ iff $\mu(f) \leq \nu(f)$ for any *convex* function $f \in C(K, \mathbb{R})$. Noting that $B(\mu) = B(\nu)$ whenever $\mu \prec \nu$, the idea is that since the values of convex functions almost by definition increase towards the boundary $\partial_e K$, probability measures on K with given barycenter that are maximal with respect to $\prec$ should be supported on $\partial_e K$ (such maximal measures always exist by a Zorn's Lemma argument). This intuition is indeed correct, *provided K is metrizable*, in which case, conversely, the condition $\mathrm{supp}(\mu) \subseteq \partial_e K$ in Theorem B.55 forces μ to be maximal. In general, an alternative way to prove the first part of Theorem B.53 would be to take some maximal μ with given barycenter μ.

The key to the generalization of Theorem B.55 to the possibly non-metrizable case, then, is to replace the assumption $\mathrm{supp}(\mu) \subseteq \partial_e K$ by maximality of μ. This is achieved by the major ***Choquet–Meyer Theorem***, which we state without proof:

Theorem B.56. *Assume the measure μ in* (B.182) *is maximal with respect to $\prec$. Then μ is uniquely determined by its barycenter φ iff K is a Choquet simplex.*

B.12 A précis of infinite-dimensional Hilbert space

The main difference between infinite-dimensional Hilbert spaces and their finite-dimensional counterparts lies in issues of convergence and completeness. Every linear subspace of a finite-dimensional Hilbert space is automatically complete (cf. Proposition B.5), and all sums one encounters are finite. In infinite dimension, $\ell_c(\mathbb{N})$ is a linear but incomplete subspace of $\ell^2(\mathbb{N})$, and similarly for $C_c(\mathbb{R}) \subset L^2(\mathbb{R})$; the expansion of some vector in terms of a basis already involves an infinite sum.

Note that in metric spaces a subset is closed iff it is sequentially complete (in that it contains all limits of Cauchy sequences); this can be seen from the fact that the metric topology is generated by ε-balls and hence by $(1/n)$-balls, $n \in \mathbb{N}$. Consequently, in Banach spaces (and hence in Hilbert spaces) H, the property of some subspace $L \subset H$ being (metrically) *complete* (in the sense that every Cauchy sequence in L converges to an element of L) is the same as L being (topologically) closed (in the sense that the set-theoretic complement L^c is open). Following tradition in functional analysis, we will henceforth speak of *closed* subspaces. We denote the (metric or topological) closure of $S \subset H$ in H by S^-.

An exhaustive way of guaranteeing that some linear subspace $L \subset H$ is closed is to exhibit it as an ***orthogonal complement*** $L = S^\perp$, where $S \subset H$ is *any* subset: we write $\psi \perp S$ iff $\langle \chi, \psi \rangle = 0$ for each $\chi \in S$, and, as in (A.29), put

$$S^\perp = \{\psi \in H \mid \psi \perp S\}. \tag{B.200}$$

We also use the ***double orthogonal complement*** $S^{\perp\perp} \equiv (S^\perp)^\perp$, *et cetera.*

Proposition B.57. *Let H be a Hilbert space.*

1. *If $S \subset H$ is any subset, $S^\perp$ is a closed linear subspace of H.*
2. *For each* closed *linear subspace $L \subset H$, one has*
$$H = L \oplus L^\perp, \tag{B.201}$$
in the sense that
$$L \cap L^\perp = \{0\}, \tag{B.202}$$
and each vector $\psi \in H$ has a unique *decomposition*
$$\psi = \psi^\parallel + \psi^\perp, \tag{B.203}$$
where $\psi^\parallel \in L$ and $\psi^\perp \in L^\perp$.
3. *For any* closed *linear subspace L one has $L^{\perp\perp} = L$.*
4. *For* any *linear subspaces L, one has*
$$L^{\perp\perp} = L^-, \tag{B.204}$$
and hence $L^- = H$ iff $\langle \psi, \varphi \rangle = 0$ for each $\varphi \in L$ implies $\psi = 0$.
5. *For any subset $S \subset H$, one has $S^{\perp\perp\perp} = S^\perp$.*
6. *For any subset $S \subset H$, the closure $[S]^-$ of the (finite) linear span $[S]$ of S is $S^{\perp\perp}$.*

Proof. 1. Linearity of $S^\perp$ follows from linearity of the inner product. If $\psi_n \in S^\perp$ and $\psi_n \to \psi$, then for $\chi \in S$ and each n, we have

$$|\langle \chi, \psi \rangle| = |\langle \chi, \psi - \psi_n \rangle| \leq \|\chi\| \|\psi - \psi_n\|. \tag{B.205}$$

Taking $n \to \infty$ gives $\langle \chi, \psi \rangle = 0$ and hence $\psi \in S^\perp$, so that $S^\perp$ is closed.

2. The proof of the infinite-dimensional case (cf. Corollary A.9 for finite dimension) relies on ***Riesz Lemma*** B.58 below, which explains why L needs to be *closed*, and also neatly identifies $\psi^\parallel$ as the unique vector in L at minimal distance to ψ. Granting this important lemma, let $\psi \in H$, we take

$$C = \psi + L \equiv \{\psi + \varphi, \varphi \in L\}. \tag{B.206}$$

Lemma B.58 yields a unique vector $\chi_0 \in C$, from which we define $\psi^\parallel = \psi - \chi_0$ and $\psi^\perp = \chi_0$ (so that $\|\psi^\parallel - \psi\| = \|\chi_0\|$ is minimal). Then $\psi^\parallel \in L$, and (B.203) holds by construction. To show that $\chi_0 \in L^\perp$, we rewrite the inequality $\|\chi_0\| \leq \|\psi + \varphi\|$ (for all $\varphi \in L$) as $\|\chi_0\| \leq \|\chi_0 + \varphi\|$, since $\psi = \chi_0 + \psi^\parallel$ and $\psi^\parallel \in L$. Putting $\varphi = -(\langle \zeta, \chi_0 \rangle / \|\zeta\|^2)\zeta$, with $\zeta \in L$ arbitrary (but nonzero), the last inequality reads $0 \leq -|\langle \zeta, \chi_0 \rangle|^2 / \|\zeta\|^2$, whence $\langle \zeta, \chi_0 \rangle = 0$ for all $\zeta \in L$, so that $\chi_0 \in L^\perp$. Uniqueness of the decomposition (B.203) follows as in Corollary A.9.

3. Trivially, $L \subseteq L^{\perp\perp}$. To prove the converse inclusion, use the previous item.
4. If $A \subseteq B$, then $B^\perp \subseteq A^\perp$ and hence $A^{\perp\perp} \subseteq B^{\perp\perp}$. With $A = L$ and $B = L^-$, this gives $L^{\perp\perp} \subseteq (L^-)^{\perp\perp} = L^-$ (where, L^- being closed, we used the previous item). Conversely, $L \subseteq L^{\perp\perp}$ and hence $L^- \subseteq L^{\perp\perp}$, since $L^{\perp\perp}$ is closed by the first item.
5. Take $L = S^\perp$ and use the third item.
6. Proceeding as in the proof of no. 1, from the continuity of the inner product we find $S^\perp = ([S]^-)^\perp$, and hence, using no. 3, finally $S^{\perp\perp} = ([S]^-)^{\perp\perp} = [S]^-$. □

Lemma B.58. *The norm assumes a unique minimum on any closed convex set $C \subset H$ (i.e., there is a unique $\chi_0 \in C$ such that $\|\chi_0\| < \|\chi\|$ for each $\chi \in C$, $\chi \neq \chi_0$).*

Proof. Let $\mu = \inf\{\|\chi\|, \chi \in C\}$, which exists, as $\|\chi\| \geq 0$. Hence there is a minimizing sequence (χ_n) in C with $\|\chi_n\| \to \mu$, which we now prove to be Cauchy (in H). Since C is convex, $\frac{1}{2}(\chi_n + \chi_m) \in C$, and therefore, $\|\chi_n + \chi_m\| \geq 2\mu$. Thus

$$0 \leq \|\chi_n - \chi_m\|^2 = 2(\|\chi_n\|^2 + \|\chi_m\|^2) - \|\chi_n + \chi_m\|^2 \leq 2(\|\chi_n\|^2 + \|\chi_m\|^2) - 4\mu^2,$$

and since $2(\|\chi_n\|^2 + \|\chi_m\|^2) \to 4\mu^2$ as $n, m \to \infty$, we must have $\|\chi_n - \chi_m\| \to 0$. Since C is closed, $\chi_n \to \chi_0$ for some $\chi_0 \in C$. To prove uniqueness, let another minimizing sequence (χ'_n) converge to $\chi'_0 \in C$. Then $\frac{1}{2}(\chi_0 + \chi'_0) \in C$, so we obtain

$$\|\chi_0 + \chi'_0\| \geq 2\mu = \|\chi_0\| + \|\chi'_0\|.$$

The inequality $\|\chi_0 + \chi'_0\| \leq \|\chi_0\| + \|\chi'_0\|$ gives $\|\chi_0 + \chi'_0\| = \|\chi_0\| + \|\chi'_0\|$, i.e. $\mathrm{Re}\langle \chi'_0, \chi_0 \rangle = \|\chi'_0\| \|\chi_0\|$. Cauchy–Schwarz gives $|\langle \chi'_0, \chi_0 \rangle| \leq \|\chi'_0\| \|\chi_0\|$ with equality iff χ'_0 and χ_0 are proportional, so the previous equality can hold only if $\chi'_0 = t\chi_0$ for some $t \geq 0$. Since χ'_0 and χ_0 both minimize the norm, we have $t = 1$. □

We now turn to the important concept of a ***basis*** of a Hilbert space; as in the previous appendix, *a basis of a Hilbert space always denotes an* ***orthonormal basis***. To define this notion, we first say that some subset $\{\upsilon_i\}_{i\in I}$ of H is ***orthonormal*** if

$$\langle \upsilon_i, \upsilon_j\rangle = \delta_{ij}; \tag{B.207}$$

this condition guarantees that the υ_i are linearly independent (and easy to calculate with!). Second, in finite dimension (where I must be finite) we may simply define a basis of H as an orthonormal set that is also a basis in the usual (linear algebra) sense. This idea remains valid for general Hilbert spaces, except that we should use Definition B.6 to define infinite sums (and Lemma B.7 to analyze them). Theorem B.61 to come gives an exhaustive account of the situation, but we first need a lemma on general orthonormal sets (that do not necessarily form a basis).

Lemma B.59. *If $\{\upsilon_i\}_{i\in I}$ is an orthonormal set in H and $c_i \in \mathbb{C}$, then the sum*

$$\psi = \sum_{i\in I} c_i \upsilon_i \tag{B.208}$$

converges in H (in the sense of Definition B.6) iff

$$\sum_{i\in I} |c_i|^2 < \infty. \tag{B.209}$$

If this is the case, the coefficients $c_i \in \mathbb{C}$ are given by

$$c_i = \langle \upsilon_i, \psi\rangle. \tag{B.210}$$

Proof. The first claim follows from Proposition B.8 and the elementary computation

$$\| \sum_{i\in G'} c_i\upsilon_i\|^2 = \sum_{i\in G'} \|c_i\upsilon_i\|^2 = \sum_{i\in G'} |c_i|^2 < \varepsilon, \tag{B.211}$$

where G' is finite, so that the sums $\sum_{i\in I} c_i\upsilon_i$ and $\sum_{i\in I} |c_i|^2$ either both exist (i.e., converge) or both do not exist. When I is countable this follows more simply by noting that $\sum_{i\in\mathbb{N}} c_i\upsilon_i$ converges iff (s_n) is a Cauchy sequence, where $s_n = \sum_{i=1}^n c_i\upsilon_i$, and computing $\|s_n - s_m\|^2 = \sum_{i=m+1}^n |c_i|^2$, where $n > m$. To prove (B.210) on the assumption that (B.208) exists, by the Cauchy–Schwarz inequality, for any $\varepsilon > 0$,

$$\begin{aligned} |\langle \upsilon_j, \psi\rangle - c_j| &= |\langle \upsilon_j, \psi - \sum_{i\in G} c_i\upsilon_i + \sum_{i\in G} c_i\upsilon_i\rangle - c_j| \\ &= |\langle \upsilon_j, \psi - \sum_{i\in G} c_i\upsilon_i\rangle \leq \|\upsilon_j\|\|\psi - \sum_{i\in G} c_i\upsilon_i\| < \varepsilon, \end{aligned}$$

where we used Definition B.6 as well as $\|\upsilon_i\| = 1$. Letting $\varepsilon \to 0$ yields (B.210). □

Lemma B.60. *Let $\{\upsilon_i\}_{i\in I}$ be an orthonormal set in H. We have* **Bessel's Inequality**

$$\sum_{i\in I} |\langle \upsilon_i, \psi\rangle|^2 \leq \|\psi\|^2 \;\; (\psi \in H). \tag{B.212}$$

Proof. For any finite $G \subset I$, a computation based on (A.2) yields

$$\sum_{i\in G} |\langle \upsilon_i, \psi\rangle|^2 = \|\psi\|^2 - \|\psi - \sum_{i\in G} \langle \upsilon_i, \psi\rangle \upsilon_i\|^2 \leq \|\psi\|^2. \tag{B.213}$$

It follows that also the supremum of the left-hand side over all finite subsets $G \subset I$ is bounded by $\|\psi\|^2$ and hence is finite. By Lemma B.7, this supremum equals $\sum_{i\in I} |\langle \upsilon_i, \psi\rangle|^2$, which gives (B.212). □

Theorem B.61. *Let $B = \{\upsilon_i\}_{i\in I}$ be an orthonormal subset of a Hilbert space H. The following conditions are equivalent (and each defines B to be a* **basis** *of H):*

1. *Any $\psi \in H$ can be written (in the sense of Definition B.6) as $\psi = \sum_{i\in I} c_i \upsilon_i$.*
2. *For each $\psi \in H$, one has* **Parseval's equality**

$$\sum_{i\in I} |\langle \upsilon_i, \psi\rangle|^2 = \|\psi\|^2. \tag{B.214}$$

3. *For any $\psi, \varphi \in H$ one has*

$$\langle \varphi, \psi\rangle = \sum_{i\in I} \langle \varphi, \upsilon_i\rangle\langle \upsilon_i, \psi\rangle. \tag{B.215}$$

4. *B is not properly contained in any other orthonormal set (i.e., B is maximal).*
5. *$B^\perp = \{0\}$.*
6. *$B^{\perp\perp} = H$.*
7. *The closure of the linear span of B is H.*

Note that (B.215) is used in almost every computation in quantum physics, in which one also typically has $\|\psi\| = 1$. In that case, (B.214) at least formally turns the $|c_i|^2 = |\langle \upsilon_i, \psi\rangle|^2$ into (Born) probabilities, as discussed throughout the main text.

Proof. Assuming (B.208) and hence (B.210), take $\varepsilon > 0$ and find $F \subset X$ (finite) so that $\|\psi - \sum_{i\in G} c_i \upsilon_i\| < \varepsilon$. By (B.213), this gives

$$\sum_{i\in G} |\langle \upsilon_i, \psi\rangle|^2 - \|\psi\|^2| < \varepsilon^2. \tag{B.216}$$

Hence (B.214) holds in the sense of Definition B.6 (with $V = \mathbb{C}$). Conversely, assuming (B.214), eq. (B.213) gives (B.208). This proves the equivalence $1 \leftrightarrow 2$.

Clearly, (B.214) is a special case of (B.215), which in turn follows from (B.208) with (B.210) and continuity of the inner product, whence $3 \to 2$ and $1 \to 3$.

Furthermore, $1 \to 5$ follows by contradiction: given (B.210), any nonzero vector $\psi \in B^\perp$ could not possibly be written as (B.208). Conversely, $5 \to 1$ most easily follows by contradiction, too. For any $\psi \in H$, the sum $\varphi = \sum_{i\in I} \langle \upsilon_i, \psi\rangle \upsilon_i$ exists in H by Lemma B.59. Continuity of the inner product yields $\langle \upsilon_j, \varphi\rangle = \langle \upsilon_j, \psi\rangle$ and hence $\langle \upsilon_j, \varphi - \psi\rangle = 0$ for each $j \in I$, whence $\varphi - \psi \in B^\perp$. If φ cannot be written in the form (B.208) we have $\varphi \neq \psi$, so $B^\perp \neq \{0\}$, which is the desired contradiction.

Finally, $4 \leftrightarrow 5$ is tautological, $5 \leftrightarrow 6$ is trivial, and $6 \leftrightarrow 7$ is a special case of Proposition B.57.6 (hence this proposition is needed only for no. 7). □

For example, if $H = \ell^2(S)$, then one may take $I = S$, with $\upsilon_x = \delta_x$. Since S is an arbitrary set, this example shows that any cardinality of I may, in principle, occur. The existence of a basis has a remarkable consequence, for which we need:

Definition B.62. *Two Hilbert spaces H_1 and H_2 are called* **isomorphic**, *written $H_1 \cong H_2$, if they are isometrically isomorphic, that is, if there is an invertible linear map $u \in B(H_1, H_2)$ such that*

$$\|u\psi\|_{H_2} = \|\psi\|_{H_1} \quad (\psi \in H_1). \tag{B.217}$$

By Theorem A.3, a specific surjective isometry $u : H_1 \to H_2$ implementing an isomorphism is automatically ***unitary***, in that it is surjective and satisfies

$$\langle u\psi, u\varphi\rangle_{H_2} = \langle \psi, \varphi\rangle_{H_1}. \tag{B.218}$$

Conversely, a unitary map is an isometric isomorphism, so that isometric isomorphism of Hilbert spaces (seen as Banach spaces) is the same as unitary isomorphism. The following theorem (due to von Neumann, who was a specialist in both Hilbert space theory and axiomatic set theory) shows that the classification of Hilbert spaces up to isomorphism reduces to the classification of sets up to bijection.

Theorem B.63. *1. Any Hilbert space has a basis.*
2. All bases of a given Hilbert space H have the same cardinality (which is then, consistently, called the **dimension** *of H).*
3. Two Hilbert spaces are isomorphic iff they have the same dimension.

Specifically, clause 2 states that if $(\upsilon_i)_{i\in I}$ and $(\upsilon'_j)_{j\in J}$ are both bases of H, then $I \cong J$ as sets (i.e., there is a bijection $I \to J$). Similarly, clause 3 states that $H_1 \cong H_2$ iff H_1 has a basis $(\upsilon_i)_{i\in I}$ and H_2 has a basis $(\upsilon'_j)_{j\in J}$ for which $I \cong J$.

Proof. 1. The general proof is, alas, based on Zorn's Lemma: the collection O of all orthonormal sets in H is ordered by inclusion and each totally (i.e. linearly) ordered subset has an upper bound, namely its union. Hence O has a maximal element, which is a basis by Theorem B.61.4. Fortunately, in case that H is (topologically) ***separable*** (in that it contains a *countable* dense subset), a basis may be constructed by the well-known ***Gram–Schmidt procedure***, as follows: let $(\psi_1, \psi_2, \ldots)$ be a countable subset of H, for simplicity already taken to be linearly independent (otherwise, remove linear combinations first), start with $\upsilon_1 = \psi^{(1)}/\|\psi^{(1)}\|$, inductively define $w_n = \psi_n - \sum_{i=1}^{n-1}\langle \upsilon_i, \psi_n\rangle \upsilon_i$, $n \in \mathbb{N}$, which already yields an orthogonal set, and finally normalize to $\upsilon_n = w_n/\|w_n\|$.
2. We only prove the case where one basis, say $\{\upsilon_i\}_{i\in I}$, is finite in somde detail. Take another basis $\{\upsilon'_j\}_{j\in J}$. From (B.214) and (B.215),

$$|I| = \sum_{i\in I}\|\upsilon_i\|^2 = \sum_{i\in I}\sum_{j\in J}|\langle \upsilon'_j, \upsilon_i\rangle|^2 = \sum_{i\in I}\sum_{j\in J}\langle \upsilon'_j, \upsilon_i\rangle\langle \upsilon_i, \upsilon'_j\rangle = \sum_{j\in J}\|\upsilon'_j\|^2 = |J|.$$

A similar computation excludes the possibility that I is countable and J is not. The general case relies on some cardinal arithmetic, which we spare the reader.

3. Let $\{\upsilon_i\}_{i\in I}$ be a basis of H and let $\{\upsilon'_j\}_{j\in j}$ be a basis of H'. Assume $I \cong J$, so that there is a bijection $b : I \to J$. Define $u : H \to H'$ and $v : H' \to H$ by linear extension of $u\upsilon_i = \upsilon'_{b(i)}$ and $v\upsilon'_j = \upsilon_{b^{-1}(j)}$, that is,

$$u\psi = \sum_{i\in I}\langle \upsilon_i, \psi\rangle \upsilon'_{b(i)} = \sum_{j\in J}\langle \upsilon_{b^{-1}(j)}, \psi\rangle \upsilon'_j; \tag{B.219}$$

$$v\psi' = \sum_{j\in J}\langle \upsilon'_j, \psi'\rangle' \upsilon_{b^{-1}(j)} = \sum_{i\in I}\langle \upsilon'_{b(i)}, \psi'\rangle' \upsilon_i, \tag{B.220}$$

where in each line the first equality sign is the definition of the map, whilst the second is a useful rewriting. These maps are well defined by Lemma B.59, e.g.,

$$\sum_{j\in J}|\langle \upsilon_{b^{-1}(j)}, \psi\rangle|^2 = \sum_{i\in I}|\langle \upsilon_i, \psi\rangle|^2 = \|\psi\|^2 < \infty, \tag{B.221}$$

so that the sums in (B.219) converges, and likewise for (B.220). Furthermore,

$$\langle u\psi, u\varphi\rangle' = \sum_{i_1,i_2}\langle \psi, \upsilon_{i_1}\rangle\langle \upsilon_{i_2}, \varphi\rangle\langle \upsilon'_{b(i_1)}, \upsilon'_{b(i_2)}\rangle' = \sum_i \langle \psi, \upsilon_i\rangle\langle \upsilon_i, \varphi\rangle = \langle \psi, \varphi\rangle,$$

where we used (B.207) for the primed basis, and (B.215). Similar computations establish $\langle v\psi', v\varphi'\rangle = \langle \psi', \varphi'\rangle'$, so that (in view of their obvious surjectivity) u and v are both unitary, as well as $uv = 1_{H'}$ and $vu = 1_H$. Thus $H \cong H'$. Conversely, if H (with basis $\{\upsilon_i\}_{i\in I}$) and H' are isomorphic, so that there is a unitary $u : H \to H'$, then $\{u\upsilon_i\}_{i\in I}$ is a basis of H', hence J even equals I. □

Corollary B.64. *If $\{\upsilon_i\}_{i\in I}$ is a basis of H, then $H \cong \ell^2(I)$.*

Proof. Define $u : H \to \ell^2(I)$ by linear extension of $u\upsilon_i = \delta_i$, where $i \in I$. □

Corollary B.65. *A Hilbert space is (topologically) separable iff it either has a countable basis, or is finite-dimensional.*

Proof. One direction of the proof is the Gram–Schmidt procedure (since the given countable dense set contains a basis). Conversely, if $\{\upsilon_i\}$ is a countable (or finite) basis of H, then the complex rational linear span of this set, i.e., the set of all finite linear combinations $\sum_i c_i\upsilon_i$ with $c_i \in \mathbb{Q} + i\mathbb{Q}$, is countable as well as dense in H. □

In particular, any finite-dimensional Hilbert space is isomorphic to $\mathbb{C}^n$ with standard inner product, and any separable Hilbert space is isomorphic to $\ell^2(\mathbb{N})$; when speaking of a separable Hilbert spaces we actually tend to think of the infinite-dimensional case. Although at first sight separability appears to be a rather restrictive condition, in fact the non-separable case only appears in some weird proofs in the theory of operator algebras (as well as in the theory of almost continuous functions in the sense of H. Bohr). Indeed, every Hilbert space naturally occurring in applications to mathematical physics (or to partial differential equations) is separable.

B.13 Operators on infinite-dimensional Hilbert space

The fact that all (infinite-dimensional) separable Hilbert spaces are isomorphic suggests that the riches of the theory are not be found in the spaces themselves, but in the operators that act on them (whose explicit form typically depends on some concrete realization of H, like $\ell^2(\mathbb{N})$, or $L^2(\mathbb{R}^d)$, etc.). The simplest operators are functionals, i.e., linear maps $f : H \to \mathbb{C}$, and the main new feature compared to the finite-dimensional case is that f is no longer *necessarily* bounded, see §B.9. The nature of bounded linear functionals, i.e., elements of the dual H^*, is totally settled by the ***Riesz–Fréchet Theorem*** (which we already know; cf. Proposition A.5 and nos. 6 and 7 in Table B.1 in §B.9), showing that little is gained by looking at them.

Theorem B.66. *Let H be a Hilbert space. The map $\psi \mapsto f_\psi$ from H to H^*, where*

$$f_\psi(\varphi) = \langle \psi, \varphi \rangle, \tag{B.222}$$

is an isometric anti-linear isomorphism $H \to H^*$.

Proof. For convenience we rewrite (B.124) for the case at hand as

$$\|f\| = \sup\{|f(\psi)|, \psi \in H, \|\psi\|_H \leq 1\}. \tag{B.223}$$

Since $|f_\psi(\varphi)| = |\langle \psi, \varphi \rangle| \leq \|\psi\|\|\varphi\|$ by Cauchy–Schwarz, it follows that $f_\psi \in H^*$ for any $\psi \in H$, with $\|f_\psi\| \leq \|\psi\|$. We may sharpen this to equality, i.e.,

$$\|f_\psi\| = \|\psi\|, \tag{B.224}$$

by choosing $f = f_\psi$ and $\varphi = \psi$ in (B.223). Hence $\psi \mapsto f_\psi$ is isometric and therefore also injective. To prove surjectivity, we find a vector ψ for which some given *nonzero* functional f equals f_ψ (of $f = 0$, then $\psi = 0$ does the job). Assume $f \neq 0$ (otherwise, $\psi = 0$ does the job). Then $\ker(f)^\perp \neq \{0\}$: namely, $\ker(f)$ is closed by continuity of f and is linear by linearity of f, whence $\ker(f)^{\perp\perp} = \ker(f)$ by Proposition B.57.3, so that (arguing by contradiction) $\ker(f)^\perp = \{0\}$ would imply $\ker(f)^{\perp\perp} = H$ and hence $\ker(f) = H$, or $f = 0$.

The remainder of the proof is the same as for Proposition A.5. □

This allows one to make the weak topology on H (or, equivalently, the weak* topology on H^*) explicit (cf. §B.9): we have $\psi_n \to \psi$ weakly iff $\langle \varphi, \psi_n - \psi \rangle \to 0$ for each $\varphi \in H$ (and similarly for nets). From the general theory, or directly from Cauchy–Schwarz, it is immediate that (at least for infinite-dimensional H) the weak topology on H is indeed weaker than the strong one (that is, strong convergence implies weak convergence), but not the other way round. A simple example is provided by any ordered countable basis $(\upsilon_n)_{n\in\mathbb{N}}$ of a separable Hilbert space, where $\upsilon_n \to 0$ weakly but not strongly for any $n \in \mathbb{N}$ (more generally, for any infinite-dimensional Hilbert space and any basis $\{\upsilon_i\}$ we have $\upsilon_i \to 0$ weakly but not strongly in the sense of convergence of nets). Nonetheless, as a corollary of Proposition B.46:

Corollary B.67. *The functional f_ψ defined by* (B.222) *is weakly continuous.*

We now move from functionals als special operators from H to $\mathbb{C}$ to operators in the usual sense, i.e., linear maps from H to itself. Once again, the main new feature compared to the finite-dimensional case is that a linear map $a : H \to H$ is no longer necessarily bounded, where (cf. Definition B.32) we recall that a is ***bounded*** if it satisfies one (and hence both) of the following equivalent conditions:

$$\|a\psi\| \leq C\|\psi\| \quad (\psi \in H); \tag{B.225}$$

$$\sup\{\|a\psi\|, \psi \in H, \|\psi\| \leq 1\} < \infty. \tag{B.226}$$

In that case, the (finite) supremum is called the ***norm*** $\|a\|$ of a, exactly as in (A.18). Using Theorem B.66 and (B.130), we therefore have

$$\|a\| = \sup\{\|a\psi\|, \psi \in H, \|\psi\| = 1\} \tag{B.227}$$

$$= \sup\{|\langle\varphi, a\psi\rangle|, \psi, \varphi \in H, \|\psi\| = \|\varphi\| = 1\}, \tag{B.228}$$

and we have the inequalities (A.20) and (A.21), as in the finite-dimensional case.

It is clear from (A.20) and (B.225) that bounded operators a are continuous, in that if $\psi_n \to \psi$, then $a\psi_n \to a\psi$. On the other hand, ***unbounded operators*** are discontinuous in this sense: for each $n \in N$ there is $\psi_n \in H$ with $\|\psi_n\| = 1$ and $\|a\psi_n\| \geq n$. The sequence $(\psi_n = \psi_n/n)$ then converges to zero, but since $\|a\tilde{\psi}_n\| \geq 1$, the sequence $(a\tilde{\psi}_n)$ does not converge to $a \cdot 0 = 0$. Thus on infinite-dimensional Hilbert spaces a sharp distinction emerges between *bounded* and *unbounded* operators.

Among the former, we will distinguish between *compact* operators and the rest, whilst among the latter, one has the *closed* operators (i.e., those with a closed graph), which are still reasonably well-behaved, and the (non-closed) rest. Yet cutting through the bounded-unbounded divide is the notion of *self-adjointness*. For any linear (not necessarily bounded) map $a : H \to H$, we say that a is ***self-adjoint*** if

$$\langle a\varphi, \psi\rangle = \langle\varphi, a\psi\rangle, \quad (\psi, \varphi \in H). \tag{B.229}$$

The remarkable ***Hellinger–Toeplitz Theorem*** then states that such maps are bounded:

Theorem B.68. *If a linear map $a : H \to H$ satisfies* (B.229)*, then it is bounded.*

Proof. The proof is based on the Closed Graph Theorem B.37. If the sequence $(\psi_n, a\psi_n)$ in $G(a) \subset H \oplus H$ converges, say to $(\psi, \varphi) \in H \oplus H$, then $\psi_n \to \psi$ and $a\psi_n \to \varphi$. Using (B.229) and continuity of the inner product, for $\chi \in H$ we have

$$\langle\chi, \varphi\rangle = \lim_n \langle\chi, a\psi_n\rangle = \lim_n \langle a\chi, \psi_n\rangle = \langle a\chi, \psi\rangle = \langle\chi, a\psi\rangle.$$

For $\chi = \varphi - a\psi$, this yields $\varphi = a\psi$, and hence $(\psi, \varphi) \in G(a)$. This means that $G(a)$ is closed, upon which the Closed Graph Theorem states that a is bounded. □

More generally, if V and W are Banach spaces, with dual spaces V^* and W*, respectively, and two linear (but not *a priori* bounded) maps $a : V \to W$ and $b : W^* \to V^*$

satisfy $\varphi(av) = (b\varphi)(v)$ for each $v \in W$ and $\varphi \in W^*$, then a and b are bounded, with $b = a^*$, as defined in (B.125). The proof is similar.

This generalization of Theorem B.68 also places the familiar adjoint a^* from Hilbert space in broader perspective: making the identification $f_\psi \leftrightarrow \psi$ of H^* with H described by the Riesz–Fréchet Theorem B.66, the Banach space definition (B.125) of the adjoint $a^* : H^* \to H^*$ of a bounded linear map $a : H \to H$ reproduces the definition (A.15) of the Hilbert space adjoint $a^* : H \to H$. Thus we also infer that (B.128) is valid for arbitrary Hilbert spaces. Note that in the Hilbert space case, boundedness of a^* may be proved more simply, as follows.

Proposition B.69. *Let $a \in B(H)$ and let $a^* : H \to H$ be its adjoint, that is,*

$$\langle a^*\psi, \varphi\rangle = \langle \psi, a\varphi\rangle \quad (\psi, \varphi \in H). \tag{B.230}$$

Then a^* is bounded, with $\|a^*\| = \|a\|$.

Proof. Eq. (B.230) gives $|\langle a^*\psi, \varphi\rangle| \leq \|a\|\|\psi\|\|\varphi\|$. Taking $\varphi = a^*\psi$ yields $\|a^*\psi\| \leq \|a\||\psi|$, and hence $\|a^*\| \leq \|a\|$. Replacing a by a^* gives the last claim. □.

Since unbounded self-adjoint operators $a : H \to H$ do not exist, von Neumann defined such operators on some (proper) linear subspace $D(a) \subseteq H$ (*always assumed to be dense in H*), called the ***domain*** of a. This affects the definition of the adjoint:

Definition B.70. *1. The adjoint a^* of an operator $a : D(a) \to H$ has domain $D(a^*) \subset H$ consisting of all $\psi \in H$ for which the functional $f^a_\psi : D(a) \to \mathbb{C}$, defined by*

$$f^a_\psi(\varphi) = \langle \psi, a\varphi\rangle \quad (\varphi \in D(a)), \tag{B.231}$$

is bounded, i.e., there is $C > 0$ such that $|f^a_\psi(\varphi)| \leq C\|\varphi\|$ for all $\varphi \in D(a)$.

2. For $\psi \in D(a^*)$, the functional f^a_ψ has a unique bounded extension $f^a_\psi : H \to \mathbb{C}$, so by Theorem B.66 there is a unique vector $\psi' \in H$ such that $f^a_\psi(\varphi) = \langle \psi', \varphi\rangle$.

3. The adjoint $a^* : D(a^*) \subset H$, then, is defined by $a^*\psi = \psi'$, or, equivalently, by

$$\langle a^*\psi, \varphi\rangle = \langle \psi, a\varphi\rangle, \ \psi \in D(a^*), \varphi \in D(a). \tag{B.232}$$

Note that, on our assumption that $D(a)$ be dense in H, i.e., $D(a)^- = H$, eq. (B.232) indeed uniquely specifies $a^*\psi$ because of Proposition B.57.4.

4. An operator $a : D(a) \to H$ is called **self-adjoint** *when $D(a^*) = D(a)$ and $a^* = a$.*

If $D(a) = H$, and a is bounded, then also $D(a^*) = H$, since $|f^a_\psi(\varphi)| \leq \|a\|\|\psi\|\|\varphi\|$, so that f^a_ψ is bounded for any $\psi \in H$. Accordingly, for $a \in B(H)$, Definition B.70 reduces to the usual definition (A.15). Furthermore, even if $D(a)$ is merely dense in H, if $a : D(a) \to H$ is bounded in the sense of (B.225) - (B.226), but now with $\psi \in D(a)$ instead of $\psi \in H$, then a has a unique extension to a a bounded operator $a : H \to H$, whose adjoint a^* may be either defined through Definition B.70 as the adjoint of $a : D(a) \to H$, or, equivalently, as the adjoint of the extension $a : H \to H$.

Here, as well as in Definition B.70.2, a general Banach space principle is at work:

Proposition B.71. *Let V and W be Banach spaces, and let V' be a dense subset of V. Any bounded linear map $a' : V' \to W$ (in the sense of Definition B.32) has a unique* bounded *linear extension $a : V \to W$, with $\|a\| = \|a'\|$.*

Proof. For $v \in V$ there is a sequence (v_n) in V' with $v_n \to v$. Since $a' : V' \to W$ is bounded and (v_n) is convergent in V' and hence Cauchy in V, also the sequence $(a'v_n)$ in W is Cauchy. Since W is assumed complete, we may define $av = \lim_n a'v_n$. This limit is easily seen to be independent of the approximating sequence to v, and the ensuing map $a : V \to W$ is clearly linear. Furthermore, since by (B.5) we have $\|v\| = \lim_n \|v_n\|$, if we assume $\|v\| = 1$ we can take v_n to have unit norm also.

Once again from (B.5), we also have $\|av\| = \lim_n \|a'v_n\| \leq \sup_n \|a'v_n\|$, whence $\|a\| \leq \|a'\|$. But for $v \in V'$, taking $v_n = v$ we have $a'v = av$, and hence the bound $\|a'v\| \leq \|a\|\|v\|$, from which $\|a'\| \leq \|a\|$, so that finally $\|a\| = \|a'\|$. □

To complete these basic definitions, we say that an (unbounded) operator $a : D(a) \to H$ is ***closed*** if its graph $G(a) = \{(\psi, a\psi), \psi \in D(a)\}$ is a closed subspace of $H \oplus H$, cf. (B.108). Note that in the Hilbert space case it is more appropriate to replace the norm (B.107) on $H \oplus H$ by the equivalent norm

$$\|(v,w)\| = \sqrt{\|v\|^2 + \|w\|^2}, \tag{B.233}$$

since this alternative norm comes from the canonical inner product on $H \oplus H$, viz.

$$\langle (v,w),(v',w')\rangle_{H\oplus H} = \langle v,v'\rangle_H + \langle w,w'\rangle_H. \tag{B.234}$$

We now prove an important property of self-adjoint operators:

Proposition B.72. *The adjoint a^* of any operator $a : D(a) \to H$ is closed. In particular, self-adjoint operators are closed.*

Proof. The proof can be elegantly given in terms of the graph $G(a)$. Defining

$$u : H \oplus H \to H \oplus H; \tag{B.235}$$
$$u(\psi_1, \psi_2) = (-\psi_2, \psi_1), \tag{B.236}$$

it is easy to verify that u is a unitary operator, and that

$$G(a^*) = u(G(a)^\perp) = (uG(a))^\perp. \tag{B.237}$$

Hence $G(a^*)$ is closed by Proposition B.57.1, and the claim follows. □

In the the context of spectral theory, we will see later what the real importance of self-adjointness (and, more generally, closedness) is. It is time for some examples.

Proposition B.73. *Let $H = \ell^2(X)$, with X countable for simplicity, and for $f \in \ell^\infty(X)$ define the* **multiplication operator** *$m_f : H \to H$ by*

$$m_f\psi = f\psi, \tag{B.238}$$

i.e., $m_f\psi(x) = f(x)\psi(x)$. *Then* m_f *is bounded, with norm, cf.* (A.107),

$$\|m_f\| = \|f\|_\infty. \tag{B.239}$$

More generally, let $H = L^2(X)$ *for some* σ-finite *Borel space* (X,Σ,μ), *and for* $f \in L^\infty(X)$, *define* m_f *in the same way. Then* m_f *is again bounded, with norm*

$$\|m_f\| = \|f\|_\infty^{\text{ess}}. \tag{B.240}$$

Finally, let $f : X \to \mathbb{R}$ *be measurable (but not necessarily essentially bounded). Then*

$$D(m_f) = \{\psi \in L^2(X) \mid f\psi \in L^2(X)\}. \tag{B.241}$$

is dense in $L^2(X)$, *and if* $f^* = f$, *the operator* $m_f : D(m_f) \to L^2(X)$ *is self-adjoint.*

Proof. On $\ell^2(X)$ we have $\|f\psi\|_2 \leq \|f\|_\infty\|\psi\|_2$, and hence $\|m_f\| \leq \|f\|_\infty$. Assume $f \neq 0$. Then $\|f\|_\infty > 0$, and for any $0 < t < \|f\|_\infty$ there is $x_t \in X$ such that $|f(x_t)| \geq t$, so that $\psi_t = 1_{\{x_t\}} \in \ell^2(X)$ satisfies $\|m_f\psi_t\|_2 = |f(x_t)| \geq t$, whence $\|m_f\| \geq t$. This holds for all $0 < t < \|f\|_\infty$, hence $\|m_f\| \geq \|f\|_\infty$, which yields (B.239).

To prove (B.240), again assume $\|f\|_\infty^{\text{ess}} > 0$ and $0 < t < \|f\|_\infty^{\text{ess}}$. Then the set $X_t = \{x \in X, |f(x)| \geq t\}$ is measurable, with $\mu(X_t) > 0$. Since (X,Σ,μ) is σ-finite, there is $X_t' \subset X_t$ with $0 < \mu(X_t') < \infty$. Take $\psi = 1_{X_t'}$, so that $\|f\psi\|_2 \geq t\|\psi\|_2$, etc.

To prove the density of $D(m_f)$, for $n \in \mathbb{N}$ define $\tilde{X}_n = \{x \in X \mid |f(x)| \leq n\}$, so that $X = \cup_n \tilde{X}_n$. For each $\psi \in L^2(X)$ we then have $1_{\tilde{X}_n}\psi \in D(m_f)$. Writing $\varphi_n = 1_{\tilde{X}_n}\psi$, we have $\langle \psi, \varphi_n \rangle = \int_{\tilde{X}_n} d\mu\, |\psi|^2$, hence $\langle \psi, \varphi_n \rangle = 0$ iff $\psi = 0$ μ-a.e. on $\tilde{X}_n$. This is true for each $n \in \mathbb{N}$ iff $\psi = 0$, so the required density follows from Proposition B.57.4

In the last claim (where $f^*(x) = \overline{f(x)}$), the domain $D(m_f^*)$ consists of all $\psi \in L^2(X)$ for which the map $\varphi \mapsto \int_X d\mu\, \overline{\psi} f\varphi$ is bounded; by Theorem B.66 this is the case iff $f\psi \in L^2(X)$, so that $D(m_f^*) = D(m_f)$. Moreover, (B.232) obviously holds for $a^* = m_f$ (if f takes complex values, then $m_f^* = m_{f^*}$, still on $D(m_f^*) = D(m_f)$). □

For quantum mechanics, a key example is $H = L^2(\mathbb{R})$ with $f(x) = x$, i.e., the position operator. It then follows from Proposition B.73 that x is self-adjoint on the domain

$$D(m_x) = \{\psi \in L^2(\mathbb{R}) \mid \int_{\mathbb{R}} dx x^2 |\psi(x)|^2 < \infty\}. \tag{B.242}$$

See also §5.11. It happens often that a given operator on some domain is not closed as it stands, but can be made so by slightly enlarging its domain. Thus an operator $a : D(a) \to H$ is ***closable*** if the closure of the graph $G(a)$ in $H \oplus H$ is the graph of a closed operator a^-, called the ***closure*** of a, i.e., $G(a)^- = G(a^-)$. The following easy lemma is very useful in proving closability (the proof is a definition chase).

Lemma B.74. *Each of the following conditions is equivalent to closability of* a:

1. *If* (ψ_n) *is a sequence in* $D(a)$ *such that* $\psi_n \to 0$, *and if its image* $(a\psi_n)$ *converges, too, then* $a\psi_n \to 0$.
2. *The domain* $D(a^*)$ *of the adjoint* a^* *(see Definition B.70) is dense in* H.

The domain $D(a^-)$ of the closure a^- of a closable operator a consists of all $\psi \in H$ for which there exists a sequence (ψ_n) in $D(a)$ such that $\psi_n \to \psi$ and $a\psi_n$ converges, so that $a^-\psi = \lim_n a\psi_n$. Finally, if a is closable, then $a^- = a^{**}$ and $(a^-)^* = a^*$.

An equality $a = b$ between unbounded operators always stands for $D(a) = D(b)$ and $a = b$. Furthermore, $a \subset b$ means $D(a) \subseteq D(b)$ and $b = a$ on $D(a)$.

Definition B.75. *Let $a : D(a) \to H$ (where $D(a)$ is dense) be an operator.*

- *If $a \subset a^*$ i.e., if $\langle a\varphi, \psi\rangle = \langle \varphi, a\psi\rangle$, $\varphi, \psi \in D(a)$, then a is called* **symmetric**.
- *If a is closable and $a^- = a^*$ (in which case the closure a^- of a is self-adjoint), then a is called* **essentially self-adjoint**.

It follows from Lemma B.74 that a symmetric operator is closable (because $D(a^*)$, containing $D(a)$, is dense). For a symmetric operator one has $a \subseteq a^- = a^{**} \subseteq a^*$, with equality at the first position when a is closed, and equality at the second position when a is essentially self-adjoint; when both equalities hold, a is self-adjoint. Conversely, an essentially self-adjoint operator is symmetric. A symmetric operator may or may not be essentially self-adjoint; we will not discuss this problem here.

As in the finite-dimensional case, the notion of the adjoint allows one to define a ***projection*** as an operator $e : H \to H$ that satisfies $e^2 = e^* = e$. However, Proposition A.8 should be slightly adapted in order to cover the infinite-dimensional case:

Proposition B.76. *There is a bijective correspondence $e \leftrightarrow L$ between:*

- *projections e on H;*
- closed *linear subspaces L of H,*

still given by (A.27) - (A.28), *where now $\{\upsilon_i\}_{i\in I}$ is a basis of L, and the latter sum must be applied to fixed $\psi \in H$ according to Definition B.6 with $V = H$, i.e.,*

$$e\psi = \sum_{i\in I}\langle \upsilon_i, \psi\rangle \upsilon_i, \quad \psi \in H. \tag{B.243}$$

Alternatively, without invoking the concept of a basis, one may use the decomposition (B.203) as proved via Lemma B.58, to define e directly by $e\psi = e\psi^{\|}$.

Proof. The linear subspace $L = eH$ is closed, since e is bounded by Theorem B.68.

Conversely, note that since L is closed, it is a Hilbert space, so that it has a basis by Theorem B.63. The sum in (B.243) then converges by Lemma B.59, and since

$$\langle \varphi, e\psi\rangle = \sum_{i\in I}\langle \upsilon_i, \psi\rangle\langle \varphi, \upsilon_i\rangle = \sum_{i\in I}\overline{\langle \upsilon_i, \varphi\rangle\langle \psi, \upsilon_i\rangle} = \overline{\langle \psi, e\varphi\rangle} = \langle e\varphi, \psi\rangle;$$

$$e^2\psi = \sum_{i\in I}\langle \upsilon_i, \psi\rangle e\upsilon_i = \sum_{i,j\in I}\langle \upsilon_i, \psi\rangle\langle \upsilon_j \upsilon_i\rangle \upsilon_j = \sum_{i\in I}\langle \upsilon_i, \psi\rangle \upsilon_i = e\psi,$$

the operator e is a projection (in the second computation we used boundedness of e to pull it through the sum). Next, (B.243) is independent of the choice of a basis of L, since if $\{\upsilon_{i'}\}_{i'\in I'}$ is another basis of L, for arbitrary $\varphi \in L$ we may compute:

$$\langle \varphi, \sum_{i\in I} \langle \upsilon_i, \psi \rangle \upsilon_i \rangle - \sum_{i'\in I'} \langle \upsilon_{i'}, \psi \rangle \upsilon_{i'} \rangle = \sum_{i\in I} \langle \varphi, \upsilon_i \rangle \langle \upsilon_i, \psi \rangle - \sum_{i'\in I} \langle \varphi, \upsilon_{i'} \rangle \langle \upsilon_{i'}, \psi \rangle$$
$$= \langle \varphi, \psi \rangle - \langle \varphi, \psi \rangle = 0, \tag{B.244}$$

where we twice used (B.215), applied to the Hilbert space L. Hence

$$\sum_{i\in I} \langle \varphi, \upsilon_i \rangle \langle \upsilon_i, \psi \rangle = \sum_{i'\in I} \langle \varphi, \upsilon_{i'} \rangle \langle \upsilon_{i'}, \psi \rangle. \tag{B.245}$$

Finally, we prove bijectivity of the correspondence $L \leftrightarrow e$:

- Given L, by Lemma B.59 (applied to the Hilbert space L), $e\psi \in L$ for any $\psi \in H$, whereas if $\psi \in L$, then $e\psi = \psi$ by Theorem B.61 and (B.210). Hence $eH = L$.
- Given e, we first note that for any $\chi \in eH = L$, by definition we have $\chi = e\psi$ for some $\psi \in H$, whence $e\chi = e^2\psi = e\psi = \chi$. Now pick a basis $\{\upsilon_i\}$ of the Hilbert space eH, so that in particular $e\upsilon_i = \upsilon_i$. For arbitrary $\varphi, \psi \in H$, writing $\varphi = e\varphi + (1-e)\varphi = \varphi^{\parallel} + \varphi^{\perp}$, so that $\varphi^{\parallel} \in L$ and hence $e\varphi^{\parallel} = \varphi^{\parallel}$, we compute

$$\langle \varphi^{\parallel}, e\psi \rangle - \sum_i \langle \varphi^{\parallel}, \upsilon_i \rangle \langle \upsilon_i, \psi \rangle = \langle \varphi^{\parallel}, \psi \rangle - \langle \varphi^{\parallel}, \psi \rangle = 0;$$
$$\langle \varphi^{\perp}, e\psi \rangle - \sum_i \langle \varphi^{\perp}, \upsilon_i \rangle \langle \upsilon_i, \psi \rangle = \langle \varphi, (1-e)e\psi \rangle - \sum_i \langle \varphi, (1-e)\upsilon_i \rangle \langle \upsilon_i, \psi \rangle = 0,$$

where is the first line we used (B.215), applied to the Hilbert space H. □

It is easy to see why the sum (B.243) cannot, in general, converge in norm without the ψ, i.e., in the original (finite-dimensional) form (A.28); it suffices to take $e = 1$ (for $H = \ell^2(\mathbb{N})$, for simplicity). Writing $e_n = \sum_{i=1}^n |\upsilon_i\rangle\langle\upsilon_i|$, where, for example, $\upsilon_i = \delta_i$, for any unit vector ψ and $m > n$, from (A.18) we have

$$\|e_m - e_n\|^2 \geq \|(e_m - e_n)\psi\|^2 = \sum_{i=n+1}^{m} |\langle \upsilon_i, \psi \rangle|^2. \tag{B.246}$$

Taking $\psi = \upsilon_j$ for any $n+1 \leq j \leq m$ shows that that $\|e_m - e_n\|^2 \geq 1$ for all m, n, so that (e_n) cannot be a Cauchy sequence in $B(H)$. This argument applies to any infinite-dimensional subspace L. Therefore, if H is infinite-dimensional we should work with at least two notions of convergence within the Banach space $B(H)$ (cf. Theorem B.33), which for simplicity we state for sequences (more generally, one should *define* the corresponding topologies in terms of convergence of nets):

- $a_n \to a$ in the ***norm topology*** (or ***uniformly***) in $B(H)$ iff $\|(a_n - a)\| \to 0$.
- $a_n \to a$ in the ***strong topology*** in $B(H)$ iff $\|(a_n - a)\psi\| \to 0$, for each $\psi \in H$.

The strong topology on $B(H)$ is also called the ***strong operator topology***, in order to distinguish it from the strong topology on H itself (which, confusingly, is another name for the norm topology) in terms of which it is defined. Similarly, the weak topology on H (cf. §B.12) defines a ***weak operator topology*** on $B(H)$, as follows:

- $a_n \to a$ ***weakly*** on $B(H)$ iff $\langle \varphi, (a_n - a)\psi \rangle \to 0$, for each $\varphi, \psi \in H$.

In decreasing strength we have 'norm - strong - weak', and we show that this trio is distinguishable on $H = \ell^2(\mathbb{N})$ (and hence on any infinite-dimensional Hilbert space):

- Let $a_n\psi(x) = 0$ for $x = 1,\ldots,n$ whilst $a_n\psi(x) = \psi(x)$ for $x > n$. In other words, if $\psi = (\psi_1, \psi_2, \ldots)$, then $a_n\psi = (0,\ldots,0,\psi_{n+1},\psi_{n+2},\ldots)$ with n zeros. Hence
$$\|a_n\psi\|^2 = \sum_{x=n+1}^{\infty} |\psi(x)|^2,$$
so that $\|a_n\psi\| \to 0$ as $n \to \infty$ in order for ψ to be in $\ell^2(\mathbb{N})$. Thus $a_n \to 0$ strongly (and hence also weakly). If (a_n) were to have a norm limit, it therefore would have to be zero, too, but since $\|a_n\| \geq \|a_n\psi\|$ for any unit vector ψ, taking e.g. $\psi = \delta_{n+1}$, we have $\|a_n\| \geq 1$ for any n and hence (a_n) cannot converge in norm.
- A slight variation on this example is $a_n\psi(x) = 0$ for $x = 1,\ldots,n$ (once again), but now $a_n\psi(x) = \psi(x-n)$ for $x > n$, or, equivalently, $a_n\psi = (0,\ldots,0,\psi_1,\psi_2,\ldots)$ with n zeros. This time, we have $\|a_n\psi\| = \|\psi\|$, so to begin with, $a_n \to 0$ strongly is excluded. However, $\langle\varphi, a_n\psi\rangle = \sum_{x=1}^{\infty} \overline{\varphi(x+n)}\psi(x)$, so $\lim_{n\to\infty}\langle\varphi, a_n\psi\rangle = 0$: to see this, take $N < \infty$ fixed and use Cauchy–Schwarz to estimate
$$\begin{aligned} |\langle\varphi, a_n\psi\rangle| &\leq \left| \sum_{x=1}^{N} \overline{\varphi(x+n)}\psi(x) + \sum_{x=N+1}^{\infty} \overline{\varphi(x+n)}\psi(x) \right| \\ &\leq \|\psi\| \left(\sum_{x=n+1}^{\infty} |\varphi(x)|^2 \right)^{1/2} + \|\varphi\| \left(\sum_{x=N+1}^{\infty} |\psi(x)|^2 \right)^{1/2}. \end{aligned} \quad \text{(B.247)}$$
Letting $N \to \infty$ and then $n \to \infty$ yields $\langle\varphi, a_n\psi\rangle \to 0$, so that $a_n \to 0$ weakly. But (a_n) has no strong limit (for if it existed, it would have to be zero, too).

It is clear from Theorem B.33 that $B(H)$ is ***sequentially complete*** in its norm topology. This is true also in the weak and strong operator topologies:

Proposition B.77. *Let (a_n) be a sequence in $B(H)$.*

1. *If $(a_n\psi)$ converges in H for each $\psi \in H$, then the operator $a: H \to H$ defined by $a\psi = \lim_n a_n\psi$ is bounded (and hence $a_n \to a$ strongly, where $a \in B(H)$).*
2. *If $(\langle\varphi, a_n\psi\rangle)$ converges in $\mathbb{C}$ for each $\varphi, \psi \in H$, then there is an operator $a \in B(H)$ such that $a_n \to a$ weakly (and hence $a_n \to a$ weakly, where $a \in B(H)$).*

It is instructive to prove this, using two results of independent interest.

Theorem B.78. *Suppose V is a Banach space, W is a normed space (not necessarily complete), X is an arbitrary set, and $\{a_x\}_{x\in X}$ is some family of operators in $B(V,W)$ indexed by X. If the family is* pointwise *bounded in that*
$$\sup\{\|a_x v\|, x \in X\} < \infty \quad (v \in V), \tag{B.248}$$
then the family is uniformly *bounded in that*
$$\sup\{\|a_x\|, x \in X\} < \infty. \tag{B.249}$$

This is the ***Principle of Uniform Boundedness*** or ***Banach–Steinhaus Theorem***.

Proof. If W is not complete, use its completion in what follows. Define $\ell^\infty(X,W)$ to be the set of all bounded functions $f : X \to W$, i.e., those function such that $\sup\{\|f(x)\|, x \in X\} < \infty$, with pointwise operations. This is easily checked to be a Banach space itself in the natural norm $\|f\|_\infty = \sup\{\|f(x)\|, x \in X\}$ (using the auxiliary functions $\tilde{f} : X \to \mathbb{C}$ defined by each $f \in \ell^\infty(X,W)$ as $\tilde{f}(x) = \|f(x)\|$, so that $\|f\|_\infty = \|\tilde{f}\|_\infty$, one may largely reduce the proof to the ordinary $\ell^\infty(X)$ case).

For fixed $v \in V$, define $f_v : X \to W$ by $f_v(x) = a_x(v)$. By assumption, $f \in \ell^\infty(X,W)$, so we may define an operator $F : V \to \ell^\infty(X,W)$ by $F(v) = f_v$. We now show that the graph $G(F)$ is closed: if $v_n \to v$ in V and $Fv_n \to g$ in $\ell^\infty(X,W)$, then since uniform convergence implies pointwise convergence, for each $x \in X$ we have

$$g(x) = \lim_n (Fv_n)(x) = \lim_n f_{v_n}(x) = \lim_n a_x v_n = a_x \lim_n v_n = a_x v = f_v(x) = (Fv)(x).$$

Thus $g = Fv$, and hence $G(F)$ closed. By Theorem B.37, F is bounded, so that:

$$\begin{aligned}\|F\| &= \sup\{\|f_v\|_\infty, v \in V, \|v\| = 1\} = \sup\{\|a_x v\|, v \in V, \|v\| = 1, x \in X\} \\ &= \sup\{\|a_x\|, x \in X\} < \infty.\end{aligned}$$ □

This gives part 1 of Proposition B.77: since $\lim_n a_n \psi$ exists, $\sup_n\{\|a_n\psi\|\} < \infty$ for each ψ, hence $\sup_n\{\|a_n\|\} < \infty$. Since $a_n\psi \to a\psi$ implies $\|a_n\psi\| \to \|a\psi\|$, cf. (B.5),

$$\|a\psi\| = \lim_n \|a_n\psi\| \leq \lim_n \|a_n\|\|\psi\| \leq \sup_n\{\|a_n\|\}\|\psi\|, \tag{B.250}$$

so taking the supremum over all unit vectors ψ gives $\|a\| < \infty$.

As to the second part, suppose $a_n \to a$ weakly. Since $(\langle\varphi, a_n\psi\rangle)$ converges for $\varphi, \psi \in H$, we have $\sup_n\{|\langle\varphi, a_n\psi\rangle|\} < \infty$. Using (B.222), this is the same as $\sup_n\{|f_{a_n\psi}(\varphi)|\} < \infty$ for each $\varphi \in H$, so using Banach–Steinhaus with $V = H^*$, $X = \mathbb{N}$, and $a_x = f_{a_n\psi}$, we find $\sup_n\{\|f_{a_n\psi}\|_{H^*}\} < \infty$. By Theorem B.66, this gives $\sup_n\{\|a_n\psi\|\} < \infty$, and hence, via a second application of Theorem (B.78), $\sup_n\{\|a_n\|\} < \infty$, or $\|a_n\| < C < \infty$ for all n, as in the case of strong limits.

This time we have to do a little more work to construct the limit operator a. This requires a second lemma, which generalizes Proposition A.23 to general Hilbert spaces. To this effect, we say that a sesquilinear form $B : H \times H$ is ***bounded*** if there is a finite constant C such that $|B(\varphi, \psi)| \leq C\|\varphi\|\|\psi\|$ for all $\varphi, \psi \in H$.

Proposition B.79. *The relation $B(\varphi, \psi) = \langle\varphi, a\psi\rangle$ provides a bijective correspondence between* bounded *(hermitian/positive) sesquilinear forms and* bounded *(self-adjoin/positive) operators $a \in B(H)$, cf. Proposition A.22.1.*

Like Proposition A.23, this is a trivial consequence of Theorem B.66.

To finish the proof of Proposition B.77.2, define $B(\varphi, \psi) = \lim_n \langle\varphi, a_n\psi\rangle$, so

$$|B(\varphi, \psi)| \leq \lim_n \|a_n\|\|\varphi\|\|\psi\| \leq \sup_n \|a_n\|\|\varphi\|\|\psi\| \leq C\|\varphi\|\|\psi\|. \tag{B.251}$$

Hence B is bounded, and Proposition B.79 gives the weak limit $a \in B(H)$. □

B.14 Basic spectral theory

In linear algebra, which in our context means the theory of operators on finite-dimensional Hilbert spaces H, the spectrum $\sigma(a)$ of an operator (i.e., a linear map) $a : H \to H$ was defined as the set of eigenvalues of a. This led to the Spectral Theorems A.10 and A.15. However, as soon as $\dim(H) = \infty$, simple examples show that even bounded operators may have no eigenvectors (and hence no eigenvalues) at all. For example, take $H = L^2(0,1)$ and $f(x) = x$, with associated (bounded) multiplication operator $a = m_f \equiv m_x$, cf. (B.238); this is just a bounded version of the position operator of quantum mechanics. Then the eigenvalue equation $a_x \psi = \lambda \psi$ implies $\int_0^1 dx\, |x-\lambda|^2 |\psi(x)|^2 = 0$, which holds iff $|x-\lambda||\psi(x)| = 0$ a.e. Since $|x-\lambda|$ is nonzero a.e. for any $\lambda \in \mathbb{C}$, this implies $\psi(x) = 0$ a.e. and hence $\psi = 0$ in $L^2(0,1)$. More generally, taking $H = L^2(\mathbb{R}^d)$ and $f \in C_b(\mathbb{R}^d)$, a similar argument shows that the multiplication operator m_f has eigenvalue $\lambda \in \mathbb{C}$ whenever the equality $f(x) = \lambda$ holds on a set of positive (Lebesgue) measure. Therefore, if f varies sufficiently, then m_f has no eigenvalues at all (e.g., in $d = 1$, $f \in C^{(1)}([0,1])$ with $f'(x) \neq 0$ a.e.).

Even amidst his magnificent *oeuvre*, covering most of mathematics, it was one of Hilbert's most prophetic insights that finite-dimensional spectral theory could not merely be rescued, but also greatly enriched, by defining the spectrum as follows:

Definition B.80. *Let H be a Hilbert space. The* **spectrum** *$\sigma(a)$ of $a \in B(H)$ consists of all $\lambda \in \mathbb{C}$ for which the operator $a - \lambda : H \to H$ is not bijective. The complement*

$$\rho(a) = \mathbb{C} \backslash \sigma(a) \tag{B.252}$$

of the spectrum in $\mathbb{C}$ is called the **resolvent** *of a, i.e., $z \in \rho(a)$ iff $a - z$ is invertible.*

Here $a - \lambda \equiv a - \lambda \cdot 1_H$, where 1_H is the unit operator on H, and by 'bijective' and 'invertible' we *a priori* mean: injective and surjective. This set-theoretic notion of invertibility is considerably strengthened by Corollary B.35, according to which the set-theoretic inverse of $a - \lambda : H \to H$, if it exists for $a \in B(H)$, is automatically in $B(H)$. Consequently, we may equivalently say that $\lambda \in \sigma(a)$ if $a - \lambda$ is not invertible *in* $B(H)$. This means that if $z \in \rho(a)$, then the equation $(a - z)\psi = \varphi$ for $\psi \in H$:

- actually *has* a solution, since $(a - z)$ is *surjective*;
- has a *unique* solution, for $(a - z)$ is *injective*;
- has a unique solution that *continuously* depends φ, as $(a - z)^{-1}$ is *bounded*.

Thus Definition B.80 becomes a special case of the following purely algebraic idea:

Definition B.81. *Let A be a (complex) algebra with unit. The* **spectrum** *$\sigma(a)$ of $a \in A$ consists of all $\lambda \in \mathbb{C}$ for which the operator $a - \lambda$ is* not *invertible in A.*

The notation (B.252) also extends to this case. This generalization is especially powerful when A is a Banach algebra, and, particularly a C*-algebra, cf. Definition C.1. The latter case actually incorporates Definition B.80:

Proposition B.82. *For any Hilbert space H, the set $B(H)$ of all bounded operators on H is a C*-algebra with unit in the operator norm* (A.18)

The proof of Proposition A.7 goes through unchanged. In a different direction:

Proposition B.83. *Let $A = C(X)$, where X is a compact Hausdorff space. Then*

$$\sigma(f) = \operatorname{ran}(f). \tag{B.253}$$

Proof. Since multiplication in $C(X)$ is pointwise, if $f - \lambda \cdot 1_X$ has an inverse, it must be $1/(f - \lambda \cdot 1_X)$. This function exists (and is continuous) iff $\lambda \notin \operatorname{ran}(f)$. □

Theorem B.84. *Let $A = B(H)$ or, more generally, a unital C^*-algebra, or, even more generally, a Banach algebra with unit 1_A (cf. Definition C.1). Then the spectrum $\sigma(a)$ of any $a \in A$ is a nonempty compact subset of $\mathbb{C}$.*

Furthermore, defining the **spectral radius** *of $a \in A$ by*

$$r(a) = \sup\{|\lambda|, \lambda \in \sigma(a)\}, \tag{B.254}$$

for general unital Banach algebras we have

$$r(a) \leq \|a\|, \tag{B.255}$$

as well as Gelfand's **spectral radius formula**

$$r(a) = \lim_{n\to\infty} \|a^n\|^{1/n}. \tag{B.256}$$

If $a \in A_{\mathrm{sa}}$ is a self-adjoint element of a unital C^*-algebra, such as $A = B(H)$, then

$$r(a) = \|a\| \quad (a^* = a). \tag{B.257}$$

Proof. The claim about the spectrum obviously follows from the following facts:

1. $\sigma(a)$ is a bounded subset of $\mathbb{C}$.
2. $\sigma(a)$ is a closed subset of $\mathbb{C}$.
3. $\sigma(a)$ is a nonempty subset of $\mathbb{C}$.

Eq. (B.255) is equivalent to the implication $|\lambda| > \|a\| \Rightarrow \lambda \in \rho(a)$. For $\lambda \neq 0$ we have $(a - \lambda) = \lambda((a/\lambda) - 1)$, so, rescaling a if necessary, we only need to show that if $\|a\| < 1$, then $1 \in \rho(a)$. Indeed, in that case the geometric series $\sum_k a^k$ for a converges absolutely and hence (A being a Banach space) converges, with

$$\sum_{k=0}^{n} a^k = (1-a)^{-1}; \tag{B.258}$$

the proof is virtually the same as for complex numbers. Thus $1 \in \rho(a)$.

Fact 2 is equivalent to the set A_* of of invertible elements in A being open in A. Indeed, for given $a \in A_*$, take a $b \in A$ for which $\|b\| < \|a^{-1}\|^{-1}$. This implies

$$\|a^{-1}b\| \leq \|a^{-1}\| \, \|b\| < 1. \tag{B.259}$$

Hence by (B.258) for $\|a\| < 1$, the operator $a + b = a(1 + a^{-1}b)$ has an inverse, namely $(1 + a^{-1}b)^{-1}a^{-1}$. Taking $\varepsilon \leq \|a^{-1}\|^{-1}$, it follows that all $c \in A$ for which $\|a - c\| < \varepsilon$ lie in A_* (which is therefore an open subset of the metric space A).

For the third claim, take $a \in A$ and define $f : \mathbb{C} \to A$ by $f(z) = z - a$. Since

$$\|f(z+\delta) - f(z)\| = \delta,$$

we see that f is continuous (take $\delta = \varepsilon$ in the definition of continuity). By part 2 of the proof, $f^{-1}(A_*)$ is open in $\mathbb{C}$. But $f^{-1}(A_*)$ is the set of all $z \in \mathbb{C}$ where $z - a$ has an inverse, so that $f^{-1}(A_*) = \rho(a)$. This set being open, its complement $\sigma(a)$ is closed. Now define

$$g : \rho(a) \to A; \tag{B.260}$$
$$z \mapsto (z-a)^{-1}. \tag{B.261}$$

For fixed $z_0 \in \rho(a)$, choose $z \in \mathbb{C}$ such that $|z - z_0| < \|(a - z_0)^{-1}\|^{-1}$. From part 2 of the proof, with a replaced by $a - z_0$ and c replaced by $a - z$, we see that $z \in \rho(a)$, as $\|a - z_0 - (a - z)\| = |z - z_0|$. Moreover, because

$$\|(z_0 - z)(z_0 - a)^{-1}\| = |z_0 - z| \, \|(z_0 - a)^{-1}\| < 1, \tag{B.262}$$

the power series

$$\frac{1}{z_0 - a} \sum_{k=0}^{n} \left(\frac{z_0 - z}{z_0 - a} \right)^k \tag{B.263}$$

is absolutely convergent and hence convergent for $n \to \infty$. By (B.258), the limit $n \to \infty$ of this power series is

$$\frac{1}{z_0 - a} \sum_{k=0}^{\infty} \left(\frac{z_0 - z}{z_0 - a} \right)^k = \frac{1}{z_0 - a} \left(1 - \left(\frac{z_0 - z}{z_0 - a} \right) \right)^{-1} = \frac{1}{z - a} = g(z). \tag{B.264}$$

Hence

$$g(z) = \sum_{k=0}^{\infty} (z_0 - z)^k (z_0 - a)^{-k-1} \tag{B.265}$$

is a norm-convergent power series. For $z \neq 0$ we write $\|g(z)\| = |z|^{-1} \|(1_A - a/z)^{-1}\|$ and observe that $\lim_{z \to \infty} 1_A - a/z = 1_A$, since $\lim_{z \to \infty} \|a/z\| = 0$. Hence we obtain $\lim_{z \to \infty} (1_A - a/z)^{-1} = 1_A$, and

$$\lim_{z \to \infty} \|g(z)\| = 0. \tag{B.266}$$

Let $\varphi \in A^*$; since φ is bounded, eq. (B.265) implies that the function $g_\varphi : z \to \varphi(g(z))$ is given by a convergent power series (i.e. is analytic), and (B.266) implies

$$\lim_{z \to \infty} g_\varphi(z) = 0. \tag{B.267}$$

Now suppose that $\sigma(a) = \emptyset$, so that $\rho(a) = \mathbb{C}$. The function g, and hence g_φ, is then defined on $\mathbb{C}$, where it is analytic and vanishes at infinity. In particular, g_φ is bounded, so that by Liouville's Theorem of elementary complex analysis it must be constant. By (B.267) this constant is zero, so that $g = 0$ by Corollary B.45. This is absurd, so that $\rho(a) \neq \mathbb{C}$, and hence $\sigma(a) \neq \emptyset$.

We now prove the spectral radius formula (B.256). For $|z| > \|a\|$ the function g, defined in (B.260) - (B.261) has a norm-convergent power series

$$g(z) = \frac{1}{z}\sum_{k=0}^{\infty}\left(\frac{a}{z}\right)^k. \tag{B.268}$$

On the other hand, we have seen that for any $z \in \rho(a)$ one may find a $z_0 \in \rho(a)$ such that the power series (B.265) converges (i.e. in norm). If $|z| > r(a)$ then $z \in \rho(a)$, so (B.265) converges for $|z| > r(a)$, uniformly in z. Therefore (by the theory of analytic functions taking values in Banach spaces), eq. (B.268) is norm-convergent for $|z| > r(a)$, too, which in turn implies that $\|a^n\|/|z|^n < 1$ for large enough n (proof by contradiction). Since this is true for all z for which $|z| > r(a)$, we must have

$$\limsup_{n\to\infty} \|a^n\|^{1/n} \leq r(a). \tag{B.269}$$

To derive a second inequality towards (B.256), we use the ***spectral mapping property*** for polynomials, which states that for any (complex) polynomial p on $\mathbb{C}$,

$$\sigma(p(a)) = p(\sigma(a)) \equiv \{p(\lambda) \mid \lambda \in \sigma(a)\}. \tag{B.270}$$

Given some polynomial p of degree n (in a variable z) and some fixed $\lambda \in \mathbb{C}$, let

$$q(z) = p(z) - \lambda = c_0 \prod_{k=1}^{n}(z - c_k), \tag{B.271}$$

for some $c_0, \ldots, c_k \in \mathbb{C}$. Hence by (A.53) - (A.55), we have $q(a) = c_0 \prod_{k=1}^{n}(a - c_k)$. Now an operator $b = b_0 \cdots b_n$ is invertible iff each factor b_k is invertible (in which case $b^{-1} = b_n^{-1} \cdots b_0^{-1}$), so $\lambda \in \sigma(p(a))$ iff some $c_k \in \sigma(a)$ (where $k > 0$, as $c_0 \neq 0$), which is true iff $q(c_k) = 0$, which holds iff $\lambda = p(c_k)$. This proves (B.270).

To conclude the proof of (B.256), we note that since $\sigma(a)$ is closed, there is $\lambda \in \sigma(a)$ for which $|\lambda| = r(a)$. Since $\lambda^m \in \sigma(a^m)$ by (B.270), one has $|\lambda^m| \leq \|a^m\|$ by (B.255). Hence $\|a^m\|^{1/m} \geq |\lambda| = r(a)$. Combining this with (B.269) yields

$$\limsup_{n\to\infty} \|a^n\|^{1/n} \leq r(a) \leq \|a^m\|^{1/m} \quad (m \in \mathbb{N}). \tag{B.272}$$

Hence the limit must exist, and $\lim_{n\to\infty} \|a^n\|^{1/n} = \inf_m \|a^m\|^{1/m} = r(a)$, i.e., (B.256).

Finally, given axiom (C.2) for C*-algebras (which include $B(H)$ by Proposition A.7 and Theorem B.33), eq. (B.257) follows from (B.256): for self-adjoint a, eq. (C.2) reads $\|a^2\| = \|a\|^2$, so if we take the limit in (B.256) along the subsequence of even numbers (as we are entitled to, given convergence), we obtain (B.257). □

We may also generalize Definition B.80 in a different direction, where we allow $a : D(a) \to H$ to be unbounded. In that case, there is room for some ambiguity, as a possible set-theoretic inverse of $a - z$, if it exists as a (necessarily linear) map $(a-z)^{-1} : H \to D(a)$ is no longer guaranteed to be bounded. By the argument preceding Definition B.81 this would, of course, be desirable, which motivates:

Definition B.85. *Let H be a Hilbert space, and let $a : D(a) \to H$ be a possibly unbounded operator (always by definition with dense domain).*

1. *The* **resolvent** $\rho(a)$ *consists of all $z \in \mathbb{C}$ for which $a - z : D(a) \to H$ has a* bounded *(linear) inverse $(a-z)^{-1} : H \to D(a)$, so that $(a-z)^{-1} \in B(H)$.*
2. *The* **spectrum** $\sigma(a) = \mathbb{C} \backslash \rho(a)$ *is the complement of the resolvent (i.e. in $\mathbb{C}$).*

This provides further motivation for requiring an unbounded operator to be closed:

Proposition B.86. *Let $a : D(a) \to H$ be a possibly unbounded operator.*

- *If a is closed, then $z \in \rho(a)$ iff $a - z$ has a* set-theoretic *inverse.*
- *If a is not closed, then $\rho(a) = \emptyset$.*

Proof. The graph $G(a^{-1})$ in $H \oplus H$ is the image of $G(a)$ under the linear homeomorphism $(\psi_1, \psi_2) \mapsto (\psi_2, \psi_1)$, hence if a is closed, then a^{-1} is closed and hence bounded (cf. Theorem B.37). Similarly, if $G(a)$ is not closed, then $G(a^{-1})$ cannot be closed either, and hence a^{-1} cannot be bounded. Likewise with $a \rightsquigarrow a - z$. □

Thus spectral theory always deals with *closed* operators a, like self-adjoint ones.

We now show that Definition B.80 is compatible with our earlier §A.4.

Proposition B.87. *Let V be a finite-dimensional vector space and let $a : V \to V$ be a linear map. Then a is injective iff it is surjective.*

Proof. This follows from the elementary fact that for any linear map $a : V \to W$ one has $\mathrm{ran}(a) \cong V/\ker(a)$. Now if $V = W$ is finite-dimensional one has $V \cong \mathbb{C}^n$ (on choice of a basis), and one may simply count dimensions to infer that

$$\dim(\mathrm{ran}(a)) = n - \dim(\ker(a)).$$

Surjectivity of a then yields injectivity and *vice versa*: we have $\dim(\mathrm{ran}(a)) = n$ iff $\dim(\ker(a)) = 0$ iff $\ker(a) = 0$. □

Note that his proposition yields the very simplest case of the ***Atiyah–Singer index theorem***, for which these mathematicians received the Abel Prize for 2004. We define the **index** of a linear map $a : V \to W$ as

$$\mathrm{index}(a) = \dim(\ker(a)) - \dim(\mathrm{coker}(a)), \tag{B.273}$$

where $\mathrm{cokern}(a) = W/\mathrm{ran}(a)$, *provided both quantities are finite*. If V and W are finite-dimensional, Proposition B.87 yields the ***baby index theorem***

$$\mathrm{index}(a) = \dim(V) - \dim(W). \tag{B.274}$$

In particular, if $V = W$, then $\text{index}(a) = 0$ for any linear map a (in general, the index theorem expresses the index of an operator in terms of topological data; in this simple situation the only such data are the dimensions of V and W).

Corollary B.88. *If a is an operator on a finite-dimensional Hilbert space, then the spectrum $\sigma(a)$ of a is the set of its eigenvalues.*

Proof. It immediately follows from Proposition B.87 that $a - z$ is invertible iff z is *not* an eigenvalue of a. □

Returning to Definition B.80, we see that if λ is an eigenvalue of a (in that, as in finite dimension, there exists a nonzero vector $\psi \in H$ for which $a\psi = \lambda\psi$), then $\lambda \in \sigma(a)$ (for $a - \lambda$) is not even injective, let alone invertible). Thus we may define:

- the ***point spectrum*** $\sigma_p(a)$ of a as the set of its eigenvalues, so that $\sigma_p(a) \subseteq \sigma(a)$;
- the ***continuous spectrum***, which (if it exists) is the remainder of $\sigma(a)$, i.e.,

$$\sigma_c(a) = \sigma(a) \backslash \sigma_p(a). \tag{B.275}$$

If $\sigma(a) = \sigma_p(a)$, we call $\sigma(a)$ ***discrete***. The example at the beginning of this section shows the opposite case, viz. $\sigma_p(a_x) = \emptyset$ and $\sigma_c(a_x) = [0,1]$. This follows from:

Proposition B.89. *Let $H = L^2(X, \Sigma, \mu)$ for some σ-finite Borel space (X, Σ, μ) such that $\mu(A) > 0$ for each open $A \subset X$, and let $f \in C(X)$. Then*

$$\sigma(m_f) = \text{ran}(f)^-. \tag{B.276}$$

Cf. Proposition B.73. More generally, let $f : X \to \mathbb{C}$ be (Borel) measurable. Then

$$\sigma(m_f) = \text{ess-ran}(f), \tag{B.277}$$

wgere the **essential range** ess-ran(f) *of f consists of all $z \in \mathbb{C}$ such that*

$$\forall \varepsilon > 0 : \mu(\{x \in X : |f(x) - z| < \varepsilon\}) > 0. \tag{B.278}$$

Proof. The second claim implies the first, for $\text{ess-ran}(f) = \text{ran}(f)^-$ if $f \in C(X)$.

To prove the second claim, we use the functions $\varphi_n = 1_{\tilde{X}_n}\psi$ from the proof of Proposition B.73, where $\psi \in H$ is arbitrary. If $0 \notin \sigma(m_f)$, then m_f is invertible, so there is $b \in B(H)$ such that $fb\varphi_n = \varphi_n$. This implies that $f(x) \neq 0$ a.e. on $\tilde{X}_n$, with $b\varphi_n = m_{1/f}\varphi_n$. Because $n \in \mathbb{N}$ is also arbitrary and $X = \cup_n \tilde{X}_n$, this gives $f(x) \neq 0$ a.e. on X, and since the linear span of the φ_n is dense in H, we obtain $b = m_{1/f}$, provided $b = m_f^{-1}$ exists (which should not surprise us, for $m_f m_g = m_{fg}$). From (B.240), with $f \rightsquigarrow 1/f$, we then obtain $\|1/f\|_\infty^{\text{ess}} = \|m_{1/f}\| < \infty$ (from $0 \in \rho(m_f)$).

The point is that $\|1/f\|_\infty^{\text{ess}} < \infty$ iff there is $\varepsilon > 0$ such that $|f(x)| \geq \varepsilon$ almost everywhere, i.e., $\mu(\{x \in X : |f(x)| < \varepsilon\}) = 0$. The negation of this condition states that $\forall \varepsilon > 0 : \mu(\{x \in X : |f(x)| < \varepsilon\}) > 0$, that is, $0 \in \text{ess-ran}(f)$. Therefore, we have shown that $0 \in \sigma(m_f)$ iff $0 \in \text{ess-ran}(f)$; if $f \in C(X)$, this is the same as $0 \in \text{ran}(f)^-$.

To finish, note that $m_f - \lambda \cdot 1_H = m_{f-\lambda}$, where $f - \lambda$ is the function $x \mapsto f(x) - \lambda$. This gives $\lambda \in \sigma(m_f)$ iff $0 \in \sigma(m_{f-\lambda})$, which is true iff $\lambda \in \text{ess-ran}(f)$. □

Corollary B.90. *If $\mu(f=\lambda)=0$ for all $\lambda\in\mathbb{C}$, then $\sigma_p(m_f)=\emptyset$.*

Thus the combination $\sigma_p(a)=\emptyset$ and $\sigma_c(a)\neq\emptyset$, which is the opposite of the finite-dimensional situation, is very well possible. To shed further light on the still somewhat mysterious idea of a continuous spectrum, we now present Weyl's theory of the spectrum. We say that a possibly unbounded operator $a: D(a)\to H$ is ***normal*** when $D(a^*)=D(a)$ and $\|a^*\psi\|=\|a\psi\|$ for each $\psi\in D(a)$; if a is bounded, this is equivalent to the familiar definition $a^*a=aa^*$. Self-adjoint operators are normal.

Theorem B.91. *Let $a: D(a)\to H$ be normal. Then $\lambda\in\sigma(a)$ iff there exists a sequence (ψ_n) of unit vectors in $D(a)$ such that*

$$\lim_{n\to\infty}\|(a-\lambda)\psi_n\|=0. \tag{B.279}$$

Of course, this is useful only as a new characterization of $\lambda\in\sigma_c(a)$; if $\lambda\in\sigma_p(a)$ one may simply take $\psi_n=\psi$ for all n, where $a\psi=\lambda\psi$. For a simple example, take

$$H = L^2(\mathbb{R}); \tag{B.280}$$
$$a = m_f \ (f\in C(\mathbb{R})), \tag{B.281}$$
$$\lambda = f(x_0) \ (x_0\in\mathbb{R}), \tag{B.282}$$

so that $\lambda\in \mathrm{ran}(f)\subset\sigma_c(m_f)=\sigma(m_f)$, and

$$\psi_n(x)=(n/\pi)^{1/4}e^{-n(x-x_0)^2/2}. \tag{B.283}$$

Then $\|\psi_n\|=1$ and $\lim_n\|(m_f-\lambda)\psi_n\|=0$, although (ψ_n) has no limit in $L^2(\mathbb{R})$.

Proof. One direction is easy by *reductio ad absurdum*: if the given sequence (ψ_n) exists yet $\lambda\in\rho(a)$, then, since $(a-\lambda)^{-1}$ would exist and would be bounded, for any sequence (φ_n) in H, $\varphi_n\to 0$ implies $(a-\lambda)^{-1}\varphi_n\to 0$, so taking $\varphi_n=(a-\lambda)\psi_n$, we find that $(a-\lambda)\psi_n\to 0$ implies $\psi_n\to 0$. Therefore, the assumption $\|\psi_n\|=1$ cannot be true, and hence $\lambda\notin\rho(\sigma(a)$, which is to say that $\lambda\in\sigma(a)$.

The converse direction requires two instructive lemmas of independent interest.

Lemma B.92. *Let $a\in B(H)$ (or, more generally, let $a: D(a)\to H$ be closed). Then*

$$\mathrm{ran}(a)^- = \ker(a^*)^\perp; \tag{B.284}$$
$$\mathrm{ran}(a)^\perp = \ker(a^*). \tag{B.285}$$

In particular, we have $\mathrm{ran}(a)^-=H$ *iff* $\ker(a^*)=\{0\}$.

Furthermore, we say that a is **norm-positive** *(a neologism!) if there exists $\alpha>0$ such that $\|a\psi\|\geq\alpha\|\psi\|$ for each $\psi\in H$ (or each $\psi\in D(a)$). Then:*

1. *If a is norm-positive, then* $\mathrm{ran}(a)$ *is closed.*
2. *The operator a is invertible iff a is norm-positive and* $\ker(a^*)=\{0\}$.
3. *A normal operator is invertible iff it is norm-positive.*

The last point provides the remainder of the proof of Theorem B.91, for if $\lambda \in \sigma(a)$, then $a - \lambda$ is *not* invertible, so for each $\varepsilon = 1/n$ there is a unit vector $\psi_n \in H$ (or $\psi \in D(a)$) such that $\|(a-\lambda)\psi_n\| < 1/n$, and hence we have our sequence (ψ_n). □

It remains to prove Lemma B.92. Eqs. (B.284) - (B.285) are easy exercises, using (B.204). For clause 1, if (φ_n) is a Cauchy sequence in $\text{ran}(a)$ converging to $\varphi \in H$, then $\varphi_n = a\psi_n$ for some $\psi_n \in D(a)$. Since $\|\psi_m - \psi_n\| \leq \alpha^{-1}\|\varphi_n - \varphi_m\|$, the sequence (ψ_n) is Cauchy, too, and if $\psi_n \to \psi$, then $\varphi_n \to a\psi = \varphi$, so $\varphi \in \text{ran}(a)$; in the unbounded case this is because a is closed. For clause 2, if a is invertible, then for $\psi \in D(a)$, we have $\|\psi\| = \|a^{-1}a\psi\| \leq \|a^{-1}\|\|a\psi\|$, since a^{-1} is bounded, and therefore a is norm-positive with (for example) $\alpha = \|a^{-1}\|^{-1}$. Moreover, invertibility implies surjectivity, i.e., $\text{ran}(a) = H$, and hence $\ker(a^*) = \{0\}$ by (B.284).

Conversely, if a is norm-positive, then it is trivially injective, and if $\ker(a^*) = \{0\}$, then $\text{ran}(a)^- = H$, again by (B.284). But since a is also norm-positive, $\text{ran}(a)^- = \text{ran}(a)$ so $\text{ran}(a) = H$ and a is surjective, too. Clause 3 now also follows, since for normal operators a we have $\ker(a) = \ker(a^*)$, so a being norm-positive implying $\ker(a^*) = \{0\}$ in any case, now also implies $\ker(a) = \{0\}$. □

The same lemma yields crucial information on spectra of self-adjoint operators.

Theorem B.93. *If $a : D(a) \to H$ is self-adjoint, then $\sigma(a) \subseteq \mathbb{R}$, and if two eigenvalues $\lambda, \lambda' \in \sigma_p(a)$ are different, then corresponding eigenvectors are orthogonal.*

Furthermore, for each $z \in \mathbb{C}$ exactly one of the following possibilities applies:

- *$z \in \rho(a)$ iff $\text{ran}(a-z) = H$;*
- *$z \in \sigma_c(a)$ iff $\text{ran}(a-z)^- = H$ but $\text{ran}(a-z) \neq H$;*
- *$z \in \sigma_p(a)$ iff $\text{ran}(a-z)^- \neq H$.*

Proof. If $a^* = a$ then $\langle\psi, a\psi\rangle$ is real, so $|\langle\psi,(a-z)\psi\rangle| \geq |\text{Im}(z)|\|\psi\|^2$ for any $z \in \mathbb{C}$. Combined with Cauchy–Schwarz, this gives the inequality

$$\|(a-z)\psi\| \geq |\text{Im}(z)|\|\psi\|. \tag{B.286}$$

Therefore, for $z \in \mathbb{C}\backslash\mathbb{R}$ the normal operator $a - z$ is norm-positive, and hence invertible by Lemma B.92.3, so that $\sigma(a) \subseteq \mathbb{R}$. Next, if $a\psi = \lambda\psi$ and $a\psi' = \lambda'\psi'$,

$$\langle\psi,\psi'\rangle = \frac{1}{\lambda-\lambda'}(\langle\lambda\psi,\psi'\rangle - \langle\psi,\lambda'\psi'\rangle) = \frac{1}{\lambda-\lambda'}\langle\psi,(a^*-a)\psi'\rangle = 0. \tag{B.287}$$

given that $\lambda, \lambda' \in \mathbb{R}$ and assuming $\lambda' \neq \lambda$ and $a^* = a$.

Furthermore, for $z \in \mathbb{C}\backslash\mathbb{R}$, we have $z \in \rho(a)$ and hence trivially $\text{ran}(a-z) = H$; conversely, the latter property states surjectivity of $a - z$, whilst (B.286) yields injectivity, so jointly, $z \in \rho(a)$. For $z \in \mathbb{R}$, assuming $\text{ran}(a-z) = H$, eq. (B.285) yields $\ker(a^* - \overline{z}) = \{0\}$, but since $a^* = a$ and $\overline{z} = z$, this is just injectivity of $a - z$, whence once more $z \in \rho(a)$. Similarly, if $z \in \mathbb{R}$, then $\text{ran}(a-z)^- \neq H$ iff $\ker(a-z) \neq \{0\}$, which yields the third case $z \in \sigma_p(a)$. The middle case is all that remains. □

This result reconfirms Corollary B.88 to the effect that continuous spectrum cannot occur if $\dim(H) < \infty$, since in that case (where linear subspaces are automatically closed) the second scenario in Theorem B.93 is impossible.

B.15 The spectral theorem

Although he did not live to see it, on Hilbert's viosnary Definition B.80 of the spectrum, part 1 of Theorem A.15 still holds *verbatim* even if H is infinite-dimensional:

Theorem B.94. *Let H be a Hilbert space, suppose $a \in B(H)$ is self-adjoint, and let $C^*(a)$ be the C*-algebra generated within $B(H)$ by a and 1_H (that is, the intersection of all C*-algebras containing a and 1_H). Then $C^*(a)$ is commutative, and there is a (necessarily isometric) isomorphism of (commutative) C*-algebras*

$$C(\sigma(a)) \stackrel{\cong}{\to} C^*(a), \; f \mapsto f(a), \tag{B.288}$$

which is unique if it is subject to the following conditions:

- *the unit function $1_{\sigma(a)} : \lambda \mapsto 1$ corresponds to the unit operator 1_H;*
- *the identity function $\mathrm{id}_{\sigma(a)} : \lambda \mapsto \lambda$ is mapped to the given operator a.*

The map $f \mapsto f(a)$ is called the ***continuous functional calculus***. In particular,

$$(tf+g)(a) = tf(a)+g(a); \tag{B.289}$$

$$(fg)(a) = f(a)g(a); \tag{B.290}$$

$$f(a)^* = f^*(a). \tag{B.291}$$

It is worth mentioning that by Theorem C.62 (cf. Appendix C) an isomorphism of C*-algebras is automatically isometric, but in this case the equality

$$\|f(a)\| = \|f\|_\infty, \tag{B.292}$$

acts as a lemma in the proof that (B.288) is an isomorphism, so we need to prove it explicitly; cf. (B.225) for the left-hand side, and (1.24) for the right-hand side.

Note that Theorem B.94 is even true for the larger class *normal* bounded operators a (which might even be *defined* by the property that $C^*(a)$ is commutative), but for applications to quantum mechanics it is sufficient to deal with the self-adjoint case (which even mathematically is not a restriction, as it implies the normal case).

Proof. We repeat (A.52) and (A.53) - (A.55), obtaining a map $f \mapsto f(a)$ defined for *polynomials* f on $\mathbb{R}$, restricted to $\sigma(a) \subset \mathbb{R}$. The *-algebra $P^*(a)$ of all polynomials in a is dense in $C^*(a)$ by definition of the latter, since one cannot have a smaller C*-algebra in $B(H)$ containing a and 1_H than the norm-closure of $P^*(a)$. In order to take advantage of this, we need the following lemma.

Lemma B.95. *For any $a \in B(H)$ and any polynomial p on $\mathbb{C}$, we have*

$$\sigma(p(a)) = p(\sigma(a)) \equiv \{p(\lambda) \mid \lambda \in \sigma(a)\}; \tag{B.293}$$

$$\|a\| = \sqrt{r(a^*a)}, \tag{B.294}$$

see (B.254). *In particular, if $a^* = a$, then $\|a\| = r(a)$, cf.* (B.257).

This is part of Theorem B.84, but we now give a direct proof of the second part. We first note that if $a^* = a$, then either $\|a\|$ or $-\|a\|$ (or both) are in $\sigma(a)$. To show this, take a sequence (ψ_n) of unit vectors in H such that $\lim_n \|a\psi_n\| = \|a\|$. Then

$$\begin{aligned} \|(a^2 - \|a\|^2)\psi_n\|^2 &= \langle (a^2 - \|a\|^2)\psi_n, (a^2 - \|a\|^2)\psi_n \rangle \\ &= \|a^2\psi_n\|^2 + \|a\|^4 - 2\|a\|^2\|a\psi_n\|^2 \\ &\leq 2\|a\|^4 - 2\|a\|^2\|a\psi_n\|^2, \end{aligned} \tag{B.295}$$

so that $\lim_n \|(a^2 - \|a\|^2)\psi_n\|^2 = 0$, and hence $\|a\|^2 \in \sigma(a^2)$ by Theorem B.91. But part 1 of the lemma gives $\sigma(a^2) = \{\lambda^2 \mid \lambda \in \sigma(a)\}$, so that $\pm\|a\| \in \sigma(a)$.

The second observation is that, for general $a \in B(H)$, if some $z \in \mathbb{C}$ has $|z| > \|a\|$, then $z \in \rho(a)$. This follows from (part 1 of) the proof of Theorem B.84. Thus we firstly have $r(a) \geq \|a\|$ ($a^* = a$), and secondly (for all a), $r(a) \leq \|a\|$.

Using Lemma B.95, we now prove that (B.292) holds for *real* polynomials $f = p$:

$$\begin{aligned} \|p(a)\| = r(p(a)) &= \sup\{|\lambda|, \lambda \in \sigma(p(a))\} = \sup\{|\lambda|, \lambda \in p(\sigma(a))\} \\ &= \sup\{|p(\lambda)|, \lambda \in \sigma(a)\} = \|p\|_\infty. \end{aligned} \tag{B.296}$$

The case of *complex* polynomials p follows from this, since, using (B.289) - (B.291),

$$\|p(a)\|^2 = \|p(a)^*p(a)\| = \||p|^2(a)\| = \||p|^2\|_\infty = \|p\|_\infty^2. \tag{B.297}$$

Thus we have proved the isometric *-algebra isomorphism $P(\sigma(a)) \cong P^*(a)$, where $P(\sigma(a))$ and $P^*(a)$ are the canonically normed vector spaces of all finite polynomials in $t \in \sigma(a)$ and in $a \in B(H)$, respectively. Neither is complete (when H is infinite-dimensional and $a \neq 0$), but given isometricity, it is easy to pass to their completions, which by Weierstrass and by definition are $C(\sigma(a))$ and $C^*(a)$, respectively. Thus for $f \in C(\sigma(a))$ we find a sequence (p_n) in $P(\sigma(a))$ such that $p_n \to f$ (from which it follows that (p_n) is Cauchy in $C(\sigma(a))$), and define

$$f(a) = \lim_n p_n(a); \tag{B.298}$$

this limit exists because $\|p_n(a) - p_m(a)\| = \|p_n - p_m\|_\infty$, so that $(p_n(a))$ is Cauchy in the Banach space $C^*(a)$. Furthermore, if $p'_n \to f$, and $f'(a) = \lim_n p'_n(a)$, then

$$\|f(a) - f'(a)\| = \lim_n \|p_n(a) - p'_n(a)\| = \lim_n \|p_n - p'_n\|_\infty = 0, \tag{B.299}$$

so $f'(a) = f(a)$. From (B.296) - (B.298) and continuity of the norm—i.e. $\|f(a)\| = \lim_n \|p_n(a)\|$, which gives d (B.292)—the map $f \mapsto f(a)$ is isometric and hence injective on $C(\sigma(a))$, and the above construction trivially makes it surjective.

Finally, the properties (B.289) - (B.291) follow from (A.53) - (A.55) by continuity. These properties also imply the uniqueness of the map $f \mapsto f(a)$ given the conditions states in the theorem, because these conditions and (A.53) - (A.55) define the map on $P(\sigma(a))$ and hence, by continuity, also on $C(\sigma(a))$. □

For a nice reformulation of Theorem B.94 in terms the Gelfand spectrum, cf. §C.4.
For later use (cf. Proposition B.98 below) we add a related result.

Lemma B.96. *If $a \in B(H)$ is self-adjoint, then*

$$\|a\| = \sup\{|\langle\psi, a\psi\rangle|, \psi \in H, \|\psi\| = 1\}. \tag{B.300}$$

In particular, if $a, b \in B(H)$ are both positive and $a \leq b$, then $\|a\| \leq \|b\|$.

Proof. Define the ***numerical range*** $\nu(a)$ of an arbitrary $a \in B(H)$ as

$$\nu(a) = \{\langle\psi, a\psi\rangle, \psi \in H, \|\psi\| = 1\}. \tag{B.301}$$

Clearly, if $\lambda \in \sigma_p(a)$, then $\lambda \in \nu(a)$. If $\lambda \in \sigma_c(a)$, then, in the notation of Theorem B.91, by Cauchy–Schwarz and normalization of ψ_n we have

$$|\langle\psi_n, (a-\lambda)\psi_n\rangle| \leq \|(a-\lambda)\psi_n\|. \tag{B.302}$$

Hence in view of (B.279) we have

$$\lim_{n\to\infty} \langle\psi_n, a\psi_n\rangle = \lambda. \tag{B.303}$$

So $\lambda \in \nu(\sigma)^-$, whence $\sigma(a) \subseteq \nu(a)^-$, and hence $r(a) \leq \sup\{|\lambda|, \lambda \in \nu(a)\}$. From Cauchy–Schwarz, in (B.301) we have $|\langle\psi, a\psi\rangle| \leq \|a\|$. If also $a^* = a$, by (B.300),

$$\|a\| = r(a) \leq \sup\{|\lambda|, \lambda \in \nu(a)\} \leq \|a\|.$$

Hence we have equalities everywhere, and (B.300) follows. □

Generalizing parts 2 and 3 of Theorem A.15 to the infinite-dimensional case requires some motivation. To this effect, note that the continuous functional calculus $a \mapsto f(a)$ is *positive*, i.e., if $f \geq 0$ pointwise, then $f(a) \geq 0$ in that $\langle\psi, f(a)\psi\rangle \geq 0$ for each $\psi \in H$. Indeed, we have $f \geq 0$ iff $f = g^*g$ for some $g \in C(\sigma(a))$, with $g^*(x) = \overline{g(x)}$ as usual, and hence, by (B.290) - (B.291), $f(a) = g(a)^*g(a)$ and therefore $\langle\psi, f(a)\psi\rangle = \|g(a)\|^2 \geq 0$. By Corollary B.17, if $\psi \in H$ is a *unit* vector, there is a *probability* measure μ_ψ on $\sigma(a)$ such that for each $f \in C(\sigma(a))$,

$$\langle\psi, f(a)\psi\rangle = \int_{\sigma(a)} d\mu_\psi f. \tag{B.304}$$

The key to the envisaged generalization of Theorem A.15 is that the integral on the right may actually be defined for a far larger class of functions than $C(\sigma(a))$; cf. (B.29). This suggests that the expression $f(a)$ on the left-hand side should similarly be generalized to a larger class of functions f. However, the L^p spaces considered in §B.6 are defined on the basis of some measure μ; since μ_ψ in (B.304) varies with ψ and $f(a)$ should be independent of ψ, it is appropriate to use the space $\mathscr{B}(\sigma(a))$ of *bounded* functions $f : \sigma(a) \to \mathbb{C}$ that are *measurable* with respect to the Borel σ-algebra on $\sigma(a)$ (which consist of the Borel sets on $\mathbb{R}$ intersected with $\sigma(a)$).

Since both boundedness and measurability are preserved under uniform limits (measurability even being preserved under pointwise limits), $\mathscr{B}(\sigma(a))$ is complete in the sup-norm, which makes it a commutative C*-algebra (under pointwise operations). Among all functions in $\mathscr{B}(\sigma(a))$, we will be particularly interested in the characteristic functions 1_A, where $A \subset \sigma(a)$ is measurable. The expressions

$$e_A = 1_A(a), \ (A \subset \sigma(a)); \tag{B.305}$$
$$e_A = e_{A\cap\sigma(a)}, \ (A \subset \mathbb{R}); \tag{B.306}$$
$$e_\lambda \equiv 1_{\{\lambda\}}(a), \ (\lambda \in \sigma_p(a)), \tag{B.307}$$

to be defined below, where A is a Borel set (and $e_\emptyset = 0$ by convention), are the ***spectral projections*** of a (which are of fundamental importance to quantum mechanics).

Lemma B.97. *Any* positive *function* $f \in \mathscr{B}(\sigma(a))$ *is a* pointwise *limit of some monotone increasing bounded sequence* (f_n) *in* $C(\sigma(a))$, *written* $f_n \nearrow f$. *That is,*

$$0 \leq f_1(x) \leq \cdots \leq f_n(x) \leq f_{n+1}(x) \leq \cdots \leq c \cdot 1_{\sigma(a)}; \tag{B.308}$$
$$f(x) = \lim_{n\to\infty} f_n(x), \ x \in \sigma(a). \tag{B.309}$$

Proof. We start with $f = 1_K$, where $K \subseteq \sigma(a)$ is compact. Then $K = \cap_n U_n$ for certain open sets U_n (this is true for any second countable space), and taking "Urysohn" functions f_n for each U_n (i.e., $f_n \in C_c(U_n)$, $0 \leq f_n(x) \leq 1$ for $x \in \sigma(a)$, and $f_n(x) = 1$ for $x \in K$), we obviously have $f_n \to 1_K$. Next, if $U \subset \sigma(a)$ is open, we have $U = \cup_n K_n$ for suitable compact K_n (since $\mathbb{R}$ and hence $\sigma(a)$ is σ-compact), so $1_{K_n} \to 1_U$. This also gives 1_C for closed sets $C = \sigma(a)\backslash U$, since $1_C = 1_{\sigma(a)} - 1_U$. Using the so-called Borel hierarchy, it can be shown that any Borel set $A \subset \sigma(a)$ can be constructed from open and closed sets in at most a countable number of steps, at each of which a countable union or intersection of sets from the previous steps is used. This gives 1_A for any Borel set, and hence also yields the simple functions $s = \sum_k c_k 1_{A_k}$ with $c_k \geq 0$. For arbitrary measurable $f \geq 0$ (not necessarily bounded and not even necessarily finite) it is a standard result in measure theory that there is a sequence (s_n) of simple functions such that $s_n \nearrow f$: to wit, define

$$A_{n,k} = \{x \in \sigma(a) \mid 2^{-n}k < f(x) \leq 2^{-n}(k+1)\}; \tag{B.310}$$
$$A_n = \{x \in \sigma(a) \mid n < f(x) < \infty\}; \tag{B.311}$$
$$s_n = n \cdot 1_{A_n} + 2^{-n} \sum_{k=1}^{2^n n-1} k 1_{A_{n,k}}. \tag{B.312}$$

Relabeling the (at most) countable number of sequences thus obtained as a single sequence then gives a positive sequence (h_n) in $C(\sigma(a))$ such that $h_n \to f$ pointwise.

A final trick turns (h_n) into a monotone increasing bounded sequence (f_n): for $m > n$, define $f_{n,m} = \min\{h_n, \ldots, h_m\}$, which is monotone *de*creasing in m and positive, and hence has a (pointwise) limit $f_n = \lim_{m\to\infty} f_{n,m}$. The ensuing sequence (f_n) is monotone *in*creasing and still converges to f. If f is bounded (as we assume by definition of $\mathscr{B}(\sigma(a))$), then (f_n) must also be bounded eventually. □

If $f \in \mathscr{B}(\sigma(a))$ and $f_n \nearrow f$ with $f_n \in C(\sigma(a))$, we would like to define $f(a)$ as $\lim_n f_n(a)$, just as in the case where $f \in C(\sigma(a))$ and $f_n \in P(\sigma(a))$. However, in the former case convergence $f_n \to f$ is merely pointwise, whereas in the latter case it was uniform, translated into norm convergence $f_n(a) \to f(a)$. Pointwise convergence of functions, then, becomes *strong* convergence of operators:

Proposition B.98. *If (a_n) is a sequence of positive operators on H for which*

$$0 \leq a_1 \leq \cdots \leq a_n \leq a_{n+1} \leq \cdots \leq c1_H, \tag{B.313}$$

where $a_i \leq a_j$ means that $\langle \psi, a_i\psi\rangle \leq \langle \psi, a_j\psi\rangle$ for each $\psi \in H$, then there exists a unique positive operator a such that $a_n \nearrow a$ strongly, i.e., for each $\psi \in H$,

$$a\psi = \lim_{n\to\infty} a_n\psi. \tag{B.314}$$

Furthermore, $a = \sup_n a_n$ with respect to the partial ordering $\leq$ on the set of positive bounded operators (that is, $a_n \leq a$ for each n, and if $a_n \leq b$ for each n, then $a \leq b$).

Proof. Recalling Proposition A.4, define a sequence of bounded quadratic forms $Q_n : H \to \mathbb{R}$ by $Q_n(\psi) = \langle \psi, a_n\psi\rangle$. Then $(Q_n(\psi))$ is a monotone increasing bounded sequence for each $\psi \in H$, so that $Q(\psi) = \lim_{n\to\infty} Q_n(\psi)$ exists. Like each Q_n, also Q satisfies (A.8) - (A.9). Since $|Q_n(\psi)| \leq c\|\psi\|^2$ and hence $|Q(\psi)| \leq c\|\psi\|^2$, it remains bounded. Hence (A.10) defines a bounded hermitian form B, upon which Proposition B.79 yields a bounded operator a, satisfying $B(\varphi,\psi) = \langle \varphi, a\psi\rangle$. Since

$$\langle \psi, a\psi\rangle = \lim_{n\to\infty} \langle \psi, a_n\psi\rangle, \tag{B.315}$$

we have $a \geq 0$. To prove (B.314), note that (B.315) gives $\langle \psi, (a-a_n)\psi\rangle \to 0$, but (B.313) implies $a - a_n \geq 0$, so that $a - a_n$ has a self-adjoint square root $\sqrt{a-a_n}$, defined by Theorem B.94 (see also Proposition B.99 below). Hence

$$\langle \psi, (a-a_n)\psi\rangle = \langle \sqrt{a-a_n}\psi, \sqrt{a-a_n}\psi\rangle = \|\sqrt{a-a_n}\psi\|^2 \to 0. \tag{B.316}$$

Now if a sequence of operators (b_n) is such that $\|b_n\| \leq C$ for all n, and $\|b_n\psi\| \to 0$, then also $\|b_n^2\psi\| \to 0$, for $\|b_n^2\psi\| \leq \|b_n\|\|b_n\psi\| \leq C\|b_n\psi\| \to 0$. This applies here, since $a_m \leq a_n$ for $m \leq n$, and hence $a - a_n \leq a - a_m$, from which $\|a - a_n\| \leq \|a - a_m\|$ (see Lemma B.96). Fixing m, this gives $\|a - a_n\| \leq C$ with $C = \|a - a_m\|$, for all $n \geq m$. So (B.316) implies $\|(a-a_n)\psi\| \to 0$, which is (B.314).

As to the final claim, eq. (B.315) is the same as $\langle \psi, a\psi\rangle = \sup_n\{\langle \psi, a_n\psi\rangle\}$. □

In this proof, we used the following generalization of Proposition A.22:

Proposition B.99. *The following conditions on $a \in B(H)$ are equivalent:*

1. *$\langle \psi, a\psi\rangle \geq 0$ for arbitrary $\psi \in H$;*
2. *$a^* = a$ and $\sigma(a) \subset \mathbb{R}^+$;*
3. *$a = c^2$ for some bounded self-adjoint operator $c \in B(H)$;*
4. *$a = b^*b$ for some bounded operator $b \in B(H)$.*

Proof. The proof is the same as in the finite-dimensional case, except that:

- In $1 \to 2$ we use (B.303) to exclude the possibility that some $\lambda < 0$ lies in $\sigma(a)$;
- In $2 \to 3$ we need Theorem B.94) to define the square root $c = \sqrt{a}$ from the function $\sqrt{\cdot} : \sigma(a) \to \mathbb{R}$ (which is well defined because $\sigma(a) \subset \mathbb{R}^+$). By (B.290) with $g = f = \sqrt{\cdot}$, we then have $\sqrt{a}\sqrt{a} = a$. □

Given some positive $f \in \mathscr{B}(\sigma(a))$, we now use Lemma B.97 to find a monotone increasing bounded sequence (f_n) in $C(\sigma(a))$ such that $f_n \nearrow f$ pointwise, and subsequently use Proposition B.98 to define $f(a)$ as the *strong* limit

$$f(a)\psi = \lim_{n\to\infty} f_n(a)\psi \ (\psi \in H). \tag{B.317}$$

Arbitrary functions f are then dealt with using (B.30) and performing the above constructing term-wise. This, then, yields $f(a)$ for any $a^* = a \in B(H)$ and $f \in \mathscr{B}(\sigma(a))$.

It is natural to ask which corner of $B(H)$ the operators $f(a)$ land in when $f \in \mathscr{B}(\sigma(a))$, much as we have shown that $f(a) \in C^*(a)$ for $f \in C(\sigma(a))$. A safe choice would be $C^*(a)^-$, i.e., the strong closure of $C^*(a)$, which by definition contains all limits of all strongly convergent nets in $C^*(a)$ (so that it certainly contains all limits (B.317)), and which is automatically a strongly closed unital *-algebra. This may seem too large, but if H is separable, it turns out to be the right choice, because these more general limits add nothing to (B.317)). For a more explicit description of $C^*(a)^-$ we need the ***commutant*** S' of any $S \subset B(H)$, which is defined by

$$S' = \{a \in B(H) \mid ab = ba \forall b \in S\}; \tag{B.318}$$

the ***bicommutant*** of S is $S'' = (S')'$. If $S^* = S$, in that $a \in S$ iff $a^* \in S$, then S' is easily seen to be a unital *-algebra within $B(H)$. Furthermore, it is obvious that $S \subset S''$, so that the passage $S \mapsto S''$ is some sort of a closure operation within $B(H)$, comparable to the closure operation $S \mapsto S^{\perp\perp}$ within H itself. Indeed, there is a striking analogue of (B.204) at the operator level, due to von Neumann (see Theorem C.127):

Theorem B.100. *If A is a unital *-algebra in $B(H)$, then*

$$A'' = A^-, \tag{B.319}$$

*where A^- is the strong closure of A in $B(H)$ (which is automatically a *-algebra).*

Corollary B.101. *Denoting the strong closure $C^*(a)^-$ of $C^*(a)$ by $W^*(a)$, we have*

$$W^*(a) = C^*(a)''. \tag{B.320}$$

Though not obvious from (B.320), the alternative description through (B.319) shows that $W^*(a)$ inherits the commutativity of $C^*(a)$; in fact $W^*(a)$ is a commutative C*-algebra, too. Moreover, by construction it is also a ***von Neumann algebra*** in that $W^*(a)'' = W^*(a)$, cf. Appendix C. Such unital *-algebras in $B(H)$ are not merely norm closed, but are also closed in at least three other natural topologies on $B(H)$, including the strong one. The situation may be summarized in the ***spectral theorem***:

Theorem B.102. *Let $a^* = a \in B(H)$. The isomorphism $C(\sigma(a)) \to C^*(a)$ of Theorem B.94 has a unique extension to a homomorphism*

$$\mathscr{B}(\sigma(a)) \to W^*(a),\ a \mapsto f(a), \tag{B.321}$$

for (B.289) - (B.291) *continue to hold. In particular, the operator e_A in* (B.305) *is a projection. Also, eq.* (B.304) *remains valid, and for each $f \in \mathscr{B}(\sigma(a))$, one has*

$$\|f(a)\| \leq \|f\|_\infty. \tag{B.322}$$

Proof. The map $a \mapsto f(a)$ is given by (B.317) and preceding discussion. Eqs. (B.289) and (B.291) easily follows by limiting arguments. Using the same trick as in the proof of Proposition B.98 it can be shown that $f(a)^2 = f(a^2)$, whence, using the identity $fg = \frac{1}{2}((f+g)^2 - f^2 - g^2)$, eq. (B.290) follows. This implies $e_A^2 = 1_A^2(a) = 1_A(a) = e_A$, whilst (B.291) gives $e_A^* = 1_A^*(a) = 1_A(a) = e_A$.

We prove (B.322) for $f \geq 0$; this implies the general case by (B.30) and the triangle equality. Writing H_1 for the set of unit vectors in H, approximating $f_n \nearrow f$, repeatedly using (B.300), the property $f(a) = \sup_n f_n(a)$ established at the end of Proposition B.98, and finally using (B.292) for each $f_n \in C(\sigma(a))$, we may estimate:

$$\begin{aligned}
\|f(a)\| &= \sup_{\psi \in H_1} \{|\langle \psi, f(a)\psi\rangle|\} \\
&= \sup_{\psi \in H_1} \sup_{n \in \mathbb{N}} \{|\langle \psi, f_n(a)\psi\rangle|\} \\
&= \sup_{n \in \mathbb{N}} \sup_{\psi \in H_1} \{|\langle \psi, f_n(a)\psi\rangle|\} \\
&= \sup_{n \in \mathbb{N}} \|f_n(a)\| = \sup_{n \in \mathbb{N}} \|f_n\|_\infty \\
&\leq \|f\|_\infty,
\end{aligned} \tag{B.323}$$

where the last inequality is a trivial consequence of the specific limit $f_n \nearrow f$.

Finally, our motivating identity (B.304) follows from the same equality for each $f_n \in C(\sigma(a))$, upon which Lebesgue's Monotone Convergence Theorem yields the right-hand side, whereas (B.315) gives the left-hand side. □

Of course, in finite dimension, Theorem B.102 coincides with Theorems A.15 and Theorem B.94. Theorem A.15 implies Theorem A.10 through (A.58) - (A.59), and, as we will now explain, in infinite dimension Theorem B.102 similarly implies a certain approximate version of Theorem A.10, namely Corollary B.104.

Lemma B.103. *If $K \subset \mathbb{R}$ is compact, any $f \in C(K)$ may be uniformly approximated by simple functions. More precisely, for each $\varepsilon > 0$ there is a decomposition $K = \bigsqcup_{i=1}^n A_i$ of K as a disjoint union of $n < \infty$ Borel sets A_i, such that for any $x_i \in A_i$,*

$$\left\| f - \sum_{i=1}^n f(x_i) 1_{A_i} \right\|_\infty < \varepsilon. \tag{B.324}$$

Proof. Since K is compact, f is actually uniformly continuous on K. This means that for $\varepsilon > 0$ there is $\delta > 0$ such that $|f(x) - f(y)| < \varepsilon$ whenever $|x - y| < \delta$. Since (B.324) just states that $|f(x) - f(x_i)| < \varepsilon$ for each $i = 1, \ldots, n$ and each $x \in A_i$, any partition for which $0 < |A_i| < \delta$ will do (where $|A| = \sup\{|x - y|, x, y \in A\}$). □

From (B.305), Lemma B.103, and Theorem B.102, we then immediately have:

Corollary B.104. *Let $a^* = a \in B(H)$. For any $f \in C(\sigma(a))$ and any $\varepsilon > 0$, there is a partition $\sigma(a) = \bigsqcup_{i=1}^n A_i$ of $\sigma(a)$ as a disjoint union of $n < \infty$ Borel sets A_i, such that for arbitrary $\lambda_i \in A_i$, one has*

$$\left\| f(a) - \sum_{i=1}^n f(\lambda_i) e_{A_i} \right\| < \varepsilon. \tag{B.325}$$

In particular, for $f(x) = x$ and $f(x) = 1$ we have

$$\left\| a - \sum_{i=1}^n \lambda_i e_{A_i} \right\| < \varepsilon; \tag{B.326}$$

$$\left\| 1_H - \sum_{i=1}^n e_{A_i} \right\| < \varepsilon. \tag{B.327}$$

If a has discrete spectrum $\sigma(a) = \sigma_p(a)$), then (B.326) - (B.327) *reduce to* (A.37) - (A.38), *where e_λ is defined by* (B.307), *and the sums converge in norm.*

Hence in this version of the spectral theorem, one approximates a by linear combinations of projections in a way that reflects the approximation of the identity function $x \mapsto x$ on $\sigma(a)$ by simple functions. Eq. (B.326) is often symbolically written as

$$a = \int_{\sigma(a)} de_\lambda \, \lambda, \tag{B.328}$$

which may also be given some direct meaning as an operator-valued Stieltjes integral, but even so, this neat expression eventually boils down to (B.326) itself.

Corollary B.105. *Let $\mathscr{P}(A) = \{e \in A \mid e^2 = e^* = e\}$, where A is a von Neumann algebra. Then A is the norm-closure of the linear span of $\mathscr{P}(A)$, and*

$$A = \mathscr{P}(A)''. \tag{B.329}$$

Proof. The first claim follows from Corollary B.104. This implies (B.329), which may also be proved directly: since $\mathscr{P}(A) \subset A$, the inclusion $\mathscr{P}(A)'' \subseteq A'' = A$ is obvious. Conversely, let $a \in A$ and assume $a^* = a$ (if not, decompose $a = a' + ia''$ with a' and a'' self-adjoint). Then $W^*(a) \subset A$, so that A contains all spectral projections of a, cf. Theorem B.102. Moreover, by Corollary B.104, a lies in the norm-closure of the linear span of $\mathscr{P}(A)$, which by Theorem B.100 in turn is contained in A''. □

B.16 Abelian *-algebras in $B(H)$

Compared with Theorem B.94, it seems a weakness of Theorem B.102 that the map $f \mapsto f(a)$ fails to be an isomorphism from $\mathscr{B}(\sigma(a))$ to $W^*(a)$. The reason is that although the map is surjective (at least when H is separable), it fails to be injective: for real-valued f one has $f(a) = 0$ iff $\langle \psi, f(a)\psi \rangle = 0$ for all $\psi \in H$, which by (B.304) is the case iff $\int_{\sigma(a)} d\mu_\psi f = 0$ for all unit vectors $\psi \in H$, which in turn is the case iff $f = 0$ a.e. with respect to μ_ψ, in other words, iff $f = 0$ in $L^\infty(\sigma(a), \mu_\psi)$.

Thus the right kind of algebra to be isomorphic to $W^*(a)$ is $L^\infty(\sigma(a), \mu)$ rather than $\mathscr{B}(\sigma(a))$, where μ is some (probability) measure on $\sigma(a)$ such that $\mu(A) = 0$ iff $\mu_\psi(A) = 0$ for all unit vectors $\psi \in H$. Indeed, in that case, since by construction

$$L^\infty(\sigma(a), \mu) \cong \mathscr{B}(\sigma(a))/\{f \mid f = 0\, \mu\text{-a.e.}\} = \mathscr{B}(\sigma(a))/\ker(f \mapsto f(a)), \quad \text{(B.330)}$$

our map $\mathscr{B}(\sigma(a)) \to W^*(a)$ descends to an isomorphism of von Neumann algebras:

$$L^\infty(\sigma(a), \mu) \stackrel{\cong}{\to} W^*(a). \quad \text{(B.331)}$$

This is quite nontrivial; let us first present a case study where everything is clear.

Proposition B.106. *Let $H = L^2(0,1) = L^2([0,1])$ (with Lebesque measure), and let $a = m_{\text{id}} \in B(H)$ (where $\text{id}(x) = x$) be the self-adjoint position operator*

$$a\psi(x) = x\psi(x). \quad \text{(B.332)}$$

Then the map $f \mapsto f(a)$ in both Theorems B.94 and B.102 is given by

$$f(a) = m_f, \quad \text{(B.333)}$$

*cf. Proposition B.73. The two *-algebras in $B(H)$ defined by a are given by*

$$C^*(a) = C([0,1]); \quad \text{(B.334)}$$
$$W^*(a) = L^\infty(0,1), \quad \text{(B.335)}$$

both realized as multplication operators (i.e., identifying f with m_f). Furthermore,

$$L^\infty(0,1)' = L^\infty(0,1). \quad \text{(B.336)}$$

More generally, let $K \subset \mathbb{R}$ be compact, let μ be a regular probability measure on K with support K, take $H = L^2(K, \mu)$ and the define a as in (B.332)*. Then:*

$$\sigma(a) = K; \quad \text{(B.337)}$$
$$C^*(a) = C(K); \quad \text{(B.338)}$$
$$W^*(a) = L^\infty(K, \mu); \quad \text{(B.339)}$$
$$f(a) = m_f; \quad \text{(B.340)}$$
$$L^\infty(K, \mu)' = L^\infty(K, \mu). \quad \text{(B.341)}$$

Proof. We just prove the case $K = [0,1]$ with $d\mu(x) = dx$; the general case is similar.

Eq. (B.333) is obvious for polynomials f, and otherwise follows from easy limiting arguments. Consequently, eq. (B.334) is an instance of Theorem B.94. Everything else then follows if we can prove that

$$C([0,1])' = L^\infty(0,1). \tag{B.342}$$

Namely, assuming (B.342), since $C([0,1]) \subset L^\infty(0,1)$ (and $A \subseteq B$ implies $B' \subseteq A'$), we automatically have $L^\infty(0,1)' \subseteq C([0,1])'$, so (B.342) implies $L^\infty(0,1)' \subseteq L^\infty(0,1)$, and since the converse inclusion is trivial from commutativity of $L^\infty(0,1)$, eq. (B.342) implies (B.336). Furthermore, since $W^*(a) = C([0,1])''$, taking the commutant of (B.342) and applying (B.336) yields (B.335).

So let us prove (B.342). The inclusion $L^\infty(0,1) \subseteq C([0,1])'$ is obvious, since $m_f m_g = m_{fg} = m_{gf} = m_g m_f$, so we need to prove the converse. Take $b \in C([0,1])'$ and define $f = b1_{[0,1]} \in L^2(0,1)$. For $\psi \in C([0,1]) \subset L^2(0,1)$, we have

$$b\psi = bm_\psi 1_{[0,1]} = m_\psi b 1_{[0,1]} = m_\psi f = m_\psi m_f 1_{[0,1]} = m_f m_\psi 1_{[0,1]} = m_f \psi, \tag{B.343}$$

so $b = m_f$ on the dense domain $C([0,1]) \subset L^2(0,1)$, with $f \in L^2(0,1)$. Now b is bounded by definition of the commutant $C([0,1])'$ and hence $\|m_f\| < \infty$. If $f \notin L^\infty(0,1)$, the proof of Proposition B.73 gives that X_t has positive measure for each $t > 0$, whence $\|m_f\| \geq t$ for all t, which is a contradiction. Hence $f \in L^\infty(0,1)$, in which case m_f extends to all of $L^2(0,1)$ by continuity. This extension must equal b, so that $b = m_f$, and hence $C([0,1])' \subseteq L^\infty(0,1)$. □

The following variation on this example turns out to be qualitatively different:

Proposition B.107. *Realizing $\ell^\infty(\mathbb{N})$ as multiplication operators on $\ell^2(\mathbb{N})$, one has*

$$\ell^\infty(\mathbb{N})' = \ell^\infty(\mathbb{N}). \tag{B.344}$$

Proof. For each $N \in \mathbb{N}$, we define a finite-dimensional subspace $\ell^2(N) \subset \ell^2(\mathbb{N})$ by

$$\ell^2(N) = \{\psi \in \ell^2(\mathbb{N}) \mid \psi(x) = 0 \forall x > N\},$$

with ensuing projection $1_N : \ell^2(\mathbb{N}) \to \ell^2(N)$, i.e., $1_N\psi(x) = \psi(x)$ for $x \leq N$ and $1_N\psi(x) = 0$ for $x > N$. If $b \in \ell^\infty(\mathbb{N})'$, we have $b : \ell^2(N) \to \ell^2(N)$, because $1_N \in \ell^\infty(\mathbb{N})$ (and hence $\psi \in \ell^2(N)$, i.e., $1_N\psi = \psi$, implies $b\psi \in \ell^2(N)$, i.e., $1_N b\psi = b\psi$). With $f_N : \mathbb{N} \to \mathbb{C}$ given by $f_N = b1_N$, define $f : \mathbb{N} \to \mathbb{C}$ by $f(x) = f_N(x)$ for any $N > x$; this is well defined, in that if $x < N < M$, then $f_N(x) = f_M(x)$. For any N and $\psi \in \ell^2(N)$, as in (B.343) we have $b\psi = m_f\psi$, which therefore holds on a dense subspace $\cup_N \ell^2(N)$ of $\ell^2(\mathbb{N})$. Again as in the previous proof, this gives

$$\|f\|_\infty = \|m_f\| = \|b\| < \infty, \tag{B.345}$$

i.e., $f \in \ell^\infty(\mathbb{N})$. Thus $b = m_f \equiv f \in \ell^\infty(\mathbb{N})$, whence $\ell^\infty(\mathbb{N})' \subseteq \ell^\infty(\mathbb{N})$. With the trivial opposite inclusion, this gives (B.344). □

Note that since a possible (discrete) position operator (B.332) would be unbounded on $\ell^2(\mathbb{N})$, a possible counterpart to (B.335), although it exists, would blast the framework of the this section (cf. §B.21). See, however, the proof of Theorem B.118.

More generally, we have:

Proposition B.108. *Let (X,Σ,μ) be a σ-finite Borel space and realize $L^\infty(X,\mu)$ as multplication operators on $L^2(X,\mu)$. Then*

$$L^\infty(X,\mu)' = L^\infty(X,\mu). \tag{B.346}$$

Proof. Writing $X = \cup_{N\in\mathbb{N}} X_N$ with $\mu(X_N) < \infty$, which holds by virtue of σ-finiteness, the proof is practically the same as for $X = \mathbb{N}$ (except for the fact that $L^2(X_N) \subset L^2(X)$ need not be finite-dimensional, but it is closed, which suffices). □

If $A \subset B(H)$ is a commutative *-algebra, we say that A is ***maximal (abelian)*** if $A \subseteq B \subset B(H)$ for some commutative *-algebra B implies $B = A$. Any *-algebra $A \subset B(H)$ is abelian iff $A \subseteq A'$ (this is trivial), and is maximally abelian iff $A' = A$. To see the nontrivial "⇒" direction, for any subsets $C \subset B(H)$ and $D \subset B(H)$ the inclusion $C \subseteq D$ implies $D' \subseteq C'$ (as is immediate from the definition of the commutant), so $B' \subseteq A'$. Since B is commutative, we also know that $B \subseteq B'$, whence $B \subseteq A'$. If $A' = A$ this gives $B \subseteq A$, so $B = A$. The condition $A' = A$, in turn, implies $A'' = A$, i.e., any maximal abelian *-algebra A in $B(H)$ is automatically a von Neumann algebra.

Corollary B.109. *In the setting of Proposition B.108, $L^\infty(X,\mu)$ is a maximal abelian *-algebra in $B(L^2(X,\mu))$, and hence a von Neumann algebra. In particular:*

- *$L^\infty(0,1)$ is a maximal abelian *-algebra in $B(L^2(0,1))$;*
- *$\ell^\infty(\mathbb{N})$ is a maximal abelian *-algebra in $B(\ell^2(\mathbb{N}))$.*

The above examples suggest a neat reformulation of the spectral theorem. This requires a few more concepts from the theory of operator algebras, cf. Appendix C.

Definition B.110. *For any *-algebra $A \subset B(H)$ and $\psi \in H$, we write $A\psi^- \subseteq H$ for the closure of the linear subspace of all vectors $a\psi$, $a \in A$. We say that ψ ($\neq 0$) is:*

- **cyclic** *for A if $A\psi^- = H$;*
- **separating** *for A if $a\psi = 0$ for $a \in A$ implies $a = 0$.*

If $a^* = a \in B(H)$, we similarly say that ψ is cyclic (separating) for a if ψ is cyclic (separating) for $A = C^*(a)$, or, equivalently, for $A = W^*(a)$.

The equivalence of the two ways of writing the last definition follows from the relation $W^*(a)\psi^- = C^*(a)\psi^-$, cf. Corollary B.101; more generally, ψ is cyclic (separating) for A iff it is cyclic (separating) for its strong closure A^-.

For example, if $A = B(H)$, any vector is cyclic for A, and none is separating. On the other hand, if $A = \mathbb{C}\cdot 1_H$, then no vector is cyclic for A and all vectors are separating. If $H = L^2(X,\mu)$ on some finite measure space, then $\psi = 1_X$ is cyclic as well as separating for $A = L^\infty(X,\mu)$. Noting (B.346), as well as the property $B(H)' = \mathbb{C}\cdot 1_H$, these examples illustrates a general phenomenon:

Lemma B.111. *If $1_H \in A$, a vector ψ is cyclic for A iff it is separating for A', and* vice versa. *In particular, if $A' = A$, then ψ is cyclic for A iff it is separating for A.*

If A is abelian, then every vector that is cyclic for A is also separating for A.

Proof. If $A\psi^- = H$ and $b\psi = 0$ for $b \in A'$, then $ba\psi = 0$ for each $a \in A$ and hence b vanishes on a dense subspace of H. Since b is bounded, $b = 0$. Conversely, let e be the projection onto $A\psi^-$; then $e \in A'$ and hence $1_H - e \in A'$. Since $1_H \in A$ we have $\psi \in A\psi^-$ and hence $e\psi = \psi$, whence $(1_H - e)\psi = 0$. If ψ is separating for A', this implies $e = 1_H$ and hence $A\psi^- = H$. Finally, A is abelian iff $A \subseteq A'$. □

Theorem B.112. *Let $a^* = a \in B(H)$, and suppose some unit vector $\psi \in H$ is cyclic for a. Then a is unitarily equivalent to the position operator* (B.332) *on $L^2(\sigma(a), \mu_\psi)$, where the probability measure μ_ψ on $\sigma(a)$ is given by* (B.304). *Furthermore, through the unitary operator $u : H \to L^2(\sigma(a), \mu_\psi)$ in question we have*

$$uf(a)u^{-1} = f; \tag{B.347}$$

$$uC^*(a)u^{-1} = C(\sigma(a)); \tag{B.348}$$

$$uW^*(a)u^{-1} = L^\infty(\sigma(a), \mu_\psi), \tag{B.349}$$

all of which being realized as multiplication operators on $L^2(\sigma(a), \mu_\psi)$.

Moreover, $L^\infty(\sigma(a), \mu_\psi)$ is maximally abelian, and hence satisfies

$$L^\infty(\sigma(a), \mu_\psi) = L^\infty(\sigma(a), \mu_\psi)'. \tag{B.350}$$

Proof. First, define u on a dense subspace of H by

$$u : C^*(a)\psi \to L^2(\sigma(a), \mu_\psi); \tag{B.351}$$

$$uf(a)\psi = f, \; f \in C(\sigma(a)). \tag{B.352}$$

It follows from (B.289) - (B.291) and (B.304) that $\|f(a)\psi\|_H = \|f\|_2$, which makes u well defined (since $f(a)\psi = g(a)\psi$ implies $f = g$), as well as isometric. In particular, u is bounded, and hence it can be extended from $C^*(a)\psi$ to H by continuity. This extension is surjective, since $C(\sigma(a))$ is dense in $L^2(\sigma(a), \mu_\psi)$, and therefore $u : H \to L^2(\sigma(a), \mu_\psi)$ is unitary. Then (B.347) - (B.348) hold by construction; the special case $f = \mathrm{id}$ yields (B.332). As in Proposition B.106, we obtain $C(\sigma(a))' = L^\infty(\sigma(a), \mu_\psi)$, which implies (B.349) - (B.350). □

Note that this proposition implies that H is separable. When does a self-adjoint (or normal) operator a have a cyclic vector? To practice, we first look at $H = \mathbb{C}^n$.

Proposition B.113. *Let $H = \mathbb{C}^n$ and let $a = \mathrm{diag}(\lambda_1, \ldots, \lambda_n)$ be a diagonal matrix. Then the following properties are equivalent:*

1. *All λ_i are distinct, i.e., $|\sigma(a)| = n$ (in words, a is non-degenerate);*
2. *The operator a has a cyclic vector;*
3. *$C^*(a)' = C^*(a)$;*
4. *$C^*(a)$ is a maximal abelian C*-subalgebra of $B(H)$.*

Proof. We first show that all λ_i are distinct iff

$$C^*(a) = D_n(\mathbb{C}), \tag{B.353}$$

i.e., the set of all diagonal matrices. To see this, first note that for any $f : \sigma(a) \to \mathbb{C}$ (and any such function is continuous, since $\sigma(a)$ is a finite subset of $\mathbb{C}$) we have

$$f(\mathrm{diag}(\lambda_1, \ldots, \lambda_n)) = \mathrm{diag}(f(\lambda_1), \ldots, f(\lambda_n)); \tag{B.354}$$

this is true by computation for polynomials in a, and these exhaust all functions on $\sigma(a)$. It follows that $C^*(a) \subseteq D_n(\mathbb{C})$. We know from (A.49) that $C^*(a) \cong C(\sigma(a))$ whether or not $\sigma(a)$ is non-degenerate, and since $\dim(C(\sigma(a))) = |\sigma(a)|$ (i.e., the number of elements of $\sigma(a)$), we obtain

$$\dim(C^*(a)) = |\sigma(a)|. \tag{B.355}$$

So if a is non-degenerate, noting that $\dim(D_n(\mathbb{C})) = n$ we must have (B.353). If, on the other hand, a is degenerate, we have $|\sigma(a)| = m < n$, so that also $\dim(C(\sigma(a))) = m < n$ and $C^*(a) \subset D_n(\mathbb{C})$ is a strict inclusion. Furthermore, by direct computation or as a special case of Proposition B.108, we have

$$D_n(\mathbb{C})' = D_n(\mathbb{C}). \tag{B.356}$$

To prove $1 \to 2$, take the cyclic vector to be

$$\psi = (1, \ldots, 1)/\sqrt{n}; \tag{B.357}$$

indeed, any vector $(z_1, \ldots, z_n)$ is equal to $\sqrt{n} \cdot \mathrm{diag}(z_1, \ldots, z_n)\psi$, and we have $\mathrm{diag}(z_1, \ldots, z_n) \in D_n(\mathbb{C}) = C^*(a)$ by (B.353). For $2 \to 1$, if H has a cyclic vector ψ for a, then by definition $C^*(a)\psi = \mathbb{C}^n$, so that $\dim(C^*(a)\psi) = n$. But also

$$\dim(C^*(a)\psi) \leq \dim(C^*(a)), \tag{B.358}$$

whether or not ψ is cyclic for a. If ψ is cyclic this gives

$$n \leq \dim(C^*(a)) \leq n \tag{B.359}$$

by (B.355), so that $\dim(C^*(a)) = n$, whence $|\sigma(a)| = n$ by (B.355).

Given this, the implication $1 \to 3$ follows from (B.356), whilst $3 \to 4$ follows from Theorem A.21. Finally, we prove $4 \to 1$: we already know that $C^*(a) \subset D_n(\mathbb{C})$, and by (B.356) and the above argument it follows that $D_n(\mathbb{C})$ is maximal. So if $C^*(a)$ is maximal, then $C^*(a) = D_n(\mathbb{C})$, and we already know from the first stage of the proof that this is equivalent to a being non-degenerate. □

With slightly more effort, an analogous result holds for general Hilbert spaces.

Proposition B.114. *A self-adjoint operator a on a separable Hilbert space H has a cyclic vector iff $W^*(a)$ is maximal abelian (i.e., $W^*(a)' = W^*(a)$).*

In other words, a has a cyclic vector iff $C^*(a)' = C^*(a)''$, cf. (B.320). As we have just seen, if $\dim(H) < \infty$, this is the case iff a is non-degenerate. Consistent with (B.349) (with $u = 1$) and (B.350), the position operator (B.332) acting on the Hilbert space $L^2(\sigma(a), \mu_\psi)$ is maximal in this sense, with $\psi = 1_{\sigma(a)}$ as a cyclic unit vector.

Proof. If ψ is cyclic for a, then (B.349) and (B.350) (along with the self-evident property $uA'u^{-1} = (uAu^{-1})'$) yield $W^*(a)' = W^*(a)$. Conversely, for any *-algebra $A \subset B(H)$, one can find unit vectors (ψ_i) such that $H = \oplus_i H_i$ with $H_i = A\psi_i^-$: start with any ψ_1, then take any $\psi_2 \in (A\psi_1^-)^\perp$ (in case this is nonzero, otherwise one was already done), etc. To show that this procedure terminates, Zorn's Lemma must be invoked (take the collection of all sets (H_i) of mutually orthogonal A-stable subspaces $H_i \subset H$ that contain a cyclic vector for A). Then $\psi = \sum_n 2^{-n}\psi_n$ is clearly separating for A. If $A' = A$, then ψ is also cyclic for A; cf. Lemma B.111. □

Thus we call a self-adjoint operator $a \in B(H)$ ***maximal*** if it has a cyclic vector.

Corollary B.115. *A maximal self-adjoint operator $a \in B(H)$ is unitarily equivalent to the position operator* (B.332) *on $L^2(\sigma(a), \mu)$, where μ is an appropriate probability measure on the spectrum $\sigma(a) \subset \mathbb{R}$. Moreover, the map $\mathscr{B}(\sigma(a)) \to W^*(a)$ in* (B.321) *induces an isomorphism* (B.331) *of von Neumann algebras.*

Proof. Take $\mu = \mu_\psi$, cf. (B.304), where ψ is cyclic (or, equivalently, separating) for a. The map $f \mapsto f(a)$ from $\mathscr{B}(\sigma(a))$ to $W^*(a)$ described in Theorem B.102 can be propelled further by conjugation with the unitary u of Theorem B.112, that is,

$$f \mapsto f(a) \mapsto uf(a)u^{-1} = m_f; \tag{B.360}$$

$$\mathscr{B}(\sigma(a)) \to B(H) \to B(L^2(\sigma(a), \mu_\psi)), \tag{B.361}$$

where the final equality in (B.360) follows from the computation

$$uf(a)u^{-1}g = uf(a)g(a)\psi = u(f \cdot g)(a)\psi = fg = m_f g, \tag{B.362}$$

where for simplicity $g \in C(\sigma(a)) \subset L^2(\sigma(a), \mu_\psi)$, the inclusion being dense. The claim then immediately follows from (B.349). □

If a is not maximal, we can still prove a weaker version of Theorem B.112, which is sometimes seen as the ultimate version of the spectral theorem. To justify this view, take $H = \mathbb{C}^n$ and let $a \in M_n(\mathbb{C})$ be self-adjoint (or, more generally, normal). By Theorem A.10, H has a basis (υ_i) of eigenvectors of a, with $a\upsilon_i = \lambda_i \upsilon_i$. This yields a unitary map $H \to \ell^2(\underline{n})$, where $\underline{n} = \{1, 2, \ldots, n\}$, defined by $u\upsilon_i = \delta_i$ (where $\delta_i(j) = \delta_{ij}$, as usual). It is easy to check that $uau^{-1} = m_\lambda$, where $\lambda : \underline{n} \to \mathbb{C}$ is defined by $\lambda(i) = \lambda_i$, and $m_\lambda \psi = \lambda\psi$, again as usual. In other words, a is unitarily equivalent to a multiplication operator (whose precise nature is left unspecified). Conversely, each multiplication operator m_f on some $L^2(X, \mu)$ is normal, and is self-adjoint if the function $f \in L^\infty(X, \mu)$ is real-valued (μ-almost everywhere).

Theorem B.116. *Any bounded self-adjoint (more generally, normal) operator on a separable Hilbert space is unitarily equivalent to a multiplication operator.*

Proof. As in the proof of Theorem B.114, decompose $H = \oplus_{i\in I} H_i$, where each H_i contains some take some separating vector ψ_i for a. Applying the proof of Theorem B.112 to each H_i then yields unitary isomorphisms $H_i \cong L^2(\sigma(a), \mu_i)$, with $\mu_i \equiv \mu_{\psi_i}$, from which, taking direct sums, we obtain a further unitary isomorphism

$$H \cong \bigoplus_{i\in I} L^2(\sigma(a), \mu_i). \tag{B.363}$$

Now take the disjoint union $X \equiv \sqcup_{i\in I}\sigma(a)$, i.e., $X = \cup_{i\in I} X_i$, where $X_i = \sigma(a)\times\{i\}$, endowed with the σ-finite measure $\mu = \sum_i \mu_i$ (so that if $A \subset X$ is given by $A = \cup_i A_i$ with $A_i \subset X_i$, we have $\mu(A) = \sum_i \mu_i(A_i)$). This gives a second isomorphism

$$\bigoplus_i L^2(\sigma(a), \mu_i) \cong L^2(X, \mu), \tag{B.364}$$

defined by mapping $\varphi_j \in L^2(\sigma(a), \mu_j)$ to the same function on X_j, extended to X by putting it zero on all other X_i, $i \neq j$. This map is obviously unitary. By Theorem B.112, the isomorphism (B.363) maps the operator a to a direct sum $\oplus_i m_{\mathrm{id}_{\sigma(a)}}$ of multiplication operators, upon which the second isomorphism (B.364) maps this direct sum to a (single) multiplication operator m_q, where the function $q : X \to \mathbb{C}$ is defined by $q(x,i) = x$ (in which $(x,i) \in X_i \subset X$, so that $x \in \sigma(a) \subset \mathbb{C}$). □

More generally, the operator $f(a)$ on H, for some $f \in \mathscr{B}(\sigma(a))$, is first mapped to $\oplus_i m_{f_i}$, where f_i is the image of f in $L^\infty(\sigma(a), \mu_i)$ in the obvious way, which in turn is mapped to a multiplication operator $m_{\hat{f}}$, where $\hat{f}(x,i) = f(x)$, analogously to the position operator $q = \widehat{\mathrm{id}}_{\sigma(a)}$ above. This leads to an isomorphism $W^*(a) \cong L^\infty(X, \mu)$, which, by the same reasoning as in the proof of Corollary B.115, also induces an isomorphism (B.331) of von Neumann algebras. See also Theorem C.140.

Finally, proposition B.114 may be generalized, to which end (and also as a result of independent interest) we extend Corollary A.20 to the infinite-dimensional case:

Theorem B.117. *Let H be separable and let $A \subset B(H)$ be an abelian von Neumann algebra. Then $A = W^*(a)$ for some self-adjoint $a \in B(H)$, i.e., A is singly generated.*

Proof. Let $\mathscr{P}(A)$ be the set of all projections in A, and let $\psi \in H$ be separating for A and hence cyclic for A' (cf. Lemma B.111 and the proof of Proposition B.114). The ensuing subset $\mathscr{P}(A)\psi = \{e\psi \mid e \in \mathscr{P}(A)\}$ may be uncountable, but since any subspace of a separable metric space is separable, there is a countable subset $\mathscr{P}_{\mathbb{N}}(A) = \{e_n, n \in \mathbb{N}\}$ of $\mathscr{P}(A)$ such that $\mathscr{P}_{\mathbb{N}}(A)\psi$ is dense in $\mathscr{P}(A)\psi$, i.e., for any $e \in \mathscr{P}(A)$ there is a subsequence e_{n_k} in $\mathscr{P}_{\mathbb{N}}(A)$ such that $\lim_{k\to\infty} e_{n_k}\psi = e\psi$. But since $\mathscr{P}(A) \subset A \subseteq A'$ and $A'\psi^- = H$, this is true not only on ψ but on a dense set of vectors $a\psi$, $a \in A'$, so that $e_{n_k} \to e$ in the strong operator topology. Thus $\mathscr{P}_{\mathbb{N}}(A)$ is strongly dense in $\mathscr{P}_{\mathbb{N}}(A)$, and by (B.329) and Theorem B.100 we have

$$\mathscr{P}_{\mathbb{N}}(A)'' = A. \tag{B.365}$$

The self-adjoint operator that does the job is now given by von Neumann's formula

$$a = \sum_n 3^{-n}(2e_n - 1_H). \tag{B.366}$$

To see this, let $C^*(e_n, n \in \mathbb{N}) \equiv C^*(e_n)_n$ be the C*-algebra generated by the projections e_n, so that by construction

$$\mathscr{P}_{\mathbb{N}}(A)'' = C^*(e_n)_n''. \tag{B.367}$$

We will show that

$$C^*(a) = C^*(e_n)_n, \tag{B.368}$$

which combined with (B.320), (B.365) and (B.367) yields the desired conclusion:

$$A = \mathscr{P}_{\mathbb{N}}(A)'' = C^*(e_n)_n'' = C^*(a)'' = W^*(a). \tag{B.369}$$

The simplest argument for (B.368) uses the Gelfand isomorphism

$$C^*(e_n)_n \cong C(X) \tag{B.370}$$

as commutative C*-algebras, cf. Theorem C.8, where the set of characters

$$X = \{x : C^*(e_n)_n \to \mathbb{C} \mid x(bc) = x(b)x(c), x(1_H) = 1\} \tag{B.371}$$

of $C^*(e_n)_n$ is equipped with the weakest topology that makes all maps

$$\hat{b} : X \to \mathbb{C}; \tag{B.372}$$
$$\hat{b}(x) = x(b),\ b \in C^*(e_n)_n, \tag{B.373}$$

continuous. This makes X a compact Hausdorff space, and the isomorphism (B.370) is given by the Gelfand transform $b \mapsto \hat{b}$. Defining $s_n \equiv 2e_n - 1_H$, we have $\|s_n\| = 1$, since $s_n\psi = \psi$ if $\psi \in e_nH$ and $s_n\psi = -\psi$ if $\psi \in (1_H - e_n)H = (e_nH)^\perp$. The series (B.366) therefore converges absolutely in $B(H)$, and hence converges, to some limit $a \in C^*(e_n)_n$. We claim that its Gelfand transform $\hat{a} \in C(X)$ separates points of X, so that by the Stone-Weierstrass Theorem B.51, the *-algebra it generates is dense in $C(X)$ (in its canonical sup-norm). Thus a likewise generates $C^*(e_n)_n$, and the proof of Theorem is ready up to the proof of the above claim, which we now give.

First, note that since by definition $C^*(e_n)_n$ is generated by the projections e_n, so that by (B.371) (and the automatic continuity this implies, i.e., $x \in C^*(e_n)_n^*$), each $x \in X$ is determined by its values on all e_n. Therefore, for each pair $x_i, x_j \in X$, $i \neq j$, there must be some $n \in \mathbb{N}$ for which $x_i(e_n) \neq x_j(e_n)$. Consequently, for each $i \neq j$, the set $N_{ij} = \{n \in \mathbb{N} \mid x_i(e_n) \neq x_j(e_n)\}$ is not empty; let $n_{ij} = \min N_{ij}$. Since for any projection e the corresponding function $\hat{e}$ can only take the values 0 or 1, each $\hat{s}_n$ must take the values ± 1, so that, with $\hat{a} = \sum_n 3^{-n}\hat{s}_n$, we have

$$\tfrac{1}{2}(\hat{a}(x_i) - \hat{a}(x_j)) = \pm 3^{-n_{ij}} + \sum_{n \in N_{ij}, n > n_{ij}} \pm 3^{-n} \neq 0, \tag{B.374}$$

since whatever the signs, the sum is always smaller than the first term. □

B.17 Classification of maximal abelian *-algebras in $B(H)$

We now prove the following classification of maximal abelian *-algebras in $B(H)$, which forms the basis of the Kadison–Singer Conjecture discussed in §2.6 and §4.3.

Theorem B.118. *If H is separable (and infinite-dimensional), and $A \subset B(H)$ is a maximal abelian *-algebra, then A is unitarily equivalent to one of the following:*

1. $L^\infty(0,1) \subset B(L^2(0,1))$ *(realized as multiplication operators);*
2. $\ell^\infty(\mathbb{N}) \subset B(\ell^2(\mathbb{N}))$ (idem);
3. $L^\infty(0,1) \oplus \ell^\infty(\mathbb{N}) \subset B(L^2(0,1) \oplus \ell^2(\mathbb{N}))$ (idem);
4. $L^\infty(0,1) \oplus D_n(\mathbb{C}) \subset B(L^2(0,1) \oplus \mathbb{C}^n)$, *for some* $n \in \mathbb{N}$ (idem),

and these possibilities are (mutually) unitarily inequivalent.

The first claim means that there is a unitary operator u from H to, say, $L^2(0,1)$, such that the map $a \mapsto uau^{-1}$ from $B(H)$ to $B(L^2(0,1))$ restricts to $uAu^{-1} = L^\infty(0,1)$, so that $A \cong L^\infty(0,1)$ as both C*-algebras and von Neumann algebras (and likewise for the other possibilities). The last claim, then, means that there is *no* unitary map from, say, $L^2(0,1)$ to $\ell^2(\mathbb{N})$ that similarly induces an isomorphism $L^\infty(0,1) \cong \ell^\infty(\mathbb{N})$

Proof. We begin with the easy part, which is the last clause. The key notion to proving the claimed inequivalence is that of an ***atomic projection*** in a von Neumann algebra $M \subset B(H)$. If we partially order projections on H by (cf. Theorem 2.50 and §C.21)

$$e \leq f \text{ iff } eH \subseteq fH, \tag{B.375}$$

we say that f is atomic if $f \neq 0$, and $0 \leq e \leq f$ implies either $e = 0$ or $e = f$. This property is preserved under unitary equivalence: if $M \subset B(H)$ and $N \subset B(H')$ and $N = uMu^{-1}$ for some unitary $u : H \to H'$ (again in the sense that $a \mapsto uau^{-1}$ is an isomorphism $M \xrightarrow{\cong} N$), then f is atomic in M iff ufu^{-1} is atomic in N. The reason is that $a \mapsto uau^{-1}$ induces an isomorphism of the pertinent posets of projections in M and N, so that all order-theoretical notions are preserved under unitary equivalence.

In the case at hand, the projections are easy to classify:

1. The nonzero projections in $L^\infty([0,1]$ are the characteristic functions on measurable subsets of $[0,1]$ *of positive Lebesgue measure*. Since any such subset properly contains another such subset, *there are no atomic projections in* $L^\infty([0,1]$.
2. The nonzero projections in $\ell^\infty(\mathbb{N})$ are the characteristic functions on $\mathbb{N}$, among which there are plenty of atomic ones, namely the one-dimensional projections δ_x, $x \in \mathbb{N}$. Thus $\ell^\infty(\mathbb{N})$ has countably many atomic projections. Moreover, each other projection majorizes an atomic one.
3. Similarly, $L^\infty(0,1) \oplus \ell^\infty(\mathbb{N})$ has has countably many atomic projections, as well as uncountably many projections that do not majorize any atomic one.
4. Since the atomic projections $D_n(\mathbb{C})$ are the one-dimensional ones (given by diagonal matrices with $n-1$ zero's and exactly one entry equal to unity), $L^\infty(0,1) \oplus D_n(\mathbb{C})$ has exactly n atomic projections, as well as uncountably many projections that do not majorize any atomic one (namely the ones in $L^\infty(0,1)$).

Any unitary equivalence between two of the entries in the list would have to preserve this fine structure of projections, and hence cannot exist.

We now prove that the list in Theorem B.118 is exhaustive. According to Theorem B.117, we only need to look at abelian von Neumann algebras $A = W^*(a)$, where a is maximal. According to Theorem B.112 and its Corollary B.115 (whilst noting that some unitary equivalence $a \cong b$ induces a unitary equivalence $W^*(a) \cong W^*(b)$), we may further restrict our attention to the case where a is the position operator on $L^2(K,\mu)$, where $K = \sigma(a) \subset \mathbb{R}$ is compact and μ is a regular probability measure (here and in what follows, this is always meant with respect to the Borel structure inherited from $\mathbb{R} \supset K$), with support equal to K, and hence

$$W^*(a) = L^\infty(K,\mu) \subset B(L^2(K,\mu)). \tag{B.376}$$

The final step is to further reduce the possibilities by exploiting equivalences.

Definition B.119. *Two measure spaces (X,Σ,μ) and (X',Σ',μ') are:*

- **equivalent** *if there is a measurable bijection $\varphi : X \to X'$ with measurable inverse, and the measures $\varphi_*\mu$ and μ' on X' are equivalent in the sense that $\varphi_*\mu(A') = 0$ iff $\mu'(A') = 0$ for each $A' \in \Sigma'$. Here $\varphi_*\mu$ is the measure on (X',Σ') defined by*

$$\varphi_*\mu(A') = \mu(\varphi^{-1}(A')) \ (A' \in \Sigma'). \tag{B.377}$$

- **isomorphic** *if there is a measurable bijection $\varphi : X \to X'$ with measurable inverse, and $\varphi_*\mu(A') = \mu'(A')$ for each $A' \in \Sigma'$.*

The ambiguity of the notation φ^{-1} in (B.377) is innocent: for general measurable maps $\varphi : X \to X'$ the set $\varphi^{-1}(A')$ can only denote the pre-image $\{x \in X \mid \varphi(x) \in A'\}$, whereas for invertible maps one might construe $\varphi^{-1}(A')$ as $\{\varphi^{-1}(x') \mid x' \in A'\}$, where φ^{-1} is the theoretic inverse φ^{-1} of φ. Of course, these sets duly coincide.

Lemma B.120. *Let K and K' be compact subsets of $\mathbb{R}$, with Σ and Σ' the Borel structures inherited from $\mathbb{R} \supset K$ and $\mathbb{R} \supset K'$, respectively (often omitted in what follows). Let μ and μ' be probability measures on K and K', respectively, and suppose that the associated measure spaces (K,Σ,μ) and (K',Σ',μ') are isomorphic.*

Then there exists a unitary operator

$$u : L^2(K,\mu) \to L^2(K,\mu')$$

such that

$$uL^\infty(K,\mu)u^{-1} = L^\infty(K',\mu'). \tag{B.378}$$

Note that u does *not* intertwine the positions operators (B.332) on $L^2(K,\mu)$ and $L^2(K',\mu')$. These operators have already done their job in reducing the situation to $L^2(K,\mu)$, and from that point onwards (B.378) is exactly what we need.

Proof. All maps appearing below are assumed Borel. The change-of-variables formula for a general (i.e., not necessarily invertible) map $\varphi : K \to K'$ reads

$$\int_{K'} d(\varphi_*\mu)\, g = \int_K d\mu\, g \circ \varphi, \tag{B.379}$$

where $g : K' \to \mathbb{C}$. Under the assumption that φ *is* invertible, this can be rewritten as

$$\int_{K'} d(\varphi_*\mu)\, f \circ \varphi^{-1} = \int_K d\mu\, f, \tag{B.380}$$

where $f : K \to \mathbb{C}$. If φ is also an isomorphism of measure spaces, this becomes

$$\int_{K'} d\mu'\, f \circ \varphi^{-1} = \int_K d\mu\, f. \tag{B.381}$$

If $\varphi_*\mu$ and μ' are equivalent and hence mutually absolutely continuous, the Radon–Nikodym derivative $d(\varphi_*\mu)/d\mu'$ exists (as does its counterpart $d(\varphi_*^{-1}\mu')/d\mu$), and using (B.137) and (B.380), one easily verifies that the operator

$$u : L^2(K,\mu) \to L^2(K,\mu'); \tag{B.382}$$

$$u\psi = \sqrt{\frac{d(\varphi_*\mu)}{d\mu'}}\, \psi \circ \varphi^{-1}, \tag{B.383}$$

is isometric. Moreover, u is unitary, because it has an inverse, given by

$$u^{-1} : L^2(K',\mu') \to L^2(K,\mu); \tag{B.384}$$

$$u^{-1}\chi = \sqrt{\frac{d(\varphi_*^{-1}\mu')}{d\mu}}\, \chi \circ \varphi, \tag{B.385}$$

We give these general expressions for later use; if $\varphi_*\mu = \mu'$, they simplify to

$$u\psi = \psi \circ \varphi^{-1}; \tag{B.386}$$

$$u^{-1}\chi = \chi \circ \varphi. \tag{B.387}$$

For $f \in L^\infty(K,\mu)$ we then have (cf. Proposition B.73)

$$u m_f u^{-1} = m_{f \circ \varphi^{-1}}. \tag{B.388}$$

We already know that the map $f \mapsto m_f$ injects $L^\infty(K,\mu)$ isometrically into $B(L^2(K,\mu))$, and analogously for $L^\infty(K',\mu')$. Furthermore, The map $f \mapsto f \circ \varphi^{-1}$ gives an isomorphism $L^\infty(K,\mu) \xrightarrow{\cong} L^\infty(K',\mu')$: the property

$$\|f \circ \varphi^{-1}\|_\infty^{\mathrm{ess}} = \|f\|_\infty^{\mathrm{ess}}, \tag{B.389}$$

which yields injectivity, may be checked either from (B.240) or from the assumed isomorphism of measures (and hence equivalence of measures, which in fact suffices for this purpose), whereras invertibility of φ gives surjectivity (since $g \in L^\infty(K',\mu')$ is the image of $f = g \circ \varphi \in L^\infty(K,\mu')$. Eq. (B.378) follows. □

The final step of the proof appeals to a deep and fundamental classification theorem in measure theory, which goes back to Kuratowski in a form that applies to general Polish (i.e., complete separable metric) spaces. This theorem implies:

Lemma B.121. *Let (K, Σ, μ) be a infinite probability space (in that infinitely many different elements of Σ have positive measure), where $K \subset \mathbb{R}$ is compact and Σ is the σ-algebra inherited from the Borel structure on $\mathbb{R}$. Then (K, Σ, μ) is isomorphic to exactly one of the following possibilities (called* **standard measure spaces***):*

1. *$K = [0,1]$ with μ equal to Lebesgue measure μ_L;*
2. *$K = \underline{\mathbb{N}}' \equiv \{2^{-n}, n \in \mathbb{N}\} \cup \{1\}$, equipped with any probability measure μ' for which $\mu'(\{2^{-n}\}) > 0$ for each $n \in \mathbb{N}$ and $\mu'(\{1\}) = 0$;*
3. *$K = [0,1]$ with $\mu = t\mu_L + (1-t)\mu'$, for some $0 < t < 1$;*
4. *$K = [0,1]$ with $\mu = t\mu_L + (1-t)\mu_n$, for some $n \in \mathbb{N}$ and $0 < t < 1$,*

where μ_n is an arbitrary strictly nonzero probability measure on the n-point set

$$\underline{n}' \equiv \{1/n, \ldots, (n-1)/n, 1\}. \tag{B.390}$$

Here we have stated the result in terms of *probability* measures μ on *compact* spaces $K \subseteq [0,1]$; this is convenient in the context of our proof. To understand the last two cases, for general measure spaces (X, Σ, μ) we say that $A \in \Sigma$ is an ***atom*** if for any $B \subset A$ we have either $\mu(B) = 0$ or $\mu(A \backslash B) = 0$ (but not both; this implies $\mu(A) > 0$, whence an equivalent definition of an atom as a set $A \in \Sigma$ having positive measure as well as the property that if some measurable subset $B \subset A$ has measure $\mu(B) < \mu(A)$, then $\mu(B) = 0$). In our case at hand (K, μ), each atom A contains a point $x \in K$ such that $\mu(A) = \mu(\{x\})$ and $\mu(A \backslash \{x\}) = 0$, so that modulo null sets we may identify each atom A with the measure-carrying point x it contains. Moreover, K can contain at most a countable set $\mathscr{A} = \{x_n\}_n$ of such points x_n. The formulae

$$\mu = \mu_a + \mu_c; \tag{B.391}$$
$$\mu_a(A) = \mu(A \cap \mathscr{A}); \tag{B.392}$$
$$\mu_c(A) = \mu(A \backslash (A \cap \mathscr{A})), \tag{B.393}$$

then give the canonical decomposition of μ into an ***atomic*** part μ_a and a ***continuous*** part μ_c. This, then, is the sense in which the last two cases of Lemma B.121 are meant. Note that characteristic functions 1_A on atoms $A \subset K$ yield atomic projections in $L^\infty(K, \mu)$, linking the two notions of atomicity that play a role in this proof.

The first entry of this lemma yields the first entry in the list in the theorem. To obtain the others, we need a few more unitary equivalences. For the second, define

$$u : L^2(\underline{\mathbb{N}}', \mu') \to \ell^2(\mathbb{N}); \tag{B.394}$$
$$u\psi(n) = \sqrt{\mu'(n)}\psi(2^{-n}), \tag{B.395}$$

and $u\psi(1)$ irrelevant. This operator is unitary and, just like in (B.378), it intertwines

$$uL^\infty(\underline{\mathbb{N}}', \mu')u^{-1} = \ell^\infty(\mathbb{N}). \tag{B.396}$$

Note that (B.394) is a special case of (B.383)). The third and fourth cases require the following construction: if $\mathscr{A} \subset K$ is the set of atoms in (K, Σ, μ), we decompose

$$K = (K \backslash \mathscr{A}) \bigsqcup \mathscr{A}, \tag{B.397}$$

as a disjoint union. For any measure μ this induces an orthogonal decomposition

$$L^2(K,\mu) = L^2(K\backslash\mathscr{A},\mu) \oplus L^2(\mathscr{A},\mu); \tag{B.398}$$
$$L^2(K\backslash\mathscr{A},\mu) = eL^2(K,\mu); \tag{B.399}$$
$$L^2(\mathscr{A},\mu) = (1_{L^2(K,\mu)} - e)L^2(K,\mu), \tag{B.400}$$

where $e = 1_{K\backslash\mathscr{A}}$ and $1_{L^2(K,\mu)} - e = 1_{\mathscr{A}}$ are projections. Using (B.391), this gives

$$L^2(K\backslash\mathscr{A},\mu) = L^2(K,\mu_c); \tag{B.401}$$
$$L^2(\mathscr{A},\mu) = L^2(\mathscr{A},\mu_a), \tag{B.402}$$

so that at the end of the day we obtain

$$L^2(K,\mu) = L^2(K,\mu_c) \oplus L^2(\mathscr{A},\mu_a). \tag{B.403}$$

This in turn induces the decomposition

$$L^\infty(K,\mu) = L^\infty(K,\mu_c) \oplus L^\infty(\mathscr{A},\mu_a); \tag{B.404}$$
$$L^\infty(K,\mu_c) = eL^\infty(K,\mu) = eL^\infty(K,\mu)e; \tag{B.405}$$
$$\begin{aligned} L^\infty(\mathscr{A},\mu_a) &= (1_{L^2(K,\mu)} - e)L^\infty(K,\mu) \\ &= (1_{L^2(K,\mu)} - e)L^\infty(K,\mu)(1_{L^2(K,\mu)} - e). \end{aligned} \tag{B.406}$$

Combined with (B.396), this shows that the third entry of the lemma yields the third entry of the theorem. To obtain the fourth and last, we need the unitary map

$$u : L^2(\underline{n}', \mu_n) \to \mathbb{C}^n; \tag{B.407}$$
$$u\psi_m = \sqrt{\mu_n(m/n)}\,\psi(m/n)\ (m = 1,\ldots,n), \tag{B.408}$$

which delivers the unitary equivalence

$$uL^\infty(\underline{n}', \mu_n)u^{-1} = D_n(\mathbb{C}). \tag{B.409}$$

Short of a proof of Lemma B.121, we have (at last!) proved Theorem B.118. □

Thus one of the remarkable novelties of infinite-dimensional Hilbert space is that even in the separable case, uniqueness of maximal abelian *-algebras is lost.

There is a different proof of Theorem B.118 that does not rely on Kuratowski's Lemma B.121, but instead is based on properties of the projection lattice $\mathscr{P}(A)$ in A. In the following outline of this proof, A is a maximal abelian *-subalgebra of $B(H)$, where H is a separable Hilbert space. Hence A is a von Neumann algebra, which is generated by its projections. This leaves three mutually exclusive possibilities:

1. A has no minimal projections;
2. A is generated by its minimal projections;
3. A has minimal projections that do not generate A.

The following lemma, whose proof we merely sketch, replaces Lemma B.121.

Lemma B.122. *If H is separable and $A \subset B(H)$, then $\mathscr{P}(A)$ contains a maximal totally ordered set $\mathscr{T}(A)$ that generates A (as a von Neumann algebra).*

Proof. This is proved in two steps. First, $\mathscr{P}(A)$ contains a *countable* subsets $\mathscr{P}_c(A)$ that generates A. Indeed, according to Lemma B.111 and Proposition B.114 (and maximality of A), H contains a unit vector ψ that is both cyclic and separating for A. Since H is separable, $\mathscr{P}(A)\psi \subset H$ has a countable dense subset, which is $\mathscr{P}_c(A)$.

The second step is trickier, namely to construct a maximal totally ordered set $\mathscr{T}(A)$ from $\mathscr{P}_c(A)$. This is done inductively. We number $\mathscr{P}_c(A) = \{e_1, e_2, \dots\}$. Starting from $\mathscr{P}_1 = \{0_H, e_1, 1_H\}$, we now construct finite totally ordered sets $\mathscr{P}_n$ of projections such that $\mathscr{P}_n \subset \mathscr{P}_{n+1}$ and e_n lies in the linear span of $\mathscr{P}_n$. Let

$$\mathscr{P}_n = \{e'_0 = 0_H, e'_1, \dots, e'_{r_n-1}, e'_{r_n} = 1_H\}, \tag{B.410}$$

where $e'_1 < \cdots < e'_{r_n}$ (where $e < f$ means $e \leq f$ and $e \neq f$), and define

$$\mathscr{P}_{n+1} = \mathscr{P}_n \cup \{e'_i + (e'_{i+1} - e'_i)e_{n+1}, i = 0, \dots, r_n - 1\}. \tag{B.411}$$

Given the total ordering in $\mathscr{P}_n$, it is easy to see that each $e'_i + (e'_{i+1} - e'_i)e_{n+1}$ is indeed a projection, and, by the same token, that $\mathscr{P}_{n+1}$ meets its specification. Let

$$\mathscr{P}_\infty = \cup_n \mathscr{P}_n, \tag{B.412}$$

which remains totally ordered but typically is infinite, and take the poset $\mathscr{P}$ of all totally ordered subsets of $\mathscr{P}(A)$ that contain $\mathscr{P}_\infty$, ordered by inclusion. Zorn's Lemma then yields a maximal element of $\mathscr{P}$, and this is our $\mathscr{T}(A)$: this maximal element is itself totally ordered, and since its linear span contains each projection $e_n \in \mathscr{P}_c(A)$, the projections in $\mathscr{T}(A)$ generate A (since the e_n already do so). □

The above trichotomy then leaves the following possibilities:

1. Let $\psi \in H$ be a unit vector that is cyclic and separating for A. Then

$$\alpha : \mathscr{T}(A) \to [0,1]; \tag{B.413}$$
$$e \mapsto \langle \psi, e\psi \rangle, \tag{B.414}$$

is an isomorphism of posets. It is easy to show that the linear span of the set of all vectors $\alpha^{-1}(t)\psi$, $t \in [0,1]$, is dense in H, and that the map

$$u\alpha^{-1}(t)\psi = 1_{(0,t)} \tag{B.415}$$

extends (by linearity and continuity) to a unitary isomorphism

$$u : H \to L^2(0,1), \tag{B.416}$$

which intertwines A with $L^\infty(0,1)$ in the sense that

$$uAu^{-1} = L^\infty(0,1). \tag{B.417}$$

2. This case relies on a general fact about von Neumann algebras M: if $e \in \mathscr{P}(M)$ is minimal, then $pMp \cong \mathbb{C}$. This implies that if $M = A$ is abelian, then for each $a \in A$ one has $ea = \lambda a$ for some $\lambda \in \mathbb{C}$. It follows that:

   - Each minimal projection e_i in $\mathscr{P}(A)$ is one-dimensional.
   - Different minimal projections are orthogonal.
   - $1_H = \sum_i e_i$ (strongly), where the sum is over all minimal projections in A.

   Since H is separable, we may assume $i \in \mathbb{N}$, so that we obtain a countable basis (υ_i) of H in which $e_i = |\upsilon_i\rangle\langle\upsilon_i|$, and hence have a unitary isomorphism

$$u : H \to \ell^2(\mathbb{N}); \tag{B.418}$$
$$\upsilon_i \mapsto \delta_i, \tag{B.419}$$

   i.e., u is defined by linear and continuous extension of (B.419). Clearly,

$$uAu^{-1} = \ell^\infty(\mathbb{N}). \tag{B.420}$$

3. The first part of the analysis in the previous item still applies, but this time, the sum $e = \sum_i e_i$ over all minimal projections in A is not equal to 1_H. If there are $n \in \mathbb{N}$ such projections, we obtain

$$eH \cong \mathbb{C}^n, \tag{B.421}$$

   and otherwise

$$eH \cong \ell^2(\mathbb{N}). \tag{B.422}$$

   We combine these in the notation

$$eH \cong \ell^2(\kappa), \tag{B.423}$$

   where $\kappa = \underline{n}$, in which case $\ell^2(\kappa) = \mathbb{C}^n$ and $\ell^\infty(\kappa) = D_n(\mathbb{C})$, or $\kappa = \mathbb{N}$. Furthermore, we have

$$(1_H - e)H \cong L^2(0,1), \tag{B.424}$$

   as in the first item. By construction, the corresponding unitary

$$u : H \to \ell^2(\kappa) \oplus L^2(0,1) \tag{B.425}$$

   then satisfies

$$uAu^{-1} = \ell^\infty(\kappa) \oplus L^\infty(0,1). \tag{B.426}$$

This finishes the alternative proof (sketch) of Theorem B.118.

B.18 Compact operators

The spectral theorem (in whatever version) on infinite-dimensional Hilbert spaces considerably simplifies for a class of well-behaved operators called ***compact***.

Definition B.123. *A linear map $a : V \to W$ between Banach spaces V,W is called* **compact** *if for some (and hence all) $d > 0$ the image $a(V_{\leq d})$ of the closed d-ball*

$$V_{\leq d} = \{v \in V : \|v\| \leq d\} \tag{B.427}$$

is pre-compact in W (i.e., its closure $a(V_{\leq d})^-$ is compact), or, equivalently, if the image (av_n) of any bounded sequence (v_n) in V has a convergent subsequence.

Before turning to Hilbert spaces, we mention two facts of general interest.

Proposition B.124. *A compact operator is bounded.*

Proof. If not, then for any $n \in \mathbb{N}$ there is some $v_n \in V_{\leq 1}$ for which $\|av_n\| \geq n$, so that (av_n) cannot possibly have a convergent subsequence. □

Proposition B.125. *A compact operator $a : V \to W$ maps weakly convergent sequences in V to norm-convergent sequences in W.*

Proof. Let (v_n) be a sequence in V that weakly converges to v. It is easy to show that if $a : V \to W$ is (norm) continuous, then it maps weakly convergent sequences in V to weakly convergent sequences in W. Therefore, the sequence (av_n) weakly converges to av. If (av_n) failed to converge to av in norm, then it would have a subsequence (av_{n_k}) such that for some $\varepsilon > 0$ and all sufficiently large k one had

$$\|av_{n_k} - av\| \geq \varepsilon. \tag{B.428}$$

However, (v_n), being weakly convergent, is bounded by Lemma B.126 below, and hence also its subsequence (v_{n_k}) must be bounded. Since a is compact, (av_{n_k}) has some norm-convergent subsequence, which necessarily converges to av (since we know this is the weak limit of the ambient sequence (av_n) and hence also of any of its subsequences, and if a norm-limit exists, the corresponding weak limit must be the same). But for large enough k this convergence flatly contradicts (B.428). □

Lemma B.126. *A weakly convergent sequence in a Banach space is bounded.*

Proof. Since $v_n \to v$ weakly, the sequence $(\varphi(v_n))$ in $\mathbb{C}$ converges to $\varphi(v)$ for each $\varphi \in V^*$, so that $\sup_n\{|\varphi(v_n)|\} < \infty$. Using the notation (B.129), this may be rewritten as $\sup_n\{|\hat{v}_n(\varphi)|\} < \infty$. Using Theorem B.78 (with $V \rightsquigarrow V^{**}$, $W = \mathbb{C}$, and $X = \mathbb{N}$), this implies $\sup_n\{\|\hat{v}_n\|\} < \infty$, and hence $\sup_n\{\|v_n\|\} < \infty$ by Proposition B.44. □

Definition B.123 simplifies if $V = W = H$ is a Hilbert space, since we have:

Proposition B.127. *If the image $a(H_{\leq 1}) \subset H$ of a linear map $a : H \to H$ is pre-compact, then this image is in fact compact (and hence a is compact).*

For the proof, call a Banach space V ***reflexive*** if $V^{**} \cong V$ (i.e. through the canonical injection $v \mapsto \hat{v}$, cf. Proposition B.44). Hilbert spaces H are reflexive, since $H^* \cong H$ by Theorem B.66. Proposition B.127 then follows from yet another lemma:

Lemma B.128. *If V is a reflexive Banach space and $a : V \to W$ is compact, then $a(V_{\leq 1})$ is compact.*

Proof. The proof relies on a corollary of the Banach–Alaoglu Theorem B.48, according to which $V_{\leq 1}$ is weakly compact if V is reflexive (indeed, by applying Banach–Alaoglu to V^* instead of V, it follows that the unit ball in V^{**} is compact in its weak*-topology; if, in addition, V is reflexive, then the inverse of the canonical injection $V \hookrightarrow V^{**}$ maps the weak*-topology on V^{**} to the weak topology on V).

So let $a : V \to W$ be compact, and let w_n be a sequence in $a(V_{\leq 1})$, say $w_n = av_n$ for some sequence (v_n) in $V_{\leq 1}$. Then since $V_{\leq 1}$ is weakly compact, v_n has a weakly convergent subsequence v_{n_k} in $V_{\leq 1}$, say $\lim_{k\to\infty} v_{n_k} = v$ *weakly*. By Proposition B.125, $\lim_{k\to\infty} av_{n_k} = av$ *in norm*. In other words, (av_n) has a norm-convergent subsequence, namely (av_{n_k}), with limit in $a(V_{\leq 1})$. Hence $a(V_{\leq 1})$ is compact. □

In view of Proposition B.127, we may as well take the following starting point:

Definition B.129. *If H is a Hilbert space, a linear map $a : H \to H$ is called* **compact** *when the image $a(H_{\leq 1})$ of the closed unit ball in H is compact.*

We write $B_0(H)$ for the set of all compact operators on H.

Theorem B.130. *The compact operators $B_0(H)$ form a C*-algebra in $B(H)$ in the operations inherited from $B(H)$. Furthermore, $B_0(H)$ is a two-sided ideal in $B(H)$.*

Unfolding this theorem, the claim consists of the following parts:

1. $B_0(H) \subset B(H)$, i.e., a compact operator is automatically bounded.
2. $B_0(H)$ is a vector space.
3. If $a, b \in B_0(H)$, then $ab \in B_0(H)$.
4. If (a_n) is a convergent sequence in $B(H)$ with limit a, i.e., $\|a_n - a\| \to 0$ for some $a \in B(H)$, and if each $a_n \in B_0(H)$, then $a \in B_0(H)$.
5. If $a \in B_0(H)$, then $a^* \in B_0(H)$.
6. If $a \in B_0(H)$ and $b \in B(H)$, then $ab \in B_0(H)$ and $ba \in B_0(H)$.

Proof. The first clause is Proposition B.124, and the second and sixth (which implies the third) are almost trivial. For the fourth, we use the following criterion for pre-compactness (in a metric space): $K \subset H$ is pre-compact iff for each $\varepsilon > 0$ it can be covered by a *finite* number of open ε-balls $B_\varepsilon(\chi_i) = \{\psi \in H : \|\psi - \chi_i\| < \varepsilon\}$, where $i = 1, \ldots, m < \infty$ (i.e., all balls have the same radius ε). Given that $\|a_n - a\| \to 0$, for each $\varepsilon > 0$ there is n such that $\|a_n - a\| < \varepsilon/2$. Since $a_n(H_{\leq 1})$ is compact, it has a finite cover with $\varepsilon/2$-balls; in other words, for each $\psi \in H_{\leq 1}$ there is an i such that $\|a_n\psi - \chi_i\| < \varepsilon/2$. Hence, as $\|\psi\| \leq 1$, we may estimate

$$\|a\psi - \chi_i\| \leq \|(a_n - a)\psi\| + \|a_n\psi - \chi_i\| \leq \|a_n - a\|\|\psi\| + \tfrac{1}{2}\varepsilon < \tfrac{1}{2}\varepsilon + \tfrac{1}{2}\varepsilon = \varepsilon.$$

So $a(H_{\leq 1})$ has a finite cover with ε-balls and hence is pre-compact. This finishes the proof from Definition B.123; from Definition B.129, invoke Proposition B.127.

To prove the fifth clause, we need a result of independent interest. We say that a linear map $a : H \to H$ is (or has) ***finite rank*** if its image is finite-dimensional.

Proposition B.131. *A bounded operator $a \in B(H)$ is compact iff it is a norm-limit of finite-rank operators.*

Proof. Since it is easy to see that finite-rank operators are compact, the "$\Leftarrow$" direction follows from clause 4 of Theorem B.130. The difficult direction is the opposite one, which we prove by contradiction (as a technical note, our proof assumes that H is separable, but the claim also holds in the non-separable case, in which it can be shown that $\text{ran}(a)$ is separable whenever a is compact).

Pick a basis (υ_i) of H (or, in the non-separable case, of ran(a)), and define e_n to be the projection onto the linear span of the first n basis vectors. Given some $a \in B_0(H)$, define $a_n = e_n a$. We show that $\|a_n - a\| \to 0$. If not, then

$$\exists \varepsilon > 0 \forall N \exists n > N : \|a_n - a\| \geq \varepsilon, \tag{B.429}$$

which in turn implies that for any $\delta > 0$ there are unit vectors ψ_n for which we have $\|(a_n - a)\psi_n\| \geq \varepsilon - \delta$. Take $\delta = \varepsilon/2$, whence

$$\exists \varepsilon > 0 \forall N \exists n > N : \|(a_n - a)\psi_n\| \geq \varepsilon/2. \tag{B.430}$$

Now a is compact, so that, noting that $\psi_n \in H_{\leq 1}$, the sequence $(a\psi_n)$ has a convergent subsequence, say with limit φ. We may then write

$$(a_n - a)\psi_n = (e_n - 1_H)(a\psi_n - \varphi + \varphi), \tag{B.431}$$

so that, for each ψ_n,

$$\|(a_n - a)\psi_n\| \leq \|(e_n - 1_H)\| \|a\psi_n - \varphi\| + \|(e_n - 1_H)\varphi\|. \tag{B.432}$$

If we now restrict the ψ_n so as to lie in the convergent subsequence in question, then the right-hand side vanishes as $n \to \infty$:

- Since $\|e_n\| = \|1_H\| = 1$ we have $\|(e_n - 1_H)\| \leq 2$;
- By construction we have $\lim_n \|a\psi_n - \varphi\| = 0$;
- For any basis of H, and any $\varphi \in H$, we have $\lim_n \|(e_n - 1_H)\varphi\| = 0$ (although $\|e_n - 1_H\|$ fails to converge to anything if H is infinite-dimensional!).

However, this contradicts (B.430). □.

We use the notation of this proof to establish the fifth clause of Theorem B.130. By the sixth, the operator $a_n^* = a^* e_n$ is compact, since any finite-rank operator such as e_n is compact and a^* is bounded. Therefore, $\|a_n^* - a^*\| = \|a_n - a\| \to 0$, so $a_n^* \to a^*$ and hence $a^* \in B_0(H)$ by clause 4. □

B.19 Spectral theory for self-adjoint compact operators

If only to establish our notation, let us begin by recalling Theorem A.10:

Theorem B.132. *Let* $\dim(H) < \infty$ *and let* $a : H \to H$ *be a self-adjoint operator. Then the eigenvalues* λ *of* a *are real (collected in the point spectrum* $\sigma_p(a) \subset \mathbb{R}$*), the eigenspaces* H_λ *corresponding to different eigenvalues* λ *are orthogonal, and we have the* **spectral resolutions**

$$a = \sum_{\lambda \in \sigma_p(a)} \lambda \cdot e_\lambda; \tag{B.433}$$

$$1_H = \sum_{\lambda \in \sigma_p(a)} e_\lambda, \tag{B.434}$$

where e_λ *is the projection onto the eigenspace*

$$H_\lambda = \{\psi \in H \mid a\psi = \lambda\psi\}. \tag{B.435}$$

This theorem is equivalent to the following alternative version:

Theorem B.133. *Let* $\dim(H) < \infty$ *and let* $a : H \to H$ *be a self-adjoint operator (i.e.,* $a^* = a$*). Then* a *is* **diagonalizable**, *in the sense that* H *has a basis* (υ_i) *consisting of eigenvectors of* a*. Furthermore, the eigenvalues* λ_i *of* a *are real.*

If a is diagonalizable, using the familiar notation $e_{\upsilon_i} = |\upsilon_i\rangle\langle\upsilon_i|$, cf. (2.7), we write

$$a\upsilon_i = \lambda_i \upsilon_i; \tag{B.436}$$

$$a = \sum_{i \in I} \lambda_i e_{\upsilon_i}. \tag{B.437}$$

To move from Theorem B.132 to Theorem B.133, pick some basis $(\upsilon_k^{(\lambda)})$ of each eigenspace H_λ. By Proposition A.8 we then have

$$e_\lambda = \sum_{k=1}^{\dim(H_\lambda)} |\upsilon_k^{(\lambda)}\rangle\langle\upsilon_k^{(\lambda)}|. \tag{B.438}$$

The totality of all $\upsilon_k^{(\lambda)}$, where $\lambda \in \sigma_p(a)$ and $k = 1, \dots, \dim(H_\lambda)$ is our basis: relabeling this set as (υ_i), eq. (B.434) becomes $1_H = \sum_i |\upsilon_i\rangle\langle\upsilon_i|$, or $\psi = \sum_i c_i \psi_i$ with $c_i = \langle\upsilon_i, \psi\rangle$ for each $\psi \in H$, which according to Theorem B.61.1 shows that (υ_i) is a basis of H (and hence $i = 1, \dots, \dim(H)$). Furthermore, (B.433) yields $a\upsilon_k^{(\lambda)} = \lambda \upsilon_k^{(\lambda)}$, or (B.436), so that each υ_i is an eigenvector of a.

Conversely, for each $\lambda \in \sigma_p(a)$, assemble all eigenvalues λ_i that are equal to λ and relabel those as $\upsilon_k^{(\lambda)}$. This yields e_λ through (B.438), and the above argument may be rerun in the opposite direction: the basis property of the (υ_i) implies (B.434), and the eigenvector property (B.436) yields (B.434) by verifying it on each basis vector $\upsilon_i \equiv \upsilon_k^{(\lambda)}$, recalling that by construction, $\lambda_i = \lambda$.

We now adapt these results to infinite dimension. We still say that an operator $a : H \to H$ is ***diagonalizable*** if H has a basis (υ_i) consisting of eigenvectors of a.

Proposition B.134. *Let $H \cong \ell^2(I)$ for some set I (i.e., H has a basis $(\upsilon_i)_{i\in I}$). Then some collection $(\lambda_i)_{i\in I}$ of complex numbers occurs as the set of eigenvalues of some bounded operator $a \in B(H)$ iff $(\lambda_i)_{i\in I}$ is bounded, i.e.,* $\sup\{|\lambda_i|, i \in I\} < \infty$.

Defining a function $\tilde{\lambda} : I \to \mathbb{C}$ by $\tilde{\lambda}(i) = \lambda_i$, we may express this as $\tilde{\lambda} \in \ell^\infty(I)$.

Proof. If $a \in B(H)$ is diagonal in some basis (υ_i), with eigenvalues (λ_i), then

$$|\lambda_i| = \|\lambda_i \upsilon_i\| = \|a\upsilon_i\| \leq \|a\|\|\upsilon_i\| = \|a\|, \tag{B.439}$$

for each $i \in I$, whence the eigenvalues are bounded. Conversely, if they are, so that $\|\tilde{\lambda}\|_\infty < \infty$, take a basis $(\upsilon_i)_{i\in I}$ of H, write $\psi = \sum_i c_i \upsilon_i$ with $\sum_i |c_i|^2 < \infty$, cf. Theorem B.61 and define $a\psi = \sum_i \lambda_i c_i \upsilon_i$. Since

$$\sum_i |\lambda_i c_i|^2 \leq \|\tilde{\lambda}\|_\infty^2 \sum_i |c_i|^2 = \|\tilde{\lambda}\|_\infty^2 \|\psi\|^2 < \infty, \tag{B.440}$$

we have $a\psi \in H$ by Lemma B.59. These estimates also prove that $\|a\psi\| \leq \|\tilde{\lambda}\|_\infty \|\psi\|$, so that a is bounded, with $\|a\| \leq \|\tilde{\lambda}\|_\infty$ (in fact, equality holds here). □

This characterization of bounded diagonalizable operators by a property of their eigenvalues may be considerably sharpened for self-adjoint compact operators.

Theorem B.135. *Let* $\dim(H) = \infty$, *and let $a \in B(H)_{\text{sa}}$. Then a is compact iff it is diagonalizable with $\tilde{\lambda} \in \ell_0(I)$, in which case the sum in* (B.437) *converges in norm.*

We recall that some function $f : I \to \mathbb{C}$ is in $\ell_0(I)$ if for each $\varepsilon > 0$ there is a *finite* subset $I_\varepsilon \subset I$ such that $|f(i)| < \varepsilon$ for all $i \notin I_\varepsilon$. If $I = \mathbb{N}$ (and in fact the proof below will produce this labeling of the basis), then the condition $\tilde{\lambda} \in \ell_0(\mathbb{N})$ just means that

$$\lim_{n\to\infty} \lambda_n = 0. \tag{B.441}$$

Before proving this, we state the infinite-dimensional analogue of Theorem B.132:

Theorem B.136. *Let* $\dim(H) = \infty$ *and let a be some bounded self-adjoint operator.*

Then a is compact iff it has the properties stated in Theorem B.132, amended by the following clarifications and addenda (cf. Definition B.6, where $X = \sigma_p(a)$):

1. *The sum in* (B.433) *converges* in norm*;*
2. *The sum in* (B.434) *converges* strongly, *i.e., for each $\psi \in H$ we have*

$$\psi = \sum_{\lambda\in\sigma_p(a)} e_\lambda \psi; \tag{B.442}$$

3. *If $\lambda \in \sigma(a)$ and $\lambda \neq 0$, then $\lambda \in \sigma_p(a)$ and* $\dim(H_\lambda) < \infty$*;*
4. *Always $0 \in \sigma(a)$, and $\sigma_p(a) \subset \mathbb{R}$ has 0 as its only accumulation point.*

The equivalence between Theorems B.135 and B.136 is a bit more subtle than in finite dimension, but the key to the proof of both is the following lemma.

Lemma B.137. *A compact self-adjoint operator a has an eigenvalue* $\lambda = \pm\|a\|$.

Note that by definition of the operator norm, one always has $|\lambda| \leq \|a\|$, whether or not a is compact, but the point about compact self-adjoint operators is firstly that *they have an eigenvalue at all*, and secondly that the above equality is saturated.

Proof. We use the fact that the norm $\psi \mapsto \|\psi\|$ is continuous on H, see (B.5), so that it attains a maximum on the compact set $a(H_{\leq 1})$. Assume that this maximum is attained at $a\psi_1$, with $\|\psi_1\| = 1$. By definition of the operator norm, this maximum must be $\|a\|$, so that $\|a\|^2 = \|a\psi_1\|^2$. Cauchy–Schwarz and $a^* = a$ then yield

$$\|a\|^2 = \langle a\psi_1, a\psi_1\rangle = \langle \psi_1, a^2\psi_1\rangle \leq \|\psi_1\|\|a^2\psi_1\| \leq \|a^2\| = \|a\|^2, \tag{B.443}$$

where we have used (C.2). In the Cauchy–Schwarz inequality (A.1) one has equality iff either $v = 0$ or $w = zv$ for some $z \in \mathbb{C}$, so that we must have $a^2\psi_1 = z\psi_1$, with $|z| = \|a\|^2$. Moreover, $z \in \mathbb{R}$, as eigenvalues must be real (which trivially follows from $a^* = a$, one does not even need Theorem B.93 here), so $a^2\psi_1 = \lambda^2\psi_1$, with either $\lambda = \|a\|$ or $\lambda = -\|a\|$. If $a\psi_1 = \lambda\psi_1$, we are ready. If not, then $\chi_1 = a\psi_1 - \lambda\psi_1 \neq 0$, in which case $a\chi_1 = a^2\psi_1 - \lambda a\psi_1 = \lambda^2\psi_1 - \lambda a\psi_1 = -\lambda\chi_1$. □

Corollary B.138. *A compact self-adjoint operator is diagonalizable.*

Proof. Using the notation of the above proof, we call the (normalized) eigenvector in question υ_1 (so either $\upsilon_1 = \psi_1$ or $\upsilon_1 = \chi_1$). Note that if $\langle\varphi, \upsilon_1\rangle = 0$, then $\langle a\varphi, \upsilon_1\rangle = \langle\varphi, a^*\upsilon_1\rangle = \langle\varphi, a\upsilon_1\rangle = \pm\lambda\langle\varphi, \upsilon_1\rangle = 0$, so that a maps the orthogonal complement $\upsilon_1^\perp = \{\varphi \in H \mid \langle\upsilon_1, \varphi\rangle = 0\}$ of υ_1 into itself. This implies that a commutes with the projection e_1 onto $\upsilon_1^\perp$, i.e., $e_1 a = ae_1$ and hence also $e_1 a = e_1 a e_1$, in which the right-hand side is essentially the restriction of a to $\upsilon_1^\perp = e_1 H$.

By Theorem B.130.6, the operator $e_1 a$ is compact, like a itself, and it is also self-adjoint. If $e_1 a = 0$ we are ready, since υ_1 plus any basis of $e_1 H$ is a basis of H that diagonalizes a. If not, we apply Lemma B.137 to the operator $e_1 a$, finding an eigenvector υ_2 with nonzero eigenvalue λ_2. A simple computation shows that $e_1\upsilon_2 = \upsilon_2$, so that $\upsilon_2 \in e_1 H$, from which we infer, in turn, that $a\upsilon_2 = \lambda_2\upsilon_2$.

So we have found two basis vectors (υ_1, υ_2) of H that are eigenvectors of a. The above procedure may then be iterated: we define e_2 as the projection onto the orthogonal complement of υ_1 and υ_2, and consider $e_2 a$. If $e_2 a = 0$ we are ready; if not, we find a third eigenvector of $e_2 a$ and hence of a in $e_2 H$, *et cetera*.

- If $H \cong \mathbb{C}^n$ is finite-dimensional, this procedure terminates after n steps, leaving a basis $\{\upsilon_1, \ldots, \upsilon_n\}$ of H that by construction consists of eigenvectors of a.
- If H is separable, the iteration procedure may be continued countably many times, leading to an ordered countable set $B = (\upsilon_1, \upsilon_2, \ldots)$ of orthogonal unit vectors that are eigenvectors of a. By construction we have $|\lambda_N| \geq |\lambda_{N+1}|$ for all $N \in \mathbb{N}$, and hence there are two scenarios: either $e_N a = 0$ for all $N > N_0 \in \mathbb{N}$ (with $e_N a \neq 0$ if $N \leq N_0$), in which case $a = 0$ on $(\upsilon_1, \ldots, \upsilon_{N_0})^\perp$, or all $|\lambda_N| > 0$.
- In general, consider the set of all orthonormal sets in H that consist of eigenvectors of a. This set is nonempty by the argument above, and is inductively ordered by inclusion, so by Zorn's Lemma it must have a maximal element B.

By Theorem B.61.5, the set B is a basis of H iff $B^{\perp} = \{0\}$. To show that this is the case, suppose $B^{\perp}$ is a nonzero Hilbert space. Define f as the projection on H with image $B^{\perp}$ and consider the self-adjoint compact operator fa. If $fa = 0$, there is at least one eigenvector of a in $B^{\perp} = fH$ (namely, with eigenvalue zero), which is a contradiction. If $fa \neq 0$, then a has an eigenvector by Lemma (B.137), and again a contradiction has been found: for in all three cases, by construction all eigenvectors were already contained in $B^{\perp\perp} = \mathrm{span}(\upsilon_1, \ldots)^{-}$. □.

Even if H is non-separable, the image of a compact operator a must nonetheless be separable. Therefore, the non-zero eigenvalues of a form a countable set, and the eigenvalue zero (which, by the same token, must occur in the non-separable case) has some uncountable multiplicity (in sharp contrast to which, each nonzero eigenvalue has finite multiplicity). Also in the separable case, the only eigenvalue that may have infinite multiplicity is zero (though in the separable case it does not necessarily occur). Theorem B.135 is now a consequence of the following lemma:

Lemma B.139. *A diagonalizable operator a is compact iff $\tilde{\lambda} \in \ell_0(I)$.*

Proof. In view of the proof and subsequent comment above, we may as well assume that $I = \mathbb{N}$. For any $\psi \in H$, the sum in (B.214) converges, so we must have $\lim_n \langle \upsilon_n, \psi \rangle = 0$, or, in other words, $\upsilon_n \to 0$ weakly. If $a \in B_0(H)$, then $a\upsilon_n \to 0$ in norm by Proposition B.125, and hence $\lambda_n \to 0$, i.e., $\tilde{\lambda} \in \ell_0(\mathbb{N})$. Conversely, if this holds, then for each $\varepsilon > 0$, the set $I_\varepsilon = \{n \in \mathbb{N} : |\lambda_n| \geq \varepsilon\}$ is finite. This implies that the operator $a_n = \sum_{m \in I_{1/n}} \lambda_m e_{\upsilon_m}$ has finite rank. Since $|\lambda_m| < \varepsilon$ whenever $m \notin I_{1/n}$,

$$\|(a_n - a)\psi\|^2 = \|\sum_{m \notin I_{1/n}} \lambda_m e_{\upsilon_m} \psi\|^2 \leq \sum_{m \notin I_{1/n}} |\lambda_m|^2 |\langle \upsilon_m, \psi \rangle|^2 \leq \varepsilon^2 \|\psi\|^2, \quad \text{(B.444)}$$

where in the last step we also used (B.213). Hence $a_n \to a$ in norm, so that a is compact by Proposition B.131. □

To finish the proof of Theorem B.135, we show that the sum in (B.437), which for general bounded diagonalizable operators converges strongly, in fact converges in norm. To put this in perspective, eq. (B.437) with $a = 1_H$ reads

$$1_H = \sum_{i \in I} e_{\upsilon_i}. \quad \text{(B.445)}$$

If I is infinite, this sum cannot converge uniformly: e.g., if we take $I = \mathbb{N}$, then

$$\lim_{N \to \infty} \left\| 1_H - \sum_{n=1}^{N} e_{\upsilon_i} \right\| = \lim_{N \to \infty} \sup \left\{ \left\| \psi - \sum_{n=1}^{N} \langle \upsilon_n, \psi \rangle \upsilon_n \right\|, \psi \in H_{\leq 1} \right\} \quad \text{(B.446)}$$

cannot be zero, as shown by taking ψ orthogonal to all $\upsilon_1, \ldots, \upsilon_N$. However, by Theorem B.61.1 the sum does converge strongly (i.e., applied to each fixed ψ). This seemingly special case even yields strong convergence of the sum in (B.437) for general diagonalizable bounded operators a, for by continuity of a we have:

$$a\psi = a\sum_{i\in I}\langle \upsilon_i,\psi\rangle \upsilon_i = \sum_{i\in I}\langle \upsilon_i,\psi\rangle a\upsilon_i = \sum_{i\in I}\lambda_i\langle \upsilon_i,\psi\rangle \upsilon_i = \sum_{i\in I}\lambda_i e_{\upsilon_i}\psi. \tag{B.447}$$

If a is compact, strong convergence of (B.437) may be strengthened to norm convergence. The argument is analogous to the proof of Lemma B.139, but for completeness and contrast we now present it for general I. Since $\tilde{\lambda}\in\ell_0(I)$, for given $\varepsilon>0$ there is a finite set $I_\varepsilon\subset I$ for which $|\lambda_i|<\varepsilon$ for all $i\notin I_\varepsilon$. For fixed $\psi\in H$, we have

$$\left\|(a-\sum_{i\in I_\varepsilon}\lambda_i e_{\upsilon_i})\psi\right\|^2 = \left\|\sum_{i\notin I_\varepsilon}\lambda_i e_{\upsilon_i}\psi\right\|^2 < \varepsilon^2\sum_{i\notin I_\varepsilon}|\langle \upsilon_i,\psi\rangle|^2 \leq \varepsilon^2\|\psi\|^2, \tag{B.448}$$

so that $\|a-\sum_{i\in I_\varepsilon}\lambda_i e_{\upsilon_i}\|<\varepsilon$. By Definition B.6, eq. (B.437) holds in norm. □

This analysis by no means contradicts Corollary B.104, including (B.327): applied to compact operators, exactly one of the subsets $A_{i_0}\subset\sigma(a)$ contains $\sigma(a)\cap U_0$, where U_0 is some neighborhood of $0\in\sigma(a)$, so that the corresponding projection $e_{A_{i_0}}$ is infinite-dimensional and all the other e_{A_i} are finite-dimensional. Thus the sum $\sum_i e_{A_i}$ in (B.327) takes a rather different form from either the sum $\sum_i e_{\upsilon_i}$ in (B.445) or the sum $\sum_\lambda e_\lambda$ in (B.434); see also the end of this section.

We now prove Theorem B.136. First, as soon as $\dim(H_\lambda)=\infty$ for some $\lambda\neq 0$, then $\tilde{\lambda}\notin\ell_0(I)$. Therefore, $\dim(H_\lambda)<\infty$ by Theorem B.135. In fact, is is easy to show directly that $\dim(\ker(a-\lambda))<\infty$ for any $a\in B_0(H)$ and $\lambda\neq 0$: since a is bounded and hence $\ker(a-\lambda)$ is closed, the latter is a Hilbert space in its own right, so if it were infinite-dimensional, any basis (u_n) of it would have the property that $u_n\to 0$ weakly and hence $au_n\to 0$ in norm (cf. the proof of the above lemma). But $au_n=\lambda u_n$, so that (au_n) cannot converge in norm as soon as $\lambda\neq 0$.

Second, take $0\neq\lambda\in\sigma(a)$. According to Theorem B.93, in order to prove that $\lambda\in\sigma_p(a)$, it suffices to show that $\mathrm{ran}(a-\lambda)$ is closed. We may assume that $\lambda\neq\lambda_i$ for all $i\in I$ (for otherwise, trivially $\lambda\in\sigma_p(a)$), which implies $\ker(a-\lambda)=\{0\}$.

Let $\psi_n=(a-\lambda)\varphi_n\in\mathrm{ran}(a-\lambda)$, with $\varphi_n\neq 0$ for all n, and suppose $\psi_n\to\psi$. We prove that (φ_n) is bounded. If not, then $\|\varphi_n\|\to\infty$, but since (φ'_n) is bounded, with $\varphi'_n=\varphi_n/\|\varphi_n\|$, and (ψ_n) converges, we have $(a-\lambda)\varphi'_n=\psi_n/\|\varphi_n\|\to 0$. Now a is compact, so $(a\varphi'_n)$ has a convergent subsequence, which together with the previous result implies that (φ'_n) itself must have a convergent subsequence (as $\lambda\neq 0$), say to φ'. Continuity of a gives $(a-\lambda)\varphi'=0$, hence $\varphi'_n\in\ker(a-\lambda)=\{0\}$. But this is impossible, as $\|\varphi'_n\|=1$ for all n. Thus knowing that (φ_n) is bounded, once again using compactness of a, we infer that $(a\varphi_n)$ has a convergent subsequence. Now

$$\varphi_n=\lambda^{-1}(a\varphi_n-(a-\lambda)\varphi_n)=\lambda^{-1}(a\varphi_n-\psi_n), \tag{B.449}$$

and since (ψ_n) converges by assumption, this implies that (φ_n) has a convergent subsequence, say with limit φ. Continuity of a then implies that

$$\psi=(a-\lambda)\varphi\in\mathrm{ran}(a-\lambda), \tag{B.450}$$

and hence $\mathrm{ran}(a-\lambda)$ is closed. Therefore, $\lambda\in\sigma_p(a)$.

To show that $0 \in \sigma(a)$, assume that a were invertible (which is to say that $0 \in \rho(a)$). Then its inverse a^{-1} would be bounded, so that $a^{-1}a = 1_H \in B_0(H)$ by Theorem B.130. But this is impossible in infinite dimension: a similar argument to the one below (B.445) shows that 1_H cannot possibly be approximated by finite-rank operators. The last claim of Theorem B.136 is the same as $\tilde{\lambda} \in \ell_0(I)$. □

Here is a nice example of compact operators, also justifying the notation $B_0(H)$.

Corollary B.140. *Let $H = \ell^2(\mathbb{N})$ and for $f \in \ell^\infty(\mathbb{N})$, define the multiplication operator m_f as usual, i.e., $m_f\psi = f\psi$. Then m_f is compact iff $f \in \ell_0(\mathbb{N})$.*

Proof. This follows from Theorem B.135, where the label set is $I = \mathbb{N}$, the basis $(\upsilon_i)_{i\in I}$ is $(\delta_n)_{n\in\mathbb{N}}$, where $\delta_n(m) = \delta_{nm}$ as usual, $m \in \mathbb{N}$, and the eigenvalues are

$$\lambda_n = f(n), \tag{B.451}$$

since obviously $m_f\delta_n = f\delta_n = f(n)\delta_n$. We already know from (B.276) that $\sigma(m_f) = \mathrm{ran}(f)^-$, which for $f \in \ell_0(\mathbb{N})$ equals $\mathrm{ran}(f)$ if $0 \in \mathrm{ran}(f)$, and

$$\mathrm{ran}(f)^- = \mathrm{ran}(f) \cup \{0\}, \tag{B.452}$$

otherwise. In the first case, $\sigma(m_f) = \sigma_p(m_f) = \mathrm{ran}(f)$, so $\sigma_c(m_f) = \emptyset$, whereas in the second case we have $\sigma_p(m_f) = \mathrm{ran}(f)$ and $\sigma_c(m_f) = \{0\}$. This also shows that in clause 4 of Theorem B.136, both possibilities $0 \in \sigma_p(a)$ and $0 \in \sigma_c(a)$ may occur, depending on a. Finally, the condition $\tilde{\lambda} \in \ell_0(I)$, which in the example $a = m_f$ reduces to (B.441), is just a restatement of the condition $f \in \ell_0(\mathbb{N})$. □

In the continuous case, for $H = L^2(X)$, say for some connected open set $X \subset \mathbb{R}^n$ with Lebesgue measure, the multiplication operator m_f defined by a function $f \in C_0(X)$ is never compact, cf. (B.276); it is the very opposite of a compact operator!

To close, in our (traditional) proof of Theorem B.136 we did not use the powerful spectral Theorem B.94. If $\dim(H) < \infty$, Theorem B.132 indeed follows from Theorem B.94: if, for $\lambda \in \mathbb{R}$, we define $1_{\{\lambda\}} \equiv \delta_\lambda : \mathbb{R} \to \mathbb{C}$ by $\delta_\lambda(x) = \delta_{\lambda x}$, then

$$\mathrm{id}_{\sigma_p(a)} = \sum_{\lambda\in\sigma_p(a)} \lambda \cdot \delta_\lambda; \tag{B.453}$$

$$1_{\sigma_p(a)} = \sum_{\lambda\in\sigma_p(a)} \delta_\lambda. \tag{B.454}$$

Now define $e_\lambda = \delta_\lambda(a)$. Then (B.290) - (B.291) give $e_\lambda^2 = e_\lambda^* = e_\lambda$, so that e_λ is a projection. Furthermore, since $\mathrm{id}_{\sigma_p(a)} \cdot \delta_\lambda = \lambda \cdot \delta_\lambda$, eq. (B.290) gives $ae_\lambda = \lambda e_\lambda$, so that $e_\lambda H \subseteq H_\lambda$. Applying the map $f \mapsto f(a)$ to (B.453) - (B.454) then yields (B.433) - (B.434), from which the equality $e_\lambda H = H_\lambda$ follows *a fortiori*.

If $\dim(H) = \infty$ and $a \in B_0(H)_{\mathrm{sa}}$, this still works for each nonzero $\lambda \in \sigma_p(a)$, and since the sum (B.453) converges uniformly in $C(\sigma(a))$, we obtain (B.433) in the same way, including its norm-convergence. Unfortunately, even if we replace $\sigma_p(a)$ by $\sigma(a)$, as we should, eq. (B.454) now fails, even pointwise, so that (B.434) still requires the kind of proof we gave (or a complicated argument based on (B.327)).

B.20 The trace

For finite-dimensional H the trace was defined by (A.77). There are (at least) two difficulties in generalizing this expression to the infinite-dimensional case in the naive way. First, not every operator has a finite trace; for example, take $a = 1_H$, so that $\mathrm{Tr}\,(1_H) = \dim(H)$. Second, Lemma A.25 is no longer valid in general: it is easy to find an operator $a \in B(H)$ and bases (υ_i) and (υ_i') of H for which

$$\sum_i \langle \upsilon_i, a\upsilon_i \rangle \neq \sum_i \langle \upsilon_i', a\upsilon_i' \rangle,$$

typically because one of these expressions converges, whereas the other diverges. For example, take $a = \sum_i (-1)^i |\upsilon_i\rangle\langle\upsilon_i|$ as a strong limit, i.e., $a\psi = \sum_i (-1)^i \langle \upsilon_i, \psi\rangle \upsilon_i$; this lies in H by Theorem B.61, from which (B.214) shows that $\|a\psi\| = \|\psi\|$. Take $\upsilon_1' = (\upsilon_1 + \upsilon_2)/\sqrt{2}$, $\upsilon_2' = (\upsilon_1 - \upsilon_2)/\sqrt{2}$, $\upsilon_3' = (\upsilon_3 + \upsilon_4)/\sqrt{2}$, $\upsilon_4' = (\upsilon_3 - \upsilon_4)/\sqrt{2}$, etc. Then $\sum_i \langle \upsilon_i, a\upsilon_i\rangle = \sum_i (-1)^i$ diverges, whereas $\sum_i \langle \upsilon_i', a\upsilon_i'\rangle = \sum_i 0 = 0$.

However, if $a \in B(H)$ is *positive*, i.e., $a \geq 0$ in the usual sense that $\langle \psi, a\psi\rangle \geq 0$ for each $\psi \in H$, then we will show that for any two bases (υ_i) and (υ_i') of H,

$$\sum_i \langle \upsilon_i, a\upsilon_i \rangle = \sum_i \langle \upsilon_i', a\upsilon_i' \rangle \tag{B.455}$$

where both sides may be infinite. Equivalently, (A.79) is valid, since any unitary operator defines and is defined by a basis transformation. To prove (B.455), we need a very useful construction of independent interest, cf. (A.73).

Lemma B.141. *Any positive operator $a \in B(H)$ has a (unique)* **square root**, *i.e., a positive operator $\sqrt{a} \in C^*(a)$ that satisfies $\sqrt{a}^2 = a$.*

Proof. This follows from Theorem B.94, since if $a \geq 0$, then $\sigma(a) \subset \mathbb{R}^+$, and hence $\sqrt{\cdot}$ is defined on $\sigma(a)$. Alternatively, one may use the following construction due to the Dutch mathematician C. Visser (which is a special case of the approach just mentioned). If necessary, first rescale a so that $\|a\| \leq 1$, take the power series for

$$\sqrt{1-x} = \sum_{k\geq 0} t_k x^k, \tag{B.456}$$

(in which $t_0 = 1$), which converges absolutely for $|x| \leq 1$, and put

$$\sqrt{a} = \sum_{k\geq 0} t_k (1_H - a)^k. \tag{B.457}$$

As in the numerical case, squaring the series and rearranging terms yields $\sqrt{a}^2 = a$. Since uniqueness will not be needed, we omit the proof. □

For $a \geq 0$, we now use (B.215) to compute

$$\sum_i \langle \upsilon_i, a\upsilon_i \rangle = \sum_i \langle \sqrt{a}\upsilon_i, \sqrt{a}\upsilon_i \rangle = \sum_{i,j} \langle \sqrt{a}\upsilon_i, \upsilon'_j \rangle \langle \upsilon'_j, \sqrt{a}\upsilon_i \rangle$$
$$= \sum_{i,j} \langle \sqrt{a}\upsilon'_j, \upsilon_i \rangle \langle \upsilon_i, \sqrt{a}\upsilon'_j \rangle = \sum_j \langle \upsilon'_j, a\upsilon'_j \rangle, \quad \text{(B.458)}$$

where each term in every sum is positive, so that rearrangements are valid. Let

$$B(H)_+ = \{a \in B(H) \mid a \geq 0\}; \quad \text{(B.459)}$$

In view of (B.458), we have a well-defined map

$$\mathrm{Tr} : B(H)_+ \to [0, \infty]; \quad \text{(B.460)}$$
$$\mathrm{Tr}\,(a) = \sum_i \langle \upsilon_i, a\upsilon_i \rangle, \quad \text{(B.461)}$$

where (υ_i) is an arbitrary basis of H, of which the result is independent by (B.455).

To drop the restriction $a \geq 0$ in the argument of the trace, for *any* $a \in B(H)$ we note that $a^*a \geq 0$, so that we may define the ***absolute value*** $|a|$ of a by

$$|a| = \sqrt{a^*a}. \quad \text{(B.462)}$$

Then $|a| \geq 0$ for all a by construction, and if $a \geq 0$, then $|a| = a$. Finally, we define the set of ***trace-class operators*** in $B(H)$, later seen to be a Banach space, as

$$B_1(H) = \{a \in B(H) \mid \mathrm{Tr}\,(|a|) < \infty\}. \quad \text{(B.463)}$$

The ***trace-norm*** of $a \in B_1(H)$, which for now is just a formula, is given by

$$\|a\|_1 = \mathrm{Tr}\,(|a|), \quad \text{(B.464)}$$

Lemma B.142. *1. For any $a \in B_1(H)$ we have*

$$\|a\| \leq \|a\|_1. \quad \text{(B.465)}$$

2. Any trace-class operator is compact, i.e., $B_1(H) \subset B_0(H)$.
3. For $b \in B(H)$ and $a \in B_1(H)$ one has (A.100), *i.e., $|\mathrm{Tr}\,(ab)| \leq \|a\|_1 \|b\|$.*
4. The trace-class operators $B_1(H)$ form a vector space with norm (B.464).

Part 4 will shortly be improved to $B_1(H)$ actually being a Banach space.

Let us note that Lemma A.28 and Proposition A.29 on the polar decomposition remain valid for infinite-dimensional Hilbert space, with essentially the same proof.

Proof. 1. By definition of the operator norm (B.227), for every $\varepsilon > 0$ there is a unit vector $\psi \in H$ such that for any $b \in B(H)$ one has $\|b\|^2 \leq \|b\psi\|^2 + \varepsilon$ (proof by contradiction). Put $b = (a^*a)^{1/4}$, and note that $\|(a^*a)^{1/4}\|^2 = \||a|\| = \|a\|$ by (C.2) and (A.93). Completing ψ to a basis (υ_i), and noting that

$$\sum_i \|(a^*a)^{1/4}\upsilon_i\|^2 = \sum_i \langle (a^*a)^{1/4}\upsilon_i, (a^*a)^{1/4}\upsilon_i \rangle = \sum_i \langle \upsilon_i, |a|\upsilon_i \rangle = \|a\|_1, \quad \text{(B.466)}$$

$$\|a\| = \|(a^*a)^{1/4}\|^2 \leq \|(a^*a)^{1/4}\psi\|^2 + \varepsilon \leq \sum_i \|(a^*a)^{1/4}\upsilon_i\|^2 + \varepsilon = \|a\|_1 + \varepsilon.$$

Since this holds for all $\varepsilon \geq 0$, one has (B.465).

2. Let $a \in B_1(H)$. Since $\sum_i \langle \upsilon_i, |a|\upsilon_i\rangle < \infty$, for each $\varepsilon > 0$ we can find n such that $\sum_{i>n} \langle \upsilon_i, |a|\upsilon_i\rangle < \varepsilon$. Let e_n be the projection onto the linear span of $\{\upsilon_i\}_{i=1,\dots,n}$. Using (C.2) in the form $\|a\|^2 = \|aa^*\|$ (which is valid by (A.22)) and (B.465)),

$$\|e_n^\perp |a|^{1/2}\|^2 = \|e_n^\perp |a| e_n^\perp\| \leq \|e_n^\perp |a| e_n^\perp\|_1 = \sum_i \langle \upsilon_i, e_n^\perp |a| e_n^\perp \upsilon_i\rangle = \sum_{i>n} \langle \upsilon_i, |a|\upsilon_i\rangle < \varepsilon,$$

for $|(e_n^\perp |a| e_n^\perp)| = e_n^\perp |a| e_n^\perp$, for if $c \geq 0$ then $b^*cb \geq 0$ for any $b, c \in B(H)$. Since $e_n^\perp = 1 - e_n$, it follows that $e_n|a|^{1/2} \to |a|^{1/2}$ in the norm topology. Since each operator $e_n|a|^{1/2}$ obviously has finite rank, $|a|^{1/2}$ and hence $|a|$ is compact. Finally, a has polar decomposition $a = u|a|$ and $B_0(H)$ is a two-sided ideal in $B(H)$.

3. We just showed that a is compact. By Theorem B.130, also a^*a is compact, and since it is self-adjoint, Theorem B.136 applies. This gives an expansion (A.101); although the sum may be infinite, this is no problem, as it is norm-convergent. Thus the computation will be analogous to the finite-dimensional case, cf. Proposition A.30, except that we cannot use (A.78), which is valid but has not been proved yet. Fortunately, this problem may be obviated using (A.94). It follows from Lemma A.28 and Proposition A.29 that $(\upsilon_i' = u\upsilon_i)$ also forms an orthonormal set, like the υ_i themselves, since the closed linear space spanned by the unit vectors υ_i is just $(\text{ran}|a|)^-$ and u is unitary from this space onto its image $(\text{ran}\, a)^-$. Taking the trace over any basis that contains the vectors υ_i', we compute

$$\begin{aligned} |\text{Tr}\,(ab)| &= |\text{Tr}\,(u|a|u^*ub)| = |\sum_i p_i \langle \upsilon_i', ub\upsilon_i'\rangle| \\ &\leq \sum_i p_i |\langle \upsilon_i', ub\upsilon_i'\rangle| \leq \sum_i p_i \|b\|\|u\|\|\upsilon_i\| = \|a\|_1 \|b\|, \end{aligned} \quad \text{(B.467)}$$

where we used $\|a\|_1 = \sum_i p_i$, which follows from (A.101) applied to $|a|$.

4. Let $a, b \in B_1(H)$, and let $a + b = u|a+b|$ be the polar decomposition. Then

$$\|a+b\|_1 = \text{Tr}\,(u^*(a+b)) = \text{Tr}\,(u^*a) + \text{Tr}\,(u^*b).$$

Applying (A.100) with $\|u^*\| \leq 1$, one has $\|a+b\|_1 \leq \|a\|_1 + \|b\|_1$. Hence $B_1(H)$ is a vector space and $\|\cdot\|_1$ satisfies the triangle inequality. The other axioms for a norm are obviously satisfied. □

Proposition B.143. *Let $H = \ell^2(\mathbb{N})$ (or even $\ell^2(X)$, for any* countable *set X), and for $f \in \ell^\infty(\mathbb{N})$, define the corresponding multiplication operator m_f by $m_f\psi = f\psi$, cf. Proposition B.73. We have seen that m_f is bounded, with norm* (B.239). *Then:*

$$m_f \in B_0(H) \text{ iff } f \in \ell_0(\mathbb{N}); \quad \text{(B.468)}$$

$$m_f \in B_1(H) \text{ iff } f \in \ell^1(\mathbb{N}); \quad \text{(B.469)}$$

$$\|m_f\|_1 = \|f\|_1. \quad \text{(B.470)}$$

Here $\ell_0(\mathbb{N})$ *consists of all* $f : \mathbb{N} \to \mathbb{C}$ *for which* $\lim_{x\to\infty} f(x) = 0$.
In particular, If $\dim(H) = \infty$ *we have* proper *inclusions*

$$B_1(H) \subset B_0(H) \subset B(H). \tag{B.471}$$

Proof. 1. For any $a \in B(H)$ we have $a \in B_0(H)$ iff $|a| \in B_0(H)$ by the polar decomposition (since $a = u|a|$ and $|a| = u^*a$ and $B_0(H)$ is a two-sided ideal in $B(H)$). In the present case, we have $|m_f| = \sqrt{m_f^* m_f} = \sqrt{m_{|f|^2}} = m_{|f|}$, whence $m_f \in B_0(H)$ iff $m_{|f|} \in B_0(H)$. Since $\sigma_p(m_{|f|}) = \{|f(x)|, x \in \mathbb{N}\}$, part 6 of Theorem B.136 applied to $a = m_{|f|}$ states that $f \in \ell_0(\mathbb{N})$.
2. This rapidly follows by computing $\mathrm{Tr}\,(|m_f|) = \mathrm{Tr}\,(m_{|f|})$ in the basis $\upsilon_x = \delta_x$, $x \in \mathbb{N}$, where $\delta_x(y) = \delta_{xy}$, as usual. □

Proposition B.144. *The map*

$$\mathrm{Tr}\ :\ B_1(H) \to \mathbb{C}; \tag{B.472}$$
$$a \mapsto \sum_i \langle \upsilon_i, a\upsilon_i \rangle, \tag{B.473}$$

where (υ_i) *is some basis of H, is well defined, (obviously) linear, and independent of the choice of basis. Furthermore,* (A.78), *i.e.,* $\mathrm{Tr}\,(ab) = \mathrm{Tr}\,(ba)$*, holds.*

Proof. Taking $a = 1_H$ in (A.100), we have $|\mathrm{Tr}\,(a)| \leq \|a\|_1 < \infty$ for $a \in B_1(H)$. Independence of the choice of basis follows by first decomposing $a = a' + ia''$, with $a' = \frac{1}{2}(a + a^*)$ and $a'' = -\frac{1}{2}i(a - a^*)$ self-adjoint, as usual, and subsequently using Theorem B.132 to write $a' = a'_+ - a'_-$, with

$$a'_\pm = \pm \sum_{\lambda \in \sigma_p(a') \cap \mathbb{R}_\pm} \lambda \cdot e_\lambda, \tag{B.474}$$

and likewise for a''. This makes a is a linear combination of four positive operators, whence the claim follows from (B.458) and the obvious linearity of (B.473).

To establish (A.78), we first note that $\mathrm{Tr}\,(au) = \mathrm{Tr}\,(ua)$ for any unitary u; this is the same as (A.79), which has just been proved. The claim then follows from the following (generally useful) lemma. □

Lemma B.145. *Any* $a \in B(H)$ *is a linear combination of at most four unitaries.*

Proof. By the previous argument, we may assume that $a^* = a$, and for convenience we also assume that $\|a\| \leq 1$. In that case, $\|a\psi\| \leq \|\psi\|$ and hence $1 - a^2 \geq 0$, so that $\sqrt{1 - a^2}$ is defined, cf. Lemma B.141. Defining the two operators

$$u_\pm = a \pm i\sqrt{1 - a^2}, \tag{B.475}$$

we find $u_\pm^* u_\pm = u_\pm u_\pm^* = 1_H$, making each $u_\pm$ unitary, and $a = \frac{1}{2}(u_+ + u_-)$. If $a \neq a^*$, the number of terms at most doubles. □

The deeper significance of the trace-class operators now emerges.

Theorem B.146. *For any Hilbert space H, we have dualities and double dualities*

$$B_0(H)^* \cong B_1(H); \tag{B.476}$$
$$B_1(H)^* \cong B(H); \tag{B.477}$$
$$B_0(H)^{**} \cong B(H); \tag{B.478}$$
$$B_1(H)^{**} \cong B(H)^*, \tag{B.479}$$

where the symbol $\cong$ *stands for* isometric isomorphism. *Explicitly:*

- *Any norm-continuous linear map* $\omega : B_0(H) \to \mathbb{C}$ *takes the form*

$$\omega(b) = \mathrm{Tr}\,(ab), \tag{B.480}$$

for some $a \in B_1(H)$ *uniquely determined by* ω, *and* vice versa, *giving a bijective correspondence between* $\omega \in B_0(H)^*$ *and* $a \in B_1(H)$ *satisfying*

$$\|\omega\| = \|a\|_1. \tag{B.481}$$

This equality remains valid if ω *is regarded as an element of* $B(H)^*$ *via* (B.479) *and the isometric embedding* $B_1(H) \hookrightarrow B_1(H)^{**}$ *(cf. Proposition B.44).*
- *Any norm-continuous linear map* $\chi : B_1(H) \to \mathbb{C}$ *takes the form*

$$\chi(a) = \mathrm{Tr}\,(ab), \tag{B.482}$$

for some $b \in B(H)$ *uniquely determined by* χ, *and* vice versa, *giving a bijective correspondence between* $\chi \in B_1(H)^*$ *and* $b \in B(H)$ *satisfying*

$$\|\chi\| = \|b\|. \tag{B.483}$$

Proof. It is clear from (A.100) that $B_1(H) \subseteq B_0(H)^*$, with $\|\omega\| \leq \|a\|_1$. For the opposite direction, we return to the projections e_n in the proof of part 2 of Lemma B.142. Taking the trace over the basis (υ_i), we have

$$\begin{aligned}\|a\|_1 &= \mathrm{Tr}\,(|a|) = \lim_n \mathrm{Tr}\,(e_n|a|e_n) = \lim_n \mathrm{Tr}\,(e_n|a|) = \lim_n \mathrm{Tr}\,(e_n u^* a)\\ &= \lim_n \omega(e_n u^*);\end{aligned} \tag{B.484}$$

since $\omega(e_n u^*) \geq 0$, we have $\omega(e_n u^*) \leq \|\omega\|\|e_n u^*\| \leq \|\omega\|$, whence $\|a\|_1 \leq \|\omega\|$ (note that the limiting procedure is necessary here, since $\omega(u^*)$ would not be defined because typically u^* is not compact). This proves (B.481).

To prove (B.476), it remains to be shown that every $\omega \in B_0(H)^*$ can be represented as (B.480). Noting that $B_0(H)$ is the norm-closure of the linear span of all operators of the sort $a = |\psi\rangle\langle\varphi|$, where $\psi, \varphi \in H$ are unit vectors, the functional ω is determined by its values on those operators. Given ω, we define a by its matrix elements $\langle\varphi, a\psi\rangle = \omega(|\psi\rangle\langle\varphi|)$. Evaluating the trace on a basis containing φ yields $\mathrm{Tr}\,(a|\psi\rangle\langle\varphi|) = \langle\varphi, a\psi\rangle$ and hence gives (B.480) on operators a of the said form, upon which the general case follows by continuity.

We now prove (B.477). As in the previous case, the inclusion $B(H) \subset B_1(H)^*$ is clear from (A.100), as is the inequality $\|\chi\| \leq \|a\|$. This time, the proof of the opposite inequality uses $a = |\psi\rangle\langle\varphi|$, in which case one easily obtains

$$\||\psi\rangle\langle\varphi|\|_1 = \|\psi\|\|\varphi\|, \tag{B.485}$$

which in the case of unit vectors equals unity. Assuming (B.482), this gives

$$|\chi(b)| = |\chi(|\psi\rangle\langle\varphi|)| = |\mathrm{Tr}\,(|\psi\rangle\langle\varphi|b)| = |\langle\varphi, b\psi\rangle| \leq \|\chi\|\||\psi\rangle\langle\varphi|\|_1 = \|\chi\|. \tag{B.486}$$

Combined with (B.228), this gives $\|b\| \leq \|\chi\|$, and hence (B.483).

Finally, as in the previous case, given χ, we find b though its matrix elements $\langle\varphi, b\psi\rangle = \chi(|\psi\rangle\langle\varphi|)$, which gives (B.482) on the special trace-class operators defined by $a = |\psi\rangle\langle\varphi|$. Noting that the linear span of such operators in dense (in the trace-norm) in $B_1(H)$, once again this gives the general case by continuity. □

Corollary B.147. *1. The vector space $B_1(H)$ is complete in the norm* (B.464).
2. $B_1(H)$ is a two-sided ideal in $B(H)$ ($a \in B(H), b \in B_1(H) \Rightarrow ab \in B_1(H) \ni ba$).

Proof. The first claim follows from (B.476) and the completeness of $B_0(H)^*$ (cf. Theorem B.33 and §B.9). The second follows from (A.100) and (A.78). □

This actually reveals a subtlety in (B.471): as a normed space, $B_0(H)$ simply inherits the norm of $B(H)$, in which it is complete. Clearly, $B_1(H)$ also inherits the norm of $B(H)$, but that is the wrong one: firstly, $B_1(H)$ is not complete in the operator norm (indeed, its completion is $B_0(H)$), and secondly, the operator norm is the wrong one for the fundamental dualities stated in Theorem B.146.

The following trace-class operators occupy the center stage in quantum theory.

Definition B.148. *A* **density operator** *is a positive operator $\rho \in B_1(H)$ such that*

$$\mathrm{Tr}\,(\rho) = 1. \tag{B.487}$$

Equivalently, ρ is a density operator iff it has a norm-convergent expansion

$$\rho = \sum_{\lambda \in \sigma_p(\rho)} \lambda \cdot e_\lambda, \tag{B.488}$$

where $\sigma_p(\rho)$ is some countable subset of $\mathbb{R}^+$ with 0 as its only possible accumulation point, the multiplicity $m_\lambda = \dim(H_\lambda)$ of each eigenvalue $\lambda > 0$ is finite, and

$$\sum_{\lambda \in \sigma_p(\lambda)} \lambda \cdot m_\lambda = 1. \tag{B.489}$$

Similarly, (2.6) holds just as in finite dimension, i.e., (B.488) is equivalent to

$$\rho = \sum_i p_i |\upsilon_i\rangle\langle\upsilon_i|, \tag{B.490}$$

where (υ_i) is a basis of H, and the coefficients (p_i) satisfy $p_i > 0$ and $\sum_i p_i = 1$. Furthermore, the p_i have 0 as their only possible accumulation point and are such that each $t > 0$ occurs in the set $\{p_i\}$ at most finitely many times. Like (B.488), also the equivalent expansion (B.490) is norm-convergent by Theorem B.136.

Definition B.149. *Let H be a separable Hilbert space. An operator $a \in B(H)$ is called a* **Hilbert–Schmidt operator** *if for some (and hence any) basis (υ_i) of H,*

$$\sum_i \|a\upsilon_i\|^2 < \infty, \tag{B.491}$$

We write $B_2(H)$ for the set of all Hilbert–Schmidt operators on H.

The argument that the sum in (B.491) is independent of the basis is based on (B.215) and is analogous to the computation (B.458), thjis time even without the complication of the square root, for we simply have $\sum_i \|a\upsilon_i\|^2 = \sum_i \langle a\upsilon_i, a\upsilon_i\rangle$, etc. For $a \in B_2(H)$, with foresight we define the expression (where (υ_i) is any basis of H):

$$\|a\|_2 = \sqrt{\mathrm{Tr}\,(a^*a)} = \left(\sum_i \|a\upsilon_i\|^2\right)^{1/2}. \tag{B.492}$$

Theorem B.150. *Let H be a separable Hilbert space.*

1. *For any $a \in B(H)$ we have*

$$\|a\| \leq \|a\|_2 \leq \|a\|_1. \tag{B.493}$$

2. *Every Hilbert–Schmidt operator is compact, and refining* (B.471) *one has*

$$B_1(H) \subset B_2(H) \subset B_0(H). \tag{B.494}$$

3. *The Hilbert–Schmidt operators $B_2(H)$ form a Hilbert space with inner product*

$$\langle a, b\rangle_2 = \mathrm{Tr}\,(a^*b), \tag{B.495}$$

and a Banach space in the ensuing norm (A.2)*, which equals (B.492). Clearly,*

$$B_2(H)^* \cong B_2(H). \tag{B.496}$$

4. *The Banach space $B_2(H)$ is a two-sided *-ideal in $B(H)$, and if $a \in B_2(H)$ and $b \in B(H)$ we have $\|ba\|_2 \leq \|b\|\|a\|_2$ and $\|ab\|_2 \leq \|b\|\|a\|_2$.*

Proof. 1. Take any unit vector $\psi \in H$ and complete it to a basis of H. This gives $\|a\psi\| \leq \|a\|_2$. Taking the supremum over all such ψ gives the first inequality. The second one will be proved in the next item.

2. With e_n from the proof of Lemma B.142.2, for $a \in B_2(H)$ we define $a_n = ae_n$, and note that because $\sum_i \|a\upsilon_i\|^2$ converges, $\|(a - a_n)\|_2^2 = \sum_{i=n+1}^{\infty} \|a\upsilon_i\|^2 \to 0$. By the previous item, $\|a_n \to a\| \to 0$. Since $a_n \in B_0(H)$, by Proposition B.131 also a is compact . For the second inequality in (B.493), Theorem B.136 yields:

$$\|a\|_2^2 = \sum_i \mu_i \leq \left(\sum_i \sqrt{\mu_i} \right)^2 = \|a\|_1^2, \tag{B.497}$$

where the $\mu_i \geq 0$ are the eigenvalues of the positive compact operator a^*a; the eigenvalues of the compact operator $|a| = \sqrt{a^*a}$ are $(\sqrt{\mu_i})$.

3. We first show that $B_2(H)$ is a vector space. For any $a, b \in B(H)$ we have

$$2(a^*a + b^*b) = (a+b)^*(a+b) + (a-b)^*(a-b), \tag{B.498}$$

so that $(a+b)^*(a+b) \leq 2(a^*a + b^*b)$ and hence $\|a+b\|_2^2 \leq 2(\|a\|_2 + \|b\|_2^2$. Therefore, if $a, b \in B_2(H)$, then $a+b \in B_2(H)$. Since $\|\lambda a\|_2 = |\lambda| \|a\|_2$, it is clear that if $a \in B_2(H)$, then $\lambda a \in B_2(H)$. Hence $B_2(H)$ is a vector space. Furthermore, because of the identity

$$a^*b = \tfrac{1}{4} \sum_{k=0}^{3} i^k (b + i^k a)^* (b + i^k a), \tag{B.499}$$

the inner product (B.495) may be rewritten as

$$\langle a, b \rangle_2 = \sum_i \langle e_i, a^*be_i \rangle = \tfrac{1}{4} \sum_{k=0}^{3} i^k \|(b + i^k a\|_2^2, \tag{B.500}$$

which shows that if $a, b \in B_2(H)$, then $\langle a, b \rangle_2 < \infty$. This reconfirms the fact that the trace in (B.495) may be computed in any basis, since this is true for each term on the right-hand side of (B.500). Sesquilinearity of (B.495) is straightforward. To prove positive definiteness, we use part 1: if $\|a\|_2 = 0$, then $\|a\| = 0$ and hence $a = 0$, since we already know that $\| \cdot \|$ is a norm.
Knowing that (B.495) is an inner product on $B_2(H)$, it immediately follows that $\| \cdot \|_2$ is a norm on $B_2(H)$, since, as already noted, $\|a\|_2 = \langle a, a \rangle_2$.
Finally, to prove completeness, we pick a basis (υ_i) in H and note that $B_2(H)$ is the closure of the linear span of all operators of the form $a = \sum_{i,j} c_{ij} |\upsilon_i\rangle\langle\upsilon_i|$. This is because of the continuity of the inclusion $B_2(H) \subset B_0(H)$ (which is true because of part 1 and the fact that $B_0(H)$ is itself the closure of this linear span). An easy calculation then gives

$$\|a\|_2 = \|\sum_{i,j} c_{ij} |\upsilon_i\rangle\langle\upsilon_i|\|_2^2 = \sum_{i,j} |c_{ij}|^2. \tag{B.501}$$

Hence $B_2(H)$ is isometrically isomorphic to the space of square-summable sequences (c_{ij}) indexed by $\mathbb{N} \times \mathbb{N}$, which by Theorem B.9 is complete in the ℓ^2-norm $\|c\|_2^2 = \sum_{i,j} |c_{ij}|^2$. Hence $B_2(H)$ is complete, too.

4. From (A.78) (proved in Proposition B.144) we have $\mathrm{Tr}\,(a^*a) = \mathrm{Tr}\,(aa^*)$, so that $a \in B_2(H)$ iff $a^* \in B_2(H)$. If $b \in B(H)$ and $a \in B_2(H)$, then $\|ba\upsilon_i\| \leq \|b\| \|a\upsilon_i\|$ and hence $\|ba\|_2 \leq \|b\| \|a\|_2$, so $ba \in B_2(H)$, and hence also $a^* \in B_2(H)$ and $a^*b^* \in B_2(H)$. Similarly, $ab \in B_2(H)$, with $\|ab\|_2 \leq \|b\| \|a\|_2$. □

B.21 Spectral theory for unbounded self-adjoint operators

Although there is hardly any distinction between bounded and unbounded self-adjoint operators in so far as the definition and elementary properties of the spectrum are concerned (cf. Definitions B.80 and B.85, Theorem B.91, and Theorem B.93), extending the various versions of the spectral theorem to the unbounded case is a highly nontrivial matter. There are many ways of accomplishing this, among which our presentation has the virtues that firstly (in contrast to von Neumann's original approach based on the Cayley transform) we stay within the realm of self-adjointness, and secondly we preserve the C*-algebraic spirit of Theorem B.94. Thirdly, our treatment is sufficiently general to cover the two main applications in quantum mechanics (viz. the Born rule and Stone's Theorem). For those applications, setting up a functional calculus for bounded Borel functions suffices, but in order to state even the defining property $\mathrm{id}_{\sigma(a)} \mapsto a$ of the functional calculus also for unbounded a (cf. Theorem B.94), unbounded *continuous* functions will also have to be incorporated (but we refrain from a further generalization to unbounded *Borel* functions).

Our approach starts from the observation that (with slight abuse of notation)

$$y : \mathbb{R} \to (-1,1); \tag{B.502}$$

$$y(x) = x(1+x^2)^{-1/2}; \tag{B.503}$$

$$y^{-1}(x) = x(1-x^2)^{-1/2}, \tag{B.504}$$

provides a homeomorphism $\mathbb{R} \cong (-1,1)$. This has an operatorial counterpart

$$a \mapsto a(1_H + a^2)^{-1/2} \equiv b; \tag{B.505}$$

$$b \mapsto b(1_H - b^2)^{-1/2} \equiv a, \tag{B.506}$$

where the notation for the square roots should be carefully disambiguated as

$$(1_H + a^2)^{-1/2} \equiv ((1_H + a^2)^{-1})^{1/2}; \tag{B.507}$$

$$(1_H - b^2)^{-1/2} \equiv ((1_H - b^2)^{1/2})^{-1}. \tag{B.508}$$

As we shall see, the operator $(1_H + a^2)^{-1}$ is bounded (and so is $1_H - b^*b$), of course), so that square roots are only taken of *bounded* operators, in which case they are defined by Lemma B.141. As in the numerical case (B.503), the correspondence $a \leftrightarrow b$ in (B.505) - (B.506) will turn out to be bijective, mapping the class of (possibly unbounded) self-adjoint operators into the class of self-adjoint pure contractions:

Definition B.151. *A* **pure contraction** *is a bounded operator* $b : H \to H$ *for which*

$$\|b\psi\| < \|\psi\| \quad (\psi \in H\backslash\{0\}). \tag{B.509}$$

If b is in addition self-adjoint, this is equivalent to $\|b\| \leq 1$ and $\ker(b \pm 1_H) = \{0\}$, i.e., $\pm 1 \notin \sigma_p(b)$; the argument is similar to the proof of Lemma B.137.

Eqs. (B.505) - (B.506) form a special case of a more general correspondence.

Theorem B.152. *The formal expressions*

$$b = a(1_H + a^*a)^{-1/2} \equiv a((1_H + a^*a)^{-1})^{1/2}; \tag{B.510}$$
$$a = b(1_H - b^*b)^{-1/2} \equiv b((1_H - b^*b)^{1/2})^{-1}, \tag{B.511}$$

make rigorous sense and define a bijective correspondence between the class of **closed operators** *a (with dense domain) and the class of (necessarily bounded)* **pure contractions** *b. This correspondence preserves the adjoint, in that*

$$b^* = a^*(1_H + aa^*)^{-1/2}; \tag{B.512}$$
$$a^* = b^*(1_H - bb^*)^{-1/2}, \tag{B.513}$$

and hence specializes to a a bijective correspondence (B.505) - (B.506) *between* **self-adjoint operators** *a and* **self-adjoint pure contractions** *b.*

The (bounded) operator b is called the **bounded transform** *of a.*

Proof. 1. *From b to a.* If b is a pure contraction, then $1_H - b^*b \geq 0$, since this means

$$\langle \psi, b^*b\psi \rangle \leq \langle \psi, \psi \rangle, \tag{B.514}$$

or $\|b\psi\|^2 \leq \|\psi\|^2$. Furthermore, $1_H - b^*b$ is injective, since $(1_H - b^*b)\psi = 0$ implies $\|\psi\|^2 = \|b\psi\|^2$, contradicting (B.509). This implies that $(1_H - b^*b)^{1/2}$ is injective, as $(1_H - b^*b)^{1/2}\psi = 0$ implies $(1_H - b^*b)\psi = 0$ and hence $\psi = 0$. Thus the inverse (B.508) exists, with domain

$$D((1_H - b^*b)^{-1/2}) = \text{ran}((1_H - b^*b)^{1/2}). \tag{B.515}$$

This domain in dense in H, since for any $c \in B(H)$ (which in our case is $c = (1_H - b^*b)^{1/2}$) we have $H = \ker(c) \oplus \ker(c)^\perp$; for $c^* = c$ we have $\ker(c) = \text{ran}(c)^\perp$ and hence $\ker(c)^\perp = \text{ran}(c)^-$, so that injectivity of c yields $H = \text{ran}(c)^-$. Hence (B.511) is well defined on

$$D(a) = \text{ran}((1_H - b^*b)^{1/2}). \tag{B.516}$$

To prove that a is closed, we write $a = bc^{-1}$, as above, and note that

$$G(a) = \{(b\psi, c\psi), \psi \in H\} = \text{ran}(v), \tag{B.517}$$

where $v : H \to H \oplus H$ is obviously defined by $v\psi = (b\psi, c\psi)$. Hence

$$\|v\psi\|^2 = \|b\psi\|^2 + \|c\psi\|^2 = \|b\psi\|^2 + \|(1_H - b^*b)^{1/2}\psi\|^2 = \|\psi\|^2, \tag{B.518}$$

so that v is an isometry. As such, $\text{ran}(v) = G(a)$ is closed.

2. *From a to b.* By definition, $D(1_H + a^*a) = D(a^*a)$, with

$$D(a^*a) = \{\psi \in D(a) \mid a^*\psi \in D(a)\}. \tag{B.519}$$

We show that $1_H + a^*a : D(a^*a) \to H$ is bijective. First, (B.237) implies

$$H \oplus H = G(a) \oplus G(a)^\perp = G(a) \oplus uG(a^*), \tag{B.520}$$

so for any $(\psi_1, \psi_2) \in H \oplus H$ there are *unique* $\varphi \in D(a)$ and $\chi \in D(a^*)$ such that

$$\psi_1 = \varphi - a^*\chi; \tag{B.521}$$
$$\psi_2 = a\varphi + \chi. \tag{B.522}$$

In particular, for $(\psi_1, \psi_2) = (\psi, 0)$ we obtain

$$\psi = (1_H + a^*a)\varphi, \tag{B.523}$$

This shows both surjectivity and injectivity, since φ is uniquely determined by ψ. Consequently, the inverse

$$(1_H + a^*a)^{-1} : H \to D(a^*a) \tag{B.524}$$

exists as a linear map, and since

$$\|(1_H + a^*a)^{-1}\psi\| = \|\varphi\| \leq \|(1_H + a^*a)\varphi\| = \|\psi\|, \tag{B.525}$$

we see that $(1_H + a^*a)^{-1}$ is bounded, with $\|(1_H + a^*a)^{-1}\| \leq 1$. A similar argument shows that $(1_H + a^*a)^{-1}$ is positive:

$$\langle \psi, (1_H + a^*a)^{-1}\psi \rangle = \langle (1_H + a^*a)\varphi, \varphi \rangle = \|\varphi\|^2 + \|a\varphi\|^2 \geq 0, \tag{B.526}$$

so that the square root (B.507) exists. As before, injectivity of $(1_H + a^*a)^{-1}$ implies injectivity of its square root, whence $\mathrm{ran}((1_H + a^*a)^{-1/2})$ is dense in H. Clearly, $(1_H + a^*a)^{-1/2}$ maps $\mathrm{ran}((1_H + a^*a)^{-1/2})$ to

$$\mathrm{ran}((1_H + a^*a)^{-1}) = D(a^*a) \subseteq D(a), \tag{B.527}$$

so that the operator b in (B.510) is defined on $\mathrm{ran}((1_H + a^*a)^{-1/2})$. We now show that b is bounded on the latter: for any $\psi \in H$ we have

$$\begin{aligned}
\|b(1_H + a^*a)^{-1/2}\psi\|^2 &= \|a(1_H + a^*a)^{-1}\psi\|^2 \\
&= \langle (1_H + a^*a)^{-1}\psi, a^*a(1_H + a^*a)^{-1}\psi \rangle \\
&\leq \langle (1_H + a^*a)^{-1}\psi, (1_H + a^*a)(1_H + a^*a)^{-1}\psi \rangle \\
&= \langle (1_H + a^*a)^{-1}\psi, \psi \rangle = \|(1_H + a^*a)^{-1/2}\psi\|^2,
\end{aligned} \tag{B.528}$$

so that b may be extended to all of H by continuity, with $\|b\| \leq 1$. Still denoting this extension by b, we have

$$b^*b = (1_H + a^*a)^{-1/2}a^*a(1_H + a^*a)^{-1/2} = 1_H - (1_H + a^*a)^{-1}, \tag{B.529}$$

from which it easily follows that b is a pure contraction: for any $\psi \neq 0$, we have

$$\|b\psi\|^2 = \langle \psi, b^*b\psi \rangle = \|\psi\|^2 - \|(1_H + a^*a)^{-1/2}\psi\|^2 < \|\psi\|^2, \tag{B.530}$$

since $\|1_H + a^*a)^{-1/2}\psi\|^2 > 0$ by injectivity of $(1_H + a^*a)^{-1/2}$.

3. *Bijectivity of the correspondence $a \leftrightarrow b$.* If a is determined by b according to (B.511), then

$$1_H + a^*a = (1_H - b^*b)^{-1}, \tag{B.531}$$

so that

$$(1_H - b^*b)^{1/2} = (1_H + a^*a)^{-1/2}, \tag{B.532}$$

whence

$$b = b(1_H - b^*b)^{-1/2}(1_H - b^*b)^{1/2} = a(1_H - b^*b)^{1/2} = a(1_H + a^*a)^{-1/2}. \tag{B.533}$$

Similarly, if b is defined by a according to (B.510), then (B.529), rewritten as

$$1_H - b^*b = (1_H + a^*a)^{-1}, \tag{B.534}$$

reproduces (B.511). To see that the domains match, in view of (B.516) we need

$$D(a) = \text{ran}((1_H + a^*a)^{-1/2}). \tag{B.535}$$

The inclusion $D(a) \supseteq \text{ran}((1_H + a^*a)^{-1/2})$ already having been established in step 2 above, we prove the opposite inclusion $\subseteq$. Indeed, for any $\psi \in D(a)$ we have

$$\psi = (1_H + a^*a)^{-1/2}(b^*a + (1_H + a^*a)^{-1/2})\psi, \tag{B.536}$$

where b is given by (B.510). This follows by taking inner products with $\varphi \in H$:

$$\begin{aligned} &\langle \varphi, (1_H + a^*a)^{-1/2}b^*a\psi \rangle + \langle \varphi, (1_H + a^*a)^{-1}\psi \rangle \\ &= \langle a^*a(1_H + a^*a)^{-1}\varphi, \psi \rangle + \langle \varphi, (1_H + a^*a)^{-1}\psi \rangle = \langle \varphi, \psi \rangle. \end{aligned} \tag{B.537}$$

4. *Self-adjointness.* Since a is closed we have $a^{**} = a$ (cf. Lemma B.74), so using a^* instead of a in part 2 above, we have

$$(1_H + aa^*)^{-1} : H \to D(aa^*) \subset D(a^*), \tag{B.538}$$

bijectively. If, in addition, $\psi \in D(a^*)$, we may compute

$$a^*\psi = a^*(1_H + aa^*)(1_H + aa^*)^{-1}\psi = (1_H + a^*a)a^*(1_H + aa^*)^{-1}\psi, \tag{B.539}$$

from which it follows that

$$a^*(1_H + aa^*)^{-1}\psi = (1_H + a^*a)^{-1}a^*\psi. \tag{B.540}$$

Similarly, for any polynomial p in one real variable we have

$$a^*p((1_H+aa^*)^{-1})\psi = p((1_H+a^*a)^{-1})a^*\psi. \tag{B.541}$$

By Weierstrass, we can find polynomials p_n such that

$$\lim_{n\to\infty} p_n((1+x)^{-1}) = (1+x)^{-1/2}, \tag{B.542}$$

for any $x \geq 0$, also cf. the proof of Lemma B.141. Hence by Theorem B.94 and closeness of a^* we obtain

$$\begin{aligned} a^*(1_H+aa^*)^{-1/2}\psi &= (1_H+a^*a)^{-1/2}a^*\psi = (a(1_H+a^*a)^{-1/2})^*\psi \\ &= b^*\psi, \end{aligned} \tag{B.543}$$

for $\psi \in D(a^*)$. Since the latter is dense, we have (B.512). Bijectivity of the correspondence $a \leftrightarrow b$ then also implies (B.513). In particular, $a^* = a$ iff $b^* = b$, which implies the last claim of the theorem. □

Though not needed in what follows, it would be a pity not to state:

Corollary B.153. *If $a : D(a) \to H$ is closed (with $D(a)^- = H$), then:*

1. *$1_H + a^*a$ is self-adjoint on $D(a^*a)$;*
2. *$(1_H + a^*a)^{-1} = \pi_1 \circ e_{G(a)} \circ \iota_1$, where:*
   - *$\iota_1 : H \to H \oplus H$ is defined by $\iota_1\psi = (\psi, 0)$;*
   - *$e_{G(a)} : H \oplus H \to H \oplus H$ is the projection onto the graph $G(a)$;*
   - *$\pi_1 : H \oplus H \to H$ is the projection $\pi_1(\psi_1, \psi_2) = \psi_1$ onto the first coordinate,*

   so that in total we duly have $\pi_1 \circ e_{G(a)} \circ \iota_1 : H \to H$.
3. *The closure of $a_{|D(a^*a)}$ is a (in other words, $D(a^*a)$ is a core for a).*

Proof. 1. Part 2 of the proof of Theorem B.152 yields positivity and hence self-adjointness of $(1_H+a^*a)^{-1}$. The claim now follows from the (easily established) fact that the inverse of an invertible self-adjoint operator is self-adjoint, too.
2. The reasoning following (B.521) - (B.522) yields $\pi_1 e_{G(a)}\iota_1(\psi) = \varphi$, where $\varphi = (1_H+a^*a)^{-1}\psi$ by (B.523). Hence

$$(1_H+a^*a)\pi_1 e_{G(a)}\iota_1 = 1_H; \tag{B.544}$$

$$\pi_1 e_{G(a)}\iota_1(1_H+a^*a) = 1_H. \tag{B.545}$$

3. This is a consequence of the fact that $\text{ran}(1_H+a^*a) = H$, cf. part 2 of the above proof, too. Indeed, we need to show that the graph of the restriction

$$G(a_{|D(a^*a)}) = \{(\psi, a\psi), \psi \in D(a^*a)\} \tag{B.546}$$

is dense in the grapg $G(a) = \{(\psi, a\psi), \psi \in D(a)\}$ within $H \oplus H$. In other words, if $\Psi \in G(a)$ satisfies $\langle \Phi, \Psi \rangle_{H\oplus H} = 0$ for each $\Phi \in G(a_{|D(a^*a)})$, then $\Psi = 0$. With $\Psi = (\psi, a\psi)$ and $\Phi = (\varphi, a\varphi)$, where $\psi \in D(a)$ and $\varphi \in D(a^*a)$, we obtain $\langle \Phi, \Psi \rangle_{H\oplus H} = \langle (1_H + a^*a)\varphi, \psi \rangle_H$, which indeed vanishes for each $\varphi \in D(a^*a)$ iff $\psi = 0$. □

To get a feeling for the constructions to follow, we first look at the bounded case.

Proposition B.154. *If* $a = a^*$ *is bounded and* b *is given by* (B.505), *then*

$$C^*(a) = C^*(b). \tag{B.547}$$

Furthermore, $\sigma(a) \subset \mathbb{R}$ *and* $\sigma(b) \subset (-1,1)$ *(both included as compact subsets) are homeomorphic via the maps* (B.503) - (B.504), *preserving eigenvalues, that is,*

$$\sigma(a) = \{\mu(1-\mu^2)^{-1/2} \mid \mu \in \sigma(b)\}; \tag{B.548}$$
$$\sigma(b) = \{\lambda(1+\lambda^2)^{-1/2} \mid \lambda \in \sigma(a)\}; \tag{B.549}$$
$$\sigma_p(a) = \{\mu(1-\mu^2)^{-1/2} \mid \mu \in \sigma_p(b)\}; \tag{B.550}$$
$$\sigma_p(b) = \{\lambda(1+\lambda^2)^{-1/2} \mid \lambda \in \sigma_p(a)\}. \tag{B.551}$$

Proof. By Theorem B.84 and Theorem B.93, $\sigma(a) \subset \mathbb{R}$ and $\sigma(b) \subseteq [-1,1]$ are compact. We now show that in fact $\sigma(b) \subset (-1,1)$; in particular, $\pm 1 \notin \sigma(b)$. For if $\pm 1 \in \sigma(b)$, then $b \mp 1$ is not invertible, so that, given that $\sqrt{1_H + a^2}$ is invertible, by (B.505) the operator $\sqrt{1_H + a^2} \pm a$ is not invertible. But since the function

$$f_{\pm}(x) = \sqrt{1+x^2} \pm x \tag{B.552}$$

is strictly positive on any compact subset of $\mathbb{R}$, and

$$\sqrt{1_H + a^2} \pm a = f_{\pm}(a), \tag{B.553}$$

the operator in question *is* invertible, with inverse $f_{\pm}(a)^{-1} = (1/f_{\pm})(a)$. Contradiction. Having thus localized $\sigma(b)$, it follows that y^{-1} in (B.504) is continuous on $\sigma(b)$, so that, with $a = y^{-1}(b)$, we have $a \in C^*(b)$ and hence $C^*(a) \subseteq C^*(b)$. Similarly, $b = y(a)$ and hence $C^*(b) \subseteq C^*(a)$, whence (B.547).

Eqs. (B.550) - (B.551) for follows from the explicit construction of the square root in the proof of Lemma B.141: if $c\psi = \lambda\psi$, then $\sqrt{c}\psi = \sqrt{\lambda}\psi$. Likewise (more trivially), if c is invertible (whence $\lambda \neq 0$), then $c^{-1}\psi = \lambda^{-1}\psi$. The same result for the full spectra follows either from the spectral mapping property (C.53), or from the following direct argument. Given (B.547), Theorem B.94 yields an isomorphism $C(\sigma(a)) \cong C(\sigma(b))$ of commutative C*-algebras, since we have

$$C(\sigma(a)) \xrightarrow[\cong]{f \mapsto f(a)} C^*(a) = C^*(b) \xleftarrow[\cong]{g(b) \leftarrow g} C(\sigma(b)). \tag{B.554}$$

Eqs. (B.548) - (B.549) then follow from the identities

$$f(a) = (f \circ y^{-1})(b), \; f \in C(\sigma(a)); \tag{B.555}$$
$$g(b) = (g \circ y)(a), \; g \in C(\sigma(b)), \tag{B.556}$$

which in turn follow from Theorem B.94. □

Now suppose a is unbounded. In that case, its bounded transform b remains bounded, but its spectrum contains at least one of the points ± 1. We abbreviate

$$\tilde{\sigma}(b) = \sigma(b) \cap (-1,1). \tag{B.557}$$

Proposition B.155. *If a and b are as in Theorem B.152, their spectra are related by*

$$\sigma(a) = \{\mu(1-\mu^2)^{-1/2} \mid \mu \in \tilde{\sigma}(b)\}; \tag{B.558}$$

$$\sigma(b) = \{\lambda(1+\lambda^2)^{-1/2} \mid \lambda \in \sigma(a)\}^-. \tag{B.559}$$

$$\sigma_p(a) = \{\mu(1-\mu^2)^{-1/2} \mid \mu \in \tilde{\sigma}_p(b)\}; \tag{B.560}$$

$$\sigma_p(b) = \{\lambda(1+\lambda^2)^{-1/2} \mid \lambda \in \sigma_p(a)\}. \tag{B.561}$$

If a is bounded this duly reduces to (and reproves) eqs. (B.548) - (B.551), since $\sigma(b) \cap (-1,1) = \sigma(b)$, and the right-hand side of (B.559) is already closed in $\mathbb{R}$.

Lemma B.156. *Let $a = a^* \in B(H)$. Then the spectrum $\sigma(a)$ according to Definition B.80 coincides with the set $\sigma(a)$ in Definition B.81, where $A = C^*(a)$.*

Proof. We must show that if $(a-\lambda)^{-1}$ exists in $B(H)$, then its exists in $C^*(a)$ (in the double sense that $(a-\lambda)^{-1}$ lies in $C^*(a)$ and is the inverse of $(a-\lambda)$ in $C^*(a)$); the converse is trivial. Using Theorem B.94 as well as the obvious invariance of the spectrum (as in Definition B.81) under isomorphism, we might as well show that if $(a-\lambda)^{-1}$ exists in $B(H)$, then the function $(\mathrm{id}_{\sigma(a)} - \lambda)^{-1}$ exists in $C(\sigma(a))$. This is the case, since, by definition of $\sigma(a)$, the antecedent holds iff $\lambda \notin \sigma(a)$. □

We apply this lemma with $a \rightsquigarrow b$ in order to prove Proposition B.155.

Proof. We know from (B.516) that $\sqrt{1_H - b^2} : H \to D(a)$ is a bijection. If $\lambda \in \rho(a)$, then both maps in the following diagram are bijections:

$$H \overset{\sqrt{1_H-b^2}}{\longrightarrow} D(a) \overset{a-\lambda}{\longrightarrow} H, \tag{B.562}$$

and this is the case iff $(a-\lambda) \circ \sqrt{1_H - b^2}$ is invertible, which, using (B.505), is true iff $b - \lambda\sqrt{1_H - b^2}$ is invertible. Hence $\lambda \in \sigma(a)$ iff $b - \lambda\sqrt{1_H - b^2} \in C^*(b)$ is not invertible in $B(H)$, or, equivalently, in $C^*(b)$. Define $g_\lambda(y) = y - \lambda\sqrt{1_H - y^2}$ in $C(\sigma(b))$, so that $g_\lambda(b) = b - \lambda\sqrt{1_H - b^2}$. Theorem B.94 (again with $a \rightsquigarrow b$) then implies that $\lambda \in \sigma(a)$ iff g_λ is not invertible in $C(\sigma(b))$, which according to (B.253) (with $\lambda = 0$) is true iff $0 \in \mathrm{ran}(g_\lambda)$. Since $g_\lambda(\pm 1) = \pm 1 \neq 0$, even if $\pm 1 \in \sigma(b)$, these values play no role, so that $0 \in \mathrm{ran}(g_\lambda)$ iff $\lambda = \mu(1-\mu^2)^{-\frac{1}{2}}$ for some $\mu \in \sigma(b) \cap (-1,1)$. This yields (B.558) for $\sigma(a)$ and $\sigma(b)$.

The claimed refinement to the point spectrum follows as in the proof of Proposition B.154. The same argument shows that any $\mu \in \sigma(b) \cap (-1,1)$ must come from $\lambda \in \sigma(a)$, and since $\sigma(b)$ must be a closed subset of $[-1,1]$, this gives (B.559). □

As an illustration, take a to be the position operator on $H = L^2(\mathbb{R})$, so that $b = m_f$ with $f(x) = x/\sqrt{1+x^2}$. Eq. (B.276) then gives $\sigma(a) = \mathbb{R}$ and $\sigma(b) = [-1,1]$.

If a is bounded, there are only two (commutative) C*-algebras to be concerned with in a spectral theorem à la Theorem B.94, viz. $C(\sigma(a))$ and $C^*(a)$. In the unbounded case, where $\sigma(a) \subseteq \mathbb{R}$ is no longer compact, already no fewer than four algebras of continuous functions are associated with the spectrum, namely (cf. §B.3):

- the set $C_c(\sigma(a))$ of all continuous functions $f : \sigma(a) \to \mathbb{C}$ *with compact support*;
- the set $C_0(\sigma(a))$ of all continuous functions $f : \sigma(a) \to \mathbb{C}$ *that vanish at infinity*;
- the set $C_b(\sigma(a))$ of all *bounded* continuous functions $f : \sigma(a) \to \mathbb{C}$;
- the set $C(\sigma(a))$ of *all* continuous functions $f : \sigma(a) \to \mathbb{C}$.

Of these, the second and the third are commutative C*-algebras in the supremum-norm; the first fails to be closed in this norm, whereas the last does not carry it (as it would be infinite on any unbounded function). We have the obvious inclusions

$$C_c(\sigma(a)) \subset C_0(\sigma(a)) \subset C_b(\sigma(a)) \subset C(\sigma(a)). \tag{B.563}$$

Each of these plays a role in spectral theory (as do measurable versions of them). On the side of the bounded operator b, on top of $C(\sigma(b))$, we have analogous function algebras, this time with inclusions

$$C_c(\tilde{\sigma}(b)) \subset C_0(\tilde{\sigma}(b)) \subset C(\sigma(b)) \subset C_b(\tilde{\sigma}(b)) \subset C(\tilde{\sigma}(b)), \tag{B.564}$$

since $C(\sigma(b))$ consists of all functions g in $C_b(\tilde{\sigma}(b))$ for which $\lim_{y\to\pm 1} g(y)$ exists, which limit is equal to zero iff $g \in C_0(\tilde{\sigma}(b))$. Since $y^{-1} : (-1,1) \to \mathbb{R}$ in (B.504) restricts to a homeomorphism $\tilde{\sigma}(b) \to \sigma(a)$ because of (B.558), the map

$$C_{\bullet}(\sigma(a)) \stackrel{\cong}{\to} C_{\bullet}(\tilde{\sigma}(b)),\ f \mapsto f \circ y^{-1}, \tag{B.565}$$

is an isomorphism for $\bullet = c, 0, b$, or blank (which is isometric for 0 and b). If $f \in C_0(\sigma(a))$, as in (B.555) (but no longer assuming a to be bounded), we may define

$$f(a) = (f \circ y^{-1})(b), \tag{B.566}$$

since $f \circ y^{-1} \in C_0(\tilde{\sigma}(b))$, and in view of (B.564), the right-hand side is defined by the continuous functional calculus for b, i.e., $g \mapsto g(b)$, where $g \in C(\sigma(b))$; the same is then true for $f \in C_c(\sigma(a))$. Let the (typically non-unital) *-algebras

$$C_c^*(b) = \{g(b) \mid g \in C_c(\tilde{\sigma}(b))\}; \tag{B.567}$$
$$C_0^*(b) = \{g(b) \mid g \in C_0(\tilde{\sigma}(b))\}, \tag{B.568}$$

be the pertinent images under this calculus. In view of (B.568), we then have

$$C_c^*(b) \subset C_0^*(b) \subset C^*(b) \subset M(C_0^*(b)) \subset M(C_c^*(b)), \tag{B.569}$$

where $M(C_0^*(b))$ and $M(C_c^*(b))$ are the multiplier algebras of $C_0^*(b)$ and $C_c^*(b)$, respectively, cf. §C.10. Note that $M(C_0^*(b))$ is a C*-algebra contained in $B(H)$, whereas $M(C_c^*(b))$ consists (partly) of unbounded operators (see below).

Lemma B.157. *The (finite) linear span $C_c^*(b)H$ of all vectors of the form $g(b)\psi$, where $g \in C_c(\tilde{\sigma}(b))$ and $\psi \in H$, is dense in H, i.e., $C_c^*(b)H^- = H$.*

This would be trivial for $C^*(b)H$, since unlike $C_c^*(b)H$ it contains the unit 1_H.

Proof. Approximate $1_{\tilde{\sigma}(b)}$ pointwise by some monotone increasing bounded sequence (f_n) with compact support, cf. Lemma B.97; for example, define

$$f_n : (-1,1) \to \mathbb{R}; \tag{B.570}$$
$$f_n(x) = 0 \ \ (x \in (-1,-1+1/n], x \in [1-1/n,1)); \tag{B.571}$$
$$f_n(x) = 1 \ \ (x \in [-1+2/n, 1-2/n]), \tag{B.572}$$

and linear interpolation elsewhere. As in (B.317), we then have $f_n(b) \to 1_H$ strongly. By definition of $C_c^*(b)$, this yields the claim. □

Theorem B.158. *Let a be a (possibly unbounded) self-adjoint operator on H.*

1. *For any $f \in C_b(\sigma(a))$, the operator $f(a)_0$, initially defined by linear extension of*

$$f(a)_0 h(a)\psi = (fh)(a)\psi = ((fh)\circ y^{-1})(b)\psi, \tag{B.573}$$

i.e., defined on the domain $C_0^*(b)H^-$ (cf. (B.565) with $\bullet = 0$), is bounded, with

$$\|f(a)\| \leq \|f\|_\infty, \tag{B.574}$$

and hence extends from $C_0^*(b)H$ to all of H by continuity; we write

$$f(a) = f(a)_0^-. \tag{B.575}$$

2. *The functional calculus $f \mapsto f(a)$ from $C_b(\sigma(a))$ to $B(H)$ thus established satisfies the algebraic rules* (B.289) - (B.291), *and one has the reassuring cases*

$$1_{\sigma(a)}(a) = 1_H. \tag{B.576}$$
$$\frac{1}{\mathrm{id}_{\sigma(a)} - z}(a) = (a-z)^{-1} \ \ (z \in \rho(a)). \tag{B.577}$$

Conceptually, what is going on here is that the homomorphism

$$C_0(\sigma(a)) \to B(H); \tag{B.578}$$
$$f \mapsto f(a), \tag{B.579}$$

as defined in (B.566), is extended to the multiplier algebra

$$M(C_0(\sigma(a))) = C_b(\sigma(a)). \tag{B.580}$$

Theorem C.77 then applies, since by Lemma B.157 the initial homomorphism is nondegenerate, immediately yielding boundedness of $f(a)$. Below we will also give an independent proof of (B.574).

Proof. The operator $f(a)_0$ is densely defined by Lemma B.157 (which *a fortiori* implies that $C_0^*(b)H$ is dense in H). To prove that $f(a)_0$ is bounded, take $\varepsilon > 0$ and hence find a compact subset $K \subset \mathbb{R}$ such that $|f(x)h(x)| < \varepsilon$ whenever $x \notin K$. Writing $\tilde{f} = f \circ y^{-1}$ etc., using (B.322) with $f \rightsquigarrow 1_{K^c}fh$ we obtain

$$\|(\widetilde{1_{K^c}fh})(b)\psi\| \leq \|(\widetilde{1_{K^c}fh})(b)\|\|\psi\| \leq \|\widetilde{1_{K^c}fh}\|_\infty\|\psi\| < \varepsilon\|\psi\|. \tag{B.581}$$

From this, using also the homomorphism property in Theorem B.102, we then find

$$\begin{aligned}\|(fh)(a)\psi\| &= \|(\widetilde{fh})(b)\psi\| \\ &= \|(\widetilde{1_K fh})(b) + (\widetilde{fh} - \widetilde{1_K fh})(b)\psi\| \\ &\leq \|(\widetilde{1_K fh})(b)\psi\| + \|(\widetilde{1_{K^c}fh})(b)\psi\| \\ &= \|(\widetilde{1_K f})(b)\tilde{h}(b)\psi\| + \|(\widetilde{1_{K^c}fh})(b)\psi\| \\ &< \|(\widetilde{1_K f})\|_\infty\|h(a)\psi\| + \varepsilon\|\psi\|, \\ &\leq \|f\|_\infty\,\|h(a)\psi\| + \varepsilon\|\psi\|,\end{aligned} \tag{B.582}$$

since

$$\|(\widetilde{1_K f})\|_\infty \leq \|\tilde{f}\|_\infty = \|f\|_\infty. \tag{B.583}$$

Since the last expression in (B.582) is independent of K, we may let $\varepsilon \to 0$, obtaining boundedness of $f(a)$ as well as (B.574).

The second claim should be obvious from (B.566) and Theorem B.94.

Eq. (B.576) is trivial. To prove (B.577), write $f(x) = (x-z)^{-1}$, where $z \in \rho(a)$ is fixed and $x \in \sigma(a)$. We have

$$f(a)_0 h(a)\psi = (fh)(a)\psi = (a-z)^{-1}h(a)\psi, \tag{B.584}$$

and hence

$$f(a)_0\varphi = (a-z)^{-1}\varphi, \tag{B.585}$$

for any $\varphi \in D(f(a)_0) = C_0^*(b)H$. So if $\varphi_n \to \varphi$ for $\varphi \in H$ and $\varphi_n \in D(f(a)_0)$, boundedness and hence continuity of the operator $(a-z)^{-1}$ implies

$$f(a)\varphi = \lim_{n\to\infty} f(a)_0\varphi_n = \lim_{n\to\infty}(a-z)^{-1}\varphi_n = (a-z)^{-1}\varphi. \qquad \square$$

To construct a (typically unbounded) operator $f(a)$ for $f \in C(\sigma(a))$ in this fashion (think of a itself, corresponding to $f = \mathrm{id}_{\sigma(a)}$), we first define

$$D(f(a)_0) = C_c^*(b)H = \mathrm{span}\{h(a)\psi \mid h \in C_c(\sigma(a)), \psi \in H\}, \tag{B.586}$$

and an operator $f(a)_0 : D(f(a)_0) \to H$ may once again be defined by (B.573); once again, the whole point is that although f may well be unbounded, h and hence fh lie in $C_c(\sigma(a))$, so that $(fh)(a)$ is defined by (B.566), and hence eventually by the continuous functional calculus for the *bounded* self-adjoint operator b.

As in the remark following Theorem B.158, from the point of view of multiplier algebras, eq. (B.573) extends the (nondegenerate) homomorphism $C_c(\sigma(a)) \to B(H)$ to the algebra $C(\sigma(a))$ of unbounded multipliers on $C_c(\sigma(a))$.

This is not the end of the construction, since $f(a)_0$ is typically not closed on the domain (B.586). However, it is a very near miss, since $f(a)_0$ is *closable*, cf. §B.13. To prove that the operator $f(a)_0$ in B.573 is closable, we use the second criterion in Lemma B.74. For $g,h \in C_c(\sigma(a))$ and $\psi, \varphi \in H$ we may compute:

$$\langle g(a)\varphi, f(a)_0 f(a)\psi\rangle = \langle \varphi, g(a)^* f(a)_0 h(a)\psi\rangle = \langle \varphi, (g^* fh)(a)\psi\rangle; \qquad \text{(B.587)}$$
$$\langle (gf^*)(a)\varphi, h(a)\psi\rangle = \langle \varphi, (gf^*)(a)^* g(a)\psi\rangle = \langle \varphi, (g^* fh)(a)\psi\rangle. \qquad \text{(B.588)}$$

Hence $D(f(a)_0^*)$ must contain $D(f(a)_0)$, and on the latter we may put

$$f(a)_0^* g(a)\varphi = (gf^*)(a)\varphi, \qquad \text{(B.589)}$$

as in (B.573). In particular, $D(f(a)_0^*)$ is dense in H, so that $f(a)_0$ is closable. Furthermore, if $f^* = f$, then $f(a)_0$ is symmetric, i.e., $f(a)_0 \subset f(a)_0^*$. Hence the closure

$$f(a) = f(a)_0^- : D(f(a)) \to H, \qquad \text{(B.590)}$$

is the operator we are looking for, where $D(f(a))$ consists of all $\psi \in H$ for which there exists a sequence (ψ_n) in $D(f(a)_0)$ such that $\psi_n \to \psi$ *and* $f(a)_0\psi_n$ converges, upon which Lemma B.74 gives

$$f(a)\psi = \lim_n f(a)_0\psi_n. \qquad \text{(B.591)}$$

What's more, if $f^* = f$, then $f(a)_0$ is *essentially self-adjoint*, i.e.,

$$f(a)_0^- = f(a)_0^*, \qquad \text{(B.592)}$$

which (by taking the adjoint) is equivalent to the property we will actually prove:

$$f(a)^* = f(a). \qquad \text{(B.593)}$$

Theorem B.159. *For real-valued $f \in C(\sigma(a))$, the operator $f(a)$ is self-adjoint.*

The proof of self-adjointness relies on ***Nelson's Lemma***:

Lemma B.160. *Let $c \subset c^*$ be densely defined and symmetric. Then c is essentially self-adjoint if there exists a continuous unitary representation$t \mapsto u_t$ of $\mathbb{R}$ on H such that $u_t : D(c) \to D(c)$ for each $t \in \mathbb{R}$, and*

$$\frac{du_t}{dt}\psi \equiv \lim_{s\to 0} \frac{u_{t+s}\psi - u_t\psi}{s} = icu_t\psi, \ \psi \in D(c). \qquad \text{(B.594)}$$

This lemma is closely related to Stone's Theorem; see Theorem 5.73 in §5.12.

Proof. The proof of Nelson's lemma relies on the following variation of Lemma 5.74 in §5.12, proved by applying the latter (or rather its proof) to the closure of a:

Lemma B.161. *Let a be symmetric. Then a is essentially self-adjoint ($a^{**} = a^*$) iff*

$$\mathrm{ran}(a+i)^- = \mathrm{ran}(a-i)^- = H. \tag{B.595}$$

Applying Lemma B.161 in the same way as Lemma 5.74 is used in the proof of self-adjointness of the generator a in Theorem 5.73, yields Lemma B.160.

For Theorem B.159, with $c = f(a)_0$ for some $f \in C(\sigma(a), \mathbb{R})$, informally define

$$u_t = \exp(itf(a)), \tag{B.596}$$

and formally define u_t as the closure of the bounded operator

$$(u_t)_0 = e_0^t(a) \tag{B.597}$$

defined by the bounded function $e^t(x) = \exp(itf(x))$ on $\sigma(T)$, cf. (B.573). The verification that $t \mapsto u_t$ defines a continuous one-parameter group of unitary operators on H is practically the same as in our proof of part 1 of Stone's Theorem, and the proof of (B.594) is almost the same as a similar step in the proof of part 3 of that theorem, so we will not repeat these here. Therefore, Lemma B.160 applies, showing that $f(a)_0$ is essentially self-adjoint. □

As an important special case of our continuous functional calculus, we have

$$\mathrm{id}_{\sigma(a)}(a) = a, \tag{B.598}$$

just as in the bounded case. Writing a_0 for the operator $(\mathrm{id}_{\sigma(a)})_0(a)$, eq. (B.573) gives $a_0\varphi = a\varphi$ for $\varphi \in D(a_0)$, cf. (B.586). Let $\psi \in D(a_0^-)$, so that there is a sequence (ψ_n) in $D(a_0)$ such that $\psi_n \to \psi$ and $(a_0\psi_n)$ converges. Since a is closed, it follows that $a_0\psi_n = a\psi_n \to a\psi$, so that $\psi \in D(a)$. Hence $a_0^- \subseteq a$. Since both operators are self-adjoint, this implies $a_0^- = a$, which proves (B.598). The proof of (B.577) is similar but easier, since $(a-z)^{-1}$ is bounded.

In similar vein, we may set up a functional calculus for bounded Borel functions of a. If $f \in \mathscr{B}(\sigma(a))$, then $f \circ y^{-1} \in \mathscr{B}(\sigma(b))$, so that $(f \circ y^{-1})(b)$ is defined, cf. Theorem B.102, and we may define $f(a)$ by (B.566). As in the continuous case, this map $f \mapsto f(a)$ yields a homomorphism $\mathscr{B}(\sigma(a)) \to B(H)$, satisfying (B.322).

What is still missing, however, is the von Neumann algebra $W^*(a)$ in which this homomorphism takes values. To close this section, we solve this issue.

If $c \in B(H)$ and a is possibly unbounded, we say that (by convention):

$$[a,c] = 0 \text{ iff } ca \subseteq ac, \tag{B.599}$$

that is, if $c \cdot D(a) \subseteq D(a)$ and $ca\psi = ac\psi$ for each $\psi \in D(a)$. We write $\{a\}'$ for the set of all $c \in B(H)$ that commute with a. If $a^* = a$, looking at the graph of a (and using the fact that a is closed), it is easy to see that $\{a\}'$ is a strongly closed unital *-subalgebra in $B(H)$. Therefore, by the bicommutant theorem, $\{a\}'$ is a von Neumann algebra. Its commutant $W^*(a)$, defined in the usual sense (B.318), i.e.,

$$W^*(a) = \{a\}''. \tag{B.600}$$

Theorem B.162. *Let a be a (possibly unbounded) self-adjoint operator on H. Then*

$$W^*(a) = W^*(b), \tag{B.601}$$

where b is the bounded transform (B.510) *of a. Consequently, if $f \in \mathscr{B}(\sigma(a))$ and the operator $f(a)$ is defined by* (B.566) *and Theorem B.102, then $f(a) \in W^*(b)$.*

Proof. We will prove a more general result of independent interest.

Definition B.163. *A closed unbounded operator $a : D(a) \to H$ is* **affiliated** *to a von Neumann algebra $A \subset B(H)$, written $a\eta A$, iff $[a,c] = 0$ for each $c \in A'$.*

For example, if $a^* = a$, then $a\eta W^*(a)$, and if $a\eta B$ for some $B = B''$, then $W^*(a) \subseteq B$.

Proposition B.164. *Let $A \subset B(H)$ be a von Neumann algebra and assume a is a self-adjoint operator on H with bounded transform b. Then $a\eta A$ iff $b \in A$.*

Proof. The first step consists in the observation that $a\eta A$ iff $[a,u] = 0$ (or, equivalently, $uau^* = a$) merely for each unitary $u \subset A'$. To see this, we strengthen Lemma B.145 (in which we replace a by c): if $c \in A'$, then c is a linear combination of at most four unitaries *in* A'. Indeed, the unitaries $u_\pm$ in the proof are constructed via the continuous functional calculus of Theorem B.94, and hence they lie in $C^*(c) \subset A'$.

The second step is to show that $[a,u] = 0$ iff $[b,u] = 0$ for any unitary u. This is a simple computation: if $uau^* = a$, then, looking at the domains in question,

$$u(1_H + a^2)^{-1}u^* = (1_H + a^2)^{-1}; \tag{B.602}$$
$$u((1_H + a^2)^{-1})^{1/2}u^* = ((1_H + a^2)^{-1})^{1/2}, \tag{B.603}$$

from which $ubu^* = b$ with b defined by (B.510). Similarly, if $bu = ub$, then $uau^* = a$, where a is defined by (B.511). Theorem B.152 therefore yields the claim. □

Theorem B.162 now follows: taking $A = W^*(a)$, so that $a\eta A$, yields $b \in W^*(a)$, and hence $W^*(b) \subseteq W^*(a)$. On the other hand, taking $A = W^*(b)$, in which case $b \in A$, gives $a\eta W^*(b)$, and hence $W^*(a) \subseteq W^*(b)$. This yields (B.601), from which the final claim follows by our definition (B.566) and Theorem B.102. □

Using this language, it can be shown that for possibly unbounded Borel functions f on $\sigma(a)$, the possibly unbounded operator $f(a)$ is affiliated to $W^*(a)$. Furthermore, there exists a Borel measure μ on $\sigma(a)$ such that the map $f \mapsto f(a)$ may also be seen as a so-called ***essential*** homomorphism from $\mathscr{B}(\sigma(a))/\mathscr{N}(\sigma(a))$ into the *-algebra of normal operators affiliated with $W^*(a)$, where $\mathscr{N}(\sigma(a))$ is the set of μ-null functions on $\sigma(a)$; this means that the algebraic properties hold after closure.

Notes

The history of functional analysis is described from various points of view by Bernkopf (1966, 1967), Birkhoff & Kreyszig (1984), Brezis & Browder (1998), Dieudonné (1981), Monna (1973), Pier (2001), Pietsch (2007), Siegmund-Schultze (2003), and Steen (1973). Apart from von Neumann (1932), the other founding books of functional analysis—coincidentally from the same year, which closed the foundational era that began around 1900—were Banach (1932) and Stone (1932).

The concept of a Hilbert space eventually emerged from Hilbert's work on quadratic forms in infinitely many variables (see especially his fourth paper on the subject, Hilbert, 1906), which in turn was inspired by his analysis of integral equations (Hilbert, 1912). From a modern point of view, Hilbert's space was the unit ball in $\ell^2(\mathbb{N})$; he did not adopt the perspective of linear spaces and operators.

An important step towards this perspective was what is now called the Riesz–Fischer Theorem from 1907; Riesz (1907a) proved the isomorphism

$$L^2([a,b]) \cong \ell^2(\mathbb{N}), \tag{B.604}$$

whereas Fischer (1907) proved the completeness of $L^2([a,b])$ and obtained Riesz's isomorphism as a corollary. Riesz (1907b) also obtained the the Riesz–Fréchet Theorem for the special case $L^2([a,b])$), independently found also by Fréchet (1907). In fact, Hilbert (1906) had already shown this (*mutatis mutandis*) for what we now call $\ell^2(\mathbb{N})$; the general case had to wait for Riesz (1934) and Löwig (1934). The latter was the first to study non-separable Hilbert spaces, including Corollary B.64. Both Riesz and Fréchet in addition played major roles in establishing another famous duality theorem, namely the one on the representation of linear functionals on continuous functions by measures (cf. Theorem B.15 etc.); see Gray (1984).

Subsequently, Schmidt (1908) developed the linear and geometric structure of $\ell^2(\mathbb{N})$, arguably the first Hilbert space studied as such, and Riesz (1913) explicitly studied linear operators on this space. Finally, it was von Neumann (1927ab, 1932) who first introduced Hilbert space and operator theory from an abstract point of view, i.e., axiomatically. For a historical analysis of this step, which was triggered by the attempts of von Neumann (originally jointly with Hilbert and Nordheim) to provide a mathematical foundation for quantum mechanics), see Rédei (2005) and Duncan & Janssen (2013); also cf. Corry (2004) on the role of Hilbert himself.

Functional analysis textbooks perused by the author include Conway (2007), Dudley (1989), Kadison & Ringrose (1983), Maurin (1972), Reed & Simon (1972), Rudin (1973), Schmüdgen (2012), and Weidmann (2000). A good place to start for contemporary beginners is Rynne & Youngson (2008), followed by the more advanced text by MacCluer (2009), which also introduces C*-algebras. A natural next step would then be Pedersen (1989), and on to operator algebras!

Since most of the material in this appendix is standard except for the last three sections, it seems pointless to give detailed notes and attributions (so that several section even lack notes), except for a few comments on unusual cases, and some supplementary material which would have distracted too much from the main text.

§B.2. ℓ^p **spaces**

Hölder's Inequality (which incorporates the claim $fg \in \ell^1$) should be clear for $p = 1$ or $p = \infty$. For $1 < p < \infty$, we use the fact that for any $s,t \in [0,\infty)$, one has

$$s^{1/p}t^{1/q} \leq \frac{s}{p} + \frac{t}{q}. \tag{B.605}$$

Using (B.605) with $s = (|f(x)|/\|f\|_p)^p$ and $t = (|g(x)|/\|g\|_q)^q$ and summing over x gives (B.15). To derive Minkowski's Inequality for $1 < p < \infty$ (the cases $p = 1$ and $p = \infty$ are obvious), define

$$h(x) = |f(x) + g(x)|^{p-1}. \tag{B.606}$$

Arguing as in part 1 above, if $f \in \ell^p$ and $g \in \ell^p$, then $f + g \in \ell^p$ and hence $h \in \ell^q$, since $h(x)^q = |h(x)|^q = |f(x) + g(x)|^p$. Now compute

$$\begin{aligned} \|f+g\|_p^p &= \sum_x |f(x)+g(x)|^p = \sum_x h(x)|f(x)+g(x)| \\ &\leq \sum_x |h(x)f(x)| + \sum_x |h(x)g(x)| = \|fh\|_1 + \|gh\|_1 \\ &\leq \|h\|_q(\|f\|_p + \|g\|_p) = \|f+g\|_p^{p-1}(\|f\|_p + \|g\|_p), \end{aligned} \tag{B.607}$$

where in the last inequality we have used (B.15). This immediately gives (B.14).

§B.4. **Basic measure theory** Standard textbooks on measure theory include Bogachev (2006), Dudley (1989), Malliavin (1995), Rudin (1986), etc.

§B.5. **Measure theory on locally compact Hausdorff spaces**

Urysohn's Lemma states that if X is a locally compact Hausdorff space and $K \subset U \subset X$ with K compact and U open, then there is a function $g \in C_c(U)$ such that $0 \leq g(x) \leq 1$ for each $x \in X$ and $g(x) = 1$ for $x \in K$. Similarly, since a locally compact Hausdorff space is completely regular, for each closed set $F \subset X$ and point $x \notin F$ there is a continuous function such that $f(x) = 0$ and $f_{|F} = 0$.

An example of a space that is locally compact Hausdorff but not σ-compact, given by Rudin ((1986), is $X = \mathbb{R}^2$ with topology given by the strange metric $d((x,y),(x',y')) = 1 + |y - y'|$ if $x \neq x'$ and $d((x,y),(x,y')) = |y - y'|$.

For a (tedious) direct proof of Theorem B.19, see Rudin (1986), Thm. 2.14. Alternatively, Theorem B.19 may be derived from Choquet theory, as mentioned in the main text, or from the Daniell–Stone construction of measures from positive functionals in a more general setting, see e.g. Bogachev, 2007, §7.8 or Dudley, 1989, §4.5. For a proof of Theorem B.22 see Malliavin (1995), Thm. 5.3.8.

The theory of finitely additive measures is exhaustively discussed in Rao & Rao (1983); for a summary see Luxemburg (1991). The notion of a semiring of subsets of X goes back to von Neumann (1950). See also Loya (2008), including a detailed proof that $\mathrm{Step}(X,\mathscr{R})$ is a (commutative) algebra.

§B.6. L^p **spaces**

An nice result "taming" $L^p(X,\Sigma,\mu)$ is ***Lusin's Theorem***, assuming μ is regular:

Theorem B.165. *Let $1 \leq p < \infty$. If the support of $f \in L^p(X)$ has finite measure, then for any $\varepsilon > 0$ there exists $g \in C_c(X)$ such that $\mu(\{x \in X \mid f(x) \neq g(x)\}) < \varepsilon$.*

§B.7. **Morphisms and isomorphisms of Banach spaces**

The ***Baire Category Theorem*** states that a *complete* metric space cannot be a countable union of nowhere dense sets (where a set in a topological space is called ***nowhere dense*** if its closure has empty interior, i.e., does not contain a non-empty open set). In other words, if (M,d) is complete and $M = \cup_n M_n$ with each M_n closed, then there is at least one $n \in \mathbb{N}$ for which M_n contains an open ball.

§B.9. **Duality**

The idea of writing (B.136) as $\lim_U f$ has the following origin.

1. Let $f : X \to K$ be any function between any pair of sets, and let F be a filter on X. Then f_*F, which consists of all $B \subset K$ for which $f^{-1}(B) \in F$, is a filter on K, called the ***push-forward*** of F by f. Moreover, if U is an ultrafilter on X, then f_*U is an ultrafilter on K. This gives a map
$$f_* : \text{Ultra}(X) \to \text{Ultra}(K). \tag{B.608}$$
If we equip Ultra(X) with the topology generated by all sets of the form
$$U_A = \{U \in \beta X \mid A \in U\}, \tag{B.609}$$
where $A \subset X$, as in the main text, and likewise Ultra(K), then f_* is continuous If X is discrete, then Ultra$(X) = \beta X$, but not otherwise.
2. We say that some filter F on a topological case X ***converges*** to $x \in X$ if $N_x \subseteq F$, where N_x is the ***neighbourhood filter*** of x, consisting of all neighbourhoods of x. This is denoted by $\lim F = x$.
3. Combining points 1 and 2, if $\lim f_*F = z$, i.e., if $N_z \subseteq f_*F$, we write
$$\lim_F f = z \ (z \in K). \tag{B.610}$$
4. As for sequences, it can be shown that filters on Hausdorff spaces have *at most* one limit, and that ultrafilters on compact spaces have *at least* one limit. Consequently, ultrafilters on compact Hausdorff spaces K have *exactly* one limit, i.e., converge to a unique point. This gives a continuous map
$$\lim : \text{Ultra}(K) \to K. \tag{B.611}$$
5. It follows that if X is any set (seen as a discrete topological space), K is a compact Hausdorff space, $f : X \to K$ is some function, and U is an ultrafilter on X, then f_*U has a unique limit $z \in K$, written $\lim_U f = z$ or $\lim f_*U = z$, or $\beta f(U) = z$, since the latter notation gives the extension βf in the diagram (B.135). Thus $\beta f = \lim \circ f_*$, as in the diagram that combines (B.608) and (B.611), viz.
$$\beta X = \text{Ultra}(X) \stackrel{f_*}{\to} \text{Ultra}(K) \stackrel{\lim}{\to} K. \tag{B.612}$$

§B.11. **Choquet's Theorem**

Our proof of Choquet's Theorem was adapted from Simon (2011) and Ebbesen (2012). For an extensive treatment of the surrounding *Choquet Theory* see e.g. Alfsen (1970), Bratteli & Robinson (1987), or Phelps (2001). For the Schläfli classification see Coxeter (1948).

§B.12. **A précis of infinite-dimensional Hilbert space**

To prove separability of $H = L^2(\mathbb{R}^d)$, note that a dense subset is given by the set of all functions of the form $1_{B_n^d} p$, where $n \in \mathbb{N}$, $B_r^d = \{x \subset \mathbb{R}^d \mid \|x\|^2 \leq r\}$ is the d-ball of radius r, and p is some polynomial on $\mathbb{R}^d$ with rational coefficients. Alternatively, take the complex rational linear span of all functions of the form 1_A, where $A \subset \mathbb{R}^d$ is a rectangle with rational coefficients (proving density in either case requires some measure theory). The latter construction has the advantage over the former that it can be generalized to Hilbert spaces $H = L^2(X)$ for which the underlying measure space (X, Σ, μ) satisfies the condition that the space of sets $A \in \Sigma$ with $\mu(A) < \infty$ is separable in the metric $d(F,G) = \mu(F \Delta G)$, where $F \Delta G = (E \cap F^c) \cup (E^c \cap F)$ is the symmetric difference. Indeed, $L^2(X)$ is separable iff this condition is satisfied.

This class includes the important case where the underlying topological space X is ***Polish*** (i.e., homeomorphic to a complete separable metric space), Σ consists of the associated Borel sets, and μ is a σ-finite regular measure. If, furthermore, μ is finite, then Lemma B.121 (in its original form for Polish spaces) applies. As in the proof of Theorem B.118, this induces Hilbert space isomorphisms like (in the second case) $L^2(X) \cong L^2(0,1)$, which do not require a choice of basis. See Royden (1988), Thm. 15.5.16 and Prop. 15.5.12, and Halmos (1974), p. 177.

§B.14. **Basic spectral theory**

Our terminology "***continuous spectrum***" $\sigma_c(a)$ for the complement of the point spectrum $\sigma_p(a)$ is not standard; many authors reserve the former term for the complement of $\sigma_p(a)$ *as well as* the so-called ***residual spectrum*** $\sigma_r(a)$, which is defined as the set of those $\lambda \in \sigma(a)$ for which $\lambda \notin \sigma_p(a)$ and $\text{ran}(a-\lambda)^- \neq H$. However, for self-adjoint operators a (which is all we need in this book, and in quantum mechanics), it follows from e.g. Theorem B.93 that $\sigma_r(a) = \emptyset$, so that at least for $a^* = a$ "our" continuous spectrum $\sigma_c(a)$ matches with the usual terminology.

The proof of (B.258) in any Banach algebra A with unit 1_A is as follows. We first show that the sum is a Cauchy sequence. Indeed, for $n > m$ one has

$$\left\| \sum_{k=0}^{n} a^k - \sum_{k=0}^{m} a^k \right\| = \left\| \sum_{k=m+1}^{n} a^k \right\| \leq \sum_{k=m+1}^{n} \|a^k\| \leq \sum_{k=m+1}^{n} \|a\|^k. \tag{B.613}$$

For $n, m \to \infty$ this converges to 0 by the theory of the geometric series. Since A is complete, the Cauchy sequence $\sum_{k=0}^{n} a^k$ converges for $n \to \infty$. Now compute

$$\sum_{k=0}^{n} a^k (1_A - a) = \sum_{k=0}^{n} (a^k - a^{k+1}) = 1_A - a^{n+1}. \tag{B.614}$$

Hence

$$\left\| 1_A - \sum_{k=0}^{n} a^k(1_A - a) \right\| = \|a^{n+1}\| \leq \|a\|^{n+1}, \tag{B.615}$$

which converges to zero when $n \to \infty$, as $\|a\| < 1$ by assumption. Thus

$$\lim_{n\to\infty} \sum_{k=0}^{n} a^k(1_A - a) = 1_A. \tag{B.616}$$

By a similar argument,

$$\lim_{n\to\infty} (1_A - a) \sum_{k=0}^{n} a^k = 1_A, \tag{B.617}$$

so that, by continuity of multiplication in a Banach algebra, one finally has

$$\lim_{n\to\infty} \sum_{k=0}^{n} a^k = (1_A - a)^{-1}. \tag{B.618}$$

To see that the closure a^- of a closable operator a is indeed closed (!), suppose $f_n \to f$ and $af_n \to g$, with (f_n) in $D(a^-)$. Since $f_n \in D(a^-)$ for fixed n, there exists $(f_{m,n})$ in $D(a)$ such that $\lim_m f_{m,n} = f_n$ and $\lim_m af_{m,n} \equiv g_n$ exists. Then clearly

$$\lim_{m,n} f_{m,n} = f, \tag{B.619}$$

and we claim that

$$\lim_{m,n} af_{m,n} = g. \tag{B.620}$$

Namely, $\|af_{m,n} - g\| \leq \|af_{m,n} - af_n\| + \|af_n - g\|$. For $\varepsilon > 0$, take n so that the second term is $< \varepsilon/2$. For that n, the vectors $a(f_{m,n} - f_n)$ converge, as $m \to \infty$, since $af_{m,n} \to g_n$ and af_n is independent of m. Also, recall that $f_{m,n} - f_n \to 0$ as $m \to \infty$. By assumption, a is closable, hence by definition one must have $a(f_{m,n} - f_n) \to 0$ in m. Hence we may find m so that $\|af_{m,n} - af_n\| < \varepsilon/2$, so that $\|af_{m,n} - g\| < \varepsilon$, and (B.620) follows. Hence $f \in D(a^-)$. Finally, since $a^- f = \lim_{m,n} af_{m,n}$ one has $a^- f = g$ by (B.620), or $a^- f = \lim_n af_n$ by definition of g. Thus a^- is closed.

§B.15. **The spectral theorem**

By (B.319), von Neumann algebas like $W^*(a)$ are complete under strong convergence of *nets* (rather than merely *sequences*), and if some net is monotone increasing (or decreasing) and bounded, the strong limit equals the supremum (or infimum), as in Proposition B.98. This yields operatorial versions of (B.40) - (B.44):

$$e_U = \sup\{f(a) \mid f \in C_c(U), 0 \leq f \leq 1_{\sigma(a)}\}; \tag{B.621}$$
$$e_K = \inf\{f(a) \mid f \in C(\sigma(a)), 0 \leq f \leq 1_{\sigma(a)}, f_{|K} = 1_K\}; \tag{B.622}$$
$$e_A = \inf\{e_U \mid U \supseteq A, U \in \mathscr{O}(\sigma(a))\}; \tag{B.623}$$
$$= \sup\{e_K \mid K \subseteq A, K \in \mathscr{K}(\sigma(a))\}, \tag{B.624}$$

where $U \in \mathscr{O}(\sigma(a))$ is open, $K \in \mathscr{K}(\sigma(a))$ is compact, and $A \subset \sigma(a)$ is Borel.

§B.16. **Abelian *-algebras in $B(H)$**

For an alternative proof of Proposition B.106, one observes that

$$\psi \to \int_0^1 f\psi = \int_0^1 b\psi = \langle\sqrt{|\psi|}, b\psi/\sqrt{|\psi|}\rangle \tag{B.625}$$

defines a *bounded* functional on $L^2(0,1)$ *seen as a dense subspace of* $L^1(0,1)$, and use the duality $L^1(0,1)^* \cong L^\infty(0,1)$. Indeed, using Cauchy–Schwarz, one has

$$\left|\int_0^1 f\psi\right| = |\langle\sqrt{|\psi|}, b\psi/\sqrt{|\psi|}\rangle| \leq \|b\|\|\sqrt{|\psi|}\|_2\|\psi/\sqrt{|\psi|}\|_2 = \|b\|\|\psi\|_1. \tag{B.626}$$

§B.17. **Classification of maximal abelian *-algebras in $B(H)$**

Theorem B.118 goes back to von Neumann (1931); for the details of the second proof see Kadison & Ringrose (1986), §9.4, or, very lucidly, Stevens (2016).

§B.20. **The trace**

The trace is often neglected in functional analysis books, except when these tend to quantum mechanics (Reed & Simon, 1972) or to operator algebras (Pedersen, 1989). Eqs. (B.476) - (B.477) and (B.496) reflect the function space dualities

$$\ell_0(\mathbb{N})^* \cong \ell^1(\mathbb{N}); \tag{B.627}$$
$$\ell^1(\mathbb{N})^* \cong \ell^\infty(\mathbb{N}); \tag{B.628}$$
$$\ell^2(\mathbb{N})^* \cong \ell^2(\mathbb{N}). \tag{B.629}$$

Similar to the ℓ^p-spaces, one has Banach spaces $B_p(H)$ residing in $B_0(H)$ for each $1 \leq p < \infty$, called ***Schatten–von Neumann ideals***, see e.g. Simon (2005).

§B.21. **Spectral theory for unbounded self-adjoint operators**

Our approach to unbounded operators via the bounded transform combines ideas from Kaufman (1978), Woronowicz (1991), Woronowicz & Napiórkowski (1992), Schmüdgen (2012), and Koliha (2014). The proof of Theorem B.159 via Lemma B.160 (due to Nelson, 1959), was suggested to the author by Nigel Higson. The last part of §B.21 was inspired by Lemma 5.2.8 in Pedersen (1989), in which we have simply replaced the Cayley transform by the bounded transform.

The idea of affiliating closed operators to von Neumann algebra goes back to von Neumann; our brief treatment is hopefully more appealing than the elaborate constructions in Kadison & Ringrose (1983), §5.6. A number of details were supplied in the M.Sc Thesis of Christian Budde (2015); see also Budde & Landsman (2016).

For general C*-algebras A, the multiplier algebra consists of all maps $m : A \to A$ for which there exists an adjoint $n \equiv m^* : A \to A$ such that $b^*m(a) = n(b)^*a$. Such maps are automatically linear and bounded, and $M(A)$ is a C*-algebra itself as a subalgebra of the Banach space $B(A)$ of all bounded linear maps on A, enriched with the adjoint $m^* = n$. See, e.g., Lance (1995), or §C.10 below. For commutative C*-algebras this reduces to the definition in the main text, which dates from Wang (1961). For unbounded multipliers see Woronowicz (1991) and Lance (1995); Woods (1979) treats the bounded case.

Appendix C
Operator algebras

This appendix provides a short course in operator algebras, building on the previous appendix. Indeed, there is surprisingly little algebra in the subject (so that there are hardly any prerequisites in that direction), and quite a lot of functional analysis, involving both operators on Hilbert space and more general Banach space theory.

Traditionally, the field of operator algebras has had two branches: C*-algebras and von Neumann algebras. Although historically speaking the latter (invented by von Neumann in 1930) preceded the former (introduced by Gelfand and Naimark in 1943), the logical order of presentation is the opposite, since von Neumann algebras turned out to be special cases of C*-algebras (with additional structure). Furthermore, for reasons in the foundations of quantum mechanics (as explained in the main text), beside von Neumann algebras we will discuss a few lesser known special cases of C*-algebras, such as *scattered* C*-algebras and AW*-algebras.

C.1 Basic definitions and examples

A C*-algebra is both an associative algebra and a Banach space, as follows:

Definition C.1. *1. A* **Banach algebra** *is a Banach space A that is simultaneously an algebra in which*

$$\|ab\| \leq \|a\|\,\|b\| \ (a,b \in A). \tag{C.1}$$

2. An **involution** *on an algebra A is a real-linear map $* : A \to A$, written $a \mapsto a^*$, such that $a^{**} = a$, $(ab)^* = b^*a^*$, and $(\lambda a)^* = \overline{\lambda} a^*$ for all $a,b \in A$ and $\lambda \in \mathbb{C}$. An algebra with involution is also called a* ***-algebra**.

3. A **C*-algebra** *is a Banach algebra A with involution in which*

$$\|a^*a\| = \|a\|^2 \ (a \in A). \tag{C.2}$$

With the same proof as (A.22), these axioms imply

$$\|a^*\| = \|a\|. \tag{C.3}$$

K. Landsman, *Foundations of Quantum Theory*,
Fundamental Theories of Physics 188, DOI 10.1007/978-3-319-51777-3

The three main examples (at least for a first orientation) are:

- The space $C_0(X)$ of all continuous functions $f : X \to \mathbb{C}$ that vanish at infinity, where X is some locally compact Hausdorff space (see §B.3). This is an algebra under pointwise operations: addition is given by $(\lambda \cdot f + g)(x) = \lambda f(x) + g(x)$, and multiplication is $(fg)(x) = f(x)g(x)$. Furthermore, it has a natural involution $f^*(x) = \overline{f(x)}$, and a natural norm $\|f\|_\infty = \sup_{x\in X}\{|f(x)|\}$, cf. (B.23). The above axioms of a C*-algebra are easily verified. Note that $C_0(X)$ has a unit (namely the function 1_X equal to 1 for any x) iff X is compact. It is of fundamental importance for physics and mathematics that $C_0(X)$ is a *commutative* C*-algebra.
- The space $B(H)$ of all bounded operators on some Hilbert space H, with obvious algebraic operations, involution given by the adjoint (see (A.15)), and the standard operator norm $\|a\| = \sup\{\|a\psi\|, \psi \in H, \|\psi\| = 1\}$. See Proposition A.7 and Theorem B.33 for the proof that $B(H)$ is a C*-algebra; it has a unit, given by the identity 1_H. If $\dim(H) > 1$, this is a highly *non-commutative* C*-algebra.
- The space $B_0(H)$ of all *compact* operators on some Hilbert space H, with operations inherited from $B(H)$; see Theorem B.130, which not merely shows that $B_0(H)$ is a C*-algebra, but also that it is a (closed) two-sided ideal in $B(H)$. It fails to have a unit whenever H is infinite-dimensional (this follows from almost any result in §B.19, such as Theorem B.135).

Definition C.2. *1. A* **homomorphism** *between C*-algebras A and B is a linear map $\varphi : A \to B$ that for all $a, b \in A$ satisfies*

$$\varphi(ab) = \varphi(a)\varphi(b); \tag{C.4}$$

$$\varphi(a^*) = \varphi(a)^*. \tag{C.5}$$

2. An **isomorphism** *between two C*-algebras is an invertible homomorphism. If A and B are isomorphic as C*-algebras in this sense, we write $A \cong B$.*

It follows from linear algebra that the set-theoretic inverse of an invertible linear map $\varphi : A \to B$ is automatically linear. It is similarly easy to show that the inverse of an invertible homomorphism is itself a homomorphism, but it is a deeper fact about C*-algebras that an isomorphism is automatically isometric (and hence has an isometric inverse); see Theorem C.62. Furthermore, if $B = \mathbb{C}$, then the property $\varphi(a^*) = \varphi(a)^*$ follows from the other conditions on a homomorphism.

The following notion, originally inspired by quantum mechanics (and turned into mathematics by von Neumann), gives a geometric flavor to operator algebras.

Definition C.3. *A* **state** *on a C*-algebra A is a bounded linear map $\omega : A \to \mathbb{C}$ that satisfies:*

1. $\omega(a^*a) \geq 0$, $a \in A$ (**positivity**);
2. $\|\omega\| = 1$ (**normalization**).

If A has a unit, the definition of a state considerably simplifies.

Lemma C.4. *Let A be a C*-algebra with unit and let $\omega : A \to \mathbb{C}$ be a linear map. Then ω is positive iff it is bounded and satisfies $\|\omega\| = \omega(1_A)$.*

The proof requires some positivity theory in C*-algebras, so we postpone it to §C.7, but as of now, we immediately infer that in the unital case we have:

Proposition C.5. *A linear map $\omega : A \to \mathbb{C}$ on a unital C*-algebra is a state iff ω is positive and satisfies $\omega(1_A) = 1$, and hence iff ω is bounded with $\|\omega\| = \omega(1_A) = 1$.*

Using the Banach–Alaoglu Theorem B.48, this implies that the ***state space*** $S(A)$ of a unital C*-algebra A, i.e., the set of all states on A, is a compact convex subset of A^* in its w^*-topology. Defining the ***pure state space*** $P(A)$ of A as the extreme boundary $\partial_e S(A)$, the Krein–Milman Theorem B.50 almost immediately implies:

Theorem C.6. *Let A be a C*-algebra with unit, having state space $S(A)$ and pure state space $P(A) = \partial_e S(A)$. Then $P(A) \neq \emptyset$ and $S(A) = \text{co}(P(A))^-$.*

In words, C*-algebras have sufficiently many pure states to approximate general states arbitrarily well, at least in the w^*-topology (of "expectation values").

The only complication in applying Theorem B.50 to $K = S(A) \subset A^*$ is that A is a complex Banach space, but the situation may be reduced to the real Banach space

$$A_{\text{sa}} = \{a \in A \mid a^* = a\}. \tag{C.6}$$

Lemma C.7. *Let A be a C*-algebra with unit. If $\omega \in S(A)$, then $\omega(a^*) = \overline{\omega(a)}$.*

Proof. Using Definition C.3.2 and eq. (C.2), for any $a^* = a$ and $t \in \mathbb{R}$ we have

$$|\omega(a+it)|^2 \leq \|a+it\|^2 = \|(a-it)(a+it)\| = \|a^2+t^2\| \leq \|a\|^2 + t^2. \tag{C.7}$$

Writing $\omega(a) = \alpha + i\beta$, where $\alpha, \beta \in \mathbb{R}$, this gives $\alpha^2 + \beta^2 + 2\beta t \leq \|a\|^2$ for all $t \in \mathbb{R}$, which forces $\beta = 0$. This proves the claim for self-adjoint a. For the general case, one uses the following decomposition of a as a sum of two self-adjoint operators:

$$a = b + ic \ (b^* = b, c^* = c); \tag{C.8}$$
$$b = \tfrac{1}{2}(a + a^*), \ c = -\tfrac{1}{2}i(a - a^*). \tag{C.9}$$

Consequently, we may restrict a state $\omega \in S(A)$ to a real-linear functional

$$\omega_{\mathbb{R}} = \omega_{|A_{\text{sa}}} : A_{\text{sa}} \to \mathbb{R} \tag{C.10}$$

that satisfies $\omega(1_A) = 1$ and $\omega(a^2) \geq 0$ for any $a \in A_{\text{sa}}$, where we used Theorem C.52 below to reformulate the positivity condition on states in terms of self-adjoint operators alone. Conversely, we may extend a state $\omega_{\mathbb{R}}$ on A_{sa} to a state ω on A by

$$\omega(a) = \omega_{\mathbb{R}}(b) + i\omega_{\mathbb{R}}(c), \tag{C.11}$$

assuming (C.8) - (C.9). We then have $\|\omega\| = \|\omega_{\mathbb{R}}\| = 1$, since obviously $\|\omega_{\mathbb{R}}\| \leq \|\omega\| = 1$ (since its sup-norm is computed on fewer operators), but also $\omega(1_A) = 1$. Thus we may regard $S(A)$ as a compact convex set in the real Banach space A^*_{sa} rather than in the complex Banach space A^*, and Theorem B.50 applies. Alternatively, one could have extended the latter to the complex case, which is possible with a similar (lack of) effort as in the procedure above.

C.2 Gelfand isomorphism

The example $A = C_0(X)$ of a commutative C*-algebra given in the previous section is more than that; as proved in the very first (1943) paper on C*-algebras by Gelfand and Naimark (despite whom one often speaks of , it is generic.

Theorem C.8. *Every commutative C*-algebra A is isomorphic to $C_0(X)$ for some locally compact Hausdorff space X, which is unique up to homeomorphism.*

The proof is technically intricate at points, but the main idea is quite simple:

1. The space X may be taken to be the ***Gelfand spectrum*** $\Sigma(A)$ of A, i.e., the set of all nonzero linear maps $\omega : A \to \mathbb{C}$ that satisfy $\omega(ab) = \omega(a)\omega(b)$ (and hence are homomorphisms $A \to \mathbb{C}$ as algebras). For example, if A is already given as $C_0(X)$, then each $x \in X$ defines $\omega_x \in \Sigma(A)$ by $\omega_x(f) = f(x)$, which is linear multiplicative (by the pointwise definition of addition and multiplication in A).
2. The ***Gelfand transform*** maps each $a \in A$ to a function $\hat{a} : \Sigma(A) \to \mathbb{C}$ by
$$\hat{a}(\omega) = \omega(a), \quad (a \in A, \ \omega \in \Sigma(A)). \tag{C.12}$$
3. The ***Gelfand topology*** is the weakest topology on $\Sigma(A)$ making all functions $\hat{a}$ continuous (i.e., the topology generated by the sets $\hat{a}^{-1}(U)$, $U \in \mathbb{C}$ open, $a \in A$). In this topology, $\Sigma(A)$ is *compact* iff A has a unit, and *locally compact* otherwise.
4. The isomorphism $A \to C_0(\Sigma(A))$, then, is just given by the Gelfand transform.

This picture becomes even more compelling from the following observation:

Lemma C.9. *For any (i.e. not necessarily commutative) C*-algebra A we have $\Sigma(A) \subset A^*$. Furthermore, for any $\omega \in \Sigma(A)$,*
$$\|\omega\| = 1, \tag{C.13}$$
and if A has a unit, 1_A, then also
$$\omega(1_A) = 1. \tag{C.14}$$

In other words, multiplicative linear functionals on A are automatically continuous (recall that A^* is the Banach space of continuous linear maps from A to $\mathbb{C}$, see §B.9).

Throughout the rest of this section we restrict all proofs to the unital case; the general case may be handled by the technique of unitization to be discussed in §C.6.

Proof. Let $\omega \in \Sigma(A)$. By multiplicativity, $\ker(\omega)$ is a two-sided ideal in A. Trivially, for any $a \in A$, we have $a - \omega(a) \cdot 1_A \in \ker(\omega)$. If this element were invertible, then $\ker(\omega)$ would contain the unit 1_A and hence would coincide with A, contradicting the definition of $\Sigma(A)$ (which requires ω to be nonzero). Hence $\omega(a) \in \sigma(a)$. By the spectral radius formula (B.255) we have $|\omega(a)| \leq \|a\|$, whence $\omega \in A^*$.

Furthermore, $\omega(1_A)^2 = \omega(1_A)$, whence $\omega(1_A) = 1$ or 0, the latter being excluded since it would imply that $\omega(a) = 0$ for all $a \in A$. This gives (C.14) (which also follows from Lemma C.4, given Lemma C.11 below), which in turn gives (C.13). □

The Gelfand topology on $\Sigma(A)$ coincides with the weak* topology inherited from A^*, which is simply the topology of pointwise convergence (i.e. $\omega_\lambda \to \omega$ iff $\omega_\lambda(a) \to \omega(a)$ for each $a \in A$), and the Gelfand transform $a \mapsto \hat{a}$ is (by abuse of notation) the image of a in A^{**} under the canonical injection $A \hookrightarrow A^{**}$ appearing in Proposition B.44, restricted (as a function on A^*) to the subset $\Sigma(A) \subset A^*$. From this perspective, continuity of $\hat{a}$ immediately follows from Proposition B.46.

This picture of the Gelfand topology also has a technical advantage, for we infer:

Lemma C.10. *If A is unital, then its Gelfand spectrum $\Sigma(A)$ is compact Hausdorff.*

Proof. By Lemma C.9, $\Sigma(A)$ lies in the unit ball of A^*, which by the Banach–Alaoglu Theorem is compact in its weak* topology. So we are ready if we show that $\Sigma(A)$ is a weak*-closed subset of A^*, which is obvious from its definition: if $\omega_\lambda \to \omega$, then for any $a \in A$ we obviously have

$$\omega(ab) = \lim_\lambda \omega_\lambda(ab) = \lim_\lambda \omega_\lambda(a)\omega_\lambda(b) = \omega(a)\omega(b). \tag{C.15}$$

We know show that the Hausdorff property of $\Sigma(A)$ is inherited from A^*. A subbasis of its weak* topology is given by sets of the form

$$U_a^\varepsilon(\varphi) = \{\rho \in A^*, |\varphi(a) - \rho(a)| < \varepsilon\}, \tag{C.16}$$

where $a \in A$, $\varphi \in A^*$, and $\varepsilon > 0$. Replacing $\rho \in A^*$ by $\rho \in \Sigma(A)$ we thus obtain a subbasis of the Gelfand topology. If ω and ω' are distinct points in $\Sigma(A)$, there exists $a \in A$ such that $\omega(a) \neq \omega'(a)$. Taking some $0 < \varepsilon < |\omega(a) - \omega'(a)|/2$, the two points in question are separated by the opens $U_a^\varepsilon(\omega)$ and $U_a^\varepsilon(\omega')$. □

It is immediate from the definition of $\Sigma(A)$ that $a \mapsto \hat{a}$ is an algebra homomorphism, since we have

$$\widehat{ab}(\omega) = \omega(ab) = \omega(a)\omega(b) = \hat{a}(\omega)\hat{b}(\omega) = (\hat{a} \cdot \hat{b})(\omega). \tag{C.17}$$

The fact that the Gelfand transform preserves the involution follows from:

Lemma C.11. *If $\omega \in \Sigma(A)$, then $\omega(a^*) = \overline{\omega(a)}$, and hence $\widehat{a^*} = (\hat{a})^*$.*

Proof. Using (C.14) and (C.2), the proof is the same as for Lemma C.7. □

The hard part of the proof of Theorem C.8 is isometricity of the Gelfand transform:

$$\|\hat{a}\|_\infty = \|a\|. \tag{C.18}$$

As always, isometricity obviously implies *injectivity*. Surprisingly, using the Stone–Weierstrass Theorem B.51, in this case isometricity also yields *surjectivity* of the map $a \mapsto \hat{a}$. Namely, if we take $X = \Sigma(A)$, and B to be the image $\hat{A}$ of A under the Gelfand transform, then the conditions on B in Theorem B.51 are easily verified. Assuming (C.18), this image is obviously closed, so that $\hat{A} = C(\Sigma(A))$. With injectivity also implied by (C.18), it follows that the Gelfand transform is an isomorphism.

It remains to prove (C.18), which conceptually is a conjunction of two equalities:

$$\|\hat{a}\|_\infty = r(a); \tag{C.19}$$
$$\|a\| = r(a) \ (a^* = a), \tag{C.20}$$

where $r(a) = \sup\{|\lambda|, \lambda \in \sigma(a)\}$ is the spectral radius of a, see Theorem B.84. These immediately yield (C.18) for self-adjoint a, from which the general case follows from (C.2), noting that a^*a is self-adjoint for any a: asuming (C.19) - (C.20) as well as the homomorphism property of the Gelfand transform, we compute

$$\|\hat{a}\|_\infty^2 = \|\hat{a}^*\hat{a}\|_\infty = \|\widehat{a^*a}\|_\infty = \|a^*a\| = \|a\|^2. \tag{C.21}$$

Since (C.20) just repeats (B.257), we already know it is true for general C*-algebras (so far, with unit). As we shall now show, (C.19) holds in any commutative Banach algebra with unit. The key is the following lemma.

Lemma C.12. *Let A be a commutative Banach algebra with unit and let $a \in A$. For any $\lambda \in \sigma(a)$ there is an element $\omega \in \Sigma(A)$ such that $\lambda = \omega(a)$.*

Granted this, and using the proof of Lemma C.9 as well as (B.253), we obtain

$$\sigma(a) = \sigma(\hat{a}), \tag{C.22}$$

for any $a \in A$. Given (B.254), this yields (C.19) and hence the Gelfand isomorphism.

There are two approaches to our crucial Lemma C.12, each having its own merits. The first and best known proof, going back to Gelfand himself, relies on the theory of (maximal) ideals in Banach algebras. It is based on the following identification:

Proposition C.13. *Let A be a commutative Banach algebra with unit. There is a bijective correspondence between $\Sigma(A)$ and the set $\mathscr{M}(A)$ of maximal ideals in A,*

$$\omega \leftrightarrow \ker(\omega). \tag{C.23}$$

This will be proved in §C.8 below, which also contains the relevant background.

It implies Lemma C.12, as follows: if $\lambda \in \sigma(a)$, then by definition $a - \lambda$ is not invertible in A, so that $J = \{(a-\lambda)b \mid b \in A\}$ is an ideal in A. By Zorn's Lemma (or Hausdorff's Maximality Theorem), applied to the partially ordered set of all proper ideals in A that contain J, ordered by inclusion), J is contained in some maximal ideal, so that $J \subseteq \ker(\omega)$ for some $\omega \in \Sigma(A)$. Since $a - \lambda \in J$ (take $b = 1_A$), from (C.14) we obtain $\omega(a) = \lambda$. Note the non-constructive nature of this argument!

The other line of proof, due to Kadison, uses a different characterization of $\Sigma(A)$:

Proposition C.14. *Let A be a commutative C*-algebra with unit. Then the Gelfand spectrum $\Sigma(A)$ coincides with the pure state space $P(A)$.*

Recall Definition 1.10 and Theorem C.6; the pure state space $P(A) = \partial_e S(A)$ of a C*-algebra A is defined as the boundary of the state space of A. The argument that instantly delivers Lemma C.12 from Proposition C.14, then, is as follows:

Proposition C.15. *Let A be a C*-algebra with unit. For any normal element $a \in A$ (i.e., $aa^* = a^*a$) and $\lambda \in \sigma(a)$, there is a pure state $\omega \in P(A)$ such that $\omega(a) = \lambda$.*

The proof of both results uses some positivity theory for C*-algebras, which is systematically developed in §C.7 below. Here, we just need that $a \in A$ is positive, written $a \geq 0$, iff $a = b^*b$ for some $b \in A$, iff a is self-adjoint with $\sigma(a) \subset [0, \infty)$.

We write $a \geq b$ or $b \leq a$ if $a - b$ is positive. Also, a linear functional $\omega : A \to \mathbb{C}$ is called positive iff $\omega(a) \geq 0$ for all $a \geq 0$, and we write $\omega \geq \varphi$ or $\varphi \leq \omega$ if $\omega - \varphi \geq 0$.

Let us note that the proofs of these results in §C.7 use some Gelfand theory, but this use is limited to Theorem C.25, which could have been proved à la Theorem C.24, whose proof *derives* the Gelfand isomorphism in the special case at hand. Therefore, the use of Propositions C.14 and C.15 in the proof of (C.18) and hence of Theorem C.8 does not render this line of proof of the latter circular.

In particular, the proof of Proposition C.14 relies on:

Lemma C.16. *If $a^* = a \in A$ there is a number $t \geq 0$ such that $t \pm a \geq 0$.*

Proof. Since $\sigma(a) \subset \mathbb{R}$ is compact (see Corollary C.27 and Theorem B.84), we have $\sigma(a) \subseteq [-t, t]$ for some $t \geq 0$. It is clear from the definition of $\sigma(a)$ that $\sigma(t \pm a) = t \pm \sigma(a)$, which yields the lemma by the criterion for positivity just stated. □

We now prove Proposition C.14.

Proof. It is clear from Lemma C.11 and eq. (C.14) that $\omega \in \Sigma(A)$ is a state. To show that ω is pure, we use the fact that for any state $\omega \in S(A)$, the expression

$$\langle b, a \rangle = \omega(b^*a) \tag{C.24}$$

defines an hermitian form on A; the easy proof again uses use Lemma C.11. Applying Cauchy–Schwarz with $b \rightsquigarrow 1_A$ and using $1_A^* = 1_A = 1_A^2$ gives

$$|\omega(a)|^2 \leq \omega(a^*a). \tag{C.25}$$

Now suppose that $\omega = \lambda\omega_1 + (1-\lambda)\omega_2$ with $\omega_i \in S(A)$ and $\lambda \in (0,1)$. Applying (C.25) (in the opposite direction) to ω_1 and ω_2 gives

$$\omega(a^*a) \geq \lambda|\omega_1(a)|^2 + (1-\lambda)|\omega_2(a)|^2. \tag{C.26}$$

On the other hand, multiplicativity of ω gives

$$\omega(a^*a) = \lambda^2|\omega_1(a)|^2 + \lambda(1-\lambda)(\omega_1(a)\overline{\omega_2(a)} + \omega_2(a)\overline{\omega_1(a)}) + (1-\lambda)^2|\omega_2(a)|^2.$$

Subtracting this from (C.26) gives the inequality $0 \geq \lambda(1-\lambda)|\omega_1(a) - \omega_2(a)|^2$, so that $\omega_1 = \omega_2$, and hence ω is pure by definition. This shows that $\Sigma(A) \subseteq P(A)$.

To prove the converse inclusion, we need another lemma.

Lemma C.17. *Let $\omega \in P(A)$ be a pure state on A. If $\tau : A \to \mathbb{C}$ is a linear functional such that $0 \leq \tau \leq \omega$, then we can find a scalar $s \in [0,1]$ such that $\tau = s\omega$.*

Proof. We assume $\tau \neq 0$ and $\tau \neq \omega$ (otherwise the claim is trivially true). By Lemma C.16, this implies $\tau(1_A) \neq 0$ and $\tau(1_A) \neq 1$. For if $\tau(1_A) = 0$, then for $a^* = a$ we find t as in Lemma C.16, so that $t \pm a \geq 0$ and hence $0 \leq \tau(t \pm a) = \pm\tau(a)$. Hence $\tau(a) = 0$ on each self-adjoint a, which forces $\tau = 0$ by the usual decomposition (C.8). If $\tau(1_A) = 1$, we apply a similar argument to the positive functional $\omega - \tau$. Therefore, $t = 1 - \tau(1_A)$ satisfies $t \in (0,1)$, and defining $\omega_1 = (\omega - \tau)/t$ and $\omega_2 = \tau/\tau(1_A)$ we obtain a decomposition $\omega = t\omega_1 + (1-t)\omega_2$. Since ω is pure, this gives $\omega_1 = \omega_2 = \omega$ and hence $\tau = \tau(1_A)\omega$. Clearly, $0 \leq \tau \leq \omega$ enforces $0 \leq \tau(1_A) \leq 1$, so the claim follows with $s = \tau(1_A)$. □

We now prove that $\omega \in P(A)$ is multiplicative on arbitrary $a \in A$, and $b \in A$ such that (for the moment) $0 \leq b \leq 1_A$. Define $\omega_b : A \to \mathbb{C}$ by $\omega_b(a) = \omega(ab)$. Then $0 \leq \omega_b \leq \omega$: taking $b = c^*c$, the first inequality $0 \leq \omega_b$ follows from

$$\omega_b(a^*a) = \omega(c^*ca^*a) = \omega((ac)^*ac) \geq 0, \tag{C.27}$$

since A is abelian, and the second is analogous, using the fact that $0 \leq b \leq 1_A$ implies $0 \leq 1_A - b \leq 1_A$. Therefore, Lemma C.17 gives $\omega_b = s\omega$ with $s = \omega_b(1_A) = \omega(b)$.

For general $0 \neq b \geq 0$, we rewrite b as $b = \|b\| \cdot (b/\|b\|)$, and use linearity of ω and the previous result to obtain multiplicativity. For general self-adjoint b we use Lemma C.53, and finally we use (C.8). □

At last, we are now in a position to prove Proposition C.15, so let $a \in A$ be normal.

Proof. Let $C^*(a)$ be the commutative C*-algebra generated by a (and hence a^*) and 1_A within A; as in Theorem C.25 below, this is the norm-closure of all polynomials in a and a^*, and $C(\sigma(a)) \cong C^*(a)$ via the map $f(\lambda, \overline{\lambda}) \mapsto f(a, a^*)$. Using Proposition C.14, define a pure state ω_λ on $C^*(a)$ by linear and multiplicative extension of $\omega_\lambda(1_A) = 1$, $\omega_\lambda(a) = \lambda$, and $\omega_\lambda(a^*) = \overline{\lambda}$, i.e., $\omega_\lambda(f(a,a^*)) = f(\lambda, \overline{\lambda})$.

Since $\|\omega_\lambda\| = 1$, Hahn–Banach (Corollary B.41, with $V \rightsquigarrow A$ and $W \rightsquigarrow C^*(a)$) yields a linear extension $\omega'_\lambda : A \to \mathbb{C}$ of ω_λ, which is in fact a state by Lemma C.4. To show that ω'_λ may be chosen to be pure also on A, let $S_\lambda(A) \subset S(A)$ be the set of all states on A that extend ω_λ. This is a nonempty weak*-closed and hence weak*-compact convex subset of $S(A)$, which by the Krein–Milman Theorem B.50 has nonempty boundary $\partial_e S_\lambda(A)$. It is easy to show that $\partial_e S_\lambda(A) \subset \partial_e S(A) = P(A)$: for $\omega \in \partial_e S_\lambda(A)$, suppose $\omega = t\omega_1 + (1-t)\omega_2$, with $t \in (0,1)$ and $\omega_i \in S(A)$. Since $\omega_{|C^*(a)} = \omega_\lambda$ is pure, we have $\omega_{1|C^*(a)} = \omega_{2|C^*(a)} = \omega_\lambda$, or $\omega_i \in S_\lambda(A)$. But ω was assumed pure in $S_\lambda(A)$, so that $\omega_1 = \omega_2 = \omega$, i.e., $\omega \in \partial_e S(A)$. Hence if we choose $\omega'_\lambda \in \partial_e S_\lambda(A)$, then the extension ω'_λ of ω_λ is also pure on A. □

The following ingredients are still missing from the proof of Theorem C.8:

- The proof the uniqueness of X up to homeomorphism (see §C.3).
- The proof of Proposition C.13 (see §C.8).
- The extension of the entire argument to the non-unital case (see §C.6).

We start with the first issue, which we fill in more broadly than needed for the proof of Theorem C.8, namely, as part of a broader picture called ***Gelfand duality*** (which will fall into place if one uses the language of category theory, see Appendix E).

C.3 Gelfand duality

Theorem C.8 is a consequence of the following two propositions.

Proposition C.18. *Let A and B be unital commutative C*-algebras. Then*

$$\varphi = \alpha^*, \tag{C.28}$$

where $\alpha^*(\omega) = \omega \circ \alpha$, establishes a bijective correspondence between unital homomorphisms $\alpha : A \to B$ and continuous maps $\varphi : \Sigma(B) \to \Sigma(A)$.

In particular, $\Sigma(A)$ and $\Sigma(B)$ are homeomorphic iff A and B are isomorpic.

Proof. Since $\alpha(ab) = \alpha(a)\alpha(b)$, if $\omega \in \Sigma(B)$ it is clear that then $\alpha^*(\omega) \in \Sigma(A)$.

Conversely, denoting the pertinent Gelfand transforms by $G_A : A \to C(\Sigma(A))$ and $G_B : A \to C(\Sigma(B))$, given $\varphi : \Sigma(B) \to \Sigma(A)$, we define $\alpha : A \to B$ by

$$\alpha = G_B^{-1} \circ \varphi^* \circ G_A, \tag{C.29}$$

where $\varphi^* : C(\Sigma(A)) \to C(\Sigma(B))$ is the pullback of φ (i.e., $\varphi^*(f) = f \circ \varphi$).

It is easy to verify that given φ, the map α defined in (C.29) returns φ through (C.28), whereas given α, the map φ defined in (C.28) returns α through (C.29). □

Proposition C.19. *For any compact Hausdorff space X, the evaluation map*

$$\text{ev} : X \to \Sigma(C(X)); \tag{C.30}$$

$$\text{ev}_x(f) = f(x), \tag{C.31}$$

is a homeomorphism, so that

$$\Sigma(C(X)) \cong X. \tag{C.32}$$

Proof. Injectivity of ev immediately follows from Urysohn's lemma (which applies because a compact Hausdorff space is normal), which implies that $C(X)$ separates points on X (i.e., for all $x \neq y$ there is an $f \in C(X)$ for which $f(x) \neq f(y)$).

To prove surjectivity, suppose there is $\omega \in \Sigma(C(X))$ such that $\omega \neq \text{ev}_x$ for all $x \in X$. Now $\ker(\omega) = \ker(\text{ev}_x)$ would imply $\omega = \text{ev}_x$ (because $\omega(f) = \lambda$ then implies $f - \lambda \cdot 1_X \in \ker(\omega)$, and hence $f(x) = \lambda$, and *vice versa*), so $\ker(\omega) \neq \ker(\text{ev}_x)$. Since $\text{ev}_x \in \Sigma(C(x))$, and $\omega \in \Sigma(C(x))$ by assumption, by Proposition C.13 both kernels are maximal ideals in $C(X)$, and hence $\ker(\omega) \subset \ker(\text{ev}_x)$ is impossible (and so is the opposite inclusion). Therefore, for each x there is a function $f_x \in \ker(\omega)$ for which $f_x(x) \neq 0$ (for otherwise $f(x) = 0$ for all $f \in \ker(\omega)$, so that $\ker(\omega) \subseteq \ker(\text{ev}_x)$). Redefining f_x by a phase if necessary, we may assume that $f_x(x) > 0$, and taking the real part of f_x if necessary, we may also assume that f is real-valued.

For each x, the set U_x where $f_x > 0$ is open, because f is continuous. This gives a covering $\{U_x\}_{x \in X}$ of X, which by compactness has a finite subcovering $\{U_{x_n}\}_{n=1,\dots,N}$. Then define the function

$$f = \sum_{n=1}^{N} f_{x_n}, \tag{C.33}$$

which is strictly positive by construction, so that it is invertible. But $\ker(\omega)$ is an ideal, so that, with all $f_{x_n} \in \ker(\omega)$ (since all $f_x \in \ker(\omega)$) also $f \in \ker(\omega)$. But an ideal containing an invertible element must contain 1_X and hence coincides with $C(X)$, contradicting the fact that $\ker(\omega)$ was maximal. Hence ev is surjective.

Finally, to prove that ev is a homeomorphism, we equip X with the topology induced by ev, in which the open sets are of the form $\mathrm{ev}^{-1}(U)$, with U open in $\Sigma(C(X))$ in the Gelfand topology. We claim that this new topology on X is weaker than the original one (this terminology includes the possibility that the two topologies in question coincide). Namely, for $f \in C(X)$ one has $\hat{f} \circ \mathrm{ev} = f$. Therefore, since the Gelfand topology on $\Sigma(C(X))$ is the weakest topology for which all Gelfand transforms $\hat{f}$ are continuous, the new topology on X is the weakest topology for which all f are continuous. But f was already continuous with respect to the given topology, so the claim follows. Without proof we now state a result from topology:

Lemma C.20. *If a set X is Hausdorff in some topology $\mathcal{O}_1(X)$ and compact in a topology $\mathcal{O}_2(X)$, and if $\mathcal{O}_1(X) \subseteq \mathcal{O}_2(X)$, then $\mathcal{O}_1(X) = \mathcal{O}_2(X)$.*

Since X is in fact compact and Hausdorff in both topologies, we conclude from this lemma that the new topology on X must coincide with the original one. □

Uniqueness of the Gelfand spectrum up to homeomorphism follows from Propositions C.18 and C.19: if A is a unital commutative C*-algebra for which $A \cong C(X)$ as well as $A \cong C(Y)$, then applying Σ and using (C.32) makes X and Y both homeomorphic to $\Sigma(A)$, and hence to each other.

With minor changes, the proof of Proposition C.19: applies also to "well-behaved" manifolds, by which we mean *second countable smooth locally compact Hausdorff manifolds*. These are the ones encountered in physics (especially in classical mechanics); we need this for Theorem 3.10 in the main text. Such manifolds admit partitions of unity subordinate to any given cover (U_λ) that are locally finite as well as countable, i.e., sequences of smooth functions $\chi_n :\to [0,1]$ such that:

1. Each $x \in X$ has an open neighbourhood U that intersects only finitely many of the sets $\mathrm{supp}(\chi_n)$;
2. For each $x \in X$ we have $\sum_n \chi_n(x) = 1$ (where the sum is finite);
3. Each set $\mathrm{supp}(\chi_n)$ is contained in some U_λ.

Furthermore, $\Sigma(C^\infty(X))$ is defined as for any complex associative algebra A, i.e., as the set of nonzero multiplicative linear maps $\omega : C^\infty(X) \to \mathbb{C}$.

Proposition C.21. *For any second countable smooth locally compact Hausdorff manifold X, the evaluation map* $\mathrm{ev} : X \to \Sigma(C^\infty(X))$ *in* (C.31) *is a bijection.*

Proof. Since X is not necessarily compact, we cannot use Urysohn's Lemma directly to prove that $C^\infty(X)$ separates points of X (so that ev is injective), but this time, if $U \subseteq X$ is open and $F \subset U$ is closed, there exists a smooth function $\chi : X \to [0,1]$ such that $\chi = 1$ on F and $\chi = 0$ on $X \backslash U$. Indeed, $\{U, X\backslash F\}$ is an open cover of X, and if $(\chi_U, \chi_{X\backslash F})$ is a partition of unity subordinate to this cover, $\chi = \chi_U$ will do.

Now for $x \neq y$, take $F = \{x\}$ and use the Hausdorff property to separate (x,y) by disjoint open sets (U,V), and we have $\chi(x) = 1$ whilst $\chi(y) = 0$.

The proof of surjectivity is the same as for $C(X)$, including the proof that $\ker(\omega)$ is a maximal ideal in $C^\infty(X)$, until the point (C.33) is reached. Here compactness is no longer available, so that we need to replace (C.33) by the expression

$$f = \sum_n c_n \chi_n f_{x_n}, \tag{C.34}$$

where (χ_n) is a smooth partition of unity subordinate to the cover (U_x), for each $n \in \mathbb{N}$, f_{x_n} is picked by no. 3 in the list of properties of a partition of unity listed above, and the coefficients c_n are chosen so that $0 < c_n < (n^2 \|\chi_n f_{x_n}\|_\infty)^{-1}$ (note that χ_n and hence $\chi_n f_{x_n}$ has compact support and is continuous, so that it is bounded). Since $\sum_n (1/n^2) < \infty$, the insertion of the c_n makes f bounded and the sum (C.34) uniformly convergent. which is necessary to pull ω through the sum so as to prove that $f \in \ker(\omega)$, as follows. Since the sup-norm is not defined on all of $C^\infty(X)$, we need a little argument here. Take $t > \|f\|_\infty$, so that $t \cdot 1_X \pm f$ nowhere vanishes and hence is invertible, so that $\omega(t \cdot 1_X \pm f) = t \pm \omega(f) \neq 0$ by multiplicativity of f, i.e., $\pm\omega(f) \neq t$. Since f and hence $\omega(f)$ is real, this gives $|\omega(f)| \leq \|f\|_\infty$. Since $\omega(f_{x_n}) = 0$, and similarly for each finite sum in (C.34), we finally obtain

$$|\omega(f)| = \left|\omega(f - \sum_{n=1}^N c_n \chi_n f_{x_n})\right| \leq \left\| f - \sum_{n=1}^N c_n \chi_n f_{x_n}) \right\|, \tag{C.35}$$

so letting $N \to \infty$ gives $\omega(f) = 0$, or $f \in \ker(\omega)$. Since f is invertible, this implies $1_X \in \ker(\omega)$ and hence $\ker(\omega) = C(X)$, contradicting $\omega \neq 0$. □

Corollary C.22. *Let X and Y be compact Hausdorff spaces. Then $\alpha(f) = f \circ \varphi$, i.e.,*

$$\alpha = \varphi^*, \tag{C.36}$$

establishes a canonical bijective correspondence between unital homomorphisms $\alpha : C(Y) \to C(X)$ (as C-algebras) and continuous maps $\varphi : X \to Y$. In particular, $C(X)$ and $C(Y)$ are isomorphic iff X and Y are homeomorphic.*

Likewise, X and Y are second countable smooth locally compact Hausdorff manifolds, eq. (C.36) *gives a canonical bijective correspondence between homomorphisms $\alpha : C^\infty(Y) \to C^\infty(X)$ (as commutative algebras) and smooth maps $\varphi : X \to Y$. In particular, $C^\infty(X)$ and $C^\infty(Y)$ are isomorphic iff X and Y are diffeomorphic.*

Proof. The passage from φ to α is obvious. We write $\mathrm{ev}_X : X \to \Sigma(C(X))$ and $\mathrm{ev}_Y : Y \to \Sigma(C(Y))$ for the bijections previously just called ev. Since these maps are invertible by the previous proposition, we may define a map $\varphi : X \to Y$ by

$$\varphi = \mathrm{ev}_Y^{-1} \circ \alpha^* \circ \mathrm{ev}_X, \tag{C.37}$$

where $\alpha^* : \Sigma(C(X)) \to \Sigma(C(Y))$ is defined by $\alpha^*(\omega) = \omega \circ \alpha$; this lies in $\Sigma(C(Y))$, because α is linear and $\alpha(fg) = \alpha(f)\alpha(g)$. Eq. (C.36) then holds by construction.

- In the compact case, we still need to prove that φ is continuous. To do so, note that a compact Hausdorff space Y is completely regular, and as such a subbase for its topology is given by sets of the form $U = f^{-1}(U')$, where $f \in C(Y)$ and $U' \in \mathcal{O}(\mathbb{C})$. Hence $\varphi^{-1}(U) = (\varphi^* f)^{-1}(U')$, and since we know that $\varphi^* f = \alpha(f) \in C(X)$, we conclude that $\varphi^{-1}(U)$ is open in X. Thus φ is continuous.
- Similarly, in the manifold case, a map $\varphi : X \to Y$ is smooth iff $\varphi^* f \in C^\infty(X)$ for each $f \in C^\infty(Y)$; using localization by bump functions à la the first part of the proof of Proposition C.21, it is enough to prove this for open sets $X \subset \mathbb{R}^n$ and $Y \subset \mathbb{R}^m$, so that $\varphi(y) = (\varphi^1(y), \ldots, \varphi^n(y))$. Knowing that $\varphi^* f$ is smooth for each $f \in C^\infty(Y)$, we simply take $f(x^1, \ldots, x^n) = x^k$ to be the k'th coordinate function. This declares each φ^k to be smooth, and therewith also φ itself. □

We now state ***Gelfand duality***, explaining its categorical interpretation in §E.1.

Theorem C.23. *1. If X is a compact Hausdorff space, then $C(X)$ is a unital commutative C*-algebra. A continuous map $\varphi : X \to Y$ induces a unital homomorphism $C(f) \equiv \varphi^* : C(Y) \to C(X)$, which behaves well under composition, in that:*

- *If φ is the identity, then so is $C(\varphi)$.*
- *If $\psi : Y \to Z$ is another continuous map, then $C(\varphi \circ \psi) = C(\psi) \circ C(\varphi)$.*

2. If A is a unital commutative C-algebra, then $\Sigma(A)$ is a compact Hausdorff space. A unital homomorphism $\alpha : A \to B$ induces a continuous function $\Sigma(\alpha) \equiv \alpha^* : \Sigma(B) \to \Sigma(A)$, which behaves well under composition in a similar way:*

- *If α is the identity, then so is $\Sigma(\alpha)$.*
- *If $\beta : B \to C$ is another unital homomorphism, then $\Sigma(\beta \circ \alpha) = \Sigma(\alpha) \circ \Sigma(\beta)$.*

3. There are canonical homeomorphisms and isomorphisms:

$$\mathrm{ev}_X : X \xrightarrow{\cong} \Sigma(C(X)); \tag{C.38}$$

$$G_A : A \xrightarrow{\cong} C(\Sigma(A)), \tag{C.39}$$

with the following "naturality" properties:

- *If $\Sigma \circ C(\varphi) : \Sigma(C(X)) \to \Sigma(C(Y))$ is the map induced by $\varphi : X \to Y$, then*

$$\Sigma \circ C(\varphi) \circ \mathrm{ev}_X = \mathrm{ev}_Y \circ \varphi; \tag{C.40}$$

- *If $C \circ \Sigma(\alpha) : C(\Sigma(A)) \to C(\Sigma(B))$ is the map induced by $\alpha : A \to B$, then*

$$C \circ \Sigma(\alpha) \circ G_A = G_B \circ \alpha. \tag{C.41}$$

Proof. The proof is an assembly of previous results and routine verifications. □

In the language of category theory, Theorem C.23 states that the categories CH of compact Hausdorff spaces (with continuous functions as arrows) and CCA_1 of commutative unital C*-algebras (with unital homomorphisms as arrows, cf. Definition C.2) are dual (i.e., contravariantly equivalent). In particular, we have an adjunction between the functors $C : \mathsf{CH} \to \mathsf{CCA}_1$ and $\Sigma : \mathsf{CCA}_1 \to \mathsf{CH}$.

C.4 Gelfand isomorphism and spectral theory

As an example of Gelfand's theory, Theorem 4.3 may be reformulated as follows:

Theorem C.24. *Let H be a Hilbert space, and let $a = a^* \in B(H)_{\mathrm{sa}}$, with associated (commutative) C^*-algebra $C^*(a)$ generated by a and 1_H. The Gelfand spectrum $\Sigma(C^*(a))$ of $C^*(a)$ is homeomorphic to $\sigma(a)$, under the mutually inverse maps*

$$\Sigma(C^*(a)) \xrightarrow{\cong} \sigma(a), \ \omega \mapsto \omega(a); \tag{C.42}$$

$$\sigma(a) \xrightarrow{\cong} \Sigma(C^*(a)), \ \lambda \mapsto \omega_\lambda : f(a) \mapsto f(\lambda). \tag{C.43}$$

In particular, the image of the map $\omega \mapsto \omega(a)$ from $\Sigma(C^*(a))$ to $\mathbb{C}$ is $\sigma(a)$, and the isomorphism $C^*(a) \to C(\sigma(a))$, $f(a) \mapsto f$, of Theorem B.94 is obtained by composing the Gelfand transform $f(a) \mapsto \widehat{f(a)}$ from $C^*(a)$ to $C(\Sigma(C^*(a)))$ with the isomorphism $C(\Sigma(C^*(a))) \xrightarrow{\cong} C(\sigma(a))$ obtained by pulling back the map (C.43).

Proof. First, we note that map (C.43) is well defined. Indeed, it follows from (B.289) that the map $\omega_\lambda : C^*(a) \to \mathbb{C}$ is linear for any $\lambda \in \sigma(a)$, whilst the following computation, which uses (B.290), implies that ω_λ multiplicative:

$$\omega_\lambda(f(a)g(a)) = \omega_\lambda(fg(a)) = (fg)(\lambda) = f(\lambda)g(\lambda) = \omega_\lambda(f(a))\omega_\lambda(g(a)). \tag{C.44}$$

Injectivity of the map $\lambda \mapsto \omega_\lambda$ holds because $\sigma(a)$ is Hausdorff, so that $f(\lambda') = f(\lambda)$ for each $f \in C(\sigma(a))$ implies $\lambda' = \lambda$. Surjectivity follows from (B.253), since

$$\sigma_{C^*(a)}(f(a)) = \sigma_{C(\sigma(a))}(f) = \mathrm{Ran}(f), \tag{C.45}$$

where we used invariance of the spectrum under isomorphisms. Consider the function $f(x) = x$, so that $f(a) = a$. It follows from (C.43) that $\omega_\lambda(a) = \lambda$. Conversely, using the same function f, for given $\omega \in \Sigma(C^*(a))$ we find $\omega_{\omega(a)} = \omega$, so that the maps in (C.42) - (C.43) are mutually inverse. It is clear from (C.42) - (C.43) dat $\omega_{\lambda_i} \to \omega_\lambda$ in the Gelfand topology on $\Sigma(C^*(a))$ (which is the topology of pointwise convergence) iff $f(\lambda_i) \to f(\lambda)$ for each $f \in C(\sigma(a))$, which is the case iff $\lambda_i \to \lambda$ on $\sigma(a)$. Hence both of our maps $\Sigma(C^*(a)) \leftrightarrow \sigma(a)$ are continuous.

The final claim is a definition chase, using the computation

$$\widehat{f(a)}(\omega_\lambda) = \omega_\lambda(f(a)) = f(\lambda).$$

□

If $\dim(H) < \infty$, one may replace this proof by using the fact that $\sigma(a)$ consists of the eigenvalues of a. If p is a polynomial, then $\omega \in \Sigma(C^*(a))$ must satisfy $\omega(p(a)) = p(\omega(a))$. The characteristic polynomial p_c of a, i.e., $p_c(x) = \prod_{i=1}^n(\lambda_i - x)$, where the λ_i are the $n = \dim(H)$ eigenvalues of a (including repetitions), satisfies $p_c(a) = 0$, so that $\omega(p_c(a)) = 0$, i.e., $\prod_{i=1}^n(\lambda_i - \omega(a)) = 0$, and hence $\omega(a) = \lambda_i$ for some i, or $\omega(a) \in \sigma(a)$. Thus (C.42) is well defined. In the opposite direction, eqs. (A.53) - (A.55) show that (C.43) is also well defined, in that indeed $\omega_\lambda \in \Sigma(C^*(a))$.

The construction of $C^*(a)$ as a C*-algebra within $B(H)$ may trivially be generalized to arbitrary unital C*-algebras A, i.e., if $a \in A$, we define $C^*(a)$ as the C*-algebra generated (within A) by a and the unit 1_A. If $a = a^*$, then $C^*(a)$ still equals the norm-closure of the algebra of all polynomials in a, and hence $C^*(a)$ is once again commutative. Defining the spectrum $\sigma(a)$ as in Definition B.81, we then have the following generalization of Theorem C.24:

Theorem C.25. *Let A be a unital C*-algebra and let $a^* = a \in A$. Then*

$$\Sigma(C^*(a)) \cong \sigma(a), \;\; \omega \leftrightarrow \omega(a); \tag{C.46}$$

$$C^*(a) \cong C(\sigma(a)), \;\; f(a) \leftrightarrow f, \tag{C.47}$$

as spaces and as (commutative) C-algebras, respectively. Under the Gelfand isomorphism* (C.47), *the Gelfand transform $\hat{a}$ of $a \in C^*(a)$ is the identity* $\mathrm{id}_{\sigma(a)} : \lambda \to \lambda$, *whereas the Gelfand transform $\widehat{1_A}$ of $1_A \in C^*(a)$ is the unit* $1_{\sigma(a)} : \lambda \to 1$.

This ***continuous functional calculus*** may be proved in exactly the same way as Theorems B.94 and C.24, with $B(H) \rightsquigarrow A$. However, these proofs did not invoke Gelfand's Theorem (but rather derived it in the special case at hand), so it may give additional insight in the situation if we reprove Theorem C.25 from Theorem C.8.

Proof. We now *assume* the isomorphism $C^*(a) \cong C(\Sigma(C^*(a)))$ via the Gelfand transform. According to (C.22) and (B.253), which imply $\sigma(\hat{a}) = \mathrm{ran}(\hat{a})$, the function $\hat{a} : \Sigma(C^*(a)) \to \mathbb{C}$ is surjective onto the spectrum $\sigma(a) \subset \mathbb{C}$. We now prove injectivity. If $\omega_1, \omega_2 \in \Sigma(C^*(a))$ and $\omega_1(a) = \omega_2(a)$, then, for all $n \in \mathbb{N}$, we have

$$\omega_1(a^n) = \omega_1(a)^n = \omega_2(a)^n = \omega_2(a^n), \tag{C.48}$$

Since also $\omega_1(1_A) = \omega_2(1_A) = 1$, we conclude by linearity that $\omega_1 = \omega_2$ on all polynomials in a. By continuity (cf. Lemma C.9) this implies that $\omega_1 = \omega_2$, since by definition the linear span of all polynomials is dense in $C^*(a)$. Using (C.12), we have therefore proved that $\hat{a}(\omega_1) = \hat{a}(\omega_2)$ implies $\omega_1 = \omega_2$, i.e., $\hat{a}$ is injective.

Since $\hat{a} \in C(\Sigma(C^*(a)))$ by Theorem C.8, $\hat{a}$ is continuous. To prove continuity of the inverse, recall that $\hat{a} : \Sigma(C^*(a)) \to \sigma(a)$ is the map $\omega \mapsto \omega(a)$, so that for $\lambda \in \sigma(a)$, the functional $\hat{a}^{-1}(\lambda) \in \Sigma(C^*(a))$ maps a to λ. By multiplicativity, $\hat{a}^{-1}(\lambda)$ then maps a^n to λ^n. Hence ny linearity and (C.14), for polynomials p in a one has

$$\hat{a}^{-1}(\lambda) : p(a) \mapsto p(\lambda). \tag{C.49}$$

Since polynomials are continuous, if $\lambda_n \to \lambda$ in $\sigma(a)$, then $p(\lambda_n) \to p(\lambda)$, so

$$(\hat{a}^{-1}(\lambda_n))(p) \to (\hat{a}^{-1}(\lambda))(p). \tag{C.50}$$

Since such polynomials $p(a)$ are dense in $C^*(a)$ by definition, and functionals in $\Sigma(C^*(a))$, being continuous, are therefore determined by their values on polynomials, we conclude that $\hat{a}^{-1}(\lambda_n) \to \hat{a}^{-1}(\lambda)$ pointwise. Since the Gelfand topology is the topology of pointwise convergence, we conclude that $\hat{a}^{-1}$ is continuous, so that $\hat{a}$ is a homeomorphism. This proves (C.46).

Finally, for compact Hausdorff spaces X and Y, a homeomorphism $\varphi : X \to Y$ induces an isomorphism $\varphi^* : C(Y) \to C(X)$ of C*-algebras, where $\varphi(f) = f \circ \varphi$ (cf. §C.3). Theorem C.8 and (C.46) give (C.47). Unfolding the latter isomorphism gives

$$C^*(a) \xrightarrow{\text{GT}} C(\Sigma(C^*(a))) \xrightarrow{(\hat{a}^{-1})^*} C(\sigma(a)), \tag{C.51}$$

where GT is the Gelfand transform and $(\hat{a}^{-1})^*$ is the pullback of the homeomorphism $\hat{a}^{-1} : \sigma(a) \to \Sigma(C^*(a))$, as in φ^* above. Following these arrows and using (C.49), one obtains the last claim. □

Corollary C.26. *Let A be a unital C*-algebra and let $a^* = a \in A$, with spectrum $\sigma(a)$. For each selfadjoint element $a \in A$ and each $f \in C(\sigma(a))$, there is an operator $f(a) \in A$, which is the obvious expression when f is a polynomial (and in general is given via the uniform approximation of f by polynomials), such that*

$$\|f(a)\| = \|f\|_\infty; \tag{C.52}$$
$$\sigma(f(a)) = f(\sigma(a)). \tag{C.53}$$

Eq. (C.53) *is called the* **spectral mapping property**. *Furthermore, the norm and spectrum of a as an element of A coincide with the norm and spectrum of a in $C^*(a)$.*

Proof. We write (C.51) in the opposite direction, i.e.,

$$C(\sigma(a)) \xrightarrow{(\omega \mapsto \omega(a))^*} C(\Sigma(C^*(a))) \xrightarrow{\hat{a} \mapsto a} C^*(a). \tag{C.54}$$

Indeed, if $\tilde{f} \in C(\Sigma(C^*(a)))$ is the image of $f \in C(\sigma(a))$ under the first arrow, then $\tilde{f}(\omega) = f(\omega(a))$, and the second arrow says that $\widehat{f(a)} = \tilde{f}$. Together these give $f(\omega(a)) = \omega(f(a))$, which by multiplicativity, linearity, and (C.14), is the case for polynomials $f = p$; the general case follows from the polynomial case by continuity.

Eq. (C.52) follows from (C.18) and the fact that also the first arrow in (C.54) is an isometry, and (C.53) follows from (C.22), with with $a \rightsquigarrow f(a)$.

To close, take $f = \text{id}_{\sigma(a)}$; then (C.52) gives $\|a\|_A = r(a)$, cf. (B.257), whilst (C.18) gives $\|a\|_{C^*(a)} = r(a)$, too. Finally, (C.47) and (B.253) show that the spectrum of a in $C^*(a)$ is $\sigma(a)$, which by definition is its spectrum in A. □

Corollary C.27. *If $a^* = a$, then $\sigma(a) \subset \mathbb{R}$.*

By Corollary C.26, we may take the spectrum of a in $C^*(a)$. By Lemma C.11, the Gelfand transform $\hat{a}$ is real-valued. Then use the last part of Theorem C.25. □

Corollary C.28. *The norm in a C*-algebra is unique (given all other structure).*

Using (B.257) for $a = a^*$, and then (C.2), for arbitrary $a \in A$ we find

$$\|a\| = \sqrt{r(a^*a)}. \tag{C.55}$$

Since the spectrum (and hence the spectral radius r) is determined by the algebraic structure, (C.55) shows that the norm is determined by the algebraic structure. □

C.5 C*-algebras without unit: general theory

In classical physics, non-compact phase spaces are described by commutative C*-algebras *without unit*. Proper ideals in C*-algebras necessarily lack a unit, too. To set the stage, we first assume that A is a Banach algebra, and form the vector space

$$\dot{A} = A \oplus \mathbb{C}, \tag{C.56}$$

and turn this into an algebra in the obvious way, i.e., by means of

$$(a+\lambda \cdot 1_{\dot{A}})(b+\mu \cdot 1_{\dot{A}}) = ab + \lambda b + \mu a + \lambda\mu \cdot 1_{\dot{A}}, \tag{C.57}$$

where we have written $a+\lambda \cdot 1_{\dot{A}}$ for (a,λ), etc. This turns the number 1 in $\mathbb{C}$ into a unit $1_{\dot{A}}$ for $\dot{A}$, and this is the point: $\dot{A}$ is unital, even if A lacks a unit. Defining

$$\|a+\lambda \cdot 1_{\dot{A}}\| = \|a\| + |\lambda|, \tag{C.58}$$

we also have a norm on $\dot{A}$, with $\|1_{\dot{A}}\| = 1$. Using (C.1), (C.57), and (C.58), we have

$$\begin{aligned}\|(a+\lambda \cdot 1_{\dot{A}})(b+\mu \cdot 1_{\dot{A}})\| &\leq \|a\|\,\|b\| + |\lambda|\,\|b\| + |\mu|\,\|a\| + |\lambda|\,|\mu| \\ &= \|a+\lambda \cdot 1_{\dot{A}}\|\,\|b+\mu \cdot 1_{\dot{A}}\|,\end{aligned}$$

so that $\dot{A}$ is a *Banach algebra with unit*. Since by (C.58) the norm of $a \in A$ in A coincides with the norm of $a + 0 \cdot 1_{\dot{A}}$ in $A \oplus \mathbb{C}$, we have shown the following:

Proposition C.29. *For every Banach algebra (with or without unit) there exists a unital Banach algebra $\dot{A}$, called the* **unitization** *of A, and an isometric (hence injective) morphism $A \to \dot{A}$, such that $\dot{A}/A \cong \mathbb{C}$.*

If A is a C*-algebra, (C.58) fails to be a C*-norm with respect to the involution

$$(a+\lambda \cdot 1_{\dot{A}})^* = a^* + \overline{\lambda} \cdot 1_{\dot{A}}, \tag{C.59}$$

since (C.2) is not satisfied. Instead, the correct norm in which $A \oplus \mathbb{C}$ is a unital C*-algebra is the one borrowed from $B(A)$, i.e., the Banach space of bounded linear maps from A to A (regarded as a Banach space), relying on an embedding $A \subset B(A)$:

Proposition C.30. *Let A be a C*-algebra (with or without unit).*

1. *The map $L : A \to B(A)$, $a \mapsto L_a$, given by*

$$L_a(b) = ab \tag{C.60}$$

   establishes an isometric isomorphism between A and $L(A) \subset B(A)$.
2. *When A has no unit, define a norm on $\dot{A} = A \oplus \mathbb{C}$ by*

$$\|a+\lambda \cdot 1_{\dot{A}}\| = \|L_a + \lambda \cdot 1_{B(A)}\|, \tag{C.61}$$

   where the right-hand side uses the operator norm in $B(A)$. With the operations (C.57) and (C.59), the norm (C.61) turns $\dot{A}$ into a C-algebra with unit.*

Proof. By (C.1) we have $\|L_a b\| = \|ab\| \leq \|a\|\,\|b\|$ for all b, so that $\|L_a\| \leq \|a\|$. On the other hand, using (C.2) and (A.22), assuming $a \neq 0$, we can write

$$\|a\| = \|aa^*\|/\|a\| = \left\| L_a \frac{a^*}{\|a\|} \right\| \leq \|L_a\|. \tag{C.62}$$

Hence

$$\|L_a\| = \|a\|. \tag{C.63}$$

Being isometric, the map L must be injective; it is clearly a homomorphism, so that we have proved the first claim of the proposition.

It is clear from (C.57) and (C.59) that the map $a + \lambda \cdot 1_{\dot{A}} \mapsto L_a + \lambda \cdot 1_{B(H)}$ is a homomorphism. Hence the norm (C.61) satisfies (C.1), for this is satisfied in the Banach algebra $B(A)$. In order to prove that the norm (C.61) satisfies (C.2), we note that if an involution on a Banach algebra A satisfies $\|a\|^2 \leq \|a^*a\|$, then A is a C*-algebra, because substituting $a \rightsquigarrow a^*$ gives $\|a^*\|^2 \leq \|aa^*\| \leq \|a\|\|a^*\|$, i.e., $\|a^*\| \leq \|a\|$, so that $\|a^*a\| \leq \|a\|^2$ and hence $\|a\|^2 = \|a^*a\|$.

Thus it suffices to show that for each $a \in A$ and $\lambda \in \mathbb{C}$ we have

$$\|L_a + \lambda \cdot 1_{\dot{A}}\|^2 \leq \|(L_a + \lambda \cdot 1_{\dot{A}})^*(L_a + \lambda \cdot 1_{\dot{A}})\|. \tag{C.64}$$

To prove (C.64), we note that by definition of the norm in $B(A)$, for given $T \in B(A)$ and $\varepsilon > 0$, there exists a $b \in A$, with $\|b\| = 1$, such that $\|T\|^2 - \varepsilon \leq \|T(b)\|^2$. Applying this with $T = L_a + \lambda \cdot 1_{\dot{A}}$, we infer that for every $\varepsilon > 0$ one has

$$\|L_a + \lambda \cdot 1_{\dot{A}}\|^2 - \varepsilon \leq \|(L_a + \lambda \cdot 1_{\dot{A}})b\|^2 = \|ab + \lambda b\|^2 = \|(ab + \lambda b)^*(ab + \lambda b)\|.$$

Here we used (C.2) in A. Using (C.60), the right-hand side may be rearranged as

$$\|L_{b^*} L_{a^* + \overline{\lambda} \cdot 1_{\dot{A}}} L_{a + \lambda \cdot 1_{\dot{A}}} b\| \leq \|L_{b^*}\|\,\|(L_a + \lambda \cdot 1_{\dot{A}})^*(L_a + \lambda \cdot 1_{\dot{A}})\|\,\|b\|. \tag{C.65}$$

Since $\|L_{b^*}\| = \|b^*\| = \|b\| = 1$ by (C.63) and (A.22), and $\|b\| = 1$ also in the last term, the inequality (C.64) follows by letting $\varepsilon \to 0$. □

Hence the C*-algebraic version of Theorem C.29, slightly supplemented, is:

Theorem C.31. *For every C*-algebra A, there is a unique unital C*-algebra $\dot{A}$ and an isometric (hence injective) morphism $A \to \dot{A}$, such that $\dot{A}/A \cong \mathbb{C}$. Moreover, any homomorphism $\alpha : A \to B$ extends to a* unital *homomorphism $\dot{\alpha} : \dot{A} \to \dot{B}$ by*

$$\dot{\alpha}(a + \lambda \cdot 1_{\dot{A}}) = \alpha(a) + \lambda \cdot 1_{\dot{B}}. \tag{C.66}$$

Proof. Uniqueness of $\dot{A}$ follows from Corollary C.28; the rest is obvious. □

This is very important, if only for the following reason:

Definition C.32. *Let A be a C*-algebra without unit. Then the spectrum $\sigma(a)$ of any $a \in A$ consists of all $\lambda \in \mathbb{C}$ for which the operator $a - \lambda$ is* not *invertible in $\dot{A}$.*

Proposition C.33. *If A has no unit, then $0 \in \sigma(a)$ for any $a \in A$.*

Proof. If $0 \notin \sigma(a)$, i.e., if a were invertible in $\dot{A}$, then $a^{-1} = b + \mu \cdot 1_{\dot{A}}$, for some $b \in A$ and $\mu \in \mathbb{C}$. Then $1_{\dot{A}} = aa^{-1} = ab + \mu a \in A$. This is a contradiction. □

The spectral theory of compact operators provides a nice illustration of this proposition: see Theorem B.136.4. At the commutative end of the operator-algebraic world, we have the obvious fact that if X is not compact, no $f \in C_0(X)$ is invertible.

The construction of $\dot{A}$ through (C.56), (C.57), (C.59), and (C.61) also works *verbatim* if A already has a unit 1_A, in which case the spectrum $\sigma(a)$ of $a \in A$ may be compared with the spectrum $\sigma(\dot{a})$ of its image $\dot{a} \equiv (a,0)$ in $\dot{A}$.

Lemma C.34. *Let A be a C*-algebra with unit, embedded in $\dot{A}$. For any $a \in A$, the spectrum $\sigma(a)$ in A is related to the spectrum $\sigma(\dot{a})$ of its image $\dot{a} \equiv (a,0)$ in $\dot{A}$ by*

$$\sigma(\dot{a}) = \sigma(a) \cup \{0\}. \tag{C.67}$$

This will be important for the proof of the fundamental Theorem C.62 below.

Proof. Suppose $0 \neq z \in \rho(a)$, so that $b \equiv (a - z \cdot 1_A)^{-1}$ exists and satisfies

$$ab - zb = ba - zb = 1_A. \tag{C.68}$$

Then $b' = b + z^{-1} \cdot (1_A - 1_{\dot{A}})$ satisfies $ab' - zb' = b'a - zb' = 1_{\dot{A}}$, so that $b' = (a - z \cdot 1_{\dot{A}})^{-1}$ exists in $\dot{A}$, and hence $z \in \rho(\dot{a})$. Conversely, if $0 \neq z \in \rho(\dot{a})$ with corresponding b' as before, then we first form $b = b' - z^{-1} \cdot (1_A - 1_{\dot{A}})$, which satisfies (C.68) but may not lie in A. If $b = b'' + \beta \cdot 1_{\dot{A}}$, where $b'' \in A$ and $\beta \in \mathbb{C}$, this is remedied by redefining $b''' = b + \beta \cdot (1_A - 1_{\dot{A}})$, which lies in A and is inverse to $a - z \cdot 1_A$. Furthermore, by the proof of Proposition C.33 with $a \rightsquigarrow \dot{a}$, we always have $0 \in \sigma(\dot{a})$. If $0 \in \sigma(a)$, then the above argument gives $\sigma(\dot{a}) = \sigma(a)$, which is a special case of (C.67). If $0 \notin \sigma(a)$, then (C.67) follows as it stands. □

To close this section, we intoduce the technique of approximate units, which will play a decisive role in the theory of ideals in C*-algebras (see §C.9). Let us first give an example. For any noncompact space X, the C*-algebra $C_0(X)$ has no unit (the unit would be 1_X, which does not vanish at infinity because it is constant). There is a certain substitute for the absentee unit, though. Taking $X = \mathbb{R}$ for simplicity, and pick a sequence of functions 1_n, $n \in \mathbb{N}$, that take the value 1 on $[-n,n]$ and vanish for $|x| > n+1$. It is clear that one does not have $1_n \to 1_{\mathbb{R}}$ in the sup-norm, but instead one has $\lim_{n\to\infty} \|1_n f - f\|_\infty = 0$ for all $f \in C_0(\mathbb{R})$. More generally, one puts:

Definition C.35. *An* **approximate unit** *in a non-unital C*-algebra A indexed by some directed set Λ is a family $\{1_\lambda\}_{\lambda \in \Lambda}$ of selfadjoint elements of A, such that*

$$\|1_\lambda\| \leq 1, \tag{C.69}$$

and, for each $a \in A$,

$$\lim_{\lambda\to\infty} \|1_\lambda a - a\| = \lim_{\lambda\to\infty} \|a 1_\lambda - a\| = 0. \tag{C.70}$$

Here the limit is meant in the sense of convergence of the nets $\lambda \mapsto \|1_\lambda a - a\|$ and $\lambda \mapsto \|a 1_\lambda - a\|$ in $\mathbb{R}$ indexed by Λ (i.e., for each open neighbourhood U of 0 in $\mathbb{R}$ there is some $\lambda_U \in \Lambda$ such that $\|1_\lambda a - a\| \in U$ for all $\lambda \geq \lambda_U$, etc.).

Proposition C.36. *Every non-unital C*-algebra A has an approximate unit $\{1_\lambda\}_{\lambda \in \Lambda}$. When A is separable, one may choose the directed set Λ countable (i.e. $\Lambda = \mathbb{N}$).*

Proof. One takes Λ to be the set of all finite subsets of A (or, if A is separable, from a countable dense subset of A), partially ordered by inclusion. Hence $\lambda \in \Lambda$ is of the form $\lambda = \{a_1, \ldots, a_n\}$, from which we build the element $b_\lambda = \sum_i a_i^* a_i$. Clearly b_λ is selfadjoint, and according to Theorem C.52 and Proposition C.51 one has $\sigma(b_\lambda) \subset \mathbb{R}^+$, so that $n^{-1} 1_{\dot{A}} + b_\lambda$ is invertible in the unitization $\dot{A}$ of A. Take

$$1_\lambda = b_\lambda (n^{-1} 1_{\dot{A}} + b_\lambda)^{-1}. \tag{C.71}$$

Since $b_\lambda^* = b_\lambda^*$ and b_λ commutes with functions of itself like $(n^{-1} 1_{\dot{A}} + b_\lambda)^{-1}$, one has $1_\lambda^* = 1_\lambda$. Although $(n^{-1} 1_{\dot{A}} + b_\lambda)^{-1}$ is computed in $\dot{A}$, so that it is of the form $c + \mu 1_{\dot{A}}$ (for some $c \in A$ and $\mu \in \mathbb{C}$), one has $1_\lambda = b_\lambda c + \mu b_\lambda$, which lies in A. Using the continuous functional calculus (i.e. Theorem C.25) with $f(t) = t/(n+t)$ on b_λ, one sees from (C.53) and the positivity of b_λ that $\sigma(1_\lambda) \subset [0,1]$. This implies (C.69) because of (B.257). Putting $c_i = 1_\lambda a_i - a_i$, a simple computation shows that

$$\sum_i c_i c_i^* = n^{-2} b_\lambda (n^{-1} 1_{\dot{A}} + b_\lambda)^{-2}. \tag{C.72}$$

We now apply (C.52) with $a \rightsquigarrow b_\lambda$ and $f(t) = n^{-2} t (n^{-1} + t)^{-2}$. Since $f \geq 0$, and f assumes its maximum at $t = 1/n$, one has $\sup_{t \in \mathbb{R}^+} |f(t)| = 1/4n$. As $\sigma(b_\lambda) \subset \mathbb{R}^+$, it follows that $\|f\|_\infty \leq 1/4n$. Therefore, by (C.52) we have

$$\|n^{-2} b_\lambda (n^{-1} 1_{\dot{A}} + b_\lambda)^{-2}\| \leq 1/4n, \tag{C.73}$$

so that $\|\sum_i c_i c_i^*\| \leq 1/4n$ by (C.72). By Lemma C.37 below this implies that $\|c_i c_i^*\| \leq 1/4n$ for each $i = 1, \ldots, n$. Since any $a \in A$ sits in some directed subset of Λ with $n \to \infty$, eq. (C.2) implies

$$\lim_{\lambda \to \infty} \|1_\lambda a - a\|^2 = \lim_{\lambda \to \infty} \|(1_\lambda a - a)^* 1_\lambda a - a\| = \lim_{\lambda \to \infty} \|c_i^* c_i\| = 0. \tag{C.74}$$

The other equality in (C.70) follows analogously. □

In this proof we used the following lemma.

Lemma C.37. *If $a, b \in A^+$ and $\|a + b\| \leq k$, then $\|a\| \leq k$.*

Proof. We first pass to the unitization $\dot{A}$ of A. By (C.83) we have $a + b \leq k 1_{\dot{A}}$, hence $0 \leq a \leq k 1_{\dot{A}} - b$ by linearity of $\leq$ (see Proposition C.51 below), which also implies that $k 1_{\dot{A}} - b \leq k 1_{\dot{A}}$, as $0 \leq b$. Hence, using $-k 1_{\dot{A}} \leq 0$ (since $k \geq 0$), we obtain $-k 1_{\dot{A}} \leq a \leq k 1_{\dot{A}}$, from which $\|a\| \leq k$ by (C.84)). □

C.6 C*-algebras without unit: commutative case

We still owe the reader a proof of Theorems C.8 and C.23 for the nonunital case.

In the commutative case, the unitization procedure has a simple topological meaning, which illustrates the general principle that the use of commutative C*-algebras often allows one to trade topological properties for algebraic ones.

The ***one-point compactification*** $\dot{X}$ of a non-compact locally compact topological space X is the set $\dot{X} = X \cup \infty$, topologized by the open sets in X plus those subsets of $X \cup \infty$ whose complement is compact in X. The injection $i : X \hookrightarrow \dot{X}$ is continuous, and any continuous function $f \in C_0(X)$ extends uniquely to a function $\dot{f} \in C(\dot{X})$ satisfying $\dot{f}(\infty) = 0$. The space $\dot{X}$ is the solution (unique up to homeomorphism) of a universal problem: if $\varphi : X \to Y$ is a map between locally compact Hausdorff spaces such that $Y \backslash f(X)$ is a point and f is a homeomorphism onto its image, then there is a unique homeomorphism $\psi : \dot{X} \to Y$ such that $\varphi = \psi \circ i$. All this is true even when X is compact, in which case ∞ is an isolated point of $\dot{X}$.

The unitization of $C_0(X)$ corresponds to the one-point compactification of X:

Lemma C.38. *Let X be a locally compact Hausdorff space. Then* $\dot{C}_0(X) \cong C(\dot{X})$.

Proof. The map $c_X : \dot{C}_0(X) \to C(\dot{X})$ given by $c_X(f + \lambda \cdot 1_{\dot{A}}) = \dot{f} + \lambda \cdot 1_X$ is obviously an injective homomorphism. To prove surjectivity, note that any $f \in C(\dot{X})$ assumes the form $f = \dot{f} + f(\infty) \cdot 1_{\dot{X}}$, where $\dot{f} = f - f(\infty) \cdot 1_{\dot{X}}$ is such that $\dot{f}_{|X} \in C_0(X)$. Thus our map is an algebraic isomorphism, which by Theorem C.62 is also isometric. □

Lemma C.39. *Let A be a commutative C*-algebra, with unitization $\dot{A}$. Then the following map $s_A : \dot{\Sigma}(A) \to \Sigma(\dot{A})$ between their Gelfand spectra is a homeomorphism:*

1. Each $\omega \in \Sigma(A)$ extends to a character $\dot{\omega} \equiv s_A(\omega)$ on $\dot{A}$ by

$$\dot{\omega}(a + \lambda 1_{\dot{A}}) = \omega(a) + \lambda. \tag{C.75}$$

2. The following functional $\omega_\infty \equiv s_A(\infty)$ on $\dot{A}$ is a character of $\dot{A}$:

$$\omega_\infty(a + \lambda 1_{\dot{A}}) = \lambda. \tag{C.76}$$

3. There are no other characters on $\dot{A}$ (i.e. except ω_∞ and $\dot{\omega}$, where $\omega \in \Sigma(A)$).

Proof. Only the third part is nontrivial: any $\omega' \in \Sigma(\dot{A})$ restricts to $\Sigma(A)$; if this restriction is zero, then $\omega' = \omega_\infty$, and if not, we have $\omega' = \dot{\omega}$ with $\omega = \omega'_{|\Sigma(A)}$. □

We are now in a position to prove Theorem C.8 also in the nonunital case. Applying the unital case of Theorem C.8 to $\dot{A}$ and using Lemma C.39, one finds

$$A \oplus \mathbb{C} = \dot{A} \cong C(\Sigma(\dot{A})) \cong C(\dot{\Sigma}(A)) \cong \dot{C}_0(\Sigma(A)) = C_0(\Sigma(A)) \oplus \mathbb{C}. \tag{C.77}$$

Keeping track of all isomorphisms, the initial $\mathbb{C}$ is duly mapped to the final $\mathbb{C}$ (as befits an isomorphism of unital C*-algebras), and A is mapped to $C_0(\Sigma(A))$. □

Next, we return to Theorem C.23. If X fails to be compact, the difficulty arises that a map $\varphi : X \to Y$ does not, in general, pull back to a morphism $\varphi^* : C_0(Y) \to C_0(X)$. For example, with Y equal to a point, any $f \in C(Y) \cong \mathbb{C}$ pulls back to a constant function on X, which does not vanish at infinity. Hence some restriction is necessary on the class of allowed maps between locally compact Hausdorff spaces.

Definition C.40. *A map $\varphi : X \to Y$ between locally compact Hausdorff spaces is* **proper** *when $\varphi^{-1}(K)$ is compact for any compact set $K \subset Y$.*

Without proof (since this is basic topology), we list some properties of proper maps.

Lemma C.41. *Let $\varphi : X \to Y$ be a map between locally compact Hausdorff spaces.*

1. *φ is proper iff it is closed and $\varphi^{-1}(pt)$ is compact for any point $pt \in Y$.*
2. *If X is compact and φ is continuous, then φ is proper.*
3. *If Y is compact and X is not, proper maps φ (trivially) do not exist.*
4. *If φ is continuous, then φ is proper iff $\dot{\varphi} : \dot{X} \to \dot{Y}$, given by $\dot{\varphi}(x) = \varphi(x)$, $x \in X$, and $\dot{\varphi}(\infty_X) = \infty_Y$, is continuous (which is automatic if X is compact, of course).*
5. *The composition of two proper maps is again proper.*

The algebraic (or "noncommutative") counterpart of a proper map is as follows.

Definition C.42. *A homomorphism $\alpha : A \to B$ between C*-algebras is called* **nondegenerate** *when $\alpha(A)B^- = B$, in other words, if $\alpha(A)B$ (i.e., the linear span of all expressions of the form $\alpha(a)b$, $a \in A$, $b \in B$) is dense in B.*

For example, any unital homomorphism between unital C*-algebras is trivially nondegenerate, and conversely, a nondegenerate homomorphism $\alpha : A \to B$ between unital C*-algebras is automatically unital. To see this, it follows from (C.4) - (C.5) that $e = \alpha(1_A)$ is a projection in B (i.e., $e^2 = e^* = e$), so that $\alpha(A)B \subseteq eB$. Since $B = eB \oplus (1_B - e)B$ as a vector space, $\alpha(A)B$ and hence eB can only be dense in B when $e = 1_B$. Similarly, using an approximate unit in B it is easy to show that nondegenerate homomorphisms $A \to B$ cannot exist if A is unital but B is not.

This is a "noncommutative" version of the third part of Lemma C.41 above.

Lemma C.43. *Let $\varphi : X \to Y$ be a continuous proper map between locally compact Hausdorff spaces. If $f \in C_0(Y)$, then $f \circ \varphi \in C_0(X)$, and the corresponding pullback $\varphi^* : C_0(Y) \to C_0(X)$ is a nondegenerate homomorphism of C*-algebras.*

Proof. Let $f \in C_0(Y)$ and $\varepsilon > 0$, giving a compact $K \subset Y$ such that $|f(y)| < \varepsilon$ for each $y \notin K$. Then $K' = \varphi^{-1}(K) \subset X$ is compact, and $|\varphi^* f(x)| < \varepsilon$ for each $x \notin K'$.

For nondegeneracy, take $g \in C_0(X)$ and $\varepsilon > 0$; these yield a compact set $L \subset X$ such that $|g(x)| < \varepsilon$ for each $x \notin L$. Then $\varphi(L) \subset Y$ is compact, so Urysohn gives us $f \in C_c(Y)$ with $0 \leq f(y) \leq 1$ for each $y \in Y$ and $f(y) = 1$ for each $y \in \varphi(L)$. Then:

$$\|(\varphi^* f) \cdot g - g\|_\infty = \sup_{x \notin L}\{|f(\varphi(x))g(x) - g(x)|\} < 2\varepsilon. \qquad \square$$

The (commutative) C*-algebraic counterpart of this lemma is as follows:

Lemma C.44. *Let $\alpha : A \to B$ be a nondegenerate homomorphism between commutative C*-algebras. If $\omega \in \Sigma(B)$, then $\omega \circ \alpha \in \Sigma(A)$, and the ensuing pullback $\alpha^* : \Sigma(B) \to \Sigma(A)$ is a continuous proper map between the two Gelfand spectra.*

Proof. Multiplicativity of $\omega \circ \alpha$ is clear, as α is a homomorphism. If $\omega \circ \alpha$ were identically zero, then (since ω is not), $\alpha(a) = 0$ for each $a \in A$, which contradicts the assumption that α be nondegenerate. Continuity of α^* follows from the fact that the Gelfand topology is the topology of pointwise convergence. Finally, in the present context, properness of α^* is most appropriately derived as follows:

1. Use (C.66) to pass to a unital homomorphism $\dot{\alpha} : \dot{A} \to \dot{B}$.
2. Theorem C.23.2 gives a continuous map $(\dot{\alpha})^* : \Sigma(\dot{B}) \to \Sigma(\dot{A})$.
3. Lemma C.46 below and continuity of s_B and s_A^{-1} make $(\alpha^*)^{\cdot}$ continuous.
4. Lemma C.41.4 then proves that α^* is proper (and continuous). □

This suggests the following generalization of Theorem C.23:

Theorem C.45. *1. If X is a locally compact Hausdorff space, then $C_0(X)$ is a unital commutative C*-algebra. A continuous proper map $\varphi : Y \to X$ induces a nondegenerate homomorphism $C_0(f) \equiv \varphi^* : C_0(X) \to C_0(Y)$, which behaves well under composition (exactly as in Theorem C.23).*
2. If A is a commutative C-algebra, then $\Sigma(A)$ is a locally compact Hausdorff space. A nondegenerate homomorphism $\alpha : A \to B$ induces a continuous proper map $\Sigma(\alpha) \equiv \alpha^* : \Sigma(B) \to \Sigma(A)$, which behaves well under composition, too.*
3. There are canonical homeomorphisms and isomorphisms,

$$\mathrm{ev}_X : X \stackrel{\cong}{\to} \Sigma(C_0(X)); \tag{C.78}$$

$$G_A : A \stackrel{\cong}{\to} C_0(\Sigma(A)), \tag{C.79}$$

with similar naturalness properties as the corresponding maps in Theorem C.23.

Categorically speaking, Theorem C.23 thus expanded states that *the category* LCHp *of locally compact Hausdorff spaces and proper continuous maps is dual to the category* CCAn *of commutative C*-algebras and nondegenerate homomorphisms.*

Proof. Parts 1 and 2 are Lemmas C.43 and C.44, respectively; correct composition of the maps in question is easily checked (as simply as in the unital case).

Eq. (C.79) has already been proved, cf. (C.77). Similarly, using Proposition C.19 (with $X \rightsquigarrow \dot{X}$) and Lemma C.39 (with $A \rightsquigarrow C_0(X)$), we have

$$X \cup \{\infty\} = \dot{X} \cong \Sigma(C(\dot{X})) \cong \Sigma(\dot{C}_0(X)) \cong \dot{\Sigma}(C_0(X)) = \Sigma(C_0(X)) \cup \omega_\infty. \tag{C.80}$$

Keeping track of the isomorphisms in question, it is easily verified that X and ∞ are mapped to $\Sigma(C_0(X))$ and ω_∞, respectively, and this proves (C.78).

Naturality follows from the unital case (Theorem C.23) and the following lemma:

Lemma C.46. *1. Let $\alpha : A \to B$ be a nondegenerate homomorphism between commutative C*-algebras. Then the following diagram commutes:*

$$\begin{array}{ccc} \dot{\Sigma}(B) & \xrightarrow{s_B} & \Sigma(\dot{B}) \\ \downarrow{\scriptstyle (\alpha^*)^{\cdot}} & & \downarrow{\scriptstyle (\dot{\alpha})^*} \\ \dot{\Sigma}(A) & \xrightarrow{s_A} & \Sigma(\dot{A}), \end{array}$$

where s_A *and* s_B *are defined in Lemma C.39,* $\dot{\alpha}$ *is defined in* (C.66)*, and* $(\alpha^*)^{\cdot} \equiv \dot{\varphi}$ *for* $\varphi = \alpha^* : \Sigma(B) \to \Sigma(A)$*, where the dot is defined as in Lemma C.41.4.*

2. *Let* $\varphi : X \to Y$ *be a proper continuous map between locally compact Hausdorff spaces. Then the following diagram commutes:*

$$\begin{array}{ccc} \dot{C}_0(Y) & \xrightarrow{c_Y} & C(\dot{Y}) \\ \downarrow{\scriptstyle (\varphi^*)^{\cdot}} & & \downarrow{\scriptstyle (\dot{\varphi})^*} \\ \dot{C}_0(X) & \xrightarrow{c_X} & C(\dot{X}), \end{array}$$

where c_X *and* c_Y *are defined in the proof of Lemma C.38,* $(\varphi^*)^{\cdot} \equiv \dot{\alpha}$ *for* $\alpha = \varphi^* : C_0(Y) \to C_0(X)$ *defined by* (C.66)*, and* $\dot{\varphi} : \dot{X} \to \dot{Y}$ *is defined in Lemma C.41.4.*

The proof is a diagram chase, but let us note that in clause 1 the role of nondegeneracy is to ensure that α^* (and hence $(\alpha^*)^{\cdot}$) is *defined* in the first place (cf. Lemma C.44). Similarly, in clause 2, the properness assumption on φ ensures that φ^* (and hence $(\varphi^*)^{\cdot}$) is *defined*. Once defined, commutativity of these diagrams is obvious.

Finally, the property that LCHp is indeed a category is trivial (as the identity maps $\mathrm{id} : X \to X$ are proper), but the corresponding fact for CCAn is not, for we need to show that the identity arrows $\mathrm{id} : A \to A$ are nondegenerate. This comes down to the property that $A^2 = A \cdot A$ is dense in A. In fact, the situation is even better:

Lemma C.47. *In any C*-algebra* A *one has* $A^2 = A$ *(and hence* $A^n = A$*,* $n \in \mathbb{N}$*).*

Proof. We prove that any self adjoint $a \in A$ takes the form

$$a = a_1 a_2, \tag{C.81}$$

for suitable $a_1, a_2 \in A$. Since the linear span of such a is A, this proves the lemma.

We assume A has no unit, for otherwise the claim is trivial. We then embed $A \subset \dot{A}$ and, for $a^* = a \in A$, consider $C^*(a) \subset \dot{A}$. We factor the identity function $t \mapsto t$ on $\sigma(a) \subset \mathbb{R}$ as $t = f_1(t) f_2(t)$ for some $f_i \in C(\sigma(a))$, so that by Corollary C.26, we have (C.81) for $a_i = f_i(a) \in C^*(a)$. By the properties of the map $f \mapsto f(a)$ mentioned in Corollary C.26, including the fact that $f(1_{\sigma(a)}) = 1_{\dot{A}}$, it follows that if $f(a) = b + \mu \cdot 1_{\dot{A}}$ for some $b \in A$ and $\mu \in \mathbb{C}$, then $f(0) = \mu$; note that $0 \in \sigma(a)$ by Proposition C.33. Consequently, imposing the additional condition $f_i(0) = 0$ enforces $a_i \in A$. □

Corollary C.48. *Each nondegenerate homomorphism* $\alpha : C_0(Y) \to C_0(X)$ *is induced by a proper continuous map* $\varphi : X \to Y$ *via* $\alpha = \varphi^*$*.*

Proof. Given (C.78), the proof is the same as for the compact case, cf. Corollary C.22. In particular, φ is given by (C.37), which map is proper because α^* is proper by Lemma C.44 and ev_X, and ev_X are homeomorphisms. □

C.7 Positivity in C*-algebras

We now turn to the important notion of *positivity*. First, we give two examples:

- An *operator* $a \in B(H)$ on a Hilbert space H is called *positive* when $\langle \psi, a\psi \rangle \geq 0$ for each $\psi \in H$. By Proposition B.99, this property is equivalent to $a^* = a$ and $\sigma(a) \subseteq \mathbb{R}^+$, or to $a = b^*b$ for some $b \in B(H)$, or to $a = c^2$ for some $c = c^* \in B(H)$.
- A *function* f on some space X is called *positive* when $f(x) \geq 0$ for all $x \in X$. This applies, in particular, to elements of the commutative C*-algebra $C_0(X)$.

These examples are not as dissimilar as they might appear at first sight: $a \in B(H)$ is positive iff its Gelfand transform $\mathrm{id}_{\sigma(a)} = \hat{a}$ is positive as a function in $C(\sigma(a))$; cf. Theorem C.24. Hence we have a notion of positivity for certain concrete C*-algebras, which we would like to generalize to arbitrary abstract C*-algebras.

Definition C.49. *An element a of a C*-algebra A is called* **positive** *when $a = a^*$ and its spectrum is positive; i.e., $\sigma(a) \subset \mathbb{R}^+$. We write $a \geq 0$ when a is positive, and A^+ for the set of all positive elements in A.*

The basic structure of A^+ is captured by the following definition.

Definition C.50. *A* **convex cone** *in a real vector space V is a subspace V^+ such that:*

1. *If $v \in V^+$ and $t \in \mathbb{R}^+$, then $tv \in V^+$.*
2. *If $v, w \in V^+$, then $v + w \in V^+$.*
3. *$V^+ \cap -V^+ = \{0\}$.*

A **linear partial ordering** *in V is a partial ordering $\leq$ in which $v \leq w$ implies $tv \leq tw$ for all $t \in \mathbb{R}^+$, as well as $v + u \leq w + u$ for all $u \in V$.*

These structures are equivalent: A convex cone $V^+ \subset V$ defines a linear partial ordering $\leq$ by $v \leq w$ if $w - v \in V^+$, and conversely, $\leq$ yields $V^+ = \{v \in V \mid 0 \leq v\}$.

Proposition C.51. *The set A^+ of all positive elements of a C*-algebra A is a convex cone in the real vector space A_{sa}, see* (C.6).

Proof. Let $a \in A^+$. Property 1 follows from $\sigma(ta) = t\sigma(a)$, which is a special case of (B.270). Since $\sigma(a) \subseteq [0, r(a)]$, we have $|c - \lambda| \leq c$ for all $\lambda \in \sigma(a)$ and all $c \geq r(a)$. Hence $\sup_{\lambda \in \sigma(a)} |c \cdot 1_{\sigma(a)} - \hat{a}(\lambda)| \leq c$ by (C.22) and Theorem C.24, i.e., $\|c \cdot 1_{\sigma(a)} - \hat{a}\|_\infty \leq c$. Gelfand transforming back to $C^*(a)$, by (C.18) this implies

$$\|c \cdot 1_A - a\| \leq c, \tag{C.82}$$

for all $c \geq \|a\|$. Inverting this, one sees that if (C.82) holds for some $c \geq \|a\|$, then $\sigma(a) \subset \mathbb{R}^+$. Use this with $a \rightsquigarrow a + b$ and $c = \|a\| + \|b\|$, so $c \geq \|a + b\|$. Then

$$\|c \cdot 1_A - (a + b)\| \leq \|(\|a\| - a)\| + \|(\|b\| - b)\| \leq c,$$

where in the last step we used the previous paragraph for $a \in A^+$ and $b \in A^+$ separately. As for a, this inequality implies $a + b \in A^+$. Finally, when $a \in A^+$ and $a \in -A^+$ it must be that $\sigma(a) = \{0\}$, hence $a = 0$ by (B.257) and Definition A.1. □

For example, when $a = a^*$ one checks the validity of the important inequalities

$$-\|a\| \cdot 1_A \leq a \leq \|a\| \cdot 1_A, \tag{C.83}$$

by taking the Gelfand transform of $C^*(a)$. This also yields the implication

$$-b \leq a \leq b \Longrightarrow \|a\| \leq \|b\|, \tag{C.84}$$

because the antecedent and (C.83) with $a \rightsquigarrow b$ yield $-\|b\| \cdot 1_A \leq a \leq \cdot \|b\| \, 1_A$, so that $\sigma(u) \subseteq [-\|b\|, \|b\|]$, hence $\|a\| \leq \|b\|$ by (B.257) and (B.254).

We now come to the central result in the theory of positivity in C*-algebras, which generalizes the cases $A = B(H)$ and $A = C_0(X)$ discussed at the beginning.

Theorem C.52. *With $A^+ = \{a \in A \mid a \geq 0\}$ as in Definition C.49, one has*

$$A^+ = \{a^2 \mid a^* = a\} \tag{C.85}$$

$$= \{a^* a | a \in A\}. \tag{C.86}$$

Proof. If $\sigma(a) \subset \mathbb{R}^+$ and $a = a^*$, then $\sqrt{a} \in A$ is defined by Corollary C.26 for $f = \sqrt{\ }$, and satisfies $\sqrt{a}^2 = a$. Hence $A^+ \subseteq \{a^2 \mid a^* = a\}$. The opposite inclusion follows from (C.53) and Corollary C.27. This proves (C.85).

Towards (C.86), the inclusion $A^+ \subseteq \{a^* a | a \in A\}$ is trivial from (C.85).

Lemma C.53. *Each selfadjoint element a has a unique decomposition*

$$a = a_+ - a_-, \tag{C.87}$$

where $a_+, a_- \in A^+$ and $a_+ a_- = 0$. Moreover, $\|a_\pm\| \leq \|a\| = \max\{\|a\|_+, \|a\|_-\}$.

Proof. Apply Corollary C.26 with $f = \mathrm{id}_{\sigma(a)} = f_+ - f_-$, where $\mathrm{id}_{\sigma(a)}(t) = t$ and $f_\pm(t) = \max\{\pm t, 0\}$. The norm property follows from (C.52). Uniqueness follows from the corresponding property in $C(\sigma(a))$, where it is obvious. □

Apply the lemma to $a = b^* b$ (noting that a is selfadjoint). Then

$$(a_-)^3 = -a_-(a_+ - a_-)a_- = -a_- a a_- = -a_- b^* b a_- = -(ba_-)^* ba_-. \tag{C.88}$$

Since $\sigma(a_-) \subset \mathbb{R}^+$ because a_- is positive, we see from (C.53) with $f(t) = t^3$ that $(a_-)^3 \geq 0$. Hence $-(ba_-)^* ba_- \geq 0$.

Lemma C.54. *If $-c^* c \in A^+$ for some $c \in A$ then $c = 0$.*

Proof. We can write $c = d + ie$, d and e selfadjoint, so that

$$c^* c = 2d^2 + 2e^2 - cc^*. \tag{C.89}$$

Now for any $a, b \in A$ one has

$$\sigma(ab) \cup \{0\} = \sigma(ba) \cup \{0\}. \tag{C.90}$$

This is because for $z \neq 0$, invertibility of $ab - z$ implies invertibility of $ba - z$; indeed,

$$(ba - z)^{-1} = z^{-1}(b(ab - z)^{-1}a - 1_A). \tag{C.91}$$

Applying (C.90) with $a \rightsquigarrow c$ and $b \rightsquigarrow c^*$, it follows that $\sigma(c^*c) \subset \mathbb{R}^-$ implies $\sigma(cc^*) \subset \mathbb{R}^-$, hence $\sigma(-cc^*) \subset \mathbb{R}^+$. By (C.85) and Proposition C.51 (applied to Definition C.50.2), eq. (C.89) then implies that $c^*c \geq 0$, i.e., $\sigma(c^*c) \subset \mathbb{R}^+$, so that the assumption $-c^*c \in A^+$ now yields $\sigma(c^*c) = 0$. Hence $c = 0$ by Proposition C.51 applied to Definition C.50.3. □

By this lemma, the last claim preceding it implies $ba_- = 0$. As

$$(a_-)^3 = -(ba_-)^*ba_- = 0, \tag{C.92}$$

we see that $(a_-)^3 = 0$, and finally $a_- = 0$ by Corollary C.26 with $f(t) = t^{1/3}$. Hence $b^*b = a_+ \in A^+$. Thus $\{a^*a | a \in A\} \subseteq A^+$, which ends the proof of Theorem C.52. □

An important consequence of (C.86) is the fact that inequalities $a_1 \leq a_2$ for selfadjoint a_1, a_2 are stable under conjugation by arbitrary elements $b \in A$, so that $a_1 \leq a_2$ implies $b^*a_1b \leq b^*a_2b$. This is because $a_1 \leq a_2$ is the same as $a_2 - a_1 \geq 0$, and hence by (C.86) there is an $a_3 \in A$ such that $a_2 - a_1 = a_3^*a_3$. But $(a_3b)^*a_3b \geq 0$, i.e., $b^*ab \leq b^*a_2b$. For example, replace a in (C.83) by a^*a, and use (C.2), yielding $a^*a \leq \|a\|^2 1_A$. Applying the above principle gives the operator inequality

$$b^*a^*ab \leq \|a\|^2 b^*b \ (a, b \in A). \tag{C.93}$$

We note that the definition of a state implies that if $a \leq b$, then $\omega(a) \leq \omega(b)$, so that

$$\omega(b^*a^*ab) \leq \|a\|^2 \omega(b^*b), \tag{C.94}$$

from (C.93). This is a key lemma for the GNS-construction (cf. Theorem C.88).

At last, we are also in a position to prove the fundamental Lemma C.4.

Proof. If ω is positive and $a^* = a$, then (C.83) in the form $\|a\| \cdot 1_A \pm a \geq 0$ gives $\omega(a) \leq \|a\|\omega(1_A)$, and hence $\omega(a) \in \mathbb{R}$. For general $a \in A$, eq. (C.8) then implies $\omega(a^*) = \overline{\omega(a)}$ (which may alternatively be proved from Lemma C.53). This, in turn, makes the form (C.24) hermitian. Cauchy–Schwarz then gives $|\omega(a)|^2 \leq \omega(a^*a)\omega(1_A)$, as in (C.25). Furthermore, if $\|a\| \leq 1$ then also $\|a^*a\| \leq 1$ by (C.2), so that (C.83) gives $\omega(a^*a) \leq \omega(1_A)$. Combining these inequalities yields $|\omega(a)| \leq \omega(1_A)$, so ω is bounded with $\|\omega\| \leq \omega(1_A)$; taking $a = 1_A$ gives equality.

Conversely, assume that $\|\omega\| = \omega(1_A) = 1$. In proving that $\omega(a) \geq 0$ whenever $a \geq 0$, we may also assume that $0 \leq a \leq 1_A$. Then (C.7) shows that $\alpha \equiv \omega(a) \in \mathbb{R}$. Also, we have $\sigma(a) \subseteq [0,1]$ and hence $\sigma(1_A - a) \subseteq [0,1]$, which in turns implies $0 \leq (1_A - a) \leq 1_A$, and hence $\|1_A - a\| \leq 1$, cf. (C.84). Then

$$1 - \alpha \leq |1 - \alpha| = |\omega(1_A - a)| \leq \|\omega\|\|1_A - a\| \leq 1, \tag{C.95}$$

whence $\alpha \geq 0$, and hence $\omega(a) \geq 0$. □

C.8 Ideals in Banach algebras

This section returns to general Banach algebras. It has two aims: it completes the (first) proof of Theorem C.8, and it prepares for the theory of ideals in C*-algebras.

Definition C.55. *Let A be a Banach algebra.*

- *A* **left ideal** *(***right ideal***) in A is a closed linear subspace J for which $a \in J$ implies $ba \in J$ ($ab \in J$) for all $b \in A$.*
- *An* **ideal** *in A is both a left and a right ideal (i.e., a closed two-sided ideal).*
- *A* **maximal ideal** *is a proper ideal $J \subset A$ (i.e., $J \neq \{0\}$ and $J \neq A$) that is not properly contained in any larger proper ideal.*

Thus *an ideal is closed by definition*. However, it is useful to know that if we omit the word 'closed' throughout Definition C.55, a *maximal* ideal $J \subset A$ (defined in the purely algebraic sense) is automatically closed. Indeed, note that the closure $\overline{J}$ of J cannot be A, since J does not contain any invertible element of A (otherwise it would coincide with A), and the set A_* of all invertible elements in A is open (see the proof of Theorem B.84). Since $J \subseteq \overline{J} \subset A$ and J is maximal, $\overline{J} = J$.

Furthermore, one often uses the fact that an ideal J that contains an invertible element a must coincide with A (since $a^{-1}a = 1_A$ must then lie in J, whence $J = A$).

In the commutative case, left and right ideals are the same as ideals. For example, if $A = C(X)$ for a compact space X, then each closed subspace $Y \subset X$ defines an ideal

$$C(X;Y) = \{f \in C(X) \mid f(x) = 0 \,\forall x \in Y\}. \tag{C.96}$$

Note that $C(X;Y)$ is indeed closed by definition of the sup-norm, and that

$$C(X;Y) \cong C_0(X \backslash Y). \tag{C.97}$$

Proposition C.83 in §C.11 shows that all ideals in $C(X)$ are of this form. It is not necessary to assume that Y is closed, but this assumption entails no loss of generality, since $C(X;Y) = C(X;\overline{Y})$, where $\overline{Y}$ is the closure of Y. We will see that $C(X;Y)$ is maximal iff Y is a point, and that all *maximal* ideals in $C(X)$ are of this form.

The next proposition is predicated on an elementary Banach space result:

Lemma C.56. *If V is a Banach space and W is a closed linear subspace of V, then the vector space quotient V/W is a Banach space in the "distance to W" norm*

$$\|\tau(v)\| = \inf_{w \in W} \|v - w\|, \tag{C.98}$$

where $\tau : V \to V/W$ is the canonical projection. Also, $\|\tau(v)\| \leq \|v\|$ for any $v \in V$.

Proof. First, (C.98) is well defined, for if $\tau(v') = \tau(v)$, i.e., $v - v' = w' \in W$, then

$$\begin{aligned}\|\tau(v')\| &= \inf\{\|v' - w\|, w \in W\} = \inf\{\|v' - w - w'\|, w \in W\}\\ &= \inf\{\|v - w\|, w \in W\} = \|\tau(v)\|.\end{aligned}$$

The axioms for a norm are easily verified, except positive definiteness: we have $\|\tau(v)\| = 0$ iff $\inf\{\|v-w\|, w \in W\} = 0$; hence there must be a sequence (w_n) in W with $v - w_n \to 0$, or $w_n \to v$. Since W is closed, $v \in W$, so that $\tau(v) = 0$. For the last claim, eq. (C.98) yields $\|[\tau(v)\| \leq \|v-w\|$ for all $w \in W$; take $w = 0$.
There seems to be no natural proof of the completeness of V/W, but here is a trick: for any Cauchy sequence $(\tau(v_n))_n$ in V/W, find a subsequence $(\tau(v_{n_k}))_k$ with $\|\tau(v_{n_{k+1}}) - \tau(v_{n_k})\| < 2^{-k}$ for all k. Using induction in k, one finds a sequence (u_k) in V with $\tau(u_k) = \tau(v_{n_k})$ and $\|u_{k+1} - u_k\| < 2^{-k}$. Hence $u_k \to u$ (since V is complete), and hence $\tau(v_{n_k}) \to \tau(u)$ by continuity of τ. Then also $(\tau(v_n))_n \to 0$. □

Proposition C.57. *If J is an ideal in a Banach algebra A, then the quotient A/J is a Banach algebra with multiplication*

$$\tau(a)\tau(b) = \tau(ab). \tag{C.99}$$

If A is unital and J is proper, A/J is unital, with unit $\tau(1_A)$ satisfying

$$\|\tau(1_A)\| = 1. \tag{C.100}$$

Proof. As far as the Banach algebra structure is concerned, first note that (C.99) is well defined: when $j_1, j_2 \in J$ one has

$$\tau(a+j_1)\tau(b+j_2) = \tau(ab + aj_2 + j_1 b + j_1 j_2) = \tau(ab) = \tau(a)\tau(b), \tag{C.101}$$

since $aj_2 + j_1 b + j_1 j_2 \in J$ by definition of an ideal, and $\tau(j) = 0$ for all $j \in J$.

To prove (C.1), observe that, by definition of the infimum, for given $a \in A$, for each $\varepsilon > 0$ there exists a $j \in J$ such that

$$\|\tau(a)\| + \varepsilon \geq \|a + j\|. \tag{C.102}$$

For if such a j would not exist, then $\|\tau(a)\| \leq \|a+j\| - \varepsilon$ for all $j \in J$, violating (C.98). On the other hand, for any $j \in J$, it is clear from (C.98) that

$$\|\tau(a)\| = \|\tau(a+j)\| \leq \|a+j\|. \tag{C.103}$$

For $a, b \in A$, choose $\varepsilon > 0$ and $j_1, j_2 \in J$ such that (C.102) holds for a, b, and estimate

$$\begin{aligned} \|\tau(a)\tau(b)\| &= \|\tau(a+j_1)\tau(b+j_2)\| = \|\tau((a+j_1)(b+j_2))\| \\ &\leq \|(a+j_1)(b+j_2)\| \leq \|a+j_1\|\,\|b+j_2\| \\ &\leq (\|\tau(a)\| + \varepsilon)(\|\tau(b)\| + \varepsilon). \end{aligned} \tag{C.104}$$

Letting $\varepsilon \to 0$ yields

$$\|\tau(a)\tau(b)\| \leq \|\tau(a)\|\,\|\tau(b)\|. \tag{C.105}$$

If A has a unit, $\tau(1_A)$ is a unit in A/J, cf. (C.99). By (C.103) with $a = 1_A$ and $j = 0$ one has $\|\tau(1_A)\| \leq \|1_A\| = 1$. On the other hand, from (C.105) and (C.99) with $b = 1_A$ and $a \in A\backslash J$, one derives $\|\tau(1_A)\| \geq 1$. Hence (C.100) follows. □

In a C*-algebra the last step is unnecessary, since a unit necessarily has norm one.

In the commutative case, a nice example (with X and Y compact, as above), is

$$C(X)/C(X;Y) \cong C(Y), \tag{C.106}$$

as two elements f,g of $C(X)$ are identified in $C(X)/C(X;Y)$ when $f-g \in C(X;Y)$, i.e., when they coincide on Y. If one looks at $C(X;Y)$ as the kernel of the restriction map $r_Y : C(X) \to C(Y)$, then $\text{ran}(r_Y) \cong C(X)/\ker(r_Y)$, which is just (C.106).

We now prove Proposition C.13, which we unfold as:

1. If $\omega \in \Sigma(A)$, then $J_\omega = \ker(\omega)$ is a maximal ideal in A;
2. $\omega_1 = \omega_2$ iff $J_{\omega_1} = J_{\omega_2}$;
3. Every maximal ideal J is of the form $J = J_\omega$, for some $\omega \in \Sigma(A)$.

For the first claim, J_ω is an ideal since ω is multiplicative. To prove maximality, suppose $J_\omega \subseteq I \subset A$ for some ideal I. Then $\omega(I)$ is an ideal in $\mathbb{C}$, so either $\omega(I) = \{0\}$ or $\omega(I) = \mathbb{C}$. In the former case, $I = J_\omega$ (since $I \subseteq \ker_\omega = J$), in the latter, $I = A$ (because for any $a \in A$ there is $b \in I$ such that $\omega(a) = \omega(b)$, whence $a - b \in \ker_\omega$ and hence $a - b \in I$, or $a \in b + I = I$). Thus J_ω is maximal.

For the second, if $\omega_1(a) = c$, then $\omega_1(a - c \cdot 1_A) = 0$ by (C.14), so if $\ker(\omega_1) = \ker(\omega_2)$, then also $\omega_2(a - c \cdot 1_A) = 0$ and hence $\omega_2(a) = c = \omega_1(a)$.

Finally, let J be maximal. Since $J \neq A$, there is a nonzero $b \in A$, $b \notin J$. Form

$$J_b = \{ba + j \mid a \in A, j \in J\}. \tag{C.107}$$

Since A is commutative, J_b is an ideal. Taking $a = 0$ gives $J \subseteq J_b$. Taking $a = 1_A$ and $j = 0$ gives $b \in J_b$, so that $J_b \neq J$. Hence $J_b = A$, as J is maximal. In particular, $1_A \in J_b$, so that $1_A = ba + j$ for some $a \in A, j \in J$. Applying $\tau : A \to A/J$ gives

$$\tau(1_A) = 1_A = \tau(ba) = \tau(b)\tau(a), \tag{C.108}$$

because of (C.99) and $\tau(J) = 0$. Hence $\tau(a) = \tau(b)^{-1}$ in A/J. Since $b \neq 0$ was arbitrary, this shows that every nonzero element of A/J is invertible. At this point it is therefore appropriate to invoke the ***Gelfand–Mazur Theorem***:

Theorem C.58. *If every nonzero element of a unital commutative Banach algebra B is invertible (i.e., if B is simple), then $B \cong \mathbb{C}$ as Banach algebras.*

Proof. Since $\sigma(b) \neq \emptyset$, for each $b \neq 0$ there is $\lambda \in \mathbb{C}$ for which $b - \lambda \cdot 1_B$ is not invertible. Hence $b - \lambda \cdot 1_B = 0$ by assumption, and $b \mapsto \lambda$ is an isomorphism. □

Hence there is an isomorphism $\psi : A/J \to \mathbb{C}$, from which we define $\omega : A \to \mathbb{C}$ by $\omega(a) = \psi(\tau(a))$. This map is clearly linear (since τ and ψ are), and nonzero (because $\omega(1_A) = 1$). Also, $\omega(a)\omega(b) = \omega(ab)$ by (C.99) and the fact that ψ is a homomorphism, so $\omega \in \Sigma(A)$. Finally, since $\ker(\tau) = J$ and ψ is an isomorphism, $J = \ker(\omega)$. This proves claim 3 above, and therefore Proposition C.13 also follows.

C.9 Ideals in C*-algebras

Definition C.55 *verbatim* applies to C*-algebras. One would expect that an ideal in a C*-algebra is required to be selfadjoint by definition, but this is unnecessary:

Proposition C.59. *Let J be an ideal in a C*-algebra A. If $a \in J$ then $a^* \in J$; in other words, every ideal in a C*-algebra is automatically selfadjoint.*

The proof (which generalizes a similar argument for compact operators, given at the end of §B.18) relies on the theory of approximate units (see §C.5).

Proof. Let $J \subset A$ be the given ideal, and put $J^* = \{a^* | a \in J\}$. Note that $j \in J$ implies $j^* j \in J \cap J^*$: it lies in J because J is an ideal, hence a left-ideal, and it lies in J^* because J^* is an ideal, hence a right-ideal. Since J is an ideal, $J \cap J^*$ is a C*-subalgebra of A. Hence by C.36 it has an approximate unit $\{1_\lambda\}$. Take $j \in J$. Using (C.2),

$$\begin{aligned} \|j^* - j^* 1_\lambda\|^2 &= \|(j - 1_\lambda j)(j^* - j^* 1_\lambda)\| \\ &= \|(j^* j - j^* j 1_\lambda)\| + \|1_\lambda\| \, \|(jj^* - jj^* 1_\lambda)\|, \end{aligned} \tag{C.109}$$

since $1_\lambda^* = 1_\lambda$. As we have seen, $j^* j \in J \cap J^*$, so that, also using (C.69), both terms vanish for $\lambda \to \infty$. Hence $\lim_{\lambda \to \infty} \|j^* - j^* 1_\lambda\| = 0$. But 1_λ lies in $J \cap J^*$, so certainly $1_\lambda \in J$, and since J is an ideal it must be that $j^* 1_\lambda \in J$ for all λ. Hence j^* is a norm-limit of elements in J; since J is closed, it follows that $j^* \in J$. □

We now turn to a C*-algebraic analogue of Proposition C.57, which is of sufficient importance to promote it to the status of a theorem:

Theorem C.60. *Let J be an ideal in a C*-algebra A. Then A/J is a C*-algebra with respect to the norm (C.98), the multiplication (C.99), and the involution*

$$\tau(a)^* = \tau(a^*). \tag{C.110}$$

The proof of this theorem uses approximate units, too. In view of Proposition C.57, all we need to prove to establish Theorem C.60 is the property (C.2). This uses:

Lemma C.61. *Let $\{1_\lambda\}$ be an approximate unit for J, and let $a \in A$. Then*

$$\|\tau(a)\| = \lim_{\lambda \to \infty} \|a - a 1_\lambda\|. \tag{C.111}$$

Proof. It is obvious from (C.98) that

$$\|a - a 1_\lambda\| \geq \|\tau(a)\|. \tag{C.112}$$

For the opposite inequality, add a unit 1_A to A if necessary, pick any $j \in J$, and write

$$\|a - a 1_\lambda\| = \|(a + j)(1 - 1_\lambda) + j(1_\lambda - 1)\| \leq \|a + j\| \, \|1 - 1_\lambda\| + \|j 1_\lambda - j\|. \tag{C.113}$$

Note that

$$\|1 - 1_\lambda\| \leq 1, \tag{C.114}$$

by Definition C.35 and the proof of Proposition C.51. The second term on the right-hand side goes to zero for $\lambda \to \infty$, since $j \in J$. Hence

$$\lim_{\lambda \to \infty} \|a - a1_\lambda\| \leq \|a + j\|. \tag{C.115}$$

For each $\varepsilon > 0$ we can choose $j \in J$ so that (C.102) holds. For this specific j, we combine (C.112), (C.115), and (C.102) to find

$$\lim_{\lambda \to \infty} \|a - a1_\lambda\| - \varepsilon \leq \|\tau(a)\| \leq \|a - a1_\lambda\|. \tag{C.116}$$

Letting $\varepsilon \to 0$ proves (C.111). □

We now prove (C.2) in A/J. Successively using (C.111), (C.2) in $\dot{A}$, (C.114), (C.111), (C.99), and (C.110), we find

$$\begin{aligned}
\|\tau(a)\|^2 &= \lim_{\lambda \to \infty} \|a - a1_\lambda\|^2 = \lim_{\lambda \to \infty} \|(a - a1_\lambda)^*(a - a1_\lambda)\| \\
&= \lim_{\lambda \to \infty} \|(1_A - 1_\lambda)a^*a(1_A - 1_\lambda)\| \leq \lim_{\lambda \to \infty} \|1 - 1_\lambda\| \, \|a^*a(1_A - 1_\lambda)\| \\
&\leq \lim_{\lambda \to \infty} \|a^*a(1_A - 1_\lambda)\| = \|\tau(a^*a)\| = \|\tau(a)\tau(a^*)\| \\
&= \|\tau(a)\tau(a)^*\|.
\end{aligned} \tag{C.117}$$

As in the proof of Proposition C.30, this implies (C.2), and hence Theorem C.60. □

We now state and prove the key result about morphisms.

Theorem C.62. *Let $\alpha : A \to B$ be a nonzero homomorphism between C*-algebras.*

1. *The homomorphism α is continuous, with norm $\|\alpha\| = 1$.*
2. *Its kernel $\ker(\alpha)$ is an ideal in A.*
3. *If α is injective, then it is isometric.*
4. *An isomorphism of C*-algebras is automatically isometric.*
5. *The range $\alpha(A)$ is a C*-subalgebra of B; in particular, $\alpha(A)$ is closed in B.*

Proof. If necessary, we first reduce the proof of the first claim to the case where A and B have units and α is unital: we do so by replacing A and B by $\dot{A}$ and $\dot{B}$, respectively (even if A and/or B was already unital in the first place, but α was not), and replacing α by the homomorphism $\dot{\alpha} : \dot{A} \to \dot{B}$ defined in (C.66). If we do so, it follows from Lemma C.34 that in the worst case the spectrum of a or $\alpha(a)$ is modified by adding 0, which does not change the spectral radius. Therefore, the move from α to $\dot{\alpha}$ makes no difference to the argument to follow, so we assume that $1_A \in A$ and $1_B \in B$, and $\alpha(1_A) = 1_B$. If $z \in \rho(a)$, so that $(a - z)^{-1}$ exists in A, then $\alpha(a - z)$ is certainly invertible in B, for (C.4) implies that $(\alpha(a - z))^{-1} = \alpha((a - z)^{-1})$. Hence $\rho(a) \subseteq \rho(\alpha(a))$, so that

$$\sigma(\alpha(a)) \subseteq \sigma(a). \tag{C.118}$$

Replacing a by a^*a this gives $r(\alpha(a^*a)) \leq r(a^*a)$, and since $\alpha(a^*a) = \alpha(a)^*\alpha(a)$, eq. (C.55) yields $\|\alpha(a)\| \leq \|a\|$, and hence $\|\alpha\| \leq 1$. This proves continuity of α.

Recalling that ideals in C*-algebras have to be closed by definition, this also implies the second claim of the theorem (whose algebraic content is trivial).

We now prove the third claim of the theorem (which trivially implies the fourth). Assume there is $b \in A$ for which $\|\alpha(b)\| \neq \|b\|$, so that by $\sigma(a) \neq \sigma(\alpha(a))$ for $a = b^*b$ by (C.55). Then (C.118) implies the strict inclusion $\sigma(\alpha(a)) \subset \sigma(a)$ (as a closed subset). By Urysohn's lemma, there is a nonzero function $f \in C(\sigma(a))$ that vanishes on $\sigma(\alpha(a))$, so that $f(\alpha(a)) = 0$ by Corollary C.26. By Lemma C.63 below, this implies $\alpha(f(a)) = 0$. If α is injective, this contradicts the property $f(a) \neq 0$, which follows from $f \neq 0$ and (C.52). Thus α must be isometric.

Combining the second claim with Theorem C.60, we see that $A/\ker(\alpha)$ is a C*-algebra. By the theory of vector spaces, we have a vector space isomorphism

$$\psi : A/\ker(\alpha) \to \alpha(A), \tag{C.119}$$

so that

$$\psi \circ \tau = \alpha. \tag{C.120}$$

Since α and τ are homomorphisms between C*-algebras, so is ψ. Since ψ is injective, it is isometric, as we have just shown. Hence $\psi(A/\ker(\alpha))$ has closed range in B. But $\psi(A/\ker(\alpha)) = \alpha(A)$, so that α has closed range in B. Since α is a morphism, its image is a *-algebra in B, which by the preceding sentence is closed in the norm of B. Hence $\alpha(A)$, inheriting all operations in B, is a C*-algebra.

Finally, we prove that for the projection $\tau : A \to A/J$ in the case at hand we have

$$\|\tau\| = 1. \tag{C.121}$$

If A has a unit, this follows from Lemma C.56 with (C.100). If not, the argument is similar, using an approximate identity (1_λ) for A: from (C.105) we obtain $\lim_\lambda \|\tau(1_\lambda\| \geq 1$, which with (C.69) gives $\sup_\lambda \|\tau(1_\lambda\| = 1$. Since $\|\tau\| \leq 1$ from Lemma C.56, this yields (C.121).

Because ψ is an isometry, it then follows from (C.120) that $\|\alpha\| = 1$. □

Here we used a nice property of the continuous functional calculus (Theorem C.25):

Lemma C.63. *If $\alpha : A \to B$ is a morphism, and $a = a^*$, then*

$$f(\alpha(a)) = \alpha(f(a)) \ \ (f \in C(\sigma(a))). \tag{C.122}$$

Here $f(a)$ and $f(\alpha(a))$ are defined through Theorem C.25, cf. (C.118).

Proof. The property is true for polynomials by (C.4), since for those functions, $f(a)$ and $f(\alpha(a))$ have their naive meaning. The general claim follows by continuity. □

Corollary C.64. *Every ideal in a C*-algebra is the kernel of some homomorphism.*

Proof. This follows from Proposition C.59, since J is the kernel of $\tau : A \to A/J$, where A/J is a C*-algebra and τ is a morphism by (C.99), and (C.110). □

C.10 Hilbert C*-modules and multiplier algebras

In §C.5 we explained the *minimal* way of adding a unit to a C*-algebra that did not have one to begin with (although the procedure even works if it does). There is also a *maximal* way, which embeds a non-unital C*-algebra in its *multiplier algebra*. In our view, this maximal extension is actually more elegant and useful than the minimal one, although the commutative case might give the opposite impression: here (as we have seen), the minimal extension corresponds to the simple one-point compactification of the Gelfand spectrum, whereas the maximal one extends the latter to its awesome Čech–Stone compactification. In topology one may doubt if the latter is indeed the neater choice, but for many noncommutative C*-algebras the multiplier algebra comes naturally. For example, the C*-algebra $B_0(H)$ of compact operators on a Hilbert space H is thereby turned into the C*-algebra $B(H)$ of bounded ones.

There are various ways of defining multiplier algebras. Although not strictly necessary, we offer the powerful entrance provided by Hilbert C*-modules, which are simultaneous generalizations of C*-algebras, Hilbert spaces, and vector bundles.

Definition C.65. **A pre-Hilbert C*-module** *over a C*-algebra A consists of:*

- *A right A-module E, i.e., a complex linear space equipped with a bilinear map* $E \times A \to A$, *written* $(\psi, a) \mapsto \psi a$ *(where* $\psi \in E$ *and* $a \in A$*) such that*

$$(\psi b)a = \psi(ba). \tag{C.123}$$

- *A map* $\langle \, , \rangle_A : E \times E \to A$, *linear in the second entry (the axioms below implying antilinearity in the first entry) that for all* $\psi, \varphi \in E$ *and* $b \in A$, *satisfies*

$$\langle \psi, \varphi \rangle_A^* = \langle \varphi, \psi \rangle_A; \tag{C.124}$$
$$\langle \psi, \varphi a \rangle_A = \langle \psi, \varphi \rangle_A a; \tag{C.125}$$
$$\langle \psi, \psi \rangle_A \geq 0; \tag{C.126}$$
$$\langle \psi, \psi \rangle_A = 0 \Leftrightarrow \psi = 0. \tag{C.127}$$

It is useful to note that (C.124) and (C.125) imply that

$$\langle \psi a, \varphi \rangle_A = a^* \langle \psi, \varphi \rangle_A. \tag{C.128}$$

Lemma C.66. *In a pre-Hilbert C*-module E over a C*-algebra A one has:*

$$\langle \psi, \varphi \rangle_A \langle \varphi, \psi \rangle_A \leq \|\varphi\|^2 \langle \psi, \psi \rangle_A; \tag{C.129}$$
$$\|\langle \psi, \varphi \rangle_A\| \leq \|\psi\| \, \|\varphi\|; \tag{C.130}$$
$$\|\psi a\| \leq \|\psi\| \, \|a\|. \tag{C.131}$$

in which the following expression (which duly defines a norm on E) occurs:

$$\|\psi\| = \|\langle \psi, \psi \rangle_A\|^{1/2}. \tag{C.132}$$

Proof. To prove (C.129), we assume $\varphi \neq 0$ (otherwise, the claim clearly holds), so that also $\|\varphi\| > 0$ by (C.127) and (C.132). Replacing φ by $\varphi/\|\varphi\|$ if necessary, (i.e., if $\|\varphi\| \neq 1$), it is then enough to show that whenever $\|\varphi\| = 1$, we have

$$\langle \psi, \varphi \rangle_A \langle \varphi, \psi \rangle_A \leq \langle \psi, \psi \rangle_A. \tag{C.133}$$

To this effect, we substitute $\varphi\langle \varphi, \psi\rangle_A - \psi$ for ψ in (C.126) and use (C.128), (C.124), and (C.125), and (C.93), the latter in form $b^*cb \leq \|c\|b^*b$ for any b and $c \geq 0$ in A. This gives (C.129). Eqs. (C.2), (C.124), and (C.129) then imply (C.130). Eq. (C.131) follows from (C.128), (C.93), (C.84), and (C.2).

Finally, (C.132) defines a norm: scaling is clear, positive definiteness follows from (C.127), and the triangle inequality is easily derived from (C.130). □

Corollary C.67. *The inner product on a pre-Hilbert C*-module is nondegenerate, in that $\psi = 0$ iff $\langle \psi, \varphi \rangle_B = 0$ for all $\varphi \in E$.*

Proof. It follows from (C.129) that for any $\psi \in E$, we have

$$\|\psi\| = \sup\{\|\langle \psi, \varphi \rangle_B\|, \varphi \in E, \|\varphi\| = 1\}. \tag{C.134}$$

. We now come to the main definition.

Definition C.68. *A* **Hilbert C*-module** *over A is a pre-Hilbert C*-module over A that is complete in the norm* (C.132). *We also say that E is a* **Hilbert** A**-module**.

The three most straightforward examples of this concept, written "$E \rightleftharpoons A$", are:

- C*-algebras themselves: $E = A$ with action $(a,b) \mapsto ab$ and inner product

$$\langle a, b \rangle_A = a^* b. \tag{C.135}$$

  By (C.2), the norm in E defined by (C.132) coincides with the original norm.
- Hilbert spaces: $E = H$ and $A = \mathbb{C}$, acting on H by the given scalar multiplication.
- Hermitian vector bundles $\mathscr{E}$ over locally compact Hausdorff spaces X: here $E = C_0(X, \mathscr{E})$ consists of the continuous cross-sections ψ of $\mathscr{E}$ vanishing at infinity, $A = C(X)$ has natural action on E given by $(\psi a)(x) = a(x)\psi(x)$, and the $C_0(X)$-valued inner product is given by the hermitian structure $< \cdot, \cdot >_{\mathscr{E}_x}$ on each fiber,

$$\langle \psi, \varphi \rangle_{C(X)} = x \mapsto < \psi(x), \varphi(x) >_{\mathscr{E}_x}. \tag{C.136}$$

  This implies a norm $\|\psi\| = \sup\{\|\psi(x)\|_{\mathscr{E}_x}, x \in X\}$, where $\|v\|^2_{\mathscr{E}_x} = < v, v >_{\mathscr{E}_x}$.

A Hilbert C*-module $E \rightleftharpoons A$ defines a C*-algebra $C^*(E, A)$ that consists of all maps $a : E \to E$ for which there exists a map $a^* : E \to E$ such that for all $\psi, \varphi \in E$,

$$\langle \psi, a\varphi \rangle_A = \langle a^* \psi, \varphi \rangle_A. \tag{C.137}$$

Such maps are called ***adjointable***. For example, if $E = A$, as in the first example above, then any element $a \in A$ defines an adjointable map simply by left multiplication (i.e., $a(b) = ab$). If A has a unit, then this is it, whereas in the nonunital case there are (many) more adjointable maps on $A \rightleftharpoons A$.

We now show that adjointable maps on a Hilbert C*-module form a C*-algebra.

Theorem C.69. *1. An adjointable map on a Hilbert A-module is automatically $\mathbb{C}$-linear, A-linear (that is, $(a\psi)b = a(\psi b)$ for all $\psi \in E$ and $b \in A$), and bounded.*
2. The adjoint of an adjointable map is unique, and the map $a \mapsto a^*$ defines an involution on the space $C^*(E,A)$ of all adjointable maps on E.
3. Equipped with this involution, and with the usual operator norm on the Banach space E, the space $C^*(E,A)$ is a C-algebra.*
4. For each $a \in C^*(E,A)$ and $\psi \in E$, the usual bound $\|a\psi\| \leq \|a\|\|\psi\|$ sharpens to

$$\langle a\psi, a\psi\rangle_A \leq \|a\|^2 \langle \psi, \psi\rangle_A. \tag{C.138}$$

Proof. The property of $\mathbb{C}$-linearity is obvious, whereas A-linearity follows from (C.128): this gives $\langle a(\psi b), \varphi\rangle_A = \langle a(\psi)b, \varphi\rangle_A$, upon which Corollary C.67 yields the claim. A similar argument shows that $a^* \in C^*(E,A)$ when $a \in C^*(E,A)$.

To prove boundedness, fix $\psi \in E$ and $a \in C^*(E,A)$, and define $T_\psi : E \to A$ by $T_\psi \varphi = \langle a^* a\psi, \varphi\rangle_A$. It is clear from (C.130) that $\|T_\psi\| \leq \|a^* a\psi\|$, so that T_ψ is bounded. On the other hand, since a is adjointable, one has $T_\psi \varphi = \langle \psi, a^* a\varphi\rangle_A$, so that, using (C.130) once again, one has $\|T_\psi \psi\| \leq \|a^* a\varphi\| \, \|\psi\|$. Since E is complete we may apply the Banach–Steinhaus Theorem B.78, which gives

$$\sup\{\|T_\psi\|, \psi \in E, \|\psi\| = 1\} < \infty. \tag{C.139}$$

It then follows from (C.132) that $\|a\| < \infty$. Uniqueness and involutivity of the adjoint are proved as for Hilbert spaces; the former follows from (C.127), the latter in addition requires (C.124). The space $C^*(E,A)$ is norm-closed, since one easily verifies from (C.137) and (C.132) that if $a_n \to a$, then a_n^* converges to a^*. As a norm-closed space of linear maps on a Banach space, $C^*(E,A)$ is a Banach algebra, so that its satisfies (C.1). To check (C.2), one infers from (C.132) and the definition (C.137) of the adjoint that $\|a\|^2 \leq \|a^* a\|$; using (C.1) and the argument leading to (A.22), one first obtains $\|a^*\| = \|a\|$, and subsequently $\|a^* a\| = \|a^2\|$.

Finally, it follows from (C.126), (C.86), and (C.137) that for fixed $\psi \in E$, the map $a \mapsto \langle \psi, a\psi\rangle_A$ from $C^*(E,A)$ to A is positive. Replacing a by $a^* a$ in (C.83) and using (C.2) and (C.137) then leads to (C.138). □

In our first example the C*-algebra $C^*(A,A)$ is usually called the ***multiplier algebra***, denoted by $M(A)$. If A has a unit, then $M(A) = A$, but in general $M(A)$ is much larger than A, and obviously it always has a unit (given by the unit operator on A).

Proposition C.70. *For any commutative C*-algebra A we have an isomorphism*

$$M(A) \xrightarrow{\cong} C_b(\Sigma(A)); \tag{C.140}$$

$$a \mapsto \hat{a}, \tag{C.141}$$

where, in terms of the Gelfand isomorphisms $A \cong C_0(\Sigma(A))$, $f \mapsto \hat{f}$, we have

$$\widehat{a(f)} = \hat{a}\hat{f}. \tag{C.142}$$

In particular, for any locally compact space X we have an isomorphism

$$M(C_0(X)) \cong C_b(X), \tag{C.143}$$

where $a \in C_b(X)$ *simply acts on* $f \in C_0(X)$ *by* $a(f) = af$.

Proof. If A is commutative, then by Theorem C.69.1, any $a \in M(A)$ satisfies

$$a(fg) = a(f)g = fa(g), \quad f,g \in A. \tag{C.144}$$

For any $f,g \in A$ and $\omega \in \Sigma(A)$ such that $\omega(f) \neq 0$ and $\omega(g) \neq 0$, the second equality in (C.144) gives $\omega(a(f))/\omega(f) = \omega(a(g))/\omega(g)$. Since $\omega \neq 0$, there is at least one $f \in A$ for which $\omega(f) \neq 0$, so that the function $\hat{a} : \Sigma(A) \to \mathbb{C}$ given by

$$\hat{a}(\omega) = \frac{\omega(a(f))}{\omega(f)} = \frac{\widehat{a(f)}(\omega)}{\hat{f}(\omega)}, \tag{C.145}$$

is well defined. Thus (C.142) holds by construction. Since $a(f) \in A$, continuity of the Gelfand transform makes $\hat{a}$ continuous. Next, we estimate

$$|\hat{a}(\omega)\hat{f}(\omega)| = |\widehat{a(f)}(\omega)| \leq \|\widehat{a(f)}\|_\infty = \|a(f)\| \leq \|a\|\|f\|, \tag{C.146}$$

where we used (C.145) and isometry of the Gelfand transform, cf. (C.18). Hence

$$|\hat{a}(\omega)| = \left|\frac{\hat{a}(\omega)\hat{f}(\omega)}{\hat{f}(\omega)}\right| \leq \frac{\|a\|}{|\hat{f}(\omega)|}, \tag{C.147}$$

for any $f \in A$, and $\omega \in \Sigma(A)$ for which $\omega(f) \neq 0$ and $\|f\| = 1$. For those, we have

$$\begin{aligned} &\inf\{|\hat{f}(\omega)|^{-1} \mid \omega \in \Sigma(A), \omega(f) \neq 0, \|f\| = 1\} = \\ &(\sup\{|\hat{f}(\omega)| \mid \omega \in \Sigma(A), \omega(f) \neq 0, \|f\| = 1\})^{-1} = \|\hat{f}\|_\infty^{-1} = 1, \end{aligned} \tag{C.148}$$

again using $\|f\| = \|\hat{f}\|_\infty$. Together with (C.147), this gives $|\hat{a}(\omega)| \leq \|a\|$, and hence

$$\|\hat{a}\|_\infty \leq \|a\|. \tag{C.149}$$

In particular, $\hat{a}$ is bounded, so that the map (C.140) - (C.141) is well defined. This map has an inverse, as clearly any function $\hat{a} \in C_b(\Sigma(A))$ defines an element of $M(C_0(\Sigma(A)))$ by multiplication, and hence defines an element $a \in M(A)$ by the inverse Gelfand transform, cf. (C.142). □

Since an isomorphism of C*-algebras is isometric, we have $\|\hat{a}\|_\infty = \|a\|$. This may also be proved directly from (C.149) and the converse inequality

$$\begin{aligned} \|a\| &= \sup\{\|a(f)\| \mid f \in A, \|f\| = 1\} = \sup\{\|\widehat{a(f)}\|_\infty \mid f \in A, \|\hat{f}\|_\infty = 1\} \\ &= \sup\{\|\hat{a}\hat{f}\|_\infty \mid f \in A, \|\hat{f}\|_\infty = 1\} \leq \|\hat{a}\|_\infty. \end{aligned} \tag{C.150}$$

Most of this argument also works for the pre-Hilbert $C_0(X)$ module $E = C_c(X)$ (whose completion is $C_0(X)$, of course), except for the inequality (C.149), which relies on boundedness of a (cf. Theorem C.69). This is lost if E fails to be complete, and we now merely obtain an isomorphism of algebras with involution:

$$M(C_c(X)) \cong C(X). \tag{C.151}$$

For a slightly different take on this, for a general C*-algebra A we define an ***unbounded multiplier*** on A (seen as a Hilbert A-module) as a closed $\mathbb{C}$-linear and A-linear map $m : D(m) \to A$, where $D(m)$ is a dense right-ideal in A (in the algebraic sense, i.e., by exception we do *not* require an ideal to be closed). In general, the set $UM(A)$ of all unbounded multipliers on A has little algebraic structure (like the set of all closed operators on a Hilbert space), but in the commutative case we have

$$UM(C_0(X)) \cong C(X), \tag{C.152}$$

under the same identification as in (C.143). This means that any unbounded multiplier on $C_0(X)$ takes the form $g \mapsto fg$ for some $f \in C(X)$, with domain

$$D(f) = \{g \in C_0(X) \mid fg \in C_0(X)\}. \tag{C.153}$$

The argument is the same as in the proof of Proposition C.70 (except for boundedness), adding that fact that $C_c(X)$ is a core for each f, in that its closure (defined as usual by the set of all $g \in C_0(X)$ for which there is a sequence (g_n) in $C_c(X)$ such that $g_n \to g$ and fg_n is Cauchy) is given by $D(f)$; then $fg_n \to fg$ (in the sup-norm).

Let us return to the bounded case, concentrating on the multiplier algebra

$$M(A) = C^*(A,A). \tag{C.154}$$

Proposition C.71. *There is an inclusion $A \hookrightarrow M(A)$, where A (seen as a subspace of $B(A)$) acts on A (seen as a Hilbert A-module) by left multiplication. Moreover, A is an* **essential ideal** *in $M(A)$, in having nonzero intersection with any other ideal.*

Proof. We first note that each map $L_a : b \mapsto ab$ $(a, b \in A)$ is adjointable, because

$$\langle c, L_a(b)\rangle_A = \langle c, ab\rangle_A = c^*ab = (a^*c)^*b = \langle a^*c, b\rangle_A = \langle L_{a^*}(c), b\rangle_A,$$

so that the adjoint of L_a is L_{a^*}. Furthermore, $L_a = 0$ iff $a = 0$, as can be seen by taking an approximate unit in A, or from Lemma C.47. Hence $A \subset M(A)$, which is a proper inclusion iff A has no unit (since $M(A)$ always has one, i.e. the unit of $B(A)$).

Now let $m \in M(A)$ and $a \in A$. Then $(m \circ a)(b) = m(ab) = m(a)b$, since $m \in C^*(A,A)$ is A-linear. Hence $ma \equiv m \circ a \in A$, since $m(a) \in A$. Since $am = (m^*a^*)^*$, this argument shows that also $am \in A$, making A an ideal in $M(A)$.

To see that this ideal is essential, we note (as a little exercise) that an ideal $J \subset B$ in a C*-algebra B is essential iff $bJ = 0$ (i.e., $bj = 0$ for each $j \in J$ and some $b \in B$) implies $b = 0$. Again by Lemma C.47, if $m(ja) = 0$ for each $j \in A$, $a \in A$, and some $b \in M(A)$, then $b(c) = 0$ for each $c \in A$, and hence $c = 0$. □

In general, one may compute $M(A)$ as follows. If A and B are C*-algebras and E is a Hilbert A-module, we say that a homomorphism $\alpha : B \to C^*(E,A)$ is ***nondegenerate*** if $\alpha(B)E^- = E$, that is, if the closed linear span of all vectors of the type $\alpha(b)\psi$, where $b \in B$ and $\psi \in E$, equals E. It can be shown (from the Cohen–Hewitt factorization theorem) that in this case one needs neither the linear span nor the closure to recover E, in that each each element of E literally factorizes:

$$E = \{\alpha(b)\psi \mid b \in B, \psi \in E\}. \tag{C.155}$$

Theorem C.72. *Suppose A and B are C*-algebras, E is a Hilbert A-module, and*

$$\alpha : B \to C^*(E,A)$$

is a nondegenerate homomorphism. If B is an ideal in a C-algebra C, then α has a unique extension to C (which is injective if B is essential in C and α is injective).*

Proof. The idea is easy: write $\varphi \in E$ as $\varphi = \alpha(b)\psi$ for some $b \in B$ and $\psi \in E$, cf. (C.155), and define the desired extension

$$\tilde{\alpha} : C \to C^*(E,A) \tag{C.156}$$

by

$$\tilde{\alpha}(c)\varphi = \alpha(cb)\psi, \tag{C.157}$$

provided this is well defined (in which case $\tilde{\alpha}$ is clearly uniquely determined by α). Adjointability then also follows, since we may define $\tilde{\alpha}(c)^* = \tilde{\alpha}(c^*)$, and compute

$$\begin{aligned}\langle\tilde{\alpha}(c)^*\alpha(b')\psi', \alpha(b)\psi\rangle_B &= \langle\alpha(c^*b')\psi', \alpha(b)\psi\rangle_B = \langle\psi', \alpha(c^*b')^*\alpha(b)\psi\rangle_B \\ &= \langle\psi', \alpha(b')^*\alpha(cb)\psi\rangle_B \\ &= \langle\alpha(b')\psi', \tilde{\alpha}(c)\alpha(b)\psi\rangle_B.\end{aligned} \tag{C.158}$$

Furthermore, it is easy to see that $\tilde{\alpha}$ is a homomorphism. Also, $\alpha(c) = 0$ for $c \in C$ implies $\alpha(cb) = 0$ for each $b \in B$; if α is injective, then $cb = 0$, and if B is an essential ideal in C, then $c = 0$, so that $\tilde{\alpha}$ is injective.

To show that (C.157) is independent of the representatives b and ψ, we estimate

$$\begin{aligned}\|\tilde{\alpha}(c)\alpha(b)\psi\| &= \lim_\lambda \|\alpha(ce_\lambda b)\psi\| = \lim_\lambda \|\alpha(ce_\lambda)\alpha(b)\psi\| \\ &\leq \lim_\lambda \|\alpha(ce_\lambda)\|\|\alpha(b)\psi\| \leq \lim_\lambda \|ce_\lambda\|\|\alpha(b)\psi\| \\ &= \|c\|\|\alpha(b)\psi\|,\end{aligned} \tag{C.159}$$

where (e_λ) is an approximate unit in C. In particular, if $\alpha(b)\psi = \alpha(b')\psi'$, then $\tilde{\alpha}(c)\alpha(b)\psi = \tilde{\alpha}(c)\alpha(b')\psi'$. □

This proof works also without (C.155); one then has a finite sum $\varphi = \sum_i \alpha(b_i)\psi_i$, and a computation similar one to the previous one shows that $\tilde{\alpha}(c)$ is bounded on the dense subspace of E consisting of such sums.

This theorem (with $B \rightsquigarrow A$ and $E \rightsquigarrow A$) explains in which sense $M(A)$ is a *maximal* unitization of A (whereas $\dot{A}$ is a *minimal* one): all we need to do is abstractly define a unitization of a non-unital C*-algebra A as a unital C*-algebra containing A as an essential ideal (cf. Proposition C.71). This incorporates both $\dot{A}$ and $M(A)$, each being distinguished by a universal property it satisfies, namely:

Corollary C.73. *For each unital C*-algebra C containing A as an essential ideal, there are unique injective homomorphisms: $C \to M(A)$ and $\dot{A} \to C$ whose restriction to A is the identity map. In other words, denoting the inclusion of A into C by ι, we have commutative diagrams*

The topological counterpart of this corollary is the construction of the one-point compactification $\dot{X}$ and of the Čech-Stone compactification βX, respectively; cf. Lemma C.38, which we may now supplement by simply *defining* βX as the Gelfand spectrum of the commutative C*-algebra $C_b(X) \cong C(\beta X)$. In this analogy, the condition on an ideal $B \subset C$ to be essential simply corresponds to a non-compact space X being a dense subspace of some compactification of it.

Corollary C.74. *Let E be some Hilbert A-module E and let $\alpha : B \to C^*(E,A)$ be an injective nondegenerate homomorphism. The unique extension $\tilde{a} : M(B) \to C^*(E,A)$ of α that exists according to Theorem C.72 maps $M(B)$ isomorphically onto*

$$Z_\alpha(E) = \{a \in C^*(E,A) \mid a\alpha(b) \in \alpha(B), \alpha(b)a \in \alpha(B) \,\forall b \in B\}. \qquad \text{(C.160)}$$

Proof. Note that $Z_\alpha(E)$ is essential in $C^*(E,A)$, as easily follows from the nondegeneracy of α. Therefore, by the argument just given (plus the abstract nonsense that shows that universal objects are unique up to isomorphism), we only need to prove that $Z_\alpha(E)$ is a maximal unitization of B. Let B be an essential ideal in C and consider the injective extension $\tilde{a} : C \to C^*(E,A)$ of α given by Theorem C.72. Then $\tilde{a}$ maps C into $Z_\alpha(E)$ by construction, as $\tilde{\alpha}(c)\alpha(b) = \alpha(bc) \in \alpha(B)$, etc. □

Corollary C.75. *A nondegenerate homomorphism $\alpha : B \to M(A)$ has a unique extension to a homomorphism $\tilde{\alpha} : M(B) \to M(A)$.*

Proof. Take $C = M(B)$ and $E = A$ in Theorem C.72.1. □

Note that two nondegenerate homomorphisms $\alpha : A \to M(B)$ and $\beta : B \to M(C)$ can be composed into a nondegenerate homomorphism $\beta \circ \alpha : A \to M(C)$, which by definition equals $\beta \circ \tilde{\alpha}$. Thus one obtains a category CAm whose objects are C*-algebras and whose arrows are nondegenerate homomorphism $\alpha : A \to M(B)$, with a full subcategory CCAm whose objects are commutative C*-algebras (with the same arrows). This leads to a neat extension of Gelfand duality (cf. Theorem C.45):

Theorem C.76. *The category* LCH *of locally compact Hausdorff spaces and continuous maps is dual to the category* CCAm *of commutative C*-algebras just defined.*

This claim may be unfolded as in Theorem C.45, omitting 'proper' on the topological side and replacing $\alpha : A \to B$ on the algebraic side by $\alpha : A \to M(B)$.

Proof. First, a continuous map $\varphi : Y \to X$ trivially induces a nondegenerate homomorphism $\varphi^* : C_0(X) \to C_b(Y)$. Second, since $\omega \in \Sigma(B)$ defines a nondegenerate homomorphism $B \to \mathbb{C}$, by Theorem C.72 it extends to a homomorphism $\tilde{\omega} : M(B) \to \mathbb{C}$. Thus the pullback $\alpha^* : \Sigma(B) \to \Sigma(A)$ of a nondegenerate homomorphism $\alpha : A \to M(B)$ is well defined (and still continuous). Part 3 of Theorem C.45 stays the same, and the pertinent naturality properties are easily verified. □

Corollary C.77. *A nondegenerate homomorphism $\alpha : C_0(X) \to B(H)$ has a unique extension to a homomorphism $\tilde{\alpha} : C_b(X) \to B(H)$.*

Proof. Taking $A = \mathbb{C}$, $E = H$, and $B = B_0(H)$, Theorem C.72.2 gives

$$M(B_0(H)) \cong B(H). \tag{C.161}$$

Combine this with the previous corollary (with $B \rightsquigarrow C_0(X)$ and $A \rightsquigarrow B_0(H)$). □

Finally, we show how to reconstruct A as a C*-algebra from A as a Hilbert A-module. The key to this is a more general construction:

Definition C.78. *The collection $C_0^*(E,A)$ of "compact" operators on a Hilbert A-module E is the C*-algebra generated (within $C^*(E,A)$) by all operators of the type $|\varphi\rangle\langle\psi|$, where $\varphi, \psi \in E$, and*

$$|\varphi\rangle\langle\psi|(\zeta) = \varphi\langle\psi,\zeta\rangle_A. \tag{C.162}$$

Such operators are easily seen to be adjointable, with adjoint

$$|\varphi\rangle\langle\psi|^* = |\psi\rangle\langle\varphi|, \tag{C.163}$$

and hence bounded, with norm majorized by $\|\psi\|\|\varphi\|$. If $E = H$ is a Hilbert space, then $C_0^*(H,\mathbb{C}) = B_0(H)$, since the maps $|\varphi\rangle\langle\psi|$ obviously generate the finite-rank operators on H, whose norm-closure is $B_0(H)$, cf. Proposition B.131. Hence the name "compact" operators, but in general elements of $C_0^*(E,A)$ need not be compact (as operators on a Banach space) at all. The next and final example is a case in point:

Proposition C.79. *If $E = A$ as a Hilbert A-module in the usual way, then*

$$C_0^*(A,A) \cong A. \tag{C.164}$$

Proof. We have $|a\rangle\langle b| = L_{ab^*}$, where $a \mapsto L_a$ is the canonical map from A to $C^*(A,A) \subset B(A)$ given by $L_a(b) = ab$, see Proposition C.30. This map is isometric, cf. (C.63), and hence injective. The map $|a\rangle\langle b| \mapsto ab^*$ from the linear span of all operators (C.162) within $C_0^*(E,A)$ to A is therefore bounded, and has dense image by Lemma C.47. Its unique continuous extension maps $C_0^*(E,A)$ onto A, see Theorem C.62.5 (or use the Cohen–Hewitt factorization theorem to conclude). □

C.11 Gelfand topology as a frame

In the traditional approach to the Gelfand isomorphism, which we have followed so far, the Gelfand spectrum $\Sigma(A)$ of a commutative unital C*-algebra A is first constructed as a set, upon which it is equipped with a natural topology $\mathcal{O}(\Sigma(A))$, i.e., the Gelfand topology. Alternatively, one may start with the latter and reconstruct $\Sigma(A)$ as a set from it. This not only gives a better conceptual understanding of Gelfand's theory (relating it, for example, to a well-known construction in algebraic geometry); it also has the technical advantage of making good sense in constructive mathematics and hence in topos theory (which the classical theory does not).

In the language of lattice theory, the topology $\mathcal{O}(X)$ of any space X is an example of a so-called ***frame*** (cf. Appendix D, compared to which we change notation so as to avoid abuse of the ubiquitous symbol X) i.e., a complete lattice L in which

$$U \wedge \bigvee S = \bigvee \{U \wedge V, V \in S\}, \tag{C.165}$$

for arbitrary elements $U \in L$ and subsets $S \subset L$. This is sometimes written in the form $U \wedge (\bigvee_\lambda V_\lambda) = \bigvee_\lambda (U \wedge V_\lambda)$, from which it is clear that the (binary) distributive law $U \wedge (V \vee W) = (U \wedge V) \vee (U \wedge W)$, which of course is implied by (C.165), is now required for arbitrary families. Indeed, the definition of a frame is primarily motivated by the example $L = \mathcal{O}(X)$, in which it should be noted that the supremum

$$\bigvee S = \bigcup S = \bigcup_\lambda \{U_\lambda \in S\}, \tag{C.166}$$

is simply given by the set-theoretic union of the elements of S, which are open sets whose union is open by definition of a topology, whereas the infimum of arbitrary families of open sets has to be doctored so as to make it open, and hence is given by

$$\bigwedge S = \bigvee \{U \in \mathcal{O}(X) \mid U \subseteq V \forall V \in S\}. \tag{C.167}$$

Frame maps, then, are defined as order-preserving maps between the underlying posets that preserve *finite* infima and *arbitrary* joins. For example, if

$$\varphi : Y \to X \tag{C.168}$$

is a continuous map, then the inverse image map

$$\varphi^{-1} : \mathcal{O}(X) \to \mathcal{O}(Y) \tag{C.169}$$

is a frame map. This also defines the category Frm of frames, whose opposite category (that has the same objects but all arrows inverted) is called the category Loc of ***locales***. Thus a locale is a frame, seen as an object in the opposite category. If no confusion arises (which, unfortunately, is rarely the case), elements of Frm are written as $\mathcal{O}(X)$, *even if they are not topologies* (and indeed there are such frames, see below), in which case the corresponding element of Loc is written as X.

In this spirit, frame maps are always written as (C.169), in which case the map in the opposite direction between the corresponding locales is (C.168). This notation suggests the right way of thinking, and we will use it whenever it is convenient.

Frames are very closely related to ***Heyting algebras***, which were originally meant to formalize the intuitionistic (propositional) logic of Brouwer, and are defined as distributive lattices L (with top $\top$ and bottom $\bot$) equipped with a binary map

$$\rightarrow : L \times L \rightarrow L, \tag{C.170}$$

playing the role of implication in logic, that satisfies the axiom

$$U \leq (V \rightarrow W) \text{ iff } (U \wedge V) \leq W. \tag{C.171}$$

Every Boolean algebra is a Heyting algebra, but not *vice versa*; in fact, a Heyting algebra is Boolean iff $\neg\neg U = U$ for all U, which is the case iff $(\neg U) \vee U = \top$ for all U (which states the law of the excluded middle denied by Brouwer). In a Heyting algebra (unlike a Boolean algebra), negation is a derived notion, defined by

$$\neg U = U \rightarrow \bot. \tag{C.172}$$

A Heyting algebra is ***complete*** when it is complete as a lattice, in that arbitrary suprema (and hence also infima) exist. The infinite distributivity law (C.165) is automatically satisfied in a complete Heyting algebra, which therefore is also a frame. Conversely, a frame may be turned into a complete Heyting algebra by defining

$$V \rightarrow W = \bigvee \{U \mid U \wedge V \leq W\}. \tag{C.173}$$

Frames and complete Heyting algebras drift apart as soon as morphisms are concerned, for although in both cases one requires maps to preserve the partial order, maps between Heyting algebras must preserve $\rightarrow$ rather than infinite suprema.

The map $X \mapsto \mathcal{O}(X)$ from topological spaces to frames (which extends to a contravariant functor in the obvious way, i.e., via (C.168) - (C.169)) is a competitor to the map $X \mapsto C_0(X)$ from topological spaces to commutative C*-algebras, and one goal of this section is to find out how these two constructions are related.

First, there is a frame-theoretic analogue of the categorical duality between locally compact Hausdorff spaces and commutative C*-algebras (cf. Theorem C.45), in which locally compact Hausdorff spaces are replaced by so-called *sober* spaces (and no restrictions on continuous maps are made), whilst the category of frames must be restricted to so-called *spatial* frames (which move is somewhat analogous to restricting C*-algebras to commutative ones). We now explain these notions.

A particularly simple frame is $\underline{2} = \{0,1\} \equiv \{\bot, \top\}$, with order $0 \leq 1$; this is just the topology $\mathcal{O}(*)$ of a singleton $*$. In agreement of the above convention, a frame map $p^{-1} : \mathcal{O}(X) \rightarrow \underline{2}$ will be written as a locale map $p : * \rightarrow X$. Such a map defines a ***point*** of the locale X (i.e., of the frame $\mathcal{O}(X)$), and we denote the set of points of X by $\mathrm{Pt}(X)$. To appreciate this definition, let us suppose that $\mathcal{O}(X)$ is the topology of some space X. Each point $x \in X$ then corresponds to a *genuine* map

$$p_x : * \to X,\ p_x(*) = x; \tag{C.174}$$

whose inverse image map $p_x^{-1} : \mathcal{O}(X) \to \underline{2}$ is frame map and hence defines a point in the above sense. Conversely, if X is sober (see below), each point of $\mathcal{O}(X)$ arise in that way. The set $\mathrm{Pt}(X)$ has a natural topology, with opens

$$\mathrm{Pt}(U) = \{p \in \mathrm{Pt}(X) \mid p(*) \in U\}, \tag{C.175}$$

where $U \in \mathcal{O}(X)$; here $p(*) \in U$ really means $p^{-1}(U) = 1$. This gives a frame map

$$U \mapsto \mathrm{Pt}(U) \tag{C.176}$$

from $\mathcal{O}(X)$ to $\mathrm{Pt}(X)$. We say $\mathcal{O}(X)$ (or the locale X) is ***spatialspatial*** if this map is an isomorphism of frames. Roughly speaking, therefore, spatial frames are just topologies (an example of a non-spatial frame is the lattice $\mathcal{O}_{\mathrm{reg}}(\mathbb{R})$ of regular open subsets of $\mathbb{R}$, i.e., of open subsets U with the property $\neg\neg U = U$, where $\neg U$ is the interior of the complement of U). This does not mean, however, that any topology $\mathcal{O}(X)$ (seen as a frame) is isomorphic to $\mathcal{O}(\mathrm{Pt}(X))$, since $\mathrm{Pt}(X)$ may not be homeomorphic to X.

Spaces X for which this *is* the case are called ***sober***; more precisely, this means that the map $x \mapsto p_x$ from X to $\mathrm{Pt}(X)$ considered above is a homeomorphism; less precisely, we may say that sober spaces X may be reconstructed from their topology $\mathcal{O}(X)$, up to homeomorphism. To give a more direct topological characterization of sobriety, call $W \in \mathcal{O}(X)$ ***meet-irreducible*** if $U \cap V \subseteq W$ (where $U, V \in \mathcal{O}(X)$) implies either $U \subseteq W$ or $V \subseteq W$. In any space X, all open sets of the form $W_x = X\backslash x^-$ are meet-irreducible, where $x \in X$ (and x^- is the closure of $\{x\}$). A space X is sober, then, iff these are the only such opens. For example, any Hausdorff space is sober (an example of a non-sober space is $X = \mathbb{N}$ with the unusual topology in which all complements of finite subsets are open, along with the empty set, of course).

The category Frm, then, has a full subcategory Spat of spatial frames, whilst likewise the category Top of topological spaces has a full subcategory Sob of sober spaces. We now have the following counterpart of Theorem C.45:

Theorem C.80. *The categories* Spat *and* Sob *are dual, in that:*

1. *If X is a sober space, then $\mathcal{O}(X)$ is a spatial frame. A continuous map $\varphi : Y \to X$ induces a frame map $\varphi^{-1} : \mathcal{O}(X) \to \mathcal{O}(Y)$ in the natural way, such that if we have another continuous map $\psi : Z \to Y$, then $(\varphi \circ \psi)^{-1} = \psi^{-1} \circ \varphi^{-1}$.*
2. *If $\mathcal{O}(X)$ is a spatial frame, then* $\mathrm{Pt}(X)$ *is a sober space. Furthermore, a locale map $\varphi : Y \to X$ (i.e., a frame map $\varphi^{-1} : \mathcal{O}(X) \to \mathcal{O}(Y)$) induces a continuous function $\varphi^* : \mathrm{Pt}(Y) \to \mathrm{Pt}(X)$ by $\varphi^*(p) = \varphi \circ p$ (i.e., $\varphi^*(p^{-1}) = p^{-1} \circ \varphi^{-1}$), which similarly behaves well under composition.*
3. *There are canonical homeomorphisms and frame maps:*

$$\mathrm{p}_X : X \overset{\cong}{\leftrightarrows} \mathrm{Pt}(\mathcal{O}(X)),\ x \mapsto p_x; \tag{C.177}$$

$$\mathrm{Pt}_X : \mathcal{O}(X) \overset{\cong}{\leftrightarrows} \mathcal{O}(\mathrm{Pt}(\mathcal{O}(X))),\ U \mapsto \mathrm{Pt}(U), \tag{C.178}$$

cf. (C.174) - (C.176), *with the correct naturality properties (cf. Theorem C.23).*

Proof. We will not give a complete proof of this, but the main points are that:

- Any locale defined by the topology of some space is spatial.
- Any space Pt(X) of points of some locale X (not necessarily a space) is sober.
- The map p_X in (C.177) is a homeomorphism by definition of sobriety (which, alternatively, could have been defined by requiring bijectivity of p_X, in which case it can be shown that this map is continuous as well as open).
- By definition of the topology on Pt(X), the map (C.176) is surjective for any locale X. If X is spatial; then for any distinct elements $U,V \in \mathcal{O}(X)$ there is a point p such that $p^{-1}(U) \neq p^{-1}(V)$, but this is the same as saying that Pt$(U) =$ Pt(V) implies $U = V$. So in that case, (C.176) is also injective. □

Our aim is to apply these ideas to Gelfand duality, specifically to an independent description of the topology $\mathcal{O}(\Sigma(A))$ of the Gelfand spectrum $\Sigma(A)$ of a commutative C*-algebra A. To put this in perspective, let A for the moment be a general C*-algebra, and recall Definition C.55 of left, right and two-sided ideals (all taken to be closed by definition). Further to these, there is another interesting notion.

Definition C.81. *A* **hereditary subalgebra** *of a C*-algebra A is a C*-subalgebra B of A with the property that $a \leq b$ for $b \in B^+$ and $a \in A^+$ implies $a \in B^+$. The set of of all hereditary subalgebras of A is denoted by $H(A)$.*

It is a simple exercise to show that there are bijective correspondences between hereditary subalgebras B of A, left ideals L of A, and right ideals R of A, given by:

$$L = \{a \in A \mid a^*a \in B^+\}; \tag{C.179}$$
$$R = \{a \in A \mid aa^* \in B^+\}; \tag{C.180}$$
$$B = L \cap L^* = R \cap R^*. \tag{C.181}$$

Furthermore, one has $I(A) \subseteq H(A)$, where $I(A)$ is the set of closed two-sided ideals in A, and likewise we write $L(A)$ and $R(A)$. If A is commutative, these ideals are two-sided, so that $L^* = L$ etc., and $L = R = B$, so that $H(A) = I(A) = L(A) = R(A)$.

Proposition C.82. *The set $H(A)$ is a complete lattice under inclusion as the partial order, with inf and sup of any subset $S \subset H(A)$ given by*

$$\bigwedge S = \bigcap S; \tag{C.182}$$
$$\bigvee S = \bigcap \{U \in H(A) \mid V \subseteq U \forall V \in S\}. \tag{C.183}$$

Moreover, if A is commutative, then $H(A) = I(A) = L(A) = R(A)$ is a frame.

Proof. The defining conditions on hereditary subalgebras of A are preserved by arbitrary intersections, which means that $H(A)$ has infima of arbitrary subsets, given by (C.182). This implies that $H(A)$ also has arbitrary suprema, given by (C.183), which is a standard formula in lattice theory. Hence $H(A)$ is a complete lattice.

The last claim follows from Corollary C.84 below (and the ensuing fact for topology). It may also be proved directly, using the fact that $H(A) = I(A)$. □

Proposition C.83. *Let X be a locally compact Hausdorff space. Then the map*

$$\mathcal{O}(X) \stackrel{\cong}{\rightarrow} H(C_0(X)); \tag{C.184}$$
$$U \mapsto C_0(U), \tag{C.185}$$

where $C_0(U)$ is seen as a subspace of $C_0(X)$, is a frame isomorphism, with inverse

$$H(C_0(X)) \stackrel{\cong}{\rightarrow} \mathcal{O}(X); \tag{C.186}$$
$$B \mapsto X\backslash F_B, \tag{C.187}$$

where, for any subset $B \subset C_0(X)$ one defines the (necessarily closed) set $F_B \subset X$ by

$$F_B = \{x \in X \mid f(x) = 0 \forall f \in B\}. \tag{C.188}$$

Proof. For any open $U \in \mathcal{O}(X)$, we may regard $f \in C_0(U)$ as an element of $C_0(X)$ by extending f to all of X through $f_{|X\backslash U} = 0$. Continuity of f is only an issue at boundary points of $U^c \equiv X\backslash U$, so take $x_0 \in \partial U^c$ (i.e., any neighbourhood of x_0 has nonempty intersection with both U^c and U). Since $f(x_0) = 0$, to prove continuity of f at x_0 we need to show that for any $\varepsilon > 0$, there is neigbourhood N of x_0 such that $|f(x)| < \varepsilon$ for each $x \in N$. Indeed, since $f \in C_0(U)$, there is a compact set $K \subset U$ such that $|f(x)| < \varepsilon$ for each $x \in U\backslash K$ (and hence also for each $x \in X\backslash K$). Then $x_0 \notin K$ (since $x_0 \in U^c$), so, we may take the open neighbourhood $N = X\backslash K$.

Since the ordering in $C_0(X)$ is pointwise, it is trivial that $C_0(U) \in H(C_0(X))$. The map (C.185) also clearly preserves the order, i.e., if $U \subseteq V$, then $C_0(U) \subseteq C_0(V)$.

Half of the proof that (C.185) and (C.187) are mutually inverse is the equality

$$C_0(U) = C_0(X; X\backslash U), \tag{C.189}$$

where for any $F \subset X$ (usually taken to be closed), we define $C_0(X;F) \subset C_0(X)$ by

$$C_0(X;F) = \{f \in C_0(X) \mid f_{|F} = 0\}. \tag{C.190}$$

To prove (C.189), we just need to prove that $C_0(X; X\backslash U) \subseteq C_0(U)$, since the opposite inclusion has been proved before Proposition C.83. Since $f \in C_0(X)$, for each $\varepsilon > 0$ and each boundary point $x \in \partial U^c$, there is an open neighbourhood N_x of x where $|f| < \varepsilon$, as well as a compact set $K \subset X$ outside which the same is true. Then $V = \cup_{x \in \partial U^c} U_x \cap U$ is open in U, so that its complement $U\backslash V$ is closed in U, and $K' = (U\backslash V) \cap K$ is compact in U. Clearly, $|f| < \varepsilon$ outside K', whence $f \in C_0(U)$.

Having proved (C.189), the other half of the proof of bijectivity of (C.184) is

$$B = C_0(X; F_B), \tag{C.191}$$

for any $B \in H(C_0(X))$. The inclusion $B \subseteq C_0(X; F_B)$ is trivial. For the converse, we exploit the fact that B is an ideal in $C_0(X)$, so that $C_0(X)/B$ is a C*-algebra by Theorem C.60. Let $\tau : C_0(X) \rightarrow C_0(X)/B$ be the canonical projection. If $f \notin B$, then $\tau(f) \neq 0$. Hence there is a character $\omega' \in \Sigma(C_0(X)/B)$, such that $\omega'(\tau(f)) \neq 0$.

Lift ω' to $\omega = \omega' \circ \tau \in \Sigma(C_0(X)) \cong X$, so that there is $x \in X$ such that $\omega(g) = g(x)$ for all $g \in C_0(X)$. Since $\tau(g) = 0$ for each $g \in B$, we have $\omega(g) = 0$, and hence $g(x) = 0$ for each $g \in B$, so that $x \in F_B$. But $f(x) \neq 0$, so $f \notin C_0(X; F_B)$, and hence we have proved the inclusion $C_0(X; F_B) \subseteq B$. □

Thus C.83 could just as well have been formulated in terms of closed sets, albeit at the cost of inverting the partial order. Also, note the isomorphism

$$C_0(X)/C_0(U) \xrightarrow{\cong} C_0(X \backslash U), \ [f] \mapsto f_{|X \backslash U}. \tag{C.192}$$

Corollary C.84. *For any commutative C*-algebra A, there is a frame isomorphism*

$$\mathcal{O}(\Sigma(A)) \cong H(A). \tag{C.193}$$

This sheds new light on maximal ideals in A as points of the Gelfand spectrum $\Sigma(A)$, cf. Proposition C.13. We need a lemma that applies to any frame $\mathcal{O}(X)$. A ***prime element*** $P \in \mathcal{O}(X)$ is an element $P \neq \top$ such that $U \wedge V \leq P$ iff $U \leq P$ or $V \leq P$. For a point $p^{-1} : \mathcal{O}(X) \to \underline{2}$, we write $\ker(p^{-1})$ for $\{U \in \mathcal{O}(X) \mid p^{-1}(U) = 0\}$.

Lemma C.85. *For any frame $\mathcal{O}(X)$ (i.e. locale X), there is a bijective correspondence between points $p^{-1} : \mathcal{O}(X) \to \underline{2}$ of X and prime elements $P \in \mathcal{O}(X)$, viz.*

$$P = \bigvee \ker(p^{-1}); \tag{C.194}$$

$$p^{-1}(U) = 0 \text{ iff } U \leq P. \tag{C.195}$$

Under this correspondence, the topology on Pt(X) *is given by the* **Zariski topology**, *whose* closed *sets F_P consist of all $Q \supseteq P$, where P is some prime element of $\mathcal{O}(X)$.*

Proof. The requirement that p^{-1} be a frame map implies the following properties of its kernel $K = \ker(p^{-1})$: $\top \notin K$, $U \wedge V \in K$ iff $U \in K$ or $V \in K$, and $\bigvee S \in K$ iff each $V \in S$ is in K. Any subset $K \subset \mathcal{O}(X)$ satisfying these properties in turn defines a point p of X whose kernel is K. Then $P = \bigvee K$ is a prime element of $\mathcal{O}(X)$, and conversely, K (and hence p) may be recovered from P as its downset $K = \downarrow P$.

The given topology on the set of prime elements is a rewriting of (C.175). □

The prime elements of $H(A)$, where A is a commutative C*-algebra, are the ***prime ideals*** in A, i.e., the proper ideals $J \subset A$ such that $J_1 J_2 \subset J$ iff $J_1 \subseteq A$ or $J_2 \subseteq A$, for any ideals J_1, J_2 of A (closed by definition, like J); note that $J_1 J_2 = J_1 \cap J_2$.

Theorem C.86. *1. The frame $H(A)$ of hereditary subalgebras of a commutative C*-algebra A is spatial, with* Pt$(H(A)) \cong \Sigma(A)$ *as topological spaces.*

2. The prime elements of $H(A)$ are the maximal ideals of A, so that, equipping the set $\mathcal{M}(A)$ of maximal ideals of A with the Zariski topology, also $\mathcal{M}(A) \cong \Sigma(A)$.

Proof. 1. Proposition C.83 bijectively relates prime elements in $H(A)$ to meet-irreducible sets in $\Sigma(A)$. The description of sobriety in terms of meet-irreducibility after (C.176), which applies because $\Sigma(A)$ is locally compact Hausdorff and hence sober, then bijectively relates these meet-irreducible sets to points of $\Sigma(A)$.
2. Proposition C.13 in turn relates points of $\Sigma(A)$ to maximal ideals of A. □

C.12 The structure of C*-algebras

Having understood the structure of commutative C*-algebras, we now turn to the general case. We already know that the algebra $B(H)$ of all bounded operators on some Hilbert space H is a C*-algebra in the obvious way (i.e., the algebraic operations are the natural ones, the involution is the operator adjoint $a \mapsto a^*$, and the norm is the operator norm of Banach space theory). Moreover, each (operator) norm-closed *-algebra in $B(H)$ is a C*-algebra. Our goal is to prove the converse:

Theorem C.87. *Each C*-algebra A is isomorphic to a norm-closed *-algebra in $B(H)$, for some Hilbert space H. Equivalently, for any C*-algebra A there exist a Hilbert space H and an injective homomorphism $\pi : A \to B(H)$.*

A homomorphism $\pi : A \to B(H)$ is called a ***representation*** of A on H. The equivalence between the two statements in the theorem follows from Theorem C.62.

Let us note that Theorems C.8 and C.87 harmonize as follows: any measure μ on X satisfying $\mu(U) > 0$ for each open $U \subset X$ leads to an injective representation of $C_0(X)$ on $L^2(X,\mu)$ by multiplication operators, that is, $\pi(f) = m_f$, cf. (B.238).

The proof of Theorem C.87 uses the elegant GNS-***construction***, named after Gelfand, Naimark, and Segal, which is important in its own right. We initially assume that A is unital. First, we call a representation π ***cyclic*** if its carrier space H contains a ***cyclic vector*** Ω for π, i.e., the closure of $\pi(A)\Omega$ coincides with H.

Theorem C.88. *Let ω be a state on a C*-algebra A. There exists a cyclic representation π_ω of A on a Hilbert space H_ω with cyclic unit vector Ω_ω such that*

$$\omega(a) = \langle \Omega_\omega, \pi_\omega(a)\Omega_\omega \rangle, \ a \in A. \tag{C.196}$$

Proof. We first give the proof in the special case that A has a unit 1_A, and $\omega(a^*a) > 0$ for all $a \neq 0$. Define a sesquilinear form $(-,-)$ on A by

$$(a,b) = \omega(a^*b). \tag{C.197}$$

This form is positive definite by definition of a state, so that we may complete A in the ensuing norm

$$\|a\|_\omega = \sqrt{\omega(a^*a)}, \tag{C.198}$$

to a Hilbert space called H_ω. For each $a \in A$, we then define a map

$$\pi_\omega(a) : A \to A; \tag{C.199}$$

$$\pi_\omega(a)b = ab. \tag{C.200}$$

Regarding A as a dense subspace of H_ω, this defines an operator $\pi_\omega(a)$ on a dense domain in H_ω. This operator is bounded, since (C.94) implies

$$\|\pi_\omega(a)\| \leq \|a\|. \tag{C.201}$$

Hence $\pi_\omega(a)$ may be extended from A to H_ω by continuity, and we obtain a map $\pi_\omega : A \to B(H_\omega)$. Simple computations show that π_ω is a representation. The special vector Ω_ω is the unit $1_A \in A$, seen as an element of H_ω: its cyclicity is obvious, and:

$$\|\Omega_\omega\|^2 = \langle \Omega_\omega, \Omega_\omega \rangle = \omega(1_A^* 1_A) = \omega(1_A) = 1; \tag{C.202}$$
$$\langle \Omega_\omega, \pi_\omega(a)\Omega_\omega \rangle = \omega(1_A^* a 1_A) = \omega(a). \tag{C.203}$$

Under our standing assumption $\omega(a^*a) > 0$ if $a \neq 0$, this not only proves Theorem C.88, but also Theorem C.87: for $\pi_\omega(a) = 0$ implies $\|\pi_\omega(a)\Omega_\omega\|^2 = 0$, whose left-hand side is precisely $(\Omega_\omega, \pi_\omega(a^*a)\Omega_\omega) = \omega(a^*a)$. Thus π_ω is faithful.

In general, a C*-algebra may lack such states, and we must adapt the proof of both theorems. The GNS-construction is easy: for an arbitrary state ω, we introduce

$$N_\omega = \{a \in A \,|\, \omega(a^*a) = 0\}. \tag{C.204}$$

If a_ω is the image of $a \in A$ in A/N_ω, we may define an inner product on the latter by

$$\langle a_\omega, b_\omega \rangle = \omega(a^*b); \tag{C.205}$$

this is well defined and positive definite, and we define the Hilbert space H_ω as the completion of A/N_ω in this inner product. Furthermore, we define

$$\pi_\omega(a) : A/N_\omega \to H_\omega; \tag{C.206}$$
$$\pi_\omega(a)b_\omega = (ab)_\omega; \tag{C.207}$$

this is well defined, because N_ω is a left ideal in A by (C.94). Finally, we define

$$\Omega_\omega = (1_A)_\omega. \tag{C.208}$$

The proof that everything works is then a simple exercise. Another way to look at the cyclic vector Ω_ω is to let ω define a linear functional $\tilde{\omega} : A/N_\omega \to \mathbb{C}$ by

$$\tilde{\omega}(a_\omega) = \omega(a); \tag{C.209}$$

this functional is continuous on $A/N_\omega \subset H_\omega$, because $|\omega(a)|^2 \leq \omega(a^*a) = \|a_\omega\|^2_{H_\omega}$, as follows from the Cauchy–Schwarz inequality for the positive semidefinite form (C.197). Hence by Riesz–Fréchet there is an implementing vector Ω_ω such that

$$\omega(a) = \langle \Omega_\omega, a_\omega \rangle. \tag{C.210}$$

Finally, when A has no unit, in defining Ω_ω we either use the GNS-construction for the unitization $\dot{A}$ and restrict $\pi_{\dot{\omega}}(\dot{A})$ to A to define $\pi_\omega(A)$, or use (C.210). □

One of the nicest feature of the GNS-construction is the link between purity of the state ω and irreducibility of the corresponding representation π_ω.

Definition C.89. *We call a representation π of a C*-algebra A on a Hilbert space H* **irreducible** *if the only closed subspaces K of H that are stable under $\pi(A)$ (in the sense that if $\psi \in K$, then $\pi(a)\psi \in K$ for all $a \in A$) are either $K = H$ or $K = \{0\}$.*

Theorem C.90. *Each of the following conditions is equivalent to irreducibility:*

1. *$\pi(A)' = \mathbb{C} \cdot 1$, where S' is the commutant of $S \subset B(H)$* (**Schur's Lemma**)*;*
2. *$\pi(A)'' = B(H)$;*
3. *Every vector in H is cyclic for $\pi(A)$.*

Furthermore, if ω is a state on A, then ω is pure iff the corresponding GNS-*representation π_ω is irreducible.*

Proof. If $\pi(A)' \neq \mathbb{C} \cdot 1$, then $\pi(A)'$ must contain a nontrivial self-adjoint element a (as it is a *-algebra), and hence also a nontrivial projection e (as the spectral projections $e_\Delta = 1_\Delta(a)$ of a, defined as in Theorem B.102, lie in $\pi(A)'$, too). But if $e \in \pi(A)'$, then eH is stable under $\pi(A)$, and hence π cannot be irreducible. Thus irreducibility implies 1. Conversely, if $\pi(A)' = \mathbb{C} \cdot 1$, then π must be irreducible by the same argument, since if not, any projection onto some proper stable subspace K for π would be an nontrivial element of $\pi(A)'$. The equivalence $1 \leftrightarrow 2$ is clear, since $(\mathbb{C} \cdot 1)' = B(H)$. Similarly, if $\varphi \in H$ would fail to be cyclic for π, then $\pi(A)\varphi^-$ would be a proper, $\pi(A)$-stable subspace of H, so that irreducibility implies 3. The converse is trivial, since if $K \subset K$ were stable for $\pi(A)$, then 3 cannot hold. □

Another useful result relates general representations to GNS-representations. We call two representations $\pi_i : A \to B(H_i)$, $i = 1, 2$, ***unitarily equivalent*** if there is a unitary $u : H_1 \to H_2$ such that $u\pi_1(a)u^* = \pi_2(a)$ (or $u\pi_1(a) = \pi_2(a)u$) for each $a \in A$.

Proposition C.91. *Let $\pi : A \to B(H)$ be a cyclic representation of H. If $\psi \in H$ is a cyclic unit vector for π, then*

$$\omega(a) = \langle \psi, \pi(a)\psi \rangle \tag{C.211}$$

is a state on A, whose GNS-*representation π_ω is unitarily equivalent to π.*

Proof. Define $u : H_\omega \to H$ first on $\pi_\omega(A)\Omega_\omega$ (which is a dense subspace of H) by

$$u\pi_\omega(a)\Omega_\omega = \pi(a)\psi. \tag{C.212}$$

Using (C.211) and (C.196), we then obtain

$$\|\pi_\omega(a)\Omega_\omega\|^2 = \omega(a^*a) = \langle \psi, \pi(a^*a)\psi \rangle = \|\pi(a)\psi\|^2. \tag{C.213}$$

This shows that u is well defined as well as isometric, so that it extends to H_ω by continuity. Its image is then the closure of $\pi(A)\psi$, which is H, since ψ is cyclic by assumption. Thus u is surjective and hence unitary. Finally, we compute

$$u\pi_\omega(a)\pi_\omega(b)\Omega_\omega = \pi(a)\pi(b)\psi = \pi(a)u\pi_\omega(b)\Omega_\omega, \tag{C.214}$$

so that $u\pi_\omega(a) = \pi(a)u$ on the dense space $\pi_\omega(A)\Omega_\omega$, and thence everywhere. □

We now take up the proof of Theorem C.87, preceded by some general remarks on direct sums of Hilbert spaces and representations. First, if $(H_1,\dots,H_n)$ is a finite family of Hilbert spaces, one may form the ***direct sum*** $H = H_1 \oplus \cdots \oplus H_n$, initially merely as a vector space, and subsequently also as a space with inner product

$$\langle (\varphi_1,\dots,\varphi_n),(\psi_1,\dots,\psi_n)\rangle = \sum_{i=1}^{n} \langle \varphi_i,\psi_i\rangle. \tag{C.215}$$

It is easy to see that H is complete in the ensuing norm

$$\|(\psi_1,\dots,\psi_n)\|^2 = \sum_{i=1}^{n} \|\psi_i\|^2. \tag{C.216}$$

Some authors write $\psi_1 \oplus \cdots \oplus \psi_n$, $\psi_1 \dot{+} \cdots \dot{+} \psi_n$, or $\psi_1 + \cdots + \psi_n$ for $(\psi_1,\dots,\psi_n)$.

Moreover, if (π_i) is a family of representations $\pi_i : A \to B(H_i)$, then one obtains a new representation $\bigoplus_i \pi_i$ of A, called the ***direct sum*** of the π_i, by

$$\bigoplus_i \pi_i(a)(\psi_1,\dots,\psi_n) = (\pi_1(a)\psi_1,\dots,\pi_n(a)\psi_n). \tag{C.217}$$

This construction works for arbitrary families of Hilbert spaces (H_x) and representations (π_x), where $x \in X$ for some index set X. First, the elements of $H = \bigoplus_x H_x$ are families $(\psi) \equiv (\psi_x)_{x\in X}$, where $\psi_x \in H_x$, such that

$$\|(\psi)\|^2 = \sup_{F\subset X} \sum_{x\in F} \|\psi_x\|^2_{H_x} < \infty, \tag{C.218}$$

where the supremum is over all finite subsets F of X, so that the sum is defined as in (B.11). In that case, the obvious linear operations (i.e., $((\psi)+(\varphi))_x = \psi_x + \varphi_x$ and $(\lambda(\psi))_x = \lambda \cdot \psi_x$) are defined within H, since for each pair $(\varphi),(\psi) \in H$ we have, from the triangle inequality for the norm in each finite direct sum $H_F = \bigoplus_{x\in F} H_x$,

$$\left(\sum_{x\in F} \|\psi_x + \varphi_x\|^2_{H_x}\right)^{1/2} \leq \left(\sum_{x\in F} \|\psi_x\|^2\right)^{1/2} + \left(\sum_{x\in F} \|\varphi_x\|^2)\right)^{1/2} \leq \|(\psi)\|^2 + \|(\varphi)\|^2.$$

The supremum over F gives $\|(\psi)+(\varphi)\|$, which is therefore finite and satisfies the triangle inequality for the norm. Similarly, the natural inner product in H is well defined, this time by the full Definition B.6, with $V = \mathbb{C}$ and $f(x) = \langle \varphi_x, \psi_x\rangle_{H_x}$, i.e.,

$$\langle (\varphi),(\psi)\rangle = \sum_{x\in X} \langle \varphi_x,\psi_x\rangle_{H_x}. \tag{C.219}$$

To see this, we apply Cauchy–Schwarz first in each H_x and then in $\ell^2(X)$ to obtain

$$|\langle (\varphi),(\psi)\rangle| \leq \sum_{x\in X} |\langle \varphi_x,\psi_x\rangle_{H_x}| \leq \sum_{x} \|\varphi_x\|\|\psi_x\| \leq \|(\varphi)\|\|(\psi)\| < \infty. \tag{C.220}$$

Finally, the proof that the direct sum Hilbert space $\bigoplus_x H_x$ is complete in the norm (C.218) is similar to the case where $H_x = \mathbb{C}$ for each x, i.e., $H = \ell^2(X)$, cf. Theorem B.9. Let $(\psi)_n$ be a Cauchy sequence in H, consisting of sequences $(\psi_x)_n \equiv \psi_x^{(n)}$ in each H_x. For each finite $F \subset X$ and $\varepsilon > 0$, we must have $\sum_{x\in F} \|\psi_x^{(n)} - \psi_x^{(m)}\| < \varepsilon$ for sufficiently large n,m so that each $(\psi_x)_n$ must be Cauchy in H_x, with limit ψ_x. The ensuing set (ψ) of vectors lies in H by the argument following (B.19), and the given Cauchy sequence $(\psi)_n$ converges to (ψ), again by the same proof as for $\ell^2(X)$.

If one has a family (π_x) of representations $\pi_x : A \to B(H_x)$, their direct sum $\pi = \bigoplus_x \pi_x$, defined by $(\pi(a)(\psi))_x = \pi_x(a)\psi_x$, is a representation of A on H. Indeed, one has $\|\pi(a)\| = \sup_x\{\|\pi_x(a)\|\}$, and since we have $\|\pi_x(a)\| \le \|a\|$ for each x, we also have $\|\pi(a)\| \le \|a\|$, so that $\pi(a) \in B(H)$, and hence π maps A into $B(H)$.

Our first use of such direct sums shows that cyclic representation are the building blocks of any representation π, at least if we require π to be ***nondegenerate*** in the sense that $\pi(a)\psi = 0$ for all $a \in A$ and $\psi \in H$ implies $\psi = 0$.

Proposition C.92. *Any nondegenerate representation $\pi : A \to B(H)$ of a C*-algebra A on a Hilbert space H is a direct sum of cyclic representations of A.*

Proof. Consider families $(\psi_x)_{x\in X}$ of nonzero vectors in H with the property that

$$\langle \pi(a)\psi_x, \pi(b)\psi_{x'}\rangle = 0, \tag{C.221}$$

for all $a,b \in A$ and all $x \neq x'$. Such families are partially ordered by inclusion, and an easy application of Zorn's Lemma shows that there is a maximal such family. For this family $(\psi_x)_{x\in X}$, we define H_x as the closure of $\pi(A)\psi_x$ in H. Since π is a homomorhism, each H_x is stable under $\pi(A)$, and hence the restriction $\pi_x(a)$ of $\pi(a)$ to H_x defines a representation of A, which is cyclic by construction. It follows that $H = \bigoplus_x H_x$ and $\pi = \bigoplus_x \pi_x$, and so the claim has been proved. □

Our second use is the proof of Theorem C.87, where we have to solve the problem of the possible lack of injectivity of π_ω in our previous preliminary proof.

Proof. To do so, we replace H_ω by the crazy Hilbert space $H_c = \bigoplus_{\omega\in P(A)} H_\omega$, where $P(A)$ is the pure state space of A. The Hilbert space H_c carries a representation $\pi = \bigoplus_{\omega\in P(A)} \pi_\omega$. The point is that if $\pi(a) = 0$, then $\pi(a^*a)\Omega_\omega = 0$ for each $\omega \in P(A)$, which by (C.196) implies $\omega(a^*a) = 0$. Proposition C.15 then gives $\sigma(a^*a) = \{0\}$, from which the spectral radius formula (C.55) gives $\|a\| = 0$, and hence $a = 0$. It follows that π is injective, and Theorem C.87 is proved. □

It should be noted that this proof relies on *shock and awe* kind of overkill (though nothing compared to the even crazier space $H_{ec} = \bigoplus_{\omega\in S(A)} H_\omega$, which is traditionally used in the above proof), in that H_c is far larger than necessary (indeed, in all but the most trivial cases, H is non-separable). For example, already for $A = M_2(\mathbb{C})$ we have $P(A) \cong S^2$, so that $H_c = \bigoplus_{\omega\in S^2} \mathbb{C}^2$; this Hilbert space is non-separable, whereas A has an injective representation on $\mathbb{C}^2$. More generally, $B_0(H)$ or $B(H)$ has an injective representation on H by definition, whereas H_c is non-separable. In the commutative case, $A = C_0(X)$ yields the non-separable $H_c = \bigoplus_{x\in X} \mathbb{C}$, although A has an injective representation on the (typically) separable space $L^2(X,\mu)$.

As a nice illustration of the GNS-construction, let us treat this example in more detail (cf. §1.5 for the simple case where X is finite). If μ is some state on $C_0(X)$, then by Theorem B.24, there is a unique probability measure μ on X such that

$$\omega(f) = \int_X d\mu\, f, \; f \in C_0(X), \tag{C.222}$$

cf. (B.39). It follows from (C.204) and (C.222) that

$$N_\omega = \left\{ f \in C_0(X) : \int_X d\mu\, |f|^2 = 0 \right\}. \tag{C.223}$$

In particular, the support of μ is X iff $N_\omega = \{0\}$, in which case $A/N_\omega = C_0(X)$. In the opposite case where ω is a pure state, i.e., $\omega = \omega_x$ for some $x \in X$, with $\omega_x(f) = f(x)$, one has $N_\omega = \{f \in C_0(X) \mid f(x) = 0\}$, so that $A/N_\omega \cong \mathbb{C}$, under the map $[f] \mapsto f(x)$. In general, from (C.206) - (C.207) we obtain

$$H_\omega = L^2(X, \mu); \tag{C.224}$$

$$\pi_\omega(f) = m_f; \tag{C.225}$$

$$\Omega_\omega = 1_X, \tag{C.226}$$

where $m_f \psi = f\psi$, cf. (B.238). Analogously to (B.331), we then obtain

$$\pi_\omega(C_0(X))'' = L^\infty(X, \mu). \tag{C.227}$$

The state ω, initially defined on the commutative C*-algebra $C_0(X)$, then has a normal extension to the commutative von Neumann algebra $L^\infty(X, \mu)$, cf. (C.222).

More generally, if A is an arbitrary commutative C*-algebra and ω is a state on A, then, writing $\Sigma(A)$ for the Gelfand spectrum of A as usual, we have

$$H_\omega \cong L^2(\Sigma(A), \mu); \tag{C.228}$$

$$\pi_\omega(f) \cong m_{\hat{f}}; \tag{C.229}$$

$$\Omega_\omega \cong 1_{\Sigma(A)}, \tag{C.230}$$

where $\hat{f} \in C_0(\Sigma(A))$ is the Gelfand transform of $f \in A$, and μ is the probabililty measure on $\Sigma(A)$ defined by

$$\omega(f) = \int_{\Sigma(A)} d\mu\, \hat{f}. \tag{C.231}$$

With this commutative case in mind, some authors would call a pair (A, ω), where A is a general C*-algebra and ω is a state on A, or, alternatively, A is a general von Neumann algebra and ω is a normal state on A, a ***non-commutative probability space***. As such, 'aordinary" probability theory (at least, on locally compact Hausdorff sample spaces) is merely the commutative case of a much more general "non-commutative probability theory".

C.13 Tensor products of Hilbert spaces and C*-algebras

If H_A and H_B are Hilbert spaces, their algebraic tensor product $H_A \otimes H_B$ typically fails to be a Hilbert space in the obvious way, since it is not complete (unless one of the factors is finite-dimensional). Similarly, the algebraic tensor product $A \otimes B$ of two C*-algebras A and B usually fails to be a C*-algebra. However, the second case is far more complicated then the first: for Hilbert spaces there is a canonical norm on the algebraic tensor product and hence a canonical completion of $H_A \otimes H_B$ into a Hilbert space $H_A \overline{\otimes} H_B$. For C*-algebras, on the other hand, there is an *embarrasment of riches*, in that there are are many norms turning the completion $A \hat{\otimes} B$ of $A \otimes B$ in some such norm into a C*-algebra. However, if A or B is *nuclear*, there is just one possibility; see below. For example, this applies of A or B is finite-dimensional.

Let us first review the (algebraic) tensor product of two vector spaces. A and B.

Proposition C.93. *Let A and B be (complex) vector spaces. There is a vector space called $A \otimes B$, in words the* **algebraic tensor product** *of A and B (over $\mathbb{C}$), and a map $p : A \times B \to A \otimes B$, such that for any vector space C and any bilinear map $\beta : A \times B \to C$, there is a unique linear map $\beta' : A \otimes B \to C$ such that $\beta = \beta' \circ p$.*

In other words, the following diagram commutes:

$$\begin{array}{ccc} A \times B & \xrightarrow{p} & A \otimes B \\ & \searrow^{\beta} & \downarrow {\scriptstyle \exists ! \beta'} \\ & & C \end{array} \tag{C.232}$$

This universal property also shows that $A \otimes B$ is unique up to isomorphism.

Proof. In preparation for an explicit construction of $A \otimes B$, define the (complex) ***free vector space*** on any non-empty set X as$C_c(X)$, where X has the discrete topology (i.e., $C_c(X)$ consists of all functions $f : X \to \mathbb{C}$ with finite support), and pointwise operations. For each $y \in X$, the delta-function $\delta_y \in C_c(X)$ is defined by $\delta_y(x) = \delta_{xy}$, so that each element f of $C_c(X)$ is a finite sum $f = \sum_i \lambda_i \delta_{x_i}$, where $\lambda_i \in \mathbb{C}$ and $x_i \in X$.

If A and B are (complex) vector spaces, $A \otimes B$ is the quotient of the free vector space $C_c(A \times B)$ on $X = A \times B$ by the equivalence relation generated by the relations:

$$\delta_{(a_1+a_2,b)} \sim \delta_{(a_1,b)} + \delta_{(a_2,b)}; \tag{C.233}$$
$$\delta_{(a,b_1+b_2)} \sim \delta_{(a,b_1)} + \delta_{(a,b_2)}; \tag{C.234}$$
$$\lambda \delta_{(a,b)} \sim \delta_{(\lambda a,b)}; \tag{C.235}$$
$$\lambda \delta_{(a,b)} \sim \delta_{(a,\lambda b)}. \tag{C.236}$$

For $a \in A, b \in B$, the image of $\delta_{(a,b)}$ in $A \otimes B$ is called $a \otimes b$, so that by construction,

$$(a_1 + a_2) \otimes b = a_1 \otimes b + a_2 \otimes b; \tag{C.237}$$
$$a \otimes (b_1 + b_2) = a \otimes b_1 + a \otimes b_2; \tag{C.238}$$
$$\lambda (a \otimes b) = (\lambda a) \otimes b = a \otimes (\lambda b). \tag{C.239}$$

Elements of the algebraic tensor product $A \otimes B$ may therefore be written as finite sums $c = \sum_i a_i \otimes b_i$, with $a_i \in A$, $b_i \in B$, subject to the above relations.

Now consider some bilinear map $\beta : A \times B \to C$. We extend β to a map

$$\tilde{\beta} : C_c(A \times B) \to C; \tag{C.240}$$

$$\tilde{\beta}\left(\sum_i \lambda_i \delta_{(a_i,b_i)}\right) = \sum_i \lambda_i \beta(a_i, b_i). \tag{C.241}$$

Since β is bilinear, it respects the above equivalence relation, so that it duly quotients to $\beta' : A \otimes B \to C$, upon which the property $\beta = \beta' \circ p$ holds by construction. Finally, since p is surjective the latter property uniquely determines β'. □

Equivalently, $A \otimes B$ is the quotient of formal sums $\sum_i (a_i, b_i)$ by the subspace consisting of those sums for which there are $\omega_A \in A^*$ and $\omega_B \in B^*$ such that $\sum_i \omega_B(a_i)\omega_B(b_i) = 0$. Similarly, it is useful to regard $A \otimes B$ as a subspace of the vector space $L(A^*, B)$ of linear maps from the dual A^* to B through the map

$$\sum_i a_i \otimes b_i : \omega_A \mapsto \sum_i \omega_A(a_i) b_i \;\; (\omega_A \in A^*); \tag{C.242}$$

this map is injective by Corollary B.45.2, since we may assume the b_i to be linearly independent. Using the canonical embedding $B \hookrightarrow B^{**}$ of Proposition B.44, this in turn yields an injection $A \otimes B \hookrightarrow L(A^* \times B^*, \mathbb{C})$, i.e., the space of bilinear maps from $A^* \times B^*$ to $\mathbb{C}$, given on arguments (ω_A, ω_B) by

$$\sum_i a_i \otimes b_i : (\omega_A, \omega_B) \mapsto \sum_i \omega_A(a_i)\omega_B(b_i). \tag{C.243}$$

Proposition C.93 turns this into an injection $A \otimes B \hookrightarrow L(A^* \otimes B^*, \mathbb{C})$, given by

$$\sum_i a_i \otimes b_i : \sum_j (\omega_A)_j \otimes (\omega_B)_j \mapsto \sum_{i,j} (\omega_A)_j(a_i)(\omega_B)_j(b_i). \tag{C.244}$$

If A and B are Hilbert spaces, we call them H_A and H_B, denote their elements by α and β, respectively, and attempt to define a sesquilinear form on $H_A \otimes H_B$ by

$$\langle \sum_j \alpha'_j \otimes \beta'_j, \sum_i \alpha_i \otimes \beta_i \rangle = \sum_{i,j} \langle \alpha'_j, \alpha_j \rangle_A \langle \beta'_j, \beta_i \rangle_B. \tag{C.245}$$

It is a non-trivial fact that this form is well defined, because representations $\sum_i \alpha_i \otimes \beta_i$ of vectors in $H_A \otimes H_B$ may not be unique. For example, if $H_A = H_B = H = \mathbb{C}^n$, and (α_i) and (α'_i) are two bases of H, then $\sum_i \alpha_i \otimes \alpha_i = \sum_i \alpha'_i \otimes \alpha'_i$ (to see this, take inner products with an arbitrary elementary tensor $\psi \otimes \varphi$, yielding the same result).

To resolve this, we note that the injection $H_A \otimes H_B \hookrightarrow L(H_A^* \times H_B^*, \mathbb{C})$ just discussed combines with the isomorphism $H^* \cong H$ of Theorem B.66 to an injection $H_A \otimes H_B \hookrightarrow \overline{L}(H_A \times H_B, \mathbb{C})$, i.e., the space of *bi-anti-linear* maps from $H_A \times H_B$ to $\mathbb{C}$. Proposition C.93 turns this into an injection $H_A \otimes H_B \hookrightarrow \overline{L}(H_A \otimes H_B, \mathbb{C})$, viz.

$$\sum_i \alpha_i \otimes \beta_i : \sum_j \alpha'_j \otimes \beta'_j) \mapsto \sum_{i,j} \langle \alpha'_j, \alpha_i \rangle_{H_A} \langle \beta'_j, \beta_i \rangle_{H_B}. \tag{C.246}$$

Consequently, if $\sum_i \alpha_i \otimes \beta_i = 0$, then the right-hand-side of (C.245) is zero, too, since it is the image of $\sum_j \alpha'_j \otimes \beta'_j$ under the zero map. Hence (C.245) is independent of the choice of representatives in the sum $\sum_i \alpha_i \otimes \beta_i$, and by hermiticity of the form, this equally well applies to the other entry $\sum_j \alpha'_j \otimes \beta'_j$.

It remains to show that (C.245) is an inner product, i.e., that it is positive definite. To see this, for some given vector $\sum_i \alpha_i \otimes \beta_i$ in $H_A \otimes H_B$ one may take the linear span H'_A of all α_i in H_A, which is a Hilbert space, and pick a basis (υ_i) in H'_A. Absorbing the scalars in the β_j, we may therefore write $\sum_i \alpha_i \otimes \beta_i = \sum_k \upsilon_k \otimes \beta''_k$, so that

$$\langle \sum_i \alpha_i \otimes \beta_i, \sum_i \alpha_i \otimes \beta_i \rangle = \sum_{k,l} \langle \upsilon_k \otimes \beta''_k, \upsilon_l \otimes \beta''_l \rangle = \sum_k \|\beta''_k\|^2_B \geq 0, \tag{C.247}$$

with equality at the end iff each $\beta''_k = 0$, and hence $\sum_i \alpha_i \otimes \beta_i = 0$.

Finally, we complete $H_A \otimes H_B$ in the norm defined by the inner product (C.245); with abuse of notation the ensuing Hilbert space is often just called $H_A \otimes H_B$, but it would be more precise to denote it by $H_A \hat{\otimes} H_B$, as we will usually do.

It is easy to show that if $(\upsilon_i^{(A)})$ and $(\upsilon_j^{(B)})$ are bases for H_A and H_B, respectively, then $(\upsilon_i^{(A)} \otimes \upsilon_j^{(B)})$ is a basis of $H_A \hat{\otimes} H_B$. Also, if (X, Σ, μ) and (X', Σ', μ') are σ-finite measure spaces with X and X' well behaved (e.g., Polish), so that the L^2-spaces are separable, one has a natural isomorphism

$$L^2(X, \Sigma, \mu) \hat{\otimes} L^2(X', \Sigma', \mu') \cong L^2(X \times X', \Sigma \times \Sigma', \mu \times \mu'), \tag{C.248}$$

obtained as the closure of the isometric (and hence bounded) map that sends the vector $\sum_i \psi_i \otimes \psi'_i$ into the function $(x, x') \mapsto \sum_i \psi_i(x) \psi'_i(x')$ on $X \times X'$. Here $\Sigma \times \Sigma'$ is the smallest σ-algebra on $X \times X'$ that contains all sets $A \times A'$, $A \in \Sigma$, $A' \in \Sigma'$, and $\mu \times \mu'$ is the familiar product measure defined on elementary measurable sets by

$$\mu \times \mu'(A \times A') = \mu(A)\mu'(A'). \tag{C.249}$$

We now turn to tensor products of C*-algebras. If A and B are C*-algebras, then the algebraic tensor product $A \otimes B$ of A and B (just seen as vector spaces) is endowed with a natural multiplication and involution, given by linear extension of

$$(a_1 \otimes b_1) \cdot (a_2 \otimes b_2) = (a_1 a_2) \otimes (b_1 b_2); \tag{C.250}$$
$$(a \otimes b)^* = a^* \otimes b^*, \tag{C.251}$$

respectively. Thus $A \otimes B$ is a *-algebra, and Proposition C.93 specializes to:

Proposition C.94. *If C is a *-algebra and if a bilinear map $\beta : A \times B \to C$ satisfies*

$$\beta(a_1 a_2, b_1 b_2) = \beta(a_1, a_2)\beta(b_1, b_2); \;\; \beta(a^*, b^*) = \beta(a, b)^*, \tag{C.252}$$

then β factors through $A \otimes B$ (now seen as a *-algebra), as in (C.232).

The proof is similar. In order to turn $A \otimes B$ (seen as a *-algebra) into a C*-algebra, we need a ***C*-norm***, i.e., a norm on $A \otimes B$ satisfying the C*-axioms (C.1) - (C.2). If such a norm exists, we denote the completion of $A \otimes B$ in that particular norm by $A \hat{\otimes} B$, where typically $\|\cdot\|$ and hence $\hat{\otimes}$ carry some label. This completion $A \hat{\otimes} B$ is a C*-algebra in the obvious way. There will be no shortage of such norms!

For example, suppose $A \subset B(H_A)$ and $B \subset B(H_B)$. For each $a \in A$, we form the operator $a \otimes 1_B$ on $H_A \otimes H_B$ (where 1_B is the unit of $B(H_B)$, which is also the unit of B if it has one). As in (C.247), we may assume that generic elements of $H_A \otimes H_B$ take the form $\sum_k \upsilon_k \otimes \beta_k$, with the υ_k orthonormal in H_A and $\beta_k \in H_B$. We then estimate

$$\left\| (a \otimes 1_B)\left(\sum_k \upsilon_k \otimes \beta_k\right) \right\|^2 = \left\| \sum_k (a\upsilon_k) \otimes \beta_k \right\|^2 \leq \sum_k \|(a\upsilon_k) \otimes \beta_k\|^2$$
$$\leq \|a\|^2 \left\| \sum_k \upsilon_k \otimes \beta_k) \right\|^2. \tag{C.253}$$

Hence $a \otimes 1_B$ is bounded on the pre-Hilbert space $H_A \otimes H_B$, and extends to a bounded operator on $H_A \overline{\otimes} H_B$ by continuity; this extension is usually called $a \otimes 1_B$, too. Similarly, any $b \in B$ defines a bounded operator $1 \otimes b$ on $H_A \overline{\otimes} H_B$, and since

$$a \otimes b = (a \otimes 1_B) \cdot (1_A \otimes b), \tag{C.254}$$

all elements $\sum_i a_i \otimes b_i$ of $A \otimes B$ extend to elements of $B(H_A \overline{\otimes} H_B)$. Now define

$$\|\sum a_i \otimes b_i\|_{\min} = \|\sum a_i \otimes b_i\|_{B(H_A \overline{\otimes} H_B)}. \tag{C.255}$$

This is clearly a C*-norm on $A \otimes B$. Moreover, it is a ***cross-norm***, in that

$$\|a \otimes b\|_{\min} = \|a\| \|b\|. \tag{C.256}$$

This construction generalizes to any two C*-algebras, since by Theorem C.87 we have injective representations $\pi_A : A \to B(H_A)$ and $\pi_B : A \to B(H_B)$ of A and B, respectively, and it is easy to verify that the norm $\|\cdot\|_{\min}$ on $A \otimes B$ and ensuing completion $A \hat{\otimes}_{\min} B$ are independent of the chosen representation. Furthermore,

$$\|c\|_{\min} = \sup\{\|\pi_A \otimes \pi_B(c)\|_{B(H_A \overline{\otimes} H_B)}\}, \tag{C.257}$$

where π_A and π_B run through all representations of A and B, respectively. The ensuing completion $A \hat{\otimes}_{\min} B$ is called the ***injective*** tensor product of A and B. Without proof (which requires more advanced methods than the elementary arguments we use in this section), we mention that, as its name suggests, $\|\cdot\|_{\min}$ is the smallest C*-norm on $A \otimes B$. This has a very important consequence:

Proposition C.95. *Any C*-norm $\|\cdot\|$ on $A \otimes B$ satisfies $\|a \otimes b\| = \|a\| \|b\|$.*

In other words, any C*-norm $\|\cdot\|$ on $A \otimes B$ is a cross-norm. To prove this from the minimality of the spatial norm, we need a lemma of wider interest.

Lemma C.96. *If $\|\cdot\|$ is any C*-norm on $A \otimes B$, then for all $a \in A$ and $b \in B$,*

$$\|a \otimes b\| \leq \|a\| \|b\|. \tag{C.258}$$

Consequently, for any C-norm on $A \otimes B$ and any $c \in A \otimes B$, we have the bound*

$$\|c\| \leq \inf \left\{ \sum_i \|a_i\| \|b_i\|, c = \sum_i a_i \otimes b_i \right\}. \tag{C.259}$$

Proof. In any C*-algebra A, if $a \geq 0$, we have $\|a\| \leq 1$ iff $a^2 \leq a$. This is trivial for $A = C(X)$, and in general can be proved within $C^*(a) \subset A$, since $C^*(a) \cong C(\sigma(a))$. Now take $a \in A$ and $b \in B$ such that $a \geq 0$, $b \geq 0$, $\|a\| \leq 1$, and $\|b\| \leq 1$, so that $(a \otimes b)^2 = a^2 \otimes b^2 \leq a \otimes b^2 \leq a \otimes b$, and hence $\|a \otimes b\| \leq 1$. For general $a \geq 0, b \geq 0$, rescaling to $a/\|a\|$ etc. gives (C.258). For general a, b altogether, we compute:

$$\|a \otimes b\|^2 = \|(a \otimes b)^*(a \otimes b)\| = \|a^*a \otimes b^*b\| \leq \|a^*a\| \|b^*b\| = \|a\|^2 \|b\|^2. \tag{C.260}$$

Eq. (C.259) then follows from the triangle inequality on the norm. □

If A and B each have a unit, there is a simpler proof: as in (C.254), we have

$$\|a \otimes b\| = \|(a \otimes 1_B)(1_A \otimes b)\| \leq \|a \otimes 1_B\| \|1_A \otimes b\| = \|a\| \|b\|, \tag{C.261}$$

where we used $\|a \otimes 1_B\| = \|a\|$ etc., which is the case because the map $a \mapsto a \otimes 1_B$ from A to $A \hat{\otimes} B$ is injective and hence is an (isometric) isomorphism onto its image.

We now prove Proposition C.95.

Proof. For any C*-norm $\|\cdot\|$, we have $\|a \otimes b\| \geq \|a \otimes b\|_{\min} = \|a\| \|b\|$, since the spatial norm is itself a cross-norm, cf. (C.256). Then (C.258) gives equality. □

In view of (C.259) and the existence of at least one C*-norm on $A \otimes B$ (namely the spatial one), it makes sense to define the ***maximal C*-norm*** on $A \otimes B$ by

$$\left\| \sum_i a_i \otimes b_i \right\|_{\max} = \sup \left\{ \left\| \sum_i a_i \otimes b_i \right\|, \|\cdot\| \text{ is a C*-norm on } A \otimes B \right\}. \tag{C.262}$$

This is clearly a C*-norm, and hence it is also a cross-norm. i.e.,

$$\|a \otimes b\|_{\max} = \|a\| \|b\|. \tag{C.263}$$

This property may be proved without using the deep result that the spatial norm is the minimal one (which in turn led to Proposition C.95); all we need is the inequality

$$\|c\|_{\min} \leq \|c\|_{\max}, \tag{C.264}$$

for any $c \in A \otimes B$, which follows from the definition of $\|\cdot\|_{\max}$, upon which (C.264) may be proved in the same way as for general C*-norms. The completion $A \hat{\otimes}_{\max} B$ of $A \otimes B$ in the norm $\|\cdot\|_{\max}$ is called the ***projective*** tensor product of A and B.

If we define representations of the pre-C*-algebra $A \otimes B$ on Hilbert spaces in the same way as for C*-algebras, i.e., as linear maps $\pi : A \otimes B \to B(H)$ that preserve the product (C.250) and the involution (C.251), we obtain

$$\|c\|_{\max} = \sup\{\|\pi(c)\|\}, \tag{C.265}$$

where $c = \sum_i a_i \otimes b_i \in A \otimes B$, and π runs through all representations of $A \otimes B$. Indeed, according to Theorem C.87 there exists an injective representation π of $A \hat{\otimes}_{\max} B$, so that $\|c\|_{\max} = \|\pi(c)\|$ for each $c \in A \hat{\otimes}_{\max} B$, and hence also of each $c \in A \otimes B$. Furthermore, any representation of $A \otimes B$ yields a cross-norm, so that (C.265) follows. This also shows that the supremum in (C.265) is actually attained.

In what follows, we restrict ourselves to the case that A and B have a unit, which suffices for our applications, but the claim is true in general (with a slightly more complicated proof, involving either approximate units or unitizations). If A and B each have a unit, so does $A \otimes B$, viz. $1_A \otimes 1_B$. ***States*** ω on $A \otimes B$ are then defined as for unital C*-algebras, i.e., as positive linear functionals (in the usual sense that $\omega(c^*c) \geq 0$ for any $c \in A \otimes B$) that map the unit $1_A \otimes A_B$ of $A \otimes B$ to 1.

Proposition C.97. *Let A and B be unital. Then each state on $A \otimes B$ is continuous with respect to the $\|\cdot\|_{\max}$-norm, and hence extends to a state on the maximal tensor product $A \hat{\otimes}_{\max} B$. Thus identifying states on $A \otimes B$ and on $A \hat{\otimes}_{\max} B$, we have*

$$S(A \otimes B) = S(A \hat{\otimes}_{\max} B). \tag{C.266}$$

Proof. Let $\omega : A \otimes B \to \mathbb{C}$ a state. Although $A \otimes B$ may not be a C*-algebra, the GNS-construction Theorem C.88 goes through as if it were. The reason is that the only delicate point, namely boundedness of $\pi_\omega(a \otimes b)$, may be proved from (C.94), just as in the usual case. Indeed, for $a \in A$, $b \in B$, and $c \in A \otimes B$, we estimate

$$\begin{aligned} \|\pi_\omega(a \otimes b)c_\omega\|^2 &= \omega(c^*(a \otimes b)^*(a \otimes b)c) = \omega(c^*(a^*a \otimes b^*b)c) \\ &\leq \|a\|^2\|b\|^2\omega(c^*c) = \|a\|^2\|b\|^2\|c_\omega\|^2 \\ &= \|a \otimes b\|_{\max}\|c_\omega\|^2, \end{aligned}$$

so that $\|\pi_\omega(a \otimes b)\| \leq \|a \otimes b\|_{\max}$, and hence $\pi_\omega(a \otimes b)$ may be extended to the completion H_ω of $(A \otimes B)/N_\omega$ by continuity. Here we used the facts that:

- $(a \otimes b)^*(a \otimes b) = a^*a \otimes b^*b$, so that the right-hand side is positive in $A \otimes B$.
- $0 \leq a^*a \leq \|a\|^2 1_A$ and $0 \leq b^*b \leq \|b\|^2 1_B$, as A and B are C*-algebras, cf. (C.83).
- If $c' \geq 0$ in $A \otimes B$, then $c^*c'c \geq 0$, as for C*-algebras, see the argument preceding (C.93). The argument is the same: $c^*c'c = c^*d^*dc = (dc)^*dc \geq 0$.

Wriiting $\Omega_\omega = (1_A \otimes 1_B)_\omega$ for the cyclic vector of H_ω, as in (C.208), for any element $c \in A \otimes B$ we obtain, using (C.265) in the final inequality, the decisive bound

$$|\omega(c)| = |\langle \Omega_\omega, \pi_\omega(c)\Omega_\omega \rangle| \leq \|\pi_\omega(c)\| \leq \|c\|_{\max}. \tag{C.267}$$

In other words, ω is continuous with respect to the $\|\cdot\|_{\max}$-norm, and since the latter is dense in $A \hat{\otimes}_{\max} B$, the state extends to the completed tensor product by continuity.

It follows from (C.267) and $\omega(1_A \otimes 1_B) = 1$ that $\|\omega\| = 1$ as a functional on $A \otimes B$ equipped with the $\|\cdot\|_{\max}$-norm, so that the in question extension has the same norm, and hence by Proposition C.5 is a state on $A\hat{\otimes}_{\max}B$. Conversely, a state on $A \otimes_{\max} B$ restricts to a state on $A \otimes B$, since the two *-algebras have the same unit and (trivially) if c is positive in the latter, then so it is in the former. □

The above proposition concerns extensions of *arbitrary* states on $A \otimes B$. However, ***product states*** on $A \otimes B$ can be extended to any completed tensor product $A\hat{\otimes}B$.

Proposition C.98. *If ω_A and ω_B are states on A and B, respectively, then the corresponding product state $\omega_A \otimes \omega_B$ on $A \otimes B$, defined as in* (C.243) *by*

$$\omega_A \otimes \omega_B \left(\sum_i a_i \otimes b_i \right) = \sum_i \omega_A(a_i)\omega_B(b_i), \tag{C.268}$$

is continuous with respect to any cross-norm $\|\cdot\|$, and hence extends to $A\hat{\otimes}B$.

Proof. Since the spatial norm is minimal among all cross-norms, it is enough to prove continuity with respect to $\|\cdot\|_{\min}$. As in the proof of Proposition C.97, we form the GNS representation $\pi_{\omega_A\otimes\omega_B}$ induced by $\omega_A \otimes \omega_B$, so that for any $c \in A \otimes B$,

$$(\omega_A \otimes \omega_B)(c) = \langle \Omega_{\omega_A\otimes\omega_B}, \pi_{\omega_A\otimes\omega_B}(c)\Omega_{\omega_A\otimes\omega_B} \rangle. \tag{C.269}$$

Now consider the representation $\pi_{\omega_A}(A) \otimes \pi_{\omega_B}(B)$ on $H_{\omega_A} \otimes H_{\omega_B}$, with cyclic vector $\Omega_{\omega_A} \otimes \Omega_{\omega_B}$. Writing $c = \sum_i a_i \otimes b_i$ as usual, a simple computation gives

$$\begin{aligned} &\langle \Omega_{\omega_A} \otimes \Omega_{\omega_B}, (\pi_{\omega_A} \otimes \pi_{\omega_B})(c)\Omega_{\omega_A} \otimes \Omega_{\omega_B} \rangle \\ &\quad = \sum_i \langle \Omega_{\omega_A}, \pi_{\omega_A}(a_i)\Omega_{\omega_A} \rangle \langle \Omega_{\omega_B}, \pi_{\omega_B}(b_i)\Omega_{\omega_B} \rangle = \sum_i \omega_A(a_i)\omega_B(b_i) \\ &\quad = (\omega_A \otimes \omega_B)(c). \end{aligned} \tag{C.270}$$

Using the same reasoning as in (the proof of) Proposition C.91 (which does not apply literally, since it is about C*-algebras), it follows from (C.270) that $\pi_{\omega_A\otimes\omega_B}(A \otimes B)$ is unitarily equivalent to $\pi_{\omega_A}(A) \otimes \pi_{\omega_B}(B)$, so that, using (C.270), analogously to (C.267) but this time using (C.257) at the end, we have

$$|(\omega_A \otimes \omega_B)(c)| \le \|\pi_{\omega_A} \otimes \pi_{\omega_B}(c)\| \le \|c\|_{\min}.$$ □

As an application, analogously to (C.248), we show that:

Proposition C.99. *For any locally compact Hausdorff spaces X, Y and any cross-norm on $C_0(X) \otimes C_0(Y)$, with completed tensor product $C_0(X)\hat{\otimes}C_0(Y)$, we have*

$$C_0(X)\hat{\otimes}C_0(Y) \cong C_0(X \times Y), \tag{C.271}$$

under the isomorphism given by continuous extension of the map $f \otimes g \mapsto fg : (x,y) \mapsto f(x)g(y)$ from the algebraic tensor product $C_0(X) \otimes C_0(Y)$ to $C_0(X \times Y)$.

Proof. We just prove the unital case, where X and Y are compact.

Let $x \in X$ and $y \in Y$, and take the corresponding evaluations maps ev_x and ev_y on $C(X)$ and $C(Y)$, respectively. These are multiplicative states, cf. Proposition C.19. Then $\mathrm{ev}_x \otimes \mathrm{ev}_x$ is a nonzero multiplicative state on $C(X) \otimes C(Y)$, and hence also on $C(X)\hat{\otimes}C(Y)$, cf. Proposition C.98. This gives an injection of $X \times Y$ into $\Sigma(C(X)\hat{\otimes}C(Y))$, i.e., the Gelfand spectrum of $C(X)\hat{\otimes}C(Y)$, cf. §C.2.

Conversely, the restriction ω_1 of any $\omega \in \Sigma(C(X)\hat{\otimes}C(Y))$ to $C(X)$, given by $\omega_1(f) = \omega(f \otimes 1_Y)$, is multiplicative, as is the restriction ω_2 of ω to $C(Y)$, defined by $\omega_2(g) = \omega(1_X \otimes g)$. Then $\omega = \omega_1 \otimes \omega_2$, with ensuing injective map $\Sigma(C(X)\hat{\otimes}C(Y)) \to X \times Y$. Thus the above injection is also a surjection, and hence a bijection, which is easily seen to be a homeomorphism. □

This can also be proved without Proposition C.98, using only the second step: if $\Sigma(C(X)\hat{\otimes}C(Y)) \neq X \times Y$, then, since $\Sigma(C(X)\hat{\otimes}C(Y))$ is closed in $X \times Y$, there are nonempty opens $U \subset X$ and $V \subset Y$ such that $(U \times V) \cap \Sigma(C(X)\hat{\otimes}C(Y)) = \emptyset$. Now take nonzero functions $f \in C_c(U)$ and $g \in C_c(V)$ such that $\omega(f \otimes g) = 0$ for all $\omega \in \Sigma(C(X)\hat{\otimes}C(Y))$. This contradicts the isometry (C.18) of the Gelfand transform.

Proposition C.100. *For any locally compact Hausdorff space X and any C*-algebra B, let $C_0(X,B)$ be the C*-algebra of all continuous functions $\tilde{f} : X \to B$ for which the function $x \mapsto \|\tilde{f}(x)\|_B$ is in $C_0(X)$, equipped with the supremum norm*

$$\|\tilde{f}\| = \sup\{\|\tilde{f}(x)\|_B, x \in X\}. \tag{C.272}$$

For any C-norm with ensuing tensor product $\hat{\otimes}$, one then has*

$$C_0(X)\hat{\otimes}B \cong C_0(X,B), \tag{C.273}$$

under continuous extension of the map from $C_0(X) \otimes B$ to $C_0(X,B)$ defined by

$$f \otimes b \mapsto (fb : x \mapsto f(x)b). \tag{C.274}$$

We just prove this for the minimal (i.e. spatial) C*-norm; the general case follows from nuclearity of $C_0(X)$, cf. Proposition C.101 below.

Proof. Take some injective representation $\pi_B : B \to B(H_B)$, and represent $C_0(X,B)$ on $\ell^2(X)\overline{\otimes}H_B$ by linear extension of $\pi : C_0(X,B) \to B(\ell^2(X)\overline{\otimes}H_B)$, as defined by

$$\pi(\tilde{f})\delta_x \otimes \varphi = \delta_x \otimes \pi_B(f(\tilde{x}))\varphi, \tag{C.275}$$

where $\tilde{f} \in C_0(X,B)$, $x \in X$, and $\varphi \in H_B$; this operator is easily seen to be bounded. In particular, an element $fb \in C_0(X,B)$, as in (C.274), is represented by

$$\pi(fb)(\delta_x \otimes \varphi) = f(x)\delta_x \otimes \pi_B(b)\varphi. \tag{C.276}$$

Denoting the representation of $C_0(X)$ on $\ell^2(X)$ through multiplication operators by π_m, i.e., $\pi_m(f)\psi(x) = f(x)\psi(x)$, where $f \in C_0(X)$ and $\psi \in \ell^2(X)$, we then have

$$\pi_m \otimes \pi_B(f \otimes b) = \pi(fb). \tag{C.277}$$

In this way, $C_0(X) \otimes B$ is faithfully represented as a subalgebra of

$$\pi(C_0(X,B)) \cong C_0(X,B), \tag{C.278}$$

and so the final step is merely to show that $C_0(X)\hat{\otimes}B$ is dense in $C_0(X,B)$. Indeed, taking X compact for simplicity (otherwise one needs a further approximation argument), for given $\tilde{f} \in C(X,B)$ and $\varepsilon > 0$, define a cover $\mathscr{U} = (U_x)_{x\in X}$ of X by

$$U_x = \{y \in X \mid \|\tilde{f}(x) - \tilde{f}(y)\| < \varepsilon\}. \tag{C.279}$$

Since X is compact, $\mathscr{U}$ has a finite subcover $\{U_{x_1}, \ldots, U_{x_n}\}$, with associated ***partition of unity*** $\{g_{x_1}, \ldots, g_{x_n}\}$, i.e., one has $g_{x_i} \in C_c(U_{x_i})$, with $0 \leq g_{x_i} \leq 1$, and

$$\sum_{i=1}^{n} g_{x_i}(x) = 1 \ (x \in X). \tag{C.280}$$

Define an approximant $g \in C(X) \otimes B$ by

$$g(x) = \sum_i g_{x_i} \otimes \tilde{f}(x_i), \tag{C.281}$$

whose image $\tilde{g} \in C(X,B)$ is given by $\tilde{g}(x) = \sum_i g_{x_i}(x)\tilde{f}(x_i)$. Then for each $x \in X$,

$$\|\tilde{g}(x) - \tilde{f}(x)\|_B = \left\| \sum_i g_{x_i}(x)(\tilde{f}(x_i) - \tilde{f}(x)) \right\|_B < \sum_i g_{x_i}(x) \cdot \varepsilon = \varepsilon, \tag{C.282}$$

so that, taking $\sup_x$, we have $\|\tilde{g} - \tilde{f}\| < \varepsilon$. This proves the claim. □

Since $C_0(X \times Y) \cong C_0(X, C_0(Y))$ under the map $f \mapsto \tilde{f}$ with $f(x,y) = (\tilde{f}(x))(y)$, the isomorphism (C.271) is a special case of (C.273).

Another case where the choice of a cross-norm does not matter—this time because no completion is even needed—is the following. Recall Corollary C.28.

Proposition C.101. *Let A be a* finite-dimensional *C*-algebra. Then for any C*-algebra B, $A \otimes B$ is complete in any C*-norm, and hence all C*-norms coincide.*

Thus $A\hat{\otimes}B = A \otimes B$, though one still needs a norm on $A \otimes B$ to make it a C*-algebra!

Proof. In view of Theorem C.163, we only need to prove this for $A = M_n(\mathbb{C})$, $n \in \mathbb{N}$. As in the previous proof, we use the spatial tensor product on $M_n(\mathbb{C}) \otimes B$, so let us faithfully represent $M_n(\mathbb{C})$ and B on $\mathbb{C}^n$ and H_B, respectively, and form the Hilbert space $\mathbb{C}^n \overline{\otimes} H_B = \mathbb{C}^n \otimes H_B$, carrying the representation $\mathrm{id} \otimes \pi_B$ of $M_n(\mathbb{C}) \otimes B$, and hence of the (alleged) completion $M_n(\mathbb{C})\hat{\otimes}_{\min}B$. Let

$$c = \sum_{i,j=1}^{n} e_{ij} b^{ij} \in M_n(\mathbb{C}) \otimes B, \tag{C.283}$$

where (e_{ij}) is the standard basis of $M_n(\mathbb{C})$ and $b^{ij} \in B$. For any such c, we have

$$\|c\|_{\min}^2 \geq \left\| \sum_{i,j=1}^{n} e_{ij} b^{ij} (\upsilon_k \otimes \varphi) \right\|_{\mathbb{C}^n \otimes H_B}^2 = \sum_i \|b^{ik}\varphi\|_{H_B}^2, \tag{C.284}$$

where $(\upsilon_1, \ldots \upsilon_n)$ is the standard basis of $\mathbb{C}^n$, $k = 1, \ldots, n$ is fixed, and $\varphi \in H_B$ is a unit vector. Taking the supremum over φ gives

$$\left\| \sum_{i,j} e_{ij} b^{ij} \right\|_{\min} \geq \|b^{ij}\|_B, \tag{C.285}$$

for each fixed pair (i, j). Hence any Cauchy sequence (c_k) in $M_n(\mathbb{C}) \otimes B$ takes the form $c_k = \sum_{i,j=1} e_{ij} b_k^{ij}$, where each (b_k^{ij}) is a Cauchy sequence in B for fixed (i, j). Then, using the fact that $\|e_{ij}\|_{M_n(\mathbb{C})} = 1$, we have

$$\|c - c_k\|_{\min} = \left\| \sum_{i,j} e_{ij} (b^{ij} - b_k^{ij}) \right\|_{\min} \leq \sum_{i,j=1} \|b^{ij} - b_k^{ij}\|_B, \tag{C.286}$$

for any $c \in M_n(\mathbb{C}) \otimes B$, as in (C.283). Taking c such that $b^{ij} = \lim_k b_k^{ij}$, it follows that $c_k \to c$ in $\|\cdot\|_{\min}$, i.e., in $M_n(\mathbb{C}) \hat{\otimes}_{\min} B$. In particular, the limit c of any Cauchy sequence in $M_n(\mathbb{C}) \otimes B$ with respect to the norm $\|\cdot\|_{\min}$ lies in $M_n(\mathbb{C}) \otimes B$, which is therefore complete already and is a C*-algebra in the spatial norm. Since the norm in a C*-algebra is unique (cf. Corollary C.28), it follows that any C*-norm on $M_n(\mathbb{C}) \otimes B$ must coincide with the spatial one $\|\cdot\|_{\min}$. □

It is also easy to show that

$$M_n(\mathbb{C}) \otimes B \cong M_n(B), \tag{C.287}$$

i.e., the $n \times n$-matrices with entries in B, with obvious operations and norm given by faithfully representing B on some Hilbert space H_B, as above, and then letting $M_n(B)$ act on $H_B^n = H_B \oplus \cdots \oplus H_B$ (i.e., n copies) in the natural way. A specific isomorphism $M_n(\mathbb{C}) \otimes B \to M_n(B)$ is then given by sending $\sum_{i,j=1}^n e_{ij} b^{ij}$ to the matrix (b^{ij}).

Finally, one of the highlights of the theory of tensor products on $A \otimes B$ is a concept that apparently makes the entire theory superfluous:

Definition C.102. *A C*-algebra A is called* **nuclear** *if for any C*-algebra B, the norms* $\|\cdot\|_{\min}$ *and* $\|\cdot\|_{\max}$ *(and consequently all C*-norms) on* $A \otimes B$ *coincide.*

The class of nuclear C*-algebras is large but not exhaustive: if H is infinite-dimensional, then $B_0(H)$ is nuclear but $B(H)$ is not, even if H is separable. However:

- Any commutative C*-algebra is nuclear (this underpins Proposition C.100).
- Any finite-dimensional C*-algebra is nuclear (cf. Proposition C.101).
- The (unique!) tensor product of any two nuclear C*-algebras is nuclear.
- Inductive limits of nuclear C*-algebras are nuclear (see §C.14).
- If $0 \to I \to A \to B \to 0$ is a short exact sequence (i.e., if $I \subset A$ is an ideal in A and $B \cong A/I$) in which two of the three C*-algebras are nuclear, the so is the third.

C.14 Inductive limits and infinite tensor products of C*-algebras

In the main text we deal with infinite quantum systems, albeit as idealizations rather than physical systems that exist in reality. Mathematically, such systems arise as infinite tensor products of C*-algebras, which in turn are special cases of ***inductive limits***, also called ***direct limits*** (categorically, these are ***colimits***, see §E.1 below, and as such they are unique op to isomorphism—in this case, of C*-algebras).

Let I be a directed set (cf. Definition D.1), typically $I = \mathbb{N}$ with the usual order. Let (A_i) a family of C*-algebras indexed by I; in case that $I = \mathbb{N}$, these will often be

$$A_n = B^n \equiv \hat{\otimes}^n_{\max} B, \tag{C.288}$$

where B is some C*-algebra and $\hat{\otimes}_{\max}$ is the projective tensor product, extended from two C*-algebras (as discussed in the previous section) to any finite number of C*-algebras in the obvious way: for any completed C*-tensor product $\hat{\otimes}$, $n \in \mathbb{N}$, and C*-algebras $(C_1, \ldots, C_n)$, we inductively define the tensor product of the latter as

$$C_1 \hat{\otimes} \cdots \hat{\otimes} C_n = (C_1 \hat{\otimes} \cdots \hat{\otimes} C_{n-1}) \hat{\otimes} C_n. \tag{C.289}$$

In general, the cartesian product $\prod_{i \in I} A_i$ consists of all functions $a : I \to \cup_i A_i$ such that $a(i) \equiv a_i \in A_i$; we often write such functions as $(a_i)_i$, where $a_i \in A_i$. The Axiom of Choice then guarantees (or, following Russell, even states) that—provided none of the A_i is empty—the set $\prod_{i \in I} A_i$ is non-empty. Since each A_i is a *-algebra, we can turn $\prod_{i \in I} A_i$ into a *-algebra in the obvious way, i.e., by defining scalar multiplication as $(\lambda \cdot a)(i) = \lambda a(i)$, with pointwise addition, multiplication, and involution. This *-algebra, denoted by $\oplus_i A_i$, is the ***algebraic direct sum*** of the A_i.

What about the norm? There are various options here, each relying on the choice of some subspace of $\oplus_i A_i$. For example, if A_0 consists of all $a \in \prod_{i \in I} A_i$ for which $\lim_i \|a_i\| = 0$, then the ***algebraic direct sum*** $\hat{\oplus}_i A_i$ of the A_i is A_0, with norm

$$\|a\| = \sup_i \|a_i\|. \tag{C.290}$$

For the inductive limit we need additional structure, namely a family of homomorphisms $\varphi_{ij} : A_i \to A_j$, defined for each $i \leq j$ in I, such that for each $i \leq j \leq k$,

$$\varphi_{ii} = \mathrm{id}_{A_i}; \tag{C.291}$$

$$\varphi_{jk} \circ \varphi_{ij} = \varphi_{ik}. \tag{C.292}$$

Such maps turn the family (A_i) into a so-called ***directed system*** of C*-algebras. For example, in case of (C.288), and assuming B has a unit 1_B (otherwise there are analogous constructions based on projections), for $n < m$, define $\varphi_{nm} : B^n \to B^m$ by

$$\varphi_{nm}(b) = b \otimes 1_B \otimes \cdots \otimes 1_B. \tag{C.293}$$

with $m - n$ units 1_B. This can be done also in the more general situation (C.289), where we assume each C_i to be unital with unit 1_i, and define

$$A_n = \hat{\otimes}_{i=1}^n C_i; \tag{C.294}$$

$$\varphi_{nm}(c) = c \otimes 1_{C_{n+1}} \otimes \cdots \otimes 1_{C_m}. \tag{C.295}$$

As a matter of central importance to the theory of quantum spin systems. one may generalize this construction in allowing more general directed sets, whilst specializing it in picking very specific C*-algebras C_i. Let $\mathbb{Z}^d \subset \mathbb{R}^d$ be the standard lattice in spatial dimension d, and let I be the set of of all finite subsets Λ of $\mathbb{Z}^d$ (so one typically writes Λ instead of i). Furthermore, take some fixed Hilbert space H, assumed finite-dimensional for simplicity (this also suffices for most applications to quantum statistical mechanics), and for each $\Lambda \in I$, define the *cartesian* product

$$H^\Lambda = \prod_{x \in \Lambda} H_x, \tag{C.296}$$

where $H_x = H$ for each x. Thus elements $\psi : \Lambda \to H$ of H^Λ are families $(\psi_x)_{x\in\Lambda}$, where $\psi_x \in H$. To define the *tensor* product

$$H_\Lambda = \otimes_{x\in\Lambda} H_x, \tag{C.297}$$

we generalize the procedure explained between (C.245) and (C.246) in the previous section. If $\dim(H_A) < \infty$ and $\dim(H_B) < \infty$, the injection

$$H_A \otimes H_B \hookrightarrow \overline{L}(H_A \times H_B, \mathbb{C}), \tag{C.298}$$

is an isomorphism, and we use this fact (with $H_A = H_B = H$) to *define* H_Λ as $\overline{L}(H^\Lambda, \mathbb{C})$, that is, the set of all anti-multi-linear maps $\hat{\psi} : H^\Lambda \to \mathbb{C}$, equipped with pointwise operations turning it into a complex vector space. Each element $\psi : \Lambda \to H$ of H^Λ itself defines such a map $\hat{\psi} \in \overline{L}(H^\Lambda, \mathbb{C})$ via

$$\hat{\psi}(\varphi) = \prod_{x\in\Lambda} \langle \varphi_x, \psi_x \rangle_H, \tag{C.299}$$

through which the inner product on H_Λ is defined by linear extension of

$$\langle \hat{\psi}, \hat{\varphi} \rangle_{H_\Lambda} = \prod_{x\in\Lambda} \langle \psi_x, \varphi_x \rangle_H. \tag{C.300}$$

In this realization of H_Λ, the elementary tensors $\otimes_{x\in\Lambda} \psi_x \in H_\Lambda$ coincide with the above elements $\hat{\psi} \in \overline{L}(H^\Lambda, \mathbb{C}) \equiv H_\Lambda$. Furthermore, if $(\upsilon_1, \ldots, \upsilon_n)$ is a basis of $H \cong \mathbb{C}^n$, then $(\otimes_{x\in\Lambda} \upsilon_{s(x)})$ is a basis of H_Λ, where $s : \Lambda \to \{1, \ldots, n\}$. Hence

$$\dim(H_\Lambda) = \dim(H)^{|\Lambda|}. \tag{C.301}$$

Furthermore, writing $\underline{n} = \{1, 2, \ldots, n\}$, and letting $\underline{n}^\Lambda$ be the set of maps ("classical spin configurations") $s : \Lambda \to \underline{n}$, there is a natural unitary isomorphism

$$H_\Lambda \cong \ell^2(\underline{n}^\Lambda). \tag{C.302}$$

Indeed, as the functions $\delta_s : t \mapsto \delta_{st}$ form a basis of ℓ^2, the map $\delta_s \mapsto \otimes_{x\in\Lambda} \upsilon_{s(x)}$ extends to a unitary from $\ell^2(\underline{n}^\Lambda)$ to H_Λ. Under this equivalence, elements of H_Λ may be interpreted as "wave-functions" whose argument is a spin configuration.

Returning to C*-algebras, having defined H_Λ, we now put

$$A_\Lambda = B(H_\Lambda). \tag{C.303}$$

To fit this into the above framework, we note that the partial order $\leq$ on I is given by $\Lambda \leq \Lambda'$ whenever $\Lambda \subseteq \Lambda'$, in which case there is a canonical embedding

$$\iota_{\Lambda\Lambda'} : A_\Lambda \hookrightarrow A_{\Lambda'}. \tag{C.304}$$

This embedding is given as in (C.293), i.e., by adding unit operators. Let $\Lambda \subset \Lambda'$ and define $\Lambda'' = \Lambda' \backslash \Lambda$. We may split $\psi' : \Lambda' \to H$ as $\psi' \mapsto (\psi'_{|\Lambda}, \psi'_{|\Lambda''})$, from which

$$H^{\Lambda'} \cong H^\Lambda \times H^{\Lambda''}. \tag{C.305}$$

As in (C.298), this gives isomorphisms

$$H_{\Lambda'} = L(H^{\Lambda'}, \mathbb{C}) \cong \overline{L}(H^\Lambda \times H^{\Lambda''}, \mathbb{C}) \cong \overline{L}(H^\Lambda \otimes H^{\Lambda''}, \mathbb{C}) \cong H_\Lambda \otimes H_{\Lambda''}. \tag{C.306}$$

This, in turn, induces an isomorphism

$$A_{\Lambda'} = B(H_{\Lambda'}) \cong B(H_\Lambda \otimes H_{\Lambda''}) \cong B(H_\Lambda) \otimes B(H_{\Lambda''}) = A_\Lambda \otimes A_{\Lambda''}, \tag{C.307}$$

which, through the embedding

$$B(H_\Lambda) \hookrightarrow B(H_\Lambda) \otimes B(H_{\Lambda''}); \tag{C.308}$$

$$a \mapsto a \otimes 1_{B(H_{\Lambda''})}, \tag{C.309}$$

gives an embedding $B(H_\Lambda) \hookrightarrow B(H_{\Lambda'})$. This, then, is the injection (C.304).

Alternatively, $B(H_\Lambda)$ may be constructed just like H_Λ itself, i.e., by starting with the set $B(H)^\Lambda$ of functions $a : \Lambda \to B(H)$. Any such a defines an operator $\hat{a}$ on H_Λ by first defining its action on elementary tensors by $\hat{a}\hat{\psi} = \otimes_{x\in\Lambda} a_x \psi_x$, and extending the result linearly to arbitrary vectors in H_Λ. We write $\hat{a} = \otimes_{x\in\Lambda} a_x$, and reconstruct $B(H_\Lambda)$ as the complex vector space spanned by all such elementary operators. The injection (C.304) is given by linear extension of the map $\hat{a} \mapsto \hat{a}'$, where $\hat{a}'_{x'} = a_x$ whenever $x' = x \in \Lambda \subset \Lambda'$, and $\hat{a}'_{x'} = 1_H$ otherwise, i.e., if $x' \in \Lambda''$.

Either way, we obtain a directed system of C*-algebras (A_Λ), where the finite subsets $\Lambda \subset \mathbb{Z}^d$ are partially ordered by inclusion, and the maps $\varphi_{\Lambda\Lambda'} : A_\Lambda \to A_{\Lambda'}$, with properties like (C.291) - (C.292), are given by the inclusions (C.304).

There is a classical counterpart to this construction, in which the local C*-algebras are given by "functions of functions", i.e.,

$$A_\Lambda^{(c)} = C(\underline{n}^\Lambda) = C(C(\Lambda, \underline{n})). \tag{C.310}$$

Since $\underline{n}^\Lambda$ is a finite discrete set, any function on it is continuous (and lies in ℓ^2, etc.). If $\Lambda \subseteq \Lambda'$, then, $s' \in \underline{n}^{\Lambda'}$ being a map $s' : \Lambda' \to \underline{n}$, the connecting homomorphisms

$$\iota^{(c)}_{\Lambda\Lambda'} : A^{(c)}_\Lambda \hookrightarrow A^{(c)}_{\Lambda'}, \tag{C.311}$$

are given quite canonically by

$$\iota^{(c)}_{\Lambda\Lambda'}(f) : s' \mapsto f(s'_{|\Lambda}). \tag{C.312}$$

Note that $C(\underline{n}^\Lambda) = \ell^2(\underline{n}^\Lambda)$ as vector spaces, so that (C.311) also gives natural maps $\ell^2(\underline{n}^\Lambda) \hookrightarrow \ell^2(\underline{n}^{\Lambda'})$, and hence, via (C.302), $H_\Lambda \hookrightarrow H_{\Lambda'}$. These are given by linear extension of the map given on basis vectors by $\otimes_{x\in\Lambda} \upsilon_{s(x)} \mapsto \sum_{s':s'_{|\Lambda}=s} \otimes_{x'\in\Lambda'} \upsilon_{s'(x')}$.

Furthermore, analogously tot (C.307), since $\Lambda' = \Lambda \cup \Lambda''$ is finite, we have

$$\begin{aligned} A^{(c)}_{\Lambda'} &= C(\underline{n}^{\Lambda'}) = C(C(\Lambda', \underline{n})) = C(C(\Lambda \cup \Lambda'', \underline{n})) \cong C(C(\Lambda, \underline{n}) \times C(\Lambda'', \underline{n})) \\ &\cong C(C(\Lambda, \underline{n})) \otimes C(C(\Lambda'', \underline{n})) = C(\underline{n}^\Lambda) \otimes C(\underline{n}^{\Lambda''}) = A^{(c)}_\Lambda \otimes A^{(c)}_{\Lambda''}. \end{aligned} \tag{C.313}$$

Given a directed system of C*-algebras (A_i, φ_{ij}), we define the ***local part*** A_{loc} of $\prod_i A_i$ as the set of all elements $a = (a_i)$ of $\prod_i A_i$ for which there is $i_0 \in I$ (depending on a) such that $a_i = \varphi_{i_0 i}(a_{i_0})$ whenever $i_0 \leq i$. This is equivalent to the seemingly stronger condition that $a_j = \varphi_{ij}(a_i)$ whenever $i_0 \leq i \leq j$, since

$$a_j = \varphi_{i_0 j}(a_{i_0}) = \varphi_{ij} \circ \varphi_{i_0 i}(a_{i_0}) = \varphi_{ij}(a_i). \tag{C.314}$$

In the example (C.288) with (C.293), this simply means that for each sequence $(a_n)_{n\in\mathbb{N}}$, there is $n_0 \in \mathbb{N}$ such that $a_n = a_{n_0} \otimes^{n-n_0} 1_B$ for each $n > n_0$. Similarly, in the example (C.303) with (C.304), for each $a = (a_\Lambda)$, where Λ is a finite subset of $\mathbb{Z}^d$ and $a_\Lambda \in A_\Lambda$ for each Λ, there is a finite subset $\Lambda_0 \subset \mathbb{Z}^d$ such that for any $\Lambda \supseteq \Lambda_0$ we have $a_\Lambda = \iota_{\Lambda_0\Lambda}(a_{\Lambda_0})$. It is easy to see that A_{loc} is a *-algebra under the (pointwise) operations inherited from $\prod_i A_i$. For each $(a_i) \in A_{\mathrm{loc}}$, the norms $\|a_i\|$ form a net in $\mathbb{R}^+$. Recall that some net $(t_i)_{i\in I}$ in $\mathbb{R}$ (which by definition is indexed by a directed set I) is said to ***converge*** to $t \in \mathbb{R}$ if for each $\varepsilon > 0$, there is $i \in I$ such that $|t - t_j| < \varepsilon$ for all $j \geq i$ (since $\mathbb{R}$ is Hausdorff, any net in $\mathbb{R}$ converges to at most one point). Because the connecting maps φ_{ij} are homomorphisms of C*-algebras, they are norm-decreasing (cf. Theorem C.62.1), i.e., $\|\varphi_{ij}(a_i)\| \leq \|a_i\|$. Thus for any $a \in A_{\mathrm{loc}}$ with associated $i_0 \in I$, the (sub)net $(\|a_i\|)_{i\geq i_0}$ lies in the interval $[0, \|a_{i_0}\|]$, and is monotone decreasing in the sense that if $j \geq i \geq i_0$, then $\|a_j\| \leq \|a_i\|$. As for sequences (which are just nets indexed by $I = \mathbb{R}$), *bounded* monotone decreasing (or increasing) nets in $\mathbb{R}$ converge, so that net $(\|a_i\|)_{i\geq i_0}$ has a limit, and this also means that $(\|a_i\|)_i$ has the same limit. Call this limit $\|a\|_0$. The map $a \mapsto \|a\|_0$ generally fails to define a norm on A_{loc}, since it may lack the property of positive definiteness, and even if it had it, the space would not be complete (at least if I is infinite, as we tacitly assume). We do have the C*-axioms $\|ab\|_0 \leq \|a\|_0\|b\|_0$ and $\|a^*a\|_0 = \|a\|_0^2$ though, since these hold for each norm $a_i \mapsto \|a_i\|$ and are preserved in the limit.

So we say that $\|a\|_0$ is a ***C*-seminorm*** on $A_{\rm loc}$, and there is a canonical procedure to turn a *-algebra with C*-seminorm into a C*-algebra:

1. Define the null space $N \subset A_{\rm loc}$ for $\|\cdot\|_0$ by $N = \{a \in A_{\rm loc} : \|a\|_0 = 0\}$;
2. Define a norm on the quotient $A_{\rm loc}/N$ by $\|a+N\| = \|a\|_0$, and complete the quotient in this norm. The result is a C*-algebra

$$A = \underline{\lim}_i A_i, \tag{C.315}$$

called the ***inductive limit*** of the directed system (A_i, φ_{ij}).

For each $i \in I$, we now define a canonical homomorphism $\varphi_i : A_i \to A$. If $a_i \in A_i$, put $a_j = \varphi_{ij}(a_i) \in A_j$ if $j \geq i$, and $a_j = 0$ otherwise. This gives an element $a \in A_{\rm loc}$ whose image in $A_{\rm loc}/N \subset A$ is $\varphi_i(a)$. A computation shows that if $i \leq j$, then $\varphi_j \circ \varphi_{ij} = \varphi_i$. Using this fact, it follows that if we put $\tilde{A}_i = \varphi_i(A_i) \subset A$, then $\tilde{A}_i \subseteq \tilde{A}_j$ whenever $i \leq j$, and hence A may be rewritten as the norm-closure of the union of the $\tilde{A}_i$, i.e.,

$$A = \overline{\bigcup_i \tilde{A}_i}^{\|\cdot\|}. \tag{C.316}$$

In the simple situation where the maps φ_{ij} are inclusions and hence isometries, as in our examples, we have $N = \{0\}$, so that $\tilde{A}_i = A_i$, and hence (C.316) simplifies to

$$A = \overline{\bigcup_i A_i}^{\|\cdot\|}. \tag{C.317}$$

As a case in point, define (A_n, φ_{nm}) as in (C.294) - (C.295). The infinite tensor product of the C_i is then defined through (C.315) and (C.295), i.e., by definition,

$$\hat{\otimes}_{i=1}^{\infty} C_i = \underline{\lim}_n \hat{\otimes}_{i=1}^{n} C_i = \overline{\bigcup_n \hat{\otimes}_{i=1}^{n} C_i}^{\|\cdot\|}. \tag{C.318}$$

Here the first equation is general, and in the second it is understood that for any $m > n$, we have $\hat{\otimes}_{i=1}^{n} C_i \subset \hat{\otimes}_{i=1}^{m} C_i$ through the embeddings (C.295).

More generally, let $(A_x)_{x\in X}$ be a family of unital C*-algebras indexed by an arbitrary set X, and let $I = \mathscr{P}_f(X)$ the set of finite subsets of X, partially ordered by inclusion. For any $F \in I$, we have a tensor product

$$A_F = \hat{\otimes}_{x\in F} A_x, \tag{C.319}$$

where once again $\hat{\otimes}$ is an arbitrary completed C*-tensor product. An explicit construction of this tensor product along the lines of (C.289) requires an ordering of F, but two such orderings give canonically isomorphic C*-algebras; if $F \subset G$, one should order G compatibly with F for the connecting homomorphisms φ_{FG} to be well defined by (C.295). This gives a directed system of C*-algebras (A_F, φ_{FG}), whose inductive limit defines the tensor product over A, i.e.,

$$\hat{\otimes}_{x\in X} A_x = \underline{\lim}_F \hat{\otimes}_{x\in F} A_x. \tag{C.320}$$

As a special case, we may rewrite our earlier algebras A_Λ and $A^{(c)}_\Lambda$ as

$$A_\Lambda = \otimes_{x\in\Lambda} B(H); \tag{C.321}$$

$$A^{(c)}_\Lambda = \otimes_{x\in\Lambda} C(\underline{n}) \cong C\left(\prod_{x\in\Lambda} \underline{n}\right), \tag{C.322}$$

cf. (C.313). Hence we have

$$\underline{\lim}_\Lambda A_\Lambda = \otimes_{x\in\mathbb{Z}^d} B(H); \tag{C.323}$$

$$\underline{\lim}_\Lambda A^{(c)}_\Lambda = \otimes_{x\in\mathbb{Z}^d} C(\underline{n}) \cong C\left(\prod_{x\in\mathbb{Z}^d} \underline{n}\right), \tag{C.324}$$

where in the last expression the infinite product $\prod_{x\in\mathbb{Z}^d} \underline{n}$ is endowed with the product topology, so that (by Tychonoff's Theorem) the space in question is compact. Thus the ensuing inductive limit may directly be expressed as the standard commutative C*-algebra $C(X)$, where $X = \prod_{x\in\mathbb{Z}^d} \underline{n}$ is compact, equipped with pointwise operations and the sup-norm. If $n = 2$ and $d = 1$, this is a model of the Cantor set.

The homomorphisms φ_i enable us to state the universal character of A:

Theorem C.103. *Let (A_i, φ_{ij}) a directed system of C*-algebras with inductive limit A. For any C*-algebra B endowed with a family homomorphisms $\beta_i : A_i \to B$ such that $\beta_j \circ \varphi_{ij} = \beta_i$, there is a unique homomorphism $\beta : A \to B$ such that $\beta_j = \beta \circ \varphi_j$.*

In other words, the following diagram commutes:

$$\begin{array}{ccccc} A_i & \xrightarrow{\varphi_{ij}} & A_j & \xrightarrow{\varphi_j} & A \\ & \beta_i \searrow & \downarrow \beta_j & \swarrow \exists!\beta & \\ & & B & & \end{array} \tag{C.325}$$

Proof. This is true almost by construction, or rather by (C.316): since β is supposed to be a homomorphism of C*-algebras, it is continuous, so it is determined by its values on the dense subalgebra $\bigcup_i \tilde{A}_i$, and hence by its values on each $\tilde{A}_i$. But these values are necessarily given by $\beta(\varphi_i(a_i)) = \beta_i(a_i)$, where $a_i \in A_i$. □

Corollary C.104. *Let $(A_x)_{x\in X}$ be a family of mutually commuting unital C*-subalgebras of a unital C*-algebra B (sharing the unit of B), such that the C*-algebra generated by all subalgebras A_x within B is equal to B. Also, let $\hat{\otimes}$ be some completed C*-tensor product such that for each finite subset $F = \{x_1, \ldots, x_n\} \subset X$, there is an injective homomorphism $\varphi_F : A_F \to B$ (where $A_F = A_{x_1} \hat{\otimes} \cdots \hat{\otimes} A_{x_n}$) satisfying*

$$\varphi_F(a_1 \otimes \cdots \otimes a_n) = a_1 \cdots a_n \ \ (a_1 \in A_{x_1}, \ldots, a_n \in A_{x_n}). \tag{C.326}$$

Then $B \cong \hat{\otimes}_{x\in X} A_x$.

Proof. In Theorem C.103, take $A_j \rightsquigarrow A_F$ and $\beta_j \rightsquigarrow \varphi_F$. In view of (C.320), this gives a homomorphism $\beta : \hat{\otimes}_{x\in X} A_x \to B$. Here, this map is an isomorphism. □

Finally, we give a result on infinite tensor products of states, needed in §8.4.

Proposition C.105. *Let $(C_i)_{i\in\mathbb{N}}$ be unital C*-algebras, and define their infinite (projective) tensor product $\hat{\otimes}_{i=1}^{\infty}C_i$ as in (C.318). For each $i\in\mathbb{N}$, let ω_i be a state on C_i. Then there is a unique state $\hat{\otimes}_{i=1}^{\infty}\omega_i$ on $\hat{\otimes}_{n=1}^{\infty}C_i$ such that for each $N\in\mathbb{N}$ and $c_i\in C_i$,*

$$\hat{\otimes}_{i=1}^{\infty}\omega_i(\varphi_n(c_1\otimes\cdots\otimes c_n))=\prod_{n=1}^{n}\omega_i(c_i) \tag{C.327}$$

Moreover, $\hat{\otimes}_{i=1}^{\infty}$ is pure iff each ω_i is pure.

Proof. We write $C^n\equiv\hat{\otimes}_{i=1}^{n}C_i$, and similarly $\hat{\otimes}_{i=1}^{n}\omega_i\equiv\omega^n$, also for $n=\infty$.

Eq. (C.327) defines ω^∞ on a dense subset $\cup_{n\in\mathbb{N}}\varphi_n(C^n)$ of C^∞, which proves uniqueness. Existence comes from Proposition C.98, according to which the map $c_1\otimes\cdots\otimes c_n\mapsto\prod_{i=1}^{n}\omega_i(c_i)$ extends to a state $\otimes_i^n\omega_i'$ on C^n, which in turn defines a state ω^n on $\varphi_n(C^n)\subset C^\infty$. Since $(\otimes_i^n\omega_i')_{|C^m}=\otimes_i^m\omega_i'$ whenever $m\leq n$, one also has $\omega^n_{|\varphi_m(C^m)}=\omega^m$, so that we may define a functional ω^∞ on $\cup_n\varphi_n(C^n)$ by its restrictions $\omega^\infty_{|C^n}=\omega^n$. Since ω^n is a state and hence satisfies $\|\omega^n\|=\omega^n(1_{\varphi_n(C^n)})=1$, so does ω^∞ (on its dense domain). Since the continuous extension of ω^∞ to C^∞ has the same norm, this extension (still called ω^∞) is a state by Proposition C.5.

One direction of the second claim is trivial: if at least one of the ω_i fails to be pure, then ω^n inherits its convex decomposition so to speak, so contrapositively we obtain that purity of ω^n implies purity of each ω_i. We first prove the opposite direction for $n<\infty$. Using Proposition C.91 and the fact that C^n is a completion of the algebraic tensor product $\otimes_{i=1}^{n}C_i$, the GNS-representation $\pi_{\omega^n}(C^n)$ is unitarily equivalent to the representation $\pi_{\omega_1}\otimes\cdots\otimes\pi_{\omega_n}$ on $H_{\omega_1}\otimes\cdots\otimes H_{\omega_n}$, and

$$(\pi_{\omega_1}\otimes\cdots\otimes\pi_{\omega_n}(C^n))''=\pi_{\omega_1}(C_1)''\overline{\otimes}\cdots\overline{\otimes}\pi_{\omega_n}(C_n)''. \tag{C.328}$$

Here, for any two von Neumann algebras A and B, $A\overline{\otimes}B$ is the smallest von Neumann algebra containing the algebraic tensor product $A\otimes B$. The main lemma behind the second claim is the nontrivial ***commutation theorem*** for von Neumann algebras:

$$(A\overline{\otimes}B)'=A'\overline{\otimes}B', \tag{C.329}$$

which we state without proof. This iterates to n von Neumann algebras. Hence

$$(\pi_{\omega_1}\otimes\cdots\otimes\pi_{\omega_n}(C^n))'=\pi_{\omega_1}(C_1)'\overline{\otimes}\cdots\overline{\otimes}\pi_{\omega_n}(C_n)', \tag{C.330}$$

so that the claim for $n<\infty$ follows from Theorem C.90.

Now take $n=\infty$, and assume each ω_i is pure. Suppose that for some $t\in(0,1)$,

$$\omega^\infty=t\omega'+(1-t)\omega'', \tag{C.331}$$

and restrict this equality to $\varphi_n(C^n)$. By the previous argument, the restriction of ω^∞ to $\varphi_n(C^n)$, which is just ω^n, is pure for any $n\in\mathbb{N}$. This gives $\omega'_{|\varphi_n(C^n)}=\omega''_{|\varphi_n(C^n)}$. This is true for each n, so that $\omega'=\omega''$. Hence ω^∞ is pure. □

C.15 Gelfand isomorphism and Fourier theory

One of the most beautiful applications of Theorem C.8 is to commutative harmonic analysis. Let G be an *abelian* locally compact Hausdorff group (e.g., $G=\mathbb{R}$, $G=\mathbb{Z}$, or $G=\mathbb{T}$). Such groups have an invariant ***Haar measure*** dx, which satisfies

$$\int_G dx\, L_y f(x) = \int_G dx\, f(x^{-1}) = \int_G dx\, f(x), \tag{C.332}$$

for any $f \in C_c(G)$ and $y \in G$, where

$$L_y f(x) = f(y^{-1}x). \tag{C.333}$$

This measure is unique up to rescaling; if G is compact, it is normalized such that $\int_G dx = 1$. For $G=\mathbb{R}$, this recovers Lebesgue measure on $\mathbb{R}$, whilst for $\mathbb{Z}$ and $\mathbb{T}$,

$$\int_{\mathbb{Z}} dx\, f(x) = \sum_{n\in\mathbb{Z}} f(n); \tag{C.334}$$

$$\int_{\mathbb{T}} dx\, f(x) = \int_0^{2\pi} \frac{d\theta}{2\pi} f(e^{i\theta}). \tag{C.335}$$

For $f,g \in C_c(G)$, the ***convolution product*** $f*g$ is defined by

$$f*g(x) = \int_G dy\, f(y)g(y^{-1}x). \tag{C.336}$$

Using (C.332), it is easy to verify that this product is commutative and associative. Also, one may define an involution on $C_c(G)$ by

$$f^*(x) = \overline{f(x^{-1})}. \tag{C.337}$$

We would now like to turn $C_c(G)$ into a commutative C*-algebra, but the obvious norms like the L^p-ones do not accomplish this. Instead, for $f \in C_c(G)$ we define an operator $\pi(f)$ on the Hilbert space $L^2(G)$ (defined with respect to Haar measure) by

$$\pi(f)\psi = f*\psi, \tag{C.338}$$

initially for $\psi \in C_c(G)$. Equivalently, we may write

$$\pi(f) = \int_G dy\, f(y) L_y, \tag{C.339}$$

where we regard L_y as an (obviously unitary) operator on $L^2(G)$, and the integral is most easily defined weakly, i.e., $\pi(f)$ is the unique bounded operator for which

$$\langle \varphi, \pi(f)\psi\rangle = \int_G dy\, f(y)\langle \varphi, L_y\psi\rangle. \tag{C.340}$$

Since L_y is unitary, this formula also shows that $|\langle\varphi,\pi(f)\psi\rangle| \leq \|f\|_1\|\varphi\|\|\psi\|$, where $\|f\|_1 = \int_G dx\,|f(x)|$. Taking $\varphi = \pi(f)\psi$ gives $\|\pi(f)\psi\| \leq \|f\|_1\|\psi\|$, whence

$$\|\pi(f)\| \leq \|f\|_1. \tag{C.341}$$

Hence $\pi(f)$ is bounded and extends from $C_c(G)$ to all of $L^2(G)$ by continuity.

Lemma C.106. *The map $f \mapsto \pi(f)$ from $C_c(G)$ to $B(L^2(G))$ is injective and satisfies*

$$\pi(f * g) = \pi(f)\pi(g); \tag{C.342}$$
$$\pi(f^*) = \pi(f)^*. \tag{C.343}$$

Proof. Eq. (C.342) follows from associativity of convolution, and (C.343) follows from the last equality in (C.332). To prove injectivity, we fix $f \in C_c(G)$, pick $\varepsilon > 0$, and find a neighbourhood U of $e \in G$ such that $y^{-1}x \in U$ implies $|f(y) - f(x)| < \varepsilon$. Then, using Urysohn's Lemma, one may find a positive function $\psi_U \in C_c(U)$ such that $\int_U \psi_U = 1$. Injectivity of π then immediately follows from the easy estimate

$$|f * \psi_U(x) - f(x)| \leq \int_G dy\,|f(y) - f(x)| \cdot |\psi_U(y^{-1}x)| < \varepsilon. \qquad \square$$

Definition C.107. *Let G be an abelian locally compact Hausdorff group. The* **group C*-algebra** *$C^*(G)$ is the norm closure of $\pi(C_c(G))$ in $B(L^2(G))$, with norm*

$$\|f\|_{C^*} = \|\pi(f)\|_{B(L^2(G))}. \tag{C.344}$$

Since $\pi(C_c(G))$ is a commutative *-algebra in $B(L^2(G))$ by Lemma C.106, it is easy to see (from joint continuity of multiplication) that its norm closure $C^*(G)$ is a commutative C*-algebra, whose Gelfand spectrum we wish to compute.

To this effect, we first define the ***dual group*** or ***character group*** $\hat{G}$ of G as

$$\hat{G} = \mathrm{Hom}(G, \mathbb{T}), \tag{C.345}$$

i.e., the set of continuous group homomorphisms from G to $\mathbb{T}$, equipped with the ***compact-open topology***. This topology is defined as the restriction to $\mathrm{Hom}(G,\mathbb{C})$ of the topology on $C(G,\mathbb{C})$ generated by the neigbourhood basis of some $\gamma \in \hat{G}$, i.e.,

$$O(\gamma, K, \varepsilon) = \{\varphi \in \hat{G} : |\gamma(x) - \varphi(x)| < \varepsilon\,\forall x \in K\}, \tag{C.346}$$

where $K \in \mathscr{K}(G)$ and $\varepsilon > 0$. The corresponding notion of convergence is uniform convergence on each compact subset of G; in particular, if G is compact, this is just uniform convergence. Equipped with this topology, it can be shown that $\hat{G}$ is itself an abelian locally compact Hausdorff group under pointwise operations, i.e.,

$$(\gamma_1\gamma_2)(x) = \gamma_1(x)\gamma_2(x); \tag{C.347}$$
$$\gamma^{-1}(x) = \overline{\gamma(x)}; \tag{C.348}$$

hence the ensuing unit $\hat{e}$ in $\hat{G}$ is the identity function $\hat{e} = 1_G$ in $\mathrm{Hom}(G, \mathbb{T})$.

Proposition C.108. *We have the following examples of dual groups:*

$$\hat{\mathbb{Z}} \cong \mathbb{T}, \quad \gamma_z(n) = z^n; \tag{C.349}$$

$$\hat{\mathbb{R}} \cong \mathbb{R}, \quad \gamma_p(x) = e^{ipx}; \tag{C.350}$$

$$\hat{\mathbb{T}} \cong \mathbb{Z}, \quad \gamma_n(z) = z^n; \tag{C.351}$$

$$\hat{\mathbb{Z}}_p \cong \mathbb{Z}_p, \quad \gamma_{[m]}([n]) = e^{2\pi imn/p}. \tag{C.352}$$

Here $\mathbb{Z}_p = \mathbb{Z}/(p \cdot \mathbb{Z})$ is the (finite) group of integers mod p.

Proof. For (C.349), any character $\gamma : \mathbb{Z} \to \mathbb{T}$ is determined by its value $\gamma(1) = z$, since for $n > 0$ we have $\gamma(n) = \gamma(1 + \cdots + 1) = \gamma(1)^n = z^n$, where the sum has n terms; for $n < 0$, we obtain the same result from $\gamma(n) = \gamma(-n)^{-1} = (z^{-n})^{-1} = z^n$.

To prove (C.350), we need to solve $\gamma(x+y) = \gamma(x)\gamma(y)$ with $\gamma(0) = 1$, where $\gamma : \mathbb{R} \to \mathbb{T}$ is continuous. To see that (C.350) gives all solutions, find $\varepsilon > 0$ for which $\int_0^\varepsilon dy\, \gamma(y) \equiv a > 0$; this is possible, since $\gamma(0) = 1$ and γ is continuous. Then

$$\int_0^\varepsilon dy\, \gamma(y)\gamma(x) = \int_0^\varepsilon dy\, \gamma(x+y) = \int_x^{\varepsilon + x} dy\, \gamma(y), \tag{C.353}$$

so that γ is differentiable, with, writing $\dot{\gamma}$ for $d\gamma/dx$,

$$a\dot{\gamma}(x) = \gamma(\varepsilon + x) - \gamma(x) = (\gamma(\varepsilon) - 1)\gamma(x). \tag{C.354}$$

Hence $\dot{\gamma}(x) = c\gamma(x)$ with $c = (\gamma(\varepsilon) - 1)/a$, so that $\gamma(x) = \exp(cx)$. Since $|\gamma(x)| = 1$, this forces $c = ip$ for some $p \in \mathbb{R}$. This also implies (C.351), since $\mathbb{T} = \mathbb{R}/\mathbb{Z}$ and hence the characters of $\mathbb{T}$ are those characters of $\mathbb{R}$ that map $\mathbb{Z}$ to 1. Similarly, (C.352) follows from (C.349): the characters on $\mathbb{Z}$ that are trivial on $p \cdot \mathbb{Z}$ take the form $\gamma(n) = z^n$ for some p-roots of unity $z = \exp(2\pi im/p)$, $m \in \{1, \ldots, p\}$. □

Theorem C.109. *Let G be an abelian locally compact Hausdorff group. Then the Gelfand spectrum $\Sigma(C^*(G))$ is homeomorphic to $\hat{G}$, and the Gelfand isomorphism*

$$C^*(G) \cong C_0(\hat{G}) \tag{C.355}$$

is given on the dense subspace $C_c(G) \subset C^*(G)$ by the generalized Fourier transform

$$\hat{f}(\gamma) = \int_G dx\, \overline{\gamma(x)} f(x). \tag{C.356}$$

Thus the Fourier transform is a special case of the Gelfand transform (which is noteworthy if only because Gelfand himself promulgated the unity of mathematics).

Proof. We will prove that each character $\gamma \in \hat{G}$ on G defines a character ω_γ on $C^*(G)$ by continuous extension (i.e., from its dense subspace $C_c(G)$ to $C^*(G)$) of

$$\omega_\gamma(f) = \hat{f}(\gamma), \tag{C.357}$$

as in (C.356), and that the map $\gamma \mapsto \omega_\gamma$ gives a homeomorphism $\hat{G} \xrightarrow{\cong} \Sigma(C^*(G))$.

It follows from simple computations that for $f,g \in C_c(G)$, one has

$$\omega_\gamma(f*g) = \omega_\gamma(f)\omega_\gamma(g); \tag{C.358}$$
$$\omega_\gamma(f^*) = \overline{\omega_\gamma(f)}. \tag{C.359}$$

To finish the proof, we need three further nontrivial facts about the map $\gamma \mapsto \omega_\gamma$:

1. It is *surjective*, i.e., if $\omega \in \Sigma(C^*(G))$, then $\omega = \omega_\gamma$ for some $\gamma \in \hat{G}$.
2. It is *injective*, in that $\hat{f}(\gamma) = \hat{f}(\gamma')$ for all $f \in C_c(G)$ implies $\gamma = \gamma'$. Moreover, the character ω_γ initially defined on $C_c(G)$ by (C.357) will be shown to satisfy

$$|\omega_\gamma(f)| \leq \|f\|_{C^*}, \tag{C.360}$$

   Thus $\omega_\gamma : C_c(G) \to \mathbb{C}$ may be extended to $C^*(G)$ by continuity in the usual way.
3. The compact-open topology on $\hat{G}$ is mapped to the Gelfand topology on $\Sigma(C^*(G))$.

To prove the first point, we restrict a character $\omega : C^*(G) \to \mathbb{C}$ to $C_c(G)$ and note that because of the bound (C.341), this restriction in turn extends to an element of $L^1(G)^*$, which we still call ω. Entry 10 in Table B.1 gives $L^1(X)^* \cong L^\infty(X)$, in the sense that any $\varphi \in L^1(X)^*$ is given by $\varphi_f(g) = \int_X fg$ for some $f \in L^\infty(X)$. Hence

$$\omega(f) = \int_G dx\, \tilde{\omega}(x) f(x), \tag{C.361}$$

where $\tilde{\omega} \in L^\infty(G)$. The multiplicative property $\omega(f*g) = \omega(f)\omega(g)$ then gives

$$\tilde{\omega}(xy) = \tilde{\omega}(x)\tilde{\omega}(y) \tag{C.362}$$

almost everywhere (a.e.) with respect to Haar measure.

To prove continuity of $\tilde{\omega}$, compare the following expressions with $f,g \in C_c(G)$:

$$\omega(f)\omega(g) = \omega(f) \int_G dx\, \tilde{\omega}(x) g(x);$$
$$\omega(f*g) = \int_G dx\, \omega(L_x f) g(x).$$

These must coincide, so if we pick some $f \in C_c(G)$ for which $\omega(f) \neq 0$ (which is possible since $C_c(G)$ is dense in $C^*(G)$ and ω is not identically zero), then we obtain

$$\tilde{\omega}(x) = \omega(L_x f)/\omega(f), \tag{C.363}$$

almost everywhere. Hence we may redefine $\tilde{\omega}$ by (C.363) for all $x \in G$. Since

$$|\omega(L_x f) - \omega(L_y f)| \leq \|L_x f - L_y f\|_{C^*} \leq \|L_x f - L_y f\|_1 \leq C\|L_x f - L_y f\|_\infty, \tag{C.364}$$

recalling that f has compact support, it follows that the function $x \mapsto \omega(L_x f)$ is continuous, whence also $\tilde{\omega}$ as redefined by (C.363) is continuous.

We now show that $\tilde{\omega}(x) \in \mathbb{T}$. If $|\tilde{\omega}(x)| > 1$, then $\tilde{\omega}$ cannot be bounded (whereas we know it lies in $L^\infty(G)$), because $\tilde{\omega}(x^n) = \tilde{\omega}(x)^n$ by (C.362). But the same is true if $|\tilde{\omega}(x)| < 1$, because using $\tilde{\omega}(x^{-1}) = \tilde{\omega}(x)^{-1}$ (which follows from (C.362) and (C.363), which gives $\tilde{\omega}(e) = 1$), the same argument applies with x^{-1} instead of x. Thus $\tilde{\omega} : G \to \mathbb{T}$ is a character $\overline{\gamma} \in \hat{G}$ (where the bar is conventional), so that (C.361) turns into (C.356). As to injectivity, if $\hat{f}(\gamma) = \hat{f}(\gamma')$ for all $f \in C_c(G)$, then

$$\int_G dx\,(\overline{\gamma(x)} - \overline{\gamma'(x)})f(x) = 0, \tag{C.365}$$

for all such f, which by standard integration theory gives $\gamma' = \gamma$ a.e. and hence everywhere, since both functions are continuous. To prove (C.360), we use a trick: take some fixed $\omega_0 \in \Sigma(C^*(G))$, so that $\omega_0(f) = \hat{f}(\gamma_0)$ for some $\gamma_0 \in \hat{G}$ by the previous step of the proof, and $\|\omega_0(f)\| \leq \|f\|_{C^*}$ for all $f \in C^*(G)$. For $\gamma \in \hat{G}$ and $f \in C_c(G)$, eqs. (C.356) and (C.347) give $\omega_\gamma(f) = \omega_0(\overline{\gamma}\gamma_0 f)$, where $\overline{\gamma}\gamma_0 f$ is the *pointwise* product of the three given functions from G to $\mathbb{C}$. Hence

$$|\omega_\gamma(f)| = |\omega_0(\overline{\gamma}\gamma_0 f)| \leq \|\pi(\overline{\gamma}\gamma_0)f\| = \|\overline{\gamma}\gamma_0 f\|_{C^*}. \tag{C.366}$$

We now denote $\overline{\gamma}\gamma_0$ by γ', which lies in $\hat{G}$, and note that for any $\gamma' \in \hat{G}$, we have

$$\langle \varphi, \pi(\gamma' f)\psi \rangle = \langle \overline{\gamma'}\varphi, \pi(f)(\overline{\gamma'}\psi)\rangle \;\; (\varphi, \psi \in L^2(G), f \in C_c(G)). \tag{C.367}$$

Taking $\varphi = \pi(\gamma' f)\psi$ and using Cauchy–Schwarz as well as $\|\overline{\gamma'}\varphi\| = \|\varphi\|$, gives

$$\|\pi(\gamma' f)\psi\| \leq \|\pi(f)\overline{\gamma'}\psi\| \;\; (\psi \in L^2(G), f \in C_c(G), \gamma' \in G). \tag{C.368}$$

Taking the sup over all $\psi \in L^2(G)$ with $\|\psi\| = 1$ (which also means $\|\overline{\gamma'}\psi\| = 1$) gives $\|\pi(\gamma' f)\| \leq \|\pi(f)\|$. Combined with (C.366) and (C.360), this gives the bound

$$|\omega_\gamma(f)| \leq \|f\|_{C^*}. \tag{C.369}$$

We now prove continuity of the map $\omega_\gamma \to \gamma$ from $\Sigma(C^*(G))$ to $\hat{G}$ (using sequences for simplicity, the argument for nets being similar). If $\omega_{\gamma_n} \to \omega_\gamma$, i.e., $\hat{f}(\gamma_n) \to \hat{f}(\gamma)$ for each $f \in C^*(G)$, and hence for each $f \in C_c(G)$, then $\gamma_n \to \gamma$ uniformly on any $K \subset \mathcal{K}(G)$. Writing $\gamma'_n = \gamma_n\overline{\gamma}$ and $g = f\overline{\gamma}$, we first notice that

$$|\gamma_n(x) - \gamma(x)| = |\gamma'_n(x) - 1|; \tag{C.370}$$
$$\hat{f}(\gamma_n) - \hat{f}(\gamma) = \hat{g}(\gamma'_n) - \hat{g}(1_G). \tag{C.371}$$

This shows that we may reduce the proof to the case $\gamma = 1_G$; otherwise, simply change γ_n to γ'_n. Thus we assume that $\hat{f}(\gamma_n) \to \hat{f}(1_G)$ for each $f \in C_c(G)$. We now pick some fixed $g \in C_c(G)$ such that $\hat{g}(1_G) = \int_G dx\, g(x) = 1$. For $\varepsilon > 0$, by uniform continuity there is a neighbourhood U of the identity $e \in G$ such that, cf. (C.364),

$$\|L_u g - g\|_1 < \varepsilon/3 \;\; (u \in U). \tag{C.372}$$

Then $\cup_{x\in G} xU$ covers G, and hence also covers each compact set $K \subset G$. Therefore, K has a finite subcover $\cup_{j\in J} x_j U$. Define $g_j = L_{x_j} g$. By invariance of the Haar measure, we have $\hat{g}_j(1_G) = 1$, so that by definition of $\omega_{\gamma_n} \to \omega_{1_G}$, we may find $N \in \mathbb{N}$ such that for each $j \in J$ and for all $n > N$, we have

$$|\hat{g}_j(\gamma_n) - 1| < \varepsilon/3. \tag{C.373}$$

Also, if $x \in K$, then $x = x_j u$ for some $j \in J$ and $u \in U$. Eq. (C.372) then implies

$$|\hat{g}_j(\gamma_n)(\gamma_n(x) - 1)| = \left| \int_G dy\, (L_u g(y) - g(y))\overline{\gamma_n(y)} \right| < \varepsilon/3. \tag{C.374}$$

Hence for any $K \in \mathscr{K}(G)$ and $x \in K$ as above, we may estimate, for all $n > N$,

$$\begin{aligned} |\gamma_n(x) - 1| \leq\ & |\gamma_n(x)(1 - \hat{g}_j(\gamma_n))| + |\hat{g}_j(\gamma_n)(\gamma_n(x) - 1)| \\ & + |\hat{g}_j(\gamma_n) - 1| < \varepsilon/3 + \varepsilon/3 + \varepsilon/3 = \varepsilon. \end{aligned} \tag{C.375}$$

Consequently, $\hat{f}(\gamma_n) \to \hat{f}(1_G)$ for each $f \in C_c(G)$ implies $\gamma_n \to 1_G$ in $\hat{G}$, as we have argued, this proves continuity of the bijection $\Sigma(C^*(G)) \to \hat{G}$ given by $\omega_\gamma \to \gamma$.

If $\Sigma(C^*(G))$ and $\hat{G}$ are compact (which is the case iff G is discrete, in which case $C^*(G)$ has a unit δ_e) we are ready, since a continuous bijection from a compact space to a Hausdorff space has a continuous inverse, and hence is a homeomorphism (in our case, both spaces are compact as well as Hausdorff). In general, continuity of the map $\gamma \mapsto \omega_\gamma$ from $\hat{G}$ to $\Sigma(C^*(G))$ almost immediately follows from the definition of the compact-open topology on $\hat{G}$: if $\gamma_n \to \gamma$ in this topology (similarly for nets), and $f \in C_c(G)$, then $\hat{f}(\gamma_n) \to \hat{f}(\gamma)$, and hence $\omega_{\gamma_n}(f) \to \omega_\gamma(f)$. A simple $\varepsilon/3$-argument then gives the same result for $f \in C^*(G)$. □

Note that local compactness of $\hat{G}$ (though provable directly) also follows from this theorem, since we know this for the Gelfand spectrum $\Sigma(C^*(G))$, cf. Theorem C.45.

Beside the Gelfand isomorphism (C.355), in which the two function spaces $C^*(G)$ and $C_0(\hat{G})$ are of a different type, there exist more symmetric versions of the generalized Fourier transform (C.356). In the setting of Banach spaces (as opposed to spaces of distributions, which would take us into the territory of locally convex topological vector spaces, and hence outside the scope of this appendix, though cf. §5.11), there are (at least) two natural possibilities. The traditional and most familiar one is provided by the Hilbert spaces $L^2(G)$ and $L^2(\hat{G})$, defined with respect to suitably normalized Haar measures dx (on G) and $d\gamma$ (on $\hat{G}$), respectively. A second, more recent possibility is to use the following two Banach spaces.

Definition C.110. *The Banach space $C_0^*(G)$ is the completion of $C_c(G)$ in the norm*

$$\|f\|_0 = \max\{\|f\|_\infty, \|\hat{f}\|_\infty\}. \tag{C.376}$$

Similarly, the Banach space $C_0^*(\hat{G})$ is the completion of $C_c(\hat{G})$ in the norm

$$\|\zeta\|_0 = \max\{\|\zeta\|_\infty, \|\check{\zeta}\|_\infty\}. \tag{C.377}$$

It follows that $C_0^*(G)$ can be norm-decreasingly injected into both $C^*(G)$ and $C_0(G)$, so that $C_0^*(G)$ is a subspace of $C_0(G)$ as well as of $C^*(G)$. By (C.341) and (C.360),

$$L^1(G) \cap C_0(G) \subset C_0^*(G), \tag{C.378}$$

and similarly for $C_0^*(\hat{G})$. Indeed, $C_0^*(G)$ (and likewise $C_0^*(\hat{G})$) could equivalently have been defied as the completion of $L^1(G) \cap C_0(G)$ in the norm (C.376).

Theorem C.111. *The Fourier transform* (C.356) *induces isometric isomorphisms*

$$L^2(G) \cong L^2(\hat{G}); \tag{C.379}$$

$$C_0^*(G) \cong C_0^*(\hat{G}), \tag{C.380}$$

such that, on suitably normalizing dx and $d\gamma$, the **Fourier inversion formula**

$$f(x) = \int_{\hat{G}} d\gamma\, \gamma(x) \hat{f}(\gamma), \tag{C.381}$$

cf. (C.356), *in both cases holds* verbatim *whenever $f \in L^1(G)$ and $\hat{f} \in L^1(\hat{G})$, in which case f and $\hat{f}$ are continuous, and* (C.356) *and* (C.381) *hold pointwise.*

The Fourier inversion formula (C.381) is actually equivalent to its special case

$$f(e) = \int_{\hat{G}} d\gamma\, \hat{f}(\gamma), \tag{C.382}$$

where $e \in G$ is the unit, since (C.381) follows by substituting $L_{x^{-1}} f$ for f and using

$$\widehat{L_{x^{-1}} f} = \gamma \hat{f}. \tag{C.383}$$

It is also important to realize that conceptually, the inversion formula (C.381) reads

$$\check{\hat{f}}(\hat{x}) = f(x^{-1}), \tag{C.384}$$

where the Fourier transform $\check{\zeta}$ for suitable $\zeta : \hat{G} \to \mathbb{C}$ is defined, as in (C.356), by

$$\check{\zeta}(\chi) = \int_{\hat{G}} d\gamma\, \overline{\chi(\gamma)} \zeta(\gamma). \tag{C.385}$$

Here $\chi : \hat{G} \to \mathbb{T}$ is some character on $\hat{G}$, i.e., $\chi \in \hat{\hat{G}}$, and we have a natural map

$$G \to \hat{\hat{G}}; \tag{C.386}$$

$$x \mapsto \hat{x}; \tag{C.387}$$

$$\hat{x}(\gamma) = \gamma(x). \tag{C.388}$$

Pontryagin duality states that (C.386) - (C.388) define an isomorphism, i.e.,

$$\hat{\hat{G}} \cong G. \tag{C.389}$$

We omit the lengthy proof of this beautiful isomorphism of topological groups (cf. the examples in Proposition C.108), and turn to the proof of Theorem C.111.

Proof. First, we (re)construct a correctly normalized Haar measure on $\hat{G}$ by defining

$$\int_{\hat{G}} : C_c(\hat{G},\mathbb{R}) \to \mathbb{R}; \tag{C.390}$$

$$\zeta \mapsto \inf\{f(e) \mid f \in C_0^*(G), \hat{f} \geq \zeta \text{ (pointwise)}\}. \tag{C.391}$$

This map takes values in $\mathbb{R}$, since if $\hat{f}$ is real, as required by $\hat{f} \geq \zeta$ in (C.391), then, noting that the Gelfand (= Fourier) transform on $C_0^*(G)$ maps the involution (C.337) into complex conjugation on $C_0(\hat{G})$, so is $f(e)$, cf.(C.337). Furthermore, $\int_{\hat{G}}$ is linear, as well as positive: if $\zeta \geq 0$ (i.e., pointwise), then also $\hat{f} \geq 0$ in $C_0(\hat{G})$, so that $f \geq 0$ in $C^*(G)$, because by Theorem C.109 the map $f \mapsto \hat{f}$ is an isomorphism, which by Theorem C.52 preserves positivity. This gives $\langle \psi, \pi(f)\psi\rangle_{L^2(G)} \geq 0$ for all $\psi \in L^2(G)$, which by a simple continuity argument (in a proof by contradiction, using the inclusion $C_0^*(G) \subset C_0(G)$) enforces $f(e) \geq 0$, and hence $\inf\{f(e)\} \geq 0$.

By Theorem B.19, there is a measure $d\gamma$ on $\hat{G}$ defining the integral $\int_{\hat{G}}$, i.e,

$$\int_{\hat{G}} d\gamma\, \zeta(\gamma) = \inf\{f(e) \mid f \in C_0^*(G), \hat{f} \geq \zeta\}, \tag{C.392}$$

where initially ζ is real-valued, upon which the integral is extended to $C_c(\hat{G})$ by complex linearity, as usual in (Lebesgue) integration. The point is that the measure $d\gamma$ is translation invariant and hence is a Haar measure on $\hat{G}$: indeed, replacing g by $L_{\gamma'} g$ amounts to replacing f (as a function that satisfies $\hat{f} \geq g$) by $\gamma' f$. Invariance then follows from $\gamma'(e) = 1$ for any character $\gamma' \in \hat{G}$, which obviously implies $(\gamma' f)(e) = \gamma'(e) f(e) = f(e)$. The Banach spaces $L^p(G)$ and $L^p(\hat{G})$ are then defined with respect to dx on G (assumed given) and $d\gamma$ on $\hat{G}$ (as above), respectively.

Furthermore, the proof uses an approximate unit (δ_U) of $C^*(G)$ that lies in $C_c(G)$ and is indexed by shrinking neighbourhoods U of $e \in G$. More precisely, take the directed set of all symmetric neighbourhoods of e (i.e., $U^{-1} = U$), ordered by reverse inclusion $\supseteq$, take positive functions $h_U \in C_c(W)$ for some neighbourhood W of e satisfying $W^2 \subset U$, normalize h_U such that $\int_G h_U * h_U^* = 1$, and define

$$\delta_U = h_U * h_U^*; \tag{C.393}$$

$$f_U = f * \delta_U \quad (f \in C^*(G)). \tag{C.394}$$

We will show that for each $f \in C^*(G)$, we have

$$\lim_U \|f_U - f\|_{C^*} = 0. \tag{C.395}$$

To this end, we first show that $\|\delta_U\|_{C^*} \leq 1$, which follows from the estimate

$$\|\pi(\delta_U)\psi\| = \left\| \int_G dy\, \delta_U(y) L_y \psi \right\| \leq \int_G dy\, \delta_U(y) \|L_y \psi\| = \|\psi\|. \tag{C.396}$$

Similar estimates give $f * \delta_U \to f$ for $f \in C_c(G)$, so that finally

$$\|f * \delta_U - f\|_{C^*} = \|f * \delta_U - g * \delta_U + g * \delta_U - g + g - f - f\|_{C^*}$$
$$\leq 2\|f - g\|_{C^*} + \|g * \delta_U - g\|_{C^*}. \tag{C.397}$$

Taking $g \in C_c(G)$, an $\varepsilon/3$ argument finishes the proof of (C.395). Moreover,

$$f_U \in C_0^*(G) \ \ (f \in C^*(G)). \tag{C.398}$$

To prove this, take $g, h \in C_c(G)$. Regarding g and h as elements of $L^2(G)$, note that

$$g * h(x) = \langle g^*, L_{x^{-1}} h \rangle_{L^2(G)}, \tag{C.399}$$

so that Cauchy–Schwarz and unitarity of $L_{x^{-1}}$ give $\|g * h\|_\infty \leq \|g\|_2 \|h\|_2$. Applying this with $g \rightsquigarrow \pi(f)g$ and $h \rightsquigarrow h$, where $f \in C^*(G)$, $g \in C_c(G)$, and $h \in C_c(G)$, yields

$$\|f * g * h\|_\infty \leq \|\pi(f)g\|_2 \|h\|_2 \leq \|f\|_{C^*} \|g\|_2 \|h\|_2; \tag{C.400}$$
$$\|f * g * h\|_2 = \|\pi(f)(g * h)\|_2 \leq \|f\|_{C^*} \|g * h\|_2. \tag{C.401}$$

Eq. (C.401) will be applied later, in the proof of (C.379). Eq. (C.400) shows that if $f_n \to f$ in $C^*(G)$ for some net (f_n) in $C_c(G)$, then $f_n * g * h \to f * g * h$ uniformly, so that $f * g * h \in C_0(G)$ and $f_n * g * h \to f * g * h$ in $C_0(G)$. Also,

$$\|\widehat{f * g * h}\|_\infty = \|\hat{f}\hat{g}\hat{h}\|_\infty \leq \|\hat{f}\|_\infty \|\hat{g}\hat{h}\|_\infty = \|f\|_{C^*} \|\hat{g}\hat{h}\|_\infty, \tag{C.402}$$

by isometry of the Gelfand transform, so that also $\widehat{f_n * g * h} \to \widehat{f * g * h}$ in $C_0(\hat{G})$. If $f_n, g, h \in C_c(G)$ then $f_n * g * h \in C_c(G) \subset C_0^*(G)$, and the above computations give $f_n * g * h \to f * g * h$ in $C_0^*(G)$. This shows that $f * g * h \in C_0^*(G)$; taking $g = h_U$ and $h = h_U^*$ yields (C.398).

We now turn to the Fourier inversion formula (C.381). Since the Gelfand transform $C^*(G) \to C_0(\hat{G})$ is an isomorphism, for any $\zeta \in C_0(\hat{G})$, we can find $f \in C^*(G)$ such that $\hat{f} = \zeta$, and we can find a net $f_U = f * \delta_U$ in $C_0^*(G)$ such that

$$\lim_U \|f_U - f\|_{C^*} = \lim_U \|\hat{f}_U - \hat{f}\|_{C_0(\hat{G})} = \lim_U \|\hat{f}_U - \hat{f}\|_\infty = 0. \tag{C.403}$$

If $\zeta \in C_c(\hat{G})$, we in addition have $\check{\hat{f}}_U \to \check{\zeta}$ in $C_0(\hat{\hat{G}})$, or, equivalently,

$$\lim_U \|\hat{f}_U - \hat{f}\|_{C^*(\hat{G})} = 0. \tag{C.404}$$

Eq. (C.403) and the fact that $\hat{\delta}_U$ is continuous, and hence uniformly continuous on every compact $K \subset \hat{G}$ (which we take such that it contains the support of $\hat{f} = \zeta$), gives $\lim_U \|\hat{\delta}_U - 1\|_\infty^{(K)} = 0$, where $\|\eta\|_\infty^{(K)}$ is the supremum of $|\eta(\gamma)|$ over all $\gamma \in K$. For $\hat{f} \in C_c(\hat{G})$, with $\hat{f}_U = \hat{\delta}_U \hat{f}$, this gives $\hat{f}_U \to \hat{f}$ in $L^1(\hat{G})$. As we trivially have $\|\hat{f}\|_{C^*(\hat{G})} \leq \|\hat{f}\|_{L^1(\hat{G})}$ (and similarly, of course, on G itself), we obtain (C.404), which together with (C.403) also yields $\hat{f}_U \to \zeta$ in $C_0^*(\hat{G})$.

Since $f_U \in C_0^*(G)$, the infimum in (C.392) is saturated, and hence the Fourier inversion formula (C.381) holds for f_U. Pontryagin duality then yields isometry, i.e., $\|\hat{f}_U\|_0 = \|f_U\|_0$. Convergence of $\hat{f}_U$ in $C_0^*(\hat{G})$ therefore yields convergence of f_U in $C_0^*(G)$, necessarily to f, since we already knew that $f_U \to f$ in $C^*(G)$, cf. (C.395). This shows that $f \in C_0^*(G)$, so that (C.381) holds for f, implying

$$\|\hat{f}\|_0 = \|f\|_0. \tag{C.405}$$

Thus the Fourier transform $\mathscr{F} : C^*(G) \to C_0(\hat{G})$ from Theorem C.109 is given by continuous extension of $\mathscr{F}(f) = \hat{f}$ as defined by (C.356), where $f \in C_c(G)$.

To prove (C.380), let $B(G)$ be the set of all $f \in C_0^*(G)$ for which $\hat{f} \in C_c(\hat{G})$, and let $B(G)^-$ be its closure in $C_0^*(G)$. Then $\mathscr{F}$ restricts to an isometric isomorphism $B(G) \to C_c(\hat{G})$, and hence also to an isometric isomorphism $B(G)^- \to C_0^*(\hat{G})$; we recall that (by definition) $C_0^*(\hat{G})$ is the completion of $C_c(\hat{G})$ in its norm $\|\cdot\|_0$.

Repeating this construction for $\hat{G}$ instead of G, and using Pontryagin duality (C.389) with the ensuing isomorphisms $C^*(\hat{\hat{G}}) \cong C^*(G)$ etc., we also have a Fourier transform $\check{\mathscr{F}} : C^*(\hat{G}) \to C_0(G)$. Since the Fourier inversion formula (C.381) holds on $C_c(\hat{G})$, we see that $\check{\mathscr{F}}$ maps $C_c(\hat{G})$ isometrically to $B(G)$ and hence by continuity maps $C_0^*(\hat{G})$ to $B(G)^-$. At the same time, $\check{\mathscr{F}}$ maps $\hat{B}(\hat{G})$ (defined, *mutatis mutandis*, like $B(G)$) to $C_c(G)$, and hence maps $\hat{B}(\hat{G})^-$ to $C_0^*(G)$. Since $\hat{B}(\hat{G})^- \subseteq C_0^*(\hat{G})$, this implies $B(G)^- = C_0^*(G)$ and $\hat{B}(\hat{G})^- = C_0^*(\hat{G})$. This proves (C.380).

Returning to (C.381), we know from the above analysis that (C.356) and (C.381) hold if $f \in C_0^*(G)$ and $\hat{f} \in C_c(\hat{G})$. If $f \in L^1(G)$, then, by Lebesgue integration theory, eq. $\mathscr{F}(f)$ remains given by (C.356). If also $\hat{f} \in L^1(\hat{G})$, then $\hat{f} \in L^1(\hat{G}) \cap C_0(\hat{G})$ and hence $\hat{f} \in C_0^*(\hat{G})$, cf. (C.378). By (C.380), there exists $\tilde{f} \in C_0^*(G)$ such that $f = \tilde{f}$ in $C^*(G)$, and hence for a.e. $x \in G$ (with respect to Haar measure), we have

$$f(x) = \lim_U f * \delta_U(x) = \lim_U \tilde{f} * \delta_U(x) = \tilde{f}(x). \tag{C.406}$$

It follows that $f = \tilde{f}$ a.e., and so the inversion formula (C.382), and hence (C.381), holds, provided (if necessary) f is replaced by its representative $\tilde{f}$.

Finally, to prove (C.379), take $f = \psi$ in (C.356) in $C_c(G)$, so that we may compute

$$\|\psi\|_2^2 = \int_G dx\, |\psi(x)|^2 = \psi * \psi^*(e) = \int_{\hat{G}} d\gamma\, \widehat{\psi * \psi^*}(\gamma) = \int_{\hat{G}} d\gamma\, |\hat{\psi}|^2 = \|\hat{\psi}\|_2^2. \tag{C.407}$$

We may therefore extend $\mathscr{F}$, initially given by $\mathscr{F}(f) = \hat{f}$, from $C_c(G)$ to its competion $L^2(G)$ in $\|\cdot\|_2$. Second, we prove surjectivity similarly to the previous part:

Pick $\zeta \in C_c(\hat{G})$, and hence $f \in C^*(G)$ with $\hat{f} = \zeta$. Then $f_U = f * \delta_U \in L^2(G)$, as follows from (C.401). Then $\hat{f}_U \to \hat{f}$ in $L^2(\hat{G})$, since analogously to the previous proof, we find that $(\hat{f}_U)$ is a Cauchy net in $L^2(\hat{G})$. By isometry of $\mathscr{F}$ (as just proved), this implies that (f_U) is a Cauchy net in $L^2(G)$. Let $f_U \to g$ in $L^2(G)$; continuity of $\mathscr{F}$ gives $\mathscr{F}(g) = \zeta$, making $\mathscr{F}$ surjective at least onto $C_c(\hat{G})$. Since $L^2(\hat{G})$ is the completion of $C_c(\hat{G})$ in the L^2-norm $\|\cdot\|_2$, the Fourier transform $\mathscr{F} : L^2(G) \to L^2(\hat{G})$ is an isometric surjection, and hence is unitary. □

We close this section with the SNAG-Theorem (named after Stone, whose Theorem 5.73 it generalizes, Naimark, Ambrose, and Godement, each of who published versions of it in 1944). This theorem uses projection-valued measures, which we have avoided so far, but which are appropriate here as well as in our application of the SNAG-Theorem to the Goldstone Theorem 10.28. Recall that the Riesz–Radon representation theorems B.19 and B.24 establish a bijective correspondence between *states* on $C_0(X)$ and *probability measures* on X. There is a similar correspondence between *representations* of $C_0(X)$ and *projection-valued measures* on X. Cf. §B.4.

Definition C.112. *Let X be a set with σ-algebra $\Sigma \subseteq \mathscr{P}(X)$, and H a Hilbert space. A* **projection-valued measure** *for (X,Σ,H) is a map $e : \Sigma \to \mathscr{P}(H)$ such that for each unit vector $\psi \in H$, the map $e^{(\psi)} : \Sigma \to [0,1]$ defined by*

$$e^{(\psi)}(A) = \langle \psi, e(A)\psi \rangle, \tag{C.408}$$

is a probability measure. Equivalently, $e(\emptyset) = 0_H$, $e(X) = 1_H$, $e(A \cap B) = e(A)e(B)$, and $e(\cup_n A_n) = \sum_n e(A_n)$ for pairwise disjoint A_n in the strong topology on $B(H)$.

The simplest example must be $H = L^2(X,\Sigma,\mu)$ with $e(A) = 1_A$, cf. §B.6.

As in (B.328), one can integrate any bounded measurable function $f : X \to \mathbb{C}$ "against" e, i.e., there is a unique operator $\int_X de\, f$ such that for any $\varepsilon > 0$ there is a finite partition $X = \bigsqcup_{i=1}^n A_i$ of X into n Borel sets A_i, such that for any $x_i \in A_i$,

$$\left\| \int_X de\, f - \sum_{i=1}^n f(x_i)e(A_i) \right\| < \varepsilon. \tag{C.409}$$

Analogously to the Riesz–Radon representation theorem, one may then prove:

Theorem C.113. *Let X be a locally compact Hausdorff space. There is a bijective correspondence between non-degenerate representations $\pi : C_0(X) \to B(H)$ and projection-valued measures e for (X,Σ,H) (where Σ is the Borel σ-algebra), viz.*

$$\pi(f) = \int_X de\, f; \tag{C.410}$$

$$e(A) = \pi(1_A), \tag{C.411}$$

where $\pi(1_A)$ is defined by extending π from $C_0(X)$ to the C^*-algebra $\mathscr{B}(X)$ of bounded Borel functions on X (cf. Theorem B.102 and Proposition B.98).

We finally need the existence of a bijective correspondence between continuous unitary representations u of G and non-degenerate representations of $C^*(G)$ given by (C.506) in §C.18 below; see the comment below Definition C.119. Combined with Theorems C.109 and C.113, we then obtain the SNAG-***Theorem***:

Theorem C.114. *There is a bijective correspondence between continuous unitary representations u of a locally compact abelian group G on some Hilbert space H and projection-valued measures $e : \mathscr{B}(\hat{G}) \to \mathscr{P}(H)$ on the dual group $\hat{G}$, such that*

$$u(x) = \int_{\hat{G}} de(\gamma)\, \gamma(x). \tag{C.412}$$

C.16 Intermezzo: Lie groupoids

Groupoids generalize groups, group actions, and equivalence relations. As such, they provide a more flexible language for dealing with symmetries than either of these. Like Lie groups, one also has Lie groupoids, which form an important tool in constructing continuous bundles of C*-algebras (see §C.19 below). These, in turn, provide the mathematical foundation of (deformation) quantization, see Chapter 7.

Definition C.115. *A* **groupoid** $G = (G_1, G_0, s, t, i, I)$ *is a small category (i.e. a category in which the underlying classes are sets, cf. §E.1) in which each arrow is invertible. Thus one has a set (of arrows) G_1 doubly fibered over some base space G_0 through source, and target maps $s,t : G_1 \to G_0$. These maps define the set*

$$G_2 = \{(x,y) \in G_1 \times G_1 \mid s(x) = t(y)\} \tag{C.413}$$

of composable pairs, on which a multiplication $m : G_2 \to G_1$ is defined, which we simply denote by $xy = m(x,y)$, subject to the axioms

$$s(xy) = s(y);\ t(xy) = t(x)\ \ (xy \in G_2); \tag{C.414}$$
$$(xy)z = x(yz)\ \ (xy \in G_2, yz \in G_2), \tag{C.415}$$

the third being well defined by virtue of the first and the second.

Furthermore, there is an object inclusion map $i : G_0 \hookrightarrow G_1,\ u \mapsto \mathrm{id}_u$, satisfying

$$s(\mathrm{id}_u) = t(\mathrm{id}_u) = u\ \ (u \in G_0); \tag{C.416}$$
$$x\mathrm{id}_{s(x)} = \mathrm{id}_{t(x)}x = x\ \ (x \in G_1). \tag{C.417}$$

Finally, what makes a (small) category a groupoid is the existence of an inverse

$$I : G_1 \to G_1,\ x \mapsto x^{-1},$$

satisfying

$$s(x^{-1}) = t(x);\ t(x^{-1}) = s(x)\ \ (x \in G_1); \tag{C.418}$$
$$x^{-1}x = \mathrm{id}_{s(x)};\ xx^{-1} = \mathrm{id}_{t(x)}\ \ (x \in G_1). \tag{C.419}$$

A **Lie groupoid** *is a groupoid for which G_1 and G_0 are manifolds, s and t are surjective submersions, and multiplication and inversion are smooth.*

We often identify u with id_u, so that $x^{-1}x = s(x)$, etc. We allow manifolds with boundary, which provide key examples; cf. Proposition C.117 below.

Proposition C.116. *In a Lie groupoid, object inclusion is an immersion, inversion is a diffeomorphism, G_2 is a closed submanifold of $G_1 \times G_1$, and for each $u \in G_0$, the fibers $s^{-1}(u)$ and $t^{-1}(u)$ are submanifolds of G_1.*

Abusing notation, G_1 is often called G. Some basic examples of Lie groupoids are:

- ***Lie groups*** G, where $G_1 = G$ and $G_0 = \{e\}$, $e \in G$ being the unit.
- ***Manifolds*** M, where $G_1 = G_0 = M$ with the obvious trivial groupoid structure $s(x) = t(x) = \mathrm{id}_x = x^{-1} = x$, and $xx = x$.
- ***Pair groupoids*** over a manifold $G_0 = M$, where $G_1 = M \times M$ and $s(x,y) = y$, $t(x,y) = x$, $(x,y)^{-1} = (y,x)$, $(x,y)(y,z) = (x,z)$, and $\mathrm{id}_x = (x,x)$.
- ***Smooth equivalence relations***, i.e. immersed submanifolds of $M \times M$.
- ***Action groupoids*** $\Gamma \ltimes M$, which are defined by a smooth (left) action $\Gamma \circlearrowright M$ of a Lie group Γ on a manifold M, where $G_1 = \Gamma \times M$, $G_0 = M$, $s(g,m) = g^{-1}m$, $t(g,m) = m$, $(g,m)^{-1} = (g^{-1}, g^{-1}m)$, and $(g,m)(h,g^{-1}m) = (gh,m)$.
- ***Vector bundles*** $\pi : E \to M$ over a manifold $G_0 = M$, with $s = t$ given by the bundle projection π, object inclusion $M \hookrightarrow E$ as the zero section, multiplication as as fiberwise addition of tangent vectors, and inverse $\xi^{-1} = -\xi$.

Any Lie groupoid G defines an associated ***tangent groupoid*** G^T, which will play a crucial role in §C.19. We first explain the (surprising) underlying differential geometry in three steps of increasing complexity. We start with the manifold $M = \mathbb{R}^n$, with tangent bundle $TM = \mathbb{R}^{2n}$. Our goal is to describe a smooth structure on

$$F = TM \sqcup (0,1] \times M \times M, \tag{C.420}$$

seen as a bundle over $[0,1]$, where (as the notation already indicates) the fibers are

$$F_0 = TM; \tag{C.421}$$
$$F_\hbar = M \times M \ (\hbar > 0). \tag{C.422}$$

Although each fiber $F_\hbar$ of this bundle is isomorphic to $\mathbb{R}^{2n}$, its smooth structure is not equal or even diffeomorphic to the usual one on $[0,1] \times \mathbb{R}^{2n}$. Instead, we define

$$\phi : [0,1] \times TM \to TM \sqcup (0,1] \times M \times M; \tag{C.423}$$
$$\phi(0,\xi) = \xi; \tag{C.424}$$
$$\phi(\hbar,\xi) = (\hbar, \exp^W(\hbar\xi)) \ (\hbar > 0), \tag{C.425}$$

where the symmetrized ("Weyl") exponential map $\exp^W : TM \to M \times M$ is given by

$$\exp^W(x,v) = (x - \tfrac{1}{2}v, x + \tfrac{1}{2}v). \tag{C.426}$$

Here the coordinates (x,v) of $\xi \in T_xM$ denote $\xi f(x) = \sum_i v^i \left(\frac{\partial f}{\partial x_i}\right)(x) \equiv \sum_i v^i \partial_i f(x)$. Like its more familiar counterpart $(x,v) \mapsto (x, x+v)$, $\exp^W$ is a diffeomorphism.

For $M = \mathbb{R}^n$, our map ϕ is a bijection, with inverse given by

$$\phi^{-1}(x,v) = (0,x,v); \tag{C.427}$$
$$\phi^{-1}(\hbar,x,y) = \left(\hbar, \frac{x+y}{2}, \frac{y-x}{\hbar}\right) \ (\hbar > 0). \tag{C.428}$$

We use this to transfer the product topology (and also the smooth structure as a manifold with boundary) from $[0,1] \times TM$ to F. Then a sequence $(\hbar_n, x_n, y_n)$ in F, where $\hbar_n \to 0$, converges iff $x_n \to x$, $y_n \to x$ for some $x \in M$, and $(y_n - x_n)/\hbar_n \to v$, in which case $(\hbar_n, x_n, y_n) \to (0, x, v)$. More abstractly, F has two key properties:

1. The map $F \to [0,1] \times M \times M$ given by

$$(x,v) \qquad \mapsto (0,x,x); \tag{C.429}$$
$$(\hbar,x,y) \mapsto (\hbar,x,y) \;\; (\hbar > 0), \tag{C.430}$$

is smooth. Indeed, as a map $[0,1] \times TM \to [0,1] \times M \times M$, this map is given by

$$(0,x,v) \mapsto (0,x,x); \tag{C.431}$$
$$(\hbar,x,v) \mapsto (\hbar, x - \tfrac{1}{2}\hbar v, y + \tfrac{1}{2}\hbar v). \tag{C.432}$$

2. For any $f \in C^\infty(M \times M)$ that vanishes on the diagonal

$$\Delta(M) = \{(x,x) \mid x \in M\} \subset M \times M, \tag{C.433}$$

the function δf on F defined by

$$\delta f(x,v) = \xi_\perp f(x,x); \tag{C.434}$$
$$\delta f(\hbar,x,y) = f(x,y)/\hbar \;\; (\hbar > 0), \tag{C.435}$$

where the tangent vector $\xi_\perp \in T_{(x,x)}(M \times M)$ has components $(-\frac{1}{2}v, \frac{1}{2}v)$, is smooth. Indeed, as a function on $[0,1] \times TM$, the pullback $\delta^* f \equiv \delta f \circ \phi$ is given by

$$\delta^* f(0,x,v) = \xi_\perp f(x,x); \tag{C.436}$$
$$\delta^* f(\hbar,x,v) = f(\hbar, x - \tfrac{1}{2}\hbar v, y + \tfrac{1}{2}\hbar v)/\hbar, \tag{C.437}$$

which is smooth given our assumptions on f.

A similar construction works for any (smooth) manifold M, except that the smooth structure on F may no longer be definable in terms of a single map ϕ. Instead, we invoke a special case of the well-known ***tubular neighbourhood theorem*** of Riemannian (or, more generally, affine) geometry, which states that M, identified with the zero section in its tangent bundle TM, has an open neighbourhood U such that the (symmetrized) exponential map $\exp^W : U \to M \times M$ is a diffeomorphism onto its image. Here $\exp^W(\xi) = (\gamma(-\frac{1}{2}), \gamma(\frac{1}{2}))$, where $\xi \in T_xM$ and γ is the unique affinely parametrized geodesic with $\gamma(0) = x$ and $\dot{\gamma}(0) = \xi$. We now replace the space $[0,1] \times TM$ used in the special case $M = \mathbb{R}^n$ by the pair of spaces

$$V_1 = \{(\hbar, \xi) \in [0,1] \times TM \mid \hbar\xi \in U\}; \tag{C.438}$$
$$V_2 = (0,1] \times M \times M, \tag{C.439}$$

with associated maps $\phi_1 : V_1 \to F$ and $\phi_2 : V_2 \to F$ defined by

$$\phi_1(0,\xi) = \xi; \tag{C.440}$$
$$\phi_1(\hbar,\xi) = \exp^W(\hbar\xi)\ (\hbar > 0); \tag{C.441}$$
$$\phi_2(\hbar,x,y) = (\hbar,x,y)\ (\hbar > 0). \tag{C.442}$$

Then ϕ_1 and ϕ_2 are injective and, writing $F_i = \varphi_i(V_i)$, we have $F = F_1 \cup F_2$, which is far from a disjoint union; let $F_{ij} = F_i \cap F_j$. Also, let $V_{ij} = \{\alpha \in V_i \mid \phi_i(\alpha) \in F_{ij}\}$, with associated maps $\phi_{ij} = \phi_j^{-1} \circ \phi_i : V_{ij} \to V_{ji}$. We now define the smooth structure of F by declaring $f : F \to \mathbb{R}$ to be smooth iff $f_i \circ \phi_i^{-1} : V_i \to \mathbb{R}$ is smooth, $i = 1,2$, where f_i is the restriction of f to F_i. These conditions are compatible on the overlap F_{ij}, since $\phi_{12}(\hbar,\xi) = \exp^W(\hbar\xi)$ is a diffeomorphism (with inverse ϕ_{21}). This smooth structure may also be defined by imposing conditions 1 and 2 above, *mutatis mutandis*. In particular, (C.434) should now read

$$\delta f(\xi) = \xi_\perp f(x,x);\ \ \xi_\perp = (-\tfrac{1}{2}\xi, \tfrac{1}{2}\xi) \in T_{(x,x)}(M \times M) \cong T_xM \oplus T_xM. \tag{C.443}$$

A more general form of the above construction, which will be used to generate a vast class of continuous bundles of C*-algebras, is as follows. Let M be a closed submanifold of another manifold G (in the above situation we take $G = M \times M$ and identify M with $\Delta(M)$), and replace TM above by the ***normal bundle***

$$N_MG = T_MG/T_MM, \tag{C.444}$$

i.e., the quotient of the restriction T_MG of the tangent bundle TG to $M \subset G$ by its subbundle $T_MM \cong TM$; hence the fiber of N_MG at $x \in M \subset G$ is T_xG/T_xM.

In the above case $G = M \times M$, one therefore has

$$N_M(M \times M) \cong TM, \tag{C.445}$$

through the isomorphism $[(\xi_1,\xi_2)] \mapsto \frac{1}{2}(\xi_2 - \xi_1)$, where $(\xi_1,\xi_2) \in T_{(x,x)}(M \times M)$ and $[(\xi_1,\xi_2)]$ is its equivalence class in the quotient $T_{(x,x)}(M \times M)/T_{(x,x)}(\Delta(M))$.

Other easy examples are Lie groups G, for which $N_MG = T_eG = \mathfrak{g}$ is just the Lie algebra of G (at least as a vector space), and $G = M$, for which $N_MG = M$.

For the bundle F, defined over $I = [0,1]$, we take the fibers and total space as

$$F_0 = N_MG; \tag{C.446}$$
$$F_\hbar = G\ (\hbar > 0); \tag{C.447}$$
$$F = N_MG \sqcup (0,1] \times G. \tag{C.448}$$

Once again, there are two equivalent ways to define a smooth structure on F. The first uses a more general version of the tubular neighbourhood theorem from differential geometry, which states that $M \subset N_MG$ (seen as its zero section) has an open neighbourhood U that is diffeomorphic to some open neighbourhood U' of $M \subset G$ via a diffeomorphism φ that maps M to itself (i.e., pointwise). Then put

$$V_1 = \{(\hbar,\xi) \in [0,1] \times N_M G \mid \hbar\xi \in U\}; \tag{C.449}$$
$$V_2 = (0,1] \times G, \tag{C.450}$$

again with associated maps $\phi_1 : V_1 \to F$ and $\phi_2 : V_2 \to F$, this time defined by

$$\phi_1(0,\xi) = \xi; \tag{C.451}$$
$$\phi_1(\hbar,\xi) = \varphi(\hbar\xi) \ (\hbar > 0); \tag{C.452}$$
$$\phi_2(\hbar,n) = (\hbar,n) \ (\hbar > 0). \tag{C.453}$$

One then proceeds exactly as above. Equivalently, we impose that:

1. The map $F \to [0,1] \times G$, defined at $\hbar = 0$ by $\xi \mapsto (0,x)$, where $\xi \in N_M G$ (where $x \in M \subset G$), and $(\hbar,n) \mapsto (\hbar,n)$ for $\hbar > 0$ and $n \in G$, is smooth.
2. For each $f \in C^\infty(G)$ that vanishes on M, the function δf on F defined by

$$\delta f(\xi) = \xi f; \tag{C.454}$$
$$\delta f(\hbar,n) = f(n)/\hbar \ (\hbar > 0), \tag{C.455}$$

is smooth (note that ξf is well defined despite the fact that $\xi \in T_M G/T_M M$ rather than $\xi \in T_M G$, since any two representatives of ξ in $T_M G$ differ by vectors in $T_M M$, which vanish on f because $f_{|M} = 0$ by assumption).

After this preparation, we are at last in a position to define tangent groupoids.

Proposition C.117. *Any Lie groupoid G over some base space $G_0 = M$ defines an associated* **tangent groupoid** *G^T, with total space $G^T = F$, cf. (C.448), with smooth structure as explained, base space $G_0^T = [0,1] \times M$, source and target projections*

$$s^T(\xi) = t^T(\xi) = (0,\pi(\xi)) \ (\hbar = 0); \tag{C.456}$$
$$s^T(\hbar,x) = (\hbar,s(x)) \ (\hbar > 0); \tag{C.457}$$
$$t^T(\hbar,x) = (\hbar,t(x)) \ (\hbar > 0), \tag{C.458}$$

where $\pi : T_M G/T_M M \to M$ is the bundle projection, and $x \in G$, multiplication

$$\xi \cdot \eta = \xi + \eta \ (\hbar = 0); \tag{C.459}$$
$$(\hbar,x) \cdot (\hbar,y) = (\hbar,xy) \ (\hbar > 0), \tag{C.460}$$

and inverse

$$\xi^{-1} = -\xi \ (\hbar = 0); \tag{C.461}$$
$$(\hbar,x)^{-1} = (\hbar,x^{-1}) \ (\hbar > 0). \tag{C.462}$$

In other words, G^T, seen as a bundle over $[0,1]$ is a "bundle of groupoids": the groupoid above $\hbar = 0$ is the normal bundle $\pi : N_M G \to M$, as in the vector bundle example above, whereas the fibers above $\hbar > 0$ are G itself.

C.17 C*-algebras associated to Lie groupoids

One may associate two C*-algebras to a Lie groupoid G, called $C^*_r(G)$ and $C^*(G)$, which coincide for abelian Lie groups, and as such generalize the construction in §C.15, cf. (C.332) and (C.336) - (C.337). We first generalize the Haar measure.

Definition C.118. *A* **Haar system** *on a Lie groupoid G is a family of measures* $(\mu^u, u \in G_0)$*, where* μ^u *is defined on the t-fiber*

$$G^u = t^{-1}(u), \tag{C.463}$$

where it is locally equivalent to Lebesgue measure, and each function

$$u \mapsto \int_{G^u} d\mu^u f \ (f \in C_c^\infty(G)) \tag{C.464}$$

on G_0 *is smooth. A Haar system is* **left-invariant** *if for each* $f \in C_c^\infty(G)$ *and* $x \in G$*,*

$$\int_{G^{t(x)}} d\mu^{t(x)}(y)\, f(y) = \int_{G^{s(x)}} d\mu^{s(x)}(y)\, f(xy). \tag{C.465}$$

It is sometimes convenient to regard μ^u as a measure on all of G but having support in G^u. Either way, *any Lie groupoid possesses a left-invariant Haar system*, briefly called a ***left Haar system***. For example, if G is a Lie group, $u \in G_0$ can only be the identity $e \in G$, so that a left-invariant Haar system is the same as a left-invariant Haar measure on G (which exists on any locally compact group). Furthermore:

1. If $G = G_0 = M$, where M is a manifold (as always), then $s(x) = t(x) = x = x^{-1}$, and the condition (C.465) is empty, so that a left-invariant Haar system is just a smooth function $\mu : M \to (0,\infty)$, i.e. $\mu(u) \equiv \mu^u$. In what follows, we simply take $\mu(u) = 1$ for each $u \in M$. More generally, whenever G^u is compact, we normalize a Haar system by imposing $\mu^u(G^u) = 1$, as in the case of groups.
2. For a pair groupoid $G = M \times M$, on the other hand, (C.465) forces the system of measures to collapse to a single measure μ on M, i.e., $\mu^u = \mu$ for each $u \in M$. For $M = \mathbb{R}^n$, we take μ to be Lebesgue measure.
3. For the tangent bundle $G = TM$ (with fiberwise addition), which is essentially a bundle of abelian groups $\mathbb{R}^n$, eq. (C.465) forces each measure μ^u on
$$t^{-1}(u) = T_uM, \tag{C.466}$$
to be translation invariant. For $M = \mathbb{R}^n$ (or, more generally, if TM is a trivial bundle), we take all μ^u to be the same and all equal to Lebesgue measure.
4. For action groupoids $\Gamma \ltimes M$, we have $t^{-1}(u) = G$, and any left-invariant Haar measure $d\gamma$ on Γ yields a left Haar system on G as $d\mu^u = d\gamma$, for each $u \in M$.
5. In case of a tangent groupoid G^T, the t-fibers are indexed by $(\hbar, u)$, where $\hbar \in [0,1]$ and $u \in M$, so that a left Haar system consists of a family $\mu^{(\hbar,u)}$. It turns out that given any left Haar system $(\mu^u, u \in M)$ on G, there exists a (suitably normalized) left Haar system $(\mu_0^u, u \in M)$ on the vector bundle N_MG such that

$$\mu^{(0,u)} = \mu_0^u \ (\hbar = 0); \tag{C.467}$$
$$\mu^{(\hbar,u)} = \hbar^{-n}\mu^u \ (\hbar > 0), \tag{C.468}$$

where $n = \dim(G) - \dim(M)$, defines a Haar system on G^T; the extra factor $\hbar^{-n}$ in (C.468) is necessary and sufficient for this Haar system to satisfy the smoothness condition on (C.464). For example, if $G = \mathbb{R}^n \times \mathbb{R}^n$ is the pair groupoid on $\mathbb{R}^n$, where each fiber $G^u \cong \mathbb{R}^n$ is endowed with Lebesgue measure $d^n x$, then the fibers $\mathbb{R}^n$ of the vector bundle $N_M G \cong T\mathbb{R}^n$ should carry exactly the same measure. To see this, in (C.464) we substitute $G \rightsquigarrow G^T$ and $u \rightsquigarrow (\hbar, y)$ $(y \in \mathbb{R}^n)$, so that for each $f \in C_c^\infty(G^T)$ the following function on $[0,1] \times \mathbb{R}^n$ should be smooth:

$$(0,y) \mapsto \int_{\mathbb{R}^n} d^n v\, f(0,y,v); \tag{C.469}$$
$$(\hbar,y) \mapsto \hbar^{-n} \int_{\mathbb{R}^n} d^n x\, f(\hbar,x,y) \ (\hbar > 0). \tag{C.470}$$

To interpret this condition, we put $f = \tilde{f} \circ \phi^{-1}$, where $\tilde{f}$ is smooth on $[0,1] \times T\mathbb{R}^n$, and ϕ^{-1} is given by (C.427) - (C.428). This transforms the above function into

$$(0,y) \mapsto \int_{\mathbb{R}^n} d^n v\, \tilde{f}(0,y,v); \tag{C.471}$$
$$(\hbar,y) \mapsto \int_{\mathbb{R}^n} d^n v\, \tilde{f}(\hbar, y - \tfrac{1}{2}\hbar v, v) \ (\hbar > 0). \tag{C.472}$$

We now define C*-algebras $C^*(G)$ and $C_r^*(G)$, which depend on the choice of a left Haar system on G, but different choices lead to isomorphic C*-algebras. We start from $C_c^\infty(G)$, on which we define a convolution product and an involution by

$$f * g(x) = \int_{G^{s(x)}} d\mu^{s(x)}(y)\, f(xy) g(y^{-1}); \tag{C.473}$$
$$f^*(x) = \overline{f(x^{-1})}. \tag{C.474}$$

We then define a C*-algebra $C^*(G)$ as the completion of $C_c^\infty(G)$ in the norm

$$\|f\| = \sup\{\|\pi(f)\|\}, \tag{C.475}$$

where the supremum is over all Hilbert space representations of $C_c^\infty(G)$ that satisfy

$$\|\pi(f)\| \leq \|f\|_1 \equiv \max\{\|f\|_1^{(s)}, \|f\|_1^{(t)}\}, \tag{C.476}$$

where the canonical L^1-norm on the right-hand side is defined by

$$\|f\|_1^{(s)} = \sup_{u \in M} \int_{G_u} d\mu_u(y)\, |f(y)|; \quad \|f\|_1^{(t)} = \sup_{u \in M} \int_{G^u} d\mu^u(y)\, |f(y)|. \tag{C.477}$$

A more tractable possibility is to limit these representations to a selected class, such as the following one. Further to the t-fiber (C.463), we denote the s-fibers of G by

$$G_u = s^{-1}(u), \tag{C.478}$$

which carries a canonical measure

$$d\mu_u(x) = d\mu^u(x^{-1}). \tag{C.479}$$

This leads to Hilbert spaces

$$H_u = L^2(G_u, \mu_u), \tag{C.480}$$

on which $C_c^\infty(G)$ can be represented through the formula

$$\pi_u(f)\psi(x) = \int_{G^u} d\mu^u(y)\, f(xy)\psi(y^{-1}) \;\; (\psi \in H_u, x \in G_u, y \in G^u). \tag{C.481}$$

Such representations automatically satisfy the bound (C.476); restricting the representations π in (C.475) to these π_u, $u \in M$, gives the ***reduced groupoid C*-algebra*** $C_r^*(G)$. In other words, $C_r^*(G)$ is the completion of $C_c^\infty(G)$ in the norm

$$\|f\|_r = \sup\{\|\pi_u(f)\|, u \in M\}. \tag{C.482}$$

One often has $C_r^*(G) = C^*(G)$, but if G is for example a non-compact and semi-simple Lie group, then the two differ (in which case $C_r^*(G)$ is a quotient of $C^*(G)$). Deferring groups to the next section, the other examples on our list are as follows.

1. For a space $G = M$, the algebraic operations are

$$f * g(x) = f(x)g(x); \tag{C.483}$$
$$f^*(x) = \overline{f(x)}, \tag{C.484}$$

from which we obtain

$$C_r^*(M) = C_0(M). \tag{C.485}$$

Indeed, $G^x = \{x\}$, so with $\mu(x) = 1$ for each $x \in M$, we obtain

$$H_x = \mathbb{C}; \tag{C.486}$$
$$\pi_x(f) = f(x), \tag{C.487}$$

and hence $\|f\|_r = \|f\|_\infty$; the completion of $C_c^\infty(M)$ in this norm is $C_0(M)$.

2. A pair groupoid $G = M \times M$, with left Haar system $\mu^u = \mu$ for all $u \in M$, gives

$$f * g(u,v) = \int_M d\mu(w)\, f(u,w)g(w,v); \tag{C.488}$$
$$f^*(u,v) = \overline{f(v,u)}, \tag{C.489}$$

which of course is reminiscent of the corresponding operations on matrices. Also,

$$H_u = L^2(M,\mu); \tag{C.490}$$
$$\pi_u(f)\psi(v) = \int_M d\mu(w)\, f(v,w)\psi(w), \tag{C.491}$$

where we wrote $x = (v,u)$ and $y = (u,w)$, and identified $\psi(v)$ with $\psi(v,u)$. With this identification, the representations π_u are the same for each u. Using the fact that $C_c^\infty(M\times M)$ is dense in $L^2(M\times M)$ and that integral operators (C.491) of Hilbert–Schmidt type are dense in the compact operators, we obtain

$$C_r^*(M\times M)\cong B_0(L^2(M)). \tag{C.492}$$

3. For a tangent bundle $G = TM$, we have, identifying T_uM with $\mathbb{R}^n$, $n = \dim(M)$,

$$f*g(u,v) = \int_{\mathbb{R}^n} d^n w\, f(u,v+w)g(u,-w); \tag{C.493}$$
$$f^*(u,v) = \overline{f(v,-u)}, \tag{C.494}$$

where we used local coordinates (u,v) on TM. Furthermore, we have

$$H_u = L^2(T_uM) = L^2(\mathbb{R}^n); \tag{C.495}$$
$$\pi_u(f)\psi(v) = \int_{\mathbb{R}^n} d^n w f(u,v+w)\psi(-w), \tag{C.496}$$

which is diagonalized by a Fourier transform $f\mapsto \hat{f}$ (cf. Theorem C.109), with

$$\hat{f}(u,p) = \int_{\mathbb{R}^n} d^n v f(u,v)e^{ipv}. \tag{C.497}$$

This map therefore gives an isomorphism

$$C^*(TM)\cong C_0(T^*M). \tag{C.498}$$

4. The (reduced) C*-algebra of an action groupoid $G = \Gamma \ltimes M$ has operations

$$f*g(\gamma,u) = \int_G d\delta\, f(\gamma\delta,u)g(\delta^{-1},\delta^{-1}\gamma^{-1}u); \tag{C.499}$$
$$f^*(\gamma,u) = \overline{f(\gamma^{-1},\gamma^{-1}u)}, \tag{C.500}$$

and the special representations π_u are given by

$$H_u = L^2(G); \tag{C.501}$$
$$\pi_u(f)\psi(\gamma) = \int_G d\delta\, f(\gamma\delta,\gamma u)\psi(\delta^{-1}). \tag{C.502}$$

This gives the (reduced) ***transformation group C*-algebra*** (see the end of §C.18)

$$C_r^*(\Gamma\ltimes M) = C_r^*(\Gamma,M). \tag{C.503}$$

5. The C*-algebra $C^*(G^T)$ of a tangent groupoid will be analyzed in §C.19.

C.18 Group C*-algebras and crossed product algebras

It can be shown that in cases 1–3 above we have $C^*(G) = C^*_r(G)$. It is useful to give a more direct and general construction of both $C^*(G)$ and $C^*_r(G)$ in the case where G is a group or an action groupoid; although the former is a special case of the latter by taking the trivial G-action on a point, we treat the group case separately first.

Let G be a Lie group, or, more generally, a locally compact group, which for simplicity we assume to be unimodular (so that it has a left Haar measure dx that is also right invariant). We turn $C_c^\infty(G)$, or, more generally, $C_c(G)$, into an algebra with involution by specializing (C.473) - (C.474) to groups, i.e. (changing $y \mapsto x^{-1}y$),

$$f * g(x) = \int_G dy\, f(y)g(y^{-1}x); \tag{C.504}$$

$$f^*(x) = \overline{f(x^{-1})}. \tag{C.505}$$

Any unitary representation u of G on a Hilbert space H (assumed strongly continuous, as always) then gives rise to a representation $u^\int$ of this *-algebra by

$$u^\int(f) = \int_G dy\, f(x)u(x), \tag{C.506}$$

in that $u^\int(f*g) = u^\int(f)u^\int(g)$ and $u^\int(f^*) = u^\int(f)^*$. Let

$$\|f\| = \sup\{\|u^\int(f)\|\}, \tag{C.507}$$

where the supremum is over all continuous unitary representations of G.

Definition C.119. *The* **group C*-algebra** $C^*(G)$ *of G is the closure of* $C_c^\infty(G)$ *or* $C_c(G)$ *in the norm* (C.507). *The* **reduced group C*-algebra** $C^*_r(G)$ *of G is the closure of* $C_c^\infty(G)$ *or* $C_c(G)$ *in the norm*

$$\|f\|_r = \|u_L^\int(f)\|, \tag{C.508}$$

where u_L is the left-regular representation $u_L(G)$ on $H = L^2(G)$, cf. (7.52).

The relationship between the two group C*-algebras is given by

$$C^*_r(G) \cong u_L^\int(C^*(G)) \cong C^*(G)/\ker\left(u_L^\int\right). \tag{C.509}$$

Definition C.120. *A unitary representation u_1 is* **weakly contained** *in u_2, if* $\|u_1^\int(f)\| \leq \|u_2^\int(f)\|$ *for all $f \in C_c(G)$. If every unitary representation of G is weakly contained in u_L, and hence* $\ker\left(u_L^\int\right) = \{0\}$ *and* $C^*_r(G) \cong C^*(G)$, *we call G* **amenable**.

It can be shown that G is amenable iff the commutative C*-algebra $C_b(G)$ of bounded continuous functions on G with sup-norm has a left-invariant state ω, i.e.,

$$\omega(L_y f) = \omega(f) \; (y \in G, f \in C_b(G)). \tag{C.510}$$

Here $L_y f(x) = f(y^{-1}x)$ as usual. This is the case, for example, for all compact groups, all abelian groups, and all solvable groups (and semi-direct products thereof, like the Euclidean group). Non-compact semi-simple Lie groups, like $SL_n(\mathbb{R})$, or the Lorentz group, are not amenable, similarly for e.g. the Poincaré group.

Bij construction, there is a bijective correspondence $u \leftrightarrow u^{\int}$ between unitary representation of G and non-degenerate representations of $C^*(G)$ (which restricts to a bijection between unitary representation of G that are weakly contained in u_L and non-degenerate representations of $C_r^*(G)$). In one direction, this is given by (C.506), whilst in the other, one first decomposes $u^{\int} \equiv \rho$ as a direct sum of cyclic representations with cyclic vectors Ω_i, and then, for each Ω_i in the sum, puts

$$u(x)\rho(f)\Omega_i = \rho(L_x f)\Omega_i. \tag{C.511}$$

Now take any C*-algebra A on which G acts, in that there is a continuous group homomorphism $\alpha : G \to \mathrm{Aut}(G)$, i.e., for each $x \in G$ we have an invertible homomorphism $\alpha_x : A \to A$ such that $\alpha_x \circ \alpha_y = \alpha_{xy}$ and $\alpha_e = \mathrm{id}_A$ (or, equivalently, $\alpha_x^{-1} = \alpha_{x^{-1}}$), and for each $a \in A$, the function $x \mapsto \alpha_x(A)$ from G to A is continuous. We turn the space $C_c(G,A)$ into a *-algebra by generalizing (C.504) - (C.505) to

$$f * g(x) = \int_G dy\, f(y)\alpha_y(g(y^{-1}x)); \tag{C.512}$$

$$f^*(x) = \alpha_x(f(x^{-1})^*). \tag{C.513}$$

We construct representations of $C_c(G,A)$ as a *-algebra from pairs $(u(G),\pi(A))$, where u is a unitary representation of G, and π is a representation of A (both defined on the same Hilbert space H) that satisfy the ***covariance condition***

$$\pi(\alpha_x(a)) = u(x)\pi(a)u(x)^*. \tag{C.514}$$

Writing $\pi \rtimes u^{\int}$ for the associated representation of $C_c(G,A)$, we put

$$\pi \rtimes u^{\int}(f) = \int_G dx\, \pi(f(x))u(x), \tag{C.515}$$

and define

$$\|f\| = \sup\{\|\pi \rtimes u^{\int}(f)\|\}, \tag{C.516}$$

where the supremum runs over all pairs $(u(G),\pi(A))$ satisfying (C.515). The closure $C^*(G,A,\alpha)$ of $C_c(G,A)$ in this norm is a C*-algebra called the ***crossed product*** or ***covariance algebra*** defined by G, A, and α. Once again, by construction there is a bijective correspondence $(u,\pi) \leftrightarrow \pi \rtimes u^{\int}$ between pairs (u,π) satisfying (C.515) and non-degenerate representations $\pi \rtimes u^{\int} \equiv \rho$ of $C^*(G,A,\alpha)$, in one direction given by (C.515), and in the other by

$$u(x)\rho(f)\Omega_i = \rho(\alpha_x(L_x f))\Omega_i; \tag{C.517}$$

$$\pi(a)\rho(f)\Omega_i = \rho(af)\Omega_i. \tag{C.518}$$

Here $\alpha_x(L_x f) \in C_c(G,A)$ is the function $y \mapsto \alpha_x(f(x^{-1}y))$, similarly $af \in C_c(G,A)$ is given by $y \mapsto af(y)$, and the cyclic vectors Ω_i are defined as in (C.511).

To construct a reduced crossed product, we take any injective representation $\pi_r(A)$ on some Hilbert space K, and from it construct a new Hilbert space

$$H = L^2(G,K) \cong L^2(G) \otimes K, \tag{C.519}$$

consisting of all measure functions $\psi : G \to K$ for which $\int_G dx \, \|\psi(x)\|_K^2 < \infty$, with

$$\langle \varphi, \psi \rangle = \int_G dx \, \langle \varphi(x), \psi(x) \rangle_K \tag{C.520}$$

as the inner product. This Hilbert space H carries a covariant pair $(u(G), \pi(A))$, viz.

$$u(y)\psi(x) = \psi(y^{-1}x); \tag{C.521}$$
$$\pi(a)\psi(x) = \pi_r(\alpha_{x^{-1}}(a))\psi(x), \tag{C.522}$$

and hence an associated representation $\pi \rtimes u^{\int}$ of $C_c(G,A)$ given by (C.515), which by continuity extends to a representation ρ_r of $C^*(G,A,\alpha)$. As in the group case, we define $C_r^*(G,A,\alpha)$ as the closure of $C_c(G,A)$ in the norm $\|f\|_r = \|\rho_r(f)\|$, or as

$$C_r^*(G,A,\alpha) = \rho_r(C^*(G,A,\alpha)). \tag{C.523}$$

If G is amenable, we once again have $C_r^*(G,A,\alpha) = C^*(G,A,\alpha)$, as for $C_r^*(G)$.

The main case of interest to us is given by a group action $G \circlearrowright Q$, as above, which gives rise to a crossed product $C^*(G,C_0(Q),\alpha) \equiv C^*(G,Q)$ through the choices

$$A = C_0(Q); \tag{C.524}$$
$$\alpha_x(\tilde{f}) = L_x \tilde{f}, \tag{C.525}$$

i.e., $\alpha_x(\tilde{f})(q) = \tilde{f}(x^{-1}q)$. The (reduced) crossed product $C^*_{(r)}(G,Q)$, then, is the same as the (reduced) C*-algebra of the action groupoid $G \ltimes Q$. Identifying the spaces $C_c(G \times Q)$ and $C_c(G, C_c(Q))$, eqs. (C.512) - (C.513) now become

$$f * g(x,q) = \int_G dy \, f(y,q) g(y^{-1}x, y^{-1}q); \tag{C.526}$$
$$f^*(x,q) = \overline{f(x^{-1}, x^{-1}q)}. \tag{C.527}$$

The obvious candidate for a faithful representation of $C_0(Q)$ comes from a measure ν on Q with support Q, so that we may take $K = L^2(Q,\nu)$ and $\pi_r(\tilde{f}) = m_{\tilde{f}}$, i..e, $\pi_r(\tilde{f})\psi = \tilde{f}\psi$, $\tilde{f} \in C_0(Q)$. Identifying $L^2(G) \otimes L^2(Q)$ with $L^2(G \times Q)$, this yields

$$u(y)\psi(x,q) = \psi(y^{-1}x, q); \tag{C.528}$$
$$\pi(\tilde{f})\psi(x,q) = \tilde{f}(x^{-1}q)\psi(x,q); \tag{C.529}$$
$$\rho_r(f)\psi(x,q) = \int_G dy \, f(y, xq)\psi(y^{-1}x, q). \tag{C.530}$$

C.19 Continuous bundles of C*-algebras

As shown on Chapter 7, continuous bundles of C*-algebras form a mathematical bridge between the classical and the quantum worlds, but they also form a beautiful structure in their own right. In what follows, I is an arbitrary locally compact Hausdorff space, but in the main text it is a subset of the unit interval $[0,1]$ that always contains 0 as an accumulation point, so one may have e.g. $I = [0,1]$ itself, or

$$I = (1/\mathbb{N}) \cup \{0\} \equiv 1/\dot{\mathbb{N}}, \tag{C.531}$$

where $\mathbb{N} = \{1,2,\ldots\}$). In physics, I plays the role of the value set for Planck's constant, but also below we generically write $\hbar \in I$, if only to avoid notational confusion with $x \in X$ (as $C_0(X)$ will be a typical fiber of the continuous bundles we study).

Definition C.121. *Let I be a locally compact Hausdorff space. A* **continuous bundle of C*-algebras** *over I consists of a C*-algebra A, a collection of C*-algebras $(A_\hbar)_{\hbar\in I}$, and surjective homomorphisms $\varphi_\hbar : A \to A_\hbar$ for each $\hbar \in I$, such that:*

1. *The function $\hbar \mapsto \|\varphi_\hbar(a)\|_\hbar$ is in $C_0(I)$ for each $a \in A$.*
2. *Writing $\|\cdot\|_\hbar$ for the norm in $A_\hbar$, the norm of any $a \in A$ is given by*

$$\|a\| = \sup_{\hbar\in I} \|\varphi_\hbar(a)\|_\hbar. \tag{C.532}$$

3. *For any $f \in C_0(I)$ and $a \in A$, there is an element $fa \in A$ such that for each $\hbar \in I$,*

$$\varphi_\hbar(fa) = f(h)\varphi_\hbar(a). \tag{C.533}$$

A **continuous (cross-) section** *of the bundle in question is a map $\hbar \mapsto a(\hbar) \in A_\hbar$, $\hbar \in I$, for which there is an $a \in A$ such that $a(\hbar) = \varphi_\hbar(a)$ for each $\hbar \in I$.*

Thus A may be identified with the space of continuous sections of the bundle: if we do so, the homomorphism $\varphi_\hbar$ is just the evaluation map at $\hbar$. The structure of A as a C*-algebra then corresponds to pointwise operations on sections. The idea is that the family $(A_\hbar)_{\hbar\in I}$ of C*-algebras is glued together by specifying a topology on the disjoint union $\sqcup_{\hbar\in I}A_\hbar$, seen as a fibre bundle over I. However, this topology is in fact given rather indirectly, namely via the specification of the space of continuous sections. This is reminiscent of Theorem C.23, which specifies the topology on a locally compact Hausdorff space X via the C*-algebra $C_0(X)$. More generally (the previous case being the trivial vector bundle $E = X \times \mathbb{C}$), the Serre–Swan Theorem about fiber bundles allows one to reconstruct the topology on a locally trivial vector bundle $E \xrightarrow{\pi} X$ from the (finitely generated projective) $C_0(X)$-module $C_0(X,E)$ of continuous sections of E. As in Definition C.121, one has maps $\varphi_x : C_0(X,E) \to E_x$ given by evaluation at x, so that (C.533) holds. However, *continuous bundles of C*-algebras need not be locally trivial*; for us, this is even the whole point!

Another way of looking at continuous bundles of C*-algebras starts from a *non-degenerate* homomorphism φ from $C_0(I)$ to the center $Z(M(A))$ of the multiplier algebra $M(A)$ of A (see §C.10); we simply write fa for $\varphi(f)a$, and similarly $C_0(I)A$.

In this notation, nondegeneracy means that $C_0(I)A$ is dense in A. Given such a nondegenerate homomorphism $\varphi : C_0(I) \to Z(M(A))$, one may define fiber algebras by

$$A_\hbar = A/(C_0(I;\hbar)\cdot A); \tag{C.534}$$
$$C_0(I;\hbar) = \{f \in C_0(I) \mid f(\hbar) = 0\}; \tag{C.535}$$

since $C_0(I;\hbar)\cdot A$ is an ideal in A, the quotient $A_\hbar$ is a C*-algebra. The projections $\varphi_\hbar : A \to A_\hbar$ are then given by the corresponding quotient maps sending $a \in A$ to its equivalence class in $A_\hbar$. In general, the function $\hbar \mapsto \|\varphi_\hbar(a)\|_\hbar$ is merely upper semicontinuous, so that one only obtains a structure equivalent to the one described in Definition C.121 if one explicitly requires the above function to be in $C_0(I)$, in which case clause 2 of Definition C.121 follows, too.

It is easy to find "trivial" examples of continuous bundles of C*-algebras: fix some C*-algebra B and take $A = C_0(I,B)$ with pointwise operations. In that case, $A_\hbar = B$ for each $\hbar \in I$, and the map $\varphi_\hbar : A \to B$ is given by $\varphi_\hbar(a) = a(\hbar)$.

It is not so easy to find nontrivial examples, even with isomorphic fibers (these were first given by Dixmier and Douady, who took the fiber algebras to be the compact operators $B_0(H)$). To connect classical to quantum, we need bundles over $I \subseteq [0,1]$ as described above, with non-isomorphic fibers, of which the fiber A_0 above $\hbar = 0$ is isomorphic to $C_0(X)$ for some (locally compact) phase space X, and hence is commutative, whereas all other fibers are noncommutative. One might say that it is the job of (deformation) quantization theory to construct such fields. Without proof, we now describe the main class of examples relevant to physics.

As we have seen, each Lie groupoid G canonically defines an associated C*-algebra $C_r^*(G)$, in which C_c^∞ functions on G endowed with a generalized convolution product (C.473) and involution (C.474) form a dense subspace. In particular,

$$C_r^*(TM) \cong C_0(T^*M); \tag{C.536}$$
$$C_r^*(M \times M) \cong B_0(L^2(M)), \tag{C.537}$$

where M is a manifold (without boundary) with tangent bundle TM and cotangent bundle T^*M. More generally, for any given Lie groupoid G one may define

$$A_0 = C_r^*(N_M G) \;\; (\hbar = 0); \tag{C.538}$$
$$A_\hbar = C_r^*(G) \;\; (\hbar > 0), \tag{C.539}$$

where $N_M G$ is the normal bundle to the embedding $M \hookrightarrow G$, cf. (C.444). Now consider the tangent groupoid G^T, which is a bundle over $[0,1]$ with fibers

$$G_0^T = N_M G \;\; (\hbar = 0); \tag{C.540}$$
$$G_\hbar^T = G \;\; (\hbar > 0), \tag{C.541}$$

The interplay between the differential geometry of the tangent groupoid and the notion of (reduced) Lie groupoid C*-algebras is described by the following lemma.

Lemma C.122. *The map $C_c^\infty(G^T) \to C_c^\infty(G_\hbar^T)$ that restricts f to $G_\hbar^T \subset G^T$ continuously extends to a surjective homomorphism $\varphi_\hbar : C_r^*(G^T) \to C_r^*(G_\hbar^T)$, $\hbar \in [0,1]$.*

Various special cases and this lemma ultimately led to the key result of the 1990s:

Theorem C.123. *For any Lie groupoid G, the fibers* (C.538) - (C.539) *merge into a continuous bundle of C*-algebras over $I = [0,1]$ with total algebra $A = C_r^*(G^T)$ and homomorphisms $\varphi_\hbar : A \to A_\hbar$ as described in Lemma C.122.*

The same result holds for the full groupoid C*-algebras $C^*(G^T)$ and $C^*(G_\hbar^T)$.

For the pair groupoid $G = \mathbb{R}^n \times \mathbb{R}^n$, as in the argument (C.469) - (C.472) we take some $\check{f} \in C_c^\infty(T\mathbb{R}^n)$, seen as a function $\tilde{f} \in C^\infty([0,1] \times T\mathbb{R}^n)$ that is independent of $\hbar$. This yields a function $\tilde{f} \circ \phi^{-1} \in C_c^\infty(G^T)$, and by construction,

$$\tilde{f} \circ \phi^{-1}(0,x,v) = \check{f}(x,v); \tag{C.542}$$

$$\tilde{f} \circ \phi^{-1}(\hbar,x,v) = \check{f}\left(\frac{x+y}{2}, \frac{y-x}{\hbar}\right) \quad (\hbar > 0). \tag{C.543}$$

By lemma C.122, the function $\varphi_0(\tilde{f} \circ \phi^{-1})$ is an element of

$$A_0 = C_r^*(T\mathbb{R}^n); \tag{C.544}$$

this element is just the function $\check{f}$. For $\hbar > 0$, we see $\varphi_\hbar(\tilde{f} \circ \phi^{-1})$ as an element of

$$A_\hbar \cong B_0(L^2(\mathbb{R}^n)), \tag{C.545}$$

through (C.490) - (C.491). Calling this element $Q_\hbar^W(\check{f})$, we have

$$Q_\hbar^W(\check{f})\psi(x) = \hbar^{-n} \int_{\mathbb{R}^n} d^n y\, f\left(\frac{x+y}{2}, \frac{y-x}{\hbar}\right) \psi(y). \tag{C.546}$$

We now use the isomorphism (C.536), implemented through the Fourier transform

$$f(x,p) = \int_{\mathbb{R}^n} d^n v \check{f}(x,v) e^{ipv}; \tag{C.547}$$

$$\check{f}(x,v) = \int_{\mathbb{R}^n} \frac{d^n p}{(2\pi)^n} f(x,p) e^{-ipv}. \tag{C.548}$$

Hence as an element of $C_0(T^*\mathbb{R}^n)$, the operator $\varphi_0(\tilde{f} \circ \phi^{-1})$ is f. From this perspective, using (C.548), eq. (C.546) may be rewritten in the more familiar form

$$Q_\hbar^W(f)\psi(x) = \int_{T^*\mathbb{R}^n} \frac{d^n p d^n y}{(2\pi\hbar)^n} e^{ip(x-y)/\hbar} \psi(y) f(\tfrac{1}{2}(x+y), p). \tag{C.549}$$

It follows that any $\check{f} \in C_c^\infty(T\mathbb{R}^n)$ defines a continuous cross-section of the continuous bundle of C*-algebra defined by $A = C^*((\mathbb{R}^n \times \mathbb{R}^n)^T)$, given by (C.547), and

$$0 \mapsto f \in C_0(T^*\mathbb{R}^n); \tag{C.550}$$

$$\hbar \mapsto Q_\hbar^W(f) \in B_0(L^2(\mathbb{R}^n)). \tag{C.551}$$

See also §7.1. These formulae were written down for the special case $M = \mathbb{R}^n$, but similar results (based on the exponential map as defined in Riemannian geometry) apply to any manifold. Moreover, as explained in §§7.2–7.4, Mackey's theory of quantization based on systems of imprimitivity and induced group representations falls squarely under the above umbrella, where G is an action groupoid.

We also employ continuous bundles of C*-algebras with non-isomorphic fibers even away from $\hbar = 0$. The construction of these fields relies on the following result, which is a special case of a more general claim; we just state the case we need, in which $I = 1/\dot{\mathbb{N}}$; continuity then imposes conditions at $\hbar = 0$ only (as I is discrete elsewhere). We identify the total space A of a (continuous) bundle of C*-algebras with the space of its (continuous) sections, as explained at the beginning of this section; thus $a \in A \subset \prod_\hbar A_\hbar$ takes the form $a = \{a_\hbar\}_{\hbar\in I}$, $a_\hbar \in A_\hbar$.

Proposition C.124. *Suppose one has a family $\{A_\hbar\}_{\hbar\in I}$ of C*-algebras over $I = 1/\dot{\mathbb{N}}$, as well as a subset $\tilde{A} \subset \prod_\hbar A_\hbar$ that satisfies the following conditions:*

1. *The set $\{\tilde{a}_\hbar \mid \tilde{a} \in \tilde{A}\}$ is dense in $A_\hbar$ for each $\hbar \in I$.*
2. *One has $\lim_{N\to\infty} \|\tilde{a}_{1/N}\| = \|\tilde{a}_0\|$ for each $\tilde{a} \in \tilde{A}$;*
3. *The set $\tilde{A}$ is a *-algebra (under pointwise operations).*

Let A consist of all $a \in \prod_\hbar A_\hbar$ for which one has

$$\lim_{N\to\infty} \|a_{1/N} - \tilde{a}_{1/N}\| = \|a_0 - \tilde{a}_0\| \quad (\tilde{a} \in \tilde{A}). \tag{C.552}$$

Regard A as a C-algebra under pointwise operations and norm* (C.532), *and define*

$$\varphi_\hbar(a) = a_\hbar. \tag{C.553}$$

Then $(A, \{A_\hbar, \varphi_\hbar\}_{\hbar\in I})$ is a continuous bundle of C-algebras (and is the unique such bundle whose space of sections contains $\tilde{A}$).*

The proof relies on the following lemma (which we state for general compact I).

Lemma C.125. *The total C*-algebra A of (sections of) a continuous bundle of C*-algebras is* **locally uniformly closed**. *That is, if $a \in \prod_\hbar A_\hbar$ is such that for every $\hbar_0 \in I$ and every $\varepsilon > 0$, there exists $b^{\hbar_0} \in A$ and a neighborhood $\mathcal{N}$ of $\hbar_0$ in which $\|a_\hbar - b^{\hbar_0}_\hbar\| < \varepsilon$ for all $\hbar \in \mathcal{N}$, then $a \in A$.*

Equivalently, if A (etc.) is a continuous bundle of C-algebras, and $a \in \prod_\hbar A_\hbar$ is such that the function $\hbar \mapsto \|a_\hbar - b_\hbar\|$ lies in $C(I)$ for each $b \in A$, then $a \in A$.*

Proof. Since I is compact, it has a finite cover $\{U_1, \ldots, U_n\}$ with associated partition of unity $\{u_i\}$. With a and ε as in the lemma, take $\hbar_i \in U_i$ and $b^{\hbar_i}$ also as in the lemma, and define $b = \sum_i u_i b^{\hbar_i}$. Then b satisfies $\sup_{\hbar\in I} \|a_\hbar - b_\hbar\| < \varepsilon$, and also $b \in A$, because of Definition C.121.3. Hence $a \in A$ by Definition C.121.2 and completeness of A.

As to the equivalent version, given $a \in \prod_\hbar A_\hbar$ and $\hbar_0 \in I$, because $\varphi_\hbar$ is surjective, there is a $b^{\hbar_0} \in A$ such that $a_{\hbar_0} = b^{\hbar_0}_{\hbar_0}$. The assumption in the second part then implies that the conditions in the first part are satisfied, such that $a \in A$. □

We are now in a position to prove Proposition C.124.

Proof. We first show that A as defined in the proposition is locally uniformly closed. With the notation of Lemma C.125 and its proof, take $\tilde{a} \in \tilde{A}$, and define the functions

$$f_{a\tilde{a}} : \hbar \mapsto \|a_\hbar - \tilde{a}_\hbar\|; \tag{C.554}$$
$$f_{b\tilde{a}} : \hbar \mapsto \|b_\hbar^{\hbar_0} - \tilde{a}_\hbar\|. \tag{C.555}$$

Since $|(\|X\| - \|Y\|)| \leq \|X - Y\|$, one obtains

$$|f_{a\tilde{a}}(\hbar) - f_{b\tilde{a}}(\hbar)| < \varepsilon, \tag{C.556}$$

for all $\hbar \in I$. By assumption, $f_{b\tilde{a}}$ is continuous, so that

$$|f_{b\tilde{a}}(\hbar) - f_{b\tilde{a}}(\hbar_0)| < \varepsilon, \tag{C.557}$$

for all $\hbar$ in some neighborhood U' of $\hbar_0$. Combining the two inequalities yields

$$|f_{a\tilde{a}}(\hbar) - f_{a\tilde{a}}(\hbar_0)| < 3\varepsilon, \tag{C.558}$$

for all $\hbar \in U'$. Hence $f_{a\tilde{a}}$ is continuous at any $\hbar_0 \in I$, so that $a \in A$ by Lemma C.125.

Using this property, it is easily shown that A is a C*-algebra, and that condition 3 in Definition C.121 is satisfied. It is clear from Definition C.121.1 and the definition of A in the proposition that A is maximal. On the other hand, according to the second part of Lemma C.125, A is minimal, so that it is unique. □

To close, let us explain to what extent we can say that a given section $(a_{1/N})_N$ of either one of our continuous bundles $A^{(c)}$ or $A^{(q)}$ "converges" to its value a_0.

Proposition C.126. *Let $(a_0, a_{1/N})$ and $(a'_0, a'_{1/N})$ be continuous cross-sections of some continuous bundle A of C*-algebras over $I = 1/\dot{\mathbb{N}}$, such that*

$$\lim_{N\to\infty} \|a'_{1/N} - a_{1/N}\| = 0. \tag{C.559}$$

Then $a'_0 = a_0$. In particular, if $(a_0, a_{1/N})$ is a continuous cross-section, then a_0 is uniquely determined by the $(a_{1/N})$ and we may symbolically write

$$a_0 = \lim_{N\to\infty} a_{1/N}. \tag{C.560}$$

Proof. The last part of Lemma C.125 states that the function defined by

$$0 \mapsto \|a_0 - a'_0\|;$$
$$1/N \mapsto \|a_{1/N} - a'_{1/N}\|,$$

is continuous on $1/\dot{\mathbb{N}}$ (i.e., continuous at 0). □

C.20 von Neumann algebras and the σ-weak topology

In this section and in §C.24 we turn to special classes of C*-algebras that are occasionally used in quantum (field) theory. Since the arguments tend to become very lengthy and technical, we will only prove some key results (e.g. von Neumann's Double Commutant Theorem), and mention other results without proof (references to which may be found in the Notes). This also applies to the next four sections.

The subject of operator algebras historically started with what we now call ***von Neumann algebras***, in honour of the founder of the subject (although, curiously, C*-algebras are not called "Gelfand–Naimark algebras"; perhaps they should!).

The first result in operator algebras was and is the ***Double Commutant Theorem***:

Theorem C.127. *Let M be a unital *-subalgebra of $B(H)$. Then the following conditions are equivalent—and, if satisfied, define M to be a* **von Neumann algebra***:*

(i) $M'' = M$;
(ii) M is closed in the weak operator topology;
(iii) M is closed in the strong operator topology.

Recall that the ***commutant*** S' of any $S \subset B(H)$ is defined by

$$S' = \{a \in B(H) \mid ab = ba \forall b \in S\}, \tag{C.561}$$

and that the ***bicommutant*** of S is $S'' = (S')'$. If $S^* = S$, in that $a \in S$ iff $a^* \in S$, then S' is easily seen to be a unital *-algebra within $B(H)$. Furthermore, it is obvious that $S \subseteq S''$, so that the passage $S \mapsto S''$ is some sort of a closure operation within $B(H)$, comparable to the closure operation $L \mapsto L^{\perp\perp}$ within H itself. Theorem C.127 shows that if S is a unital *-algebra, the algebraic closure operation $S \mapsto S''$ coincides with two topological closure operations. To this effect, recall also that:

- The ***weak operator topology*** on $B(H)$ may be defined by saying that $a_\lambda \to a$ (where (a_λ) is some net in M) iff $\langle \varphi, (a_\lambda - a)\psi \rangle \to 0$ for all $\varphi, \psi \in H$;
- The ***strong operator topology*** on $B(H)$ yields convergence $a_\lambda \to a$ of some net (a_λ) iff $\|(a_\lambda - a)\psi\| \to 0$ for each $\psi \in H$.

Proof. The essence of the proof is already contained in the finite-dimensional case $H = \mathbb{C}^n$, where the nontrivial claim in Theorem C.127 is:

If M is a unital *-subalgebra of $M_n(\mathbb{C})$, then $M'' = M$.

In fact, all we need to prove is $M'' \subseteq M$, since the converse inclusion is obvious. The idea is to take n arbitrary (and hence possibly linearly independent) vectors $\upsilon_1, \dots, \upsilon_n$ in H, and, given $a \in M''$, find some $b \in M$ such that $a\upsilon_i = b\upsilon_i$ for all $i = 1, \dots, n$. Hence $a = b$, so $a \in M$. To this end, we start with a single vector $\upsilon \in H$.

Form the linear subspace $M\upsilon = \{m\upsilon \mid m \in M\}$ of H, with associated projection e (i.e. $ew = w$ if $w \in M\upsilon$ and $ew = 0$ if $w \in (M\upsilon)^\perp$). Then $e \in M'$, and hence $a \in M''$ commutes with e. Since $1_H \in M$, we have $\upsilon \in M\upsilon$, so $\upsilon = e\upsilon$, and we compute $a\upsilon = ae\upsilon = ea\upsilon \in M\upsilon$. Hence $a\upsilon = b\upsilon$, for some $b \in M$.

Now run the same argument with the following substitutions:

- $H \rightsquigarrow H^n = H \oplus \cdots \oplus H$ (with n terms).
- $M \rightsquigarrow M_n = \{\text{diag}(m,\ldots,m) \mid m \in M\}$.
- $\upsilon \rightsquigarrow \mathsf{v} = \oplus_i \upsilon_i \equiv (\upsilon_1,\ldots,\upsilon_n)$.

We then have $(M_n)'' = (M'')_n$, so for any matrix $\mathsf{a} = \text{diag}(a,\ldots,a)$ in $(M'')_n$, the previous argument yields a matrix $\mathsf{b} = \text{diag}(b,\ldots,b) \in M_n$ such that $\mathsf{a}\upsilon = \mathsf{b}\upsilon$. But this is $a\upsilon_i = b\upsilon_i$ for all $i = 1,\ldots,n$, so that $a = b$ and hence $M'' \subseteq M$.

If H is infinite-dimensional, the above proof may be adapted by taking the closure of $M\upsilon$ in H, which gives (3) $\Rightarrow$ (1). Finally, (1) $\Rightarrow$ (2) $\Rightarrow$ (3) is trivial. □

Corollary C.128. *Let M be a unital *-subalgebra of $B(H)$. Then the closures of M in the strong and weak topologies coincide with each other and with M''.*

Corollary C.129. *A von Neumann algebra is norm-closed, i.e., is a C*-algebra.*

Since $S''' = S'$, the commutant of any self-adjoint set $S^* = S \subset B(H)$ is a von Neumann algebra. As a case in point, take a (strongly continuous) unitary group representation $u : G \to B(H)$. Then $u(x)^* = u(x^{-1})$, so $u(G)'$ is a von Neumann algebra. In fact, any von Neumann algebra M takes this form, since one may take G to be the group of all unitaries in M (and u its defining representation). Furthermore, the bicommutant A'' of any C*-algebra $A \subset B(H)$ is a von Neumann algebra. An important example of this construction is the abelian von Neumann algebra $W^*(a) = C^*(a)''$ generated by a self-adjoint operator $a = a^* \in B(H)$, cf. (B.320).

Although the weak and strong topologies on M appear in the fundamental double commutant theorem, the most important topology on a von Neumann algebra (besides the norm topology) is the so-called the ***σ-weak topology*** (sometimes called the ***ultraweak topology***). This topology corresponds to the following convergence:

- One has $a_\lambda \to a$ σ-weakly iff $\text{Tr}\,(b(a_\lambda - a)) \to 0$ for each $b \in B_1(H)$.

To begin with, as far as Theorem C.127 is concerned this topology is at least on a par with the weak and the strong ones:

Theorem C.130. *Let M be a unital *-subalgebra of $B(H)$. Then $M'' = M$ (i.e. M is a von Neumann algebra) iff M is closed in the σ-weak operator topology.*

This one is a bit more technical, so we just sketch the proof.

Proof. Define a new Hilbert space $H^\infty = H \otimes \ell^2$, whose elements v are infinite sequences of vectors $(\upsilon_1, \upsilon_2, \ldots)$ in H with $\sum_i \|\upsilon_i\|^2 < \infty$. The inner product is

$$\langle \mathsf{v}, \mathsf{v}' \rangle_{H^\infty} = \sum_i \langle \upsilon_i, \upsilon_i' \rangle_H. \tag{C.562}$$

The obvious (diagonal) embedding of $B(H)$ in $B(H^\infty)$, whose image is denoted by $B(H)_\infty$, restricts to $M \subset B(H)$, with image $M_\infty \subset B(H^\infty)$. Then the σ-weak topology on $B(H)$ is the relative weak topology on $B(H)_\infty$ (i.e., the weak topology on $B(H^\infty)$ restricted to $B(H)_\infty$), so that Theorem C.130 follows from Theorem C.127. □

This brings us to an important refinement of Theorem C.127, called ***Kaplansky's Density Theorem*** (which should actually be seen as a lemma for numerous results):

Theorem C.131. *Let $A \subset B(H)$ be a C*-algebra (or a *-algebra). Then the unit ball of A is dense in the unit ball of A'' in the weak, strong, and σ-weak topologies.*

The real significance of the σ-weak topology comes from ***Sakai's Theorem***:

Theorem C.132. *A C*-algebra $M \subset B(H)$ is a von Neumann algebra iff M is the (Banach) dual of a unique Banach space M_* (called the* **predual** *of M).*

We turn to the proof below. For example, by Theorem B.146, the predual of $B(H)$ is

$$B(H)_* \cong B_1(H). \tag{C.563}$$

In the commutative case, entry 10 in Table B.1 in §B.9 gives

$$L^\infty(X,\mu)_* \cong L^1(X,\mu); \tag{C.564}$$

the fact that $L^\infty(X,\mu)$, acting on $H = L^2(X,\mu)$ as multiplication operators, is a von Neumann algebra was established in §B.16. In the first example, the σ-weak topology on $B(H)$ obviously coincides with the weak*-topology defined by $B(H)_*$.

In general, there is a canonical embedding $M_* \hookrightarrow M^*$, $\check{\varphi} \mapsto \varphi$, with $\varphi(a) = a(\check{\varphi})$, cf. §B.9. Proposition B.46 then shows that the image of M_* in M^* consists precisely of the weak*-continuous functionals on M (recall that the weak*-topology on M is the topology of pointwise convergence, seeing M as the dual of M_*). If we now identify $\check{\varphi}$ with φ, we have the following generalization of the observation just made:

Theorem C.133. *Let $M \subset B(H)$ be a von Neumann algebra. The predual M_* of M (seen as a subspace of M^*) coincides with the space of σ-weakly continuous functionals on M, and hence the σ-weak topology on M coincides with the weak*-topology in its role as the dual Banach space of M_*.*

σ-weakly continuous functionals on a von Neumann algebra M are called ***normal***.

Proof. Identifying $\check{\varphi}$ with φ, we introduce the following spaces:

$$M^\perp = \{\varphi \in B(H)_* \mid \varphi(a) = 0 \, \forall a \in M\};$$
$$M^{\perp\perp} = \{a \in B(H) \mid \varphi(a) = 0 \, \forall \varphi \in M^\perp\}.$$

Having proved the theorem for $M = B(H)$, i.e., (C.563), the key is to show that

$$M^{\perp\perp} = M; \tag{C.565}$$
$$M_* \cong B(H)_*/M^\perp, \tag{C.566}$$

where (C.566) denotes an isometric isomorphism of normed spaces. Since the right-hand side of (C.566) is a Banach space, so is the left-hand side. This yields the first claim. Combining (C.566) with (C.565) and the duality $B(H) = B_1(H)^*$, we have

$$M_*^* \cong (B(H)_*/M^\perp)^* = M^{\perp\perp} = M.$$

This is the second claim. The first equality sign is true, because if Y is a closed subspace of a Banach space Y, then $(X/Y)^* = \{\varphi \in X^* \mid \varphi \restriction Y = 0\}$.

For the remainder of the theorem, recall that $a_\lambda \to a$ σ-weakly in M whenever $\varphi(a_\lambda - a) \to 0$ for all $\varphi \in B(H)_*$. By (C.566), this is equivalent to $a_\lambda \to a$ in the weak*-topology, since a possible component of φ in $M^\perp$ drops out.

We next prove (C.565). The inclusion $M \subset M^{\perp\perp}$ is trivial. For the converse, pick $a \notin M$; since M is a von Neumann algebra, it is σ-weakly closed, so its complement M^c in $B(H)$ is σ-weakly open. Hence there are $\varphi \in B(H)_*$ and $\varepsilon > 0$ such that the open neighbourhood $\mathscr{O}(a) = \{b \in B(H) : |\varphi(a) - \varphi(b)| < \varepsilon\}$ of a entirely lies in M^c. So $|\varphi(a) - \varphi(b)| \geq \varepsilon$ for all $b \in M$. This implies $\varphi(b) = 0$ by linearity in b. Hence $|\varphi(a)| \geq \varepsilon$, so $a \notin M^{\perp\perp}$, hence $M^{\perp\perp} \subset M$.

For (C.566), first note that $M^\perp$ is a norm-closed subspace of $B(H)_* = B_1(H)$, which is a Banach space in the trace-norm (which coincides with the norm inherited from $B(H)^*$, since the injection $B_1(H) \hookrightarrow B(H)^*$ is an isometry). Hence the quotient $B(H)_*/M^\perp$ is a Banach space in the canonical norm $\|\dot{\varphi}\| = \inf\{\|\varphi + \psi\| \mid \psi \in M^\perp\}$, where $\dot{\varphi}$ is the image of $\varphi \in B(H)_*$ under the canonical projection, and the norm is the one in $B(H)^*$. Let $\varphi^\upharpoonright = \varphi \upharpoonright M$ be the restriction of $\varphi \in B(H)_*$ to M. It is clear that the map $\varphi^\upharpoonright \mapsto \dot{\varphi}$ is well defined and is a linear bijection from M_* to $B(H)_*/M^\perp$. In fact, this map is isometric. First, one trivially has

$$\|\varphi^\upharpoonright\| = \sup\{|\varphi(a)| \mid a \in M_u\} = \inf_{\psi \in M^\perp} \sup\{|\varphi(a) + \psi(a)| \mid a \in M_u\}, \quad \text{(C.567)}$$

since $\psi(a) = 0$. But this is clearly majorized by

$$\|\dot{\varphi}\| = \inf_{\psi \in M^\perp} \sup\{|\varphi(a) + \psi(a)|, a \in B(H)_1\}, \quad \text{(C.568)}$$

since now the supremum is taken over a larger set. Hence $\|\varphi^\upharpoonright\| \leq \|\dot{\varphi}\|$.

Conversely, for any $\varphi \in B(H)_*$ with $\|\dot{\varphi}\| = 1$, by Corollary B.41 there exists an $a \in B(H)$ with $\check{a} \in M^{\perp\perp}$, $\varphi(a) = 1$ and $\|a\| = 1$. From (C.565), one then has $\|\varphi^\upharpoonright\| \geq |\varphi(a)| = 1 = \|\dot{\varphi}\|$. This finishes the proof of Theorem C.133. □

Half of Theorem C.132 evidently follows from Theorem C.133. The converse ('if') implication uses a refinement of the GNS-construction, where the state ω is assumed to be σ-weakly continuous. In that case, using the theory of σ-weakly closed ideals of von Neumann algebras, it can be shown that $\pi_\omega(M)$ coincides with $\pi_\omega(M)''$ and hence is a von Neumann algebra. Since normal pure state on a von Neumann algebra may not exist (for example, take $M = L^\infty(0,1)$), the 'crazy' Hilbert space H_c in the proof of Theorem C.87 must be replaced by the perhaps even crazier direct sum $H_{ec} = \bigoplus_{\omega \in S_n(M)} H_\omega$, where this time the sum is over all *normal* states on M. Similarly, in Lemma C.15 one should now have a normal state instead of a pure state. Otherwise, the proof that M has a faithful representation as a von Neumann algebra on a Hilbert space essentially follows the proof of Theorem C.87.

Finally, uniqueness of the predual follows from Corollary C.139 below. □

Corollary C.134. *Let $M \subset B(H)$ be a von Neumann algebra. Each normal functional $\varphi \in M_*$ on M is of the form $\varphi(a) = \mathrm{Tr}\,(ba)$, for some $b \in B_1(H)$. In particular, φ is a normal state iff b is a density operator.*

C.21 Projections in von Neumann algebras

General C*-algebras need not have any nontrivial projections; think of $C_0([0,1])$. On the other hand, von Neumann algebras are generated by their projections:

Theorem C.135. *Let $\mathscr{P}(M) = \{p \in M \mid e^2 = e^* = e\}$, where M is a von Neumann algebra. Then M is the norm-closure of the linear span of $\mathscr{P}(M)$, and $M = \mathscr{P}(M)''$.*

This is Corollary B.105. In addition, $\mathscr{P}(M)$ is not just a set.

Proposition C.136. *The set $\mathscr{P}(M)$ of projections in a von Neumann algebra M is a complete lattice under the partial ordering $e \leq f$ iff $ef = fe = e$.*

Proof. Since $e \leq f$ in $M \subset B(H)$ iff $eH \subseteq fH$, the supremum $e \vee f$ is the projection on $\overline{eH + fH}$, whilst the infimum $e \wedge f$ is the projection on $eH \cap fH$. For arbitrary families $(e_\lambda)_{\lambda \in \Lambda}$ of projections, $\vee_\lambda e_\lambda$ equals the projection on the closure of the linear span of all subspaces $H_\lambda \equiv e_\lambda H$, whereas $\wedge_\lambda e_\lambda \equiv e$ is the projection on their intersection. To show that the latter lies in M (provided all the e_λ do, of course), note that each unitary $u \in M'$ satisfies $uH_\lambda = H_\lambda$ for all λ, so that also $u(\cap_\lambda H_\lambda) = \cap_\lambda H_\lambda$. Hence $eu = ue$ and so $e \in M'' = M$ (since each element of a von Neumann algebra is a linear combination of at most four unitaries in it; the proof is similar to Lemma B.145). Finally, by de Morgan's Law we have $\vee_\lambda e_\lambda = (\wedge_\lambda e_\lambda^\perp)^\perp$, with $f^\perp = 1 - f$ for any $f \in \mathscr{P}(M)$. Hence also $\vee_\lambda e_\lambda \in M$. □

This is nice in itself, but is also implies a very important result about maps between von Neumann algebras. Recall that a (purely algebraic) isomorphism between C*-algebras (seen as *-algebras) is automatically isometric and hence norm-continuous; see Theorem C.62. An even better result holds for von Neumann algebras:

Theorem C.137. *A (purely algebraic) isomorphism $\varphi : M \to N$ between von Neumann algebras (seen as *-algebras) is an isomorphism of Banach spaces as well as a homeomorphism with respect the σ-weak topologies on M and N.*

This theorem only seems to have rather difficult proofs. One, based on Proposition C.136, is based on the following result. First, we say that a map $\varphi : M \to N$ of von Neumann algebras is ***completely additive*** if for any family (e_λ) in $\mathscr{P}(M)$,

$$\varphi(\vee_\lambda e_\lambda) = \vee_\lambda \varphi(e_\lambda). \tag{C.569}$$

Lemma C.138. *Let $\varphi : M \to N$ be a homomorphism of von Neumann algebras.*

1. *φ is σ-weakly continuous iff it is completely additive.*
2. *If φ is a (purely algebraic) isomorphism, then it is completely additive.*

The proof of claim 2 is easy, as is the implication from σ-weak continuity to completely additivity in claim 1. The converse implication, however, is quite difficult. In any case, Theorem C.137 now follows, so that we may speak of isomorphisms between von Neumann algebras without any ambiguity.

Corollary C.139. *If two von Neumann algebras are algebraically isomorphic, then their preduals M_* and N_* are isomorphic as Banach spaces. In particular (take $M = N$), the predual of a von Neumann algebra is unique (up to isometric isomorphism).*

A second proof of Theorem C.137 uses Theorem C.132 (and hence provides no non-circular proof of Corollary C.139), as follows.

Proof. Since φ is isometric by Corollary C.129 and Theorem C.62, it induces a dual isomorphism (of Banach spaces) $\varphi^* : N^* \to M^*$, with the property that $M \cong (\varphi^*(N_*))^*$ under the map

$$a \mapsto (\varphi^*(\omega) \mapsto \omega(\varphi(a))) \ (a \in M, \omega \in N_*). \tag{C.570}$$

Uniqueness of the predual then yields $\varphi^*(N_*) \cong M_*$, which in turn implies that φ preserves pointwise convergent nets: if $\omega'(a_\lambda) \to \omega'(a)$ for all $\omega' \in M_*$, then $\omega(\varphi(a_\lambda)) \to \omega(\varphi(a))$ for all $\omega \in N_*$. Hence φ is σ-weakly continuous. □

Theorem C.137 shows that the notion of isomorphism to be used in the classification of von Neumann algebras M is unambiguous. There are two totally different cases of von Neumann algebras (only $M = \mathbb{C}$ falls in both classes):

- ***Abelian*** von Neumann algebras, which equal their center ($M \cap M' = M$);
- ***Factors***, which have trivial center ($M \cap M' = \mathbb{C} \cdot 1$).

A factor has no nontrivial decomposition $M = M_1 \oplus M_2$, whereas an abelian von Neumann algebra (except $M = \mathbb{C}$) does have such a decomposition (typically even many of them). Using von Neumann's technique of *direct integrals*, which generalizes direct sums (and will not be reviewed here), the classification of general von Neumann algebras may be reduced to these two cases. We start with the first class.

We know that if (X, Σ, μ) is some σ-finite Borel space with associated Hilbert space $L^2(X, \mu)$, then the commutative C*-algebra $L^\infty(X, \mu)$ is mapped isometrically into $B(L^2(X, \mu))$ via $f \mapsto m_f$, see Proposition B.73 and especially (B.240). If we denote the image of this map by $L^\infty(X, \mu)$ also, then $L^\infty(X, \mu)'' = L^\infty(X, \mu)$ by (B.346), so $L^\infty(X, \mu) \subset B(L^2(X, \mu))$ is an abelian von Neumann algebra. In general:

Theorem C.140. *Let $M \subset B(H)$ be an abelian von Neumann algebra, Then*

$$M \cong L^\infty(X, \mu), \tag{C.571}$$

for some locally compact space X and probability measure μ on X.

If H is separable, this follows from Theorems B.116 (including the remarks after its proof) and B.117 in §B.16. The proof for arbitrary Hilbert space is quite technical and will be omitted, but the idea is to find an abelian C*-algebra A for which $M = A''$, upon which $X = \Sigma(A)$, and the measure μ is constructed such that $\mu(\Delta) = 0$ iff $\mu_\psi(\Delta) = 0$ for all unit vectors $\psi \in H$, with μ_ψ defined similarly to (B.304). In general, one cannot take $A = M$, since $\Sigma(M)$ may not support such measures. Thus we have a complete and satisfactory characterization of abelian von Neumann algebras, including their projections: these are simply the (equivalence classes of) characteristic functions 1_A, where $A \in \Sigma$ is a Borel set in X (modulo null sets).

The advantage of this approach is that there are often simple models for X; we know from the classification of maximal abelian von Neumann algebras on separable Hilbert space in §B.17 that $X = [0,1]$ with (Lebesgue measure) and $X = \mathbb{N}$ (with counting measure) are enough in that case. However, the pair (X,μ) lacks intrinsic uniqueness properties. Thus it also makes sense to apply Theorem C.8 to abelian von Neumann algebras, so that $M \cong C(X)$. Since by Theorem C.135, M has plenty of projections, which as elements of $C(X)$ are realized by characteristic functions 1_A, where $A \subset X$, the space X must have lots of ***clopen*** (i.e. closed and open) sets.

It can be shown that X arises as the Gelfand spectrum of some abelian von Neumann algebra iff it is *hyperstonean*, where we say that a compact Hausdorff X is:

- ***Stone*** if the only connected subsets are points (equivalently, a Stone space is compact, T_0, and has a basis of clopen sets).
- ***stonean*** if it is Stone and the closure of each open set is open.
- ***hyperstonean*** if it is stonean, and for any nonzero $f \in C(X,\mathbb{R}^+)$ there exists a completely additive positive measure μ such that $\mu(f) > 0$.

This replaces the classification of abelian von Neumann algebras up to isomorphism by the classification of hyperstonean spaces up to homeomorphism, which is hardly an improvement (the only other area of mathematics where such wacky spaces appear is algebraic logic). However, we do obtain a nice relationship between the projection lattice of an abelian von Neumann algebra and its Gelfand spectrum (at this point please recall Theorem D.5 and surrounding text in Appendix D).

Theorem C.141. *The projection lattice $\mathscr{P}(M)$ of a von Neumann algebra M is Boolean iff M is abelian, in which case there is a homeomorphism*

$$\Sigma(M) \cong \mathscr{S}(\mathscr{P}(M)) \tag{C.572}$$

between the Gelfand spectrum of M (as a commutative C-algebra) and the Stone spectrum of $\mathscr{P}(M)$ (as a Boolean lattice). Hence we have isomorphisms*

$$M \cong C(\mathscr{S}(\mathscr{P}(M))); \tag{C.573}$$

$$\mathscr{O}(\Sigma(M)) \cong \mathrm{Idl}(\mathscr{P}(M)), \tag{C.574}$$

as (commutative) C-algebras and as frames, respectively.*

Proof. In the commutative case, the lattice operations in $\mathscr{P}(M)$ are given by

$$e \wedge f = ef; \tag{C.575}$$

$$e \vee f = e + f - ef; \tag{C.576}$$

$$e^{\perp} = 1_M - e, \tag{C.577}$$

as may be verified by embedding $M \subset B(H)$ and using the proof of Proposition C.136; eq. (C.577) is true for any M. One then finds that M is distributive, since

$$e \wedge (f \vee g) = ef + eg - efg = (e \wedge f) \vee (e \wedge g), \tag{C.578}$$

and similarly with $\vee$ and $\wedge$ swapped. Since $\mathscr{P}(M)$ is orthomodular for arbitrary von Neumann algebras M and is distributive if M is abelian, it follows that $\mathscr{P}(M)$ is Boolean. Conversely, if $\mathscr{P}(M)$ is Boolean, we may compute

$$(e\wedge(e\wedge f)^{\perp})^{\perp}=(e\wedge(e^{\perp}\vee f^{\perp}))^{\perp}=((e\wedge e^{\perp})\vee(e\wedge f^{\perp}))^{\perp}=(e\wedge f^{\perp})^{\perp}=e^{\perp}\vee f,$$

and since $f\leq g\vee f$ for any g, this implies $f\leq(e\wedge(e\wedge f)^{\perp})^{\perp}$. Now $f\leq g^{\perp}$ implies $fg=gf=0$, so

$$f(e\wedge(e\wedge f)^{\perp})=(e\wedge(e\wedge f)^{\perp})f=0. \tag{C.579}$$

If $g\leq e$, then $e\wedge g=g$, hence $e\wedge g^{\perp}+g=e\wedge(1_M-g)+g=e$. So $g=e\wedge f$ gives

$$e-(e\wedge f)=e\wedge(e\wedge f)^{\perp}. \tag{C.580}$$

Using (C.579) - (C.580) finally yields

$$ef=((e\wedge f)+e-(e\wedge f))f=(e\wedge f)f+(e\wedge(e\wedge f)^{\perp})f=e\wedge f, \tag{C.581}$$

and since $e\wedge f=f\wedge e$, we find $ef=fe$ for any two projections $e,f\in\mathscr{P}(M)$. Hence M is abelian by Theorem C.135.

If we now realize the Gelfand spectrum $\Sigma(M)$ as the multiplicative state space of M, and realize the Stone spectrum $\mathscr{S}(\mathscr{P}(M))$ as the space $\mathrm{Pt}(\mathscr{P}(M))$ of points of $\mathscr{P}(M)$, then a homeomorphism $\Sigma(M)\cong\mathrm{Pt}(\mathscr{P}(M))$ arises as follows:

- First, the restriction $\varphi:\mathscr{P}(M)\to\mathbb{C}$ of any multiplicative state $\varphi:M\to\mathbb{C}$ must be $\{0,1\}$-valued. Using (C.575) - (C.577), it is then easy to show that the ensuing map $\varphi:\mathscr{P}(M)\to\{0,1\}$ is a homomorphism of Boolean lattices.
- *Vice versa*, by Corollary B.104 a point $\varphi:\mathscr{P}(M)\to\{0,1\}$ extends by continuity to a map $\varphi:M\to\mathbb{C}$. Since φ must preserve $\perp$, this map is nonzero. By continuity, multiplicativity in general follows from multiplicativity on projections, which follows by running the previous point backward (or from Theorem C.168).

Finally, (C.573) and (C.574) follow from (C.572) and the Gelfand isomorphism (Theorem C.8) and eq. (D.35), respectively. See also Theorem C.168 below. □

Note that (C.574) is a special case of Corollary C.84, for if M is a commutative von Neumann algebra, and $H(M)$ its frame of heriditary subalgebras, we have

$$H(M)\cong\mathrm{Idl}(\mathscr{P}(M)); \tag{C.582}$$
$$J\mapsto\{e\in\mathscr{P}(M)\mid Me\subseteq J\}, \tag{C.583}$$

whose inverse maps an ideal $I\subset\mathscr{P}(M)$ to the norm-closure of $\bigcup_{e\in I}Me$ in M. In particular, if J is σ-weakly closed, then $J=Me$ for a unique projection $e\in\mathscr{P}(M)$, in which case the right-hand side of (C.583) is just the principal ideal $\downarrow e$. To see this special case, we quote a useful result about arbitrary von Neumann algebras:

Proposition C.142. *Let I be a σ-weakly closed left (right) ideal in a von Neumann algebra M. Then there is a unique projection $e\in\mathscr{P}(M)$ such that $I=Me$ ($I=eM$).*

Indeed, e is the σ-weak limit of any approximate identity in I.

C.22 The Murray–von Neumann classification of factors

After this analysis of abelian von Neumann algebras, we now turn to their opposites, viz. factors. The main tool in the classification of factors, introduced by Murray and von Neumann, is a new partial ordering $\lesssim$ on the projection lattice $\mathscr{P}(M)$, which is defined for general von Neumann algebras M. Unlike the familiar partial ordering $\leq$ (see Proposition C.136), $\lesssim$ gives a total ordering on $\mathscr{P}(M)$ if M is a factor.

Definition C.143. *Let $\mathscr{P}(M)$ be the projection lattice of a von Neumann algebra M. We say that $e \sim f$ in $\mathscr{P}(M)$ iff there exists $u \in M$ such that $u^*u = e$ and $uu^* = f$. Subsequently, we write $e \lesssim f$ if there is $e' \in \mathscr{P}(M)$ with $e \sim e'$ and $e' \leq f$.*

It is easy to show that $\sim$ is an equivalence relation. The operator u in this definition is unitary from eH to fH, vanishes on $(eH)^{\perp}$, and has range fH. Such an operator is therefore a ***partial isometry*** (cf. Definition A.27), with initial projection e and final projection f. It follows that a *necessary* condition for $e \sim f$ is that $\dim(eH) = \dim(fH)$, but (unless $M = B(H)$) this is by no means *sufficient*, since the unitary u that maps eH to fH *is required to lie in M*. For example, if $H = \mathbb{C} \oplus \mathbb{C}$, then $e = \text{diag}(1,0)$ is equivalent to $f = \text{diag}(0,1)$ with respect to $M = M_2(\mathbb{C})$, but not with respect to $M = D_2(\mathbb{C}) = \mathbb{C} \oplus \mathbb{C}$ (i.e., the diagonal 2×2 matrices).

To see how natural this definition is, consider a unitary representation u of a group G on H. If $H_i \subset H$ is stable under $u(G)$, $i = 1,2$, then the restrictions u_i of u to H_i are unitarily equivalent precisely when $e_1 \sim e_2$ with respect to $M = u(G)'$ (where e_i is the projection onto H_i). Furthermore, u_1 is unitarily equivalent to a subrepresentation of u_2 iff $e_1 \lesssim e_2$. More generally, if $N \subset B(H)$ is a von Neumann algebra, with stable subspaces H_i, $i = 1,2$, then the restrictions N_i to H_i are unitarily equivalent iff $e_1 \sim e_2$ with respect to $M = N'$, et cetera.

One may compare projections in M with sets and compare $\leq$, $\sim$, and $\lesssim$ with $\subseteq$ (inclusion), $\cong$ (isomorphism), and $\hookrightarrow$ (the existence of an injective map), respectively. The Schröder–Bernstein Theorem of set theory (which von Neumann knew well) states that if $X \hookrightarrow Y$ and $Y \hookrightarrow X$, then $X \cong Y$. Similarly, it can be shown that:

Proposition C.144. *If $e \lesssim f$ and $f \lesssim e$, then $e \sim f$.*

The special role of factors with respect to the partial ordering $\lesssim$ now emerges.

Proposition C.145. *If M is a factor, then $\lesssim$ is a total ordering (i.e., $e \lesssim f$ or $f \lesssim e$).*

The property of a factor that leads to this result is:

Lemma C.146. *Let M be a factor. For any nonzero projections $e, f \in \mathscr{P}(M)$, there are nonzero projections $e', f' \in \mathscr{P}(M)$ such that $e' \leq e$, $f' \leq f$, and $e' \sim f'$.*

The first step in the Murray—von Neumann classification of factors is as follows:

Definition C.147. *A projection e in M is called* **finite** *if $f \sim e$ and $f \leq e$ for some $f \in \mathscr{P}(M)$ implies $f = e$, and* **minimal** *if $f \leq e$, $f \in \mathscr{P}(M)$, implies $f = e$ or $f = 0$.*

Accordingly, a factor M is called **finite** *iff 1_M is finite,* **semifinite** *iff 1_M majorizes a finite projection, and* **purely infinite** *iff all nonzero projections are infinite.*

For $M = B(H)$, which is evidently a factor, a projection e is (in)finite iff $\dim(eH)$ is (in)finite, so that $B(H)$ is finite iff H is finite-dimensional, and semifinite otherwise. Surprisingly, we will see that finite factors different from $M_n(\mathbb{C})$ exist, as do semifinite factors different from $B(\ell^2)$. Even purely infinite factors (initailly defined as what was left out from the previous two cases) turn out to exist (even in physics).

We first rephrase Definition C.147 in terms of generalized traces.

Definition C.148. *A* **trace** *on a von Neumann algebra M is a map*

$$\mathrm{tr} : M_+ \to [0,\infty] \tag{C.584}$$

satisfying

$$\mathrm{tr}(\lambda \cdot a + b) = \lambda \cdot \mathrm{tr}(a) + \mathrm{tr}(b) \;\; (a,b \in M_+, \lambda \geq 0; \tag{C.585}$$
$$\mathrm{tr}(aa^*) = \mathrm{tr}(a^*a) \;\; (a \in M). \tag{C.586}$$

Equivalently, $\mathrm{tr}(uau^*) = \mathrm{tr}(a)$ *for all $a \in M_+$ and unitary $u \in M$ (so that $uau^* \in M^+$).*

A trace is **finite** *if* $\mathrm{tr}(a) < \infty$ *for all $a \in M_+$,* **semifinite** *if for any $a \in M_+$ there is a nonzero $b \leq a$ in M_+ for which* $\mathrm{tr}(b) < \infty$, *and* **infinite** *otherwise.*

The usual trace Tr is a trace tr on $B(H)$ in this new sense, which is finite iff $\dim(H)$ is finite. As we will see, other factors admit other traces. The following result could have been used as a definition of (semi)finite and purely infinite factors.

Proposition C.149. *A factor is (semi)finite iff it admits a faithful σ-weakly continuous (semi)finite trace, and is purely infinite otherwise.*

It can be shown that a finite trace on a factor is automatically σ-weakly continuous, so a factor is finite iff it admits a faithful finite trace. Hence we recover the fact that $B(H)$ is finite iff $\dim(H) < \infty$, and semifinite otherwise. For a completely different kind of trace, defined on factors remote from $B(H)$, we turn to discrete groups G. For these, Haar measure is simple the counting measure, so that $L^2(G) = \ell^2(G)$, and convolution (C.504) and involution (C.505), initially defined on $C_c(G)$, are given by

$$f * g(x) = \sum_{y \in G} f(xy^{-1})g(y); \quad f^*(x) = \overline{f(x^{-1})}. \tag{C.587}$$

According to Definition C.119, the reduced group C*-algebra $C_r^*(G)$ is the norm-closure of the *-algebra in $B(L^2(G))$ containing all operators

$$u_L^{\int}(f)\psi(x) = \sum_{y \in G} f(y)\psi(y^{-1}x) \;\; (f \in C_c(G)). \tag{C.588}$$

Thus $C_r^*(G)$ is realized as a concrete C*-algebra of operators on $B(\ell^2(G))$, so that, following von Neumann himself, we may form the ***group!von Neumann algebra***

$$W^*(G) = C_r^*(G)''. \tag{C.589}$$

Theorem C.150. *The group von Neumann algebra $W^*(G)$ of a countable group is a factor iff all nontrivial conjugacy classes in G (i.e., all except $\{e\}$) are infinite.*

In that case, we say that G has (or "is") ***icc***, i.e., has ***infinite conjugacy classes***.

Proof. From (C.587), for $f \in C_c(G)$ we have $f * g = g * f$ for each $g \in C_c(G)$ iff $f(yxy^{-1}) = f(x)$ for all $x, y \in G$. In other words, f lies in the center of $C_c(G) \subset W^*(G)$ iff f is constant on each conjugacy class of G. If G is icc, this implies that f can have support only at e, i.e., $f = \lambda \cdot \delta_e$, $\lambda \in \mathbb{C}$. Noting that δ_e is the unit in the algebra $C_c(G)$, this proves the claim, except for the fact that we should extend this argument from $C_c(G)$ to $W^*(G)$, which by Theorem C.127 is its strong closure.

The key to this extension is the fact that one has $f * g(x) = \langle R_{x^{-1}} f^*, g \rangle$ for f and g in $C_c(G)$, where $R_x f(y) = f(yx)$ and the inner product is in $\ell^2(G)$. Hence

$$|f * g(x)| = |\langle R_{x^{-1}} f^*, g \rangle| \leq \|R_{x^{-1}} f\|_2 \|g\|_2 = \|f\|_2 \|g\|_2, \tag{C.590}$$

so that the sum in (C.587) is actually defined and converges (absolutely) for $f, g \in \ell^2(G)$. This also shows that if f_n strongly converges to some $a \in B(\ell^2(G))$, i.e., $\|f_n * \psi - a\psi\| \to 0$ for each $\psi \in \ell^2(G)$, then $a\psi = f * \psi$, where $f \in \ell^2(G)$ is the limit of (f_n) seen as a sequence in $\ell^2(G)$. Hence $W^*(G) \subset \ell^2(G)$, and the above computation of the center of $W^*(G)$ remains valid: we have $f \in W^*(G) \cap W^*(G)'$ iff f is constant on each conjugacy class of G. Conversely, any f that is constant on some finite conjugacy class (different from $\{e\}$) and zero elsewhere is central without being a multiple of the unit. □

Whether or not G has icc, we have a map $\mathrm{tr} : W^*(G) \to \mathbb{C}$, defined by

$$\mathrm{tr}(f) = f(e), \tag{C.591}$$

which satisfies (C.585) - (C.586) and hence defines a finite trace on $W^*(G)$. Also,

$$\mathrm{tr}(f) = \langle \delta_e, f * \delta_e \rangle, \tag{C.592}$$

so this trace is σ-weakly continuous.

Corollary C.151. *If G has icc, $W^*(G)$ is a finite factor non-isomorphic to any $B(H)$.*

Since G must obviously be infinite for it to have icc, $W^*(G)$ is infinite-dimensional, and hence $W^*(G) \not\cong M_n(\mathbb{C})$ for any $n \in \mathbb{N}$. Furthermore, if H is infinite-dimensional, then $B(H)$ does not admit any σ-weakly continuous finite faithful trace:

Proposition C.152. *Any two nonzero σ-weakly continuous (semi)finite traces* $\mathrm{tr}, \mathrm{tr}'$ *on a (semi)finite factor are proportional, i.e.,* $\mathrm{tr}' = \lambda \,\mathrm{tr}$ *for some* $\lambda \in \mathbb{R}^+$.

See also Theorem C.155 below. Consequently, since Tr and tr are both σ-weakly continuous, and $\mathrm{Tr}(1_H) = \infty$ on $B(H)$, whereas $\mathrm{tr}(1_{\ell^2(G)}) = 1$ on $W^*(G)$, we conclude that $W^*(G) \not\cong B(H)$ for any H. Note also that (still assuming that G has icc), all projections in $W^*(G)$ are finite, and $W^*(G)$ has no minimal projections (see below), whereas $B(H)$ has both finite and infinite projections, and also has plenty of minimal projections, namely those with one-dimensional range.

Do such "icc" groups actually exist? In fact, there are infinitely many of them: each free group on $n > 1$ generators is an example. Another example is the group $\mathfrak{S}_\infty$ of finite permutations of $\mathbb{N}$. A ***j-cycle*** is a cyclic permutation of j objects (called the *carrier* of the cycle in question). Any element p of $\mathfrak{S}_\infty = \cup_n \mathfrak{S}_n$ is a finite product of j-cycles with disjoint carriers, and for each $j \in \mathbb{N}$, the number of j-cycles in such a decomposition of p is uniquely determined by p. Two permutations in $\mathfrak{S}_\infty$, then, are conjugate iff they have the same number of j-cycles, for all $j \in \mathbb{N}$

We present the ***type classification*** of factors due to Murray and von Neumann.

Definition C.153. *A factor M is said to be of type:*

- I *if it has at least one minimal projection, subdivided into:*
  - *Type* I_n *($n \in \mathbb{N}$) if M is finite and 1_M is the sum of n minimal projections.*
  - *Type* I_∞ *if M is type* I *and semifinite but not finite.*
- II *if it has no minimal projections, but has some nonzero finite projection, with:*
  - *Type* II_1 *if M is type* II *and finite.*
  - *Type* II_∞ *if M is type* II *and semifinite but not finite.*
- III *if all nonzero projections are infinite.*

A nice understanding of these types arises from a construction similar to the trace.

Definition C.154. *A* **dimension function** *on a von Neumann algebra M is a function $d : \mathscr{P}(M) \to [0,\infty]$ such that $d(e) < \infty$ iff e is finite, $d(e+f) = d(e) + d(f)$ if $ef = 0$ (i.e., $eH \perp fH$), and $d(e) = d(f)$ if $e \sim f$.*

Paraphrasing results in Murray and von Neumann's great series of papers, we have:

Theorem C.155. *For any von Neumann algebra M, the restriction of a trace to $\mathscr{P}(M)$ is a dimension function. If $M \subset B(H)$ is a factor, with H separable, then:*

1. *Any σ-weakly continuous trace on M restricts to a completely additive dimension function with the additional property that $d(e) = d(f)$ if* and only if *$e \sim f$.*
2. *Any dimension function with this additional property arises from a σ-weakly continuous trace, and hence is completely additive, and unique up to scaling.*
3. *In that case, the dimension function d induces an isomorphism between $\mathscr{P}(M)/\sim$ and some subset of $[0,\infty]$. Suitably scaling d, this subset must be one of:*
   - $\{0,1,2,\ldots,n\}$, *for some $n \in \mathbb{N}$ (type* I_n*).*
   - $\mathbb{N} \cup \infty$ *(type* I_∞*).*
   - $[0,1]$ *(type* II_1*).*
   - $[0,\infty]$ *(type* II_∞*).*
   - $\{0,\infty\}$ *(type* III*).*

We may now strengthen the few examples we had so far in the following way:

Corollary C.156.
- *If* $\dim(H) = n$, *then $B(H)$ is a factor of type* I_n.
- *If* $\dim(H) = \infty$, *then $B(H)$ is a factor of type* I_∞.
- *Let G be icc. Then $W^*(G)$ is a factor of type* II_1.

C.23 Classification of hyperfinite factors

Throughout this section we assume that our von Neumann algebras $M \subseteq B(H)$ act on a *separable* Hilbert space H. We say that M is ***hyperfinite*** if $M = (\cup_n M_n)''$, for a family of finite-dimensional von Neumann subalgebras $M_n \subset M$ with $M_n \subset M_{n+1}$. For example, $M = B(H)$ is hyperfinite. If G is a group such that $G = \cup_n G_n$ for finite subgroups $G_n \subset G_{n+1}$, as is the case e.g. for the (icc) group $\mathfrak{S}_\infty = \cup_n \mathfrak{S}_n$ of finite permutations of $\mathbb{N}$, then the associated von Neumann algebra $W^*(G)$ is hyperfinite.

Murray and von Neumann partly classified hyperfinite factors, as follows:

Theorem C.157. *Let $M \subset B(H)$ be a hyperfinite factor.*

- *If M is type* I_n, *then* $M \cong M_n(\mathbb{C})$.
- *If M is type* I_∞, *then* $M \cong B(\ell^2)$.
- *If M is type* II_1, *then* $M \cong W^*(\mathfrak{S}_\infty)$.

The unique hyperfinite II_1-factor $W^*(\mathfrak{S}_\infty)$, which turns out to be isomorphic to $W^*(G)$ for *any* finitely generated icc group G, is usually called R. Similarly (and trivially), $B(\ell^2) \cong B(H')$ for any separable infinite-dimensional Hilbert space H'.

An example of a hyperfinite II_∞ factor is also quickly found, viz. $M = R \otimes B(\ell^2)$, but Murray and von Neumann were unable to *classifiy* such factors. About type III, they knew almost nothing, except for a couple of examples from ergodic theory. Between 1971–1975, Connes made two decisive steps forward in this area:

1. Dividing type III factors into III_λ, $\lambda \in [0,1]$, by means of a new invariant.
2. Completely classifying *hyperfinite* type II_∞ and type III factors, as follows:
   - There is a unique hyperfinite II_∞ factor, namely $R \otimes B(\ell^2)$.
   - There is a unique hyperfinite III_1 factor (Connes and Haagerup).
   - There is a unique hyperfinite III_λ factor for each $\lambda \in (0,1)$.
   - There is an infinite family of hyperfinite III_0 factors, completely classified by the so-called *flow of weights* introduced by Connes and Takesaki.

We list III_1 separately from III_λ for $\lambda \in (0,1)$ for two reasons: first, "hyperfinite III_1" turns out to be *the* factor occurring in quantum field theory and quantum statistical mechanics of infinite systems, whereas III_λ for $\lambda \in (0,1)$ seems artificial) and second, the proof of uniqueness of the hyperfinite III_1 factor is much more difficult.

An important technical tool of Connes was his own profound discovery that a von Neumann algebra $M \subset B(H)$ is hyperfinite iff it is ***injective***, in that there exists a σ-weakly continuous ***conditional expectation*** $E : B(H) \to M$, that is, a linear map $E : B(H) \to B(H)$ such that $E(a) \in M$ and $E(a^*) = E(a)^*$ for all $a \in B(H)$, $E^2 = E$, and $\|E\| = 1$. It follows that $E(abc) = aE(b)c$ for all $a, c \in M$, $b \in B(H)$. The equivalence of hyperfiniteness and injectivity implies, for example, that if $M = N \otimes B(\ell^2)$ is hyperfinite, then so is N. Another crucial tool was the ***Tomita–Takesaki theory***, which we briefly summarize (this theory was paralleled by simultaneous and independent work in mathematical physics by the German-Dutch mathematical physics trio Haag–Hugenholtz–Winnink, which among other things allowed a direct definition of thermal equilibrium states in infinite volume, see §9.6.

Definition C.158. *A von Neumann algebra $M \subset B(H)$ is in* **standard form** *if H contains a unit vector Ω that is cyclic and separating for M.*

Recall that Ω is *separating* for M if $a\Omega \neq 0$ for all nonzero $a \in M$, and that Ω is *cyclic* for M iff it is separating for M'. Any von Neumann algebra can be brought into standard form. For separable H, this follows by picking an injective density operator ρ on H, whose associated state $\omega(a) = \mathrm{Tr}(\rho a)$ is faithful (in that $\omega(a^*a) > 0$ for all nonzero $a \in M$), and passing to the GNS-representation $\pi_\omega(M) \cong M$. For example, $M = B(H)$ acting on H is not in standard form, but acting on $B_2(H)$ by *left* multiplication it is, where $B_2(H)$ is the Hilbert space of Hilbert–Schmidt operators on H with the familiar inner product $\langle a, b\rangle = \mathrm{Tr}(a^*b)$. If $\rho \in B_1(H)$ is an injective density operator on H, then $\Omega = \sqrt{\rho} \in B_2(H)$ brings M into standard form. In this case, $M' \cong B(H)^{op}$ (where the suffix "op" means that multiplication is done in the opposite order, i.e. ab in $B(H)^{op}$ is equal to ba in $B(H)$), which acts on $B_2(H)$ by *right* multiplication. If $H = \mathbb{C}^n$, one simply has $B(H) = B_2(H) = M_n(\mathbb{C})$.

Let $M \subset B(H)$ be in standard form. Tomita introduced the (unbounded) *antilinear* operator S as the closure of the operator S_0 having domain $D(S_0) = M\Omega$ and action

$$S_0(a\Omega) = a^*\Omega. \tag{C.593}$$

This domain is dense because Ω is cyclic for M, the action is well defined since Ω is separating for M, and S_0 indeed turns out to be closable, with closure S. Any closed operator a has a polar decomposition $a = v|a|$, where v is a partial isometry and $|a| = \sqrt{a^*a}$. We write the polar decomposition of the above operator S as

$$S = J\Delta^{1/2}, \tag{C.594}$$

where J is an *antilinear* partial isometry, and $\Delta = S^*S$. Since S is injective with dense range, J is actually anti-unitary, satisfying $J^* = J$ and $J^2 = 1$. Furthermore, $\Delta \geq 0$, so that Δ^{it} is well defined for $t \in \mathbb{R}$: writing $\Delta = \exp(h)$ for the self-adjoint operator $h = \log\Delta$, we have $\Delta^{it} = \exp(ith)$. We then have the ***Tomita–Takesaki Theorem***:

Theorem C.159. *Let $M \subset B(H)$ be a von Neumann algebra in standard form. Then:*

- $M' = JMJ \equiv \{JaJ \mid a \in M\}$.
- *For each $t \in \mathbb{R}$ and $a \in M$, the operator $\alpha_t(a) = \Delta^{it}a\Delta^{-it}$ lies in M.*
- *The map $t \mapsto \alpha_t$ is a group homomorphism from $\mathbb{R}$ to* $\mathrm{Aut}(M)$ *(i.e., the group of all automorphisms of M), which is continuous, in that for each $a \in M$ the function $t \mapsto \alpha_t(a)$ from $\mathbb{R}$ to M (with σ-weak topology) is continuous.*

The image of $\mathbb{R}$ in $\mathrm{Aut}(M)$ by α is called the ***modular group*** of M associated with the cyclic and separating vector Ω (or rather, with the associated σ-weakly continuous faithful state ω). Simple examples show that the modular group explicitly depends on the vector Ω. In his thesis, Connes analyzed the dependence of α on Ω, and showed it was innocent. To state the simplest version of his result, assume that H contains two different vectors Ω_1 and Ω_2, each of which is cyclic and separating for M. We write $\alpha_t^{(i)}$ for the modular group derived from Ω_i, $i = 1, 2$.

Theorem C.160. *There is a family U_t of unitary operators in M ($t \in \mathbb{R}$), such that*

$$\alpha_t^{(1)}(a) = U_t \alpha_t^{(2)}(a) U_t^*; \tag{C.595}$$

$$U_{t+s} = U_s \alpha_s^{(2)}(U_t). \tag{C.596}$$

Proof. The proof of this theorem is Connes's favourite (as he declared in an interview), so we present it in some detail. It is based on the following idea. Extend M to $\mathrm{Mat}_2(M)$, i.e., the von Neumann algebra of 2×2 matrices with entries in M, and let $\mathrm{Mat}_2(M)$ act on $H_2 = H \oplus H$ in the obvious way. Subsequently, let $\mathrm{Mat}_2(M)$ act on $H_4 = H \oplus H \oplus H \oplus H = H_2 \oplus H_2$ by simply doubling the action on H_2. The vector $\Omega = (\Omega_1, 0, 0, \Omega_2) \in H_4$ is then cyclic and separating for $\mathrm{Mat}_2(M)$, with corresponding modular operator $\Delta = \mathrm{diag}(\Delta_1, \Delta_4, \Delta_3, \Delta_2)$. Here Δ_1 and Δ_2 are just the operators on H originally defined by Ω_1 and Ω_2, respectively, and Δ_3 and Δ_4 are certain operators on H. Denoting elements of $\mathrm{Mat}_2(M)$ by

$$\mathbf{a} = \begin{pmatrix} a_{11} & a_{12} \\ a_{21} & a_{22} \end{pmatrix}, \tag{C.597}$$

we then have

$$\Delta^{it} \begin{pmatrix} \mathbf{a} & \mathbf{0} \\ \mathbf{0} & \mathbf{a} \end{pmatrix} \Delta^{-it} = \begin{pmatrix} \tilde{\alpha}_t^{(1)}(\mathbf{a}) & \mathbf{0} \\ \mathbf{0} & \tilde{\alpha}_t^{(2)}(\mathbf{a}) \end{pmatrix}; \tag{C.598}$$

$$\tilde{\alpha}_t^{(1)}(\mathbf{a}) = \begin{pmatrix} \Delta_1^{it} a_{11} \Delta_1^{-it} & \Delta_1^{it} a_{12} \Delta_4^{-it} \\ \Delta_4^{it} a_{21} \Delta_1^{-it} & \Delta_4^{it} a_{22} \Delta_4^{-it} \end{pmatrix}; \tag{C.599}$$

$$\tilde{\alpha}_t^{(2)}(\mathbf{a}) = \begin{pmatrix} \Delta_3^{it} a_{11} \Delta_3^{-it} & \Delta_3^{it} a_{12} \Delta_2^{-it} \\ \Delta_2^{it} a_{21} \Delta_3^{-it} & \Delta_2^{it} a_{22} \Delta_2^{-it} \end{pmatrix}. \tag{C.600}$$

But by Theorem C.159, the right-hand side of (C.598) must be of the form $\mathrm{diag}(\mathbf{b}, \mathbf{b})$ for some $\mathbf{b} \in \mathrm{Mat}_2(M)$, so that $\tilde{\alpha}_t^{(1)}(\mathbf{a}) = \tilde{\alpha}_t^{(2)}(\mathbf{a})$. This allows us to replace $\Delta_4^{it} a_{22} \Delta_4^{-it}$ in (C.599) by $\Delta_2^{it} a_{22} \Delta_2^{-it}$. We then put $U_t = \Delta_1^{it} \Delta_4^{-it}$, which, unlike either Δ_1^{it} or Δ_4^{-it}, lies in M, because each entry in $\tilde{\alpha}_t^{(1)}(\mathbf{a})$ must lie in M if all the a_{ij} do, and here we have taken $a_{12} = 1$. All claims of the theorem may then be verified using elementary computations with 2×2 matrices. For example, combining

$$\begin{pmatrix} a & 0 \\ 0 & 0 \end{pmatrix} = \begin{pmatrix} 0 & 1 \\ 0 & 0 \end{pmatrix} \begin{pmatrix} 0 & 0 \\ 0 & a \end{pmatrix} \begin{pmatrix} 0 & 0 \\ 1 & 0 \end{pmatrix} \tag{C.601}$$

with the property $\tilde{\alpha}_t^{(1)}(\mathbf{ab}) = \tilde{\alpha}_t^{(1)}(\mathbf{a}) \tilde{\alpha}_t^{(1)}(\mathbf{b})$, we recover (C.595). Using the identity

$$\begin{pmatrix} 0 & U_t \\ 0 & 0 \end{pmatrix} = \begin{pmatrix} 0 & 1 \\ 0 & 0 \end{pmatrix} \begin{pmatrix} 0 & 0 \\ 0 & U_t, \end{pmatrix}, \tag{C.602}$$

evolving each side to time s yields (C.596). A proof from *The Book*! □

We say that an automorphism $\gamma : M \to M$ is ***inner*** if there exists a unitary element $u \in M$ such that $\gamma(a) = uau^*$ for all $a \in M$. The inner automorphisms of M form a normal subgroup $\mathrm{Inn}(M)$ of the group $\mathrm{Aut}(M)$ of all automorphisms, with quotient $\mathrm{Out}(M) = \mathrm{Aut}(M)/\mathrm{Inn}(M)$. Theorem C.160 shows that the image $\pi(\alpha(\mathbb{R}))$ of the modular group in $\mathrm{Out}(M)$ under the canonical projection $\pi : \mathrm{Aut}(M) \to \mathrm{Out}(M)$ is independent of Ω, and invariants of this image will be invariants of M itself.

Such invariants are trivial if M is a factor of type I or II, since in that case $\pi(\alpha(\mathbb{R})) = \{e\}$; to see this in the finite case (i.e., type I_n or type II_1), take a finite trace τ on M and check that $\Delta = 1$ for $\pi_\tau(M) \cong M$. For the semifinite but not finite case (i.e., type I_∞ or type II_∞), a slight generalization of the GNS-construction leads to the same conclusion. To find invariants for type III factors, we therefore need to extract information from the modular group $t \mapsto \alpha_t$ up to inner automorphisms.

Definition C.161. *Let $\alpha : \mathbb{R} \to \mathrm{Aut}(M)$ be a continuous action of $\mathbb{R}$ on M, defining:*

$$M^\alpha = \{x \in M \mid \alpha_t(x) = x \forall t \in \mathbb{R}\}; \tag{C.603}$$
$$M_e = \{x \in M \mid xe = ex = x\} \ (e \in \mathscr{P}(M^\alpha)). \tag{C.604}$$

- *The* **Arveson spectrum** $\mathrm{sp}(\alpha)$ *of α consists of all $p \in \mathbb{R}$ for which there is a sequence (x_n) in M with $\|x_n\| = 1$ and $\lim_{n\to\infty} \|\alpha_t(x_n) - e^{ipt}x_n\| = 0 \forall t \in \mathbb{R}$.*
- *For each $e \in \mathscr{P}(M^\alpha)$, the map $\alpha_t : M \to M$ restricts to $\alpha_t^e : M_e \to M_e$, defining a (group) homomorphism $\alpha^e : \mathbb{R} \to \mathrm{Aut}(M_e)$, $t \mapsto \alpha_t^e$. The* **Connes spectrum** *of α is $\Gamma(\alpha) = \exp(\Gamma'(\alpha)) \subset \mathbb{R}_*^+$, where $\Gamma'(\alpha) = \bigcap_{0 \neq e \in P(M^\alpha)} \mathrm{sp}(\alpha^e) \subset \mathbb{R}$.*

The Connes spectrum $\Gamma(\alpha)$ is a closed subgroup of $\mathbb{R}_*^+$, which has the great virtue that if $\pi(\alpha(\mathbb{R})) = \pi(\alpha'(\mathbb{R}))$, then $\Gamma(\alpha) = \Gamma(\alpha')$. So if α is the modular group of M with respect to some state ω, then $\Gamma(\alpha)$ is independent of ω, and may therefore be called $\Gamma(M)$. This invariant can also be defined through the usual spectrum of self adjoint operators on Hilbert space. To this effect, Connes defined and proved

$$S(M) = \bigcap_\omega \sigma(\Delta_\omega) = \bigcap_{0 \neq e \in \mathscr{P}(M^\alpha)} \sigma(\Delta_{\varphi_e}), \tag{C.605}$$

where the first intersection is over all σ-weakly continuous faithful states ω on M, whereas in the second one takes *a fixed* σ-weakly continuous faithful state φ on M, and restricts it to $\varphi_e = \varphi_{|M_e}$. Furthermore, Δ_ω denotes the operator Δ on H_ω, defined with respect to the usual cyclic unit vector Ω_ω of the GNS-construction, etc. If M is a type I or II factor, one has $S(M) = \{1\}$, whereas $0 \in S(M)$ iff M is type III.

Connes showed that $\Gamma(M) = S(M) \cap \mathbb{R}_*^+$, and the known classification of closed subgroups of $\mathbb{R}_*^+$ yields his path-breaking parametrization of type III factors:

Definition C.162. *Let M be a type* III *factor. Then M is said to be of type:*

- III_0 *if $\Gamma(M) = \{1\}$;*
- III_λ, *where $\lambda \in (0,1)$, if $\Gamma(M) = \lambda^{\mathbb{Z}}$;*
- III_1 *if $\Gamma(M) = \mathbb{R}_*^+$.*

The unique hyperfinite III_1 factor appears throughout algebraic quantum field theory, where it plays the role of a universal algebra of localized observables.

C.24 Other special classes of C*-algebras

There are many other special classes of C*-algebras apart from von Neumann algebras and commutative C*-algebras. The classes we consider here contain both commutative and non-commutative C*-algebras; in the spirit of (exact) Bohrification, whenever possible we try to characterize them through properties of their (maximal) commutative subalgebras. Like the von Neumann algebras already studied, each class in this section is sandwiched between the ***finite-dimensional*** C*-algebras, i.e. those C*-algebras that are finite-dimensional as a vector space (which it contains), and the *real rank zero* C*-algebras defined below (in which it is contained).

Finite-dimensional C*-algebras admit a straightforward classification:

Theorem C.163. *Every finite-dimensional C*-algebra A is isomorphic to a direct sum of matrix algebras, i.e., $A \cong \oplus_k M_{n_k}(\mathbb{C})$, where $n_k \in \mathbb{N}$, and the sum is finite.*

Proof. Let A be a finite-dimensional C*-algebra, and take the injective representation $\pi = \bigoplus_{\omega \in P(A)} \pi_\omega$ on $H_c = \bigoplus_{\omega \in P(A)} H_\omega$, where $P(A)$ is the pure state space of A; cf. the last stage of the proof of Theorem C.87. The proof now unfolds:

1. Since H_ω is the closure of $\pi_\omega(A)\Omega_\omega$, it must be finite-dimensional.
2. Since each ω is pure, by Theorem C.90 we must have $\pi_\omega(A)'' = B(H_\omega)$.
3. By Theorem C.127, $\pi_\omega(A)''$ equals the weak or strong closure of $\pi_\omega(A)$, but since this algebra is finite-dimensional by step 1, these closures coincide with $\pi_\omega(A)$, and hence $\pi_\omega(A) = B(H_\omega) \cong M_n(\mathbb{C})$, where $n = \dim(H_\omega)$.
4. One can find an injective subrepresentation π_i of π using only a finite number of pure states (proof by contradiction to $\dim(A) < \infty$), so that $\pi_i(A) \cong A$. □

The real rank of a C*-algebra A is a non-commutative generalization of the (Lebesgue) ***covering dimension*** of a non-empty space X, defined as follows. First say that $\dim(X) \leq n$ iff every open cover of X has an open refinement $\mathscr{U}$ for which every $x \in X$ is contained in at most $n+1$ elements of $\mathscr{U}$. We then say that $\dim(X) = n$ iff $\dim(X) \leq n$ but $\dim(X) \nleq n-1$ (such n need not exist).

If X is a compact Hausdorff space, then $\dim(X) = n$ iff n is the smallest integer n such that for every $f \in C(X, \mathbb{R}^{n+1})$ and $\varepsilon > 0$, there is $g \in C(X, \mathbb{R}^{n+1})$ such that $g(x) \neq 0$ for all x and $\|f - g\|_\infty < \varepsilon$, where $\|f\|_\infty = \sup_{x \in X}\{|f(x)|\}$. If no such n exists, we say that $\dim(X) = \infty$. If $g : X \to \mathbb{R}^{n+1}$ is described by its coordinates $(g_1, \ldots, g_{n+1})$, then $g(x) \neq 0$ iff $\sum_{k=1}^{n+1} g_k(x)^2 > 0$, or equivalently, $\sum_k g_k^2$ is invertible in $C(X)$. We may replace the usual norm $\|\mathbf{v}\|$ in $\mathbb{R}^{n+1}$ by the equivalent max-norm, i.e., $\|\mathbf{v}\| = \max_i\{|v_i|\}$, where $\mathbf{v} = (v_1, \ldots, v_{n+1})$. If we do so, we may generalize the covering dimension to possibly noncommutative unital C*-algebras, as follows.

Let $A^n = A \oplus \cdots \oplus A$ (with n terms) be the C*-algebra $A \times \cdots \times A$ with pointwise operations and norm $\|(a_1, \ldots, a_n)\| = \max_i\{\|a_i\|\}$. Let $Q(A^n)$ be the set of all *self-adjoint* elements $(a_1, \ldots, a_n)$ in A^n for which $\sum_i a_i^2$ is invertible (i.e. in A). The ***real rank*** $\mathrm{rr}(A)$ of a unital C*-algebra A is defined as the smallest integer n for which $Q(A^{n+1})$ is dense in A_{sa}^{n+1}, i.e., if for every $\mathbf{a} \in A_{\mathrm{sa}}^{n+1}$ and $\varepsilon > 0$, there is $\mathbf{b} \in Q(A^{n+1})$ such that $\|\mathbf{a} - \mathbf{b}\| < \varepsilon$. If no such n exists, we define $\mathrm{rr}(A) = \infty$. If A has no unit, we define its real rank as $\mathrm{rr}(A) = \mathrm{rr}(\dot{A})$, i.e., as the real rank of its unitization.

Taking $A = C(X)$, it follows from the previous paragraph that

$$\mathrm{rr}(C(X)) = \dim(X). \tag{C.606}$$

Now $\dim(X) = 0$ iff X has a basis of clopen sets, and if X is compact Hausdorff, then $\dim(X) = 0$ iff X is a Stone space. Hence from (C.606) we immediately have:

Proposition C.164. *If A is a commutative C*-algebra, $\mathrm{rr}(A) = 0$ iff $\Sigma(A)$ is Stone.*

This makes dimension zero somewhat pathological. On the other hand, for non-commutative C*-algebras real rank zero is ubiquitous. Note that if $a = a^*$ and a^2 is invertible, then its inverse is positive, too, and has a square-root which inverts a. Thus A has real rank zero iff its invertible self-adjoint elements are dense in A_{sa}.

Proposition C.165. *Any von Neumann algebra has real rank zero.*

Proof. For $a \in A_{\mathrm{sa}}$ and $\varepsilon > 0$, with $A \subseteq B(H)$, use Theorem B.102 to define

$$b = (\mathrm{id}_{\sigma(a)} + (\tfrac{1}{2}\varepsilon \cdot 1_{\sigma(a)} - \mathrm{id}_{\sigma(a)}) \cdot 1_{[-\varepsilon/2,\varepsilon/2]})(a). \tag{C.607}$$

Using (B.322), we may then compute

$$\begin{aligned} \|a - b\| &\leq \|\tfrac{1}{2}\varepsilon \cdot 1_{\sigma(a)} - \mathrm{id}_{\sigma(a)} \cdot 1_{[-\varepsilon/2,\varepsilon/2]}\|_\infty \\ &\leq \|\tfrac{1}{2}\varepsilon \cdot 1_{\sigma(a)}\|_\infty + \|\mathrm{id}_{\sigma(a)} \cdot 1_{[-\varepsilon/2,\varepsilon/2]}\|_\infty \\ &\leq \tfrac{1}{2}\varepsilon + \tfrac{1}{2}\varepsilon = \varepsilon. \end{aligned} \tag{C.608}$$

Writing (C.607) as $b = f(a)$, the function $f \in \mathscr{B}(\sigma(a))$ satisfies $f(x) = x$ if $x \notin [-\varepsilon/2, \varepsilon/2]$ and $f(x) = \frac{1}{2}\varepsilon$ if $x \in [-\varepsilon/2, \varepsilon/2]$; either way, $f(x) \neq 0$. Hence f is invertible in $\mathscr{B}(\sigma(a))$, and therefore $b = f(a)$ is invertible in $W^*(a)$ and in $B(H)$.□

We now turn to classes of C*-algebras that are sandwiched between the finite-dimensional ones at the lower end and those with real rank zero at the upper end.

Definition C.166. *Let A be a unital C*-algebra. Then A is said to be:*

1. **Finite-dimensional** *if it is finite-dimensional as a vector space.*
2. **AF** *(Approximately Finite-dimensional) if it is the norm-closure of the union of some (not necessarily countable) directed set of finite-dimensional C*-subalgebras.*
3. **Scattered** *if every $a \in A_{\mathrm{sa}}$ has countable spectrum.*
4. *A* **W*-algebra** *if it is the dual of a Banach space, and a* **von Neumann algebra** *if it is a represented W*-algebra, i.e., $A \subseteq B(H)$; this can always be achieved.*
5. **Monotone complete** *if every upward directed bounded subset in A_{sa} (under the usual order $\leq$) has a least upper bound (i.e. supremum).*
6. **AW*** *if for each nonempty subset $S \subset A$ there is $e \in \mathscr{P}(A)$ so that $R(S) = eA$.*
7. **Rickart** *if for each $a \in A$ there is a projection $e \in \mathscr{P}(A)$ so that $R(a) = eA$.*
8. **Real rank zero** *if its invertible self-adjoint elements are dense in A_{sa}.*

Here a subset S of a poset P is ***upward directed*** if for each $x, y \in S$ there is $z \in S$ such that $x \leq z$ and $y \leq z$ (for example, this is true in a complete lattice).

Furthermore, the ***right-annihilator*** $R(S)$ of $S \subset A$ is defined as

$$R(S) = \{a \in A \mid ba = 0 \forall b \in S\}, \tag{C.609}$$

and $R(a) \equiv R(\{a\})$; in the presence of an involution, equivalent definitions may be given in terms of the left-annihilator. In all cases, the projection e is unique. Since Rickart himself already showed that A is Rickart iff for each nonempty *countable* subset $S \subset A$ there is $e \in \mathscr{P}(A)$ so that $R(S) = eA$, the difference between Rickart and AW* lies in the countability assumption on S in the former but not in the latter.

It is known that if a C*-algebra A has a faithful representation on a separable Hilbert space, then it is a Rickart C*-algebra iff it is an AW*-algebra, but otherwise these classes are different. Similarly, an AW*-algebra is a W*-algebra iff it has a separating family of normal states, where normality of functionals on AW*-algebras is defined as in Definition 4.11, i.e. through complete additivity on orthogonal familes of projections, which always have an upper bound (cf. Theorem C.169 below). This is the case in all examples relevant to mathematical physics, but set-theoretically the class of AW*-algebras has higher cardinality than the class of W*-algebras it contains. It is generally believed that a C*-algebra is Rickart iff it is monotone σ-complete, and that it is AW* iff it is monotone complete, but there are neither proofs of nor counterexamples to these claims. We have the inclusions:

$$\begin{aligned} W^* &\subset \text{ monotone complete} \subseteq AW^* \subset \text{ Rickart} \subset \text{ real rank zero;} \\ AF &\subset \text{ real rank zero;} \\ \text{scattered} &\subset \text{ real rank zero.} \end{aligned}$$

Scattered C*-algebras may alternatively be characterized as those C*-algebras on which every state is a w^*-convergent convex sum of pure states; this condition is far stronger than what the Krein–Milman theorem gives, namely that every state is a w^*-limit of some net consisting of finite convex sums of pure states. For example, for any Hilbert space the compact operators $B_0(H)$ form a scattered C*-algebra (extending the definition of the latter to the non-unital case as appropriate).

Two kinds of results are of interest for Bohrification: one is the topological characterization of the commutative case of each class, the other is the characterization of the class itself through properties of its commutative subalgebras. Without proof we state what is known in this respect.

Theorem C.167. *Let A be a commutative unital C*-algebra. Then A is:*

1. *Finite-dimensional iff $\Sigma(A)$ is finite (with discrete topology).*
2. *AF iff $\Sigma(A)$ is a Stone space.*
3. *Scattered iff $\Sigma(A)$ is scattered.*
4. *A W*-algebra or a von Neumann algebra iff $\Sigma(A)$ is hyperstonean.*
5. *Monotone complete iff $\Sigma(A)$ is stonean.*
6. *AW* iff $\Sigma(A)$ is stonean.*
7. *Rickart iff $\Sigma(A)$ is σ-stonean.*
8. *Real rank zero iff $\Sigma(A)$ is a Stone space.*

Here we used the convention that a ***Stone space*** is a zero-dimensional compact Hausdorff space (equivalently, it is compact Hausdorff and totally disconnected in the sense that the only connected subsets are points). A ***(σ-) stonean space*** is a Stone space with the additional property that $\mathrm{Clopen}(\Sigma(A))$ is a (σ-) complete lattice (equivalently, a stonean space is a compact Hausdorff space that is extremally disconnected in that the closure of each open set is open). Furthermore, a space is ***hyperstonean*** if it is stonean, and for any nonzero $f \in C(X, \mathbb{R}^+)$ there exists a completely additive positive measure μ such that $\mu(f) > 0$. In particular, in the commutative case the classes *AF* and real rank zero coincide, as do AW* and monotone complete algebras. A space X is called ***scattered*** if each non-empty closed subset $C \subset X$ contains an isolated point (i.e., a point $x \in C$ with an open neighbourhood U such that $U \cap C = \{x\}$). If X is scattered, then it is totally disconnected. An example of a compact scattered space is $(1/\mathbb{N}) \cup \{0\}$ with the relative topology from $\mathbb{R}$.

This leads to the following generalization and extension of Theorem C.141.

Theorem C.168. *Let A be a commutative unital C*-algebra. The projections $\mathscr{P}(A)$ in A form a Boolean lattice, which is related to the Gelfand spectrum $\Sigma(A)$ through*

$$\mathscr{P}(A) \cong \mathrm{Clopen}(\Sigma(A)). \tag{C.610}$$

If A is also AF, then its Gelfand spectrum $\Sigma(A)$ is a Stone space, and we have

$$\Sigma(A) \cong \mathscr{S}(\mathscr{P}(A)); \tag{C.611}$$

$$\mathscr{O}(\Sigma(A)) \cong \mathrm{Idl}(\mathscr{P}(A)); \tag{C.612}$$

$$A \cong C(\mathscr{S}(\mathscr{P}(A))), \tag{C.613}$$

as topological spaces, frames, and (commutative) C-algebras, respectively.*

Conversely, for any Boolean lattice L the C-algebra $C(\mathscr{S}(L))$ is AF, and*

$$L \cong \mathscr{P}(C(\mathscr{S}(L))). \tag{C.614}$$

Proof. Using the Gelfand isomorphism $A \cong C(\Sigma(A))$, eq. (C.610) follows from

$$\mathscr{P}(C(X)) \cong \mathrm{Clopen}(X), \tag{C.615}$$

where X is some compact Hausdorff space. Indeed, if $e^2 = e^* = e \in C(X)$, then e must be $\{0,1\}$-valued, so it must be $e = 1_U$ for some $U \subset X$, viz. $U = e^{-1}(\{1\})$. Since $e \in C(X)$ is continuous, U must be clopen. Conversely, for each $U \in \mathrm{Clopen}(X)$, the function $1_U \in C(X)$ is a projection, and the maps $U \mapsto 1_U$ and $e \mapsto e^{-1}(\{1\})$ are each other's inverse. Theorem D.5 then implies that $\mathscr{P}(A)$ is Boolean.

If $A \cong C(X)$ is *AF*, then $C(X) = (\cup_\lambda A_\lambda)^-$ is the norm-closure of the union of finite-dimensional C*-algebras A_λ, which union by the Stone–Weierstrass theorem separates points of X. Since each A_λ is the linear span of its projections, the finite-dimensional projections $\cup_\lambda \mathscr{P}(A_\lambda)$ already separate points in X, and this in turn implies that X is ***totally separated***, i.e., for each $x \neq y \in X$, there is $U \in \mathrm{Clopen}(X)$ such that $x \in U$ and $y \notin U$. Since a compact Hausdorff space is zero-dimensional (and hence Stone) iff it is totally separated, X is a Stone space.

Again using Theorem C.8, we only need to prove (C.611) in the special case

$$X \cong \mathscr{S}(\mathscr{P}(C(X))), \tag{C.616}$$

where X is a Stone space; this follows from (C.615) and Theorem D.5. Eq. (C.612) follows from (D.35), whilst (C.613) is immediate from (C.611) and Theorem C.8.

Finally, using Theorem D.5 we see that (C.614) reduces to (C.615), so we only need to prove that $C(X)$ is AF for any Stone space X. This is just the above proof of the converse ran backwards: since X is totally separated, for each $x \neq y$ we find $U \in \text{Clopen}(X)$ separating x and y, so that also the associated projection 1_U separates x and y, and hence $\mathscr{P}(C(X))$ separates X. Taking Λ to label the finite subsets of $\mathscr{P}(C(X))$, and A_λ to be the finite-dimensional C*-algebra generated by $\lambda \in \Lambda$, by Stone–Weierstrass we have $C(X) = (\cup_\lambda A_\lambda)^-$. Hence $C(X)$ is AF. □.

Theorem C.169. *The claim that a unital C*-algebra lies in class $\mathscr{X}$ iff each of its maximal abelian *-subalgebras lies in class $\mathscr{X}$ is true for the following classes:*

1. *Finite-dimensional C*-algebras.*
2. *Scattered C*-algebras.*
3. *von Neumann algebras.*
4. *AW*-algebras.*
5. *Rickart C*-algebras.*

The claim is false for AF-algebras, true for monotone complete C*-algebras iff these coincide with AW*-algebras, false for real rank zero C*-algebras, and unknown for W*-algebras, which we therefore state as a *conjecture*:

A C-algebra is a W*-algebra iff each maximal abelian *-subalgebra is a W*-algebra.*

Proposition C.170. *For any C*-algebra A and any projections $e, f \in \mathscr{P}(A)$, we have $ef = e$ iff $e \leq f$ (with partial ordering $\leq$ as defined in A_{sa} via A^+, cf. §C.7).*

Proof. As explained above (C.93), if $a_1 \leq a_2$, then $b^* a_1 b \leq b^* a_2 b$, so $e \leq f$ implies

$$(1_A - f)e(1_A - f) \leq (1_A - f)f(1_A - f) = 0. \tag{C.617}$$

However, since $e^2 = e^* = e$, with $c = e(1_A - f)$ we have $(1_A - f)e(1_A - f) = c^*c$, and hence $(1_A - f)e = 0$ (as $c^*c \geq 0$), or $e = fe$. Taking adjoints gives $ef = e$, and consequently $ef = fe$. Conversely, if $ef = e$, we have $ef = fe$ and hence

$$(f - e)^2 = f - 2e + e = f - e. \tag{C.618}$$

Of course, $f - e = (f - e)^*$, so that (C.618) makes $f - e$ a projection. Since any projection lies in A^+, we have $f - e \geq 0$, and hence $e \leq f$. □

The set of projections $\mathscr{P}(A)$ in a C*-algebra is always a poset in the order $\leq$, but it is not automatically a lattice. It is a σ-complete lattice if A is Rickart, and hence also in all "lower" classes, including von Neumann algebras (cf. Proposition C.136), where $\mathscr{P}(A)$ is even a complete lattice.

C.25 Jordan algebras and (pure) state spaces of C*-algebras

Let A be a unital C*-algebra. As we know, the ***state space*** $S(A)$ is the set of all states on A, seen as a compact convex set in the w^*-topology inherited from the embedding $S(A) \subset A^*$ (note that $S(A)$ fails to be compact if A lacks a unit). To see which information $S(A)$ carries about A, we need to impoverish A as follows.

Definition C.171. *A* **Jordan algebra** *is a real* commutative *(but generally* non-associative*) algebra A whose product* $\circ$ *satisfies (writing* $a^2 = a \circ a$*):*

$$a \circ (b \circ a^2) = (a \circ b) \circ a^2. \tag{C.619}$$

A **JB-algebra** *is a Jordan algebra that is also a (real) Banach space such that:*

$$\|a \circ b\| \leq \|a\| \|b\|; \tag{C.620}$$
$$\|a\|^2 \leq \|a^2 + b^2\|. \tag{C.621}$$

Given (C.620), axiom (C.621) is equivalent to $\|a^2\| \leq \|a^2 + b^2\|$ and $\|a^2\| = \|a\|^2$.

It is easy to see that the self-adjoint part A_{sa} of any C*-algebra A is a JB-algebra if we put $a \circ b = \frac{1}{2}(ab + ba)$, cf. (5.14). If A and B are unital C*-algebras, we say that a linear map $\varphi : A_{sa} \to B_{sa}$ is a ***Jordan homomorphism*** if it preserves $\circ$; to this effect it clearly suffices that $\varphi(a^2) = \varphi(a)^2$ for each a. If φ in addition is bijective, then it is called a ***Jordan isomorphism***; in that case its inverse is necessarily linear and also preserves te Jordan product $\circ$. A Jordan***Jordan automorphism*** of a C*-algebra A is a Jordan isomorphism $A_{sa} \to A_{sa}$. Of course, we may complexify $\varphi : A_{sa} \to B_{sa}$ so as to obtain a $\mathbb{C}$-linear map $\varphi_{\mathbb{C}} : A \to B$ that equally well satisfies $\varphi_{\mathbb{C}}(a^2) = \varphi_{\mathbb{C}}(a)^2$, this time for all $a \in A$ (rather than all $a \in A_{sa}$). However, the conceptual point here is that quantum-mechanical observables are supposed to be self-adjoint, and that the Jordan product (but not the ordinary associative product) always preserves self-adjointness. Generalizing Proposition 5.19, we then have the key result:

Theorem C.172. *Let A and B be unital C*-algebras. There is a bijective correspondence between Jordan isomorphisms* $\varphi : A_{sa} \to B_{sa}$ *and affine homeomorphisms* $f : S(B) \to S(A)$*, given by* $f = \varphi^*$ *(i.e.* $f(\omega)(a) = \omega(\varphi(a))$*). In particular, each affine homeomorphism of* $S(A)$ *is induced by a Jordan automorphism of A.*

The proof is similar to Proposition 5.19; generalizing Lemma 5.20 we now have:

Lemma C.173. *Let A and B be unital C*-algebras. Then* $f = \varphi^*$ *gives a bijective correspondence between affine bijections* $f : S(B) \to S(A)$ *and unital positive linear bijections* $\varphi : A_{sa} \to B_{sa}$*. Moreover, if* $\varphi : A_{sa} \to B_{sa}$ *is a unital linear bijection, then* φ *is positive iff* φ *is isometric iff* φ *is a Jordan isomorphism.*

Most of the proof is practically the same as for Lemma 5.20 (so we omit it), expect for the last equivalence between invertible unital isometries and Jordan isomorphisms, which is deeper and relies on ***Kadison's inequality*** $\varphi(a^*a) \geq \varphi(a)^*\varphi(a)$ for positive unital linear maps φ between C*-algebras and normal operators a.

A similar result is Hamhalter's generalization of ***Dye's Theorem*** to AW*-algebras:

Theorem C.174. *Let A and B be AW*-algebras and let $\mathsf{N} : \mathscr{P}(A) \to \mathscr{P}(B)$ be an isomorphism of the corresponding orthocomplemented projection lattices that in addition preserves arbitrary suprema. If A has no summand isomorphic to either $\mathbb{C}^2$ or $M_2(\mathbb{C})$, then there is a unique Jordan isomorphism $\mathsf{J} : A_{\mathrm{sa}} \to B_{\mathrm{sa}}$ that extends N (and hence Jordan isomorphisms are characterized by their values on projections).*

This generalizes Corollary 5.22 in the main text, but has a much more difficult proof.

Proof. If $e, f \in \mathscr{P}(A)$ are orthogonal, then so are $\mathsf{N}(e)$ and $\mathsf{N}(f)$, so that

$$\mathsf{N}(e+f) = \mathsf{N}(e) + \mathsf{N}(f). \tag{C.622}$$

Gleason's Theorem for AW*-algebras then gives a Jordan homomorphism

$$\mathsf{J}_{(e,f)} : AW^*(e,f)_{\mathrm{sa}} \to B_{\mathrm{sa}}, \tag{C.623}$$

where $AW^*(e,f)$ is the AW*-algebra generated by e, f, and the unit 1_A, which in particular preserves all ***Jordan triple products***

$$\{a,b,c\} = (a \circ b) \circ c + a \circ (b \circ c) - b \circ (a \circ c), \tag{C.624}$$

which in terms of the usual operator product equals $\frac{1}{2}(abc + cba)$. This implies

$$\mathsf{N}((1_A - 2e) f (1_A - 2e)) = (1_B - 2\mathsf{N}(e))\mathsf{N}(f)(1_B - 2\mathsf{N}(e)), \tag{C.625}$$

which (in the second major step of the proof, after the application of Gleason's Theorem) is necessary and sufficient for φ to extend to a Jordan isomorphism. □

The structure of Jordan isomorphisms may be inferred from the following remarkable result, in which a linear map $\varphi : A \to B$ between C*-algebras is called an ***anti-homomorphism*** of $\varphi(a^*) = \varphi(a)^*$ as usual, but $\varphi(ab) = \varphi(b)\varphi(a)$.

Theorem C.175. *If $\varphi : A_{\mathrm{sa}} \to B(H)_{\mathrm{sa}}$ is a Jordan homomorphism (where A is a C*-algebra and H is a Hilbert space), there exist three mutually orthogonal projections e_1, e_2, e_3 in the center $\varphi(A)' \cap \varphi(A)''$ of the von Neumann algebra $\varphi(A)''$, such that:*

1. $e_1 + e_2 + e_3 = 1_H$;
2. *The map $a \mapsto \varphi_{\mathbb{C}}(a)e_1$ from A to $B(e_1H)$ is a homomorphism (of C*-algebras).*
3. *The map $a \mapsto \varphi_{\mathbb{C}}(a)e_2$ from A to $B(e_2H)$ is an anti-homomorphism* (ibid.*).*
4. *The map $a \mapsto \varphi_{\mathbb{C}}(a)e_3$ from A to $B(e_3H)$ is both a homomorphism and an anti-homomorphism of C*-algebras (so that $\varphi_{\mathbb{C}}(A)e_3$ is commutative).*

If in addition $a \mapsto \varphi_{\mathbb{C}}(a)e_1$ is not an anti-homomorphism and $a \mapsto \varphi_{\mathbb{C}}(a)e_2$ is not a homomorphism, then e_1, e_2, and e_3 are uniquely determined by these conditions.

Like the previous theorem, the proof of this one exceeds the scope of this book.

Corollary C.176. *Let $\mathsf{J} : B(H)_{\mathrm{sa}} \to B(H)_{\mathrm{sa}}$ be a Jordan isomorphism. Then $\mathsf{J}_{\mathbb{C}} : B(H) \to B(H)$ is either a homomorphism or an anti-homomorphism of C*-algebras.*

Proof. The center of $B(H)$ is trivial, so either $e_1 = 1_H$ or $e_2 = 1_H$. □

The ***pure state space*** $P(A) = \partial_e S(A)$ is the extreme boundary of the state space $S(A)$. According to the Krein–Milman Theorem B.50, $P(A)$ is not empty, and

$$S(A) = (\mathrm{co}\partial_e P(A))^-, \tag{C.626}$$

see (B.165) for notation. In order to recover $S(A)$ from $P(A)$, the latter obviously needs more structure than just that of a set. First, it inherits the w^*-topology from A^*, but it turns out that we need to equip $P(A)$ with the more refined w^*-*uniformity*.

In general, a ***uniform structure*** on a set X (also called an ***entourage uniformity***) is a nonempty filter $\mathscr{U}$ on $X \times X$ (i.e.,a collection $\mathscr{U} \subset \mathscr{P}(X \times X)$ of subsets of $X \times X$ such that $U \in \mathscr{U}$ and $U \subset V$ imply $V \in \mathscr{U}$, and $U \in \mathscr{U}$ and $V \in \mathscr{U}$ imply $U \cap V \in \mathscr{U}$) satisfying the following conditions:

1. Each $U \in \mathscr{U}$ contains the diagonal $\Delta_X = \{(x,x) \mid x \in X\}$;
2. If $U \in \mathscr{U}$, then $U^T \in \mathscr{U}$, where $U^T = \{(y,x) \mid (x,y) \in U\}$;
3. If $U \in \mathscr{U}$, then there is some $V \in \mathscr{U}$ such that $V^2 \subseteq U$, where

$$V^2 = \{(x,z) \mid \exists y \in X : (x,y) \in V, (y,z) \in V\}. \tag{C.627}$$

A set with a uniformity is called a ***uniform space***. If X and Y are uniform spaces, a function $f : X \to Y$ is ***uniformly continuous*** if $f^{-1}(V) \in \mathscr{U}_X$ whenever $V \in \mathscr{U}_Y$.

The w^*-unformity $\mathscr{U}_{w^*}$ on A^*, where A is any Banach space, is the smallest one containing all subsets of the type

$$\{(\varphi, \varphi') \in A \times A : |\varphi(a) - \varphi'(a)| < \varepsilon\}, \tag{C.628}$$

where $a \in A$ and $\varepsilon > 0$; this implies that $U \in \mathscr{U}_{w^*}$ iff U contains some such subset.

Second, $P(A)$ carries a natural ***transition probability***, cf. Definition 1.17 and (2.43). For $\omega, \omega' \in P(A)$, this function $\tau : P(A) \times P(A) \to [0,1]$ is defined by

$$\tau(\omega, \omega') = \inf\{\omega(a) \mid a \in A, 0 \le a \le 1_A, \omega'(a) = 1\}. \tag{C.629}$$

This definition, and the following result, are valid even if A has no unit.

Proposition C.177. *Let A be C*-algebra and define τ by* (C.629). *Then*

$$\tau(\omega, \omega') = 1 - \tfrac{1}{4}\|\omega - \omega'\|^2, \tag{C.630}$$

and the following dichotomy applies:

1. *If ω and ω' are equivalent (in the sense that the corresponding* GNS*-representations π_ω and $\pi_{\omega'}$ are unitarily equivalent), so that we may assume that the associated cyclic vectors Ω_ω and $\Omega_{\omega'}$ lie in the same Hilbert space, we have*

$$\tau(\omega, \omega') = \mathrm{Tr}\,(e_{\Omega_\omega} e_{\Omega_{\omega'}}) = |\langle \Omega_\omega, \Omega_{\omega'} \rangle|^2. \tag{C.631}$$

2. *If ω and ω' are inequivalent (in that π_ω and $\pi_{\omega'}$ are inequivalent), then*

$$\tau(\omega, \omega') = 0. \tag{C.632}$$

Proof. We first show that (C.630) yields (C.631) and (C.632). In the first case,

$$\begin{aligned}\|\omega-\omega'\| &= \sup\{|\omega(a)-\omega'(a)|, a\in A, \|a\|=1\}\\ &= \sup\{|\langle\Omega_\omega,\pi_\omega(a)\Omega_\omega\rangle-\langle\Omega_{\omega'},\pi_\omega(a)\Omega_{\omega'}\rangle|, a\in A, \|a\|=1\}\\ &= \sup\{|\mathrm{Tr}((e_{\Omega_\omega}-e_{\Omega_{\omega'}})\pi_\omega(a))|, a\in A, \|a\|=1\}\\ &= \sup\{|\mathrm{Tr}((e_{\Omega_\omega}-e_{\Omega_{\omega'}})a)|, a\in \pi_\omega(A), \|a\|=1\}\\ &= \sup\{|\mathrm{Tr}((e_{\Omega_\omega}-e_{\Omega_{\omega'}})a)|, a\in B(H_\omega), \|a\|=1\}\\ &= \|e_{\Omega_\omega}-e_{\Omega_{\omega'}}\|_1, \end{aligned} \tag{C.633}$$

where $\|\cdot\|_1$ is the trace norm on $B_1(H_\omega)$. In the fifth step we used the fact that the map $a\mapsto \mathrm{Tr}(ba)$ is σ-weakly continuous for any $b\in B_1(H_\omega)$, so that we may replace the supremum over $a\in\pi_\omega(A)$ by the supremum over a in the σ-weak closure of $\pi_\omega(A)$ which by the Theorem C.130 is $\pi_\omega(A)''$, which in turn is $B(H_\omega)$ because $\pi_\omega(A)$ is irreducible (since ω is pure, cf. Theorem C.90). The last step then follows from Theorem B.146. To compute the last expression in (C.633), we assume that Ω_ω and $\Omega_{\omega'}$ are not proportional (if they are, then $\omega=\omega'$, so that (C.630) reduces to $1=1$, and hence holds). We may then work in the 2-dimensional Hilbert space spanned by $\Omega_\omega\equiv(1,0)$ and $\Omega_{\omega'}=(c_1,c_2)$, with $|c_1|^2+|c_2|^2=1$. In that case,

$$(e_{\Omega_\omega}-e_{\Omega_{\omega'}})^2 = |c_2|^2\cdot 1_2; \tag{C.634}$$

$$|e_{\Omega_\omega}-e_{\Omega_{\omega'}}| = \sqrt{(e_{\Omega_\omega}-e_{\Omega_{\omega'}})^2} = |c_2|\cdot 1_2; \tag{C.635}$$

$$\|e_{\Omega_\omega}-e_{\Omega_{\omega'}}\|_1 = \mathrm{Tr}(|e_{\Omega_\omega}-e_{\Omega_{\omega'}}|) = 2|c_2|. \tag{C.636}$$

Using (C.633), this gives

$$1-\tfrac{1}{4}\|\omega-\omega'\|^2 = 1-\tfrac{1}{4}\|e_{\Omega_\omega}-e_{\Omega_{\omega'}}\|_1^2 = 1-|c_2|^2 = |c_1|^2 = |\langle\Omega_\omega,\Omega_{\omega'}\rangle|^2. \tag{C.637}$$

To deal with the second case, we use the following version of Schur's Lemma:

Lemma C.178. *Let π_ω and $\pi_{\omega'}$ be irreducible representations of some C*-algebra A, and let $w: H_\omega\to H_{\omega'}$ be an intertwiner, i.e., a bounded linear map that satisfies*

$$w\pi_\omega(a) = \pi_{\omega'}(a)w \;\; (a\in A). \tag{C.638}$$

- *If π_ω and $\pi_{\omega'}$ are equivalent, then w is either zero or invertible.*
- *If π_ω and $\pi_{\omega'}$ are inequivalent, then w is zero.*

Proof. The proof is the same as for group representations: taking the adjoint of (C.638), it follows that $w^*w\in\pi_\omega(A)'$ and $ww^*\in\pi_{\omega'}(A)'$, so by Theorem C.90 (i.e. the mother of all Schur's lemma's) we have $w^*w=\lambda\cdot 1_{H_\omega}$ and $ww^*=\mu\cdot 1_{H_{\omega'}}$, for some $\lambda,\mu\in\mathbb{R}^+$ (since w^*w and ww^* are positive operators). Moreover, since $w=\lambda ww^*w=\mu w$, in fact we have $\lambda=\mu$ whenever $w\neq 0$. If $\lambda>0$, then the operator $(\lambda)^{-1/2}w: H_\omega\to H_{\omega'}$ is a unitary intertwiner, so π_ω and $\pi_{\omega'}$ are equivalent. If $\lambda=0$, then $w^*w=0$ and hence $w=0$, since $\|w^*w\|=\|w\|^2$. □

Continuing the proof of (C.632), we form the direct sum

$$\pi(A) = \pi_\omega(A) \oplus \pi_{\omega'}(A); \tag{C.639}$$
$$H = H_\omega \oplus H_{\omega'}. \tag{C.640}$$

The second case of Lemma C.178 then gives

$$(\pi_\omega(A) \oplus \pi_{\omega'}(A))' = \pi_\omega(A)' \oplus \pi_{\omega'}(A)', \tag{C.641}$$

whose right-hand side consist of operators $\lambda \cdot 1_{H_\omega} \oplus \mu \cdot 1_{H_{\omega'}}$ $(\lambda, \mu \in \mathbb{C})$, so that

$$(\pi_\omega(A) \oplus \pi_{\omega'}(A))'' = \pi_\omega(A)'' \oplus \pi_{\omega'}(A)'' = B(H_\omega) \oplus B(H_{\omega'}). \tag{C.642}$$

Once again using Theorem C.130, a computation à la (C.633) therefore gives

$$\begin{aligned}\|\omega - \omega'\| &= \sup\{|\mathrm{Tr}((e_{\Omega_\omega} - e_{\Omega_{\omega'}})a)|, a \in B(H_\omega) \oplus B(H_{\omega'}), \|a\| = 1\} \\ &= \sup\{|\mathrm{Tr}(e_{\Omega_\omega} a) - \mathrm{Tr}(e_{\Omega_{\omega'}} a')|, a \in B(H_\omega), a' \in B(H_{\omega'}), \|a \oplus a'\| = 1\} \\ &= \sup\{|\mathrm{Tr}(e_{\Omega_\omega} a)|, a \in B(H_\omega), \|a\| = 1\} \\ &+ \sup\{|\mathrm{Tr}(e_{\Omega_{\omega'}} a')|, a' \in B(H_{\omega'}), \|a'\| = 1\} \\ &= \|e_{\Omega_\omega}\|_1 + \|e_{\Omega_{\omega'}}\|_1 = 1 + 1 = 2,\end{aligned} \tag{C.643}$$

since the trace may be computed in a basis of $H_\omega \oplus H_{\omega'}$ consisting of a basis of H_ω and a basis of $H_{\omega'}$, and $\|a \oplus a'\| = \max\{\|a\|, \|a'\|\}$ for $a \in B(H_\omega)$ and $a' \in B(H_{\omega'})$.

Finally, we prove that (C.629) and (C.630) coincide. If ω and ω' are equivalent,

$$\tau(\omega, \omega') = \inf\{\mathrm{Tr}(e_{\Omega_\omega} \pi_\omega(a)) \mid a \in A, 0 \leq a \leq 1_A, \mathrm{Tr}(e_{\Omega_{\omega'}} \pi_\omega(a)) = 1\} \tag{C.644}$$

and, as in (C.633), Theorem C.130 allows us to replace the infimum over $a \in A$ by the one over $a \in B(H_\omega)$. The claim then follows from Theorem 2.12 and eq. (C.631).

Similarly, if ω and ω' are inequivalent, eq. (C.642) and Theorem C.130 give

$$\tau(\omega, \omega') = \inf\{\mathrm{Tr}(e_{\Omega_\omega} a) \mid a \in B(H_\omega) \oplus B(H_{\omega'}), 0 \leq a \leq 1_H, \mathrm{Tr}(e_{\Omega_{\omega'}} a) = 1\},$$

and notice that the infimum zero is reached by $a = 0 \cdot 1_{H_\omega} \oplus 1_{H_{\omega'}}$. □

The final result of this appendix, then, is the "pure" counterpart of Theorem C.172:

Theorem C.179. *Let A and B be unital C*-algebras. There is a bijective correspondence $f = \varphi^*$ between Jordan isomorphisms $\varphi : A_{\mathrm{sa}} \to B_{\mathrm{sa}}$ and bijections $f : P(B) \to P(A)$ that preserve transition probabilities and are w^*-uniformly continuous along with their inverse. In particular, $\varphi : A_{\mathrm{sa}} \to A_{\mathrm{sa}}$ is a Jordan automorphism of A iff $\varphi^* : P(A) \to P(A)$ has the properties just stated for f.*

The proof of this theorem is far more difficult than Theorem C.172, so we omit it.

If $A \cong C(X)$ and $B \cong C(Y)$ are commutative, we obtain a variation on Corollary C.22 featuring *uniform* homeomorphisms. Also, we see from Wigner's Theorem 5.4.1 that for $A = B(H)$ it is enough to consider normal pure states, in which case also the (uniform) continuity condition on f is superfluous.

Notes

As already mentioned in the Introduction, the theory of operator algebras on Hilbert spaces was created by von Neumann, partly in collaboration with his assistant Murray (von Neumann, 1930, 1931, 1938, 1940, 1949; Murray & von Neumann, 1936, 1937, 1943, reprinted in von Neumann, 1961). His motivation for doing so certainly included quantum mechanics, but also functional analysis, measure theory, ergodic theory, and representation theory, all of which fields in turn benefited from their interaction with operator algebras. Von Neumann (and Murray) studied what they called "rings of operators", which are now deservedly called von Neumann algebras. John von Neumann (1903–1957) was one of the greatest mathematicians in history, especially considering the totality of his oeuvre in pure and applied mathematics (including numerical mathematics, computer science, and mathematical economics). His work in mathematical physics, notably on the mathematical structure of quantum mechanics, in some sense forms a bridge between the two.

Von Neumann was a Hungarian prodigy; he wrote his first mathematical paper at the age of seventeen. Except for this first paper, his early work was in set theory and the foundations of mathematics. In the Fall of 1926, he moved to Göttingen to work with Hilbert. Around 1920, Hilbert had initiated his *Beweistheory*, an approach to the foundations of mathematics whose specific technical goals were not achieved because of Gödel's work, but whose overall view of mathematics (i.e. as an activity whose correctness is to be established purely syntactically and whose meaning is a semantic matter to be distinguished from its syntax) still reigns. However, at the time that von Neumann arrived, Hilbert was also interested in quantum mechanics. Apart from his broad interest in general (mathematical) physics (for example, his Sixth Problem from 1900 called for the mathematical axiomatization of physics), Hilbert was specifically attracted to quantum mechanics because Göttingen was, next to Copenhagen, a leading center for research in this area. Indeed, Heisenberg's (1925) paper initiating quantum mechanics (at least in its preliminary guise of "matrix mechanics") was followed by the *Dreimännerarbeit* of Born, Heisenberg, and Jordan (1926), and all three were in Göttingen at the time. Born was one of the few physicists of his day to be familiar with the concept of a matrix; in previous research he had even used infinite matrices. Born turned to his former teacher Hilbert for mathematical advice. Aided by his assistants Nordheim and von Neumann, Hilbert thus ran a seminar on the mathematical structure of quantum mechanics, and the three wrote a joint paper on the subject (which is now exclusively of historical value).

It was von Neumann (1927ab) who, at the age of 23, discovered the mathematical structure of quantum mechanics. In this process, he defined the abstract concept of a Hilbert space, which previously had only appeared in examples that went back to the work of Hilbert and his pupils on integral equations, spectral theory, and infinite-dimensional quadratic forms. Hilbert's famous memoirs on integral equations had appeared between 1904 and 1906; in 1908, his student Schmidt had defined the space ℓ^2 in the modern sense, and F. Riesz had studied the space of all continuous linear maps on ℓ^2 in 1912. Various examples of L^2-spaces had emerged around the same time (with hindsight, Hilbert himself mainly worked with the unit ball of ℓ^2).

However, the abstract notion of a Hilbert space was missing until von Neumann provided it. In particular, von Neumann saw that Schrödinger's wave functions were unit vectors in a Hilbert space of L^2 type, and that Heisenberg's observables were linear operators on a different Hilbert space, of ℓ^2 type. A unitary transformation between these spaces provided the the mathematical equivalence between wave mechanics and matrix mechanics. Moreover, von Neumann developed the spectral theory of bounded as well as unbounded normal operators on a Hilbert space. This work culminated in his book *Mathematische Grundlagen der Quantenmechanik* (1932).

Despite the tremendous prestige of von Neumann, initially few mathematicians recognized the importance of his subsequent theory of operator algebras. For example, after a lecture by von Neumann on operator algebras in the weekly mathematics colloquium at Harvard sometime in the 1930s, G. H. Hardy, one of the leading mathematicians of his time, is reported to have said:[1]

> "He is quite clearly a brilliant man, but why does he waste his time on this stuff?"

Fortunately, among those who did study operator algebras were Gelfand & Naimark (1943), who linked the subject to Gelfand's earlier work on (commutative) Banach algebras and in doing so created the theory of C*-algebras. This, in turn, was picked up by Segal (1947ab), who thereby also restored the link with quantum theory.

A survey of von Neumann's mathematical work is given in Oxtoby et al (1958), which contains a biographical introduction by von Neumann's friend and colleague Ulam, and some of von Neumann's correspondence is collected in Rédei (2005b), which also contains a short mathematical biography. One of the most insightful documents about von Neumann is the rare manuscript Vonneumann (1987) by his brother Nicholas, of which the author got a copy from von Neumann's only PhD student Israel Halperin, who visited Cambridge on a peace mission in the early 1990s.[2] Politically, von Neumann was a controversial figure because of his enthusiastic contributions to nuclear weapons and the arms race between the USA and the Soviet Union; see Heims (1980) and Macrae (1992) for different perspectives on this. A substantial scholarly scientific biography of von Neumann remains to be written.

The history of operator algebras (i.e. von Neumann algebras and C*-algebras, which terms were probably introduced by Dieudonné and Segal, respectively) has been described in Kadison (1982), Doran & Belfi (1986), and Doran (1994).

Leading textbooks on operator algebras, written by some of the original contributors, are Neumark (1968), Sakai (1971), Dixmier (1977, 1981), Pedersen (1979), Kadison & Ringrose (1983, 1986), and Takesaki (2002, 2003a, 2003b). See also Murphy (1990), Li (1992), Davidson (1996), Blackadar (2006), and the remarkable lectures on von Neumann algebras by algebraic topologist Lurie (2011). Connes (1994), written by arguably the greatest contemporary mathematician working in operator algebras, also provides innumerable fascinating insights into the subject.

---

[1] Reported by G.D. Birkhoff (who overheard Hardy saying this) to his son, Garrett Birkhoff, who in turn mentioned it to G.C. Rota, who wrote it down in the Introduction to Stern (1991).

[2] According to Rhodes (1996, pp. 245–246), Halperin was a spy for the Soviet Union, although his evidence seems limited to the fact Halperin was arrested in 1946 *suspected* of espionage, having Klaus Fuchs in his address book.

§C.1.**Basic definitions and examples**

As in the notes to the previous appendix, we only comment on results whose origins are less well known or which are less standard by themselves, the rest belonging to the foundations of the field as described in the textbooks just mentioned. Once again, for this reason not all sections in this appendix come with notes.

§C.2. **Gelfand isomorphism**

The implication $\omega \in \Sigma(A) \Rightarrow \omega(a) \in \sigma(a)$ $(a \in A)$ in the proof of Lemma C.9 also holds in the oppositie direction (given that A is a Banach algebra with unit and $\omega : A \to \mathbb{C}$ is linear); this is the ***Gleason-Kahane-Zelazko Theorem*** (Sourour, 1994). A recent monograph about $C(X)$ is Groenewegen & van Rooij (2016), following up on earlier books like Semadeni (1971) and Gillman & Jerison (1976).

§C.3.**Gelfand duality**

Proposition C.19 is due to Gelfand & Kolmogorov (1939). In the spirit of the proof of the Stone–Weierstrass Theorem B.51 in §B.10, let us give an alternative proof of this proposition (Simon, 2011), which is based on Proposition C.14 and Corollary B.17. These identify $\Sigma(C(X))$ with the set $\partial_e M_1^+(X)$ of extreme completely regular probability measures on X, provided we identify the latter with the corresponding functionals on $C(X)$, as in (B.39). That is, we must prove that the map $x \mapsto \delta_x$ (i.e., the Dirac measure at x, which, seen as a functional on $C(X)$, is just the evaluation map ev_x) is a bijection.

Proof. We first show that a measure $\mu \in \partial_e M_1^+(X)$ must satisfy $\mu(A) = 1$ or $\mu(A) = 0$ for any $A \in \Sigma$. For if there is some $C \in \Sigma$ for which $0 < \mu(C) < 1$, we have a nontrivial convex decomposition $\mu = t\mu_1 + (1-t)\mu_2$, namely $t = \mu(C)$, $\mu_1(A) = \mu(A|C)$ (i.e., $\mu(A \cap C)/\mu(C)$), and $\mu_2(A) = \mu(A \backslash C)/\mu(X \backslash C)$. From this, we show that $\mathrm{supp}(\mu)$ is a point. Indeed, if both x and $y \neq x$ would lie in $\mathrm{supp}(\mu)$, we could separate these with disjoint open sets $x \in U$ and $y \in V$. This would leave four (im)possibilities:

- $\mu(U) = \mu(V) = 1$ would imply $\mu(X) \geq 2$, contradicting $\mu(X) = 1$;
- $\mu(U) = 0$ would make $U^c \cap \mathrm{supp}(\mu)$ a proper closed subset of $\mathrm{supp}(\mu)$ whose open complement has measure zero, contradicting the definition of $\mathrm{supp}(\mu)$; this applies to all four cases $\mu(V) = 0$, $\mu(V) = 1$, $\mu(U) = 1$, and $\mu(V) = 0$.

Thus $\mathrm{supp}(\mu) = \{x\}$ for some $x \in X$, i.e., $\mu = \delta_x$, so that $\partial_e M_1^+(X) \subseteq X$. Finally, we also have $X \subseteq \partial_e M_1^+(X)$, since $\delta_x = t\mu_1 + (1-t)\mu_2$ forces

$$\mathrm{supp}(\mu_1) = \mathrm{supp}(\mu_2) = \{x\}, \tag{C.645}$$

and hence $\mu_1 = \mu_2 = \delta_x$. □

In the unital/compact case, categorical Gelfand duality was first established in Negrepontis (1969, 1971), and was reproved in a different way by Johnstone (1982). Our proof of is taken from Landsman (2004), with some improvements in the non-unital case due to Brandenburg (2015), but it should be considered "folklore".

In the smooth case, Corollary C.22 is often called ***Milnor's exercise***. The result even holds without the second countability assumption on the manifold X, but with a completely different proof (Mrĉun, 2005). See also Burtscher (2009).

§C.6. **C*-algebras without unit: commutative case**

For proper maps see e.g. Bourbaki (1989), §I.10.

§C.10. **Hilbert C*-modules and multiplier algebras**

The theory of Hilbert C*-modules goes back to Kaplansky, Paschke, and Rieffel. See Lance (1995) and Raeburn & Williams for textbook coverage, and Landsman (1998a) for applications to mathematical physics (e.g. constrained quantization).

Theorem C.76 is due to An Huef, Raeburn, & Williams (2010).

The ***Cohen–Hewitt Factorization Theorem*** à la Fell & Doran (1988), Theorem v.9.2, adapted to C*-algebras, states that if A and B are C*-algebras and $\alpha : A \to B)$ is a homomorphism, then $\{\alpha(a)b \mid a \in A, b \in B\}$ is a closed linear subspace of B. Consequently, if α is nondegenerate, then each element $c \in B$ factors as $c = \alpha(a)b$. In particular, taking $B = A$ and α to be the identity, we see that Lemma C.47 may be sharpened to the claim that any $c \in A$ takes the form $c = ab$ for suitable $a, b \in A$.

§C.11. **Gelfand topology as a frame**

Our treatment of frames and locales has been borrowed from Mac Lane & Moerdijk (1992), where also the details of the proof of Theorem C.80 may be found. See also Picado & Pultr (2012). Hereditary subalgebras are discussed e.g. in Pedersen (1979) and Blackadar (2006).

The fact that $H(A)$ forms a complete lattice was noted by Akemann & Bice (2014), who also pursued the analogy with open sets, though not in a frame-theoretic setting. The theory is still disappointing in various ways, most notably in the fact that $H(A)$ fails to be a frame unless A is commutative. Also, Theorem C.86 has (so far) been proved by conventional means, i.e., via the Gelfand isomorphism; it would be preferable to prove it purely algebraically (and if possible constructively).

From a localic point of view, the Gelfand transform $\hat{a} : \Sigma(A) \to \mathbb{C}$ of $a \in A$ should primarily be described as the corresponding frame map $\hat{a}^{-1} : \mathcal{O}(\mathbb{C}) \to \mathcal{O}(\Sigma(A))$, and hence, using Corollary C.84, as a frame map

$$\hat{a}^{-1} : \mathcal{O}(\mathbb{C}) \to H(A). \tag{C.646}$$

Denoting the hereditary subalgebra generated by a by H_a, i.e., the closure of $a \cdot A$, for $U \in \mathcal{O}(\mathbb{C})$) we obtain a nice formula whose use remains to be established:

$$\hat{a}^{-1}(U) = \bigcap_{z \in \mathbb{C} \setminus U} H_{a-z}. \tag{C.647}$$

A direct proof of the last claim of Proposition C.82 uses the property $H(A) = I(A)$ (in the commutative case), the identification of $I \wedge J$ with $(IJ)^{-}$ (i.e., the closure of the linear span of all ab, $a \in I$, $b \in J$, which follows by taking an approximate unit in I or J), and the identification of $\bigvee S$ with the closure of the linear span of $\bigcup S$.

§C.13. **Tensor products of Hilbert spaces and C*-algebras**

For the proof of (C.248) see Reed & Simon (1972), Theorem II.10.

For tensor products of C*-algebra we mainly relied on Lance (1982), Li (1992), Wegge-Olsen (1993), and Takesaki (2002), by one of the founders of the theory.

Tensor products of Banach spaces and Hilbert spaces were first studied by Schatten (1946) and Schatten & von Neumann (1946, 1948). The subject was subsequently taken up by Grothendieck (1955) for locally convex spaces, and hence involves two of the greatest mathematicians of the twentieth century. Nuclearity of C*-algebras is a vast and important field, to which Takesaki (2003) is a good introduction.

Yet another expression for the maximal C*-norm on $A \otimes B$ arises if we say that two representations $\pi_A : A \to B(H)$ and $\pi_B : B \to B(H)$ on the same Hilbert space H ***commute*** if $\pi_A(a)\pi_B(b) = \pi_B(b)\pi_A(a)$ for all $a \in A$ and $b \in B$. Such a pair defines a representation $\pi_A \otimes \pi_B$ of $A \otimes B$ by

$$\pi_A \otimes \pi_B(c) = \sum_i \pi_A(a_i) \otimes \pi_B(b_i), \tag{C.648}$$

which makes sense because $(a,b) \mapsto \pi(a)\pi(b)$ is bilinear and hence (by universality of $\otimes$) factors through $A \otimes B$. This gives a third formula for $\|\cdot\|_{\max}$, namely

$$\|c\|_{\max} = \sup\{\|\pi_A \otimes \pi_B(c)\|_{B(H_A \overline{\otimes} H_B)}\}, \tag{C.649}$$

where π_A and π_B run through all *commuting* representations of A and B. Indeed, the restrictions of any representation of $A \otimes B$ to A and B define commuting representations, so that although at first sight the expression (C.649) appears to majorize (C.265), it must be equal to it in view of the equality of (C.265) and (C.263).

The name ***projective tensor product*** for $A \hat{\otimes}_{\max} B$, where A and B are C*-algebras, is actually confusing, since if A and B are regarded as Banach algebras, their projective tensor product is usually defined as the completion of $A \otimes B$ in the norm

$$\|c\|_{\text{proj}} = \inf\left\{\sum_i \|a_i\|\|b_i\|, c = \sum_i a_i \otimes b_i\right\}, \tag{C.650}$$

cf. (C.259), which is defined for any two Banach algebras A and B. This may not be a C*-norm, and hence $A \hat{\otimes}_{\text{proj}} B$ may not be a C*-algebra. However, for any Banach algebra C with involution, one may canonically construct a C*-algebra C* (sic) and a homomorphism $\varphi : C \to C^*$ of involutive Banach algebras, with the universal property that for any morphism $\beta : C \to D$, where D is a C*-algebra, there is a unique homomorphism $\beta' : C^* \to D$ of C*-algebras such that $\beta = \beta' \circ \varphi$. This C*-algebra C*, which by the usual argument is unique up to isomorphism, is called the ***C*-envelope*** of C. An explicit construction is obtained by completing C in the norm

$$\|c\| = \sup\{\|\pi(c)\|\}, \tag{C.651}$$

where the supremum runs over all representations of C on Hilbert spaces; it is finite since $\|\pi(c)\| \leq \|c\|$ for each $c \in C$, see Dixmier (1977), §1.3.7 and §2.7. It is easy to see that $\|\cdot\|_{\text{proj}}$ is a cross-norm on $A \otimes B$, and that one has a bijective correspondence between representations of $A \otimes B$ that satisfy $\|\pi(a \otimes b)\| \leq \|a\|\|b\|$ and representations of $A \hat{\otimes}_{\text{proj}} B$. The point, then, is that one has $A \hat{\otimes}_{\max} B = (A \hat{\otimes}_{\text{proj}} B)^*$.

C*-algebras (with homomorphism) and $\hat{\otimes}_{\max}$ form a ***monoidal category*** (also called a ***tensor category***), with commutative C*-algebras as a full subcategory CCA. The map $X \mapsto C_0(X)$ then defines a duality *as monoidal categories* between the category LCHp of locally compact Hausdorff spaces and proper continuous maps (with cartesian product as a tensor product) and the category CCAn of commutative C*-algebras and nondegenerate homomorphisms (with its unique C*-algebraic tensor product, for example realized as $\hat{\otimes}_{\max}$). Cf. Theorem C.45. See Hofmann (1970).

§C.14. **Inductive limits and infinite tensor products of C*-algebras**

For inductive limits of C*-algebras see in particular Sakai (1971); they were originally a Japanese invention (Takeda). Infinite tensor products of operator algebras (which partly motivated inductive limits) go back to von Neumann (1938). Bounded monotone nets converge under very general conditions; see McArthur (1970).

§C.15. **Gelfand isomorphism and Fourier theory**

For details on the Haar measure and for the proof of local compactness of $\hat{G}$ see Weil (1965), §27. Our approach to the Fourier transform is largely taken from Deitmar & Echterhoff (2009), where complete proofs may be found (though we sometimes followed a slightly different approach). In particular, these authors introduced the Banach spaces $C_0^*(G)$ and $C_0^*(\hat{G})$, whose use forms a marked improvement over older and less elegant treatments, as in e.g. Rudin (1962) or Folland (1995).

Eq. (C.379) is often called ***Plancherel's Theorem***.

We may add a third entry to the 'symmetric' isomorphisms (C.379) - (C.380). The ***Bruhat space*** $\mathscr{S}(G)$ of rapidly decreasing functions on G is defined by

$$A(G) = \{f \in L^\infty(G) \mid \exists K \in \mathscr{K}(G) \forall n > 0 \exists C_n > 0 \forall k > 0 : \|f_{|G\backslash K^k}\|_\infty \leq C_n k^{\ \ n}\};$$
$$\mathscr{S}(G) = \{f \in L^\infty(G) \mid f \in A(G), \hat{f} \in A(\hat{G})\}.$$

For $G = \mathbb{R}$ this recovers the usual test functions $\mathscr{S}(\mathbb{R})$ (cf. Definition 5.64), where the condition $f \in A(\mathbb{R})$ gives rapid decrease whereas $\hat{f} \in A(\mathbb{R})$ gives smoothness. Pontryagin duality then yields an isomorphism $\mathscr{S}(G) \cong \mathscr{S}(\hat{G})$ (Osborne, 1975).

The author originally learnt the SNAG-Theorem from Barut & Raçka (1977), whose proof (due to K. Maurin) is quite different; the argument given above was inspired by the treatment of projection-valued spectral measures in Conway (2007, Ch. 9, §1), who calls them *spectral measures*. Conway also proves our Theorem C.113 as his Theorem 1.14, albeit for the case where X is compact; passage to the locally compact case may be done through unitization, as in §C.6. The need for π to be non-degenerate may then be traced back to (our) Lemma C.43.

§C.16. **Intermezzo: Lie groupoids**

For introductions to Lie groupoids see Moerdijk & Mrčun (2003) or Mackenzie (2005), who also described the link with symplectic geometry. For their use in noncommutative geometry and mathematical physics cf. Connes (1994) and Landsman (1998a, 2006b), respectively. The tangent groupoid was invented by Connes, with further contributions by Hilsum & Skandalis (1987), Weinstein (1989) and Landsman (1998a). See also Connes (1994), Landsman (2003), Higson (2010), and van Erp (2010) for applications of the tangent groupoid to index theory.

§C.17. **C*-algebras associated to Lie groupoids**

C*-algebras associated to locally compact groupoids (with Haar system) were first studied in detail by Renault (1980). Originally in the setting of foliation theory, the Lie (i.e. smooth) case was pioneered by Connes (1994), who noted in particular that Lie groupoids carry an intrinsic Haar system, and gave many interesting examples. The uniqueness of $C^*(G)$ for Lie groupoids G, i.e., the independence of the underlying left Haar system (up to isomorphism) is proved in Paterson (1999).

§C.18. **Group C*-algebras and crossed product algebras**

The *locus classicus* is Pedersen (1979), but Williams (2007) may even be better.

§C.19. **Continuous bundles of C*-algebras**

The bundles studied in this section were originally introduced by Fell (1961) and their theory was further developed by Dixmier & Douady (1963); see also Dixmier (1977), Fell & Doran (1988), and, for a modern treatment, Raeburn & Williams (1998). Lemma C.125 was part of Dixmier's definition of a continuous field of C*-algebras, before it was recast into the rather more appealing Definition C.121 by Kirchberg & Wassermann (1995) and Blanchard (1996). Theorem C.123 is due to Landsman & Ramazan (2001); see also Landsman (1998a) for a detailed discussion. Aastrup, Nest, & Schrohe (2006) discuss applications to manifolds with boundary.

§C.20. **von Neumann algebras and the σ-weak topology**

There are many other topologies on von Neumann algebras, se e.g. Takesaki (2002), Chapter II. In any case, we only scratch the surface of the subject.

§C.21.**Projections in von Neumann algebras**

The first part of the proof of Theorem C.141 is taken from Rédei (1998), Prop. 4.16. The remainder is adapted from Heunen, Landsman, & Spitters (2012). The details of the proof of Theorem C.140 may be found in Takesaki (2002), Thm. III.1.18; see also Dixmier (1981), Ch. 7 and Lurie (2011), lectures 13–17.

§C.23. **Classification of hyperfinite factors**

This material, which is a high point in modern mathematics, is explained in great detail in Takesaki (2003ab). See also Wright (1989) for the uniqueness of the hyperfinite III_1 factor. In his review **MR1030046 (91a:46059)** of the latter book for *Mathematical Reviews* in 1991, E. Størmer wrote:

> 'At the time of writing this review, by far the deepest and most difficult proof in von Neumann algebra theory is the one of Connes and Haagerup on the uniqueness of the injective factor of type III_1 with separable predual.'

The applications of C*-algebras and von Neumann algebras to quantum field theory are reviewed in Haag (1992), where the identification of the unique hyperfinite III_1 factor with local algebras of observables may be found in §V.6. This book also explains the relationship between Tomita–Takesaki theory and quantum statistical mechanics, as do Bratteli & Robinson (1981). It should be mentioned that the Tomita–Takesaki theory, including the modular group (i.e. of time translations) has a classical analogue in Poisson geometry (Weinstein, 1997), which somewhat softens the spectacular claim by Connes & Rovelli (1994) that time has a quantum-mechanical (or non-commutative) origin related to thermodynamics.

§C.24. **Other special classes of C*-algebras**

The classic reference on AW*-algebras and Rickart C*-algebras is Berberian (1972). For monotone complete C*-algebras see the monograph by Saitô & Maitland Wright (2015b). Real rank zero was introduced by Brown & Pedersen (1991), who also proved that the definition of real rank zero in the main text may be replaced by an equivalent property that is often taken as the definition:

Proposition C.180. *Let A be a unital C*-algebra. Then* $\mathrm{rr}(A) = 0$ *iff the set of self-adjoint elements with finite spectrum is dense in* A_{sa}.

See Davidson (1996), Theorem V.7.3, for a streamlined proof.

Scattered C*-algebras were independently introduced by Jensen (1977) and Huruya (1978). The results in the main text are due to Kusuda (2011).

Theorem C.167.1 should be obvious. No. 2 is due to Kusuda (2011), no. 3 may be found in Takesaki (2002), §III.1, no. 4 is (a restatement of) Theorem 2.3.7 in Saitô & Maitland Wright (2015b), no. 5 is Theorem 1.7.1 in Berberian (1972), no. 6 is Theorem 1.8.1 in the same reference, no. 7 is from Saitô & Maitland Wright (2015a), and finally no. 8 may be found in Blackadar (1994), §6.1.3.

Theorem C.169.1 is Exercise 4.6.12 in Kadison & Ringrose (1983); it should be hidden from students that the AMS published two volumes with the answers to all their exercises! No. 2 is in Kusuda (2011), no. 3 is in Pedersen (1972), no. 4 is (a restatement of) Theorem 8.2.5 in Saitô & Maitland Wright (2015b), and no. 5 easily follows from Corollary 2.7 in Saitô & Maitland Wright (2015a). See also Lindenhovius (2016), where results of this kind are used to study the invariant $\mathscr{C}(A)$.

§C.25. **Jordan algebras and (pure) state spaces of C*-algebras**

Theorem C.172 is Corollary 4.20 in Alfsen & Shultz (2001), based on Kadison (1951). See also Roberts & Roepstorff (1969). Theorem C.174 is due to Hamhalter (2015); the second step in the proof had been given earlier by Heunen & Reyes (2014). A complete proof of Lemma C.173 may be found in Bratteli & Robinson (1997), Theorem 3.2.3. In particular, Kadison's inequality is Proposition 3.2.4 in the same book. Theorem C.175 is the culmination of a long chain of argument, starting with Jacobson & Rickart (1950) and ending with Thomsen (1982). See also Bratteli & Robinson (1987), Theorem 3.2.3.

The formula (C.629) was proposed by Mielnik (1968, 1969). Otherwise, case 1 of Proposition C.177 is due to Roberts & Roepstorff (1969), who state case 2 without proof, referring to Glimm & Kadison (1960). Theorem C.179 is due to Shultz (1982). A completely different proof of the last claim, based on a reconstruction of A from $P(A)$, appears in Landsman (1998a), §I.3. Both authors add further structure to $P(A)$ to make it an invariant for A as a C*-algebra, viz. an orientation and a Poisson structure, respectively. The notion of an orientation was originally introduced by Alfsen & Shultz in order to make $S(A)$ a complete invariant for A; see their final work Alfsen & Shultz (2001, 2003).

Appendix D
Lattices and logic

In this appendix we collect some basic material from the theory of lattices, including Stone's representation theorem for Boolean lattices and the connection between Boolean (Heyting) lattices and classical (intuitionistic) propositional logic. In preparation for Appendix E, we also provide an introduction to first-order logic.

D.1 Order theory and lattices

One hopes that the reader has seen some of the following concepts before!

Definition D.1. *1. A* **preorder** *on a set X is a subset $R \subset X \times X$ (i.e., a* **relation** *on X), where we write $x \leq y$ or $y \geq x$ iff $(x,y) \in R$, such that $x \leq x$, and $x \leq y$ and $y \leq z$ imply $x \leq z$. A preorder is a* **partial order** *if in addition $x \leq y$ and $y \leq x$ imply $x = y$. A set with a partial order is called a* **poset** *(for* **p***artially* **o***rdered* **set***). A a poset (or preorder) is* **directed** *if every pair $\{x,y\}$ has an upper bound, i.e., some z for which $x \leq z$ and $y \leq z$. A poset may have a largest element (also called a* **top element***) denoted by 1 or $\top$ that satisfies $x \leq \top$ for each $x \in X$, and/or a smallest element (also called a* **bottom element***) 0 or $\perp$ that satisfies $\perp \leq x$ for each $x \in X$. For $x, z \in X$, the* **order interval** *$[x,z]$ is defined by*

$$[x,z] = \{y \mid x \leq y \leq z\}. \tag{D.1}$$

An **atom** *in a poset with 0 is an element $x \neq 0$ for which $[0,x] = \{0,x\}$.* In other words, x is an atom if $x \neq 0$, and $0 \leq y \leq x$ implies $y = 0$ or $y = x$. Thus x is an atom iff x ***covers*** 0, where we say that x covers y if $x \neq y$ and $[y,x] = \{y,x\}$.

A **homomorphism** *between posets is a map that preserves $\leq$. As usual, an* **isomorphism** *is an invertible (i.e. bijective) homomorphism, such that the inverse also preserves the given structure (which, in this case, is $\leq$).*

Thus a bijection $\varphi : X \to Y$ between posets X and Y is an isomorphism when $\varphi(x) \leq \varphi(y)$ iff $x \leq y$). In some cases, the inverse of a bijective homomorphism automatically preserves the relevant structure.

K. Landsman, *Foundations of Quantum Theory*,
Fundamental Theories of Physics 188, DOI 10.1007/978-3-319-51777-3

2. *A* **lattice** *is a poset in which for any two elements x, y, there exists:*
   - *an element $x \vee y$, called the a* **supremum (sup)** *of x and y, such that*

$$x \leq x \vee y; \tag{D.2}$$
$$y \leq x \vee y, \tag{D.3}$$

   and if $x \leq z$ and $y \leq z$ for some z, then $x \vee y \leq z$;
   - *an element $x \wedge y$, called the* **infimum (inf)** *of x and y, such that*

$$x \geq x \wedge y; \tag{D.4}$$
$$y \geq x \wedge y, \tag{D.5}$$

   and if $x \geq z$ and $y \geq z$ for some z, then $x \wedge y \geq z$.

   Suprema and infima are unique (if they exist). Equivalently, a lattice may be defined algebraically (rather than order-theoretically) as a set equipped with two idempotent, commutative, and associative binary operations $\vee, \wedge$ that satisfy

$$x \vee (y \wedge x) = x; \tag{D.6}$$
$$x \wedge (y \vee x) = x. \tag{D.7}$$

   The corresponding partial ordering is then defined by $x \leq y$ if $x \wedge y = x$.
3. *A* **homomorphism** *between lattices is a map that preserves $\vee$ and $\wedge$.*
   In this case, an isomorphism of lattices may be defined as a bijective homomorphism, which automatically preserves $\vee$ and $\wedge$, (and similarly in all other cases below). One may also consider ***order homomorphisms*** between lattices just regarded as posets. This is a weaker notion: a lattice homomorphism is an order homomorphism, but not necessarily the other way round. However, an order isomorphism between lattices turns out to be the same as a lattice isomorphism.
4. *A* **complete lattice** *is a poset X in which* each *subset S of X has a supremum $\bigvee S$ (i.e., $x \leq \bigvee S$ for each $x \in S$, and $x \leq z$ for each $x \in S$ implies $\bigvee S \leq z$), as well as an infimum $\bigwedge S$ (i.e., $x \geq \bigwedge S$ for each $x \in S$, and $x \geq z$ for each $x \in S$ implies $\bigvee S \geq z$).* Clearly, taking S finite makes a complete lattice (merely) a lattice. A complete lattice X has a largest element $0 = \bigvee X$ and a smallest element $1 = \bigwedge X$.
5. *A lattice is* **distributive** *if either one (and hence both) of the following equivalent properties holds:*

$$x \vee (y \wedge z) = (x \vee y) \wedge (x \vee z); \tag{D.8}$$
$$x \wedge (y \vee z) = (x \wedge y) \vee (x \wedge z). \tag{D.9}$$

6. *A* **frame** *is a complete lattice X which is "infinitely distributive" in that*

$$x \wedge \bigvee S = \bigvee \{x \wedge y, y \in S\}, \tag{D.10}$$

   for arbitrary subsets $S \subset X$. A frame is clearly distributive. Frame homomorphism by definition preserve finite infima and arbitrary suprema.

7. *A* **Heyting algebra** *is a lattice X with top* $\top$ *and bottom* $\perp$*, equipped with a map* $\dashrightarrow: X \times X \to X$*, called* **(material) implication** *that satisfies*

$$x \leq (y \dashrightarrow z) \text{ iff } (x \wedge y) \leq z. \tag{D.11}$$

A Heyting algebra is automatically distributive. Negation is *defined* by

$$\neg x \equiv (x \dashrightarrow \perp). \tag{D.12}$$

A Heyting algebra is ***complete*** when it is complete as a lattice, in that arbitrary suprema (and hence also infima) exist. *In that case,* (D.10) *is satisfied, so that a complete Heyting algebra is a frame. Conversely, a frame becomes a complete Heyting algebra if we define the implication arrow* $\dashrightarrow$ *by*

$$y \dashrightarrow z = \bigvee \{x \in X \mid x \wedge y \leq z\}. \tag{D.13}$$

However, frames and complete Heyting algebras drift apart as soon as morphisms are concerned, for although in both cases one requires maps to preserve the partial order, maps between Heyting algebras must preserve $\dashrightarrow$ rather than $\bigvee$.

8. *An* **orthocomplementation** *on a lattice (poset) X with* 0 *and* 1 *is a map*

$$\perp: X \to X, \;\; x \to x^{\perp}, \tag{D.14}$$

that satisfies:

$$x^{\perp\perp} = x; \tag{D.15}$$

$$x \leq y \text{ iff } y^{\perp} \leq x^{\perp}; \tag{D.16}$$

$$x \wedge x^{\perp} = 0 \;\; (x \wedge x^{\perp} \text{ exists and equals } 0); \tag{D.17}$$

$$x \vee x^{\perp} = 1 \;\; (x \vee x^{\perp} \text{ exists and equals } 1). \tag{D.18}$$

A lattice (poset) with an orthocomplementation is called **orthocomplemented**. *A* **homomorphism** *of orthocomplemented lattices (posets) is an lattice (order) morphism that also preserves the orthocomplementation, as well as 0 or 1.*

9. *A lattice is called* **modular** *if* $x \leq z$ *implies* $x \vee (y \wedge z) = (x \vee y) \wedge z$ *for each* y *(i.e., if distributivity holds merely if* $x \leq z$*).*
   Hence modularity is a weakening of the following property:
10. *A distributive orthocomplemented lattice is called* **Boolean**. *A* **homomorphism** *between Boolean lattices is just a homomorphism of orthocomplemented lattices. An* **isomophism** *of Boolean lattices is a map that preserves* $\vee$*,* $\wedge$*, and* $\perp$*, i.e., an invertible homomophism.*
11. *An orthocomplemented lattice (poset) is called* **orthomodular** *if* $x \leq z$ *implies (that* $x \vee z^{\perp}$ *and hence* $x^{\perp} \wedge z$ *exist and that)*

$$x \vee (x^{\perp} \wedge z) = z. \tag{D.19}$$

That is, the modularity axiom holds for $y = x^{\perp}$ (note that $x \vee (x^{\perp} \wedge z) = z$ exists because $x \leq x \vee z^{\perp}$). For lattices this axiom is equivalent to each of:

- $x \leq z$ *and* $x^{\perp} \wedge z = 0$ *imply* $x = z$.
- *xCy iff yCx, where xCy if* $x = (x \wedge y) \vee (x \wedge y^{\perp})$ (i.e., x and y are ***compatible***). In a Boolean lattice any two elements are compatible, reconfirming the fact that orthomodularity is a weakening of modularity and hence of Booleanity. *A* **homomorphism** *between orthomodular lattices (posets) is a map* φ *that preserves 0 and* $\perp$ *(and hence preserves 1), and satisfies* $\varphi(x \vee y) = \varphi(x) \vee \varphi(y)$ *(if* $x \leq y^{\perp}$*). An* **isomorphism** *between orthomodular lattices (posets) is an invertible homomorphism, which is automatically an order isomorphism.*

Every Boolean algebra is a Heyting algebra, but not *vice versa*; a Heyting algebra is Boolean iff one and hence both of the following equivalent conditions hold:

$$\neg\neg x = x \qquad (x \in X); \tag{D.20}$$

$$(\neg x) \vee x = \top \qquad (x \in X), \tag{D.21}$$

which state the ***law of the excluded middle*** (famously denied by Brouwer).

The following result will be used implicitly throughout the main text.

Proposition D.2. *An order isomorphism of a lattice preserves all suprema and infima that exist. Hence in a complete lattice all suprema and infima are preserved.*

An important source of orthocomplemented lattices is provided by (possibly infinite-dimensional) complex vector spaces V with inner product, cf. Definition A.1: the elements of X are the ***orthoclosed*** subspaces $L \subset V$, i.e., those subspaces for which $L^{\perp\perp} = L$, where $L^{\perp\perp} = (L^{\perp})^{\perp}$, and orthocomplementation is defined by

$$L^{\perp} = \{v \in V \mid \forall w \in L : \langle v, w \rangle = 0\}, \tag{D.22}$$

and the partial ordering is given by inclusion. This yields

$$L \wedge M = L \cap M; \tag{D.23}$$

$$L \vee M = (L+M)^{\perp\perp} = (L^{\perp} \cap M^{\perp})^{\perp}, \tag{D.24}$$

where $L+M$ is the linear span of L and M. We have the ***Amemiya–Araki Theorem:***

Theorem D.3. *The lattice of orthoclosed subspaces of an inner product space V is orthomodular iff V is* complete *in the norm* (A.2) *associated to the inner product.*

A space X is called ***totally disconnected*** if it has no other connected subspaces than its points (so any larger subspace $\neq X$ is the union of two proper clopen sets).

Definition D.4. *A* **Stone space** *is a totally disconnected compact Hausdorff space.*

Any finite set (with the discrete topology) is a Stone space. The best-known example of an infinite Stone space is the Cantor set $\{0,1\}^{\mathbb{N}}$ with product topology, which in addition is metrizable and has no isolated points (these properties even characterize the Cantor set up to homeomorphism). ***Stone's Representation Theorem*** reads:

Theorem D.5. *A lattice L is Boolean iff it is isomorphic to the lattice* Clopen(X) *of all clopen subsets of some Stone space X (partially ordered by set-theoretic inclusion), where X is uniquely determined by L up to homeomorphism.*

Thus the lattice operations in Clopen(X) are simply geven set-theoretically by

$$U \vee W = U \cup W; \tag{D.25}$$
$$U \wedge V = U \cap W, \tag{D.26}$$

with orthocomplementation given by set-theoretic complementation (the theorem is obviously predicated on the fact that such a lattice is Boolean). The space X is called the ***Stone spectrum*** of L, generically denoted by $\mathscr{S}(L)$. Just like Gelfand duality, Theorem D.5 extends to a categorical duality theorem in an obvious way.

The Stone spectrum $\mathscr{S}(L)$ of L has the following canonical realizations:

1. Consider the space $\mathrm{Pt}(L) = \mathrm{Hom}(L, \underline{2})$, where $\underline{2} = \{0,1\}$ is seen as a Boolean lattice ordered by $0 \leq 1$ (and $0 \neq 1$), with topology inherited from the product topology on 2^L. That is, the basic opens in $\mathrm{Pt}(L)$ are the sets
$$U_x = \{\varphi \in \mathrm{Pt}(L) \mid \varphi(x) = 1\}, \tag{D.27}$$
where $x \in L$, and similarly with $1 \rightsquigarrow 0$. This is a Stone space, with isomorphism
$$L \xrightarrow{\cong} \mathrm{Clopen}(\mathrm{Pt}(L)); \tag{D.28}$$
$$x \mapsto U_x. \tag{D.29}$$
2. Generalizing the case of a power set (cf. Definition B.49), a ***filter*** in a (Boolean) lattice L is a nonempty subset $F \subset L$ such that $x, y \in F$ implies $x \wedge y \in F$, and $y \geq x \in F$ implies $y \in F$ (whence $1 \in F$). A filter F is ***proper*** if $F \neq L$, which is the case iff $0 \notin F$. An ***ultrafilter*** is a filter that is maximal in the set of all proper filters, ordered by inclusion. Ultrafilters (i.e. maximal filters) in a Boolean lattice are the same as ***prime filters***, which are filters for which $x \vee y \in F$ implies $x \in F$ or $y \in F$. More generally, in a distributive lattice with 0 any maximal filter is prime, and the presence of an orthocomplementation also gives the converse inclusion. Moreover, a filter F in a Boolean lattice is maximal (and hence prime) iff for any $x \in L$ either $x \in F$ or $x^{\perp} \in F$ (but not both). For $x \in L$, let
$$U'_x = \{F \in \mathscr{U}(L) \mid x \in F\}, \tag{D.30}$$
where $\mathscr{U}(L)$ is the set of all ultrafilters on L. One has $U'_x \cap U'_y = U'_{x \wedge y}$, as well as $U'_x \cup U'_y = U'_{x \vee y}$, $U'_x \subseteq U'_y$ if $x \leq y$, and subsets $U'_x \subset \mathscr{U}(L)$ form the basis of a topology on $\mathscr{U}(L)$ whose open sets are sets $U' \subseteq \mathscr{U}(L)$ with the property that for each $F \in U'$ there is $x \in L$ with $F \in U'_x \subseteq U'$. This topology makes $\mathscr{U}(L)$ a Stone space, whose basis of clopen sets is given by the U'_x, $x \in L$, with isomorphism

$$L \overset{\cong}{\to} \text{Clopen}(\mathscr{U}(L)); \tag{D.31}$$
$$x \mapsto U'_x. \tag{D.32}$$

3. Instead of *filters*, one may consider the dual notion of *ideals*, obtained by reversing the order (and hence swapping $\wedge$ and $\vee$). Thus an ***ideal*** in L is a subset $I \subseteq L$ such that $x, y \in I$ implies $x \vee y \in I$, and $y \leq x \in I$ implies $y \in I$ (whence $0 \in I$). An ideal I is ***proper*** if $I \neq L$, which is the case iff $1 \notin I$. A ***maximal ideal*** is an ideal that is maximal in the set of all proper ideals, ordered by inclusion. In a Boolean lattice, maximal ideals coincide with ***prime ideals***, which are ideals I that do not contain 1, and where $x \wedge y \in I$ implies $x \in I$ or $y \in I$. In a distributive lattice with 0 any maximal ideal is prime. The (set-theoretic) complement of a maximal ideal is a maximal filter (i.e. an ultrafilter), so that an ideal I in a Boolean lattice is maximal (and hence prime) iff for any $x \in L$ either $x \in I$ or $x^{\perp} \in I$ (but not both). The space $\mathscr{I}(L)$ of all maximal (i.e. prime) ideals in L is topologized by basic opens $U''_x = \{I \in \mathscr{I}(L) \mid x \notin I\}$, and so this time the desired isomorphism is

$$L \overset{\cong}{\to} \text{Clopen}(\mathscr{I}(L)); \tag{D.33}$$
$$x \mapsto U''_x. \tag{D.34}$$

4. Finally, the set $\text{Idl}(L)$ of *all* ideals in a (Boolean) lattice L is a frame if it is partially ordered by inclusion (cf. §C.11). One may realize the points of the frame $\text{Idl}(L)$ as its prime elements (cf. Lemma C.85), which are simply the prime (and hence maximal) ideals in L considered above. Hence $\text{Pt}(\text{Idl}(L))$ forms a model of the Stone spectrum X of L, too. The advantage of this realization is that it gives a direct description of the topology of X (seen as a frame), namely as

$$\mathscr{O}(X) \cong \text{Idl}(L). \tag{D.35}$$

The relationship between the first three approaches is that for any $\varphi \in \text{Pt}(L)$, the set $\varphi^{-1}(\{1\})$ is a maximal filter in L, whose complement $\varphi^{-1}(\{0\})$ is a maximal ideal. This can be shown to give homeomorphisms $\text{Pt}(L) \cong \mathscr{U}(L) \cong \mathscr{I}(L)$, under which the opens U_x, U'_x, and U''_x are mapped to each other. The (contravariant) functorial nature of the Stone spectrum comes out particularly clearly in the first description: given a homomorphism $h : L \to L'$, we immediately obtain a map $h^* : \text{Pt}(L') \to \text{Pt}(L)$ by pullback (i.e., $h^*\varphi = \varphi \circ h$). In this description the isomorphism $X \overset{\cong}{\to} \text{Pt}(\text{Clopen}(X))$ is given by $x \mapsto \varphi_x$, where $\varphi_x(U) = 1_U(x)$, with $U \in \text{Clopen}(X)$. In the second description, the isomorphism $X \overset{\cong}{\to} \mathscr{U}(\text{Clopen}(X))$ is given by $x \mapsto \{U \in \text{Clopen}(X) \mid x \in U\}$, which also gives the isomorphism $X \overset{\cong}{\to} \mathscr{I}(\text{Clopen}(X))$ of the third description as $x \mapsto \{U \in \text{Clopen}(X) \mid x \notin U\}$.

Eq. (D.35) follows from Theorem D.5, which implies an isomorphism of frames

$$\mathscr{O}(X) \overset{\cong}{\to} \text{Idl}(\text{Clopen}(X)); \tag{D.36}$$
$$U \mapsto \{V \in \text{Clopen}(X) \mid V \subseteq U\}, \tag{D.37}$$

with inverse $I \mapsto \bigcup_{U \in I} U$. However, by itself, eq. (D.35) may also be taken as a constructive version of Stone's Representation Theorem; the next, non-constructive step (relying on Zorn's Lemma) then gives the points of X from $\mathrm{Idl}(L)$, cf. §C.11.

To close this brief introduction to lattice theory, we present a general construction of free distributive lattices, possibly with relations, which will be needed for the theory of the constructive Gelfand spectrum in §12.2. The main advantage of this construction is that it can be performed in any topos, as will indeed be done in §12.4.

Definition D.6. *The* **free distributive lattice** $\mathscr{L}_S$ *on a set* S *is the set of* irredundant *finite subsets* $\{A_1,\ldots,A_n\}$ *of the finite power set* $\mathscr{P}_f$ *of* S*, i.e.,* $A_i \subset S$*,* $|A_i| < \infty$*,* $n \in \mathbb{N}$*, and no* A_i *is a proper subset of any* A_j*, with lattice operations inductively generated (using distributivity) from the following singleton cases:*

$$\{\{s\}\} \vee \{\{t\}\} = \{\{s\},\{t\}\}; \tag{D.38}$$
$$\{\{s\}\} \wedge \{\{t\}\} = \{\{s,t\}\}. \tag{D.39}$$

For $\{A_1,\ldots,A_n\} \in \mathscr{L}_S$ as above, and similarly $\{B_1,\ldots,B_m\} \in \mathscr{L}_S$, these rules imply

$$\{A_1,\ldots,A_n\} \vee \{B_1,\ldots,B_m\} = \{A_1,\ldots,A_n,B_1,\ldots,B_m\}_{\mathrm{ir}}; \tag{D.40}$$
$$\{A_1,\ldots,A_n\} \wedge \{B_1,\ldots,B_m\} = \{A_i \cup B_j \mid i = 1,\ldots,n, j = 1,\ldots,m\}_{\mathrm{ir}}, \tag{D.41}$$

where the subscript ir means that redundancies in the above sense have been removed by deleting any set on the list that properly contains some other set on the list. The motivation for this rule is that, using distributivity, any element x of a distributive lattice can be brought into the ("normal") form $x = x_1 \vee \cdots \vee x_n$, where each $x_i = y_i^{(1)} \wedge \cdots \wedge y_i^{(m_i)}$ is a finite meet. We then identify A_i with $\{y_i^{(1)},\cdots,y_i^{(m_i)}\}$, so that $x_i = \bigwedge A_i$, and identify $\{A_1,\ldots,A_n\}$ with $x_1 \vee \cdots \vee x_n$. If we allow empty sets (as we do), then $\mathscr{L}_S$ has both a bottom element $\bot = \bigvee \emptyset$ and a top element $\top = \bigwedge \emptyset$.

Consequently, an equivalent description of $\mathscr{L}_S$ is to first define the set Σ of all formal expressions inductively defined by the rules: (i) $S \subset \Sigma$, $\bot \in \Sigma$, and $\top \in \Sigma$; (ii) if $x \in \Sigma$ and $y \in \Sigma$, then $x \vee y \in \Sigma$ and $x \wedge y \in \Sigma$. Secondly, we quotient Σ by the equivalence relation generated by all of the basic identities in a distributive lattice, i.e., the commutativity, associativity, idempotency, and distributivity laws for $\vee$ and $\wedge$, the rules $x \vee \bot = x$ and $x \wedge \top = x$, and the absorption law $x \vee (x \wedge y) = x$. The lattice operations on the quotient are the ones inherited from concatenation on Σ.

As in most free constructions, the map $S \mapsto \mathscr{L}_S$ is left adjoint to the forgetful functor from the category of distributive lattices into Sets. One has a canonical map $i : S \to \mathscr{L}_S$, given by $i(s) = \{s\}$, with the universal property that any function $f : S \to L$ from S to some distributive lattice L factors through $\mathscr{L}_S$, i.e., there is a unique lattice homomorphism $g : \mathscr{L}_S \to L$ such that $f = g \circ i$. Indeed, g may be inductively generated from the special case $g(\{\{s\}\}) = f(s)$ using the rules (D.38) - (D.39).

One may enrich this construction by introducing a congruence $\sim$ on $\mathscr{L}_S$, e.g., one generated by relations $x_i = y_i$, $i \in I$. In that case, the ensuing quotient $\mathscr{L}_S/\sim$ exists, and is universal for homomorphisms $f : \mathscr{L}_S \to M$ of distributive lattices that satisfy $f(x_i) = f(y_i)$, i.e., if $p : \mathscr{L}_S \to \mathscr{L}_S/\sim$ is the canonical projection, there is a unique homomorphisms of distributive lattices $g : (\mathscr{L}_S/\sim) \to M$ such that $f = g \circ p$.

D.2 Propositional logic

The topos-theoretical approach to quantum logic discussed in Chapter 12 uses an advanced version of an elementary construction in algebraic logic that relates classical propositional logic to Boolean algebras (or lattices), and similarly relates intuitionistic propositional logic to Heyting algebras. Tough easy to state, these relationships are conceptually quite deep, based as they are on a separation between syntax and semantics that is decidedly "modern", reflecting a view on the nature of mathematics that would have been completely foreign to e.g. Newton and Euler or even Gauss, not to speak of Euclid and Archimedes, notwithstanding their use of the axiomatic-deductive method that has been a defining property of (real) mathematics since its birth in Plato's Academy. As expressed by Boole himself, this modern view is:

> 'They who are acquainted with the present state of the theory of Symbolic Algebra, are aware that the validity of the processes of analysis does not depend upon the interpretation of the symbols which are employed, but solely upon the laws of their combination.' (Boole, 1847, Preface)

The formalization of mathematics starts with ***propositional logic***, whose notation consists of the following groups of symbols in terms of which a ***theory*** is defined:

1. ***Purely logical symbols*** $\neg$, $\wedge$, $\vee$, and $\rightarrow$ (which, *because of the axioms they will be subject to*, may later be interpreted as *not, and, or, implies*, respectively).
2. ***Non-logical symbols*** $p_1, p_2, \ldots$ (also written $p, p', \ldots$ or $p, q, \ldots$), which denote ***atomic propositions*** (being the simplest examples of propositions, see below). The set $\Sigma = \{p_1, \ldots\}$ (at most countable) is called the ***signature*** of a theory.

As in arithmetic, there is some ambiguity to be dispelled. This may be done either by introducing ***brackets*** (,), subject to obvious rules we omit, or by conventions to the effect that $\neg$ "binds" symbols more strongly that $\vee$ and $\wedge$, which in turn "bind" more strongly than $\rightarrow$. For example, $\neg\alpha \vee \delta \rightarrow \beta \wedge \gamma$ is the same as $((\neg\alpha) \vee \delta) \rightarrow (\beta \wedge \gamma)$.

In propositional logic (unlike in first-order logic), ***well-formed formulae*** and ***propositions*** coincide; typically denoted by Greek letters $\alpha, \beta, \ldots$, both are defined as expressions in the above symbols that (iteratively) arise in the following way:

i) Each non-logical symbol $p_1, p_2, \ldots$ present in the signature Σ is a proposition.
ii) If α and β are proposition, then so are $\alpha \wedge \beta$, $\alpha \vee \beta$, $\alpha \rightarrow \beta$, and $\neg\alpha$.

Also here one may use brackets in the obvious way, e.g., if α is $p_1 \rightarrow p_2$, and β is $p_1 \wedge p_3$, then $(p_1 \rightarrow p_2) \rightarrow (p_1 \wedge p_3)$ is the same as $\alpha \rightarrow \beta$.

For example, one may check that the following expression is a valid proposition:

$$(p_1 \rightarrow (p_2 \rightarrow p_3)) \rightarrow ((p_1 \rightarrow p_2) \rightarrow (p_1 \rightarrow p_3)). \quad \text{(D.42)}$$

A final informal symbol we use is $\equiv$, as in $\alpha \equiv \beta$, which has no logical meaning, but states that α is the same as β (e.g., for $\alpha \equiv (p_1 \rightarrow (p_2 \rightarrow p_3))$, consider $\neg\alpha$).

The notion of a (propositional) *theory* will be picked up later, but we now interrupt the construction of the syntax of propositional logic and discuss its ***semantics***. In its most elementary form, this means that there is a ***valuation*** on Σ, i.e.,

$$V : \Sigma \to \{0,1\}, \tag{D.43}$$

also called a ***truth function***, where 0 = false and 1 = true; one often writes $\alpha = 1$ for $V(\alpha) = 1$ (i.e., α is true, and $\alpha = 0$ if α is false (this formally introduces a new symbol "=", which however is foreign to propositional logic). Let B_Σ be the set of all propositions (i.e., well-formed formulae) on the given signature Σ. With abuse of notation (justified by the property $\Sigma \subset B_\Sigma$), V uniquely extends to a function

$$V : B_\Sigma \to \{0,1\}, \tag{D.44}$$

as follows. First, each $V(p_i)$ is fixed by the given function (D.43). Second, the value of V on compound expressions is (iteratively) determined through the use of ***truth tables***, which formalize the everyday meaning of the symbols $\neg, \wedge, \vee, \to$:

α	$\neg\alpha$
0	1
1	0

α	β	$\alpha \wedge \beta$
0	0	0
0	1	0
1	0	0
1	1	1

α	β	$\alpha \vee \beta$
0	0	0
0	1	1
1	0	1
1	1	1

α	β	$\alpha \to \beta$
0	0	1
0	1	1
1	0	0
1	1	1

The first table should be read as follows: if α is false, then $\neg\alpha$ is true, and if α is true, than $\neg\alpha$ is false. Similarly, the second table means that if α and β are both false, then so is $\alpha \wedge \beta$, etc. For example, to see if $\gamma \equiv p_1 \wedge (\neg p_2)$ is true or false given the valuation $p_1 = p_2 = 0$, we first look at the truth table for $\neg$ with $\alpha \equiv p_2$, inferring from the first row that $\neg p_2 = 1$ als $p_2 = 0$. We subsequently inspect the table for $\wedge$ with $\alpha \equiv p_1$ and $\beta \equiv \neg p_2$. Since $p_1 = p_2 = 0$ is the same as $p_1 = 0$ and $\neg p_2 = 1$, we look at the second row, obtaining $\gamma = 0$. Another example, just involving the implication symbol $\to$, is (D.42), given e.g. $p_1 = 1$, $p_2 = 0$, and $p_3 = 1$. This is settled through the following steps, each of which involves the table for $\to$:

1. Taking $\alpha \equiv p_2 = 0$ and $\beta \equiv p_3 = 1$, the second row gives $(p_2 \to p_3) = 1$.
2. Taking $\alpha \equiv p_1 = 1$ and $\beta \equiv (p_2 \to p_3) = 1$, row 4 yields $(p_1 \to (p_2 \to p_3)) = 1$.
3. Similarly, $(p_1 \to p_2) = 0$ and $(p_1 \to p_3) = 1$.
4. From these, the second row gives $((p_1 \to p_2) \to (p_1 \to p_3)) = 1$.
5. Finally, taking $\alpha \equiv (p_1 \to (p_2 \to p_3)) = 1$ and $\beta \equiv ((p_1 \to p_2) \to (p_1 \to p_3)) = 1$ in the fourth row gives

$$(p_1 \to (p_2 \to p_3)) \to ((p_1 \to p_2) \to (p_1 \to p_3)) = 1. \tag{D.45}$$

The proposition in (D.42) is actually rather special, in that *all* truth values for the atomic propositions (p_1, p_2, p_3) it contains make it true (as is easily checked).

Definition D.7. *A proposition φ that is true whatever the (un)truth of the atomic propositions it contains, is called a* **tautology**, *denoted by* $\models \varphi$.

For example, $\alpha \to \alpha$ is a tautology for any proposition α; this follows from the truth table for $\to$ by replacing β by α, in which case only the first and the fourth rows are consistent (both yielding 1). Introducing a new logical symbol $\leftrightarrow$ by stipulating that $\alpha \leftrightarrow \beta$ is the same as $(\alpha \to \beta) \wedge (\beta \to \alpha)$, then one easily proves:

Theorem D.8. *The proposition $\alpha \leftrightarrow \beta$ is a tautology iff α and β are either both true or both false for each joint truth value of the atomic propositions they contain.*

Here α and β need not contain the same atomic propositions, but if they do, this proposition says that $\alpha \leftrightarrow \beta$ is a tautology iff α and β have the same truth table.

Here and in what follows, one should distinguish theorems *about* logic from theorems *within* logic. The former are themselves derived from logical rules that can be formalized, as first done by Hilbert and his school in "meta-mathematics". The latter is what we now turn to, motivated by the above semantic intermezzo. The syntax of any logical system, such as propositional logic, is completed by stating axioms and deduction rules that enable one to prove ***theorems***. In the case of propositional logic, these are propositions (i.e., expressions correctly formed from rules i) and ii) above) that can be derived from the axioms and deduction rules in a finite number of steps, starting with (some of) the axioms and applying (some of) the deduction rules to the previous step of the proof. The axioms are considered to be theorems, too. Theorems are often denoted by φ, and to show that a proposition φ is indeed a theorem we write $\vdash \varphi$. Thus the question if $\vdash \varphi$ holds is purely syntactic, and hence is independent of the truth-value of the atomic propositions p_i in φ.

This is a baby version of the fundamental idea of Boole mentioned above, that the possible meaning of mathematical symbols should not affect the validity of mathematical reasoning about them, Nonetheless, there is a consistency requirement (on the axioms and deduction rules) that one should not be able to derive φ if φ is semantically false under some truth assignment to the atomic propositions it contains. In other words, *a theorem must be true for any truth assignment to the pertinent atomic propositions*, or, then again, *a theorem within propositional logic must be a tautology*, symbolically: $\vdash \varphi$ implies $\vDash \varphi$ (meta-mathematically). This is the ***soundness*** condition on any logical system. Conversely, one would like to prove as many true propositions as possible. Optimally, this is expressed by the ***completeness*** condition that $\vDash \varphi$ imply $\vdash \varphi$. If both hold, i.e., if a system is sound as well as complete, one has $\vdash \varphi$ iff $\vDash \varphi$: in (other) words, *a proposition is a theorem iff it is a tautology.*

Achieving this should be the goal of our axioms and deduction rules. This can indeed be done in propositional logic (and also in first-order logic, on a suitable interpretation of $\vDash$, see §D.4). Even this requirement does not fix the axioms and the deduction rules, although it clearly makes any two such systems equivalent, in the sense that each leads to the same theorems (namely the tautologies). In particular, one can switch between axioms and deduction rules (matters like this were first systematically sorted out by Hilbert and his school, notably Bernays and Ackermann, partly motivated by the *Principia Mathematica* of Russell and Whitehead).

One particularly convenient choice has just a single *deduction rule*, namely:

- ***Modus ponens***: if $\vdash \alpha$ and $\vdash \alpha \to \beta$, then $\vdash \beta$.

Even so, the *axioms* of propositional logic may be stated in many different ways. Although it is even possible to use a single logical symbol (namely the ***Sheffer stroke*** $|$, called NAND in computer science, where $\alpha|\beta$ means $\neg(\alpha \wedge \beta)$), we proceed less radically and initially use two symbols. To this end, it is easy to show that

$$\alpha \wedge \beta \leftrightarrow \neg(\alpha \rightarrow \neg\beta) \tag{D.46}$$

$$\alpha \vee \beta \leftrightarrow \neg\alpha \rightarrow \beta \tag{D.47}$$

are tautologies, so that in principe the symbols $\vee$ and $\wedge$ are superfluous, in that $\alpha \wedge \beta$ may be regarded as an abbreviation of $\neg(\alpha \rightarrow \neg\beta)$, and likewise, $\alpha \vee \beta$ stands for $\neg\alpha \rightarrow \beta$. A possible choice for the axioms that regulate $\neg$ and $\rightarrow$, is:

$$\vdash \beta \rightarrow (\alpha \rightarrow \beta); \tag{D.48}$$

$$\vdash (\beta \rightarrow (\gamma \rightarrow \delta)) \rightarrow ((\beta \rightarrow \gamma) \rightarrow (\beta \rightarrow \delta)); \tag{D.49}$$

$$\vdash (\neg\alpha \rightarrow \neg\beta) \rightarrow ((\neg\alpha \rightarrow \beta) \rightarrow \alpha). \tag{D.50}$$

The third axiom axiom settles the use of $\neg$ and, jointly, with *modus ponens*, justifies ***proof by contradiction*** or ***reductio ad absurdum***: suppose one has established

$$\vdash \neg\alpha \rightarrow \beta; \tag{D.51}$$

$$\vdash \neg\alpha \rightarrow \neg\beta, \tag{D.52}$$

then (D.50) and *modus ponens* yield $(\neg\alpha \rightarrow \beta) \rightarrow \alpha$. Axiom (D.48) and *modus ponens* then yield α. Furthermore, as another proof technique (i.e. a theorem *about* propositional calculus) one can prove the ***deduction theorem***:

Theorem D.9. *If α and $(\gamma_1,\ldots,\gamma_n)$ imply β, then $(\gamma_1,\ldots,\gamma_n)$ imply $\vdash \alpha \rightarrow \beta$.*

Introducing an external implication symbol $\Rightarrow$, such statements are often written:

$$(\alpha,\gamma_1,\ldots,\gamma_1) \vdash \beta \Rightarrow (\gamma_1,\ldots,\gamma_1) \vdash \alpha \rightarrow \beta. \tag{D.53}$$

Writing the external "and" as a comma, one can similarly prove the rules

$$\beta \rightarrow \gamma, \gamma \rightarrow \delta \Rightarrow \beta \rightarrow \delta; \tag{D.54}$$

$$\beta \rightarrow (\gamma \rightarrow \delta), \gamma \Rightarrow \beta \rightarrow \delta. \tag{D.55}$$

As already mentioned, the central result about propositional logic is:

Theorem D.10. *For any proposition φ, one has $\vdash \varphi$ iff $\vDash \varphi$.*

Proof. We only prove the easy direction. Axioms are tautologies, and *modus ponens* preserves truth, in that $\vDash \alpha$ and $\vDash \alpha \rightarrow \beta$ imply $\vDash \beta$, as follows from the fourth row of the truth table for $\alpha \rightarrow \beta$. Hence each step in a proof preserves tautologies. □

Nonetheless, the notions of *theorem* and *tautology* are quite different conceptually: the first is defined syntactically, whereas the latter is defined semantically.

At the other end of the spectrum, we mention an axiom system that involves all four logical connectives (whilst keeping *modus ponens* as the only deduction rule):

$$\vdash (\beta \wedge \gamma) \to \beta; \tag{D.56}$$
$$\vdash (\beta \wedge \gamma) \to \gamma; \tag{D.57}$$
$$\vdash \beta \to (\gamma \to (\beta \wedge \gamma)); \tag{D.58}$$
$$\vdash \beta \to (\beta \vee \gamma); \tag{D.59}$$
$$\vdash \gamma \to (\beta \vee \gamma); \tag{D.60}$$
$$\vdash (\beta \to \delta) \to ((\gamma \to \delta) \to ((\beta \vee \gamma) \to \delta)); \tag{D.61}$$
$$\vdash \beta \to (\gamma \to \beta); \tag{D.62}$$
$$\vdash (\beta \to (\gamma \to \delta)) \to ((\beta \to \gamma) \to (\beta \to \delta)); \tag{D.63}$$
$$\vdash \neg\beta \to (\beta \to \gamma); \tag{D.64}$$
$$\vdash (\beta \to \gamma) \to ((\beta \to \neg\gamma) \to \neg\beta); \tag{D.65}$$
$$\vdash \neg\neg\beta \to \beta. \tag{D.66}$$

We now describe the relationship between propositional logic and Boolean algebras. Define an equivalence relation $\sim$ on the set B_Σ of propositions by

$$\varphi \sim \psi \text{ iff } \psi \vdash \varphi \text{ and } \varphi \vdash \psi, \tag{D.67}$$

where, as in (D.53), the notation $\psi \vdash \varphi$ means that φ can be derived from ψ, which is the case iff $\vdash \psi \leftrightarrow \varphi$. The ensuing set of equivalence classes

$$L_\Sigma = B_\Sigma / \sim \tag{D.68}$$

is called the (classical) ***Lindenbaum (–Tarski) algebra*** for the given signature Σ.

Theorem D.11. *The set L_Σ defined by* (D.68) *is partially ordered by*

$$[\psi] \leq [\varphi] \text{ iff } \psi \vdash \varphi. \tag{D.69}$$

In this ordering, the ensuing poset is a Boolean algebra, with operations

$$[\psi] \vee [\varphi] = [\psi \vee \varphi]; \tag{D.70}$$
$$[\psi] \wedge [\varphi] = [\psi \wedge \varphi]; \tag{D.71}$$
$$[\psi]^\perp = [\neg\psi]. \tag{D.72}$$

Furthermore, the bottom and top elements of L_Σ are the equivalence classes of any contradiction and any tautology, respectively. The Boolean algebra L_Σ thus obtained is the free Boolean algebra $\mathscr{B}_\Sigma$ on the set Σ, and hence any valuation (D.44) - (D.43)*, induces a homomorphism of Boolean algebras*

$$V : \mathscr{B}_\Sigma \to \{0, 1\}. \tag{D.73}$$

Here the ***free Boolean algebra*** $\mathscr{B}_\Sigma$ on a set Σ is defined as usual, namely as "the" Boolean algebra (unique up to isomorphism), along with an injection $\iota : \Sigma \to \mathscr{B}_\Sigma$, such that any map $g : \Sigma \to A$, where A is some Boolean algebra, factors through ι (i.e., there is a unique homomorphism $f : \mathscr{B}_\Sigma \to A$ such that $g = f \circ \iota$).

Constructions like this become more interesting for propositional ***theories***, in which (beyond specifying the signature Σ) further axioms are added to whatever system for which Theorem D.10 holds. Let us call the list of such axioms $\mathscr{T}$, where we assume that the theory is ***consistent***, in that no contradiction can be derived from $\mathscr{T}$ (in propositional logic—as opposed to predicate logic—this question is decidable). We also assume that $\mathscr{T}$ contains no tautologies (which would add no new theorems). We now write $\mathscr{T} \vdash \varphi$ if φ can be derived (in a finite number of steps) from $\mathscr{T}$ and the basic axioms and deduction rule(s). Unless $\mathscr{T}$ is empty, the set of theorems will be larger now (e.g., any member of $\mathscr{T}$ itself, say $\vdash p_1$, is trivially a theorem of $\mathscr{T}$). In order to preserve Theorem D.10, now in the form

$$\mathscr{T} \vdash \varphi \text{ iff } \mathscr{T} \vDash \varphi, \tag{D.74}$$

we should define the right-hand side appropriately. Call a valuation (D.43), or, equivalently, the corresponding homomorphism (D.73), a ***binary model*** of $\mathscr{T}$ if

$$V(\alpha) = 1, \tag{D.75}$$

for each $\alpha \in \mathscr{T} \subset B_\Sigma$ (by soundness this is already the case for the axioms of propositional logic *per se*). We then say that $\mathscr{T} \vDash \varphi$ iff $V(\varphi) = 1$ (i.e., φ is true) in any binary model of $\mathscr{T}$. On this definition of $\mathscr{T} \vDash$, eq. (D.74), and hence Theorem D.10 (with $\mathscr{T}$ added to the axioms), holds. Moreover, for $\alpha, \beta \in B_\Sigma$, define

$$\alpha \sim_{\mathscr{T}} \beta \text{ iff } \mathscr{T} \vdash (\alpha \leftrightarrow \beta), \tag{D.76}$$

where the right-hand side stands for $(\mathscr{T}, \alpha) \vdash \beta$ and $(\mathscr{T}, \beta) \vdash \alpha$. Then define

$$L_{(\Sigma,\mathscr{T})} = B_\Sigma / \sim_{\mathscr{T}}, \tag{D.77}$$

and (partially) order $L_{(\Sigma,\mathscr{T})}$ by $[\psi] \leq [\varphi]$ iff $(\mathscr{T}, \psi) \vdash \varphi$; as before, this is equivalent with $\mathscr{T} \vdash (\psi \to \varphi)$. This construction obviously generalizes (D.67), etc. Then Theorem D.11 holds (*mutatis mutandis*) for $L_{(\Sigma,\mathscr{T})}$. In particular, $L_{(\Sigma,\mathscr{T})}$ is a Boolean algebra, which can also be shown to have the following universal property.

A ***model*** of $\mathscr{T}$ in some Boolean algebra B is a map $V : \Sigma \to B$ whose unique extension $V : B_\Sigma \to B$ makes the axioms of $\mathscr{T}$ true, i.e. $V(\varphi) = \top$ for each $\varphi \in \mathscr{T}$ (where $\top$ is the top element of B). Note that $\alpha \mapsto [\alpha]$ is a model of $\mathscr{T}$ in $L_{(\Sigma,\mathscr{T})}$.

Theorem D.12. *For each model $V : \Sigma \to B$ of $\mathscr{T}$, there is a unique homomorphism $V' : L_{(\Sigma,\mathscr{T})} \to B$ of Boolean algebras such that $V(\alpha) = V'([\alpha])$ for each $\alpha \in B_\Sigma$.*

D.3 Intuitionistic propositional logic

In view of its importance for quantum mechanics and topos theory, we now briefly discuss the intuitionistic version of the preceding material on (classical) propositional logic. Intuitionism in mathematics originated with the Dutch mathematician L.E.J. Brouwer (1881–1966), who was also one of the most important early contributors to the field of (algebraic) topology. Brouwer held a rather subjective view of mathematics (sometimes even tending towards solipsism), in which mathematics primarily resided within the mind of the "creative subject" (perhaps the right translation of Brouwers' "scheppend" is: "creating" rather than "creative"). Any means of communication supposedly weakened this effort, so that Brouwer saw the formalization of mathematics (including logic) as secondary and even potentially dangerous; he openly (and polemically) opposed his views to the "formalism" he attributed to Hilbert, with whom he also fell out personally. A more technical consequence of Brouwer's intuitionism was an emphasis on explicit *constructions*, rejecting not only proofs by contradiction, but even the abstract existence of mathematical objects in general (as claimed by the so-called *Platonic* philosophy of mathematics).

Brouwer's lasting influence on logic is partly due to his student Arend Heyting (1989–1980), who was less radical than his teacher and formalized (!) intuitionistic logic analogously to its classical counterpart. In fact, the system (D.56) - (D.65), with *modus ponens*, gives axioms for ***intuitionistic propositional logic***, which therefore differs from classical propositional logic exclusively by the absence of the law of the excluded middle (D.66). It is customary in intuitionistic logic to use the purely logical symbols $\wedge, \vee, \rightarrow$ and $\perp$, in terms of which negation is defined by

$$\neg\alpha \equiv \alpha \rightarrow \perp. \tag{D.78}$$

In that case, axiom (D.65) is simply replaced by

$$\vdash \perp \rightarrow \alpha, \tag{D.79}$$

and in the presence of (D.56) - (D.64) with (D.79), the axiom that makes the system classical may now be formulated as the validity of *reductio ad absurdum*, i.e.,

$$\vdash ((\alpha \rightarrow \perp) \rightarrow \perp) \rightarrow \alpha, \tag{D.80}$$

which is therefore denied in intuitionistic logic. Similarly, classical rules like:

$$\alpha \vee \neg\alpha; \tag{D.81}$$
$$\neg\neg\alpha \vee \neg\alpha; \tag{D.82}$$
$$(\neg\alpha \rightarrow \neg\beta) \rightarrow (\beta \rightarrow \alpha); \tag{D.83}$$
$$(\alpha \rightarrow \beta) \vee (\beta \rightarrow \alpha); \tag{D.84}$$
$$\neg(\neg\alpha \wedge \neg\beta) \rightarrow (\alpha \vee \beta); \tag{D.85}$$
$$\neg(\neg\alpha \vee \neg\beta) \rightarrow (\alpha \wedge \beta), \tag{D.86}$$

are invalid in intuitionistic logic, as is, of course, (D.66). Fortunately, as theorems of intuitionistic propositional logic one does have:

$$\vdash \alpha \to \neg\neg\alpha; \tag{D.87}$$
$$\vdash \neg\neg\neg\alpha \leftrightarrow \neg\alpha; \tag{D.88}$$
$$\vdash (\alpha \to \beta) \to (\neg\beta \to \neg\alpha); \tag{D.89}$$
$$\vdash \neg\alpha \vee \neg\beta \to \neg(\alpha \wedge \beta); \tag{D.90}$$
$$\vdash \neg(\alpha \vee \beta) \to (\neg\alpha \wedge \neg\beta); \tag{D.91}$$
$$\vdash (\alpha \to \beta) \to (\neg\beta \to \neg\alpha). \tag{D.92}$$

More generally, Gödel's ***negative translation*** of classical (propositional) logic into intuitionistic (propositional) logic establishes the fact that if one puts $\neg\neg$ in front of atomic propositions and recursively replaces $\alpha \vee \beta$ by $\neg(\neg\alpha \wedge \neg\beta)$, which changes nothing classically, the ensuing proposition is intuitionistically valid. In this sense, intuitionistic logic is *stronger* than classical logic, although at first sight it looks *weaker* (as is has fewer axioms). Also more generally, one often sees that classical results whose proofs apparently rely on intuitionistically invalid reasoning are classically equivalent to intuitionistically valid results. (e.g. Gelfand duality).

A natural (and complete) semantics for intuitionistic propositional logic is given by Heyting algebras (replacing the Boolean algebras of the classical case). Let I_Σ denote the set of all propositions (i.e., well-formed formulae) on some signature Σ built from the letters $p \in \Sigma$ and the symbols $\wedge, \vee, \to$ and $\perp$, where in formation rule i) preceding (D.42) we also declare $\perp$ to be a proposition, and we omit $\neg\alpha$ at the end of rule ii), as it is a special case of the preceding part with (D.78). If H is a Heyting algebra, we may then extend any function $V : \Sigma \to H$ to a function

$$V : I_\Sigma \to H \tag{D.93}$$

by recursively using the following rules, where $\bullet$ is $\wedge$, $\vee$, or $\to$ in I_Σ and $\dashrightarrow$ in H:

$$V(\perp) = \perp; \tag{D.94}$$
$$V(\alpha \bullet \beta) = V(\alpha) \bullet V(\beta). \tag{D.95}$$

Then each axiom φ of intuitionistic propositional logic is valid, in that

$$V(\varphi) = \top. \tag{D.96}$$

Moreover, if Γ is some finite set of propositions, then

$$\Gamma \vdash \varphi \text{ implies } V\left(\bigwedge \Gamma\right) \leq V(\varphi). \tag{D.97}$$

In particular, suppose we a theory $\mathscr{T}$. As in the classical case, we call a valuation (D.93) a ***model*** of $\mathscr{T}$ if (D.96) holds for each $\varphi \in \mathscr{T}$. It then follows from (D.97) that each model V of $\mathscr{T}$ is ***sound*** in that for all propositions φ one has the rule:

$$\mathscr{T} \vdash \varphi \text{ implies } V(\varphi) = \top. \tag{D.98}$$

That is, φ is ***true*** in the given model. As in Theorem D.10, soundness and completeness of Heyting algebra semantics of intuitionistic propositional logic are then jointly expressed by the following result (where $\vdash$ denotes derivability using only the intuitionistically valid axioms (D.56) - (D.64) with (D.79), and *modus ponens*):

Theorem D.13. *For any theory $\mathscr{T}$ in intuitionistic propositional logic, $\mathscr{T} \vdash \varphi$ holds iff $\mathscr{T} \vDash \varphi$, i.e., $V(\varphi) = \top$ for all Heyting algebra models $V : I_\Sigma \to H$.*

The classical construction of the Lindenbaum algebra may also be copied by defining L_Σ and $L_{(\Sigma,\mathscr{T})}$ as (D.67) - (D.68), where this time the symbol $\vdash$ defining $\sim$ through (D.67) or (D.76) is the one using the intuitionistically valid axioms only. It follows that any Heyting algebra model $V : I_\Sigma \to H$ factors through a homomorphism $L_\Sigma \to H$ of Heyting algebras, just as in the classical case (cf. Theorem D.11).

Kripke models are special Heyting algebra models, which already form a complete semantics for intuitionistic propositional logic. For any poset X, the set

$$\text{Upper}(X) = \mathscr{O}(X) \tag{D.99}$$

of all upper subsets U of X (i.e. $y \leq x \in U$ implies $y \in U$), which by definition coincides with the set $\mathscr{O}(X)$ of open sets in the Alexandrov topology on X, is a Heyting algebra in the partial order defined by inclusion, with $\vee = \cup$, $\wedge = \cap$, and

$$U \dashrightarrow V = \{x \in X \mid (\uparrow x) \cap U \subseteq V\}. \tag{D.100}$$

Given a valuation $V : \Sigma \to \text{Upper}(X)$ with associated Heyting algebra homomorphism $V : I_\Sigma \to \text{Upper}(X)$, for any $x \in X$ and $\varphi \in I_\Sigma$ we write $x \Vdash \varphi$ iff $x \in V(\varphi)$, and say that x ***forces*** φ. Then $V(\varphi) = \top$ iff $x \Vdash \varphi$ for all $x \in X$, and we have:

$$x \Vdash \varphi \text{ and } y \geq x \text{ imply } y \Vdash \varphi; \tag{D.101}$$
$$x \Vdash \bot \text{ for no } x \in X; \tag{D.102}$$
$$x \Vdash \varphi \wedge \psi \text{ iff } x \Vdash \varphi \text{ and } x \Vdash \psi; \tag{D.103}$$
$$x \Vdash \varphi \vee \psi \text{ iff } x \Vdash \varphi \text{ or } x \Vdash \psi; \tag{D.104}$$
$$x \Vdash \varphi \to \psi \text{ iff for all } y \geq x : y \Vdash \varphi \text{ implies } y \Vdash \psi; \tag{D.105}$$
$$x \Vdash \neg\varphi \text{ iff for all } y \geq x,\ y \Vdash \varphi \text{ is false.} \tag{D.106}$$

Hence these are *properties* of any homomorphism $V : I_\Sigma \to \text{Upper}(X)$; originally, (D.101) - (D.105), which imply (D.106), were taken to be *axioms* extending a binary "forcing" relation $x \Vdash p$ on $X \times \Sigma$ to $X \times I_\Sigma$. In topos theory, generalizations of the rules (D.101) - (D.106), once again theorems rather than axioms, will provide the Kripke–Joyal semantics of the (intuitionistic) internal logic op toposes (cf. §E.5).

D.4 First-order (predicate) logic

Propositional logic lacks the structure to describe arithmetic (not to speak of set theory), because it has neither variables—as we shall see, the symbols p_i are not variables but *predicate symbols*—nor quantification symbols like 'there exists' ($\exists$) and 'for all' ($\forall$). This defect is remedied by the formalism of ***predicate logic***, also called ***first-order logic***, which was essentially introduced by Frege and was adopted by Hilbert's school as a universal language for mathematics (as they knew it), in which for example the ***Zermelo–Fraenkel*** (ZF) axioms for set theory may be formulated as a foundation of mathematics (against competitors like the *Principia Mathematica* system of Russell and Whitehead, and others). A simple mathematical theory that can be formalized using classical first-order logic is ***Peano Arithmetic*** (PA).

- The notation of a first-order theory consists of *symbols* from two groups:
  1. The ***purely logical symbols*** are the familiar symbols $\neg, \wedge, \vee, \to$ from propositional logic (or some logically independent subset thereof, such as $\neg$ and $\to$), supplemented by the equality sign $=$ and the quantification symbols $\forall$ and $\exists$ (the latter is in fact superfluous in the classical system discussed here, since, the combination $\exists_x$ defined below is the same as $\neg\forall_x\neg$).
  2. Unlike the ones above, the ***non-logical symbols*** (comprising the ***signature*** of the theory) depend on the field of mathematics to be formalized (such as set theory or arithmetic), but the general format is as follows. One has:
     a. ***Variables*** $a, b, c, \ldots, x, y, z, x_1, x_2, \ldots$, assumed countable many at most. For example, in PA these variables may be thought of as denoting natural numbers, whereas in ZF they will be sets, but of course such *interpretations* do not form part of the syntax! This warning also applies to the next items. In ***many-sorted theories*** the variables are ***sorted***, in that there is a set $\{A, B, \ldots\}$ of ***sorts***, and each variable $x \equiv x_A$ belongs to one of these sorts.
     b. ***Constants***, arbitrarily formatted. For example, PA has just one constant, called $\mathbf{0}$, to be interpreted as the number zero. Also ZF has just a single (even superfluous) constant $\emptyset$, to be interpreted as the empty set.
     c. ***Function symbols*** $f, g, \ldots$. Each such symbol has an ***arity*** $a(f)$, which is a natural number indicating the number of variables it has (as formalized below). Formally, one allows $a(f) = 0$, in which case f is also a constant. PA has three function symbols, viz. S, $+$, and $\times$, with arities $a(S) = 1$, $a(+) = 2$, and $a(\times) = 2$ (these will be interpreted as the successor function $n \mapsto n+1$, addition, and multiplication, respectively). Perhaps surprisingly (especially in the light of category theory), ZF *has no function symbols*: in set theory, functions $f: X \to Y$ are defined as special subsets of $X \times Y$.
     d. ***Predicate symbols*** $P, \ldots$, coming with an arity $a(P) \in \mathbb{N}$, too. These will play a role in the construction of formulae, see below (some authors count $=$ as a predicate symbol with arity 2, instead of as a purely logical symbol). PA has no predicate symbols. ZF has one predicate symbol $\in$, with arity 2.

- According to rules we are about to state, from these symbols one subsequently constructs ***terms***, ***formulae***, and ***sentences*** (or ***closed formulae***). Sentences are at least candidates for theorems, in that one may attempt to prove them (and may either succeed or fail, the latter even in two possible ways: the sentence may be *false*, in that its negation can be proved, or it may be *undecidable*, in that neither it nor its negation can be proved—it was Hilbert's outspoken intention to exclude the last possibility, which however was famously shown to be unavoidable by Gödel). For example, both $x^2 = 1$ and $\exists_x(x^2 = 1)$ are formulae in PA, but only the latter is a sentence, which is even a theorem. The rules, then, are as follows.

  1. ***Term formation*** is done by iterating the steps:
     a. Any variable x_i is a term.
     b. Any constant is a term.
     c. Any function symbol f and any set of $k = a(f)$ terms $(t_1,\ldots,t_k)$ jointly yield a term $f(t_1,\ldots,t_k)$; if $a(f) = 0$ this reduces to the previous case.

     In PA, this means that $S(t)$ is a term, and that $t_1 + t_2 \equiv +(t_1,t_2)$ and $t_1 \times t_2 \equiv \times(t_1,t_2)$ are terms (provided t, t_1, and t_2 are terms). For example, the constant $\mathbf{0}$ is a terms, and hence $S(\mathbf{0})$ is a term, which one calls $\mathbf{1}$. Similarly, $S^n(\mathbf{0})$ is a term called $\mathbf{n}$, where e.g. $S^2(\mathbf{0}) \equiv S(S(\mathbf{0}))$, etc.). From these, we can make terms $\mathbf{n}+\mathbf{m}$, or $\mathbf{n} \times x_i$, and subsequently $(\mathbf{n}+\mathbf{m}) \times (\mathbf{n} \times x_i)$, enz.
     In ZF, the only terms are $\emptyset$ and the variables (as ZF lacks function symbols).
  2. ***Formulae*** are (once again iteratively) constructed from terms using the equality sign $=$ and the predicate symbols, according to the following rules:
     a. If t_1 and t_2 are terms, then $t_1 = t_2$ is a formula.
     b. Any predicate symbol P and any set of $k = a(P)$ terms $(t_1,\ldots,t_{a(P)})$ jointly yield a formula $P(t_1,\ldots,t_{a(P)})$; if $a(P) = 0$, then P is a formula by itself.
     c. As in propositional logic: if φ and ψ are formulae, then so are $\neg\varphi$, $\varphi \vee \psi$, $\varphi \wedge \psi$, and $\varphi \to \psi$. What is new to first-order logic is that also $\exists_x\varphi$ and $\forall_x\varphi$ are formulae, for any variable x (which may or may not occur in φ).

     In PA, the expression $t_1 = t_2$ is a formula (provided t_1 and t_2 are terms).
     In ZF, the expressions $t_1 \in t_2$ and $t_1 = t_2$ are formulae (if t_1 and t_2 are terms).
  3. A variable x occurring in a formula φ is called ***bound*** if it only occurs via $\forall_x\psi_i(x)$ and/or $\exists_x\psi_i(x)$, where ψ is a subformula of φ; otherwise, x is ***free***. A formula containing at least one free variable is called ***open***; if x occurs freely in an open formula φ, the latter is sometimes called $\varphi(x)$, and analogously $\varphi(x_1,\ldots,x_n)$. If all variables in a formula are bound (or if it contains no variables at all), then it is said to be ***closed***. A ***sentence*** is a closed formula.

- ***Axioms*** are, syntactically speaking, special cases of formulae. As for propositional logic, we may either keep $\wedge$ and $\vee$, and add adding (D.46) - (D.47) as axioms, or, equivalently, we may see $\alpha \wedge \beta$ and $\alpha \vee \beta$ as abbreviations for $\neg(\alpha \to \neg\beta)$ and $\neg\alpha \to \beta$, respectively. Similarly, we may see $\exists$ as a derived symbol, in that $\exists_x\varphi$ is an abbreviation for $\neg\forall_x\neg\varphi$.
  As in propositional logic, the axioms for predicate logic come in two groups: *purely logical axioms* and *domain specific axioms*. We will state the latter for the theories PA and ZF in §D.5 below, and now discuss the former (common to both).

From propositional logic, we adopt (D.48) - (D.50), where $\alpha, \beta, \gamma, \delta$ are arbitrary formulae. These are also *Axioms 1–3* of predicate logic, to which one adds:

Axiom 4 : $\vdash (\forall_x \varphi(x)) \rightarrow \varphi(t)$ for any term t (unless x occurs freely in φ through a subformula $\forall_y \psi$ where y occurs in t); some authors write $\varphi(x/t)$ for $\varphi(t)$.

Axiom 5 : $\vdash (\forall_x(\varphi \rightarrow \psi)) \rightarrow (\forall_x \varphi \rightarrow \forall_x \psi)$.

Axiom 6 : $\vdash \forall_x(x = x)$.

Axiom 7 : $\vdash \forall_x \forall_y((x = y) \rightarrow (\varphi(x) \rightarrow \varphi(y)))$ for each formula φ that contains the variable x *freely* and contains y either *freely* or not at all.

- The only two deduction rules of predicate logic (for formulae φ, ψ) are:

  1. ***Modus ponens***: $\vdash (\varphi \rightarrow \psi)$ and $\vdash \varphi$ imply $\vdash \psi$.
  2. ***Universal generalization***: $\vdash \varphi(x)$ implies $\vdash \forall_x \varphi(x)$.

  These rules also apply to theories, provided that in the second, $\mathscr{T} \vdash \varphi(x)$ implies $\mathscr{T} \vdash \forall_x \varphi(x)$ *provided no formula in* $\mathscr{T}$ *used in the proof of* φ *freely contains* x.
- A ***theorem*** is a sentence φ that can be proved form the axioms using the deduction rules in a finite number of steps. In that case, we write $\vdash \varphi$
- A ***theory*** $\mathscr{T}$ is a set of formulae (assumed contradiction-free, although for e.g. ZF this cannot been proved within ZF because of Gödel's Incompleteness Theorem).
- An ***interpretation*** of a theory $\mathscr{T}$ consists of a *nonempty* set M (the ***carrier*** of the interpretation), elements $[[c]]_M \in M$ for each constant c, functions $[[f]]_M : M^{a(f)} \rightarrow M$ for each function symbol f of arity $a(f)$, and subsets $[[P]]_M \subset M^{(a(P)}$ for each predicate symbol P of arity $a(P)$. The interpretation $[[\varphi]]_M$ of a formula φ then follows by giving the logical symbols $\neg, \wedge, \vee, \rightarrow$, and $=$ their usual meanings of "not", "and", "or", "implies", and "is equal to", whereas the range of each variable x occurring in $\forall_x$ or $\exists_x$ in taken to be M. If a sentence φ is true in this interpretation, we write $M \models \varphi$. If each axiom of $\mathscr{T}$ is true, then we call the given interpretation a ***model*** of $\mathscr{T}$ (so that in a model, $\mathscr{T} \vdash \varphi$ implies $M \models \varphi$).

Gödel's Completeness Theorem (to be contrasted with his *in*completeness theorem, which roughly states that any first-order theory that incorporates PA contains undecidable sentences) generalizes Theorem D.10 and eq. (D.74) to first-order logic:

Theorem D.14. *A first-order theory* $\mathscr{T}$ *is consistent iff it has a model. In that case, a sentence* φ *of* $\mathscr{T}$ *is a theorem iff it is true in all models of* $\mathscr{T}$.

Propositional logic is a special case of predicate logic, namely by assuming no variables, constants, and function symbols, and taking the atomic propositions $(p_1, \ldots)$ to be predicate symbols with arity zero (or else $\{0,1\}$-valued variables). The rules of term formation in predicate logic then show that propositional logic has no terms, so that step 2.a above is empty, and step 2.b only yields the p_i. These may be turned into compound expressions by the original uses of propositional logic, which in this case coincide with the rules of predicate logic (since there are no variables, $\exists_x \varphi$ and $\forall_x \varphi$ are both equivalent to φ). Finally, formulae coincide with sentences, since in the absence of variables, all formulae are closed.

As a transition to the next appendix, we continue our discussion on intuitionistic logic started in §D.3. The propositional fragment of first-order intuitionistic logic is still given by (D.56) - (D.65), in which the connectives $\wedge, \vee, \rightarrow$, and $\neg$ (or $\perp$) are independent. The equality sign = is treated with suspicion in intuitionism, and hence is omitted, whilst $\exists$ can no longer be defined in terms of $\forall$ through the classical identification of $\exists_x$ with $\neg\forall_x\neg$. Instead, it is regulated by the two axioms

$$\vdash(\forall_x(\varphi \rightarrow \psi)) \rightarrow (\exists_x\varphi \rightarrow \exists_x\psi); \tag{D.107}$$
$$\vdash\varphi(t) \rightarrow \exists_x\varphi(x), \tag{D.108}$$

subject to the same proviso as Axiom 4 of the classical case, plus a deduction rule:

- ***$\exists$-elimination***: $\vdash \exists_x\varphi$ implies $\vdash \varphi$ (provided x is not free in φ).

This will be the logic on which the topos theory of the next chapter is based. Scary examples of intuituitionistically *invalid* rules involving $\forall$ and $\exists$ include:

$$\neg\forall_x\neg\varphi(x) \leftrightarrow \exists_x\varphi(x); \tag{D.109}$$
$$\forall_x\neg\neg\varphi(x) \leftrightarrow \forall_x\varphi(x); \tag{D.110}$$
$$\neg\neg\exists_x\varphi(x) \leftrightarrow \exists_x\neg\neg\varphi(x); \tag{D.111}$$
$$(\varphi \rightarrow \exists_x\psi(x)) \rightarrow \exists_x(\varphi \rightarrow \psi(x)), \tag{D.112}$$

whereas useful intuituitionistically *valid* theorems containing $\forall$ and $\exists$ are, e.g.,

$$\neg\exists_x\varphi(x) \leftrightarrow \forall_x\neg\varphi(x); \tag{D.113}$$
$$\neg\neg\forall_x\varphi(x) \leftrightarrow \forall_x\neg\neg\varphi(x); \tag{D.114}$$
$$\neg\neg\exists_x\varphi(x) \leftrightarrow \neg\forall_x\neg\varphi(x). \tag{D.115}$$

Gödel's ***negative translation*** of classical logic to intuitionistic logic extends to first-order logic: if, further to the manipulations mentioned after (D.92), one also replaces $\exists_x\varphi(x)$ by $\neg\forall_x\neg\varphi(x)$, then theorems φ of classical first-order logic are turned into theorem of intuitionstic first-order logic. Although we will not use it, we mention that the notion of a ***Kripke model*** also extends from propositional to predicate intuitionistic logic: compared to a classical model carried by a set M, as described above, we now have a *family* of (classical!) sets (M_p) indexed by some poset P, in which constants, functions, and predicate symbols are similarly interpreted as families $[[c]]_{M_x} \in M_x$, $([[f]]_{M_x} : M_x^{(a(f)} \rightarrow M_x)$, and $([[P]]_{M_x} \subset M_x^{(a(P)})$, such that if $x \leq y$, then $M_x \subseteq M_y$, $[[c]]_{M_x} = [[c]]_{M_y}$, $G([[f]]_{M_x}) \subseteq G([[f]]_{M_y})$ (where $G(f)$ is the graph of f), and $[[P]]_{M_x} \subset [[P]]_{M_y}$. Further to the forcing rules (D.101) - (D.106) for intuitionistic propositional logic, there are additional ones for $\exists$ and $\forall$, viz.

$$x \Vdash \exists\varphi(x) \text{ if there exists } m \in M_x \text{ such that } x \Vdash [[\varphi]]_{M_x}(m); \tag{D.116}$$
$$x \Vdash \forall_x\varphi(x) \text{ if for all } y \geq x \text{ and all } m \in M_y \text{ one has } y \Vdash [[\varphi]]_{M_x}(m). \tag{D.117}$$

We will revisit these rules in topos theory, see §E.5; indeed, Kripke models for intuitionistic predicate logic emerge much more naturally in categorical language.

D.5 Arithmetic and set theory

Completing our running examples (for classical first-order logic), we now give the theories PA and ZF, starting with the axioms of ***Peano Arithmetic***:

PA1 $\vdash \forall_x(\neg(S(x) = \mathbf{0}))$;
PA2 $\vdash \forall_x\forall_y(S(x) = S(y) \to x = y)$;
PA3 $\vdash \forall_x(x + \mathbf{0} = x)$;
PA4 $\vdash \forall_x\forall_y(x + S(y) = S(x + y))$;
PA5 $\vdash \forall_x(x \times \mathbf{0} = \mathbf{0})$;
PA6 $\vdash \forall_x\forall_y(x \times S(y) = (x \times y) + x)$;
PA7 $(\varphi(0) \wedge (\forall_x(\varphi(x) \to \varphi(S(x))))) \to \forall_x\varphi(x)$, for any formula $\varphi(x)$.

Thinking of the variables in question as natural numbers (which is what Peano himself still did), these axioms obviously capture their properties pretty well and may require no further explanation (except perhaps the last one, which enables the proof technique of induction). The point, however, is that the axioms only form a *syntax*; the natural numbers $\mathbb{N}$ (as a set) themselves form a *model* of PA in the general sense discussed in the previous section (though by no means the only possible model, and hence $\mathbb{N}$ is called the ***standard model*** of PA). In particular, this means that the set $\mathbb{N}$ is assumed to be known (e.g. via ZF, see below), upon which the interpretation $[[\varphi]]_{\mathbb{N}}$ of some formula φ in PA is determined by the rules given earlier. In particular:

- The constant $\mathbf{0}$ is interpreted as the number zero.
- The function (symbol) S is interpreted as $S(x) = x + 1$, whilst the functions $+$ and $\times$ are interpreted as addition and multiplication, respectively.
- The range of all variables is taken to be $\mathbb{N}$, i.e., $\forall_x$ means "for all $x \in \mathbb{N}$", and $\exists_x$ means "there exists $x \in \mathbb{N}$".

According to the general definition, a sentence φ of PA is then called ***true*** in the given model (i.e., in the natural numbers) if $[[\varphi]]_{\mathbb{N}}$ is true, in which case we write $\mathbb{N} \models \varphi$. For example, $[[\forall_x\forall_y(x+y=y+x)]]_{\mathbb{N}}$ means that for all natural numbers $x, y \in \mathbb{N}$, one has $x+y=y+x$ (which is true, isn't it). Another example is $\mathbf{1}+\mathbf{1}=\mathbf{2}$, which abbreviates $S(\mathbf{0})+S(\mathbf{0}) = S(S(\mathbf{0}))$. The interpretation of $[[\mathbf{1}+\mathbf{1}=\mathbf{2}]]_{\mathbb{N}}$ is given by $1+1=2$ (which once again is true!). In particular, the above axioms of PA are true in this interpretation. The key conceptual point here is that (following Hilbert) one interprets a theory in a domain that is supposed to be known and consistent, so that it has its own methods of proof (for otherwise the semantic entailment symbol $\models$ would be undefined). In this particular case, the domain is ZF set theory (or at least its lower echelons); see the comments to axioms **ZF7** below.

It is quite instructive to see the crucial role of the seemingly technical axiom **PA7**. Suppose we try to define a model of PA in the set $\mathbb{Q}^+$ of positive rational numbers (including zero), so that $\forall_x$ means "for all $x \in \mathbb{Q}^+$", and $\exists_x$ stands for "there exists $x \in \mathbb{Q}^+$"; the number zero (as the interpretation of the constant $\mathbf{0}$) and the functions S, $+$, and $\times$ have their usual meaning, however. Then all of **PA1–PA6** hold, but **PA7** fails, and hence the given interpretation of PA in $\mathbb{Q}^+$ is not a model of PA.

The axioms of ZF are a trifle more complex than those of PA, but then they are supposed to describe all of mathematics! We use the following abbrevations:

$$\forall_{x,y} \equiv \forall_x \forall_y; \tag{D.118}$$
$$\alpha \leftrightarrow \beta \equiv (\alpha \to \beta) \wedge (\beta \to \alpha); \tag{D.119}$$
$$x \neq y \equiv \neg(x = y); \tag{D.120}$$
$$x \notin y \equiv \neg(x \in y). \tag{D.121}$$

Other notation of ZF will be explained in the text following the axioms, which are:

ZF1 $\vdash \forall_{x,y}((\forall_z(z \in x \leftrightarrow z \in y)) \leftrightarrow x = y)$ (***Extensionality***)

ZF2 $\vdash \forall_x \exists_y \forall_z(((z \in x) \wedge \varphi(z)) \leftrightarrow z \in y)$ (***Separation***)

ZF3 $\vdash \neg \exists_x x \in \emptyset$ (***Empty set***)

ZF4 $\vdash \forall_{v,w} \exists_y \forall_z(z \in y \leftrightarrow (z = v) \vee (z = w))$ (***Pairing***)

ZF5 $\vdash \forall_x \exists_y \forall_z(z \in y \leftrightarrow \exists_{w \in x} z \in w)$ (***Union***)

ZF6 $\vdash \forall_x \exists_y \forall_z(z \in y \leftrightarrow z \subset x)$ (***Power set***)

ZF7 $\vdash \exists_x(\emptyset \in x \wedge \forall_y(y \in x \to y^+ \in x))$ (***Infinity***)

ZF8 $\vdash \forall_u((\forall_{x \in u} \exists!_z \varphi(x,z)) \to \exists_y \forall_z(z \in y \leftrightarrow \exists_{x \in u} \varphi(x,z)))$ (***Replacement***)

ZF9 $\vdash \forall_{v \neq \emptyset} \exists_{x \in v} \forall_y(y \in x \to y \notin v)$ (***Regularity***)

AC $\vdash \forall_u \exists_w((w \subset \mathscr{P}(u) \times u) \wedge (\forall_{x \in \mathscr{P}(u)}(x \neq \emptyset \to \exists!_{y \in x} < x, y > \in w)))$ (***Choice***)

In **ZF2** and **ZF8**, $\varphi(\cdot)$ is an arbitrary formula with at least the specified free variables, so that these axioms are more properly thought of as *axiom schemes*.

These axioms have been the subject of entire monographs, but we will be brief here. All intuition about the axioms comes from "naive" sets, although the whole point should be that the axioms stand on their own, and circumvent the problem of defining sets conceptually (as Frege and Cantor desperately tried to do, much as Euclid tried to define in vain what a point is, before he was was liberated by Hilbert). The axioms may be put into two groups: Axioms **ZF1**, **ZF3**, **ZF9**, and **AC** are concerned with *given* sets, whereas nos. **ZF2**, **ZF4**, **ZF5**, **ZF6**, **ZF7**, and **ZF8** regulate the way *new* sets may be constructed from old ones. Here are some comments on the axioms one by one (which should, however, be seen as a whole).

ZF1 states that a set is determined by its members (which themselves are sets!).

ZF2 is a correct version of the naive idea of Cantor, Dedekind, and Frege that every property (or predicate) defines a set. If we look at a predicate as a formula $\varphi(z)$ stating that z has a certain property, the naive idea of these gentlemen was that $y = \{z \mid \varphi(z)\}$ is a set. This idea would be secured by the axiom

$$\exists_y \forall_z(\varphi(z) \leftrightarrow z \in y), \tag{D.122}$$

which however leads to Russell's Paradox (in which $\varphi(z) \equiv z \notin z$).

The crucial difference between **ZF2** and this naive version is that in ZF one restricts set formation to those z that satisfy $\varphi(z)$ *and are a member of some set* x *that is already given*. By **ZF1**, the set y defined by **ZF2** is unique; it is written as

$$y \equiv \{z \in x \mid \varphi(z)\}. \tag{D.123}$$

This notation introduces the familiar brackets $\{\cdots\}$ from naive set theory, which are therefore derived concepts not belonging to the notation of ZF. This is also true for most of the other symbols from naive set theory (except $\in$, which is a predicate symbol in ZF). For example, for arbitrary "sets" x and v (which so far are really just variables in ZF), we introduce $x \cap v$ as a name (i.e., an abbreviation) for the set y defined by taking $\varphi(z)$ in **ZF2** to be $z \in v$. Using the notation (D.123), this *defines* the symbol $\cap$ (for "intersection") by

$$x \cap v \equiv \{z \in x \mid z \in v\}. \tag{D.124}$$

ZF3 states that $\emptyset$, which was the only constant in ZF, has no elements. According to **ZF1** this set is unique, so that $\emptyset$ may be thought of as *the* empty set. In particular, **ZF3** implies that there are sets in the first place (instead of defining it as a constant, one could alternatively introduce the symbol $\emptyset$ at this stage).
An equivalent form of **ZF3** is: $\vdash \forall_x \neg(x \in \emptyset)$, also written as $\forall_x x \notin \emptyset$.

ZF4 states that for given sets v and w, there exists a set y with exactly those two members. We write this y as $y = \{v, w\}$, which uses brackets $\{\cdots\}$ consistently: in **ZF2** take $\varphi(z)$ to be $(z = v) \vee (z = w)$ and take x to be the y just considered. This may be iterated, so that we may write $\{x_1, \ldots, x_n\}$ for the set y that satisfies

$$\forall_{x_1,\ldots,x_n} \exists_y \forall_z (z \in y \leftrightarrow (z = x_1) \vee \cdots \vee (z = x_n)); \tag{D.125}$$

this set is unique by **ZF1**. Using the notation from **ZF2** we may then write

$$\{x_1, \ldots, x_n\} \equiv \{z \in y \mid (z = x_1) \vee \cdots \vee (z = x_n)\}. \tag{D.126}$$

ZF5 postulates the existence of a set y whose elements are the elements of x. In this axiom, the generic notation

$$\exists_{w \in x} \psi \equiv \exists_w ((w \in x) \wedge \psi), \tag{D.127}$$

is used, where ψ is some formula, which in **ZF5** is $z \in w$. We write $y = \cup x$, which *defines* the symbol $\cup$, i.e.,

$$\cup x \equiv \{z \in y \mid \exists_{w \in x} z \in w\}, \tag{D.128}$$

where $y = \cup x$ is the set whose existence is guaranteed by **ZF5**. In the special case $x = \{x_1, \ldots, x_n\}$, we write

$$x_1 \cup \cdots \cup x_n \equiv \cup \{x_1, \ldots, x_n\}. \tag{D.129}$$

ZF6 calls for each x to have a power set y. The notation

$$z \subset x \equiv \forall_y (y \in z \rightarrow y \in x), \tag{D.130}$$

defines the symbol $\subset$; note that $z = x$ is allowed, so that $\subset \equiv \subseteq$. As usual, the set y is unique due to **ZF1**, and is denoted by $\mathscr{P}(x)$, whose elements are therefore the subsets z of x. We may write this à la (D.123) as (y being the set from **ZF6**):

$$\mathscr{P}(x) \equiv \{z \in y \mid z \subset x\}. \tag{D.131}$$

ZF7 postulates the existence of a set y whose elements are

$$\emptyset, \emptyset^+ = \{\emptyset\}, \{\emptyset\}^+ = \{\emptyset, \{\emptyset\}\}, \{\emptyset, \{\emptyset\}\}^+ = \{\emptyset, \{\emptyset\}, \{\emptyset, \{\emptyset\}\}\}, \dots \tag{D.132}$$

in which the notation

$$y^+ \equiv \bigcup \{y, \{y\}\} = y \cup \{y\}, \tag{D.133}$$

is underwritten by **ZF5**. Hence the elements of y^+ are the elements of y, supplemented with the single element y. Following von Neumann, the sets in (D.132) are called $\dot{0}, \dot{1}, \dot{2}, \dot{3}, \dots$, respectively, where $\dot{0}$ is identified with the empty set, and $n > 0$ is realized in a very specific way. Thus **ZF7** states the existence of a set containing $\dot{0}, \dot{1}, \dot{2}, \dot{3}, \dots$. The intersection of all sets with this property is the smallest set containing $\dot{0}, \dot{1}, \dot{2}, \dot{3}, \dots$; this is the smallest infinite set, called ω. In the standard model of ZF (see below), ω is (a copy of) the set $\mathbb{N}$ of natural numbers.

ZF8, in which φ should not contain y, states that if some formula $\varphi(x,z)$ assigns exactly one z to any given x, then these z form a set, provided the variables x form a set (i.e., u). Such a formula φ is really a function f so that $f(x) = z$, and hence this axioms states that the image of any set under some function is again a set. Using the notation (D.123), we then have

$$f(u) = \{z \in y \mid \exists_{x \in u} \varphi(x,z)\}. \tag{D.134}$$

ZF9 is the most contrived axiom in ZF, stating that every nonempty set v contains some element x disjoint from x. Its formulation uses the generic abbreviation

$$\forall_{v \neq \emptyset} \psi \equiv \forall_v ((\exists_z z \in v) \rightarrow \psi) \tag{D.135}$$

Using the symbol $\cap$ from (D.124), one easily checks that $\forall_y (y \in x \rightarrow y \notin v)$ is the same as $x \cap v = \emptyset$, in terms of which **ZF9** reads

$$\vdash \forall_{v \neq \emptyset} \exists_{x \in v} (x \cap v = \emptyset). \tag{D.136}$$

This implies $x \notin x$, which avoids all kinds of paradoxes (though not Russell's, which was taken care of by **ZF2**). Moreover, **ZF9** enables transfinite induction.

AC warrants the choice of an element of each nonempty subset of any set. Indeed, rewriting the expression $\exists!_{y \in x} (x,y) \in w$ as $\exists!_{y \in u} ((x,y) \in w \wedge y \in x)$, **AC** reads

$$\vdash \forall_u \exists_w ((w \subset \mathscr{P}(u) \times u) \wedge (\forall_{x \in \mathscr{P}(u)} (x \neq \emptyset \rightarrow \exists!_{y \in u} (x,y) \in w \wedge y \in x))). \quad \text{(D.137)}$$

As we shall see shortly, this shows that there exists a function

$$f : \mathscr{P}(u) \longrightarrow u \quad \text{(D.138)}$$

that maps $x \in \mathscr{P}(u)$ to $f(x) \in u$, such that $\forall_{x \in \mathscr{P}(u)} (x \neq \emptyset \rightarrow f(x) \in x)$. Although $\exists_f$ is undefined in (first-order) ZF, one may therefore *informally* rewrite **AC** as

$$\forall_u \exists_{f : \mathscr{P}(u) \longrightarrow u} \forall_{x \in \mathscr{P}(u), x \neq \emptyset} f(x) \in x. \quad \text{(D.139)}$$

We now formally define ***functions***, which, as already noted, are curiously absent in ZF (which lacks function symbols). This relies on the following theorem of ZF:

$$\vdash \forall_{u,v} \forall_{x,y} ((x \in u) \wedge (y \in v)) \rightarrow \{\{x\}, \{x,y\}\} \in \mathscr{P}(\mathscr{P}(u \cup v)). \quad \text{(D.140)}$$

We now introduce the abbreviation

$$< x,y > = \{\{x\}, \{x,y\}\}, \quad \text{(D.141)}$$

which by (D.140) is an element of the double power set $\mathscr{P}(\mathscr{P}(u \cup v))$ (assuming that $x \in u$ and $y \in v$); this notation makes $< x,y >$ an *ordered* pair, as opposed to $\{x,y\} = \{y,x\}$. The ***(cartesian) product*** of two sets u and v is now defined as the set

$$u \times v \equiv \{z \in \mathscr{P}(\mathscr{P}(u \cup v)) \mid \exists_{x \in u} \exists_{y \in v} z = < x,y >\}, \quad \text{(D.142)}$$

i.e., in **ZF2** we substitute $x \rightsquigarrow \mathscr{P}(\mathscr{P}(u \cup v))$ as well as

$$\varphi(z) \rightsquigarrow \exists_{x \in u} \exists_{y \in v} z = < x,y >, \quad \text{(D.143)}$$

and denote the (unique) set y thus defined by $u \times v$. Informally, one often writes

$$u \times v = \{< x,y > \mid x \in u, y \in v\}. \quad \text{(D.144)}$$

We are now in a position to define functions in ZF set theory:

Definition D.15. *A* **function** $f : u \rightarrow v$ *is a subset* $G_f \subset u \times v$ *for which*

$$\forall_{x \in u} \exists!_{y \in v} < x,y > \in G_f. \quad \text{(D.145)}$$

Here $\exists!_{y \in v} \psi(y)$ abbreviates

$$\exists_y ((y \in v) \wedge (\forall_z (\psi(z) \leftrightarrow z = y))), \quad \text{(D.146)}$$

cf. (D.127)), which yields (D.145) upon the substitution $\psi(y) \rightsquigarrow < x,y > \in G_f$.

More generally, one has

$$\exists!_y \psi(y) \equiv \exists_y \forall_z (\psi(z) \leftrightarrow z = y). \quad \text{(D.147)}$$

Hence in ZF set theory a function f is defined by (or even identified with) its graph G_f, which closes the historical circle: Newton clearly looked at what we now call functions through their graphs, upon which Euler began to assign some value $f(x) \in v$ to $x \in u$ (though always through some concrete prescription). The 19th century brought the abstract idea of a function as a map between sets, which, as we just saw, ZF set theory replaced by the view that a function is defined by its graph.

Compared to the standard interpretation of PA in the natural numbers, which was a special case of the general notion of a model described in §D.4, the ***standard model of*** ZF is unusual, in that its carrier is not a set (but a so-called *class*), called the ***set-theoretic universe*** (or ***cumulative hierarchy***) V, whose construction was first given by none other than von Neumann, whose name already pervaded this book. We will not go into the details of this construction except by noting that—much as the natural numbers may be built from zero by repeated use of the successor function S—the universe V is constructed from the empty set $\emptyset$ by "repeated" use of:

- The successor operation $V \mapsto V^+ \equiv V \cup \{V\}$, cf. **ZF7**.
- The union operation $V \mapsto \cup V$, cf. **ZF5**.
- The power set operation $V \mapsto \mathscr{P}(V)$, cf. **ZF6**.

However, what is really meant here by "repeated" defies imagination (and may drive one crazy); fortunately, most of mathematics only uses the lower echelons of V.

Furthermore, interpreting the constant $\emptyset$ by the usual empty set (with the same name), the interpretation ε of $\in$ in V needs to be defined. This is done as follows:

1. There exists no set Z such that $Z \varepsilon \emptyset$.
2. One has $Z \varepsilon V^+$ iff $Z \varepsilon V$ or $Z = V$.
3. One has $Z \varepsilon \bigcup V$ iff there exists $W \varepsilon V$ with $Z \varepsilon W$.
4. One has $Z \varepsilon \mathscr{P}(V)$ iff $Z \subseteq V$, where we say $Z \subseteq V$ iff for all $Y \varepsilon Z$ one has $Y \varepsilon V$.

Here V, Y, and Z are sets in V. Applying these rules "iteratively" (see, however, the above comment on "repeated"), for all sets X and Y in V, it can in principle be established whether or not $X \varepsilon Y$, so that the symbol ε is defined within V. Having access to the universe V, ε, and the empty set $\emptyset$, one may then define the interpretation $[[\varphi]]_{\mathsf{V}}$ of some formula φ of ZF in V by the following rules (cf. PA and $\mathbb{N}$):

1. The range of all variables is V, i.e., $\forall_x \varphi(x)$ means that $\varphi(V)$ holds for all $V \in \mathsf{V}$.
2. The constant $\emptyset$ is interpreted as the empty set.
3. The predicate symbol $\in$ is interpreted as the membership relation ε.

A sentence φ in ZF is then ***true***, denoted by $\mathsf{V} \vDash \varphi$, if $[[\varphi]]_{\mathsf{V}}$ is true. For example, all axioms of ZF are true in this interpretation (which is by no means trivial!).

In particular, in this model we interpret $\dot{n}$ (see the explication of **ZF7** above) as the n-fold iteration of the successor operation to $\emptyset$, i.e., $\dot{n} = \emptyset^{+\cdots+}$ (with n pluses), seen as an element of V, and recover the standard model of the natural numbers (and hence the carrier of the standard interpretation of PA) as $\mathbb{N} = \cup_n \dot{n}$, which is the intersection of all sets in V that contain all sets $\dot{n}$ (for any finite n).

Notes

The "modernist" transformation of mathematics led by Hilbert, including its complete prehistory and aftermath, is delightfully described in Gray (2008). The revolutionary nature of Hilbert's views, which started with his influential book *Grundlagen der Geometrie* from 1899, is nowhere clearer than from his correspondence with Frege (cf. Gabriel et al, 1980), who, though one of the fathers of the formalisation of mathematics (specifically through first-order logic), infuriated Hilbert by stating that the latter did not bother to define the notions of "point" or "line" because Hilbert assumed these to be familiar to his readers. But no, quite to the contrary:

> 'Hier liegt wohl der Cardinalpunkt des Misverständnisses (...) Ich will nichts als bekannt voraussetzen (...) Wenn ich unter meinen Punkten irgendwelche Systeme von Dingen, z.B. das System: Liebe, Gesetz, Schornsteinfeger ..., denke und dann meine sämtlichen Axiome als Beziehungen zwischen diesen Dingen annehme, so gelten meine Sätze, z.B. der Pythagoras, auch von diesen Dingen.' (Hilbert to Frege, 29-12-1899).[1]

This may be an exaggeration, however. Einstein probably came closer to the truth:

> 'An dieser Stelle nun taucht ein Rätsel auf, das Forscher aller Zeiten so viel beunruhigt hat. Wie ist es möglich, daß die Mathematik, die doch ein von aller Erfahrung unabhängiges Produkt des menschlichen Denkens ist, auf die Gegenstände der Wirklichkeit so vortrefflich paßt? Kann denn die menschliche Vernunft ohne Erfahrung durch bloßes Denken Eigenschaften der wirklichen Dinge ergründen?
>
> Hierauf ist nach meiner Ansicht kurz zu antworten: Insofern sich die Sätze der Mathematik auf die Wirklichkeit beziehen, sind sie nicht sicher, und insofern sie sicher sind, beziehen sie sich nicht auf die Wirklichkeit.' (Einstein, 1921).[2]

The great irony is that Hilbert's call for abstraction, which at first sight decoupled mathematics from its origins in physics and other applications, in fact very rapidly led to the deepest applications of mathematics to physics so far, such as the use of (pseudo) Riemannian geometry in general relativity, and the use of Hilbert (!) spaces and operator algebras in quantum mechanics. In the present book, a high point of this paradox is the use of Grothendieck toposes (cf. Appendix E) in quantum mechanics (see Chapter 12), especially because Grothendieck himself almost made a sport of extreme abstraction, partly motivated by internal mathematical needs in algebraic geometry, but undoubtedly also by his indignation about the use of (mathematical) physics for military purposes (which put him diametrically against von Neumann).

---

[1] This is surely the central point of the misunderstanding (...) I do not want to assume anything as known (...) If I interpret my notions by arbitrary things, for example, by the system: love, law, chimney sweeper, and subsequently interpret my axioms as relations between these things, then my theorems, like the one of Pythagoras, hold about these things. (Translation by the author)

[2] At this point an enigma presents itself, which in all ages has agitated inquiring minds. How can it be that mathematics, being after all a product of human thought which is independent of experience, is so admirably appropriate to the objects of reality? Is human reason, then, without experience, merely by taking thought, able to fathom the properties of real things?

In my opinion the answer to this question is, briefly, this: as far as the propositions of mathematics refer to reality, they are not certain; and as far as they are certain, they do not refer to reality. (Translation: Sonja Bargmann)

§D.1. **Order theory and lattices**

For lattice theory in general and Stone's Theorem see Givant & Halmos (2009), Davey & Priestley (2002), and Johnstone (1982). For (D.36) - (D.37) see Theorem 33 in Chapter 35 of Givant & Halmos (2009).

§D.2. **Propositional logic**

Halmos & Givant (1998) is an elementary exposition of the connection between Boolean lattices and logic. Other useful (propositional as well as first-order) logic texts include Bell & Machover (1977), Johnstone (1987), Kaye (2007), and Mendelson (2010).

§D.3. **Intuitionistic propositional logic**

Key writings on intuitionism (at least from the Dutch school) include Brouwer (1907, 1918, 1975), Heyting (1956) and Troelstra & van Dalen (1988). See also Dummett (2000) for a view from abroad. Our treatment of Kripke models for intuitionisistic propositional logic is taken from Goldblatt (1984) and Palmgren (2009).

§D.4. **First-order (predicate) logic**

For the history of first-order logic see Grattan-Guinness (2000) and Mancosu, Zach, & Badesa (2004), plus innumerable books about Frege, Russell, Hilbert, etc. It is regrettable that the close companionship of mathematics and philosophy at the time, whose cross-fertilization has given us both the modern foundations of mathematics on the one hand and analytic philosophy on the other, has not lasted.

§D.5. **Arithmetic and set theory** For PA see e.g. Kaye (1991), which focuses on non-standard models. The bible of ZF set theory is Jech (2006).

Appendix E
Category theory and topos theory

This appendix gives a brief introduction to category theory, moving towards the particular categories that are of interest to quantum theory (viz. categories of presheaves and sheaves) as quickly as possible (but not more quickly). However, even the basic setup of category theory is already relevant for e.g. the conceptually most satisfactory formulation of Gelfand duality, as described below Theorem C.23 (see also Theorem C.45), and likewise of Stone duality, see Theorem D.5. Otherwise, this material will only be used in Chapter 12 on quantum logic. We omit most proofs.

Categories were originally introduced by Eilenberg & Mac Lane (1945) in order to define *natural transformations*, through which they formalized (and explained) the intuition that certain isomorphism in mathematics are "natural" or "canonical" (like the one between the second dual V^{**} of a finite-dimensional vector space V and V itself, as opposed to the isomorphism between V^* and V). Natural transformations are predicated on *categories* and *functors*, i.e. maps between categories, which are analogous to continuous functions between topological spaces, and in turn give rise to new categories, similarly to functions giving rise to function spaces in functional analysis. Initially meant to organize certain fields of mathematics in a systematic way (such as algebraic topology and homological algebra), categories soon became objects of study in their own right. As such, the basic vocabulary of category theory is completed by defining *adjoint functors* (invented by Kan in 1958) and *(co)limits*.

Toposes are categories with enough structure to support the interpretation of first-order (and even higher-order) intuitionistic logic, similar to set theory providing semantics for classical predicate logic, which in turn generalizes the relationship between propositional logic and Boolean algebra, cf. §D. In this respect, the presence of a *truth object* (i.e. *subobject classifier*) partly explains their potential relevance to quantum mechanics. However, toposes were introduced in the 1960s by Grothendieck from a completely different motivation, namely algebraic geometry, and were originally seen by him as generalizations of topological spaces. This aspect plays an equally important role for quantum mechanics, and hence we quote:

> 'A startling aspect of topos theory is that it unifies two seemingly wholly distinct mathematical subjects: on the one hand, topology and algebraic geometry, and on the other hand, logic and set theory.' (Mac Lane & Moerdijk, 1992, p. 1).

K. Landsman, *Foundations of Quantum Theory*,
Fundamental Theories of Physics 188, DOI 10.1007/978-3-319-51777-3

E.1 Basic definitions

The definition of a category emphasizes the idea that one is at least as interested in the maps between objects as in the objects themselves. The only complication (which we ignore) is the uses of ***classes***; categories are often too big to be sets, and hence they require an axiomatization of mathematics different from standard ZF set theory (such as von Neumann–Bernays–Gödel set theory or algebraic set theory).

Definition E.1. *A* **category** $\mathsf{C} = (\mathsf{C}_1, \mathsf{C}_0, i, s, t, m)$ *consists of:*

- *A class* C_0 *of* **objects**.
- *A class* C_1 *of* **arrows** *(also called* **morphisms***).*
- *Maps* $s : \mathsf{C}_1 \to \mathsf{C}_0$ *(the* **source map***),* $t : \mathsf{C}_1 \to \mathsf{C}_0$ *(the* **target map***),* $i : \mathsf{C}_0 \to \mathsf{C}_1$ *(the* **identities map***), and* $m : \mathsf{C}_2 = \mathsf{C}_1 \times_{\mathsf{C}_0} \mathsf{C}_1 \to \mathsf{C}_1$ *(***multiplication***), where*

$$\mathsf{C}_1 \times_{\mathsf{C}_0} \mathsf{C}_1 = \{(f,g) \in \mathsf{C}_1 \times \mathsf{C}_1 \mid s(f) = t(g)\}, \tag{E.1}$$

such that, writing $fg \equiv m(f,g)$ *and* $\mathrm{id}_x \equiv i(x)$,

$$s(fg) = s(g); \tag{E.2}$$
$$t(fg) = t(f); \tag{E.3}$$
$$(fg)h = f(gh); \tag{E.4}$$
$$s(\mathrm{id}_x) = t(\mathrm{id}_x) = x; \tag{E.5}$$
$$f\mathrm{id}_{s(f)} = \mathrm{id}_{t(f)} f = f. \tag{E.6}$$

Note that (E.4) is well defined by virtue of (E.2) - (E.3). We often write $x \xrightarrow{f} y$ or $f : x \to y$ or, even better in principle but cumbersome in practice (see below), $y \xleftarrow{f} x$, when $f \in \mathsf{C}_1$ satisfies $s(f) = x$ and $t(f) = y$, and interpret f as an arrow from x to y, so that id_x is an arrow from x to x. Composition $f \circ g \equiv fg$ of arrows is defined whenever $s(f) = t(g)$ (so that on paper the preferred direction of an arrow is from right to left!). Arrow composition is associative whenever defined, and each $i(x)$ acts as an identity under this composition operation. The class of all arrows from x to y in a category C is sometimes written as $\mathrm{Hom}_{\mathsf{C}}(x,y)$, or simply as $\mathrm{Hom}(x,y)$, when C is unambiguous. A category is called ***small*** if both C_0 and C_1 are sets (otherwise, a category is called ***large***), and ***locally small*** if for each $x, y \in \mathsf{C}_0$ the class $\mathrm{Hom}_{\mathsf{C}}(x,y)$ is a set (although C_1 itself may be a proper class). All categories used in this book are locally small (though not necessarily small). Here are some examples.

- Sets has sets as objects and functions as arrows. Sets is a large category, but it follows from the ZF axioms for set theory that it is locally small (as promised).
- Any set X with a preorder $\leq$ (and hence any poset) defines a category X where $X_0 = X$ and $\mathrm{Hom}(x,y)$ contains a unique arrow iff $x \leq y$, being empty otherwise.
- A small category in which each arrow is invertible is called a ***groupoid***; see §C.16. In particular, any group may be seen as a category with just a single object.

Categories come with an intrinsic notion of isomorphism: one calls two objects $x,y \in \mathsf{C}_0$ ***isomorphic***, written $x \cong y$, when there exist arrows $f : x \to y$ and $g : y \to x$ such that $fg = \mathrm{id}_y$ and $gf = \mathrm{id}_x$. For example, two sets are in bijective correspondence iff they are isomorphic objects in Sets, two topological spaces are homeomorphic iff they are isomorphic in the category of topological spaces and continuous maps, and two C*-algebras are isomorphic in the sense of Definition C.2 iff they are isomorphic in CA, where we define the following useful categories of C*-algebras:

- CA, which has C*-algebras as objects and homomorphims as arrows.
- CAm, again having C*-algebras as objects, but now with nondegenerate homomorphims into the multiplier algebra as arrows (cf. Theorem C.76 etc.).
- CAn, with C*-algebras and nondegenerate homomorphims (cf. Definition C.42).
- CA_1, with *unital* C*-algebras as objects and *unital* homomorphims as arrows.
- CCA, CCAm, CCAn, and CCA_1, i.e. the full subcategories of CA, CAm, CAn, and CCA_1, respectively, in which the objects are *commutative* C*-algebras.

Here the notion of a ***subcategory*** $\mathsf{C} \subset \mathsf{D}$ is the obvious one, i.e. $\mathsf{C}_0 \subset \mathsf{D}_0$, $\mathsf{C}_1 \subseteq \mathsf{D}_1$, and C is a category by itself (in particular, C is closed under the maps s,t,i,m). We say that C is a ***full subcategory*** of D if $\mathrm{Hom}_{\mathsf{C}}(x,y) = \mathrm{Hom}_{\mathsf{D}}(x,y)$ for all $x,y \in \mathsf{C}_0$.

We now define the "canonical" maps between categories (which, in the spirit of the subject, are often more important than the underlying categories themselves!).

Definition E.2. *Let* C *and* D *be categories. A* **covariant functor** *or simply* **functor** $F : \mathsf{C} \to \mathsf{D}$ *consists of a pair of maps* $F_i : \mathsf{C}_i \to \mathsf{D}_i$, $i = 0,1$, *such that:*

$$i_{\mathsf{D}} \circ F_0 = F_1 \circ i_{\mathsf{C}}; \tag{E.7}$$
$$s_{\mathsf{D}} \circ F_1 = F_0 \circ s_{\mathsf{C}}; \tag{E.8}$$
$$t_{\mathsf{D}} \circ F_1 = F_0 \circ t_{\mathsf{C}}; \tag{E.9}$$
$$F_1(fg) = F_1(f)F_1(g) \quad (f,g \in \mathsf{C}_2), \tag{E.10}$$

where $i_{\mathsf{D}} : \mathsf{D}_0 \to \mathsf{D}_1$ *is the inclusion map in* D, *etc.*

A **contravariant functor** $F : \mathsf{C} \to \mathsf{D}$ *is a pair* $F_i : \mathsf{C}_i \to \mathsf{D}_i$, $i = 0,1$, *such that:*

$$i_{\mathsf{D}} \circ F_0 = F_1 \circ i_{\mathsf{C}}; \tag{E.11}$$
$$s_{\mathsf{D}} \circ F_1 = F_0 \circ t_{\mathsf{C}}; \tag{E.12}$$
$$t_{\mathsf{D}} \circ F_1 = F_0 \circ s_{\mathsf{C}}; \tag{E.13}$$
$$F_1(fg) = F_1(g)F_1(f) \quad (f,g \in \mathsf{C}_2). \tag{E.14}$$

It follows that F_0 is determined by F_1, since i is injective, but nonetheless it is useful to keep them apart. The use of contravariant functors may be avoided by introducing the ***opposite category*** C^{op} of C, which has the same objects and arrows as C, but the latter going in the opposite direction (i.e. $s_{\mathsf{C}^{\mathrm{op}}} = t_{\mathsf{C}}$, etc.). For example, if $\mathsf{C} = \mathsf{X}$ is a preorder, in the category X^{op} the partial order is reversed. A contravariant functor $F : \mathsf{C} \to \mathsf{D}$ is then obviously the same thing as a covariant functor $F : \mathsf{C} \to \mathsf{D}^{\mathrm{op}}$, or, equivalently, $F : \mathsf{C}^{\mathrm{op}} \to \mathsf{D}$. This is very important for us, because Gelfand duality is based on *contravariant* functors and hence on *opposite* categories; see below.

Definition E.3. *A* **natural transformation** *between two functors $F : \mathsf{C} \to \mathsf{D}$ and $G : \mathsf{C} \to \mathsf{D}$ (that are either both covariant or both contravariant) is a map $\tau : \mathsf{C}_0 \to \mathsf{D}_1$, written $x \mapsto \tau_x$, such that $s_{\mathsf{D}}(\tau_x) = F_0(x)$ and $t_{\mathsf{D}}(\tau_x) = G_0(x)$— in other words, τ is a collection of maps $\tau_x : F_0(x) \to G_0(x)$ indexed by $x \in \mathsf{C}_0$—such that the following diagram commutes for all arrows $f : x \to y$:*

$$\begin{array}{ccc} F_0(x) & \xrightarrow{\tau_x} & G_0(x) \\ {\scriptstyle F_1(f)}\downarrow & & \downarrow{\scriptstyle G_1(f)} \\ F_0(y) & \xrightarrow{\tau_y} & G_0(y) \end{array} \qquad \text{(E.15)}$$

Two functors F and G as above are called **naturally isomorphic***, written $F \cong G$, when there exists a natural transformation τ between them for which all arrows τ_x are invertible (i.e., are isomorphisms).*

It follows that if F and G are naturally isomorphic, then $F_0(x) \cong G_0(x)$ for all $x \in \mathsf{C}_0$, but this condition is not sufficient by itself to render F and G naturally isomorphic, for the isomorphisms τ_x between $F(x)$ and $G(x)$ must be compatible with the arrows, as expressed by the diagram in the above definition; this is even the whole point!

Definition E.3 clarifies the idea that the double dual V^{**} of any finite-dimensional vector space V is isomorphic to V in a "natural" way: namely, the functor $**$ from the category of finite-dimensional vector spaces (over $\mathbb{C}$) to itself (with linear maps as arrows) is naturally isomorphic to the identity functor through the natural transformation whose components $\tau_V : V \to V^{**}$ are given by the "Gelfand transform" $v \mapsto \hat{v}$, where $\hat{v}(\theta) = \theta(v)$ for $\theta \in V^*$. In contrast, the dual V^* is isomorphic to V in an "unnatural" way, in that any isomorphism depends on the choice of a basis.

Definition E.4. *Two categories C, D are called* **equivalent***, written $\mathsf{C} \simeq \mathsf{D}$, when there exist (covariant) functors $F : \mathsf{C} \to \mathsf{D}$ and $G : \mathsf{D} \to \mathsf{C}$ such that $F \circ G \cong \mathrm{id}_{\mathsf{D}}$ and $G \circ F \cong \mathrm{id}_{\mathsf{C}}$. Similarly, C and D are called* **dual** *when there exist* contravariant *functors with the same properties, i.e., if C and D^{op} are equivalent.*

Here id_{C} is the identity functor on C, etc. Spelling out what this means, using Definition E.3, yields the commutative diagrams

$$\begin{array}{ccc} G_0 \circ F_0(x) & \xrightarrow{\tau_x} & x \\ {\scriptstyle G_1 \circ F_1(f)}\downarrow & & \downarrow{\scriptstyle f} \\ G_0 \circ F_0(y) & \xrightarrow{\tau_y} & y \end{array} \qquad \text{(E.16)}$$

for all $f : x \to y$ in C_1, where each τ_x is invertible, and also for all $f' : x' \to y'$ in D_1,

$$\begin{array}{ccc} F_0 \circ G_0(x') & \xrightarrow{\tau'_{x'}} & x' \\ {\scriptstyle F_1 \circ G_1(f')}\downarrow & & \downarrow{\scriptstyle f'} \\ F_0 \circ G_0(y') & \xrightarrow{\tau'_{y'}} & y' \end{array} \qquad \text{(E.17)}$$

We are now in a position to give a categorical (re)formulation of Gelfand duality. Further to the categories of commutative C*-algebras CCA_1, CCAn, and CCAm defined earlier in this section, this involves the following categories of spaces:

- CH, i.e. the category of compact Hausdorff spaces and continuous maps.
- LCHp, with locally compact Hausdorff spaces and *proper* continuous maps.
- LCH, the category of locally compact Hausdorff spaces and continuous maps.

Theorem E.5. *There are categorical equivalences (i.e., dualities if 'op' is omitted):*

$$\mathsf{CCA}_1 \simeq \mathsf{CH}^{\mathrm{op}}; \tag{E.18}$$

$$\mathsf{CCAn} \simeq \mathsf{LCHp}^{\mathrm{op}}; \tag{E.19}$$

$$\mathsf{CCAm} \simeq \mathsf{LCH}^{\mathrm{op}}. \tag{E.20}$$

Proof. In the proof of Theorem C.23, the maps ev_X provide a natural isomorphism between the functors $\mathrm{id}_{\mathsf{CH}}$ and $\Sigma \circ C$ from CH to itself, whilst the maps G_A perform the same job for the functors $\mathrm{id}_{\mathsf{CCA}_1}$ and $C \circ \Sigma$ from CCA_1 to itself; the naturality properties (C.40) and (C.41) precisely express commutativity of the above diagrams. Likewise for the other two cases, which restate Theorems C.45 and C.76. □

Similarly, Stone's Theorem D.5 is best seen categorically, stating that the category of Boolean lattices (with homomorphisms preserving $\vee$, $\wedge$, and $\perp$ as arrows) is dual to the category of Stone spaces (as a full subcategory of CH). With hindsight, Stone's Theorem (which predated category theory) was the first such duality result.

Definition E.4 may be *strengthened* by replacing the isomorphisms

$$F \circ G \cong \mathrm{id}_{\mathsf{D}}; \tag{E.21}$$

$$G \circ F \cong \mathrm{id}_{\mathsf{C}}, \tag{E.22}$$

by equalities, i.e.,

$$F \circ G = \mathrm{id}_{\mathsf{D}}; \tag{E.23}$$

$$G \circ F = \mathrm{id}_{\mathsf{C}}. \tag{E.24}$$

In that case, the categories C and D are called ***isomorphic***. However, this is less relevant than the following *weakening* of the first two conditions, called an ***adjunction***:

Definition E.6. *Two functors $F : \mathsf{C} \to \mathsf{D}$ and $G : \mathsf{D} \to \mathsf{C}$ form an* **adjoint pair** *if there exist natural transformations η from id_{C} to $G \circ F$ (called the* **unit** *of the adjunction), and ε from $F \circ G$ to id_{D} (called the* **counit** *of the adjunction), such that the following diagrams of natural transformarions (i.e. the* **triangle identities***) commute:*

$$\begin{array}{ccc} F & \xrightarrow{F_0 \circ \eta} & FGF \\ & {\scriptstyle \mathrm{id}} \searrow & \downarrow {\scriptstyle \varepsilon \circ F_0} \\ & & F \end{array} \qquad\qquad \begin{array}{ccc} G & \xrightarrow{\eta \circ G_0} & GFG \\ & {\scriptstyle \mathrm{id}} \searrow & \downarrow {\scriptstyle G_0 \circ \varepsilon} \\ & & G. \end{array} \tag{E.25}$$

We write $F \dashv G$, and say that F is **left-adjoint** *to G, or that G is* **right-adjoint** *to F.*

It is easy to see that if they exist, left or right adjoints are unique up to isomorphism. If we assume that C is locally small (in that all classes $\mathrm{Hom}_{\mathsf{C}}(x,y)$ and $\mathrm{Hom}_{\mathsf{D}}(x',y')$ are sets), then the above definition states that the functors $\mathrm{Hom}_{\mathsf{D}}(F(-),-)$ and $\mathrm{Hom}_{\mathsf{C}}(-,G(-))$, both defined from $\mathsf{C}^{\mathrm{op}} \times \mathsf{D}$ to Sets, are naturally isomorphic. In other words, for each $x \in \mathsf{C}_0$ and $y' \in \mathsf{D}_0$, we have a bijection:

$$\mathrm{Hom}_{\mathsf{D}}(F(x),y') \cong \mathrm{Hom}_{\mathsf{C}}(x,G(y')) \tag{E.26}$$

that is natural in both variables x and y' (i.e., for each $y' \in \mathsf{D}_0$, the functors $\mathrm{Hom}_{\mathsf{D}}(F(-),y')$ and $\mathrm{Hom}_{\mathsf{C}}(-,G(y'))$ from C^{op} to Sets are naturally isomorphic, and for each $x \in \mathsf{C}_0$, the functors $\mathrm{Hom}_{\mathsf{D}}(F(x),-)$ and $\mathrm{Hom}_{\mathsf{C}}(x,G(-))$ from D to Sets are naturally isomorphic). Indeed, the natural bijection (E.26) is given by

$$\left(F(x) \xrightarrow{f'} y'\right) \mapsto \left(x \xrightarrow{\eta_x} GF(x) \xrightarrow{G(f')} G(y')\right); \tag{E.27}$$

$$\left(x \xrightarrow{f} G(y')\right) \mapsto \left(F(x) \xrightarrow{F(f)} FG(y') \xrightarrow{\varepsilon_{y'}} y'\right). \tag{E.28}$$

This may even be interesting if $\mathsf{C} = \mathsf{D}$, and hence $F : \mathsf{C} \to \mathsf{C}$ and $G : \mathsf{C} \to \mathsf{C}$. For example, a Heyting algebra H (seen as a posetal category) is home to an adjunction

$$(-) \wedge y \dashv y \dashrightarrow (-), \tag{E.29}$$

for any fixed $y \in H$, where, writing (E.29) as $F \dashv G$ as usual, we put

$$F_0(x) = x \wedge y; \tag{E.30}$$

$$G_0(x) = (y \dashrightarrow x). \tag{E.31}$$

Definition E.4 of an equivalence of categories involves an adjunction $F \dashv G$ whose unit and counit are both natural *isomorphisms*, as opposed to mere natural *transformations*, as in Definition E.6 of an adjunction. In that case, G is an inverse to F up to isomorphism of objects (which still falls short of an exact inverse, which as mentioned would lead to the less important notion of isomorphism of categories). But even for an adjunction, one may regard G as a weak kind of inverse to F, which allows one to move between categories in the direction opposite to F.

Other than equivalences of categories, the traditional examples of adjunctions yield left adjoints to so-called ***forgetful functors***, which strip some class of mathematical objects of (some of) its structure. For example, if Grp is the category of groups and homomorphisms, the forgetful functor G : Grp $\to$ Sets sends a given group to its underlying set; this functor has a left adoint F : Sets $\to$ Grp that assigns the free group on a set X to X. Similarly for vector spaces, Boolean algebras, etc.

We now move on to *limits* and *colimits*, whose general definition we precede by a few special cases. These abstract the corresponding constructions from Sets (and hence pave the way for topos theory, which resembles set theory in various ways), so that for the right "feeling" we switch to labeling objects in a category by capitals.

Definition E.7. *Let* C *be a category (for simplicity assumed to be locally small).*

- *A* **product** *of a pair* $X, Y \in \mathsf{C}_0$ *is an object* $X \times Y \in \mathsf{C}_0$, *with arrows* $p_1 : X \times Y \to X$ *and* $p_2 : X \times Y \to Y$, *such that for all arrows* $q_1 : Z \to X$ *and* $q_2 : Z \to Y$, *there is a unique arrow* $Z \to X \times Y$ *making the following diagram commute:*

$$\begin{array}{ccccc} & & Z & & \\ & \swarrow^{q_1} & \downarrow \exists! & \searrow^{q_2} & \\ X & \xleftarrow[p_1]{} & X \times Y & \xrightarrow[p_2]{} & Y \end{array} \tag{E.32}$$

If each pair of objects in C_0 *has a product,* C *is said to have* **binary products**. The next part of the definition relies on the following fact about products, which is easy to prove: given $f : X \to X'$ and $g : Y \to Y'$ in C_1, there is a unique arrow

$$f \times g : X \times Y \to X' \times Y' \tag{E.33}$$

such that the following diagram commutes:

$$\begin{array}{ccccc} X & \xleftarrow{p_1} & X \times Y & \xrightarrow{p_2} & Y \\ {\scriptstyle f}\downarrow & & \downarrow{\scriptstyle \exists! f \times g} & & \downarrow{\scriptstyle g} \\ X' & \xleftarrow{p_1'} & X' \times Y' & \xrightarrow{p_2'} & Y' \end{array} \tag{E.34}$$

- *A* **function space** *or* **exponential** *of a pair* $Y, Z \in \mathsf{C}_0$ *in a category* C *with binary products is an object* $Z^Y \in \mathsf{C}_0$ (which in Sets is the set of all functions $g : Y \to Z$) *with an* **evaluation map** $\mathrm{ev} : Z^Y \times Y \to Z$ (which in Sets is $(g, y) \mapsto g(y) \subset Z$), *such that for each* $f : X \times Y \to Z$ *there is a unique arrow* $\tilde{f} : X \to Z^Y$ (which in Sets is $\tilde{f}(x)(y) = f(x, y)$) *making the following diagram commute:*

$$\begin{array}{ccc} X \times Y & \xrightarrow{f} & Z \\ & \searrow_{\tilde{f} \times \mathrm{id}_Y} & \uparrow{\scriptstyle \mathrm{ev}} \\ & & Z^Y \times Y \end{array} \tag{E.35}$$

- *A* **terminal object** *is an object* $\mathbf{1} \in \mathsf{C}_0$ *such that for each* $X \in \mathsf{C}_0$, *there is a unique arrow* $X \to \mathbf{1}$ *(in other words,* $\mathrm{Hom}_{\mathsf{C}}(X, \mathbf{1})$ *contains precisely one element).*
- *A category* C *having a terminal object, binary products, and function spaces for all objects, is called* **cartesian closed**.

The relationship between products and function spaces is just the adjunction

$$(-) \times Y \dashv (-)^Y, \tag{E.36}$$

for each $Y \in \mathsf{C}_0$, where the left-hand side denotes the following functor:

$$(-)\times Y : \mathsf{C} \to \mathsf{C}; \tag{E.37}$$
$$X \mapsto X \times Y; \tag{E.38}$$
$$(f : X \to X') \mapsto (f \times \mathrm{id}_Y : X \times Y \to X' \times Y). \tag{E.39}$$

Here $f \times \mathrm{id}_Y$ is a special case of (E.33), whilst the right-hand side of (E.36) is

$$(-)^Y : \mathsf{C} \to \mathsf{C}; \tag{E.40}$$
$$Z \mapsto Z^Y; \tag{E.41}$$
$$(g : Z \to Z') \mapsto (\widetilde{g \circ \mathrm{ev}} : Z^Y \to (Z')^Y), \tag{E.42}$$

where the arrow $\widetilde{g \circ \mathrm{ev}}$ is defined as in the text above (E.35), in which we substitute $X \rightsquigarrow Z^Y$, $Z \rightsquigarrow Z'$, and $f \rightsquigarrow g \circ \mathrm{ev}$; note that the latter is an arrow $Z^Y \times Y \to Z'$.

As in (E.26), the adjunction (E.36) gives a bijection

$$\mathrm{Hom}_{\mathsf{C}}(X \times Y, Z) \cong \mathrm{Hom}_{\mathsf{C}}\left(X, Z^Y\right), \tag{E.43}$$

which of course is precisely the correspondence $f \leftrightarrow \tilde{f}$; the counit of (E.36) is $\varepsilon = \mathrm{ev}$ (i.e., its component at Z is $\mathrm{ev} : Z^Y \times Y \to Z$), whereas the unit (at Z) is the map $\tilde{f} : X \to Z^Y$ corresponding to $f : X \times Y \to Z$ on the choices $X \rightsquigarrow Z$, $Z \rightsquigarrow Z \times Y$, and $f : Z \times Y \to Z \times Y$ being the identity arrow.

The following construction, generalizing binary products, is very important.

Definition E.8. *The* **pullback** *of two arrows $f : X \to Z$ and $g : Y \to Z$ consists of two arrows $p : P \to X$ and $q : P \to Y$ such that the following square commutes,* and *has the universal property that for any arrows $p' : P' \to X$ and $q' : P' \to Y$ with $fp' = gq'$, there is a unique arrow $h : P' \to P$ such that the entire diagram commutes:*

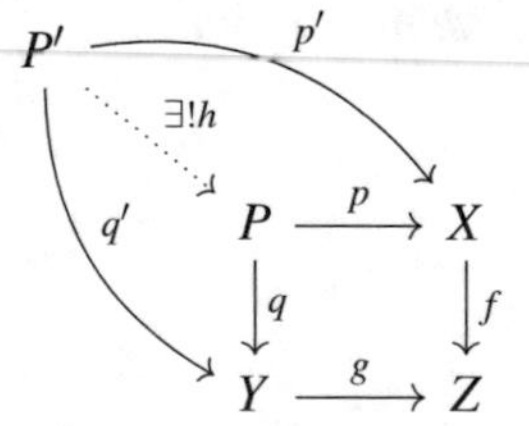

(E.44)

One says that q is a pullback of f over g, whilst p is a pullback of g over f.

In the category Sets, pullbacks coincide with fibered products, that is,

$$P = X \times_Z Y \equiv \{(x,y) \in X \times Y \mid f(x) = g(y)\}, \tag{E.45}$$

where p and q are the projections on the first and the second coordinate, respectively. In particular, taking Z to be a singleton reproduces binary products as special cases of pullbacks. This can be done in all categories C with a terminal object.

At last, we turn to limits and colimits in a category. A (finite) ***diagram*** in a category C is a functor $D : \mathsf{J} \to \mathsf{C}$, where J is some (finite) category. In case that J is empty, we say that there is a unique functor D into C; even this is interesting!

The diagram just consisting of two objects $X, Y \in \mathsf{C}_0$ corresponds to $J_0 = \{0,1\}$ with $0 \neq 1$ and only identity arrows. The next case is an arrow $f : X \to Y$, obtained from $J_0 = \{0,1\}$ as a poset, i.e., $0 \leq 1$. Finally, consider $J_0 = \{0,1,2\}$ with nontrivial arrows $0 \to 1$ and $2 \to 1$; this defines a diagram

$$Y \xrightarrow{g} Z \xleftarrow{f} X. \tag{E.46}$$

For any $C \in \mathsf{C}_0$, let $D_C : \mathsf{J} \to \mathsf{C}$ be the constant functor that sends all $j \in \mathsf{J}_0$ to C, and all arrows in J to id_C. A ***cone*** over a diagram $D : \mathsf{J} \to \mathsf{C}$ is an object $C \in \mathsf{C}_0$ (called the ***vertex*** of the cone) with a natural transformation from D_C to D, i.e., a collection of arrows $c_j : C \to D_j \equiv D_0(j)$ indexed by $j \in J_0$, such that for each arrow $\chi : j \to k$ in J_1, with induced arrow $J_1(\chi) : D_j \to D_k$, the following triangle commutes:

$$\begin{array}{ccc} C & \xrightarrow{c_k} & D_k \\ & \searrow_{c_j} & \uparrow J_1(\chi) \\ & & D_j \end{array} \tag{E.47}$$

A cone over the empty diagram is just a loose object C. A cone over our two-object diagram without arrows is $X \leftarrow C \to Y$, whereas a cone over (E.46) is a commuting square as in (E.44). A ***limit*** of a diagram $D : \mathsf{J} \to \mathsf{C}$ is a ***universal cone*** over D, i.e., a cone $(C, \{c_j : C \to D_j\}_{j \in \mathsf{J}_0})$ such that for any other cone $(C', \{c'_j : C' \to D_j\}_j)$ for the same diagram there is a unique arrow $h : C' \to C$ such that

$$c_j \circ h = c'_j \ (j \in \mathsf{J}_0). \tag{E.48}$$

A more elegant way of phrasing this is via the category $\mathsf{Cone}(D)$, whose objects are cones over D, and whose arrows are arrows $h : C' \to C$ in C_1 satisfying (E.48). A limit of D, then, is just a terminal object in $\mathsf{Cone}(D)$. Either way, it is clear from the universal property that any two limits of a given diagram must be isomorphic. Despite this lack of uniqueness, the typical notation for a limit of a diagram D is

$$C = \varprojlim{}_j D_j. \tag{E.49}$$

It should now be clear that a terminal object is a limit over the unique diagram over the empty category, a (binary) product is a limit over a two-object diagram obtained from $J_0 = \{0,1\}$ with only identities, and finally a pullback is a limit over the diagram (E.46) obtained via $J_0 = \{0,1\}$ seen as a posetal category.

Especially in connection with topos theory, the following fact is quite useful:

Proposition E.9. *A category has all finite limits (i.e. limits based on finite diagrams) iff it has all pullbacks and has a terminal object.*

Replacing C by its opposite category C^{op}, we obtain the ***colimit*** $C = \varinjlim{}_j D_j$ of a diagram, which is defined as a limit of the same diagram seen in C^{op}, so that in all definitions all arrows are reversed. Thus terminal objects are replaced by ***initial objects***, products become ***coproducts***, and pullbacks are turned into ***pushouts***.

E.2 Toposes and functor categories

The last ingredient we need for the definition of a topos is a categorical abstraction of subsets $X \subset Y$ and their characteristic functions 1_X, i.e. a *subobject classifier.*

Definition E.10. *1. An arrow $m : X \to Y$ in any category* C *is a* **monomorphism** *(or briefly a* **mono***) if for any $g,h : Z \to X$, the equality $mg = mh$ implies $g = h$. Similarly, abstracting surjectivity rather than injectivity, $e : X \to Y$ is called an* **epimorphism** *or an* **epi** *if for any $g,h : Y \to Z$, the equality $ge = he$ implies $g = h$.*

2. Two monomorphisms $m : X \to Y$ and $m' : X' \to Y$ are **equivalent** *if there is a (necessarily unique) isomorphism $h : X \to X'$ such that $m = m'h$.*

3. A **subobject** *of Y is an equivalence class of monomorphisms $m : X \to Y$. The class of all subobjects of Y (which is not necessarily a set) is called* Sub(Y).

4. A **subobject classifier** *in a category* C *is a mono $t : \mathbf{1} \to \Omega$ such that all pullbacks of t exist in* C, *and for any mono $m : X \to Y$ there is a* unique *arrow $\chi_m : Y \to \Omega$ (called the* **characteristic function** *or* **classifying map** *of m, or, loosely, of X) that makes the following diagram a pullback (and hence makes it commutative):*

$$\begin{array}{ccc} X & \xrightarrow{\exists!} & \mathbf{1} \\ {\scriptstyle m}\downarrow & & \downarrow{\scriptstyle t} \\ Y & \xrightarrow{\chi_m} & \Omega \end{array} \tag{E.50}$$

It follows that the object $\mathbf{1}$ is terminal in C (which of course constrains C to have a terminal object in the first place); Ω is often called the ***truth object*** of C.

Proposition E.11. *If a locally small category* C *has a subobject classifier, then for any $Y \in \mathsf{C}_0$, the class* Sub(Y) *is a set, and the map $m \mapsto \chi_m$ induces a bijection*

$$\mathrm{Sub}(Y) \cong \mathrm{Hom}_{\mathsf{C}}(Y, \Omega). \tag{E.51}$$

Proof. It follows from the definition of a pullback that equivalent monos $m : X \to Y$ and $m' : X' \to Y$ yield the same arrow χ_m, so that the map $m \mapsto \chi_m$ from monos to arrows passes to equivalence classes, i.e., we have a map $[m] \mapsto \chi_m$. The universal part in the definition of a pullback (i.e., monos with the same classifying maps are isomorphic) makes the latter map injective, whereas surjectivity follows from the general fact that the pullback (namely m) of any arrow (namely χ) over a mono (namely t) is a mono, where we see (E.50) as a pullback *for given χ and t.* □

For example, in Sets a mono is an injective function (and an epi is surjective), so that any mono into Y originates in some set that is isomorphic to some subset of Y. Any singleton $\mathbf{1} = * = \{\emptyset\}$ serves as a terminal object, and Sets has a truth object

$$\Omega = \underline{2} = \{0, 1\}, \tag{E.52}$$

with subobject classifier $t(*) = 1$; if $X \subset Y$, and m is the inclusion map, then $\chi_m = 1_X$ is just the characteristic function of X, and $\mathrm{Sub}(X) \cong \mathscr{P}(X)$ is the power set of X.

The haunting name "truth object" for Ω might explain some of the fascination logicians and quantum physicists have felt for topos theory, which we now define:

Definition E.12. *A* **topos** *is a cartesian closed category (i.e., having a terminal object, binary products, and function spaces) with pullbacks and a subobject classifier.*

More precisely, this defines an ***elementary topos***. It follows from Proposition E.9 that a topos has all finite limits, and it can be shown that it also has all finite colimits.

It should be clear that Sets is a topos; indeed, in our presentation the presence of the necessary ingredients of a topos within Sets partly motivated these ingredients. More generally, all toposes relevant to this book are of the following sort. We first note that for any two categories C and D we obtain a new category $[\mathsf{C},\mathsf{D}]$ whose objects are (covariant) functors from C to D, and whose arrows are natural transformations between such functors. It is often natural to consider *contravariant* functors, giving the category $[\mathsf{C}^{\mathrm{op}},\mathsf{D}]$. If D = Sets, such functors are called ***presheaves*** on C. The category $[\mathsf{C}^{\mathrm{op}},\mathsf{Sets}]$ is often denoted by $\mathsf{Sets}^{\mathsf{C}^{\mathrm{op}}}$. An important special case is

$$\mathsf{C} = \mathcal{O}(X), \tag{E.53}$$

i.e. the topology of some space X (seen as a posetal category), with slight abuse of notation, functors $F : \mathcal{O}(X)^{\mathrm{op}} \to \mathsf{Sets}$ are called ***presheaves on*** X.

Theorem E.13. *For any small category* C, *the category* $[\mathsf{C}^{\mathrm{op}},\mathsf{Sets}]$ *is a topos.*

Proof. We focus on the subobject classifier; the remainder following from the fact that limits in $[\mathsf{C}^{\mathrm{op}},\mathsf{Sets}]$ (including pullbacks and function spaces) are computed pointwise, i.e., if $D : \mathsf{J} \to [\mathsf{C}^{\mathrm{op}},\mathsf{Sets}]$ is a diagram, then for each $C \in \mathsf{C}_1$ we obtain a diagram D_C in Sets defined by $D_C(j) = D(j)(C)$. Since Sets has all limits, we obtain limits $\mathscr{C}_C$ for each D_C. These form a single functor $\mathscr{C}$, which is a limit of D.

The simplest example is the ***terminal object*** in $[\mathsf{C}^{\mathrm{op}},\mathsf{Sets}]$, which comes out as $\mathbf{1}_0(C) = *$ for each $C \in \mathsf{C}_1$, where $*$ is some arbitrary (but fixed) singleton.

To discuss the ***truth object*** in $[\mathsf{C}^{\mathrm{op}},\mathsf{Sets}]$, we need a few definitions.

Definition E.14. *1. In any small category* C, *a* **sieve** *on an object $C \in \mathsf{C}_0$ is a set S of arrows with target C such that if $f \in S$, then $fh \in S$ whenever fh is defined.*

2. The **maximal sieve** $S^{(C)}_{\max}$ *on C consists of all arrows with target C.*

3. The **pullback sieve** f^*S *(on D) over an arrow $f : D \to C$ consists of all arrows*

$$f^*S = \{h : X \to D \mid fh \in S\}. \tag{E.54}$$

4. We denote the set of all sieves on C by $\mathrm{Sieves}(C)$.

Clearly, if $\mathrm{id}_C \in S$, then $S = S^{(C)}_{\max}$. We will show that the truth object in $[\mathsf{C}^{\mathrm{op}},\mathsf{Sets}]$ is

$$\Omega_0(C) = \mathrm{Sieves}(C); \tag{E.55}$$

$$\Omega_1(f) = f^*. \tag{E.56}$$

The ***subobject classifier*** in $[\mathsf{C}^{\mathrm{op}}, \mathsf{Sets}]$, then, is the natural transformation

$$t : \mathbf{1} \to \Omega; \tag{E.57}$$

$$t_C : \mathbf{1}_0(C) \to \mathrm{Sieves}(C); \tag{E.58}$$

$$t_C(*) = S^{(C)}_{\max}. \tag{E.59}$$

To understand this, we need the ***Yoneda Lemma*** E.15 below. In preparation, for any (fixed) $C \in \mathsf{C}_0$, we define a functor $y_C : \mathsf{C}^{\mathrm{op}} \to \mathsf{Sets}$ by

$$(y_C)_0(D) = \mathrm{Hom}_{\mathsf{C}}(D, C); \tag{E.60}$$

$$(y_C)_1\left(D \xrightarrow{f} D'\right) = (g \mapsto gf), \tag{E.61}$$

the latter being a map from $\mathrm{Hom}_{\mathsf{C}}(D', C)$ to $\mathrm{Hom}_{\mathsf{C}}(D, C)$. This is often written as

$$y_C = \mathrm{Hom}_{\mathsf{C}}(-, C), \tag{E.62}$$

and the functors y_C are called ***representable presheaves***. Since $f : C \to C'$ induces a natural transformation $y_C \to y_{C'}$ in the obvious way, i.e., its component τ_D at D is the map $g \mapsto fg$ from $\mathrm{Hom}_{\mathsf{C}}(D, C)$ to $\mathrm{Hom}_{\mathsf{C}}(D, C')$, the map $C \mapsto y_C$ extends to a functor $y : \mathsf{C} \to [\mathsf{C}^{\mathrm{op}}, \mathsf{Sets}]$, called the ***Yoneda embedding***.

Lemma E.15. *For any $F \in [\mathsf{C}^{\mathrm{op}}, \mathsf{Sets}]$, any $D \in \mathsf{C}_0$, and $x \in F_0(C)$, the map*

$$\tau^{(x)}_D : \mathrm{Hom}_{\mathsf{C}}(D, C) \to F_0(D); \tag{E.63}$$

$$\left(D \xrightarrow{f} C\right) \mapsto F_1(f)(x), \tag{E.64}$$

where $F_1(f)$ duly maps $F_0(C)$ to $F_0(D)$, forms the component at D of a natural transformation $\tau^{(x)}$ from y_C to F, and the ensuing map $x \mapsto \tau^{(x)}$ gives a bijection

$$F_0(C) \cong \mathrm{Hom}_{[\mathsf{C}^{\mathrm{op}}, \mathsf{Sets}]}(y_C, F). \tag{E.65}$$

Recall that by definition of a functor category, the right-hand side of (E.65) consists precisely of the set $\mathrm{Nat}(y_C, F)$ of natural transformations from y_C to F.

Lemma E.16. *For any $C \in \mathsf{C}_0$ and $S \in \mathrm{Sieves}(C)$, the presheaf $X^{(S)}$ defined by*

$$X^{(S)}_0(D) = \mathrm{Hom}_{\mathsf{C}}(D, C) \cap S; \tag{E.66}$$

$$X^{(S)}_1\left(D \xrightarrow{f} D'\right) = \{gf \mid g \in X^{(S)}_0(D')\}, \tag{E.67}$$

defines a subobject $m : X^{(S)} \to y_C$, and the ensuing map $S \mapsto X^{(S)}$ yields a bijection

$$\mathrm{Sieves}(C) \xrightarrow{\cong} \mathrm{Sub}(y_C). \tag{E.68}$$

More generally, if X and Y (generalizing $X^{(S)}$ and y_C above) are presheaves on C with $X_0(D) \subseteq Y_0(D)$ for all D, then the equivalence class of X is a subobject of Y.

The proof of Lemma E.16 below uses the converse fact: any subobject of Y has a representative X' for which X' is a ***subfunctor*** of Y, i.e., $X'_0(D) \subseteq Y_0(D)$ for all D, and X'_1 is the restriction of Y_1, as in (E.70). below. To see this, suppose one has a mono $m : X \to Y$, so that each component $m_D : X_0(C) \to Y_0(D)$ of m is an injective function. We now define a presheaf X' on C by

$$X'_0(D) = m_D(X_0(D)) \subseteq Y_0(D); \tag{E.69}$$
$$X'_1(f) = Y_1(f)_{|X'_0(D')} \ (f : D \to D'). \tag{E.70}$$

Furthermore, we define a natural transformation $m' : X' \to Y$, whose components $m'_D : X'_0(D) \to Y_0(D)$ are given by set-theoretic inclusion. The natural transformation $h : X \to X'$, defined through its components $h_D = \tilde{m}_D$ (where $\tilde{m}'_D$ is m_D, but seen as a map from $X_0(D)$ to $X'_0(D)$ rather than to $Y_0(D)$) then renders m and m' isomorphic.

Proof. The map $S \mapsto X^{(S)}$ has an inverse $X \mapsto S_X$, where $S_X \in \text{Sieves}(C)$ is given by

$$S_X = \bigcup_{D \in \mathsf{C}_0} X_0(D). \qquad \square$$

Combining (E.55) - (E.56) with Lemma E.15 applied to $F = \Omega$, gives

$$\text{Hom}_{[\mathsf{C}^{\text{op}},\mathsf{Sets}]}(y_C, \Omega) \cong \text{Sieves}(C), \tag{E.71}$$

so that Lemma E.16 yields a bijective correspondence between arrows from y_C to Ω as defined in (E.55) - (E.56), and subobjects of y_C. At D, diagram (E.50) is

$$\begin{array}{ccc} \text{Hom}_{\mathsf{C}}(D,C) \cap S & \xrightarrow{\exists!} & * \\ {\scriptstyle m_D}\downarrow & & \downarrow{\scriptstyle t_D} \\ \text{Hom}_{\mathsf{C}}(D,C) & \xrightarrow{(\chi_m)_D} & \text{Sieves}(D) \end{array}, \tag{E.72}$$

where m_D is the inclusion map, $t_D(*) = S_{\max}(D)$, and $(\chi_m)_D(f) = f^*S$. Commutativity of this diagram follows from the fact that if $f \in \text{Hom}_{\mathsf{C}}(D,C) \cap S$, then f^*S is the maximal sieve on D, as trivially follows from the definition of a sieve. The pullback condition is then easy to verify from Lemma E.16.

If we replace y_C by any presheaf Y, the classifying map χ_m is given by

$$(\chi_m)_D : Y_0(D) \to \text{Sieves}(D); \tag{E.73}$$
$$x \mapsto \{f : D' \to D \mid Y_1(f)(x) \in X_0(D')\}, \tag{E.74}$$

noting that $Y_i(f)$ maps $Y_0(D)$ into $Y_0(Z)$, and $X_0(Z) \subseteq Y_0(Z)$, since we assume that X represents a subobject of Y such that $X_0(D) \subseteq Y_0(D)$. This generalizes the previous case where $Y = y_C$. To show that χ_m is unique, we write down (E.50) at D:

$$\begin{array}{ccc} X_0(D) & \xrightarrow{\exists!} & * \\ {\scriptstyle m_D}\downarrow & & \downarrow{\scriptstyle t_D} \\ Y_0(D) & \xrightarrow{(\chi_m)_D} & \mathsf{Sieves}(D) \end{array} \quad . \tag{E.75}$$

Also, the condition that χ_m be a natural transformation implies that the diagram

$$\begin{array}{ccc} Y_0(D) & \xrightarrow{(\chi_m)_D} & \mathsf{Sieves}(D) \\ {\scriptstyle Y_1(f)}\downarrow & & \downarrow{\scriptstyle f^*} \\ Y_0(D') & \xrightarrow{(\chi_m)_{D'}} & \mathsf{Sieves}(D') \end{array} \tag{E.76}$$

commutes for any $f : D' \to D$. Then (E.75) with $D \rightsquigarrow D'$ implies that for any $y \in Y_0(D')$, we have $y \in X_0(D')$ iff $(\chi_m)_{D'}(y) = S_{\max}(D')$. In particular, we may take $y = Y_1(f)(x)$ for $x \in Y_0(D)$, so that $Y_1(f)(x) \in X_0(D')$ iff $\chi_{D'}(Y_1(f)(x)) = S_{\max}(D')$. Commutativity of the diagram (E.76) gives $(\chi_m)_{D'} \circ Y_1(f) = f^* \circ (\chi_m)_D$, so that $Y_1(f)(x) \in X_0(D')$ iff $f^*((\chi_m)_D(x)) = S_{\max}(D')$, which in turn is the case if and only if $f \in (\chi_m)_D(x)$. Hence we finally obtain

$$f \in (\chi_m)_D(x) \text{ iff } Y_1(f)(x) \in X_0(D'), \tag{E.77}$$

which is the definition (E.73) - (E.74) of χ_m, and renders it unique (given m).

Finally, the universal property of (E.50) follows from Proposition E.11: since if X' in (E.50) is like P' in (E.44), then $m' : X' \to Y$ is the pullback of χ over t, and hence m' must be equivalent to m. But we know (cf. Definition E.10.2) that an equivalence between mono's is unique. This closes the proof of Theorem E.13. □

Refining presheaves, we also introduce the category $\mathsf{Sh}(X)$ of ***sheaves*** on X, which is the full subcategory of $[\mathcal{O}(X)^{\mathrm{op}}, \mathsf{Sets}]$ defined by the following condition.

Definition E.17. *A presheaf $F : \mathcal{O}(X)^{\mathrm{op}} \to \mathsf{Sets}$ on X is a* **sheaf** *if for any open $U \in \mathcal{O}(X)$, any open cover $U = \cup_j U_j$ of U, and any family $\{s_j \in F_0(U_j)\}$ such that*

$$F_1(U_{jk} \leq U_j)(s_j) = F_1(U_{jk} \leq U_k)(s_k), \tag{E.78}$$

for all j,k, there is a unique $s \in F(U)$ such that $s_j = F(U_j \leq U)(s)$ for all j.

Here $U_{jk} = U_j \cap U_k$, and $F_1(V \leq W) : F_0(W) \to F_0(V)$ is the arrow part of the functor F. If F is a sheaf on X, then for each open $U = \cup_{j \in J} U_j$, it has the continuity property

$$F_0(U) = \varprojlim{}_j \, F_0(U_j), \tag{E.79}$$

where the limit is defined with respect to the diagram $D : \mathsf{J} \to \mathsf{Sets}$ where J is the posetal category whose objects are $j \in J$, and $(i, j) \in J \times J$ provided $U_{ij} \neq \emptyset$, ordered by $i \leq (i, j)$ and $j \leq (i, j)$, with $D(i) = F(U_i)$ and $D(i, j) = F(U_{ij})$, etc.

A key example of a sheaf on X is the ***sheaf of continuous functions***, where

$$F_0(U) = C(U, \mathbb{R}). \tag{E.80}$$

If $U \leq V$, then the associated map $F_1(U \leq V) : C(V,\mathbb{R}) \to C(U,\mathbb{R})$ is simply given by restriction. Sheaves may be defined far more generally (as done by Grothendieck), namely on a *site* (which is a category equipped with a so-called *Grothendieck topology*), but sheaves on a space are all we need in this book.

Analogously to Theorem E.13, $\mathsf{Sh}(X)$ is a topos, whose truth object is the sheaf

$$\Omega_0(U) = \mathcal{O}(U); \tag{E.81}$$

$$\Omega_1(U \leq V) = (-) \cap U, \tag{E.82}$$

i.e., if $W \in \mathcal{O}(V)$, then $\Omega_1(W) = W \cap U \in \mathcal{O}(U)$. With the terminal object in $\mathsf{Sh}(X)$ being borrowed from $[\mathcal{O}(X), \mathsf{Sets}]$, i.e., $\mathbf{1}_0(U) = *$, its subobject classifier is

$$t_U(*) = U. \tag{E.83}$$

In fact, let X be a poset equipped with its intrinsic ***Alexandrov topology***, whose open sets are the ***upper sets***, i.e. those $U \subseteq X$ for which $x \subset U$ and $x \leq y$ implies $y \in U$. Examples of opens are ***up-sets*** $U = \uparrow x = \{y \in X \mid x \leq y\}$, which form a basis of the Alexandrov topology; in fact, $\uparrow x$ is the smallest open set containing x. For any $x \in X$, we write $\mathrm{Upper}(x)$ for the set of all upper sets containing $\uparrow x$.

Proposition E.18. *If X is a poset, the category $[X, \mathsf{Sets}]$ of functors $F : X \to \mathsf{Sets}$ (where X is seen as a category defined by the underlying poset) is isomorphic to the category $\mathsf{Sh}(X)$ of sheaves on X (equipped with the Alexandrov topology), i.e.,*

$$[X, \mathsf{Sets}] \cong \mathsf{Sh}(X). \tag{E.84}$$

Note that $[X, \mathsf{Sets}]$ consists of presheaves on X^{op} (in which $x \leq y$ iff $y \leq x$ in X).

Proof. This isomorphism is given by mapping a functor $\underline{F} : X \to \mathsf{Sets}$ to a sheaf $F : \mathcal{O}(X)^{\mathrm{op}} \to \mathsf{Sets}$, by defining the latter on a basis of the Alexandrov topology as

$$F(\uparrow x) = \underline{F}(x), \tag{E.85}$$

extended to general Alexandrov opens by (E.79). *Vice versa*, a sheaf F on X immediately defines $\underline{F}$ by reading (E.85) from right to left. □

Corollary E.19. *If X is a poset, the subobject classifier in $[X, \mathsf{Sets}]$ is given by*

$$\Omega_0(x) = \mathrm{Upper}(x); \tag{E.86}$$

$$\Omega_1(x \leq y) = (-) \cap (\uparrow y); \tag{E.87}$$

$$t_x(*) = \uparrow x. \tag{E.88}$$

Proof. If C is a poset X, then a sieve on $x \in X$ is a ***lower subset*** of $\downarrow x$ (i.e., if $y \in S$ then $y \leq x$, and if also $z \leq y$, then $z \in S$). Recalling the comment after (E.84), the claim then follows from (E.55) - (E.59). Alternatively, using Proposition E.18, the claim also follows from (E.86) - (E.88). □

E.3 Subobjects and Heyting algebras in a topos

There are numerous connections between topos theory and intuitionistic logic, most of which generalize links between set theory and classical logic. The beginning of algebraic logic was Boole's work, which in modern parlance structured the power set $\mathscr{P}(X)$ of any set as a Boolean lattice, and hence provided a semantics for classical propositional logic, cf. §D.2. From a categorical view, $\mathscr{P}(X)$ is the set $\mathrm{Sub}(X)$, cf. (E.52) and subsequent text. This generalizes to any topos in which $\mathrm{Sub}(X)$ is a set (rather than a proper class), except for the decisive difference that $\mathrm{Sub}(X)$ is no longer a Boolean lattice but a *Heyting algebra*, making topos logic intuitionistic.

Proposition E.20. *For any object X in a locally small topos T, the set $\mathrm{Sub}(X)$ of subobjects of X is a Heyting algebra with respect to the partial ordering $\leq$ defined by $[m : U \to X] \leq [m' : V \to X]$ iff there is $h : U \to V$ such that $m'h = m$.*

It is easy to show from Definition E.10.1 that $\leq$ is well defined, and since it is defined "on the nose", i.e., at the level of representatives of the equivalence classes in question, in what follows we will use mono's rather than their equivalence classes.

Proof. Since we only need this result for presheaf toposes, we just list the pertinent operations, and omit the verification of the details (which is left to the reader).

- The ***bottom element*** $\perp$ of $\mathrm{Sub}(X)$ is the unique arrow $\mathbf{0} \to X$, where $\mathbf{0}$ is the ***initial object*** in T (any category with finite colimits has such an object, denoted by $\emptyset$, whose defining property is that for any X there is a unique arrow $\emptyset \to X$; as the notation suggests, in Sets the empty set is an initial object).
- The ***top element*** $\top$ of $\mathrm{Sub}(X)$ is the identity arrow id_X at X.
- The ***inf*** of $m : U \to X$ and $m' : V \to X$ is their pullback, i.e., abusing notation,

$$
\begin{array}{ccc}
U \wedge V & \xrightarrow{p} & U \\
{\scriptstyle q}\downarrow & & \downarrow{\scriptstyle m} \\
V & \xrightarrow{m'} & X,
\end{array}
\tag{E.89}
$$

  so that the desired arrow $U \wedge V \to X$ is $mp = m'q$ (which is indeed a mono).
- The ***sup*** of $m : U \to X$ and $m' : V \to X$ is more complicated. In any topos T, arrows have an epi-mono factorization $f = me$, where m is mono (and as such is unique up to isomorphism), called the ***image*** of f, and e is epi. Furthermore, T has finite colimits including coproducts. Reversing all arrows in (E.32) gives

$$
\begin{array}{ccccc}
 & & X & & \\
 & {\scriptstyle m}\nearrow & \uparrow{\scriptstyle \exists! f} & \nwarrow{\scriptstyle m'} & \\
U & \xleftarrow[p_2]{} & U+V & \xrightarrow[p_1]{} & V
\end{array}
\tag{E.90}
$$

  The sup $U \vee V$, then, is "the" image of the arrow f in this diagram.

- Finally, ***implication*** $\dashrightarrow$ is defined in terms of an ***equalizer***, which may be constructed as a pullback, as follows: taking $Y = Z = X$ and $q_1 = q_2 = \mathrm{id}_X$ in (E.32) gives a unique arrow $\Delta_X : X \to X \times X$, called the ***diagonal***; in Sets it is $\Delta_X(x) = (x,x)$. Furthermore, if we have two arrows $f,g : X \to Y$, taking $Z \rightsquigarrow X$, $X \rightsquigarrow Y$, $q_1 \rightsquigarrow f$, and $q_2 \rightsquigarrow g$ in (E.32) gives a unique arrow $(f,g) : X \to Y \times Y$, which in Sets is of course given by $(f,g)(x) = (f(x),g(x))$.
  The equalizer of f and g, then, is the arrow $e : E \to X$ in the pullback

$$\begin{array}{ccc} E & \xrightarrow{p} & Y \\ {\scriptstyle e}\downarrow & & \downarrow{\scriptstyle \Delta_Y} \\ X & \xrightarrow{(f,\,g)} & Y \times Y. \end{array} \tag{E.91}$$

The equalizer indeed deserves its name, because the map p equals both fe and ge; in Sets, $E \subseteq X$ may simply be taken to be the subset on which f and g coincide.

We return to our monos $m : U \to X$ and $m' : V \to X$, with inf $U \wedge V$: the mono $(U \dashrightarrow V) \to X$ is the equalizer of the classifying maps $\chi_U, \chi_{U \wedge V} : X \to \Omega$. □

Recall that in Sets we may identify $\mathrm{Sub}(X)$ with the power set $\mathscr{P}(X)$, so that

$$\bot = \emptyset; \tag{E.92}$$
$$\top = X. \tag{E.93}$$

For $U, V \subseteq X$, the above constructions reduce to the well-known expressions

$$U \leq V \text{ iff } U \subseteq V; \tag{E.94}$$
$$U \wedge V = U \cap V; \tag{E.95}$$
$$U \vee V = U \cup V; \tag{E.96}$$
$$U \dashrightarrow V = U^c \cup V, \tag{E.97}$$

where for comparison below we may rewrite the right-hand side of (E.97) as

$$U^c \cup V = \{x \in X \mid x \in U \to x \in V\}. \tag{E.98}$$

The (derived) expression (D.12) for negation then equates $\neg$ with complementation:

$$\neg U = U^c = \{x \in X \mid x \notin U\}. \tag{E.99}$$

In a presheaf topos $[\mathsf{C}^{\mathrm{op}}, \mathrm{Sets}]$, one obtains similar expressions for $\bot$ and $\top$, viz.

$$\bot_0(C) = \emptyset; \tag{E.100}$$
$$\top_0(C) = X(C), \tag{E.101}$$

where the functor $\bot$ is the initial object in $[\mathsf{C}^{\mathrm{op}}, \mathrm{Sets}]$. The logical connectives resemble the set-theoretic case, too, except for the last ones: if U and V are representatives of subobjects of X such that $U_0(C) \subseteq X_0(C)$ and $V_0(C) \subseteq X_0(C)$, we have:

$$U \leq V \text{ iff } U_0(C) \subseteq V_0(C) \text{ for all } C; \tag{E.102}$$

$$(U \wedge V)_0(C) = U_0(C) \cap V_0(C); \tag{E.103}$$

$$(U \vee V)_0(C) = U_0(C) \cup V_0(C); \tag{E.104}$$

$$(U \dashrightarrow V)_0(C) = \{x \in X_0(C) \mid \forall D \xrightarrow{f} C : X_1(f)(x) \in U_0(D) \Rightarrow X_1(f)(x) \in V_0(D)\}; \tag{E.105}$$

$$\neg U_0(C) = \{x \in X_0(C) \mid \forall D \xrightarrow{f} C : X_1(f)(x) \in U_0(D) \Rightarrow X_1(f)(x) \notin V_0(D)\}. \tag{E.106}$$

This Heyting algebra is Boolean iff $\neg\neg U = U$ for each U, so we are interested in

$$\neg\neg U_0(C) = \{x \in X_0(C) \mid \forall D \xrightarrow{g} C \exists E \xrightarrow{f} D : X_1(gf)(x) \in U_0(E)\}. \tag{E.107}$$

It can be shown that $\text{Sub}(X)$ is Boolean for each object X iff $\text{Sub}(\Omega)$ is Boolean. In order to settle this, we specialize (E.107) to subfunctors $m : U \to \Omega$, which gives

$$\neg\neg U_0(C) = \{S \in \text{Sieves}(C) \mid \forall D \xrightarrow{g} C \exists E \xrightarrow{f} D : (gf)^* S \in U_0(E)\}. \tag{E.108}$$

For example, if $\mathsf{C} = X^{\text{op}}$ is a posetal category, this expression becomes

$$\neg\neg U_0(x) = \{S \in \text{Upper}(x) \mid \forall y \geq x \exists z \geq y : S \cap (\uparrow z) \in U_0(z)\}, \tag{E.109}$$

which is clearly an additional property of $S \in U_0(x)$; examples abound in Chapter 12. Thus the (propositional) logic of $\text{Sub}(X)$ may be genuinely intuitionistic (and given our examples, this conclusion especially applies to quantum logic).

Although X is an object in a topos, $\text{Sub}(X)$ is a Heyting algebra in ordinary set theory. This is called an ***external*** description of X. Alternatively, one may study a topos using so-called ***internal*** reasoning. We will develop the logical foundation of internal reasoning (at least to some extent) in the next section, and for the moment just look at a special example, namely Heyting algebras *within some given topos*.

Definition E.21. *Let* T *be a topos (more generally, a category with all finite limits).*

- *A* **preorder** *on an object* $X \in \mathsf{T}_0$ *in* T *is a mono* $m_\leq : R \to X \times X$ *for which:*
  1. *The diagrammatic version of* reflexivity *(in set theory:* $x \leq x$ *) holds, as follows. The diagonal* $\Delta_X \equiv \Delta : X \to X \times X$ *factors through* $m_\leq$*, i.e. there is an arrow* $X \to R$ *such that the following diagram commutes:*

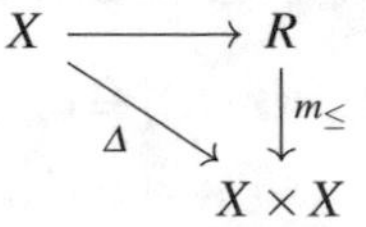

(E.110)

  2. *The diagrammatic version of* transitivity *(in set theory:* $x \leq y$ *and* $y \leq z$ *imply* $x \leq z$*) holds, as follows. First, define* P *as the pullback*

$$\begin{array}{ccc} P & \xrightarrow{\;p\;} & R \\ {\scriptstyle q}\big\downarrow & & \big\downarrow{\scriptstyle p_1\circ m_{\leq}} \\ R & \xrightarrow{p_2\circ m_{\leq}} & X, \end{array} \qquad (E.111)$$

where $p_1, p_2 : X \times X \to X$ are the arrows in (E.32)*, and p, q are defined as in* (E.44)*. The arrows $p_1 \circ m_{\leq} \circ p : P \to X$ and $p_2 \circ m_{\leq} \circ q : P \to X$, then yield an arrow $P : X \to X \times X$ via* (E.32)*, which must factor through $m_{\leq}$, too.*

- *A* **partial order** *on X is a preorder that is* antisymmetric *(in set theory: $x \leq y$ and $y \leq x$ imply $x = y$), in the following sense. First, define the* **twist map**

$$\tau : X \times X \to X \times X \qquad (E.112)$$

by taking $Z \rightsquigarrow X \times X$, $Y \rightsquigarrow X$, $q_1 \rightsquigarrow p_2$ and $q_2 \rightsquigarrow p_1$ in (E.32)*; in set theory, this would be $\tau(x,y) = (y,x)$. This enables us to reverse the order by defining a monic*

$$m_{\geq} : R \to X \times X; \qquad (E.113)$$

$$m_{\geq} = \tau \circ m_{\leq}, \qquad (E.114)$$

with associated pullback

$$\begin{array}{ccc} P' & \xrightarrow{\;p'\;} & R \\ {\scriptstyle q'}\big\downarrow & & \big\downarrow{\scriptstyle m_{\leq}} \\ R & \xrightarrow{\;m_{\geq}\;} & X \end{array} \qquad (E.115)$$

The arrow $m_{\leq} \circ p' = m_{\geq} \circ q' : P' \to X$, then, must factor through $\Delta : X \to X \times X$.

- *A* **lattice** *in T is a partial order on some object X for which there are arrows*

$$\wedge : X \times X \to X; \qquad (E.116)$$

$$\vee : X \times X \to X, \qquad (E.117)$$

such that:

1. *The arrow $m_{\leq} : R \to X \times X$ is an equalizer of the arrows $\wedge : X \times X \to X$ and $p_1 : X \times X \to X$ (in set theory this expresses the property $x \leq y$ iff $x \wedge y = x$).*
2. *One has $\wedge \circ \Delta = \mathrm{id}_X$ and $\vee \circ \Delta = \mathrm{id}_X$ (i.e., $x \wedge x = x$ and $x \vee x = x$).*
3. *The following square (stating that $x \wedge (y \vee x) = (x \wedge y) \vee x = x$) commutes:*

$$\begin{array}{ccc} X & \xrightarrow{\;\;\wedge\;\;} & X \times X \\ {\scriptstyle p_1}\big\uparrow & & \big\uparrow{\scriptstyle \mathrm{id}_X \times \vee} \\ X \times X & \xrightarrow{\;\;c\;\;} & X \times X \times X \\ {\scriptstyle p_1}\big\downarrow & & \big\downarrow{\scriptstyle \wedge \times \mathrm{id}_X} \\ X & \xleftarrow[\;\;\vee\;\;]{} & X \times X. \end{array} \qquad (E.118)$$

Here the middle arrow c is the composition

$$X \times X \xrightarrow{\Delta \times \mathrm{id}_X} (X \times X) \times X \xrightarrow{\cong} X \times (X \times X) \xrightarrow{\mathrm{id}_X \times \tau} X \times X \times X. \qquad \text{(E.119)}$$

- *Let* **1** *be 'the" terminal object in* T*, with associated arrow* $X \to X \times \mathbf{1}$ *from* (E.32), *with* $Z \rightsquigarrow X$, $q_1 \rightsquigarrow \mathrm{id}_X$*, and* $Y \rightsquigarrow \mathbf{1}$*. A* **top element** *in an internal lattice is an arrow* $\top : \mathbf{1} \to X$ *such that the following composite arrow is the identity* id_X*:*

$$X \xrightarrow{\cong} X \times \mathbf{1} \xrightarrow{\mathrm{id}_X \times \top} X \times X \xrightarrow{\wedge} X. \qquad \text{(E.120)}$$

A **bottom** *element is an arrow* $\perp : \mathbf{1} \to X$ *for which the following arrow is* id_X*:*

$$X \xrightarrow{\cong} X \times \mathbf{1} \xrightarrow{\mathrm{id}_X \times \perp} X \times X \xrightarrow{\vee} X. \qquad \text{(E.121)}$$

- *A* **Heyting algebra** *in* T *is a lattice X with* $\top$ *and* $\perp$*, endowed with an arrow*

$$\dashrightarrow : X \times X \to X, \qquad \text{(E.122)}$$

such that the monos m_1 *and* m_2 *in the double pullback diagram*

$$\begin{array}{ccccc} P_1 & \longrightarrow & R & \longleftarrow & P_2 \\ {\scriptstyle m_1}\downarrow & & \downarrow{\scriptstyle m_\leq} & & \downarrow{\scriptstyle m_2} \\ X \times X \times X & \xrightarrow{\wedge \times \mathrm{id}_X} & X \times X & \xrightarrow{\mathrm{id}_X \times \dashrightarrow} & X \times X \times X \end{array} \qquad \text{(E.123)}$$

are equivalent (and hence define the same subobject of $X \times X \times X$*).*

The reader may check that in Sets these definitions reduce to the usual ones; as one can clearly see, finding diagrammatic versions of familiar definitions is an art!

The most important example of an internal Heyting algebra in a topos is Ω.

Theorem E.22. *The truth object* Ω *in a topos* T *with subobject classifier* $t : \mathbf{1} \to \Omega$*, is a Heyting algebra in the partial ordering* $m_\leq : R \to \Omega \times \Omega$ *defined as the equalizer of the projection* $p_1 : \Omega \times \Omega \to \Omega$ *and the classifying map* $\chi_{(t,t)} : \Omega \times \Omega \to \Omega$ *of the product arrow* $(t,t) : \mathbf{1} \to \Omega \times \Omega$ *derived from* $t : \mathbf{1} \to \Omega$*. In particular:*

1. *The inf arrow* $\wedge : \Omega \times \Omega \to \Omega$ *equals the classifying map* $\chi_{(t,t)}$ *of* (t,t)*.*
2. *The sup arrow* $\vee : \Omega \times \Omega \to \Omega$ *is the classifying map* $\chi_\cup$ *of the arrow (see below)*

$$(t, \mathrm{id}_\Omega) \cup (\mathrm{id}_\Omega, t) : (\mathbf{1} \times \Omega) \cup (\Omega \times \mathbf{1}) \to \Omega \times \Omega;. \qquad \text{(E.124)}$$

3. *The implication arrow* $\dashrightarrow : \Omega \times \Omega \to \Omega$ *is the classifying map* $\chi_{m_\leq}$ *of* $m_\leq$*.*
4. *The top element* $\top : \mathbf{1} \to \Omega$ *coincides with the subobject classifier* $t : \mathbf{1} \to \Omega$*.*
5. *The bottom element* $\perp : \mathbf{1} \to \Omega$ *is equal to the classifying map* $\chi_\mathbf{0}$ *of* $\mathbf{0} \to \mathbf{1}$*.*
6. *The negation arrow* $\neg : \Omega \to \Omega$ *equals the classifying map* $\chi_\perp$ *of* $\perp : \mathbf{1} \to \Omega$*.*

For every object $Y \in \mathsf{T}_0$*, this structure makes* $\mathrm{Hom}_{\mathsf{T}}(Y, \Omega)$ *an external Heyting algebra (i.e., in* Sets*), such that* (E.51) *is an isomorphism of Heyting algebras.*

We omit the proof of Theorem E.22, which is a straightforward verification.

In no. 1, the arrow (t,t) is a special case of the arrow (f,g) defined just before (E.91). In no. 2, we need the following construction, applied to the arrows

$$(t,\mathrm{id}_\Omega):(\mathbf{1}\times\Omega)\to(\Omega\times\Omega); \quad \text{(E.125)}$$
$$(\mathrm{id}_\Omega,t):(\Omega\times\mathbf{1})\to(\Omega\times\Omega), \quad \text{(E.126)}$$

To define maps like the one in (E.124) in general, recall the the coproduct diagram

$$\begin{array}{ccccc} X & \longrightarrow & X+Y & \longleftarrow & Y \\ & \searrow & \downarrow \exists! & \swarrow & \\ & & Z & & \end{array} \quad \text{(E.127)}$$

which is just the opposite of the product diagram (E.32). In particular, for any given mono's $m_1 : X\to Z$ and $m_2 : Y\to Z$, we obtain a unique map

$$(m_1,m_2):X+Y\to Z. \quad \text{(E.128)}$$

The image of the latter in the sense of its epi-mono factorization $(m_1,m_2)=me$, i.e.

$$(m_1,m_2):X+Y\xrightarrow{e}X\cup Y\xrightarrow{m}Z, \quad \text{(E.129)}$$

is the mono denoted by $m\cup m' : X\cup Y\to Z$ (which is called m in the above diagram).

In no. 5, $\mathbf{0}$ is the initial object in T. Note that the truth arrow $t:\mathbf{1}\to\Omega$ is the same as the classifying map χ_{id_1} of the identity arrow $\mathrm{id}_\mathbf{1}:\mathbf{1}\to\mathbf{1}$, so that all arrows in Theorem E.22 are classifying maps.

In the presheaf topos $[\mathsf{C}^{\mathrm{op}},\mathsf{Sets}]$, where C is any category, products are taken pointwise, and also, set-theoretic intersection commutes with pullback of sheaves:

$$f^*(S\cap S')=f^*(S)\cap f^*(S') \ \ (f:D\to C;\ S,S'\in\mathrm{Sieves}(C)). \quad \text{(E.130)}$$

These facts imply that the component at $\wedge_C$ of the natural transformation $\wedge$ is just

$$\wedge_C(S,S')=S\cap S' \ \ (S,S'\in\mathrm{Sieves}(C)), \quad \text{(E.131)}$$

which in turns implies that if R is taken to be a subfunctor of $\Omega\times\Omega$, so that

$$(m_\leq)_C:R_0(C)\hookrightarrow\mathrm{Sieves}(C)\times\mathrm{Sieves}(C) \quad \text{(E.132)}$$

is the inclusion map, we have $(S,S')\in R_0(C)$ iff $S\subseteq S'$. We also find

$$\dashrightarrow_C(S,S')=\{f:D\to C\mid f^*S\subseteq f^*S'\}; \quad \text{(E.133)}$$
$$\neg_C(S)=\{f:D\to C\mid f\notin S\}, \quad \text{(E.134)}$$

which are easily checked to be natural in C. Finally, the top element $\top_C\in\mathrm{Sieves}(C)$ is the maximal sieve, and similarly the bottom element $\bot_C$ is the empty sieve.

E.4 Internal frames and locales in sheaf toposes

As we have seen in §D.1 as well as in §C.11, a *complete* Heyting algebra is the same qua lattice structure as a *frame*, except that maps between frames are defined differently: a frame map is required to preserve order and arbitrary suprema, whereas a Heyting algebra frame map preserves order and implication. Furthermore, one has *locales*, which are frames, too, except that maps go in the opposite direction. Hence if Frm is the category of frames (within Sets), then the category Loc of locales is

$$\mathsf{Loc} = \mathsf{Frm}^{\mathrm{op}}. \tag{E.135}$$

We also recall the bizarre (but wonted) notation X for an object in Loc that is *the same* as the object denoted by $\mathcal{O}(X)$ in Frm, where nothing is implied about the spatiality of the frames in question (i.e., it is not necessarily the case that there is an actual space X of which the given frame called $\mathcal{O}(X)$ is the topology). In the same spirit, *frame* maps are written $f^* : \mathcal{O}(Y) \to \mathcal{O}(X)$ or $f^{-1} : \mathcal{O}(Y) \to \mathcal{O}(X)$, the corresponding *locale* map being $f : X \to Y$ (which is *the same map between the same objects*), once again, even if no space X in the usual sense is around.

In any case, in order to define internal frames, locales, or complete Heyting algebras in a topos, one must define completeness of internal lattices. This is difficult diagrammatically, but it can be done through the internal language of §E.5, e.g. by

$$\models \forall_S \exists x\, (S \subseteq \downarrow x) \wedge \forall_y\, (S \subseteq \downarrow y \to x \leq y), \tag{E.136}$$

where $S \subseteq X$ and $x, y \in X$ (technically, S is a variable of type Ω^X, and x and y are variables of type X, see §E.5). We may avoid this, however, since due to the identification (E.84) in Chapter 12 we can work in a sheaf topos $\mathsf{Sh}(X)$, where *internal* frames have a simple *external* description, as follows: there is an equivalence

$$\mathsf{Frm}_{\mathsf{Sh}(X)} \simeq (\mathsf{Frm}_{\mathsf{Sets}}/\mathcal{O}(X))^{\mathrm{op}} \tag{E.137}$$

between the category of internal frames in $\mathsf{Sh}(X)$ and the category of frame maps in Sets with domain $\mathcal{O}(X)$, where the arrows between two such maps

$$\pi_Y^{-1} : \mathcal{O}(X) \to \mathcal{O}(Y); \tag{E.138}$$

$$\pi_Z^{-1} : \mathcal{O}(X) \to \mathcal{O}(Z); \tag{E.139}$$

are the frame maps

$$\varphi^{-1} : \mathcal{O}(Z) \to \mathcal{O}(Y) \tag{E.140}$$

that satisfy

$$\varphi^{-1} \circ \pi_Z^{-1} = \pi_Y^{-1}. \tag{E.141}$$

This looks more palpable in terms of the "virtual" underlying spaces (i.e. locales): If (E.138) - (E.140) are seen as inverse images of maps $\pi_Y : Y \to X$, $\pi_Z : Z \to X$, and $\varphi : Y \to Z$, then the condition (E.141) corresponds to the equality $\pi_Z \circ \varphi = \pi_Y$.

To explain the equivalence (E.137), we underline locales in $\mathsf{Sh}(X)$, writing $\underline{Y}$ etc.; the corresponding internal frame is denoted by $\mathscr{O}(\underline{Y})$ (which is the same object in $\mathsf{Sh}(X)$ as $\underline{Y}$). The external description of $\underline{Y}$ in Sets, then, is a continuous map

$$\pi : Y \to X, \tag{E.142}$$

where Y is a locale in Sets (in which X was a a space to begin with), with frame

$$\mathscr{O}(Y) = \mathscr{O}(\underline{Y})(X). \tag{E.143}$$

Also here, the notation $\pi : Y \to X$ is purely symbolic, and stands for a frame map

$$\pi^{-1} : \mathscr{O}(X) \to \mathscr{O}(Y), \tag{E.144}$$

from which one may reconstruct $\underline{Y}$ as the sheaf

$$\mathscr{O}(\underline{Y}) : U \mapsto \{V \in \mathscr{O}(Y) \mid V \leq \pi^{-1}(U)\} \ \ (U \subset \mathscr{O}(X)). \tag{E.145}$$

The frame maps (E.138) - (E.140) yield an internal frame map $\underline{\varphi}^{-1} : \mathscr{O}(\underline{Z}) \to \mathscr{O}(\underline{Y})$ in $\mathsf{Sh}(X)$, which is a natural transformation, by defining its components as

$$\underline{\varphi}^{-1}(U) : \downarrow \pi_Z^{-1}(U) \to \downarrow \pi_Y^{-1}(U); \tag{E.146}$$
$$S \mapsto \varphi^{-1}(S). \tag{E.147}$$

As an application, the Dedekind real numbers $\mathbb{R}$ can be axiomatized by what is called a *geometric propositional theory* $\mathbb{T}$. In any topos T (with natural numbers object), such a theory determines a certain frame $\mathscr{O}(\mathbb{T})_{\mathsf{T}}$, whose "points" are defined as frame maps $\mathscr{O}(\mathbb{T}) \to \Omega$, where Ω is the subobject classifier in T (more precisely, the object of points of $\mathscr{O}(\mathbb{T})$ in T is the subobject of $\Omega^{\mathscr{O}(\mathbb{T})}$ consisting of frame maps). If $\mathbb{T}_{\mathbb{R}}$ is the theory axiomatizing $\mathbb{R}$, in Sets one simply has the frame

$$\mathscr{O}(\mathbb{T}_{\mathbb{R}})_{\mathsf{Sets}} = \mathscr{O}(\mathbb{R}), \tag{E.148}$$

whose points are $\mathbb{R}$. More generally, if $\mathbb{T}$ is some geometric propositional theory, and X is a space with associated sheaf topos $\mathsf{Sh}(X)$, then the internal frame $\mathscr{O}(\mathbb{T})_{\mathsf{Sh}(X)}$ is given by the sheaf (E.145) defined by taking the frame map (E.144) to be the inverse image map $\pi_{\mathbb{T}}^{-1} \equiv \pi_{\mathbb{T}}^{-1} : \mathscr{O}(X) \to \mathscr{O}(X \times \mathscr{O}(\mathbb{T})_{\mathsf{Sets}})$ of the projection $\pi_{\mathbb{T}} : X \times \mathscr{O}(\mathbb{T})_{\mathsf{Sets}} \to X$ onto the first component. Using (E.148), this yields the *frame* of Dedekind real numbers $\mathscr{O}(\mathbb{R}) \equiv \mathscr{O}(\mathbb{T}_{\mathbb{R}})$ in a sheaf topos $\mathsf{Sh}(X)$ as the sheaf

$$\mathscr{O}(\mathbb{R})_{\mathsf{Sh}(X)} : U \mapsto \mathscr{O}(U \times \mathbb{R}). \tag{E.149}$$

The Dedekind real numbers *object*, on the other hand, is given by the sheaf

$$(\mathbb{R})_{\mathsf{Sh}(X)} : U \mapsto C(U, \mathbb{R}). \tag{E.150}$$

Using (E.85), such results may immediately be transferred to $\mathsf{T}(A)$, see §12.1.

E.5 Internal language of a topos

The ***internal language*** (also called ***Mitchell–Bénabou language***) of a given topos T looks like a first-order language, except that it is ***typed*** (i.e., many-sorted), in that each term σ has a certain type, written $\sigma : X$, indexed by the objects X of T. For example, formulae (by definition) have type Ω. In addition, symbols, terms, and formulae have a list $\mathrm{FV}(\sigma)$ of free variables. Furthermore, the internal language has a canonical model in which it may be interpreted, whose carrier is T itself. We often make no difference in notation between σ as an element of the internal language of T and its interpretation $[[\sigma]]$ in T, which is some arrow in T; the two are so closely interwoven that making such a difference would be very artificial. Here are the rules.

- ***Constants*** c of type X correspond to arrows $c : \mathbf{1} \to X$ (and so in Sets are elements of X) and have no free variables, i.e. $\mathrm{FV}(c) = \emptyset$. Here and in what follows, we write 'corresponds to' in the following sense: for each arrow $c : \mathbf{1} \to X$ there is a constant c of type X, and the interpretation of this constant is this arrow.
- Logically interesting constants are the subobject classifier $t : \mathbf{1} \to \Omega$, which as in Theorem E.22 we often write as $\top$, and its antipode $\bot : \mathbf{1} \to \Omega$, defined as the classifying map for the mono $\mathbf{0} \to \mathbf{1}$ (where $\mathbf{0}$ is the initial object in T).
- ***Variables*** x of type X correspond to the identity $\mathrm{id}_X : X \to X$, with $\mathrm{FV}(x) = \{x\}$.
- ***Function symbols*** f of type Y correspond to arrows $f : X \to Y$. Thus in addition to its type, f has a ***source*** (namely X). Arities are unnecessary here; we may take $Y = Z \times \cdots \times Z$, with n terms, and say that f has source Z with arity n, but this is superfluous. Similarly, predicate symbols P would be function symbols of type Ω^X, and hence they are redundant (clearly, even constants and variables are special cases of function symbols, but it is useful to keep them apart).
- ***Terms*** are built by iteratively applying the following formation rules:
  1. Constants and variables are terms of the given type.
  2. If $\tau : X$ is a term of type X, and $f : X \to Y$ is a function symbol, then $f(\tau)$ is a term of type Y, and $\mathrm{FV}(f(\tau)) = \mathrm{FV}(\tau)$. Furthermore, $[[f(\tau)]] = f \circ \tau = f\tau$.
  3. If we have n terms $\tau_i : X_i$ $(i = 1, \ldots, n)$, with $\mathrm{FV}(\tau_1) = \cdots = \mathrm{FV}(\tau_n) \equiv F$, then $(\tau_1, \ldots, \tau_n)$ is a term of type $X_1 \times \cdots \times X_n$ and $\mathrm{FV}(\tau_1, \ldots, \tau_n) = F$. If τ_i has interpretation $\tau_i : Y \to X_i$, then $(\tau_1, \ldots, \tau_n) : Y \to X_1 \times \cdots \times X_n$ is the corresponding product arrow, as defined (for $n = 1$) before (E.91).
  4. One may add free variables to terms; if $\tau : Z$ with interpretation $\tau : X \to Z$ has a single free variable $x : X$, and we add a free variable y, then the interpretation of the revised term τ' with $\mathrm{FV}(\tau') = \{x, y\}$ is $\tau' : X \times Y \xrightarrow{p_1} X \xrightarrow{\tau} Z$ (etc.).
  5. From $\tau : X$ with $\mathrm{FV}(\tau) = \{z_1, \ldots, z_n\}$ with $z_i : Z_i$, and n terms $\sigma_i : Z_i$, all having the same free variables $\mathrm{FV}(\sigma_i) = \{y_1, \ldots, y_m\}$, with $y_j : Y_j$, we can form a new term $\tau(\sigma_1, \ldots, \sigma_n)$ of type X (i.e. the same type τ had), with free variables
  $$\mathrm{FV}(\tau(\sigma_1, \ldots, \sigma_n)) = \{y_1, \ldots, y_m\}. \tag{E.151}$$
  As the notation suggests, the interpretation of $\tau(\sigma_1, \ldots, \sigma_n)$ is $\tau \circ (\sigma_1, \ldots, \sigma_n)$.

- A ***formula*** is a term of type Ω. A ***sentence*** is a formula without free variables, which is therefore interpreted as an arrow $\varphi : \mathbf{1} \to \Omega$. The rules for formulae are:
  1. Let φ be a formula with $\mathrm{FV}(\varphi) = \{x, y\}$, with $x : X$ and $y : Y$. As in first-order logic, we may write φ as $\varphi(x, y)$. Then $\{x \mid \varphi(x, y)\}$ is a term of type Ω^X, with
$$\mathrm{FV}(\{x \mid \varphi(x,y)\}) = \{y\}. \tag{E.152}$$
This rule implements the isomorphism (sometimes called ***λ-conversion***)
$$\mathrm{Hom}_{\mathrm{T}}(X \times Y, \Omega) \cong \mathrm{Hom}_{\mathrm{T}}\left(Y, \Omega^X\right), \tag{E.153}$$
which follows from the existence of exponentials in a topos. Indeed, (E.153) turns the interpretation $\varphi : X \times Y \to \Omega$ into an arrow $\{x \mid \varphi(x,y)\} : Y \to \Omega^X$. Similarly, from $\varphi : X \times Y \to \Omega$ we obtain a term $\{(x,y) \mid \varphi(x,y)\}$ of type $X \times Y$, which is none other than the subobject classified by φ.
Taking $Y = \mathbf{1}$ to be the terminal object and using (E.51), we see that
$$\mathrm{Hom}_{\mathrm{T}}(X, \Omega) \cong \mathrm{Sub}(X) \cong \mathrm{Hom}_{\mathrm{T}}\left(\mathbf{1}, \Omega^X\right), \tag{E.154}$$
which shows that Ω^X plays the role the power set $\mathscr{P}(X)$ of X plays in Sets.
  2. If $\sigma : Y$ and $\tau : Y$ are terms with the same free variables, then $\sigma = \tau$ is a formula having the same set of free variables as τ and σ. If $\sigma : X \to Y$ and $\tau : X \to Y$, then the interpretation $[[\sigma = \tau]] : X \to \Omega$ is the composite arrow
$$X \xrightarrow{(\sigma,\tau)} Y \times Y \xrightarrow{=_Y} \Omega, \tag{E.155}$$
where $=_Y$ is the classifying map of the diagonal $\Delta_Y : Y \to Y \times Y$.
  3. If $\tau : Y$ and $\sigma : \Omega^Y$ are terms with the same free variables, then $\tau \in \sigma$ is a formula with the same free variables. If $\tau : X \to Y$ and $\sigma : X \to \Omega^Y$, then
$$[[\tau \in \sigma]] : X \xrightarrow{(\tau,\sigma)} Y \times \Omega^Y \xrightarrow{\mathrm{ev}} \Omega. \tag{E.156}$$
  4. As in first-order (or propositional) logic, new formulae may be made from old ones using the logical connectives $\wedge$, $\vee$, $\to$, and $\neg$. To interpret such composites, it is convenient to assume that their components have the same free variables, which can always be achieved using rule 4 for term-building above (i.e, by adding free variables). So let $\varphi : X \to \Omega$ and $\psi : X \to \Omega$ be (interpretations of) formulae, and let $\bullet$ be either $\wedge$, $\vee$, or $\to$. We then define
$$[[\varphi \bullet \psi]] : X \xrightarrow{(\varphi,\psi)} \Omega \times \Omega \xrightarrow{\bullet} \Omega, \tag{E.157}$$
where the arrow $\bullet : \Omega \times \Omega \to \Omega$ is defined from the Heyting algebra structure on Ω described in Theorem E.22. Similarly, negation is given by
$$[[\neg\varphi]] : X \xrightarrow{\varphi} \Omega \xrightarrow{\neg} \Omega. \tag{E.158}$$

5. If a formula $\varphi(x,y)$ contains x freely, as well as other free variables collectively called y, then $\exists_x \varphi(x,y)$ is a formula, whose interpretation we now give, after a bit of preparation. First, consider the commutative diagram

$$\begin{array}{ccccc} P & \longrightarrow & Z & \longrightarrow & \mathbf{1} \\ {\scriptstyle f^*m}\downarrow & & \downarrow{\scriptstyle m} & & \downarrow{\scriptstyle t} \\ X & \xrightarrow{f} & Y & \xrightarrow{\chi_m} & \Omega, \end{array} \tag{E.159}$$

where m is a mono, so that its equivalence class defines an element of $\mathrm{Sub}(Y)$. Taking the pullback of either m and f, or, equivalently, of t and $\chi_m \circ f$, we obtain a monic $f^*m : P \to X$, whose equivalence class is an element of $\mathrm{Sub}(Y)$. Consequently, any arrow $f : X \to Y$ induces a map

$$f^* : \mathrm{Sub}(Y) \to \mathrm{Sub}(X), \tag{E.160}$$

which is a homomorphism of (external) Heyting algebras (i.e. in Sets). For example, in Sets, where $\mathrm{Sub}(X)$ may be identified with $\mathscr{P}(X)$ (see comment after (E.52)), the map $f^* : \mathscr{P}(Y) \to \mathscr{P}(X)$ is simply the inverse image f^{-1} of f. If we regard the lattices $\mathrm{Sub}(X)$ and $\mathrm{Sub}(Y)$ as posetal categories, the map f^* has both a left-adjoint and a right-adjoint, denoted by

$$\exists_f : \mathrm{Sub}(X) \to \mathrm{Sub}(Y); \tag{E.161}$$
$$\forall_f : \mathrm{Sub}(X) \to \mathrm{Sub}(Y). \tag{E.162}$$

To justify this suggestive notation, replace X by $X \times Y$ and take $f : X \times Y \to Y$ to be p_2 (i.e., projection on the second space). Hence this gives maps

$$\exists_{p_2} : \mathrm{Sub}(X \times Y) \to \mathrm{Sub}(Y); \tag{E.163}$$
$$\forall_{p_2} : \mathrm{Sub}(X \times Y) \to \mathrm{Sub}(Y). \tag{E.164}$$

In Sets, we identify the Heyting algebras $\mathrm{Sub}(X \times Y)$ and $\mathrm{Sub}(Y)$ (now Boolean) with $\mathscr{P}(X \times Y)$ and $\mathscr{P}(Y)$, respectively, and obtain (on $A \subset X \times Y$):

$$\exists_{p_2}(A) = \{y \in Y \mid \exists_{x \in X} : (x,y) \in A\}; \tag{E.165}$$
$$\forall_{p_2}(A) = \{y \in Y \mid \forall_{x \in X} : (x,y) \in A\}. \tag{E.166}$$

Returning to a general topos, given $\varphi : X \times Y \to \Omega$, the diagram

$$\begin{array}{ccccccc} \mathbf{1} & \longleftarrow & \{(x,y) \mid \varphi(x,y)\} & \longrightarrow & \exists_{p_2}(\{(x,y) \mid \varphi(x,y)\}) & \longrightarrow & \mathbf{1} \\ {\scriptstyle t}\downarrow & & \downarrow & & \downarrow & & \downarrow{\scriptstyle t} \\ \Omega & \xleftarrow{\varphi} & X \times Y & \xrightarrow{p_2} & Y & \xrightarrow{[[\exists_x \varphi(x,y)]]} & \Omega \end{array}$$

defines the interpretation $[[\exists_x \varphi(x,y)]]$ (with innocent abuse of notation in applying the map $\exists_{p_2}$). The interpretation of $\forall_x \varphi(x,y)$ via $\forall_{p_2}$ is quite similar.

We now define the (semantic) notion of ***truth*** for sentences in the internal language of a topos; this is a far-reaching categorical generalization of the idea initially studied in the straightforward context of propositional logic, cf. §D.2.

Definition E.23. *1. A sentence φ in the internal language of* T *is* **true**, *written* $\Vdash \varphi$, *if its interpretation* $[[\varphi]]$ *coincides with the subobject classifier* $t : \mathbf{1} \to \Omega$.
2. An open formula $\varphi(x)$ *is true if its interpretation* $[[\varphi(x)]] : X \to \Omega$ *factors through* t, *or, equivalently, if (the interpretation of)* $\{x \mid \varphi(x)\}$, *seen as the subobject of* X *classified by* φ *(as explained between* (E.153) *and* (E.154)*), is* X *itself.*

The two clauses of this definition are actually equivalent, since no. 1 is obviously a special case of no. 2 by omitting the free variable x (and hence taking $X = \mathbf{1}$), but also, the second reduces to the first, because $\varphi(x)$ is true iff $\forall_x \varphi(x)$ is true.

As a refinement of this concept of truth, for $[[\varphi(x)]] : X \to \Omega$ as above, which we simply write as $\varphi : X \to \Omega$, take an arrow $f : Y \to X$. By definition:

$$Y \Vdash \varphi(f) \text{ means } \Vdash \varphi \circ f. \tag{E.167}$$

If φ is a sentence (i.e. $X = \mathbf{1}$), this means that $Y \Vdash \varphi(f)$ iff $\varphi = \chi_Y$ (in other words, φ classifies $Y \to \mathbf{1}$). There are (at least) two applications of this idea:

- The notion of ***partial truth*** states that φ is true ***at stage*** Y if $Y \Vdash \varphi(f)$.
- We say that a set $\mathscr{G} \subset \mathsf{T}_0$ of objects ***generates*** T if for every pair of parallel arrows $f : X \to Y$ and $h : X \to Y$, the property $fg = hg$ for all arrows $g : G \to X$ from all objects $G \in \mathscr{G}$ implies $f = g$. For example, any singleton generates Sets, and for any category C, it can be shown that the functors y_C generate the presheaf topos $[\mathsf{C}^{\mathrm{op}}, \mathsf{Sets}]$, as C runs through C_0. In general, it then follows that $\Vdash \varphi$ holds iff $G \Vdash \varphi$ for all $G \in \mathscr{G}$ (and, implicitly, all arrows $G \to X$).

Both play a role in Chapter 12, in the case where $\mathsf{T} = [\mathsf{C}^{\mathrm{op}}, \mathsf{Sets}]$ for a poset C. By the Yoneda Lemma E.15, any arrow $\alpha' : y_C \to X$ bijectively corresponds to some element $\alpha \in X_0(C)$. In that case, we write $C \Vdash \varphi(\alpha)$ for $C \Vdash \varphi(\alpha')$, which by (E.167) that the arrow $\varphi \circ \alpha' : y_C \to \Omega$ factors through the subobject classifier $t : \mathbf{1} \to \Omega$.

Kripke–Joyal semantics unfolds the expression $Y \Vdash \varphi(f)$ by looking at the formula φ in terms of its constituent terms. As one sees in Chapter 12, since this procedure may be used iteratively, it is extremely useful for computational purposes.

Although more than we need (which is the posetal case), we now give the rules for the validity of $C \Vdash \varphi(\alpha)$ in an arbitrary presheaf topos $[\mathsf{C}^{\mathrm{op}}, \mathsf{Sets}]$, as just mentioned; the posetal case follows in that $f : D \to C$ can only mean $D \leq C$.

We use the following notation:

- In clauses 1–4 below, we assume $\varphi : X \to \Omega$, and also $\psi : X \to \Omega$ (as already noted, this can always be achieved by adding free variables to φ and/or ψ).
- In 5–6, we assume $\varphi : X \times Y \to \Omega$ so as to accommodate the free variable $y : Y$.
- In 7 and 8, we have $\tau : X \to Y$, with $\sigma : X \to Y$ in no. 7, and $\sigma : X \to \Omega^Y$ in 8.

We then have the following ***forcing rules***, which generalize the ones given at the end of §D.3, and should be seen as theorems of categorical logic and topos theory:

1. $C \Vdash \varphi(\alpha) \wedge \psi(\alpha)$ iff $C \Vdash \varphi(\alpha)$ and $C \Vdash \psi(\alpha)$.
2. $C \Vdash \varphi(\alpha) \vee \psi(\alpha)$ iff $C \Vdash \varphi(\alpha)$ or $C \Vdash \psi(\alpha)$.
3. $C \Vdash \varphi(\alpha) \to \psi(\alpha)$ iff $D \Vdash \varphi(\alpha f)$ implies $D \Vdash \psi(\alpha f)$ for each $f : D \to C$.
4. $C \Vdash \neg\varphi(\alpha)$ iff no arrow $f : D \to C$ exists such that $D \Vdash \psi(\alpha f)$ holds.
5. $C \Vdash \exists_y \varphi(y)(\alpha)$ iff there exists $\beta \in Y_0(C)$ such that $C \Vdash \varphi(\alpha, \beta)$, where

$$\varphi(\alpha,\beta) : C \to \Omega; \tag{E.168}$$
$$\varphi(\alpha,\beta) \equiv \varphi \circ (\alpha', \beta'). \tag{E.169}$$

is obtained by combining the maps $\alpha' : y_C \to X$ and $\beta' : y_C \to Y$ into

$$(\alpha', \beta') : y_C \to X \times Y. \tag{E.170}$$

If φ has no free variables except y, then $C \Vdash \exists_y \varphi(y)$ iff there is $\beta \in Y_0(C)$ such that $C \Vdash \varphi(\beta)$.

6. $C \Vdash \forall_y \varphi(y)(\alpha)$ iff $D \Vdash \varphi(\alpha f, \beta)$ for each $f : D \to C$ and each $\beta \in Y_0(D)$. Here the arrow $f : D \to C$ induces a natural transformation $f' : y_D \to y_C$, yielding $\alpha f \equiv \alpha' \circ f' : y_D \to X$, which combines with $\beta' : y_D \to Y$ to

$$(\alpha f, \beta) : y_D \to X \times Y. \tag{E.171}$$

Similarly to the previous case, If φ has no free variables except y, we have

$$C \Vdash \forall_y \varphi(y) \text{ iff } D \Vdash \varphi(\beta), \tag{E.172}$$

for each $f : D \to C$ and each $\beta \in Y_0(D)$.

7. $C \Vdash (\tau = \sigma)(\alpha)$ iff $\tau \circ \alpha' = \sigma \circ \alpha'$.
8. $C \Vdash (\tau \in \sigma)(\alpha)$ iff the arrow

$$(\sigma \circ \alpha', \tau \circ \alpha') : y_C \to \Omega^Y \times Y \tag{E.173}$$

factors through the subobject of $\Omega^Y \times Y$ that is classified by the evaluation map $\mathrm{ev} : \Omega^Y \times Y \to \Omega$. As a special case, take $Y \rightsquigarrow \mathbf{1}$ and hence $\tau : X \to \mathbf{1}$, so that

$$\sigma : X \to \Omega^{\mathbf{1}} \cong \Omega \tag{E.174}$$

corresponds to a subobject $S \to X$ (i.e. classified by $\sigma \equiv \chi$). The above subobject of $\Omega^{\mathbf{1}} \times \mathbf{1} \cong \Omega$ is then simply given by the truth arrow $t : \mathbf{1} \to \Omega$. Writing $x \in S$ for $\tau \in \sigma$ (where $x : X$ is a variable of type X), we therefore obtain the rule:

9. $C \Vdash (x \in S)(\alpha)$ iff $\sigma \circ \alpha : y_C \to \Omega$ factors through t (in other words, the subobject of y_C classified by $\sigma \circ \alpha$ is y_C itself).

Notes

The standard introduction to category theory by one of its founders is Mac Lane (1998); see also the book by his student Awodey (2010), as well as the lecture notes by van Oosten (2002) and Cheng (2002). A nice book, which studies set theory from the point of view of category theory, is Lawvere & Rosebrugh (2003). At high-school level, see also Lawvere & Schanuel (1997) or (informally) Cheng (2015).

Toposes were invented by Grothendieck in the early 1960s as part of his rebuilding of algebraic geometry; see Artin, Grothendieck, & Verdier (1972). The history and philosophy of category theory (including topos theory) has been described by Krömer (2007) and by Marquis (2009); for categorical logic see also Marquis & Reyes (2012) and Bell (2005). According to a leading category & topos theorist:

> 'category theory was the objective form of dialectical materialism (...) set theory was considered to be essentially bourgeois since it is founded on the relationship of belonging.' (Marquis & Reyes, 2012, p. 30).

Books on topos theory and categorical logic we used include (in increasing order of scope and sophisitication): Goldblatt (1984), Bell (1988), Borceux (1994), Mac Lane & Moerdijk (1992), and last but not least, the encyclopedic Johnstone (2002).

§E.1. **Basic definitions**

von Neumann–Bernays–Gödel set theory is discusses in some detail in Mendelson (2010); for algebraic set theory see Joyal & Moerdijk (1995). Category theorists also typically rely on the notion of a *Grothendieck Universe*, see e.g. Mac Lane (1998, §1.6), Marquis (2009, §5.5), and Krömer (2007, Ch. 6).

§E.2. **Toposes and functor categories**

An axiomatization of Grothendieck's toposes (and certain generalizations thereof) equivalent to Definition E.12 was given in 1970 by Lawvere and Tierney (it seems to have been customary among the pioneers of topos theory, who also include Joyal, not to publish their findings too lavishly and in fact no joint paper by Lawvere & Tierney recording their definition seems to exist at least in the open literature).

§E.3. **Subobjects and Heyting algebras in a topos**

See Mac Lane & Moerdijk (1992), §§I.8, IV.8, and Borceux (1994), §1.2.

§E.4. **Internal frames and locales in sheaf toposes**

The external description of internal locales in sheaf toposes originates with Joyal & Tierney (1984); see also Johnstone (2002), §C1.6.

§E.5. **Internal language of a topos**

More details and proofs of the Kripke–Joyal semantics for the internal language of a topos may be found in Bell (1988), Ch. 4, Mac Lane & Moerdijk (1992), §IV.6, Borceux (1994), §6.6, and Johnstone (2002), §D1.2.

For an analysis of the notion of partial truth (as defined here) applied to quantum mechanics (differently from our Chapter 12), see Butterfield (2002).

References

1. Aarens, J.F. (1970). Quasi states on C*-algebras. *Transactions of the American Mathematical Society* 149, 601–625.
2. Aarens, J.F. (1991). Quasi-states and quasi-measures. *Advances in Mathematics* 86, 41–67.
3. Aastrup, J., Nest, R., Schrohe, E. (2006). A continuous field of C*-algebras and the tangent groupoid for manifolds with boundary. *Journal of Functional Analysis* 237, 482–506.
4. Abraham, R., J.E. Marsden (1985). *Foundations of Mechanics*, 2nd ed. Redwood City: Addison Wesley.
5. Abramsky, S., Brandenburger, A. (2011). The sheaf-theoretic structure of non-locality and contextuality. *New Journal of Physics* 13, 113036.
6. Acín, A., Fritz, T., Leverrier, A., Sainz, A.B. (2015). A combinatorial approach to nonlocality and contextuality. *Communications in Mathematical Physics* 334, 533–628.
7. Aczel, P, (2006). Aspects of general topology in constructive set theory. *Annals of Pure and Applied Logic* 137, 3–29.
8. Adelman, M., Corbett, J. (1995). A sheaf model for intuitionistic quantum mechanics. *Applied Categorical Structures* 3, 79–104.
9. Aguirre, A., Tegmark, M. (2011). Born in an infinite universe: a cosmological interpretation of quantum mechanics. *Physical Review D* 84, 105002.
10. Akemann, C.A., Bice, T. (2014). Hereditary C*-subalgebra lattices. `arXiv:1410.0093`.
11. Alfsen, E.M. (1970). *Compact Convex Sets and Boundary Integrals*. Springer: Berlin.
12. Alfsen, E.M., Shultz, F.W. (2001). *State Spaces of Operator Algebras*. Basel: Birkhäuser.
13. Alfsen, E.M., Shultz, F.W. (2003). *Geometry of State Spaces of Operator Algebras*. Basel: Birkhäuser.
14. Allahverdyan, A.E., Balian, R., Nieuwenhuizen, Th.M. (2013). Understanding quantum measurement from the solution of dynamical models. *Physics Reports* 525, 1–166.
15. Altland, A., Simons, B. (2010). *Condensed Matter Field Theory*. Cambridge: Cambridge University Press.
16. Amir, D. (1986). *Characterizations of Inner Product Spaces*. Basel: Birkhäuser.
17. An Huef, A., Raeburn, I., Williams, D.P. (2010). Functoriality of Rieffel's generalised fixed-point algebras for proper actions. *Proceedings of Symposia in Pure Mathematics* 81, 9-26.
18. Anderson, J. (1979). Extensions, restrictions, and representations of states on C*-algebras. *Transactions of the American Mathematical Society* 249, 303–329.
19. Anderson, P.W. (1972). More is different. *Science* 177 (4047), 393–396.
20. Appleby, D.M. The Bell–Kochen–Specker Theorem. *Studies in History and Philosophy of Modern Physics* 36, 1–28.
21. Araki, H. (1974). On the equivalence of the KMS condition and the variational principle for quantum lattice systems. *Communications in Mathematical Physics* 38, 1–10.
22. Araki, H. (1975). On uniqueness of KMS states of one-dimensional quantum lattice systems. *Communications in Mathematical Physics* 44, 1–7.

K. Landsman, *Foundations of Quantum Theory*,
Fundamental Theories of Physics 188, DOI 10.1007/978-3-319-51777-3

23. Araki, H. (1987). Bogoliubov automorphisms and Fock representations of canonical anti-commutation relations. *Contemporary Mathematics* 62, 23–165.
24. Araki, H., Matsui, T. (1985). Ground states of the XY-model. *Communications in Mathematical Physics* 101, 213–245.
25. Arends, F. (2009). A lower bound on the size of the smallest Kochen–Specker vector system. Master's thesis, Oxford University. `www.cs.ox.ac.uk/people/joel.ouaknine/download/arends09.pdf`.
26. Arens, R. (1961). The analytic-functional calculus in commutative topological algebras. *Pacific Journal of Mathematics* 11, 405–429.
27. Arndt, M., Dörre, S., Eibenberger, S., Haslinger, P., Rodewald, J., Hornberger, K., Nimmrichter, S., Mayor, M. (2015). Matter-wave interferometry with composite quantum objects. `arXiv:1501.07770`.
28. Artin, M., Grothendieck, A., Verdier, J.-L. (1972). *Théorie de Topos et Cohomologie Étale des Schémas (SGA4)*. Berlin: Springer.
29. Awodey, S. (2010). *Category Theory*. Second Edition. Oxford: Oxford University Press.
30. Bacciagaluppi, G. (1993). Separation theorems and Bell inequalities in algebraic quantum mechanics. *Proceedings of the Symposium on the Foundations of Modern Physics (Cologne, 1993)*, pp. 29–37. Busch, P., Lahti, P.J., Mittelstaedt, P., eds. Singapore: World Scientific.
31. Bacciagaluppi, G. (2014). Insolubility from no-signalling. *International Journal of Theoretical Physics* 53, 3465–3474.
32. Bach, A. (1997). *Indistinguishable Classical Particles*. Berlin: Springer.
33. Baez, J. (1987). Bell's inequality for C*-algebras. *Letters in Mathematical Physics* 13, 135–136.
34. Baker, D.J., Halvorson, H., Swanson, N. (2015). The conventionality of parastatistics. *British Journal for the Philosophy of Science* 66, 929–976.
35. Banach, S. (1932). *Théorie des Opérations Linéaires*. Warszawa: Instytut Matematyczny Polskiej Akademii Nauk, New York: Chelsea.
36. Banaschewski, B., Mulvey, C.J. (2000a). The spectral theory of commutative C*-algebras: The constructive spectrum. *Quaestiones Mathematicae* 23, 425–464.
37. Banaschewski, B., Mulvey, C.J. (2000b). The spectral theory of commutative C*-algebras: The constructive Gelfand–Mazur theorem. *Quaestiones Mathematicae* 23, 465–488.
38. Banaschewski, B., Mulvey, C.J. (2006). A globalisation of the Gelfand duality theorem. *Annals of Pure and Applied Logic* 137, 62–103.
39. Bargmann, V. (1954). On unitary ray representations of continuous groups. *Annals of Mathematics* 59, 1–46.
40. Bargmann, V. (1964). Note on Wigner's Theorem on symmetry operations. *Journal of Mathematical Physics* 5, 862–868.
41. Barrett, J., Kent, A. (2004). Non-contextuality, finite precision measurement and the Kochen–Specker theorem. *Studies in History and Philosophy of Modern Physics* 35, 151–176.
42. Barut, A.O., Rączka, R. (1977). *Theory of Group Representations and Applications*. Warszawa: PWN.
43. Bassi, A., Ghirardi, G.C. (2000). A general argument against the universal validity of the superposition principle. *Physics Letters A* 275, 373–381.
44. Bassi, A., Ghirardi, G.C. (2003). Dynamical reduction models. *Physics Reports* 379, 257–426.
45. Bassi, A., Lochan, K., Satin, S., Singh, T.P., Ulbricht, H. (2013). Models of wave-function collapse, underlying theories, and experimental tests. *Reviews of Modern Physics* 85, 471–527.
46. Bassi, A., Ghirardi, G.C. (2007). The Conway-Kochen argument and relativistic GRW models. *Foundations of Physics* 37, 169–185.
47. Batterman, R. (2002). *The Devil in the Details: Asymptotic Reasoning in Explanation, Reduction, and Emergence*. Oxford: Oxord University Press.
48. Batterman, R. (2005). Response to Belot's "Whose devil? Which details?". *Philosophy of Science* 72, 154–163.

49. Batterman, R. (2011). Emergence, singularities, and symmetry breaking. *Foundations of Physics* 41, 1031–1050.
50. Bayen, F., Flato, M., Fronsdal, C., Lichnerowicz, A., Sternheimer, D. (1978). Deformation theory and quantization I, II. *Annals of Physic (N.Y.)* 110, 61–110, 111–151.
51. Bedau, M.A.,Humphreys, P., eds. (2008). *Emergence*. Cambridge (Mass.): MIT Press.
52. Beebee, H. (2003). Local Miracle Compatibilism. *Noûs* 37, 258–277.
53. Beebee, H. (2013). *Free Will: An Introduction*. New York: Palgrave Macmillan.
54. Bell, J.L. (1988). *Toposes and Local Set Theories: An Introduction*. Oxford: Clarendon Press.
55. Bell, J.L. (2005). The development of categorical logic. *Handbook of Philosophical Logic, Vol. 12*, pp. 279–360. Gabbay, D.M., Guenthner, F., eds. New York: Springer.
56. Bell, J.L., Machover, M. (1977). *A Course in Mathematical Logic*. Amsterdam: North-Holland.
57. Bell, J.S. (1964). On the Einstein–Podolsky–Rosen Paradox. *Physics* 1, 195–200.
58. Bell, J.S. (1966). On the problem of hidden variables in quantum mechanics. *Reviews of Modern Physics* 38, 447–452.
59. Bell, J.S. (1975). On wave packet reduction in the Coleman-Hepp model. *Helvetica Physica Acta* 48, 93–98.
60. Bell, J.S. (1976). The theory of local beables. *Epistemological Letters* 9, 11–24.
61. Bell, J.S. (1990a). La nouvelle cuisine, *Between Science and Technology*, pp. 97–115. Sarlemijn, A., Kroes, P., eds. Amsterdam: Elsevier.
62. Bell, J.S. (1990b). Against "Measurement". *Physics World* 3, 33–41.
63. Bell, M., Gottfried, K., Veltman, M. (2001). *John S. Bell on the Foundations of Quantum Mechanics*. Singapore: World Scientific.
64. Beller, M. (1999). *Quantum Dialogue: The Making of a Revolution*. Chicago: Chicago University Press.
65. Belinfante, F.J. (1973). *A Survey of Hidden-Variable Theories*. Oxford: Pergamon Press.
66. Belot, G. (2005). Whose devil? Which details? *Philosophy of Science* 72, 128–153.
67. Beltrametti, E.G.,Bugajski, S. (1995). A classical extension of quantum mechanics, *Journal of Physics A* 28, 3329–3343.
68. Benaceraff, P., Putnam, H. (1983). *Philosophy of Mathematics: Selected Readings, Second Edition*. Cambridge: Cambridge University Press.
69. Bennett, J. (1984). Counterfactuals and temporal direction. *The Philosophical Review* 93, 57–91.
70. Berberian, S.K. (1972). *Baer *-Rings*. Berlin: Springer-Verlag.
71. Berglund, N. (2011). Kramers' law: Validity, derivations and generalizations. arXiv:1106.5799.
72. Bernkopf, M.(1966). The development of function spaces with particular reference to their origins in integral equation theory. *Archive for History of Exact Sciences* 3, 1–96.
73. Berezin, F.A. (1975). General concept of quantization. *Communications in Mathematical Physics* 40, 153–174.
74. Berg, B. van den, Heunen, C. (2012). Noncommutativity as a colimit. *Applied Categorical Structures* 20, 393–414.
75. Berg, B. van den, Heunen, C. (2014). Extending obstructions to noncommutative functorial spectra. *Theory and Applications of Categories* 29, 457–474.
76. Bernkopf, M.(1967). A history of infinite matrices. *Archive for History of Exact Sciences* 4, 308–358.
77. Berzi, V. (1979). Arveson spectral theory and C*-algebraic Goldstone-type theorems. *Reports on Mathematical Physics* 16, 293–304.
78. Berzi, V. (1981). Remarks on the Goldstone theorem in the C*-algebraic approach. *Letters in Mathematical Physics* 5, 373–377.
79. Birkhoff, G. & Kreyszig, E. (1984). The establishment of functional analysis. *Historia Mathematica* 11, 258–321.
80. Birkhoff, G., von Neumann, J. (1936). The logic of quantum mechanics. *Annals of Mathematics* 37, 823–843.

81. Blackadar, B. (1994). Projections in C*-algebras. *C*-algebras: 1943–1993, Contemporary Mathematics* Vol. 167, pp. 131–150. Doran, R.S., ed. Providence: American Mathematical Society.
82. Blackadar, B. (2006). *Operator Algebras: Theory of C*-Algebras and von Neumann Algebras*. Berlin: Springer.
83. Blanchard, E. (1996). Déformations de C*-algebras de Hopf. *Bulletin de la Société mathématique de la France* 124, 141–215.
84. Blanchard, P., Fröhlich, J., Schubnel, B. (2016). A "garden of forking paths" - the quantum mechanics of histories of events. `arXiv:1603.09664`.
85. Blanco, A., Turnšek, A. (2006). On maps that preserve orthogonality in normed spaces. *Proceedings of the Royal Society of Edinburgh* A136, 709–716.
86. Bogachev, V.I. (2007). *Measure Theory*, Vol. 2. Berlin: Springer.
87. Bogoliubov, N.N. (1958). On a new method in the theory of superconductivity. *Nuovo Cimento* 7, 794–805.
88. Bohr, N. (1913). On the constitution of atoms and molecules (Part I). *Philosophical Magazine* 26, 1–25.
89. Bohr, N. (1928). The quantum postulate and the recent development of atomic theory ("Como Lecture"). *Nature suppl.* April 14, 580–590.
90. Bohr, N. (1935). Can quantum-mechanical description of physical reality be considered complete? *Physical Review* 48, 696–702.
91. Bohr, N. (1948). On the notions of causality and complementarity. *Dialectica* 2, 312–319.
92. Bohr, N. (1949). Discussion with Einstein on epistemological problems in atomic physics. *Albert Einstein: Philosopher-Scientist*, pp. 201–241. Schlipp, P.A., ed. La Salle: Open Court.
93. Bohr, N. (1958). *Atomic Physics and Human Knowlegde*. New York: Wiley.
94. Bohr, N. (1976). *Collected Works. Vol. 3: The Correspondence Principle (1918–1923)*. Rosenfeld, J., Nielsen, R., eds. Amsterdam: North-Holland.
95. Bohr, N. (1985). *Collected Works. Vol. 6: Foundations of Quantum Physics* I *(1926–1932)*. Kalckar, J., ed. Amsterdam: North-Holland.
96. Bohr, N. (1996). *Collected Works. Vol. 7: Foundations of Quantum Physics* II *(1933–1958)*. Kalckar, J., ed. Amsterdam: North-Holland.
97. Bokulich, A.(2008). *Reexamining the Quantum-Classical Relation: Beyond Reductionism and Pluralism*. Cambridge: Cambridge University Press.
98. Bona, P. (1980). A solvable model of particle detection in quantum theory. *Acta Facultatis Rerum Naturalium Universitatis Comenianae Physica* XX, 65–94.
99. Bona, P. (1988). The dynamics of a class of mean-field theories. *Journal of Mathematical Physics* 29, 2223–2235.
100. Bona, P. (1989). Equilibrium states of a class of mean-field theories. *Journal of Mathematical Physics* 30, 2994–3007.
101. Bona, P. (2000). Extended quantum mechanics. *Acta Physica Slovaca* 50, 1–198.
102. Bongaarts, P. (2015). *Quantum Theory: A Mathematical Approach*. Heidelberg: Springer.
103. Bonolis, L. (2004). From the rise of the group concept to the stormy onset of group theory in the new quantum mechanics: A saga of the invariant characterization of physical objects, events and theories. *Nuovo Cimento Rivista* 27. DOI: `10.1393/ncr/i2004-10006-4`.
104. Boole, G. (1847). *The Mathematical Analysis of Logic: Being an Essay Towards a Calculus of Deductive Reasoning*. London: Henderson & Spalding.
105. Borceux, F. (1994). *Handbook of Categorical Algebra 3: Categories of Sheaves*. Cambridge: Cambridge University Press.
106. Borchers, H.J. (2000). On revolutionizing quantum field theory with Tomita's modular theory. *Journal of Mathematical Physics* 41, 3604–3673.
107. Born, M. (1926a). Zur Quantenmechanik der Stoßvorgänge. *Zeitschrift für Physik* 37, 863–867.
108. Born, M. (1926b). Quantenmechanik der Stoßvorgänge. *Zeitschrift für Physik* 38, 803–827.
109. Born, M., Heisenberg, W., Jordan, P. (1926). Zur Quantenmechanik II. *Zeitschrift für Physik* 36, 557–615.

110. Botet, R., & Julien, R. (1982). Large-size critical behavior of infinitely coordinated systems. *Physical Review* B28, 3955–3967.
111. Botet, R., Julien, R., & Pfeuty, P. (1982). Size scaling for infinitely coordinated systems. *Physical Review Letters* 49, 478–481.
112. Bourbaki, N. (1989). *General Topology: Chapters 1–4*. Berlin: Springer.
113. Brandenburg, M. (2015). Non-unital Gelfand duality. Unpublished.
114. Bratteli, O., Robinson, D.W. (1987). *Operator Algebras and Quantum Statistical Mechanics. Vol. I: C*- and W*-Algebras, Symmetry Groups, Decomposition of States*. 2nd Ed. Berlin: Springer.
115. Bratteli, O., Robinson, D.W. (1981). *Operator Algebras and Quantum Statistical Mechanics. Vol. II: Equilibrium States, Models in Statistical Mechanics*. Berlin: Springer.
116. Braunstein, S.L., Caves, C.M. (1990). Wringing out better Bell inequalities. *Annals of Physics* 202, 22–56.
117. Breuer, T., Amann, A., Landsman, N.P. (1993). Robustness in quantum measurements. *Journal of Mathematical Physics* 34, 5441–5450.
118. Brezis, H., Browder, F. (1998). Partial difference equations in the 20th century. *Advances in Mathematics* 135, 76–144.
119. Bricmont, J. (2016). *Making Sense of Quantum Mechanics*. Switzerland: Springer International Publishing.
120. Broad, C.D. (1925). *The Mind and its Place in Nature*. London: Routledge & Kegan Paul.
121. Brock, S. (2003). *Niels Bohr's Philosophy of Quantum Physics: In the Light of the Helmholtzian Tradition of Theoretical Physics*. Berlin: Logos Verlag.
122. Brody, D.C., Hughston, L.P. (2001). Geometric quantum mechanics. *Journal of Geometry and Physics* 38, 19–53.
123. Brouwer, L.E.J. (1907). *Over de Grondslagen der Wiskunde*. PhD Thesis, University of Amsterdam. Amsterdam: Maas & Van Suchtelen.
124. Brouwer, L.E.J. (1918). *Begründung der Mengenlehre unabhängig vom logischen Satz vom ausgeschlossenen Dritten*. Amsterdam: Müller.
125. Brouwer, L.E.J. (1975). *Collected Works, Vol. 1: Philosophy and Foundations of Mathematics*. Amsterdam: North-Holland.
126. Brown, H. (1986). The insolubility proof of the quantum measurement problem. *Foundations of Physics* 16, 857–870.
127. Brown, H. (1992). Bell's other theorem and its connection with non-locality. *Bell's Theorem and the Foundations of Modern Physics*, pp. 104–116. van der Merwe, A., Selleri, F., Tarozzi, G., eds. Singapore: World Scientific.
128. Brown, H., Svetlichny, G. (1990). Nonlocality and Gleason's lemma. Part I. Deterministic theories. *Foundations of Physics* 20, 1379–1387.
129. Brown, H., Timpson, C. (2015). Bell on Bell's Theorem: The changing face of nonlocality. `arXiv:1501.03521`.
130. Brown, L.G., Pedersen, G.K. (1991). C*-algebras of real rank zero. *Journal of Functional Analysis* 99, 131–149.
131. Brunner, C., Cavalcanti, D., Scarani, V., Wehner, S. (2014). Bell nonlocality. *Reviews of Modern Physics* 86, 419–478.
132. Bub, J. (1997). *Interpreting the Quantum World*. Cambridge: Cambridge University Press.
133. Bub, J. (2004). Why the quantum? *Studies in History and Philosophy of Modern Physics* 35, 241–266.
134. Bub, J. (2011a). Is von Neumann's 'no hidden variables' proof silly?, *Deep Beauty: Mathematical Innovation and the Search for Underlying Intelligibility in the Quantum World*, pp. 393–408. Halvorson, H., ed. Cambridge: Cambridge University Press.
135. Bub, J. (2011b). Quantum probabilities: an information-theoretic interpretation. *Probabilities in Physics*, pp. 231–262. Hartmann, S., Beisbart, C., eds. Oxford: Oxford University Press.
136. Bub, J., Clifton, R. (1996). A uniqueness theorem for 'no collapse' interpretations of quantum mechanics. *Studies in History and Philosophy of Modern Physics* 27, 181–219.
137. Buchholz, D., Doplicher, S., Longo, R., Roberts, J.E. (1992). A new look at Goldstones theorem. *Reviews in Mathematical Physics* 4 (Special Issue), 49–83.

138. Budde, C. (2015). *Operator Algebras and Unbounded Self-adjoint Operators*. MSc Thesis, Radboud University Nijmegen. www.math.ru.nl/~landsman/Budde.pdf.
139. Budde, C., Landsman, N.P. (2016). A bounded transform approach to self-adjoint operators: Functional calculus and affiliated von Neumann algebras. *Annals of Functional Analysis* 7, 411–420.
140. Bugajski, S., Motyka, Z. (1981).Generalized Borel law and quantum probabilities. *International Journal of Theoretical Physics* 20, 262–268.
141. Burtscher, A. (2009). *Isomorphisms of Algebras of Smooth and Generalized Functions*. Diplomarbeit, Universität Wien.
142. Busch, P. (2003). Quantum states and generalized observables: a simple proof of Gleason's Theorem. *Physical Review Letters* 91, 120403.
143. Busch, P., Grabowski, M., Lahti, P.J. (1998). *Operational Quantum Physics*, 2nd corrected ed. Berlin: Springer.
144. Busch, P., Lahti, P.J., Mittelstaedt, P. (1996). *The Quantum Theory of Measurement. Second Edition*. Berlin: Springer-Verlag.
145. Busch, P., Lahti, P.J., Pellonpää, J.-P., Ylinen, K. (2016). *Quantum Measurement*. Heidelberg: Springer.
146. Busch, P., Shimony, A. (1996). Insolubility of the quantum measurement problem for unsharp observables. *Studies in History and Philosophy of Modern Physics* 27, 397–404.
147. Butterfield, J. (1992a). David Lewis meets John Bell. *Philosophy of Science* 59, 26–43.
148. Butterfield, J. (1992b). Bell's theorem: what it takes. *British Journal for the Philosophy of Science* 43, 41–83.
149. Butterfield, J. (2002). Topos theory as a framework for partial truth. *In the Scope of Logic, Methodology and Philosophy of Science, Vol. I*, pp. 307–330. Gärdenfors, P. et al, eds. Dordrecht: Kluwer Academic Publishers.
150. Butterfield, J. (2011). Less is different: Emergence and reduction reconciled. *Foundations of Physics* 41, 1065–1135.
151. Butterfield, J., & Bouatta, N. (2011). Emergence and reduction combined in phase transitions. arXiv:1104.1371.
152. Butterfield, J., Isham, C.J. (1999). A topos perspective on the Kochen-Specker theorem. II. Conceptual aspects and classical analogues. *International Journal of Theoretical Physics* 38, 827–859 (1999).
153. Butterfield, J., Isham, C.J. (2002). A topos perspective on the Kochen-Specker theorem. IV. Interval valuations. *International Journal of Theoretical Physics* 41, 613–639.
154. Cabello, A., Estebaranz, J.M., Alcaine, G.C. (1996). Bell–Kochen–Specker theorem: A proof with 18 vectors. *Physics Letters A* 183–187.
155. Camilleri, K. (2007). Bohr, Heisenberg, and the divergent views of complementarity. *Studies in History and Philosophy of Modern Physics* 38, 514–528.
156. Callender, C. (2011). Thermodynamic asymmetry in time. *The Stanford Encyclopedia of Philosophy (Fall 2011 Edition)*, Zalta, E.N., ed. plato.stanford.edu/archives/fall2011/entries/time-thermo/.
157. Callender, C. (2016). *What Makes Time Special*. To appear.
158. Camilleri, K. (2009a). A history of entanglement: Decoherence and the interpretation problem. *Studies in History and Philosophy of Modern Physics* 40, 290–302.
159. Camilleri, K. (2009b). *Heisenberg and the Interpretation of Quantum Mechanics: The Physicist as Philosopher*. Cambridge: Cambridge University Press.
160. Camilleri, K., Schlosshauer, M. (2015). Niels Bohr as philosopher of experiment: Does decoherence theory challenge Bohr's doctrine of classical concepts? *Studies in History and Philosophy of Modern Physics* 49, 73–83.
161. Campanino, M., Klein, A., Perez, J.F. (1991). Localization in the ground state of the Ising model with a random transverse field. *Communications in Mathematical Physics* 135, 499-515.
162. Caruana, L. (1995). John von Neumann's 'Impossibility Proof' in a historical perspective. *Physis* 32, 109–124.

163. Casazza, P.G., Fickus, M., Tremain, J.C., Weber, E. (2005). The Kadison-Singer Problem in mathematics and engineering. `arXiv:math/0510024`.
164. Casazza, P.G., Tremain, J.C. (2016). Consequences of the Marcus/Spielman/Srivastava solution of the Kadison-Singer Problem. *New Trends in Applied Harmonic Analysis*, pp. 191–213. Aldroubi, A. et al, eds. Basel: Birkhäuser.
165. Caspers, M. (2008). *Gelfand Spectra of C*-algebras in Topos Theory*. MSc Thesis, Radboud Universiteit Nijmegen. `www.math.ru.nl/~landsman/scriptieMartijn.pdf`.
166. Caspers, M., Heunen, C., Landsman, N.P., Spitters, B. (2009). Intuitionistic quantum logic of an n-level system. *Foundations of Physics* 39, 731–759.
167. Cassinelli, G., De Vito, E., Lahti, P.J., Levrero, A. (2004). *The Theory of Symmetry Actions in Quantum Mechanics. Lecture Notes in Physics* 654. Berlin: Springer-Verlag.
168. Cassinelli, G., Lahti, P.J. (1989). The measurement statistics interpretation of quantum mechanics: Possible values and possible measurement results of physical quantities. *Foundations of Physics* 19, 873–890.
169. Cassinello, A., Sánchez-Gómez, J.L. (1996). On the probabilistic postulate of quantum mechanics. *Foundations of Physics* 26, 1357–1374.
170. Cator, E., Landsman, N.P. (2014). Constraints on determinism: Bell versus Conway–Kochen. *Foundations of Physics* 44, 781–791.
171. Caves, C.M., Fuchs, C.A., Schack, R. (2002a). Unknown quantum states: the quantum de Finetti representation. *Journal of Mathematical Physics* 43, 4537–4559.
172. Caves, C.M., Fuchs, C.A., Schack, R. (2002b). Quantum probabilities as Bayesian probabilities. *Physical Review A* 65, 022305.
173. Caves, C.M., Fuchs, C.A., Manne, K.K., Renes, J.M. (2004). Gleason-type derivations of the quantum probability rule for generalized measurements. *Foundations of Physics* 34, 193–209.
174. Caves, C., Schack, R. (2005). Properties of the frequency operator do not imply the quantum probability postulate. *Annals of Physics (N.Y.)* 315, 123–146.
175. Cederquist, J., Coquand, T. (2000). Entailment relations and distributive lattices. *Lecture Notes in Logic* 13, 127–139.
176. Cesi, F. (1989). An algorithm to study tunneling in a wide class of one-dimensional multiwell potentials. I.*Journal of Physics* A22, 1027–1052.
177. Cesi, F., Rossi, G.C., Testa, M. (1991). Non-symmetric double well and Euclidean functional integral. *Annals of Physics* 206, 318–333.
178. Chayes, L., Crawford, N., Ioffe, D., Levit, A. (2008). The phase diagram of the quantum Curie–Weiss model. *Journal of Statistical Physics* 133, 131–149.
179. Cheng, E. (2002). *Category Theory*. University of Cambridge Lecture Notes. `cheng.staff.shef.ac.uk/catnotes/categorynotes-cheng.pdf`.
180. Cheng, E. (2015). *Cakes, Custard and Category Theory: Easy Recipes for Understanding Complex Maths*. London: Profile Books.
181. Clauser, J.F., Horne, M.A, Shimony, A., Holt, R.H. (1969). Proposed experiment to test local hidden-variable theories. *Physical Review Letters* 23, 880–884.
182. Claverie, P., Jona-Lasinio, G. (1986). Instability of tunneling and the concept of molecular structure in quantum mechanics: The case of pyramidal molecules and the enantiomer problem. *Physical Review* A33, 2245–2253.
183. Clifton, R. (1993). Getting contextual and nonlocal elements-of-reality the easy way. *American Journal of Physics* 61, 443–447.
184. Clifton, R., Kent, A. (2000). Simulating quantum mechanics by non-contextual hidden variables. *Proceedings of the Royal Society of London A* 456, 2101–2114.
185. Clifton, R., Redhead, M., J. Butterfield, J. (1990). Nonlocal influences and possible worlds—A Stapp in the wrong direction. *The British Journal for the Philosophy of Science* 41, 5–58.
186. Clifton, R., Redhead, M., J. Butterfield, J. (1991). Generalization of the Greenberger–Horne–Zeilinger algebraic proof of nonlocality. *Foundations of Physics* 21, 149–184.
187. Coecke, B. (2002). Quantum logic in intuitionistic perspective. *Studia Logica* 70, 411–440.
188. Colbeck, R., Renner, R. (2011). No extension of quantum theory can have improved predictive power. *Nature Communications* 2:411.

189. Colbeck, R., Renner, R. (2012a). Is a system's wave function in one-to-one correspondence with its elements of reality? *Physical Review Letters* 108, 150402.
190. Colbeck, R., Renner, R. (2012b). The completeness of quantum theory for predicting measurement outcomes. arXiv:1208.4123.
191. Combes, J. M., Duclos, P., Seiler, R. (1983). Convergent expansions for tunneling. *Communications in Mathematical Physics* 92, 229–245.
192. Connes, A. (1994). *Noncommutative Geometry*. San Diego: Academic Press.
193. Connes, A., Marcolli, M. (2008). *Noncommutative Geometry, Quantum Fields, and Motives*. New Delhi: Hindustan Book Agency.
194. Connes, A., Rovelli, C. (1994). Von Neumann algebra automorphisms and time-thermodynamics relation in generally covariant quantum theories. *Classical and Quantum Gravity* 11, 2899-2917.
195. Conway, J.B. (2007). *A Course in Functional Analysis*. Second Edition. New York: Springer.
196. Conway, J.H. (2009). Six video lectures on the Free Will Theorem. paw.princeton.edu/issues/2009/07/15/pages/6596/index.xml.
197. Conway, J.H., Kochen, S. (2006). The Free Will Theorem. *Foundations of Physics* 36, 1441–1473.
198. Conway, J.H., Kochen, S. (2009). The Strong Free Will Theorem. *Notices of the American Mathematical Society* 56, 226–232.
199. Cooke, R., Keane, M., Moran, W. (1985). An elementary proof of Gleason's theorem. *Mathematical Proceedings of the Cambridge Philosophical Society* 98, 117–128.
200. Coquand, T. (2005). About Stone's notion of a spectrum. *Journal of Pure and Applied Algebra* 197, 141–158.
201. Coquand, T., Spitters, B. (2005). Formal topology and constructive mathematics: The Gelfand and Stone–Yosida representation theorems. *Journal of Universal Computer Science* 11, 1932–1944.
202. Coquand, T., Spitters, B. (2009). Constructive Gelfand duality for C*-algebras. *Mathematical Proceedings of the Cambridge Philosophical Society* 147, 339–344.
203. Corry, L. (2004). *David Hilbert and the Axiomatization of Physics (1898–19018): From* Grundlagen der Geometrie *zur* Grundlagen der Physik. Dordrecht: Kluwer Academic Publishers.
204. Courant, R., Hilbert, D. (1924). *Methoden der mathematischen Physik*, Vols. I, II. Berlin: Springer-Verlag.
205. Coxeter, H.S.M. (1948). *Regular Polytopes*. London: Methuen and Co.
206. Dam, W. van, Hayden, P. (2003). Universal entanglement transformations without communication. *Physical Review* A67, 060302(R).
207. Danieri, A., Loinger, A., Prosperi, G.M. (1962). Quantum theory of measurement and ergodic condition. *Nuclear Physics* 33, 297–319.
208. Darrigol, O. (1992). *From c-Numbers to q-Numbers*. Berkeley: University of California Press.
209. Davey, B.A., Priestley, H.A. (2002). *Introduction to Lattices and Order*, 2nd ed. Cambridge: Cambridge University Press.
210. Davidson, K.R. (1996). *C*-Algebras by Example*. Providence: American Mathematical Society.
211. Davies, E.B. (1976). *Quantum Theory of Open Systems*. London: Academic Press.
212. Dawid, R., Thébault, K. (2015). Many worlds: decoherent or incoherent? *Synthese* 192, 1559–1580.
213. De Gennes, P. (1963). Collective motions of hydrogen bonds. *Solid State Communications* 1, 132–137.
214. De Mola, D. (2016). *Compatibilism and Actual Miracles*. MSc Thesis, Radboud University Nijmegen. www.math.ru.nl/~landsman/Davide.pdf.
215. Deitmar, A. (2005). *A First Course in Harmonic Analysis*. New York: Springer-Verlag.
216. Deitmar, A., Echterhoff, S. (2009). *Principles of Harmonic Analysis*. New York: Springer-Verlag.

217. Dennett, D. (1984). I could not have done otherwise, so what? *Journal of Philosophy* 81, 553–565.
218. DeWitt, B.S., Graham, N. (1973). *The Many Worlds Interpretation of Quantum Mechanics*. Princeton: Princeton University Press.
219. Diaconis, P., Freedman, D. (1980). Finite exchangeable sequences. *The Annals of Probability* 8, 745–764.
220. Dieks, D. (2016a). Niels Bohr and the formalism of quantum mechanics. To appear in Faye & Folse (2017). `philsci-archive.pitt.edu/12312/`.
221. Dieks, D. (2016b). Von Neumann's impossibility proof: Mathematics in the service of rhetorics. `philsci-archive.pitt.edu/12443/`.
222. Dieudonné, J. (1981). *History of Functional Analysis*. Amsterdam: North-Holland.
223. Dijksterhuis, E.J. (1924). *Val en Worp: Een bijdrage tot de Geschiedenis der Mechanica van Aristoteles tot Newton*. Groningen: Noordhoff.
224. Dirac, P.A.M. (1925). The fundamental equations of quantum mechanics. *Proceedings of the Royal Society (London)* A109, 642–653.
225. Dirac, P.A.M. (1926). On the theory of quantum mechanics. *Proceedings of the Royal Society (London)* A 112, 661–677.
226. Dirac, P.A.M. (1927). The physical interpretation of the quantum dynamics. *Proceedings of the Royal Society (London)* A 113, 621–641.
227. Dirac, P.A.M. (1930). *The Principles of Quantum Mechanics*. Oxford: Clarendon Press.
228. Dirac, P.A.M. (1947). *The Principles of Quantum Mechanics, 3d ed.*. Oxford: Clarendon Press.
229. Dixmier, J. (1977). *C*-algebras*. Amsterdam: North Holland.
230. Dixmier, J. (1981). *Von Neumann Algebras*. Amsterdam: North-Holland.
231. Dixmier, J., Douady, A. (1963). Champs continus d'espaces hilbertiens et de C*-algèbres. *Bulletin de la Société mathématique de la France* 91, 227–284.
232. Doran, R.S. (1994). *C*-algebras: 1943–1993*. *Contemporary Mathematics* Vol. 167. Providence: American Mathematical Society.
233. Doran, R.S., Belfi, V. (1986). *Characterization of C*-algebras*. New York: Marcel Dekker.
234. Döring, A. (2005). Kochen–Specker Theorem for von Neumann algebras. *International Journal of Theoretical Physics* 44, 139–160.
235. Döring, A. (2012). Topos-based logic for quantum systems and bi-Heyting algebras. *Logic and Algebraic Structures in Quantum Computing*, pp. 151–173. Chubb, J., Eskandarian, A., Harizanov, V., eds. Cambridge: Cambridge University Press.
236. Döring, A., (2014). Two new complete invariants of von Neumann algebras. `arXiv:1411.5558`.
237. Döring, A., Harding, J. (2010). Abelian subalgebras and the Jordan structure of a von Neumann algebra. `arXiv:1009.4945`.
238. Döring, A., Isham, C.J. (2008a). A topos foundation for theories of physics. I. Formal languages for physics. *Journal of Mathematical Physics* 49, 053515.
239. Döring, A., Isham, C.J. (2008b). A topos foundation for theories of physics. II. Daseinisation and the liberation of quantum theory. *Journal of Mathematical Physics* 49, 053516.
240. Döring, A., Isham, C.J. (2008c). A topos foundation for theories of physics. III. The representation of physical quantities with arrows. *Journal of Mathematical Physics* 49, 053517.
241. Döring, A., Isham, C.J. (2008d). A topos foundation for theories of physics. IV. Categories of systems. *Journal of Mathematical Physics* 49, 053518.
242. Döring, A., Isham, C.J. (2010). "What is a Thing?": Topos theory in the foundations of physics. *Lecture Notes in Physics* 813, 753–937.
243. Doyle, B. (2011). *Free Will: The Scandal in Philosophy*. Cambridge (MA): I-Phi Press.
244. Dowker, J. S. (1972). Quantum mechanics and field theory on multiply connected and on homogeneous spaces. *Journal of Physics* A5, 936–943.
245. Drühl, K., Haag, R., Roberts, J.E. (1970). Parastatistics. *Communications in Mathematical Physics* 18, 204–226.
246. Dudley, R.M. (1989). *Real Analysis and Probability*. Pacific Grove: Wadsworth & Brooks/Cole.

247. Duffield, N.G. (1990). Classical and thermodynamic limits for generalized quantum spin systems. *Communications in Mathematical Physics* 127, 27–39.
248. Duffield, N.G., Werner, R.F. (1992a). Classical Hamiltonian dynamics for quantum Hamiltonian mean-field limits. *Stochastics and Quantum Mechanics: Swansea, Summer 1990*, pp. 115–129. Truman, A., Davies, I.M., eds. Singapore: World Scientific.
249. Duffield, N.G., Werner, R.F. (1992b). On mean-field dynamical semigroups on C*-algebras. *Reviews in Mathematical Physics* 4, 383–424.
250. Duffield, N.G., Werner, R.F. (1992c). Local dynamics of mean-field quantum systems. *Helvetica Physica Acta* 65, 1016–1054.
251. Duffield, N.G., Roos, H., , Werner, R.F. (1992). Macroscopic limiting dynamics of a class of inhomogeneous mean field quantum systems. *Annales de l' Institut Henri Poincaré - Physique Théorique* 56, 143–186.
252. Duffner, E., Rieckers, A. (1988). On the global quantum dynamics of multilattice systems with nonlinear classical effects. *Zeitschrift für Naturforschung* A43, 521–532.
253. Duistermaat, J.J., Kolk, J. (2000). *Lie Groups*. Berlin: Springer.
254. Dummett, M. (2000). *Elements of Intuitionism, 2nd Ed.*. Oxford: Clarendon Press.
255. Duncan, A., Janssen, M. (2013). (Never) Mind your p's and q's: Von Neumann versus Jordan on the foundations of quantum theory. *European Physical Journal* H38, 175–259.
256. Dutta, A., Divakaran, U., Sen, D., Chakrabarti, B.K., Rosenbaum, T.F., Aeppli, G. (2015). *Quantum Phase Transitions in Transverse Field Spin Models: From Statistical Physics to Quantum Information*. Cambridge: Cambridge University Press.
257. Dvurečenskij, A. (1993). *Gleason's Theorem and its Applications*. Dordrecht: Kluwer Academic Publishers.
258. Earman, J. (1986). *A Primer on Determinism*. Dordrecht: D. Reidel.
259. Earman, J. (2003). Rough guide to spontaneous symmetry breaking. *Symmetries in Physics: Philosophical Reflections*, pp. 335–346. Brading, K., & Castellani, E., eds. Cambridge: Cambridge University Press.
260. Earman, J. (2004). Curie's Principle and spontaneous symmetry breaking. *International Studies in Philosophy of Science* 18, 173–198.
261. Earman, J. (2010). Understanding permutation invariance in quantum mechanics, unpublished preprint. See also `www.youtube.com/watch?v=xciuUhnsx1k`.
262. Ebbesen, A.M. (2012). *Convex Sets and their Integral Representations*. BSc Thesis. Copenhagen: University of Copenhagen.
263. Eckart, W. (1926). Operator calculus and the solution of the equations of quantum dynamics. *Physical Review* 28, 711–728.
264. Eddington, A.S. (1923). *The Mathematical Theory of Relativity*. Cambridge: Cambridge University Press.
265. Eilers, M., Horst, E. (1975). The theorem of Gleason for nonseparable Hilbert spaces. *International Journal of Theoretical Physics* 13, 419–424.
266. Einstein, A. (1921). *Geometrie und Erfahrung (erweiterte Fassung des Festvortrages gehalten an der Preussischen Akademie der Wissenschaften zu Berlin am 27. Januar 1921)*. Berlin: Springer. Repr. Einstein, A. (1982). *Mein Weltbild*, pp. 119–127. Frankfurt/M: Ullstein.
267. Einstein, A. (1954). *Ideas and Opinions*. New York: Bonanza Books.
268. Einstein, A., Born, M. (2005). *Briefwechsel 1916-1955*. München: Langen Müller.
269. Emch, G.G. (2007). Quantum statistical physics. *Handbook of the Philosophy of Science. Vol. 2: Philosophy of Physics, Part B*, pp. 1075–1182. Butterfield, J., Earman, J. eds. Amsterdam: North-Holland.
270. Emch, G.G., Whitten-Wolfe, B. (1976). Mechanical quantum measuring process. *Helvetica Physica Acta* 49, 45–55.
271. Enter, A.C.D. van, Fernandez, R., Sokal, A.D. (1993). Regularity properties and pathologies of position-space renormalization-group transformations: Scope and limitations of Gibbsian theory. *Journal of Statistical Physics* 72, 879–1167.
272. Erp, E. van (2010). The Atiyah-Singer index formula for subelliptic operators on contact manifolds. Part I, II. *Annals of Mathematics* 171, 1647–1681, 1683–1706.

273. Evans, D.E., Kawahigashi, Y. (1998). *Quantum Symmetries on Operator Algebras*. Oxford: Oxford University Press.
274. Fannes, M., Pule, J.V., Verbeure, A. (1982). Goldstone theorem for Bose systems. *Letters in Mathematical Physics* 6, 385–389.
275. Fannes, M., Spohn, H., Verbeure, A. (1980). Equilibrium states for mean field models. *Journal of Mathematical Physics* 21, 355–358.
276. Farhi, E., Goldstone, J., Gutmann, S. (1989). How probability arises in quantum mechanics. *Annals of Physics (N.Y.)* 192, 368–382.
277. Faye, J., Folse, J. (2017). *Niels Bohr and Philosophy of Physics: Twenty First Century Perspectives*. London: Bloomsbury.
278. Fell, J.M.G. (1961). The structure of algebras of operator fields. *Acta Mathematica* 106, 223–280.
279. Fell, J.M.G., Doran, R.S. (1988). *Representations of *-Algebras, Locally Compact Groups, and Banach *-Algebraic Bundles, Vol. 1*. Boston: Academic Press.
280. Fine, A. (1970). Insolubility of the quantum measurement problem. *Physical Review* D2, 2783–2787.
281. Fine, A. (1974). On the completeness of quantum theory. *Synthese* 29, 257–289.
282. Fine, T.L. (1973). *Theories of Probability*. New York: Academic Press.
283. Finkelstein, D. (1965). The logic of quantum physics. *Transactions of the New York Academy of Science* 25, 621–637.
284. Fine, A. (1974). On the completeness of quantum theory. *Synthese* 29, 257–289.
285. Fine, A. (1982). Hidden variables, joint probability, and the Bell inequalities. *Physical Review Letters* 48, 291–295.
286. Firby, P.A. (1973). Lattices and compactifications, II. *Proceedings of the London Mathematical Society* 27, 51–60.
287. Fischer, E. (1907). Sur la convergence en moyenne. *Comptes Rendus de l'Académie des Sciences* 144, 1022–1024.
288. Fischer, J.M. (1994). *The Metaphysics of Free Will*. Oxford: Blackwell.
289. Flori, C. (2013). *A First Course in Topos Quantum Theory*. Heidelberg: Springer.
290. Folland, G.B. (1995). *A Course in Abstract Harmonic Analysis*. Boca Raton: CRC Press.
291. Folse, H.J. (1985). *The Philosophy of Niels Bohr*. Amsterdam: North-Holland.
292. Fraassen, B. van (1991). *Quantum Mechanics: An Empiricist View*. Oxford: Oxford University Press.
293. Fraser, J.D. (2016). Spontaneous symmetry breaking in finite systems. *Philosophy of Science* 83, 585–605.
294. Fraser, D., Koberinski, A. (2016). The Higgs mechanism and superconductivity: A case study of formal analogies. *Studies in History and Philosophy of Modern Physics*, 55, 72–91.
295. Fréchet, M. (1907). Sur les opérations linéaires (Troisième note). *Transactions of the American Mathmatical Society* 8, 433-446.
296. Freed, D.S. (2012). On Wigner's Theorem. *Geometry & Topology Monographs* 18, 83–89.
297. Freed, D.S., Moore, G. (2012). Twisted equivariant matter. `arXiv:1208.5055`.
298. Freire, O. (2009). Quantum dissidents: Research on the foundations of quantum theory circa 1970 *Studies in History and Philosophy of Modern Physics* 40, 280–289.
299. French, S., Krause, D. (2006). *Identity in Physics: A Historical, Philosophical, and Formal Analysis*. Oxford: Clarendon Press.
300. Fritz, T. (2016). Beyond Bell's Theorem II: Scenarios with arbitrary causal structure. *Communications in Mathematical Physics* 341, 391–434.
301. Fröhlich, J., Schubnel, B. (2013). Quantum probability theory and the foundations of quantum mechanics. `arXiv:1310.1484`.
302. Fulton, W. (1997). *Young Tableaux*. Cambridge: Cambridge University Press.
303. Gaal, S.A. (1973). *Linear Analysis and Representation Theory*. Heidelberg: Springer.
304. Gabriel, G., et al, eds. (1980). *Gottlob Frege: Philosophical and Mathematical Correspondence*. London: Blackwell.
305. Garg, A. (2000). Tunnel splittings for one-dimensional potential wells revisited, *American Journal of Physics* 68, 430–437.

306. Gelfand, I.M., Kolmogorov, A.N. (1939). On rings of continuous functions on topological spaces. *Doklady Akademy Nauka SSSR* 22, 11–15.
307. Gelfand, I.M., Naimark, M.A. (1943). On the imbedding of normed rings into the ring of operators in Hilbert space. *Sbornik: Mathematics* 12, 197–213.
308. Georgii, H.O. (2011).*Gibbs Measures and Phase Transitions. Second Edition*. Berlin: De Gruyter.
309. Gerisch, T. (1993). Internal symmetries and limiting Gibbs states in quantum lattice mean-field models. *Physica* A197, 284–300.
310. Gilder, L. (2008). *The Age of Entanglement: When Quantum Physics Was Reborn*. New York: A. Knopf.
311. Gillies, D. 2000), *Philosophical Theories of Probability*. Cambridge: Cambridge University Press.
312. Ghirardi, G.C., Romano, R. (2013). About possible extensions of quantum theory. *Foundations of Physics* 43, 881–894.
313. Gill, R.D., Keane, M.S. (1996). A geometric proof of the Kochen–Specker no-go theorem. *Journal of Physics A* 29 L289–L291.
314. Gillman, L., Jerison, M. (1976). *Rings of Continuous Functions*. New York: Springer.
315. Givant, S., Halmos, P.R. (2009). *Introduction to Boolean Algebras*. New York: Springer.
316. Gleason, A.M. (1957). Measures on the closed subspaces of a Hilbert space. *Journal of Mathematics and Mechanics* 6, 885–893.
317. Glimm, J. (1960). A Stone-Weierstrass Theorem for C*-Algebras. *Annals of Mathematics* 72, 216–244.
318. Glimm, J., Kadison, R.V. (1960). Unitary operators in C*-algebras. *Pacific Journal of Mathematics* 1, 227–232.
319. Goldblatt, R. (1984). *Topoi: The Categorical Analysis of Logic*. Amsterdam: North-Holland.
320. Goldstein, S. (2013). Bohmian mechanics. *The Stanford Encyclopedia of Philosophy (Spring 2013 Edition)*, Zalta, E.N. ed. `plato.stanford.edu/archives/spr2013/entries/qm-bohm/`.
321. Goldstein, S., Tausk, D.V., Tumulka, R., Zanghi, N. (2010). What does the free will theorem actually prove? *Notices of the AMS* December 2010, 1451–1453.
322. Goldstone, J., Salam, A., Weinberg, S. (1962). Broken symmetries. *Physical Review* 127, 965–970.
323. Navarro González, J.A., Sancho de Salas, J.B. (2003). *C^∞-Differentiable Spaces*. New York: Springer.
324. Goodman, R., Wallach, N.R. (2000). *Representations and Invariants of the Classical Groups*. Cambridge: Cambridge University Press.
325. Gould, E. (2009). *Proofs of the Kochen-Specker Theorem in 3 Dimensions*. BSc Thesis, Worcester Polytechnic Institute. `www.wpi.edu/Pubs/E-project/Available/E-project-051509-073137/unrestricted/MQPReport.pdf`.
326. Grattan-Guinness, I. (2000). *The Search for Mathematical Roots 1870–1940: Logics, Set Theories, and the Foundations of Mathematics from Cantor through Russell to Gödel*. Princeton: Princeton University Press.
327. Gray, J.D. (1984). The shaping of the Riesz representation theorem: A chapter in the history of analysis. *Archive for History of Exact Sciences* 31, 127–187.
328. Gray, J.D. (2008). *Plato's Ghost: The Modernist Transformation of Mathematics*. Princeton: PrincetonUniversity Press.
329. Gray, R.M. (2009). *Probability, Random Processes, and Ergodic Properties*. New York: Springer.
330. Graffi, S., Grecchi, V., Jona-Lasinio, G. (1984). Tunnelling instability via perturbation theory. *Journal of Physics A* 17, 2935–2944.
331. Greenberger, D.M., Horne, M.A., Shimony, A., Zeilinger, (1990). A. Bell's theorem without inequalities. *American Journal of Physics* 58, 1131–1143.
332. Groenewegen, G.L.M., van Rooij, A.C.M. (2016). *Spaces of Continuous Functions*. Atlantis Press/Springer.

333. Groenewold, H.J. (1946). On the principles of elementary quantum mechanics. *Physica* 12, 405–460.
334. Groote, H.F. de (2011). Classical and quantum observables. *Deep Beauty: Mathematical Innovation and the Search for Underlying Intelligibility in the Quantum World*, pp. 239–269. Halvorson, H., ed. Cambridge: Cambridge University Press.
335. Grothendieck, A. (1955). Produits tensoriels topologiques et espaces nucléaires. *Memoirs of the American Mathematical Society* 16.
336. Grothendieck, A. (1986). *Récoltes et Semailles*. `http://lipn.univ-paris13.fr/~duchamp/Books&more/Grothendieck/RS/pdf/RetS.pdf`.
337. Grübl, G. (2003). The quantum measurement problem enhanced. *Physics Letters A* 316, 153–158.
338. Guillemin, V., Sternberg, S. (1984). *Symplectic Techniques in Physics*. Cambridge: Cambridge University Press.
339. Gustafson, S.J., Sigal, I.M. (2003). *Mathematical Concepts of Quantum Mechanics*. Berlin: Springer-Verlag.
340. Haag, R. (1992). *Local Quantum Physics: Fields, Particles, Algebras*. Heidelberg: Springer-Verlag.
341. Haag, R. (2010). Some people and some problems met in half a century of commitment to mathematical physics. *The European Physics Journal H* 35, 263–307.
342. Haag, R., Hugenholtz, N.M., Winnink, M. (1967). On the equilibrium states in quantum statistical mechanics. *Communications in Mathematical Physics* 5, 215–236.
343. Haag, R., Kadison, R.V., Kastler, D. (1970). Nets of C*-algebras and classification of states. *Communications in Mathematical Physics* 16, 81–104.
344. Hacking, I. (1975).*The Emergence of Probability*. Cambridge: Cambridge University Press.
345. Hacking, I. (1990). *The Taming of Chance*. Cambridge: Cambridge University Press.
346. Hacking, I. (2001). *Probability and Inductive Logic*. Cambridge: Cambridge University Press.
347. Hájek, A., Hitchcock, C., eds. (2016). *The Oxford Handbook of Probability and Philosophy*. Oxford: Oxford University Press.
348. Hall, B.C. (2013). *Quantum Theory for Mathematicians*. New York: Springer.
349. Halmos, P.R. (1958). *Finite-dimensional Vector Spaces*. Princeton: Van Nostrand.
350. Halmos, P.R. (1970). Finite-dimensional Hilbert Spaces. *The American Mathematical Monthly* 77, 457–464.
351. Halmos, P.R. (1974). *Measure Theory*. New York: Springer.
352. Halmos, P.R., Givant, S. (1998). *Logic as Algebra*. Washington (D.C.): The Mathematical Association of America.
353. Hamhalter, J. (1993). Pure Jauch-Piron states on von Neumann algebras. Annales de l'IHP Physique théorique 58, 173–187.
354. Hamhalter, J. (2004). *Quantum Measure Theory*. Dordrecht: Kluwer Academic Publishers.
355. Hamhalter, J. (2011). Isomorphisms of ordered structures of abelian C*-subalgebras of C*-algebras, *Journal of Mathematical Analysis and Applications* 383, 391–399.
356. Hamhalter, J. (2015). Dye's Theorem and Gleason's Theorem for AW*-algebras. *Journal of Mathematical Analysis and Applications* 422, 1103–1115.
357. Hamhalter, J. Turilova, E. (2013). Structure of associative subalgebras of Jordan operator algebras. *Quarterly Journal of Mathematics* 64, 397–408.
358. Hamilton, J., Isham, C.J., Butterfield, J. (2000). Topos perspective on the Kochen–Specker Theorem: III. Von Neumann Algebras as the base category. *International Journal of Theoretical Physics* 39, 1413–1436.
359. Hanche-Olsen, H., Størmer, E. (1984). *Jordan Operator Algebras*. Boston: Pitman.
360. Hänggi, P., Talkner, P., Borkovec, M. (1990). Reaction-rate theory: fifty years after Kramers, *Reviews of Modern Physics* 62, 251–341
361. Harrell, E.W. (1980). Double wells. *Communications in Mathematical Physics* 75, 239–261.
362. Hartle, J.B. (1968). Quantum mechanics of individual systems. *American Journal of Physics* 36, 704–712.

363. Heisenberg, W. (1925). Über quantentheoretische Umdeutung kinematischer und mechanischer Beziehungen. *Zeitschrift für Physik* 33, 879–893.
364. Heisenberg, W. (1926). Mehrkörperproblem und Resonanz in der Quantenmechanik. *Zeitschrift für Physik* 38, 411–426.
365. Heisenberg, W. (1927). Über den anschaulichen Inhalt der quantentheoretischen Kinematik und Mechanik. *Zeitschrift für Physik* 43, 172–198.
366. Heisenberg, W. (1928). Zur Theorie des Ferromagnetismus. *Zeitschrift für Physik* 49, 619–636.
367. Heisenberg, W. (1955). The development of the interpretation of quantum theory. *Niels Bohr and the Development of Physics*, pp. 12–29. Pauli, W., Rosenfeld, L., Weiskopf, V., eds. New York: McGraw Hill.
368. Heisenberg, W. (1958). *Physics and Philosophy: The Revolution in Modern Science*. London: Allen & Unwin.
369. Heisenberg, W. (1967). Quantum theory and its interpretation. *Niels Bohr*, pp. 94–108. Rozental, S., ed. Amsterdam: North-Holland.
370. Heisenberg, W. (2004). *Ordnung der Wirklichkeit*. München: Piper.
371. Havlicek, H., Krenn, G., Summhammer, J., Svozil, K. (2001). Colouring the rational quantum sphere and the Kochen–Specker theorem. *Journal of Physics A: Mathematical and General* 34, 3071–3077.
372. Heims, S.J. (1980). *John von Neumann and Norbert Wiener: From Mathematics to the Technologies of Life and Death*. Cambridge (Mass.): MIT Press.
373. Heinosaari, T., Ziman, M. (2008). Guide to mathematical concepts of quantum theory. *Acta Physica Slovaca* 58, 487–674.
374. Helffer, B. (1988). *Semi-classical Analysis for the Schrödinger Operator and Applications*. Heidelberg: Springer.
375. Helffer, B., & Sjöstrand, J. (1985). Puits multiples en limite semi-classique. II. Interaction moléculaire. Symétries. Perturbation. *Ann. Inst. H. Poincar Phys. Théor.* 42, 127–212.
376. Hempel, C. (1965). On the idea of emergence. Hempel, C., *Aspects of Scientific Explanation and Other Essays in the Philosophy of Science*, pp. 258–264. New York: The Free Press.
377. Hemmen, L. van (1978). Linear fermion systems, molecular field models, and the KMS condition. *Fortschritte der Physik* 26, 397–439.
378. Hemmick, D.L., Shakur, A.M. (2012). *Bell's Theorem and Quantum Realism: Reassessment in the Light of the Schrödinger Paradox*. Heidelberg: Springer.
379. Hendry, J. (1984). *The Creation of Quantum Mechanics and the Bohr-Pauli Dialogue*. Dordrecht: Reidel.
380. Henry, S. (2014a). Localic Metric spaces and the localic Gelfand duality. `arXiv:1411.0898`.
381. Henry, S. (2014b). Constructive Gelfand duality for non-unital commutative C*-algebras. `arXiv:1412.2009`.
382. Henry, S. (2015). A geometric Bohr topos. `arXiv:1502.01896`.
383. Hensen, B. et al. (2015). Experimental loophole-free violation of a Bell inequality using entangled electron spins separated by 1.3 km. *Nature* 526, 682–686.
384. Hepp, K. (1972). Quantum theory of measurement and macroscopic observables. *Helvetica Physica Acta* 45, 237–248.
385. Hepp, K., Lieb, E. (1974). Phase transitions in reservoir driven open systems with applications to lasers and superconductors. *Helvetica Physica Acta* 46, 573–602.
386. Hermann, G. (1935). Die naturphilosphischen Grundlagen der Quantenmechanik. *Abhandlungen der Fries'schen Schule* 6, 75–152.
387. Hermens, R. (2009). *Quantum Mechanics: From Realism to Intuitionism*. MSc Thesis, Radboud University Nijmegen. `philsci-archive.pitt.edu/5021/`.
388. Hermens, R. (2014). ConwayKochen and the finite precision loophole. *Foundations of Physics* 44, 1038–1048.
389. Hermens, R. (2016). *Philosophy of Quantum Probability: An Empiricist Study of its Formalism and Logic*. PhD Thesis, Rijksuniversiteit Groningen.

390. Heugten, J. van, Wolters, S. (2016). Obituary for a flea. *Proceedings of the Nagoya Winter Workshop 2015: Reality and Measurement in Algebraic Quantum Theory*, to appear. Ozawa, M., ed. arXiv:1610.06093.
391. Heunen, C. (2009). *Categorical Quantum Models and Logics*. PhD Thesis, Radboud University Nijmegen.
392. Heunen, C. (2014a). Characterizations of categories of commutative C*-algebras. *Communications in Mathematical Physics* 331, 215–238.
393. Heunen, C. (2014b). The many classical faces of quantum structures. arXiv:1412.2177.
394. Heunen, C., Landsman, N.P., Spitters, B. (2008). The principle of general tovariance. *Proc. XVI International Fall Workshop on Geometry and Physics (Lisbon, 2007)*, pp. 93–102. Fernandes, R.L., Picken, R., eds. Melville: American Physical Society.
395. Heunen, C., Landsman, N.P., Spitters, B. (2009). A topos for algebraic quantum theory. *Communications in Mathematical Physics* 291, 63–110.
396. Heunen, C., Landsman, N.P., Spitters, B. (2012). Bohrification of operator algebras and quantum logic. *Synthese*, 186, 719–752.
397. Heunen, C., Landsman, N.P., Spitters, B., Wolters, S. (2012). The Gelfand spectrum of a noncommutative C*-algebra: a topos-theoretic approach. *Journal of the Australian Mathematical Society* 90, 32–59.
398. Heunen, C., Lindenhovius, A.J. (2015). Domains of commutative C*-subalgebras. *Proceedings of the 30th annual ACM/IEEE symposium on Logic in Computer Science*, pp. 450–461.
399. Heunen, C., Reyes, M.L. (2014). Active lattices determine AW*-algebras. *Journal of Mathematical Analysis and Applications* 416, 289–313.
400. Heyting, A. (1956). *Intuitionism: An Introduction*. Amsterdam: North-Holland.
401. Hewitt, E., Savage, L.J. (1955). Symmetric measures on Cartesian products. *Transactions of the American Mathematical Society* 80, 470–501.
402. Heywood, P., Redhead, M. (1983). Nonlocality and the Kochen–Specker paradox. *Foundations of Physics* 13, 481–499.
403. Higgs, P.W. (1964a). Broken symmetries, massless particles and gauge fields. *Physics Letters* 12, 132–133.
404. Higgs, P.W. (1964b). Broken symmetries and the masses of gauge bosons. *Physical Review Letters* 13, 508–509.
405. Higson, N. (2010). The tangent groupoid and the Index Theorem. *Clay Mathematics Proceedings* 11, 241–256.
406. Hilbert, D. (1906). Grundzüge einer allgemeinen Theorie der linearen Integralgleichungen, Vierte Mitteilung. *Nachrichten von der Gesellschaft der Wissenschaften zu Göttingen, Mathematisch-Physikalische Klasse*, 157–228.
407. Hilbert, D. (1912). *Grundzüge einer allgemeinen Theorie der linearen Integralgleichungen*. Leipzig: Teubner.
408. Hilbert, D. (1925). Über das Unendliche. *Mathematische Annalen* 95, 161–190. English translation in Benaceraff & Putnam (1983), pp. 183–201.
409. Hilsum, M., Skandalis, G. (1987). Morphismes K-orientés d'espaces de feuilles et fonctorialité en théorie de Kasparov. *Annales scientifiques de l'École normale supérieure (4e série)* 20, 325–390.
410. Hislop, P.D., & Sigal, I.M. (1996). *Introduction to Spectral Theory*. New York: Springer.
411. Hofer-Szabó, G. (2015). On the relation between the probabilistic characterization of the common cause and Bell's notion of local causality. *Studies in History and Philosophy of Modern Physics* 49, 32–41.
412. Hofmann, K.H. (1970). *The Duality of Compact Semigroups and C*-Bigebras*. Berlin: Springer-Verlag.
413. Holevo, A.S. (1982). *Probabilistic and Statistical Aspects of Quantum Theory*. Amsterdam: North-Holland.
414. Hooft, G. 't (2007). The free-will postulate in quantum mechanics. arXiv:quant-ph/0701097.
415. Hooft, G. 't (2016). *The Cellular Automaton Interpretation of Quantum Mechanics*. Cham: Springer.

416. Hooker, C.A. (2004). Asymptotics, reduction and emergence. *British Journal for the Philosophy of Science* 55, 435–479.
417. Hörmander, L. (1966). *An Introduction to Complex Analysis in Several Variables.* New York: D. Van Nostrand.
418. Hornberger, K., Gerlich, S., Haslinger, P., Nimmrichter, S., Arndt, M. (2012). Colloquium: Quantum interference of clusters and molecules. *Reviews of Modern Physics* 84, 157–173.
419. Horsch, P., von der Linden, W. (1988). Spin-correlations and low lying excited states of the spin-1/2 Heisenberg antiferromagnet on a square lattice. *Zeitschrift für Physik B* 72, 181–193.
420. Horvathy, P.A., Morandi, G., Sudarshan, E.C.G. (1989). Inequivalent quantizations in multiply connected spaces. *Nuovo Cimento* D11, 201–228.
421. Howard, D. (2004). Who Invented the Copenhagen Interpretation? *Philosophy of Science* 71, 669–682.
422. Howson, C. (1995). Theories of probability. *The British Journal for the Philosophy of Science* 46, 1–32.
423. Huang, Y.-F., Li, C.-F., Zhang, Y.-S., Pan, J.-W., Guo, G.-C. (2003). Experimental test of the Kochen–Specker theorem with single photons. *Physical Review Letters* 90, 250401.
424. Hudson, R.L.,Moody, G.R. (1976). Locally normal symmetric states and an analogue of de Finetti's theorem. *Zeitschrift für Wahrscheinlichkeitstheorie und verwandte Gebiete* 33, 343–351.
425. Hunziker, W. (1972). A note on symmetry operations in quantum mechanics. *Helvetica Physica Acta* 45, 233–236.
426. Huruya, T. (1978). A spectral characterization of a class of C*-algebras. *Science Reports of Niigata University Series A* 15, 21–24.
427. Inwagen, P. van (1975). The incompatibility of Free Will and Determinism. *Philosophical Studies* 27, 185–199.
428. Inwagen, P. van (2008). How to think about the problem of free will. *Journal of Ethics* 12, 337–341.
429. Ioffe, D., Levit, A. (2013). Ground states for mean field models with a transverse component. *Journal of Statistical Physics* 151, 1140–1161.
430. Isham, C.J. (1984). Topological and global aspects of quantum theory. *Relativity, Groups and Topology II*, pp. 1059–1290. DeWitt, B.S., Stora, R., eds. Amsterdam: North-Holland.
431. Isham, C.J. (1997). Topos theory and consistent histories: The internal logic of the set of all consistent sets. *International J. of Theoretical Physics* 36, 785–814.
432. Isham, C.J., Butterfield, J. (1998). Topos perspective on the Kochen–Specker theorem. I. Quantum states as generalized valuations. *International Journal of Theoretical Physics* 37, 2669–2733.
433. Ismael, J.T. (2016). *How Physics Makes Us Free*. Oxford: Oxford University Press.
434. Israel, R.B. (1979). *Convexity in the Theory of Lattice Gases*. Princeton: Princeton University Press.
435. Janssen, H. (2008). *Reconstructing Reality*. M.Sc Thesis, Radboud University Nijmegen. `www.math.ru.nl/~landsman/scriptieHanneke.pdf`.
436. Jarrett, J.P. (1984). On the physical significance of the locality conditions in the Bell arguments. *Noûs* 18, 569–589.
437. Jacobson, N., Rickart, C.E. (1950). Jordan homomorphisms of rings. *Transactions of the American Mathematical Society* 69, 479–502. *Physical Review Letters* 86, 2490–2493.
438. Jammer, M. (1966). *The Conceptual Development of Quantum Mechanics*. New York: McGraw-Hill.
439. Jammer, M. (1974). *The Philosophy of Quantum Mechanics*. New York: Wiley.
440. Jauch, J.M. (1964). The problem of measurement in quantum mechanics. *Helvetica Physics Acta* 37, 293–316.
441. Jech, T. *Set Theory: The Third Millennium Edition*. Berlin: Springer.
442. Jensen, H.E. (1977). Scattered C*-algebras. *Mathematica Scandinavica* 41, 308–314.
443. Johnson, S. (2001). *Emergence: The Connected Lives of Ants, Brains, Cities, and Software*. New York: Scribner.

444. Johnstone, P.T. (1982) *Stone Spaces*. Cambridge: Cambridge University Press.
445. Johnstone, P.T. (1983). The point of pointless topology. *Bulletin (New Series) of the American Mathematical Society* 8, 41–53.
446. Johnstone, P.T. (1987) *Notes on Logic and Set Theory*. Cambridge: Cambridge University Press.
447. Johnstone, P.T. (2002) *Sketches of an Elephant: A topos theory compendium, Vols. 1, 2*. Oxford: Clarendon Press.
448. Jona-Lasinio, G., Martinelli, F., Scoppola, E. (1981a). New approach to the semiclassical limit of quantum mechanics. *Communications in Mathematical Physics* 80, 223–254.
449. Jona-Lasinio, G., Martinelli, F., Scoppola, E. (1981b). The semiclassical limit of quantum mechanics: A qualitative theory via stochastic mechanics. *Physics Reports* 77, 313–327.
450. Joos, E., Zeh, H.D. (1985). The emergence of classical properties through interaction with the environment. *Zeitschrift für Physik B* 59, 223–243.
451. Joos, E., Zeh, H.D., Kiefer, C., Giulini, D., Kupsch, J., Stamatescu, I.-O. (2003). *Decoherence and the Appearance of a Classical World in Quantum Theory. Second Edition*. Berlin: Springer-Verlag.
452. Jordan, P. (1927). Über quantummechanische Darstellung von Quantensprungen. *Zeitschrift für Physik* 40, 661–666.
453. Jordan, P., von Neumann, J. (1935). On inner products in linear, metric spaces. *Annals of Mathematics* 36, 719–723.
454. Jordan, P., von Neumann, J., Wigner, E.P. (1934). On an algebraic generalization of the quantum mechanical formalism. *Annals of Mathematics* 35, 29–64.
455. Joyal, A., Moerdijk, I. (1995). *Algebraic Set Theory*. Cambridge: Cambridge University Press.
456. Joyal, A., Tierney, M. (1984). An extension of the Galois theory of Grothendieck. *Memoirs of the American Mathematical Society* 51.
457. Kadison, R.V. (1951). Isometries of Operator Algebras. *Annals of Mathematics* (2nd series) 54, 325–338.
458. Kadison, R.V. (1965). Transformation of states in operator theory and dynamics. *Topology* 3, 177-198.
459. Kadison, R.V. (1982). Operator algebras: The first forty years. *Proceedings of Symposia in Pure Mathematics* 38(1), pp. 1–18. Providence: American Mathematical Society.
460. Kadison, R.V., Ringrose, J.R. (1983). *Fundamentals of the Theory of Operator Algebras. Vol. 1: Elementary Theory*. New York: Academic Press.
461. Kadison, R.V., Ringrose, J.R. (1986). *Fundamentals of the Theory of Operator Algebras. Vol. 2: Advanced Theory*. New York: Academic Press.
462. Kadison, R.V., Singer, I.M. (1959). Extensions of pure states. *American Journal of Mathematics* 81, 383–400.
463. Kaiser, D. (2010) *How the Hippies Saved Physics: Science, Counterculture, and the Quantum Revival*. New York: W.W Norton.
464. Kallenberg, O. (2005). *Probabilistic Symmetries and Invariance Principles*. New York: Springer.
465. Kalmbach, G. (1998). *Quantum Measures and Spaces*. Dordrecht: Springer.
466. Kaltenbaek, R. et al (2015). Macroscopic quantum resonators (MAQRO): 2015 Update. `arXiv:1503.02640`.
467. Kapitan, T. (2002). A master argument for incompatibilsm? *The Oxford Handbook of Free Will*, pp. 127–157. Kane, R., ed. Oxford: Oxford University Press.
468. Kaplan, T.A., Horsch, P., Von der Linden, W. (1989). Order parameter in quantum antiferromagnets. *Journal of the Physical Society of Japan* 58, 3894–3898
469. Kaplan, T.A., Von der Linden, W., Horsch, P. (1990). Spontaneous symmetry breaking in the Lieb-Mattis model of antiferromagnetism *Physical Review B* 42, 4663–4669.
470. Karevski, D. (2006). *Ising Quantum Chains*. `arXiv:hal-00113500`.
471. Kastler, D., Robinson, D.W., Swieca, A. (1966). Conserved currents and associated symmetries: Goldstone's theorem. *Communications in Mathematical Physics* 2, 108–120.

472. Kaufman, W.E. (1983). Closed operators and pure contractions in Hilbert space. *Proceedings of the American Mathematical Society* 87, 83–87.
473. Kaye, R. (1991). *Models of Peano Arithmetic*. Oxford: Clarendon Press.
474. Kaye, R. (2007). *The Mathematics of Logic*. Cambridge: Cambridge University Press.
475. Kim, J. (1999). Making sense of Emergence. *Philosophical Studies* 95, 3–36.
476. Kingman, J.F.C. (1978). Uses of exchangeability. *The Annals of Probability* 6, 183–197.
477. Kirchberg, E., Wassermann, S. (1995). Operations on continuous bundles of C*-algebras. *Mathematische Annalen* **303**, 677–697
478. Kirillov, A.A. (1976). *Elements of the Theory of Representations.* Heidelberg: Springer.
479. Knapp, A.W. (1988). *Lie Groups, Lie Algebras, and Cohomology*. Princeton: Princeton University Press.
480. Kochen, S., Specker, E. (1967). The problem of hidden variables in quantum mechanics. *Journal of Mathematics and Mechanics* 17, 59–87.
481. Koliha, J.J. (2014). On Kaufman's theorem. *Journal of Mathematical Analysis and Applications* 411, 688–692.
482. Koma, T., Tasaki, H. (1993). Symmetry breaking in Heisenberg antiferromagnets. *Communications in Mathematical Physics* 158, 191–214.
483. Koma, T., Tasaki, H. (1994). Symmetry breaking and finite-size effects in quantum many-body systems. *Journal of Statistical Physics* 76, 745–803.
484. König, R., & Mitchison, G. (2009). A most compendious and facile quantum de Finetti theorem. *Journal of Mathematical Physics* 50, 012105.
485. Kovachy, T., et al (2015). Quantum superposition at the half-metre scale. *Nature* 528, 530–533.
486. Kraus, K. (1983). *States, Effects, and Operations*. Heidelberg: Springer-Verlag.
487. Krishnaprasad, P.S., Marsden, J.E. (1987). Hamiltonian structure and stability for rigid bodies with flexible attachments. *Archive for Rational Mechanics and Analysis* 98, 137–158.
488. Krömer, R. (2007). *Tool and Object: A History and Philosophy of Category Theory*. Basel: Birkhäuser.
489. Kubo, R. (1957). Statistical-mechanical theory of irreversible processes. I. General theory and simple applications to magnetic and conduction problems. *Journal of the Physical Society of Japan* 12, 570–586.
490. Kusuda, M. (2011). C*-algebras in which every C*-subalgebra is AF. *The Quarterly Journal of Mathematics* 63, 675–680.
491. Laidlaw, M.G., DeWitt-Morette, C.M. (1971). Feynman functional integrals for systems of indistinguishable particles. *Physical Review* D3, 1375–1378.
492. Lance, E.C. (1982). Tensor products and nuclear C*-algebras. *Operator Algebras and Applications, Proceedings of Symposia in Pure Mathematics* 38(1), pp. 379–399. Providence: American Mathematical Society.
493. Lance, E.C. (1995). *Hilbert C*-Modules*. Cambridge: Cambridge University Press.
494. Landau, L.J., Perez, J.F., Wreszinski, W.F. (1981). Energy gap, clustering, and the Goldstone theorem in statistical mechanics. *Journal of Statistical Physics* 26, 755–766.
495. Landsman, N.P. (1991). Algebraic theory of superselection sectors and the measurement problem in quantum mechanics. *International Journal of Modern Physics* A6, 5349–5372.
496. Landsman, N.P. (1995). Observation and superselection in quantum mechanics. *Studies in History and Philosophy of Modern Physics* 26, 45–73.
497. Landsman, N.P. (1996). Classical behaviour in quantum mechanics: a transition probability approach. *International Journal of Modern Physics* B10, 1545–1554.
498. Landsman, N.P. (1997). Poisson spaces with a transition probability. *Reviews in Mathematical Physics* 9, 29–57.
499. Landsman, N.P. (1998a). *Mathematical Topics Between Classical and Quantum Mechanics*. New York: Springer-Verlag.
500. Landsman, N.P. (1998b). Strict quantization of coadjoint orbits. *Journal of Mathematical Physics* 39, 6372–6383.

501. Landsman, N.P., Ramazan, B. (2001). Quantization of Poisson algebras associated to Lie algebroids. *Groupoids in Analysis, Geometry, and Physics, Contemporary Mathematics* 282, pp. 159–192.
502. Landsman, N.P. (1999). Quantum mechanics on phase space. *Studies in History and Philosophy of Modern Physics* 30, 287–305.
503. Landsman, N.P. (2002) Getting even with Heisenberg. *Studies in History and Philosophy of Modern Physics* 33, 297–325.
504. Landsman, N.P. (2003). Deformation quantization and the Baum–Connes conjecture. *Communications in Mathematical Physics* 237, 87–103.
505. Landsman, N.P. (2004). *Lecture Notes on C*-algebras and K-Theory* (University of Amsterdam), unpublished.
506. Landsman, N.P. (2006a). When champions meet: Rethinking the Bohr–Einstein debate. *Studies in History and Philosophy of Modern Physics* 37, 212–242.
507. Landsman, N.P. (2006b). Lie groups and Lie algebroids in physics and noncommutative geometry. *Journal of Geometry and Physics* 56, 24–54.
508. Landsman, N.P. (2007). Between classical and quantum. *Handbook of the Philosophy of Science. Vol. 2: Philosophy of Physics, Part A*, pp. 417–553. Butterfield, J., Earman, J. eds. Amsterdam: North-Holland.
509. Landsman, N.P. (2008). Macroscopic observables and the Born rule. *Reviews in Mathematical Physics* 20, 1173–1190.
510. Landsman, N.P. (2009). The Born rule and its interpretation. *Compendium of Quantum Physics*, pp. 64–70. Greenberger, D., Hentschel, K., Weinert, F., eds. Dordrecht: Springer.
511. Landsman, N.P. (2013). Spontaneous symmetry breaking in quantum systems: Emergence or reduction? *Studies in History and Philosophy of Modern Physics* 44, 379–394.
512. Landsman, N.P. (2015). On the Colbeck–Renner Theorem. *Journal of Mathematical Physics* 56, 122103.
513. Landsman, N.P. (2016a). Quantization and superselection III: Mutliply connected spaces and indistinguishable particles. *Reviews in Mathematical Physics* 28, 1650019.
514. Landsman, N.P. (2016b). Bohrification: From classical concepts to commutative algebras. To appear in Faye & Folse (2017). arXiv:1601.02794.
515. Landsman, N.P. (2016c). On the notion of free will in the Free Will Theorem. *Studies in History and Philosophy of Modern Physics*. DOI: 10.1016/j.shpsb.2016.11.001.
516. Landsman, N.P., Lindenhovius, A.J. (2016). Symmetries in exact Bohrification. *Proceedings of the Nagoya Winter Workshop 2015: Reality and Measurement in Algebraic Quantum Theory*, to appear. Ozawa, M., ed.
517. Landsman, N.P., Reuvers, R. (2013). A flea on Schrödinger's Cat. *Foundations of Physics* 43, 373–407.
518. Landsman, N.P., van Weert, Ch.G. (1987). Real-and imaginary-time field theory at finite temperature and density. *Physics Reports* 145, 141–249.
519. Landau, L.D., Lifshitz, E.M. (1977). *Quantum Mechanics: Non-relativistic Theory*. 3d Ed. Oxford: Pergamon Press.
520. Lanford, O.E., Ruelle, D. (1969). Observables at infinity and states with short range correlations in statistical mechanics. *Communications in Mathematical Physics* 13, 194–215.
521. Lawvere, F.W., Schanuel, S.H. (1997). *Conceptual Mathematics: A First Introduction to Categories*. Cambridge: Cambridge University Press.
522. Lawvere, F.W., Rosebrugh, R. (2003). *Sets for Mathematics*. Cambridge: Cambridge University Press.
523. Leggett, A.J. 2002). Testing the limits of quantum mechanics: motivation, state of play, prospects. *Journal of Physics: Condensed Matter* 14, R415–R451.
524. Leegwater, G. (2016). An impossibility theorem for parameter independent hidden variable theories. *Studies in History and Philosophy of Modern Physics* 54, 18–34.
525. Leifer, M. (2014). Is the quantum state real? An extended review of ψ-ontology theorems. *Quanta* 3, 67–155.
526. Leinaas, J.M., Myrheim, J. (1977). On the theory of identical particles. *Nuovo Cimento B* 37, 1–23.

527. Lewis, D. (1973). *Counterfactuals*. Cambridge: Harvard University Press.
528. Lewis, D. (1979). Counterfactual dependence and time's arrow. *Noûs* 13, 455–476.
529. Lewis, D. (1980). A subjectivist's guide to objective chance. *The University of Western Ontario Series in Philosophy of Science* 15, 267–297.
530. Lewis, D. (1981). Are we free to break the laws? *Theoria* 47, 112–121.
531. Lewis, D. (2000). Causation as Influence. *Journal of Philosophy* 97, 182–97.
532. Li, Bing-Ren (1992). *Introduction to Operator Algebras*. Singapore: World Scientific.
533. Liang, Y.C., Spekkens, R.W., Wiseman H.M. (2011). Specker's parable of the overprotective seer: A road to contextuality, nonlocality and complementarity. *Physics Reports* 506, 1–39.
534. Lieb, E.H. (1973). The classical limit of quantum spin systems. *Communications in Mathematical Physics* 62, 327–340.
535. Lieb, E.H., Schultz, T., Mattis, D. (1961). Two soluble models of an antiferromagnetic chain. *Annals of Physics* 16, 407–466.
536. Lindenhovius, B. (2015). Classifying finite-dimensional C*-algebras by posets of their commutative C*-subalgebras. `arXiv:1501.03030`.
537. Lindenhovius, B. (2016). $\mathscr{C}(A)$. PhD Thesis, Radboud University Nijmegen. `www.math.ru.nl/~landsman/Lindenhovius.pdf`.
538. Liu, C., Emch, G.G. (2005). Explaining quantum spontaneous symmetry breaking. *Studies in History and Philosophy of Modern Physics* 36, 137–163.
539. London, F., Bauer, E. (1939). *La Théorie de l'Observation en Mécanique Quantique*. Paris: Hermann.
540. Loya, P. (2008). *Lebesgue's Remarkable Theory of Measure and Integration with Probability*. Lecture Notes. Binghamton: State University of New York.
541. Ludwig, G. (1983). *Foundations of Quantum Mechanics, Vol. I*. Berlin: Springer-Verlag.
542. Lurie, J. (2011). *Math261y: von Neumann algebras*. Harvard University lecture notes available at `www.math.harvard.edu/~lurie/261y.html`.
543. Lusanna, L., Valtancoli, P. (1996a). Dirac's Observables for the Higgs Model: I) the Abelian case. `arXiv:hep-th/9606078`.
544. Lusanna, L., Valtancoli, P. (1996b). Dirac's Observables for the Higgs Model: II) the non-Abelian SU(2) case. `arXiv:hep-th/9606079`.
545. Luxemburg, W.A.J. (1991). Integration with respect to finitely additive measures. *Positive Operators, Riesz Spaces, and Economics*, pp. 109–150. Aliprantis, C.D. Border, K.C., Luxemburg, W.A.J., eds. Berlin: Springer.
546. Lyubich, Y.I. (1988). *Introduction to the Theory of Banach Representations of Groups*. Basel: Birkhäuser.
547. Mac Lane, S. (1998). *Categories for the Working Mathematician, 2nd ed.* Berlin: Springer.
548. Mac Lane, S., Moerdijk, I. (1992). *Sheaves in Geometry and Logic: A First Introduction to Topos Theory*. New York: Springer-Verlag.
549. MacCluer, B.D. (2009). *Elementary Functional Analysis*. New York: Springer.
550. Maeda, S. (1980). *Lattice Theory and Quantum Logic* (in Japanese). Tokyo: Maki-Shoten.
551. Maeda, S. (1990). Probability measures on projections in von Neumann algebras. *Reviews in Mathematical Physics* 1, 235–290.
552. Mack, B. (1993). *The Lost Gospel: The Book of Q & Christian Origins*. New York: HarperCollins.
553. Mackenzie, K.C.H. (2005). *An Introduction to Lie Groupoids and Lie Algebroids*. Cambridge: Cambridge University Press.
554. Mackey, G.W. (1957). Quantum mechanics and Hilbert space. *American Mathematical Monthly* 64, 45–57.
555. Mackey, G.W. (1958). Unitary representations of group extensions. I. *Acta Mathematica* 99, 265–311.
556. Mackey, G.W. (1963). *The Mathematical Foundations of Quantum Mechanics*. New York: Benjamin.
557. Mackey, G.W. (1968). *Induced Representations of Groups and Quantum Mechanics*. New York: Benjamin.

558. Mackey, G.W. (1974). Ergodic theory and its significance for statistical mechanics and probability theory. *Advances in Mathematics* 12, 178–268.
559. Mackey, G.W. (1978). *Unitary Group Representations in Physics, Probability, and Number Theory*. New York: Benjamin.
560. Mackey, G.W. (1992). *The Scope and History of Commutative and Noncommutative Harmonic Analysis*. Providence: American Mathematical Society.
561. Macrae, N. (1992). *John von Neumann: The Scientific Genius Who Pioneered the Modern Computer, Game Theory, Nuclear Deterrence, and Much More*. Providence: American Mathematical Society.
562. Maeda, S. (1990). Probability measures on projections in von Neumann algebras. *Reviews in Mathematical Physics* 1, 235–290.
563. Malliavin, P. (1995). *Integration and Probability*. NewYork: Springer.
564. Mancosu, P., Zach, R., Badesa, C. (2004). The development of mathematical logic from Russell to Tarski: 1900–1935. *The Development of Modern Logic*, pp. 318–470. Haaparanta, L., ed. New York: Oxford University Press.
565. Marcus, A., Spielman, D.A., Srivastava, N. (2014a). Interlacing families II: Mixed characteristic polynomials and the Kadison-Singer Problem. arXiv:1306.3969.
566. Marcus, A., Spielman, D.A., Srivastava, N. (2014b). Ramanujan Graphs and the solution of the Kadison–Singer Problem. arXiv:1408.4421.
567. Marquis, J.-P. (2009). *From a Geometrical Point of View: A Study of the History and Philosophy of Category Theory*. New York: Springer.
568. Marquis, J.-P., Reyes, G.E. (2012). The history of categorical logic: 1963–1977. *Handbook of the History of Logic, Vol. 6*, pp. 689–800. Gabbay, D.M., Kanamori, A., Woods, J., ed. Amsterdam: Elsevier.
569. Marsden, J.E., Ratiu, T.S. (1994). *Introduction to Mechanics and Symmetry*. New York: Springer-Verlag.
570. Martin, P.C., Schwinger, J. (1959). Theory of many-particle systems. I. *Physical Review* 115, 1342–1373.
571. Mattis, D.C. (1965). *The Theory of Magnetism: An Introduction to the Study of Cooperative Phenomena*. New York: Harper & Row.
572. Maudlin, T. (1995). Three measurement problems. *Topoi* 14, 7–15.
573. Maurin, K. (1972). *Methods of Hilbert Spaces. Second Edition Revised*. Warszawa: PWN.
574. McArthur, C.W. (1970). Convergence of monotone nets in ordered topological vector spaces. *Studia Mathematica* 34, 1–16.
575. McEvoy, P. (2001). *Niels Bohr: Reflections on Subject and Object*. Microanalytix.
576. McKenna, M., Coates, D.J. (2015). Compatibilism. *The Stanford Encyclopedia of Philosophy (Summer 2015 Edition)*. Zalta, E.N., ed. plato.stanford.edu/archives/sum2015/entries/compatibilism/.
577. McLaughlin, B.P. (2008). The rise and fall of British Emergentism. *Emergence*, pp. 19–59. Bedau, M.A., & Humphreys, P., eds. Cambridge (Mass.): MIT Press.
578. Mehra, J., Rechenberg, H. (1982). *The Historical Development of Quantum Theory. Vol. 1: The Quantum Theory of Planck, Einstein, Bohr, and Sommerfeld: Its Foundation and the Rise of Its Difficulties*. New York: Springer-Verlag.
579. Mehra, J., Rechenberg, H. (2000). *The Historical Development of Quantum Theory. Vol. 6: The Completion of Quantum Mechanics 1926–1941. Part 1: The Probabilistic Interpretation and the Empirical and Mathematical Foundation of Quantum Mechanics, 1926-1936*. New York: Springer-Verlag.
580. Mendelson, E. (2010). *An Introduction to Mathematical Logic: Fifth Edition*. Boca Raton: CRC Press.
581. Mendivil, F. (1999). Function algebras and the lattices of compactifications. *Proceedings of the American Mathematical Society* 127, 1863–1871.
582. Menzies, P. (2014). Counterfactual Theories of Causation. *The Stanford Encyclopedia of Philosophy (Spring 2014 Edition)*, Zalta, E.N., ed. plato.stanford.edu/archives/spr2014/entries/causation-counterfactual/.

583. Messiah, A.M., Greenberg, O.W. (1964). Symmetrization postulate and Its experimental foundation, *Physical Review* B136, 248–267.
584. Meyer, D.A. (1999). Finite precision measurement nullifies the Kochen–Specker theorem. *Physical Review Letters* 83, 3751–3754.
585. Mielnik, B. (1968). Geometry of quantum states. *Communications in Mathematical Physics* 9, 55–80.
586. Mielnik, B. (1969). Theory of filters. *Communications in Mathematical Physics* 15, 1–46.
587. Mill, J.S. (1952) [1843]. *A System of Logic, 8th ed.* London: Longmans, Green, Reader, and Dyer.
588. Mittelstaedt, P. (2004). *The Interpretation of Quantum Mechanics and the Measurement Process*. Cambridge: Cambridge University Press.
589. Molnár, J. (2002). Orthogonality preserving transformations on indefinite inner product spaces: generalization of Uhlhorn's version of Wigner's theorem. *Journal of Functional Analysis* 194, 248–262.
590. Moerdijk, I., Mrčun, J. (2003). *Introduction to Foliations and Lie Groupoids.* Cambridge: Cambridge University Press.
591. Monna, A.F. (1973). *Functional Analysis in Historical Perspective*. Utrecht: Oosthoek.
592. Moore, G.E. (1912). *Ethics*. New York: Henry Holt and Company.
593. Morandi, G. 1992). *The Role of Topology in Classical and Quantum Mechanics*. New York: Springer-Verlag.
594. Moretti, V. (2013). *Spectral Theory and Quantum Mechanics*. Mailand: Springer-Verlag.
595. Morchio, G., Strocchi, F. (1987). Mathematical structures for long-range dynamics and symmetry breaking. *Journal of Mathematical Physics* 28, 622–635.
596. Morchio, G., Strocchi, F. (2007). Quantum mechanics on manifolds and topological effects. *Letters in Mathematical Physics* 82, 219–236.
597. Moulay, E. (2014). Non-collapsing wave functions in an infinite universe. *Results in Physics* 4,164–167.
598. Moyal, J.E. (1949). Quantum mechanics as a statistical theory. *Mathematical Proceedings of the Cambridge Philosophical Society*, 45, 99–124.
599. Mrĉun, J. (2005). On isomorphisms of algebras of smooth functions. *Proceedings of the American Mathematical Society* 133, 3109–3113.
600. Muller, F.A. (1997a). The equivalence myth of quantum mechanics—Part I. *Studies in History and Philosophy of Modern Physics* 28, 35–61.
601. Muller, F.A. (1997b). The equivalence myth of quantum mechanics—Part II. *Studies in History and Philosophy of Modern Physics* 28, 219–247.
602. Murdoch, D. (1987). *Niels Bohr's Philosophy of Physics*. Cambridge: Cambridge University Press.
603. Murphy, G.J. (1990). *C*-algebras and Operator Theory*. San Diego: Academic Press.
604. Murray, F.J., Neumann, J. von (1936). On rings of operators. *Annals of Mathematics* 37, 116–229.
605. Murray, F.J., Neumann, J. von (1937). On rings of operators II. *Transactions of the American Mathematical Society* 41, 208–248.
606. Murray, F.J., Neumann, J. von (1943). On rings of operators IV. *Annals of Mathematics* 44, 716–808.
607. Muynck, W.M. de (2002) *Foundations of Quantum Mechanics: an Empiricist Approach.* Dordrecht: Kluwer Academic Publishers.
608. Naaijkens, P. (2013). Quantum spin systems on infinite lattices. `arXiv:1311.2717`.
609. Nachtergaele, B. (2007). Quantum spin systems after DLS 1978. *Proceedings of Symposia in Pure Mathematics* 76.1, pp. 47–68.
610. Negrepontis, J.W. (1969). *Applications of the Theory of Categories to Analysis*. PhD Thesis (McGill University).
611. Negrepontis, J.W. (1971). Duality in analysis from the point of view of triples. *Journal of Algebra* 19, 228–253.
612. Nelson, E. (1959). Analytic vectors. *Annals of Mathematics* 70, 572–614.

613. Neumann, J. von (1927a). Mathematische Begründung der Quantenmechanik. *Nachrichten von der Gesellschaft der Wissenschaften zu Göttingen, Mathematisch-Physikalische Klasse*, 1–57.
614. Neumann, J. von (1927b). Wahrscheinlichkeitstheoretischer Aufbau der Quantenmechanik. *Nachrichten von der Gesellschaft der Wissenschaften zu Göttingen, Mathematisch-Physikalische Klasse* 273–291.
615. Neumann, J. von (1930). Zur Algebra der Funktionaloperationen und Theorie der normalen Operatoren. *Mathematische Annalen* 102, 370–427.
616. Neumann, J. von (1931). Über Funktionen von Funktionaloperatoren. *Annals of Mathematics* 32, 191–226.
617. Neumann, J. von (1932). *Mathematische Grundlagen der Quantenmechanik.* Berlin: Springer–Verlag. English translation (1955): *Mathematical Foundations of Quantum Mechanics.* Princeton: Princeton University Press.
618. Neumann, J. von (1936). On an algebraic generalization of the quantum mechanical formalism (Part I). *Sbornik: Mathematics (N.S.)* 1, 415–484.
619. Neumann, J. von (1938). On infinite direct products. *Compositio Mathematica* 6, 1–77.
620. Neumann, J. von (1940). On rings of operators III. *Annals of Mathematics* 41, 94–161.
621. Neumann, J. von (1949). On rings of operators V. Reduction theory. *Annals of Mathematics* 50, 401–485.
622. Neumann, J. von (1950). *Functional Operators. I. Measures and Integrals*. Annals of Mathematics Studies, no. 21. Princeton: Princeton University Press.
623. Neumann, J. von (1961). *Collected Works, Vol. III: Rings of Operators*. Taub, A.H., ed. Oxford: Pergamon Press.
624. Neumann, J. von (1981). Continuous geometries with a transition probability. Edited by Halperin, I.S. *Memoirs of the American Mathematical Society* 252, 1–210. (MS from 1937).
625. Neumann, J. von, Wigner, E.P. (1928). Zur Erklärung einiger Eigenschaften der Spektren aus der Quantenmechanik des Drehelektrons. *Zeitschrift für Physik* A49, 73–94.
626. Neumark, M.A. *Normierte Algebren*. Berlin: VEB Deutscher Verlag der Wissenschaften.
627. Newton, I. (1999). *The Principia: Mathematical Principles of Natural Philosophy. A New Translation by I. Bernard Cohen and Anne Whitman. Preceded by a Guide to Newton's Principia by I. Bernard Cohen.* Berkeley: University of California Press.
628. Niederer, U.H., O'Raifeartaigh, L.O. (1974). Realizations of the unitary representations of the inhomogeneous space-time groups I. *Fortschritte der Physik* 22, 111–129.
629. Norsen, T. (2009). Local causality and completeness: Bell vs. Jarrett. *Foundations of Physics* 39, 273–294.
630. Norton, J.D. (1993). General covariance and the foundations of general relativity: eight decades of dispute. *Reports on Progress in Physics* 56, 791–861.
631. Norton, J.D. (1995). Did Einstein stumble? The debate over general covariance. *Erkenntnis* 42, 223–245.
632. Norton, J.D. (2012). Approximation and idealization: Why the difference matters. *Philosophy of Science* 79, 207–232.
633. Ochs, W. (1977). On the strong law of large numbers in quantum probability theory. *Journal of Philosophical Logic* 6, 473–480.
634. Ochs, W. (1980). Concepts of convergence for a quantum law of large numbers. *Reports on Mathematical Physics* 17, 127–143.
635. O'Connor, T., Wong, H.Y. (2012). Emergent Properties. *The Stanford Encyclopedia of Philosophy (Spring 2012 Edition)*, Zalta, E.N., ed. `plato.stanford.edu/archives/sum2015/entries/properties-emergent/`.
636. Olson, M.P. (1971). The selfadjoint operators of a von Neumann algebra form a conditionally complete lattice. *Proceedings of the American Mathematical Society* 28, 537–544.
637. Oosten, J. van (2002). *Basic Category Theory.* Lecture Notes, Utrecht University. `www.staff.science.uu.nl/~ooste110/syllabi/catsmoeder.pdf`.
638. Osborne, M.S. (1975). On the Schwartz–Bruhat space and the Paley–Wiener Theorem for locally compact abelian groups. *Journal of Functional Analysis* 19, 40–49.

639. Oxtoby, J.C., Pettis, B.J., Price, G.B. (1958). John von Neumann: 1903–1957. *Bulletin of the American Mathematical Society* 64, No. 3, Part 2.
640. Paffenholz, A. (2010). *Polyhedral Geometry and Linear Optimization*. Lecture notes, Freie Universität Berlin. www.mathematik.tu-darmstadt.de/~paffenholz/daten/preprints/ln.pdf.
641. Pais, A. (1986). *Inward Bound: Of Matter and Forces in the Physical World*. Oxford: Oxford University Press.
642. Palmgren, E. (2009). Semantics of intuitionistic propositional logic. Lecture Notes, Uppsala University.
643. Palomaki, T.A., Teufel, J.D., Simmonds, R.W., Lehnert, K.W. (2013). Entangling mechanical motion with microwave fields. *Science* 342 , no. 6159, 710–713.
644. Pauli, W. (1927). Über Gasentartung und Paramagnetismus. *Zeitschrift für Physik* 41, 81–102.
645. Pauli, W. (1933). *Die allgemeinen Prinzipien der Wellenmechanik*. 2. Auflage, Band XXIV, Teil 1. *Handbuch der Physik*. Geiger, H., Scheel, K, eds. Neu herausgegeben und mit historischen Anmerkungen versehen von N. Straumann in 1990. English translation (1980): *General Principles of Quantum Mechanics*. Berlin: Springer–Verlag.
646. Pavicic, M., Merlet, J.P., McKay, B., Megill, N.D. (2005). Kochen-Specker vectors. *Journal of Physics A: Mathematical and General* 38, 1577–1592.
647. Pedersen, G.K. (1972). Operator algebras with weakly closed abelian subalgebras. *Bulletin of the London Mathematical Society* 4, 171–175.
648. Pedersen, G.K. (1979). *C*-algebras and their Automorphism Groups*. London: Academic Press.
649. Pedersen, G.K. (1989). *Analysis Now*, 2nd ed. New York: Springer-Verlag.
650. Perelomov, A.M. (1972). Coherent states for arbitrary Lie groups. *Communications in Mathematical Physics* 26, 222–236.
651. Perelomov, A. (1986). *Generalized Coherent States and their Applications*. Berlin: Springer.
652. Peres, A. (1995). *Quantum Theory: Concepts and Methods*. Dordrecht: Kluwer Academic Publishers.
653. Pfeuty, P. (1970). The one-dimensional Ising model with a transverse field. *Annals of Physics* 47, 79–90.
654. Phelps, R. (2001). *Lectures on Choquet's Theorem, Second Edition*. Berlin: Springer-Verlag.
655. Philipse, H. (1999). *Heidegger's Philosophy of Being: A Critical Interpretation*. Princeton: Princeton University Press.
656. Picado, J., Pultr, A. (2012). *Frames and Locales: Topology Without Points*. Basel: Birkhäuser.
657. Pier, J. (2001). *Mathematical Analysis During the 20th Century*. New York: Oxford University Press.
658. Pietsch, W. (2007). *History of Banach Spaces and Linear Operators*. Basel: Birkhäuser.
659. Pitowsky, I. (1989). *Quantum Probability - Quantum Logic*. Berlin: Springer-Verlag.
660. Pitowsky, I. (1994). George Boole's 'Conditions of possible experience' and the quantum puzzle. *British Journal for the Philosophy of Science* 45, 95–125.
661. Plato, J. von (1994). *Creating Modern Probability*. Cambridge: Cambridge University Press.
662. Popescu, S., Rohrlich, D. (1994). Quantum nonlocality as an axiom. *Foundations of Physics* 24, 379–385.
663. Popper, K.R. (1938). A set of independent axioms for probability. *Mind* 47(186), 275–77.
664. Popper, K.R. (1968). Birkhoff and von Neumann's interpretation of quantum mechanics. *Nature* 219, 682–685.
665. Powers, R.T. (1967). Representations of uniformly hyperfinite algebras and their associated von Neumann rings. *Annals of Mathematics* 86, 138–171.
666. Price, H. (1996). *Time's Arrow and Archimedes' Point: New Directions for the Physics of Time*. Oxford: Oxford University Press.
667. Pulmannová, S., Stehlková, B. (1986). Strong law of large numbers and central limit theorem on a Hilbert space logic. *Reports on Mathematical Physics* 23, 99–107.
668. Raeburn, I., Williams, D.P. (1998). *Morita Equivalence and Continuous-Trace C*-Algebras*. Providence: American Mathematical Society.

669. Raggio, G.A. (1981). *States and composite systems in W*-algebras quantum mechanics*. PhD Thesis, ETH Zürich.
670. Raggio, G.A. (1988). A remark on Bell's inequality and decomposable normal states. *Letters in Mathematical Physics* 15, 27–29.
671. Raggio, G.A., Werner, R.F. (1989). Quantum statistical mechanics of general mean field systems. *Helvetica Physica Acta* 62, 980–1003.
672. Raggio, G.A., Werner, R.F. (1991). The Gibbs variational principle for inhomogeneous mean field systems. *Helvetica Physica Acta* 64, 633–667.
673. Rao, K.P.S.B., Rao, M.B (1983). *Theory of Charges: A Study of Finitely Additive Measures*. London: Academic Press.
674. Raynaud, G. (2014). *Fibered Contextual Quantum Physics*. PhD Thesis, University of Birmingham.
675. Reece, G. (1973). The theory of measurement in quantum mechanics. *International Journal of Theoretical Physics* 7, 81–117.
676. Reed, M., Simon, B. (1972). *Methods of Modern Mathematical Physics. Vol I. Functional Analysis*. New York: Academic Press.
677. Reed, M., Simon, B. (1978). *Methods of Modern Mathematical Physics. Vol IV. Analysis of Operators*. New York: Academic Press.
678. Rédei, M. (1998). *Quantum Logic in Algebraic Approach*. Dordrecht: Kluwer Academic Publishers.
679. Rédei, M. (2005a). John von Neumann on mathematical and axiomatic physics. *The Role of Mathematics in the Physical Sciences*, pp. 43–54. Boniolo, G. et al, eds. Dordrecht: Springer.
680. Rédei, M. (2005b). *John von Neumann: Selected Letters*. Providence: American Mathematical Society.
681. Renault, J. (1980). *A Groupoid Approach to C*-algebras*. Berlin: Springer.
682. Rényi, A. (1955). On a new axiomatic theory of probability. *Acta Mathematica Academiae Scientiarum Hungaricae* 6, 286–335.
683. Rhodes, R. (1995). *Dark Sun: The Making of the Hydrogen Bomb*. New York: Touchstone.
684. Richman, F., Bridges, D. (1999). A constructive proof of Gleason's theorem. *Journal of Functional Analysis* 162, 287–312.
685. Rieckers, A. (1984). On the classical part of the mean field dynamics for quantum lattice systems in grand canonical representations. *Journal of Mathematical Physics* 25, 2593–2601.
686. Rieffel, M.A. (1989). Deformation quantization of Heisenberg manifolds. *Communications in Mathematical Physics* 122, 531–562.
687. Rieffel, M.A. (1990). Lie group convolution algebras as deformation quantizations of linear Poisson structures. *American Journal of Mathematics* 112, 657–686.
688. Rieffel, M.A. (1994). Quantization and C^*-algebras. *Contemporary Mathematics* 167, 66–97.
689. Riesz, F. (1907a). Sur les systèmes orthogonaux des functions. *Comptes Rendus de l'Académie des Sciences* 144, 615–619.
690. Riesz, F. (1913). *Les Systèmes d'Équations Linéaires à une Infinité d'Inconnues*. Paris: Gauthiers–Villars.
691. Riesz, F. (1934). Zur Theorie des Hilbertschen Raumes. *Acta Scientiarum Mathematicarum (Szeged)* 7, 34–38.
692. Roberts, B. (2016). Three myths about time reversal in quantum theory. `philsci-archive.pitt.edu/12305/`.
693. Roberts, S. (2015). *Genius at Play: The Curious Mind of John Horton Conway*. New York: Bloomsbury.
694. Roberts, J.E. and G. Roepstorff (1969). Some basic concepts of algebraic quantum theory. *Communications in Mathematical Physics* 11, 321–338.
695. Royden, H.L. (1988), *Real Analysis*, 3d Ed. Englewood Cliffs: Prentice-Hall.
696. Rubakov, R. (2002). *Classical Theory of Gauge Fields*. Princeton: Princeton University Press.
697. Rudin, W. (1962). *Fourier Analysis on Groups*. New York: Wiley Interscience.

698. Rudin, W. (1973). *Functional Analysis*. New York: McGraw–Hill.
699. Rudin, W. (1986). *Real and Complex Analysis*. New York: McGraw–Hill.
700. Ruelle, D. (1969). *Statistical Mechanics: Rigorous Results*. New York: Benjamin.
701. Ruetsche, L. (2011). *Interpreting Quantum Theories.* Oxford: Oxford University Press.
702. Ruijsenaars, S.N.M. (1978). On Bogoliubov transformations. II. The general case. *Annals of Physics* 116, 105–134.
703. Rynne,B.P., Youngson, M.A. (2008). *Linear Functional Analysis*. London: Springer.
704. Sachdev, S. (2011). *Quantum Phase Transitions*, 2nd ed. Cambridge: Cambridge University Press.
705. Saitô, K., Maitland Wright, J.D. (2015a). On defining AW*-algebras and Rickart C*-algebras. arXiv:1501.02434.
706. Saitô, K., Maitland Wright, J.D. (2015b). *Monotone Complete C*-algebras and Generic Dynamics.* London: Springer.
707. Sakai, S. (1971). *C*-Algebras and W*-Algebras*. Berlin: Springer-Verlag.
708. Saunders, J.T. (1968). The temptations of "powerlessness". *American Philosophical Quarterly* 5, 100–108.
709. Saunders, S. (2013). Indistinguishability. *Handbook of Philosophy of Physics*, pp. 340–380. Batterman, R., ed. Oxford: Oxford University Press.
710. Schatten, R. (1946). The cross-space of linear transformations. *Annals of Mathematics* 47, 73–84.
711. Schatten, R., Neumann, J. von (1946). The cross-space of linear transformations. II. *Annals of Mathematics* 47, 608–630.
712. Schatten, R., Neumann, J. von (1948). The cross-space of linear transformations. III. *Annals of Mathematics* 49, 557–582.
713. Scheibe, E. (1973). *The Logical Analysis of Quantum Mechanics*. Oxford: Pergamon Press.
714. Scheibe, E. (2001). *Between Rationalism and Empiricism: Selected Papers in the Philosophy of Physics*. Falkenburg, B., ed. New York: Springer.
715. Schlosshauer, M. (2007). *Decoherence and the Quantum-to-Classical Transition*. Berlin: Springer.
716. Schlosshauer, M. (2011). *Elegance and Enigma: The Quantum Interviews*. Berlin: Springer.
717. Schmidt, E. (1908). Über die Auflösung linearer Gleichungen mit unendlich vielen Unbekannten. *Rendiconti del Circolo Matematico di Palermo* 25, 53–77.
718. Schmüdgen, K. (2012). *Unbounded Self-adjoint Operators on Hilbert Space*. Dordrecht: Springer.
719. Scholz, E. (2006). Introducing groups into quantum theory (1926–1930). *Historia Mathematica* 33, 440–490.
720. Schottenloher, M. (2013). The unitary group in its strong topology. arXiv:1309.5891.
721. Schrödinger, E. (1926a). Quantisierung als Eigenwertproblem; Erste Mitteilung. *Annalen der Physik* 79, 361–376.
722. Schrödinger, E. (1926b). Quantisierung als Eigenwertproblem; Zweite Mitteilung. *Annalen der Physik* 79, 489–527.
723. Schrödinger, E. (1926c). Über das Verhältnis der Heisenberg–Born–Jordansche Quantenmechanik zu der meinen. *Annalen der Physik* 79, 734–756.
724. Schrödinger, E. (1926d). Quantisierung als Eigenwertproblem; Dritte Mitteilung. *Annalen der Physik* 80, 437–490.
725. Schrödinger, E. (1926e). Quantisierung als Eigenwertproblem; Vierte Mitteilung. *Annalen der Physik* 81, 109–139.
726. Schrödinger, E. (1935). Die gegenwärtige Situation in der Quantenmechanik. *Die Naturwissenschaften* 23, 807–812, 823–828, 844–849.
727. Schroeck, F.E., Jr. (1996). *Quantum Mechanics on Phase Space*. Dordrecht: Kluwer Academic Publishers.
728. Schulman, L.S. (1981). *Techniques and Applications of Path Integrals*. New York: Wiley.
729. Seevinck, M.P. (2012). Challenging the gospel: Grete Hermann on von Neumann's no-hidden-variables proof. Slides available at mpseevinck.ruhosting.nl/seevinck/Aberdeen_Grete_Hermann2.pdf.

730. Seevinck, M.P., Uffink, J. (2011). Not throwing out the baby with the bathwater: Bell's condition of local causality mathematically 'sharp and clean'. *Explanation, Prediction, and Confirmation* 2, 425–450.
731. Segal, I.E. (1947a). Postulates for general quantum mechanics. *Annals of Mathematics* 48, 930–948.
732. Segal, I.E. (1947b). Irreducible representations of operator algebras. *Bulletin of the American Mathematical Society* 53, 73–88.
733. Semadeni, Z. (1971). *Banach Spaces of Continuous Functions, Vol. 1.* Warszawa: PWN.
734. Sewell, G. L. (1986). *Quantum Theory of Collective Phenomena*. New York: Oxford University Press.
735. Sewell, G. L. (2002). *Quantum Mechanics and its Emergent Macrophysics*. Princeton: Princeton University Press.
736. Sewell, G.L. (2005). On the mathematical structure of quantum measurement theory. *Reports on Mathematical Physics* 56, 271–290.
737. Shale, D., Stinespring, W.F. (1964). States on the Clifford Algebra. *Annals of Mathematics* 80, 365–381.
738. Shale, D., Stinespring, W.F. (1965). Spinor representations of infinite orthogonal groups. *Journal of Mathematics and Mechanics* 14, 315–322.
739. Shimony, A. (2013). Bell's Theorem. *The Stanford Encyclopedia of Philosophy (Winter 2013 Edition)*, Zalta E.N., ed. `plato.stanford.edu/archives/win2013/entries/bell-theorem/`.
740. Shultz, F.W. (1982). Pure states as dual objects for C*-algebras. *Communications in Mathematical Physics* 82, 497–509.
741. Siegmund-Schultze, R. (2003). The origins of functional analysis. *A History of Analysis*, pp. 385–408. Jahnke, H. N., ed. Providence: American Mathemematical Society.
742. Silberstein, M. (2002). Reduction, emergence and explanation. *The Blackwell Guide to the Philosophy of Science*, pp. 80–107. Machamer, P.K., & Silberstein, M., eds. Oxford: Blackwell.
743. Simms, D.J. (1968). *Lie Groups and Quantum Mechanics*. Berlin: Springer-Verlag.
744. Simms, D.J. (1971). A short proof of Bargmann's criterion for the lifting of projective representations of Lie groups. *Reports on Mathematical Physics* 2, 283–287.
745. Simon, B. (1976). Quantum dynamics: from automorphism to Hamiltonian. *Studies in Mathematical Physics: Essays in Honor of Valentine Bargmann*, pp. 327–349. Lieb, E., Simon, B., Wightman, A.S., eds. Princeton: Princeton University Press.
746. Simon, B. (1980). The classical limit of quantum partition functions. *Communications in Mathematical Physics* 71, 247–276.
747. Simon, B. (1985). Semiclassical analysis of low lying eigenvalues. IV. The flea on the elephant. *Journal of Functional Analysis* 63, 123–136.
748. Simon, B. (1993). *The Statistical Mechanics of Lattice Gases. Vol. I.* Princeton: Princeton University Press.
749. Simon, B. (1996). *Representations of Finite and Compact Groups*. Providence: American Mathematical Society.
750. Simon, B. (2005). *Trace Ideals and Their Applications*. Second Edition. Providence: American Mathematical Society.
751. Simon, B. (2011). *Convexity: An Analytic Viewpoint*. Cambridge: Cambridge University Press.
752. Souriau, J.-M. (1967). Quantification géométrique. Applications. *Annales de l'Institut Henri Poincaré* A6, 311–341.
753. Sourour, A.R. (1994). The Gleason-Kahane-Zelazko Theorem and its generealizations. *Banach Center Publications* 30, 327–331.
754. Spehner, D. (2009). *Models for Quantum Measurements*. Lecture Notes. `www-fourier.ujf-grenoble.fr/~spehner/notes_de_cours_q_meas6.pdf`
755. Spehner, D., Haake, F. (2008). Quantum measurements without macroscopic superpositions. *Physical Review A* 77, 052114.

756. Spitters, B., Vickers, S., Wolters, S. (2014). Gelfand spectra in Grothendieck toposes using geometric mathematics. *Electronic Proceedings in Theoretical Computer Science* 158, 77—107.
757. Stairs, A. (1983). Quantum logic, realism, and value definiteness. *Philosophy of Science* 50, 578–602.
758. Steen, L.A. (1973). Highlights in the history of spectral theory. *American Mathematical Monthly* 80, 359–381.
759. Stein, D.L., Newman, C.M. (2013). *Spin Glasses and Complexity*. Princeton: Princeton University Press.
760. Stephan, A. (1992). Emergence - a systematic view of its historical facets. *Emergence or Reduction?: Essays on the Prospects of Nonreductive Physicalism*, pp. 25–48. Beckermann, A., Flohr, H., Kim, J., eds. Berlin: De Gruyter.
761. Stern, M. (1991). *Semimodular Lattices*. Stuttgart: Teubner.
762. Stevens, M. (2016). *The Kadison–Singer Property*. Heidelberg: Springer.
763. Stöltzner, M. (2014). Higgs models and other stories about mass generation. *Journal for General Philosophy of Science* 45, 369–386.
764. Stone, M.H. (1932). *Linear Transformations in Hilbert space and their Applications to Analysis.* Providence: AMS.
765. Stone, M.H. (1937). Applications of the theory of Boolean rings to topology. *Transactions of the American Mathematical Society* 41, 375–481.
766. Størmer, E. (1969). Symmetric states of infinite tensor products of C*-algebras. *Jornal of Functional Analysis* 3, 48–68.
767. Strocchi, F. (2008). *Symmetry Breaking, Second Edition.* Heidelberg: Springer.
768. Strocchi, F. (2012). Spontaneous symmetry breaking in quantum systems. *Scholarpedia* 7, 11196. DOI: `10.4249/scholarpedia.11196`.
769. Suzuki, S., Inoue, J.-i., & Chakrabarti, B.K. (2013). *Quantum Ising Phases and Transitions in Transverse Ising Models*, 2nd ed. Heidelberg: Springer.
770. Takesaki, M. (2002). *Theory of Operator Algebras. Vol. I.* New York: Springer-Verlag.
771. Takesaki, M. (2003a). *Theory of Operator Algebras. Vol. II.* New York: Springer-Verlag.
772. Takesaki, M. (2003b). *Theory of Operator Algebras. Vol. III.* New York: Springer-Verlag.
773. Takhtajan, L.A. (2008). *Quantum Theory for Mathematicians*. Providence: American Mathematical Society.
774. Tanona, S. (2013). Decoherence and the Copenhagen cut. *Synthese* 190, 3625–3649.
775. Tao, T. (2013). Real stable polynomials and the Kadison-Singer problem. Blog: `terrytao.wordpress.com/2013/11/04/real-stable-polynomials-and-the-kadison-singer-problem/`.
776. Thirring, W. (1968). On the mathematical structure of the BCS model. II. *Communications in Mathematical Physics* 7, 181–199.
777. Thirring, W. (2002). *Quantum Mathematical Physics: Atoms, Molecules, and Large Systems.* Berlin: Springer-Verlag.
778. Thirring, W., Wehrl, A. (1967). On the mathematical structure of the BCS model. I. *Communications in Mathematical Physics* 4, 303–314.
779. Thomsen, K. (1982). Jordan-morphisms in *-algebras. *Proceedings of the American Mathematical Society* 86, 283–286.
780. Troelstra, A.S., van Dalen, D. (1988). *Constructivism in Mathematics: An Introduction. Vols. 1, 2.* Amsterdam: North-Holland.
781. Tuynman, G.M., W.A.J.J. Wiegerinck (1987). Central extensions in physics. *Journal of Geometry and Physics* 4, 207–258.
782. Uhlhorn, U. (1963). Representation of symmetry transformations in quantum mechanics. *Arkiv för Fysik* 23, 307–340.
783. Uijlen, S., Westerbaan, B. (2015). A Kochen–Specker system has at least 22 vectors. *Next Generation Computing* 34, 3–23.
784. Unnerstall, T. (1990a). Phase-spaces and dynamical descriptions of infinite mean-field quantum systems. *Journal of Mathematical Physics* 31, 680–688.

785. Unnerstall, T. (1990b). Schrödinger dynamics and physical folia of infinite mean-field quantum systems. *Communications in Mathematical Physics* 130, 237–255.
786. Van Wesep, R.A. (2006). Many worlds and the appearance of probability in quantum mechanics. *Annals of Physics (N.Y.)* 321, 2438–2452.
787. Varadarajan, V.S. (1985). *Geometry of Quantum Theory* (2nd ed.). New York: Springer-Verlag.
788. Vickers, S. (2007). Locales and toposes as spaces. *Handbook of Spatial Logics*, pp. 429–496. Aiello, M., Pratt-Hartmann, I., van Benthem, J., eds. Dordrecht: Springer.
789. Vidal, J., Palacios, G., & Mosseri, R. (2004). Entanglement in a second-order quantum phase transition. *Physical Review* A69, 022107.
790. Vihvelin, K. (2013). *Causes, Laws, and Free Will: Why Determinism Doesn't Matter*. Oxford: Oxford University Press.
791. Visser, C. (1937). Note on linear operators. *Proceedings of the Royal Academy of Sciences at Amsterdam* 40, 270–272.
792. Vonneumann, N. (1987). *John von Neumann as Seen by his Brother*. Typescript.
793. Waegell, M., Aravind, P.K. (2012). Proofs of the Kochen-Specker theorem based on a system of three qubits. *Journal of Physics A* 45, 405301.
794. Wallace, D. (2012). *The Emergent Multiverse: Quantum Theory According to the Everett Interpretation*. Oxford. Oxford University Press.
795. Walleczek, J. (2016). The super-indeterminism in orthodox quantum mechanics does not implicate the reality of experimenter free will. `arXiv:1603.03400`.
796. Walter, H. (2001). *Neurophilosophy of Free Will*. Cambridge (MA): MIT Press.
797. Wayne, A., Arciszewski, M. (2009). Emergence in physics. *Philosophy Compass* 4/5, 846–858.
798. Weaver, N. (2004). The Kadison-Singer problem in discrepancy theory. *Discrete Mathematics* 278, 227–239.
799. Weidmann, J. (2000). *Lineare Operatoren in Hilberträumen, Vol. I*. Stuttgart: Teubner.
800. Weil, A. (1965). *L'Intégration Dans Les Groupes Topologiques*. Deuxième Édition. Paris: Hermann.
801. Wegge-Olsen, N.E. (1993). *K-theory and C*-algebras*. Oxford: Oxford University Press.
802. Weinstein, A. (1983). The local structure of Poisson manifolds. *Journal of Differential Geometry* 18, 523–557.
803. Weinstein, A. (1989). Blowing up realizations of Heisenberg–Poisson manifolds. *Bulletin des Sciences Mathématiques* (2) 113, 381–406.
804. Weinstein, A. (1997). The modular automorphism group of a Poisson manifold. *Journal of Geometry and Physics* 23, 379–394.
805. Werner, R.F. (1989). Quantum states with Einstein-Podolsky-Rosen correlations admitting a hidden-variable model. *Physical Review* A40, 4277–4281.
806. Werner, R.F., Wolf, M.M. (2001). Bell inequalities and entanglement. *Quantum Information & Computation* 1, 1–25.
807. Wigner, E.P. (1931). *Gruppentheorie und ihre Anwendung auf die Quantenmechanik der Atomspektren*. Wiesbaden: Vieweg.
808. Wigner, E.P. (1939). Unitary representations of the inhomogeneous Lorentz group. *Annals of Mathematics* 40, 149–204.
809. Weyl, H. (1918). *Raum · Zeit · Materie: Vorlesungen über allgemeine Relativitätstheorie*. Berlin: J. Springer.
810. Weyl, H. (1927). Quantenmechanik und Gruppentheorie. *Zeitschrift für Physik* 46, 1–46.
811. Weyl, H. (1928). *Gruppentheorie und Quantenmechanik*. Leipzig: Hirzel.
812. Wezel, J. van (2007). *Quantum Mechanics and the Big World*. PhD Thesis. `openaccess.leidenuniv.nl/handle/1887/11468`.
813. Wezel, J. van (2008). Quantum dynamics in the thermodynamic limit. *Physical Review B* 78, 054301.
814. Wezel, J. van (2010). Broken time translation symmetry as a model for quantum state reduction. *Symmetry* 2, 582–608.

815. Wezel, J. van, Brink, J. van den, (2007). Spontaneous symmetry breaking in quantum mechanics. *American Journal of Physics* 75, 635–638.
816. Wezel, J. van, Brink, J. van den, Zaanen, J. (2005). An intrinsic limit to quantum coherence due to spontaneous symmetry breaking. *Physical Review Letters* 94, 230401.
817. Wheeler, J.A., Zurek, W.H. (1983). *Quantum Theory and Measurement*. Princeton: Princeton University Press.
818. Wigner, E.P. (1931). *Gruppentheorie und ihre Anwendung auf die Quantenmechanik der Atomspektren*. Braunschweig: Vieweg. English translation (1959): *Group Theory and its Application to the Quantum Mechanics of Atomic Spectra*. New York: Academic Press.
819. Wigner, E.P. (1963). The problem of measurement. *American Journal of Physics* 31, 6–15.
820. Willard, S. (1970). *General Topology*. Reading: Addison-Wesley Publishing Company.
821. Williams, D.P. (2007). *Crossed Products of C*-Algebras*. Providence: American Mathematical Society.
822. Winnink, M. (1972). Some general properties of thermodynamic states in an algebraic approach. *Statistical Mechanics and Field Theory*, pp. 311–333. Sen, R.N., Weil, C., eds. Jerusalem: Keter Press.
823. Wiseman, H. (2014). The two Bell's Theorems of John Bell. *Journal of Physics A* 47, 424001.
824. Wittgenstein, L. (2009/1953). *Philosophical Investigations*. Anscombe, G.E.M., Hacker, P.M.S., Schulte, J., eds. Chisester: Wiley–Blackwell.
825. Wolters, S.A.M. (2013a). *Quantum Toposophy*. PhD Thesis, Radboud University Nijmegen. www.math.ru.nl/~landsman/Wolters.pdf.
826. Wolters, S.A.M. (2013b). A comparison of two topos-theoretic approaches to quantum theory. *Communications in Mathematical Physics* 317, 3–53.
827. Woronowicz, S.L. (1991). Unbounded elements affiliated with C*-algebras and noncompact quantum groups. *Communications in Mathematical Physics* 136, 399–432.
828. Woronowicz, S. L., Napiórkowski, K. (1992). Operator theory in the C*-algebra framework. *Reports in Mathematical Physics* 31, 353–371.
829. Wreszinski, W.F. (1987). Charges and symmetries in quantum theories without locality. *Fortschritte der Physik* 35, 379–413.
830. Wright, S. (1989). *Uniqueness of the Injective* III_1 *Factor*. Berlin: Springer.
831. Wüthrich, C. (2011). Can the world be shown to be indeterministic after all? *Probabilities in Physics*, pp. 365–389. Beisbart, C., Hartmann, S., eds. Oxford: Oxford University Press.
832. Yu, S., Oh, C.H. (2012). State-independent proof of Kochen–Specker Theorem with 13 rays. *Physical Review Letters* 108, 030402.
833. Zalamea, F. (2016). *Chasing Individuation: Mathematical Description of Physical Systems*. PhD Thesis, Université Sorbonne Paris Cité.
834. Zeh, H.D. (1970). On the interpretation of measurement in quantum theory. *Foundations of Physics* 1, 69–76.
835. Zimba, J., Penrose, R. (1993). On Bell non-locality without probabilities: more curious geometry. *Studies in History and Philosophy of Science A* 24, 697–720.
836. Zinkernagel, H. (2016). Niels Bohr on the wave function and the classical/quantum divide. *Studies in History and Philosophy of Modern Physics* 53, 9–19.
837. Zurek, W.H. (1981). Pointer basis of quantum apparatus: Into what mixture does the wave packet collapse? *Physical Review* D24, 1516–1525.
838. Zurek, W.H. (2003). Decoherence, einselection, and the quantum origins of the classical. *Reviews of Modern Physics* 75, 715–775.

Index

K. Landsman, *Foundations of Quantum Theory*,
Fundamental Theories of Physics 188, DOI 10.1007/978-3-319-51777-3

Printed in the United States
By Bookmasters